TEXTBOOK OF NEPHRO-ENDOCRINOLOGY

Dr. Singh wishes to dedicate this book to his wife, Ritu, his daughter, Anika and son, Vikrum and our respective parents.

Dr. Williams wishes to dedicate this book to his family for the many hours that they gave up being with him so that the work could be completed. Specifically, these include his wife, Dee Dee and his children: Jeffrey, Christopher, Jonathan, Tarryn, Megan and Brenya and their spouses and children.

Textbook of Nephro-Endocrinology

Edited by

Ajay K. Singh
*Renal Division,
Brigham and Women's Hospital and Harvard Medical School,
Boston, Massachusetts, USA*

Gordon H. Williams
*Endocrinology, Diabetes, and Hypertension,
Brigham and Women's Hospital and Harvard Medical School,
Boston, Massachusetts, USA*

AMSTERDAM • BOSTON • HEIDELBERG • LONDON • NEW YORK
• OXFORD • PARIS • SAN DIEGO • SAN FRANCISCO • SINGAPORE
• SYDNEY • TOKYO
Academic Press is an imprint of Elsevier

Academic Press is an imprint of Elsevier
525 B Street, Suite 1900, San Diego, CA 92101-4495, USA
30 Corporate Drive, Suite 400, Burlington, MA 01803, USA
32 Jamestown Road, London NW1 7BY, UK
360 Park Avenue South, New York, NY 10010-1710, USA

First edition 2009

Copyright © 2009 Elsevier Inc. All rights reserved

No part of this publication may be reproduced, stored in a retrieval system or transmitted in any form or by any means electronic, mechanical, photocopying, recording or otherwise without the prior written permission of the publisher

Permissions may be sought directly from Elsevier's Science & Technology Rights Department in Oxford, UK: phone (+44) (0) 1865 843830; fax (+44) (0) 1865 853333; email: permissions@elsevier.com. Alternatively visit the Science and Technology Books website at www.elsevierdirect.com/rights for further information

Notice
No responsibility is assumed by the publisher for any injury and/or damage to persons or property as a matter of products liability, negligence or otherwise, or from any use or operation of any methods, products, instructions or ideas contained in the material herein.

Medicine is an ever-changing field. Standard safety precautions must be followed, but as new research and clinical experience broaden our knowledge, changes in treatment and drug therapy may become necessary or appropriate. Readers are advised to check the most current product information provided by the manufacturer of each drug to be administered to verify the recommended dose, the method and duration of administrations, and contraindications. It is the responsibility of the treating physician, relying on experience and knowledge of the patient, to determine dosages and the best treatment for each individual patient. Neither the publisher nor the authors assume any liability for any injury and/or damage to persons or property arising from this publication.

Library of Congress Cataloging in Publication Data
A catalog record for this book is available from the Library of Congress

British Library Cataloguing in Publication Data
A catalogue record for this book is available from the British Library

ISBN: 978-0-12-373870-7

Printed and bound in the United States of America
09 10 11 12 13   10 9 8 7 6 5 4 3 2 1

For information on all Academic Press publications
visit our website at elsevierdirect.com

Working together to grow
libraries in developing countries

www.elsevier.com | www.bookaid.org | www.sabre.org

ELSEVIER   BOOK AID International   Sabre Foundation

Contents

List of Contributors vii

Preface xi

PART I: THE KIDNEY AS AN ENDOCRINE ORGAN 1

Section I: Erythropoietin 1

1 Erythropoietin: An Historical Overview of Physiology, Molecular Biology
and Gene Regulation
David R. Mole, Peter J. Ratcliffe. 3

2 Erythropoiesis: The Roles of Erythropoietin and Iron
Herbert Y. Lin 19

3 Extra-Hematopoietic Action of Erythropoietin
Zheqing Cai, Gregg L. Semenza. 27

4 Development of Recombinant Erythropoietin and Erythropoietin Analogs
Iain C. Macdougall. 35

5 Erythropoietin. Anemia and Kidney Disease
Ajay K. Singh, Tejas Patel, Shona Pendse, Sairam Keithi-Reddy. 47

Section II: Vitamin D, PTH and Novel Regulators of Phosphate 59

6 Vitamin D and the Kidney: Introduction and Historical Perspective
Tejas Patel, Ajay K. Singh. 61

7 Vitamin D: Molecular Biology and Gene Regulation
Adriana S. Dusso, Alex J. Brown. 67

8 Molecular Biology of Parathyroid Hormone
Peter A. Friedman. 93

9 Endocrine Regulation of Phosphate Homeostasis
Harald Jüppner, Anthony A. Portale 103

Section III: Renin–Angiotensin 125

10 The History of the Renin–Angiotensin System
Joel Menard. 127

11 Molecular Biology of Renin and Regulation of its Gene
Timothy L. Reudelhuber, Daniel F. Catanzaro. 133

12 Physiology and Regulation of the Renin–Angiotensin–Aldosterone System
Robert M. Carey, Shetal H. Padia. 145

13	The Renin–Angiotensin–Aldosterone System and the Kidney *Benjamin Ko, George Bakris*	165
14	The Renin–Angiotensin System and the Heart *Aaron J. Trask, Carlos M. Ferrario*	179
15	Renin–Angiotensin Blockade: Therapeutic Agents *Domenic A. Sica*	187

PART II: THE KIDNEY AS A HORMONAL TARGET — 201

Section IV: Antidiuretic Hormone — 201

16	Vasopressin in the Kidney: Historical Aspects *Lynn E. Schlanger, Jeff M. Sands*	203
17	Molecular Biology and Gene Regulation of Vasopressin *Swasti Tiwari, Carolyn A. Ecelbarger*	223
18	Vasopressin Antagonists in Physiology and Disease *Tomas Berl, Robert W. Schrier*	247
19	Diabetes Insipidus and SIADH *Michael L. Moritz, Juan Carlos Ayus*	259

Section V: The Atrial Natriuretic Peptides — 285

20	ANP, BNP and CNP: Physiology and Pharmacology of the Cardiorenal Axis *Candace Y. W. Lee, John C. Burnett Jr*	287

Section VI: Aldosterone — 307

21	Aldosterone: History and Introduction *John Coghlan, James F. Tait*	309
22	Aldosterone Receptors and Their Renal Effects: Molecular Biology and Gene Regulation *Celso E. Gomez-Sanchez, Elise P. Gomez-Sanchez, Mario Galigniana*	327
23	Aldosterone and its Cardiovascular Effects *Rajesh Garg, Gail K. Adler*	347
24	Regulation of Aldosterone Production *William E. Rainey, Wendy B. Bollag, Carlos M. Isales*	359

Section VII: Endocrine Disorders in Renal Failure — 381

25	Insulin Resistance and Diabetes in Chronic Renal Disease *Donald C. Simonson*	383
26	Growth Hormone *John D. Mahan*	409
27	Sexual Dysfunction in Men and Women with Chronic Kidney Disease *Biff F. Palmer*	427
28	Thyroid Status in Chronic Renal Failure Patients – a Non-Thyroidal Illness Syndrome *Victoria S. Lim, Manish Suneja*	439
29	Metabolic Acidosis of Chronic Kidney Disease *Jeffrey A. Kraut, Glenn T. Nagami*	455
30	Pregnancy and the Kidney *Chun Lam, S. Ananth Karumanchi*	481

INDEX — 515

List of Contributors

Gail K. Adler
Division of Endocrinology, Diabetes and Hypertension, Brigham and Women's Hospital, Harvard Medical School, Boston, MA 02115, USA

Juan Carlos Ayus
Director of Clinical Research, Renal Consultants of Houston, Houston, TX, USA

George Bakris
University of Chicago, Department of Medicine, Sections of Nephrology and Endocrinology, Diabetes and Metabolism, Hypertensive Diseases Unit, Chicago, IL, USA

Tomas Berl
University of Colorado at Denver and Health Sciences Center, Division of Renal Diseases and Hypertension, 4200 East Ninth Avenue, BRB 423, Denver, CO 80262, USA

Wendy B. Bollag
Medical College of Georgia, The Institute of Molecular Medicine and Genetics, 1120 15th Street, Augusta, Georgia 30912, USA

Alex J. Brown
The Renal Division, Washington University School of Medicine, St Louis, Missouri, USA

John C. Burnett Jr
Cardiorenal Research Laboratory, Division of Cardiovascular Diseases, Departments of Medicine and Physiology and Division of Clinical Pharmacology, Department of Molecular Pharmacology & Experimental Therapeutics, Mayo Clinic and Mayo Clinic College of Medicine, Rochester, MN, USA

Zheqing Cai
Division of Cardiology, Department of Medicine, The Johns Hopkins University School of Medicine, Baltimore, MD 21205, USA

Robert M. Carey
Division of Endocrinology and Metabolism, Department of Medicine, University of Virginia Health System, Charlottesville, VA, USA

Daniel F. Catanzaro
Division of Cardiovascular Pathophysiology, Department of Medicine, Weill Cornell Medical College, New York, NY 10021, USA

John Coghlan
210 Clarendon St, East Melbourne Victoria 3002, Australia

Adriana S. Dusso
The Renal Division, Washington University School of Medicine, St Louis, Missouri, USA

Carolyn A. Ecelbarger
Department of Medicine, Division of Endocrinology and Metabolism, Georgetown University, Washington DC, 20007, USA

Carlos M. Ferrario
The Hypertension & Vascular Research Center, Department of Physiology & Pharmacology, Wake Forest University School of Medicine, Winston-Salem, North Carolina, 27157, USA

Peter A. Friedman
Department of Pharmacology, University of Pittsburgh School of Medicine, Pittsburgh, PA 15217, USA

Mario Galigniana
Fundacion Leloir, Buenos Aires, Argentina

Rajesh Garg
Division of Endocrinology, Diabetes and Hypertension, Brigham and Women's Hospital, Harvard Medical School, Boston, MA 02115, USA

Celso E. Gomez-Sanchez
Endocrinology, G.V. Sonny Montgomery VA Medical Center, Jackson and University of Mississippi Medical Center, Jackson, MS 39216, USA

Elise P. Gomez-Sanchez
Endocrinology, G.V. Sonny Montgomery VA Medical Center, Jackson and University of Mississippi Medical Center, Jackson, MS 39216, USA

Carlos M. Isales
Medical College of Georgia, The Institute of Molecular Medicine and Genetics, 1120 15th Street, Augusta, Georgia 30912, USA

Harald Jüppner
Endocrine Unit and Pediatric Nephrology Unit, Massachusetts General Hospital and Harvard Medical School, Boston, MA 02114, USA

S. Ananth Karumanchi
Beth Israel Deaconess Medical Center and Harvard Medical School, Boston, MA, USA

Sairam Keithi-Reddy
Renal Division, Brigham and Women's Hospital and Harvard Medical School, 75 Francis Street, Boston, MA 02115, USA

Benjamin Ko
University of Chicago, Department of Medicine, Sections of Nephrology and Endocrinology, Diabetes and Metabolism, Hypertensive Diseases Unit, Chicago, IL, USA

Jeffrey A. Kraut
UCLA Membrane Biology Laboratory, Division of Nephrology, Medical and Research Services, VA Greater Los Angeles Health Care System; UCLA David Geffen School of Medicine, Los Angeles, California, 90073, USA

Chun Lam
Merck Research Laboratories, Rahway, NJ, USA

Candace Y.W. Lee
Cardiorenal Research Laboratory, Division of Cardiovascular Diseases, Departments of Medicine and Physiology and Division of Clinical Pharmacology, Department of Molecular Pharmacology & Experimental Therapeutics, Mayo Clinic and Mayo Clinic College of Medicine, Rochester, MN, USA

Victoria S. Lim
Division of Nephrology, Department of Medicine, University of Iowa College of Medicine, Iowa City, Iowa 52242, USA

Herbert Y. Lin
Massachusetts General Hospital, Program in Membrane Biology/Division of Nephrology, Center for Systems Biology, Department of Medicine, Boston, Massachusetts, 02214, USA

Iain C. Macdougall
Department of Renal Medicine, King's College Hospital, London, UK

John D. Mahan
Children's Hospital, 700 Children's Drive, Columbus, OH 43205, USA

Joel Menard
Professor of Public Health, Faculté de Médecine Paris Descartes, Laboratoire SPIM, 15 rue de l'médecine, 75006 Paris, France

David R. Mole
Henry Wellcome Building for Molecular Physiology, University of Oxford, Headington Campus, Roosevelt Drive, Oxford OX3 7BN, UK

Michael L. Moritz
Division of Nephrology, Department of Pediatrics, Children's Hospital of Pittsburgh of UPMC, The University of Pittsburgh School of Medicine, Pittsburgh, PA, USA

Glenn T. Nagami
UCLA Membrane Biology Laboratory, Division of Nephrology, Medical and Research Services, VA Greater Los Angeles Health Care System; UCLA David Geffen School of Medicine, Los Angeles, California, 90073, USA

Shetal H. Padia
Division of Endocrinology and Metabolism, Department of Medicine, University of Virginia Health System, Charlottesville, VA, USA

Biff F. Palmer
Department of Medicine, Division of Nephrology, University of Texas Southwestern Medical Center, 5323 Harry Hines Blvd, Dallas Texas, 75390, USA

Tejas Patel
Renal Division, Brigham and Women's Hospital and Harvard Medical School, 75 Francis Street, Boston, MA 02115, USA

Shona Pendse
Renal Division, Brigham and Women's Hospital and Harvard Medical School, 75 Francis Street, Boston, MA 02115, USA

Anthony A. Portale
Department of Pediatrics, Division of Pediatric Nephrology, University of California San Francisco, San Francisco, CA 94143, USA

William E. Rainey
Medical College of Georgia, Department of Physiology, 1120 15th Street, Augusta, Georgia 30912, USA

Peter J. Ratcliffe
Henry Wellcome Building for Molecular Physiology, University of Oxford, Headington Campus Roosevelt Drive, Oxford OX3 7BN, UK

Timothy L. Reudelhuber
Clinical Research Institute of Montreal (IRCM), 110 Pine avenue west, Montreal, Quebec H2W 1R7, Canada

Jeff M. Sands
Emory University, School of Medicine, Atlanta, GA 30322, USA

Lynn E. Schlanger
Emory University, School of Medicine, Atlanta, GA 30322, USA

Robert W. Schrier
University of Colorado at Denver and Health Sciences Center, Division of Renal Diseases and Hypertension, 4200 East Ninth Avenue, BRB 423, Denver, CO 80262, USA

Gregg L. Semenza
Vascular Program, Institute for Cell Engineering, McKusick-Nathans Institute of Genetic Medicine, Departments of Medicine, Pediatrics, Oncology and Radiation Oncology, The Johns Hopkins University School of Medicine, Baltimore, MD 21205, USA

Ajay K. Singh
Renal Division, Brigham and Women's Hospital and Harvard Medical School, 75 Francis Street, Boston, MA 02115, USA

Domenic A. Sica
Clinical Pharmacology and Hypertension, Virginia Commonwealth University Health System, Richmond, Virginia, 23298-0160, USA

Donald C. Simonson
Division of Endocrinology, Diabetes and Hypertension, Department of Medicine, Brigham and Women's Hospital, Harvard Medical School, Boston, MA, USA

Manish Suneja
Division of Nephrology, Department of Medicine, University of Iowa College of Medicine, Iowa City, Iowa 52242, USA

James F. Tait
Granby Court, Granby Road, Harrogate, N. Yorkshire, UK

Swasti Tiwari
Department of Medicine, Division of Endocrinology and Metabolism, Georgetown University, Washington DC, 20007, USA

Aaron J. Trask
The Hypertension & Vascular Research Center, Department of Physiology & Pharmacology, Wake Forest University School of Medicine, Winston-Salem, North Carolina, 27157, USA

Preface

During the last quarter century, dramatic increases in our understanding of the relationship of the renal and endocrine systems have occurred. While many of these advances have been documented in original and review articles and in some standard medical, renal or endocrine textbooks, a compilation of them in one text has not been available. Of interest, it was 25 years ago in 1983 that Dr. Michael Dunn edited what we believe was the last text on this subject. One of the editors of this text also served as a chapter author in Dunn's book. Some subjects are similar between Dunn's and the present text. Both have chapters on the renin–angiotensin system, aldosterone, antidiuretic hormone, parathyroid hormone and Vitamin D, insulin, thyroid hormone, female sex hormones and erythropoietin. In addition, the current textbook has chapters on atrial natriuretic peptides, growth hormone, acid–base balance and pregnancy.

During the past quarter century, there have been major increases in the tools needed to understand human physiology and pathophysiology. There has been substantial growth of bench tools, e.g., molecular biology, confocal microscopy and genetic manipulations of mice, available to understand fundamental mechanisms. Likewise there have been advances in the development of clinical tools unheard of 25 years ago, e.g., human genetics, high resolution imaging, advance statistics and bioinformatics. Because of these two facts, what could be written concerning the interaction of a hormone and the kidney in a single chapter 25 years ago now requires an entire section. Thus, instead of single chapters on the Renin–Angiotensin System, Aldosterone, Antidiuretic hormone, and Erythropoietin, the current text has entire sections devoted to these specific subjects.

Similar to Dunn's textbook, the current one divides the subject matter into hormones produced by the kidney and hormones that act on the kidney. In addition the current textbook has a section on hormonal derangements and/or effects in individuals with chronic renal insufficiency. While the focus is on the actions and effects in humans, individual chapters draw on relevant preclinical data to more clearly understand the effects in humans. Each section begins with an historical introduction to the subject matter and then provides in depth discussions of it in one or more following chapters. The section on hormones and renal insufficiency discusses insulin/diabetes, growth hormone, sex steroids, thyroid hormone, acid–base disturbances and pregnancy. There are some subjects they could potential fit under the umbrella of the theme of this book that are not included. We apologize for any such omissions. However, space and time considerations limited our ability to include them.

The chapters are written to enlighten the novice and extend the knowledge base of the established investigator. None of the chapters are meant to be comprehensive of their subject matter as in many cases there are entire books written on the various topics. However, we believe the information contained herein will be of value to our audience of master's or PhD trainees, medical students, students in other biomedical professional disciplines, scientists in industry, practicing clinical investigators and administrators in the broad fields of nephrology and endocrinology.

The editors wish to thank several individuals for their untiring devotion to the production of this book. First, Barbara D. Smith who spent countless hours administratively organizing the book, encouraging editors and authors to stick to the timelines established for its on time delivery to the publisher and in proof reading some of the chapters. Also to Fran Hodge for her diligent efforts to ensure that authors were keep appraised of the status of their chapters. Finally, we thank Megan Wickline at Elsevier for her patience and expert organizational skills that allowed for a successful conclusion to this project.

Ajay K. Singh, MD
Gordon H. Williams, MD

PART I
The Kidney as an Endocrine Organ

Section I. Erythropoietin

CHAPTER 1

Erythropoietin: An Historical Overview of Physiology, Molecular Biology and Gene Regulation

DAVID R. MOLE AND PETER J. RATCLIFFE

Henry Wellcome Building for Molecular Physiology, University of Oxford, Headington Campus, Roosevelt Drive, Oxford, UK

Contents

I.	Introduction	3
II.	Hormonal regulation of erythropoiesis	4
III.	Identification of the site of erythropoietin production	5
IV.	Assays of erythropoietin	5
V.	Isolation and characterization of erythropoietin	5
VI.	Erythropoietin effector mechanisms	6
VII.	Regulation of erythropoiesis by hypoxia	6
VIII.	Regulatory elements of erythropoietin (EPO) gene	7
IX.	Erythropoietin – the paradigm for gene regulation by hypoxia	8
X.	Hypoxia inducible-factor (HIF)	10
XI.	The elusive nature of the oxygen sensor	10
XII.	Degradation of HIF by the ubiquitin-proteosomal pathway	12
XIII.	Disruption of the oxygen-sensing pathway in cancer	14
XIV.	Disruption of the oxygen-sensing pathway in hereditary polycythemia	15
XV.	Pharmacological manipulation of HIF	16
XVI.	Summary	16
	References	16

I. INTRODUCTION

Although generally ascribed to the 19th century physicians, Richard Bright, Robert Christison and Pierre Rayer, the link between kidney disease and anemia was first described 18 centuries earlier by the Greek physician Aretaeus the Cappadocian who noted: 'In all the species there are present paleness, difficulty of breathing, occasional cough; they are torpid, with much languor'. In recent years, studies into the regulation of red blood cell production by the renal hormone erythropoietin have not only confirmed this link, but have also provided effective therapeutic strategies for renal anemia and fundamental molecular insights into mechanisms of oxygen sensing and signaling that underlie oxygen homeostasis throughout the animal kingdom (see Table 1.1 for summary).

However, following Aretaeus's description, many years were to pass before key discoveries in the 17th and 18th centuries defined the importance of oxygen and oxygen transport systems to living organisms, and provided the fundamental platform necessary for modern understanding. Indeed, the description of the blood circulation by William Harvey (1578–1657) in *De Motu Cordis et Sanguinis in Animalibus* in 1628 left open the question of its purpose. Richard Lower (1631–1691), working in Oxford with Robert Hooke (1635–1702), noted that whereas the blood leaving the heart for the lungs was blue, that returning from the lungs to the heart was red. Lower mixed blood with air in a glass vessel and noted the same color change, concluding that: 'Nitrous spirit of the air, vital to life is mixed with the blood during transit through the lungs'. Furthermore, by means of the vacuum pump specially contrived by himself and Robert Hooke, Robert Boyle (1627–1691) was able to obtain 'air' from blood in 1670.

For centuries, it had been recognized that there was some active part in the air. The Chinese had called it 'yin'. The Italian polymath, Leonardo da Vinci (1452–1519), had stated that the air was not completely consumed in respiration or combustion and had claimed that there were two gases in the air. Robert Boyle had shown that a component of air was depleted by living animals. However, the nature of this 'spirit of the air' was to remain elusive for another 100 years, in part delayed by the erroneous phlogiston theory of combustion. By heating mercuric oxide to release a gas that supported combustion and respiration, Priestley and Scheele identified the essential 'dephlogistated air' or 'fire air', although by publishing first, in 1774, Priestley is commonly credited with

TABLE 1.1 Erythropoietin timeline

Date	Event
1st century AD	Aretaeus described anemia in chronic kidney disease
1590	First description of the effects of altitude on the human body by Father Joseph De Acosta
1628	Discovery of the circulation of the blood by William Harvey
1774	Discovery of oxygen by Priestley and Scheele
1837	First measurement of blood oxygenation
1862/4	Description of the oxygen transport function of hemoglobin by Hoppe-Sayer/Stokes
1863/1878	Description of the effects of altitude on blood concentration by Jourdanet/Bert
1906	Hormonal regulation of erythropoiesis first postulated by Carnot and Deflandre
1953	Definitive proof of the existence of erythropoietin provided by Erslev
1956	First bioassay of erythropoietin activity
1974	Direct proof that erythropoietin is produced in the adult kidney demonstrated by Erslev
1977	Purification, sequencing and cloning of erythropoietin
1979	Radioimmunoassay for erythropoietin developed
1986	First clinical use of recombinant erythropoietin
1989	Molecular identification and cloning of erythropoietin receptor
1991	Identification of the hypoxia response element in the 3′ erythropoietin enhancer
1992	Identification of hypoxia-inducible factor (HIF)
1993	Demonstration of the universal nature of the oxygen-sensing mechanism
1995	Biochemical purification, molecular identification and cloning of cDNA encoding HIF
1999	Demonstration of von Hippel-Lindau (VHL) dependent proteasomal degradation of HIF-α subunits
2001	Oxygen-sensing process defined as oxygen-dependent prolyl hydroxylation by non-heme (FeII)-dependent dioxygenases

the discovery. However, it was Lavoisier who overturned the phlogiston theory and, in 1777, coined the term oxygen, correctly describing the chemistry of combustion and concluding that biological energy metabolism was essentially the same process. Perhaps because of Lavoisier's untimely end, under the guillotine during the French Revolution, the term did not come into general use until it was popularized in the book 'The Botanic Garden' by Erasmus Darwin, grandfather of Charles Darwin, 'The enamour'd oxygene. The common air of the atmosphere appears by the analysis of Dr Priestley and other philosophers to consist of ... about one-fourth of pure vital air fit for the support of animal life and of combustion called oxygene'.

The first consistent measurements of oxygen in blood were performed by Gustav Magnus in 1837. In showing that there was more oxygen in arterial than venous blood, he confirmed the role of the blood circulation in delivering oxygen to the tissues. He also showed that blood contained more oxygen than could be accounted for by simple solubility and that the uptake of oxygen by the blood could be blocked by carbon monoxide suggesting a specific carrier mechanism. It was Hoppe-Seyer, in 1862, and Stokes, in 1864, who demonstrated the reversible binding of oxygen to the pigmented hemoglobin in the red cells that accounted for the color change and facilitated the transport of oxygen by the blood. By independently demonstrating the production of red cells in the bone marrow, in a process initially termed hematopoiesis, or latterly and more specifically erythropoiesis, Ernst Neumann and Giulio Bizzozero, in the 1870s, set the scene for subsequent studies into the regulation of this process.

II. HORMONAL REGULATION OF ERYTHROPOIESIS

It was only a few years after the concept of hormones was first suggested by Henri Brown-Sequard in 1889, that the idea of hormonal regulation of erythropoiesis was first formulated by Carnot and Deflandre in 1906 (Carnot and Deflandre, 1906). Their experiments involved injecting serum from rabbits, rendered anemic by venesection, into normal rabbits leading to an increased concentration of red blood cells in the recipients within 1–2 days. They concluded that the transferred serum contained a hematopoietic factor that they termed 'haematopoïetine'. Subsequently, haematopoïetine would be substituted by the more specific term 'erythropoietin'. However, initially, the existence of haematopoïetine was doubted for many years because, with a few exceptions, most investigators failed to reproduce the results of Carnot and Deflandre. Interest in the possibility of a humoral factor controlling erythropoiesis was rekindled following observations in parabiotic animal pairs in 1950. Kurt Reissmann and Gerhard Ruhenstroth-Bauer observed that induction of anemia or hypoxemia in one of the parabiotic animals would induce erythrocytic hyperplasia and reticulocytosis in the partner. Allan Erslev (1919–2003) is generally credited with providing definitive proof of the existence of erythropoietin in 1953, by transfusing large quantities of plasma from anemic rabbits. Plasma from anemic animals (but not control animals) generated a significant reticulocytosis in the recipient animal which, after repeated dosing, resulted in a rise in the hematocrit (Erslev, 1953). With remarkable foresight, he also postulated: 'Conceivably isolation and purification of this factor would provide an agent useful in the treatment of conditions associated with erythropoietic depression, such as chronic infection and chronic renal disease'.

III. IDENTIFICATION OF THE SITE OF ERYTHROPOIETIN PRODUCTION

The clinical observation that patients suffering from hypoxemia to the lower portion of the body due to a patent ductus arteriosus showed generalized erythroid hyperplasia, suggested a link between the lower part of the body and the stimulation of erythropoiesis. This was consistent with the observation that patients suffering from significant renal impairment were frequently anemic. The important role of the kidney in erythropoietin production became apparent when Leon Jacobson (1911–1992) and Eugene Goldwasser showed that nephrectomized rats failed to respond to venesection or cobalt chloride with the normal increase in erythropoietin activity, while the response was intact in rats subjected to hypophysectomy, thyroidectomy, splenectomy, adrenalectomy and gonadectomy (Jacobson et al., 1957). Nevertheless, the failure of attempts to extract erythropoietin from the kidney led to doubt that the kidney was the direct source of erythropoietin. Instead, an alternative hypothesis was advanced in which the kidney secreted an enzyme (erythrogenin) that cleaved erythropoietin from a plasma protein. However, erythropoietin could not be reliably generated by the addition of kidney extract to normal plasma. The erythrogenin concept was finally disproved by Erslev, as late as 1974, by the demonstration of erythropoietin activity in isolated serum-free perfused kidneys from hypoxic rabbits (Erslev, 1974). Further confirmation that the kidney produces erythropoietin directly came with the isolation of erythropoietin mRNA from the kidneys of hypoxic rodents (Beru et al., 1986). While mRNA studies confirmed the results of organ ablation studies in showing that the main sites of erythropoietin synthesis were the kidney in the adult and the liver in fetal and neonatal life, the spleen, lung, bone marrow, brain and testes were all shown to express small amounts of erythropoietin mRNA. The translational efficiency and potential function of erythropoietin produced in these sites is not known. For instance, it is unlikely that erythropoietin produced in the brain enters the systemic circulation because of the blood–brain barrier. The expression of erythropoietin receptor (EpoR) in the brain and the ability of erythropoietin to protect the brain from ischemic insult has led to the assumption that erythropoietin may act as a paracrine neuroprotective factor, though the physiological function of such an action is unclear (Sakanaka et al., 1998).

Within the kidneys of hypoxic rats, erythropoietin activity was mainly found in the cortex and not in the medulla. Again, this result was borne out by later mRNA studies, which demonstrated erythropoietin expression in the interstitium of the renal cortex and showed co-localization with fibroblast markers implicating this cell lineage in renal erythropoietin production (Maxwell et al., 1993a).

IV. ASSAYS OF ERYTHROPOIETIN

Early erythropoietin research was hampered by the low concentration of the hormone in the fluids and tissues to be studied, particularly in the basal state, which made its detection and quantitation unreliable. The first assays of erythropoietin activity utilized the rate of incorporation of radioactive iron-59 into hemoglobin as a measure of erythropoiesis in starved rats that had been injected with the material under test (Jacobson et al., 1956). These assays were rendered more sensitive by using 'ex-hypoxic' polycythemic mice to reduce the rate of background erythropoiesis. Initial standardization employed the 'cobalt unit' in which one unit produced the same erythrogenic response in the test animals as 5 micromol cobalt chloride. Later reference standards included preparations of sheep plasma, human urinary erythropoietin and, in 1992, a fully glycosylated purified recombinant human erythropoietin.

In vivo bioassays were costly, time-consuming and lacked precision and sensitivity. Several more sensitive bioassay methods using cell culture were described. These methods were generally applicable to purified erythropoietin samples, but were often affected by non-specific inhibitors present in crude samples. Such methods were eventually replaced by radioimmunoassay in the late 1970s and early 1980s. Today, there are many commercially available enzyme-linked immunoassay kits available for the determination of erythropoietin levels.

V. ISOLATION AND CHARACTERIZATION OF ERYTHROPOIETIN

Armed with the early bioassays of erythropoietic activity, researchers next turned their attention to the biochemical purification of erythropoietin. Early attempts at partial purification of erythropoietin from anemic rabbit serum proved remarkably informative. The erythropoietic activity was found to have an electrophoretic mobility similar to alpha-2 globulin, to be heat stable and to stain for carbohydrate. Erythropoietin was therefore deduced to be a glycoprotein. These studies also showed that erythropoietin contained hexose, hexosamine and sialic acid and that erythropoietic activity was lost upon removal of neuraminic acid.

In a mammoth effort, first ovine erythropoietin was purified over a million-fold from anemic-sheep plasma and then human erythropoietin was purified from 2550 liters of urine from patients with aplastic anemia (Miyake et al., 1977). The purified human erythropoietin was subjected to tryptic digestion and the resulting fragments separated and sequenced. The partial amino acid sequences obtained enabled DNA probes to be made, which were then used to probe both genomic and cDNA libraries to identify and subsequently clone the erythropoietin gene (Jacobs et al., 1985; Lin et al., 1985). Expression of the erythropoietin cDNA in Chinese hamster ovary cells resulted in production of biologically active erythropoietin (Lin et al., 1985).

The human erythropoietin gene is a single copy gene containing five exons and is located on the long arm of chromosome 7 (7q11–q22). It encodes a 193-amino acid prohormone, from which a 27-residue leader sequence, as well as the carboxy-terminal arginine are cleaved prior to secretion. The resulting 165 amino acid, mature human erythropoietin is an acidic glycoprotein with a molecular mass of 30.4 kDa that contains two bisulfide bridges. Circulating erythropoietin has several glycosylation isoforms with 40% of the molecule consisting of carbohydrate comprising three tetra-antennary N-linked (Asn^{24}, Asn^{38} and Asn^{83}) and one small O-linked (Ser^{126}) glycans. The N-linked glycans are essential for the biological activity of erythropoietin and contain terminal sialic acid residues that protect the whole molecule from removal by galactose receptors expressed on hepatocytes. The addition of further N-linked glycans to recombinant erythropoietin by site-directed mutagenesis has been used to prolong *in vivo* activity of the molecule (Egrie and Browne, 2001).

The cloning of the human erythropoietin gene and the production of recombinant human erythropoietin led very quickly to its clinical application in the treatment of the anemia of chronic kidney disease (Winearls et al., 1986; Eschbach et al., 1987), thereby fulfilling Erslev's earlier prediction.

VI. ERYTHROPOIETIN EFFECTOR MECHANISMS

It is now nearly 50 years since self-replicating hematopoietic stem cells were first demonstrated in the bone marrow (Till et al., 1961). Derived from these pluripotent stem cells, the erythroid lineage comprises 'the burst-forming unit erythroid' (BFU-E) and the more differentiated 'colony-forming unit erythroid' (CFU-E). Each BFU-E can generate 50–200 erythroblasts when exposed to high concentrations of erythropoietin. The CFU-Es are more sensitive to the effects of erythropoietin with the number of colonies of erythroblasts derived from each increasing in response to more modest erythropoietin concentrations. In 1992, Koury and Bondurant demonstrated that erythropoietin promotes red cell formation by preventing apoptosis in these cell lineages (Koury and Bondurant, 1992).

Although evidence for a cell-surface erythropoietin receptor was first provided in 1974, it was not until 15 years later that the murine erythropoietin receptor was first cloned and characterized as belonging to the cytokine class I receptor family (D'Andrea et al., 1989). The receptor exists as a homodimer and crystallization studies reveal a conformational change on binding of erythropoietin that leads to activation of Janus kinase 2 (JAK2), which interacts with the cytoplasmic region of the receptor. The affinity of erythropoietin analogs for the receptor decreases with increasing glycosylation (see Chapter 4).

VII. REGULATION OF ERYTHROPOIESIS BY HYPOXIA

The first description of the effects of altitude on the human body is generally attributed to Father Joseph De Acosta (1539–1600), a Jesuit priest who made observations during highland expeditions following the Spanish conquest of South America. However, techniques for measuring hemoglobin concentration and red cell counts were not available until the mid-19th century, so it was not until this time that the effects of altitude on the blood were first described. During the attempted French colonization of Mexico, Denis Jourdanet (1815–1892) observed the blood of patients at altitude to be thick, dark and to flow slowly and, on measuring the number of red blood corpuscles, found them to be raised despite the patients having the symptoms of anemia (Jourdanet, 1863). His protégé, Paul Bert (1833–1886), Professor of Physiology at the Sorbonne in Paris, and later to become governor-general of French Indo-China, noted the same phenomenon in animals living at altitude (Bert, 1878). Both workers felt that these changes were acquired over generations. However, on traveling from Bordeaux to Morococha, in Peru, at 4500 m above sea level, Viault noted an increase in erythrocytes within 23 days. Following this, it was Mabel Fitzgerald (1872–1973) (a colleague of J.S. Haldane on the 1911 expedition to Pike's Peak, Colorado to study breathing responses at altitude) who first clearly described the great sensitivity of this response, illustrating that relatively minor reductions in barometric pressure at modest altitude were associated with a discernable elevation of hematocrit (FitzGerald, 1913).

These results suggested a link between physiological hypoxia and erythropoiesis. However, there is also a link between pathological hypoxia and polycythemia. For example, in conditions such as chronic lung disease or cyanotic heart disease, which result in systemic hypoxemia, the hematocrit is frequently raised.

Later recognition of the role of erythropoietin in regulating erythropoiesis (see above) was rapidly followed by studies of the effect of hypoxia on erythropoietin itself, confirming the great sensitivity of erythropoietin production to hypoxia. For instance, in the 1960s, Faura and colleagues demonstrated increased erythropoietin activity in response to hypoxia in lowlanders taken to Morococha in Peru, while previous animal work had shown a link between hypoxia and erythropoietic activity. Abbrecht and Littell showed the rapid and transient nature of the erythropoietin response during a high altitude expedition to Colorado (4360 m) with erythropoietin peaking at 1–3 days after arrival at the new altitude and falling close to baseline by day 10 (Abbrecht and Littell, 1972).

Given that the teleological purpose of erythropoietin is to maintain the blood hemoglobin concentration in the normal range and that the main function of hemoglobin is to carry and deliver oxygen from the lungs to the respiring tissues, it is appropriate that the sensed parameter in the feedback loop is tissue oxygenation. However, given that the kidney has no direct role in erythropoiesis or oxygen transport, it is not immediately apparent why the kidney is so well suited to sensing oxygen.

A possible explanation for this paradox lies in the unusual anatomical and physiological characteristics of the highly specialized blood supply to the kidney. In order to maintain the osmotic gradient generated by the loop of Henle, the arterial and venous blood vessels supplying the renal tissues run countercurrent and in close contact. This leads to shunt diffusion of oxygen between the arterial and venous circulation and the generation of an oxygen gradient throughout the renal parenchyma (Bauer and Kurtz, 1989). Consequently, oxygen tensions fall with increasing distance from the renal surface reaching levels below 10 mmHg within the medulla. Furthermore, the glomerulus and tubule are supplied by the same network of vessels and, because the renal oxygen demand, arising from the work of tubular reabsorption, varies in proportion with the glomerular filtration rate, the oxygen gradient is little affected by the rate of blood flow. Therefore, as the oxygen concentration of the blood (as determined by hemoglobin concentration) falls, so the hypoxic regions of the kidney propagate outwards and increasing numbers of peritubular fibroblasts are recruited to produce erythropoietin in a 'march' effect (Koury et al., 1989). The induction of erythropoietin in this manner shows a strikingly high gain response, with even modest changes in hemoglobin leading to large changes in erythropoietin over several orders of magnitude.

VIII. REGULATORY ELEMENTS OF ERYTHROPOIETIN (EPO) GENE

The cloning of the erythropoietin gene provided the opportunity to study the regulatory pathway by examining the effects of non-coding 'cis-acting' sequences on expression. Transgenic studies in mice demonstrated a role for long-range sequences, lying both 5′ and 3′ to the gene, in directing erythropoietin expression to the kidney and liver, respectively. The mechanisms by which these enhancer sequences interact with local sequences at the erythropoietin promoter to direct tissue-specific expression are still unclear. At the promoter itself, the 5′ flanking sequence contains a GATA binding site and it has been proposed that the fall in GATA-4 in hepatocytes during the transition from fetal to adult life may underlie the switch in erythropoietin production after birth. Nevertheless, transgenic studies demonstrate that distant 3′ sequences are also needed for proper suppression of hepatic erythropoietin expression in adult mice (Figure 1.1). In contrast, GATA-2 appears to inhibit erythropoietin gene expression. In addition, nuclear factor κB (NF-κB) also binds to the erythropoietin 5′ promoter and it has been proposed that enhanced NF-κB and GATA-2 activity might be responsible for the suppression of erythropoietin production seen during systemic inflammation (Ebert and Bunn, 1999).

An important finding in the transgenic mouse studies was that local sequences, including a hypersensitive site lying 3′ to the erythropoietin gene, were sufficient to direct oxygen-sensitive expression, at least in the liver. Further analysis of these sequences was facilitated by the use of erythropoietin producing hepatoma cell lines that demonstrated the same dynamic hypoxic response as the whole organism, and could be studied in tissue culture. Using these cell lines, reporter assays, in which 3′ erythropoietin flanking sequences were fused to DNA encoding a heterologous protein, defined a transcriptional enhancer, lying 3′, responsible for hypoxic regulation of erythropoietin expression. (For review, see Stockmann and Fandrey, 2006.)

These studies defined a specific point of interaction between the oxygen-sensitive signaling system (at least as manifest in the erythropoietin-producing hepatoma cells) and the erythropoietin gene locus. Unexpectedly, however, transfection studies of the 3′ erythropoietin enhancer also demonstrated oxygen-regulated activity after introduction into a wide variety of cells, regardless of whether the cells produced erythropoietin, or were derived from an erythropoietin-producing organ. Thus, it became clear that the highly specific and sensitive response to hypoxia that was manifest in the erythropoietin-producing tissues was, in fact, a general property of mammalian cells, irrespective of whether they produced erythropoietin and therefore predicted to have many other functions (Maxwell et al., 1993b).

Studies of proteins binding to erythropoietin 3′ sequences revealed a trans-acting nuclear factor, termed hypoxia-inducible factor-1 (HIF-1), that was induced by hypoxia in a process requiring uninterrupted protein synthesis and which bound to the erythropoietin gene enhancer at a site critically required for oxygen-dependent transcriptional activation (Semenza and Wang, 1992).

Taken together, these findings prompted a search for genes other than erythropoietin that contained similar cis-acting sequences (termed hypoxia response elements) and which responded to HIF-1.

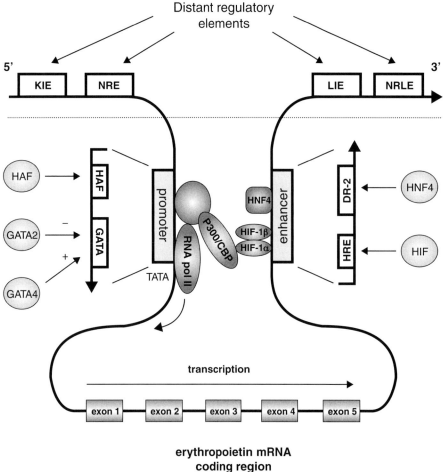

FIGURE 1.1 Regulatory elements of the erythropoietin gene. This schematic representation of the erythropoietin gene shows the promoter region, the five exons and the 3′ erythropoietin enhancer. Additional cis-acting regulatory are required for tissue-specific and developmental regulation. The kidney inducible element (KIE) confers expression in interstitial peritubular cells. The negative regulatory element (NRE) suppresses transcription. Two cis-acting regulatory elements 3′ to the erythropoietin coding sequence, the liver-inducible element (LIE) and the negative regulatory liver element (NRLE) promote and enhance gene expression in the liver. At the promoter, 5′ sequences contain a 'GATA' motif that regulates opposing effects of diverse GATA factors. While the erythropoietin promoter is a weak promoter, it does contribute to hypoxic regulation through a hypoxia-associated factor (HAF) binding site. However, the 3′ enhancer containing a hypoxia-responsive element (HRE) that binds hypoxia-inducible factor (HIF) is responsible for the majority of the fold-induction of the erythropoietin gene in response to hypoxia. In addition, the direct repeat of two steroid hormone receptor half-sites (DR2) confers regulation through binding of hepatic nuclear factor 4 (HNF-4).

IX. ERYTHROPOIETIN – THE PARADIGM FOR GENE REGULATION BY HYPOXIA

Unexpectedly, the first such genes to be identified were those encoding the enzymes phosphoglycerate kinase and lactate dehydrogenase, genes which, though modestly inducible by hypoxia, show a much lower amplitude of induction than erythropoietin. It is now clear that HIF target genes are involved in a wide range of cellular and systemic responses to hypoxia whose dynamics differ quite markedly from those of erythropoietin production (Wenger, 2002) (Figure 1.2).

HIF targets included glucose transporters and key enzymes in the glycolytic pathway such as GLUT1 (glucose uptake), 6-phosphofructo-1-kinase L, enolase, pyruvate kinase, in addition to lactate dehydrogenase A. Hence, during conditions of oxygen deficiency, under which oxidative phosphorylation cannot proceed, HIF coordinately upregulates the less efficient glycolytic pathway and facilitates conversion of the resultant pyruvate to lactate for export to the liver. In addition, in hypoxia, HIF also upregulates pyruvate dehydrogenase kinase (PDK), which phosphorylates and inactivates the pyruvate dehydrogenase enzyme complex that converts pyruvate to acetyl-coenzyme A, thereby inhibiting pyruvate entry into the tricarboxylic acid (TCA) cycle and hence directly suppressing oxidative metabolism (Brahimi-Horn and Pouyssegur, 2007).

Also important in limiting oxygen demand are cell-based decisions involving cell proliferation and apoptosis through the expression of genes encoding apoptotic regulators such as B-cell lymphoma-2 family members (Bcl-2) and the cell cycle regulators p21 and p27, which are themselves regulated either directly or indirectly by the HIF transcriptional cascade.

Increased oxygen delivery and erythropoiesis requires not only stimulation of the bone marrow by erythropoietin, but also coordinated iron provision. HIF contributes to this, through processes that include enhanced gastrointestinal uptake through downregulation of the iron-regulatory hormone, hepcidin, in addition to upregulation of transferrin, the iron carrier protein, and transferrin receptor mediating cellular iron uptake.

During localized hypoxia, vasomotor tone is subject to control by HIF-mediated transcriptional regulation of factors such as endothelin 1, inducible nitric-oxide synthase (iNOS), endothelial nitric-oxide synthase (eNOS), and heme oxygenase-1 (HO-1), endothelin-1 and atrial natriuretic peptide (ANP), though it remains unclear how these responses are coordinated in the overall regulation of vascular tone.

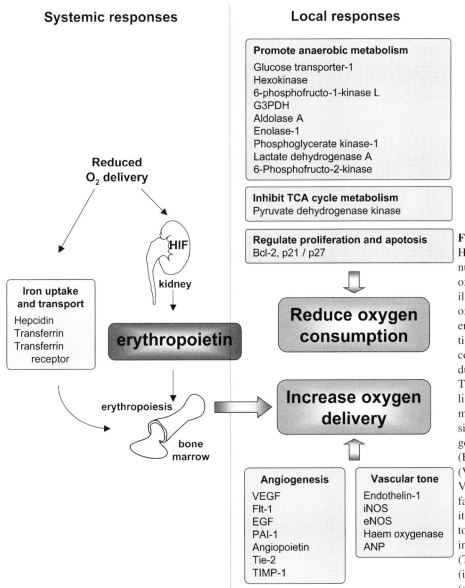

FIGURE 1.2 Transcriptonal targets of HIF. A representation of the increasing number of genes regulated by the HIF oxygen-sensing transcriptional pathway is illustrated. Broadly, these act to increase oxygen delivery systemically by promoting erythropoiesis and iron delivery and at the tissue level by promoting angiogenesis and controlling vascular tone, in addition to reducing oxygen consumption by inhibiting TCA (tricarboxylic acid) cycle metabolism, promoting anaerobic glycolysis and modulating cell proliferation and apoptosis. Glyceraldehyde-3-phosphate dehydrogenase (G3PHD), B-cell CLL/lymphoma 2 (Bcl-2), vascular endothelial growth factor (VEGF), fms-related tyrosine kinase 1/VEGF receptor (Flt-1), epidermal growth factor (EGF), plasminogen activator inhibitor-1 (PAI-1), endothelium-specific receptor tyrosine kinase 2 (Tie-2), tissue inhibitor of matrix metalloproteinase-1 (TIMP-1), inducible nitric-oxide synthase (iNOS), endothelial nitric-oxide synthase (eNOS), atrial natriuretic peptide (ANP).

In the adult, the vasculature is usually quiescent, with endothelial cells being among the longest-lived cells outside the nervous system. However, during the physiological processes of growth and development, wound healing and proliferation, as well as pathological conditions arising in neoplasia, ischemia and inflammation, localized oxygen demand may exceed supply, leading to new blood vessel growth (angiogenesis) directed at restoring this balance. Both vascular endothelial growth factor (VEGF) and its receptor Flt-1 are transcriptionally activated by HIF and this alone is capable of initiating angiogenesis in quiescent vessels. However, for an efficient vasculature to be formed, a more coordinated response is necessary, involving other growth factors such as angiopoietin and its receptor Tie-2, fibroblast growth factor (FGF) and platelet derived growth factor (PDGF), in addition to the balanced control of matrix metalloproteinases and tissue inhibitors of matrix metalloproteinases (TIMPs). Again, all these processes appear to be directly or indirectly responsive to HIF (Pugh and Ratcliffe, 2003). Though the detail of how responses are coordinated is unclear, activation of HIF does appear to be sufficient to generate an effective angiogenic response that can enhance or restore oxygen delivery.

Inflammatory tissues that are injured by infectious processes, trauma or other causes are characterized by low glucose levels, high lactate concentration with resultant low pH and, frequently, extreme degrees of hypoxia. This results from a combination of changes in metabolic activity, an increased diffusion distance, disruption of blood flow through phagocytic plugging and damage to capillaries or, in the case of an abscess, complete avascularity of certain regions. Myeloid cells, key effectors of the innate immune response, have evolved several HIF-dependent survival strategies to cope with this hypoxia, including enhanced glycolysis, inhibition of NF-κB dependent apoptosis and enhanced diapedesis of neutrophils into hypoxic regions (Cramer et al., 2003). Within nephrology, this hypoxic upregulation of the immune system is of importance not only in the modulation of autoimmune inflammation, but also in the regulation of alloimmune inflammation, following the ischemic insult sustained during transplantation.

X. HYPOXIA INDUCIBLE-FACTOR (HIF)

Using DNA-affinity chromatography, HIF-1 was purified to homogeneity allowing the determination of partial amino acid sequence and identification of the encoding cDNAs. This revealed that HIF-1 is a heterodimer of the novel HIF-1α subunit (120 kDa) and a HIF-1β subunit (91–94 kDa) previously identified as the aryl hydrocarbon nuclear receptor translocator (ARNT). In common with many transcription factors, these proteins have distinct functional domains. Each protein was found to contain both a basic-helix-loop-helix (bHLH) motif (residues 17–70), common to many transcription factors, and a PAS domain, defined by its presence in the Drosophila Per and Sim proteins and in the mammalian ARNT and AHR proteins (Wang et al., 1995). PAS domains contain two internal homology units, the A and B repeats (Figure 1.3) and are implicated in protein–protein interactions. Residues 1–166 of HIF-1α are sufficient for heterodimerization with HIF-1β, but residues 1–390 are required to effect proper DNA binding. Consistent with the role of the basic-helix-loop-helix (bHLH) domain in DNA binding of transcription factors, deletion of the basic region (residues 4–27) was found to abrogate DNA binding.

HIF-1β or ARNT was first identified from mutational and complementation studies, on mouse hepatoma (Hepa-1) cells, as essential for the transcriptional response to certain environmental hydrocarbons termed the xenobiotic response. It is expressed constitutively and, with alternative dimerization partners, it functions in a range of transcriptional systems. Thus, in the xenobiotic response, ARNT forms a heterodimer with another basic helix-loop-helix PAS protein, the aryl hydrocarbon receptor (AHR), which then binds the xenobiotic responsive element, a control sequence for genes such as *CYP1A1*, a cytochrome P450 reductase that can convert aryl hydrocarbons to toxic or carcinogenic metabolites.

In contrast, both HIF-1α and a closely related protein HIF-2α, identified bioinformatically as a homologue of HIF-1α, were found to be highly and specifically regulated by hypoxia, a process shown to be mediated through rapid degradation in the presence of oxygen (Salceda and Caro, 1997) and dependent on a central oxygen dependent degradation domain. Further regulatory domains were identified within the C-terminal portion of HIF-1α and HIF-2α, but not HIF-1β, that mediate hypoxia inducible transactivation in a manner that is independent of protein level and involves regulated binding to p300/CBP. Thus, these studies indicated that oxygen specifically regulates at least two independent properties of HIF-α subunits.

XI. THE ELUSIVE NATURE OF THE OXYGEN SENSOR

Although it was known that erythropoietin levels could be increased several hundred-fold within hours of hypoxic stimulation, the exact nature of the oxygen-sensing mechanism that regulated HIF remained elusive and the subject of much debate.

Interestingly, in mammals, it had long been observed that certain transition metal ions, such as cobalt(II), manganese (II) and nickel(II) ions, induced erythropoietin gene expression both *in vivo* and in hepatoma cell lines (Goldberg et al., 1988). More recently, it had been demonstrated that HIF-1α itself and non-erythropoietin HIF target genes were also induced by cobalt. Moreover, the efficacy of cobalt was inversely related to the availability of iron, suggesting that the metal might be competing at an oxygen-sensing iron center.

Two broad categories of process were proposed as putative oxygen sensors, either direct sensing of molecular dioxygen or indirect sensing through its effects on levels of a number of metabolites, with proposals including the products of oxidative metabolism such as ATP or heme, or reactive, oxygen-derived free radicals.

Since ATP is a major product of oxidative respiration and many systems respond to alterations in its level, it was initially attractive as a possible mediator of oxygen sensing in mammalian cells. However, early work on erythropoietin production and later studies on the HIF system clearly indicated that the effects of hypoxia were not mimicked by metabolic inhibitors of ATP production such as cyanide.

Another appealing model was that of a heme protein sensor. In addition to the effects of transition metal ions on the regulatory mechanism of erythropoietin production, it was shown that carbon monoxide could influence the system. Based on the inhibition of hypoxia-inducible but not cobalt-inducible production of erythropoietin in Hep3B cells by high concentrations of carbon monoxide (Goldberg et al., 1988), it was proposed that the oxygen sensor was a heme protein and in which cobalt substitution might generate a constitutive 'de-oxy' form. However, the observation that the HIF system was also stimulated by highly selective iron chelators, desferrioxamine and hydroxypyridinones, was difficult to accommodate within the heme-sensing model. Heme-iron is not known to be affected by these chelators and, in fact, does not freely interchange with cobalt. It was

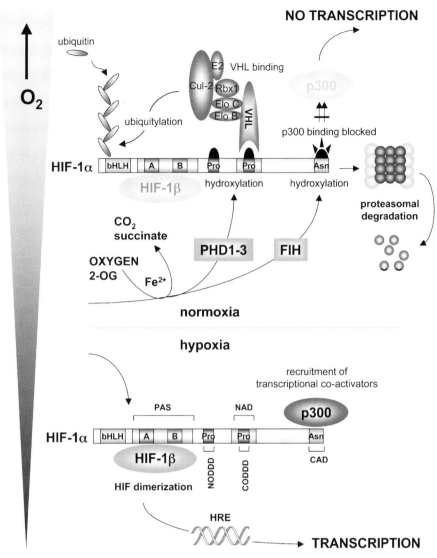

FIGURE 1.3 Oxygen-dependent regulation of HIF signaling. The figure shows the domain structure for HIF-1α. The basic helix-loop-helix (bHLH) domains and the PAS domains (PAS domain = PER, AHR, ARNT, SIM domain; PER = periodic circadian protein, AHR = aryl-hydrocarbon receptor, ARNT = aryl-hydrocarbon receptor nuclear translocator, SIM = single-minded protein, A and B = A and B domains) are responsible for dimerization and DNA binding. The C-terminal portion of HIF-1α and HIF-2α contains the regulatory domains: the amino-terminal oxygen-dependent degradation domain (NODDD) and the carboxy-terminal oxygen-dependent degradation domain (CODDD), responsible for regulating HIF-α stability and the amino-terminal activation domain (NAD) and the carboxy-terminal activation domain (CAD) involved in regulating transactivating ability. When oxygen is available, enzymatic hydroxylation of specific residues within these regulatory domains is effected by the PHD (prolyl hydroxylase domain) and FIH (factor inhibiting HIF) enzymes. Oxygen is rate limiting for these modifications, which also require divalent iron (Fe^{2+}) and 2-oxoglutarate (2-OG) as co-factors. Hydroxylation of a specific asparagine (Asn) residue within the CAD blocks recruitment of the co-activator p300, while hydroxylation of specific proline (Pro) residues in the NODDD and CODDD allows binding of the VHL E3 ubiquitin ligase complex. This complex comprises a specific recognition component, the von Hippel-Lindau protein (VHL) and elongin B (Elo B), elongin C (Elo C), Rbx and cullin-2 (Cul-2). Recognition of hydroxylated proline residues leads to covalent attachment of a polyubiquitin chain that then targets the HIF-α subunit for proteosomal destruction. Under hypoxic conditions HIF-α subunits dimerize with HIF-1β, bind DNA sequences containing hypoxia response elements (HREs) and recruit transcriptional co-activators such as p300 to effect transcription of target genes.

therefore necessary to propose that the sensing molecule was itself turning over rapidly to allow incorporation of cobalt into the nascent protein, or to permit removal of chelatable iron to affect heme synthesis and incorporation.

Despite uncertainties in the heme protein hypothesis, the perturbation of the system by iron chelators and transition metals strongly suggested that the oxygen-sensing process involved some form of iron center. Pharmacological studies in which redox active chemicals have been found to perturb the regulation of erythropoietin or HIF led to the proposal of different types of redox-sensing mechanism, in which a metabolite of oxygen, such as reactive oxygen species, might be sensed. Several sources of such reactive oxygen species (ROS) have been suggested including the mitochondrial chain, specific NAD(P)H oxidases or the possibility of direct attack by 'metal-catalyzed' oxidation (a local Fenton reaction) at the protein surface of the target molecule. However, conflicting results were obtained in different studies. Both the very large number of biological interfaces of oxygen in cells and even greater complexity of intracellular redox chemistry make for extreme difficulty in distinguishing direct from indirect responses to pharmacological or even genetic probes. Together with methodological difficulty and controversy over the measurement of intracellular metabolites that might be postulated to be involved in oxygen sensing, this creates substantial problems for such an 'outside-in' approach to probe the oxygen-sensing process, leading several groups to favor an 'inside-out' approach focusing on regulatory sequences in HIF-α.

The existence of distinct regulatory domains mediating the effects of oxygen on HIF defined specific polypeptide sequences that must interact with the upstream oxygen-sensitive signaling system. However, though these sequences were studied in detail, the nature of the signal transducing interaction was not immediately apparent. Since HIF-α subunits are heavily phosphorylated and perturbation of protein phosphatase/kinase pathways can modulate HIF activity, it was generally expected that the oxygen-sensitive pathway would involve oxygen-regulated phosphorylation of specific HIF-α residues. Unexpectedly, however, no sites of oxygen-regulated phosphorylation were defined, suggesting the operation of a different mode of signal transduction.

XII. DEGRADATION OF HIF BY THE UBIQUITIN-PROTEOSOMAL PATHWAY

Further progress into the regulation of HIF by oxygen was made by linking oxygen sensing to two related fields of research. The first of these was the ubiquitin-proteosome pathway of protein degradation. Biochemical inhibitors of this pathway were found to induce HIF-1 DNA binding when added to normoxic cells and to prevent the degradation of the HIF-1 complex in cells transferred from hypoxia to normoxia. Furthermore, cells with a temperature-dependent mutation of ubiquitylation manifested rapid accumulation of HIF-1α at the non-permissive temperature (Salceda and Caro, 1997). Hence, it was demonstrated that HIF-1α is degraded by oxygen regulated ubiquitin mediated proteosomal degradation.

The ubiquitin-proteosome system plays an important role in a broad array of basic cellular processes, including cell cycle regulation, modulation of immune and inflammatory responses and control of signal transduction pathways, development and differentiation. In outline, degradation of a protein by the ubiquitin system involves two successive steps (for review see Hershko et al., 2000):

1. covalent attachment of multiple ubiquitin molecules to the substrate
2. degradation of the tagged protein by the 26S proteasome.

Conjugation of ubiquitin proceeds via three stages: activation, transfer and ligation. In the ligation step, the first ubiquitin moiety is attached to an ε-amine group of an internal lysine residue. Subsequent chain elongation proceeds by cross-linking to lysine 48 of the ubiquitin molecule, although attachments to other lysine residues have been observed and may have separate distinct functions. More recently, other, ubiquitin-like molecules (e.g. NEDD8, SUMO-1) have been identified which target substrates to other processes within the cell.

Within the pathway, specificity and control are conferred through recognition of substrate by the E3 ubiquitin ligase complex, as E1, E2s and the proteosome are all constitutively active. However, it is rare that a single protein is targeted by a specific E3 ligase and, in most cases, an E3 recognizes a subset of proteins that contains similar structural motifs. Furthermore, some proteins are recognized by two different E3 enzymes, via distinct recognition motifs. Some of these motifs are encoded within the protein itself and the proteins which harbor them are degraded constitutively. Stability of other proteins may be regulated, depending on the state of oligomerization or on post-translational modification, such as phosphorylation.

The second field of research to provide important genetic insight into the regulation of HIF evolved from observations on the von Hippel-Lindau (VHL) syndrome and, in particular, the recognition that the von Hippel-Lindau tumor suppressor gene product (pVHL) provides the specific E3 ligase recognition component targeting HIF-α subunits for destruction in the presence of oxygen (Maxwell et al., 1999). Enhanced glucose metabolism and angiogenesis are classical features of cancer, involving upregulation of HIF target genes. While partially attributable to the hypoxic microenvironment, genetic alterations also contribute to these effects, with VHL disease providing a striking example.

Affected individuals, bearing a germ line mutation in the VHL tumor suppresser gene, develop tumors resulting from somatic loss or inactivation of the remaining wild-type allele. A cardinal feature of these tumors is a rich supply of blood vessels (e.g. hemangioblastomas, affecting the CNS and retina, and renal clear cell carcinomas) (Kaelin, 2002). Many of these tumors over-express hypoxia-inducible genes, such as VEGF and, more rarely, erythropoietin, providing early evidence of a connection between the hypoxic response and pVHL.

The VHL gene encodes a 213 amino acid protein (pVHL) and was first isolated in 1993 (Latif et al., 1993). Although the primary sequence did not immediately suggest a function, protein association experiments defined a series of pVHL interacting molecules, including elongins B and C, and CUL2. This complex showed homology to specific yeast ubiquitin ligases referred to as SCF complexes (Skp1/Cdc53/F-box protein). In such complexes, the target protein destined for polyubiquitination and destruction is recognized or bound by the F-box protein, suggesting an analogous, specific recognition, role for pVHL.

The definitive link between pVHL and HIF-α regulation was established through the study of renal carcinoma cell lines deficient in pVHL (Maxwell et al., 1999). In a series of such cell lines, HIF-α subunits were found to be constitutively stabilized and HIF was activated, leading to increased transcription of reporter plasmids containing hypoxia response elements (HREs) and dysregulation of a wide range of hypoxically regulable native genes. Re-expression of wild-type pVHL in stable transfectants restored the normal pattern of oxygen-dependent instability demonstrating the critical function of pVHL in HIF proteolysis. A physical interaction between pVHL and HIF-α was demonstrated by co-immunoprecipitation after blockade of HIF degradation by proteosomal inhibitors. Further experiments formally confirmed the function of pVHL as part of an ubiquitin ligase complex targeting HIF-α subunits.

These findings provided a new focus for analysis of the oxygen-sensitive pathway through studies of the interaction between HIF-α polypeptides and pVHL. Further studies defined two short HIF-α sub-sequences that interact with pVHL corresponding to independently active proteolytic degradation domains in HIF-α. These sequences interact directly with the β-domain of pVHL which, in turn, provides a link, through an interaction between its α-domain and elongin C, to the other components of the multi-ubiquitin ligase complex (Ivan et al., 2001; Jaakkola et al., 2001; Masson et al., 2001).

Immunoprecipitation experiments using whole cell extracts revealed that capture of HIF-α by pVHL is suppressed by treatment of cells with iron chelators and cobaltous ions (Maxwell et al., 1999) and (when oxygen is excluded from the cell extraction buffers) by hypoxia (Ivan et al., 2001; Jaakkola et al., 2001), indicating that regulation of this interaction accurately reflects the properties of the oxygen-sensitive pathway.

Further studies showed that the HIF-α/pVHL interaction could be reproduced *in vitro* using recombinant HIF-α and pVHL, but that the interaction required that the HIF-α polypeptide was pre-incubated with a cell extract in the presence of iron and oxygen. Temperature sensitivity and heat inactivation suggested that this process involved an enzymatic modification and mutational analysis together with mass spectrometry and functional testing of modified HIF-α polypeptides showed that the critical modification was hydroxylation of a specific prolyl residue (P564 in human HIF-1α (Ivan et al., 2001; Jaakkola et al., 2001). Further studies showed that each of the two independent HIF-α degradation domains contains a site of prolyl hydroxylation and defined a common LxxLxP motif at the hydroxylation site (Masson et al., 2001). Recognition of each hydroxyproline by pVHL then targets the HIF-α molecule for rapid degradation by the ubiquitin-proteosome pathway (see Figure 1.3).

The mechanism by which addition of a single oxygen atom to a prolyl residue within a HIF-α degradation motif governs recognition by pVHL was subsequently studied by X-ray crystallography of a hydroxylated HIF-α peptide bound to the VCB (pVHL, elongins B and C). These studies revealed a single, well-defined hydroxyproline-binding pocket on the surface of the pVHL β-domain. Highly specific discrimination between hydroxylated and non-hydroxylated HIF is achieved by an optimized hydrogen-bonding network between VHL residues, in the floor of the binding pocket, and the HIF-α hydroxyproline residue that would be denied to proline.

Genetic analysis in model organisms defined the critical prolyl hydroxylase and led to the identification of three homologous and closely related mammalian enzymes termed PHD1, 2 and 3 (*p*rolyl *h*ydroxylase *d*omain) that catalyze hydroxylation of prolyl residues within the two independent degradation domains of human HIF-α subunits (Epstein et al., 2001).

When assayed using synthetic peptides, kinetic analysis of HIF-α prolyl hydroxylation gives an apparent K_m for oxygen (concentration of substrate that gives half maximal activity) of 230–250 μM. Even though, when assayed using longer (and more physiological) HIF-α polypeptides, K_m values for oxygen are lower, they are still well above the oxygen concentration in tissues (believed to be 10–30 μM). Consequently, cellular availability of oxygen will limit the rate of HIF-α hydroxylation, with the reaction rate varying essentially linearly over the physiological range of tissue oxygen tensions. Because, in the presence of an intact pVHL/ubiquitin/proteosome pathway, the hydroxylation step is rate limiting for HIF degradation, this allows HIF-α levels to reflect oxygen concentration in a graded manner that is essential for the system to act as an oxygen sensor.

Following the discovery of prolyl hydroxylation in the regulation of HIF-α stability, a further enzymatic hydroxylation, that of an asparagine residue in the C-terminal of HIF-α subunits, was defined by mass spectrometry and demonstrated to direct non-proteolytic regulation of HIF transactivation (Lando et al., 2002). In the presence of oxygen, hydroxylation of this asparaginyl residue blocks the interaction of HIF-α with the CH1 domain of p300 and CBP co-activators impeding its ability to initiate transcription. It was rapidly recognized, using bioinformatic analysis, that a molecule originally defined as a HIF interacting protein that inhibited HIF transactivation by unknown mechanisms (FIH, *f*actor *i*nhibiting *H*IF) was, in fact, the HIF asparaginyl hydroxylase that targets this C-terminal asparagine residue. Thus, in the presence of oxygen, HIF is inactivated by a dual mechanism involving both proteolytic degradation and transcriptional inactivation of remaining protein HIF-α protein (Lando et al., 2003; Schofield and Ratcliffe, 2004).

Although to date only a single HIF asparaginyl hydroxylase has been identified, the presence of multiple HIF prolyl hydroxylases raises questions about their possible redundancy and diverse roles. Inactivation of each PHD individually and in combination using small interfering RNA has demonstrated that while all three contribute to the regulation of HIF, there are differences both in their relative importance and in effects on HIF-1α versus HIF-2α. Thus, while PHD2 appears to be the most important enzyme in setting basal levels of HIF-1α in normoxic cells, PHD3 has a more substantial effect on HIF-2α particularly in hypoxic cells. The importance of PHD2 under basal conditions is strongly supported by the contrast between embryonic lethality seen in PHD2 knockout mice and the relatively minor phenotypes of those lacking PHD1 and PHD3. The three PHD enzymes are conserved in mammals and a number of other vertebrate species and show differential patterns of organ expression, intracellular localization and inducibility by exogenous stimuli, suggesting that they have distinct functions (Schofield and Ratcliffe, 2004). Whether these are discreet functions in the regulation of HIF or other hydroxylation substrates is currently unclear.

Within a complex organism, levels of oxygenation are markedly heterogeneous. For example, the renal medulla has a considerably lower oxygen tension than the renal cortex. Cells within these different environments have adapted their HIF response to avoid inappropriate overactivation of this pathway. How this is achieved is incompletely understood, although may in part be due to hypoxic inducibility of PHD2 and PHD3 expression (Epstein et al., 2001). These two PHD isoforms are under transcriptional regulation of HIF itself, providing a negative feedback loop in which HIF induces the mechanism of its own destruction, permitting HIF levels to accommodate to chronic changes in oxygen availability, while still responding to rapid perturbations.

Enzymatic hydroxylation of residues within proteins directly connects the availability of molecular oxygen to protein function by an entirely novel mechanism of signal transduction, although such modifications are common in extracellular proteins and have structural roles, for instance in the formation of collagens. To date all prolyl and asparaginyl hydroxylases identified have been members of the superfamily of 2-oxoglutarate (α-ketoglutarate) dependent dioxygenases. In addition to oxygen, these enzymes also require 2-oxoglutarate as a co-substrate and non-heme iron (Fe^{2+}) as a co-factor. Furthermore, ascorbate (vitamin C) is required for full enzymatic activity, possibly to maintain the catalytic iron center in the reduced state, or to facilitate the availability of iron in a more general way (Schofield and Ratcliffe, 2004). This raises important questions as to what extent HIF-hydroxylase activity is affected by physiological or pathological variation in iron or ascorbate availability. Experiments in tissue culture certainly suggest that limiting iron availability could be pathologically important, particularly in cancer cells where rapid growth often depletes cells of iron and HIF-α levels are commonly raised even in the presence of oxygen (Gerald et al., 2004; Kaelin, 2005). These co-factor requirements also raise the intriguing possibility that physiological or pathophysiological changes in their levels in an appropriate cellular compartment might provide other links to the availability of oxygen. In theory, at least, cellular disposition of Kreb's cycle intermediates and the redox status of iron/ascorbate might also be affected by oxygen with the HIF hydroxylases effectively integrating a number of signals. Such influences appear likely to underlie at least some of the effects of pharmacological and genetic perturbations of oxygen radical metabolism on the HIF system.

XIII. DISRUPTION OF THE OXYGEN-SENSING PATHWAY IN CANCER

Inadequate tissue oxygen supply is a central component of both neoplastic and non-neoplastic ischemic vascular disease. For instance, the essential physiological requirement of achieving a balance between oxygen supply and demand is not only integral to organized growth and development, but is recapitulated during the disorganized proliferation seen in the development of cancer. Many pieces of evidence now link HIF to important aspects of tumor behavior. HIF activation is commonly seen in many types of cancer and there is a high degree of concordance between HIF target genes and those upregulated in cancerous tissue. For instance, the classical features of upregulated glycolysis and angiogenesis are driven, at least in part, by activation of HIF pathway in cancer. In many cancers, the degree of HIF-α immunostaining correlates with tumors that are more aggressive and is seen as an independent marker of prognosis (Semenza, 2003).

Microenvironmental tumor hypoxia is clearly an important mechanism of HIF induction in tumors and HIF immunostaining is often most intense in hypoxic perinecrotic regions of these growths. Nevertheless, other mechanisms clearly contribute to neoplastic upregulation of HIF and it has become clear that a large number of pathways involving oncogenic activation or tumor suppressor inactivation are linked to HIF activation, most probably reflecting the fundamental importance of linking growth to processes entraining an oxygen supply (Semenza, 2003).

These links include inactivation of PTEN or p53 pathways and activation of ras, src and myc pathways as well as stimulation by exogenous growth factors such as epidermal growth factor, insulin, insulin-like growth factors 1 and 2, angiotensin II, thrombin and PDGF. Though the mechanisms underlying some of these links remain poorly understood, evidence to date indicates that they involve both increased HIF translational and impairment of oxygen-dependent degradation.

The most direct link between genetic mutation in cancer and activation of HIF is in VHL-associated renal cell carcinoma (RCC) (see above). In hereditary VHL disease, affected individuals develop multiple cystic lesions in their kidneys and have a very high lifetime risk of (often multiple) RCC. These tumors are associated with somatic loss or inactivation of the remaining wild-type allele in individuals who are heterozygous for germline mutation in the VHL tumor suppressor gene, in accordance with the classic Knudson 'two-hit' hypothesis. Non-familial tumors may result from somatic loss or inactivation of both alleles within the same cell, a process that accounts for in excess of 80% of sporadic RCC. However, on average it takes longer to accrue two 'hits' within the same cell, accounting for the lower prevalence and older age distribution of sporadic tumors. As outlined above, loss of functional pVHL in tumor cells blocks oxygen-dependent degradation of both HIF-α subunits. This direct influence on the HIF pathway has led to intense investigation of the causal role of HIF activation in VHL-associated RCC.

Genetic manipulation, involving both upregulation and downregulation of HIF-α (particularly HIF-2α), in RCC cell lines, and assessment of growth as tumor xenografts has indicated that upregulation of HIF-2α is both necessary and sufficient to promote growth, at least in these assays. This evidence for a causal role of HIF upregulation in RCC is backed by functional genetic studies indicating, not only that all RCC-associated VHL mutations result in HIF upregulation, both also that the severity of this HIF dysregulation correlates with risk of RCC in familial VHL disease. Interestingly, however, it appears that the two main HIF-α isoforms (HIF-1α and HIF-2α) have quite different actions in the pathogenesis of VHL-associated RCC. In the kidneys of patients with VHL disease, selective HIF-1α upregulation is seen in the earliest dysplastic lesions, whereas HIF-2α is more abundant in more advanced dysplastic lesions. Concordant with this, many VHL-negative renal cancer cell-lines preferentially express HIF-2α and, while overexpression of HIF-2α in renal cancer xenografts enhances their growth, overexpression of HIF-1α inhibits xenograft growth, at least in some RCC cell lines. Most probably, this reflects differences in transcriptional selectivity between HIF-1α and HIF-2α, with HIF-2α preferentially activating pro-tumorigenic genes in RCC (Kaelin, 2002).

It is interesting that erythropoietin itself appears to be principally an HIF-2α target gene. Renal erythropoietin is normally produced by the renal interstitial fibroblasts that express HIF-2α rather than HIF-1α. Dysregulated erythropoietin production and erythrocytosis is also, however, associated with RCC, particularly in the later stages of disease. In this situation, erythropoietin mRNA is produced by the (epithelial) RCC cells themselves (Da Silva et al., 1990). As described above, this is associated with a switch from HIF-1α to HIF-2α expression in the neoplastic epithelium that might be related, in some way, to the epithelial-mesenchymal transition that occurs during RCC progression.

Since the recognition of the importance of HIF in VHL-associated RCC, several other genetic tumor syndromes have been connected with dysregulation of the HIF pathway and, interestingly, most of these include a predisposition to renal neoplasia.

Thus, certain nuclear genes encoding mitochondrial enzymes can also act as tumor suppressors. Succinate dehydrogenase deficiency (SDHD) resulting from germline mutation in genes encoding subunits B, C or D is associated with the formation of highly vascular tumors such as pheochromocytoma, paraganglioma of the carotid body (also an organ with a specialized role in oxygen sensing), papillary thyroid carcinoma and renal cell carcinoma. Similarly, germline mutation in fumarate hydratase (FH) has been linked to a syndrome predisposing affected individuals to cutaneous and uterine leiomyomas and renal cell cancer. In each case, individuals are heterozygous for the enzyme loss, while tumor cells have mutated or inactivated their second copy, resulting in high levels of upstream TCA cycle intermediates, namely succinate and fumarate, which are able competitively to inhibit 2-oxoglutarate dependent dioxygenases including the HIF hydroxylases, and result in elevated levels of HIF-α within tumor cells. Nevertheless, though it is an attractive possibility that HIF activation contributes causally to tumor development, as in VHL disease, this has not yet been assessed directly (Ratcliffe, 2007).

Tuberose sclerosis is well known to nephrologists, because of multiple renal angiomyolipomas and cysts, but clear-cell renal cell carcinomas are also observed. This and the other familial hamartoma syndromes, Peutz–Jeghers syndrome, Cowden syndrome and Bannayan–Riley–Ruvalcaba syndrome are linked mechanistically through the TSC1 and TSC2 tumor suppressors and their effects on the mTOR (mammalian target of rapamycin) pathway. Phosphorylation of TSC2 by AMPK, consequent to energy starvation, leads to inhibition of mTOR signaling. mTOR itself affects HIF-α levels independent of oxygen and may provide a means of coupling nutrient availability to oxygen sensing (Brugarolas and Kaelin, 2004). Inhibition of HIF also provides a possible explanation for some of the useful anti-cancer properties of the mTOR inhibitor and immunosuppressant agent rapamycin, in particular its actions on tumor angiogenesis.

XIV. DISRUPTION OF THE OXYGEN-SENSING PATHWAY IN HEREDITARY POLYCYTHEMIA

Though most causes of erythrocytosis are either associated with appropriate depression of serum erythropoietin levels (e.g. polycythemia rubra vera) or an obvious defect in oxygen delivery (e.g. hypoxemia, or high affinity hemoglobins), some individuals have been identified with congenital erythrocytosis and unexplained inappropriate high erythropoietin levels, suggesting a defect in the oxygen-sensing process.

A genetic analysis of some of these syndromes has indeed revealed defects in the oxygen-sensing pathways outlined above. Thus, individuals affected by a rare form of hereditary polycythemia, first described in patients from western Russia (Ang et al., 2002) and termed 'Chuvash polycythemia', have been found to be homozygous for a specific mutation in the VHL tumor suppressor gene. Unlike cancer associated VHL mutations, the protein derived from this gene mutation remains partially active. While only mild activation of the HIF pathways ensues, this is sufficient to generate excessive erythropoiesis and elevation of the hematocrit. Paradoxically, these patients do not develop tumors, as seen in the VHL syndrome, raising the possibility that pVHL has a second function that is important in tumor suppression. More recently, families have also been described with different types of inactivating mutation in the PHD2 enzyme, leading to familial erythrocytosis.

XV. PHARMACOLOGICAL MANIPULATION OF HIF

The central role of HIF in a wide range of common pathologies, including cancer and ischemia, has made it an attractive target for pharmacological manipulation. To date, such approaches have focused on downregulation of the HIF pathway in cancer and upregulation in ischemic/hypoxic/anemic diseases, where the aim has been to promote erythropoiesis, angiogenesis or cytoprotective responses in different settings.

While pharmacological small molecule strategies for inhibition of transcriptional systems such as HIF are still regarded as difficult to achieve, the enzymatic basis of regulation by hydroxylation presents a classical small molecule target for upregulation of HIF using small molecule inhibitors of the HIF hydroxylases. Inhibition of these enzymes by 2-oxoglutarate analogs was rapidly shown to block the interaction of HIF-α with VHL, stabilize HIF-α and upregulate HIF target genes. Increasing numbers of such agents are being developed (Hewitson and Schofield, 2004) and have been shown to stimulate erythropoietin production and erythropoiesis and to afford protection in models of renal and other organ ischemia.

Despite this promise, achieving the specificity needed for safe clinical intervention is likely to be challenging (Hewitson and Schofield, 2004). In addition to known 2-oxoglutarate dioxygenases such as procollagen hydroxylases, bioinformatic predictions suggest that there may be up to fifty or so related and largely uncharacterized enzymes in the human genome with potential to generate 'off-target' effects from non-specific inhibitors (Elkins et al., 2003). The possibility of unwanted actions is increased still further when the pleiotropic actions of HIF are considered. This is well illustrated by the ability of HIF activation to increase both erythropoiesis and angiogenesis. Promotion of erythropoiesis, leading to polycythemia might be an unwanted side effect of treatments aimed at enhancing angiogenesis, while safe treatment of anemia might be confounded by promotion of angiogenesis. However, dissociating these actions may simply be a matter of delivering the drug to the required organ. For example, systemic hypoxia at altitude predominantly affects erythropoiesis, with little in the way of systemic angiogenesis, whereas localized hypoxia, in healing or neoplastic tissues incapable of expressing erythropoietin, stimulates localized angiogenesis. Nevertheless, careful development will be needed to determine whether and how safe therapeutic translation of these findings can be achieved.

XVI. SUMMARY

Following classical insights into the fundamental importance of oxygen in living systems, physiological studies in the 20th century demonstrated a circulating hormone, erythropoietin, that was produced by the kidneys, in response to tissue hypoxia and stimulated red blood cell production in the bone marrow to complete a feedback loop contributing to oxygen homeostasis. The advent of molecular biology led to identification of the erythropoietin gene in the 1980s, underpinning the immense therapeutic advance provided by recombinant erythropoietin in the treatment of renal anemia. In addition, study of the pathways underlying regulation of erythropoietin gene has led to the discovery of a highly conserved, widespread oxygen-sensing mechanism, regulated by novel signaling pathways involving the post-translation hydroxylation of specific amino acid residues in the transcription factor HIF. This reaction is catalyzed by a series of non-heme iron dioxygenases whose absolute requirement for molecular oxygen underlies the oxygen-sensing process.

Conflict of interest statement

PJ Ratcliffe is a founding scientist of the company ReOx Ltd.

References

Abbrecht, P. H., and Littell, J. K. (1972). Plasma erythropoietin in men and mice during acclimatization to different altitudes. *J Appl Physiol* **32**: 54–58.

Ang, S. O., Chen, H., Hirota, K. et al. (2002). Disruption of oxygen homeostasis underlies congenital Chuvash polycythemia. *Nat Genet* **32**: 614–621.

Bauer, C., and Kurtz, A. (1989). Oxygen sensing in the kidney and its relation to erythropoietin production. *Annu Rev Physiol* **51**: 845–856.

Bert, P. (1878). La Pression Barometrique. Recherches de physiologie experimentale. Libraire de l'Acadamie de Medicine, Paris.

Beru, N., McDonald, J., Lacombe, C., and Goldwasser, E. (1986). Expression of the erythropoietin gene. *Mol Cell Biol* **6**: 2571–2575.

Brahimi-Horn, M. C., and Pouyssegur, J. (2007). Hypoxia in cancer cell metabolism and pH regulation. *Essays Biochem* **43**: 165–178.

Brugarolas, J., and Kaelin, W. G. (2004). Dysregulation of HIF and VEGF is a unifying feature of the familial hamartoma syndromes. *Cancer Cell* **6**: 7–10.

Carnot, P., and Deflandre, C. (1906). Sur l'activité hémopoiétique du sérum au cours de la régénération du sang. *C. R. Acad. Sci. Paris* **143**: 384–386.

Cramer, T., Yamanishi, Y., Clausen, B. E. et al. (2003). HIF-1alpha is essential for myeloid cell-mediated inflammation. *Cell* **112**: 645–657.

D'Andrea, A. D., Lodish, H. F., and Wong, G. G. (1989). Expression cloning of the murine erythropoietin receptor. *Cell* **57**: 277–285.

Da Silva, J. L., Lacombe, C., Bruneval, P. et al. (1990). Tumor cells are the site of erythropoietin synthesis in human renal cancers associated with polycythemia. *Blood* **75**: 577–582.

Ebert, B. L., and Bunn, H. F. (1999). Regulation of the erythropoietin gene. *Blood* **94**: 1864–1877.

Egrie, J. C., and Browne, J. K. (2001). Development and characterization of novel erythropoiesis stimulating protein (NESP). *Nephrol Dial Transplant* **16**: Suppl 33–13.

Elkins, J. M., Hewitson, K. S., McNeill, L. A. et al. (2003). Structure of factor-inhibiting hypoxia-inducible factor (HIF) reveals mechanism of oxidative modification of HIF-1 alpha. *J Biol Chem* **278**: 1802–1806.

Epstein, A. C., Gleadle, J. M., McNeill, L. A. et al. (2001). C. elegans EGL-9 and mammalian homologs define a family of dioxygenases that regulate HIF by prolyl hydroxylation. *Cell* **107**: 43–54.

Erslev, A. (1953). Humoral regulation of red cell production. *Blood* **8**: 349–357.

Erslev, A. J. (1974). In vitro production of erythropoietin by kidneys perfused with a serum-free solution. *Blood* **44**: 77–85.

Eschbach, J. W., Egrie, J. C., Downing, M. R., Browne, J. K., and Adamson, J. W. (1987). Correction of the anemia of end-stage renal disease with recombinant human erythropoietin. Results of a combined phase I and II clinical trial. *N Engl J Med* **316**: 73–78.

FitzGerald, M. P. (1913). The changes in the breathing and the blood at various high altitudes. *Philosoph Trans Roy Soc Lond Series B* **203**: 351–371.

Gerald, D., Berra, E., Frapart, Y. M. et al. (2004). JunD reduces tumor angiogenesis by protecting cells from oxidative stress. *Cell* **118**: 781–794.

Goldberg, M. A., Dunning, S. P., and Bunn, H. F. (1988). Regulation of the erythropoietin gene: evidence that the oxygen sensor is a heme protein. *Science* **242**: 1412–1415.

Hershko, A., Ciechanover, A., and Varshavsky, A. (2000). Basic Medical Research Award. The ubiquitin system. *Nat Med* **6**: 1073–1081.

Hewitson, K. S., and Schofield, C. J. (2004). The HIF pathway as a therapeutic target. *Drug Discov Today* **9**: 704–711.

Ivan, M., Kondo, K., Yang, H. et al. (2001). HIFalpha targeted for VHL-mediated destruction by proline hydroxylation: implications for O2 sensing. *Science* **292**: 464–468.

Jaakkola, P., Mole, D. R., Tian, Y. M. et al. (2001). Targeting of HIF-alpha to the von Hippel-Lindau ubiquitylation complex by O2-regulated prolyl hydroxylation. *Science* **292**: 468–472.

Jacobs, K., Shoemaker, C., Rudersdorf, R. et al. (1985). Isolation and characterization of genomic and cDNA clones of human erythropoietin. *Nature* **313**: 806–810.

Jacobson, L. O., Goldwasser, E., Fried, W., and Plzak, L. (1957). Role of the kidney in erythropoiesis. *Nature* **179**: 633–634.

Jacobson, L. O., Plzak, L., Fried, W., and Goldwasser, E. (1956). Plasma factor(s) influencing Red Cell Production. *Nature* **177**: 1240.

Jourdanet, D. (1863). De l'anemie des altitudes et de l'anemie en general dans les rapports avec la pression de l'atmosphere. J-B Balliere et fils, Paris.

Kaelin, W. G. (2005). Proline hydroxylation and gene expression. *Annu Rev Biochem* **74**: 115–128.

Kaelin, W. G. (2002). Molecular basis of the VHL hereditary cancer syndrome. *Nat Rev Cancer* **2**: 673–682.

Koury, M. J., and Bondurant, M. C. (1992). The molecular mechanism of erythropoietin action. *Eur J Biochem* **210**: 649–663.

Koury, S. T., Koury, M. J., Bondurant, M. C., Caro, J., and Graber, S. E. (1989). Quantitation of erythropoietin-producing cells in kidneys of mice by in situ hybridization: correlation with hematocrit, renal erythropoietin mRNA, and serum erythropoietin concentration. *Blood* **74**: 645–651.

Lando, D., Gorman, J. J., Whitelaw, M. L., and Peet, D. J. (2003). Oxygen-dependent regulation of hypoxia-inducible factors by prolyl and asparaginyl hydroxylation. *Eur J Biochem* **270**: 781–790.

Lando, D., Peet, D. J., Whelan, D. A., Gorman, J. J., and Whitelaw, M. L. (2002). Asparagine hydroxylation of the HIF transactivation domain a hypoxic switch. *Science* **295**: 858–861.

Latif, F., Tory, K., Gnarra, J. et al. (1993). Identification of the von Hippel-Lindau disease tumor suppressor gene. *Science* **260**: 1317–1320.

Lin, F. K., Suggs, S., Lin, C. H. et al. (1985). Cloning and expression of the human erythropoietin gene. *Proc Natl Acad Sci USA* **82**: 7580–7584.

Masson, N., Willam, C., Maxwell, P. H., Pugh, C. W., and Ratcliffe, P. J. (2001). Independent function of two destruction domains in hypoxia-inducible factor-alpha chains activated by prolyl hydroxylation. *Embo J* **20**: 5197–5206.

Maxwell, P. H., Osmond, M. K., Pugh, C. W. et al. (1993). Identification of the renal erythropoietin-producing cells using transgenic mice. *Kidney Int* **44**: 1149–1162.

Maxwell, P. H., Pugh, C. W., and Ratcliffe, P. J. (1993). Inducible operation of the erythropoietin 3′ enhancer in multiple cell lines: evidence for a widespread oxygen-sensing mechanism. *Proc Natl Acad Sci USA* **90**: 2423–2427.

Maxwell, P. H., Wiesener, M. S., Chang, G. W. et al. (1999). The tumour suppressor protein VHL targets hypoxia-inducible factors for oxygen-dependent proteolysis. *Nature* **399**: 271–275.

Miyake, T., Kung, C. K., and Goldwasser, E. (1977). Purification of human erythropoietin. *J Biol Chem* **252**: 5558–5564.

Pugh, C. W., and Ratcliffe, P. J. (2003). Regulation of angiogenesis by hypoxia: role of the HIF system. *Nat Med* **9**: 677–684.

Ratcliffe, P. J. (2007). Fumarate hydratase deficiency and cancer: activation of hypoxia signaling. *Cancer Cell* **11**: 303–305.

Sakanaka, M., Wen, T.-C., Matsuda, S. et al. (1998). In vivo evidence that erythropoietin protects neurons from ischemic damage. *Proc Natl Acad Sci* **95**: 4635–4640.

Salceda, S., and Caro, J. (1997). Hypoxia-inducible factor 1alpha (HIF-1alpha) protein is rapidly degraded by the ubiquitin-proteasome system under normoxic conditions. Its stabilization by hypoxia depends on redox-induced changes. *J Biol Chem* **272**: 22642–22647.

Schofield, C. J., and Ratcliffe, P. J. (2004). Oxygen sensing by HIF hydroxylases. *Nat Rev Mol Cell Biol* **5**: 343–354.

Semenza, G. L. (2003). Targeting HIF-1 for cancer therapy. *Nat Rev Cancer* **3**: 721–732.

Semenza, G. L., and Wang, G. L. (1992). A nuclear factor induced by hypoxia via de novo protein synthesis binds to the human erythropoietin gene enhancer at a site required for transcriptional activation. *Mol Cell Biol* **12**: 5447–5454.

Stockmann, C., and Fandrey, J. (2006). Hypoxia-induced erythropoietin production: a paradigm for oxygen-regulated gene expression. *Clin Exp Pharmacol Physiol* **33**: 968–979.

Wang, G. L., Jiang, B. H., Rue, E. A., and Semenza, G. L. (1995). Hypoxia-inducible factor 1 is a basic-helix-loop-helix-PAS heterodimer regulated by cellular O2 tension. *Proc Natl Acad Sci USA* **92**: 5510–5514.

Wenger, R. H. (2002). Cellular adaptation to hypoxia: O2-sensing protein hydroxylases, hypoxia-inducible transcription factors, and O2-regulated gene expression. *Faseb J* **16**: 1151–1162.

Winearls, C. G., Oliver, D. O., Pippard, M. J., Reid, C., Downing, M. R., and Cotes, P. M. (1986). Effect of human erythropoietin derived from recombinant DNA on the anaemia of patients maintained by chronic haemodialysis. *Lancet* **2**: 1175–1178.

Further reading

Till, J. E., and Mc, C. E. (1961). A direct measurement of the radiation sensitivity of normal mouse bone marrow cells. *Radiat Res* **14**: 213–222.

CHAPTER **2**

Erythropoiesis: The Roles of Erythropoietin and Iron

HERBERT Y. LIN

Massachusetts General Hospital, Program in Membrane Biology/Division of Nephrology, Center for Systems Biology, Department of Medicine, Boston, Massachusetts, 02114, USA

Contents

I. Erythropoiesis: an overview	19
II. Role of erythropoietin in erythropoiesis	21
III. Role of iron in erythropoiesis	22
References	24

I. ERYTHROPOIESIS: AN OVERVIEW

A. Hematopoiesis

The cellular and cell-derived components of the blood are generated in the process called hematopoiesis. In the adult, hematopoiesis occurs in the bone marrow, while some hematopoiesis occurs during early embryonic development in the yolk sac and the fetal liver (Palis et al., 1999; Buza-Vidas et al., 2007). A small population of self-renewing hematopoietic stem cell (HSC) precursors gives rise to differentiated progeny cell lineages that become progressively committed towards a single cell type such as the mature erythrocyte (Ross and Li, 2006). A simultaneous expansion in the number of progeny cells occurs at each step of differentiation. After the hematopoietic stem cells divide, daughter cells either remain undifferentiated (thus renewing the HSC population of cells) or they become differentiated. The end result is the generation of larger numbers of mature cells from a relatively few numbers of hematopoietic stem cells.

The lineage relationships among differentiating progeny of hematopoietic precursor cells have been established with great certainty (Graf, 2002). All of the intermediate cell types from multipotential hematopoietic progenitor stem cells to mature progeny cells have been characterized (Akashi et al., 2000; Adolfsson et al., 2005; Loose and Patient, 2006; Loose et al., 2007).

The characterization includes specific cell-surface markers (Ney, 2006; Socolovsky, 2007) to identify different cell types.

An early stem cell precursor, called the common hemangioblast, is the common progenitor for both hematopoietic and endothelial cell populations. This has been shown experimentally in both the zebrafish (Vogeli et al., 2006) and in mice (Ueno and Weissman, 2006). The precursor to the common hemangioblast is thought to arise from early mesoderm cells that express the marker Brachyury (T), which then initiates the expression of the transmembrane tyrosine kinase receptor FlK-1, marking these cells for eventual contribution to the hematopoietic lineages (Fehling et al., 2003). In the mouse, subsequent expression of the marker SCL signals the establishment of the common hemangioblast lineage that will give rise to both the hematopoietic and endothelial lineages. Expression of the transcription factor GATA-1 in the hemangioblast descendents signals the commitment of subsequent progeny towards the hematopoiesis pathway (Endoh et al., 2002).

The committed (but still multipotential) hematopoietic stem cell (HSC) then gives rise to two major progenitor cell lines: the common myeloid progenitor (CMP) and the common lymphoid progenitor (CLP) (Figure 2.1). The common lymphoid progenitor gives rise eventually to mature B cells, T cells, NK cells and lymphoid dendritic cells. The common myeloid progenitor gives rise to further intermediate precursor populations: the megakaryocytic/erythrocytic progenitors (MEPs) and the common granulocyte myeloid precursor (CGMP). The CGMP gives rise to mature basophils, eosinophils, neutrophils, macrophages, myeloid dendritic cells and mast cells.

The MEPs can differentiate into two cell lineages, either megakaryoblasts (MKP) or the committed erythropoietic progenitor (see Figure 2.1). Megakaryoblasts eventually diffferentiate into thrombocytes or platelets, while the committed erythropoietic progenitor, now unipotential, follow a series of well-described differentiation steps that eventually give rise to mature red cells (Socolovsky, 2007; Loose et al., 2007).

B. Erythropoiesis

During erythropoiesis, gene expression microarray profiling studies have shown that expression of genes that are not specific for the erythroid lineage is restricted in a

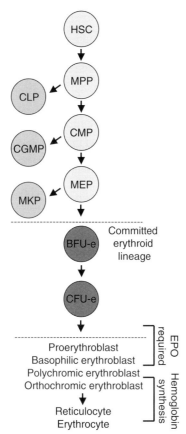

FIGURE 2.1 Schematic representation of hematopoiesis and erythropoiesis. Hematopoietic stem precursors differentiate into progeny cell lineages, as indicated by the arrows. Stages of erythropoiesis that require erythropoietin (EPO) and the stages when hemoglobin synthesis occur are indicated. HSC, hematopoietic stem cell; MPP, myeloid precursor progenitor; CLP, common lymphoid precursor; CMP, common myeloid precursor; CGMP, common granulocyte/myeloid precursor; MEP, common megakaryocyte/erythroid progenitor; MKP, megakaryocyte precursor; BFU-E, burst-forming unit-erythroid; CFU-E, colony-forming unit-erythroid.

progressive manner as progenitor cells differentiate from the hematopoietic stem cells stage to the mature red cell (Ney, 2006). A key cell fate decision is made by MEPs either to become megakaryoblasts or committed erythrocyte progenitors. The erythrocyte progenitor is committed to becoming an erythrocyte and it undergoes a series of expansion and differentiation steps on its way to becoming a mature erythrocyte.

The first truly committed erythrocyte progenitor is a cell lineage called the burst-forming unit-erythroid (BFU-E) (Iscove et al., 1974). The BFU-E has been functionally defined in classic colony-formation assays as a cell that gives rise to a burst of approximately 500 mature red blood cells in 6–10 days when grown in erythropoietin (EPO) supplemented semisolid medium. A direct descendent of the BFU-E is the later stage colony-forming unit-erythroid (CFU-E) (Stephenson and Axelrad, 1971), a more mature cell which gives rise to about 8–32 red cells in 2–3 days in culture with EPO. The CFU-e differentiates into the classic erythroblasts: proerythroblasts, basophilic erythroblasts, polychromic (or polychromatophilic) erythroblasts and orthochromic erythroblasts (see Figure 2.1). The next stage in differentiation is the reticulocyte (also known as the polychromatic erythrocyte), which finally gives rise to the mature enucleated erythrocyte. Overall, approximately 200 billion erythrocytes are newly generated each day, equal to the number of erythrocytes that become senescent.

1. Growth Factors Important for Hematopoiesis and Erythropoiesis

Pluripotent hematopoietic stem cell precursors respond to signals from a variety of specific growth factors (Gregory and Eaves, 1977; Graf, 2002). The commitment of hematopoietic precursor cells to the erythroid cell lineage involves the coordinated action of many growth factors: stem cell factor (SCF), erythropoietin (EPO), interleukins (IL) including IL-3, IL-4, IL-9, insulin-like growth factor (IGF)-1 and granulocyte-macrophage colony stimulating factor (GM-CSF) (Ross and Li, 2006; Buza-Vidas et al., 2007). The molecular mechanisms of action of these growth factors on hematopoiesis have been well characterized. In general, the roles of these growth factors includes generation of signals that leads to survival and differentiation of their target cells. It has been shown that these growth factors are somewhat interchangeable, i.e. they can somewhat compensate for the absence of each other (Wu et al., 1995; Loose et al., 2007). The effectiveness of these growth factors at each point in the differentiation pathway is dependent on the expression pattern of their respective receptors in the hematopoietic cell lineages. Once specific cell lineages are established, each major lineage has its own unique set of growth factor requirements in order to generate mature progeny cells.

For erythropoiesis to progress from the MEP stage to the committed erythrocyte progenitor/BFU-E stage, the growth factors SCF, EPO, IL-3, IL-4, IL-9, IGF-1 and GM-CSF are important (Loose and Patient, 2006; Loose et al., 2007). To progress from the BFU-E to the CFU-E stage requires the specific actions of EPO, GM-CSF and IL-3. At this stage, IL-3, GM-CSF and EPO can be somewhat substituted for each other (Socolovsky, 2007). However, starting at the late CFU-E stage, there is an absolute requirement for EPO that cannot be subserved by the other growth factors, thus making EPO the central hormone for erythropoiesis (Wu et al., 1995; Jelkmann, 2007). Interestingly, beyond this stage of erythroid development, EPO is not important for the terminal stages of differentiation of the mature erythroblasts and the number of EPO receptors per cell beyond the late CFU-E stage rapidly declines (Wickrema et al., 1992).

II. ROLE OF ERYTHROPOIETIN IN ERYTHROPOIESIS

A. EPO Receptor (see Figure 2.1)

Erythropoietin (EPO) is a glycosylated protein hormone that is produced by kidney peritubular cells in response to hypoxia (Jelkmann, 2007). Native human EPO was first isolated in 1977 (Miyake et al., 1977) and the human gene was cloned in 1985 (Jacobs et al., 1985; Lin et al., 1985). The EPO gene is under the regulation of the hypoxia inducible factor 2α (HIF-2α) (Jelkmann, 2007). During hypoxia, HIF-2α is upregulated and stimulates the expression of a variety of hypoxic genes including EPO. EPO binds to its receptor that is found to be expressed on the surface of red cell precursors.

Upon EPO binding, a series of intracellular signaling events occurs that ultimately leads to the prevention of programmed cell death or apoptosis at the late CFU-e stage of erythropoiesis (Wu et al., 1995). EPO receptor expression is tightly restricted to cells between the late BFU-e and CFU-e cell stages. In mice that lack the EPO receptor, erythroid precursor cells do not progress beyond the CFU-e stage, implying that EPO signaling is critical at this stage of erythropoiesis (Wu et al., 1995). This limited expression pattern allows the restriction of EPO's actions to this particular point in erythropoiesis and explains EPO's specific effects in increasing the number of erythrocytes but not of other hematopoietic cell types. Other growth factors important for erythropoiesis such as GM-CSF, IL-3 or SCF are promiscuous and are also required for the proliferation and differentiation of other hematopoietic cell lineages (Jelkmann, 2007; Socolovsky, 2007).

EPO receptors were first cloned in 1989 and found to be a member of the cytokine superfamily of receptors (D'Andrea et al., 1989). There are several known splice variants of the EPO receptor (Jelkmann, 2007). EPO receptors were initially thought to be expressed only on erythropoietic precursor cells, but they are now known to be expressed in other tissues, including neuronal cells and endothelial cells (Rossert and Eckardt, 2005). EPO receptors are single transmembrane receptors that are constitutively associated with tyrosine kinases. EPO receptors homodimerize upon binding to EPO at the cell surface. The signal transduction cascade initiated upon binding of EPO to its receptors has been well-studied (Richmond et al., 2005; Menon et al., 2006; Wojchowski et al., 1999).

B. Intracellular Signaling Mediators of EPO

EPO receptors are constitutively associated with the JAK2 family tyrosine protein kinase molecules (Witthuhn et al., 1993). EPO receptors are thought to exist at the cell surface as preformed dimers that are inactive in the absence of EPO. Upon ligand binding, the two EPO receptors shift into an activated mode and their associated JAK2 kinases are brought into close proximity (Constantinescu et al., 2001) (Figure 2.2). The JAK2 kinases transphosphorylate each other and become active. The activated JAK2 kinases then phosphorylate the distal cytoplasmic domain of the EPO receptors at 8 tyrosine residues. The phosphotyrosine residues of the EPO receptors serve as the docking sites for a variety of intracellular signaling mediators that contain Src homology 2 (SH2) domains, including STAT5, Ras/MAP,

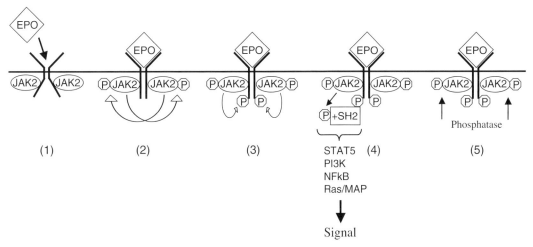

FIGURE 2.2 Schematic representation of the signaling mechanism of erythropoietin (EPO) via the EPO receptor. (1) EPO receptors exist as inactive preformed dimers on the cell surface with constitutively attached JAK2 kinases. (2) Binding of EPO triggers a conformational change in the EPO receptors that allows the transphosphorylation and activation of the JAK2 kinases. (3) Activated JAK2 kinases phosphorylate the EPO receptors on tyrosine residues, creating docking sites for SH2 containing proteins. (4) Signaling molecules containing SH2 domains dock with phosphorylated EPO receptors and become phosphorylated and activated by JAK2. Activated signaling molecules then generate signals promoting erythropoiesis. (5) Termination of signal transduction occurs when phosphatases dephosphorylate EPO receptors and the EPO-bound receptor complexes undergo ligand-induced receptor-mediated endocytosis (not shown).

PI3K, NFkB (Wojchowski et al., 1999). The translocation of these signaling mediators from the cytoplasm to the EPO receptor brings them into close proximity to the activated JAK2 kinases, allowing them to become phosphorylated by the JAK2 kinases (Figure 2.2).

In the case of the transcription factor STAT5, it binds to phosphorylated tyrosine 343 of the activated EPO receptor. Upon phosphorylation by JAK2 kinase, activated STAT5 protein then migrates to the nucleus and either activates or suppresses gene transcription that promotes the survival and proliferation of the CFU-e cell (Loose et al., 2007; Socolovsky, 2007). After STAT5 enters the nucleus, it binds to the cis-acting elements and enhances the transcription of several genes, including Bcl-Xl, an anti-apoptotic molecule in the Bcl-2 family (Silva et al., 1999). The anti-apoptotic effect of Bcl-Xl allows the CFU-e cells to survive and proliferate and to continue their irreversible differentiation into erythroblasts.

Other signaling molecules that are activated by the presence of EPO include PI-3-kinase, phospholipase Cγ, protein kinase C, Grb2 and Shc and Ras/MAP kinase (Constantinescu et al., 1999; Wojchowski et al., 1999). The genes whose expression levels are modulated by EPO signaling including the signaling molecules SOCS1 and JUNB (Buchse et al., 2006). It is the sum total of the action of EPO that leads to the survival of the CFU-e and allows continued differentiation and maturation of erythroblasts.

EPO receptor signaling is eventually downregulated by several mechanisms. One is the activity of a SH-PTP1 phosphatase which dephosphorylates JAK2 kinases and inactivates them (Klingmuller et al., 1995). Another is the inhibitory activity of the cytokine inducible SH2-containing proteins SOCS/CIS, which is activated by STAT5 and negatively feed back on STAT5 activity and on EPO receptor phosphorylation activity (Matsumoto et al., 1997; Sasaki et al., 2000; Peltola et al., 2004). Finally, ligand mediated receptor internalization leads to degradation of ligand and to recycling of receptors for further rounds of signaling (Beckman et al., 1999; Walrafen et al., 2005).

III. ROLE OF IRON IN ERYTHROPOIESIS

Approximately 200 billion new erythrocytes are generated each day, requiring 20–25 mg of iron for their corresponding hemoglobin production (Hentze et al., 2004; Andrews and Schmidt, 2007). Erythroblasts acquire iron from the serum through iron-bound transferrin-mediated uptake through the transferrin 1 receptor (Andrews and Schmidt, 2007). Since the body only absorbs 1–2 mg of new iron per day, the source for 90–95% of the iron used by developing red cell precursors is from the recycling of old and senescent red cells, which are phagocytosed by specialized reticuloendothelial macrophages whose purpose is to recover the iron from the hemoglobin for recycling and re-use (Brittenham, 1994).

Heme-oxygenase 1 liberates Fe from heme in the cytosol (Andrews and Schmidt, 2007). Thus, approximately 20–25 mg of iron are recovered and recirculated daily via macrophages from damaged red blood cells. Compared to the amount of iron that is absorbed from the duodenum on a daily basis, the recycling of iron by macrophages represents a quantitatively significant source of iron for the bone marrow.

A. Hepcidin and Regulation of Serum Iron levels

Iron is normally in homeostasis, i.e. there is no net gain or loss of iron from the body. To accomplish this goal, the intake and loss of iron in the body is a highly regulated process which prevents anemia or hemochromatosis. Each day there is normally 1 or 2 mg of iron that is absorbed or shed by sloughing of cells from the gastrointestinal tract. Free iron is taken up by DMT1 and heme is taken up by an unknown heme carrier. Inside cells, Fe is bound to ferritin as a storage form. Iron is exported by ferroportin from the basal lateral intestinal cells and is bound by transferrin for transport in the bloodstream (Andrews and Schmidt, 2007). When serum iron level is high, iron absorption from the GI tract and iron recycling from the macrophages is decreased. When serum iron level is low, iron absorption from the GI tract and iron release from reticuloendothelial macrophages is increased.

In the last several years, exciting work in the iron field has shown that the secreted liver peptide, hepcidin, is the major regulator of baseline iron homeostasis (Ganz, 2007). Hepcidin was originally discovered as LEAP-1 and found to be secreted in the liver (Krause et al., 2000). Overexpression leads to anemia in mice. Hemochromatosis occurs in KO mice and humans with mutations in hepcidin.

Hepcidin that is made in the liver (Figure 2.3) is processed from a precursor form by the protease furin (Valore and Ganz, 2008). Active hepcidin is a 25 amino acid (aa) peptide with four internal cysteine disulfide bonds. Smaller forms that are 22 or 20 aa are not active (Nemeth and Ganz, 2006). Hepcidin is freely filtered at the glomerulus and can be measured in urine (Nemeth and Ganz, 2006). In patients with renal failure, hepcidin levels are elevated (Tomosugi et al., 2006), in part due to lack of clearance by the impaired kidney. Hemodialysis can somewhat reduce hepcidin levels in end stage renal disease (ESRD) patients (Tomosugi et al., 2006). There is an emerging correlation between serum hepcidin-25 levels and the degree of anemia in a variety of patient disorders (Kemna et al., 2008), while prohepcidin levels do not correlate well with iron levels in many iron related disorders (Luukkonen and Punnonen, 2006; Roe et al., 2007). There is an ELISA for serum hepcidin 25, the active form of hepcidin (Ganz et al., 2008).

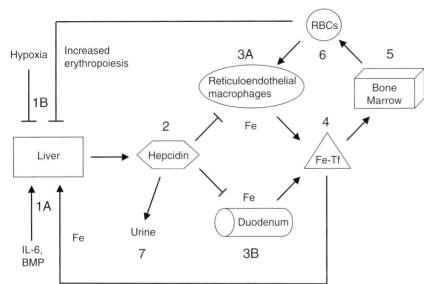

FIGURE 2.3 Schematic representation of iron metabolism. (1A) Liver produces hepcidin in reponse to increased Fe or increased inflammatory mediators. Bone morphogenetic protein (BMP) signaling is important for hepcidin expression. (1B) Hepcidin production is decreased by low iron diet, increased erythropoiesis and in response to hypoxia. (2) Circulating hepcidin blocks iron release from (3A) reticuloendothelial macrophages and blocks iron absorption from the (3B) duodenum. (4) Iron in the serum is transported by transferrin to the (5) bone marrow, where it is used by the (6) red blood cells to form hemoglobin. The iron from senescent RBCs is recycled by (3A) reticuloendothelial macrophages. Hepcidin is filtered by the kidney and appears in the urine (7).

Hepcidin acts by binding to the iron export protein ferroportin (Donovan et al., 2000) and causes its downregulation and removal from the cell surface (Nemeth et al., 2004). The downregulation of ferroportin activity leads to decreased iron release into the bloodstream from the duodenum and to decreased iron release from reticuloendothelial macrophages. Too little hepcidin leads to supranormal activity of ferroportin and to the accumulation of iron in the body, resulting in the pathologic condition known as hemochromatosis (Pietrangelo, 2007). Abnormally elevated hepcidin levels leads to underactivity of ferroportin and to iron deficiency anemia, such as seen in inflammation and renal failure and in anemia of chronic disease (Nemeth and Ganz, 2006; Ganz, 2007).

There are three identified regulators of hepcidin expression: 1. iron; 2. inflammation; 3. erythropoiesis itself (Nemeth and Ganz, 2006; Kemna et al., 2008).

The iron regulation of hepcidin expression is likely under the control of the BMP co-receptor, hemojuvelin (Babitt et al., 2006). Hemojuvelin (HJV) is a GPI-anchored cell surface protein encoded by the gene HFE2 that is mutated in chromosome 1q-linked juvenile hemochromatosis, a severe early onset form of hemochromatosis (Papanikolaou et al., 2004). HJV knockout mice have very low hepcidin levels and develop early onset hemochromatosis (Huang et al., 2005; Niederkofler et al., 2005). Inhibition of BMP signaling also blocks iron induction of hepcidin in zebrafish (Yu et al., 2007) and in mice (Wang et al., 2005). The mechanism of iron regulation of hepcidin expression is not entirely clear but is thought to involve iron-laden transferrin (Ganz, 2007).

Proinflammatory mediators such as IL-6 can stimulate hepcidin expression, providing a logical explanation for the anemia seen in inflammatory diseases (Nemeth and Ganz, 2006). There is evidence that blockade of the BMP signaling pathway can also inhibit IL-6 mediated hepcidin upregulation (Babitt et al., 2007; Yu et al., 2007). Recent data also suggest that lowering BMP signaling in mice will lower hepcidin expression and lead to increased serum iron levels (Babitt et al., 2007; Yu et al., 2007) and may potentially be useful in treating patients with anemia of end-stage renal disease (ESRD) who have elevated hepcidin levels (Tomosugi et al., 2006).

Erythropoiesis itself can inhibit hepcidin expression (Andrews and Schmidt, 2007; Ganz, 2007). EPO stimulated erythropoiesis leads to decreased hepcidin by the liver by an unknown mechanism. A possible mediator has been postulated to be the BMP-like growth factor GDF15. In thalassemia patients who have increased but defective erythropoietic activity, elevated GDF15 has been implicated in inhibiting hepatic hepcidin production (Tanno et al., 2007).

B. Iron and Regulation of Hemoglobin Production

Although iron is not absolutely required for erythropoiesis, iron profoundly influences erythropoiesis. Iron is required for heme, an essential component of hemoglobin production. Severe iron deficiency leads to lower hemoglobin production and results in small (microcytic) red blood cells with central pallor and poor stability (Koury and Ponka, 2004). Low hemoglobin levels can lead to hypoxia which leads to stimulation of EPO production by the kidney peritubular cells and to increased erythropoiesis. However, unless iron stores are repleted, in the setting of continuing inadequate iron supply, defective erythropoiesis will continue (Goodnough, 2007).

Heme is a protoporphyrin which binds to Fe and is the essential oxygen carrying component of functional hemoglobin in red cells and myoglobin in muscle cells. The rate of heme synthesis regulates the supply of iron via the transferrin-receptor uptake pathway in erythroid precursor cells (Koury and Ponka, 2004). Iron is rapidly incorporated into the heme molecule and heme is then complexed with globin chains to form hemoglobin. Under normal physiologic conditions, there are essentially no free iron, heme or globin chains in red cell precursors. Free heme that is uncommitted will inhibit further iron acquisition by the cells and will inhibit further heme synthesis.

The availability of heme regulates the synthesis of globin at both the transcriptional and translational levels. Heme deficiency leads to a decrease in globin mRNA levels and hemin treatment of erythroid precursors leads to the rapid accumulation of globlin mRNA (Ponka, 1997). The molecular mechanism of this transcriptional control is explained by heme-mediated upregulation of NF-E2 (an erythroid transcription factor) binding activity (Sassa and Nagai, 1996).

The translation of globin in intact reticulocytes is dependent on the availability of heme (Chen and London, 1995). The molecular mechanism of iron regulation of globin translational synthesis is through the effect of heme on heme-regulated inhibitor (HRI), a cyclic adenosine monophosphate (AMP)-independent protein kinase (Koury and Ponka, 2004). HRI plays an important role in the translation of globin synthesis in erythroid precursors by regulating the α-subunit of eukaryotic initiation factor 2 (eIF-2α). When free heme is absent, HRI is active and phosphorylates eIF-2α at Ser51. When phosphorylated at Ser51, the GDP-exchange factor eIF-2B is blocked and translation initiation by eIF-2 is inhibited, resulting in translational blockade of globin protein synthesis. When free heme is present, it binds to HRI at two sites. Binding of heme to site 1 is stable (i.e. HRI is a hemoprotein) and does not affect HRI activity *per se*, but binding of heme to the second site leads to rapid downregulation of HRI activity (Chefalo et al., 1998). When HRI activity is decreased, there is release of the translational blockade of globin synthesis and globin is made which then complexes to the free heme to form hemoglobin.

References

Adolfsson, J., Mansson, R. et al. (2005). Identification of Flt3+ lympho-myeloid stem cells lacking erythro-megakaryocytic potential a revised road map for adult blood lineage commitment. *Cell* **121**(2): 295–306.

Akashi, K., Traver, D. et al. (2000). A clonogenic common myeloid progenitor that gives rise to all myeloid lineages. *Nature* **404**(6774): 193–197.

Andrews, N. C., and Schmidt, P. J. (2007). Iron homeostasis. *Annu Rev Physiol* **69**: 69–85.

Babitt, J. L., Huang, F. W. et al. (2006). Bone morphogenetic protein signaling by hemojuvelin regulates hepcidin expression. *Nat Genet* **38**(5): 531–539.

Babitt, J. L., Huang, F. W. et al. (2007). Modulation of bone morphogenetic protein signaling in vivo regulates systemic iron balance. *J Clin Invest* **117**(7): 1933–1939.

Beckman, D. L., Lin, L. L. et al. (1999). Activation of the erythropoietin receptor is not required for internalization of bound erythropoietin. *Blood* **94**(8): 2667–2675.

Brittenham, G. M. (1994). New advances in iron metabolism, iron deficiency, and iron overload. *Curr Opin Hematol* **1**(2): 101–106.

Buchse, T., Prietzsch, H. et al. (2006). Profiling of early gene expression induced by erythropoietin receptor structural variants. *J Biol Chem* **281**(12): 7697–7707.

Buza-Vidas, N., Luc, S. et al. (2007). Delineation of the earliest lineage commitment steps of haematopoietic stem cells: new developments, controversies and major challenges. *Curr Opin Hematol* **14**(4): 315–321.

Chefalo, P. J., Oh, J. et al. (1998). Heme-regulated eIF-2alpha kinase purifies as a hemoprotein. *Eur J Biochem* **258**(2): 820–830.

Chen, J. J., and London, I. M. (1995). Regulation of protein synthesis by heme-regulated eIF-2 alpha kinase. *Trends Biochem Sci* **20**(3): 105–108.

Constantinescu, S. N., Ghaffari, S. et al. (1999). The erythropoietin receptor: structure, activation and intracellular signal transduction. *Trends Endocrinol Metab* **10**(1): 18–23.

Constantinescu, S. N., Keren, T. et al. (2001). Ligand-independent oligomerization of cell-surface erythropoietin receptor is mediated by the transmembrane domain. *Proc Natl Acad Sci USA* **98**(8): 4379–4384.

D'Andrea, A. D., Lodish, H. F. et al. (1989). Expression cloning of the murine erythropoietin receptor. *Cell* **57**(2): 277–285.

Donovan, A., Brownlie, A. et al. (2000). Positional cloning of zebrafish ferroportin1 identifies a conserved vertebrate iron exporter. *Nature* **403**(6771): 776–781.

Endoh, M., Ogawa, M. et al. (2002). SCL/tal-1-dependent process determines a competence to select the definitive hematopoietic lineage prior to endothelial differentiation. *Embo J* **21**(24): 6700–6708.

Fehling, H. J., Lacaud, G. et al. (2003). Tracking mesoderm induction and its specification to the hemangioblast during embryonic stem cell differentiation. *Development* **130**(17): 4217–4227.

Ganz, T. (2007). Molecular control of iron transport. *J Am Soc Nephrol* **18**(2): 394–400.

Ganz, T., Olbind, G. et al. (2008). Immunoassay for human serum hepcidin. *Blood*. In press.

Goodnough, L. T. (2007). Erythropoietin and iron-restricted erythropoiesis. *Exp Hematol* **35**(4 Suppl 1): 167–172.

Graf, T. (2002). Differentiation plasticity of hematopoietic cells. *Blood* **99**(9): 3089–3101.

Gregory, C. J., and Eaves, A. C. (1977). Human marrow cells capable of erythropoietic differentiation in vitro: definition of three erythroid colony responses. *Blood* **49**(6): 855–864.

Hentze, M. W., Muckenthaler, M. U. et al. (2004). Balancing acts: molecular control of mammalian iron metabolism. *Cell* **117**(3): 285–297.

Huang, F. W., Pinkus, J. L. et al. (2005). A mouse model of juvenile hemochromatosis. *J Clin Invest* **115**(8): 2187–2191.

Iscove, N. N., Sieber, F. et al. (1974). Erythroid colony formation in cultures of mouse and human bone marrow: analysis of the requirement for erythropoietin by gel filtration and affinity chromatography on agarose-concanavalin A. *J Cell Physiol* **83**(2): 309–320.

Jacobs, K., Shoemaker, C. et al. (1985). Isolation and characterization of genomic and cDNA clones of human erythropoietin. *Nature* **313**(6005): 806–810.

Jelkmann, W. (2007). Erythropoietin after a century of research: younger than ever. *Eur J Haematol* **78**(3): 183–205.

Kemna, E. H., Tjalsma, H. et al. (2008). Hepcidin: from discovery to differential diagnosis. *Haematologica* **93**(1): 90–97.

Klingmuller, U., Lorenz, U. et al. (1995). Specific recruitment of SH-PTP1 to the erythropoietin receptor causes inactivation of JAK2 and termination of proliferative signals. *Cell* **80**(5): 729–738.

Koury, M. J., and Ponka, P. (2004). New insights into erythropoiesis: the roles of folate, vitamin B12, and iron. *Annu Rev Nutr* **24**: 105–131.

Krause, A., Neitz, S. et al. (2000). LEAP-1, a novel highly disulfide-bonded human peptide, exhibits antimicrobial activity. *FEBS Lett* **480**(2–3): 147–150.

Lin, F. K., Suggs, S. et al. (1985). Cloning and expression of the human erythropoietin gene. *Proc Natl Acad Sci USA* **82**(22): 7580–7584.

Loose, M., and Patient, R. (2006). Global genetic regulatory networks controlling hematopoietic cell fates. *Curr Opin Hematol* **13**(4): 229–236.

Loose, M., Swiers, G. et al. (2007). Transcriptional networks regulating hematopoietic cell fate decisions. *Curr Opin Hematol* **14**(4): 307–314.

Luukkonen, S., and Punnonen, K. (2006). Serum pro-hepcidin concentrations and their responses to oral iron supplementation in healthy subjects manifest considerable inter-individual variation. *Clin Chem Lab Med* **44**(11): 1361–1362.

Matsumoto, A., Masuhara, M. et al. (1997). CIS, a cytokine inducible SH2 protein, is a target of the JAK-STAT5 pathway and modulates STAT5 activation. *Blood* **89**(9): 3148–3154.

Menon, M. P., Fang, J. et al. (2006). Core erythropoietin receptor signals for late erythroblast development. *Blood* **107**(7): 2662–2672.

Miyake, T., Kung, C. K. et al. (1977). Purification of human erythropoietin. *J Biol Chem* **252**(15): 5558–5564.

Nemeth, E., and Ganz, T. (2006). Regulation of iron metabolism by hepcidin. *Annu Rev Nutr* **26**: 323–342.

Nemeth, E., Tuttle, M. S. et al. (2004). Hepcidin regulates cellular iron efflux by binding to ferroportin and inducing its internalisation. *Science* **306**(5704): 2090–2093.

Ney, P. A. (2006). Gene expression during terminal erythroid differentiation. *Curr Opin Hematol* **13**(4): 203–208.

Niederkofler, V., Salie, R. et al. (2005). Hemojuvelin is essential for dietary iron sensing, and its mutation leads to severe iron overload. *J Clin Invest* **115**(8): 2180–2186.

Palis, J., Robertson, S. et al. (1999). Development of erythroid and myeloid progenitors in the yolk sac and embryo proper of the mouse. *Development* **126**(22): 5073–5084.

Papanikolaou, G., Samuels, M. E. et al. (2004). Mutations in HFE2 cause iron overload in chromosome 1q-linked juvenile hemochromatosis. *Nat Genet* **36**(1): 77–82.

Peltola, K. J., Paukku, K. et al. (2004). Pim-1 kinase inhibits STAT5-dependent transcription via its interactions with SOCS1 and SOCS3. *Blood* **103**(10): 3744–3750.

Pietrangelo, A. (2007). Hemochromatosis: an endocrine liver disease. *Hepatology* **46**(4): 1291–1301.

Ponka, P. (1997). Tissue-specific regulation of iron metabolism and heme synthesis: distinct control mechanisms in erythroid cells. *Blood* **89**(1): 1–25.

Richmond, T. D., Chohan, M. et al. (2005). Turning cells red: signal transduction mediated by erythropoietin. *Trends Cell Biol* **15**(3): 146–155.

Roe, M. A., Spinks, C. et al. (2007). Serum prohepcidin concentration: no association with iron absorption in healthy men; and no relationship with iron status in men carrying HFE mutations, hereditary haemochromatosis patients undergoing phlebotomy treatment, or pregnant women. *Br J Nutr* **97**(3): 544–549.

Ross, J., and Li, L. (2006). Recent advances in understanding extrinsic control of hematopoietic stem cell fate. *Curr Opin Hematol* **13**(4): 237–242.

Rossert, J., and Eckardt, K. U. (2005). Erythropoietin receptors: their role beyond erythropoiesis. *Nephrol Dial Transplant* **20**(6): 1025–1028.

Sasaki, A., Yasukawa, H. et al. (2000). CIS3/SOCS-3 suppresses erythropoietin (EPO) signaling by binding the EPO receptor and JAK2. *J Biol Chem* **275**(38): 29338–29347.

Sassa, S., and Nagai, T. (1996). The role of heme in gene expression. *Int J Hematol* **63**(3): 167–178.

Silva, M., Benito, A. et al. (1999). Erythropoietin can induce the expression of bcl-x(L) through Stat5 in erythropoietin-dependent progenitor cell lines. *J Biol Chem* **274**(32): 22165–22169.

Socolovsky, M. (2007). Molecular insights into stress erythropoiesis. *Curr Opin Hematol* **14**(3): 215–224.

Stephenson, J. R., and Axelrad, A. A. (1971). Separation of erythropoietin-sensitive cells from hemopoietic spleen colony-forming stem cells of mouse fetal liver by unit gravity sedimentation. *Blood* **37**(4): 417–427.

Tanno, T., Bhanu, N. V. et al. (2007). High levels of GDF15 in thalassemia suppress expression of the iron regulatory protein hepcidin. *Nat Med* **13**(9): 1096–1101.

Tomosugi, N., Kawabata, H. et al. (2006). Detection of serum hepcidin in renal failure and inflammation by using ProteinChip System. *Blood* **108**(4): 1381–1387.

Ueno, H., and Weissman, I. L. (2006). Clonal analysis of mouse development reveals a polyclonal origin for yolk sac blood islands. *Dev Cell* **11**(4): 519–533.

Valore, E. V., and Ganz, T. (2008). Posttranslational processing of hepcidin in human hepatocytes is mediated by the prohormone convertase furin. *Blood Cells Mol Dis* **40**(1): 132–138.

Vogeli, K. M., Jin, S. W. et al. (2006). A common progenitor for haematopoietic and endothelial lineages in the zebrafish gastrula. *Nature* **443**(7109): 337–339.

Walrafen, P., Verdier, F. et al. (2005). Both proteasomes and lysosomes degrade the activated erythropoietin receptor. *Blood* **105**(2): 600–608.

Wang, R. H., Li, C. et al. (2005). A role of SMAD4 in iron metabolism through the positive regulation of hepcidin expression. *Cell Metab* **2**(6): 399–409.

Wickrema, A., Krantz, S. B. et al. (1992). Differentiation and erythropoietin receptor gene expression in human erythroid progenitor cells. *Blood* **80**(8): 1940–1949.

Witthuhn, B. A., Quelle, F. W. et al. (1993). JAK2 associates with the erythropoietin receptor and is tyrosine phosphorylated and activated following stimulation with erythropoietin. *Cell* **74**(2): 227–236.

Wojchowski, D. M., Gregory, R. C. et al. (1999). Signal transduction in the erythropoietin receptor system. *Exp Cell Res* **253**(1): 143–156.

Wu, H., Liu, X. et al. (1995). Generation of committed erythroid BFU-E and CFU-E progenitors does not require erythropoietin or the erythropoietin receptor. *Cell* **83**(1): 59–67.

Yu, P. B., Hong, C. C. et al. (2007). Dorsomorphin inhibits BMP signals required for embryogenesis and iron metabolism. *Nat Chem Biol.*

Further reading

Wojchowski, D. M., Menon, M. P. et al. (2006). Erythropoietin-dependent erythropoiesis: New insights and questions. *Blood Cells Mol Dis* **36**(2): 232–238.

CHAPTER 3

Extra-Hematopoietic Action of Erythropoietin

ZHEQING CAI[1] AND GREGG L. SEMENZA[2,3,4]

[1]Division of Cardiology, Department of Medicine, The Johns Hopkins University School of Medicine, Baltimore, MD 21205, USA
[2]Vascular Program, Institute for Cell Engineering, The Johns Hopkins University School of Medicine, Baltimore, MD 21205, USA
[3]McKusick-Nathans Institute of Genetic Medicine, The Johns Hopkins University School of Medicine, Baltimore, MD 21205, USA
[4]Departments of Pediatrics Medicine, Oncology and Radiation Oncology, The Johns Hopkins University School of Medicine, Baltimore, MD 21205, USA

Contents

I.	Introduction	27
II.	The EPO receptor (EPOR)	27
III.	Regulation of EPOR expression	28
IV.	Action of EPO	29
V.	Mechanisms of EPO action	30
VI.	Summary	31
	References	31

I. INTRODUCTION

Erythropoietin (EPO) is primarily produced in the kidney. In response to hypoxia, renal tubular and interstitial cells synthesize EPO and release it into blood and then EPO is distributed into various tissues through blood flow. Oxygen-regulated transcription of the *EPO* gene is mediated by the hypoxia-inducible transcription factors HIF-1 (Semenza and Wang, 1992; Cai et al., 2003) and HIF-2 (Rankin et al., 2007). In response to hypoxia, plasma EPO levels increase in parallel to renal EPO levels with one hour lag time (Schuster et al., 1987). The biological effects of EPO are mediated by binding to the EPO receptor (EPOR). EPOR was first suspected in spleen cells infected with the anemia strain of Friend virus (Krantz and Goldwasser, 1984). Later Sawyer et al. (1987) identified EPOR in Friend virus-infected erythroid cells. EPOR is expressed in almost all tissues, including brain, heart, kidney and liver. EPO not only promotes hematopoiesis but also mediates protection against ischemia and other forms of tissue injury, promotes neovascularization and exerts anti-inflammatory effects.

In addition to endocrine production of EPO by the kidney, hypoxic tissues also produce EPO as an autocrine or paracrine factor. Locally-produced EPO differs from plasma EPO in size and activity. For example, brain EPO is smaller and more active (Masuda et al., 1994). The differences may be caused by different degrees of sialylation. Moreover, EPO production is tissue-specific. In response to hypoxia, EPO mRNA peaks early but decays faster in the kidney than in the brain (Chikuma et al., 2000). In the heart, there is no clear evidence of EPO production. However, EPO mediates potent cardioprotective effects (Cai et al., 2003; Calvillo et al., 2003; Parsa et al., 2003). A large body of experimental data indicates that EPO acts as an endocrine factor (hormone) to protect the body from injury. In addition, EPO functions as an autocrine and paracrine factor to prevent cell death that would otherwise be induced by ischemia/hypoxia. Here we will focus on the action of EPO as a hormone.

II. THE EPO RECEPTOR (EPOR)

EPO specifically binds to its receptor on plasma membrane, leading to activation of anti-apoptotic signaling pathways. In addition to the hematopoietic system, EPOR is expressed in a wide variety of tissues in animal models and in humans (Table 3.1).

A. Heart and Vasculature

Human umbilical vein endothelial cells possess abundant EPOR. They were the first non-hematopoietic cells shown to express EPOR on their cell surface (Anagnostou et al., 1990). Both human umbilical cord vascular endothelium and placenta had strong immunostaining with antibody against EPOR (Anagnostou et al., 1994). Abundant EPOR has been found in endothelial cells and smooth muscle

TABLE 3.1 Distribution of EPOR and Action of EPO

Organs	heart	brain	kidney	liver	lung	stomach
EPOR in Cell types	CM: + EC: + FB: +	NE: + AS: + EC: +	TU: + ME: + PO: +	+	+	+
EPO Action	AA AI NV	AA AI SCD	AA	AA	AA RE	GMCP
Signaling pathways	PI3K Erk1/2	PI3K Erk1/2 STAT-5 NF-κB	PI3K			

Cell types CM: cardiomyocytes, EC: endothelial cells, FB: fibroblasts, NE: neurons, AS: astrocytes, TU: tubular cells, ME: mesangial cells, PO: podocytes.
Actions AA: anti-apoptosis, AI: anti-inflammation, NV: neovascularization, SCD: stem cell differention, RE: recruitment of endothelial progenitor cells, GMCP: gastric mucosal cell proliferation.

cells of blood vessels in many tissues (Genc et al., 2004; Morakkabati et al., 1996). In rat hearts, EPOR is predominantly expressed in interstitial cells, including endothelial cells and fibroblasts, but weak expression of EPOR is found in cardiomyocytes (van der Meer et al., 2004). In contrast to rat hearts, human hearts show clear EPOR in both cardiomyocytes and endothelial cells (Depping et al., 2005).

B. Brain

As in the cardiovascular system, the nervous system also expresses EPOR. In cultured rat cortical neurons, EPOR protein was observed by immunostaining and its mRNA was identified by reverse transcriptase-polymerase chain reaction (RT-PCR) (Morishita et al., 1997). Moreover, EPOR mRNA expression has been demonstrated in cultured human neurons, astrocytes and microglia (Nagai et al., 2001). Interestingly, rat oligodendrocytes express EPOR, but human oligodendrocytes do not (Sugawa et al., 2002). The differences may be due to different culture conditions or may reflect species-specific regulatory mechanisms. Mouse EPOR has been detected in the hippocampus, capsula interna, cortex and midbrain areas, as determined by binding of radioiodinated EPO. EPOR mRNA expression in these areas has been confirmed by RT-PCR (Digicaylioglu et al., 1995). In adult human brain, EPOR mRNA has been detected in biopsies of the temporal cortex, hippocampus, cerebellum and amygdala (Marti et al., 1996). However, only weak expression of EPOR was observed in neurons and astrocytes (Juul et al., 1999). Strong staining of EPOR was shown in endothelial cells throughout the brain (Brines et al., 2000). In rat peripheral nerve, EPOR is present in cell bodies of the dorsal root ganglia (DRG), axons and Schwann cells (Campana and Myers, 2001).

C. Kidney and Other Organs

Abundant EPOR is expressed in the kidney. EPOR mRNA has been detected in renal cortex, medulla and papilla in rat and human by RT-PCR. Various types of renal cells, including mesangial, proximal tubular and medullary collecting duct cells, contain the EPOR mRNA. These cells show a high to intermediate affinity for binding of ^{125}I-EPO (Westenfelder et al., 1999). EPOR is also expressed in fetal liver, but its levels decrease with liver maturation (Juul et al., 1998). EPOR is found in other organs, including the lung (Wu et al., 2006), uterus and ovary (Yasuda et al., 2001) and stomach (Okada et al., 1996).

III. REGULATION OF EPOR EXPRESSION

As in the case of renal EPO expression, EPOR expression is regulated by tissue oxygen levels, an effect that is mediated by the transcription factor HIF-1 (Manalo et al., 2005). Exposure of cultured hippocampal neurons to hypoxia increases EPOR mRNA levels. The increased expression of EPOR mRNA is also found in the hippocampus of rats subjected to hypoxia (Lewczuk et al., 2000). EPOR mRNA regulation by hypoxia is not limited to the hippocampus (Marti et al., 1996). Moreover, brain ischemia or hemorrhage has similar effects on EPOR mRNA expression as hypoxia (Bernaudin et al., 1999). In humans, ischemic stroke induces EPOR expression in neurons, astrocytes and endothelial cells (Sirén et al., 2001b). The change may be related to oxygen deprivation. In failing hearts resulting from ischemia and reperfusion injury, EPOR protein expression is upregulated in the myocardium (van der Meer et al., 2004). However, there was no change in EPOR expression in peripheral nerve after injury (Campana and Myers, 2001).

IV. ACTION OF EPO

EPO promotes cell growth, differentiation, migration and survival. Among the acute effects of EPO, anti-apoptosis is predominant. Apoptosis is one of the major forms of cell death. EPO prevents apoptosis following injury in various organs including heart, brain, kidney and liver. The late phase effects of EPO include anti-inflammation, recruitment of EPC and neovascularization.

A. Anti-Apoptosis

A large body of evidence shows that EPO is a survival factor that increases cellular tolerance to injury. Pretreatment of rats with recombinant human (rh) EPO increases the recovery of cardiac function and decreases cell apoptosis after ischemia-reperfusion (Cai et al., 2003). In myocardial infarction models, rhEPO given at the time of coronary artery ligation or after reperfusion also improves cardiac function and reduces infarct size (Calvillo et al., 2003; Parsa et al., 2003). Although the protective effect is generated by administration of pharmacologic doses of exogenous EPO, endogenous EPO may also play a role in preventing apoptosis in the heart following ischemia-reperfusion. When the EPOR gene was deleted in the non-hematopoietic cells, increased infarct size and apoptosis were found in the myocardium after ischemia-reperfusion injury (Tada et al., 2006).

The protective effect of EPO is not limited to the heart. Many studies have reported that rhEPO protects the brain in various injury models. Pretreatment with rhEPO reduces cell death in cultured hippocampal neurons at the onset of hypoxia (Lewczuk et al., 2000) and increases survival of cultured rat cortical neurons after oxygen and glucose deprivation (Ruscher et al., 2002). Moreover, the protective effect of rhEPO was observed in gerbil brain exposed to transient bilateral carotid occlusion, even when it was administered systemically after reperfusion (Calapai et al., 2000). Brines et al. (2000) reported that rhEPO can penetrate the blood–brain barrier through blood endothelial cells and that systemic administration of EPO before or up to 6 h after middle cerebral artery occlusion reduces infarct volume by approximately 50–75%. Furthermore, rhEPO ameliorates brain damage caused by concussive injury, experimental autoimmune encephalomyelitis and kainite (Brines et al., 2000). There are two types of EPOR in rat brain endothelial cells. One of them may transport EPO through the capillary endothelium so that it enters the brain, probably by endocytosis (Yamaji et al., 1996).

EPO-mediated protection has been reported in many other organs. The EPO derivative darbepoetin alfa protects pig tubular cells and mouse mesangial cells in the kidney against pro-apoptotic prostaglandin D2 synthase and hypoxia (Fishbane et al., 2004). Importantly, the compound reduces apoptosis in the podocytes of glomeruli and improves proteinuria in mice with anti-glomerular antibody-induced glomerulonephritis (Echigoya et al., 2005; Logar et al., 2007). Furthermore, continuous activation of EPOR inhibits mesangial expansion and ameliorates albuminuria in diabetic kidney (Menne et al., 2007). Treatment with rhEPO also protects the eyes, liver and lungs against ischemia-reperfusion injury (Junk et al., 2002; Abdelrahman et al., 2004; Wu et al., 2006).

B. Anti-Inflammation

Myocardial reperfusion is followed by an inflammatory response, which worsens the apoptosis initiated by ischemia. The inflammatory process in the heart is induced by activation of complement and cytokines and the release of reactive oxygen species (ROS) (Frangogiannis et al., 2000). Cytokine activation and increased ROS levels are implicated in cardiac dysfunction after myocardial infarction (Entman et al., 1991; Jordan et al., 1999). EPO pretreatment decreases ROS production in cardiomyocytes and inhibits polymorphonuclear neutrophil infiltration following ischemia-reperfusion (Rui et al., 2005). Smith et al. (1988) reported that the extent of polymorphonuclear neutrophil accumulation in the myocardium is directly related to the severity of ischemia-reperfusion injury. EPO has been found to reduce the production of pro-inflammatory cytokines, including tumor necrosis factor-α, interleukin-6 and intercellular adhesion molecule-1, and promote the expression of anti-inflammatory cytokine interleukin-10 after ischemia-reperfusion (Liu et al., 2006; Li et al., 2006). EPO anti-inflammatory effects have also been demonstrated in the brain. In a mouse closed-head injury model, EPO decreased focal inflammation and neuronal death and improved recovery of cognitive function (Yatsiv et al., 2005). Systemic EPO pretreatment inhibits perihematomal inflammation in rat brain (Lee et al., 2006).

C. Recruitment of EPC

EPO is a potent factor in recruiting endothelial progenitor cells (EPC) to injured tissue (Heeschen et al., 2003). These cells are derived from bone marrow stem cells but active in peripheral tissues, possessing the ability to mature into endothelial cells. rhEPO significantly increases stem and progenitor cells in bone marrow and EPC in peripheral blood. In patients with myocardial infarction, serum EPO levels are associated with circulating EPC number and function (Shintani et al., 2001). Moreover, the circulating EPC levels are inversely related to the risk of death in patients with coronary artery disease (Werner et al., 2005). In the mice with deleted EPOR in non-hematopoietic cells, the mobilization of EPC and their recruitment to the pulmonary endothelium were impaired (Satoh et al., 2006). Hypoxia caused accelerated pulmonary hypertension in these mice, indicating that EPO inhibits pathogenic hypoxia-induced remodeling of pulmonary arterioles.

D. Neovascularization

Systemic EPO treatment promotes neovascularization in rodent ischemic hind limb (Nakano et al., 2007). The neovascularization occurs by angiogenesis and vasculogenesis. Angiogenesis is the sprouting of new vessels from existing ones, whereas vasculogenesis refers to the *de novo* formation of blood vessels from EPC. EPO may promote both forms of neovascularization. Increased capillary density was observed in hearts subjected to myocardial infarction and followed by EPO treatment 3 weeks later (van der Meer et al., 2005). EPO has also been shown to increase the levels of vascular endothelial growth factor (VEGF) in the brain and the density of cerebral microvessels in the stroke boundary (Wang et al., 2004). The role of EPC in neovascularization is further confirmed in transgene-rescued EpoR-knockout mice, which express EPOR exclusively in hematopoietic cells. These mice demonstrate decreased capillary density in skeletal muscles due to defective activation of the VEGF/VEGFR2 axis (Nakano et al., 2007).

V. MECHANISMS OF EPO ACTION

EPOR belongs to the cytokine receptor superfamily. When EPO binds to its receptor, it induces dimerization of the receptor and autophosphorylation of Janus-tyrosine-kinase-2 (Jak2). The receptor activation stimulates several downstream signaling pathways (Figure 3.1), including phosphatidylinositol-3-kinase (PI3K), mitogen-activated protein kinases (MAPKs) and members of the signal transducer and activator of transcription (STAT) family. Protein tyrosine phosphatase terminates EPO receptor activation.

A. PI3K

Many studies have reported that the PI3K pathway is involved in EPO-mediated anti-apoptosis. In isolated rat hearts, rhEPO pretreatment stimulates the association of EPOR with PI3K and activates the serine-threonine protein kinase Akt. The inhibition of PI3K with wortmannin blocks the effects of EPO on the recovery of cardiac function and cell apoptosis after ischemia-reperfusion (Cai and Semenza, 2004). When EPO is administered at the point of reperfusion in rat and dog myocardial infarction models, the activation of PI3K is also required for EPO-mediated reduction in infarct size (Bullard et al., 2005; Hirata et al., 2005). Moreover, activation of the PI3K/Akt pathway is also implicated in neuron protection against apoptosis. Systemic administration of rhEPO dramatically reduces brain infarct size and

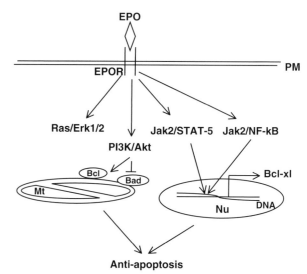

FIGURE 3.1 Signaling pathways activated by EPO. EPO binds to its receptor at the plasma membrane, leading to activation of survival signaling pathways, including PI3K/Akt, Ras/Erk1/2 and Jak2/STAT-5. PI3K/Akt and Ras/Erk1/2 phosphorylate Bcl-2 family proteins and thereby inhibit apoptosis. Activated STAT-5 increases gene expression of the anti-apoptotic protein Bcl-xL. PI3K, phosphatidylinositol-3-kinase; Akt or PKB, protein kinase B; Jak2, Janus-tyrosine kinase-2; STAT-5, signal transducer and activator of transcription-5; NF-κB; nuclear factor-κ of B cells; Erk1/2, extracellular-regulated kinase 1/2; Ras, small GTPase; PM, plasma membrane; Nu, nucleus; Mt, mitochondria.

cell apoptosis through Akt activation after middle cerebral artery occlusion (Sirén et al., 2001a). Activated Akt phosphorylates Bcl-2 family proteins, leading to inactivation of pro-apoptotic Bad and activation of anti-apoptotic Bcl-xL (Kretz et al., 2005; Zhande and Karsan, 2007). Moreover, PI3K inhibition abolishes neurogenin 1 gene expression and cell proliferation and differentiation in adult neural progenitor cells after rhEPO treatment (Wang et al., 2006a). PI3K signaling pathways may be involved in neural progenitor cell migration. rhEPO increases endothelial cell survival through PI3K activation (Zhande and Karsan, 2007) and the stimulated endothelial cells secrete matrix metalloproteinase 2 and 9 to promote neural progenitor cell migration (Wang et al., 2006b).

B. Erk1/2

Extracellular-regulated kinase (Erk) 1/2 is implicated in EPO-mediated cardioprotection against ischemia-reperfusion injury in several models. In rat and rabbit models, rhEPO activates Erk1/2 and induces cardioprotection (Parsa et al., 2003; van der Meer et al., 2004). The EPO-mediated activation of Erk1/2 is blocked by its inhibitors in isolated neonatal rabbit and rat hearts (Shi et al., 2004; Bullard et al., 2005). However, it is not clear whether the Erk1/2 signaling pathway is required for EPO-mediated protection in the *in vivo* animal models. Erk1/2 may also be involved in neuronal protection. rhEPO activates Erk1/2 in the developing rat brain and prevents N-methyl-D-aspartate receptor antagonist neurotoxicity (Dzietko et al., 2004). Moreover, constitutive overexpression of human EPO protects rat brain against focal ischemia through activation of Erk1/2 (Kilic et al., 2005). The EPO-mediated Erk1/2 activation may be required for the production of neurotrophic factors in the brain which promote cell survival and differentiation (Lee et al., 2004; Park et al., 2006).

C. Stat-5

Although an increasing body of evidence shows that EPO-mediated PI3K and Erk1/2 signaling pathways are involved in neuronal protection in the brain, STAT-5 was first found to be activated after EPO stimulation and its activation leads to gene expression of the anti-apoptotic Bcl-xl protein (Silva et al., 1999). In a transient global cerebral ischemia model, rhEPO decreased neuronal damage through STAT-5 activation and Bcl-xl gene expression in the CA1 region of the hippocampus (Zhang et al., 2007). Like STAT-5, NF-κB acts as downstream of Jak2 and the Jak2/NF-κB signaling pathway plays an important role in EPO-mediated protection in the neurons subjected to N-methyl-D-aspartate and excessive nitric oxide (Digicaylioglu and Lipton, 2001).

VI. SUMMARY

Plasma EPO is increased in response to hypoxia. The increased EPO is mainly produced by the kidney, although many tissues synthesize EPO that acts locally. EPOR is extensively expressed in various tissues, including endothelial cells, cardiomyocytes and neuronal cells. rhEPO has been shown to protect the heart and the brain against ischemia or ischemia-reperfusion injury through anti-apoptosis, anti-inflammation, EPC recruitment and neovascularization. Moreover, recent studies, by using transgenic EPOR rescued mice, indicate that endogenous EPO plays a role in preventing tissue injury from hypoxia or ischemia-reperfusion. The protective effect of EPO is likely mediated by survival signaling pathways including PI3K, Erk1/2, and STAT-5.

References

Abdelrahman, M., Sharples, E. J., McDonald, M. C. et al. (2004). Erythropoietin attenuates the tissue injury associated with hemorrhagic shock and myocardial ischemia. *Shock* **22**: 63–69.

Anagnostou, A., Lee, E. S., Kessimian, N., Levinson, R., and Steiner, M. (1990). Erythropoietin has a mitogenic and positive chemotactic effect on endothelial cells. *Proc Natl Acad Sci USA* **87**: 5978–5982.

Anagnostou, A., Liu, Z., Steiner, M. et al. (1994). Erythropoietin receptor mRNA expression in human endothelial cells. *Proc Natl Acad Sci* **91**: 3974–3978.

Bernaudin, M., Marti, H. H., Roussel, S. et al. (1999). A potential role for erythropoietin in focal permanent cerebral ischemia in mice. *J Cereb Blood Flow Metab* **19**: 643–651.

Brines, M. L., Ghezzi, P., Keenan, S. et al. (2000). Erythropoietin crosses the blood–brain barrier to protect against experimental brain injury. *Proc Natl Acad Sci USA* **97**: 10526–10531.

Bullard, A. J., Govewalla, P., and Yellon, D. M. (2005). Erythropoietin protects the myocardium against reperfusion injury in vitro and in vivo. *Basic Res Cardiol* **100**: 397–403.

Cai, Z., Manalo, D. J., Wei, G. et al. (2003). Hearts from rodents exposed to intermittent hypoxia or erythropoietin are protected against ischemia-reperfusion injury. *Circulation* **108**: 79–85.

Cai, Z., and Semenza, G. L. (2004). Phosphatidylinositol-3-kinase signaling is required for erythropoietin-mediated acute protection against myocardial ischemia/reperfusion injury. *Circulation* **109**: 2050–2053.

Calapai, G., Marciano, M. C., Corica, F. et al. (2000). Erythropoietin protects against brain ischemic injury by inhibition of nitric oxide formation. *Eur J Pharmacol* **401**: 349–356.

Calvillo, L., Latini, R., Kajstura, J. et al. (2003). Recombinant human erythropoietin protects the myocardium from ischemia-reperfusion injury and promotes beneficial remodelling. *Proc Natl Acad Sci USA* **100**: 4802–4806.

Campana, W. M., and Myers, R. R. (2001). Erythropoietin and erythropoietin receptors in the peripheral nervous system: changes after nerve injury. *FASEB J* **15**: 1804–1806.

Chikuma, M., Masuda, S., Kobayashi, T., Nagao, M., and Sasaki, R. (2000). Tissue-specific regulation of erythropoietin production in the murine kidney, brain and uterus. *Am J Physiol Endocrinol Metab* **279**: E1242–E1248.

Depping, R., Kawakami, K., Ocker, H. et al. (2005). Expression of the erythropoietin receptor in human heart. *J Thorac Cardiovasc Surg* **130**: 877–878.

Digicaylioglu, M., Bichet, S., Marti, H. H. et al. (1995). Localization of specific erythropoietin binding sites in defined areas of the mouse brain. *Proc Natl Acad Sci USA* **92**: 3717–3720.

Digicaylioglu, M., and Lipton, S. A. (2001). Erythropoietin-mediated neuroprotection involves cross-talk between Jak2 and NF-kappaB signalling cascades. *Nature* **412**: 641–647.

Dzietko, M., Felderhoff-Mueser, U., Sifringer, M. et al. (2004). Erythropoietin protects the developing brain against N-methyl-D-aspartate receptor antagonist neurotoxicity. *Neurobiol Dis* **15**: 177–187.

Echigoya, M. H., Obikane, K., Nakashima, T., and Sasaki, S. (2005). Glomerular localization of erythropoietin receptor mRNA and protein in neonatal and mature mouse kidney. *Nephron Exp Nephrol* **100**: e21–e29.

Entman, M. L., Michael, L., Rossen, R. D. et al. (1991). Inflammation in the course of early myocardial ischemia. *FASEB J* **5**: 2529–2537.

Fishbane, S., Ragolia, L., Palaia, T., Johnson, B., Elzein, H., and Maesaka, J. K. (2004). Cytoprotection by darbepoetin/epoetin alfa in pig tubular and mouse mesangial cells. *Kidney Int* **65**: 452–458.

Frangogiannis, N. G., Smith, C. W., and Entman, M. L. (2000). The inflammatory response in myocardial infarction. *Cardiovasc Res* **53**: 31–47.

Genc, S., Koroglu, T. F., and Genc, K. (2004). Erythropoietin and the nervous system. *Brain Res* **1000**: 19–31.

Heeschen, C., Aicher, A., Lehmann, R. et al. (2003). Erythropoietin is a potent physiologic stimulus for endothelial progenitor cell mobilization. *Blood* **102**: 1340–1346.

Hirata, A., Minamino, T., Asanuma, H. et al. (2005). Erythropoietin just before reperfusion reduces both lethal arrhythmias and infarct size via the phosphatidylinositol-3 kinase-dependent pathway in canine hearts. *Cardiovasc Drugs Ther* **19**: 33–40.

Jordan, J. E., Zhao, Z. Q., and Vinten-Johansen, J. (1999). The role of neutrophils in myocardial ischemia-reperfusion injury. *Cardiovasc Res* **43**: 860–878.

Junk, A. K., Mammis, A., Savitz, S. I. et al. (2002). Erythropoietin administration protects retinal neurons from acute ischemia-reperfusion injury. *Proc Natl Acad Sci USA* **99**: 10659–10664.

Juul, S. E., Yachnis, A., Rojiani, A. M., and Christensen, R. D. (1999). Immunohistochemical localization of erythropoietin and its receptor in the developing human brain. *Pediatr Dev Pathol* **2**: 148–158.

Juul, S. E., Yachnis, A. T., and Christensen, R. D. (1998). Tissue distribution of erythropoietin and erythropoietin receptor in the developing human fetus. *Early Hum Dev* **52**: 235–249.

Kilic, E., Kilic, U., Soliz, J., Bassetti, C. L., Gassmann, M., and Hermann, D. M. (2005). Brain-derived erythropoietin protects from focal cerebral ischemia by dual activation of ERK-1/-2 and Akt pathways. *FASEB J* **19**: 2026–2028.

Krantz, S. B., and Goldwasser, E. (1984). Specific binding of erythropoietin to spleen cells infected with the anemia strain of Friend virus. *Proc Natl Acad Sci USA* **81**: 7574–7578.

Kretz, A., Happold, C. J., Marticke, J. K., and Isenmann, S. (2005). Erythropoietin promotes regeneration of adult CNS neurons via Jak2/Stat3 and PI3K/AKT pathway activation. 1. *Mol Cell Neurosci* **29**: 569–579.

Lee, S. M., Nguyen, T. H., Park, M. H. et al. (2004). EPO receptor-mediated ERK kinase and NF-kappaB activation in erythropoietin-promoted differentiation of astrocytes. *Biochem Biophys Res Commun* **320**: 1087–1095.

Lee, S. T., Chu, K., Sinn, D. I. et al. (2006). Erythropoietin reduces perihematomal inflammation and cell death with eNOS and STAT3 activations in experimental intracerebral hemorrhage. *J Neurochem* **96**: 1728–1739.

Lewczuk, P., Hasselblatt, M., Kamrowski-Kruck, H. et al. (2000). Survival of hippocampal neurons in culture upon hypoxia: effect of erythropoietin. *Neuroreport* **11**: 3485–3488.

Li, Y., Takemura, G., Okada, H. et al. (2006). Reduction of inflammatory cytokine expression and oxidative damage by erythropoietin in chronic heart failure. *Cardiovasc Res* **71**: 684–694.

Liu, X., Xie, W., Liu, P. et al. (2006). Mechanism of the cardioprotection of rhEPO pretreatment on suppressing the inflammatory response in ischemia-reperfusion. *Life Sci* **78**: 2255–2264.

Logar, C. M., Brinkkoetter, P. T., Krofft, R. D., Pippin, J. W., and Shankland, S. J. (2007). Darbepoetin alfa protects podocytes from apoptosis in vitro and in vivo. *Kidney Int* **72**: 489–498.

Manalo, D. J., Rowan, A., Lavoie, T. et al. (2005). Transcriptional regulation of vascular endothelial cell responses to hypoxia by HIF-1. *Blood* **105**: 659–669.

Marti, H. H., Wenger, R. H., Rivas, L. A. et al. (1996). Erythropoietin gene expression in human, monkey and human brain. *Eur J Neurosci* **8**: 666–676.

Masuda, S., Okano, M., Yamagishi, K., Nagao, M., Ueda, M., and Sasaki, R. (1994). A novel site of erythropoietin production: oxygen-dependent production in cultured rat astrocytes. *J Biol Chem* **269**: 19488–19493.

Menne, J., Park, J. K., Shushakova, N., Mengel, M., Meier, M., and Fliser, D. (2007). The continuous erythropoietin receptor activator affects different pathways of diabetic renal injury. *J Am Soc Nephrol* **18**: 2046–2053.

Morakkabati, N., Gollnick, F., Meyer, R., Fandrey, J., and Jelkmann, W. (1996). Erythropoietin induces Ca2+ mobilization and contraction in rat mesangial and aortic smooth muscle cultures. *Exp Hematol* **24**: 392–397.

Morishita, E., Masuda, S., Nagao, M., Yasuda, Y., and Sasaki, R. (1997). Erythropoietin receptor is expressed in rat hippocampal and cerebral cortical neurons, and erythropoietin prevents in vitro glutamate-induced neuronal death. *Neuroscience* **76**: 105–116.

Nagai, A., Nakagawa, E., Choi, H. B., Hatori, K., Kobayashi, S., and Kim, S. U. (2001). Erythropoietin and erythropoietin receptors in human CNS neurons, astrocytes, microglia, and oligodendrocytes grown in culture. *J Neuropathol Exp Neurol* **60**: 386–392.

Nakano, M., Satoh, K., Fukumoto, Y. et al. (2007). Important role of erythropoietin receptor to promote VEGF expression and angiogenesis in peripheral ischemia in mice. *Circ Res* **100**: 662–669.

Okada, A., Kinoshita, Y., Maekawa, T. et al. (1996). Erythropoietin stimulates proliferation of rat-cultured gastric mucosal cells. *Digestion* **57**: 328–332.

Park, M. H., Lee, S. M., Lee, J. W. et al. (2006). ERK-mediated production of neurotrophic factors by astrocytes promotes neuronal stem cell differentiation by erythropoietin. *Biochem Biophys Res Commun* **339**: 1021–1028.

Parsa, C. J., Matsumoto, A., Kim, J. et al. (2003). A novel protective effect of erythropoietin in the infarcted heart. *J Clin Invest* **112**: 999–1007.

Rankin, E. B., Biju, M. P., Liu, Q. et al. (2007). Hypoxia-inducible factor-2 (HIF-2) regulates hepatic erythropoietin in vivo. *J Clin Invest* **117**: 1068–1077.

Rui, T., Feng, Q., Lei, M. et al. (2005). Erythropoietin prevents the acute myocardial inflammatory response induced by ischemia/reperfusion via induction of AP-1. *Cardiovasc Res* **65**: 719–727.

Ruscher, K., Freyer, D., Karsch, M. et al. (2002). Erythropoietin is a paracrine mediator of ischemic tolerance in the brain: evidence from an in vitro model. *J Neurosci* **22**: 10291–10301.

Satoh, K., Kagaya, Y., Nakano, M. et al. (2006). Important role of endogenous erythropoietin system in recruitment of endothelial progenitor cells in hypoxia-induced pulmonary hypertension in mice. *Circulation* **113**: 1442–1450.

Sawyer, S. T., Krantz, S. B., and Luna, J. (1987). Identification of the receptor for erythropoietin by cross-linking to Friend virus-infected erythroid cells. *Proc Natl Acad Sci USA* **84**: 3690–3694.

Schuster, S. J., Wilson, J. H., Erslev, A. J., and Caro, J. (1987). Physiologic regulation and tissue localization of renal erythropoietin messenger RNA. *Blood* **70**: 316–318.

Semenza, G. L., and Wang, G. L. (1992). A nuclear factor induced by hypoxia via de novo protein synthesis binds to the human erythropoietin gene enhancer at a site required for transcriptional activation. *Mol Cell Biol* **12**: 5447–5454.

Shi, Y., Rafiee, P., Su, J., Pritchard, K. A., Tweddell, J. S., and Baker, J. E. (2004). Acute cardioprotective effects of erythropoietin in infant rabbits are mediated by activation of protein kinases and potassium channels. *Basic Res Cardiol* **99**: 173–182.

Shintani, S., Murohara, T., Ikeda, H. et al. (2001). Mobilization of endothelial progenitor cells in patients with acute myocardial infarction. *Circulation* **103**: 2776–2779.

Silva, M., Benito, A., Sanz, C. et al. (1999). Erythropoietin can induce the expression of bcl-x(L) through Stat5 in erythropoietin-dependent progenitor cell lines. *J Biol Chem* **274**: 22165–22169.

Sirén, A. L., Fratelli, M., Brines, M. et al. (2001a). Erythropoietin prevents neuronal apoptosis after cerebral ischemia and metabolic stress. *Proc Natl Acad Sci USA* **98**: 4044–4049.

Sirén, A. L., Knerlich, F., Poser, W., Gleiter, C. H., Bruck, W., and Ehrenreich, H. (2001b). Erythropoietin and erythropoietin receptor in human ischemic/hypoxic brain. *Acta Neuropathol* **101**: 271–276.

Smith, E. F., Egan, J. W., Bugelski, P. J., Hillegass, L. M., Hill, D. E., and Griswold, D. E. (1988). Temporal relation between neutrophil accumulation and myocardial reperfusion injury. *Am J Physiol* **255**: H1060–H1068.

Sugawa, M., Sakurai, Y., Ishikawa-Ieda, Y., Suzuki, H., and Asou, H. (2002). Effects of erythropoietin on glial cell development; oligodendrocyte maturation and astrocyte proliferation. *Neurosci Res* **44**: 391–403.

Tada, H., Kagaya, Y., Takeda, M. et al. (2006). Endogenous erythropoietin system in non-hematopoietic lineage cells plays a protective role in myocardial ischemia/reperfusion. *Cardiovasc Res* **71**: 466–477.

van der Meer, P., Lipsic, E., Henning, R. H. et al. (2004). Erythropoietin improves left ventricular function and coronary flow in an experimental model of ischemia-reperfusion injury. *Eur J Heart Fail* **6**: 853–859.

van der Meer, P., Lipsic, E., Henning, R. H. et al. (2005). Erythropoietin induces neovascularization and improves cardiac function in rats with heart failure after myocardial infarction. *J Am Coll Cardiol* **46**: 125–133.

Wang, L., Zhang, Z., Wang, Y., Zhang, R., and Chopp, M. (2004). Treatment of stroke with erythropoietin enhances neurogenesis and angiogenesis and improves neurological function in rats. *Stroke* **35**: 1732–1737.

Wang, L., Zhang, Z. G., Zhang, R. L. et al. (2006a). Neurogenin 1 mediates erythropoietin enhanced differentiation of adult neural progenitor cells. *J Cereb Blood Flow Metab* **26**: 556–564.

Wang, L., Zhang, Z. G., Zhang, R. L. et al. (2006b). Matrix metalloproteinase 2 (MMP2) and MMP9 secreted by erythropoietin-activated endothelial cells promote neural progenitor cell migration. *J Neurosci* **26**: 5996–6003.

Werner, N., Kosiol, S., Schiegl, T. et al. (2005). Circulating endothelial progenitor cells and cardiovascular outcomes. *N Engl J Med* **353**: 999–1007.

Westenfelder, C., Biddle, D. L., and Baranowski, R. L. (1999). Human, rat, and mouse kidney cells express functional erythropoietin receptors. *Kidney Int* **55**: 808–820.

Wu, H., Ren, B., Zhu, J. et al. (2006). Pretreatment with recombined human erythropoietin attenuates ischemia-reperfusion-induced lung injury in rats. *Eur J Cardiothorac Surg* **29**: 902–907.

Yamaji, R., Okada, T., Moriya, M. et al. (1996). Brain capillary endothelial cells express two forms of erythropoietin receptor mRNA. *Eur J Biochem* **239**: 494–500.

Yasuda, Y., Fujita, Y., Musha, T. et al. (2001). Expression of erythropoietin in human female reproductive organs. *Ital J Anat Embryol* **106**(2 Suppl 2): 215–222.

Yatsiv, I., Grigoriadis, N., Simeonidou, C. et al. (2005). Erythropoietin is neuroprotective, improves functional recovery, and reduces neuronal apoptosis and inflammation in a rodent model of experimental closed head injury. *FASEB J* **19**: 1701–1703.

Zhande, R., and Karsan, A. (2007). Erythropoietin promotes survival of primary human endothelial cells through PI3K-dependent, NF-kappaB-independent upregulation of Bcl-xL. *Am J Physiol Heart Circ Physiol* **292**: H2467–H2474.

Zhang, F., Wang, S., Cao, G., Gao, Y., and Chen, J. (2007). Signal transducers and activators of transcription 5 contributes to erythropoietin-mediated neuroprotection against hippocampal neuronal death after transient global cerebral ischemia. *Neurobiol Dis* **25**: 45–53.

Further reading

Fantacci, M., Bianciardi, P., Caretti, A. et al. (2006). Carbamylated erythropoietin ameliorates the metabolic stress induced in vivo by severe chronic hypoxia. *Proc Natl Acad Sci USA* **103**: 17531–17536.

Lipsic, E., Schoemaker, R. G., van der Meer, P., Voors, A. A., van Veldhuisen, D. J., and van Gilst, W. H. (2006). Protective effects of erythropoietin in cardiac ischemia: from bench to bedside. *J Am Coll Cardiol* **48**: 2161–2167.

Marrero, M. B., Venema, R. C., Ma, H., Ling, B. N., and Eaton, D. C. (1998). Erythropoietin receptor-operated Ca2+ channels: activation by phospholipase C-gamma 1. *Kidney Int* **53**: 1259–1268.

Schmeding, M., Neumann, U. P., Boas-Knoop, S., Spinelli, A., and Neuhaus, P. (2007). Erythropoietin reduces ischemia-reperfusion injury in the rat liver. *Eur Surg Res* **39**: 189–197.

Sepodes, B., Maio, R., Pinto, R. et al. (2006). Recombinant human erythropoietin protects the liver from hepatic ischemia-reperfusion injury in the rat. *Transpl Int* **19**: 919–926.

Zhang, Q., Moe, O. W., Garcia, J. A., and Hsia, C. C. (2006). Regulated expression of hypoxia-inducible factors during postnatal and postpneumonectomy lung growth. *Am J Physiol Lung Cell Mol Physiol* **290**: L880–L889.

CHAPTER 4

Development of Recombinant Erythropoietin and Erythropoietin Analogs

IAIN C. MACDOUGALL
Department of Renal Medicine, King's College Hospital, London, UK

Contents

I. Introduction 35
II. History of recombinant human erythropoietin 35
III. Biosimilar EPOs 36
IV. Potential strategies for modifying erythropoietin to create new EPO analogs 37
V. Darbepoetin alfa 38
VI. Continuous erythropoietin receptor activator (C.E.R.A.) 41
VII. Small molecule ESAs 43
VIII. Other strategies for stimulating erythropoiesis 44
IX. Conclusions 44
References 44

I. INTRODUCTION

The ability to stimulate erythropoiesis with therapeutic agents has probably had the greatest impact in the field of nephrology. Ever since it was recognized that red cell production was controlled by the hormone erythropoietin (EPO), and that this hormone was produced *de novo* in the kidney in response to hypoxia, there was a clear rationale for administering erythropoietin replacement therapy (Jelkmann, 2007). When this dream became reality in the late 1980s, the true impact of this treatment was realized. Dialysis patients who were heavily transfusion-dependent and who, without regular blood transfusions could barely achieve hemoglobin levels above 6–7 g/dl, were rendered transfusion-independent, with hemoglobin concentrations of around 11–12 g/dl (Winearls et al., 1986; Eschbach et al., 1987; Macdougall et al., 1990). This was truly one of the major breakthroughs in nephrology, if not in the whole of medicine, within the last two or three decades.

This chapter will discuss the history of recombinant human erythropoietin and the way this therapeutic field has evolved over the last 20 years, along with novel strategies for stimulating erythropoiesis.

II. HISTORY OF RECOMBINANT HUMAN ERYTHROPOIETIN

Erythropoiesis is a complex physiological process that maintains homeostasis of oxygen levels in the body. It is primarily regulated by erythropoietin (EPO), a 30.4 kDa, 165-amino acid hematopoietic growth factor (Lai et al., 1986). In the presence of EPO, erythroid cells in the bone marrow proliferate and differentiate. In its absence, these progenitor cells undergo apoptosis. The presence of a humoral factor like EPO was first suggested by Carnot and Deflandre in 1906, following a series of elegant experiments in which they injected blood from anemic rabbits into donor rabbits and observed a significant increase in red cell production (Carnot and Deflandre, 1906). It was, however, not until the 1950s that Erslev and others conclusively demonstrated the presence of EPO (Erslev, 1953) and it took nearly another 30 years before human erythropoietin was isolated from the urine of patients with aplastic anemia (Miyake et al., 1977).

The next major development was the successful isolation and cloning of the human EPO gene in 1983 (Lin et al., 1985), which allowed for the development of recombinant human erythropoietin as a clinical therapeutic. The original recombinant human erythropoietins (epoetin alfa and epoetin beta) were synthesized in cultures of transformed Chinese hamster ovary (CHO) cells that carry cDNA encoding human erythropoietin. The amino acid sequence of both epoetins is therefore identical and the major difference between these products lies in their glycosylation pattern. There are also slight differences in the sugar profile between recombinant human EPO and endogenous EPO (Skibeli et al., 2001), but the amino acid sequence is identical. Erythropoietin exerts its mechanism of action via binding to the EPO receptor on the surface of erythroid progenitor cells. The erythropoietin receptor undergoes a conformational change following dimerization, which involves activation of the JAK2/STAT-5 intracellular signaling pathway (Rossert and Eckardt, 2005). The metabolic fate of erythropoietin has been debated for years, but it now appears likely that this is partly mediated via internalization of the EPO receptor complex, with

TABLE 4.1 Benefits of correction of anemia in chronic kidney disease (CKD) patients

↑Quality-of-life	↑Sexual function
↑Exercise capacity	↑Endocrine function
↓Cardiac output	↑Immune function
↓Angina	↑Muscle metabolism
↓Left ventricular hypertrophy	↓Hospitalizations
↓Bleeding tendency	↓Transfusions
↑Brain/cognitive function nutrition	
↓Depression	
↑Sleep patterns	

subsequent lysosomal degradation (Gross and Lodish, 2006). The latter has strong relevance for the design of new erythropoietin analogs, since the circulating half-life of recombinant human erythropoietin following intravenous administration is fairly short, at around 6–8 h (Macdougall et al., 1991). Thus, in the first clinical trials of recombinant human erythropoietin in hemodialysis patients, the drug was administered three times weekly to coincide with the dialysis sessions (Winearls et al., 1986; Eschbach et al., 1987).

The half-life of subcutaneously-administered recombinant human EPO is much longer, at around 24 h (Macdougall et al., 1991) and this characteristic, along with the recognition that the high peak levels after intravenous administration are not necessary for its biological action, means that a lower dose of drug may be administered subcutaneously to achieve the same effect as that seen following IV administration. In a randomized controlled trial, Kaufman et al. (1998) demonstrated that the dosage requirements following subcutaneous administration were approximately 30% lower than following IV administration. This was confirmed in a later meta-analysis by Besarab et al. (2002).

In the early clinical trials, the huge benefits of EPO administration were seen. In addition to transfusion independence, patients became aware of significantly improved energy levels, greater exercise capacity and generally improved quality of life (Keown, 1991). Cardiac benefits, such as a reduction in left ventricular hypertrophy, were also described (Macdougall et al., 1990; Teruel et al., 1991), as were objective measures of exercise performance (Mayer et al., 1988; Macdougall et al., 1990). A wide range of other secondary benefits were reported (Table 4.1), including improvements in cognitive function, skeletal muscle function and immune function. Several serious adverse events were seen in the early days of erythropoietin therapy, including severe hypertension, hypertensive encephalopathy and seizures (Edmunds et al., 1989), but these side effects are not commonly seen today, almost certainly due to the fact that anemia is treated earlier and with more cautious increments in hemoglobin. The only other serious adverse effect of EPO therapy was highlighted in 2002 by Casadevall et al. (2002), who described a case series of 13 patients who developed antibody-mediated pure red cell aplasia caused by the formation of antibodies against recombinant human erythropoietin. The mechanism behind this effect has been debated, but factors such as inadequate cold storage facilities, subcutaneous route of administration, and leachates from the rubber plungers of the syringes (acting as immune adjuvants) may all have contributed (Boven et al., 2005). Although this side effect was devastating, it is extremely rare and, worldwide, only approximately 300 cases have been seen. Most of these were with a particular formulation of epoetin alfa manufactured outside the USA, under the trade name of Eprex (Erypo in Germany).

The majority of chronic kidney disease (CKD) patients respond to erythropoietin therapy, although a minority show a more sluggish response, which may be due to iron insufficiency, inflammation, or a number of more minor factors (Johnson et al., 2007) (Table 4.2).

The ability to boost hemoglobin levels without blood transfusions has also generated considerable debate over the optimal target hemoglobin for patients receiving erythropoietin therapy (see Chapter 5). Anemia guidelines discussing this issue were first published in 1997, and suggested that a target hemoglobin range of 11–12 g/dl was appropriate (NKF DOQI Guidelines, 1997). A series of other anemia guidelines from Europe, Canada, Australia and the UK then followed and some of these were revised. The latest US KDOQI Guidelines update on target hemoglobin once again suggests that CKD patients receiving EPO therapy should generally target hemoglobin levels of between 11 and 12 g/dL (NKF KDOQI Anemia Guidelines update, 2007). Three large randomized controlled trials (Besarab et al., 1998; Drueke et al., 2006; Singh et al., 2006), along with a *Lancet* meta-analysis (Phrommintikul et al., 2007) suggested that there was likely harm in targeting hemoglobin levels above 13 g/dl, due to an increased risk of cardiovascular events. There continues to be debate over whether the actual hemoglobin level is dangerous, or whether the means to achieve this hemoglobin level is the cause for concern but, at the present time, full correction of anemia is not advised in CKD patients (NKF KDOQI Anemia Guidelines update, 2007).

III. BIOSIMILAR EPOS

Since the patents for epoetin alfa and epoetin beta have now expired in several countries, and because the market for recombinant human EPO is so lucrative, copies of the

TABLE 4.2 Causes of a poor response to ESA therapy

Major	Minor
Iron deficiency	Blood loss
Infection/inflammation	Hyperparathyroidism
Underdialysis	Aluminum toxicity
	B12/folate deficiency
	Hemolysis
	Marrow disorders, e.g. MDS
	Hemoglobinopathies
	Angiotensin-converting enzyme (ACE) inhibitors
	Carnitine deficiency
	Obesity (SC EPO)
	Anti-EPO antibodies (PRCA)

established EPO preparations are now beginning to appear on the market. These products are named 'biosimilars' in the European Union and 'follow on biologics' in the USA (Schellekens, 2008). Biosimilar EPOs, by definition, are those that have been through the EU regulatory process. In addition, outside the EU and the USA, 'copy' epoetins are already produced by companies other than the innovators and used clinically as anti-anemic drugs. For example, a CHO cell-derived recombinant human erythropoietin produced in Havana, Cuba, has been shown to have therapeutic efficacy (Pérez-Oliva et al., 2005). All recombinant proteins are, however, associated with a number of issues that distinguish them from traditional drugs and their generics. Recombinant proteins are highly complex at the molecular level and biological manufacturing processes are highly elaborate, involving cloning, selection of a suitable cell line, fermentation, purification and formulation. In addition, the therapeutic properties of recombinant proteins are highly dependent on each step of the manufacturing process. Since the manufacturing process will be different from that used by the innovator company, there have been serious concerns about the safety, efficacy and consistency of both biosimilar and 'copy' epoetin products (Schellekens, 2008).

Other epoetins that have recently become available or are still being developed include epoetin omega (Acharya et al., 1995; Bren et al., 2002; Sikole et al., 2002) and epoetin delta (Spinowitz and Pratt, 2006; Kwan and Pratt, 2007; Martin, 2007). As with all recombinant human erythropoietins, these products share the same amino acid sequence as for epoetin alfa and epoetin beta, as well as the endogenous hormone. The cell culture conditions, however, vary. With epoetin omega, baby hamster kidney (BHK) cell cultures have been used for the manufacture of this product, which has been used clinically in some Eastern European, Central American, and Asian countries.

Epoetin delta is another recombinant erythropoietin that has been used for treating patients with chronic kidney disease. Again, patent issues have prevented its availability in the USA, but not in Europe. Epoetin delta (Dynepo) was approved by the European Medicines Agency (EMEA) in 2002 and first marketed in Germany in 2007. Epoetin delta is synthesized in human fibrosarcoma cell cultures (line HT-1080). The product is also called gene activated EPO because the expression of the native human EPO gene is activated by transformation of the cell with the cytomegalovirus promoter (The Court Service, 2002). In contrast to CHO or BHK cell-derived recombinant human erythropoietin, epoetin delta does not possess N-glycolylneuraminic acid (Neu5Gc) since, in contrast to other mammals including great apes, humans are genetically unable to produce Neu5Gc due to an evolutionary mutation (Tangvoranuntakul et al., 2003). The implications of a lack of Neu5Gc residues in synthetic recombinant erythropoietin, if any, are not clear at the present time.

IV. POTENTIAL STRATEGIES FOR MODIFYING ERYTHROPOIETIN TO CREATE NEW EPO ANALOGS

The major limitation of recombinant human erythropoietin is its short duration of action and thus the patient needs to receive one to three injections per week. Given the lucrative nature of the anemia market, several companies have investigated means of modifying the erythropoietin molecule to create longer-acting EPO receptor agonists. Some of the strategies that have been employed in this process are summarized in Table 4.3 (Macdougall, 2008).

The first strategy to be investigated was the creation of a hyperglycosylated analog of EPO. The rationale for this is described in more detail below (see Darbepoetin alfa), but the addition of extra sialic acid residues to the erythropoietin molecule was found to confer a greater metabolic stability *in vivo*. Another strategy that has been used for prolonging the duration of action of other therapeutic proteins such as G-CSF and interferon-alfa, is pegylation of the protein. This is the strategy that was adopted in the creation of C.E.R.A. (see below) and the circulating half-life of this molecule is considerably enhanced compared to native or recombinant erythropoietin. Solid phase peptide synthesis and branched precision polymer constructs were used to create Synthetic

TABLE 4.3 EPO-receptor agonists

Protein-based ESA therapy
Epoetin (alfa, beta, delta, omega)
Biosimilar EPOs (epoetin zeta)
Darbepoetin alfa
C.E.R.A. (methoxy polyethylene glycol epoetin beta)
Synthetic erythropoiesis protein (SEP)
EPO fusion proteins
- EPO–EPO
- GM–CSF–EPO
- Fc–EPO
- CTNO 528

Small molecule ESAs
Peptide-based (e.g. Hematide)
Non-peptide based

Erythropoiesis Protein, the erythropoietic effect of which has been shown to vary in experimental animals depending on the number and type of the attached polymers (Chen et al., 2005).

Another strategy that has been adopted is the fusion of EPO with other proteins. These recombinant EPO fusion proteins may contain additional peptides at the carboxy-terminus to increase the *in vivo* survival of the molecule (Lee et al., 2006). Large EPO fusion proteins, of molecular weight 76 kDa, have been designed from cDNA encoding two human EPO molecules linked by small flexible polypeptides (Sytkowski et al., 1999; Dalle et al., 2001). A single subcutaneous administration of this compound to mice increased red cell production within 7 days at a dose at which epoetin was ineffective (Sytkowski et al., 1999). Another dimeric fusion protein incorporating both EPO and GM-CSF has been created, with the rationale that GM-CSF is required for early erythropoiesis. This EPO–GM–CSF complex proved to be able to stimulate erythropoiesis in cynomolgus monkeys (Coscarella et al., 1998a), but was later found to induce anti-erythropoietin antibodies, causing severe anemia (Coscarella et al., 1998b). Yet another approach is the genetic fusion of EPO with the Fc region of human immunoglobulin G (Fc–EPO) (Way et al., 2005). This molecular modification promotes recycling out of the cell upon endocytosis via the Fc recycling receptor (Capon et al., 1989; Yeh et al., 1992), again providing an alternative mechanism for enhancing circulating half-life. The same effect may be achieved by fusing EPO with albumin.

Another molecule currently undergoing development is CTNO 528, which is an EPO-mimetic antibody fusion protein with an enhanced serum half-life but no structural similarity to erythropoietin (Bugelski et al., 2005). Rats treated with a single subcutaneous dose of CTNO 528 showed a more prolonged reticulocytosis and hemoglobin rise compared to treatment with epoetin or darbepoetin alfa. Phase I studies in healthy volunteers showed a similar effect following a single intravenous administration of CTNO 528, with a peak reticulocyte count occurring after 8 days and the maximum hemoglobin concentration being seen after 22 days. None of the 24 subjects in this study developed antibodies against the molecule (Franson et al., 2005).

Interestingly, an Fc–EPO fusion protein has been successfully administered in a phase I trial to human volunteers as an aerosol, with a demonstrable increase in erythropoietin levels associated with an increase in reticulocyte counts (Dumont et al., 2005). In addition to the erythropoietin derivatives administered by aerosol inhalation, other delivery systems for erythropoietin have been investigated, including ultrasound-mediated transdermal uptake (Mitragotri et al., 1995) and orally via liposomes to rats (Maitani et al., 1999). Mucoadhesive tablets containing erythropoietin and an absorption enhancer (Labrasol) for oral administration have been studied in rats and dogs (Venkatesan et al., 2006). Theoretically, this preparation is designed to allow the tablet to reach the small intestine intact. Preliminary experiments in beagle dogs were conducted with intrajejunal administration of a single tablet containing 100 IU/kg of recombinant human erythropoietin, with a corresponding increase in reticulocytes 8 days after administration (Venkatesan et al., 2006). It is too early to say whether this strategy could have any clinical relevance in the treatment of anemia in CKD patients.

V. DARBEPOETIN ALFA

Darbepoetin alfa, initially termed Novel Erythropoiesis Stimulating Protein (NESP), and now marketed under the trade name of Aranesp, is a second-generation EPO analog. Its development arose from the recognition that the higher isoforms (those with a greater number of sialic acid residues) of recombinant human erythropoietin were more potent biologically *in vivo* due to a longer circulating half-life than the lower isomers (those with a lower number of sialic acid residues) (Egrie et al., 1993) (Figure 4.1). Since the majority

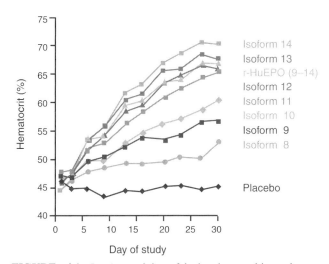

FIGURE 4.1 *In vivo* activity of isolated recombinant human erythropoietin isoforms in mice injected three times per week intraperitoneally. Redrawn from Egrie et al. (1993).

of sialic acid residues are attached to the three N-linked glycosylation chains of the erythropoietin molecule, attempts were made to synthesize EPO analogs with a greater number of N-linked carbohydrate chains. This was achieved using site-directed mutagenesis, in order to change the amino acid sequence at sites not directly involved in binding to the erythropoietin receptor (Egrie and Browne, 2001; Elliott et al., 2003). Thus, five amino acid substitutions were implemented (Ala30Asn, His32Thr, Pro87Val, Trp88Asn, Pro90Thr), allowing darbepoetin alfa to carry a maximum of 22 sialic acid residues, compared with recombinant or endogenous erythropoietin which support a maximum of 14 sialic acid residues. The additional N-linked carbohydrate chains increased the molecular weight of epoetin from 30.4 kDa to 37.1 kDa and the carbohydrate contribution to the molecule correspondingly increased from 40% to around 52% (Egrie and Browne, 2001; Elliott et al., 2003).

These molecular modifications to erythropoietin conferred a greater metabolic stability *in vivo* and this was confirmed in a single-dose pharmacokinetic study performed in EPO-naïve patients undergoing continuous ambulatory peritoneal dialysis (Macdougall et al., 1999). Following a single IV injection of darbepoetin alfa, the mean terminal half-life was approximately threefold longer compared to a single IV injection of epoetin alfa (25.3 h versus 8.5 h, respectively) and the AUC was more than twofold greater (291 ± 8 versus 138 ± 8 ng.h/ml), as well as a threefold lower clearance (1.6 ± 0.3 versus 4.0 ± 0.3 ml/h/kg), which was biphasic. The volume of distribution was similar for the two molecules (52.4 ± 2.0 and 48.7 ± 2.1 ml/kg, respectively). The mean terminal half-life in patients given darbepoetin alfa subcutaneously was approximately 49 h, which is around twice that following IV administration and the C_{max} averaged about 10% of the intravenous volume. The T_{max} averaged 54 ± 5 h, with a mean bioavailability of 37% (Macdougall et al., 1999).

More recent studies have estimated a longer half-life for subcutaneously (SC)-administered darbepoetin alfa (Macdougall et al., 2007a). These pharmacokinetic studies have employed longer sampling periods, up to 28 days, and they suggest that the half-life of SC darbepoetin alfa may be around 70 h. Two studies by Padhi and colleagues (2005, 2006) conducted in patients with chronic renal insufficiency (CRI) used sampling times of 648 to 672 h to estimate the SC half-life. The first study, a pilot, was conducted in a subset of five patients from an open-label, multicenter investigation of QM SC administration of darbepoetin alfa. These patients had been receiving darbepoetin alfa Q2W and had stable hemoglobin levels between 10.0 and 12.0 g/dl. They were switched to QM darbepoetin alfa at a dose equal to the total dose received in the previous month. Pharmacokinetic analysis was performed between 6 and 672 h after administration of the first QM darbepoetin alfa dose. Absorption after SC injection was slow in all patients, with peak concentrations of 0.75–6.29 ng/ml reached at 34–58 h post-dose, respectively, followed by a generally monophasic decline. The mean terminal half-life of darbepoetin alfa was 73 h (range 39.9–115.0 h, consistent with the variability range expected for all ESAs) (Padhi and Jang, 2005). The second study was a single-dose, open-label study of SC darbepoetin alfa in 20 adult patients with CRI. The extended sampling period was 672 h. Peak concentrations of darbepoetin alfa were reached in a median of 36.0 h (range 12.0–72.0 h), with a mean terminal half-life of 69.6 h (95% CI, 54.9–84.4 h) (Padhi et al., 2006).

The half-life of darbepoetin alfa was also investigated in peritoneal dialysis (PD) patients receiving a range of darbepoetin alfa doses. Tsubakihara and colleagues (2004) performed pharmacokinetic analyses in patients receiving PD and patients with chronic renal insufficiency (CRI) following single doses of SC darbepoetin alfa. Darbepoetin alfa was administered to 32 PD patients at 20, 40, 90, or 180 μg (8 patients per treatment group) and to 32 patients with CRI (same dose groups and patient allocation). Serum darbepoetin alfa concentrations were followed for 336 h for patients receiving the 20, 40 or 90 μg doses, or 672 h for patients receiving the 180 μg dose. The mean terminal half-life in the different dose groups ranged from 64.7 to 91.4 h in the PD patients and from 73.6 to 104.9 h in CRI patients, but was not dose-dependent. This study also showed that there was no effect of differing levels of renal function on the half-life of darbepoetin alfa (Tsubakihara et al., 2004).

The more prolonged half-life of darbepoetin alfa compared to either epoetin alfa or epoetin beta has translated into less frequent dosing, with most patients receiving injections once-weekly or once-every-other-week.

A. Intravenous Administration – Hemodialysis Patients

Results from two studies support the conclusion that IV darbepoetin alfa is clinically efficacious in maintaining hemoglobin levels without a need to increase the dose in hemodialysis (HD) patients when administered at longer intervals compared with epoetin alfa. In a 28-week, randomized study, Nissenson et al. (2002) assigned HD patients receiving stable therapy with IV epoetin alfa TIW to continue treatment or to switch to IV darbepoetin alfa QW. There was no statistically or clinically significant change in mean hemoglobin levels from baseline to the evaluation period (the final 8 weeks of the study). During the evaluation period, 49% of patients in the epoetin alfa group versus 44% of patients in the darbepoetin alfa group required a dose change in order to maintain hemoglobin levels within the 9–13 g/dl target range. The mean dose during the evaluation period did not differ statistically from baseline values in either treatment group. Safety profiles were comparable between the two treatments, with similar rates of adverse events (Nissenson et al., 2002). Locatelli and colleagues showed that hemoglobin levels were maintained in HD patients over 30 weeks of treatment with QW or Q2W darbepoetin alfa. Importantly, this study also demonstrated that there was no significant dose increase with the extension of the darbepoetin alfa interval out to Q2W and that the treatment was well tolerated at both dosing schedules (Locatelli et al., 2005).

A prospective, multicenter, 24-week study determined the bioequivalent dose of darbepoetin alfa given IV QW in stable HD patients who had previously received epoetin alfa SC or IV and who had hemoglobin levels between 10.8 and 13 g/dl (Bock, 2005). Using the European label recommended 1 μg darbepoetin alfa to 200 IU epoetin alfa conversion ratio, subjects previously stable on epoetin alfa BIW or TIW were converted to darbepoetin alfa QW, and subjects previously stable on epoetin alfa QW were converted to darbepoetin alfa Q2W. The dose of darbepoetin alfa was subsequently adjusted to maintain hemoglobin levels within ±1 g/dl of the baseline value. In the 100 study completers, hemoglobin was well maintained. The dose of darbepoetin alfa was 45.6 μg and 25.8 μg at baseline for the QW and Q2W groups, respectively, and 31.5 μg and 21.4 μg at the end of the study, respectively (Bock, 2005).

Although PD patients are more likely to receive ESA therapy via the SC route, the efficacy and safety of IV darbepoetin alfa at various dosing frequencies was also investigated in these patients. In one study, PD patients either naïve to ESAs or previously receiving epoetin (alfa or beta not specified) were treated with darbepoetin alfa Q2W. Once stable, patients could extend the dosing interval out to QM. All patients received darbepoetin alfa for up to 28 weeks to achieve and maintain hemoglobin levels between 11 and 13 g/dl. Hemoglobin in ESA-naïve patients increased from 8.15 to 11 g/dl over the first 10 weeks of darbepoetin alfa therapy, and all patients' hemoglobin levels were successfully maintained within the target range regardless of whether darbepoetin alfa was dosed Q2W or QM (Hiramatsu et al., 2005). It should be noted, however, that only stable patients were included in this study and QM administration may not be appropriate for an unselected dialysis population.

B. Subcutaneous Administration – Hemodialysis Patients

A number of studies support the efficacy of SC darbepoetin alfa administered to HD and PD patients who were switched from more frequent epoetin therapy (alfa or beta) (Locatelli et al., 2003; Mann et al., 2005, 2007; Vanrenterghem et al., 2002; Barril et al., 2005). In these studies, the patients receiving SC epoetin therapy were switched to darbepoetin alfa: epoetin BIW/TIW to darbepoetin alfa QW, and epoetin QW to darbepoetin alfa Q2W. Hemoglobin levels were successfully maintained within the target range (for the majority of studies, 10–13 g/dl) and without the need for darbepoetin alfa dose increases (Locatelli et al., 2003; Mann et al., 2005, 2007; Vanrenterghem et al., 2002; Barril et al., 2005).

Administration of darbepoetin alfa SC was also shown to be effective in PD patients not previously treated with an ESA. As part of a larger study, Macdougall and colleagues administered a range of darbepoetin alfa doses TIW or QW to 47 PD patients (Macdougall et al., 2003a). The investigators found that overall, both 0.45 and 0.75 μg/kg/week doses increased mean hemoglobin levels ≥1 g/dl/4 weeks and that there was no apparent difference in efficacy between the TIW and QW regimens. The individual patients who achieved a hemoglobin rate of rise ≥1 g/dl/4 weeks continued darbepoetin alfa treatment for up to 52 weeks with hemoglobin levels being maintained between 10 and 13 g/dl (Macdougall et al., 2003a). The safety profile of darbepoetin alfa was similar to that expected for this patient population, thus demonstrating that PD patients can safely achieve a hemoglobin response within a month of initiating darbepoetin alfa therapy.

C. Subcutaneous Administration – Pre-Dialysis CKD Patients

In addition to dialysis patients, studies have also investigated the efficacy and safety of Q2W and QM dosing with darbepoetin alfa in CRI patients (Toto et al., 2004; Suranyi et al., 2003; Disney et al., 2005; Ling et al., 2005; Agarwal et al., 2006; Sarac et al., 2004). Two, large, 24-week studies by Suranyi et al. and Toto et al. examined the administration of darbepoetin alfa in CRI patients not previously receiving an ESA. Within approximately 5 weeks of initiating SC, *de novo* Q2W darbepoetin alfa, 95% to 97% of these patients achieved hemoglobin levels between 11 and 13 g/dl. These results were supported by a chart review that compared *de novo* Q2W or QM epoetin alfa therapy with *de novo* Q2W or QM darbepoetin alfa. The proportion of patients achieving a mean hemoglobin level ≥11 g/dl within 100 days was recorded. Of the patients dosed with Q2W or QM darbepoetin alfa, 66.7% and 80.0%, respectively, achieved the hemoglobin target. In comparison, of the patients with Q2W or QM epoetin alfa, 53.8% and 50.0%, respectively achieved the hemoglobin target (Sarac et al., 2004).

Three recent studies have examined the efficacy of SC QM darbepoetin alfa in maintaining hemoglobin levels in CRI patients following extension of the dosing interval from previous Q2W darbepoetin alfa. In the study by Disney and colleagues, 83% of patients who received at least one QM dose of darbepoetin alfa (the modified intent-to-treat population; mITT) and 95% of the patients who completed the study achieved a target hemoglobin level of ≥ 10 g/dl (Disney et al., 2005). Likewise, Ling et al. reported that the hemoglobin target of 10–12 g/dl was achieved in 79% of the mITT population and in 85% of those patients completing the study following extension of the darbepoetin alfa dosing interval to QM (Ling et al., 2005). Finally, the study by Agarwal et al. further confirmed the efficacy of QM darbepoetin alfa by showing that following a switch from Q2W darbepoetin alfa dosing, hemoglobin levels could be maintained ≥ 11 g/dl in 76% of the mITT population and in 85% of patients who completed the study (Agarwal et al., 2006).

In a setting representative of current nephrology practice, nephrologists at centers participating in the Aranesp Registry Group in the Netherlands enrolled patients in a registry to investigate the feasibility of administering darbepoetin alfa QM (van Buren et al., 2005). Nephrologists were first informed of the possibility of dosing darbepoetin alfa QM and then their patients' treatments were monitored for a 12-month period. The patients had CRI, were not receiving renal replacement therapy and were currently receiving or about to initiate QM darbepoetin alfa therapy. Of 108 patients completing the 12-month follow-up period, 66% were found to be receiving QM darbepoetin alfa, and mean hemoglobin levels were maintained at approximately 12 g/dl throughout the study. Fifty-nine percent of the patients who had ever received a QM dose of darbepoetin alfa preferred this regimen over any other and 31% had no preference. These results are interesting, but clearly the potential for selection bias in this study is considerable, since nephrologists are far more likely to select patients for QM dosing if they are stable and likely to manage with less frequent dosing (van Buren et al., 2005). Unfortunately, all studies of QM darbepoetin alfa are non-randomized and uncontrolled and, until better quality studies are available, the use of the QM dosing frequency for darbepoetin alfa should be restricted to highly selected stable patients.

VI. CONTINUOUS ERYTHROPOIETIN RECEPTOR ACTIVATOR (C.E.R.A.)

Even more recently, a third-generation erythropoietic molecule, Continuous Erythropoietin Receptor Activator (C.E.R.A.), was created by integrating a large polymer chain into erythropoietin, thus increasing the molecular weight to twice that of epoetin at approximately 60 kDa (Macdougall, 2005). This methoxy-polyethylene glycol polymer chain is integrated via amide bonds between the N-terminal amino group or the ε-amino group of lysine (predominantly lysine-52 or lysine-45), using a single succinimidyl butanoic acid linker.

Evidence is accumulating that C.E.R.A. has very different receptor binding characteristics and pharmacokinetic properties compared with both epoetin and darbepoetin alfa (Macdougall et al., 2003b). It has a much lower affinity for the erythropoietin receptor compared with the natural ligand, leading to a reduced specific activity *in vitro*. However, since the elimination half-life is so prolonged, C.E.R.A. has increased erythropoietic activity *in vivo*.

Pre-clinical and then clinical studies of C.E.R.A. will be discussed in turn.

A. Effects of C.E.R.A. *In Vitro* and in Animal Models

The erythropoietic activities of C.E.R.A. and epoetin were compared *in vitro* by measuring their effect on the proliferation of a human acute myeloid leukemia cell line (UT-7) that expresses the erythropoietin receptor. Across the dose range 0.003–3 U/ml, epoetin stimulated greater proliferation of UT-7 cells compared with C.E.R.A. (Haselbeck et al., 2002). However, *in vivo* studies in normothycemic mice comparing identical amounts of protein across the dose range 60–1000 ng protein/animal have shown that C.E.R.A. was more effective than epoetin at stimulating bone marrow precursor cells and increasing reticulocyte count (Haselbeck et al., 2002). At a dose of 1000 ng, C.E.R.A. increased the mean reticulocyte count by 14%, compared with 9% for epoetin.

Pre-clinical studies in various animal models have investigated the pharmacodynamic and pharmacokinetic properties of IV and SC C.E.R.A. administered in both single and multiple doses, across the dose range 0.75–20 μg/kg. In mice, a single SC injection of C.E.R.A. 20 μg/kg increased the mean reticulocyte count by 13%, compared with 7.8% in response to a comparable dose of epoetin beta (Fishbane et al., 2003). The median duration of the response was approximately 3 days longer with C.E.R.A. compared with epoetin. In addition, a single SC or IV administration of C.E.R.A. 2.5 μg/kg in mice elicited a greater reticulocyte response than multiple doses of epoetin 2.5 μg/kg, in terms of the magnitude and duration of response. Further studies in mice showed that SC administration of C.E.R.A. once weekly (1.25 μg/kg and 5 μg/kg) or once every 2 weeks (5 μg/kg) produced a greater reticulocyte response compared with epoetin (1.25 μg/kg) three times weekly (Fishbane et al., 2003). Moreover, approximately equal numbers of reticulocytes were produced with C.E.R.A. 1.25 μg/kg administered once every 2 weeks as with epoetin 1.25 μg/kg administered three times weekly. Pharmacokinetic studies in animals showed that C.E.R.A. has a longer half-life and lower systemic clearance than epoetin (Macdougall et al., 2003b).

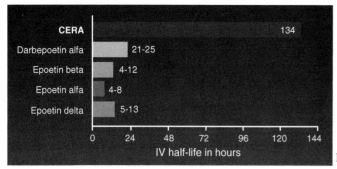

FIGURE 4.2 IV half-lives of ESA therapy.

From the results of these pre-clinical studies, it appears that C.E.R.A. has receptor binding and pharmacokinetic properties that give rise to more potent stimulation of erythropoiesis *in vivo* than epoetin, with regard to both the magnitude and duration of response. These findings suggested the potential for C.E.R.A. to be administered at extended dosing intervals.

B. C.E.R.A. in Healthy Subjects

Four phase I studies were conducted in healthy subjects to investigate the pharmacokinetic and pharmacodynamic properties of C.E.R.A. In two single ascending dose (SAD) studies, subjects were randomized to receive single IV doses of C.E.R.A. (0.4–3.2 μg/kg) or placebo [n = 38] or single SC doses of C.E.R.A. (0.1–3.2 μg/kg) or placebo [n = 70]. In two multiple ascending dose (MAD) studies, subjects were randomized to receive three IV doses of C.E.R.A. (0.4–3.2 μg/kg) or placebo [n = 61] once every 3 weeks or four SC doses of C.E.R.A. (0.4–3.2 μg/kg) or placebo [n = 48] once every 2 weeks. The half-life of C.E.R.A. (Macdougall et al., 2006a), administered IV or SC, was observed to be considerably longer than that previously reported for epoetin (alfa or beta) (Halstenson et al., 1991) or darbepoetin alfa (Macdougall et al., 1999) (Figure 4.2). The pharmacokinetics of C.E.R.A. were apparently unaffected by repeated dosing.

In the MAD studies, the clearance of both IV and SC C.E.R.A. was low (IV 27.6–44.6 ml/h, SC 97–347 ml/h) (Dougherty et al., 2004). No accumulation was observed when steady state was achieved with the different frequencies tested. The prolonged half-life and low clearance seen with both IV and SC C.E.R.A. in healthy subjects supported the data from animal studies, suggesting that it might be possible to administer C.E.R.A. at extended dosing intervals.

C. Effects of C.E.R.A. in Patients with CKD Anemia

Four phase II dose-finding studies investigated the feasibility of C.E.R.A. for the correction of anemia and maintenance of hemoglobin (Hb) levels at extended administration intervals in more than 350 patients with CKD. In two studies, one in CKD patients receiving dialysis (n = 61) (de Francisco et al., 2003) and one in CKD patients not yet receiving dialysis (n = 65) (Provenzano et al., 2007), SC C.E.R.A. was administered for the correction of anemia. All patients were aged ≥18 years, with Hb 8–11 g/dl and were ESA-naïve.

The first study, conducted in patients receiving dialysis treatment, examined escalating doses of C.E.R.A. in three patient groups (de Francisco et al., 2003). After a 4-week run-in period, patients were randomized to receive one of three C.E.R.A. doses (0.15, 0.30 or 0.45 μg/kg/week). Once-weekly, once every 2 weeks and once every 3 weeks administration schedules were assessed in each dose group. The second study, conducted in patients not receiving dialysis, also investigated escalating doses of C.E.R.A. in three patient groups (Provenzano et al., 2007). After a 2-week run-in period, patients were randomized to receive one of three C.E.R.A. doses (0.15, 0.30 or 0.60 μg/kg/week). Again, once-weekly, once every 2 weeks and once every 3 weeks administration schedules were assessed in each dose group. For both studies, individual dose adjustments were permitted according to defined Hb criteria after the initial 6-week period. Patients were followed for a total of 12 weeks (patients receiving dialysis) or 18 weeks (patients not receiving dialysis), respectively.

In these studies, there was a statistically significant dose response to C.E.R.A. treatment and the Hb response was independent of the frequency of administration. These results suggested that C.E.R.A. was capable of correcting anemia when administered to ESA-naïve CKD patients at extended dosing intervals.

Two phase II, multicenter, dose-finding studies were conducted to determine the efficacy of C.E.R.A. for the maintenance of Hb levels in adult patients with renal anemia (Hb 10–13 g/dl) and receiving dialysis treatment. In one study, 91 patients previously maintained on three-times weekly IV epoetin alfa were switched to IV C.E.R.A. (Besarab et al., 2004). After a 2-week run-in period, patients were randomized to one of three C.E.R.A. doses based on their previous epoetin dose and data on exposure to C.E.R.A. in healthy subjects. Once-weekly and once every 2 weeks administration schedules were assessed in each dose group. Patients were followed for a total of 19 weeks.

A statistically significant ($P < 0.0001$) dose-dependent Hb response was observed. At the highest dose studied, IV C.E.R.A. maintained stable Hb levels within ±1.5 g/dl from baseline in the highest percentage of patients when administered once every 2 weeks. These results suggested that IV C.E.R.A. administered once every 2 weeks may maintain stable Hb levels in dialysis patients.

In the second maintenance study, 137 patients previously maintained on three-times weekly SC epoetin treatment were switched to SC C.E.R.A. (Locatelli et al., 2004). After a 2-week run-in period, patients were randomized to one of three C.E.R.A. doses based on their previous epoetin dose and data on exposure to C.E.R.A. from healthy subjects. Once-weekly, once every 3 weeks and once every 4 weeks administration schedules were assessed in each dose group. Patients were followed for a total of 19 weeks; those in the once every 4 weeks group were followed for 21 weeks.

There was a statistically significant ($P < 0.001$) dose-dependent response to C.E.R.A. in the three treatment groups which was independent of the frequency of administration (up to once every 4 weeks). The results suggested that SC C.E.R.A. administered once every 4 weeks may maintain stable Hb levels in dialysis patients.

Six phase III studies of C.E.R.A. were conducted to further investigate the reduced dosing frequency of this new agent (Klinger et al., 2007; Macdougall et al., 2008; Levin et al., 2007; Sulowicz et al., 2007; Canaud et al., 2006; Spinowitz et al., 2008). Two of these (AMICUS, ARCTOS) investigated the correction of anemia in ESA-naïve patients, with an initial dosing frequency of once every two weeks, with a second phase randomization to either continue once every two-week dosing or switch to once every four-week administration. The AMICUS study (Klinger et al., 2007) investigated intravenously-administered C.E.R.A. in hemodialysis patients, while the ARCTOS study (Macdougall et al., 2008) investigated subcutaneously-administered C.E.R.A. in pre-dialysis CKD patients. Both studies were randomized and controlled, using a comparator drug (epoetin alfa for AMICUS, and darbepoetin alfa for ARCTOS).

The four maintenance phase studies were MAXIMA, PROTOS, STRIATA and RUBRA. MAXIMA compared once every two-week administration of C.E.R.A. with injections once every four weeks in hemodialysis patients (Levin et al., 2007) while the PROTOS study had an almost identical protocol, but with subcutaneously-administered C.E.R.A. (Sulowicz et al., 2007). STRIATA also utilized once every two-week dosing of C.E.R.A. administered intravenously in hemodialysis patients (Canand et al., 2006) and RUBRA recruited patients on dialysis to receive either intravenous or subcutaneous C.E.R.A., once every two weeks, in dialysis patients using pre-filled syringes (Spinowitz et al., 2008).

All the phase III studies met their primary non-inferiority end-point with the comparator drug, confirming the hypothesis that C.E.R.A. may be administered less frequently to achieve either correction of anemia or maintenance of the target hemoglobin in both pre-dialysis and dialysis-dependent CKD patients.

D. Safety and Tolerability of C.E.R.A.

In the studies in CKD patients reported to date, C.E.R.A. has generally been well tolerated with no unexpected safety concerns. Available information indicates that the incidence of adverse events was in accordance with that expected for this study population (Dougherty and Beyer, 2005).

VII. SMALL MOLECULE ESAs

A. Peptide-Based ESAs

Just over 10 years ago, several small bisulfide-linked cyclic peptides composed of around 20 amino acids were identified by random phage display technology that were unrelated in sequence to erythropoietin but still bound to the EPO receptor (Livnah et al., 1996; Wrighton et al., 1996). These small peptides were able to induce the same conformational change in the EPO receptor that leads to JAK2 kinase/STAT-5 intracellular signaling (Wrighton et al., 1996), as well as other intracellular signaling mechanisms, resulting in stimulation of erythropoiesis both *in vitro* and *in vivo*. The first peptide to be investigated (EPO-mimetic peptide-1; EMP-1) (Wrighton et al., 1996) was not potent enough to be considered as a potential therapeutic agent in its own right, but the potency of these peptides could be greatly enhanced by covalent peptide dimerization with a PEG linker. Thus, another EMP was selected for the development of *Hematide*, a pegylated synthetic dimeric peptidic ESA, that was found to stimulate erythropoiesis in experimental animals (Fan et al., 2006). The half-life of Hematide in monkeys ranges from 14 h to 60 h depending on the dose administered. Further studies in rats using quantitative whole-body auto radioluminography have shown that the primary route of elimination for the peptide is the kidney (Woodburn et al., 2007a).

A phase I study in healthy volunteers showed that single injections of Hematide cause a dose-dependent increase in reticulocyte counts and hemoglobin concentrations (Stead et al., 2006). Phase II studies have demonstrated that Hematide can correct the anemia associated with chronic kidney disease (Macdougall et al., 2006b), as well as maintaining the hemoglobin in dialysis patients who are already receiving conventional ESAs (Besarab et al., 2007). Doses in the range of 0.025–0.05 mg/kg appear to be therapeutically optimal in this patient population (Macdougall et al., 2006b) and, at the time of writing, four phase III studies are actively recruiting. Hematide may be administered either intravenously or subcutaneously and dosing once a month is effective (Macdougall et al., 2006b).

The potential advantages of this new agent are greater stability at room temperature, lower immunogenicity compared to conventional ESAs and a much simpler (and cheaper) manufacturing process, avoiding the need for cell lines and genetic engineering techniques. Antibodies against

Hematide do not cross-react with erythropoietin and, similarly, anti-EPO antibodies do not cross-react with Hematide (Woodburn et al., 2007b). This has two major implications: first, even if a patient were to develop anti-Hematide antibodies, these should not neutralize the patient's own endogenous erythropoietin and the patient should not develop pure red cell aplasia. Secondly, patients with antibody-mediated pure red cell aplasia should be able to respond to Hematide therapy by an increase in their hemoglobin concentration, since Hematide is not neutralized by anti-erythropoietin antibodies. This latter hypothesis has already been confirmed in animals (Woodburn et al., 2007b). Thus, rats receiving regular injections of recombinant human erythropoietin were shown to develop anti-EPO antibodies. Injections of Hematide were able to 'rescue' these animals and restore their hemoglobin concentration, in contrast to the vehicle-treated group (Woodburn et al., 2007b). A clinical trial examining this issue in patients with antibody-mediated pure red cell aplasia is also underway, and preliminary data suggest that the first 10 patients recruited were able to attain hemoglobin concentrations greater than 11 g/dl without the need for red cell transfusions (Macdougall et al., 2007b).

Other peptide-based ESAs are in pre-clinical development. A compound made by AplaGen in Germany has linked a peptide to a starch residue, again demonstrating prolongation of the circulating half-life of the peptide (Pötgens et al., 2006). Indeed, altering the molecular weight of the starch moiety has been shown to alter the pharmacological properties of the compound.

B. Non-Peptide-Based ESAs

Several non-peptide molecules capable of mimicking the effects of EPO have also been identified following screens of small molecule non-peptide libraries for molecules with EPO-receptor-binding activity (Qureshi et al., 1999). One such compound was found, but this bound to only a single chain of the erythropoietin receptor and was not biologically active. The compound was ligated to enable it to interact with both domains of the EPO receptor and this second molecule was shown to stimulate erythropoiesis. Further development of non-peptide EPO mimetics could lead to the production of an orally-active ESA in the future.

VIII. OTHER STRATEGIES FOR STIMULATING ERYTHROPOIESIS

The focus of this chapter is agents that act via stimulation of the erythropoietin receptor, inducing a conformational change and homodimerization of the receptor, followed by activation of the JAK2/STAT5 intracellular signaling pathway. There are, however, a number of other strategies that are under investigation to enhance erythropoiesis, although none is as yet in clinical trials. These include prolyl hydroxylase inhibition, GATA inhibition, hemopoietic cell phosphatase inhibition and, finally, EPO gene therapy. These strategies have been briefly reviewed elsewhere (Macdougall et al., 2008), but there are many hurdles to overcome before agents targeting these various mechanisms can be a viable therapeutic option for the treatment of CKD anemia. The attraction of the first three strategies is the potential for orally-active treatment, but the efficacy and safety of these agents in humans are not yet ascertained. Likewise, EPO gene therapy has its own problems, notably avoidance of immunogenicity and oncogenicity, as well as incorporating a mechanism to control the levels of EPO production.

IX. CONCLUSIONS

As the molecular mechanisms controlling red cell production have been elucidated, so too have new targets and strategies been developed for stimulating erythropoiesis and treating anemia. Following the introduction of recombinant human erythropoietin in the late 1980s, attempts were made to modify the molecule and produce longer-acting erythropoietic agents such as darbepoetin alfa and C.E.R.A. Other modifications to the molecule, such as the production of fusion proteins, are being explored, as is the concept that smaller molecules such as peptides, or even non-peptides, may be able to bind to and activate the erythropoietin receptor, and the first such molecule (Hematide) is in phase III clinical trials. All these molecules are EPO receptor agonists, activating the erythropoietin receptor, but other strategies, mainly still in pre-clinical development, include the development of orally-active agents for the treatment of anemia, which act by inhibition of prolyl hydroxylase, GATA or hemopoietic cell phosphatase. EPO gene therapy also has many hurdles to overcome before this can become clinical reality.

References

Acharya, V. N., Sinha, D. K., Almeida, A. F., and Pathare, A. V. (1995). Effect of low dose recombinant human omega erythropoietin (rHuEPO) on anaemia in patients on hemodialysis. *J Assoc Physicians India* **43**: 539–542.

Agarwal, A. K., Silver, M. R., Reed, J. E. et al. (2006). An open-label study of darbepoetin alfa administered once monthly for the maintenance of haemoglobin concentrations in patients with chronic kidney disease not receiving dialysis. *J Intern Med* **260**: 577–585.

Barril, G., Montenegro, J., Ronco, C. et al. (2005). Aranesp (darbepoetin alfa) administered every 2 weeks (Q2W) effectively maintains haemoglobin in patients receiving peritoneal dialysis: results of a pooled analysis of eight European studies. Paper presented at 42nd ERA-EDTA Congress, Istanbul, Turkey.

Besarab, A., Kline Bolton, W., Browne, J. K. et al. (1998). The effects of normal as compared with low hematocrit values in patients with cardiac disease who are receiving hemodialysis and epoetin. *N Engl J Med* **339**: 584–590.

Besarab, A., Reyes, C. M., and Hornberger, J. (2002). Meta-analysis of subcutaneous versus intravenous epoetin in maintenance treatment of anemia in hemodialysis patients. *Am J Kidney Dis* **40**: 439–446.

Besarab, A., Bansal, V., Fishbane, S. et al. on behalf of the BA16285 study group. (2004). Intravenous CERA (Continuous Erythropoietin Receptor activator) administered once weekly or once every 2 weeks maintains haemoglobin levels in haemodialysis patients with chronic renal anaemia. *Abstract Book of the XLI Congress of the ERA-EDTA 230*: (Abstract M047).

Besarab, A., Zeig, S., Geronemus, R. et al. (2007). Hematide, a synthetic peptide-based erythropoiesis stimulating agent, maintains hemoglobin in hemodialysis patients previously treated with epoetin alfa (EPO). Natl Kidney Found Clin Meet 2007, Abs 24.

Bock, A; EFIXNES Study Investigators. (2005). Conversion from weekly epoetin to bi-weekly darbepoetin-α maintains hemoglobin and permits dose reductions at higher doses. Results from a prospective study. Paper presented at 42nd ERA-EDTA Congress, Istanbul, Turkey.

Boven, K., Stryker, S., Knight, J. et al. (2005). The increased incidence of pure red cell aplasia with an Eprex formulation in uncoated rubber stopper syringes. *Kidney Int* **67**: 2346–2353.

Bren, A., Kandus, A., Varl, J., Buturovic, J., Ponikvar, R., and Kveder, R. (2002). A comparison between epoetin omega and epoetin alfa in the correction of anemia in hemodialysis patients: a prospective, controlled crossover study. *Artif Organs* **26**: 91–97.

Bugelski, P., Nesspor, B. T., Spinka-Doms, T. et al. (2005). Pharmacokinetics and pharmacodynamics of CTNO528, a novel erythropoiesis receptor agonist in normal & anemic rats. *Blood* **106**: 146b.

Canaud, B., Braun, J., Locatelli, F. et al. (2006). Intravenous (IV) C.E.R.A. (continuous erythropoietin receptor activator) administered once every 2 weeks maintains stable haemoglobin levels in patients with chronic kidney disease on dialysis [abstract no SP425]. *Nephrol Dial Transplant* **21**: iv 157.

Capon, D. J., Chamow, S. M., Mordenti, J. et al. (1989). Designing CD4 immunoadhesins for AIDS therapy. *Nature* **337**: 525–531.

Carnot, P., and Deflandre, C. L. (1906). Sur l'activite hemopoietique du serum au cours de la regeneration du sang. *C R Seances Acad Sci* **143**: 384–386.

Casadevall, N., Nataf, J., Viron, B. et al. (2002). Pure red-cell aplasia and antierythropoietin antibodies in patients treated with recombinant erythropoietin. *N Engl J Med* **346**: 469–475.

Chen, S. Y., Cressman, S., Mao, F., Shao, H., Low, D. W., and Beilan, H. S. (2005). Synthetic erythropoietic proteins: tuning biological performance by site-specific polymer attachment. *Chem Biol* **12**: 371–383.

Coscarella, A., Liddi, R., Bach, S. et al. (1998a). Pharmacokinetic and immunogenic behavior of three recombinant human GM-CSF-EPO hybrid proteins in cynomolgus monkeys. *Mol Biotechnol* **10**: 115–122.

Coscarella, A., Liddi, R., Di Loreto, M. et al. (1998b). The rhGM-CSF-EPO hybrid protein MEN 11300 induces anti-EPO antibodies and severe anaemia in rhesus monkeys. *Cytokine* **10**: 964–969.

Dalle, B., Henri, A., Rouyer-Fessard, P., Bettan, M., Scherman, D., and Beuzard, Y. (2001). Dimeric erythropoietin fusion protein with enhanced erythropoietic activity in vitro and in vivo. *Blood* **97**: 3776–3782.

de Francisco, A. L., Sulowicz, W., and Dougherty, F. C. (2003). Subcutaneous CERA (Continuous Erythropoietin Receptor Activator) has potent erythropoietic activity in dialysis patients with chronic renal anemia: an exploratory multiple-dose study. *J Am Soc Nephrol* **14**: 27A–28A (Abstract SA-FC124).

Disney, A., de Jersey, P., Kirkland, G. et al. (2005). Aranesp (darbepoetin alfa) administered once monthly (QM) maintains hemoglobin (Hb) levels in chronic kidney disease (CKD) patients. Paper presented at 42nd ERA-EDTA Congress, Istanbul, Turkey.

Dougherty, F. C., and Beyer, U. (2005). Safety and tolerability profile of Continuous Erythropoietin Receptor Activator (CERA) with extended dosing intervals in patients with chronic kidney disease on dialysis. *Nephrology* **10**: A313 (Abstract W-PO40130).

Dougherty, F. C., Reigner, B., Jordan, P., and Pannier, A. (2004). CERA (Continuous Erythropoietin Receptor Activator): Dose-response, pharmacokinetics and tolerability in phase I multiple ascending dose studies. *J Clin Oncol* **22**(Suppl 15): 14S (Abstract 6692).

Drueke, T. B., Locatelli, F., Clyne, N. et al. (2006). Normalization of hemoglobin level in patients with chronic kidney disease and anemia. *N Engl J Med* **335**: 2071–2084.

Dumont, J. A., Bitonti, A. J., Clark, D., Evans, S., Pickford, M., and Newman, S. P. (2005). Delivery of an erythropoietin-Fc fusion protein by inhalation in humans through an immunoglobulin transport pathway. *J Aerosol Med* **18**: 294–303.

Edmunds, M. E., Walls, J., Tucker, B. et al. (1989). Seizures in haemodialysis patients treated with recombinant human erythropoietin. *Nephrol Dial Transplant* **4**: 1065–1069.

Egrie, J. C., and Browne, J. K. (2001). Development and characterization of novel erythropoiesis stimulating protein (NESP). *Br J Cancer* **84**: 3–10.

Egrie, J. C., Grant, J. R., Gillies, D. K., Aoki, K. H., and Strickland, T. W. (1993). The role of carbohydrate on the biological activity of erythropoietin. *Glycoconjugate J* **10**: 263.

Elliott, S., Lorenzini, T., Asher, S., Aoki, K., Brankow, D., and Buck, L. (2003). Enhancement of therapeutic protein in vivo activities through glycoengineering. *Nat Biotechnol* **21**: 414–421.

Erslev, A. (1953). Humoral regulation of red cell production. *Blood* **8**: 349–357.

Eschbach, J. W., Egrie, J. C., Downing, M. R., Browne, J. K., and Adamson, J. W. (1987). Correction of the anemia of end-stage renal disease with recombinant human erythropoietin. Results of a combined phase I and II clinical trial. *N Engl J Med* **316**: 73–78.

Fan, Q., Leuther, K. K., Holmes, C. P., Fong, K. L., Zhang, J., and Velkovska, S. (2006). Preclinical evaluation of Hematide, a novel erythropoiesis stimulating agent, for the treatment of anemia. *Exp Hematol* **34**: 1303–1311.

Fishbane, S., Tare, N., Pill, J., and Haselbeck, A. (2003). Preclinical pharmacodynamics and pharmacokinetics of CERA (Continuous Erythropoietin Receptor Activator), an innovative erythropoietic agent for anemia management in patients with kidney disease. *J Am Soc Nephrol* **14**: 27A (Abstract SA-FC 123).

Franson, K. L., Burggraaf, Bouman-Trio, E. A. et al. (2005). A phase I, single and fractionated, ascending dose study evaluating the safety, pharmacokinetics, pharmacodynamics, and immunogenicity of an erythropoietic mimetic antibody fusion protein, CTNO528 in healthy male subjects. *Blood* **106**: 146b.

Gross, A. W., and Lodish, H. F. (2006). Cellular trafficking and degradation of erythropoietin and novel erythropoiesis stimulating protein (NESP). *J Biol Chem* **281**: 2024–2032.

Halstenson, C. E., Macres, M., Katz, S. A. et al. (1991). Comparative pharmacokinetics and pharmacodynamics of epoetin alfa and epoetin beta. *Clin Pharmacol Ther* **50**: 702–712.

Haselbeck, A., Bailon, P., Pahlke, W. et al. (2002). The discovery and characterization of CERA, an innovative agent for the treatment of anemia. *Blood* **100**: 227A (Abstract 857).

Hiramatsu, M., Kubota, J., Akizawa, T., Koshikawa, S., and the KRN 321 Study Group. (2005). Intravenous administration of KRN321 (darbepoetin alfa) once every 4 weeks improves and maintains haemoglobin concentrations in peritoneal dialysis patients. Paper presented at 42nd ERA-EDTA Congress, Istanbul, Turkey.

Jelkmann, W. (2007). Recombinant EPO production – points the nephrologist should know. *Nephrol Dial Transplant* **22**: 2749–2753.

Johnson, D. W., Pollock, C. A., and Macdougall, I. C. (2007). Erythropoiesis-stimulating agent hyporesponsiveness. *Nephrology (Carlton)* **12**: 321–330.

Kaufman, J. S., Reda, D. J., Fye, C. L. et al. (1998). Subcutaneous compared with intravenous epoetin in patients receiving hemodialysis. Department of Veterans Affairs Cooperative Study Group on Erythropoietin in Hemodialysis patients. *N Engl J Med* **339**: 578–583.

Keown, P. A. (1991). Quality of life in end-stage renal disease patients during recombinant human erythropoietin therapy. *Contrib Nephrol* **88**: 81–86.

Klinger, M., Arias, M., Vargemezis, V. et al. (2007). Efficacy of intravenous methoxy polyethylene-glycol epoetin beta administered every 2 weeks compared with epoetin administered 3 times weekly in patients treated by hemodialysis or peritoneal dialysis: a randomized trial. *Am J Kidney Dis* **50**: 989–1000.

Kwan, J. T., and Pratt, R. D. (2007). Epoetin delta, erythropoietin produced in a human cell line, in the management of anaemia in pre-dialysis chronic kidney disease patients. *Curr Med Res Opin* **23**: 307–311.

Lai, P. H., Everett, R., Wang, F. F., Arakawa, T., and Goldwasser, E. (1986). Structural characterization of human erythropoietin. *J Biol Chem* **261**: 3116–3121.

Lee, D. E., Son, W., Ha, B. J., Oh, M. S., and Yoo, O. J. (2006). The prolonged half-lives of new erythropoietin derivatives via peptide addition. *Biochem Biophys Res Commun* **339**: 380–385.

Levin, N. W., Fishbane, S., Canedo, F. V. et al. (2007). Intravenous methoxy polyethylene glycol-epoetin beta for haemoglobin control in patients with chronic kidney disease who are on dialysis: a randomised non-inferiority trial (MAXIMA). *Lancet* **370**: 1415–1421.

Lin, F. K., Suggs, S., and Lin, C. H. (1985). Cloning and expression of the human erythropoietin gene. *Proc Natl Acad Sci USA* **82**: 7580–7584.

Ling, B., Walczyk, M., Agarwal, A., Carroll, W., Liu, W., and Brenner, R. (2005). Darbepoetin alfa administered once monthly maintains hemoglobin concentrations in patients with chronic kidney disease. *Clin Nephrol* **63**: 327–334.

Livnah, O., Stura, E. A., Johnson, D. L., Middleton, S. A., Mulcahy, L. S., and Wrighton, N. C. (1996). Functional mimicry of a protein hormone by a peptide agonist: the EPO receptor complex at 2.8 A. *Science* **273**: 464–471.

Locatelli, F., Canaud, B., Giacardy, F., Martin-Malo, A., Baker, N., and Wilson, J. (2003). Treatment of anaemia in dialysis patients with unit dosing of darbepoetin alfa at a reduced dose frequency relative to recombinant human erythropoietin (rHuEpo). *Nephrol Dial Transplant* **18**: 362–369.

Locatelli, F., Villa, G., Arias, M. et al. (2004). CERA (Continuous Erythropoietin Receptor Activator) maintains hemoglobin levels in dialysis patients when administered subcutaneously up to once every 4 weeks. on behalf of the BA16286 study group. *J Am Soc Nephrol* **15**: 543A (Abstract SU-PO051).

Locatelli, F., Villa, G., Backs, W., and Pino, M.D. (2005). A phase 3 multicentre, randomised, double-blind, non-inferiority trial to evaluate the efficacy of Aranesp (darbepoetin alfa) once every 2 weeks (Q2W) vs Aranesp once weekly (QW) in patients on haemodialysis. Paper presented at 42nd ERA-EDTA Congress, Istanbul, Turkey.

Macdougall, I. C. (2005). C.E.R.A. (Continuous Erythropoietin Receptor Activator): a new erythropoiesis-stimulating agent for the treatment of anemia. *Curr Hematol Rep* **4**: 436–440.

Macdougall, I. C., Lewis, N. P., Saunders, M. J. et al. (1990). Long-term cardiorespiratory effects of amelioration of renal anaemia by erythropoietin. *Lancet* **335**: 489–493.

Macdougall, I. C., Roberts, D. E., Coles, G. A., and Williams, J. D. (1991). Clinical pharmacokinetics of epoetin (recombinant human erythropoietin). *Clinical Pharmacokinetics* **20**: 99–113.

Macdougall, I. C., Gray, S. J., Elston, O. et al. (1999). Pharmacokinetics of novel erythropoiesis stimulating protein compared with epoetin alfa in dialysis patients. *J Am Soc Nephrol* **10**: 2392–2395.

Macdougall, I. C., Matcham, J., and Gray, S. J. (2003a). Correction of anaemia with darbepoetin alfa in patients with chronic kidney disease receiving dialysis. *Nephrol Dial Transplant* **18**: 576–581.

Macdougall, I. C., Bailon, P., Tare, N. et al. (2003b). CERA (Continuous Erythropoietin Receptor Activator) for the treatment of renal anemia: an innovative agent with unique receptor binding characteristics and prolonged serum half-life. *J Am Soc Nephrol* **14**: 769A (Abstract SU-PO1063).

Macdougall, I. C., Robson, R., Opatrna, S. et al. (2006a). Pharmacokinetics and pharmacodynamics of intravenous and subcutaneous continuous erythropoietin receptor activator (C.E.R.A.) in patients with chronic kidney disease. *Clin J Am Soc Nephrol* **1**: 1211–1215.

Macdougall, I.C., Tucker, B., Yaqoob, M. et al. (2006b). Hematide, a synthetic peptide-based erythropoiesis stimulating agent, achieves correction of anemia and maintains Hb in patients with CKD not on dialysis. Ann Meet Am Soc Nephrol 39, Abs F-FC079.

Macdougall, I. C., Padhi, D., and Jang, G. (2007a). Pharmacology of darbepoetin alfa. *Nephrol Dial Transplant* **22**(Suppl. 4): iv2–iv9.

Macdougall, I. C., Casadevall, N., Froissart, M. et al. (2007b). Treatment of erythropoietin antibody-mediated pure red cell aplasia with a novel synthetic peptide-based erythropoietin receptor agonist. Ann Meet Am Soc Nephrol **40**: Abs SU–FC061.

Macdougall, I. C. (2008). Novel erythropoiesis-stimulating agents: A new era in anemia management. *Clin J Am Soc Nephrol* **3**: 200–207.

Macdougall, I. C., Walker, R., Provenzano, R. et al. (2008). C.E.R.A. corrects anemia in patients with chronic kidney disease not on dialysis: results of a randomized clinical trial. *Clin J Am Soc Nephrol* **3**: 337–347.

Maitani, Y., Moriya, H., Shimoda, N., Takayama, K., and Nagai, T. (1999). Distribution characteristics of entrapped recombinant human erythropoietin in liposomes and its intestinal absorption in rats. *Int J Pharm* **185**: 13–22.

Mann, J., Kessler, M., Villa, G. et al. (2005). Darbepoetin alfa (DA) every 2 weeks (Q2W) is effective in dialysis patients (pts) on high prior weekly (QW) doses of rHuEPO, a sub-analysis of combined data from 8 prospective multicenter studies. Paper presented at ASN 38th Annual Renal Week Meeting, Philadelphia.

Mann, J., Kessler, M., Villa, G. et al. (2007). Darbepoetin alfa once every 2 weeks for treatment of anemia in dialysis patients: a combined analysis of eight multicenter trials. *Clin Nephrol* **67**: 140–148.

Martin, K. J. (2007). Epoetin delta in the management of renal anaemia: results of a 6-month study. on behalf of the Epoetin Delta 3001 Study Group. *Nephrol Dial Transplant* **22**: 3052–3054.

Mayer, G., Thum, J., Cada, E. M. et al. (1988). Working capacity is increased following recombinant human erythropoietin treatment. *Kidney Int* **34**: 525–528.

Mitragotri, S., Blankschtein, D., and Langer, R. (1995). Ultrasound-mediated transdermal protein delivery. *Science* **269**: 850–853.

Miyake, T., Kung, C. K., and Goldwasser, E. (1977). Purification of human erythropoietin. *J Biol Chem* **252**: 5558–5564.

National Kidney Foundation (2007). KDOQI clinical practice guideline and clinical practice recommendations for anemia in chronic kidney disease – 2007 update of hemoglobin target. *Am J Kidney Dis* **50**: 471–530.

Nissenson, A. R., Swan, S. K., Lindberg, J. S. et al. (2002). Randomized, controlled trial of darbepoetin alfa for the treatment of anemia in hemodialysis patients. *Am J Kidney Dis* **40**: 110–118.

NKF-DOQI clinical practice guidelines for the treatment of anemia of chronic renal failure (1997). National Kidney Foundation-Dialysis Outcomes Quality Initiative. *Am J Kidney Dis* **30** (4 Suppl 3): S192–240.

Padhi, D., and Jang, G. (2005). Pharmacokinetics (PK) of Aranesp (darbepoetin alfa) in patients with chronic kidney disease (CKD) (Abstract MP 165). Paper presented at 42nd ERA-EDTA Congress, Istanbul, Turkey.

Padhi, D., Ni, L., Cooke, B., Marino, R., and Jang, G. (2006). An extended terminal half-life for darbepoetin alfa: results from a single-dose pharmacokinetic study in patients with chronic kidney disease not receiving dialysis. *Clin Pharmacokinet* **45**: 503–510.

Pérez-Oliva, J. F., Casanova-González, M., García-García, I. et al. (2005). Comparison of two recombinant erythropoietin formulations in patients with anemia due to end-stage renal disease on hemodialysis: A parallel, randomized, double blind study. the Bioequivalence Study of Erythropoietin Group. *BMC Nephrology* **6**: 1–11.

Phrommintikul, A., Haas, S. J., Elsik, M. et al. (2007). Mortality and target haemoglobin concentrations in anaemic patients with chronic kidney disease treated with erythropoietin: a meta-analysis. *Lancet* **369**: 381–388.

Pötgens, A., Haberl, U., Rybka, A. et al. (2006). An optimized supravalent EPO mimetic peptide with unprecedented efficacy. *Ann Hematol* **85**: 643.

Provenzano, R., Besarab, A., Macdougall, I. C. et al. (2007). The continuous erythropoietin receptor activator (C.E.R.A.) corrects anemia at extended administration intervals in patients with chronic kidney disease not on dialysis: results of a phase II study. and BA 16528 Study Investigators. *Clin Nephrol* **67**: 306–317.

Qureshi, S. A., Kim, R. M., Konteatis, Z., Biazzo, D. E., Motamedi, H., and Rodrigues, R. (1999). Mimicry of erythropoietin by a nonpeptide molecule. *Proc Natl Acad Sci USA* **96**: 12156–12161.

Rossert, J., and Eckardt, K.-U. (2005). Erythropoietin receptors: their role beyond erythropoiesis. *Nephrol Dial Transplant* **20**: 1025–1028.

Sarac, E., Veres, Z., Tallam, S., Barton, D., and Gemmel, D. (2004). Hemoglobin responses following *de novo* darbepoetin alfa (ARANESP) vs epoetin alfa (PROCRIT) administration in anemic chronic kidney disease (CKD) patients. Paper presented at ASN 37th Renal Week Meeting, St Louis.

Schellekens, H. (2008). The first biosimilar epoetin: but how similar is it? *Clin J Am Soc Nephrol* **3**: 174–178.

Sikole, A., Spasovski, G., Zafirov, D., and Polenakovic, M. (2002). Epoetin omega for treatment of anemia in maintenance hemodialysis patients. *Clin Nephrol* **57**: 237–245.

Singh, A. K., Szczech, L., Tang, K. L. et al. (2006). Correction of anemia with epoetin alfa in chronic kidney disease. *N Engl J Med* **335**: 2085–2098.

Skibeli, V., Nissen-Lie, G., and Torjesen, P. (2001). Sugar profiling proves that human serum erythropoietin differs from recombinant human erythropoietin. *Blood* **98**: 3626–3634.

Spinowitz, B. S., and Pratt, R. D. (2006). Epoetin delta is effective for the management of anaemia associated with chronic kidney disease. *Curr Med Res Opin* **22**: 2507–2513.

Spinowitz, B., Coyne, D. W., Lok, C. E. et al. (2008). C.E.R.A. maintains stable control of hemoglobin in patients with chronic kidney disease on dialysis when administered once every two weeks. *Am J Nephrol* **28**: 280–289.

Stead, R. B., Lambert, J., Wessels, D., Iwashita, J. S., Leuther, K. K., and Woodburn, K. W. (2006). Evaluation of the safety and pharmacodynamics of Hematide, a novel erythropoietic agent, in a phase 1, double-blind placebo-controlled, dose-escalation study in healthy volunteers. *Blood* **108**: 1830–1834.

Sulowicz, W., Locatelli, F., Ryckelynck, J.-P. et al. (2007). Once-monthly subcutaneous C.E.R.A. maintains stable hemoglobin control in patients with chronic kidney disease on dialysis and converted directly from epoetin one to three times weekly. *Clin J Am Soc Nephrol* **2**: 637–646.

Suranyi, M. G., Lindberg, J. S., Navarro, J., Elias, C., Brenner, R. M., and Walker, R. (2003). Treatment of anemia with darbepoetin alfa administered de novo once every other week in chronic kidney disease. *Am J Nephrol* **23**: 106–111.

Sytkowski, A. J., Lunn, E. D., Risinger, M. A., and Davis, K. L. (1999). An erythropoietin fusion protein comprised of identical repeating domains exhibits enhanced biological properties. *J Biol Chem* **274**: 24773–24778.

Tangvoranuntakul, P., Gagneux, P., Diaz, S., Bardor, M., Varki, N., and Varki, A. (2003). Human uptake and incorporation of an immunogenic nonhuman dietary sialic acid. *Proc Natl Acad Sci USA* **100**: 12045–12050.

Teruel, J. L., Pascual, J., Jiménez, M. et al. (1991). Hemodynamic changes in hemodialyzed patients during treatment with recombinant human erythropoietin. *Nephron* **58**: 135–137.

The Court Service – Court of Appeal – Civil Judgement. (2002). In TKT's technology, those cells were designated as R223 cells. Neutral Citation Number: EWCA Civ. 1096. www.hmcourtsservice.gov.uk/judgmentsfiles/j1329/Kirin_v_Hoechst.htm.

Toto, R. D., Pichette, V., Navarro, J. et al. (2004). Darbepoetin alfa effectively treats anemia in patients with chronic kidney disease with *de novo* every-other-week administration. *Am J Nephrol* **24**: 453–460.

Tsubakihara, Y., Hiramatsu, M., Iino, Y., Akizawa, T., Koshikawa, S., and the KRN321 Study Group. (2004). The pharmacokinetics of KRN321 (darbepoetin alfa) after subcutaneous (SC) administration: a comparison between peritoneal dialysis and predialysis chronic renal failure (CRF) patients in Japan. Paper presented at 41st ERA-EDTA Congress, Lisbon, Portugal.

van Buren, M., van Manen, J.G., Bakker-de Bruin, Y., Boeschoten, E.W. on behalf of the Aranesp registry group (2005). Feasibility to increase the dosing interval of darbepoetin alfa (Aranesp®) to once monthly (QM) in patients with chronic renal insufficiency (CRI). Paper presented at 42nd ERA-EDTA Congress, Istanbul, Turkey.

Vanrenterghem, Y., Barany, P., Mann, J. F. et al. (2002). Randomized trial of darbepoetin alfa for treatment of renal anemia at a reduced dose frequency compared with rHuEPO in dialysis patients. *Kidney Int* **62**: 2167–2175.

Venkatesan, N., Yoshimitsu, J., Ohashi, Y. et al. (2006). Pharmacokinetic and pharmacodynamic studies following oral administration of erythropoietin mucoadhesive tablets to beagle dogs. *Int J Pharm* **310**: 46–52.

Way, J. C., Lauder, S., Brunkhorst, B., Kong, S. M., Qi, A., and Webster, G. (2005). Improvement of Fc-erythropoietin structure and pharmacokinetics by modification at a disulfide bond. *Protein Eng Des Sel* **18**: 111–118.

Winearls, C. G., Oliver, D. O., Pippard, M. J., Reid, C., Downing, M. R., and Cotes, P. M. (1986). Effect of human erythropoietin derived from recombinant DNA on the anaemia of patients maintained by chronic haemodialysis. *Lancet* **ii**: 1175–1178.

Woodburn, K., Leuther, K., Holmes, C. et al. (2007a). Renal excretion is the primary route of elimination for the erythropoiesis stimulating agent Hematide as assessed by quantitative whole-body autoradioluminography in Sprague Dawley rats. *Nephrol Dial Transplant* **22**: (**Suppl 6**): Abs SaP333.

Woodburn, K. W., Fan, Q., Winslow, S. et al. (2007b). Hematide is immunologically distinct from erythropoietin and corrects anemia induced by antierythropoietin antibodies in a rat pure red cell aplasia model. *Exp Hematol* **35**: 1201–1208.

Wrighton, N. C., Farrell, F. X., Chang, R., Kashyap, A. K., Barbone, F. P., and Mulcahy, L. S. (1996). Small peptides as potent mimetics of the protein hormone erythropoietin. *Science* **273**: 458–464.

Yeh, P., Landais, D., Lemaitre, M. et al. (1992). Design of yeast-secreted albumin derivatives for human therapy: biological and antiviral properties of a serum albumin-CD4 genetic conjugate. *Proc Natl Acad Sci USA* **89**: 1904–1908.

CHAPTER 5

Erythropoietin, Anemia and Kidney Disease

AJAY K. SINGH, TEJAS PATEL, SHONA PENDSE AND SAIRAM KEITHI-REDDY
Renal Division, Brigham and Women's Hospital and Harvard Medical School, 75 Francis Street, Boston, MA 02115, USA

Contents

I. Introduction	49
II. Erythropoietin pathophysiology in CKD patients	49
III. Anemia of chronic kidney disease	51
IV. Anemia and erythropoietin treatment in children with CKD	55
V. Conclusions	56
References	56

I. INTRODUCTION

The introduction of recombinant human erythropoietin (rHuEPO) for the treatment of the anemia of chronic kidney disease (CKD) has transformed the lives of millions of patients (Pendse and Singh, 2005). Sales of recombinant erythropoietin are estimated to exceed $22 billion (Cotter et al., 2006) and newer analog such as darbepoietin and continuous erythropoietin receptor activator (CERA) have been introduced (Macdougall, 2006, 2008; Del Vecchio and Locatelli, 2008). In the era prior to the discovery and introduction of recombinant erythropoietin, severe anemia in dialysis and non-dialysis CKD patients resulted in fatigue and the need for frequent blood transfusions (Fried, 1978). Iron overload was a common complication because patients required multiple blood transfusions. In the past 20 years, since the introduction of recombinant erythropoietin (erythropoiesis stimulating agents (ESA) is the optimal term for commercially available erythropoietic agents), the average hemoglobin of CKD patients has steadily increased to an average hemoglobin (Hgb) above 10 g/dl (United States Renal Data System, 2006). Indeed, in 1991, the mean Hgb in hemodialysis patients was 9.7 g/dl, whereas in 2005, the Hgb level was approximately 12 g/dl (United States Renal Data System, 2006). The widespread use of ESAs has resulted in a major reduction in blood transfusions and in antibody sensitization and infections (Wish and Coyne, 2007). However, while the advantages of ESAs in treating anemia have been widely recognized, the potential risks of ESAs in completely correcting CKD anemia have been debated quite vigorously (Singh and Fishbane, 2008; Carrera and Macdougall, 2008). The publication of studies suggesting harm with the targeting of higher hemoglobin levels has generated considerable controversy (Singh et al., 2006; Drueke et al., 2006). These studies have resulted in a revision of National Kidney Foundation Kidney Disease Quality Initiative (K-DOQI) guidelines (KDOQI, 2006, 2007) and in a Food and Drug Administration (FDA) black box warning (www.fda.gov). Furthermore, concerns about Hgb cycling (Fishbane and Berns, 2007; Singh et al., 2008) and its relationship to outcome have emerged (Yang et al., 2007; Brunelli et al., 2008). There is also renewed appreciation that the treatment of anemia with erythropoietin requires concomitant treatment of iron deficiency (Auerbach et al., 2008), even among inflamed patients on dialysis (Coyne et al., 2007; Singh et al., 2007; Kapoian et al., 2008).

II. ERYTHROPOIETIN PATHOPHYSIOLOGY IN CKD PATIENTS

Erythropoietin is a glycoprotein hormone that regulates the survival and subsequent production of red cell precursors in the red cell lineage (Bunn, 2006), although the pleiotropic effects of erythropoietin are also well recognized (Sasaki, 2003), including in cancer (Milano and Schneider, 2007) and cardiac (Lipsic et al., 2006), brain (Grasso et al., 2007) and renal ischemia (Sharples et al., 2005). Pluripotent hematopoetic stem cells differentiate into lineage-committed blood forming erythroid (BFU-E) and colony forming erythroid (CFU-E) progenitor cells (Figure 5.1) (Testa, 2004). CFU-Es mature into early and late erythroblasts that subsequently mature into reticulocytes and erythrocytes. Although some of the molecular events remain controversial, there is consensus that erythropoietin protects erythroblasts from programmed cell death, facilitates maturation and stimulates proliferation of erythroblasts (reviewed in Koury et al., 2002; Wojchowski et al., 2006). Erythropoietin represses apoptosis through the induction of Bcl, a member of the Bcl-2 family

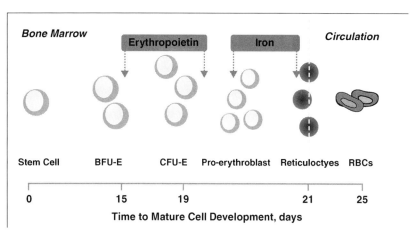

FIGURE 5.1 Steps in erythropoiesis in the bone marrow.

(Silva et al., 1996; Gregoli and Bondurant, 1997). Epo receptor and GATA-1 are also upregulated (Aispuru et al., 2008). In addition, it has been hypothesized that the Fas/FasL system, also under the influence of erythropoietin, negatively regulates erythropoiesis through its effects on mature erythroblasts (De Maria et al., 1999; Liu et al., 2006). Data from studies in astronauts in space flight suggest that erythropoietin also appears to modulate survival of young circulating red cells (a process termed neocytolysis) (Rice and Alfrey, 2005; Risso et al., 2007). Neocytolysis comprises of hemolysis of red cells that occurs in the setting of epo deficiency. It is initiated by a fall in erythropoietin levels, so this hormone remains the major regulator of red cell mass both with anemia and with red cell excess. Uremia, childbirth (the emergence of newborns from a hypoxic uterine environment) and the descent of polycythemic high-altitude dwellers to sea level may be dependent on erythropoietin-driven neocytolysis (Rice et al., 1999; Rice and Alfrey, 2000).

Erythropoietin is produced by the liver and kidney (Lacombe et al., 1988, 1991; Eckardt et al., 1994; Göpfert et al., 1997; Jelkmann, 2001). At birth, erythropoietin production switches largely from the liver (where erythropoietin is produced by both hepatocytes and interstitial fibroblastoid cell) to the kidney (in adults the liver may contribute approximately 15% of total erythropoietin production). In the kidney, erythropoietin is produced by peritubular cells in the renal cortex and, to a lesser extent, in the outer medulla (Lacombe et al., 1988). Regulation of erythropoietin synthesis in the kidneys relies on a feedback loop in response to hypoxia (this is described in more detail elsewhere in the book). The expression of erythropoietin mRNA and protein is regulated primarily at the transcriptional level through the hypoxia-inducible transcription factor HIF (Blanchard et al., 1993; Eckardt and Kurtz, 2005; Stockmann and Fandrey, 2006). Under normoxic situations, erythropoietin expression in the kidney (as well as the liver) is inhibited by the oxygen-dependent degradation of the alpha-subunit of HIF, a process that is mediated by prolyl hydroxylation. Three HIF-α prolyl hydroxylases (PHDs) have been cloned and characterized. PHD2 plays the dominant biologic role (Bunn, 2006). It has been hypothesized that the same cell types that synthesize erythropoietin also function in sensing hypoxia – the kidney plays a 'critmeter' role (Donelly, 2001; Donnelly, 2003). Further, in this critmeter role, it has been postulated that blood flow to these erythropoietin producing cells influences oxygen tension – raising the possibility that renal hemodynamics may also influence erythropoietin synthesis (Dunn et al., 2007).

In patients with progressive kidney damage, there is a reduction in the kidneys' ability to produce erythropoietin (McGonigle et al., 1984). However, there is no close correlation between the degree of renal impairment and the erythropoietin level (Radtke et al., 1979; Naets et al., 1986; Ross et al., 1994). Other important causes of anemia in CKD patients are summarized in Table 5.1. Iron deficiency is an important cause of anemia since it is progressive and may be further exacerbated among anemic CKD patients once treatment with exogenous recombinant erythropoietin is begun. In healthy individuals, serum levels of erythropoietin range between 1 and 27 mu/ml (mean 6.2 ± 4.3 mu/ml) (Garcia et al., 1990), whereas in CKD patients, serum levels of erythropoietin in patients have been reported to be in the 4.2 and 102 mu/ml (mean 29.5 ± 4.0 mu/ml). Thus, even though erythropoietin deficiency is the primary cause of the anemia of CKD, the uremic state may contribute to the anemia by creating a state of erythropoietin resistance.

TABLE 5.1 Causes of anemia in dialysis and non-dialysis CKD patients

Epo deficiency
Iron deficiency
Inflammation
Blood loss
Folate/B12 deficiency
Renal osteodystrophy
Hemoglobinopathies

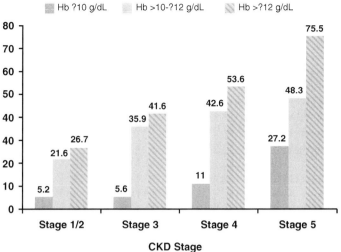

FIGURE 5.2 Prevalence of anemia stratified by different levels of anemia and at different stages of CKD.

III. ANEMIA OF CHRONIC KIDNEY DISEASE

Anemia is a common complication in CKD and increases in prevalence as kidney disease worsens (Figure 5.2). The World Health Organization defines anemia for males as a hemoglobin level <13.0 g/dl and in women as a hemoglobin level <12.0 g/dl (NKF-DOQI, 1997). Measurement of hemoglobin, rather than hematocrit (HCT), is preferred for assessing anemia, because it provides as an absolute level generally unaffected by shifts in plasma water that may occur in patients on dialysis (NKF-DOQI, 1997, 2001). Furthermore, since the HCT relies on the mean corpuscular volume (MCV) for calculation (Henry, 1996). HCT – MCV/erythrocyte count, or more precisely MCV (in fL) = (HCT [in l/l] × 1000)/(rbc count [in millions/μl]) factors affecting the accuracy of MCV such as storage at room temperature (Britten et al., 1969) and hyperglycemia (Holt et al., 1982), both affecting the MCV, may result in imprecision in the HCT value. The anemia of CKD is generally normocytic and normochromic (Pendse and Singh, 2005).

The differential diagnosis of anemia of CKD is shown in Table 5.2. The National Kidney Foundation (NKF) Kidney Disease Quality Initiative (KDOQI) recommends that an anemia work-up in CKD patients should be initiated when the Hgb < 11 g/dL (Hct is <33% in pre-menopausal females and pre-pubertal patients) and an Hgb <12 g/dL (Hct is <37%) in adult males and post-menopausal females (KDOQI, 2006, 2007). Work-up in patients should include evaluation of the red blood cell indices, the reticulocyte count and iron parameters, including serum iron, total iron binding capacity (TIBC) (the transferrin saturation or TSAT is the serum iron × 100 divided by TIBC) and serum ferritin (KDOQI, 2006, 2007). A test for occult blood in stool should also be performed. Since the central problem is usually an absolute or relative deficiency of erythropoietin, treatment

TABLE 5.2 Differential diagnosis of anemia in CKD patients

a. Microcytic (MCV < 80 fl)
- Iron-deficiency anemia
- Thalassemia
- Anemia of chronic disease
- Sideroblastic anemia (hereditary, lead)

b. Normocytic (MCV 80–100 fl)
- Nutritional anemias, including iron-deficiency
- Anemia of renal insufficiency
- Hemolytic anemias

Red cell intrinsic etiologies including membranopathies, enzymopathies, hemoglobinopathies;

Red cell extrinsic causes including drug, virus and lymphoid associated anemias and transfusion reactions

Microangiopathic anemias (TTP/HUS)

Infection associated
- Anemia of chronic disease
- Bone marrow disorders

Aplastic anemia

Pure red cell aplasia

Ineffective erythropoiesis

c. Macrocytic (MCV > 100 fl)
- Drug-induced
- Nutritional

Vitamin B12 deficiency

Folate deficiency
- Hemolytic anemia
- Clonal hematologic disorders
- Alcohol
- Liver disease
- Cold agglutinins

TABLE 5.3 Milestones in the development of erythropoietin

1878	Bert and Jourdaner propose that symptoms of anemia result from hypoxia
1906	Carnot and Deflandre demonstrate that serum from an anemic donor rabbit injected into normal rabbits results in increased erythropoeisis
1953	Erslev demonstrates plasma from anemic rabbits contains a factor that stimulates erythropoeisis
1977	Goldwasser's laboratory purifies erythropoietin from large quantities of urine obtained from patients with aplastic anemia
1985	Epo gene cloned by Lin working at Amgen
1989	Eschbach publishes study in *New England Journal of Medicine* demonstrating efficacy of erythropoietin in the treatment of CKD anemia
1989	Food and Drug Administration (FDA) approves the use of erythropoietin in the treatment of the anemia of chronic renal failure

with erythropoietin should be begun. The availability of erythropoietin has allowed correction or partial correction of anemia in the majority of CKD patients. Key milestones in the development of recombinant erythropoietin are shown in Table 5.3. Treatment with erythropoietin depends on achieving and maintaining an adequate erythropoietin level. This may be challenging in patients who have erythropoietin resistance (Pendse and Singh, 2005). While there is no consensus on how to define quantitatively erythropoietin resistance, one commonly used definition is 'a failure to achieve and/or maintain target hemoglobin levels at an erythropoietin dose of 450 IU/kg/week when administered intravenously, or 300 IU/kg/week when administered subcutaneously' (Drüeke, 2001; Kwack and Balakrishnan, 2006; van der Putten et al., 2008). These erythropoietin resistant patients represent approximately one-third of all hemodialysis patients, at least in the USA. Causes of erythropoietin resistance are shown in Table 5.4. Patients usually have a high ferritin, low TSAT and an Hgb of <11 g/dl. Recent observations from a randomized controlled study, the DRIVE study, support the empiric use of iron in these patients (Coyne et al., 2007; Singh et al., 2007; Kapoian et al., 2008).

The consequences and potential benefit of anemia correction have been well studied and include: improved exercise tolerance (Mayer et al., 1988; Teehan et al., 1989; Robertson et al., 1990; Braumann et al., 1991), cardiovascular effects (Cannella et al., 1990; Macdougall et al., 1990; Pascual et al., 1991; Wizemann et al., 1992, 1993; Harnett et al., 1995), cognitive abnormalities (Wolcott et al., 1989), effects on menstruation and sexual activity (Schaefer et al., 1989; Sobh et al., 1992) and quality of life (Alexander et al., 2007; Abu-Alfa et al., 2008; Weisbord and Kimmel, 2008). There are also substantial observational data to support a relationship between anemia and mortality and cardiovascular risk (Ma et al., 1999; Li et al., 2004; Robinson et al., 2005; Roberts et al., 2006; Xue et al., 2002). In studies on dialysis patients, hemoglobin levels of 10.0 g/dl or less are associated with a higher risk of death when those compared to levels >10 g/dl (Ma et al., 1999; Li et al., 2004; Robinson et al., 2005; Roberts et al., 2006; Xue et al., 2002). In these studies, generally, an HCT level between 33% and 36% was associated with the lowest risk of death. Left ventricular hypertrophy has also been proposed as a major consequence of the anemia of CKD (Levin et al., 1999; Levin, 2002).

Several forms of erythropoietin are available commercially including epoetin-alfa, epoetin-beta, darbepoetin (discussed in detail in Chapter 4). A central question in the correction of CKD anemia is the optimal target hemoglobin concentration and this issue, in particular, has been a source of much controversy.

A. Target Hemoglobin in CKD Patients

The FDA recommended an HCT of 30–33% when erythropoietin was launched in 1989. This was largely based on the phase III (Eschbach et al., 1989) and earlier phase I and II trials (Eschbach et al., 1987) conducted by Amgen. Most recently, the epoetin label has a black box warning with a recommendation to achieve and maintain the hemoglobin

TABLE 5.4 Causes of erythropoietin resistance

- Absolute or functional iron-deficiency anemia
- Hyperparathyroidism
- Aluminum toxicity
- Concurrent infection
- Systemic inflammation (elevated acute phase proteins i.e. CRP, fibrinogen)
- Vitamin B12 or folate deficiency
- Hemolysis
- Bone marrow disorders
- Hemoglobinopathies
- Carnitine deficiency
- Angiotensin-converting enzyme inhibitor therapy
- Anti-erythropoietin antibodies
- Gender (females have less response to erythropoietin, and thus need higher doses, than do males)
- Advanced age
- Mode of dialysis (patients on hemodialysis need higher doses of erythropoietin than those on peritoneal dialysis)
- Inadequate dialysis (decreased dose or adequacy)
- Chemical or biological contamination of water used for dialysis

level in the 10–12 g/dl range (www.fda.gov). Data from the phase III study and from several studies evaluating quality of life have supported a target HCT of 36%. Many studies have demonstrated clinical benefits with a higher HCT (Tonelli et al., 2004), although the quality of some of these studies has been debated (Singh, 2008), although most recently the FDA has indicated that the QOL studies have major methodologic flaws (US Food and Drug Administration, 2007). Large observational studies have provided evidence to support complete correction of anemia in CKD patients. Regidor et al. (2006) studied the records of 58 058 hemodialysis patients at a large dialysis organization. The all cause and cardiovascular mortality hazard ratios were found to increase progressively for hemoglobin levels less than 11.5–12.0 g/dl. Notably, however, no favorable effect on mortality was observed beyond a hemoglobin level of 13.0 g/dl. Similarly, Locatelli et al. (2004) studied 4591 subjects enrolled in the Dialysis Outcomes and Practice Patterns Study and found hemoglobin level to be an important predictor of risk for mortality and hospitalizations. However, while observational studies can point to associations and help formulate hypotheses they not provide insights regarding causality. For this, randomized controlled studies are essential.

B. Randomized Controlled Trials

Randomized controlled studies have been published thus far and are discussed in a recent meta-analysis (Phrommintikul et al., 2007). With the exception of four studies, the Normal HCT Study (Besarab et al., 1998), the Canada-Europe Study (Parfrey et al., 2005), CHOIR (Singh et al., 2006) and CREATE (Drueke et al., 2006), observations have been limited by small sample sizes. The Normal HCT study was a randomized controlled study of hemodialysis patients with established heart disease comparing an HCT target of 42% to 30% (Besarab et al., 1998). The study was stopped by the Data Safety Monitoring Board because of concern for increased mortality in the patients randomized to the higher HCT group. However, the patients in the higher HCT group also had a very nearly statistically significantly higher rate of non-fatal myocardial infarction (MI) or death. Several explanations were entertained to explain these findings. These included the possibility that high HCTs in the higher hemoglobin group resulted in hemoconcentration and therefore a higher rate of thrombosis, greater use of iron and a lower dialysis adequacy in the higher HCT group. In this study, of eight categories of quality of life studied using the SF-36 instrument, only one showed improvement in relation to higher HCT levels. The Canada-Europe study also randomized hemodialysis patients to a higher versus lower hemoglobin (hemoglobin values of 13.0 versus 11.0 g/dl, respectively) (Kapoian et al., 2008). However, in contrast to the Normal HCT Study, Parfrey et al. (2005) selected patients that were not at high risk of cardiovascular disease by excluding patients with symptomatic heart disease as well as those with left ventricular dilatation at baseline. Moreover, they enrolled incident dialysis patients. The primary endpoint was a change in left ventricular volume index (LVVI). However, changes in parameters of heart failure, stroke and quality of life were also measured. No significant benefit in either of the cardiac structural or functional parameters was observed in the high versus low hemoglobin groups. However, a statistically significantly higher rate of cerebrovascular accident in the higher hemoglobin group was observed on secondary analysis. Quality of life did show a difference in the high versus lower hemoglobin group with respect to the Vitality score, which improved over time in patients randomized to the higher hemoglobin. The 6-minute walk test and two other measures of quality of life did not improve significantly in the high hemoglobin group in this study.

The CHOIR and CREATE studies have recently provided great insight into the issue of the target hemoglobin level in CKD patients not receiving dialysis (Singh et al., 2006; Drueke et al., 2006). CHOIR was an open-label, randomized trial that studied 1432 patients with CKD: 715 patients randomized to receive epoetin alfa targeted to achieve a hemoglobin of 13.5 g/dl and 717 were randomized to receive epoetin alfa targeted to achieve a hemoglobin of 11.3 g/dl (Singh et al., 2006). The median study duration was 16 months. The primary endpoint was a composite of death, myocardial infarction, congestive heart failure (CHF), hospitalization (excluding hospitalization during which renal replacement therapy occurred) and stroke. Two hundred and twenty-two composite events occurred: 125 events among the high hemoglobin group and 97 events among the low hemoglobin group ($P = 0.03$, hazard ratio of 1.34; with 95 percent confidence interval of 1.03 and 1.74). The higher rate of composite events was explained largely by a higher rate of death (48% higher risk, $P = 0.07$) or CHF hospitalization (41%, $P = 0.07$). Although neither death nor CHF hospitalization were statistically significantly higher in the higher versus lower hemoglobin group, the study was not powered for this purpose. Among other secondary endpoints, quality of life showed improvement in both groups but was not significantly better in the higher versus lower hemoglobin groups. Notably, more subjects in the high hemoglobin group experienced at least one serious adverse event compared to the low hemoglobin group. The Cardiovascular Risk Reduction by Early Anemia Treatment with Epoetin beta (CREATE) study enrolled approximately 600 patients (Drueke et al., 2006). Subjects were randomized to an early anemia correction or a late anemia correction group (Macdougall, 2006). The early anemia correction group received epoetin beta therapy immediately for a target hemoglobin 13–15 g/dl. The late anemia correction group did not receive treatment until their hemoglobin was <10.5 g/dl; their target hemoglobin was 10.5–11.5 g/dl. The study showed that 'complete correction' was not associated with a statistically significantly higher rate of the first cardiovascular event (58 events in the high hemoglobin group versus 47 events in the low hemoglobin group; hazard ratio of 0.78, 95% confidence interval, 0.53 to 1.14; $P = 0.20$). However, left ventricular mass index remained stable in both groups, but dialysis was required in more patients in the higher versus lowed hemoglobin group (127 versus 111, $P = 0.03$). On the other hand, unlike CHOIR, in CREATE a quality of life benefit, at least in year 1 of the study, was observed for the higher versus lower hemoglobin group.

In recent meta-analyses, an aggregate of randomized studies points to increased risk with targeting an Hgb of >12 g/dl (KDOQI, 2007; Phrommintikul et al., 2007). Indeed, in a subgroup analysis where pre-dialysis and dialysis trials were disaggregated, Phrommintikul and co-workers reported that the results showing harm for higher hemoglobin levels were similar. Phrommintikul et al. suggest that the 'combined data is [sic] the best possible summation of the current database regarding this issue.' Separately, Strippoli et al. (2004), in an editorial, point out that the increased mortality observed in the higher hemoglobin group is independent of the kidney disease stage (pre-dialysis versus dialysis, interaction 0.95).

C. Role of ESAs on Kidney Outcomes

There are conflicting data on whether ESA therapy confers renal protection (Rossert and Froissart, 2006; Rossert et al., 2006; Singh, 2007). The rationale for a renoprotective role is that treatment increases the Hgb and thereby oxygen delivery to vulnerable tissues which, in turn, reduces tissue hypoxia. Recent studies in ischemia–reperfusion models support an anti-ischemic role for ESAs (Sharples et al., 2005; Sharples and Yaqoob, 2006a, 2006b; Imamura et al., 2007). It has also been hypothesized that the anti-apoptotic properties of erythropoietin protect renal tubular cells from undergoing apoptosis in response to hypoxia. Aside from an anti-ischemic effect of ESAs, clinical studies support a renoprotective effect of anemia correction (Rossert and Froissart, 2006; Singh, 2007). *Post-hoc* analysis of the Reduction in Endpoints in Non-insulin-dependent Diabetes Mellitus with the Angiotensin II Antagonist Losartan (RENAAL) study suggests that a higher hemoglobin is associated with a slower rate of progression of type 2 diabetic kidney disease (Mohanram et al., 2004). As well, a study by Gouva et al. (2004), although limited in its design and sample size, also suggests that anemia treatment may have an ameliorative effect in CKD progression.

In the CHOIR (Singh et al., 2006) and CREATE (Drueke et al., 2006) trials, kidney failure was studied as a prespecified endpoint. CREATE showed a higher risk of developing end-stage renal disease (ESRD) requiring dialysis in patients randomized to the higher Hgb concentration. In the higher Hgb group, 127 patients developed ESRD compared to 111 patients in the low Hgb group. There was a statistically significant difference in the time to initiation of dialysis. The higher Hgb group had a shorter time to dialysis compared to the lower Hgb group ($P = 0.03$). CHOIR showed a trend to a higher rate of dialysis in the higher Hgb group. The ACORD study, published recently (in 2007), enrolled patients with early diabetic nephropathy and was unable to show a benefit in retarding progression of kidney disease in patients targeted to a higher Hgb level, although the study was underpowered (Ritz et al., 2007). Earlier studies in animal models have examined whether ESA treatment affects kidney function (Garcia et al., 1988; Lafferty et al., 1991). In both studies from Brenner's laboratory, erythropoietin treatment in rats with renal ablation was associated with an acceleration in glomerular injury because of worsened systemic and glomerular hypertension. The authors stated: 'The accelerated progression associated with normalization of HCT suggests that efforts to fully correct anemia in predialysis patients with advancing renal disease may prove harmful to residual renal function' (Lafferty et al., 1991).

D. Erythropoietin Treatment and the Risk of Hypertension and Graft Thrombosis

Epoetin therapy has been implicated in the pathogenesis of hypertension, thrombosis and atherosclerosis in CKD patients. Several authors in both non-dialysis CKD and children have reported exacerbation of hypertension on dialysis (Bianchetti et al., 1991; Scharer et al., 1993; Brandt et al., 1999). However, many of the studies excluded children with moderate or uncontrolled hypertension either before or during the trial period. On the other hand, in the Normal HCT (Besarab et al., 1998) and Canada-Europe (Parfrey et al., 2005) studies in hemodialysis patients, and the CHOIR (Singh et al., 2006) and CREATE (Drueke et al., 2006) studies in non-dialysis CKD patients, blood pressure was not higher in the higher hemoglobin-higher epoetin treated arm compared to the lower Hgb-lower epoetin dose arm. Vaziri and co-workers have explored the mechanism for hypertension (Ni et al., 1998; Vaziri, 1999, 2001; Wang and Vaziri, 1999). Vaziri et al. have suggested that erythropoietin-induced hypertension is a nitric oxide-mediated complication (Ni et al., 1998; Wang and Vaziri, 1999). A recent FDA analysis of several studies in the aggregate does point to a pro-thrombotic effect of ESAs (www.fda.gov). In fact, a higher rate of graft thrombosis in the higher HCT-higher epoetin arm was one of the reasons that the Data Safety Monitoring Committee for the Normal HCT Study (Besarab et al., 1998) prematurely halted the study. Some studies have associated ESAs with increased levels of proinflammatory cytokines as well as increasing platelet activation. However, given the cross-sectional design of these studies, caution is needed in interpreting these studies until prospective or randomized data become available.

E. Relationship Between Erythropoietin Dose and Adverse Outcomes in CKD Patients

There are emerging data that raise the possibility that high doses of epoetin, particularly when used in epo-hyporesponsive patients, may be associated with toxicity. The first epidemiologic evidence for an association between epoetin dose and toxicity emerged in a study in prevalent dialysis patients by Zhang et al. (2004). These investigators reported that patients with lower Hgb levels (9–11 g/dl) exposed to high doses of epoetin appeared to be at high risk. More recently, in a *post-hoc* analysis of the CHOIR study, Szczech et al. (2008) have published data that suggest a similar finding to that of Zhang et al., namely that the use of high doses of epo in epo-hyporesponsive patients accounts for adverse outcome rather than the achieved Hgb concentration. It is important to be cautious in interpreting these data since they are observational in nature. In particular, residual confounding and/or confounding by indication need to be excluded.

IV. ANEMIA AND ERYTHROPOIETIN TREATMENT IN CHILDREN WITH CKD

Anemia is a major complication in children with chronic kidney disease (CKD) and largely reflects erythropoietin deficiency. Anemia in children is defined as Hgb levels less than 5th percentile of normal adjusted for age and sex (NKF/KDOQI, 2006; Filler et al., 2007). Between the age of 1 and 19 years, anemia is defined as an Hgb level <12.1–13.5 g/dl for boys and <11.4–11.5 g/dl for girls. For 3-day neonates anemia is defined as a Hgb < 13.5 g/dl, for a 2-week neonate Hgb < 12.5 g/dl and for a 6-month infant Hgb < 11.5 g/dl. The prevalence of anemia in pediatric patients based on KDOQI data approximates 36% (Wong et al., 2006). The benefits of anemia correction in children have been associated with improvements in quality of life, scholastic performance, growth and nutrition.

Initial reports of successful treatment of anemia with EPO use were described in children on peritoneal dialysis (Sinai-Trieman et al., 1989), however, three out of five patients treated in the study had significant exacerbation of hypertension. In a subsequent study, 14 children received almost double the dose of the previous study at 300 U/kg/wk and maintained at 100 U/kg/wk. In these children, too, hypertension was an important side effect (Offner et al., 1990). Following these reports, epoetin was routinely used in the management of anemia in children at a dose of 150 U/kg/wk. In a study from Switzerland in 18 patients aged 5–18 years on regular dialysis, packed cell volume increased by 5% on ESA (Bianchetti et al., 1991). However, during treatment, plasma potassium increased significantly, more vigorous antihypertensive measures were required in eight patients, iliofemoral thrombosis occurred in one patient 10 days after renal transplant. While these data suggested that epoetin ameliorates anemia associated with CKD in children, concerns were raised regarding hyperkalemia, arterial hypertension and possibly thrombosis. In spite of these concerns, higher doses are being used in children based on the observational data on 52 children from the North American Pediatric Renal Transplant Cooperative Study (NAPRTCS) 2004 registry report that suggested that younger children tend to require higher doses of EPO with infants requiring the highest dose ranging from 275 to 350 U/kg per week (Port et al., 2004).

Studies have shown improvement in quality of life and physical fitness and activity in patients on epoetin (Morris et al., 1993; Van Damme-Lombaerts et al., 1994). This is often attributed to increase in Hgb and reduction in left ventricular hypertrophy. Reduced requirement for blood transfusions is another important benefit of treating CKD anemia with erythropoietin, particularly in children, since it reduces the likelihood of antibody sensitization and thus increasing the chances of transplantation. Growth retardation is an important problem in children. A study in 47 children showed that catch-up growth during pre-dialysis care was independently associated with the initial Hgb levels and erythropoietin therapy (Boehm et al., 2007). This study suggested early initiation of epoetin in children.

Hypertension is the most common side effect of ESA treatment in children and often responds to decrease in dose. In one study, children aged 4 months to 21 years were assigned to either high-dose (450 U/kg/wk) or low-dose (150 U/kg/wk). ESA treatment was associated with a significant increase in diastolic blood pressure by week 12 compared with baseline (Brandt et al., 1999). The investigators reported a non-significant trend between increasing Hgb levels and increasing systolic and diastolic blood pressures despite stable or lower ESA dose.

Several studies point to a higher risk of thrombosis with the use of epoetin in children (Van Geet et al., 1990). This is attributed to thrombocytosis and an increase in HCT. While some animal models suggest an association of inflammation and further endothelial activation by epoetin, this has not been tested in human subjects. In children, it is often recommended that heparin dose be increased to avoid clotting problems during dialysis in patients on epoetin therapy and that the use of epoetin be deferred for at least 2 weeks after surgery for A-V fistula in children to avoid fistula thrombosis (Van Damme-Lombaerts and Herman, 1999). Hyperkalemia and hyperphosphatemia are described in association with the use of epoetin, but have been attributed to better appetite.

A. Hemoglobin Targets in Children

Despite remarkable success in the management of anemia with the introduction of iron and ESAs, variation in the normal Hgb with age in children introduces a significant challenge in standardizing the target Hgb. Quality of life is an important measure in children since it determines the range of activities a child can pursue, especially in terms of physical performance, exercise tolerance and school attendance. Based on largely the adult experience and smaller cross-over studies in children, a minimum Hgb level of 11 g/dL has been recommended in children by the European Pediatric Peritoneal Dialysis Working Group (EPPWG) (Schroder, 2003).

V. CONCLUSIONS

Erythropoietin deficiency is an important cause of CKD anemia in both adults and children. Erythropoietin treatment has transformed the lives of millions of CKD patients. Several forms of recombinant erythropoietin are currently available and treatment is effective. Recent evidence has raised concerns about the safety of erythropoietin but, so far, the precise mechanism underlying this risk of adverse outcome has not been delineated.

References

Abu-Alfa, A. K., Sloan, L., Charytan, C. et al. (2008). The association of darbepoetin alfa with hemoglobin and health-related quality of life in patients with chronic kidney disease not receiving dialysis. *Curr Med Res Opin* **24**(4): 1091–1100. Epub 2008 Mar 6.

Aispuru, G. R., Aguirre, M. V., Aquino-Esperanza, J. A., Lettieri, C. N., Juaristi, J. A., and Brandan, N. C. (2008). Erythroid expansion and survival in response to acute anemia stress: the role of EPO receptor, GATA-1, Bcl-x(L) and caspase-3. *Cell Biol Int* Epub 2008 Apr 10.

Alexander, M., Kewalramani, R., Agodoa, I., and Globe, D. (2007). Association of anemia correction with health related quality of life in patients not on dialysis. *Curr Med Res Opin* **23**(12): 2997–3008.

Auerbach, M., Coyne, D., and Ballard, H. (2008). Intravenous iron: from anathema to standard of care. *Am J Hematol* Jan 29. [Epub ahead of print].

Besarab, A., Bolton, W. K., Browne, J. K. et al. (1998). The effects of normal as compared with low HCT values in patients with cardiac disease who are receiving hemodialysis and epoetin. *N Engl J Med* **27**: 584–590.

Bianchetti, M. G., Hammerli, I., Roduit, C., Neuhaus, T. J., Leumann, E. P., and Oetliker, O. H. (1991). Epoetin alfa in anaemic children or adolescents on regular dialysis. *Eur J Pediatr* **150**: 509–512.

Blanchard, K. L., Fandrey, J., Goldberg, M. A., and Bunn, H. F. (1993). Regulation of the erythropoietin gene. *Stem Cells* **11** (Suppl 1): 1–7.

Boehm, M., Riesenhuber, A., Winkelmayer, W. C., Arbeiter, K., Mueller, T., and Aufricht, C. (2007). Early erythropoietin therapy is associated with improved growth in children with chronic kidney disease. *Pediatr Nephrol* **22**: 1189–1193.

Brandt, J. R., Avner, E. D., Hickman, R. O., and Watkins, S. L. (1999). Safety and efficacy of erythropoietin in children with chronic renal failure. *Pediatr Nephrol* **13**: 143–147.

Braumann, K. M., Nonnast-Daniel, B., Boning, D., Bocker, A., and Frei, U. (1991). Improved physical performance after treatment of renal anemia with recombinant human erythropoietin. *Nephron* **58**: 129–134.

Britten, G. M., Brecher, G., Johnson, C. A., and Elashoff, R. M. (1969). Stability of blood in commonly used anticoagulants. *Am J Clin Pathol* **52**: 690–694.

Brunelli, S. M., Joffe, M. M., Israni, R. K. et al. (2008). History-adjusted marginal structural analysis of the association between hemoglobin variability and mortality among chronic hemodialysis patients. *Clin J Am Soc Nephrol* **3**(3): 777–782. Epub 2008 Mar 12.

Bunn, H. F. (2006). New agents that stimulate erythropoiesis. *Blood* **109**(3): 868–873. Epub 2006 Oct 10.

Cannella, G., La Canna, G., Sandrini, M. et al. (1990). Renormalization of high cardiac output and of left ventricular size following long-term recombinant human erythropoietin treatment of anemic dialyzed uremic patients. *Clin Nephrol* **34**: 272–278.

Carrera, F., and Macdougall, I. C. (2008). Hemoglobin targets: the jury is still out. *Clin Nephrol* **69**(1): 8–9.

Cotter, D., Thamer, M., Narasimhan, K., Zhang, Y., and Bullock, K. (2006). Translating epoetin research into practice: the role of government and the use of scientific evidence. *Health Aff (Millwood)* **25**(5): 1249–1259.

Coyne, D. W., Kapoian, T., Suki, W. et al. (2007). DRIVE Study Group. Ferric gluconate is highly efficacious in anemic hemodialysis patients with high serum ferritin and low transferrin saturation: results of the Dialysis Patients' Response to IV Iron with Elevated Ferritin (DRIVE) Study. *J Am Soc Nephrol* **18**(3): 975–984. Epub 2007 Jan 31.

De Maria, R., Testa, U., Luchetti, L. et al. (1999). Apoptotic role of Fas/Fas ligand system in the regulation of erythropoiesis. *Blood* **93**(3): 796–803.

Del Vecchio, L., and Locatelli, F. (2008). New erythropoiesis-stimulating agents: how innovative are they? *Contrib Nephrol* **161**: 255–260.

Donnelly, S. (2001). Why is erythropoietin made in the kidney? The kidney functions as a critmeter *Am J Kidney Dis* **38**(2): 415–425.

Donnelly, S. (2003). Why is erythropoietin made in the kidney? The kidney functions as a 'critmeter' to regulate the HCT. *Adv Exp Med Biol* **543**: 73–87.

Drüeke, T. (2001). Hyporesponsiveness to recombinant human erythropoietin. *Nephrol Dial Transplant* **16**(Suppl 7): 25–28.

Drüeke, T. B., Locatelli, F., Clyne, N. et al. (2006). CREATE Investigators. Normalization of hemoglobin level in patients with chronic kidney disease and anemia. *N Engl J Med* **355** (20): 2071–2084.

Dunn, A., Lo, V., and Donnelly, S. (2007). The role of the kidney in blood volume regulation: the kidney as a regulator of the HCT. *Am J Med Sci* **334**(1): 65–71. Review. Erratum in: *Am J Med Sci* 2008 **335**(1), 79.

Eckardt, K. U., and Kurtz, A. (2005). Regulation of erythropoietin production. *Eur J Clin Invest* **35**(Suppl 3): 13–19.

Eckardt, K. U., Pugh, C. W., Meier, M., Tan, C. C., Ratcliffe, P. J., and Kurtz, A. (1994). Production of erythropoietin by liver cells in vivo and in vitro. *Ann NY Acad Sci* **718**: 50–60. discussion 61-3.

Eschbach, J. W., Egrie, J. C., Downing, M. R., Browne, J. K., and Adamson, J. W. (1987). Correction of the anemia of end-stage renal disease with recombinant human erythropoietin: Results of a combined phase I and II clinical trial. *N Engl J Med* **316**: 73–78.

Eschbach, J. W., Abdulhadi, M. H., Browne, J. K. et al. (1989). Recombinant human erythropoietin in anemic patients with end-stage renal disease. Results of a phase III multicenter clinical trial. *Ann Intern Med* **111**: 992–1000.

Filler, G., Mylrea, K., Feber, J., and Wong, H. (2007). How to define anemia in children with chronic kidney disease? *Pediatr Nephrol* **22**(5): 702–707. Epub 2007 Jan 10.

Fishbane, S., and Berns, J. S. (2007). Evidence and implications of haemoglobin cycling in anaemia management. *Nephrol Dial Transplant* **22**(8): 2129–2132. Epub 2007 Jun 25.

Fried, W. (1978). Hematological complications of chronic renal failure. *Med Clin North Am* **62**(6): 1363–1379.

Garcia, D. L., Anderson, S., Rennke, H. G., and Brenner, B. M. (1988). Anemia lessens and its prevention with recombinant human erythropoietin worsens glomerular injury and hypertension in rats with reduced renal mass. *Proc Natl Acad Sci USA* **85** (16): 6142–6146.

Garcia, M. M., Beckman, B. S., Brookins, J. W. et al. (1990). Development of a new radioimmunoassay for EPO using recombinant erythropoietin. *Kidney Int* **38**: 969–975.

Göpfert, T., Eckardt, K. U., Geb, B., and Kurtz, A. (1997). Oxygen-dependent regulation of erythropoietin gene expression in rat hepatocytes. *Kidney Int* **51**(2): 502–506.

Gouva, C., Nikolopoulos, P., Ioannidis, J. P., and Siamopoulos, K. C. (2004). Treating anemia early in renal failure patients slows the decline of renal function: a randomized controlled trial. *Kidney Int* **66**(2): 753–760.

Grasso, G., Sfacteria, A., Meli, F. et al. (2007). The role of erythropoietin in neuroprotection: therapeutic perspectives. *Drug News Perspect* **20**(5): 315–320.

Gregoli, P. A., and Bondurant, M. C. (1997). The roles of Bcl-X(L) and apopain in the control of erythropoiesis by erythropoietin. *Blood* **90**(2): 630–640.

Harnett, J. D., Foley, R. N., Kent, G. M., Barre, P. E., Murray, D., and Parfrey, P. S. (1995). Congestive heart failure in dialysis patients: Prevalence, incidence, prognosis and risk factors. *Kidney Int* **47**: 884–890.

Henry, J. B. (1996). *Methods hematology: basic methodology, in clinical diagnosis and management by laboratory methods*, 19th edn. Saunders, Philadelphia, pp. 578–625.

Holt, J. T., DeWandler, M. J., and Aevan, D. A. (1982). Spurious elevation of electronically determined mean corpuscular volume and HCT caused by hyperglycemia. *Am J Clin Pathol* **77**: 561–567.

Imamura, R., Isaka, Y., Ichimaru, N., Takahara, S., and Okuyama, A. (2007). Carbamylated erythropoietin protects the kidneys from ischemia-reperfusion injury without stimulating erythropoiesis. *Biochem Biophys Res Commun* **16**: 786–792. Epub 2006 Dec 22.

Jelkmann, W. (2001). The role of the liver in the production of thrombopoietin compared with erythropoietin. *Eur J Gastroenterol Hepatol* **13**(7): 791–801.

Kapoian, T., O'Mara, N. B., Singh, A. K. et al. (2008). Ferric gluconate reduces epoetin requirements in hemodialysis patients with elevated ferritin. *J Am Soc Nephrol* **19**(2): 372–379. Epub 2008 Jan 23.

KDOQI; National Kidney Foudnation. (2006). II. Clinical practice guidelines and clinical practice recommendations for anemia in chronic kidney disease in adults. *Am J Kidney Dis* **47**(5 Suppl 3): S16–85.

KDOQI. (2007). KDOQI Clinical Practice Guideline and Clinical Practice Recommendations for anemia in chronic kidney disease: 2007 update of hemoglobin target. *Am J Kidney Dis* **50**(3): 471–530.

Koury, M. J., Sawyer, S. T., and Brandt, S. J. (2002). New insights into erythropoiesis. *Curr Opin Hematol* **9**(2): 93–100.

Kwack, C., and Balakrishnan, V. S. (2006). Managing erythropoietin hyporesponsiveness. *Semin Dial* **19**(2): 146–151.

Lacombe, C., Da Silva, J. L., Bruneval, P. et al. (1988). Peritubular cells are the site of erythropoietin synthesis in the murine hypoxic kidney. *J Clin Invest* **81**(2): 620–623.

Lacombe, C., Da Silva, J. L., Bruneval, P. et al. (1991). Erythropoietin: sites of synthesis and regulation of secretion. *Am J Kidney Dis* **18**(4 Suppl 1): 14–19.

Lafferty, H. M., Garcia, D. L., Rennke, H. G., Troy, J. L., Anderson, S., and Brenner, B. M. (1991). Anemia ameliorates progressive renal injury in experimental DOCA-salt hypertension. *J Am Soc Nephrol* **1**(10): 1180–1185.

Levin, A. (2002). Anemia and left ventricular hypertrophy in chronic kidney disease populations: a review of the current state of knowledge. *Kidney Int Suppl* **80**: 35–38.

Levin, A., Thompson, C. R., Ethier, J. et al. (1999). Left ventricular mass index increase in early renal disease: impact of decline in hemoglobin. *Am J Kidney Dis* **34**(1): 125–134.

Li, S., Foley, R. N., and Collins, A. J. (2004). Anemia, hospitalization, and mortality in patients receiving peritoneal dialysis in the United States. *Kidney Int* **65**(5): 1864–1869.

Lipsic, E., Schoemaker, R. G., van der Meer, P., Voors, A. A., van Veldhuisen, D. J., and van Gilst, W. H. (2006). Protective effects of erythropoietin in cardiac ischemia: from bench to bedside. *J Am Coll Cardiol* **48**(11): 2161–2167. Epub 2006 Nov 9.

Liu, Y., Pop, R., Sadegh, C., Brugnara, C., Haase, V. H., and Socolovsky, M. (2006). Suppression of Fas-FasL coexpression by erythropoietin mediates erythroblast expansion during the erythropoietic stress response in vivo. *Blood* **108**(1): 123–133. Epub 2006 Mar 9.

Locatelli, F., Pisoni, R. L., Combe, C. et al. (2004). Anaemia in haemodialysis patients of five European countries: association with morbidity and mortality in the Dialysis Outcomes and Practice Patterns Study (DOPPS). *Nephrol Dial Transplant* **19** (1): 121–132.

Ma, J. Z., Ebben, J., Xia, H., and Collins, A. J. (1999). HCT level and associated mortality in hemodialysis patients. *J Am Soc Nephrol* **10**(3): 610–619.

Macdougall, I. C. (2006). Recent advances in erythropoietic agents in renal anemia. *Semin Nephrol* **26**(4): 313–318.

Macdougall, I. C. (2008). Novel erythropoiesis-stimulating agents: a new era in anemia management. *Clin J Am Soc Nephrol* **3**(1): 200–207. Epub 2007 Dec 12.

Macdougall, I. C., Lewis, N. P., Saunders, M. J. et al. (1990). Long-term cardiorespiratory effects of amelioration of renal anaemia by erythropoietin. *Lancet* **335**: 489–493.

Mayer, G., Thum, J., Cada, E. M., Stummvoll, H. K., and Graf, H. (1988). Working capacity is increased following recombinant human erythropoietin treatment. *Kidney Int* **34**: 525–528.

McGonigle, R. J., Wallin, J. D., Shadduck, R. K., and Fisher, J. W. (1984). Erythropoietin deficiency and inhibition of erythropoiesis in renal insufficiency. *Kidney Int* **25**: 437–444.

Milano, M., and Schneider, M. (2007). EPO in cancer anemia: benefits and potential risks. *Crit Rev Oncol Hematol* **62**(2): 119–125. Epub 2007 Jan 2.

Mohanram, A., Zhang, Z., Shahinfar, S., Keane, W. F., Brenner, B. M., and Toto, R. D. (2004). Anemia and end-stage renal disease in patients with type 2 diabetes and nephropathy. *Kidney Int* **66**(3): 1131–1138.

Morris, K. P., Sharp, J., Watson, S., and Coulthard, M. G. (1993). Non-cardiac benefits of human recombinant erythropoietin in end stage renal failure and anaemia. *Arch Dis Child* **69**: 580–586.

Naets, J. P., Garcia, J. F., Tousaaint, C., Buset, M., and Waks, D. (1986). Radioimmunoassay of erythropoietin in chronic uraemia of anephric patients. *Scand J Haematol* **37**: 390–394.

Ni, Z., Wang, X. Q., and Vaziri, N. D. (1998). Nitric oxide metabolism in erythropoietin-induced hypertension: effect of calcium channel blockade. *Hypertension* **32**(4): 724–729.

NKF/KDOQI. (2006). Guidelines clinical practice recommendations for anemia in chronic kidney disease in children. *Am J Kidney Dis* **47**, S86–S107.

NKF-DOQI. (1997). Clinical practice guidelines for the treatment of anemia of chronic renal failure. National Kidney Foundation-Dialysis Outcomes Quality Initiative. Am J Kidney Dis. 303 (4 Suppl 3), S192–240.

Offner, G., Hoyer, P. F., Latta, K., Winkler, L., Brodehl, J., and Scigalla, P. (1990). One year's experience with recombinant erythropoietin in children undergoing continuous ambulatory or cycling peritoneal dialysis. *Pediatr Nephrol* **4**: 498–500

Parfrey, P. S., Foley, R. N., Wittreich, B. H., Sullivan, D. J., Zagari, M. J., and Frei, D. (2005). Double-blind comparison of full and partial anemia correction in incident hemodialysis patients without symptomatic heart disease. *J Am Soc Nephrol* **16**(7): 2180–2189. Epub 2005 May 18.

Pascual, J., Teruel, J. L., Moya, J. L., Liano, F., Jimenez-Mena, M., and Ortuno, J. (1991). Regression of left ventricular hypertrophy after partial correction of anemia with erythropoietin in patients on hemodialysis: a prospective study. *Clin Nephrol* **35**: 280–287.

Pendse, S., and Singh, A. K. (2005). Complications of chronic kidney disease: anemia, mineral metabolism, and cardiovascular disease. *Med Clin North Am* **89**(3): 549–561.

Phrommintikul, A., Haas, S. J., Elsik, M., and Klum, H. (2007). Mortality and target haemoglobin concentrations in anemia patients with chronic kidney disease treated with erythropoietin: a meta-analysis. *Lancet* **369**: 381–388.

Port, R. E., Kiepe, D., Van Guilder, M., Jelliffe, R. W., and Mehls, O. (2004). Recombinant human erythropoietin for the treatment of renal anaemia in children: no justification for bodyweight-adjusted dosage. *Clin Pharmacokinet* **43**: 57–70.

Radtke, H. W., Claussner, A., Erbes, P. M., Scheuermann, E. H., Schoepp, W., and Koch, K. M. (1979). Serum erythropoietin concentration in chronic renal failure: Relationship to degree of anemia and excretory renal function. *Blood* **54**: 877–884.

Regidor, D. L., Kopple, J. D., Kovesdy, C. P. et al. (2006). Associations between changes in hemoglobin and administered erythropoiesis-stimulating agent and survival in hemodialysis patients. *J Am Soc Nephrol* **17**(4): 1181–1191.

Rice, L., and Alfrey, C. P. (2000). Modulation of red cell mass by neocytolysis in space and on Earth. *Pflugers Arch* **441**(2–3 Suppl): R91–R94.

Rice, L., and Alfrey, C. P. (2005). The negative regulation of red cell mass by neocytolysis: physiologic and pathophysiologic manifestations. *Cell Physiol Biochem* **15**(6): 245–250.

Rice, L., Alfrey, C. P., Driscoll, T., Whitley, C. E., Hachey, D. L., and Suki, W. (1999). Neocytolysis contributes to the anemia of renal disease. *Am J Kidney Dis* **33**(1): 59–62.

Risso, A., Turello, M., Biffoni, F., and Antonutto, G. (2007). Red blood cell senescence and neocytolysis in humans after high altitude acclimatization. *Blood Cells Mol Dis* **38**(2): 83–92. Epub 2006 Dec 26.

Ritz, E., Laville, M., Bilous, R. W. et al. (2007). Anemia Correction in Diabetes Study Investigators. Target level for hemoglobin correction in patients with diabetes and CKD: primary results of the Anemia Correction in Diabetes (ACORD) Study. *Am J Kidney Dis* **49**(2): 194–207. Erratum 49(4), 562.

Roberts, T. L., Foley, R. N., Weinhandl, E. D., Gilbertson, D. T., and Collins, A. J. (2006). Anaemia and mortality in haemodialysis patients: interaction of propensity score for predicted anaemia and actual haemoglobin levels. *Nephrol Dial Transplant* **21**(6): 1652–1662.

Robertson, H. T., Haley, N. R., Guthrie, M., Cardenas, D., Eschbach, J. W., and Adamson, J. W. (1990). Recombinant erythropoietin improves exercise capacity in anemic hemodialysis patients. *Am J Kidney Dis* **15**: 325–332.

Robinson, B. M., Joffe, M. M., Berns, J. S., Pisoni, R. L., Port, F. K., and Feldman, H. I. (2005). Anemia and mortality in hemodialysis patients: accounting for morbidity and treatment variables updated over time. *Kidney Int* **68**(5): 2323–2330. Erratum in: 68 (6), 2934.

Ross, R. P., McCrea, J. B., and Besarab, A. (1994). Erythropoietin response to blood loss in hemodialysis patients is blunted but preserved. *Am Soc Artific Int Org J* **40**: M880–M885.

Rossert, J., and Froissart, M. (2006). Role of anemia in progression of chronic kidney disease. *Semin Nephrol* **26**(4): 283–289.

Rossert, J., Levin, A., Roger, S. D. et al. (2006). Effect of early correction of anemia on the progression of CKD. *Am J Kidney Dis* **47**(5): 738–750.

Sasaki, R. (2003). Pleiotropic functions of erythropoietin. *Intern Med* **42**(2): 142–149.

Schaefer, R. M., Kokot, F., and Heidland, A. (1989). Impact of recombinant erythropoietin on sexual function in hemodialysis patients. *Contrib Nephrol* **76**: 273–282.

Scharer, K., Klare, B., Braun, A., Dressel, P., and Gretz, N. (1993). Treatment of renal anemia by subcutaneous erythropoietin in children with preterminal chronic renal failure. *Acta Paediatr* **82**: 953–958.

Schroder, C. H. (2003). The management of anemia in pediatric peritoneal dialysis patients. Guidelines by an ad hoc European committee. *Pediatr Nephrol* **18**: 805–809.

Sharples, E. J., and Yaqoob, M. M. (2006a). Erythropoietin and acute renal failure. *Semin Nephrol* **26**(4): 325–331.

Sharples, E. J., and Yaqoob, M. M. (2006b). Erythropoietin in experimental acute renal failure. *Nephron Exp Nephrol* **104**(3): e83–e88.

Sharples, E. J., Thiemermann, C., and Yaqoob, M. M. (2005). Mechanisms of disease: cell death in acute renal failure and emerging evidence for a protective role of erythropoietin. *Nat Clin Pract Nephrol* **1**(2): 87–97.

Silva, M., Grillot, D., Benito, A., Richard, C., Nunez, G., and Fernández-Luna, J. L. (1996). Erythropoietin can promote erythroid progenitor survival by repressing apoptosis through Bcl-XL and Bcl-2. *Blood* **88**(5): 1576–1582.

Sinai-Trieman, L., Salusky, I. B., and Fine, R. N. (1989). Use of subcutaneous recombinant. *Pediatrics* **114**: 550–554.

Singh, A. K. (2007). Does correction of anemia slow the progression of chronic kidney disease? *Nat Clin Pract Nephrol* **3**(12): 638–639. Epub 2007 Oct 16.

Singh, A.K. (2008). Anemia and erythropoietin: The FDA and CRAC. http://www.medscape.com/viewarticle/561804 Accessed June 29, 2008.

Singh, A. K., and Fishbane, S. (2008). The optimal hemoglobin in dialysis patients – a critical review. *Semin Dial* **21**(1): 1–6.

Singh, A. K., Szczech, L., Tang, K. L. et al. (2006). CHOIR Investigators. Correction of anemia with epoetin alfa in chronic kidney disease. *N Engl J Med* **355**(20): 2085–2098.

Singh, A. K., Coyne, D. W., Shapiro, W., and Rizkala, A. R. (2007). Predictors of the response to treatment in anemic hemodialysis patients with high serum ferritin and low transferrin saturation. DRIVE Study Group. *Kidney Int* **71**(11): 1163–1171. Epub 2007 Mar 28.

Singh, A.K., Milford, E., Fishbane, S., and Keithi-Reddy, S.R. (2008). Managing anemia in dialysis patients: hemoglobin cycling and overshoot. *Kidney Int* Mar 12 [Epub ahead of print].

Sobh, M. A., Abd el Hamid, I. A., Atta, M. G., and Refaie, A. F. (1992). Effect of erythropoietin on sexual potency in chronic haemodialysis patients: A preliminary study. *Scand J Urol Nephrol* **26**: 181–185.

Stockmann, C., and Fandrey, J. (2006). Hypoxia-induced erythropoietin production: a paradigm for oxygen-regulated gene expression. *Clin Exp Pharmacol Physiol* **33**: 968–979.

Strippoli, G. F., Craig, J. C., Manno, C., and Schena, F. P. (2004). Hemoglobin targets for the anemia of chronic kidney disease: a meta-analysis of randomized, controlled trials. *J Am Soc Nephrol* **15**: 3154–3165.

Szczech, L.A., Barnhart, H., Inrig, J.K. et al. (2008). Epoetin alfa dose and achieved hemoglobin are associated with outcomes among patients with anemia and chronic kidney disease: a secondary analysis of the CHOIR (Correction of Hemoglobin and Outcomes in Renal Insufficiency) Trial. *Kidney Int* in press.

Teehan, B., Sigler, M. H., Brown, J. M. et al. (1989). Hematologic and physiologic studies during correction of anemia with recombinant human erythropoietin in predialysis patients. *Transplant Proc* **21**: 63–66.

Testa, U. (2004). Apoptotic mechanisms in the control of erythropoiesis. *Leukemia* **18**(7): 1176–1199.

Tonelli, M., Owen, W. F., Jindal, K., Winkelmayer, W. C., and Manns, B. (2004). Regarding impact of epoetin alfa on clinical end points in patients with chronic renal failure: a meta-analysis. *Kidney Int* **66**(4): 1712. author reply 1712–1713.

United States Renal Data System. Annual Data Report 2006.

Van Damme-Lombaerts, R., and Herman, J. (1999). Erythropoietin treatment in children with renal failure. *Pediatr Nephrol* **13**: 148–152.

Van Damme-Lombaerts, R., Broyer, M., Businger, J., Baldauf, C., and Stocker, H. (1994). A study of recombinant human erythropoietin in the treatment of anaemia of chronic renal failure in children on haemodialysis. *Pediatr Nephrol* **8**: 338–342.

van der Putten, K., Braam, B., Jie, K. E., and Gaillard, C. A. (2008). Mechanisms of disease: erythropoietin resistance in patients with both heart and kidney failure. *Nat Clin Pract Nephrol* **4**(1): 47–57.

Van Geet, C., Van Damme-Lombaerts, R., Vanrusselt, M., de Mol, A., Proesmans, W., and Vermylen, J. (1990). Recombinant human erythropoietin increases blood pressure, platelet aggregability and platelet free calcium mobilisation in uraemic children: a possible link? *Thromb Haemost* **64**: 7–10.

Vaziri, N. D. (1999). Mechanism of erythropoietin-induced hypertension. *Am J Kidney Dis* **33**(5): 821–828.

Vaziri, N. D. (2001). Cardiovascular effects of erythropoietin and anemia correction. *Curr Opin Nephrol Hypertens* **10**: 633–637.

Wang, X. Q., and Vaziri, N. D. (1999). Erythropoietin depresses nitric oxide synthase expression by human endothelial cells. *Hypertension* **33**(3): 894–899.

Weisbord, S. D., and Kimmel, P. L. (2008). Health-related quality of life in the era of erythropoietin. *Hemodial Int* **12**(1): 6–15.

Wish, J. B., and Coyne, D. W. (2007). Use of erythropoiesis-stimulating agents in patients with anemia of chronic kidney disease: overcoming the pharmacological and pharmacoeconomic limitations of existing therapies. *Mayo Clin Proc* **82**(11): 1371–1380.

Wizemann, V., Kaufmann, J., and Kramer, W. (1992). Effect of erythropoietin on ischemia tolerance in anemic hemodialysis patients with confirmed coronary artery disease. *Nephron* **62**: 161–165.

Wizemann, V., Schafer, R., and Kramer, W. (1993). Follow-up of cardiac changes induced by anemia compensation in normotensive hemodialysis patients with left-ventricular hypertrophy. *Nephron* **64**: 202–206.

Wojchowski, D. M., Menon, M. P., Sathyanarayana, P. et al. (2006). Erythropoietin-dependent erythropoiesis: new insights and questions. *Blood Cells Mol Dis* **36**(2): 232–238. Epub 2006 Mar 9.

Wolcott, D. L., Marsh, J. T., La Rue, A., Carr, C., and Nissenson, A. R. (1989). Recombinant human erythropoietin treatment may improve quality of life and cognitive function in chronic hemodialysis patients. *Am J Kidney Dis* **14**: 478–485.

Wong, H., Mylrea, K., Feber, J., Drukker, A., and Filler, G. (2006). Prevalence of complications in children with chronic kidney disease according to KDOQI. *Kidney Int* **70**(3): 585–590. Epub 2006 Jun 21.

Xue, J. L., St Peter, W. L., Ebben, J. P., Everson, S. E., and Collins, A. J. (2002). Anemia treatment in the pre-ESRD period and associated mortality in elderly patients. *Am J Kidney Dis* **40**: 1153–1161.

Yang, W., Israni, R. K., Brunelli, S. M., Joffe, M. M., Fishbane, S., and Feldman, H. I. (2007). Hemoglobin variability and mortality in ESRD. *J Am Soc Nephrol* **18**(12): 3164–3170. Epub 2007 Nov 14.

Zhang, Y., Thamer, M., Stefanik, K., Kaufman, J., and Cotter, D. J. (2004). Epoetin requirements predict mortality in hemodialysis patients. *Am J Kidney Dis* **44**(5): 866–876.

Further reading

http://www.fda.gov/cder/drug/advisory/RHE.htm.

www.fda.gov/ohrms/dockets/ac/08/briefing/2008-4345b2-01-FDA.pdf Accessed June 29, 2008.

NKF-K/DOQI. (2001). Clinical practice guidelines for anemia of chronic kidney disease: update 2000. *Am J Kidney Dis* **37** (1 Suppl 1), S182–238.

US Food and Drug Administration (CDER) http://www.fda.gov/ohrms/dockets/ac/07/briefing/2007-4315b1-00-index.htm Accessed June 29, 2008.

Xue, J. L., St Peter, W. L., Ebben, J. P., Everson, S. E., and Collins, A. J. (2002). Anemia treatment in the pre-ESRD period and associated mortality in elderly patients. *Am J Kidney Dis* **40**: 1153–1161.

Section II. Vitamin D, PTH and Novel Regulators of Phosphate

CHAPTER **6**

Vitamin D and the Kidney: Introduction and Historical Perspective

TEJAS PATEL AND AJAY K. SINGH

Renal Division, Brigham and Women's Hospital and Harvard Medical School, 75 Francis Street, Boston, MA 02115, USA

Contents

I. Introduction 63
II. Vitamin D 63
III. Cinacalcet 65
References 66

I. INTRODUCTION

The discovery of vitamin D, the elucidation of the associated pathophysiology and then, more recently, the discovery and application of drugs that modulate the parathyroid hormone (PTH)-Vitamin D-mineral physiology have their origins in a fascinating series of scientific experiments that led to the cure for rickets. The pleotropic effects of vitamin D have been elucidated more recently. Vitamin D is now known to play a central role in protecting against osteoporosis, risk of fracture and falls, cancer, infection, psoriasis, multiple sclerosis, hypertension and diabetes (DeLuca, 1988; Wolf, 2004; www.beyonddiscovery.org).

II. VITAMIN D

A. Etiology of Rickets

Rickets was first described in detail by Daniel Whistler and Francis Glisson in the 17th century (Whistler, 1645; Glisson, 1650), but did not receive much attention until the industrial revolution. During that time, many families went from working outdoors to the ill-ventilated, dark and unhygienic environment of factories. The smog-polluted cities of Europe made exposure to the sun even more difficult. As a result, rickets became endemic in the factory-filled polluted cities of Europe (Hess, 1929). In the 1800s, Armand Trousseau and Frantisek Chvostek further characterized the effects of severe calcium depletion, describing the physical signs of latent tetany, spasms of the extremities due to hypocalcemia. Rickets is primarily a disorder of the young, resulting from insufficient uptake of dietary calcium. Afflicted children showed poor muscle tone with retarded sitting and walking function, pigeon breasts and bowed legs (Figure 6.1) and high mortality. Clinical manifestations also include weakness, bone deformity, muscle spasms and seizures.

B. Search for a Remedy for Rickets

In 1822, it was first observed that rickets was endemic in cities but not in small towns, implicating sun exposure as a remedy. In 1827, D. Scheutte reported cure from rickets from intake of cod-liver oil. The concept of micronutrient deficiency leading to clinical disease was yet in its infancy and his proposal went largely ignored at this time. It was not until the early part of the 20th century that a Dutch physician, Christiaan Eijkman (who discovered vitamin B_1 or thiamin) and a British scientist, Sir Frederick Hopkins (who discovered the amino acid tryptophan), proposed and convincingly proved the concept of micronutrient or 'accessory food factors'. For their pioneering work, both were jointly awarded the 1929 Nobel Prize for Medicine or Physiology (http://nobelprize.org/nobel_prizes/medicine/laureates/1929/).

C. Experimental Evidence of the Role of Vitamin D in Rickets

In 1919, Sir Edward Mellanby showed that diet-controlled dogs raised indoors developed rickets (Mellanby and Cantag, 1919). In 1922, Elmer McCollum demonstrated that aerated and heated cod-liver oil deficient in vitamin A was able to cure rickets. He coined the term 'vitamin D' (www.beyonddiscovery.org). By definition, vitamins are organic compounds required as a nutrient in small quantity by an organism. A compound is called a vitamin when it cannot be synthesized in sufficient quantities by an organism and thus must be obtained from the diet. Unlike vitamins A, B and C, vitamin D is not a true vitamin as it can be produced in adequate amounts by the body when skin is exposed to solar rays. Thus the term 'vitamin D' is a misnomer.

In 1923, Harry Goldblatt and Katherine Soames irradiated 7-dehydrocholestrol in the skin of animals that

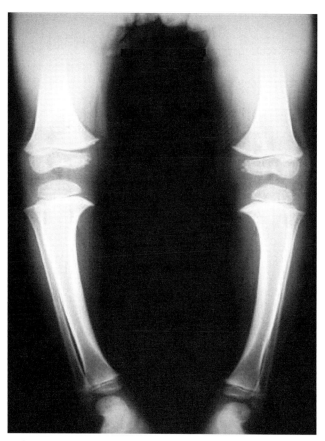

FIGURE 6.1 Roentgenogram of tibia-fibula demonstrating classic bowing of the legs in rickets.

FIGURE 6.2 Adolf Windaus discovered the structure of vitamin D3.

could produce the anti-rachitic vitamin D (Gloldblatt and Soames, 1923). Hess and Weinstock confirmed the dictum that 'light equals vitamin D' (Hess and Weinstock, 1924). They excised a small portion of skin, irradiated it with ultraviolet light and then fed it to groups of rachitic rats. The skin that had been irradiated protected against rickets; unirradiated skin did not. Thus a connection between diet and photochemistry was found (Hess, 1922). They also showed radiologic improvement of rachitic children after exposure to sunlight. Steenbock and Black found that rat food, which was irradiated with ultraviolet light, also had anti-rachitic properties (http://vitamind.ucr.edu/history.html). Soon, irradiation was used as a method to fortify the food. However, concerns surrounding the safety of radiation remained and precluded widespread use.

D. Discovery of the Structure of Vitamin D

In 1927, Adolf Otto Reinhold Windaus (1876–1959, Figure 6.2), a German chemist and a widely recognized leader on sterol chemistry, established the empirical formula of the 4-ring structure of cholesterol. He showed that purified cholesterol did not have anti-rachitic properties. Windaus and Hess tested 30 different steroid preparations with more than one double bond from various plant sources for anti-rachitic activity upon radiation. They came across ergosterol, a fungal steroid from ergot. In collaboration with Hess, Rosenheim and Webster, he concluded that ergosterol was the pro-vitamin D, convertible to vitamin D by ultraviolet irradiation. The irradiated product was named vitamin D_2 or calciferol, which possessed great anti-rachitic properties. Since ergosterol does not occur in animals, how can the animal obtain vitamin D from sunlight? This question was answered by Windaus and Bock in 1937. They identified the structure of vitamin D_3 produced from irradiated 7-dehydrocholesterol (called cholecalciferol) and showed that this was structurally identical to the anti-rachitic component in cod-liver oil. For his contribution to the constitution of sterols and their connection with vitamins, Windaus received the Nobel Prize in Chemistry in 1928 (Wolf, 2004). He is considered one of the most important contributors to our early understanding of the structure and function of vitamin D.

In 1931, Askew and colleagues identified the chemical structure of vitamin D_2 (ergocalciferol), which is derived from the precursor molecule ergosterol. In 1968, Hector DeLuca and colleagues at the University of Wisconsin isolated the active substance identified as 25-hydroxyvitamin D_3. They also proved that the liver produced the substance. The British team identified a second metabolite. Its chemical structure was not characterized as 1,25-dihydroxyvitamin D_3 until 1971. Subsequently, Michael Holick and others demonstrated the exact sequence of steps in the production of cholecalciferol in the skin (Holick et al., 1980). Identifying and isolating activated vitamin D accelerated the manufacturing of the molecule in large quantity at low cost. It offered a cure for rickets without the untoward side effects of irradiation. In 1975, Haussler et al. demonstrated, by radiolabel isotope, that active vitamin D binds to the nucleus (vitamin D receptor, VDR) of the principal cells of the small intestine (thus proving the mechanism proposed by Nicolaysen) (Brumbaugh and Haussler, 1975). Full-length cDNA coding of human VDR was cloned and confirmed that it is a member of nuclear receptor family (Baker et al., 1988).

E. Role of Vitamin D in Calcium Regulation

Researchers at the University of Wisconsin and Cambridge University discovered that a hormone produced by the parathyroid gland is critical in maintaining adequate levels of vitamin D. It was learnt that vitamin D suppresses parathyroid hormone. It was confirmed that calcium was tightly regulated by interplay between the parathyroid hormone and vitamin D. In the 1950s, Swedish scientist, Arvid Carlsson, showed that vitamin D removes calcium from the bones when the calcium levels are low in the blood. Independently, Nicolaysen showed that calcium uptake from food is governed by an unknown 'endogenous factor' that 'alerts' the intestines to the body's calcium needs. The role of kidney in the conversion of 25, vitamin D to 1,25-dihydroxyvitamin D_3 was elucidated (DeLuca and Schnoes, 1983). The enzyme responsible for conversion was identified as 1α-hydroxylase and was localized to the proximal convoluted tubules (Brunette et al., 1978). The pathway was thus identified – liver changes vitamin D_3 to 25-hydroxyvitamin D_3, the major circulating form of the vitamin. The kidneys then convert 25-hydroxyvitamin D_3 to 1,25-dihydroxyvitamin D_3, the active form of the vitamin. Although the kidney is the exclusive site of 1,25 vitamin D_3 production, evidence of extra-renal production was reported (Lambert et al., 1982).

F. Role of Vitamin D in Kidney Disease

The observation that patients with renal dysfunction had abnormalities in calcium regulation was central to invoking a role for vitamin D deficiency in patients with kidney failure (Foley et al., 1996). Oral treatment with 1,25 vitamin D_3 resulted in partial improvement of hypocalcemia. However, it was discovered that vitamin D can directly suppress parathyroid hormone synthesis (Russell et al., 1986). Oral 1,25 vitamin D_3 did not reach the parathyroid hormone due to small intestinal degradation of the compound. Administration of intravenous 1,25 vitamin D has been more effective in suppressing parathyroid hormone in dialysis patients (Slatopolsky et al., 1984). There is indirect evidence that, compared to the oral form, intravenous 1,25 vitamin D_3 confers better survival in dialysis patients (Teng et al., 2003).

G. FDA Approval of Vitamin D

In 1994, the US Food and Drug Administration (FDA) approved vitamin D treatment of psoriasis – the first disease for which vitamin D was approved. Subsequently, oral calcitriol (Rocaltrol) and injectable activated vitamin D (Zemplar®, Abbott Laboratory) were approved for treatment of secondary hyperparathyroidism in chronic renal failure (www.fda.gov/cder/foi/nda/98/20819; www.fda.gov/cder/foi/nda/98/021068a). At present, vitamin D_3 is included in multiple vitamin supplements available over the counter in the USA and Europe.

H. Role of Vitamin D in Other Disease States

Over the past decade, vitamin D deficiency has been associated with risk of hypertension (Hermann and Ruschitzka, 2008), type 2 diabetes mellitus (Palomer et al., 2008), multiple sclerosis (Brown, 2006) and left ventricular hypertrophy (Achinger and Ayus, 2005). High dose vitamin D has been shown to retard the dedifferentiation of malignant leukemia cells. Its role in regulating the immune system was also demonstrated in a series of experiments by DeLuca et al. and Holick et al. Holick showed that topical 1,25 vitamin D_3 dramatically improved psoriasis (Holick et al., 1987). However, the causality has yet to be firmly established. The story of vitamin D in pathogenesis of infection is interesting. It is known that tuberculosis reactivates in people who migrate from an endemic area to a developed western country. The etiology was unclear. Recently, a team of researchers from the UK demonstrated, in an elegant experiment, the critical role of vitamin D in protecting the host against tuberculosis that includes the induction of antimicrobial peptide ll-37 (Martineau et al., 2007). Important milestones are outlined in Table 6.1.

Since the early description of rickets almost half a millennium ago, we have come a long way in discovering the pleotropic effects of vitamin D. However, much work remains to be done to unravel its precise role in human health and in chronic disease states. Overall, Windaus would not be disappointed the way things have turned out for vitamin D.

III. CINACALCET

In the 1980s, the existence of some kind of calcium receptor mechanism on parathyroid cells was hypothesized and had some success in rat and bovine parathyroid cells (Gylfe et al., 1986). By 1990, several characteristics of the calcium receptor were known: this receptor could be activated by a variety of inorganic and organic cations; the receptor was coupled to the formation of inositol trisphosphate and the mobilization of intracellular $Ca2+$ from a non-mitochondrial pool. In 1993, Ed Brown et al. cloned the DNA encoding the bovine parathyroid calcium receptor (Brown et al., 1993). There is

TABLE 6.1 Milestones in the discovery of Vitamin D

1600s	Whistler and Glisson describe rickets
1827	D. Scheutte reports cure of rickets by cod-liver oil; goes largely unnoticed
Early 1900s	Independently, Christiaan Eijkman and Sir Frederick Hopkins demonstrate that whole foods contain unknown constituents essential to health
1913	Huldschinsky cures children of rickets by artificial ultraviolet light
1919	Sir Edward Mellanby induces rickets in dogs and then cures the disease by feeding them cod-liver oil
1920–23	Harry Goldblatt, Katherine Soames, Mildred Weinstock and Alfred Hess independently discover that irradiating certain food with ultraviolet light renders them anti-rachitic
1922	Chick and coworkers show that rickets could be cured by whole milk or cod-liver oil
1922	McCollum demonstrates that a different substance in cod-liver oil carries an anti-rachitic property (not vitamin A). He calls the substance vitamin D
1920s	Adolf Windaus, Rosenheim and Webster conclude that erogsterol is the most likely parent substance of vitamin D in food
1931	Askew defines the chemical makeup of the form of vitamin D found in irradiated foods (later called ergocalciferol)
1936	Adolf Windaus reports the chemical structure of vitamin D3 produced in the skin (now known as cholecalciferol) and identifies the structure of its parent molecule, 7-dehydrocholesterol
1968	DeLuca and colleagues isolate and identify 25-hydroxyvitamin D3. They demonstrate that the substance is produced in the liver
1970s	Regulation of calcium by vitamin D is identified
1971	Molecular structure of the active 1,25 vitamin D3 is reported
1975	Haussler confirms the discovery of vitamin D receptor (VDR) in pig small intestine
1980s	Michael Holick and others show that vitamin D inhibits skin cell growth
1980s	Vitamin D modulates immunity
1988	Human VDR is cloned
1994	United States Food and Drug Administration approves vitamin D for the treatment of psoriasis
1998	United States Food and Drug Administration approves Paricalcitol for secondary hyperparthyroidism in chronic renal failure
1998	United States Food and Drug Administration approves Calcitriol for secondary hyperparthyroidism in chronic renal failure
2000-	Association studies linking vitamin D deficiency to hypertension, diabetes mellitus type 2, cardiovascular disease, left ventricular hypertrophy, multiple sclerosis and mortality
2007	Confirmatory evidence showing modulation of host immunity by vitamin D to Mycobacterium tuberculosis

conflict of opinion about the importance of this study on the development of the first calcimimetic, Cinacalcet (Stewart, 2004; Nemeth, 2006). Nevertheless, Cinacalcet (Sensipar® in the USA and Australia and Mimpara® in Europe; Amgen, Thousand Oaks, CA, USA) became the first and the only drug available thus far to treat refractory secondary hyperparathyroidism in patients with chronic kidney disease and to treat hypercalcemia in patients with parathyroid carcinoma.

References

Achinger, S. G., and Ayus, J. C. (2005). The role of vitamin D in left ventricular hypertrophy and cardiac function. *Kidney Int*, pp. S37–S42.

Baker, A. R., McDonnell, D. P., Hughes, M. et al. (1988). Cloning and expression of full-length cDNA encoding human vitamin D receptor. *Proc Natl Acad Sci USA* **85**: 3294–3298.

Brown, S. J. (2006). The role of vitamin D in multiple sclerosis. *Ann Pharmacother* **40**: 1158–1161.

Brown, E. M., Gamba, G., Riccardi, D. et al. (1993). Cloning and characterization of an extracellular Ca(2+)-sensing receptor from bovine parathyroid. *Nature* **366**: 575–580.

Brumbaugh, P. F., and Haussler, M. R. (1975). Specific binding of 1alpha, 25-dihydroxycholecalciferol to nuclear components of chick intestine. *J Biol Chem* **250**: 1588–1594.

Brunette, M. G., Chan, M., Ferriere, C., and Roberts, K. D. (1978). Site of 1,25(OH)2 vitamin D3 synthesis in the kidney. *Nature* **276**: 287–289.

DeLuca, H. F. (1988). The vitamin D story: a collaborative effort of basic science and clinical medicine. *FASEB J* **2**: 224–236.

DeLuca, H. F., and Schnoes, H. K. (1983). Vitamin D: recent advances. *Annu Rev Biochem* **52**: 411–439.

Foley, R. N., Parfrey, P. S., Harnett, J. D. et al. (1996). Hypocalcemia, morbidity, and mortality in end-stage renal disease. *Am J Nephrol* **16**: 386–393.

Glisson, F. (1650). De Rachitide sive morbo puerili, qui vulgo The Rockets diciteur, London 1–416.

Gloldblatt, H., and Soames, K. M. (1923). Studies on the fat-soluble growth-promoting factor. *Biocehm J* **17**: 446–453.

Gylfe, E., Larsson, R., Johansson, H. et al. (1986). Calcium-activated calcium permeability in parathyroid cells. *FEBS Lett* **205**: 132–136.

Hermann, M., and Ruschitzka, F. (2008). Vitamin D and hypertension. *Curr Hypertens Rep* **10**: 49–51.

Hess, A. (1922). Influence of light on the prevention of rickets. *Lancet* **2**: 1222.

Hess, A. (1929). Rickets, including osteomalacia and tetany. In *The history of rickets*. Philadelphia: Lea and Febiger.

Hess, A. F., and Weinstock, M. (1924). Antirachitic properties imparted to lettuce and to growing wheat by ultraviolet irradiation. *Proc Soc Exp Biol Med* **22**: 5.

Holick, M. F., MacLaughlin, J. A., Clark, M. B. et al. (1980). Photosynthesis of previtamin D3 in human skin and the physiologic consequences. *Science* **210**: 203–205.

Holick, M. F., Smith, E., and Pincus, S. (1987). Skin as the site of vitamin D synthesis and target tissue for 1,25-dihydroxyvitamin D3. Use of calcitriol (1,25-dihydroxyvitamin D3) for treatment of psoriasis. *Arch Dermatol* **123**: 1677–1683.

Lambert, P. W., Stern, P. H., Avioli, R. C. et al. (1982). Evidence for extrarenal production of 1 alpha, 25-dihydroxyvitamin D in man. *J Clin Invest* **69**: 722–725.

Martineau, A. R., Wilkinson, K. A., Newton, S. M. et al. (2007). IFN-gamma- and TNF-independent vitamin D-inducible human suppression of mycobacteria: the role of cathelicidin LL-37. *J Immunol* **178**: 7190–7198.

Mellanby, E., and Cantag, M. (1919). Experimental investigation on rickets. *Lancet* **196**: 407–412.

Nemeth, E. F. (2006). Misconceptions about calcimimetics. *Ann NY Acad Sci* **1068**: 471–476.

Palomer, X., Gonzalez-Clemente, J. M., Blanco-Vaca, F., and Mauricio, D. (2008). Role of vitamin D in the pathogenesis of type 2 diabetes mellitus. *Diabetes Obes Metab* **10**: 185–197.

Russell, J., Lettieri, D., and Sherwood, L. M. (1986). Suppression by 1,25(OH)2D3 of transcription of the pre-proparathyroid hormone gene. *Endocrinology* **119**: 2864–2866.

Slatopolsky, E., Weerts, C., Thielan, J., Horst, R., Harter, H., and Martin, K. J. (1984). Marked suppression of secondary hyperparathyroidism by intravenous administration of 1,25-dihydroxy-cholecalciferol in uremic patients. *J Clin Invest* **74**: 2136–2143.

Stewart, A. F. (2004). Translational implications of the parathyroid calcium receptor. *N Engl J Med* **351**: 324–326.

Teng, M., Wolf, M., Lowrie, E., Ofsthun, N., Lazarus, J. M., and Thadhani, R. (2003). Survival of patients undergoing hemodialysis with paricalcitol or calcitriol therapy. *N Engl J Med* **349**: 446–456.

Whistler, D. (1645). Morbo puerili Anglorum, quem patrio idiomate indigenae vocant The Rickets. *Lugduni Batavorum*, pp. 1–13.

Wolf, G. (2004). The discovery of vitamin D: the contribution of Adolf Windaus. *J Nutr* **134**: 1299–1302.

Further reading

http://www.beyonddiscovery.org/content/view.article.asp?a=414 (Accessed June 9, 2008).

http://nobelprize.org/nobel_prizes/medicine/laureates/1929/ (Accessed June 9, 2008).

http://vitamind.ucr.edu/history.html (Accessed June 9, 2008).

http://www.fda.gov/cder/foi/nda/98/20819_Zemplar_approv.pdf (Accessed June 9, 2008).

http://www.fda.gov/cder/foi/nda/98/021068a_appltr_prntlbl.pdf (Accessed June 9, 2008).

Xue, J. L., St Peter, W. L., Ebben, J. P., Euerson, S. E., and Collins, A. J. (2002). Anemia treatment in the Pre-ESRD period and associated mortality in elderly patients. *Am J Kidney Dis* **40**: 1153–1161.

CHAPTER 7

Vitamin D: Molecular Biology and Gene Regulation

ADRIANA S. DUSSO AND ALEX J. BROWN
The Renal Division, Washington University School of Medicine, St Louis, Missouri, USA

Contents

I.	Vitamin D	69
II.	The 1,25-dihydroxyvitamin D/vitamin D receptor complex	70
III.	Relevance of 1,25-dihydroxyvitamin D/VDR actions in health and in kidney disease	80
IV.	Concluding remarks	85
	References	86

I. VITAMIN D

Vitamin D is essential for bone health, as has been recognized since the 1930s when vitamin D fortification of milk eradicated rickets, a skeletal disorder characterized by under-mineralized bone (Dusso et al., 2005). In recent years, strong epidemiological associations in the normal population between vitamin D deficiency and higher risk of fractures and osteoporosis, and also of cardiovascular disease, hypertension, diabetes, tuberculosis, autoimmune disease and cancer (Holick, 2007) have unraveled novel vitamin D biological actions critical in preventing these devastating disorders (Dusso et al., 2005). However, vitamin D, which can be either obtained from the diet or formed from 7-dehydrocholesterol in the skin via a non-enzymatic, ultraviolet light-dependent reaction, has little intrinsic activity. To exert its plethora of biological effects, vitamin D must be metabolized (Figure 7.1). The first step in vitamin D bioactivation is its hydroxylation in carbon 25, which occurs primarily in the liver and renders the most abundant metabolite in circulation, 25-hydroxyvitamin D. Serum levels of 25-hydroxyvitamin D directly correlate with vitamin D concentrations, thus constituting the best indicator of vitamin D status. The kidney proximal convoluted tubule is the principal site for the final and most critical step in vitamin D bioactivation: 1α-hydroxylation of 25-hydroxyvitamin D to the potent calcitropic steroid hormone 1,25-dihydroxyvitamin D (1,25(OH)$_2$D or calcitriol). However, numerous non-renal sources of calcitriol have been identified throughout the body. Under normal physiological conditions, non-renal calcitriol production appears to serve cell-specific autocrine/paracrine functions in the microenvironment where it is generated, rather than contribute to serum calcitriol levels. The cytochrome p450 containing enzymes that catalyze 25- and 1α-hydroxylations are CYP2R1 and CYP27B1 (Dusso et al., 2005).

Similar to other steroid hormones, most calcitriol actions are mediated by a high affinity, intracellular receptor, the vitamin D receptor (VDR), which acts as a ligand-activated transcription factor either to increase or suppress the rate of transcription of vitamin D target genes by RNA-polymerase II (Dusso et al., 2005). Vitamin D maintenance of a healthy skeleton requires calcitriol/VDR regulation of the expression of numerous genes involved not only in calcium and phosphate fluxes but also in the hormonal interactions between the kidney, the parathyroid gland, the intestine and bone. These genes control transepithelial calcium transport, the expression of the calcium sensing receptor (CaSR) (Brown et al., 1999; Canaff and Hendy, 2002) and, consequently, cellular responses to extracellular calcium, phosphate absorption and resorption, the coupling of bone forming and resorbing activities, and the tight loops between the rates of renal calcitriol synthesis and catabolism, parathyroid cell growth and the synthesis and secretion of parathyroid hormone (PTH), and bone synthesis of the phosphatonin FGF23 (Barthel et al., 2007). Figure 7.2 summarizes the central role of the calcitriol/VDR complex in coordinating with PTH and FGF23 the numerous hormonal feed back loops required for the maintenance of calcium and phosphate homeostasis and skeletal health in normal individuals. A decrease in serum calcium is sensed by the parathyroid glands which, to restore calcium balance, rapidly enhance the secretion and/or synthesis of parathyroid hormone (PTH). Elevations in serum PTH induce calcium resorption from bone and stimulate renal 1α-hydroxylase activity to increase calcitriol production. Increased serum calcitriol induces intestinal calcium absorption and, upon calcium normalization, both calcium and calcitriol close the loop by suppressing PTH synthesis and renal 1α-hydroxylase activity. Calcitriol also elevates

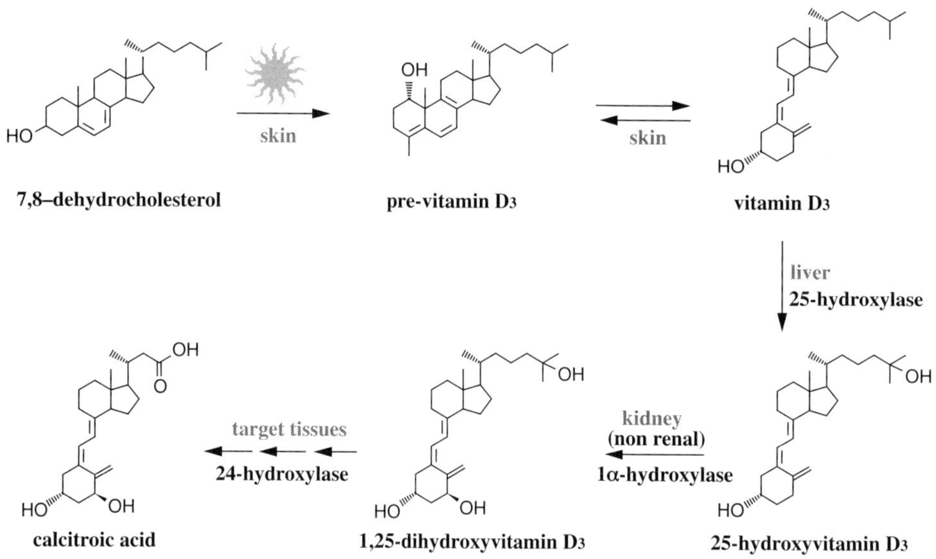

FIGURE 7.1 Vitamin D bioactivation and calcitriol catabolism.

serum phosphate levels by promoting intestinal absorption, renal reabsorption and bone resorption. Calcitriol also induces the synthesis of the phosphaturic hormone FGF23 in bone (Barthel et al., 2007). FGF23 in turn suppresses renal calcitriol synthesis through direct inhibition of 1α-hydroxylase expression and lowers serum phosphate levels by inhibiting renal phosphate reabsorption (Schiavi and Kumar, 2004). Thus, whereas calcitriol-induced increases in serum calcium and phosphate ensure appropriate bone mineralization (Dusso et al., 2005), calcitriol-coordinated regulation of PTH and FGF23 prevents the excesses of these minerals that could lead to over-mineralization of bone or ectopic calcification (Barthel et al., 2007).

Vitamin D actions unrelated to mineral homeostasis or skeletal health and critical in disease prevention require calcitriol/VDR regulation of the expression of numerous genes with potent effects on proliferation (Eelen et al., 2007), differentiation and function in many cell types including skin, pancreas and muscle as well as the hematopoietic, immune and nervous systems (Dusso et al., 2005).

Comprehensive studies of mineral and skeletal homeostasis in mice lacking the VDR and/or 1α-hydroxylase have further validated the obligatory role of VDR in calcitriol signaling and revealed not only which VDR responses are the most pathophysiologically relevant, but also the existence of $1,25(OH)_2D_3$-independent actions of the VDR as well as VDR-independent actions of $1,25(OH)_2D_3$. In addition to the evidence from transgenic mouse models, kidney disease underscores the pathophysiological repercussions of defective renal calcitriol synthesis causing systemic calcitriol deficiency and reduced calcitriol/VDR complex formation. These patients not only develop secondary hyperparathyroidism and the resulting disorders in skeletal remodeling known as renal osteodystrophy, but also cardiovascular disorders, hypertension, diabetes, as well as immunological and neuromuscular abnormalities that greatly enhance morbidity and mortality rates (Andress, 2006). Importantly, calcitriol/analog therapy, even at doses insufficient to correct either high serum PTH or mineral abnormalities, increases survival (Teng et al., 2003, 2005) thereby supporting the relevance of non-classical calcitriol/VDR actions and their distinct sensitivity to therapy.

II. THE 1,25-DIHYDROXYVITAMIN D/VITAMIN D RECEPTOR COMPLEX

A. Gene Targets and Biological Actions

Most of $1,25(OH)_2D_3$ biological activities require the VDR, an ancient member of the superfamily of nuclear receptors for steroid hormones. The biological relevance of the VDR in mediating the myriad of gene regulatory effects of calcitriol in mineral and skeletal homeostasis is underscored by

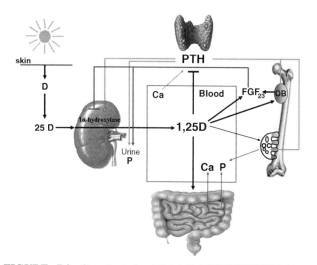

FIGURE 7.2 Coordinated calcitriol (1,25D)/PTH/FGF23 interactions control calcium, phosphate and skeletal homeostasis.

the phenotypes of hypocalcemia, severe skeletal growth disorders, secondary hyperparathyroidism and variable hypophosphatemia that are similar to those in vitamin D deficiency rickets, in patients with hereditary hypocalcemic vitamin D-resistant rickets (HVDRR, a syndrome which results from loss-of-function mutations in the VDR gene causing functional defects in the VDR) and also in transgenic mice in which the VDR has been ablated (Li et al., 1998). Intriguingly, however, clinical evidence from HVDRR patients and studies in the VDR-ablated transgenic models suggested a 'dispensable' role for vitamin D in bone metabolism. Bone mineralization defects in these patients were completed resolved by calcium infusion. Similarly, in the VDR-null mice, both the parathyroid and skeletal phenotypes of rickets were rescued through feeding a diet high in calcium, phosphate and lactose, which allows normalization of serum calcium by facilitating vitamin D-independent intestinal calcium absorption. These findings confirmed that the induction of intestinal calcium absorption is the dominant biological action of the calcitriol/VDR complex for skeletal health, especially when there is a physiologic stress from the requirement of bone mineralization under conditions of low calcium availability.

Recent comprehensive studies by Panda and collaborators using a double 1α-hydroxylase- and VDR-null mice model, challenged the concept of a 'dispensable' role of vitamin D in the calcium/PTH/bone axis (Panda et al., 2004). They conclusively demonstrated that not only for calcium absorption, but also for longitudinal bone growth, osteoblast and osteoclast activity, both calcitriol and the VDR are essential. They confirmed the findings in the VDR-null mice that bone mineralization can occur by providing a high ambient calcium concentration, which can also very effectively suppress PTH secretion. However, to prevent parathyroid hyperplasia, both calcium and calcitriol are necessary. In addition, more recent reports using a double 1α-hydroxylase and PTH null mice model also revealed that calcitriol uses intestinal and renal, but not skeletal, mechanisms to elevate serum calcium and also that calcitriol-induced increases in endochondral and appositional bone are independent of endogenous PTH (Xue et al., 2006). Clearly, although calcium and calcitriol have discrete and overlapping functions in mineral and skeletal homeostasis, calcium cannot entirely substitute for calcitriol. Indeed, calcitriol/VDR-regulated genes are involved in the induction of intestinal and renal transepithelial calcium transport, in bone anabolism and catabolism, and in the tight interactions between the parathyroid gland and the kidney that mediate the maintenance of PTH/calcitriol/calcium axis, and also between bone and the kidney in the control of the FGF23/calcitriol/phosphate axis for phosphate and bone metabolism.

As indicated earlier, the calcitriol/VDR complex also regulates the expression of genes unrelated to mineral or skeletal homeostasis. The complex controls cell proliferation, differentiation and survival in numerous cell types (Eelen et al., 2007) including cells of the immune system (Adorini et al., 2004), regulates specific cell function such as the suppression of the renin–angiotensin system (Li et al., 2002), the enhancement of glucose-mediated insulin secretion (Zeitz et al., 2003), the induction of the expression of the natriuretic peptide receptor (Chen et al., 2007) and induction of cytochromes p450 that mediate xenobiotic detoxification (Thompson et al., 2002). Indeed, VDR-null mice elicited uterine hypoplasia (Yoshizawa et al., 1997), mammary gland hyperplasia (Zinser and Welsh, 2004), hypertension (Li et al., 2002), enhanced sensitivity to skin carcinogens (Zinser et al., 2002), defective myocyte development (Endo et al., 2003) and impaired Th1 responses due to defective macrophage cytokine synthesis (O'Kelly et al., 2002). The non-classical calcitriol/VDR actions suggest that calcitriol deficiency in kidney disease could partially account for the development of hypertension, diabetes, immunological disorders and neuromuscular defects, important comorbid conditions in this patient population (Andress, 2006). The next section presents our current understanding of the basic mechanisms underlying VDR mediation of calcitriol regulation of gene transcription.

B. Structure-Function

Like the other members of the steroid receptor family, the VDR acts as a ligand-activated transcription factor (Brown et al., 1999). Figure 7.3 depicts the two critical domains of the VDR molecule involved in VDR control of gene transcription: the classical zinc finger motif for DNA binding and a multifunctional C-terminal domain for ligand binding, heterodimerization with the VDR partner, the retinoid X receptor (RXR) and the interface for recruiting of VDR-interacting nuclear proteins (co-regulators), which markedly enhance or suppress the rate of gene transcription by the VDR. The VDR molecule lacks, however, the long N-terminal extension known as activation function 1 (AF1), typical of the traditional steroid hormone receptors (Whitfield et al., 1999).

The DNA-binding domain (DBD), extending from amino acid 22 to 110 in the human VDR, is highly conserved among nuclear steroid receptors. The DBD is organized into two zinc-nucleated modules, the zinc finger DNA binding motifs, that are responsible for high affinity interaction with specific DNA sequences in the promoter region of 1,25 $(OH)_2D_3$ target genes, called vitamin D-responsive elements (VDREs). Natural mutations in the zinc finger region of the human VDR result in defective DNA binding and the most severe clinical phenotypes of vitamin D resistance (Haussler et al., 1998), thus providing compelling evidence not only for the obligatory role of the VDR in vitamin D actions, but for the absolute requirement for DNA binding as a mechanism for VDR signaling. Four nuclear localization

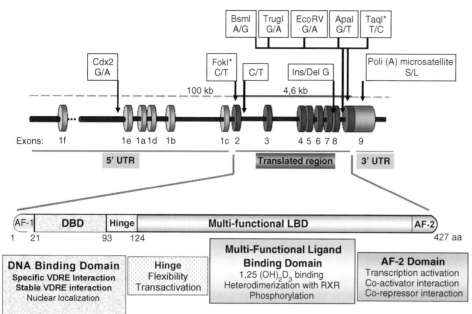

FIGURE 7.3 The VDR gene, its polymorphisms and functional domains of the encoded VDR.

signals (NLS), short stretches of basic amino acids, exist in the DBD of the VDR. Two of them, at amino acids 49–50 and 102–104, show co-identity in contacting DNA and as NLS. Point mutations in these NLS cause both defective cytoplasmic-to-nuclear translocation and the phenotype of HVDRR (Hewison et al., 1993; Barsony et al., 1997). Interestingly, basic residues at equivalent position are also important for the nuclear translocation of the VDR protein partner, the retinoid X receptor, RXR (Prufer et al., 2000). PKC phosphorylation of the VDR at serine 51, in close vicinity with this dual NLS-DNA binding sequence, markedly impairs nuclear localization and DNA binding affinity (Hsieh et al., 1993).

The hinge region between the DBD and the ligand binding domain is thought to make the VDR flexible. The function of the hinge region in gene regulation by the VDR is still unfolding (Shaffer et al., 2005).

The ligand binding domain (LBD) is located in the carboxyterminal portion of the VDR molecule and mediates the high affinity binding of $1,25(OH)_2D_3$ (Kd = 10^{-10}–10^{-11} M). In contrast, $25(OH)D_3$, the precursor of calcitriol and the most abundant metabolite in the circulation, binds the VDR nearly 100 times less avidly (Brumbaugh and Haussler, 1974; Mellon and DeLuca, 1979). The A ring containing the 1α-hydroxyl group is a critical portion of the $1,25(OH)_2D_3$ molecule responsible for high affinity VDR binding (Peleg et al., 1996). Calcitriol is anchored into the highly hydrophobic ligand pocket of the VDR molecule through six hydrogen bonds with the 1α-hydroxyl, the 3β-hydroxyl and the 25-hydroxyl. However, as will be discussed below, ligand binding affinity is not an absolute predictor of the VDR transcriptional activity conferred by the ligand.

Three different domains of the ligand binding domain of the VDR molecule operate as dimerization surfaces for the selective association between the VDR and the RXR, which induces a VDR conformation that is essential for VDR transactivating function. An interplay between ligand binding and heterodimerization domains was suggested by two natural mutations (I314S and R391C) in the LBD of the VDR that confer the phenotype of vitamin D resistance by significantly impairing both VDR-RXR heterodimerization and ligand retention (Haussler et al., 1997). Two additional domains of the LBD serve as adaptor surfaces for nuclear proteins necessary for VDR-co-regulator interactions (Carlberg, 2004): one is the RXR-heterodimerization domain containing residue 246, which is highly conserved among nuclear receptors, and forms part of the binding interface with transcriptional co-activators. Its alteration severely compromises transactivation. The second region is the activation function 2 (AF2) domain, which undergoes a dramatic conformational shift upon ligand binding (Figure 7.4) that allows the recruitment of VDR-interacting proteins including components of the transcription initiation complex, RNA polymerase II and nuclear transcriptional co-regulators that promote chromatin remodeling and control gene transcription. Removal of the AF2 domain eliminates $1,25(OH)_2D_3$-VDR transcriptional activity with little effect on ligand binding or heterodimeric DNA binding.

The next section presents the dynamic changes in the VDR molecule induced by calcitriol binding that are responsible for the activation of the calcitriol-bound VDR to translocate from the cytosol to the nucleus either to increase or suppress the rate of transcription of vitamin D responsive genes by RNA-polymerase II (Brown et al., 1999).

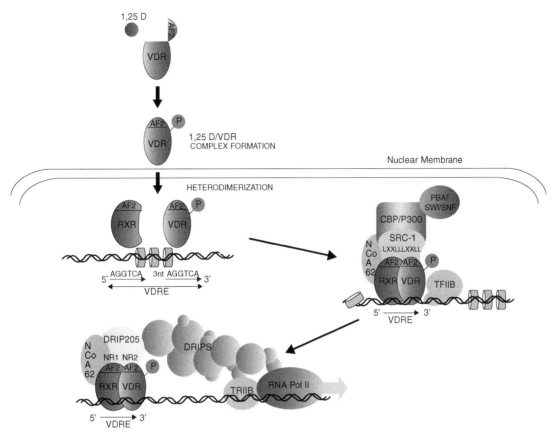

FIGURE 7.4 Calcitriol/VDR regulation of gene expression (see color plate section).

C. Regulation of Gene Expression

This section addresses the basic molecular mechanisms involved in:

1. ligand/VDR complex formation
2. VDR heterodimerization with RXR
3. DNA binding
4. gene transactivation and transrepression,

with special focus on recent findings concerning the modulation of each step under physiological conditions and in kidney disease.

1. Ligand/VDR complex formation

Intracellular levels of both ligand and VDR determine the magnitude of ligand/VDR complex formation in a vitamin D target cell. As indicated previously, both calcitriol and its precursor, 25-hydroxyvitaminD, can bind and activate the VDR, although with distinct affinities. How do these lipophylic molecules enter target cells from the circulation? The calcitriol synthesized in the kidney by mitochondrial 1α-hydroxylase and the 25-hydroxyvitamin D produced by the liver are transported in the blood by carrier proteins. Vitamin D binding protein (DBP) is the main carrier. However, calcitriol also binds albumin and lipoproteins (Brown et al., 1999). Recently, the prior dogma that the free form of calcitriol and other steroid hormones trigger biological responses after entering target cells by simple diffusion through the plasma membrane has been challenged. The uptake of 25-hydroxyvitamin D by renal proximal tubular cells occurs through megalin-mediated endocytosis (Nykjaer et al., 1999), thus raising the possibility that calcitriol entrance to target cells could also involve megalin-mediated endocytosis, as recently demonstrated for sex steroids in reproductive tissues (Hammes et al., 2005). Megalin is expressed in several epithelial cell types (Verroust and Christensen, 2002) including parathyroid cells (Segersten et al., 2002) and osteoblasts (van Driel et al., 2006), whose responses to calcitriol are critical for mineral and skeletal homeostasis. In kidney disease, a reduction in parathyroid or osteoblast megalin levels, similar to that demonstrated in renal proximal tubular cells after 5/6 nephrectomy in rats, could contribute to the resistance to calcitriol therapy in these patients. The expression of 1α-hydroxylase in parathyroid cells (Segersten et al., 2002; Ritter et al., 2006) and osteoblasts (van Driel et al., 2006) also suggests that endogenously produced $1,25(OH)_2D_3$ can bind and activate the VDR. $1,25(OH)_2D_3$ production, however, is contingent upon the adequacy of both circulating levels of

25-hydroxyvitamin D and cellular uptake, the latter susceptible to the same limitations in kidney disease as those outlined for calcitriol entrance. In fact, impaired uptake of 25-hydroxyvitamin D was demonstrated in peripheral blood mononuclear cells from hemodialysis patients compared to that in monocytes from normal individuals (Gallieni et al., 1995). Differential cellular uptake of the less calcemic calcitriol analogs compared to calcitriol could contribute to their tissue-specific potency. The LDL receptor was shown to mediate calcitriol uptake by human fibroblasts (Teramoto et al., 1995), supporting the existence of an active component in the cellular uptake of vitamin D metabolites. However, the more rapid uptake of the 22-oxa-calcitriol analog that binds DBP poorly (Dusso et al., 1991b) supports either the 'free' hypothesis or a facilitated endocytosis of the alternative carrier.

The concentration of ligand available for VDR binding is also dependent on the rate of its inactivation within the cell. 24-hydroxylase (CYP24A1) is the enzyme responsible for 25-hydroxyvitamin D and calcitriol degradation. CYP24A1 is induced by calcitriol in every vitamin D responsive tissue through a classical VDR-mediated mechanism. Calcitriol induction of its own degradation operates as a fine local tuning for cellular activation of the VDR and also as a highly efficient mechanism to maintain serum calcitriol levels within the narrow limits required for normal calcium homeostasis, despite the large seasonal variations in serum 25-hydroxyvitamin D levels. Reduced $1,25(OH)_2D_3$ clearance and signs of vitamin D intoxication in the 24-hydroxylase null mice support a critical role for ligand inactivation in the *in situ* control of the VDR responses to vitamin D (St-Arnaud et al., 2000). Conversely, calcitriol induction of its own degradation is a limitation for therapy and suggests an advantage of local calcitriol production over exogenous administration.

25-hydroxyvitamin D, calcitriol and less calcemic calcitriol analogs differ markedly in their VDR binding affinity. However, once inside the cell, cell-specific variations in the expression of intracellular vitamin D binding proteins (IDBPs) that mediate the delivery of the ligand to and from the VDR (Adams et al., 2004) modulate ligand-VDR association/dissociation rates. This influences both the half-life of the VDR molecule, which is protected from proteosomal degradation (Masuyama and MacDonald, 1998) through ligand binding (Wiese et al., 1992) and the degree of VDR activation. Physical interactions of liganded VDR with SUG1, a component of the proteosome complex, targets the VDR for ubiquitination and subsequent proteolysis (Masuyama and MacDonald, 1998), thus providing an additional *in situ* control of transcriptional responses to $1,25(OH)_2D_3$. In fact, reduced recruitment of SUG1 to the VDR by $1,25(OH)_2D_3$ analogs *in vitro* accounts for their higher transactivating potency (Jaaskelainen et al., 2000). It is unclear at present what directs the liganded VDR to the proteosome or to the transcriptional pre-initiation complex. Differential expression of parathyroid IDBPs may contribute to VDR-binding saturation for the calcitriol analog 22-oxacalcitriol at much lower concentration of the ligand compared to other tissues with similar VDR content (Koike et al., 1998).

Upon ligand binding, the repositioning of helix 12 in the AF2 carboxy terminus of the VDR ligand binding domain imparts a major conformational change in the 3-dimensional structure of the VDR that acts as a switch to activate the VDR. Despite the prediction that the structure of the ligand would induce unique conformational changes in the VDR molecule, thereby altering its biological function, crystallographic analysis could not attribute the unusual actions of vitamin D analogs *in vivo* and *in vitro* to their ability to induce a differential VDR conformation (Tocchini-Valentini et al., 2001; Vanhooke et al., 2004). This ligand-induced activation step appears to be required for the recruitment by the VDR of motor proteins (Racz and Barsony, 1999), that rapidly translocate cytoplasmic VDR along microtubules (Barsony and McKoy, 1992) to the nuclear membrane where importin facilitates nuclear entrance. In human peripheral blood monocytes, disruption of microtubular integrity is sufficient to abolish $1,25(OH)_2D_3$ induction of 24-hydroxylase gene transcription (Kamimura et al., 1995).

2. VDR HETERODIMERIZATION WITH RXR

Ligand binding also facilitates a high affinity heterodimeric interaction of VDR with RXR leading to the recognition by the VDR-RXR heterodimers of specific VDREs upstream of vitamin D target genes and serine phosphorylation of the VDR. Liganded VDR also influences the RXR to pivot its AF2 region into the 'closed' or active position (see Figure 7.4), which greatly reduces the affinity of RXR for its 9-cis retinoic acid ligand and enhances VDR/RXR capacity to bind co-activators. The RXR is a transcriptionally active partner of the VDR, as demonstrated by inhibition of VDR-mediated gene transcription through mutagenesis of the AF2 region of the RXR (Thompson et al., 2001) and by blocking RXR recruitment of co-activators using RXR specific blocking peptides (Bettoun et al., 2003).

Kidney disease provides *in vivo* evidence of the relevance of intracellular RXR levels in VDR-mediated transcription. In unilaterally nephrectomized rats, a reduction in renal content of a 50 kDa RXR isoform results in a reduction of the binding of the endogenous VDR/RXR heterodimer to the VDRE of the mouse osteopontin promoter (Sawaya et al., 1997). A similar reduction of parathyroid RXR content in these rats could explain reduced RXR/VDR suppression of PTH (MacDonald et al., 1994) with high serum PTH levels in the absence of hypocalcemia or hypophosphatemia (Sawaya et al., 1997).

3. DNA BINDING

High resolution crystal structures show the DBD of the VDR bound to the major groove of the hexameric VDRE (Shaffer and Gewirth, 2002), the most common VDRE type designated DR3, which contains direct repeats of two variable six-base half elements with the general consensus sequence AGGTCA, separated by a spacer of three nucleotides. This sequence directs the VDR-RXR heterodimer to the promoter region of $1,25(OH)_2D_3$-regulated genes, with the RXR binding the 5' half site and the VDR occupying the 3' half site (Haussler et al., 1998). Another type of VDRE, the IP9, consists of two inverted palindromic sequences separated by nine base pairs (Schrader et al., 1997). The VDREs of genes suppressed by the VDR, such as rat and chicken parathyroid hormone and mouse osteocalcin, are similar to the DR3 sequence found in genes in which transcription is induced by vitamin D. This finding raised important questions regarding the mechanisms determining whether gene transcription will be induced or suppressed by $1,25(OH)_2D_3$ (Darwish and DeLuca, 1996).

Multiple factors impair VDR/RXR binding to DNA in kidney disease including the accumulation of uremic toxin, increases in nuclear levels of calreticulin or the activation of cytokine signaling pathways. The contribution of uremic toxins to calcitriol resistance has been well documented (Hsu and Patel, 1995). Ultrafiltrate from uremic plasma causes a dose dependent inhibition of VDR/RXR binding to VDRE and calcitriol/VDR-transactivating function (Patel et al., 1995).

Increases in nuclear calreticulin also impair calcitriol/VDR actions. Calreticulin is a cytosolic protein that binds integrins in the plasma membrane and the DNA-binding domain of nuclear receptors, including the VDR, thus interfering with receptor mediated transactivation (Sela-Brown et al., 1998). Hypocalcemia, commonly present in renal failure and caused by either low serum calcitriol levels or hyperphosphatemia, enhances nuclear levels of parathyroid calreticulin. *In vitro* studies demonstrated that increases in nuclear calreticulin inhibit VDR/RXR binding to VDRE in a dose-dependent manner and totally abolish calcitriol suppression of PTH gene transcription (Sela-Brown et al., 1998). However, there is no report demonstrating that kidney disease increases nuclear levels of calreticulin in the parathyroid glands of patients or experimental animals to account for the resistance to calcitriol/VDR suppression of PTH.

In human monocytes and macrophages, cytokine activation markedly inhibits VDR binding to DNA and calcitriol/VDR-mediated gene transcription. Activation by the cytokine gamma interferon of its signaling molecule, Stat1, induces physical interactions between Stat1 and the DNA-binding domain of the VDR that impair VDR/RXR binding to VDRE and consequently, calcitriol transactivation of the 24-hydroxylase and osteocalcin genes (Vidal et al., 2002). Higher levels of inflammatory cytokines after hemodialysis could partially contribute to calcitriol resistance.

This interaction, however, does not compromise calcitriol induction of macrophage immune function, since it prolongs the half-life and transcriptional signaling of Stat 1, the most potent macrophage activating cytokine.

4. GENE TRANSACTIVATION/TRANSREPRESSION

The recruitment by the VDR/RXR heterodimer of nuclear co-regulators is a critical step in vitamin D regulation of gene expression. The co-regulators that interact within the AF-2 region of the nuclear receptors in general include the yeast SWI/SNIF complex, whose mammalian counterpart is the multisubunit Brahma/SWI12-related gene 1 associated factor (BAF), or the closely related polybromo- and BAF containing complex (PBAF) (Becker and Horz, 2002), that functions to remodel chromatin in an ATP-dependent manner, by sliding nucleosomes along DNA, a process common for transactivation or transrepression by the VDR.

a. Transactivation. In VDR-mediated transactivation, the recruitment of the p300/CBP co-activator complexes that contain the p160 class of nuclear receptor interacting proteins SRC1, SRC2 and SRC3 functions to re-acetylate nucleosomal histones via histone acetyl transferase (HAT) activity intrinsic to the co-activators. Histone acetylation appears to reduce internucleosomal interaction in chromatin, converting it to transcriptionally active euchromatin. Acetylation by HATs continues until co-activators undergo self-acetylation and dissociate from VDR (Shang et al., 2000). This allows the recruitment of a second complement of transcriptional co-activators, the DRIP/TRAP complex that consists of approximately 15 proteins. DRIP205 interacts directly with the VDR at a similar or overlapping site to that vacated by the p160 class of co-activators, thus supporting the sequential model of events shown in Figure 7.4. DRIP/TRAP recruitment builds a bridge with the basal transcriptional machinery that favors the assembly of the pre-initiation complex (PIC) to potentiate VDR induction of gene expression (Rachez and Freedman, 2000; Jurutka et al., 2001). DRIP 205 binding to the VDR appears to displace the transcription factor IIB (TFIIB) and delivers it to RNA Pol II facilitating PIC formation. Once transcriptional initiation has occurred and the DRIP205 mediator has departed the VDR/RXR interface for co-activator binding, the promoter DNA and its nucleosomes in the VDRE region would be exposed to the repressive actions of histone deacetylases. Evidence from chromatin immunoprecipitation assays indicates rapid oscillatory binding of p160 and DRIP co-activators to the VDR (Pike et al., 2007) which, taken together with the protection from proteosome-mediated degradation of liganded VDR, suggests that the next event in the VDR transcriptional cycle would be rebinding of SRC-1 and DRIP205 and sustained transcriptional activation. Alternatively, TRIP1 (the mammal ortholog of SUG1) could occupy the vacant VDR/RXR interface, with the subsequent recruitment of the proteosomal machinery. In gene transactivation by the glucocorticoid receptor, the linkage of less than four ubiquitin molecules precludes the receptor from proteosomal degradation (Masuyama and MacDonald, 1998). A similar mechanism for the VDR would allow four transcription cycles to occur before directing the VDR for proteosomal degradation. The dissociation of both TRIP1 and the ligand would lead to a re-opening of the AF2 to the inactive position, leading to re-association of co-repressors with the unliganded-VDR-RXR heterodimer. Undoubtedly, this cycle is subjected to many variations depending upon the specific target cell and gene regulated by the VDR. There are multiple variations in the sequences of the natural DR3 VDREs and there are indications that these sequence differences may allosterically confer the VDRE-bound heterodimer with distinct conformations that favor the recruitment of different sets of co-activators (Staal et al., 1996).

An additional modulator of VDR-transactivating potency is VDR phosphorylation, which can be either stimulatory when catalyzed by casein kinase 2 at serine 208 (Jurutka et al., 1996) or inhibitory, as indicated earlier for PKC phosphorylation of serine 51. A novel VDR phosphorylation, catalyzed by a yet unidentified kinase, appears to mediate high affinity heterodimeric binding to the VDRE (Jurutka et al., 2002).

Although the DR3 VDRE is the prevalent sequence for genes transactivated by liganded VDR, the calcitriol/VDR complex induces gene expression from 'non-consensus' VDREs such as in the case of RANKL, key to calcitriol induction of osteoclastogenesis (Zella et al., 2007), or indirectly by regulating the expression of gene-specific enhancers or repressors. For example, calcitriol transactivation of the FGF23 gene appears to depend upon the upregulation of GATA-3, an enhancer of FGF23 transcription (Barthel et al., 2007). The calcitriol-mediated induction of the cyclin dependent kinase inhibitor p27 involves calcitriol suppression of Skp2, which targets p27 for ubiquitination and proteosomal degradation, thus prolonging p27 half-life (Huang and Hung, 2006). Interestingly, the epidermal growth factor receptor (EGFR) gene, critical in the control of cell growth, apoptosis and survival, that possesses consensus VDREs and is induced by calcitriol in osteoblasts (Gonzalez et al., 2002), is suppressed by the sterol in cancer cells through mechanisms that involve competition by the VDR for a SP1 site mandatory for EGFR transactivation (McGaffin and Chrysogelos, 2005).

The exact set of co-modulators recruited by the activated VDR-RXR heterodimer may constitute another tissue-specific permutation influencing the nature of transcriptional activation. Many co-activators, including SRC1, GRIP1 (SRC-2) and ACTR (SRC-3), have distinct tissue distributions (Gehin et al., 2002). Indeed, recent studies suggest a bifunctional role for the VDR co-modulator NCoA62/Skip. It can promote transcriptional activation or repression, in a cell-specific manner, depending upon the expression of other co-regulator molecules (Leong et al., 2004). The co-activator p300 and the co-repressors NCoR and SMRT interact with the same N-terminal region of the Skip molecule. The relative levels of expression of the nuclear co-repressor NCoR and co-activator CBP/p300 in CV-1 and P19 cells dictate whether Skip activates or represses VDR/RXR-dependent transcription (Leong et al., 2004). Similar to Skip, a novel ATP-dependent chromatin remodeling complex containing the Williams' syndrome transcription factor, WSTF, potentiates ligand-induced VDR action in both gene transactivation and repression (Kato et al., 2004).

An additional critical component is $1,25(OH)_2D_3$ modulation of the transcriptional competence of target cells through regulation, in a cell-specific manner, of the relative expression of genes of two important co-regulator families, TIF2 from the p160 co-activators and SMRT among co-repressors (Picard et al., 2002; Dunlop et al., 2004).

NCoA62/Skip also mediates a link between transcriptional regulation by the VDR (and other nuclear receptors) and RNA splicing by the spliceosome (Zhang et al., 2003). Skip physically interacts with components of the splicing machinery and nuclear matrix associated proteins. In fact, expression of a dominant negative Skip interfered with appropriate splicing of transcripts derived from $1,25(OH)_2D_3$-VDR transactivation (Zhang et al., 2003).

b. Transrepression. VDR activation can also reduce the transcription rates of many target genes; those of particular relevance to chronic kidney disease will be discussed below. Perhaps as fascinating as the physiologic and pathologic consequences of the negative regulation of these genes is the diversity of the mechanisms by which the VDR represses their transcription. In general, for transcriptional repression by the VDR, VDR/RXR binding to a negative VDRE recruits co-repressors of the family of histone deacetylases. These molecules prevent chromatin exposure and, consequently, the binding of proteins (TATA binding protein) mandatory to initiate the transcription of the target gene by RNA-polymerase II (Rachez and Freedman, 2000; Jurutka et al., 2001).

The best-known example of transrepression by liganded VDR is the control of PTH. Parathyroid glands express high levels of VDR and studies *in vivo* and in cell culture demonstrated reduction of PTH gene transcription with $1,25(OH)_2D_3$ treatment. The mechanism by which $1,25(OH)_2D_3$ represses PTH remains controversial. Initial studies revealed a VDR binding site at around 100 bp upstream of the transcription start site in the human PTH gene (Demay et al., 1992) that mediated transrepression, but RXR binding to this site could not be demonstrated (Mackey et al., 1996). On the other hand, the negative VDREs of the rat and chicken PTH gene promoters bind the VDR-RXR heterodimer and resemble the DR3 of positive VDREs (Russell et al., 1999). The demonstration that changing two bases of the negative VDRE of the chicken PTH gene converts it to a positive element indicates that the DNA sequence itself may alter the conformation of the heterodimer, leading to replacement of co-repressors with co-activators. More recently, an additional/alternative mechanism has been proposed in which the activated VDR interacts with a positive regulator of PTH gene transcription (VDIR or *VD*R *i*nteracting *r*epressor), displacing co-activators from VDIR and recruiting co-repressors to this site, which is located at around 500 bp from the transcription start site of the human PTH gene (Kim et al., 2007). VDIR also appears to be involved in the transrepression of the 1α-hydroxylase gene.

The repression of the 1α-hydroxylase gene by 1,25(OH)$_2$D$_3$ has been controversial, as the negative regulation has not been universally observed. However, recent reports have implicated a mechanism similar to that described for PTH, involving activation of the 1α-hydroxylase gene by VDIR binding to its element and displacement of VDIR-bound co-activators by liganded VDR (Murayama et al., 2004). More recently, evidence was presented for VDR-RXR binding to a more distal element of the 1α-hydroxylase promoter that conferred negative regulation in the presence of 1,25(OH)$_2$D$_3$ (Turunen et al., 2007). Negative regulation via this distal element was observed in embryonic kidney (HEK 293) cells but not in breast cancer (MCF-7) cells. The inability of 1,25(OH)$_2$D$_3$ to repress the 1α-hydroxylase gene has been reported for other cells including macrophages and dendritic cells. The basis for the cell-specificity requires further investigation.

The expression of several genes critical in bone remodeling is distinctly suppressed by calcitriol. Specifically:

1. VDR/RXR binding to a negative VDRE in osteoblasts mediates calcitriol suppression of the expression of Runt-related transcription factor (Runx2 or Cbfa1). Cbfa1 is a key regulator of osteoblast formation during development and continues to regulate gene expression in mature osteoblasts (Drissi et al., 2002)
2. The endopeptidase PHEX plays a key role in phosphate metabolism and bone mineralization, as PHEX mutations lead to hypophosphatemic rickets. Calcitriol suppresses PHEX gene expression indirectly by influencing the activity or the expression of a transcription factor (PAP110) that is a positive regulator of basal PHEX gene transcription (Kream et al., 1995; Hines et al., 2004)
3. The Col1a1 gene (Kream et al., 1995) mediates the synthesis of type I collagen, the major protein component of bone matrix, acting as a scaffold for mineralization. However, details are lacking concerning the inhibitory complex and it site of action
4. Bone sialoprotein (BSP) is a mineralized tissue-specific protein that is highly expressed during the initial formation of bone. Repression of this gene by 1,25(OH)$_2$D$_3$ was reported to be mediated by activated VDR homodimers acting on an element near the transcriptional start site of the rat BSP gene (Kim et al., 1996)
5. Osteoprotegerin (OPG), a decoy receptor that attenuates the osteoclastogenic actions of the 1,25(OH)$_2$D$_3$-inducible RANKL. To enhance the effectiveness of RANKL, 1,25(OH)$_2$D$_3$ reduces OPG expression in osteoblasts by accelerating degradation of OPG mRNA and by repression of the OPG gene. The mechanism for transrepression involves inactivation of a stimulatory AP-1 site by promoting dephosphorylation of the AP-1 transcription factor c-Jun (Kondo et al., 2004).

The PTH-related polypeptide (PTH-rP) is also negatively regulated by liganded VDR. Although PTHrP is not believed to control calcium homeostasis, it does play a role in development and is secreted by certain cancer cells causing hypercalcemia. Like PTH, it stimulates the renal 1α-hydroxylase and is subject to feedback by 1,25(OH)$_2$D$_3$. The mechanism of transrepression appears to differ, however. The negative VDRE in the PTH-rP promoter binds monomeric VDR (no RXR) in the absence of ligand. 1,25(OH)$_2$D$_3$ appears to produce a conformation change that induces VDR interaction with Ku antigen, a DNA-dependent kinase that phosphorylates the VDR, leading to the release of VDR from the element (Okazaki et al., 2003).

1,25(OH)$_2$D$_3$ negatively regulates numerous genes unrelated to mineral homeostasis. Important for kidney disease progression is calcitriol suppression of renin expression. Indeed, VDR-ablated mice develop hypertension and cardiac hypertrophy (Xiang et al., 2005). Calcitriol repression of renin gene transcription does not require binding of the VDR to DNA or to RXR. It involves two regions of the promoter, one near the transcription start site and a second more distal site 2700 bp upstream. The distal element confers much greater repression and involves interference with the strong positive cAMP response element (CRE) via direct binding of the liganded VDR to the CRE binding protein (CREB) (Yuan et al., 2007).

The expression of several genes responsible for calcitriol immunomodulatory properties are suppressed through non-classical transrepression mechanisms. Specifically:

1. Inhibition of interferon-γ (IFN-γ) expression by 1,25(OH)$_2$D$_3$ in macrophages and dendritic cells prevents further antigen presentation and recruitment of T-cells. Transrepression of the IFN-γ gene involves, in part, binding of the liganded VDR-RXR complex to a negative VDRE-like element in the promoter. In addition, 1,25(OH)$_2$D$_3$ may also interfere with an enhancer element crucial for high level IFN-γ expression (Cippitelli and Santoni, 1998)
2. Inhibition by 1,25(OH)$_2$D$_3$ of interleukin-2 (IL-2), an autocrine growth factor for lymphocytes, prevents further activation and proliferation. Repression of IL-2 transcription is indirect. Liganded VDR-RXR binds to the distal site for nuclear factor of activated T-cells (NFAT) in the IL-2 promoter preventing activation by NFAT. In addition, 1,25(OH)$_2$D$_3$ treatment also disrupts formation of the active NFAT-AP1 complex (Takeuchi et al., 1998)

3. Granulocyte-macrophage colony-stimulating factor (GMCSF) is a T-cell cytokine that activates mature granulocytes and macrophages, part of the inflammatory response to infection. Repression of GMCSF occurs via a unique mechanism in which liganded VDR monomers bind to a negative regulatory element in the GMCSF promoter (Towers et al., 1999)
4. $1,25(OH)_2D_3$ suppresses Fas ligand (FasL) expression. The Fas/FasL is the major pathway that mediates activation-induced cell death of T-cells. Calcitriol transrepression of FasL involves binding of liganded VDR-RXR to a non-canonical c-myc site in the FasL promoter (Cippitelli et al., 2002)
5. Interleukin-12 (IL-12) is the major cytokine for stimulating formation of the pro-inflammatory Th1. By suppressing IL-12, $1,25(OH)_2D_3$ shifts the distribution away from Th1 to the anti-inflammatory Th2. Repression of IL-12 by $1,25(OH)_2D_3$ is accomplished by interfering with NF-κB activation and binding to the enhancer element in the p40 subunit of IL-12 (D'Ambrosio et al., 1998)
6. RelB is a member of the Rel/NFκB family of transcription factors that are crucial for inflammatory and immune responses. RelB heterodimers have low affinity for IκBs and are, therefore, primarily nuclear and contribute to the constitutive NFκB activity of immune cells. $1,25(OH)_2D_3$ represses RelB transcription by binding of the liganded VDR-RXR complex to a negative response element in the RelB promoter (Dong et al., 2003).

In summary, the mechanisms by which the activated VDR reduces target gene expression vary greatly. Often the liganded VDR-RXR complex binds to a negative element and recruits co-repressors and this action may be determined by the DNA sequence of the element. In at least one case, VDR monomers (or dimers with non-RXR partners) can bind to a negative element. However, in many instances, the liganded VDR, with or without RXR, interferes with the actions or expression of enhancers of gene expression.

5. NON-GENOMIC ACTIONS OF CALCITRIOL

An additional level of complexity in the cell- and promoter-specific VDR/RXR-nuclear co-regulator interactions responsible for VDR regulation of gene expression is imparted by cell-specific VDR, RXR or co-regulator phosphorylation. Signaling pathways leading to activation of kinases and phosphatases are activated by the rapid, non-genomic actions of a variety of hormones, including calcitriol itself (Baran et al., 1992; Baran, 1994; Farach-Carson and Davis, 2003), although this potential modulatory action of calcitriol remains controversial. Indeed, numerous studies demonstrated that non-genomic actions may not be critical for $1,25(OH)_2D_3$-mediated gene activation (Farach-Carson et al., 1993; Jurutka et al., 1993; Norman et al., 1993; Khoury et al., 1994, 1995; Zhou and Norman, 1995) or inhibition of cell proliferation (Norman et al., 1993; Hedlund et al., 1996). However, as indicated, $1,25(OH)_2D_3$ non-genomic stimulation of protein kinases including MAP kinases (Beno et al., 1995; Song et al., 1998), could potentially influence the VDR-mediated effects. $1,25(OH)_2D_3$ can rapidly stimulate phosphoinositide metabolism (Lieberherr et al., 1989; Bourdeau et al., 1990; Morelli et al., 1993), cytosolic calcium levels (Lieberherr, 1987; Hruska et al., 1988; Sugimoto et al., 1988; Lucas et al., 1989; Morelli et al., 1993), cyclic GMP levels (Guillemant and Guillemant, 1980; Vesely and Juan, 1984), PKC (Sylvia et al., 1996) and the opening of chloride channels (Zanello and Norman, 1996) all of which could affect not only subcellular localization, but the trafficking of VDR, RXR and critical co-regulators.

The depicted complexity in the interactions of the VDR/RXR with VDRE and co-regulators raises the possibility that, in kidney disease, vitamin D resistance could also result from decreased expression of essential co-activator or co-repressor molecules, or from defective recruitment of these molecules by the VDR. Megalin was shown to modulate VDR-mediated transactivation through sequestration of a component of the VDR-transcriptional complex (May et al., 2003). Thus, the reduction in megalin expression induced by kidney disease could partially account for abnormalities in calcitriol/VDR transcriptional activity. Uremia-induced activation of VDR-unrelated signaling pathways may also interfere with the recruitment by the VDR of co-regulator molecules to the transcription pre-initiation complex or with their transactivating function.

6. VDR POLYMORPHISMS

Finally, the quantity and quality of VDR-mediated transactivation or tranrepression can be affected by polymorphic variants in the human VDR gene that occur naturally in the human population (Morrison et al., 1994; Koshiyama et al., 1995), with substantial differences between races and ethnic groups (Uitterlinden et al., 2004a). Their expression associates with decreased bone density (Eisman, 1999), susceptibility to infections, autoimmune diseases and cancer (Uitterlinden et al., 2004b) and, in the population of kidney disease patients, with the propensity to hyperparathyroidism (Carling et al., 1995; Gomez Alonso et al., 1998), resistance to vitamin D therapy (Kontula et al., 1997), the severity of secondary hyperparathyroidism (Fernandez et al., 1997), bone loss after renal transplantation (Akiba et al., 1997; Nagaba et al., 1998; Giannini et al., 2002), time undergoing hemodialysis before parathyroidectomy is needed (Borras et al., 2003) and parathyroid responsiveness to serum calcium (Yokoyama et al., 2001). The top panel of Figure 7.3 depicts the human VDR gene, which maps to chromosome 12q13 and consists of at least 14 exons. Exons 1a to 1f are located at the 5' non-coding end of the genes whereas exons 2-9 encode the actual VDR. Although the polymorphisms most frequently studied are located in the intron separating

(Chanard et al., 1976; Juttmann et al., 1978) suggests that intestinal calcitriol/VDR-complex formation may be normal and that the removal of uremic toxins restores VDR-mediated calcium absorption (Chanard et al., 1976).

1,25(OH)$_2$D$_3$ also increases active phosphate transport through stimulation of the expression of the Na/Pi co-transporter (Yagci et al., 1992) and changes in the composition of the enterocyte plasma membrane (Kurnik and Hruska, 1985) that increase fluidity and phosphate uptake. Little is known, however, concerning the molecular mechanisms involved in the extrusion of phosphate across the basolateral membrane into the circulation. As with intestinal calcium absorption, phosphate absorption is decreased in renal failure and could be enhanced by administration of calcitriol of 1α-hydroxy vitamin D$_3$.

The calcitriol/VDR complex also activates the transcription of the gene encoding for Cyp3A4 that converts lithocholic acid (LCA) to its 6α-isoform (Araya and Wikvall, 1999). LCA is the product of the degradation by gut bacteria of one of the major bile acids, chenodeoxycholic acid. LCA is a carcinogen that accumulates in the colon due to ineffective reabsorption. Calcitriol induction of Cyp3A4 converts LCA in a molecule than can be transported out of the intestine, thereby eliminating a potential contributor to colon cancer. It is unknown whether kidney disease impairs this important VDR detoxifying mechanism.

B. Parathyroid Glands

The vitamin D endocrine system is a potent modulator of parathyroid function. Whereas vitamin D deficiency results in parathyroid hyperplasia and increased PTH synthesis and secretion, 1,25(OH)$_2$D$_3$ administration inhibits PTH synthesis and parathyroid cell growth, *in vivo* and *in vitro*, thus rendering 1,25(OH)$_2$D$_3$ and its analogs the therapy of choice in treating the secondary hyperparathyroidism of chronic kidney disease (Dusso et al., 2004). The demonstration that normalization of serum calcium in the VDR- and 1α-hydroxylase-null mice (Kollenkirchen et al., 1991) corrects the high serum PTH levels suggests that neither the VDR nor 1,25(OH)$_2$D$_3$ are essential, but are cooperative with calcium in controlling PTH synthesis. However, compelling evidence demonstrates that direct actions of calcitriol on parathyroid cells do indeed regulate PTH. In addition to the described transrepression of the PTH gene by the 1,25(OH)$_2$D$_3$-VDR complex, 1,25(OH)$_2$D$_3$ regulates both parathyroid levels of VDR and the response of the parathyroid gland to calcium. 1,25(OH)$_2$D$_3$-induced increases in parathyroid VDR result from increases in mRNA levels, possibly secondary to increases in serum calcium (Brown et al., 1995) and/or to calcitriol induction of C/EBPβ, a transactivator of the VDR gene in renal cells and in osteoblasts (Dhawan et al., 2005) and also from prolonging VDR half-life through ligand-dependent protection of the VDR from proteosomal degradation (Wiese et al., 1992). In fact, in rats with kidney failure, a strong direct correlation exists between serum 1,25(OH)$_2$D$_3$ levels and parathyroid VDR-protein content (Denda et al., 1996). Furthermore, prophylactic 1,25(OH)$_2$D$_3$/analog administration prevents the decrease in parathyroid VDR expression that accompanies the progression of kidney disease. 1,25(OH)$_2$D$_3$ modulation of the response of the parathyroid glands to calcium involves direct calcitriol/VDR induction of the expression of the calcium sensing receptor (CaSR) gene (Brown et al., 1996), by interactions with VDREs in both of the CaSR promoters (Canaff and Hendy, 2002). In rats, parathyroid CaSR mRNA levels were decreased by 40% by vitamin D deficiency (Brown et al., 1996) and enhanced by calcitriol treatment in a time and dose-dependent manner. Upregulation of parathyroid CaSR by calcitriol treatment in hemodialysis patients could explain the decrease in the set point for PTH suppression by serum calcium (Delmez et al., 1989) and, also, the higher CaSR levels in surgically removed parathyroid glands from patients receiving calcitriol compared to untreated patients (Shiraishi et al., 2001). The strong association between vitamin D deficiency, high serum PTH and increased fracture risk in the normal population suggests that PTH gene suppression could also result from VDR activation by calcitriol synthesized endogenously by parathyroid 1α-hydroxylase since, as stated previously, serum levels of 25-hydroxyvitamin D, but not calcitriol levels, are low in vitamin D deficiency. The impact of endogenously produced calcitriol to induce VDR and CaSR cannot be assessed easily but could be inferred from the correction of PTH by vitamin D supplementation in kidney disease patients. Eighty percent of these patients are vitamin D deficient (Andress, 2006). In kidney disease, both parathyroid levels of the CaSR and VDR decrease with the progressive severity of the hyperplasia (Kifor et al., 1996), rendering nodular hyperplasia, the most severe form of secondary hyperparathyroidism, insensitive to high calcium and calcitriol suppression of either PTH or cellular growth.

To identify the mechanisms mediating calcitriol/VDR suppression of parathyroid cell growth in kidney disease, it was necessary to characterize the pathogenesis of parathyroid hyperplasia. Recent studies demonstrated that increases in parathyroid expression of the growth promoter TGFα and its receptor, the epidermal growth factor receptor (EGFR) and TGFα upregulation of its own expression generate a feedforward growth loop of EGFR activation that causes the aggressive parathyroid growth induced by kidney disease and aggravated by low calcium or high P intake, within one week after 5/6 nephrectomy in rats (Dusso et al., 2001; Cozzolino et al., 2001). In closer agreement with the features of human secondary hyperparathyroidism (Gogusev et al., 1996), in established secondary hyperparathyroidism in rats, exclusive increases in TGFα and TGFα self upregulation, with unchanged parathyroid EGFR levels, determine the severity of parathyroid cell growth.

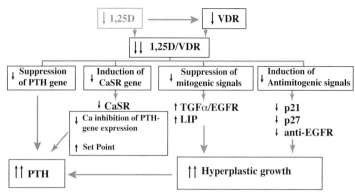

FIGURE 7.5 Calcitriol (1,25D)/VDR control of parathyroid function in kidney disease.

Calcitriol arrests parathyroid hyperplasia not only by suppressing mitogenic signals but also by inducing antimitogenic signals, including:

1. prevention of the increases in parathyroid expression of both components of the highly mitogenic TGFα/EGFR growth loop (Cozzolino et al., 2001; Dusso et al., 2001)
2. enhancement of parathyroid expression of the cyclin dependent kinase inhibitors p21 and p27 (Cozzolino et al., 2001; Dusso et al., 2001; Tokumoto et al., 2002), known suppressors of cellular growth, and
3. downregulation of cell membrane and nuclear growth signals from TGFα-activated EGFR through sequestering ligand-activated EGFR in the early endosomal compartment (Cordero et al., 2002), a phenomenon compatible with calcitriol-induced autophagy, as demonstrated in breast cancer cells (Hoyer-Hansen et al., 2005).

The time of exposure to calcitriol required to suppress parathyroid expression of TGFα and EGFR *in vivo* and/or TGFα/EGFR growth signals *in vitro* (Cordero et al., 2002) suggests a genomic VDR-mediated mechanism rather than rapid, non-genomic calcitriol actions. It is unclear at present whether these novel antiproliferative properties of the calcitriol/VDR complex involve direct transcriptional regulation.

The critical role of $1,25(OH)_2D_3$ in the control of parathyroid cell growth emerged from studies in the 1α-hydroxylase null mice. As stated earlier, normalization of serum calcium corrected serum PTH levels but could not suppress parathyroid hyperplasia (Panda et al., 2004). A novel, uncharacterized VDR-independent mechanism appears partially to mediate the antiproliferative properties of high serum $1,25(OH)_2D_3$ in the parathyroid glands, since calcium normalization effectively arrests parathyroid growth in the VDR-null mouse, which have supraphysiological circulating levels of $1,25(OH)_2D_3$. Similarly, intra-parathyroid (percutaneous) injection therapy with the $1,25(OH)_2D_3$ analog 22-oxacalcitriol regressed parathyroid hyperplasia and induced apoptosis in patients with nodular secondary hyperparathyroidism resistant to intravenous 22-oxacalcitriol (Shiizaki et al., 2003). It is unclear whether this is attributable to the very high concentration of 22-oxacalcitriol acting on the diminished VDR levels or due to a VDR-independent mechanism. High concentrations of calcitriol/analog could also trigger the increases in cytosolic calcium required to induce macroautophagy, as demonstrated for the calcitriol analog EB1089 in breast cancer cells (Hoyer-Hansen et al., 2007) and for calcitriol adjuvant effects to ionizing radiation in cancer therapy (Demasters et al., 2006).

Figure 7.5 summarizes the multiple direct effects of low serum calcitriol on the parathyroid glands leading to parathyroid hyperplasia and secondary hyperparathyroidism. Early interventions with vitamin D supplementation or low doses of calcitriol or its less calcemic analogs should prevent the decreases in VDR and CaSR responsible for the reduced sensitivity of hyperplastic parathyroid glands to control PTH synthesis and cell growth in response to calcium, calcitriol or analog therapy. Indeed, the reduction in VDR triggering the onset of calcitriol resistance is the most serious challenge for the efficacious treatment of secondary hyperparathyroidism. Recent studies identified the pathogenesis for the strong association between the severity of parathyroid cell growth and the resistance to calcitriol/analog therapy. A direct cause–effect relationship exists between EGFR activation and the reduction in VDR. The use of EGFR-tyrosine kinase inhibitors, highly specific inhibitors of EGFR activation, demonstrated that inhibition of TGFα activation of the EGFR was sufficient not only to arrest growth but also to prevent VDR reduction and reverse the resistance to calcitriol suppression of PTH (Arcidiacono et al., in press in the J. Am. Soc. Nephrol. and summarized in Figure 7.6). More significantly, the characterization of the molecular mechanisms by which the activation of the TGFα/EGFR growth loop reduces VDR expression has shed some light into the determinants of the switch from the mild growth of diffuse glands to the aggressive growth in the nodules of human secondary hyperparathyroidism and also into novel antiproliferative properties of calcitriol. TGFα activation of the EGFR induces the synthesis of LIP, a potent mitogen and inducer of transformation (Lamb et al., 2003; Baldwin et al., 2004) and, also, the dominant negative isoform of the transcription factor C/EBPβ, an inducer of VDR gene expression (Dhawan et al., 2005). Increases in LIP antagonize C/EBPβ transactivation of the VDR gene, thus reducing VDR content. Since contrary to LIP, C/EBPβ arrest cellular growth,

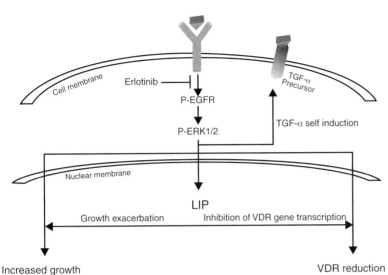

FIGURE 7.6 Parathyroid EGFR activation mediates the association between the severity of hyperplastic growth and the reduction of VDR content.

and the higher the C/EBP LIP ratios in a cell the lower the proliferation rates, calcitriol induction of C/EBPβ in the parathyroid glands should counteract both LIP mitogenic and transforming properties and LIP antagonism on VDR expression. It is unclear at present whether calcitriol induction of C/EBPβ expression involves transcriptional or translational regulation. As will be discussed below, appropriate therapeutic interventions to control the EGFR/VDR antagonism will affect not only the rate of progression of secondary hyperparathyroidism but also that of lesions in the renal parenchyma, left ventricular hypertrophy and vascular disorders.

C. Bone

The essential role of vitamin D in the development and maintenance of a mineralized skeleton has been known since 1920. However, only recently, transgenic mouse models, null for 1α-hydroxylase and/or the VDR, have extended calcitriol/VDR actions in bone beyond the traditional view as a promoter of intestinal absorption of calcium and phosphate and as a stimulator of osteoclastic bone resorption. Indeed, except for cartilage and skeletal mineralization, normalization of serum calcium through enhanced intestinal absorption cannot entirely substitute for a defective $1,25(OH)_2D_3$/VDR in skeletal homeostasis. Specifically, whereas growth plate development requires coordinated calcium and $1,25(OH)_2D_3$ actions, optimal osteoblastic bone formation, basal and PTH-induced osteoclastogenesis, osteoclastic bone resorption and the normal coupling of bone remodeling (Panda et al., 2004) demand both $1,25(OH)_2D_3$ and the VDR. Reduced osteoblast number, mineral apposition rates and bone volume in the 1α-hydroxylase null mice *in vivo* and decreased production of mineralized colonies *ex vivo* suggest that the $1,25(OH)_2D_3$/VDR system may exert an 'anabolic' effect necessary to sustain bone-forming activity, an effect unmasked only when the defective $1,25(OH)_2D_3$/VDR system exists in the presence of normal PTH. Calcitriol induction of LRP5 (Barthel et al., 2007), a Wnt co-receptor that is required for osteoblast proliferation and function, could contribute to osteoblastogenesis and the induction of calcitonin, osteopontin and alkaline phosphatase to mature osteoblast activity. A defective control of the RANKL/RANK interactions by an altered $1,25(OH)_2D_3$/VDR system contributes to abnormal coupling in bone turnover (Khosla, 2001; Boyle et al., 2003). Osteoblasts express a surface ligand RANKL (receptor activator of NFκB ligand), which can bind either RANK or an osteoblast-derived soluble decoy receptor osteoproteregin (OPG). The binding of RANKL to RANK induces a signaling cascade that results in differentiation and maturation of osteoclasts. $1,25(OH)_2D_3$ as well as PTH and prostaglandins stimulate RANKL expression (Kitazawa et al., 2003), but $1,25(OH)_2D_3$ also inhibits OPG production (Kondo et al., 2004) with a corresponding increase in osteoclastogenesis and osteoclast activity. The expression of 1-hydroxylase in osteoblasts (van Driel et al., 2006) and in monocyte/macrophages (Dusso et al., 1991a), the precursors of ostoclasts, suggests that VDR activation by endogenous calcitriol production could contribute to the association between vitamin D deficiency, osteoporosis and increased fracture risk in the normal population.

A recent relevant finding of calcitriol actions in skeletal homeostasis has been calcitriol induction of osteoblast FGF23 gene expression, an important counterbalance to vitamin D that could prevent calcitriol induction of hyperphosphatemia and ectopic calcification. FGF23 acts on the kidney to reduce phosphate reabsorption, via the sodium-dependent phosphate transporter type 2a, and to repress renal 1α-hydroxylase expression, thereby operating as a reciprocal hormone to calcitriol in phosphate metabolism and completing an integrated loop in calcium and phosphate homeostasis. Thus, as summarized in Figure 7.2, the calcitriol/VDR complex operates as a master integrator of calcium and phosphate homeostasis through its cross-regulation of PTH and FGF23, the genesis of osteoblasts and osteoclasts, as well as calcium and phosphate acquisition through its regulation of additional anabolic and catabolic genes to achieve proper mineral and skeletal metabolism.

In kidney disease, the development of secondary hyperparathyroidism has been considered the major determinant of the disorders in skeletal remodeling known as renal osteodystrophy, characterized by skeletal frailty, increased risk of fracture, osteopenia, osteoporosis and cardiovascular calcification. The emerging critical PTH-independent role of calcitriol in bone anabolism (Panda et al., 2004), together with the high impact of reduced bone formation on vascular calcification at low PTH levels (London et al., 2004; Xue et al., 2006), suggest that the survival advantage conferred by active vitamin D therapy, at doses unable to control secondary hyperparathyroidism (Teng et al., 2005), could result in part from calcitriol induction of bone formation. The high FGF23 levels associated with kidney disease could further aggravate secondary hyperpathyroidism and calcification through direct inhibition of parathyroid 1α-hydroxylase, since parathyroid cells express FGF receptor and klotho, both of which are required for FGF23 signaling. Reduced parathyroid calcitriol production will result in decreased autocrine VDR activation and PTH suppression by circulating 25-hydroxyvitamin D.

D. Kidney

The most important endocrine effect of $1,25(OH)_2D_3$ in the kidney is a tight control of its own homeostasis through simultaneous suppression of 1α-hydroxylase and stimulation of 24-hydroxylase and, very likely, through its ability to induce megalin expression in the proximal tubule (Liu et al., 1998). Since megalin also mediates protein reabsorption by renal proximal tubular cells (Gekle, 2005) and renal megalin levels decrease rapidly after the onset of kidney disease in rats (Takemoto et al., 2003), calcitriol/analog induction of renal megalin could partially account for the efficacy of vitamin D analogs in correcting proteinuria at early stages of kidney disease (Agarwal et al., 2005). Since proteinuria is a recognized marker of renal disease progression and cardiovascular complications (Jerums et al., 1997), calcitriol induction of megalin expression could also contribute to the survival advantage of active vitamin D therapy.

$1,25(OH)_2D_3$ involvement in the renal handling of calcium and phosphate continues to be controversial due to the simultaneous effects suppressing serum PTH, inducing bone synthesis of the phosphatonin FGF23, and on intestinal calcium and phosphate absorption, which affects the filter load of both ions. $1,25(OH)_2D_3$ enhances renal calcium reabsorption and calbindin expression and accelerates PTH-dependent calcium transport in the distal tubule (Friedman and Gesek, 1993), the main determinant of the final excretion of calcium into the urine and the site with the highest VDR content. As in the intestine, the epithelial calcium channel TRPV5 is an important target in $1,25(OH)_2D_3$ mediated calcium reabsorption. Several putative VDR binding sites have been located in the human promoter of the renal epithelial calcium channel and decreases in circulating levels of $1,25(OH)_2D_3$ concentrations resulted in a marked decline in the expression of the channel at the protein and mRNA levels (Hoenderop et al., 2001). Calcitriol also inhibits the expression of SBPRY (B-box and SPRY domain-containing protein), a TRPV5 and TRPV6 interacting protein, and an inhibitor of TRVP5 calcium channel activity at the cell surface (van de Graaf et al., 2006).

$1,25(OH)_2D_3$ also induces Npt2c, a renal sodium–phosphate co-transporter that facilitates renal absorption of phosphate. Although this gene plays a minor role in renal phosphate reabsorption in rodents, it underlies the human phosphate wasting disorder of hereditary hyperphosphatemic rickets with hypercalciuria (Bergwitz et al., 2006; Lorenz-Depiereux et al., 2006).

Two critical non-classical renal actions of calcitriol in the kidney could mediate the reported beneficial effects of vitamin D therapy in the cardiovascular system: suppression of renin and induction of the type A natriuretic peptide receptor (NPR-A) in cells of the inner medullary collecting duct (Chen et al., 2007), key in determining urinary sodium concentration (Zeidel, 1993). Calcitriol induces NPR-A expression through a transcriptional mechanism that involves a VDRE upstream from the gene start site.

Calcitriol inhibition of signaling from activated EGFR could be critical in preventing the progressive deterioration of the renal parenchyma in kidney disease. Renal EGFR activation occurs upon nephron reduction, prolonged renal ischemia (Terzi et al., 2000) and prolonged exposure to angiotensin II (Lautrette et al., 2005) and is the cause of accelerated progression of renal lesions, including not only tubular hyperplasia, but also mononuclear cell infiltration, tubulointerstitial inflammation and interstitial fibrosis. These effects of angiotensin II are independent of hypertension (Lautrette et al., 2005) and result from EGFR transactivation from G protein-coupled receptor upon angiotensin binding (Figure 7.7). The findings and the antagonistic cross-talk delineated in hyperplastic parathyroid glands between the degree of EGFR activation and the reduction of VDR levels causing vitamin D resistance underscore the importance of implementing therapeutic maneuvers that effectively prevent the onset of resistance to calcitriol antiproliferative, antifibrotic and anti-inflammatory actions. Indeed, combination therapy with angiotensin-converting enzyme inhibitors and calcitriol analog was proven more effective in reducing macrophage infiltration and chemokine production compared to monotherapy in rat kidney disease (Mizobuchi et al., 2007). It is unclear at present whether calcitriol anti-EGFR signaling could mediate the renoprotective effect for $1,25(OH)_2D_3$ therapy in the development of glomerulosclerosis and the progression of albuminuria that involve PTH-independent antiproliferative actions

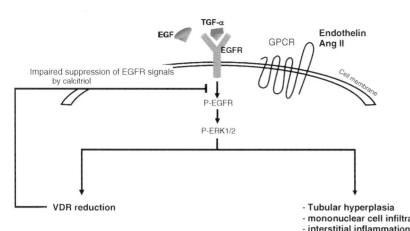

FIGURE 7.7 EGFR transactivation, calcitriol resistance and the progression of renal and cardiovascular lesions in kidney disease.

(Schwarz et al., 1998) or 1,25(OH)$_2$D$_3$-induced decreases in podocyte loss and podocyte hypertrophy (Kuhlmann et al., 2004) that also contribute to less pronounced albuminuria and glomerulosclerosis.

E. Cardiovascular System

Although hypervitaminosis D results in vascular damage due to calcinosis, several clinical and animal studies demonstrate a role of vitamin D in cardiovascular health. Vitamin D deficiency in rats associates with hypertension and cardiac hypertrophy (Weishaar and Simpson, 1987). *In vitro*, calcitriol reverses endothelin I-induced hypertrophy of neonatal cardiac myocytes (Wu et al., 1996). More recently, vitamin D deficiency was also associated with arterial stiffening (London et al., 2007). Arterial compliance correlates with healthy endothelial and arterial mechanical physiology and inversely with calcification. Furthermore, calcitriol/analog therapy reduces the mortality rates in advanced kidney disease attributable to extremely high incidence of cardiovascular disease. These findings have prompted the characterization of the mechanisms underlying vitamin D actions. Multiple actions of the calcitriol/VDR complex appear to contribute to cardiovascular health (see reviews by Andress, 2006; Towler, 2007; Wu-Wong, 2007). As indicated earlier, one mechanism could be the maintenance of normal bone formation and skeletal mineralization since there is an inverse relationship between osteoblast dependent bone formation and arterial calcification (Weber et al., 2004; Coen et al., 2005). Other mechanisms also described, that are independent of calcitriol control of vascular calcium load, and related to:

1. calcitriol suppression of the renin-angiotensin II-aldosterone system (Li et al., 2002)
2. the regulation of atrial natriuretic peptide and its receptor (Chen et al., 2007). Activation of this system mediates vasodilation, increased urinary excretion of sodium and water and suppression of myocardial hypertrophy and fibrosis (Knowles et al., 2001)
3. the control of inflammation that restores endothelial cell dysfunction (Wu-Wong et al., 2006b)
4. inhibition of smooth muscle cell proliferation (Wu-Wong et al., 2006a).

Although hypothetical, calcitriol/VDR inhibition of signals downstream from EGFR transactivation (see Figure 7.7) could counteract the well-known deleterious effects of angiotensin II and endothelin I in left ventricular hypertrophy and in vascular physiology. Similar to parathyroid cells and osteoblasts, aortic endothelial cells and vascular smooth muscle cells express 1α-hydroxylase activity, suggesting that defective autocrine/paracrine calcitriol production could mediate the association between vitamin D deficiency and increased risk of cardiovascular disease in the normal population and underscoring the importance of vitamin D supplementation both in healthy individuals and kidney disease patients.

IV. CONCLUDING REMARKS

The essential role of vitamin D in calcium and phosphate homeostasis often overshadows its more subtle, but critical, functions in other physiologic systems. Many of these non-classical actions are disrupted in chronic kidney disease and contribute to morbidity and mortality, as evidenced by the increased survival in patients receiving active vitamin D therapy. Calcitriol and its analogs can activate and repress genes that control progression of renal disease, vascular calcification and secondary hyperparathyroidism, with overall beneficial effects on outcomes. It is also important to consider that many of the target cells for vitamin D are capable of producing their own calcitriol and under controlled conditions in response to natural cues. Thus, the often poor vitamin D status of chronic kidney disease patients should be corrected to allow these systems to function properly. Continued basic and clinical research on the integrated network of vitamin D actions should allow the development of improved strategies to enhance the survival of chronic kidney disease patients.

Acknowledgments

This work was supported in part by grants DK62713 from the National Institutes of Health (to ASD) and CeDAR (Center for vitamin D receptor Activator Research) grant (to ASD) from Massachusetts General Hospital.

References

Adams, J. S., Chen, H., Chun, R. et al. (2004). Response element binding proteins and intracellular vitamin D binding proteins: novel regulators of vitamin D trafficking, action and metabolism. *J Steroid Biochem Mol Biol* **89–90**: 461–465.

Adorini, L., Penna, G., Giarratana, N. et al. (2004). Dendritic cells as key targets for immunomodulation by Vitamin D receptor ligands. *J Steroid Biochem Mol Biol* **89–90**: 437–441.

Agarwal, R., Acharya, M., Tian, J. et al. (2005). Antiproteinuric effect of oral paricalcitol in chronic kidney disease. *Kidney Int* **68**: 2823–2828.

Akiba, T., Ando, R., Kurihara, S., Heishi, M., Tazawa, H., and Marumo, F. (1997). Is the bone mass of hemodialysis patients genetically determined? *Kidney Int Suppl* **62**: S69–S71.

Andress, D. L. (2006). Vitamin D in chronic kidney disease: a systemic role for selective vitamin D receptor activation. *Kidney Int* **69**: 33–43.

Arai, H., Miyamoto, K. I., Yoshida, M. et al. (2001). The polymorphism in the caudal-related homeodomain protein Cdx-2 binding element in the human vitamin D receptor gene. *J Bone Miner Res* **16**: 1256–1264.

Araya, Z., and Wikvall, K. (1999). 6alpha-hydroxylation of taurochenodeoxycholic acid and lithocholic acid by CYP3A4 in human liver microsomes. *Biochim Biophys Acta* **1438**: 47–54.

Baldwin, B. R., Timchenko, N. A., and Zahnow, C. A. (2004). Epidermal growth factor receptor stimulation activates the RNA binding protein CUG-BP1 and increases expression of C/EBPbeta-LIP in mammary epithelial cells. *Mol Cell Biol* **24**: 3682–3691.

Baran, D. T. (1994). Nongenomic actions of the steroid hormone 1 alpha,25-dihydroxyvitamin D3. *J Cell Biochem* **56**: 303–306.

Baran, D. T., Sorensen, A. M., Shalhoub, V., Owen, T., Stein, G., and Lian, J. (1992). The rapid nongenomic actions of 1 alpha,25-dihydroxyvitamin D3 modulate the hormone-induced increments in osteocalcin gene transcription in osteoblast-like cells. *J Cell Biochem* **50**: 124–129.

Barsony, J., and McKoy, W. (1992). Molybdate increases intracellular 3′,5′-guanosine cyclic monophosphate and stabilizes vitamin D receptor association with tubulin-containing filaments. *J Biol Chem* **267**: 24457–24465.

Barsony, J., Renyi, I., and McKoy, W. (1997). Subcellular distribution of normal and mutant vitamin D receptors in living cells. Studies with a novel fluorescent ligand. *J Biol Chem* **272**: 5774–5782.

Barthel, T. K., Mathern, D. R., Whitfield, G. K. et al. (2007). 1,25-Dihydroxyvitamin D3/VDR-mediated induction of FGF23 as well as transcriptional control of other bone anabolic and catabolic genes that orchestrate the regulation of phosphate and calcium mineral metabolism. *J Steroid Biochem Mol Biol* **103**: 381–388.

Becker, P. B., and Horz, W. (2002). ATP-dependent nucleosome remodeling. *Annu Rev Biochem* **71**: 247–273.

Beno, D. W. A., Brady, L. M., Bissonnette, M., and Davis, B. H. (1995). Protein kinase C and mitogen-activated protein kinase are required for 1,25-dihydroxyvitamin D_3-stimulated Egr induction. *J Biol Chem* **270**: 3642–3647.

Bergwitz, C., Roslin, N. M., Tieder, M. et al. (2006). SLC34A3 mutations in patients with hereditary hypophosphatemic rickets with hypercalciuria predict a key role for the sodium-phosphate cotransporter napi-iic in maintaining phosphate homeostasis. *Am J Hum Genet* **78**: 179–192.

Bettoun, D. J., Burris, T. P., Houck, K. A. et al. (2003). Retinoid X receptor is a nonsilent major contributor to vitamin D receptor-mediated transcriptional activation. *Mol Endocrinol* **17**: 2320–2328.

Borras, M., Torregrossa, V., Oliveras, A. et al. (2003). BB genotype of the vitamin D receptor gene polymorphism postpones parathyroidectomy in hemodialysis patients. *J Nephrol* **16**: 116–120.

Bouillon, R., Okamura, W. H., and Norman, A. W. (1995). Structure-function relationships in the vitamin D endocrine system. *Endocrine Rev* **16**: 200–257.

Bouillon, R., Van Cromphaut, S., and Carmeliet, G. (2003). Intestinal calcium absorption: molecular vitamin D mediated mechanisms. *J Cell Biochem* **88**: 332–339.

Bourdeau, A., Atmani, F., Grosse, B., and Lieberherr, M. (1990). Rapid effects of 1,25-dihydroxyvitamin D3 and extracellular Ca^{2+} on phospholipid metabolism in dispersed porcine parathyroid cells. *Endocrinology* **127**: 2738–2743.

Boyle, W. J., Simonet, W. S., and Lacey, D. L. (2003). Osteoclast differentiation and activation. *Nature* **423**: 337–342.

Brickman, A. S., Coburn, J. W., and Norman, A. W. (1972). Action of 1,25-dihydroxycholecalciferol, a potent, kidney-produced metabolite of vitamin D, in uremic man. *N Engl J Med* **287**: 891–895.

Brown, A. J., Dusso, A., and Slatopolsky, E. (1999). Vitamin D. *Am J Physiol* **277**: F157–F175.

Brown, A. J., Finch, J., and Slatopolsky, E. (2002). Differential effects of 19-nor-1,25-dihydroxyvitamin D(2) and 1,25-dihydroxyvitamin D(3) on intestinal calcium and phosphate transport. *J Lab Clin Med* **139**: 279–284.

Brown, A. J., Zhong, M., Finch, J. et al. (1996). Rat calcium-sensing receptor is regulated by vitamin D but not by calcium. *Am J Physiol* **270**: F454–F460.

Brown, A. J., Zhong, M., Finch, J., Ritter, C., and Slatopolsky, E. (1995). The roles of calcium and 1,25-dihydroxyvitamin D3 in the regulation of vitamin D receptor expression by rat parathyroid glands. *Endocrinology* **136**: 1419–1425.

Brumbaugh, P. F., and Haussler, M. R. (1974). 1 Alpha,25-dihydroxycholecalciferol receptors in intestine. I. Association of 1 alpha,25-dihydroxycholecalciferol with intestinal mucosa chromatin. *J Biol Chem* **249**: 1251–1257.

Calkhoven, C. F., Muller, C., and Leutz, A. (2002). Translational control of gene expression and disease. *Trends Mol Med* **8**: 577–583.

Canaff, L., and Hendy, G. N. (2002). Human calcium-sensing receptor gene. Vitamin D response elements in promoters P1 and P2 confer transcriptional responsiveness to 1,25-dihydroxyvitamin D. *J Biol Chem* **277**: 30337–30350.

Carlberg, C. (2004). Ligand-mediated conformational changes of the VDR are required for gene transactivation. *J Steroid Biochem Mol Biol* **89–90**: 227–232.

Carling, T., Kindmark, A., Hellman, P. et al. (1995). Vitamin D receptor genotypes in primary hyperparathyroidism. *Nat Med* **1:** 1309–1311.

Chanard, J. M., Drueke, T., Zingraff, J., Man, N. K., Russo-Marie, F., and Funck-Brentano, J. L. (1976). Effects of haemodialysis on fractional intestinal absorption of calcium in uraemia. *Eur J Clin Invest* **6:** 261–264.

Chen, S., Olsen, K., Grigsby, C., and Gardner, D. G. (2007). Vitamin D activates type A natriuretic peptide receptor gene transcription in inner medullary collecting duct cells. *Kidney Int* **72:** 300–306.

Cippitelli, M., Fionda, C., Di Bona, D. et al. (2002). Negative regulation of CD95 ligand gene expression by vitamin D3 in T lymphocytes. *J Immunol* **168:** 1154–1166.

Cippitelli, M., and Santoni, A. (1998). Vitamin D3: a transcriptional modulator of the interferon-gamma gene. *Eur J Immunol* **28:** 3017–3030.

Coen, G., Mantella, D., Manni, M. et al. (2005). 25-hydroxyvitamin D levels and bone histomorphometry in hemodialysis renal osteodystrophy. *Kidney Int* **68:** 1840–1848.

Colin, E. M., Weel, A. E., Uitterlinden, A. G. et al. (2000). Consequences of vitamin D receptor gene polymorphisms for growth inhibition of cultured human peripheral blood mononuclear cells by 1,25-dihydroxyvitamin D3. *Clin Endocrinol (Oxf)* **52:** 211–216.

Cordero, J. B., Cozzolino, M., Lu, Y. et al. (2002). 1,25-Dihydroxyvitamin D down-regulates cell membrane growth- and nuclear growth-promoting signals by the epidermal growth factor receptor. *J Biol Chem* **277:** 38965–38971.

Cozzolino, M., Lu, Y., Finch, J., Slatopolsky, E., and Dusso, A. S. (2001). p21WAF1 and TGF-alpha mediate parathyroid growth arrest by vitamin D and high calcium. *Kidney Int* **60:** 2109–2117.

D'Ambrosio, D., Cippitelli, M., Cocciolo, M. G. et al. (1998). Inhibition of IL-12 production by 1,25-dihydroxyvitamin D3. Involvement of NF-kappaB downregulation in transcriptional repression of the p40 gene. *J Clin Invest* **101:** 252–262.

Darwish, H. M., and DeLuca, H. F. (1996). Analysis of binding of the 1,25-dihydroxyvitamin D3 receptor to positive and negative vitamin D response elements. *Arch Biochem Biophys* **334:** 223–234.

Delmez, J. A., Tindira, C., Grooms, P., Dusso, A., Windus, D. W., and Slatopolsky, E. (1989). Parathyroid hormone suppression by intravenous 1,25-dihydroxyvitamin D. A role for increased sensitivity to calcium. *J Clin Invest* **83:** 1349–1355.

Demasters, G., Di, X., Newsham, I., Shiu, R., and Gewirtz, D. A. (2006). Potentiation of radiation sensitivity in breast tumor cells by the vitamin D3 analogue, EB 1089, through promotion of autophagy and interference with proliferative recovery. *Mol Cancer Ther* **5:** 2786–2797.

Demay, M. B., Kiernan, M. S., DeLuca, H. F., and Kronenberg, H. M. (1992). Sequences in the human parathyroid hormone gene that bind the 1,25-dihydroxyvitamin D3 receptor and mediate transcriptional repression in response to 1,25-dihydroxyvitamin D3. *Proc Natl Acad Sci USA* **89:** 8097–8101.

Denda, M., Finch, J., Brown, A. J., Nishii, Y., Kubodera, N., and Slatopolsky, E. (1996). 1,25-dihydroxyvitamin D3 and 22-oxacalcitriol prevent the decrease in vitamin D receptor content in the parathyroid glands of uremic rats. *Kidney Int* **50:** 34–39.

Dhawan, P., Peng, X., Sutton, A. L. et al. (2005). Functional cooperation between CCAAT/enhancer-binding proteins and the vitamin D receptor in regulation of 25-hydroxyvitamin D3 24-hydroxylase. *Mol Cell Biol* **25:** 472–487.

Dong, X., Craig, T., Xing, N. et al. (2003). Direct transcriptional regulation of RelB by 1alpha,25-dihydroxyvitamin D3 and its analogs: physiologic and therapeutic implications for dendritic cell function. *J Biol Chem* **278:** 49378–49385.

Drissi, H., Pouliot, A., Koolloos, C. et al. (2002). 1,25-(OH)2-vitamin D3 suppresses the bone-related Runx2/Cbfa1 gene promoter. *Exp Cell Res* **274:** 323–333.

Dunlop, T. W., Vaisanen, S., Frank, C., and Carlberg, C. (2004). The genes of the coactivator TIF2 and the corepressor SMRT are primary 1alpha, 25(OH)2D3 targets. *J Steroid Biochem Mol Biol* **89–90:** 257–260.

Dusso, A. S., Brown, A. J., and Slatopolsky, E. (2005). Vitamin D. *Am J Physiol Renal Physiol* **289:** F8–F28.

Dusso, A. S., Finch, J., Brown, A. et al. (1991a). Extrarenal production of calcitriol in normal and uremic humans. *J Clin Endocrinol Metab* **72:** 157–164.

Dusso, A. S., Negrea, L., Gunawardhana, S. et al. (1991b). On the mechanisms for the selective action of vitamin D analogs. *Endocrinology* **128:** 1687–1692.

Dusso, A. S., Pavlopoulos, T., Naumovich, L. et al. (2001). p21WAF1 and transforming growth factor-alpha mediate dietary phosphate regulation of parathyroid cell growth. *Kidney Int* **59:** 855–865.

Dusso, A. S., Thadhani, R., and Slatopolsky, E. (2004). Vitamin D receptor and analogs. *Semin Nephrol* **24:** 10–16.

Eelen, G., Gysemans, C., Verlinden, L. et al. (2007). Mechanism and potential of the growth-inhibitory actions of vitamin D and analogs. *Curr Med Chem* **14:** 1893–1910.

Eisman, J. A. (1999). Genetics of osteoporosis. *Endocr Rev* **20:** 788–804.

Ellgaard, L., and Frickel, E. M. (2003). Calnexin, calreticulin, and ERp57: teammates in glycoprotein folding. *Cell Biochem Biophys* **39:** 223–247.

Endo, I., Inoue, D., Mitsui, T. et al. (2003). Deletion of vitamin D receptor gene in mice results in abnormal skeletal muscle development with deregulated expression of myoregulatory transcription factors. *Endocrinology* **144:** 5138–5144.

Erben, R. G., Soegiarto, D. W., Weber, K. et al. (2002). Deletion of deoxyribonucleic acid binding domain of the vitamin D receptor abrogates genomic and nongenomic functions of vitamin D. *Mol Endocrinol* **16:** 1524–1537.

Fang, Y., van Meurs, J. B., Bergink, A. P. et al. (2003). Cdx-2 polymorphism in the promoter region of the human vitamin D receptor gene determines susceptibility to fracture in the elderly. *J Bone Miner Res* **18:** 1632–1641.

Farach-Carson, M. C., Abe, J., Nishii, Y., Khoury, R., Wright, G. C., and Norman, A. W. (1993). 22-Oxacalcitriol: dissection of 1,25 (OH)2D3 receptor-mediated and Ca2+ entry-stimulating pathways. *Am J Physiol* **265:** F705–F711.

Farach-Carson, M. C., and Davis, P. J. (2003). Steroid hormone interactions with target cells: cross talk between membrane and nuclear pathways. *J Pharmacol Exp Ther* **307:** 839–845.

Fernandez, E., Fibla, J., Betriu, A., Piulats, J. M., Almirall, J., and Montoliu, J. (1997). Association between vitamin D receptor gene polymorphism and relative hypoparathyroidism in patients with chronic renal failure. *J Am Soc Nephrol* **8:** 1546–1552.

Friedman, P. A., and Gesek, F. A. (1993). Vitamin D3 accelerates PTH-dependent calcium transport in distal convoluted tubule cells. *Am J Physiol* **265**: F300–F308.

Gallieni, M., Kamimura, S., Ahmed, A. et al. (1995). Kinetics of monocyte 1 alpha-hydroxylase in renal failure. *Am J Physiol* **268**: F746–F753.

Gardiner, E. M., Esteban, L. M., Fong, C. et al. (2004). Vitamin D receptor B1 and exon 1d: functional and evolutionary analysis. *J Steroid Biochem Mol Biol* **89–90**: 233–238.

Gehin, M., Mark, M., Dennefeld, C., Dierich, A., Gronemeyer, H., and Chambon, P. (2002). The function of TIF2/GRIP1 in mouse reproduction is distinct from those of SRC-1 and p/CIP. *Mol Cell Biol* **22**: 5923–5937.

Gekle, M. (2005). Renal tubule albumin transport. *Annu Rev Physiol* **67**: 573–594.

Giannini, S., D'Angelo, A., Nobile, M. et al. (2002). The effects of vitamin D receptor polymorphism on secondary hyperparathyroidism and bone density after renal transplantation. *J Bone Miner Res* **17**: 1768–1773.

Gogusev, J., Duchambon, P., Stoermann-Chopard, C., Giovannini, M., Sarfati, E., and Drueke, T. B. (1996). De novo expression of transforming growth factor-alpha in parathyroid gland tissue of patients with primary or secondary uraemic hyperparathyroidism. *Nephrol Dial Transplant* **11**: 2155–2162.

Gomez Alonso, C., Naves Diaz, M. L., Diaz-Corte, C., Fernandez Martin, J. L., and Cannata Andia, J. B. (1998). Vitamin D receptor gene (VDR) polymorphisms: effect on bone mass, bone loss and parathyroid hormone regulation. *Nephrol Dial Transplant* **13**: 73–77.

Gonzalez, E. A., Disthabanchong, S., Kowalewski, R., and Martin, K. J. (2002). Mechanisms of the regulation of EGF receptor gene expression by calcitriol and parathyroid hormone in UMR 106-01 cells. *Kidney Int* **61**: 1627–1634.

Guillemant, J., and Guillemant, S. (1980). Early rise in cyclic GMP after 1,25-dihydroxycholecalciferol administration in the chick intestinal mucosa. *Biochem Biophys Res Commun* **93**: 906–911.

Hammes, A., Andreassen, T. K., Spoelgen, R. et al. (2005). Role of endocytosis in cellular uptake of sex steroids. *Cell* **122**: 751–762.

Hartenbower, D. L., Coburn, J. W., Reddy, C. R., and Norman, A. W. (1974). Calciferol metabolism and intestinal calcium transport in the chick with reduced renal function. *J Lab Clin Med* **83**: 38–45.

Haussler, M. R., Haussler, C. A., Jurutka, P. W. et al. (1997). The vitamin D hormone and its nuclear receptor: molecular actions and disease states. *J Endocrinol* **154**(Suppl): S57–S73.

Haussler, M. R., Whitfield, G. K., Haussler, C. A. et al. (1998). The nuclear vitamin D receptor: biological and molecular regulatory properties revealed. *J Bone Miner Res* **13**: 325–349.

Hedlund, T. E., Moffatt, K. A., and Miller, G. J. (1996). Stable expression of the nuclear vitamin D receptor in the human prostatic carcinoma cell line JCA-1: evidence that the antiproliferative effects of 1 alpha, 25-dihydroxyvitamin D3 are mediated exclusively through the genomic signaling pathway. *Endocrinology* **137**: 1554–1561.

Hewison, M., Rut, A. R., Kristjansson, K. et al. (1993). Tissue resistance to 1,25-dihydroxyvitamin D without a mutation of the vitamin D receptor gene. *Clin Endocrinol (Oxf)* **39**: 663–670.

Hines, E. R., Kolek, O. I., Jones, M. D. et al. (2004). 1,25-dihydroxyvitamin D3 down-regulation of PHEX gene expression is mediated by apparent repression of a 110 kDa transfactor that binds to a polyadenine element in the promoter. *J Biol Chem* **279**: 46406–46414.

Hoenderop, J. G., Chon, H., Gkika, D. et al. (2004). Regulation of gene expression by dietary Ca2+ in kidneys of 25-hydroxyvitamin D3-1 alpha-hydroxylase knockout mice. *Kidney Int* **65**: 531–539.

Hoenderop, J. G., Muller, D., Van Der Kemp, A. W. et al. (2001). Calcitriol controls the epithelial calcium channel in kidney. *J Am Soc Nephrol* **12**: 1342–1349.

Holick, M. F. (2007). Vitamin D deficiency. *N Engl J Med* **357**: 266–281.

Hoyer-Hansen, M., Bastholm, L., Mathiasen, I. S., Elling, F., and Jaattela, M. (2005). Vitamin D analog EB1089 triggers dramatic lysosomal changes and Beclin 1-mediated autophagic cell death. *Cell Death Differ* **12**: 1297–1309.

Hoyer-Hansen, M., Bastholm, L., Szyniarowski, P. et al. (2007). Control of macroautophagy by calcium, calmodulin-dependent kinase kinase-beta, and Bcl-2. *Mol Cell* **25**: 193–205.

Hruska, K. A., Bar-Shavit, Z., Malone, J. D., and Teitelbaum, S. (1988). Ca2+ priming during vitamin D-induced monocytic differentiation of a human leukemia cell line. *J Biol Chem* **263**: 16039–16044.

Hsieh, J. C., Jurutka, P. W., Nakajima, S. et al. (1993). Phosphorylation of the human vitamin D receptor by protein kinase C. Biochemical and functional evaluation of the serine 51 recognition site. *J Biol Chem* **268**: 15118–15126.

Hsu, C. H., and Patel, S. R. (1995). Altered vitamin D metabolism and receptor interaction with the target genes in renal failure: calcitriol receptor interaction with its target gene in renal failure. *Curr Opin Nephrol Hypertens* **4**: 302–306.

Huang, Y. C., and Hung, W. C. (2006). 1,25-dihydroxyvitamin D3 transcriptionally represses p45Skp2 expression via the Sp1 sites in human prostate cancer cells. *J Cell Physiol* **209**: 363–369.

Huhtakangas, J. A., Olivera, C. J., Bishop, J. E., Zanello, L. P., and Norman, A. W. (2004). The vitamin D receptor is present in caveolae-enriched plasma membranes and binds $1\alpha,25(OH)_2$-vitamin D_3 *in vivo* and *in vitro*. *Mol Endocrinol* **18**: 2660–2671.

Jaaskelainen, T., Ryhanen, S., Mahonen, A., DeLuca, H. F., and Maenpaa, P. H. (2000). Mechanism of action of superactive vitamin D analogs through regulated receptor degradation. *J Cell Biochem* **76**: 548–558.

Jerums, G., Panagiotopoulos, S., Tsalamandris, C., Allen, T. J., Gilbert, R. E., and Comper, W. D. (1997). Why is proteinuria such an important risk factor for progression in clinical trials? *Kidney Int Suppl* **63**: S87–S92.

Ji, Y., and Studzinski, G. P. (2004). Retinoblastoma protein and CCAAT/enhancer-binding protein beta are required for 1,25-dihydroxyvitamin D3-induced monocytic differentiation of HL60 cells. *Cancer Res* **64**: 370–377.

Jundt, F., Raetzel, N., Muller, C. et al. (2005). A rapamycin derivative (everolimus) controls proliferation through downregulation of truncated CCAAT enhancer binding protein {beta} and NF-{kappa}B activity in Hodgkin and anaplastic large cell lymphomas. *Blood* **106**: 1801–1807.

Jurutka, P. W., Hsieh, J. C., Nakajima, S., Haussler, C. A., Whitfield, G. K., and Haussler, M. R. (1996). Human vitamin D receptor phosphorylation by casein kinase II at Ser-208 potentiates transcriptional activation. *Proc Natl Acad Sci USA* **93**: 3519–3524.

Jurutka, P. W., MacDonald, P. N., Nakajima, S. et al. (2002). Isolation of baculovirus-expressed human vitamin D receptor: DNA responsive element interactions and phosphorylation of the purified receptor. *J Cell Biochem* **85:** 435–457.

Jurutka, P. W., Terpening, C. M., and Haussler, M. R. (1993). The 1,25-dihydroxy-vitamin D3 receptor is phosphorylated in response to 1,25-dihydroxy-vitamin D3 and 22-oxacalcitriol in rat osteoblasts, and by casein kinase II, in vitro. *Biochemistry* **32:** 8184–8192.

Jurutka, P. W., Whitfield, G. K., Hsieh, J. C., Thompson, P. D., Haussler, C. A., and Haussler, M. R. (2001). Molecular nature of the vitamin D receptor and its role in regulation of gene expression. *Rev Endocr Metab Disord* **2:** 203–216.

Juttmann, J. R., Hagenouw-Taal, J. C., Lameyer, L. D., Ruis, A. M., and Birkenhager, J. C. (1978). Intestinal calcium absorption, serum phosphate, and parathyroid hormone in patients with chronic renal failure and osteodystrophy before and during hemodialysis. *Calcif Tissue Res* **26:** 119–126.

Kamimura, S., Gallieni, M., Zhong, M., Beron, W., Slatopolsky, E., and Dusso, A. (1995). Microtubules mediate cellular 25-hydroxyvitamin D3 trafficking and the genomic response to 1,25-dihydroxyvitamin D3 in normal human monocytes. *J Biol Chem* **270:** 22160–22166.

Kasinath, B. S., Mariappan, M. M., Sataranatarajan, K., Lee, M. J., and Feliers, D. (2006). mRNA translation: unexplored territory in renal science. *J Am Soc Nephrol* **17:** 3281–3292.

Kato, S., Fujiki, R., and Kitagawa, H. (2004). Vitamin D receptor (VDR) promoter targeting through a novel chromatin remodeling complex. *J Steroid Biochem Mol Biol* **89–90:** 173–178.

Khosla, S. (2001). Minireview: the OPG/RANKL/RANK system. *Endocrinology* **142:** 5050–5055.

Khoury, R., Ridall, A. L., Norman, A. W., and Farach-Carson, M. C. (1994). Target gene activation by 1,25-dihydroxyvitamin D3 in osteosarcoma cells is independent of calcium influx. *Endocrinology* **135:** 2446–2453.

Khoury, R. S., Weber, J., and Farach-Carson, M. C. (1995). Vitamin D metabolites modulate osteoblast activity by Ca+2 influx-independent genomic and Ca+2 influx-dependent nongenomic pathways. *J Nutrit* **125:** 1699S–1703S.

Kifor, O., Moore, F. D., Wang, P. et al. (1996). Reduced immunostaining for the extracellular Ca2+-sensing receptor in primary and uremic secondary hyperparathyroidism. *J Clin Endocrinol Metab* **81:** 1598–1606.

Kim, M. S., Fujiki, R., Murayama, A. et al. (2007). 1Alpha,25(OH)2D3-induced transrepression by vitamin D receptor through E-box-type elements in the human parathyroid hormone gene promoter. *Mol Endocrinol* **21:** 334–342.

Kim, R. H., Li, J. J., Ogata, Y., Yamauchi, M., Freedman, L. P., and Sodek, J. (1996). Identification of a vitamin D3-response element that overlaps a unique inverted TATA box in the rat bone sialoprotein gene. *Biochem J* **318:** 219–226.

Kitazawa, S., Kajimoto, K., Kondo, T., and Kitazawa, R. (2003). Vitamin D3 supports osteoclastogenesis via functional vitamin D response element of human RANKL gene promoter. *J Cell Biochem* **89:** 771–777.

Knowles, J. W., Esposito, G., Mao, L. et al. (2001). Pressure-independent enhancement of cardiac hypertrophy in natriuretic peptide receptor A-deficient mice. *J Clin Invest* **107:** 975–984.

Koike, N., Hayakawa, N., Kumaki, K., and Stumpf, W. E. (1998). In vivo dose-related receptor binding of the vitamin D analogue [3H]-1,25-dihydroxy-22-oxavitamin D3 (OCT) in rat parathyroid, kidney distal and proximal tubules, duodenum, and skin, studied by quantitative receptor autoradiography. *J Histochem Cytochem* **46:** 1351–1358.

Kollenkirchen, U., Fox, J., and Walters, M. R. (1991). Normocalcemia without hyperparathyroidism in vitamin D-deficient rats. *J Bone Miner Res* **6:** 273–278.

Kondo, T., Kitazawa, R., Maeda, S., and Kitazawa, S. (2004). 1alpha,25 dihydroxyvitamin d3 rapidly regulates the mouse osteoprotegerin gene through dual pathways. *J Bone Miner Res* **19:** 1411–1419.

Kontula, K., Valimaki, S., Kainulainen, K., Viitanen, A. M., and Keski-Oja, J. (1997). Vitamin D receptor polymorphism and treatment of psoriasis with calcipotriol. *Br J Dermatol* **136:** 977–978.

Koshiyama, H., Sone, T., and Nakao, K. (1995). Vitamin-D-receptor-gene polymorphism and bone loss. *Lancet* **345:** 990–991.

Koyama, H., Inaba, M., Nishizawa, Y. et al. (1994). Potentiated 1,25 (OH)2D3-induced 24-hydroxylase gene expression in uremic rat intestine. *Am J Physiol* **267:** F926–F930.

Kream, B. E., Harrison, J. R., Krebsbach, P. H. et al. (1995). Regulation of type I collagen gene expression in bone. *Connect Tissue Res* **31:** 261–264.

Kuhlmann, A., Haas, C. S., Gross, M. L. et al. (2004). 1,25-Dihydroxyvitamin D3 decreases podocyte loss and podocyte hypertrophy in the subtotally nephrectomized rat. *Am J Physiol Renal Physiol* **286:** F526–F533.

Kurnik, B. R., and Hruska, K. A. (1985). Mechanism of stimulation of renal phosphate transport by 1,25-dihydroxycholecalciferol. *Biochim Biophys Acta* **817:** 42–50.

Lamb, J., Ramaswamy, S., Ford, H. L. et al. (2003). A mechanism of cyclin D1 action encoded in the patterns of gene expression in human cancer. *Cell* **114:** 323–334.

Lautrette, A., Li, S., Alili, R. et al. (2005). Angiotensin II and EGF receptor cross-talk in chronic kidney diseases: a new therapeutic approach. *Nat Med* **11:** 867–874.

Leong, G. M., Subramaniam, N., Issa, L. L. et al. (2004). Ski-interacting protein, a bifunctional nuclear receptor coregulator that interacts with N-CoR/SMRT and p300. *Biochem Biophys Res Commun* **315:** 1070–1076.

Li, Y. C., Amling, M., Pirro, A. E. et al. (1998). Normalization of mineral ion homeostasis by dietary means prevents hyperparathyroidism, rickets, and osteomalacia, but not alopecia in vitamin D receptor-ablated mice. *Endocrinology* **139:** 4391–4396.

Li, Y. C., Kong, J., Wei, M., Chen, Z. F., Liu, S. Q., and Cao, L. P. (2002). 1,25-Dihydroxyvitamin D(3) is a negative endocrine regulator of the renin-angiotensin system. *J Clin Invest* **110:** 229–238.

Lieberherr, M. (1987). Effects of vitamin D3 metabolites on cytosolic free calcium in confluent mouse osteoblasts. *J Biol Chem* **262:** 13168–13173.

Lieberherr, M., Grosse, B., Duchambon, P., and Drueke, T. (1989). A functional cell surface type receptor is required for the early action of 1,25-dihydroxyvitamin D3 on the phosphoinositide metabolism in rat enterocytes. *J Biol Chem* **264:** 20403–20406.

Liu, W., Yu, W. R., Carling, T. et al. (1998). Regulation of gp330/megalin expression by vitamins A and D. *Eur J Clin Invest* **28:** 100–107.

London, G. M., Guerin, A. P., Verbeke, F. H. et al. (2007). Mineral metabolism and arterial functions in end-stage renal disease:

potential role of 25-hydroxyvitamin D deficiency. *J Am Soc Nephrol* **18**: 613–620.

London, G. M., Marty, C., Marchais, S. J., Guerin, A. P., Metivier, F., and de Vernejoul, M. C. (2004). Arterial calcifications and bone histomorphometry in end-stage renal disease. *J Am Soc Nephrol* **15**: 1943–1951.

Lorenz-Depiereux, B., Benet-Pages, A., Eckstein, G. et al. (2006). Hereditary hypophosphatemic rickets with hypercalciuria is caused by mutations in the sodium-phosphate cotransporter gene SLC34A3. *Am J Hum Genet* **78**: 193–201.

Lucas, P. A., Roullet, C., Duchambon, P., Lacour, B., and Drueke, T. (1989). Rapid stimulation of calcium uptake by isolated rat enterocytes by 1,25(OH)2D3. *Pflugers Arch Eur J Physiol* **413**: 407–413.

MacDonald, P. N., Ritter, C., Brown, A. J., and Slatopolsky, E. (1994). Retinoic acid suppresses parathyroid hormone (PTH) secretion and PreproPTH mRNA levels in bovine parathyroid cell culture. *J Clin Invest* **93**: 725–730.

Mackey, S. L., Heymont, J. L., Kronenberg, H. M., and Demay, M. B. (1996). Vitamin D receptor binding to the negative human parathyroid hormone vitamin D response element does not require the retinoid x receptor. *Mol Endocrinol* **10**: 298–305.

Masuyama, H., and MacDonald, P. N. (1998). Proteasome-mediated degradation of the vitamin D receptor (VDR) and a putative role for SUG1 interaction with the AF-2 domain of VDR. *J Cell Biochem* **71**: 429–440.

May, P., Bock, H. H., and Herz, J. (2003). Integration of endocytosis and signal transduction by lipoprotein receptors. *Sci STKE* **2003**: PE12.

McGaffin, K. R., and Chrysogelos, S. A. (2005). Identification and characterization of a response element in the EGFR promoter that mediates transcriptional repression by 1,25-dihydroxyvitamin D3 in breast cancer cells. *J Mol Endocrinol* **35**: 117–133.

Mellon, W. S., and DeLuca, H. F. (1979). An equilibrium and kinetic study of 1,25-dihydroxyvitamin D3 binding to chicken intestinal cytosol employing high specific activity 1,25-dehydroxy[3H-26,27] vitamin D3. *Arch Biochem Biophys* **197**: 90–95.

Mizobuchi, M., Morrissey, J., Finch, J. L. et al. (2007). Combination therapy with an angiotensin-converting enzyme inhibitor and a vitamin d analog suppresses the progression of renal insufficiency in uremic rats. *J Am Soc Nephrol* **18**: 1796–1806.

Morelli, S., de Boland, A. R., and Boland, R. L. (1993). Generation of inositol phosphates, diacylglycerol and calcium fluxes in myoblasts treated with 1,25-dihydroxyvitamin D3. *Biochem J* **289**: 675–679.

Morrison, N. A., Qi, J. C., Tokita, A. et al. (1994). Prediction of bone density from vitamin D receptor alleles. *Nature* **367**: 284–287.

Murayama, A., Kim, M. S., Yanagisawa, J., Takeyama, K., and Kato, S. (2004). Transrepression by a liganded nuclear receptor via a bHLH activator through co-regulator switching. *Embo J* **23**: 1598–1608.

Nagaba, Y., Heishi, M., Tazawa, H., Tsukamoto, Y., and Kobayashi, Y. (1998). Vitamin D receptor gene polymorphisms affect secondary hyperparathyroidism in hemodialyzed patients. *Am J Kidney Dis* **32**: 464–469.

Nemere, I., Dormanen, M. C., Hammond, M. W., Okamura, W. H., and Norman, A. W. (1994). Identification of a specific binding protein for 1 alpha,25-dihydroxyvitamin D3 in basal-lateral membranes of chick intestinal epithelium and relationship to transcaltachia. *J Biol Chem* **269**: 23750–23756.

Nemere, I., Farach-Carson, M. C., Rohe, B. et al. (2004). Ribozyme knockdown functionally links a 1,25(OH)2D3 membrane binding protein (1,25D3-MARRS) and phosphate uptake in intestinal cells. *Proc Natl Acad Sci USA* **101**: 7392–7397.

Nemere, I., Schwartz, Z., Pedrozo, H., Sylvia, V. L., Dean, D. D., and Boyan, B. D. (1998). Identification of a membrane receptor for 1,25-dihydroxyvitamin D_3 which mediates the rapid activation of protein kinase C. *J Bone Min Res* **13**: 1353–1359.

Nemere, I., Yoshimoto, Y., and Norman, A. W. (1984). Calcium transport in perfused duodena from normal chicks: enhancement within fourteen minutes of exposure to 1,25-dihydroxyvitamin D3. *Endocrinology* **115**: 1476–1483.

Norman, A. W., Bouillon, R., Farach-Carson, M. C. et al. (1993). Demonstration that 1 beta,25-dihydroxyvitamin D3 is an antagonist of the nongenomic but not genomic biological responses and biological profile of the three A-ring diastereomers of 1 alpha,25-dihydroxyvitamin D3. *J Biol Chem* **268**: 20022–20030.

Norman, A. W., Song, X., Zanello, L., Bula, C., and Okamura, W. H. (1999). Rapid and genomic biological responses are mediated by different shapes of the agonist steroid hormone, 1alpha,25 (OH)2vitamin D3. *Steroids* **64**: 120–128.

Nykjaer, A., Dragun, D., Walther, D. et al. (1999). An endocytic pathway essential for renal uptake and activation of the steroid 25-(OH) vitamin D3. *Cell* **96**: 507–515.

Ogg, C. S. (1968). The intestinal absorption of 47Ca by patients in chronic renal failure. *Clin Sci* **34**: 467–471.

Okazaki, T., Nishimori, S., Ogata, E., and Fujita, T. (2003). Vitamin D-dependent recruitment of DNA-PK to the chromatinized negative vitamin D response element in the PTHrP gene is required for gene repression by vitamin D. *Biochem Biophys Res Commun* **304**: 632–637.

O'Kelly, J., Hisatake, J., Hisatake, Y., Bishop, J., Norman, A., and Koeffler, H. P. (2002). Normal myelopoiesis but abnormal T lymphocyte responses in vitamin D receptor knockout mice. *J Clin Invest* **109**: 1091–1099.

O'Kelly, J., Uskokovic, M., Lemp, N., Vadgama, J., and Koeffler, H. P. (2006). Novel Gemini-vitamin D3 analog inhibits tumor cell growth and modulates the Akt/mTOR signaling pathway. *J Steroid Biochem Mol Biol* **100**: 107–116.

Panda, D. K., Miao, D., Bolivar, I. et al. (2004). Inactivation of the 25-hydroxyvitamin D 1alpha-hydroxylase and vitamin D receptor demonstrates independent and interdependent effects of calcium and vitamin D on skeletal and mineral homeostasis. *J Biol Chem* **279**: 16754–16766.

Patel, S. R., Ke, H. Q., Vanholder, R., Koenig, R. J., and Hsu, C. H. (1995). Inhibition of calcitriol receptor binding to vitamin D response elements by uremic toxins. *J Clin Invest* **96**: 50–59.

Peleg, S., Liu, Y. Y., Reddy, S., Horst, R. L., White, M. C., and Posner, G. H. (1996). A 20-epi side chain restores growth-regulatory and transcriptional activities of an A ring-modified hybrid analog of 1 alpha,25- dihydroxyvitamin D3 without increasing its affinity to the vitamin D receptor. *J Cell Biochem* **63**: 149–161.

Picard, F., Gehin, M., Annicotte, J. et al. (2002). SRC-1 and TIF2 control energy balance between white and brown adipose tissues. *Cell* **111**: 931–941.

Pike, J. W., Meyer, M. B., Watanuki, M. et al. (2007). Perspectives on mechanisms of gene regulation by 1,25-dihydroxyvitamin D3 and its receptor. *J Steroid Biochem Mol Biol* **103:** 389–395.

Prufer, K., Racz, A., Lin, G. C., and Barsony, J. (2000). Dimerization with retinoid X receptors promotes nuclear localization and subnuclear targeting of vitamin D receptors. *J Biol Chem* **275:** 41114–41123.

Rachez, C., and Freedman, L. P. (2000). Mechanisms of gene regulation by vitamin D(3) receptor: a network of coactivator interactions. *Gene* **246:** 9–21.

Racz, A., and Barsony, J. (1999). Hormone-dependent translocation of vitamin D receptors is linked to transactivation. *J Biol Chem* **274:** 19352–19360.

Ramji, D. P., and Foka, P. (2002). CCAAT/enhancer-binding proteins: structure, function and regulation. *Biochem J* **365:** 561–575.

Recker, R. R., and Saville, P. D. (1971). Calcium absorption in renal failure: its relationship to blood urea nitrogen, dietary calcium intake, time on dialysis, and other variables. *J Lab Clin Med* **78:** 380–388.

Ritter, C. S., Armbrecht, H. J., Slatopolsky, E., and Brown, A. J. (2006). 25-Hydroxyvitamin D(3) suppresses PTH synthesis and secretion by bovine parathyroid cells. *Kidney Int* **70:** 654–659.

Russell, J., Ashok, S., and Koszewski, N. J. (1999). Vitamin D receptor interactions with the rat parathyroid hormone gene: synergistic effects between two negative vitamin D response elements. *J Bone Miner Res* **14:** 1828–1837.

Sawaya, B. P., Koszewski, N. J., Qi, Q., Langub, M. C., Monier-Faugere, M. C., and Malluche, H. H. (1997). Secondary hyperparathyroidism and vitamin D receptor binding to vitamin D response elements in rats with incipient renal failure. *J Am Soc Nephrol* **8:** 271–278.

Schiavi, S. C., and Kumar, R. (2004). The phosphatonin pathway: new insights in phosphate homeostasis. *Kidney Int* **65:** 1–14.

Schrader, M., Kahlen, J. P., and Carlberg, C. (1997). Functional characterization of a novel type of 1 alpha,25- dihydroxyvitamin D3 response element identified in the mouse c-fos promoter. *Biochem Biophys Res Commun* **230:** 646–651.

Schwarz, U., Amann, K., Orth, S. R., Simonaviciene, A., Wessels, S., and Ritz, E. (1998). Effect of 1,25 (OH)2 vitamin D3 on glomerulosclerosis in subtotally nephrectomized rats. *Kidney Int* **53:** 1696–1705.

Segersten, U., Correa, P., Hewison, M. et al. (2002). 25-hydroxyvitamin D(3)-1alpha-hydroxylase expression in normal and pathological parathyroid glands. *J Clin Endocrinol Metab* **87:** 2967–2972.

Sela-Brown, A., Russell, J., Koszewski, N. J., Michalak, M., Naveh-Many, T., and Silver, J. (1998). Calreticulin inhibits vitamin D's action on the PTH gene in vitro and may prevent vitamin D's effect in vivo in hypocalcemic rats. *Mol Endocrinol* **12:** 1193–1200.

Shaffer, P. L., and Gewirth, D. T. (2002). Structural basis of VDR-DNA interactions on direct repeat response elements. *Embo J* **21:** 2242–2252.

Shaffer, P. L., McDonnell, D. P., and Gewirth, D. T. (2005). Characterization of transcriptional activation and DNA-binding functions in the hinge region of the vitamin D receptor. *Biochemistry* **44:** 2678–2685.

Shang, Y., Hu, X., DiRenzo, J., Lazar, M. A., and Brown, M. (2000). Cofactor dynamics and sufficiency in estrogen receptor-regulated transcription. *Cell* **103:** 843–852.

Shiizaki, K., Hatamura, I., Negi, S. et al. (2003). Percutaneous maxacalcitol injection therapy regresses hyperplasia of parathyroid and induces apoptosis in uremia. *Kidney Int* **64:** 992–1003.

Shiraishi, K., Tsuchida, M., Wada, T. et al. (2001). 22-Oxacalcitriol upregulates p21(WAF1/Cip1) in human parathyroid glands. A preliminary report. *Am J Nephrol* **21:** 507–511.

Song, X., Bishop, J. E., Okamura, W. H., and Norman, A. W. (1998). Stimulation of phosphorylation of mitogen-activated protein kinase by $1\alpha,25$-dihydroxyvitamin D_3 in promyelocytic NB4 leukemic cells: a structure-function study. *Endocrinology* **139:** 457–465.

Staal, A., van Wijnen, A. J., Birkenhager, J. C. et al. (1996). Distinct conformations of vitamin D receptor/retinoid X receptor-alpha heterodimers are specified by dinucleotide differences in the vitamin D- responsive elements of the osteocalcin and osteopontin genes. *Mol Endocrinol* **10:** 1444–1456.

St-Arnaud, R., Arabian, A., Travers, R. et al. (2000). Deficient mineralization of intramembranous bone in vitamin D-24- hydroxylase-ablated mice is due to elevated 1,25-dihydroxyvitamin D and not to the absence of 24,25-dihydroxyvitamin D. *Endocrinology* **141:** 2658–2666.

Sugimoto, T., Ritter, C., Ried, I., Morrissey, J., and Slatopolsky, E. (1988). Effect of 1,25-dihydroxyvitamin D3 on cytosolic calcium in dispersed parathyroid cells. *Kidney Int* **33:** 850–854.

Sylvia, V. L., Schwartz, Z., Ellis, E. B. et al. (1996). Nongenomic regulation of protein kinase C isoforms by the vitamin D metabolites 1 alpha,25-(OH)2D3 and 24R,25-(OH)2D3. *J Cell Physiol* **167:** 380–393.

Takemoto, F., Shinki, T., Yokoyama, K. et al. (2003). Gene expression of vitamin D hydroxylase and megalin in the remnant kidney of nephrectomized rats. *Kidney Int* **64:** 414–420.

Takeuchi, A., Reddy, G. S., Kobayashi, T., Okano, T., Park, J., and Sharma, S. (1998). Nuclear factor of activated T cells (NFAT) as a molecular target for 1alpha,25-dihydroxyvitamin D3-mediated effects. *J Immunol* **160:** 209–218.

Teng, M., Wolf, M., Lowrie, E., Ofsthun, N., Lazarus, J. M., and Thadhani, R. (2003). Survival of patients undergoing hemodialysis with paricalcitol or calcitriol therapy. *N Engl J Med* **349:** 446–456.

Teng, M., Wolf, M., Ofsthun, M. N. et al. (2005). Activated injectable vitamin d and hemodialysis survival: a historical cohort study. *J Am Soc Nephrol* **16:** 1115–1125.

Teramoto, T., Endo, K., Ikeda, K. et al. (1995). Binding of vitamin D to low-density-lipoprotein (LDL) and LDL receptor-mediated pathway into cells. *Biochem Biophys Res Commun* **215:** 199–204.

Terzi, F., Burtin, M., Hekmati, M. et al. (2000). Targeted expression of a dominant-negative EGF-R in the kidney reduces tubulointerstitial lesions after renal injury. *J Clin Invest* **106:** 225–234.

Thompson, P. D., Jurutka, P. W., Whitfield, G. K. et al. (2002). Liganded VDR induces CYP3A4 in small intestinal and colon cancer cells via DR3 and ER6 vitamin D responsive elements. *Biochem Biophys Res Commun* **299:** 730–738.

Thompson, P. D., Remus, L. S., Hsieh, J. C. et al. (2001). Distinct retinoid X receptor activation function-2 residues mediate transactivation in homodimeric and vitamin D receptor heterodimeric contexts. *J Mol Endocrinol* **27:** 211–227.

Tocchini-Valentini, G., Rochel, N., Wurtz, J. M., Mitschler, A., and Moras, D. (2001). Crystal structures of the vitamin D receptor complexed to superagonist 20-epi ligands. *Proc Natl Acad Sci USA* **98:** 5491–5496.

Tokumoto, M., Tsuruya, K., Fukuda, K., Kanai, H., Kuroki, S., and Hirakata, H. (2002). Reduced p21, p27 and vitamin D receptor in the nodular hyperplasia in patients with advanced secondary hyperparathyroidism. *Kidney Int* **62**: 1196–1207.

Towers, T. L., Staeva, T. P., and Freedman, L. P. (1999). A two-hit mechanism for vitamin D3-mediated transcriptional repression of the granulocyte-macrophage colony-stimulating factor gene: vitamin D receptor competes for DNA binding with NFAT1 and stabilizes c-Jun. *Mol Cell Biol* **19**: 4191–4199.

Towler, D. A. (2007). Calciotropic hormones and arterial physiology: 'D'-lightful insights. *J Am Soc Nephrol* **18**: 369–373.

Turunen, M. M., Dunlop, T. W., Carlberg, C., and Vaisanen, S. (2007). Selective use of multiple vitamin D response elements underlies the 1 alpha,25-dihydroxyvitamin D3-mediated negative regulation of the human CYP27B1 gene. *Nucleic Acids Res* **35**: 2734–2747.

Uitterlinden, A. G., Fang, Y., Van Meurs, J. B., Pols, H. A., and Van Leeuwen, J. P. (2004a). Genetics and biology of vitamin D receptor polymorphisms. *Gene* **338**: 143–156.

Uitterlinden, A. G., Fang, Y., van Meurs, J. B., van Leeuwen, H., and Pols, H. A. (2004b). Vitamin D receptor gene polymorphisms in relation to Vitamin D related disease states. *J Steroid Biochem Mol Biol* **89–90**: 187–193.

van Abel, M., Hoenderop, J. G., van der Kemp, A. W., van Leeuwen, J. P., and Bindels, R. J. (2003). Regulation of the epithelial Ca2+ channels in small intestine as studied by quantitative mRNA detection. *Am J Physiol Gastrointest Liver Physiol* **285**: G78–G85.

Van Cromphaut, S. J., Dewerchin, M., Hoenderop, J. G. et al. (2001). Duodenal calcium absorption in vitamin D receptor-knockout mice: functional and molecular aspects. *Proc Natl Acad Sci USA* **98**: 13324–13329.

van de Graaf, S. F., van der Kemp, A. W., van den Berg, D., van Oorschot, M., Hoenderop, J. G., and Bindels, R. J. (2006). Identification of BSPRY as a novel auxiliary protein inhibiting TRPV5 activity. *J Am Soc Nephrol* **17**: 26–30.

van Driel, M., Koedam, M., Buurman, C. J. et al. (2006). Evidence for auto/paracrine actions of vitamin D in bone: 1alpha-hydroxylase expression and activity in human bone cells. *Faseb J* **20**: 2417–2419.

Vanhooke, J. L., Benning, M. M., Bauer, C. B., Pike, J. W., and DeLuca, H. F. (2004). Molecular structure of the rat vitamin D receptor ligand binding domain complexed with 2-carbon-substituted vitamin D3 hormone analogues and a LXXLL-containing coactivator peptide. *Biochemistry* **43**: 4101–4110.

Verroust, P. J., and Christensen, E. I. (2002). Megalin and cubilin–the story of two multipurpose receptors unfolds. *Nephrol Dial Transplant* **17**: 1867–1871.

Vesely, D. L., and Juan, D. (1984). Cation-dependent vitamin D activation of human renal cortical guanylate cyclase. *Am J Physiol* **246**: E115–E120.

Vidal, M., Ramana, C. V., and Dusso, A. S. (2002). Stat1-vitamin D receptor interactions antagonize 1,25-dihydroxyvitamin D transcriptional activity and enhance stat1-mediated transcription. *Mol Cell Biol* **22**: 2777–2787.

Walling, M. W., Kimberg, D. V., Wasserman, R. H., and Feinberg, R. R. (1976). Duodenal active transport of calcium and phosphate in vitamin D-deficient rats: effects of nephrectomy, Cestrum diurnum, and 1alpha,25-dihydroxyvitamin D3. *Endocrinology* **98**: 1130–1134.

Weber, K., Kaschig, C., and Erben, R. G. (2004). 1 Alpha-hydroxyvitamin D2 and 1 alpha-hydroxyvitamin D3 have anabolic effects on cortical bone, but induce intracortical remodeling at toxic doses in ovariectomized rats. *Bone* **35**: 704–710.

Weishaar, R. E., and Simpson, R. U. (1987). Vitamin D3 and cardiovascular function in rats. *J Clin Invest* **79**: 1706–1712.

Whitfield, G. K., Jurutka, P. W., Haussler, C. A., and Haussler, M. R. (1999). Steroid hormone receptors: evolution, ligands, and molecular basis of biologic function. *J Cell Biochem*, pp. 110–122.

Whitfield, G. K., Remus, L. S., Jurutka, P. W. et al. (2001). Functionally relevant polymorphisms in the human nuclear vitamin D receptor gene. *Mol Cell Endocrinol* **177**: 145–159.

Wiese, R. J., Uhland-Smith, A., Ross, T. K., Prahl, J. M., and DeLuca, H. F. (1992). Up-regulation of the vitamin D receptor in response to 1,25-dihydroxyvitamin D3 results from ligand-induced stabilization. *J Biol Chem* **267**: 20082–20086.

Wong, R. G., Norman, A. W., Reddy, C. R., and Coburn, J. W. (1972). Biologic effects of 1,25-dihydroxycholecalciferol (a highly active vitamin D metabolite) in acutely uremic rats. *J Clin Invest* **51**: 1287–1291.

Wu, J., Garami, M., Cheng, T., and Gardner, D. G. (1996). 1,25(OH)2 vitamin D3, and retinoic acid antagonize endothelin-stimulated hypertrophy of neonatal rat cardiac myocytes. *J Clin Invest* **97**: 1577–1588.

Wu-Wong, J. R. (2007). Vitamin D receptor: a highly versatile nuclear receptor. *Kidney Int* **72**: 237–239.

Wu-Wong, J. R., Nakane, M., and Ma, J. (2006a). Effects of vitamin D analogs on the expression of plasminogen activator inhibitor-1 in human vascular cells. *Thromb Res* **118**: 709–714.

Wu-Wong, J. R., Noonan, W., Ma, J. et al. (2006b). Role of phosphorus and vitamin D analogs in the pathogenesis of vascular calcification. *J Pharmacol Exp Ther* **318**: 90–98.

Xiang, W., Kong, J., Chen, S. et al. (2005). Cardiac hypertrophy in vitamin D receptor knockout mice: role of the systemic and cardiac renin-angiotensin systems. *Am J Physiol Endocrinol Metab* **288**: E125–E132.

Xue, Y., Karaplis, A. C., Hendy, G. N., Goltzman, D., and Miao, D. (2006). Exogenous 1,25-dihydroxyvitamin D3 exerts a skeletal anabolic effect and improves mineral ion homeostasis in mice that are homozygous for both the 1alpha-hydroxylase and parathyroid hormone null alleles. *Endocrinology* **147**: 4801–4810.

Yagci, A., Werner, A., Murer, H., and Biber, J. (1992). Effect of rabbit duodenal mRNA on phosphate transport in Xenopus laevis oocytes: dependence on 1,25-dihydroxy-vitamin-D3. *Pflugers Arch* **422**: 211–216.

Yamamoto, H., Miyamoto, K., Li, B. et al. (1999). The caudal-related homeodomain protein Cdx-2 regulates vitamin D receptor gene expression in the small intestine. *J Bone Miner Res* **14**: 240–247.

Yokoyama, K., Shigematsu, T., Kagami, S. et al. (2001). Vitamin D receptor gene polymorphism detected by digestion with Apa I influences the parathyroid response to extracellular calcium in Japanese chronic dialysis patients. *Nephron* **89**: 315–320.

Yoshizawa, T., Handa, Y., Uematsu, Y. et al. (1997). Mice lacking the vitamin D receptor exhibit impaired bone formation, uterine hypoplasia and growth retardation after weaning. *Nat Genet* **16**: 391–396.

Yuan, W., Pan, W., Kong, J. et al. (2007). 1,25-dihydroxyvitamin D3 suppresses renin gene transcription by blocking the activity of the cyclic AMP response element in the renin gene promoter. *J Biol Chem* **282:** 29821–29830.

Zanello, L. P., and Norman, A. W. (1996). 1α, 25(OH)$_2$-vitamin D$_3$-mediated stimulation of outward anionic currents in osteoblast-like ROS 17/2. 8 cells. *Biochem Biophys Res Commun* **225:** 551–556.

Zanello, L. P., and Norman, A. W. (2003). Rapid modulation of osteoblast ion channel responses by 1alpha,25(OH)2-vitamin D3 requires the presence of a functional vitamin D nuclear receptor. *Proc Natl Acad Sci USA* **101:** 1589–1594.

Zeidel, M. L. (1993). Hormonal regulation of inner medullary collecting duct sodium transport. *Am J Physiol* **265:** F159–F173.

Zeitz, U., Weber, K., Soegiarto, D. W., Wolf, E., Balling, R., and Erben, R. G. (2003). Impaired insulin secretory capacity in mice lacking a functional vitamin D receptor. *Faseb J* **17:** 509–511.

Zella, L. A., Kim, S., Shevde, N. K., and Pike, J. W. (2007). Enhancers located in the vitamin D receptor gene mediate transcriptional autoregulation by 1,25-dihydroxyvitamin D3. *J Steroid Biochem Mol Biol* **103:** 435–439.

Zhang, C., Dowd, D. R., Staal, A. et al. (2003). Nuclear coactivator-62 kDa/Ski-interacting protein is a nuclear matrix-associated coactivator that may couple vitamin D receptor-mediated transcription and RNA splicing. *J Biol Chem* **278:** 35325–35336.

Zhou, L. X., Nemere, I., and Norman, A. W. (1992). 1,25-Dihydroxyvitamin D3 analog structure-function assessment of the rapid stimulation of intestinal calcium absorption (transcaltachia). *J Bone Min Res* **7:** 457–463.

Zhou, L. X., and Norman, A. W. (1995). 1 alpha,25(OH)2-vitamin D3 analog structure-function assessment of intestinal nuclear receptor occupancy with induction of calbindin-D28K. *Endocrinology* **136:** 1145–1152.

Zinser, G. M., Sundberg, J. P., and Welsh, J. (2002). Vitamin D(3) receptor ablation sensitizes skin to chemically induced tumorigenesis. *Carcinogenesis* **23:** 2103–2109.

Zinser, G. M., and Welsh, J. (2004). Accelerated mammary gland development during pregnancy and delayed postlactational involution in vitamin D3 receptor null mice. *Mol Endocrinol* **18:** 2208–2223.

CHAPTER **8**

Molecular Biology of Parathyroid Hormone

PETER A. FRIEDMAN
Department of Pharmacology, University of Pittsburgh School of Medicine, Pittsburgh, PA 15217, USA

Contents

I. Introduction 95
II. Biosynthesis and metabolism 95
III. PTH receptors 97
IV. Isoforms 98
V. Physiological actions of PTH 99
References 102

I. INTRODUCTION

Parathyroid hormone (PTH) regulates extracellular calcium and phosphate homeostasis in a negative feedback manner. Decreases of calcium or increases of phosphate stimulate PTH release which, through direct and indirect effects on kidney and bone, restores normal calcium and phosphate balance. In chronic kidney disease, this process is disrupted and, if untreated, leads to secondary hyperparathyroidism and renal osteodystrophy. This chapter presents an overview of the components of these regulatory systems, including the measurement of PTH, and their behavior in both normal conditions and when impaired.

II. BIOSYNTHESIS AND METABOLISM

PTH is synthesized in the chief cells of the parathyroid gland. It is initially translated as a 115–amino-acid product called preproparathyroid hormone. This single-chain peptide is sequentially converted to proparathyroid hormone by cleavage of 25 amino-terminal residues (Figure 8.1). As the peptide moves through the Golgi complex, biologically active PTH (PTH[1–84]) is formed by cleavage of the six amino acids of the pro sequence. The mature, 84-amino acid protein, PTH(1–84), resides within dense core secretory granules until it is discharged into the circulation. Neither preproparathyroid hormone nor proparathyroid hormone appears in plasma. Further details of the molecular synthesis and processing of PTH are available (Jüppner et al., 2001).

Once released into the general circulation PTH(1–84) is rapidly cleared and metabolized.

In addition to the full-length peptide, the parathyroids also generate a number of shorter fragments. PTH(7–84) is major proteolytic product of PTH resulting from the action of cathepsins B and H in the parathyroids. The mechanism of regulation is presently unknown. However, it is clear that plasma calcium levels determine the relative proportion of intact PTH and amino-truncated peptide fragments (Mayer et al., 1979). Amino-truncated PTH metabolites are also formed from intact, circulating PTH by hepatic Kupffer cells (Segre et al., 1981; Bringhurst et al., 1982; Nguyen-Yamamoto et al., 2002). Notably, PTH(7–84) and other peptide fragments were long assumed to be inactive degradation products that were inert byproducts of the catabolism of PTH(1–84) and represented a means to reduce the amount of active protein. Such an interpretation was based on the competitive inhibition by PTH(7–34) of PTH binding and action (McKee et al., 1988; Divieti et al., 2002) and the inability of PTH(7–34) or PTH(7–84) to activate the type 1 PTH receptor (PTH1R) (Slatopolsky et al., 2000). More recent work shows that the concentration of PTH(7–84) and other amino-truncated PTH fragments increases significantly during renal failure, representing as much as 95% of circulating peptide (Brossard et al., 1996). The massive accumulation of such fragments in renal failure arises, in part, because they are cleared from the circulation predominantly by the kidneys and only minimally by the liver (Martin et al., 1979), whereas intact PTH is eliminated principally by extra-renal mechanisms. The clearance of amino-truncated PTH fragments is considerably slower than that for peptides with an intact amino-terminus. The clinical significance of these peptides is of some considerable debate and uncertainty. In isolated kidney and bone cells, it is now apparent that PTH(7–84), along with a series of synthetic shorter PTH fragments, promote PTH receptor downregulation without concomitant activation (Sneddon et al., 2003, 2004). The effect of PTH fragments on receptor endocytosis is governed by the cellular abundance of the adapter protein NHERF1 (Na/H exchange regulatory factor). Notably, this

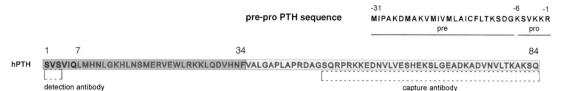

FIGURE 8.1 Sequence of pre-pro PTH sequence (above) and full-length human PTH(1–84) (below). Indicated sites in PTH(1–84) show location of detection and capture epitopes used in second-generation 'sandwich' immunometric assays. Specific deletion of the first six amino-terminal residues of PTH produces a peptide fragment that is recognized by first-generation PTH immunoassays because the detection antibody recognized an epitope in between residues 15 and 34. In second-generation assays the detection epitope is at positions 1–3 (John et al., 1999; Gao et al., 2001) and only full-length (bioactive) PTH is measured.

selective action of PTH(7–84) is not due to competitive displacement of PTH(1–84) and suggests that amino-truncated PTH fragments may promote hormone resistance by causing receptor internalization from the plasma membrane (Sneddon et al., 2003; Wheeler et al., 2007). PTH(7–84) has been shown also to antagonize the effects of PTH(1–84) on bone in experimental renal failure (Langub et al., 2003).

A. Regulation

PTH secretion is tightly regulated by changes of extracellular calcium (Chattopadhyay et al., 1996) by a classical negative feedback mechanism that is mediated by calcium-sensing receptor (CaSR), a G-protein-coupled receptor on parathyroid cells (Brown et al., 1993; Brown and MacLeod, 2001). Occupancy of the CaSR by Ca^{2+} inhibits PTH secretion, whereas reduced CaSR occupancy promotes hormone secretion (Figure 8.2). CaSR regulation of PTH release is discussed further in Physiological Actions of PTH.

The amount of preformed PTH is limited and, as noted earlier, its degradation rapid. The availability of sufficient cellular PTH to meet changing demands on extracellular mineral ion homeostasis is met by the coordinated regulation of PTH formation by controlling gene expression. PTH synthesis and secretion are regulated by extracellular Ca^{2+}, vitamin D and phosphate (Silver and Levi, 2005). Reductions of extracellular calcium increase the synthesis of PTH mRNA, whereas elevations of calcium decrease PTH formation. In a reciprocal manner, increases of serum phosphate enhance PTH synthesis, while decreases of phosphate reduce its formation. The regulation of PTH synthesis by serum phosphate occurs independently of calcium and vitamin D. Whether such control by Ca^{2+} and phosphate is exerted at transcriptional (Okazaki et al., 1988, 1991) or post-transcriptional (Moallem et al., 1998) levels remains controversial. $1,25(OH)_2D_3$ directly regulates PTH transcription (Silver et al., 1986). 22-oxacalcitriol, a vitamin D analog exhibiting few effects on serum calcium, also suppresses PTH formation (Fan et al., 2000).

More PTH is secreted and less is hydrolyzed during periods of hypocalcemia. In this setting, PTH(7–84) release is augmented. PTH synthesis also increases and the gland hypertrophies in prolonged hypocalcemia.

Secretion of PTH(1–84) is favored at normal or low calcium levels, whereas amino-truncated PTH fragments are preferentially secreted at elevated extracellular calcium concentrations (Rakel et al., 2005; Rubin et al., 2006). PTH (1–84) has a half-life in plasma of about 2 minutes. The liver, primarily, and kidneys, secondarily, account for about 90% of its clearance (Bringhurst et al., 1988). By contrast, the half-life of PTH(7–84) is some 10 minutes or longer (Martin et al., 1979) and unlike PTH(1–84), PTH(7–84) is cleared primarily by the kidneys (Martin et al., 1979). Thus, formation of PTH(7–84) in chronic kidney disease is elevated and elimination is reduced resulting in profound accumulation of carboxy-terminal PTH fragments in the peripheral circulation (Monier-Faugere et al., 2001; Coen et al., 2002). The level of these fragments may approach or exceed that of PTH(1–84) (Brossard et al., 1996; Coen et al., 2002; Nguyen-Yamamoto et al., 2002). The physiological consequences and recommended interventions for such situations

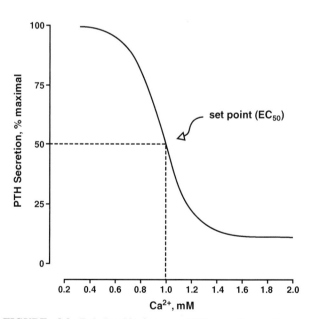

FIGURE 8.2 Relationship between PTH secretion and serum free calcium concentration (Ca^{2+}). PTH secretion is normalized to percent of maximum. The dashed line shows half-maximal PTH secretion, which occurs at approximately 1 mM Ca^{2+}. This is referred to as the set point.

are presently conflicting and uncertain (Spiechowicz et al., 2004; Chang et al., 2005) as it remains ambiguous whether low- or high-turnover bone disease can be explicitly related to the absolute or relative abundance of PTH(1–84) and carboxy-terminal PTH fragments, i.e. peptides lacking the amino terminus.

B. Circulating Forms of PTH

As noted earlier, metabolism of PTH generates smaller, non-PTH(1–84) fragments, e.g. PTH(7–84) that also circulate in blood. In fact, these peptide fragments represent 80% of circulating PTH in normal individuals and up to 95% in patients with chronic kidney disease (D'Amour, 2006). Reliable measurements of circulating PTH are essential to the diagnosis and clinical management of patients with renal disease.

C. Measurement

Assays for the determination of serum PTH have evolved considerably since the introduction of a radioimmunoassay based on the principle of competitive inhibition of binding of [^{131}I]-labeled PTH to specific hormone antibodies (Berson et al., 1963), one of the first developed radioimmunoassays (RIAs).

The presence of PTH fragments was not appreciated nor measured by the original RIAs for PTH. These assays recognized only a single epitope within the middle or carboxy regions of the peptide and therefore could not discern the presence or absence of the amino terminus. The development of two-site immunometric assays was thought to overcome the limitations of the earlier assays. First-generation immunometric assays employed an immobilized 'capture' antibody and a second, higher affinity antibody for detection. These assays were based on saturation rather than competitive binding as in RIAs. The capture antibody generally recognized an epitope in the carboxy region of PTH(39–84), while the detection antibody was in or near the amino terminus. Several observations, however, suggested the presence of anomalous peptides when these assays were employed. Compared with healthy subjects, patients with chronic kidney disease exhibited significant increases in the non-suppressible fraction of PTH, i.e. the portion that is not attenuated by increasing serum calcium when parathyroid function is measured dynamically and serum PTH is calculated as a function of serum Ca^{2+} using a four-parameter curve fitting procedure (Brown, 1983). Precise analytical characterization of circulating PTH unambiguously confirmed that first-generation immunometric assays failed to detect the presence of these PTH fragments (Brossard et al., 1996; Lepage et al., 1998). The terminology describing first-generation immunometric assays lends to some present confusion because they are promoted as measuring 'intact PTH' while, in fact, they detect both PTH(1–84) as well as amino-truncated PTH fragments.

Second-generation immunometric PTH assays now employ a detection antibody that is within the first three amino acids. These new assays distinguish between bioactive PTH (PTH[1–84] + amino-terminal peptides) and other circulating PTH forms (John et al., 1999; Gao et al., 2001). The ability to measure whole PTH separately from large amino-truncated PTH fragments is argued to increase the accuracy of laboratory testing of parathyroid status and distinguishing between high- and low-turnover bone status in patients with renal failure (Monier-Faugere et al., 2001; Donadio et al., 2007). Direct comparisons of various assays individually or as ratios as a function of serum calcium do not seem to provide improved predictive behavior (Rubin et al., 2006) and the significance of employing ratios of bioactive PTH to amino-truncated fragments for therapeutic intervention has been questioned (Coen et al., 2002; Tsuchida et al., 2006; Goodman, 2007).

III. PTH RECEPTORS

PTH actions are mediated by the canonical type I PTH receptor. This G-protein-coupled receptor is abundantly expressed in kidney and bone, where it transduces signals from PTH and the PTH-related peptide (PTHrP). Two other PTH receptors have been identified. The type 2 PTH receptor (PTH2R), which exhibits about 50% sequence identity with the PTH1R, responds virtually uniquely to PTH and not PTHrP (Usdin et al., 1995, 1997). The structural determinants that account for the ability of the PTH2R to discriminate between PTH and PTHrP have been identified (Gardella et al., 1996). Histidine at position 5 (Ile in PTH) and Phe at position 23 (Trp in PTH) account for the differential binding affinity and activation. In the kidney, PTH2R expression is thought to be limited to glomerular and other vascular cells and is not expressed on tubular epithelial cells. It has not been implicated in renal calcium or phosphate transport. The 39-amino acid tuberoinfundibular neuropeptide (TIP39) is the cognate PTH2R ligand.

A third, non-mammalian, PTH receptor (PTH3R) has been identified in zebrafish (Rubin and Jüppner, 1999). It shares higher homology with the PTH1R than with the PTH2R. However, the PTH3R displays greater affinity for PTHrP than for hPTH and exhibits more robust activation of adenylyl cyclase.

RT-PCR and *in situ* hybridization were used to define the nephron sites of mRNA expression (Lee et al., 1996; Riccardi et al., 1996; Yang et al., 1997). General consensus pointed to conspicuous PTH1R expression in the glomerulus, PCT and PST, CAL and DCT. Some differences may also be noteworthy. Riccardi and Hebert additionally reported PTH1R expression in the rat CCD (Riccardi et al., 1996), a finding not confirmed by Yang et al. (Yang et al., 1997). Immunoelectron and immunofluorescence microscopy localization detected PTH1R on basolateral (but also apical)

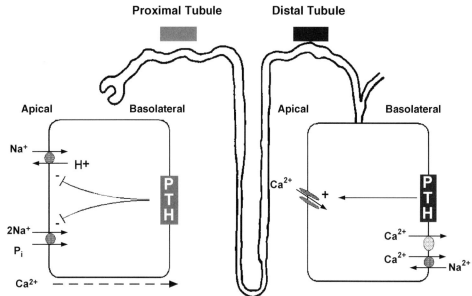

FIGURE 8.3 PTH actions on kidney tubule ion transport. PTH1Rs are found primarily on cells of the proximal tubule and distal convoluted tubule. Prototype proximal and distal tubule cells showing the principal transport pathways regulated by PTH are illustrated. In proximal tubule cells, PTH inhibits Na/H exchange and Na–Pi cotransport. Calcium absorption denoted by the dashed line proceeds through the paracellular pathway. In distal tubule cells PTH promotes cellular calcium entry through apical membrane calcium channels. Basolateral calcium efflux is elevated concomitantly, primarily through the Na/Ca exchanger. Additional details are given in the text.

membranes of proximal convoluted tubule cells (Amizuka et al., 1997). Other studies employing immunofluorescence reported both luminal and basolateral PTH1R expression in proximal tubules and thick ascending limbs but only on basolateral surfaces of distal convoluted tubules (Ba et al., 2003). The physiological actions of PTH are generally ascribed to the presence of PTH1R on proximal and distal tubules (Figure 8.3).

IV. ISOFORMS

Strictly speaking, as far as is presently known, only a single functional form of the PTH1R is expressed on the cell surface. Nonetheless, PTH1R variants using alternative upstream transcription start sites generate different splicing patterns resulting in receptors that are widely expressed as well as kidney-specific expression (Jobert et al., 1996; Joun et al., 1997). An entirely different, truncated PTH1R lacking exon 14, which encodes most of the 7th transmembrane domain, was identified and found more abundantly in a patient with pseudohypoparathyroidism type Ib (Ding et al., 1995). The significance of such a construct is not known, but it might function in a dominant negative manner similar to calcitonin receptors lacking the comparable exon (Seck et al., 2005).

A. Topology of PTH Binding

Initiation of the biological events entrained by PTH begins with its engagement with the PTH1R. A variety of experimental approaches including generation of chimeric or mutant receptors and hybrid ligands, NMR of the ligand with selected receptor domains, photoaffinity scanning and bimolecular FRET of dynamic interactions have been applied to analyze the PTH interface with the PTH1R. The carboxy-terminal helix of PTH binds with high affinity to the amino-terminal extracellular PTH1R domain, and a low-affinity interaction between the amino-terminal portion of PTH with the juxtamembrane (J) region of the receptor, the latter being essential for PTH1R activation (Carter et al., 1999; Vilardaga et al., 2001). This arrangement is shown schematically in Figure 8.4. Photochemical crosslinking has been applied to examine static interactions of PTH with the PTH1R (Gardella and Juppner, 2000). Dynamic PTH interactions with the PTH1R reveal a two-step mechanism in which PTH rapidly binds to the N-domain of the PTH1R. This is followed by a slower binding step to the J-domain that is coupled to PTH1R activation (Castro et al., 2005).

B. Signaling

The PTH1R is a member of Family B of GPCRs. These receptors signal primarily through adenylyl cyclase, using cAMP as a second messenger, or phospholipase C (PLC), where intracellular Ca^{2+} and inositol phosphates serve as second messengers. Considerable cell-specific signaling of the PTH1R has been described. In some cells, the PTH1R activates both adenylyl cyclase and PLC. In other instances, however, the PTH1R couples only to a single pathway. For example, in vascular smooth muscle cells, PTH stimulates adenylyl cyclase but not PLC (Wu et al., 1993; Maeda et al., 1996), whereas in keratinocytes (Whitfield et al., 1992; Orloff et al., 1992, 1995), cardiac myocytes (Rampe et al., 1991; Schlüter et al., 1995) and lymphocytes (Whitfield et al., 1971; Atkinson et al., 1987; Klinger et al., 1990), PLC but not adenylyl cyclase is activated. Thus, PTH actions may be

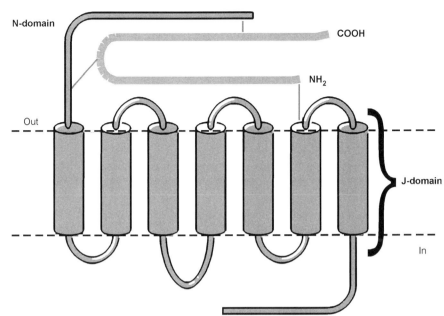

FIGURE 8.4 Schematic representation of the PTH1R occupied by PTH (light grey hairpin loop). The amino-terminus of PTH interacts with the J-domain of the PTH1R, while the mid-region is stabilized by the receptor N-domain. Further details are provided in the text.

mediated by the cAMP → protein kinase A (PKA) pathway and/or PLC → protein kinase C (PKC) route.

A variety of explanations has been advanced to explain the remarkable differences in cell-specific signaling. A major breakthrough in reconciling these disparities came from the work of Mahon and Segre (Mahon et al., 2002), who discovered that the Na/H exchanger regulatory factor (NHERF) interacts with the PTH1R thereby determining the signaling pathway. Two forms of NHERF, NHERF1 and NHERF2, have been identified (Donowitz et al., 2005). Both are capable of interacting with the PTH1R and promoting the receptor signaling switch. NHERF1 would seem to be the more biologically relevant form because, in its absence, mice exhibit profound renal mineral ion wasting and stunted growth and increased fractures (Shenolikar et al., 2002). In the kidney, NHERF1 is expressed in apical membrane microvilli. Although NHERF2 can be detected weakly in the microvilli, it is expressed predominantly in collecting duct principal cells, as well as in the glomerulus and in other renal vascular structures (Wade et al., 2001, 2003).

NHERF proteins interact with the PTH1R through its carboxy-terminal PDZ-binding domain. Insofar as the kidney expresses a number of PDZ proteins, it is likely that transient multiprotein complexes are formed that regulate PTH signaling and actions in a highly precise manner.

V. PHYSIOLOGICAL ACTIONS OF PTH

PTH exerts a range of actions that involve direct and indirect effects on mineral ion homeostasis, acid–base balance, intermediary metabolism and skeletal integrity.

A. Mineral ion Homeostasis

The most prominent renal actions of PTH are its effects on the extracellular concentrations of calcium, phosphate and, perhaps, magnesium. The spatial and vectorial effects of PTH on calcium and phosphate absorption proceed in different nephron segments and are generally regulated in opposite directions. Phosphate transport occurs largely, if not entirely, in proximal tubules, where its absorption proceeds through a cellular pathway (see Figure 8.3). Calcium absorption occurs both in proximal and distal nephron segments. However, it is only distal calcium transport that is under hormonal and pharmacological regulation. Furthermore, proximal calcium absorption, which is the site of the bulk of its recovery, occurs primarily through the lateral intercellular spaces forming the paracellular pathway. Thus, transport of calcium and phosphate is compartmentalized both at the tissue (proximal versus distal nephron) and cellular (transcellular versus paracellular) levels. Given the limited solubility of calcium phosphate and its propensity for precipitation, it is likely that evolutionary processes contributed to their discrete and spatially separated sites of transport.

PTH augments renal calcium absorption while depressing that of phosphate. In this manner, PTH actions are balanced and do not contribute to distortions of the calcium–phosphate product (Ca × P). PTH also promotes renal magnesium absorption (de Rouffignac and Quamme, 1994). Hyperparathyroidism has little effect on serum magnesium levels (Hanna et al., 1961) and the hypomagnesemia accompanying parathyroidectomy has been attributed to rapid deposition of magnesium in newly formed bone. Vitamin D and calcium supplements given to correct serum calcium in the postoperative period may also contribute to hypomagnesemia (Ahmed and Sutton, 2000). Thus, the role of PTH in management of serum magnesium is ambivalent.

Remarkably modest PTH levels are needed to maintain calcium balance, at least in mice. In animals harboring a phosphorylation-deficient form of the PTH1R, a 70% reduction of PTH had no effect on serum calcium (Bounoutas et al., 2006).

PTH secretion is regulated by the extracellular Ca^{2+}-sensing receptor (CaSR) as mentioned briefly earlier. Reductions of serum Ca^{2+} stimulate PTH secretion (see Figure 8.2). The inverse relation between the fall of Ca^{2+} and the increment of PTH release is quite steep and exhibits several distinct characteristics. The slope (3.5–4) assures small reductions of Ca^{2+} are robustly addressed, helping to defend extracellular calcium. The point at which half-maximal PTH release occurs is referred to as the 'set-point'. This occurs at a serum Ca^{2+} concentration that is somewhat less (1.0 mM) than normal serum Ca^{2+} (1.2–1.3 mM). Thus, there is some tonic stimulation of PTH release and the Ca^{2+} is poised in the linear range, where further decreases of Ca^{2+} bring about large increases in PTH secretion.

With persistent hypocalcemia, additional compensatory mechanisms are activated to fortify the short-term actions of PTH. In particular, PTH induces activation of the renal 1α-hydroxylase, thereby increasing the synthesis and release of the biologically active form of vitamin D (calcitriol), $1,25(OH)_2$vitamin D, which directly stimulates intestinal calcium absorption. Release of calcium from bone into the extracellular fluid is also enhanced. Prolonged and severe hypocalcemia activates bone remodeling units which, at the expense of skeletal integrity, restores circulating calcium levels. Persistent hypocalcemia accompanied by hyperphosphatemia, or vitamin D deficiency promotes parathyroid hyperplasia thereby enhancing the capacity of the parathyroid glands to synthesize and secrete PTH, thereby normalizing calcium.

When serum calcium rises, PTH secretion is inhibited and renal calcium absorption decreases. However, PTH secretion is not entirely suppressed even at high serum calcium concentrations. Thus, there is a non-suppressible basal component of secretion. Concomitantly, the reduction of circulating PTH promotes renal phosphate conservation and both the decreased PTH and the increased phosphate reduce $1,25(OH)_2$vitamin D production, thereby reducing intestinal calcium absorption. Finally, bone remodeling is suppressed. This negative feedback system assures the integrated physiologic regulation and appropriate responses to positive or negative excursions of serum calcium.

B. Na/H Exchange and pH

In addition to regulating extracellular mineral ion homeostasis, PTH exerts important inhibitory effects on the portion of proximal tubule sodium absorption that is mediated by Na/H exchange, which may be as much as 60%. PTH directly inhibits Na/H exchange by as much as 50% (Kahn et al., 1985). Moreover, it should be recalled that proximal tubule phosphate absorption is mediated by the Npt2a sodium–phosphate co-transporter, which is inhibited by PTH. Thus, PTH suppresses proximal sodium absorption mediated by both Na/H exchange and by Na–P co-transport. Nonetheless, PTH has little conspicuous effect on sodium excretion or volume diuresis (Agus et al., 1971). The reason for this is likely to be twofold. First, it is generally possible only to demonstrate natriuretic effects of PTH during volume depletion. Second, it is likely that the sodium and water rejected at more proximal sites are largely absorbed downstream. The absence of distal sites of phosphate absorption likely explains the phosphaturia that accompanies PTH administration and is consistent with the interpretation of the modest hormone effects on sodium and water excretion. Some degree of parathyroid insufficiency may also be necessary to demonstrate a natriuretic effect of PTH administration.

C. Regulation of Vitamin D Synthesis

As mentioned earlier, PTH promotes vitamin D synthesis by activating 1α-hydroxylase, the rate-limiting enzyme in the synthesis of $1,25(OH)_2$vitamin D. PTH rapidly increases $1,25(OH)_2D$ production by a cyclic AMP-dependent pathway. The circulating concentration of $1,25(OH)_2D$ primarily reflects its synthesis in the kidney, however, 1α-hydroxylase activity is also present in keratinocytes, macrophages and osteoblasts. 1α-hydroxylase activity is regulated not only by PTH, but also by calcium, phosphate and $1,25(OH)_2D$ itself. Hypocalcemia *per se* can directly activate the 1α-hydroxylase, in addition to affecting it indirectly through PTH action (Bland et al., 1999). Hypophosphatemia greatly augments 1α-hydroxylase activity (Haussler and McCain, 1977; Yoshida et al., 2001). Finally, $1,25(OH)_2D$ itself controls 1α-hydroxylase activity by a negative-feedback mechanism that involves a direct action on the kidney as well as by inhibition of PTH secretion. The plasma half-life of $1,25(OH)_2D$ is estimated to be between 3 and 5 hours in humans (Bailie and Johnson, 2002).

24-Hydroxylase is the enzyme responsible for the first step of $1,25(OH)_2D$ catabolism and PTH enhances its breakdown by diminishing its stability (Zierold et al., 2001). Thus, PTH promotes accumulation of $1,25(OH)_2D$ by simultaneously increasing its formation and decreasing its degradation.

Conversely, high calcium, phosphate and vitamin D intake suppresses 1α-hydroxylase activity. Regulation is both acute and chronic, with the latter due to changes in protein synthesis.

D. Intermediary Metabolism

1. GLUCONEOGENESIS

Although the liver is the principal site of gluconeogenesis, the kidney contributes importantly to this metabolic process

(Friedman and Torretti, 1978; Schoolwerth et al., 1988). PTH promotes both hepatic (Hruska et al., 1979) and renal gluconeogenesis (Wang and Kurokawa, 1984). The hepatic effects appear to be restricted to PTH(1–84) insofar as single intravenous injections of amino-truncated PTH fragments failed to increase hepatic venous glucose concentrations. This finding may result from concurrent differential activation of glycogenolysis by full-length PTH and PTH fragments. It should be borne in mind that the concentrations of PTH necessary to demonstrate an effect on hepatic gluconeogenesis is considerably above normal circulating levels and thus may not obtain under physiological conditions.

PTH-stimulated renal gluconeogenesis is attended by conspicuous segmental and internephron heterogeneity. PTH primarily stimulates gluconeogenesis in cortical S_1 and S_2 proximal tubules (Wang and Kurokawa, 1984). Although juxtamedullary S1 proximal tubules exhibit the highest rate of gluconeogenesis, it is unaffected by PTH.

2. Ammoniagenesis

Gluconeogenesis is linked to ammoniagenesis because both are stimulated by acidosis and by PTH. Moreover, L-glutamine, which is the major gluconeogenic precursor is also a substrate for ammoniagenesis. These and other observations raised the possibility that gluconeogenesis and ammoniagenesis are metabolically and functionally linked. This may be the case in acidosis but not under non-acidotic conditions, where inhibition of the gluconeogenic enzyme phosphoenolpyruvate carboxykinase (PEPCK) failed to blunt ammoniagenesis. Hence, the two processes appear to proceed through independent metabolic mechanisms under physiological conditions but may involve convergent pathways in acidosis.

E. Disordered PTH Regulation in Chronic Kidney Disease

Advanced chronic kidney disease, CKD3–5 (K/DOQI, 2003), is almost invariably accompanied by secondary hyperparathyroidism in individuals who are untreated. The primary factors involved in the pathogenesis of the secondary hyperparathyroidism of CKD include progressively reduced glomerular filtration rate, phosphate retention and elevated PTH. With time and further decline of renal function, this non-productive cycle results in impaired production of 1,25(OH)$_2$D, abnormal parathyroid growth and function and skeletal resistance to PTH. The hallmarks of secondary hyperparathyroidism due to CKD include reductions in renal 1,25(OH)$_2$D production with a decline in its serum levels, decreases in intestinal calcium absorption that result in hypocalciuria, a tendency toward mild or overt hypocalcemia and, at later stages, hyperphosphatemia.

Accumulation of amino-terminally-truncated PTH fragments in secondary hyperparathyroidism may account for reductions in bone formation and turnover. Because first-generation immunometric assays detect both PTH(1–84) and various peptide fragments, while second-generation immunometric PTH assays exclusively detect PTH(1–84), the difference in plasma PTH levels as determined by first- and second-generation PTH assays has been used as an index to estimate the concentration of amino-truncated fragments. It should be underscored that such analyses do not measure the concentration of PTH(7–84), which cannot be determined directly. Rather, the values most probably reflect the plasma level of a variety of amino-terminally-truncated fragments, one of which is PTH(7–84) (D'Amour, 2006). Using this approach, it has been proposed that low-turnover (or 'adynamic') renal osteodystrophy in some patients undergoing dialysis is due to the accumulation of PTH(7–84) (Monier-Faugere et al., 2001).

Given the limited data from clinical studies of bone histology to validate such biochemical findings, a fundamental role for amino-terminally-truncated peptides in general, and for PTH(7–84) in particular, in the pathogenesis of low-turnover renal osteodystrophy is uncertain. Additional information about vitamin D nutritional status, recent treatment with vitamin D and the ambient serum calcium concentration at the time PTH is measured should be obtained. The preponderance of evidence suggests that measurements of PTH using either first- or second-generation immunometric assays provides similar accuracy for predicting bone turnover in dialysis patients.

These caveats notwithstanding, recent findings have heightened interest in the potential importance of PTH fragments that have been considered to be biologically inactive. Studies in several experimental models indicate that synthetic PTH(7–84) and other amino-terminally-truncated PTH fragments can counteract the classical biological actions of PTH(1–84) and PTH(1–34) in regulating serum calcium levels, actions that are mediated by signal-transduction through the PTH1R. Both PTH(7–84) and other synthetic amino-terminally-truncated peptides have been shown to modify calcium efflux from neonatal mouse calvariae and to influence osteoclast differentiation *in vitro* (Divieti et al., 2001, 2002; Nguyen-Yamamoto et al., 2001). Osteocytes derived from genetically modified mice lacking the PTH1R exhibit high-affinity binding for PTH(19–84) that is presumably mediated by a putative carboxy-terminal PTH receptor (C-PTHR) as described previously in osteoblast-like cells (Divieti et al., 1998, 2002, 2005). Moreover, binding of the carboxyl-terminal portion of PTH to the putative C-PTHR in osteocyte-like cells promotes apoptosis (Divieti et al., 2001; Bringhurst, 2002). It is thus plausible that carboxy-terminal PTH fragments serve a physiological function by interacting with a discrete C-PTHR to alter bone cell metabolism and to offset or modify signal transduction by the PTH1R. Such a mechanism could account for the tissue resistance to PTH that characterizes patients with advanced renal failure and contribute to alterations in bone metabolism among those with chronic kidney disease.

The effect of different molecular species of PTH on signal transduction and trafficking of the PTH1R in renal tubular epithelial cells is governed in a cell-specific manner by the presence and abundance of the scaffolding protein NHERF (Na/H exchanger regulatory factor) (Mahon et al., 2002; Sneddon et al., 2003). NHERF1 is also expressed in osteoclasts (Khadeer et al., 2003), but it is not yet clear whether NHERF1 is expressed in osteoblasts or in osteocytes. NHERF1 might conceivably prevent or promote vascular calcification in renal disease (Vattikuti and Towler, 2004). Were PTH(1–84) to promote skeletal mineral accretion, while concurrently limiting vascular calcification, the accumulation of PTH fragments might contribute to the calcific vasculopathy of end-stage renal failure (Shao et al., 2003). Expression of NHERF1 prevents PTH(7–84)-dependent PTH1R internalization. Thus, if NHERF1 expression is downregulated in renal failure, PTH(7–84) would contribute to the blunted PTH sensitivity of uremia by allowing enhanced PTH1R endocytosis (Vattikuti and Towler, 2004). Conversely, if NHERF1 expression is upregulated in this setting, ectopic calcification may be blunted. Nevertheless, variations in the level of expression of NHERF1 in distinct populations of bone cells represent a potentially important mechanism to account for changes in PTH1R-mediated signal transduction that are mediated by amino-truncated PTH fragments as described earlier.

Acknowledgment

Original studies described herein were supported by NIH grants DK54171 and DK069998.

References

Agus, Z. S. et al. (1971). Mode of action of parathyroid hormone and cyclic adenosine 3′,5′-monophosphate on renal tubular phosphate reabsorption in the dog. *J Clin Invest* **50**: 617–626.

Ahmed, A., and Sutton, R. A. L. (2000). Disorders of magnesium metabolism. Seldin, D. W. and Giebisch, G. editors. *The Kidney. Physiology and Pathophysiology* **vol. 2**: Lippincott Williams & Wilkins, Philadelphia, pp. 1731–1748.

Amizuka, N. (1997). Cell-specific expression of the parathyroid hormone (PTH)/PTH-related peptide receptor gene in kidney from kidney-specific and ubiquitous promoters. *Endocrinology* **138**: 469–481.

Atkinson, M. J. (1987). Parathyroid hormone stimulation of mitosis in rat thymic lymphocytes is independent of cyclic AMP. *J Bone Min Res* **2**: 303–309.

Ba, J. (2003). Calcium-sensing receptor regulation of PTH-inhibitable proximal tubule phosphate transport. *Am J Physiol Renal Physiol* **285**: F1233–F1243.

Bailie, G. R., and Johnson, C. A. (2002). Comparative review of the pharmacokinetics of vitamin D analogues. *Semin Dial* **15**: 352–357.

Berson, S. A. (1963). Immunoassay of bovine and human parathyroid hormone. *Proc Natl Acad Sci USA* **49**: 613–617.

Bland, R. (1999). Constitutive expression of 25-hydroxyvitamin D_3-1a-hydroxylase in a transformed human proximal tubule cell line: Evidence for direct regulation of vitamin D metabolism by calcium. *Endocrinology* **140**: 2027–2034.

Bounoutas, G. S. (2006). Impact of impaired receptor internalization on calcium homeostasis in knock-in mice expressing a phosphorylation-deficient parathyroid hormone (PTH)/PTH-related peptide receptor. *Endocrinology* **147**: 4674–4679.

Bringhurst, F. R. (2002). PTH receptors and apoptosis in osteocytes. *J Musculoskelet Neuronal Interact* **2**: 245–251.

Bringhurst, F. R. (1982). Metabolism of parathyroid hormone by Kupffer cells: analysis by reverse-phase high-performance liquid chromatography. *Biochemistry* **21**: 4252–4258.

Bringhurst, F. R. (1988). Peripheral metabolism of PTH: fate of biologically active amino terminus in vivo. *Am J Physiol* **255**: E886–E893.

Brossard, J. H. (1996). Accumulation of a non-(1–84) molecular form of parathyroid hormone (PTH) detected by intact PTH assay in renal failure: importance in the interpretation of PTH values. *J Clin Endocrinol Metab* **81**: 3923–3929.

Brown, E. M. (1983). Four-parameter model of the sigmoidal relationship between parathyroid hormone release and extracellular calcium concentration in normal and abnormal parathyroid tissue. *J Clin Endocrinol Metab* **56**: 572–581.

Brown, E. M. (1993). Cloning and characterization of an extracellular Ca^{2+}-sensing receptor from bovine parathyroid. *Nature* **366**: 575–580.

Brown, E. M., and MacLeod, R. J. (2001). Extracellular calcium sensing and extracellular calcium signaling. *Physiol Rev* **81**: 239–297.

Carter, P. H. (1999). Studies of the N-terminal region of a parathyroid hormone-related peptide (1–36) analog: receptor subtype-selective agonists, antagonists, and photochemical cross-linking agents. *Endocrinology* **140**: 4972–4981.

Castro, M. (2005). Turn-on switch in parathyroid hormone receptor by a two step parathyroid hormone binding mechanism. *Proc Natl Acad Sci USA*.

Chang, J. M. (2005). 7–84 parathyroid hormone fragments are proportionally increased with the severity of uremic hyperparathyroidism. *Clin Nephrol* **63**: 351–355.

Chattopadhyay, N. (1996). The calcium-sensing receptor: a window into the physiology and pathophysiology of mineral ion metabolism. *Endocrine Rev* **17**: 289–307.

Coen, G. (2002). PTH 1–84 and PTH '7–84' in the noninvasive diagnosis of renal bone disease. *Am J Kidney Dis* **40**: 348–354.

D'Amour, P. (2006). Circulating PTH molecular forms: what we know and what we don't. *Kidney Int*, pp. S29–S33.

de Rouffignac, C., and Quamme, G. (1994). Renal magnesium handling and its hormonal control. *Physiol Rev* **74**: 305–322.

Ding, C. (1995). Identification of an alternatively spliced form of PTH/PTHrP receptor mRNA in immortalized renal tubular cells. *J Bone Min Res* **10**(Suppl 1): S484.

Divieti, P. (2005). Receptors specific for the carboxyl-terminal region of parathyroid hormone on bone-derived cells: determinants of ligand binding and bioactivity. *Endocrinology* **146**: 1863–1870.

Divieti, P. (2001). Receptors for the carboxyl-terminal region of PTH(1–84) are highly expressed in osteocytic cells. *Endocrinology* **142**: 916–925.

Divieti, P. (2002). Human PTH-(7–84) inhibits bone resorption in vitro via actions independent of the type 1 PTH/PTHrP receptor. *Endocrinology* **143**: 171–176.

Divieti, P. (1998). Conditionally immortalized murine osteoblasts lacking the type 1 PTH/PTHrP receptor. *J Bone Min Res* **13**: 1835–1845.

Donadio, C. (2007). Parathyroid hormone and large related C-terminal fragments increase at different rates with worsening of renal function in chronic kidney disease patients. A possible indicator of bone turnover status? *Clin Nephrol* **67**: 131–139.

Donowitz, M. (2005). NHERF family and NHE3 regulation. *J Physiol (Lond)* **567**: 3–11.

Fan, S. L. (2000). Potent suppression of the parathyroid glands by hydroxylated metabolites of dihydrotachysterol$_2$. *Nephrol Dial Transplant* **15**: 1943–1949.

Friedman, P. A., and Torretti, J. (1978). Regional glucose metabolism in the cat kidney in vivo. *Am J Physiol* **234**: F415–F423.

Gao, P. (2001). Development of a novel immunoradiometric assay exclusively for biologically active whole parathyroid hormone 1–84: implications for improvement of accurate assessment of parathyroid function. *J Bone Min Res* **16**: 605–614.

Gardella, T. J., and Juppner, H. (2000). Interaction of PTH and PTHrP with their receptors. *Rev Endocr Metab Disord* **1**: 317–329.

Gardella, T. J. (1996). Converting parathyroid hormone-related peptide (PTHrP) into a potent PTH-2 receptor agonist. *J Biol Chem* **271**: 19888–19893.

Goodman, W. G. (2007). Comments on plasma parathyroid hormone levels and their relationship to bone histopathology among patients undergoing dialysis. *Semin Dial* **20**: 1–4.

Hanna, S. (1961). Magnesium metabolism in parathyroid disease. *Br Med J* **2**: 1253–1256.

Haussler, M. R., and McCain, T. A. (1977). Basic and clinical concepts related to vitamin D metabolism and action. *N Engl J Med* **297**: 974–983.

Hruska, K. A. (1979). Effect of intact parathyroid hormone on hepatic glucose release in the dog. *J Clin Invest* **64**: 1016–1023.

Jobert, A. S. (1996). Expression of alternatively spliced isoforms of the parathyroid hormone (PTH)/PTH-related peptide receptor messenger RNA in human kidney and bone cells. *Mol Endocrinol* **10**: 1066–1076.

John, M. R. (1999). A novel immunoradiometric assay detects full-length human PTH but not amino-terminally truncated fragments: Implications for PTH measurements in renal failure. *J Clin Endocrinol Metab* **84**: 4287–4290.

Joun, H. (1997). Tissue-specific transcription start sites and alternative splicing of the parathyroid hormone (PTH)/PTH-related peptide (PTHrP) receptor gene: a new PTH/PTHrP receptor splice variant that lacks the signal peptide. *Endocrinology* **138**: 1742–1749.

Jüppner, H. W. (2001). Parathyroid hormone and parathyroid hormone-related peptide in the regulation of calcium homeostasis and bone development. DeGroot, L. J. and Jameson, J. L. editors. *Endocrinology* **vol. 2**: W.B. Saunders Company, Philadelphia, pp. 969–998.

K/DOQI. (2003). Clinical practice guidelines for bone metabolism and disease in chronic kidney disease. Am J Kidney Dis. 42, S1–S201.

Kahn, A. M. (1985). Parathyroid hormone and dibutyryl cAMP inhibit Na$^+$/H$^+$ exchange in renal brush border vesicles. *Am J Physiol* **248**: F212–F218.

Khadeer, M. A. (2003). Na$^+$-dependent phosphate transporters in the murine osteoclast: cellular distribution and protein interactions. *Am J Physiol Cell Physiol* **284**: C1633–C1644.

Klinger, M. (1990). Effect of parathyroid hormone on human T cell activation. *Kidney Int* **37**: 1543–1551.

Langub, M. C. (2003). Administration of PTH-(7–84) antagonizes the effects of PTH-(1–84) on bone in rats with moderate renal failure. *Endocrinology* **144**: 1135–1138.

Lee, K. C. (1996). Localization of parathyroid hormone parathyroid hormone-related peptide receptor mRNA in kidney. *Am J Physiol* **270**: F186–F191.

Lepage, R. (1998). A non-(I-84) circulating parathyroid hormone (PTH) fragment interferes significantly with intact PTH commercial assay measurements in uremic samples. *Clin Chem* **44**: 805–809.

Maeda, S. (1996). Cell-specific signal transduction of parathyroid hormone (PTH)-related protein through stably expressed recombinant PTH/PTHrP receptors in vascular smooth muscle cells. *Endocrinology* **137**: 3154–3162.

Mahon, M. J. (2002). Na$^+$/H$^+$ exchanger regulatory factor 2 directs parathyroid hormone 1 receptor signalling. *Nature* **417**: 858–861.

Martin, K. J. (1979). The peripheral metabolism of parathyroid hormone. *N Engl J Med* **301**: 1092–1098.

Mayer, G. P. (1979). Effects of plasma calcium concentration on the relative proportion of hormone and carboxyl fragments in parathyroid venous blood. *Endocrinology* **104**: 1778–1784.

McKee, R. L. (1988). The 7–34 fragment of human hypercalcemia factor is a partial agonist/antagonist for parathyroid hormone-stimulated cAMP production. *Endocrinology* **122**: 3008–3010.

Moallem, E. (1998). RNA-Protein binding and post-transcriptional regulation of parathyroid hormone gene expression by calcium and phosphate. *J Biol Chem* **273**: 5253–5259.

Monier-Faugere, M. C. (2001). Improved assessment of bone turnover by the PTH-(1–84)/large C-PTH fragments ratio in ESRD patients. *Kidney Int* **60**: 1460–1468.

Nguyen-Yamamoto, L. (2001). Synthetic carboxyl-terminal fragments of parathyroid hormone (PTH) decrease ionized calcium concentration in rats by acting on a receptor different from the PTH/PTH-related peptide receptor. *Endocrinology* **142**: 1386–1392.

Nguyen-Yamamoto, L. (2002). Origin of parathyroid hormone (PTH) fragments detected by intact-PTH assays. *Eur J Endocrinol* **147**: 123–131.

Okazaki, T. (1988). 5'-flanking region of the parathyroid hormone gene mediates negative regulation by 1,25-(OH)$_2$ vitamin D$_3$. *J Biol Chem* **263**: 2203–2208.

Okazaki, T. (1991). Negative regulatory elements in the human parathyroid hormone gene. *J Biol Chem* **266**: 21903–21910.

Orloff, J. J. (1992). Analysis of PTHrP binding and signal transduction mechanisms in benign and malignant squamous cells. *Am J Physiol* **262**: E599–E607.

Orloff, J. J. (1995). Further evidence for a novel receptor for amino-terminal parathyroid hormone-related protein on keratinocytes and squamous carcinoma cell lines. *Endocrinology* **136**: 3016–3023.

Rakel, A. (2005). Overproduction of an amino-terminal form of PTH distinct from human PTH(1–84) in a case of severe primary hyperparathyroidism: influence of medical treatment and surgery. *Clin Endocrinol* **62**: 721–727.

Rampe, D. (1991). Parathyroid hormone: an endogenous modulator of cardiac calcium channels. *Am J Physiol* **261**: H1945–H1950.

Riccardi, D. (1996). Localization of the extracellular Ca^{2+}-sensing receptor and PTH/PTHrP receptor in rat kidney. *Am J Physiol* **271**: F951–F956.

Rubin, D. A., and Jüppner, H. (1999). Zebrafish express the common parathyroid hormone/parathyroid hormone-related peptide receptor (PTH1R) and a novel receptor (PTH3R) that is preferentially activated by mammalian and fugufish parathyroid hormone-related peptide. *J Biol Chem* **274**: 28185–28190.

Rubin, M. R. (2006). An N-terminal molecular form of parathyroid hormone (PTH) distinct from hPTH(1 84) is overproduced in parathyroid carcinoma. *Clin Chem* **53**: 1470–1476.

Schlüter, K. D. (1995). Parathyroid hormone induces protein kinase C but not adenylate cyclase in adult cardiomyocytes and regulates cyclic AMP levels via protein kinase C-dependent phosphodiesterase activity. *Biochem J* **310**: 439–444.

Schoolwerth, A. C. (1988). Renal gluconeogenesis. *Min Electrolyte Metab* **14**: 347–361.

Seck, T. (2005). The Δe13 isoform of the calcitonin receptor forms a six-transmembrane domain receptor with dominant-negative effects on receptor surface expression and signaling. *Mol Endocrinol* **19**: 2132–2144.

Segre, G. V. (1981). Metabolism of parathyroid hormone by isolated rat Kupffer cells and hepatocytes. *J Clin Invest* **67**: 449–457.

Shao, J. S. (2003). Teriparatide (human parathyroid hormone (1–34)) inhibits osteogenic vascular calcification in diabetic low density lipoprotein receptor-deficient mice. *J Biol Chem* **278**: 50195–50202.

Shenolikar, S. (2002). Targeted disruption of the mouse NHERF-1 gene promotes internalization of proximal tubule sodium-phosphate cotransporter type IIa and renal phosphate wasting. *Proc Natl Acad Sci USA* **99**: 11470–11475.

Silver, J., and Levi, R. (2005). Regulation of PTH synthesis and secretion relevant to the management of secondary hyperparathyroidism in chronic kidney disease. *Kidney Int Suppl*, pp. S8–S12.

Silver, J. (1986). Regulation by vitamin D metabolites of parathyroid hormone gene transcription in vivo in the rat. *J Clin Invest* **78**: 1296–1301.

Slatopolsky, E. (2000). A novel mechanism for skeletal resistance in uremia. *Kidney Int* **58**: 753–761.

Sneddon, W. B. (2004). Ligand-selective dissociation of activation and internalization of the parathyroid hormone receptor. Conditional efficacy of PTH peptide fragments. *Endocrinology* **145**: 2815–2823.

Sneddon, W. B. (2003). Activation-independent parathyroid hormone receptor internalization is regulated by NHERF1 (EBP50). *J Biol Chem* **278**: 43787–43796.

Spiechowicz, U. (2004). Function of parathyroid glands in kidney transplant patients – diagnostic value of CAP and CIP. *Ann Transplant* **9**: 33–36.

Tsuchida, T. (2006). Serum levels of 1–84 and 7–84 parathyroid hormone in predialysis patients with chronic renal failure measured by the intact and bio-PTH assay. *Nephron Clin Pract* **102**: c108–c114.

Usdin, T. B. (1995). Identification and functional expression of a receptor selectively recognizing parathyroid hormone, the PTH2 receptor. *J Biol Chem* **270**: 15455–15458.

Vattikuti, R., and Towler, D. A. (2004). Osteogenic regulation of vascular calcification: an early perspective. *Am J Physiol Endocrinol Metab* **286**: E686–E696.

Vilardaga, J. P. (2001). Analysis of parathyroid hormone (PTH)/secretin receptor chimeras differentiates the role of functional domains in the PTH/PTH-related peptide (PTHrP) receptor on hormone binding and receptor activation. *Mol Endocrinol* **15**: 1186–1199.

Wade, J. B. (2003). Localization and interaction of NHERF isoforms in the renal proximal tubule of the mouse. *Am J Physiol Cell Physiol* **285**: C1494–C1503.

Wade, J. B. (2001). Differential renal distribution of NHERF isoforms and their colocalization with NHE3, ezrin, and ROMK. *Am J Physiol Cell Physiol* **280**: C192–C198.

Wang, M.-S., and Kurokawa, K. (1984). Renal gluconeogenesis: axial and internephron heterogeneity and the effect of parathyroid hormone. *Am J Physiol* **246**: F59–F66.

Wheeler, D. G. (2007). Role of NHERF-1 and the cytoskeleton in the regulation of the traffic and membrane dynamics of G-protein-coupled receptors. *J Biol Chem* **282**: 25076–25087.

Whitfield, J. F. (1992). Parathyroid hormone stimulates protein kinase C but not adenylate cyclase in mouse epidermal keratinocytes. *J Cell Physiol* **150**: 299–303.

Whitfield, J. F. (1971). The roles of calcium and cyclic AMP in the stimulatory action of parathyroid hormone on thymic lymphocyte proliferation. *J Cell Physiol* **78**: 355–368.

Wu, S. (1993). Effects of N-terminal, midregion, and C-terminal parathyroid hormone-related peptides on adenosine $3',5'$-monophosphate and cytoplasmic free calcium in rat aortic smooth muscle cells and UMR-106 osteoblast-like cells. *Endocrinology* **133**: 2437–2444.

Yang, T. X. (1997). Expression of PTHrP, PTH/PTHrP receptor, and Ca^{2+}-sensing receptor mRNAs along the rat nephron. *Am J Physiol* **272**: F751–F758.

Yoshida, T. (2001). Dietary phosphorus deprivation induces 25-hydroxyvitamin D3 1α-hydroxylase gene expression. *Endocrinology* **142**: 1720–1726.

Zierold, C. (2001). Parathyroid hormone regulates 25-hydroxyvitamin D_3-24-hydroxylase mRNA by altering its stability. *Proc Natl Acad Sci USA* **98**: 13572–13576.

Further reading

Usdin, T. B. (1997). Evidence for a parathyroid hormone-2 receptor selective ligand in the hypothalamus. *Endocrinology* **138**: 831–834.

CHAPTER 9

Endocrine Regulation of Phosphate Homeostasis

HARALD JÜPPNER[1] AND ANTHONY A. PORTALE[2]

[1]Endocrine Unit and Pediatric Nephrology Unit, Massachusetts General Hospital and Harvard Medical School, Boston, MA 02114, USA
[2]Department of Pediatrics, Division of Pediatric Nephrology, University of California San Francisco, San Francisco, CA 94143, USA

Contents

I. Introduction 105
II. Phosphate homeostasis 106
III. Renal phosphate transport 107
IV. Role of phosphate in the regulation of renal vitamin D metabolism 112
V. Mouse models with renal defects of phosphate transport 113
VI. Disorders with an abnormal regulation of renal phosphate transport 113
References 119

I. INTRODUCTION

Two important regulators of calcium homeostasis, vitamin D and parathyroid hormone (PTH), were discovered decades ago and numerous investigations exploring their actions *in vivo* and *in vitro* have been conducted (Jüppner et al., 2005; Potts and Gardella, 2007). We now have detailed knowledge of the mechanisms that lead to the formation of biologically active vitamin D metabolites and of their role in promoting intestinal calcium (and phosphate absorption) and in regulating bone metabolism (St-Arnaud and Demay, 2003; Demay et al., 2007; Holick, 2007). Likewise, a large body of literature is available exploring the biological roles of PTH, particularly regarding its actions on kidney and bone, which are mediated through the PTH/PTHrP receptor, a G protein-coupled receptor that uses at least two distinct second messenger systems cAMP/PKA and Ca^{++}/IP3/PKC (Gardella et al., 2002; Jüppner et al., 2005) (Figure 9.1).

In kidney, PTH actions in the proximal tubule are different from those in the distal tubule. In the proximal tubule, PTH decreases the expression of two sodium–phosphate co-transporters, NPT2a (*SLC34A1*) and NPT2c (*SLC34A3*), thereby decreasing the reabsorption of phosphate and thus enhancing renal phosphate excretion (Forster et al., 2006; Miyamoto et al., 2007; Tenenhouse, 2007); in this part of the nephron, PTH furthermore enhances expression of the renal 25-hydroxyvitamin D 1α-hydroxylase (*CYP27B1*), the enzyme that increases the production of the biologically active metabolite of vitamin D, 1,25-dihydroxyvitamin D (1,25 $(OH)_2D$) (see below). PTH thus enhances indirectly the intestinal absorption of calcium and phosphate. In the distal tubule, PTH increases the expression of TRPV5, a member of the transient receptor potential (TRP) cation channel subfamily V, thereby enhancing the reabsorption of calcium and thus minimizing renal calcium excretion (Mensenkamp et al., 2007; Schoeber et al., 2007).

For several decades, the parathyroid glands have served as a source to extract and purify PTH for study of its physiologic effects and mechanism of action. However, there was no readily available tissue source to extract and purify a hormonal principal that was thought to primarily regulate phosphate homeostasis. A potential tissue source of such a hormonal principal, termed 'phosphatonin', were rare tumors that often cause profound renal phosphate-wasting leading to the development of muscle weakness and osteomalacia (Econs and Drezner, 1994; Drezner, 2002). Surgical removal of such tumors resulted in normalization of phosphate homeostasis and reversal of muscle weakness and osteomalacia. Such tumors are usually small and difficult to find, which made it impossible to extract and purify quantities sufficient to obtain primary amino acid sequence information. Despite these difficulties, considerable advances have been made over the past several years in the identification and characterization of several proteins involved in the regulation of phosphate homeostasis, primarily as a result of the molecular definition of rare genetic disorders characterized by either hypo- or hyperphosphatemia. These efforts resulted in the identification of PHEX (*ph*osphate-regulating *e*nzyme with homologies to endopeptidases on the *X*-chromosome) (Consortium, 1995), fibroblast growth factor 23 (FGF23) (White et al., 2000), NPT2c (sodium-dependent phosphate co-transporter type IIc; NaPi-IIc) (Bergwitz et al., 2006; Lorenz-Depiereux et al., 2006a; Ichikawa et al., 2006), dentin

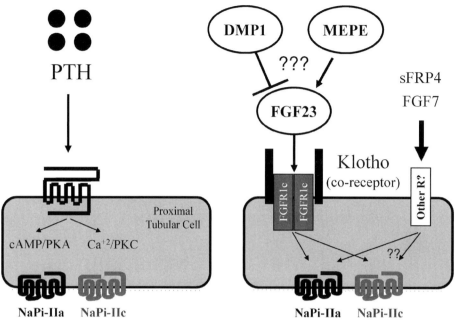

FIGURE 9.1 Regulation of phosphate handling in the proximal renal tubules. PTH-dependent actions are mediated through the PTH/PTHrP receptor and FGF23-dependent actions are mediated through FGFR1c with Klotho as a necessary co-receptor.

matrix protein 1 (DMP1) (Lorenz-Depiereux et al., 2006b; Feng et al., 2006) and GALNT3 (UDP-N-acetyl-alpha-D-galactosamine:polypeptide N-acetylgalactosaminyltransferase 3) (Topaz et al., 2004; Ichikawa et al., 2005) and helped define further the roles of the fibroblast growth factor receptor 1 (FGFR1) (White et al., 2005; Farrow et al., 2006) and Klotho, a transmembrane protein that, in addition to its role in phosphate regulation, affects sensitivity to insulin and appears to be involved in aging (Kuro-o et al., 1997; Ichikawa et al., 2007). Furthermore, additional proteins have been proposed to have roles in the regulation of phosphate homeostasis; these include soluble frizzled-related protein (sFRP4) (Jan de Beur et al., 2002; Berndt et al., 2005), matrix extracellular phosphoglycoprotein (MEPE) (Rowe et al., 2000, 2004; Martin et al., 2007) and fibroblast growth factor 7 (FGF7) (Carpenter et al., 2004). Molecular genetic studies, rather than traditional protein purification or expression cloning strategies thus have provided most of the important new insights into the regulation of phosphate homeostasis. These advances and the role of some of the identified proteins in more common disorders such as chronic kidney disease will be reviewed in this chapter.

II. PHOSPHATE HOMEOSTASIS

Ingested phosphate (Pi) is absorbed by the small intestine through a process involving the sodium-dependent phosphate co-transporter NPT2b (Hilfiker et al., 1998), deposited in bone, and filtered by the kidney where it is reabsorbed and excreted in amounts that are determined by the specific requirements of the organism. Thus, the kidney is a major determinant of Pi homeostasis by virtue of its ability to increase or decrease its Pi reabsorptive capacity to accommodate Pi need.

Pi accounts for approximately 0.6% of body weight at birth and about 1% of body weight, or 600 to 700 g, in the adult (Nordin, 1976). Approximately 85% of body Pi is in the skeleton and in teeth, approximately 15% is in soft tissue and the remainder ($\approx 0.3\%$) is in extracellular fluid. Pi is an important constituent of bone mineral and thus, in growing individuals, the balance of Pi must be positive to meet the needs of skeletal growth and consolidation; in the adult, Pi balance is zero. Pi deficiency results in osteomalacia in both children and adults.

In the plasma, Pi exists in two forms, an organic form consisting principally of phospholipids and phosphate esters and an inorganic form (Marshall, 1976). Of the total plasma Pi concentration of approximately 14 mg/dl (4.52 mM), about 4 mg/dl (1.29 mM) is inorganic Pi. Of this, about 10–15% is protein bound and the remainder, which is freely filtered by the renal glomerulus, exists principally as Pi complexed with sodium, calcium or magnesium or as the undissociated or 'free' Pi ions HPO_4^{2-} and $H_2PO_4^{-}$, which are present in plasma in a ratio of 4:1 at physiologic pH. In clinical settings, only the inorganic orthophosphate form of Pi is routinely measured.

In the adult in zero Pi balance, net intestinal absorption of Pi (dietary Pi minus fecal Pi) is approximately 60–65% of dietary intake. To satisfy the demands of rapid growth of bone and soft tissue, net intestinal absorption of Pi in infants is higher than in the adult and can exceed 90% of

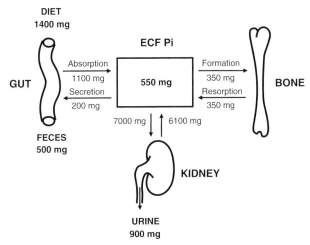

FIGURE 9.2 Phosphate fluxes between body pools in the normal human adult in zero phosphate balance. (Reprinted with permission from Portale, 1999.)

dietary intake (Rowe et al., 1984; Giles et al., 1987). Metabolic balance studies in healthy adult humans reveal that over the customary range of dietary Pi, net intestinal absorption is a linear function of intake (Wilkinson, 1976), with no indication of saturation. Thus, Pi deficiency results primarily from decreased availability of Pi rather than from changes in the intrinsic capacity of intestinal Pi transport. A small amount of Pi is secreted into the intestinal lumen in digestive fluids. Absorbed Pi enters the extracellular Pi pool which is in equilibrium with the bone and soft tissue Pi pools. Pi is filtered at the glomerulus and is reabsorbed to a large extent by the renal tubules. In subjects in zero Pi balance, the amount of Pi excreted by the kidney is equal to the net amount absorbed by the intestine and, in growing children, the amount excreted is less than the net amount absorbed due to deposition of Pi in bone. An overall schema of Pi homeostasis is depicted in Figure 9.2.

Plasma Pi concentration and overall Pi homeostasis are regulated primarily by renal tubular reabsorption of Pi. In response to a decrease in the extracellular Pi concentration, urine excretion of Pi decreases promptly. This response reflects both an acute decrease in the filtered Pi load and an adaptive increase in intrinsic proximal tubule Pi reabsorption induced by hypophosphatemia or decreased dietary Pi intake. Hypophosphatemia also is a potent stimulus for the renal synthesis of $1,25(OH)_2D$ (Tanaka and DeLuca, 1973; Baxter and DeLuca, 1976; Gray and Napoli, 1983), and the resulting increase in serum $1,25(OH)_2D$ concentration can stimulate intestinal absorption of Pi and calcium and their mobilization from bone. With the increase in plasma calcium concentration, PTH release is suppressed which leads to a further decrease in renal Pi excretion but an increase in calcium excretion. These homeostatic adjustments result in an increase in extracellular Pi concentration toward normal values, with little change in serum calcium concentration. Conversely, in response to an increase in plasma Pi concentration, production of $1,25(OH)_2D$ is decreased and release of PTH is stimulated. The effect of hyperphosphatemia on bone, kidney and intestine are opposite to those occurring with hypophosphatemia, the net result being a decrease in Pi concentration toward normal values.

In healthy subjects ingesting typical diets, the serum Pi concentration exhibits a circadian rhythm, characterized by a rapid decrease in early morning to a nadir shortly before noon, a subsequent increase to a plateau in late afternoon and a further small increase to a peak shortly after midnight (Markowitz et al., 1981; Portale et al., 1987). The amplitude of the rhythm (nadir to peak) is approximately 1.2 mg/dl (0.39 mM) or 30% of the 24-hour mean level. With restriction or supplementation of dietary Pi, a substantial decrease or increase, respectively, is observed in serum Pi concentrations during the late morning, afternoon and evening, but little or no change is observed in the morning fasting Pi concentration (Portale et al., 1987). To minimize the impact of changes in dietary Pi on the serum Pi concentration, specimens for analysis should be obtained in the morning fasting state. Specimens obtained in the afternoon are more likely to be affected by diet and thus may be more useful to monitor the effect of manipulation of dietary Pi on serum Pi concentrations, as in patients with renal insufficiency receiving phosphorus-binding agents to treat hyperphosphatemia.

Other factors can also affect the serum Pi concentration. Presumably as a result of Pi movement into cells, the serum Pi concentration can be decreased acutely by intravenous infusion of glucose or insulin, ingestion of carbohydrate rich meals, acute respiratory alkalosis or by infusion or endogenous release of epinephrine. Serum Pi concentration can be increased acutely by metabolic acidosis and by intravenous infusion of calcium (Suki and Rouse, 1996).

There are substantial effects of age on the fasting serum Pi concentration. In infants in the first 3 months of life, Pi levels are highest (4.8–7.4 mg/dl, mean 6.2 mg/dl [2 mM]) and decrease at age 1–2 years to 4.5–5.8 mg/dl (mean 5.0 mg/dl [1.6 mM]). In mid-childhood, values range from 3.5 to 5.5 mg/dl (mean 4.4 mg/dl [1.42 mM]) and decrease to adult values by late adolescence (Greenberg et al., 1960; Arnaud et al., 1973). In adult males, serum Pi is ≈3.5 mg/dl (1.13 mM) at age 20 years and decreases to ≈3.0 mg/dl (0.97 mM) at age 70 (Greenberg et al., 1960; Keating et al., 1969). In women, the values are similar to those of men until after the menopause, when they increase slightly from ≈3.4 mg/dl (1.1 mM) at age 50 years to 3.7 mg/dl (1.2 mM) at age 70.

III. RENAL PHOSPHATE TRANSPORT

A. Physiology and Tubular Localization

The proximal tubule is the major site of Pi reabsorption, with approximately 70% of the filtered load reclaimed in the proximal convoluted tubule and approximately 10% in the

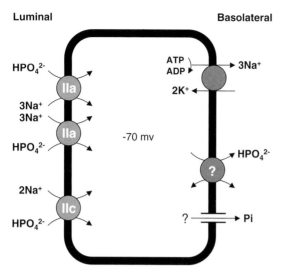

FIGURE 9.3 Location of identified Na$^+$-dependent Pi transporters in the proximal tubule cell. Available data indicate that most proximal tubule-Pi reabsorption occurs via type IIa and type IIc cotransporters (NPT2a and NPT2c), which are localized in the brush border membrane and are the major target for physiologic regulation of renal Pi reabsorption. (Modified from Murer et al., 1999.)

proximal straight tubule. In addition, a small but variable portion (<10%) of filtered Pi is reabsorbed in the distal segments of the nephron (Suki et al., 2000).

Clearance studies in humans and experimental animals have shown that when the filtered load of Pi is progressively increased, Pi reabsorption increases until a maximum tubular reabsorptive rate for Pi, or TmP, is reached, after which Pi excretion increases in proportion to its filtered load. The measurement of TmP varies among individuals and within the same individual, due in part to variation in glomerular filtration rate (GFR). Thus, the ratio, TmP/GFR, or the maximum tubular reabsorption of Pi per unit volume of GFR, is the most reliable quantitative estimate of the overall tubular Pi reabsorptive capacity and can be considered to reflect the quantity of sodium-dependent phosphate (Na/Pi) co-transporters available per unit of kidney mass (Bijvoet, 1976). The serum Pi concentration at which Pi reabsorption is maximal is called the 'theoretical renal Pi threshold'; this value is equal to the ratio, TmP/GFR, and closely approximates the normal fasting serum Pi concentration. Thus, the renal reabsorptive capacity for Pi is the principal determinant of the serum Pi concentration.

B. Cellular and Molecular Aspects

Transepithelial Pi transport is essentially unidirectional and involves uptake across the brush border membrane, translocation across the cell and efflux at the basolateral membrane (Figure 9.3). Pi uptake at the apical cell surface is the rate-limiting step in the overall Pi reabsorptive process and the major site of its regulation. It is mediated by Na$^+$-dependent Pi transporters that depend on the basolateral membrane-associated Na$^+$/K$^+$-dependent ATPase. Na/Pi co-transport is either electrogenic (NPT2a) or electroneutral (NPT2c) and sensitive to changes in pH, with 10- to 20-fold increases observed when the pH is raised from 6 to 8.5. Little is known about the translocation of Pi across the cell except that Pi anions rapidly equilibrate with intracellular inorganic and organic Pi pools. There are few data regarding the mechanisms involved in the efflux of Pi at the basolateral cell surface. It has been proposed that, in the proximal tubule, a Na$^+$-dependent electroneutral anion exchanger is at least partially responsible for Pi efflux (Barac-Nieto et al., 2002).

Three classes of Na/Pi co-transporters have been identified by expression and homology cloning. The type I Na/Pi co-transporter (NPT1, *SLC17A1*) is expressed predominately in brush border membranes (BBM) of proximal tubule cells (Custer et al., 1993). The NPT1 transporters are approximately 465 amino acids in length with seven to nine membrane spanning segments. NPT1 mediates high affinity Na/Pi co-transport, although its pH profile differs significantly from that of the pH-dependence of Na/Pi co-transport in isolated renal brush border membrane vesicles. NPT1 exhibits broad substrate specificity and mediates the transport of Cl$^-$ and organic anions as well as Pi. Conditions that physiologically regulate proximal tubule phosphate transport, such as dietary phosphorus or PTH, do not alter type I Na–Pi co-transporter protein or mRNA expression. Thus, the precise physiological role of NPT1 will thus require further study. The human gene encoding the type I Na–Pi co-transporter (*SLC17A1*) is located on chromosome 6p21.3-p23.

The type II family of Na/Pi co-transporters, whose cDNA shares only approximately 20% homology with that of NPT1 (Werner et al., 1991; Magagnin et al., 1993), is comprised of three highly homologous isoforms: type IIa (NPT2a, *SLC34A1*) and type IIc (NPT2c, *SLC34A3*) (Segawa et al., 2002; Ohkido et al., 2003), which are expressed exclusively in the brush border membrane of the renal proximal tubule, and type IIb (NPT2b, *SLC34A2*), which is expressed in several tissues, including small intestine and lung, but not in kidney, and it is thought to be responsible for intestinal absorption of Pi (Hilfiker et al., 1998). Human NPT2a and human NPT2c are comprised of 635 and 599 amino acids, respectively; both proteins are predicted to have eight membrane-spanning segments. The human genes encoding NPT2a and NPT2c are located on chromosomes 5q35 and 9q34 (Hartmann et al., 1996). NPT2a-mediated Na/Pi co-transport is electrogenic and involves the inward flux of three Na$^+$-ions and one Pi-anion (preferentially divalent) (Forster et al., 1999). The type IIb Na/Pi co-transporter is also electrogenic, whereas the Npt2c isoform mediates the electroneutral transport of two Na$^+$-ions with one divalent Pi-anion.

In the mouse, Npt2a and Npt2c are detected exclusively in the brush border membrane of proximal tubular cells. At the mRNA level, NPT2c is approximately one order of magnitude less abundant than NPT2a. The relative abundance of

Npt2c protein is significantly higher in kidneys of 22-day-old rats than in those of 60-day-old rats, suggesting that Npt2c has a particularly important role during early postnatal development (Segawa et al., 2002). However, several different homozygous and compound heterozygous mutations in *SLC34A3*, the gene encoding NPT2c, have been found in patients affected by hereditary hypophosphatemic rickets (HHRH) (Bergwitz et al., 2006; Lorenz-Depiereux et al., 2006a; Ichikawa et al., 2006) (see below), indicating that this co-transporter has a more prominent role than initially thought. Hybrid depletion studies suggested that Npt2c accounts for approximately 30% of Na/Pi co-transport in kidneys of Pi-deprived adult mice (Ohkido et al., 2003).

Type III Na/Pi co-transporters are cell surface retroviral receptors [gibbon ape leukemia virus (Glvr-1, Pit-1, *SLC20A1*) and murine amphotropic virus (Ram-1, Pit-2, *SLC20A2*)] that mediate high affinity, electrogenic Na^+-dependent Pi transport when expressed in oocytes and in mammalian cells (Kavanaugh et al., 1994; Collins et al., 2004). Glvr-1 and Ram-1 show no sequence similarity to NPT1 or NPT2a. Both Glvr-1 and Ram-1 proteins are widely expressed in mammalian tissue; in the kidney they are localized at the basolateral membrane where they serve as 'housekeeping' Na/Pi co-transporters to maintain cellular Pi homeostasis.

C. Regulation of NPT2a and NPT2c

Regulation of renal Pi reabsorption has been the subject of intense investigation. Dietary Pi intake, PTH and FGF23 are thought to be the major regulators of renal Pi handling, although many hormonal and non-hormonal factors are also known to regulate this process. Both NPT2a and NPT2c are targets of regulation. PTH, FGF23 and dietary Pi regulate renal Pi reabsorption primarily by inducing alterations in the abundance of NPT2a protein in the brush border membrane of proximal tubular cells, accomplished either by insertion of existing transporters into the membrane or retrieval of transporters from the membrane with subsequent lysosomal degradation. PTH, FGF23 and dietary Pi also regulate the abundance of NPT2c protein in the brush border membrane (Miyamoto et al., 2007), however, in contrast to NPT2a, NPT2c does not appear to undergo lysosomal degradation, but instead may get recycled and re-inserted into the apical membrane. The post-transcriptional mechanisms underlying membrane trafficking of NPT2a and NPT2c proteins are complex and involve interaction of the transporters with various scaffolding and signalling proteins such as NHERF1, NHERF2, PDZK1 (NHERF3), PDZK2 (NHERF4) and Shank2E (Forster et al., 2006; Villa-Bellosta et al., 2007).

1. Regulation by Dietary Phosphate

Dietary Pi intake is a key determinant of renal Pi handling. Pi deprivation elicits an increase in Pi reabsorption and brush border membrane Na/Pi co-transport. Acute exposure to low dietary Pi induces an increase in transport V_{max} as well as in NPT2a protein but not NPT2a mRNA (Pfister et al., 1998a). These findings are consistent with data demonstrating that low dietary Pi has no effect on NPT2a promoter activity (Hilfiker et al., 1998). The acute increase in renal tubule Pi transport induced by Pi deprivation is mediated by microtubule-dependent recruitment of existing NPT2a protein to the apical membrane (Lotscher et al., 1997). In contrast, exposure to high dietary Pi leads to the internalization of cell surface NPT2a protein into the endosomal compartment by a microtubule-independent mechanism (Lotscher et al., 1997). Internalized NPT2a protein is then delivered to the lysosome, by a microtubule-dependent process, for degradation (Pfister et al., 1998a).

A Pi response element (PRE) was identified in the mouse *NPT2a* promoter by DNA footprint analysis (Kido et al., 1999). The PRE was shown to bind a mouse transcription factor, TFE3 and the expression of TFE3 is increased in the kidney in response to Pi deprivation (Kido et al., 1999). On the basis of these results, it was suggested that TFE3 participates in the transcriptional regulation of the NPT2a gene by dietary Pi. NPT2c is also regulated by dietary Pi. Dietary Pi restriction induces an increase in NPT2c immunoreactive protein in the apical membrane of proximal tubule cells, whereas feeding a high Pi diet induces a decrease in NPT2c protein (Segawa et al., 2005; Miyamoto et al., 2007). Internalization of the type IIc transporter was slightly delayed relative to that of the type IIa transporter after acute exposure to high dietary Pi; internalized NPT2c is, however, not degraded in the lysosomes, in contrast to NPT2a protein, which is degraded.

2. Parathyroid Hormone (PTH)

PTH is a major hormonal regulator of renal Pi reabsorption. PTH acts directly on proximal tubular cells and inhibits brush border membrane sodium-dependent phosphate co-transport through mechanisms that involve internalization of cell surface NPT2a protein (Kempson et al., 1995) and its subsequent lysosomal degradation (Pfister et al., 1998b); in contrast, recent evidence suggests that NPT2c can be recycled and re-inserted into the brush border membrane (Blaine et al., 2007).

The PTH-mediated signaling pathways involved in these processes include cAMP/PKA- and Ca^{2+}/PLC/PKC-dependent mechanisms. PTH binding to the PTH/PTHrP receptors on the basolateral membrane activates protein kinase A (PKA) and/or protein kinase C (PKC) signaling pathways, whereas its binding to apical receptors appears to activate only PKC (Traebert et al., 2000). The extracellular signal-regulated kinase/mitogen-activated protein kinase (ERK/MAPK) pathway also participates in PTH-induced signaling (Lederer et al., 2000), and recent studies have shown that PKA- and PKC-dependent signaling pathways converge on the ERK/MAPK pathway to internalize NPT2a protein

(Bacic et al., 2003a). Although the downstream targets for ERK/MAPK-mediated phosphorylation remain unknown, changes in the phosphorylation state of NPT2a are not associated with its PTH-induced internalization (Jankowski et al., 2001). Rather, it has been postulated that the phosphorylation of proteins that associate with NPT2a may determine its regulation.

AKAP79, an A kinase anchoring protein (Khundmiri et al., 2003), and RAP, a receptor-associated protein (Bacic et al., 2003b), have been shown to participate in the PTH-mediated retrieval of NPT2a from the plasma membrane of proximal tubular cells. In opossum kidney (OK) cells, AKAP79 associates with NPT2a and the regulatory and catalytic subunits of PKA, and this process is necessary for PKA-dependent inhibition of sodium-dependent phosphate co-transport (Khundmiri et al., 2003). In RAP-deficient mice, PTH-induced internalization of NPT2a is significantly delayed whereas regulation by dietary Pi is not affected (Bacic et al., 2003b). In contrast, NHERF-1, which associates with NPT2a and MAP-1 in mouse kidney (Pribanic et al., 2003), does not appear to play a role in PTH-induced internalization of NPT2a (Lederer et al., 2003).

Much less is known about the PTH-dependent regulation of NPT2c. It disappears from the surface of the BBM in response to PTH at a much slower rate than does NPT2a and NPT2c does not seem to undergo lysosomal degradation (Miyamoto et al., 2007; Tenenhouse, 2007; Segawa et al., 2007a; Virkki et al., 2007); recent preliminary evidence suggests that NPT2c may be recycled and re-inserted into the BBM (Blaine et al., 2007). In rats on a low phosphate diet, NPT2a and NPT2c undergo different regulation by PTH. PTH(1–34) failed to decrease NPT2a expression in brush border membrane vesicles from rats on a low phosphate diet, consistent with the blunted phosphaturic effect of PTH observed in hypophosphatemic humans or rodents (Marcinkowski et al., 1997). In contrast, PTH was able to efficiently reduce the expression of NPT2c (Segawa et al., 2007a). Furthermore, whereas NPT2a is expressed in segments S1 through S3 of the proximal tubule, NPT2c is expressed only in the S1 segment.

3. Fibroblast Growth Factor 23 (FGF-23)

Fibroblast growth factor 23 (FGF-23) was isolated through several independent approaches, including the homology-based search of genomic databases (Yamashita et al., 2000) and the sequence analysis of a large number of cDNAs from tumors responsible for oncogenic osteomalacia (Shimada et al., 2001). However, its identification through a positional cloning approach to determine the cause of autosomal dominant hypophosphatemic rickets (ADHR) (White et al., 2000) provided first compelling evidence for involvement of FGF-23 in the regulation phosphate homeostasis (see below).

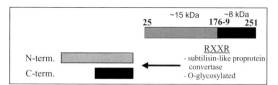

FIGURE 9.4 Schematic depiction of the mature FGF-23 protein and the sizes of the two fragments derived from cleavage by a subtilisin-like proprotein convertase (SPC); threonine at position 178 within the RXXR motif undergoes O-linked glycosylation.

The human *FGF-23* gene consists of 3 exons that span 10 kb of genomic sequence and encodes a 251 amino acid precursor protein comprising a hydrophobic amino acid sequence (residues 1 through 24), which likely serves as a leader sequence (Figure 9.4). Unlike most other fibroblast growth factors, FGF-23 thus appears to be efficiently secreted into the circulation. FGF-23 is most closely related to fibroblast growth factors 21, 19 and 15, but also shares limited homology with other fibroblast growth factors (White et al., 2000; Kato et al., 2006). Particularly important structural features of FGF-23 include an RXXR site for cleavage by subtilisin-like proprotein convertases (residues 176 through 179; note the mutations that cause ADHR affect amino acid residues 176 and 179) and the threonine residue at position 178, which appears to undergo O-linked glycosylation (Yamazaki et al., 2002; White et al., 2006). When expressed in different cell lines, cleavage at the RXXR site generates protein fragments of about 32 kDa and 12 kDa.

Across different species, FGF-23 is most conserved within the amino-terminal portion of the molecule, while the portion C-terminal of the RXXR site shows considerable amino acid sequence variation. This sequence variation is relevant for the species-specific detection of FGF-23 in serum or plasma. One of three currently available immunometric assays uses antibodies directed against conserved amino acid sequences and thus detects human and rodent FGF-23 equally well (Yamazaki et al., 2002), whereas the other two assays use a detection antibody that is directed against an epitope within the C-terminal portion of human FGF23, which shows virtually no homology with the rodent homolog (Jonsson et al., 2003; Imel et al., 2006).

The mRNA encoding human or rodent FGF-23 could not be detected by Northern blot analysis in normal tissues. Low mRNA levels were, however, identified by reverse transcriptase (RT)-PCR in heart, liver, thymus, small intestine, brain and osteoblasts (White et al., 2000; Shimada et al., 2001), and by *in situ* hybridization in bone cells (Riminucci et al., 2003). Through the use of reporter genes fused to the endogenous mouse Fgf-23 promoter, FGF-23 was furthermore shown to be expressed in bone cells, most prominently in osteocytes (Sitara et al., 2004; Liu et al., 2006).

FGF-23 binds, albeit with relatively low affinity, to most of the different splice variants of the known FGF receptors, when these are expressed in a murine bone marrow-derived pro-B cell line (BaF3 cells) (Yu et al., 2005). However, in the presence of Klotho, a membrane bound protein with β-glucuronidase activity, FGF-23 can bind to FGFR1(IIIc) with high affinity (Urakawa et al., 2006; Kurosu et al., 2006). Consistent with an important role of Klotho in mediating the actions of FGF-23, Klotho-null mice and *Fgf23*-null mice have significantly overlapping phenotypes (Kuro-o, 2006; Razzaque and Lanske, 2007; Liu et al., 2007). Furthermore, the Klotho-null animals show markedly elevated serum levels of biologically active FGF-23 (Urakawa et al., 2006). Thus, FGF-23 acts through known FGFRs but only in those tissues in which Klotho is also expressed, including kidney (Kato et al., 2000; Li et al., 2004).

Administration of FGF-23 in mice elicits a decrease in serum Pi that is associated with increased renal Pi excretion (Shimada et al., 2001), a reduction in brush border membrane Na/Pi co-transport (Saito et al., 2003) and decreased renal Npt2a expression (Bai et al., 2003). In addition, rats receiving intrahepatic injection of FGF-23 cDNA develop hypophosphatemia and a significant decrease in both brush border membrane Na/Pi co-transport and NPT2c protein abundance (Segawa et al., 2003). Furthermore, mice transplanted with cell lines stably expressing FGF-23 develop hypophosphatemia due to increased urinary phosphate excretion, and these animals show increased alkaline phosphatase activity and low serum concentration of $1,25(OH)_2D$, a marked increase of unmineralized osteoid and significant widening of growth plates leading to deformities of weight-bearing bones (Shimada et al., 2001, 2002). Similarly, transgenic mice expressing FGF-23 under the control of different promoters develop hypophosphatemia, renal phosphate-wasting and abnormal bone development (Bai et al., 2004; Shimada et al., 2004a; Larsson et al., 2004). It remains uncertain, however, whether the severe secondary hyperparathyroidism observed when FGF-23 is overexpressed in transgenic animals contributes significantly to the changes in mineral ion homeostasis (Bai et al., 2004; Larsson et al., 2004).

Findings opposite to those in transgenic animals were observed in mice homozygous for ablation of the Fgf-23 gene (*Fgf23*-null). These animals develop hyperphosphatemia and increased serum $1,25(OH)_2D$ concentrations and they die prematurely, partly due to renal failure secondary to calcifications of glomerular capillaries (Shimada et al., 2004b; Sitara et al., 2004; Liu et al., 2006). *Fgf23*-null animals furthermore showed reduced bone turnover, diminished osteoblast and osteoclast number and activity and increased osteoid volume. The latter mineralization defect was particularly unexpected, since the increase in serum phosphorous concentration, combined with a normal or slightly increased serum calcium level, provided no explanation for the profound mineralization defect (Shimada et al., 2004b; Liu et al., 2006).

4. OTHER HORMONAL REGULATORS

Other hormones also contribute to the regulation of proximal tubular Pi transport. Growth hormone, insulin-like growth factor-I, insulin, $1,25(OH)_2D$ and thyroid hormone all stimulate Pi reabsorption, whereas PTH-related peptide, calcitonin, atrial natriuretic factor, epidermal growth factor, transforming growth factor-β and glucocorticoids inhibit Pi reclamation (for review see Murer et al., 2003). The increase in brush border membrane Na/Pi co-transport induced by thyroid hormone is associated with an increase in NPT2a mRNA (Alcalde et al., 1999), whereas both hypercalcemia (Levi et al., 1995) and epidermal growth factor (Arar et al., 1999) decrease NPT2a mRNA abundance. Neither thyroid hormone nor hypercalcemia has an effect on NPT2a promoter-reporter gene expression (see Hilfiker et al., 1998), suggesting that transcriptional mechanisms are not involved.

Administration of $1,25(OH)_2D$ to vitamin D-deficient rats elicits an increase in brush border membrane Na/Pi co-transport that is accompanied by an increase in renal NPT2a mRNA and protein abundance (Taketani et al., 1998). While these results are consistent with direct effects of $1,25(OH)_2D$ on NPT2a-mediated renal Na/Pi co-transport, the effects may result from a $1,25(OH)_2D$-dependent decrease in PTH levels. However, the finding that $1,25(OH)_2D$ increased the activity of a NPT2a promoter-luciferase reporter gene construct suggests a direct effect of the vitamin D hormone on NPT2a gene transcription (Taketani et al., 1998). In vitamin D receptor (VDR)-null mice, in which serum levels of PTH and $1,25(OH)_2D$ are greatly increased, the abundance of NPT2a protein in renal BBM vesicles was significantly decreased, whereas the abundance of NPT2c protein was unaffected (Segawa et al., 2004). This finding suggests that $1,25(OH)_2D$ has little direct effect on NPT2c expression.

Other hormones that regulate renal Pi handling include stanniocalcin, 5-hydroxytryptamin (5-HT) and secreted frizzled-related protein 4 (sFRP4). Stanniocalcin is a peptide hormone that counteracts hypercalcemia and stimulates Pi reabsorption in bony fish, and is also produced by humans. Infusion of stanniocalcin in rats stimulates renal Pi reabsorption and brush border membrane Na/Pi co-transport (Wagner et al., 1997), suggesting a role for stanniocalcin in the maintenance of Pi homeostasis in mammals as well as fish. 5-HT is synthesized in the kidney, and locally generated 5-HT was shown to interfere with PTH-mediated inhibition of renal sodium-dependent phosphate co-transport (Hafdi et al., 1996). These findings suggested that 5-HT is a paracrine modulator of renal Pi transport. Fibroblast growth factor 7 (FGF-7), produced by a hypophosphatemic osteomalacia-causing tumor, was recently shown to inhibit phosphate uptake in OK cells, thus suggesting that FGF-7 can also cause phosphaturia and may be responsible for osteomalacia in those patients

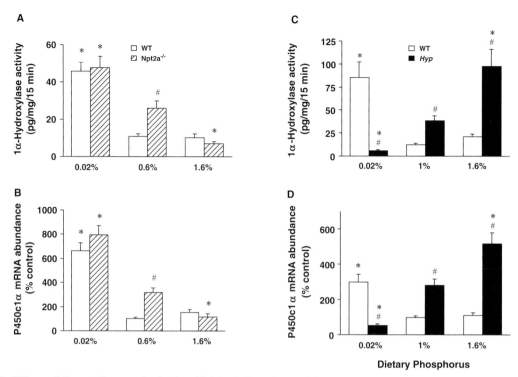

FIGURE 9.5 Effects of dietary Pi on renal mitochondrial 1α-hydroxylase activity (A, C) and renal P450c1α mRNA abundance (B, D) in wild-type mice (clear bars) and either $Npt2^{-/-}$ (A, B) or Hyp (C, D) mice (filled bars). Mice were fed either the low (0.02%), control (0.6–1%), or high (1.6%) Pi diet for 4–5 days. Renal mitochondria were prepared and incubated with 25-hydroxyvitamin D_3 to determine 1α-hydroxylase activity and renal total RNA was prepared to estimate the abundance of P450C1α mRNA, relative to β-actin mRNA, by ribonuclease protection assay. Bars depict mean ± SEM. Compared with the 0.6% or 1% Pi diet, within each species, $P < 0.05$. # Compared with wild-type animals, within each diet group, $P < 0.05$. (A, B modified from Tenenhouse et al., 2001; C, D from Azam et al., 2003.)

who have no elevation in circulating FGF23 levels (Carpenter et al., 2004).

IV. ROLE OF PHOSPHATE IN THE REGULATION OF RENAL VITAMIN D METABOLISM

Dietary Pi intake and serum Pi concentration are critically important determinants of the renal metabolism of 1,25 $(OH)_2D$. Tanaka and DeLuca (1973) first demonstrated that rats maintained on a low Pi diet synthesized predominantly 1,25$(OH)_2D$, and this effect was not diminished by parathyroidectomy. They further observed an inverse relationship between the synthesis of 1,25$(OH)_2D$ and the serum Pi concentration. Numerous studies in animals have demonstrated that hypophosphatemia induced by dietary Pi restriction induces an increase in the production rate of 1,25$(OH)_2D$ measured *in vivo* (Hughes et al., 1975; Gray, 1981) and in 1α-hydroxylase activity measured *in vitro* (Baxter and DeLuca, 1976; Gray and Napoli, 1983; Lobaugh and Drezner, 1983). Studies in healthy human subjects are consistent with those in experimental animals. Thus, restriction of dietary Pi induces an increase (Gray et al., 1977; Insogna et al., 1983; Maierhofer et al., 1986; Portale et al., 1986, 1987) and supplementation of dietary Pi (Portale et al., 1986, 1987, 1989) a decrease in the serum concentration of 1,25 $(OH)_2D$ and in its *in vivo* production rate (Portale et al., 1986). It is of interest that the stimulation of 1,25$(OH)_2D$ production by Pi depletion is abolished by hypophysectomy and is restored by administration of growth hormone or insulin-like growth factor I (IGF-1) (Gray, 1981; Gray et al., 1983; Gray and Garthwaite, 1985; Halloran and Spencer, 1988; Grieff et al., 1996), indicating that an intact growth hormone/IGF-1 axis is required for the effect; however, the molecular mechanisms are unknown.

Studies in intact mice revealed that restriction of dietary Pi induces a rapid, sustained, ≈6–8-fold increase in renal mitochondrial 1α-hydroxylase activity (Figure 9.5a, c), which can be attributed to an ≈7-fold increase in Vmax of the enzyme, with no measurable change in its Km (Zhang et al., 2002). The increase in 1α-hydroxylase activity can be attributed to an increase in the renal abundance of 1α-hydroxylase (P450C1α CYP27B1) mRNA (Figure 9.5b, d) and protein (Yoshida et al., 2001; Zhang et al., 2002). Immunohistochemical analysis demonstrated that the increase in 1α-hydroxylase expression occurs exclusively in the proximal renal tubule and is due, at least in part, to increased transcription of *CYP27B1* (Zhang et al., 2002). Degradation of 1,25$(OH)_2D$ to calcitroic acid and cholacalcioic acid via the 24-oxidation pathway is initiated by the enzyme, 25-hydroxyvitamin D-24-hydroxylase (24-hydroxylase, CYP24). Pi restriction induces a significant decrease in renal 24-hydroxylase activity (Wu et al., 1997) and mRNA abundance (Wu et al., 1996; Zhang et al., 2002).

V. MOUSE MODELS WITH RENAL DEFECTS OF PHOSPHATE TRANSPORT

A. Effect of Npt2a and NPT2c Gene Disruption

The critical role of NPT2a in the maintenance of Pi homeostasis was clearly demonstrated in mice in which the *Npt2a* gene (lower case refers to the mouse gene) was knocked out by targeted mutagenesis (Beck et al., 1998). Mice that are null for *Npt2a* exhibit decreased renal Pi reabsorption, an ≈80% loss of brush border membrane Na/Pi, hypophosphatemia, an appropriate adaptive increase in the renal synthesis (Tenenhouse et al., 2001) and serum concentration of 1,25(OH)$_2$D (Beck et al., 1998; Tenenhouse et al., 2001) and associated hypercalcemia, hypercalciuria and hypoparathyroidism, and an age-dependent skeletal phenotype (Beck et al., 1998; Gupta et al., 2001). Npt2c protein abundance was shown to be increased significantly in the BBM of Npt2a null mice (Tenenhouse et al., 2003) and Npt2c likely accounts for residual sodium-dependent co-transport in the mutant mice. Dietary Pi intake and PTH were without effect on renal BBM sodium-dependent co-transport in *Npt2a*-null mice (Hoag et al., 1999; Zhao and Tenenhouse, 2000), demonstrating that NPT2a is a major regulator of renal Pi handling.

Preliminary studies have shown that *Npt2c*-null mice have, at different ages, only a small increase in blood ionized calcium and some increase in urinary calcium excretion, but no hypophosphatemia and no increase in urinary phosphate excretion (Segawa et al., 2007b), suggesting that this co-transporter may be of limited functional significance in rodents. However, FGF-23 levels are suppressed in Npt2c-null mice and the combined ablation of Npt2a and Npt2c has a more severe phenotype than the ablation of Npt2a alone, suggesting that the Npt2c is likely to have more significant role than suggested by the Npt2c-null animals.

B. The Hyp Mouse, a Model for Human XLH

In contrast to the response to Pi depletion in wild-type and *Npt2a*-null mice, the hypophosphatemia that results from renal Pi wasting in *Hyp* mice, the murine homolog of human x-linked hypophosphatemia (XLH), is associated with inappropriately normal or low serum 1,25(OH)$_2$D concentrations (Meyer et al., 1980; Tenenhouse and Jones, 1990). Renal 1α-hydroxylase activity also is inappropriately low for the degree of hypophosphatemia (Lobaugh and Drezner, 1983), and Pi restriction induces a paradoxical decrease in serum concentration of 1,25(OH)$_2$D (Meyer et al., 1980; Tenenhouse and Jones, 1990) and in renal 1α-hydroxylase activity and mRNA abundance (Figure 9.5c, d) (Yamaoka et al., 1986; Azam et al., 2003). The disordered regulation of vitamin D metabolism can be attributed at least in part to an increase in circulating FGF-23, which has been shown to suppress renal 1α-hydroxylase expression (Bai et al., 2004; Saito et al., 2004; Kolek et al., 2005; Liu et al., 2007). Consistent with these findings in mice, most patients with XLH have increased serum levels of FGF-23 (Yamazaki et al., 2002; Jonsson et al., 2003; Weber et al., 2003), thus providing a potential mechanism to explain why their 1,25(OH)$_2$D concentrations are inappropriately low given the attendant hypophosphatemia (Scriver et al., 1978; Delvin and Glorieux, 1981). The regulation of 24-hydroxylase also is abnormal in *Hyp* mice, with renal enzyme activity being twofold higher in *Hyp* than in wild-type littermates, which may also be related to increased FGF-23 levels; paradoxically 24-hydroxylase activity increases further with Pi restriction (Tenenhouse and Jones, 1990). In *Hyp* mice, Pi restriction induced a significant increase in renal 24-hydroxylase mRNA abundance (Roy et al., 1994; Roy and Tenenhouse, 1996; Azam et al., 2003) in contrast to the decrease observed in wild-type mice (Azam et al., 2003). These findings suggest that the decrease in the serum concentration of 1,25(OH)$_2$D in phosphorus-restricted *Hyp* mice can be attributed to an increase in its renal catabolism as well as to a reduction in its renal synthesis (Azam et al., 2003). Nevertheless, current data demonstrate that regulation of both 1α-hydroxylase and 24-hydroxylase gene expression by Pi is disordered in *Hyp* mice at the level of renal 1α-hydroxylase enzyme activity and mRNA expression, and suggest that loss of *Phex* function and the increase in FGF-23 levels give rise to disordered transcriptional regulation of both genes.

VI. DISORDERS WITH AN ABNORMAL REGULATION OF RENAL PHOSPHATE TRANSPORT

A. Genetic Hypophosphatemic Disorders with Increased Urinary Phosphate Excretion, Normal or Low 1,25(OH)$_2$D and Increased FGF-23 Levels

1. AUTOSOMAL DOMINANT HYPOPHOSPHATEMIC RICKETS (ADHR)

In 1971, Bianchine et al. first described a small family in which the affected individuals presented with hypophosphatemia, inappropriately normal or diminished 1,25(OH)$_2$D concentrations and rickets or osteomalacia. In this family, the disease was transmitted from an affected male to his son suggesting an autosomal dominant trait for this hypophosphatemic disorder. Such a mode of transmission effectively excluded XLH, the most frequently encountered renal phosphate-wasting disorder, which leads to hypophosphatemia and rickets in affected children and osteomalacia in affected adults (Holm et al., 2003; Portale and Miller, 2003; White et al., 2006). The apparent autosomal dominant mode of inheritance in the family by Bianchine suggested that the disorder was caused by a genetic mutation that is different from that in patients affected by XLH. Subsequently, Econs and McEnery (1997) described a large kindred with multiple affected individuals, which confirmed the autosomal dominant mode of inheritance and led to the term autosomal dominant hypophosphatemic rickets (ADHR) (Table 9.1). In addition to the differences in inheritance pattern, ADHR frequently shows, in contrast to XLH, incomplete penetrance and a variable age of disease onset. In fact, while most patients with XLH present as toddlers or during early childhood, some patients develop ADHR only as adolescents or during adulthood, and thus present with bone pain, muscle weakness and insufficiency fractures, but not with deformities of the lower extremities (Econs and McEnery, 1997; Econs et al., 1997; White et al., 2006; Imel et al., 2007).

To identify the molecular genetic basis of ADHR, Econs and his colleagues pursued a positional cloning strategy and their genome-wide linkage scan led to the identification of a locus on chromosome 12p13.3 (Econs et al., 1997) and, subsequently, to the identification of a heterozygous missense mutation in *FGF-23*, in which an arginine residue at position 176 is changed to glutamine (R176Q) (White et al., 2000). The same heterozygous R176Q mutation was identified also in one unrelated kindred with several affected family members, and in two other additional heterozygous missense mutations (R179W and R179Q) were identified in two other unrelated ADHR families (White et al., 2000; Kruse et al., 2001; Negri et al., 2004). The clustering of these heterozygous missense mutations that alter arginine residues within the RXXR motif suggested that these mutations result in a gain of function of the mutant FGFl-23.

2. X-linked Hypophosphatemia (XLH)

XLH is the most frequent, inherited phosphate-wasting disorder (frequency: 1 in 20 000). As with ADHR, XLH is characterized by hypophosphatemia due to renal phosphate wasting, inappropriately normal or low circulating 1,25 $(OH)_2D$ concentrations and rickets or osteomalacia (see Table 9.1). This disorder is caused by inactivating mutations in PHEX, a gene located on Xp22.1 (Consortium, 1995; Holm et al., 1997). PHEX, which is expressed in kidney, bone and other tissues, shows significant peptide sequence homology to the M13 family of zinc metallopeptidases, which include neutral endopeptidase neprilysin (NEP), endothelin converting enzyme 1 (ECE-1) and 2 (ECE-2) and the Kell antigen. All of these are type II integral membrane glycoproteins that have endopeptidase activity and consist of a short N-terminal cytoplasmic domain, a single transmembrane hydrophobic region and a large extracellular domain. The substrate(s) for PHEX remains to be established. An initial report had suggested that FGF-23 is a substrate, but these findings were not confirmed by others (Bowe et al., 2001; Benet-Pages et al., 2004). PHEX appears to bind matrix extracellular phosphoglycoprotein (MEPE), thereby preventing its cleavage and thus the generation of the ASARM fragment, which may act as an inhibitor of bone matrix mineralization and of dentin matrix protein 1 (DMP1) formation (Rowe et al., 2000, 2004; Martin et al., 2007). DMP1 is produced primarily by osteocytes and its ablation in mice or homozygous inactivating mutations in humans leads to increased urinary phosphate excretion and rickets/osteomalacia. FGF-23 levels are elevated in Dmp1-null animals, and they are elevated or inappropriately normal in patients with DMP1 mutations (see below). These findings are similar to those in patients with XLH (Yamazaki et al., 2002; Jonsson et al., 2003; Weber et al., 2003) and in *Hyp* mice (Aono et al., 2003; Liu et al., 2005). Genetic ablation of FGF-23 in male *Hyp* mice, i.e. animals that are null for FGF-23 and Phex, led to blood phosphate levels that are indistinguishable from those in mice lacking FGF-23 alone (Sitara et al., 2004; Liu et al., 2006), indicating that FGF-23 resides genetically down-stream of PHEX.

3. Autosomal Recessive Hypophosphatemia (ARHP)

Hypophosphatemic rickets in consanguineous kindreds has been reported, suggesting an autosomal recessive form of hypophosphatemia (ARHP) (Perry and Stamp, 1978; Scriver et al., 1981; Bastepe et al., 1999). Affected individuals in such kindreds showed clinical findings that were similar to those observed in patients affected by ADHR or XLH, including rickets, skeletal deformities and dental defects, and affected individuals develop osteosclerotic bone lesions and enthesopathies later in life (see Table 9.1). Hypophosphatemia, which results from increased renal phosphate excretion, is accompanied by normal or low $1,25(OH)_2D$ levels and high alkaline phosphatase activity. Patients affected by ARHP have FGF-23 levels that are either elevated or inappropriately normal for the low level of serum phosphorous (Lorenz-Depiereux et al., 2006b; Feng et al., 2006).

ARHP is caused by homozygous mutations in the gene encoding DMP1. DMP1 belongs to the SIBLING protein family, which includes osteopontin, matrix extracellular phosphoglycoprotein, bone sialoprotein II and dentin sialoprotein; the genes encoding these proteins are all clustered on chromosome 4q21. DMP1 is a bone and teeth specific protein (George et al., 1993), which is involved in the regulation of transcription in undifferentiated osteoblasts (Narayanan et al., 2003; George et al., 2007). During the early phase of osteoblast maturation, DMP1 undergoes phosphorylation and subsequently is exported into the extracellular matrix where it regulates the nucleation of

TABLE 9.1 Biochemical findings in several inherited hypo- and hyperphosphatemic disorders and underlying genetic defects

	FGF-23	TRP	1,25(OH)$_2$D	PTH	Serum Calcium	Urinary Calcium	Mutant gene
Hypophosphatemic disorders							
XLH	Increased/inappropriately normal	Low	Low/inappropriately normal	Normal/increased	Normal	Normal	PHEX
ADHR	Increased/inappropriately normal	Low	Low/inappropriately normal	Normal/increased	Normal	Normal	FGF23
ARHP	Increased/inappropriately normal	Low	Low/inappropriately normal	Normal/increased	Normal	Normal	DMP1
HHRH	Low/normal	Low	High	Low	Normal/increased	High	NPT2c
Hyperphosphatemic disorders							
Different forms of tumoral calcinosis	Intact: low C-terminal: very high	High	Normal-high	Low	Normal/increased	Increased	FGF23 or GALNT3 (Polypeptide GalNAc transferase)
	Extremely high (intact and C-terminal)	High	Normal-high	Elevated	Normal/increased	Increased	Klotho
Pseudohypoparathyroidism type Ia (PHP-Ia) or Ib (PHP-Ib)	Normal-increased	Elevated	Low-normal	High	Low	Low	PHP-Ia: GNAS exons encoding Gsα PHP-Ib: microdeletions within or up-stream of GNAS

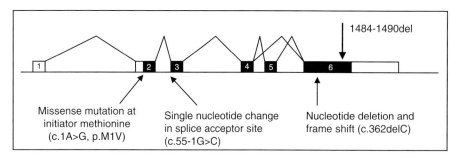

FIGURE 9.6 Structure of the DMP1 gene and location of homozygous mutations that were identified as the cause of an autosomal recessive form of hypophosphatemia (ARHP) (Lorenz-Depiereux et al., 2006a; Feng et al., 2006).

hydroxyapatite. DMP1 normally undergoes post-translational modifications that yield a 94 kDa mature protein, which is rapidly cleaved into a 37 kDa and a 57 kDa fragment. Preliminary studies suggested that the 57 kDa DMP1 fragment alone is sufficient to reverse the phenotype of *Dmp1*-null animals and to suppress FGF-23 secretion (Lu et al., 2007a) (see below).

Of the several different DMP1 mutations identified thus far, one mutation alters the translation initiation codon (M1V), two mutations are located in different intron-exon boundaries and three are frame-shift mutations within exon 6 (Figure 9.6). These mutations appear to be inactivating, suggesting that the loss of DMP1 results in hypophosphatemia. Accordingly, *Dmp1*-null mice show hypophosphatemia, osteomalacia and severe defects in dentine, bone and cartilage (Ye et al., 2005; Feng et al., 2008). Furthermore, FGF-23 levels in serum and expression in osteocytes are greatly increased in these animals (Feng et al., 2006). These findings suggest that DMP1 plays a role in the regulation of FGF-23 expression, thereby indirectly regulating phosphate homeostasis. Given the established importance of DMP1 in osteoblast function, loss of DMP1 actions in osteoblasts and extracellular matrix may also contribute to the phenotype of patients with ARHP. Consistent with this hypothesis, a high calcium/phosphate diet capable of rescuing osteomalacia in VDR null mice does not seem to prevent bone and dentine mineralization defect in Dmp1 null mice (Lu et al., 2007b).

4. Osteoglophonic Dysplasia (OGD)

Osteoglophonic dysplasia (OGD) is an autosomal dominant disorder characterized by skeletal abnormalities and frequently by changes in mineral ion homeostasis. Affected patients present with craniosynostosis, prominent superorbital ridges, mild facial hypoplasia, rhizomelic dwarfism and non-ossifying bone lesions, and most of them seem to develop hypophosphatemia due to renal phosphate-wasting associated with inappropriately normal $1,25(OH)_2D$ levels (Beighton, 1989). White et al. recently identified several different heterozygous missense mutations in the FGFR1 gene that are all located within or close to the receptor's membrane-spanning domain. These mutations affect amino acid residues that are highly conserved across species and seem to lead to constitutive receptor activation (White et al., 2005; Farrow et al., 2007). Heterozygous activating FGFR1 mutations are also the cause of patients with Pfeiffer syndrome (Muenke et al., 1994) and a patient with the skeletal findings of Jackson–Weiss syndrome (Roscioli et al., 2000). However, the mutations leading to these disorders are located between the second and third putative Ig domains of FGFR1, a location similar to that of mutations in FGFR2 and in FGFR3 that cause the Crouzon syndrome and hypo- or achondrodysplasia, respectively; note that patients affected by these other disorders do not develop hypophosphatemia. Surprisingly, some patients with OGD were reported to have elevated FGF-23 levels (White et al., 2005; Farrow et al., 2007). This could indicate that the skeletal lesions develop because the constitutive activation of the FGFR1 leads to an upregulation of FGF-23 secretion in the metaphyseal growth plate and, consequently, enhanced renal phosphate excretion. Consistent with this view, patients with more radiographic evidence for lesions appear to develop more profound hypophosphatemia.

5. Linear Nevus Sebaceous Syndrome (LNSS)/ Epidermal Nevus Syndrome (ENS)

Linear nevus sebaceous syndrome (LNSS), also known as epidermal nevus syndrome (ENS) or Schimmelpenning–Feuerstein–Mims syndrome, is a rare, highly variable congenital sporadic disorder that can feature, besides papillomatous epidermal hyperplasia and excess sebaceous glands, brain and eye abnormalities, focal or generalized skeletal disease and, in rare cases, hypophosphatemia leading to the development of rickets (Hoffman et al., 2005; Heike et al., 2005). Two recent reports have described increased FGF-23 concentrations in two patients, implying that this phosphaturic hormone contributes to renal phosphate wasting. While the skin lesions appeared to be the source of FGF-23 in one patient (Hoffman et al., 2005), the bone lesions were thought to secret this hormone in the other patient (Heike et al., 2005).

6. Fibrous Dysplasia (FD)

Fibrous dysplasia (FD) refers to a disorder characterized by fibrous skeletal lesions and associated localized mineralization defects. Some patients with FD develop systemic hypophosphatemia due to renal phosphate-wasting, which may lead to the development of generalized rickets/osteomalacia (Weinstein et al., 2001). When these skeletal findings occur in conjunction with abnormal skin pigmentation, premature sexual development and hyperthyroidism, the disease is referred to as McCune–Albright syndrome (MAS). FD and MAS are caused by heterozygous activating, post-zygotic mutations in exon 8 of *GNAS*, the gene encoding the alpha-subunit of the stimulatory G protein (Gsα) (Weinstein et al., 1991; Schwindinger et al., 1992). R201H or R201C are the most frequently encountered mutations, but other less frequent mutations have been described, including R201S, R201G and R201L (Weinstein, 2006). Recent studies have shown that FD can be associated with increased FGF-23 serum levels, which are inversely correlated with serum phosphate and 1,25(OH)$_2$D levels (Riminucci et al., 2003; Kobayashi et al., 2006). The mRNA encoding FGF-23 as well as the FGF-23 protein were found in the dysplastic lesions, including osteoblasts and fibrous cells (Riminucci et al., 2003), thus raising the possibility that FGF-23 plays an important role in the pathogenesis of increased renal phosphate excretion that is often seen in patients affected by FD/MAS. Treatment with bisphosphates was shown to reduce serum FGF-23 levels which, in turn, led to a reduction in renal phosphate excretion (Yamamoto et al., 2005). Although the mechanism(s) leading to the reduction in circulating FGF-23 remains to be explored, it appears plausible that the production of FGF-23 by osteoblasts and/or osteocytes is at least partially regulated by cAMP-dependent mechanisms.

B. Genetic Hypophosphatemic Disorders with Increased Urinary Phosphate Excretion and Increased 1,25(OH)$_2$D and Low FGF-23 Levels

1. NaPi-IIa Mutations Associated with Increased Urinary Phosphate Excretion, Nephrolithiasis and Osteoporosis

Two different heterozygous mutations (A48P and V147M) in *SLC34A1*, the gene encoding the sodium-dependent phosphate co-transporter NaPi-IIa, have been reported in patients with urolithiasis or osteoporosis and persistent idiopathic hypophosphatemia due to decreased tubular phosphate reabsorption (Prié et al., 2002). When expressed in *Xenopus laevis* oocytes, the mutant NaPi-IIa showed impaired function and dominant negative properties, when co-expressed with the wild-type co-transporter. However, these *in vitro* findings were not confirmed in another study using oocytes and OK cells, raising the concern that the identified NaPi-IIa mutations alone cannot explain the findings in the described patients (Virkki et al., 2003). On the other hand, additional heterozygous NaPi-IIa variations, either in-frame deletions or missense changes, have recently been identified upon analysis of a large cohort of hypercalciuric stone-forming kindreds; however, these genetic variations do not seem to cause functional abnormalities (Lapointe et al., 2006).

2. NaPi-IIc Mutations are the Cause of Hereditary Hypophosphatemic Rickets with Hypercalciuria (HHRH)

The homozygous ablation of *Npt2a* in mice (*Npt2a$^{-/-}$*) results, as expected, in increased urinary phosphate excretion leading to hypophosphatemia, an appropriate increase in the serum levels of 1,25(OH)$_2$D leading to hypercalcemia, hypercalciuria, decreased serum parathyroid hormone levels and increased serum alkaline phosphatase activity (Beck et al., 1998). These biochemical features are typically observed in patients with hereditary hypophosphatemic rickets with hypercalciuria (HHRH), a presumably autosomal recessive disorder affecting renal tubular phosphate reabsorption (Tieder et al., 1985, 1987). Most HHRH patients have rickets, short stature and increased renal phosphate clearance (TmP/GFR is usually 2 to 4 standard deviations below the age-related normal range). Serum concentration of 1,25(OH)$_2$D is increased leading to an increase in the gastrointestinal absorption of calcium and phosphorus and hypercalciuria with normal serum calcium concentrations. Parathyroid function is suppressed, and urinary cyclic AMP excretion within normal limits. Long-term phosphate supplementation as the sole therapy leads to reversal of the clinical and biochemical abnormalities, although TmP/GFR remains decreased (Tieder et al., 1985, 1987). Unlike HHRH patients, *Npt2a*-null mice do not have rickets or osteomalacia. Instead, they have poorly developed trabecular bone and retarded secondary ossification and, in older animals, there is a dramatic reversal and eventual overcompensation of the skeletal phenotype. Consistent with these phenotypic differences, mutations in *SLC34A1*, the gene encoding the sodium–phosphate co-transporter NPT2a were excluded in the affected members of several unrelated kindreds, including the one in whom this syndrome was first described (Jones et al., 2001; Tieder et al., 1985, 1987). Recent studies have led to the identification of homozygous or compound heterozygous mutations in *SLC34A3*, the gene encoding the sodium–phosphate co-transporter NPT-2c, in patients affected by HHRH. These findings indicate that in humans NPT-2c has a more important role in phosphate homeostasis than previously thought.

C. Genetic Hyperphosphatemic Disorders

1. Tumoral Calcinosis (TC) with/without Hyperphosphatemia

At least three variants of tumoral calcinosis (TC) with hyperphosphatemia have been described; an autosomal dominant form (Lyles et al., 1985) and two distinct autosomal recessive forms (Topaz et al., 2004). Patients with the autosomal dominant form of this disorder usually have elevated serum 1,25

(OH)₂D levels, but other findings typical of tumoral calcinosis may not always be present. The teeth are hypoplastic with short, bulbous roots and almost complete obliteration of pulp cavities, but have fully developed enamel of normal color. The molecular defect in this form of TC is not known.

The autosomal recessive forms of tumoral calcinosis with hyperphosphatemia are severe, sometimes fatal disorders characterized by inappropriately diminished renal phosphate excretion and often massive calcium deposits in the skin and subcutaneous tissues. Recently, Topaz et al. mapped the gene causing one form of the disease to 2q24-q31 and revealed homozygous or compound heterozygous mutations in *GALNT3* (Topaz et al., 2004), the gene that encodes a glycosyltransferase responsible for initiating mucin-type O-glycosylation. While the serum concentrations of intact FGF-23 are normal or low, the levels of carboxyl-terminal FGF-23 can be greatly increased (Topaz et al., 2004; Ichikawa et al., 2005). These findings suggest that defective post-translational modifications of FGF-23 are most likely responsible for the abnormal regulation of phosphate homeostasis. Another form of TC is caused by homozygous FGF-23 mutations, and patients affected by this form of the disease also showed dramatically elevated circulating concentration of carboxyl-terminal FGF-23, while the concentration of the intact hormone is within normal limits (Benet-Pages et al., 2004; Larsson et al., 2005; Araya et al., 2005).

The combination of hyperostosis with hyperphosphatemia was first described in 1970 (Melhem et al., 1970). Affected patients can have recurrent painful swelling of long bones, features of tumoral calcinosis and increased serum phosphate levels, yet normal renal function and usually normal serum calcium, 1,25(OH)₂D and PTH concentrations (Narchi, 1997). Most cases appear to be sporadic, but consanguineous parents were described for some patients, implying that the disease can be recessive; the underlying molecular defect is not yet known. Recently, *GALNT3* mutations were also identified in the recessive form of this disease, indicating that one of the two forms of tumoral calcinosis and hyperostosis with hyperphosphatemia are allelic variants (Frishberg et al., 2004). As in patients with the autosomal recessive forms of tumoral calcinosis, the carboxyl-terminal FGF-23 concentration is significantly increased (Frishberg et al., 2004).

A patient with tumoral calcinosis has recently been identified as having a homozygous inactivating Klotho mutation, who showed hyperphosphatemia due to insufficient urinary phosphate excretion, despite having greatly increased serum FGF-23 levels (Ichikawa et al., 2007). This patient further showed increased 1,25(OH)₂D levels, hypercalcemia and hyperparathyroidism. As outlined above, the Klotho gene was first discovered when determining the cause of the premature aging phenotype of a transgenic mouse strain (Kuro-o et al., 1997; Kuro-o, 2006), and Klotho acts as a necessary co-receptor for FGF-23 (Urakawa et al., 2006; Kurosu et al., 2006). Furthermore, the homozygous ablation of *Klotho* in the mouse results in a phenotype similar to that of the FGF23-null mouse, with both animals lacking the phosphaturic actions of FGF-23 and thus being hyperphosphatemic (Kuro-o, 2006; Razzaque and Lanske, 2007). The finding of a loss-of-function Klotho mutation in a patient with tumoral calcinosis underscores the importance of Klotho in the FGF-23-dependent regulation of phosphate and vitamin D homeostasis in humans.

Another form of tumoral calcinosis without hyperphosphatemia appears to be caused by mutations in SAMD9, a gene of unknown function, which encodes a TNF-alpha responsive protein (Chefetz et al., 2008).

2. Hypoparathyroidism and Pseudohypoparathyroidism

Biochemical features of hypoparathyroidism include hypocalcemia and hyperphosphatemia and low or inappropriately normal levels of PTH. Hypoparathyroidism can occur as part of complex congenital defects, as part of pluriglandular autoimmune disorders or mitochondrial disorders, or as a solitary endocrinopathy. Familial occurrences of isolated hypoparathyroidism have been described with autosomal dominant, autosomal recessive and X-linked recessive modes of inheritances (reviewed in Bastepe and Jüppner, 2005; Jüppner and Thakker, 2008). Patients affected by type I pseudohypoparathyroidism (PHP) display similar laboratory abnormalities as those observed in patients with hypoparathyroidism, namely hypocalcemia and hyperphosphatemia. However, PTH levels are increased, as evidence for resistance towards this hormone in the proximal tubules; in contrast, the actions of PTH in the distal tubules remain intact (reviewed in Jan de Beur and Levine, 2001; Weinstein et al., 2004; Bastepe and Jüppner, 2005). As a result, renal phosphate excretion is impaired, while the PTH-dependent reabsorption of calcium in the distal tubules is enhanced thus minimizing the urinary losses of calcium. In contrast to patients with hypoparathyroidism in whom the application of biologically active, exogenous PTH leads to normal or even enhanced excretion of urinary cyclic AMP and phosphate, there is a subnormal or no response in type I PHP (Jan de Beur and Levine, 2001; Weinstein et al., 2004; Bastepe and Jüppner, 2005). Patients with PHP-Ia show, besides resistance toward PTH and often other peptide hormones, developmental defects that are referred to as Albright's hereditary osteodystrophy.

PHP-Ia is caused by heterozygous inactivating mutations within exons 1 through 13 of *GNAS* located on chromosome 20q13.3, which encode the stimulatory G protein (Gsα). These mutations were shown to lead to an ≈50% reduction in Gsα activity/protein in readily accessible tissues, like erythrocytes and fibroblasts, and explain, at least partially, the resistance towards PTH and other hormones that mediate their actions through G-protein-coupled receptors

(Jan de Beur and Levine, 2001; Weinstein et al., 2004; Bastepe and Jüppner, 2005). However, a similar reduction in Gsα activity/protein is also found in patients with pseudohypoparathyroidism (pPHP), who show the same physical appearance as individuals with PHP-Ia, but lack endocrine abnormalities. Subsequent studies showed that hormonal resistance in PHP-Ia is parentally imprinted, i.e. the disease occurs only if the defective gene is inherited from a female affected by either PHP-Ia or pPHP; pPHP occurs only if the defective gene is inherited from a male affected by either form of the two disorders (Davies and Hughes, 1993; Wilson et al., 1994). Mutations in the gene encoding Gsα could not be detected in most patients with PHP-Ib, a disorder in which affected individuals show PTH-resistant hypocalcemia and hyperphosphatemia, but lack the developmental defects observed in PHP-Ia; a significant number of PHP-Ib patients also show increased TSH levels (Levine, 1996; Bastepe et al., 2001a; Jan de Beur and Levine, 2001; Weber et al., 2003; Weinstein et al., 2004; Bastepe and Jüppner, 2005). As with PHP-Ia, the autosomal dominant form of PHP-Ib is a paternally imprinted disorder; it appears to be caused in most familial cases by a 3-kb deletion located between two 391-bp repeats about 200 kb upstream of *GNAS* (Bastepe et al., 2003; Bastepe and Jüppner, 2005). Affected individuals always inherit the microdeletion from their mother and they always show an associated loss of *GNAS* methylation that is restricted exon A/B of the maternal allele (Liu et al., 2000; Bastepe et al., 2001b). Less frequently, autosomal dominant PHP-Ib is caused by maternally inherited microdeletions within *GNAS*, which are associated with broader methylation changes (Bastepe et al., 2005). Although similar or indistinguishable broad methylation changes are also observed in most patients with a sporadic form of PHP-Ib, no deletions or point mutations have yet been identified in these individuals (Linglart et al., 2007). Taken together, these findings suggest that several different deletions upstream or within the *GNAS* locus lead to indistinguishable clinical and laboratory findings.

Surprisingly, although FGF-23 is likely to be the most important phosphate-regulating hormone, its levels are normal or only mildly elevated in the few patients with hypoparathyroidism or pseudohypoparathyroidism, who were investigated so far. FGF-23 is thus unable to normalize serum phosphate levels in patients with these disorders.

D. Acquired Hypophosphatemic Disorders with Elevated FGF-23 Levels

1. Tumor-Induced Osteomalacia (TIO)

TIO (also referred to as oncogenic osteomalacia, OOM) is a rare disorder characterized by hypophosphatemia due to inappropriately increased urinary phosphate excretion, a low circulating $1,25(OH)_2D$ concentration and osteomalacia that develop in previously healthy individuals (Drezner, 2002). Thus, there are considerable similarities between TIO and ADHR, XLH and ARHP. TIO is caused by usually small, often difficult to locate tumors, most frequently hemangiopericytomas, some of which were reported to express PHEX (Lipman et al., 1998; John et al., 2001). The clinical and biochemical abnormalities resolve rapidly after the removal of the tumor. FGF-23 mRNA and protein were found to be markedly overexpressed in tumors that cause TIO, suggesting that FGF-23 promotes the excessive excretion of phosphate (Shimada et al., 2001; White et al., 2001). Recent studies have shown that FGF23-producing tumors can be identified by using a combination of different imaging techniques (Hesse et al., 2007a, 2007b) and by using selective venous catherization of vessels draining the tumor to measure FGF-23 (Takeuchi et al., 2004; van Boekel et al., 2008).

2. Chronic Kidney Disease (CKD)

FGF-23 levels are significantly increased in patients with stage V chronic kidney disease (CKD) and also in earlier stages of CKD (Jonsson et al., 2003; Larsson et al., 2003; Imanishi et al., 2004; Gutierrez et al., 2005). Patients with end-stage renal disease (ESRD) on hemodialysis or peritoneal dialysis have the highest FGF-23 levels, followed by patients with CKD stage 3 or 4 and individuals with a functioning renal graft. Post-renal transplantation, some patients had normal FGF-23 levels, but the mean for this group was slightly above the normal range. Similarly, in rodents treated with adenine to induce renal failure, FGF-23 concentrations were increased, which normalized upon treatment with an oral phosphate binder (Nagano et al., 2006). These elevations in FGF-23 may contribute the impaired production of $1,25(OH)_2D$ in affected patients and thus the development of secondary hyperparathyroidism. It is furthermore conceivable that elevations in serum FGF-23 levels may provide first evidence for impaired renal phosphate excretion in the earlier stages of CKD and these measurements might be helpful in deciding when to start therapy with oral phosphate binders. It remains to be determined whether extremely elevated FGF-23 levels lead to ectopic activation of FGF receptors and thereby contribute to morbidity in chronic renal failure.

References

Alcalde, A. I., Sarasa, M., Raldua, D. et al. (1999). Role of thyroid hormone in regulation of renal phosphate transport in young and aged rats. *Endocrinology* **140**: 1544–1551.

Aono, Y., Shimada, T., Yamazaki, Y. et al. (2003). The neutralization of FGF-23 ameliorates hypophosphatemia and rickets in Hyp mice. Meeting of the American Society for Bone and Mineral Research, Minneapolis, Minnesota, p. 1056.

Arar, M., Zajicek, H. K., Elshihabi, I., and Levi, M. (1999). Epidermal growth factor inhibits Na-Pi cotransport in weaned and suckling rats. *Am J Physiol* **276**: F72–F78.

Araya, K., Fukumoto, S. et al. (2005). A novel mutation in fibroblast growth factor 23 gene as a cause of tumoral calcinosis. *J Clin Endocrinol Metab* **90:** 5523–5527.

Arnaud, S. B., Goldsmith, R. S., Stickler, G. B., McCall, J. T., and Arnaud, C. D. (1973). Serum parathyroid hormone and blood minerals: interrelationships in normal children. *Pediatr Res* **7:** 485–493.

Azam, N., Zhang, M. Y., Wang, X., Tenenhouse, H. S., and Portale, A. A. (2003). Disordered regulation of renal 25-hydroxyvitamin D-1alpha-hydroxylase gene expression by phosphorus in X-linked hypophosphatemic (hyp) mice. *Endocrinology* **144:** 3463–3468.

Bacic, D., Schulz, N., Biber, J., Kaissling, B., Murer, H., and Wagner, C. A. (2003a). Involvement of the MAPK-kinase pathway in the PTH-mediated regulation of the proximal tubule type IIa Na+/Pi cotransporter in mouse kidney. *Pflugers Arch* **446:** 52–60.

Bacic, D., Capuano, P., Gisler, S. M. et al. (2003b). Impaired PTH-induced endocytotic down-regulation of the renal type IIa Na+/Pi-cotransporter in RAP-deficient mice with reduced megalin expression. *Pflugers Arch* **446:** 475–484.

Bai, X. Y., Miao, D., Goltzman, D., and Karaplis, A. C. (2003). The autosomal dominant hypophosphatemic rickets R176Q mutation in FGF23 resists proteolytic cleavage and enhances in vivo biological potency. *J Biol Chem* **278:** 9843–9849.

Bai, X., Miao, D., Li, J., Goltzman, D., and Karaplis, A. (2004). Transgenic mice overexpressing human fibroblast growth factor 23(R176Q) delineate a putative role for parathyroid hormone in renal phosphate wasting disorders. *Endocrinology* **145:** 5269–5279.

Barac-Nieto, M., Alfred, M., and Spitzer, A. (2002). Basolateral phosphate transport in renal proximal-tubule-like OK cells. *Exp Biol Med (Maywood)* **227:** 626–631.

Bastepe, M., and Jüppner, H. (2005). The GNAS locus and pseudohypoparathyroidism. *Hormone Res* **63:** 65–74.

Bastepe, M., Shlossberg, H., Murdock, H., Jüppner, H., and Rittmaster, R. (1999). A Lebanese family with osteosclerosis and hypophosphatemia. *J Bone Miner Res* **14:** S558.

Bastepe, M., Lane, A. H., and Jüppner, H. (2001a). Paternal uniparental isodisomy of chromosome 20q (patUPD20q) – and the resulting changes in *GNAS1* methylation – as a plausible cause of pseudohypoparathyroidism. *Am J Hum Genet* **68:** 1283–1289.

Bastepe, M., Pincus, J. E., Sugimoto, T. et al. (2001b). Positional dissociation between the genetic mutation responsible for pseudohypoparathyroidism type Ib and the associated methylation defect at exon A/B: evidence for a long-range regulatory element within the imprinted *GNAS1* locus. *Hum Mol Genet* **10:** 1231–1241.

Bastepe, M., Fröhlich, L. F., Hendy, G. N. et al. (2003). Autosomal dominant pseudohypoparathyroidism type Ib is associated with a heterozygous microdeletion that likely disrupts a putative imprinting control element of GNAS. *J Clin Invest* **112:** 1255–1263.

Bastepe, M., Fröhlich, L. F., Linglart, A. et al. (2005). Deletion of the NESP55 differentially methylated region causes loss of maternal GNAS imprints and pseudohypoparathyroidism type Ib. *Nat Genet* **37:** 25–27.

Baxter, L. A., and DeLuca, H. F. (1976). Stimulation of 25-hydroxyvitamin D3-1alpha-hydroxylase by phosphate depletion. *J Biol Chem* **251:** 3158–3161.

Beck, L., Karaplis, A. C., Amizuka, N., Hewson, A. S., Ozawa, H., and Tenenhouse, H. S. (1998). Targeted inactivation of *Ntp2* in mice leads to severe renal phosphate wasting, hypercalciuria, and skeletal abnormalities. *Proc Natl Acad Sci USA* **95:** 5372–5377.

Beighton, P. (1989). Osteoglophonic dysplasia. *J Med Genet* **26:** 572–576.

Benet-Pages, A., Lorenz-Depiereux, B., Zischka, H., White, K., Econs, M., and Strom, T. (2004). FGF23 is processed by proprotein convertases but not by PHEX. *Bone* **35:** 455–462.

Bergwitz, C., Roslin, N., Tieder, M. et al. (2006). SLC34A3 mutations in patients with hereditary hypophosphatemic rickets with hypercalciuria (HHRH) predict a key role for the sodium-phosphate co-transporter NaPi-IIc in maintaining phosphate homeostasis and skeletal function. *Am J Human Genet* **78:** 179–192.

Berndt, T., Bielesz, B., Craig, T. et al. (2005). Secreted frizzled-related protein-4 reduces sodium-phosphate co-transporter abundance and activity in proximal tubule cells. *Pflugers Arch* Epub ahead of print.

Bianchine, J. W., Stambler, A. A., and Harrison, H. E. (1971). Familial hypophosphatemic rickets showing autosomal dominant inheritance. *Birth Defects Orig Artic Ser* **7:** 287–295.

Bijvoet, O. (1976). The importance of the kidney in phosphate homeostasis. In: (Avioli, L., Bordier, P., Fleisch, H., Massry, S., eds) *Phosphate metabolism, kidney and bone,* Armour-Montagu, Paris.

Blaine, J., Breusegem, S., Giral, H., Barry, N., and Levi, M. (2007). Differential regulation of renal NaPiIIa and NaPi-IIc trafficking by PTH. Renal Week, ASN, San Francisco, pp. SA-FC103.

Bowe, A., Finnegan, R., Jan de Beur, S. et al. (2001). FGF-23 inhibits renal tubular phosphate transport and is a PHEX substrate. *Biochem Biophys Res Commun* **284:** 977–981.

Carpenter, T., Ellis, B., Insogna, K., Philbrick, W., Sterpka, J., and Shimkets, R. (2004) FGF7 – an inhibitor of phosphate transport derived from oncogenic osteomalacia-causing tumors. *J Clin Endocrinol Metab* **90:** 1012–1020.

Chefetz, I., Amitai, D. B., Browning, S. et al. (2008). Normophosphatemic familial tumoral calcinosis is caused by deleterious mutations in SAMD9, encoding a TNF-alpha responsive protein. *J Invest Dermatol* **128:** 1423–1429.

Collins, J. F., Bai, L., and Ghishan, F. K. (2004). The SLC20 family of proteins: dual functions as sodium-phosphate cotransporters and viral receptors. *Pflugers Arch* **447:** 647–652.

Consortium. (1995). A gene (PEX) with homologies to endopeptidases is mutated in patients with X-linked hypophosphatemic rickets. The HYP Consortium. *Nat Genet* **11:** 130–136.

Custer, M., Meier, F., Schlatter, E. et al. (1993). Localization of NaPi-1, a Na-Pi cotransporter, in rabbit kidney proximal tubules. I. mRNA localization by reverse transcription/polymerase chain reaction. *Pflugers Arch* **424:** 203–209.

Davies, A. J., and Hughes, H. E. (1993). Imprinting in Albright's hereditary osteodystrophy. *J Med Genet* **30:** 101–103.

Delvin, E. E., and Glorieux, F. H. (1981). Serum 1,25-dihydroxyvitamin D concentration in hypophosphatemic vitamin D-resistant rickets. *Calcif Tissue Int* **33:** 173–175.

Demay, M. B., Sabbagh, Y., and Carpenter, T. O. (2007). Calcium and vitamin D: what is known about the effects on growing bone. *Pediatrics* **119**(Suppl 2): S141–S144.

Drezner, M. K. (2002). Phosphorus homeostasis and related disorders. In: *Principles in bone biology,* 2nd edn, (Bilezikian, J.P., Raisz, L.G., Rodan, G.A., eds) Academic Press, New York, pp. 321–338.

Econs, M., and Drezner, M. (1994). Tumor-induced osteomalacia – unveiling a new hormone. *N Engl J Med* **330**: 1679–1681.

Econs, M., and McEnery, P. (1997). Autosomal dominant hypophosphatemic rickets/osteomalacia: clinical characterization of a novel renal phosphate-wasting disorder. *J Clin Endocrinol Metab* **82**: 674–681.

Econs, M., McEnery, P., Lennon, F., and Speer, M. (1997). Autosomal dominant hypophosphatemic rickets is linked to chromosome 12p13. *J Clin Invest* **100**: 2653–2657.

Farrow, E., Davis, S., Mooney, S. et al. (2006). Extended mutational analyses of FGFR1 in osteoglophonic dysplasia. *Am J Med Genet* **140**: 537–539.

Feng, J. Q., Ward, L. M., Liu, S. et al. (2006). Loss of DMP1 causes rickets and osteomalacia and identifies a role for osteocytes in mineral metabolism. *Nat Genet* **38**: 1310–1315.

Feng, J. Q., Scott, G., Guo, D. et al. (2008). Generation of a conditional null allele for Dmp1 in mouse. *Genesis* **46**: 87–91.

Forster, I. C., Loo, D. D., and Eskandari, S. (1999). Stoichiometry and Na+ binding cooperativity of rat and flounder renal type II Na+-Pi cotransporters. *Am J Physiol* **276**: F644–F649.

Forster, I. C., Hernando, N., Biber, J., and Murer, H. (2006). Proximal tubular handling of phosphate: a molecular perspective. *Kidney Int* **70**: 1548–1559.

Frishberg, Y., Araya, K., Rinat, C. et al. (2004). *Hyperostosis-hyperphosphatemia syndrome caused by mutations in GALNT3 and associated with augmented processing of FGF-23*. American Society of Nephrology, Philadelphia, pp. F-P0937.

Gardella, T. J., Jüppner, H., Bringhurst, F. R., and Potts, J. T. (2002). Receptors for parathyroid hormone (PTH) and PTH-related peptide. In: (Bilezikian, J., Raisz, L., Rodan, G., eds) *Principles of bone biology,* Academic Press, San Diego, pp. 389–405.

George, A., Sabsay, B., Simonian, P. A., and Veis, A. (1993). Characterization of a novel dentin matrix acidic phosphoprotein. Implications for induction of biomineralization. *J Biol Chem* **268**: 12624–12630.

George, A., Ramachandran, A., Albazzaz, M., and Ravindran, S. (2007). DMP1 – a key regulator in mineralized matrix formation. *J Musculoskelet Neuronal Interact* **7**: 308.

Giles, M. M., Fenton, M. H., Shaw, B. et al. (1987). Sequential calcium and phosphorus balance studies in preterm infants. *J Pediatr* **110**: 591–598.

Gray, R. W. (1981). Control of plasma 1, 25-(OH)2-vitamin D concentrations by calcium and phosphorus in the rat: effects of hypophysectomy. *Calcif Tissue Int* **33**: 485–488.

Gray, R. W., and Garthwaite, T. L. (1985). Activation of renal 1,25-dihydroxyvitamin D3 synthesis by phosphate deprivation: evidence for a role for growth hormone. *Endocrinology* **116**: 189–193.

Gray, R. W., and Napoli, J. L. (1983). Dietary phosphate deprivation increases 1,25-dihydroxyvitamin D3 synthesis in rat kidney in vitro. *J Biol Chem* **258**: 1152–1155.

Gray, R. W., Wilz, D. R., Caldas, A. E., and Lemann, J. (1977). The importance of phosphate in regulating plasma 1, 25-(OH)2-vitamin D levels in humans: studies in healthy subjects in calcium-stone formers and in patients with primary hyperparathyroidism. *J Clin Endocrinol Metab* **45**: 299–306.

Gray, R. W., Garthwaite, T. L., and Phillips, L. S. (1983). Growth hormone and triiodothyronine permit an increase in plasma 1,25 (OH)2D concentrations in response to dietary phosphate deprivation in hypophysectomized rats. *Calcif Tissue Int* **35**: 100–106.

Greenberg, B. G., Winters, R. W., and Graham, J. B. (1960). The normal range of serum inorganic phosphorus and its utility as a discriminant in the diagnosis of congenital hypophosphatemia. *J Clin Endocrinol Metab* **20**: 364–379.

Grieff, M., Zhong, M., Finch, J., Ritter, C. S., Slatopolsky, E., and Brown, A. J. (1996). Renal calcitriol synthesis and serum phosphorus in response to dietary phosphorus restriction and anabolic agents. *Am J Kidney Dis* **28**: 589–595.

Gupta, A., Tenenhouse, H. S., Hoag, H. M. et al. (2001). Identification of the type II Na(+)-Pi cotransporter (Npt2) in the osteoclast and the skeletal phenotype of Npt2-/- mice. *Bone* **29**: 467–476.

Gutierrez, O., Isakova, T., Rhee, E. et al. (2005). Fibroblast growth factor-23 mitigates hyperphosphatemia but accentuates calcitriol deficiency in chronic kidney disease: New insight into secondary hyperparathyroidism. *J Am Soc Nephrol* **16**: 2205–2215.

Hafdi, Z., Couette, S., Comoy, E., Prie, D., Amiel, C., and Friedlander, G. (1996). Locally formed 5-hydroxytryptamine stimulates phosphate transport in cultured opossum kidney cells and in rat kidney. *Biochem J* **320**: 615–621.

Halloran, B. P., and Spencer, E. M. (1988). Dietary phosphorus and 1,25-dihydroxyvitamin D metabolism: influence of insulin-like growth factor I. *Endocrinology* **123**: 1225–1229.

Hartmann, C. M., Hewson, A. S., Kos, C. H. et al. (1996). Structure of murine and human renal type II Na+-phosphate cotransporter genes (Npt2 and NPT2). *Proc Natl Acad Sci USA* **93**: 7409–7414.

Heike, C., Cunningham, M., Steiner, R. et al. (2005). Skeletal changes in epidermal nevus syndrome: does focal bone disease harbor clues concerning pathogenesis? *Am J Med Genet A* **139**: 67–77.

Hesse, E., Moessinger, E., Rosenthal, H. et al. (2007a). Oncogenic osteomalacia: exact tumor localization by co-registration of positron emission and computed tomography. *J Bone Miner Res* **22**: 158–162.

Hesse, E., Rosenthal, H., and Bastian, L. (2007b). Radiofrequency ablation of a tumor causing oncogenic osteomalacia. *N Engl J Med* **357**: 422–424.

Hoag, H. M., Martel, J., Gauthier, C., and Tenenhouse, H. S. (1999). Effects of Npt2 gene ablation and low-phosphate diet on renal Na(+)/phosphate cotransport and cotransporter gene expression. *J Clin Invest* **104**: 679–686.

Hoffman, W., Jüppner, H., Deyoung, B., O'dorisio, M., and Given, K. (2005). Elevated fibroblast growth factor-23 in hypophosphatemic linear nevus sebaceous syndrome. *Am J Med Genet A* **134**: 233–236.

Holick, M. F. (2007). Vitamin D deficiency. *N Engl J Med* **357**: 266–281.

Holm, I. A., Huang, X., and Kunkel, L. M. (1997). Mutational analysis of the PEX gene in patients with X-linked hypophosphatemic rickets. *Am J Hum Genet* **60**: 790–797.

Holm, I. A., Econs, M. J., and Carpenter, T. O. (2003). Familial hypophosphatemia and related disorders. In: (Glorieux, F.H., Pettifor, J.M., Jüppner, H., eds) *Pediatric Bone: Biology and Diseases,* Academic Press, San Diego, pp. 603–631.

Hughes, M. R., Brumbaugh, P. F., Hussler, M. R., Wergedal, J. E., and Baylink, D. J. (1975). Regulation of serum 1alpha,25-dihydroxyvitamin D3 by calcium and phosphate in the rat. *Science* **190**: 578–580.

Ichikawa, S., Lyles, K., and Econs, M. (2005). A novel GALNT3 mutation in a pseudoautosomal dominant form of tumoral calcinosis: evidence that the disorder is autosomal recessive. *J Clin Endocrinol Metab* **90**: 2420–2423.

Ichikawa, S., Sorenson, A. H., Imel, E. A., Friedman, N. E., Gertner, J. M., and Econs, M. J. (2006). Intronic deletions in the SLC34A3 gene cause hereditary hypophosphatemic rickets with hypercalciuria. *J Clin Endocrinol Metab* **91**: 4022–4027.

Ichikawa, S., Imel, E. A., Kreiter, M. L. et al. (2007). A homozygous missense mutation in human KLOTHO causes severe tumoral calcinosis. *J Clin Invest* **117**: 2684–2691.

Imanishi, Y., Inaba, M., Nakatsuka, K. et al. (2004). FGF-23 in patients with end-stage renal disease on hemodialysis. *Kidney Int* **65**: 1943–1946.

Imel, E., Peacock, M., Pitukcheewanont, P. et al. (2006). Sensitivity of fibroblast growth factor 23 measurements in tumor-induced osteomalacia. *J Clin Endocrinol Metab* **91**: 2055–2061.

Imel, E. A., Hui, S. L., and Econs, M. J. (2007). FGF23 concentrations vary with disease status in autosomal dominant hypophosphatemic rickets. *J Bone Mineral Res* **22**: 520–526.

Insogna, K. L., Broadus, A. E., and Gertner, J. M. (1983). Impaired phosphorus conservation and 1,25 dihydroxyvitamin D generation during phosphorus deprivation in familial hypophosphatemic rickets. *J Clin Invest* **71**: 1562–1569.

Jan de Beur, S. M., and Levine, M. A. (2001). Pseudohypoparathyroidism: clinical, biochemical, and molecular features. In: (Bilezikian, J.P., Markus, R., Levine, M.A., eds) *The parathyroids: basic and clinical concepts*, Academic Press, New York, pp. 807–825.

Jan de Beur, S., Finnegan, R., Vassiliadis, J. et al. (2002). Tumors associated with oncogenic osteomalacia express genes important in bone and mineral metabolism. *J Bone Min Res* **17**: 1102–1110.

Jankowski, M., Hilfiker, H., Biber, J., and Murer, H. (2001). The opossum kidney cell type IIa Na/P(i) cotransporter is a phosphoprotein. *Kidney Blood Press Res* **24**: 1–4.

John, M., Wickert, H., Zaar, K. et al. (2001). A case of neuroendocrine oncogenic osteomalacia associated with a PHEX and fibroblast growth factor-23 expressing sinusidal malignant schwannoma. *Bone* **29**: 393–402.

Jones, A., Tzenova, J., Frappier, D. et al. (2001). Hereditary hypophosphatemic rickets with hypercalciuria is not caused by mutations in the Na/Pi cotransporter NPT2 gene. *J Am Soc Nephrol* **12**: 507–514.

Jonsson, K., Zahradnik, R., Larsson, T. et al. (2003). Fibroblast growth factor 23 in oncogenic osteomalacia and X-linked hypophosphatemia. *N Engl J Med* **348**: 1656–1662.

Jüppner, H., and Thakker, R. (2008). Genetic disorders of calcium and phosphate homeostasis. In: (Pollak, M., eds) *The kidney*, W. B. Saunders Company, Philadelphia.

Jüppner, H., Gardella, T., Brown, E., Kronenberg, H., and Potts, J. (2005). Parathyroid hormone and parathyroid hormone-related peptide in the regulation of calcium homeostasis and bone development. In: *Endocrinology*, 5th edn, (DeGroot, L., Jameson, J., eds) W.B. Saunders Company, Philadelphia, pp. 1377–1417.

Kato, Y., Arakawa, E., Kinoshita, S. et al. (2000). Establishment of the anti-Klotho monoclonal antibodies and detection of Klotho protein in kidneys. *Biochem Biophys Res Commun* **267**: 597–602.

Kato, K., Jeanneau, C., Tarp, M. et al. (2006). Polypeptide GalNAc-transferase T3 and familial tumoral calcinosis. Secretion of fibroblast growth factor 23 requires O-glycosylation. *J Biol Chem* **281**: 18370–18377.

Kavanaugh, M. P., Miller, D. G., Zhang, W. et al. (1994). Cell-surface receptors for gibbon ape leukemia virus and amphotropic murine retrovirus are inducible sodium-dependent phosphate symporters. *Proc Natl Acad Sci USA* **91**: 7071–7075.

Keating, F. R., Jones, J. D., Elveback, L. R., and Randall, R. V. (1969). The relation of age and sex to distribution of values in healthy adults of serum calcium, inorganic phosphorus, magnesium, alkaline phosphatase, total proteins, albumin, and blood urea. *J Lab Clin Med* **73**: 825–834.

Kempson, S. A., Lotscher, M., Kaissling, B., Biber, J., Murer, H., and Levi, M. (1995). Parathyroid hormone action on phosphate transporter mRNA and protein in rat renal proximal tubules. *Am J Physiol* **268**: F784–F791.

Khundmiri, S. J., Rane, M. J., and Lederer, E. D. (2003). Parathyroid hormone regulation of type II sodium-phosphate cotransporters is dependent on an A kinase anchoring protein. *J Biol Chem* **278**: 10134–10141.

Kido, S., Miyamoto, K., Mizobuchi, H. et al. (1999). Identification of regulatory sequences and binding proteins in the type II sodium/phosphate cotransporter NPT2 gene responsive to dietary phosphate. *J Biol Chem* **274**: 28256–28263.

Kobayashi, K., Imanishi, Y., Koshiyama, H. et al. (2006). Expression of FGF23 is correlated with serum phosphate level in isolated fibrous dysplasia. *Life Sci* **78**: 2295–2301.

Kolek, O. I., Hines, E. R., Jones, M. D. et al. (2005). 1alpha,25-Dihydroxyvitamin D3 upregulates FGF23 gene expression in bone: the final link in a renal-gastrointestinal-skeletal axis that controls phosphate transport. *Am J Physiol Gastrointest Liver Physiol* **289**: G1036–G1042.

Kruse, K., Woelfel, D., and Strom, T. (2001). Loss of renal phosphate wasting in a child with autosomal dominant hypophosphatemic rickets caused by a FGF23 mutation. *Horm Res* **55**: 305–308.

Kuro-o, M. (2006). Klotho as a regulator of fibroblast growth factor signaling and phosphate/calcium metabolism. *Curr Opin Nephrol Hypertens* **15**: 437–441.

Kuro-o, M., Matsumura, Y., Aizawa, H. et al. (1997). Mutation of the mouse klotho gene leads to a syndrome resembling ageing. *Nature* **390**: 45–51.

Kurosu, H., Ogawa, Y., Miyoshi, M. et al. (2006). Regulation of fibroblast growth factor-23 signaling by Klotho. *J Biol Chem* **281**: 6120–6123.

Lapointe, J. Y., Tessier, J., Paquette, Y. et al. (2006). NPT2a gene variation in calcium nephrolithiasis with renal phosphate leak. *Kidney Int* **69**: 2261–2267.

Larsson, T., Nisbeth, U., Ljunggren, O., Jüppner, H., and Jonsson, K. (2003). Circulating concentration of FGF-23 increases as renal function declines in patients with chronic kidney disease, but does not change in response to variation in phosphate intake in healthy volunteers. *Kidney Int* **64**: 2272–2279.

Larsson, T., Marsell, R., Schipani, E. et al. (2004). Transgenic mice expressing fibroblast growth factor 23 under the control of the α1

(I) collagen promoter exhibit growth retardation, osteomalacia and disturbed phosphate homeostasis. *Endocrinology* **145**: 3097–3104.

Larsson, T., Davis, S., Garringer, H. et al. (2005). Fibroblast growth factor-23 mutants causing familial tumoral calcinosis are differentially processed. *J Clin Endocrinol Metab* **146**: 3883–3891.

Lederer, E. D., Sohi, S. S., and McLeish, K. R. (2000). Parathyroid hormone stimulates extracellular signal-regulated kinase (ERK) activity through two independent signal transduction pathways: role of ERK in sodium-phosphate cotransport. *J Am Soc Nephrol* **11**: 222–231.

Lederer, E. D., Khundmiri, S. J., and Weinman, E. J. (2003). Role of NHERF-1 in regulation of the activity of Na-K ATPase and sodium-phosphate co-transport in epithelial cells. *J Am Soc Nephrol* **14**: 1711–1719.

Levi, M., Shayman, J. A., Abe, A. et al. (1995). Dexamethasone modulates rat renal brush border membrane phosphate transporter mRNA and protein abundance and glycosphingolipid composition. *J Clin Invest* **96**: 207–216.

Levine, M. A. (1996). Pseudohypoparathyroidism. In: (Bilezikian, J.P., Raisz, L.G., Rodan, G.A., eds) *Principles of bone biology*, Academic Press, New York, pp. 853–876.

Li, S. A., Watanabe, M., Yamada, H., Nagai, A., Kinuta, M., and Takei, K. (2004). Immunohistochemical localization of Klotho protein in brain, kidney, and reproductive organs of mice. *Cell Struct Funct* **29**: 91–99.

Linglart, A., Bastepe, M., and Jüppner, H. (2007). Similar clinical and laboratory findings in patients with symptomatic autosomal dominant and sporadic pseudohypoparathyroidism type Ib despite different epigenetic changes at the GNAS locus. *Clin Endocrinol (Oxf.)* **67**: 822–831.

Lipman, M., Panda, D., Bennett, H. et al. (1998). Cloning of human PEX cDNA. Expression, subcellular localization, and endopeptidase activity. *J Biol Chem* **273**: 13729–13737.

Liu, J., Litman, D., Rosenberg, M., Yu, S., Biesecker, L., and Weinstein, L. (2000). A GNAS1 imprinting defect in pseudohypoparathyroidism type IB. *J Clin Invest* **106**: 1167–1174.

Liu, S., Brown, T., Zhou, J. et al. (2005). Role of matrix extracellular phosphoglycoprotein in the pathogenesis of X-linked hypophosphatemia. *J Am Soc Nephrol* **16**: 1645–1653.

Liu, S., Zhou, J., Tang, W., Jiang, X., Rowe, D. W., and Quarles, L. D. (2006). Pathogenic role of Fgf23 in Hyp mice. *Am J Physiol Endocrinol Metab* **291**: E38–E49.

Liu, S., Gupta, A., and Quarles, L. D. (2007). Emerging role of fibroblast growth factor 23 in a bone-kidney axis regulating systemic phosphate homeostasis and extracellular matrix mineralization. *Curr Opin Nephrol Hypertens* **16**: 329–335.

Lobaugh, B., and Drezner, M. K. (1983). Abnormal regulation of renal 25-hydroxyvitamin D-1 alpha-hydroxylase activity in the X-linked hypophosphatemic mouse. *J Clin Invest* **71**: 400–403.

Lorenz-Depiereux, B., Benet-Pages, A., Eckstein, G. et al. (2006a). Hereditary hypophosphatemic rickets with hypercalciuria is caused by mutations in the sodium/phosphate cotransporter gene SLC34A3. *Am J Human Genet* **78**: 193–201.

Lorenz-Depiereux, B., Bastepe, M., Benet-Pages, A. et al. (2006b). DMP1 mutations in autosomal recessive hypophosphatemia implicate a bone matrix protein in the regulation of phosphate homeostasis. *Nat Genet* **38**: 1248–1250.

Lotscher, M., Kaissling, B., Biber, J., Murer, H., and Levi, M. (1997). Role of microtubules in the rapid regulation of renal phosphate transport in response to acute alterations in dietary phosphate content. *J Clin Invest* **99**: 1302–1312.

Lu, Y., Liu, S., Yu, S. et al. (2007a). The 57 kDa C-terminal fragment of Dentin Matrix Protein 1 (DMP1) retains all biological activity: osteocytic regulation of Pi homeostasis through FGF23. 29. Annual Meeting of American Society Bone and Mineral Research, Honolulu, Hawai.

Lu, Y., Xie, Y., Zhang, S. et al. (2007b). DMP1-targeted Cre expression in odontoblasts and osteocytes. *J Dent Res* **86**: 320–325.

Lyles, K., Burkes, E., Ellis, G., Lucas, K., Dolan, E., and Drezner, M. (1985). Genetic transmission of tumoral calcinosis: autosomal dominant with variable clinical expressivity. *J Clin Endocrinol Metab* **60**: 1093–1096.

Magagnin, S., Werner, A., Markovich, D. et al. (1993). Expression cloning of human and rat renal cortex Na/Pi cotransport. *Proc Natl Acad Sci USA* **90**: 5979–5983.

Marcinkowski, W., Smogorzewski, M., Zhang, G., Ni, Z., Kedes, L., and Massry, S. G. (1997). Renal mRNA of PTH-PTHrP receptor, [Ca2+]i and phosphaturic response to PTH in phosphate depletion. *Miner Electrolyte Metab* **23**: 48–57.

Markowitz, M., Rotkin, L., and Rosen, J. F. (1981). Circadian rhythms of blood minerals in humans. *Science* **213**: 672–674.

Marshall, R.W. (ed.) (1976). *Plasma fractions*, Churchill Livingstone, New York.

Martin, A., David, V., Laurence, J. S. et al. (2007). Degradation of MEPE, DMP1 & release of SIBLING ASARM-peptides (minhibins): ASARM-peptide(s) are directly responsible for defective mineralization in HYP. *Endocrinology*.

Melhem, R., Najjar, S., and Khachadurian, A. (1970). Cortical hyperostosis with hyperphosphatemia: a new syndrome? *J Pediatr* **77**: 986–990.

Mensenkamp, A. R., Hoenderop, J. G., and Bindels, R. J. (2007). TRPV5, the gateway to Ca2+ homeostasis. *Handb Exp Pharmacol*, pp. 207–220.

Meyer, R. A., Gray, R. W., and Meyer, M. H. (1980). Abnormal vitamin D metabolism in the X-linked hypophosphatemic mouse. *Endocrinology* **107**: 1577–1581.

Miyamoto, K., Ito, M., Tatsumi, S., Kuwahata, M., and Segawa, H. (2007). New aspect of renal phosphate reabsorption: the type IIc sodium-dependent phosphate transporter. *Am J Nephrol* **27**: 503–515.

Muenke, M., Schell, U., Hehr, A. et al. (1994). A common mutation in the fibroblast growth factor receptor 1 gene in Pfeiffer syndrome. *Nat Genet* **8**: 269–274.

Murer, H., Forster, I., Hernando, N., Lambert, G., Traebert, M., and Biber, J. (1999). Posttranscriptional regulation of the proximal tubule NaPi-II transporter in response to PTH and dietary P(i). *Am J Physiol* **277**: F676–F684.

Murer, H., Hernando, N., Forster, I., and Biber, J. (2003). Regulation of Na/Pi transporter in the proximal tubule. *Annu Rev Physiol* **65**: 531–542.

Nagano, N., Miyata, S., Abe, M. et al. (2006). Effect of manipulating serum phosphorus with phosphate binder on circulating PTH and FGF23 in renal failure rats. *Kidney Int* **69**: 531–537.

Narayanan, K., Ramachandran, A., Hao, J. et al. (2003). Dual functional roles of dentin matrix protein 1. Implications in biomineralization and gene transcription by activation of intracellular Ca2+ store. *J Biol Chem* **278**: 17500–17508.

Narchi, H. (1997). Hyperostosis with hyperphosphatemia: evidence of familial occurrence and association with tumoral calcinosis. *Pediatrics* **99**: 745–748.

Negri, A. L., Negrotti, T., Alonso, G., and Pasqualini, T. (2004). [Different forms of clinical presentation of an autosomal dominant hypophosphatemic rickets caused by a FGF23 mutation in one family]. *Medicina (B Aires)* **64**: 103–106.

Nordin, B.E.C. (ed.) (1976). *Nutritional considerations,* Churchill Livingstone, New York.

Ohkido, I., Segawa, H., Yanagida, R., Nakamura, M., and Miyamoto, K. (2003). Cloning, gene structure and dietary regulation of the type-IIc Na/Pi cotransporter in the mouse kidney. *Pflugers Arch* **446**: 106–115.

Perry, W., and Stamp, T. (1978). Hereditary hypophosphataemic rickets with autosomal recessive inheritance and severe osteosclerosis. A report of two cases. *J Bone Joint Surg* **60B**: 430–434.

Pfister, M. F., Hilfiker, H., Forgo, J., Lederer, E., Biber, J., and Murer, H. (1998a). Cellular mechanisms involved in the acute adaptation of OK cell Na/Pi-cotransport to high- or low-Pi medium. *Pflugers Arch* **435**: 713–719.

Pfister, M. F., Ruf, I., Stange, G. et al. (1998b). Parathyroid hormone leads to the lysomal degradation of the renal type II Na/P$_i$ cotransporter. *Proc Natl Acad Sci USA* **95**: 1909–1914.

Portale, A. A. (1999). Calcium and phosphorus. In: (Barratt, T.M., Avner, E.D., Harmon, W.E., eds) *Pediatric nephrology,* Lippincott Williams and Wilkins, Baltimore, pp. 191–213.

Portale, A. A., and Miller, W. L. (2003). Rickets due to hereditary abnormalities of vitamin D synthesis and action. In Pediatric bone: biology and diseases. Academic Press, San Diego, pp. 583–602.

Portale, A. A., Halloran, B. P., Murphy, M. M., and Morris, R. C. (1986). Oral intake of phosphorus can determine the serum concentration of 1,25-dihydroxyvitamin D by determining its production rate in humans. *J Clin Invest* **77**: 7–12.

Portale, A. A., Halloran, B. P., and Morris, R. C. (1987). Dietary intake of phosphorus modulates the circadian rhythm in serum concentration of phosphorus. Implications for the renal production of 1,25-dihydroxyvitamin D. *J Clin Invest* **80**: 1147–1154.

Portale, A. A., Halloran, B. P., and Morris, R. C. (1989). Physiologic regulation of the serum concentration of 1,25-dihydroxyvitamin D by phosphorus in normal men. *J Clin Invest* **83**: 1494–1499.

Potts, J. T., and Gardella, T. J. (2007). Progress, paradox, and potential: parathyroid hormone research over five decades. *Ann NY Acad Sci* **1117**: 196–208.

Pribanic, S., Gisler, S. M., Bacic, D. et al. (2003). Interactions of MAP17 with the NaPi-IIa/PDZK1 protein complex in renal proximal tubular cells. *Am J Physiol Renal Physiol* **285**: F784–F791.

Prié, D., Huart, V., Bakouh, N. et al. (2002). Nephrolithiasis and osteoporosis associated with hypophosphatemia caused by mutations in the type 2a sodium-phosphate cotransporter. *N Engl J Med* **347**: 983–991.

Razzaque, M. S., and Lanske, B. (2007). The emerging role of the fibroblast growth factor-23-klotho axis in renal regulation of phosphate homeostasis. *J Endocrinol* **194**: 1–10.

Riminucci, M., Collins, M., Fedarko, N. et al. (2003). FGF-23 in fibrous dysplasia of bone and its relationship to renal phosphate wasting. *J. Clin. Invest* **112**: 683–692.

Roscioli, T., Flanagan, S., Kumar, P. et al. (2000). Clinical findings in a patient with FGFR1 P252R mutation and comparison with the literature. *Am J Med Genet* **93**: 22–28.

Rowe, J., Rowe, D., Horak, E. et al. (1984). Hypophosphatemia and hypercalciuria in small premature infants fed human milk: evidence for inadequate dietary phosphorus. *J Pediatr* **104**: 112–117.

Rowe, P., de Zoysa, P., Dong, R. et al. (2000). MEPE, a new gene expressed in bone marrow and tumors causing osteomalacia. *Genomics* **67**: 54–68.

Rowe, P., Kumagai, Y., Gutierrez, G. et al. (2004). MEPE has the properties of an osteoblastic phosphatonin and minhibin. *Bone* **34**: 303–319.

Roy, S., and Tenenhouse, H. S. (1996). Transcriptional regulation and renal localization of 1,25-dihydroxyvitamin D3-24-hydroxylase gene expression: effects of the Hyp mutation and 1,25-dihydroxyvitamin D3. *Endocrinology* **137**: 2938–2946.

Roy, S., Martel, J., Ma, S., and Tenenhouse, H. S. (1994). Increased renal 25-hydroxyvitamin D3-24-hydroxylase messenger ribonucleic acid and immunoreactive protein in phosphate-deprived Hyp mice: a mechanism for accelerated 1,25-dihydroxyvitamin D3 catabolism in X-linked hypophosphatemic rickets. *Endocrinology* **134**: 1761–1767.

Saito, H., Kusano, K., Kinosaki, M. et al. (2003). Human fibroblast growth factor-23 mutants suppress Na+-dependent phosphate co-transport activity and 1alpha,25-dihydroxyvitamin D3 production. *J Biol Chem* **278**: 2206–2211.

Saito, H., Maeda, A., Ohtomo, S. et al. (2005). Circulating FGF-23 is regulated by 1a,25-dihydroxyvitamin D3 and phosphorus in vivo. *J Biol Chem* **280**: 2543–2549.

Schoeber, J. P., Hoenderop, J. G., and Bindels, R. J. (2007). Concerted action of associated proteins in the regulation of TRPV5 and TRPV6. *Biochem Soc Trans* **35**: 115–119.

Schwindinger, W., Francomano, C., and Levine, M. (1992). Identification of a mutation in the gene encoding the alpha subunit of the stimulatory G protein of adenylyl cyclase in McCune-Albright syndrome. *Proc Natl Acad Sci USA* **89**: 5152–5156.

Scriver, C. R., Reade, T. M., DeLuca, H. F., and Hamstra, A. J. (1978). Serum 1,25-dihydroxyvitamin D levels in normal subjects and in patients with hereditary rickets or bone disease. *N Engl J Med* **299**: 976–979.

Scriver, C., Reade, T., Halal, F., Costa, T., and Cole, D. (1981). Autosomal hypophosphataemic bone disease responds to 1,25-(OH)2D3. *Arch Dis Child* **56**: 203–207.

Segawa, H., Kaneko, I., Takahashi, A. et al. (2002). Growth-related renal type II Na/Pi cotransporter. *J Biol Chem* **277**: 19665–19672.

Segawa, H., Kawakami, E., Kaneko, I. et al. (2003). Effect of hydrolysis-resistant FGF23-R179Q on dietary phosphate regulation of the renal type-II Na/Pi transporter. *Pflugers Arch* **446**: 585–592.

Segawa, H., Kaneko, I., Yamanaka, S. et al. (2004). Intestinal Na-P(i) cotransporter adaptation to dietary P(i) content in vitamin D receptor null mice. *Am J Physiol Renal Physiol* **287**: F39–47.

Segawa, H., Yamanaka, S., Ito, M. et al. (2005). Internalization of renal type IIc Na-Pi cotransporter in response to a high-phosphate diet. *Am J Physiol Renal Physiol* **288**: F587–F596.

Segawa, H., Yamanaka, S., Onitsuka, A. et al. (2007a). Parathyroid hormone-dependent endocytosis of renal type IIc Na-Pi cotransporter. *Am J Physiol Renal Physiol* **292**: F395–F403.

Segawa, H., Onitsuka, A., Kuwahata, M. et al. (2007b). *The role of type IIc sodium-dependent phosphate transporter (Npt2c), which is involved in hereditary hypophosphatemic rickets with hypercalciuria (HHRH).* ASBMR, Honululu, p. 1115.

Shimada, T., Mizutani, S., Muto, T. et al. (2001). Cloning and characterization of FGF23 as a causative factor of tumor-induced osteomalacia. *Proc Natl Acad Sci USA* **98**: 6500–6505.

Shimada, T., Muto, T., Urakawa, I. et al. (2002). Mutant FGF-23 responsible for autosomal dominant hypophosphatemic rickets is resistant to proteolytic cleavage and causes hypophosphatemia in vivo. *Endocrinology* **143**: 3179–3182.

Shimada, T., Urakawa, I., Yamazaki, Y. et al. (2004a). FGF-23 transgenic mice demonstrate hypophosphatemic rickets with reduced expression of sodium phosphate cotransporter type IIa. *Biochem Biophys Res Commun* **314**: 409–414.

Shimada, T., Kakitani, M., Yamazaki, Y. et al. (2004b). Targeted ablation of Fgf23 demonstrates an essential physiological role of FGF23 in phosphate and vitamin D metabolism. *J. Clin. Invest* **113**: 561–568.

Sitara, D., Razzaque, M. S., Hesse, M. et al. (2004). Homozygous ablation of fibroblast growth factor-23 results in hyperphosphatemia and impaired skeletogenesis, and reverses hypophosphatemia in Phex-deficient mice. *Matrix Biol* **23**: 421–432.

St-Arnaud, R. and Demay, M. B. (2003). Vitamin D biology. In Pediatric bone: biology and diseases. Academic Press, San Diego, pp. 193–216.

Suki, W. N. and Rouse, D. (eds) (1996). *Renal transport of calcium, magnesium, and phosphate,* W.B. Saunders, Philadelphia.

Suki, W. N. Lederer, E. D. and Rouse, D. (eds) (2000). *Renal transport of calcium, magnesium, and phosphate,* W.B. Saunders, Philadelphia.

Taketani, Y., Segawa, H., Chikamori, M. et al. (1998). Regulation of type II renal Na+-dependent inorganic phosphate transporters by 1,25-dihydroxyvitamin D3. Identification of a vitamin D-responsive element in the human NAPi-3 gene. *J Biol Chem* **273**: 14575–14581.

Takeuchi, Y., Suzuki, H., Ogura, S. et al. (2004). Venous sampling for fibroblast growth factor-23 confirms preoperative diagnosis of tumor-induced osteomalacia. *J Clin Endocrinol Metab* **89**: 3979–3982.

Tanaka, Y., and DeLuca, H. F. (1973). The control of 25-hydroxy-vitamin D metabolism by inorganic phosphorus. *Arch Biochem Biophys* **154**: 566–574.

Tenenhouse, H. S. (2007). Phosphate transport: molecular basis, regulation and pathophysiology. *J Steroid Biochem Mol Biol* **103**: 572–577.

Tenenhouse, H. S., and Jones, G. (1990). Abnormal regulation of renal vitamin D catabolism by dietary phosphate in murine X-linked hypophosphatemic rickets. *J Clin Invest* **85**: 1450–1455.

Tenenhouse, H. S., Martel, J., Gauthier, C., Zhang, M. Y., and Portale, A. A. (2001). Renal expression of the sodium/phosphate cotransporter gene, Npt2, is not required for regulation of renal 1 alpha-hydroxylase by phosphate. *Endocrinology* **142**: 1124–1129.

Tenenhouse, H., Martel, J., Gauthier, C., Segawa, H., and Miyamoto, K. (2003). Differential effects of Npt2a gene ablation and X-linked Hyp mutation on renal expression of Npt2c. *Am J Physiol Renal Physiol* **285**: F1271–F1278.

Tieder, M., Modai, D., Samuel, R. et al. (1985). Hereditary hypophosphatemic rickets with hypercalciuria. *N Engl J Med* **312**: 611–617.

Tieder, M., Modai, D., Shaked, U. et al. (1987). 'Idiopathic' hypercalciuria and hereditary hypophosphatemic rickets. Two phenotypical expressions of a common genetic defect. *N Engl J Med* **316**: 125–129.

Topaz, O., Shurman, D., Bergman, R. et al. (2004). Mutations in GALNT3, encoding a protein involved in O-linked glycosylation, cause familial tumoral calcinosis. *Nat Genet* **36**: 579–581.

Traebert, M., Volkl, H., Biber, J., Murer, H., and Kaissling, B. (2000). Luminal and contraluminal action of 1-34 and 3-34 PTH peptides on renal type IIa Na-P(i) cotransporter. *Am J Physiol Renal Physiol* **278**: F792–F798.

Urakawa, I., Yamazaki, Y., Shimada, T. et al. (2006). Klotho converts canonical FGF receptor into a specific receptor for FGF23. *Nature* **444**: 770–774.

van Boekel, G., Ruinemans-Koerts, J., Joosten, F., Dijkhuizen, P., van Sorge, A., and de Boer, H. (2008). Tumor producing fibroblast growth factor 23 localized by two-staged venous sampling. *Eur J Endocrinol* **158**: 431–437.

Villa-Bellosta, R., Barac-Nieto, M., Breusegem, S. Y., Barry, N. P., Levi, M., and Sorribas, V. (2007). Interactions of the growth-related, type IIc renal sodium/phosphate cotransporter with PDZ proteins. *Kidney Int* **73**: 456–466.

Virkki, L., Forster, I., Hernando, N., Biber, J., and Murer, H. (2003). Functional characterization of two naturally occurring mutations in the human sodium-phosphate cotransporter type IIa. *J Bone Miner Res* **18**: 2135–2141.

Virkki, L. V., Biber, J., Murer, H., and Forster, I. C. (2007). Phosphate transporters: a tale of two solute carrier families. *Am J Physiol Renal Physiol* **293**: F643–F654.

Wagner, G. F., Vozzolo, B. L., Jaworski, E. et al. (1997). Human stanniocalcin inhibits renal phosphate excretion in the rat. *J Bone Miner Res* **12**: 165–171.

Weber, T., Liu, S., Indridason, O., and Quarles, L. (2003). Serum FGF23 levels in normal and disordered phosphorus homeostasis. *J Bone Miner Res* **18**: 1227–1234.

Weinstein, L. S. (2006). G(s)alpha mutations in fibrous dysplasia and McCune-Albright syndrome. *J Bone Miner Res* **21**(Suppl. 2): P120–P124.

Weinstein, L. S., Shenker, A., Gejman, P. V. et al. (1991). Activating mutations of the stimulatory G protein in the McCune-Albright syndrome. *N Engl J Med* **325**: 1688–1695.

Weinstein, L., Yu, S., Warner, D., and Liu, J. (2001). Endocrine manifestations of stimulatory G protein alpha-subunit mutations and the role of genomic imprinting. *Endocr Rev* **22**: 675–705.

Weinstein, L. S., Liu, J., Sakamoto, A., Xie, T., and Chen, M. (2004). Minireview: GNAS: normal and abnormal functions. *Endocrinology* **145**: 5459–5464.

Werner, A., Moore, M. L., Mantei, N., Biber, J., Semenza, G., and Murer, H. (1991). Cloning and expression of cDNA for a Na/Pi cotransport system of kidney cortex. *Proc Natl Acad Sci USA* **88**: 9608–9612.

White, K. E., Evans, W. E., O'Riordan, J. L. H. et al. ADHR Consortium (2000). Autosomal dominant hypophosphataemic rickets is associated with mutations in FGF23. *Nat Genet* **26**: 345–348.

White, K., Jonsson, K., Carn, G. et al. (2001). The autosomal dominant hypophosphatemic rickets (ADHR) gene is a secreted polypeptide overexpressed by tumors that cause phosphate wasting. *J Clin Endocrinol Metab* **86**: 497–500.

White, K., Cabral, J. et al. (2005). Mutations that cause osteoglophonic dysplasia define roles for FGFR1 in bone elongation. *Am J Hum Genet* **76**: 361–367.

White, K., Larsson, T., and Econs, M. (2006). The roles of specific genes implicated as circulating factors involved in normal and disordered phosphate homeostasis: Frp-4, MEPE, and FGF23. *Endocr Rev* **27**: 221–241.

Wilkinson, R. (ed.) (1976). *Absorption of calcium, phosphorus and magnesium,* Churchill Livingstone, New York.

Wilson, L. C., Oude-Luttikhuis, M. E. M., Clayton, P. T., Fraser, W. D., and Trembath, R. C. (1994). Parental origin of Gsa gene mutations in Albright's hereditary osteodystrophy. *J Med Genet* **31**: 835–839.

Wu, S., Finch, J., Zhong, M., Slatopolsky, E., Grieff, M., and Brown, A. J. (1996). Expression of the renal 25-hydroxyvitamin D-24-hydroxylase gene: regulation by dietary phosphate. *Am J Physiol* **271**: F203–F208.

Wu, S., Grieff, M., and Brown, A. J. (1997). Regulation of renal vitamin D-24-hydroxylase by phosphate: effects of hypophysectomy, growth hormone and insulin-like growth factor I. *Biochem Biophys Res Commun* **233**: 813–817.

Yamamoto, T., Imanishi, Y., Kinoshita, E. et al. (2005). The role of fibroblast growth factor 23 for hypophosphatemia and abnormal regulation of vitamin D metabolism in patients with McCune-Albright syndrome. *J Bone Miner Metab* **23**: 231–237.

Yamaoka, K., Seino, Y., Satomura, K., Tanaka, Y., Yabuuchi, H., and Haussler, M. R. (1986). Abnormal relationship between serum phosphate concentration and renal 25-hydroxycholecalciferol-1-alpha-hydroxylase activity in X-linked hypophosphatemic mice. *Miner Electrolyte Metab* **12**: 194–198.

Yamashita, T., Yoshioka, M., and Itoh, N. (2000). Identification of a novel fibroblast growth factor, FGF-23, preferentially expressed in the ventrolateral thalamic nucleus of the brain. *Biochem Biophys Res Commun* **277**: 494–498.

Yamazaki, Y., Okazaki, R., Shibata, M. et al. (2002). Increased circulatory level of biologically active full-length FGF-23 in patients with hypophosphatemic rickets/osteomalacia. *J Clin Endocrinol Metab* **87**: 4957–4960.

Ye, L., Mishina, Y., Chen, D. et al. (2005). Dmp1-deficient mice display severe defects in cartilage formation responsible for a chondrodysplasia-like phenotype. *J Biol Chem* **280**: 6197–6203.

Yoshida, T., Yoshida, N., Monkawa, T., Hayashi, M., and Saruta, T. (2001). Dietary phosphorus deprivation induces 25-hydroxyvitamin D(3) 1alpha-hydroxylase gene expression. *Endocrinology* **142**: 1720–1726.

Yu, X., Ibrahimi, O. A., Goetz, R. et al. (2005). Analysis of the biochemical mechanisms for the endocrine actions of fibroblast growth factor-23. *Endocrinology* **146**: 4647–4656.

Zhang, M. Y., Wang, X., Wang, J. T. et al. (2002). Dietary phosphorus transcriptionally regulates 25-hydroxyvitamin D-1alpha-hydroxylase gene expression in the proximal renal tubule. *Endocrinology* **143**: 587–595.

Zhao, N., and Tenenhouse, H. S. (2000). Npt2 gene disruption confers resistance to the inhibitory action of parathyroid hormone on renal sodium-phosphate cotransport. *Endocrinology* **141**: 2159–2165.

Further reading

Brodehl, J., Gellissen, K., and Weber, H. P. (1982). Postnatal development of tubular phosphate reabsorption. *Clin Nephrol* **17**: 163–171.

Hilfiker, H., Hattenhauer, O., Traebert, M., Forster, I., Murer, H., and Biber, J. (1998a). Characterization of a murine type II sodium-phosphate cotransporter expressed in mammalian small intestine. *Proc Natl Acad Sci USA* **95**: 14564–14569.

Hilfiker, H., Hartmann, C. M., Stange, G., and Murer, H. (1998b). Characterization of the 5'-flanking region of OK cell type II Na-Pi cotransporter gene. *Am J Physiol* **274**: F197–F204.

Maierhofer, W. J., Gray, R. W., and Lemann, J. (1984). Phosphate deprivation increases serum 1,25-(OH)2-vitamin D concentrations in healthy men. *Kidney Int* **25**: 571–575.

Section III. Renin–Angiotensin

CHAPTER 10

The History of the Renin–Angiotensin System

JOEL MENARD
Professor of Public Health, Faculté de Médecine Paris Descartes, Laboratoire SPIM, 15 rue de l'Ecole de Médecine, 75006 Paris, France

Contents

I.	Introduction	129
II.	The 20th century	129
III.	The 21st century	131
IV.	The dream to be normotensive and drug free	132
V.	Conclusion	132
	References	133

I. INTRODUCTION

The history of the renin–angiotensin system begins with the studies of Tigerstedt and Bergman. The number of publications related to this system was relatively few until the mid-20th century. However, as shown in Figure 10.1, the number of references listed in PubMed increased exponentially between 1966 and 2006, making any formal review cumbersome. Instead, this overview will take a hypothetical voyage around the world, stopping at various cities where, throughout the years, momentous contributions have been made. The reader can readily find references to supplement this approach in two previous reviews (Menard, 1993; Menard and Patchett, 2001).

II. THE 20TH CENTURY

Our journey begins in Buenos-Aires, home of Houssay, Braun-Menendès, Fasciolo, Taquini and Leloir, where many bright and well-educated Argentinean scientists left their imprint on the physiology and pathophysiology of the renin–angiotensin system by their detailed *in vitro* and *in vivo* studies before spreading out around the globe. The next stop is Sao-Paulo, to look at the history of Feireira and his discovery of bradykinin. From there, we fly to Cleveland, where Skeggs, Gould, Goldblatt and Haas worked throughout their careers and where Page, Corcoran, Masson, MacCubbin, Dustan, Bumpus, Khairallah, Tarazi and others made seminal observations of this system over several decades in the middle of the last century. Goldblatt and Page had little contact, until Page became the second investigator in the world (the first was working in Belgium) to write to Goldblatt inquiring about the material used to clip dog renal arteries. Concurrently, in the USA, Page made seminal contributions to the study of the renin–angiotensin system. He was a leader for the field, as were Pickering in the UK, Milliez and Hatt in France, Bartorelli and Zanchetti in Italy, Genest in Canada, Gross in Germany, Brod in Czechoslovakia, Smirk and Simpson in New Zealand and Strasser in Switzerland. In 1958, Page laid out the roadmap for the study of the renin–angiotensin system for the next 50 years. He theorized: 'The possibility of operating on renin, i.e. to prevent or control the rate at which angiotensin I is formed, or on the converting enzyme to prevent or control angiotensin II formation, is worth considering. A small molecule antagonist of angiotensin (not a peptide) might be found if we had some angiotensin for the pharmacologists to use in a screening test'. In the 1960s, however, nothing was evident, as outlined by the discussions at the first two consecutive international meetings organized in 1965, in Sienna by Bartorelli and Zanchetti and in Paris by Milliez, Meyer and Tcherdakoff. Verniory and Potlieve (Brussels) said that: 'their observations suggest that renin is not a mandatory participant of the mechanisms by which a stenosis of renal artery increases blood pressure. The importance of the juxta-glomerular apparatus in hypertension development no longer seems to be predominant as suggested by Goldblatt and Goorgmaghtig. In many circumstances, high blood pressure occurs without the juxta-glomerular's or the adrenal gland's participation'. According to Genest (Montreal), 'research results of our laboratory demonstrate that the renin–angiotensin system does not seem to play an important role in the occurrence or maintenance of arterial hypertension' and also 'our studies demonstrate a close link between sodium regulation and renin in humans, and contribute to a better understanding of the renin–angiotensin system pathophysiology'. From Glasgow, Brown, Lever and Robertson declared: 'Plasma renin is inversely correlated to plasma sodium, independently on aetiology and severity of hypertension, its

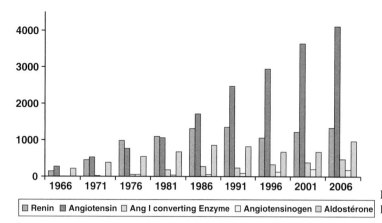

FIGURE 10.1 Renin–angiotensin–aldosterone system in PubMed (number of papers/year every 5 years for 40 years).

complications and its treatments'. And from New York, Laragh started developing his concepts on the renin–angiotensin system: 'Our studies demonstrate a homogeneous relationship between angiotensin, aldosterone and sodium balance. In this relationship, the blood level of angiotensin finally interacts with available sodium ions to maintain vascular tone regulation' (Milliez and Tcherdakoff, 1965).

For decades, many brilliant investigators were privileged to study in Boston. They were educated in endocrinology with Thorn, renal physiology with Barger, cardiology with Braunwald and immunology with Haber. They had immediate access to both basic research and clinical research – the specific element that has boosted many scientific and medical contributions in the field – an approach now known as 'translational medicine'. Fascinating, elegant studies of the adrenal glands and kidneys were performed by Williams and Hollenberg, uncovering the identity of the modulator and the non-modulator in hypertensives. The Pfeffers moved from studies on coronary artery ligature in rats to the management of large-scale randomized controlled trials that established angiotensin-converting enzyme inhibitors and angiotensin II receptor-type 1 antagonists as the major treatments for post-myocardial infarction and congestive heart failure. Brenner started out analyzing glomerular filtration rate in Wistar-Munich rats and subsequently shared his key discoveries with nephrologists worldwide, triggering major efforts to reduce proteinuria and microalbuminuria and to slow the decrease in glomerular filtration rate in most renal diseases. The desired results were accomplished using a converting enzyme inhibitor, an angiotensin II receptor antagonist, or the combination in tandem. Dzau uncovered the components of tissue renin–angiotensin systems and their function in vascular and cardiac tissue, while Ingelfinger was studying kidney angiotensinogen. Investigators associated with the Framingham Heart Study began to use this same cohort to increase our understanding of the role of the renin–angiotensin system as a cardiovascular risk factor.

Substantial contributions were being made by the Laragh and Sealey group in New York, some of whom made critical independent observations regarding the regulation of the renin–angiotensin system and the role of interrupting it in treating cardiovascular disease. The contributions made by Brunner, Buhler, Weber, Case, Atlas, Maack, Catanzaro and others have provided the underpinnings of a rational approach to hypertension treatment, based on a solid methodology established both in the clinic and in the laboratory. At the same time, blood pressure measurements were being refined by Pickering Jr, the cardiac mass was being explored by Devereux and observational epidemiologic approaches and their limitations were being confronted by Alderman.

At several laboratories in the southern USA, Guyton, Hall, Young and Cowley were refining their integrated view of the physiology of system regulation and analyzing the close fit between modelling and experimentation – with one focus being the renin–angiotensin system, the kidney and volume homeostasis. Carey was refining the role of renin and several of its active peptides in regulating renal function as O. Smithies was developing the technology required to prepare precisely modified genes in mice as tools to understand the molecular mechanisms underlying the renin–angiotensin system's physiologic effects. Inagami was increasing our understanding of the biochemical structure of this system. In Canada, Genest and his many collaborators were expanding our knowledge of how this system is modified in disease states and had a major role in the training of many young European investigators, from Germany to France and from Spain to Czechoslovakia.

At the National Institutes of Health, Catt and his colleagues were teasing out the action of various angiotensins on adrenal and vascular function and the effects of the level of salt intake on this activity. He has assumed the same responsibilities in the training of young and motivated investigators such as Mendelsohn, Capponi and the author, as those assumed at Cleveland, Boston, New York, Montreal or San Francisco. Halfway across the world in Japan, Yamamoto, his research group and many other skillful Japanese investigators were using rat models of disease to further our understanding of the renin–angiotensin system and other vasoactive factors in causing cardiovascular diseases. In Australia, beginning in the 1960s and extending for four decades, Coghlan, Denton, Funder, Johnston, Morgan, Mendelsohn, Whitworth, Campbell, Chalmers, MacMahon and colleagues have contributed substantially to our knowledge of the renin–angiotensin system. They used their easy access to sheep in the most appropriate way to analyze sodium balance and aldosterone and contributed to major observational and interventional studies. To mention just one of the prolific numbers of insights provided by Australian investigators, Skinner established the link between the perfusion pressure of the kidney, as described by Selkurt, and renin release, as investigated by the aforementioned Cleveland Clinic group.

A number of groups in Europe also contributed to our knowledge of the renin–angiotensin system during the latter third of the 20th century. In Edinburgh, Mullins was developing elegant models of knockout mice. The team composed of Brown, Lever and Robertson and colleagues set up a Blood Pressure Unit – not a Hypertension Unit – in Glasgow. Even now, Fraser is linking the generations, with Connell and Dominiczak, whereas the clinical pharmacology developed by 'Sir' Colin Dollery in London has been finely applied to the renin–angiotensin system by Reid and his colleagues. MacGregor, Edwards and Stewart further expanded our knowledge particularly related to RAS interactions with mineralocorticoids and inhibitors of the renin–angiotensin system. In Rotterdam, the complex metabolism of angiotensins was being analyzed by Schalekampf, Derkx, Deinum and Danser. In Lausanne, other groups were expanding our knowledge base in several related areas. Brunner, Nussberger and Burnier were evaluating the regulation of angiotensins in conjunction with the regulation of sodium in the proximal tubule, whereas Rossier and his colleagues were doing the same in the distal tubule. The connection between animal models and clinical pharmacology they established has been an extremely effective legacy, thanks to the methodological and conceptual efforts of Nussberger, Burnier, Weber, Pedrazini and Biollaz, who worked with Brunner. In Basel, at the top of the Ciba-Geigy building, in 1986, there was a splendid library. Some pages of the journals in that library were well worn, as the books automatically opened at certain pages. It was here that the author, working with Hofbauer, Wood and de Gasparo, found the spirit and work of Gross that helped them develop new therapeutic agents that modified the activity of the renin–angiotensin system, such as benazepril, valsartan and aliskiren.

In Paris, the Broussais Hospital has closed but, fortunately, at Georges Pompidou European Hospital a new generation of researchers – Jeunemaitre, Azizi, Plouin and Alhenc-Gelas – has carried forth the undertakings of Corvol and Menard. Laurent and Simon have extended the work of Safar on arterial functions. In parallel Geneviève Nguyen, at the College of France, is assessing the relevance of the prorenin/renin receptor.

We arrive in Berlin. Before Ganten and Unger, is the shadow of Gross and the Heidelberg trio Taugner, Hackenthal and Ritz. Close to Ganten, as a star among American and Europe as a clinical investigator, Luft reminds us that dedicated researchers should take their research, but not themselves, seriously. This journey ends in Berlin because Ganten has imbued all of us with his political vision and scientific dreams. The large contribution of many Italian scientists, around Zanchetti, Bianchi and Mantero, Spanish investigators with Ruilope and of a small country such as Denmark that has generated Bing, Poulsen, Leyssac and many others are not forgotten.

This abbreviated reminiscence has omitted many worthy names and countries; yet it is hoped that the message of this approach is clear: when a wide variety of techniques, access to large numbers of patients and a strong commitment to research freedom and education are found at the same place and when there is an effective link between experimental, clinical and/or epidemiological research, the city and the institution become as renowned for their contribution to renin–angiotensin–aldosterone system research as the scientists themselves.

III. THE 21ST CENTURY

A. The Ambiguity of the Renin–Angiotensin–Aldosterone System

The clinical benefits of blocking this system have been carefully described, but the absence of aldosterone and complete inhibition of the renin–angiotensin system appear to be dangerous (Campbell, 2002; Sung, 2003; Dluhy and Williams, 2004). Hypoaldosteronism, due to an aldosterone synthesis deficiency (White, 2004) or a mutation of the mineralocorticoid receptor, has now been analyzed in greater detail, with its various degrees of severity and multiple genotypes (Bonny and Rossier, 2002; Geller et al., 2006). The pharmacological blockade of aldosterone at the site of its receptor was found to be beneficial for patients with congestive heart failure in the RALES Study (Pitt et al., 1999), but the extension of anti-aldosterone therapy in general populations has been shown to induce adverse effects (Juurlink et al., 2004; Koch et al., 2005).

Conversely, higher aldosterone levels in the general population appear to be associated with the incidence of hypertension (Vasan et al., 2004) and the neutralization of aldosterone effects in the heart, blood vessels and kidneys is beneficial (Brilla and Weber, 1992; Rocha et al., 2002) (see Chapter 23 for more details). As Chapters 13–15 will demonstrate, inhibition of the renin–angiotensin system is beneficial as it delays renal insufficiency, after decreasing proteinuria or microalbuminuria but, if really complete, the pharmacological blockade requires the maintenance of salt intake to prevent renal insufficiency (Griffiths et al., 2001; Richer-Giudicelli et al., 2004). The most impressive finding is the discovery of myriad adverse effects on fetal kidney function of RAS inhibitor use during pregnancy, in parallel with those of knockouts of genes encoding angiotensinogen, renin, angiotensin I-converting enzyme or angiotensin II type 1 receptors (Gomez and Norwood, 1995). The confluence of genetics, pathology and clinical observation has demonstrated that tubular dysgenesis, inducing death *in utero*, or early renal failure, is a disease with multiple genotypes and phenotypes, which may arise from diverse mutations in one of several RAS genes (Lacoste et al., 2006). Gavras and Brunner once asked whether the renin system was necessary (Brunner and Gavras, 1980): the answer is 'yes', for maintenance of renal function.

B. The Discovery of New Monogenic Forms of Hypertension

Whereas the genetic analysis of the renin–angiotensin system showed the complexity of the effects of polymorphisms on biological and clinical phenotypes, the contribution of groups working on genetics, especially that of Lifton, first in Utah and then in Yale, has been particularly fruitful on the discovery of genes responsible for monogenic forms of renal or adrenal hypertension. The benefits of the research in molecular genetics (Dluhy et al., 2001; Lifton et al., 2001; Wilson et al., 2001) are likely to be amplified. Once more, the conjunction of careful clinical characterization with cell physiology and animal experiments has led to discovery of the genetic and biochemical basis of an increasing number of monogenic hypertensive diseases, involving the renin–angiotensin system directly or indirectly, first described by skilled clinicians or biologists, such as Liddle, Laidlaw, Gordon, Bartter, Ulick, New and others, in the last century. Additionally, the mutated proteins have contributed to our understanding of the normal physiology and of the response to drugs (Frazier et al., 2004; Lalioti et al., 2006). This research has contributed to reinforce the importance of sodium and potassium balance in the pathogenesis of hypertension. It is also possible that these genetic discoveries have been initially made at the place where they were expected and that unexpected findings will occur soon from many candidates inside and especially outside of the renin–angiotensin–aldosterone system. The very elegant work of the Nobel Prize winner, Oliver Smithies and his group, has created mice with 0–4 copies of the renin–angiotensin–aldosterone system genes by using homologous recombination. It confirmed the importance of each component of the system and of the kinetics of this unique enzymatic cascade.

One of the most extraordinary of these recently described genetic diseases is a new 'pregnancy-induced' form of hypertension. This disease is the result of a gain of function mutation secondary to the substitution of leucine for serine at codon 810 (S810L) in the mineralocorticoid receptor (MR). With this mutation, the interaction between the receptor and progesterone and cortisone is modified, resulting in an increased blood pressure early in life and especially during pregnancy (Geller et al., 2000; Rafestin-Oblin et al., 2003). In addition, molecular modeling has allowed the development of highly potent renin inhibitors (Wood et al., 2003) (see Chapter 15 for details). Finally, analysis of the crystal structure of the ligand-binding domain of MR, when it is associated with an agonist, has transformed our vision of the interactions between agonists, antagonists and MR (Rafestin-Oblin et al., 2003).

Although all scientists stress that interactions between genes and environment underlie chronic diseases, the optimal use of this approach remains unclear. Behind the elegance of experiments based on gene suppression, gene titration and conditional gene overexpression in non-physiologic sites, a question remains: are these animal models created to validate what is already known from human physiology and pharmacology and medicine, or are they opening up new ways of looking at unknown human disorders (Takahashi et al., 2003; Wintermantel et al., 2005)? The transplantation of angiotensin II type 1a receptor knockout kidneys to normal mice or of normal kidneys to angiotensin II type 1a receptor knockout mice illustrates this conundrum perfectly (Crowley et al., 2007). The result is consistent with Guyton's concepts and provides a supplementary means of confirming what is partly known from the renal transplantation experiments of Dahl in the USA, Bianchi in Italy, or Rettig in Germany. However, step-by-step, through such experiments, we progress within the kidney to investigate the specific role of the tubular angiotensin II type 1 receptors, which may turn out to be more important than the vascular receptors (*ibid*). The road is long if we imagine that each mutation identified in association studies will lead to translational research to investigate the functional impact of each newly discovered mutated protein (Hadchouel and Jeunemaitre, 2006; Ring et al., 2007).

IV. THE DREAM TO BE NORMOTENSIVE AND DRUG FREE

The use of siRNA or active immunization to provide benefits similar to those currently provided by drugs, but without risk and maintaining the individual disease free over decades, remains the most fascinating challenge to renin–angiotensin–aldosterone system researchers. These possibilities are already raising issues that must be publicly addressed by both researchers and society (Brown et al., 2004; Ambuhl et al., 2007; Menard, 2007).

V. CONCLUSION

Hundreds of meetings and thousands of scientific papers a year: it is not surprising that the history of the renin–angiotensin system can no longer be described objectively. Early on, history is easy to summarize. Our common memory has preserved Bright, Tigerstedt, Volhard and Fahr. Many nameless, now anonymous investigators made significant contributions in their time, but their names together with the journals and proceedings in which they published have been lost. Any reader who has made an important contribution but was not mentioned in this overview should not feel slighted. The author learned very late in his professional life that Goldblatt's experiment had previously been performed in Volhard's laboratory (Heidland et al., 2001) and that Conn's syndrome had essentially been described before Conn's discovery (Kucharz, 1991). The real challenge for the reader is to go to PubMed or other search engines and use 'the renin–angiotensin system' as a query and determine how to synthesize the information provided by the thousands of references identified. Even at a time when the chapter is closed, the author would like to add his own highlights of 2007, the connexin40 of the juxta-glomerular cells as a major and new player structure in renin release (Krattinger et al., 2007; Wagner et al., 2007) and, for 2008, the TASK channels deletion in mice which causes hyperaldosteronism, as if there was no end to the story.

In the following chapters Reudelhuber and Catanzaro will provide details of the molecular biology and gene regulation of the renin–angiotensin system and Carey and Padia will review its physiology and regulation. Bakris will update our understanding of its role in the kidney and Ferrario and Trask will do the same for the heart. Finally, Sica will review the pharmacologic and therapeutic aspects of blocking this system in treating human diseases. We anticipate that the organization of the massive amount of data in this way will facilitate the ability of the reader to understand more clearly the complex actions of the renin–angiotensin system.

References

Ambuhl, P. M., Tissot, A. C., Fulurija, A. et al. (2007). A vaccine for hypertension based on virus-like particles: preclinical efficacy and phase I safety and immunogenicity. *J Hypertens* **25**: 63–72.

Bonny, O., and Rossier, B. C. (2002). Disturbances of Na/K balance: pseudohypoaldosteronism revisited. *J Am Soc Nephrol* **13**: 2399–2414.

Brilla, C. G., and Weber, K. T. (1992). Reactive and reparative myocardial fibrosis in arterial hypertension in the rat. *Cardiovasc Res* **26**: 671–677.

Brown, M. J., Coltart, J., Gunewardena, K., Ritter, J. M., Auton, T. R., and Glover, J. F. (2004). Randomized double-blind placebo-controlled study of an angiotensin immunotherapeutic vaccine (PMD3117) in hypertensive subjects. *Clin Sci (Lond)* **107**: 167–173.

Brunner, H. R., and Gavras, H. (1980). Is the renin system necessary? *Am J Med* **69**: 739–745.

Campbell, D. J. (2002). Renin-angiotensin system inhibition: how much is too much of a good thing? *Intern Med J* **32**: 616–620.

Crowley, S. D., Gurley, S. B., and Coffman, T. M. (2007). AT(1) receptors and control of blood pressure: the kidney and more. *Trends Cardiovasc Med* **17**: 30–34.

Dluhy, R. G., Anderson, B., Harlin, B., Ingelfinger, J., and Lifton, R. (2001). Glucocorticoid-remediable aldosteronism is associated with severe hypertension in early childhood. *J Pediatr* **138**: 715–720.

Dluhy, R. G., and Williams, G. H. (2004). Aldosterone – villain or bystander? *N Engl J Med* **351**: 8–10.

Frazier, L., Turner, S. T., Schwartz, G. L., Chapman, A. B., and Boerwinkle, E. (2004). Multilocus effects of the renin-angiotensin-aldosterone system genes on blood pressure response to a thiazide diuretic. *Pharmacogenomics J* **4**: 17–23.

Geller, D. S., Farhi, A., Pinkerton, N. et al. (2000). Activating mineralocorticoid receptor mutation in hypertension exacerbated by pregnancy. *Science* **289**: 119–123.

Geller, D. S., Zhang, J., Zennaro, M. C. et al. (2006). Autosomal dominant pseudohypoaldosteronism type 1: mechanisms, evidence for neonatal lethality, and phenotypic expression in adults. *J Am Soc Nephrol* **17**: 1429–1436.

Gomez, R. A., and Norwood, V. F. (1995). Developmental consequences of the renin-angiotensin system. *Am J Kidney Dis* **26**: 409–431.

Griffiths, C. D., Morgan, T. O., and Delbridge, L. M. (2001). Effects of combined administration of ACE inhibitor and angiotensin II receptor antagonist are prevented by a high NaCl intake. *J Hypertens* **19**: 2087–2095.

Hadchouel, J., and Jeunemaitre, X. (2006). Life and death of the distal nephron: WNK4 and NCC as major players. *Cell Metab* **4**: 335–337.

Heidland, A., Gerabek, W., Sebekova, K., Volhard, F., and Fahr, T. (2001). Achievements and controversies in their research in renal disease and hypertension. *J Hum Hypertens* **15**: 5–16.

Juurlink, D. N., Mamdani, M. M., Lee, D. S. et al. (2004). Rates of hyperkalemia after publication of the Randomized Aldactone Evaluation Study. *N Engl J Med* **351**: 543–551.

Koch, E., Otarola, A., and Kirschbaum, A. (2005). A landmark for popperian epidemiology: refutation of the randomised Aldactone evaluation study. *J Epidemiol Commun Hlth* **59**: 1000–1006.

Krattinger, N., Capponi, A., Mazzolai, L. et al. (2007). Connexion 40 regulates renin production and blood pressure. *Kidney Int* **72**: 814–822.

Kucharz, E. J. (1991). Forgotten description of primary hyperaldosteronism. *Lancet* **337**: 1490.

Lacoste, M., Cai, Y., Guicharnaud, L. et al. (2006). Renal tubular dysgenesis, a not uncommon autosomal recessive disorder leading to oligohydramnios: Role of the renin-angiotensin system. *J Am Soc Nephrol* **17**: 2253–2263.

Lalioti, M. D., Zhang, J., Volkman, H. M. et al. (2006). Wnk4 controls blood pressure and potassium homeostasis via regulation of mass and activity of the distal convoluted tubule. *Nat Genet* **38**: 1124–1132.

Lifton, R. P., Gharavi, A. G., and Geller, D. S. (2001). Molecular mechanisms of human hypertension. *Cell* **104**: 545–556.

Menard, J. (2007). A vaccine for hypertension. *J Hypertens* **25**: 41–46.

Menard, J. (1993). Anthology of the renin–angiotensin system: a one hundred reference approach to angiotensin II antagonists. *J Hypertens Suppl* **11**: S3–S11.

Menard, J., and Patchett, A. (2001). Angiotensin-converting enzyme inhibitors. In Advances in protein chemistry. Drug discovery and design. (Scolnick EM, ed.). West Point, Pennsylvania, 56, p. 14–75.

Milliez, P., and Tcherdakoff, Ph. (eds) (1965). International Club on arterial hypertension. L'Expansion Scientifique. Paris, p. 524.

Pitt, B., Zannad, F., Remme, W. J. et al. (1999). The effect of spironolactone on morbidity and mortality in patients with severe heart failure. Randomized Aldactone Evaluation Study Investigators. *N Engl J Med* **341**: 709–717.

Rafestin-Oblin, M. E., Souque, A., Bocchi, B., Pinon, G., Fagart, J., and Vandewalle, A. (2003). The severe form of hypertension caused by the activating S810L mutation in the mineralocorticoid receptor is cortisone related. *Endocrinology* **144**: 528–533.

Richer-Giudicelli, C., Domergue, V., Gonzalez, M. et al. (2004). Haemodynamic effects of dual blockade of the renin-angiotensin system in spontaneously hypertensive rats: influence of salt. *J Hypertens* **22**: 619–627.

Ring, A. M., Cheng, S. X., Leng, Q. et al. (2007). WNK4 regulates activity of the epithelial Na^+ channel in vitro and in vivo. *Proc Natl Acad Sci USA* **104**: 4020–4024.

Rocha, R., Martin-Berger, C. L., Yang, P., Scherrer, R., Delyani, J., and McMahon, E. (2002). Selective aldosterone blockade prevents angiotensin II/salt-induced vascular inflammation in the rat heart. *Endocrinology* **143**: 4828–4836.

Takahashi, N., Hagaman, J. R., Kim, H. S., and Smithies, O. (2003). Minireview: computer simulations of blood pressure regulation by the renin-angiotensin system. *Endocrinology* **144**: 2184–2190.

Vasan, R. S., Evans, J. C., Larson, M. G. et al. (2004). Serum aldosterone and the incidence of hypertension in nonhypertensive persons. *N Engl J Med* **351**: 33–41.

Wagner, C., de Wit, C., Kurtz, L., Grünberger, C., Kurtz, A., and Schweda, F. (2007). Connexin40 is essential for the pressure control of renin synthesis and secretion. *Circ Res* **100**: 556–563.

White, P. C. (2004). Aldosterone synthase deficiency and related disorders. *Mol Cell Endocrinol* **217**: 81–87.

Wilson, F. H., Disse-Nicodeme, S., Choate, K. A. et al. (2001). Human hypertension caused by mutations in WNK kinases. *Science* **293**: 1107–1112.

Wintermantel, T. M., Berger, S., Greiner, E. F., and Schutz, G. (2005). Evaluation of steroid receptor function by gene targeting in mice. *J Steroid Biochem Mol Biol* **93**: 107–112.

Wood, J. M., Maibaum, J., Rahuel, J. et al. (2003). Structure-based design of aliskiren, a novel orally effective renin inhibitor. *Biochem Biophys Res Commun* **308**: 698–705.

Further reading

Davies, L. A., Hu, C., Guagliardo, N. A. et al. (2008). TASK channel deletion in mice causes primary hyperaldosteronism. *Proc Natl Acad Sci USA* **105**: 2203–2208.

Heitzmann, D., Derand, R., Jungbauer, S. et al. (2008). Invalidation of TASK1 potassium channels disrupts adrenal gland zonation and mineralocorticoid homeostasis. *EMBO J* **27**: 179–187.

Nguyen, G. (2007). The (pro)renin receptor: pathophysiological roles in cardiovascular and renal pathology. *Curr Opin Nephrol Hypertens* **16**: 129–133.

Sung, N. S., Crowley, W. F., Genel, M. et al. (2003). Central challenges facing the national clinical research enterprise. *J Am Med Assoc* **289**: 1278–1287.

CHAPTER 11

Molecular Biology of Renin and Regulation of its Gene

TIMOTHY L. REUDELHUBER[1] AND DANIEL F. CATANZARO[2]

[1] Clinical Research Institute of Montreal (IRCM), 110 Pine avenue west, Montreal, Quebec H2W 1R7, Canada
[2] Division of Cardiovascular Pathophysiology, Department of Medicine, Weill Cornell Medical College, New York, NY 10021, USA

Contents

I. Introduction 135
II. Production and activation of renin 135
III. Renin gene structure and regulation 138
IV. Renin gene mutation and disease 141
V. Future perspectives 142
References 142

I. INTRODUCTION

Renin, a member of the aspartyl protease family of enzymes, is released into the circulation from the juxtaglomerular (JG) cells of the kidney where it cleaves angiotensinogen primarily of hepatic origin to form angiotensin I (Ang I). Ang I is further processed to angiotensin II (Ang II) by angiotensin-converting enzyme (ACE) abundantly present both in the plasma and on the surface of endothelial cells. Ang II causes vasoconstriction through its effects on the vasculature and sodium retention both through its direct effects on the renal tubules and via its effects on adrenal synthesis and secretion of aldosterone. Modulation of renin–angiotensin–aldosterone system (RAAS) activity is therefore a critical determinant of blood pressure and fluid volume.

In humans, the release of renin from JG cells represents the rate-limiting step in the RAAS cascade. Because the circulating concentration of angiotensinogen (roughly 1.3 μM) is very close to the K_m of renin for angiotensinogen (1.15 μM) (Do et al., 1987; Campbell et al., 1991a), the rate of generation of angiotensin peptides is directly affected by changes in both renin and substrate levels. Renin, itself, is first synthesized as an enzymatically inactive precursor, prorenin. Conversion of prorenin to active renin occurs before secretion from JG cells which release both prorenin and renin into the circulation. The secretion of prorenin is thought to be constitutive whereas that of renin is regulated by physiological stimuli. However, as has become increasingly apparent, physiological stimuli affecting renin secretion also impact on renin gene expression and hence prorenin secretion. Overall, the activity of the RAAS hinges on the level of renin gene expression and on the efficiency with which prorenin is converted to active renin. The goal of this chapter is to review what is known about these two biological processes.

II. PRODUCTION AND ACTIVATION OF RENIN

A. Tissue Origin of Active Renin

Active renin in the plasma comes almost entirely from the kidneys. Total nephrectomy of both humans and animals causes plasma renin levels to disappear (Sealey and Rubattu, 1989). In addition to active renin secreted from JG cells, prorenin is also present in the plasma at levels 5–10 times that of active renin (Hsueh and Baxter, 1991). After nephrectomy, circulating prorenin declines, but does not disappear, suggesting that it is secreted from sites outside of the kidneys (Campbell et al., 1991a; Hsueh and Baxter, 1991; Hosoi et al., 1992). Indeed, expression of the renin gene has been detected in a number of extrarenal tissues including the adrenal and pituitary glands, the eye and reproductive tissues (Ganong et al., 1989; Mulrow, 1989; Griendling et al., 1993; Paul et al., 1993; Brandt et al., 1994) and it is therefore likely that these tissues contribute to prorenin found in the circulation.

While the renal JG cells release both prorenin and renin, they do so by different mechanisms: the secretion of prorenin is only limited by its rate of synthesis and this type of secretion has been called constitutive. Renin, on the other hand, is only generated once prorenin enters the lysosome-like secretory granules of the JG cell (Taugner et al., 1987). Renin is stored in these dense secretory granules (Figure 11.1) and is subsequently released in response to various physiological stimuli. It is estimated that about one fourth of the prorenin made in JG cells is sorted to secretory granules where it is converted to renin (Pratt et al., 1988). Release of JG cell granules thereby provides a 'rapid response' mechanism to

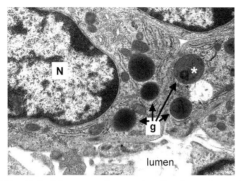

FIGURE 11.1 Electron micrograph of a juxtaglomerular cell from a 5-day-old mouse kidney. g, dense core secretory granule containing rennin, N, nucleus. The lumen of the afferent arteriole is indicated. Note the inclusions visible in some granules (asterisk) that are characteristic of autophagy. Original magnification ×15 000.

increase active renin content in the circulation while the modulation of renin gene expression provides a more long-term modulation of renin and prorenin production.

B. Physiological Control of Renin Levels

Because only the JG cells of the kidneys produce active renin, it is not surprising that the mechanisms that regulate normal renin secretion and renin gene expression reside primarily in the kidneys. JG cells repress the release of renin in response to high salt in the distal tubule (through mechanisms mediated by the macula densa) as well as increases in Ang II and/or the concomitant rise in blood pressure. In contrast, renin release increases in response to sympathetic stimulation, low blood pressure or blockade of Ang II signaling or production as well as decreases in distal tubular sodium. The physiological regulation of renin release will be dealt with in more detail in Chapter 12.

Defects in the physiological control of renin release may also play an important part in the contribution of the RAAS to disease. Although only about 10% of all hypertensive patients have elevated plasma renin levels, even a normal plasma renin level in a hypertensive patient is inappropriate because the high blood pressure should suppress renin secretion. Therefore, hypertensive patients should have low plasma renin levels if their RAAS is responding normally, and normal as well as high plasma renins should be considered an abnormality in a hypertensive patient's ability to suppress plasma renin in response to their elevated blood pressure. The dependence of an individual's blood pressure on plasma renin is reflected by their responsiveness to anti-renin drugs (Buhler et al., 1972). Conversely, in hypertensives with low plasma renin levels, anti-renin drugs have relatively little effect (Laragh, 2001).

C. Ontogeny and Plasticity of Renin Expression

In the fetal kidneys of mammals, renin-expressing cells are found dispersed throughout the walls of large renal arteries,

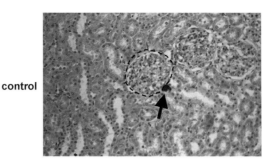
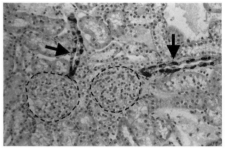

FIGURE 11.2 Recruitment of juxtaglomerular cells. The number of cells stained in rat kidney with an antibody for renin (solid arrows) increases dramatically along the afferent arteriole 2 weeks after partial occlusion of the renal artery (2-kidney, 1 clip Goldblatt model of hypertension). The glomeruli are outlined with hatched lines. Experimental details are described in Mercure et al., 1998.

in branching arterioles and even in the renal interstitium before the appearance of blood vessels (Reddi et al., 1998). This distribution is broader and differs from that seen after birth when renin is expressed primarily in the JG cells. By transplanting renin-expressing cells from genetically marked animals into control mice, Gomez and colleagues provided evidence that JG cells are derived from precursor cells that retain the ability to reversibly switch between two identities: a vascular smooth muscle cell and the epithelioid-like JG cell (Sequeira Lopez et al., 2001). Differentiation into the JG cell occurs when there is prolonged high demand for renin secretion or when the normal repression of renin secretion (angiotensin II, salt or pressure) is lost (Figure 11.2). The JG cells can revert to the smooth muscle phenotype when the stimulus for renin secretion is withdrawn. This same group also used a sophisticated approach involving targeted insertion into the renin gene and cross-breeding with a marker mouse strain to perform a 'cell fate' map designed to tag cells that at one time during development had expressed the renin gene (Sequeira Lopez et al., 2004). Their results demonstrate that renin-expressing cells are precursors to a number of cell types including smooth muscle cells in a number of renal vessels as well as mesangial and epithelial cells. Moreover, these investigators demonstrated that, in conditions of high renin demand, these cell types were able to reactivate a renin-expressing phenotype (Sequeira Lopez et al., 2004). Recently, Gomez and co-workers have identified a protein, Zis, which contains zinc finger domains that are characteristics of splicing factors and

RNA binding proteins (Karginova et al., 1997). Zis was identified by differential display on the basis of its downregulation following 10 days culture of JG cells. In addition, these investigators have shown that chromatin remodeling may play a significant role in reactivating renin gene expression in the JG cell precursor pool (Pentz et al., 2007). It will be important to determine the contributions of Zis and chromatin remodeling in these cells as recruitment/derecruitment of JG cells may contribute significantly to the overall level of renin production (Everett et al., 1990; Gomez et al., 1990).

D. Mechanisms of Proteolytic and Non-Proteolytic Activation

Prorenin, present in the circulation at 5–10 times the levels of active renin (Hsueh and Baxter, 1991) is the biochemical precursor of renin. In humans, the conversion of prorenin to active renin occurs by the proteolytic removal of a 43 amino acid prosegment from the amino-terminal end of prorenin. This activation is irreversible and there is no evidence suggesting that prorenin can be proteolytically activated once it is released into the circulation. Unlike some other proteases, renin is not able to activate itself by autoproteolysis, but rather requires the participation of another protease to remove the prosegment. While the enzyme that carries out this processing in JG cells has never been identified with certainty, it is clear that the activation only occurs once prorenin enters the acidic environment of the secretory granules (Taugner et al., 1987). The mechanism by which this processing is restricted to secretory granules is still not entirely clear, but this targeted activation ensures that active renin secretion is regulated, since the granules are stored until the cell receives a signal for their release. One strong candidate for the human prorenin processing enzyme in JG cells is cathepsin B. While cathespin B is an acid hydrolase primarily found in lysosomes, JG cell secretory granules have many of the characterisitics of 'secretory' lysosomes including content of vesicular inclusions and numerous lysosomal enzymes such as acid phosphatase, β-glucuronidase, arylsulfatase as well as cathepsins B, D, H and L (Matsuba et al., 1989) and visual evidence that they are capable of autophagy (see Figure 11.1 and Taugner et al., 1988). Cathepsin B also displays the proper cleavage specificity on prorenin, cleaving at a single site after a pair of basic amino acids (lysine and arginine) in the prosegment to release renin with the same amino-terminus as that isolated from human kidney preparations (Wang et al., 1991; Jutras and Reudelhuber, 1999). Although no other strong candidate prorenin processing enzymes have been proposed for the JG cell to date, several other proteases have been shown to activate human prorenin *in vitro* or in tissue culture models. Trypsin, plasmin as well as tissue and plasma kallikreins can all correctly process prorenin *in vitro* (Hsueh and Baxter, 1991; Kikkawa et al., 1998), but there is no compelling evidence for the physiological relevance of these enzymes in active renin generation *in vivo*. However, two members of the proprotein/prohormone convertase (PC) family of enzymes, PC1/3 and PC5/6A, have been shown both accurately to cleave human prorenin and to require the environment of secretory granules to do so (Benjannet et al., 1992; Mercure et al., 1996). While neither enzyme is expressed in JG cells, they might contribute to local production of active renin in extrarenal tissues such as the adrenal gland (Reudelhuber et al., 1994).

Proteolytic activation of mouse Ren2 prorenin in submandibular glands also appears to occur after a pair of basic amino acids and at a position analogous to that of human prorenin (Misono et al., 1982). In contrast, rat prorenin has been reported to be processed after a threonine residue seven amino acids downstream from the site used by human and mouse Ren2 prorenins (Kim et al., 1991). Rat and mouse prorenins both subsequently undergo an additional internal cleavage to produce 'two-chain' renins held together by a disulfide bridge (Misono et al., 1982; Campbell et al., 1991b; Kim et al., 1991). Thus, while the proteolytic activation of human, mouse and rat prorenins all require the activity of a second protease to remove an N-terminal prosegment, some species differences may exist in the processing proteases and the subcellular environments required for this activation event.

While proteolytic activation of prorenin is irreversible, there is significant evidence that prorenin can undergo a reversible conformational change that results in its activation with the prosegment still attached. *In vitro*, acidification (Derkx et al., 1987) or prolonged storage in the cold (Pitarresi et al., 1992) of purified prorenin both result in its partial and reversible activation. In addition, even though the prosegment is thought to block the active site of renin, prolonged incubation of prorenin with active site-directed inhibitors eventually results in all of the prorenin being bound by the inhibitor (Heinrikson et al., 1989). One interpretation of this phenomenon is that prorenin continually undergoes a transient unfolding of the prosegment and that active site-directed inhibitors either trap the 'open' conformation or perhaps even induce it. This concept has led to the proposal of a two-step model to explain activation of prorenin (Figure 11.3) and has stimulated interest in the possibility that conditions might exist *in vivo* that would mimic the effects of cold, acid or active site-directed inhibitors and result in a local unfolding of prorenin in tissues. This hypothesis received support with the description of a cell-surface 'receptor' ((P)RR) capable not only of binding both prorenin and renin with nanomolar affinity, but of triggering intracellular signaling events (Nguyen et al., 2002). Moreover, binding of prorenin to (P)RR has been reported to result in full activation of prorenin with the prosegment attached (Nabi et al., 2006). Transgenic mouse models have provided further support to the potential contribution of prorenin to the generation of angiotensin peptides within tissues. Reudelhuber and colleagues (Methot et al., 1999) reported that full-length (intact) human prorenin attached was capable of

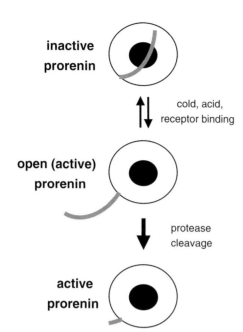

FIGURE 11.3 Schematic representation of the proposed steps in prorenin activation. The stippled area represents the prosegment that prevents access to the active site represented by a solid black circle. Several conditions are thought to cause the prosegment to unfold transiently while proteolytic cleavage of the prosegment generates active renin irreversibly. Adapted from Derkx et al., 1987.

generating angiotensin I when expressed with human angiotensinogen in the pituitaries of transgenic mice. Moreover, this same group demonstrated that the human prorenin released from the livers of transgenic mice could be taken up from the circulation and used by the heart to make angiotensin peptides (Prescott et al., 2002). Thus, even though prorenin has no enzymatic activity in the circulation, these results raise the possibility that it might have some biological or pathophysiological role in certain tissues, either through local activation possibly occurring on cell surface receptors, or by triggering intracellular signaling cascades. This question is also clinically relevant since prorenin levels increase with age, in pregnancy, following pharmacological inhibition of the RAAS and have been reported to increase prior to the appearance of microvascular disease in diabetics (Danser et al., 1998). However, definitive evidence for the physiological role of circulating prorenin is still lacking.

III. RENIN GENE STRUCTURE AND REGULATION

A. Genomic Structure in Humans and Laboratory Animals

The genes encoding human, rat and mouse renins have been cloned and characterized (Hardman et al., 1984; Hobart et al., 1984; Holm et al., 1984; Miyazaki et al., 1984; Fukamizu et al., 1988). The human and rat genomes contain only one renin gene. In humans, the renin gene is located on chromosome 1 at the q32 band (Cohen-Haguenauer et al., 1989) while, in rats, it is found on chromosome 13 at band q13 (Mori et al., 1992). In contrast, some strains of mice have two copies of the renin gene. In mice that carry only one renin gene (e.g. C57Bl/6), the gene is designated $Ren1^c$, whereas in strains with two renin genes (e.g. DBA2 and 129J), the genes are designated $Ren1^d$ and $Ren2$ (Dickinson et al., 1984). Although all three genes are highly homologous resulting in 97% sequence similarity at the amino acid levels, $Ren2$ lacks the asparagine-linked glycosylation consensus sequences that are encoded by the other two genes (Sigmund and Gross, 1991). Although $Ren1^d$ and $Ren2$ are expressed in similar tissues, their relative expression levels vary. In the kidney, $Ren1^d$ may be the predominant form (Kim et al., 1999), although earlier studies suggested approximately equal accumulation of $Ren1^d$ and $Ren2$-derived mRNAs (Field and Gross, 1985). Expression at other sites also differs, for example in submandibular gland, $Ren2$ is highly androgen dependent and therefore more abundant in male mice (Mesterovic et al., 1983; Catanzaro et al., 1985; Field and Gross, 1985).

B. Promoters and Enhancers in the Renin Gene

JG cells make up roughly 1/10 000 of the cell mass in the kidney (Taugner et al., 1984). The anatomical relationship of JG cells to the kidney architecture appears to be critical in maintaining their renin-expressing phenotype. When kidney tissue is dispersed and the cells placed in culture, JG cells rather rapidly lose the ability to secrete renin (Della and Kurtz, 1995). The paucity of JG cells in the kidney and the difficulty of growing differentiated JG cells in culture have proven to be major impediments to identifying the mechanisms regulating renin gene expression (Johns et al., 1987). An approach that has been successfully used to generate immortalized cell lines that express a particular gene is targeted tumorigenesis in transgenic mice. Targeted tumorigenesis utilizes the promoter and enhancer sequences specific for the cell type that is to be isolated linked to an oncogene, commonly the SV40 T-antigen. Targeted tumorigenesis in transgenic mice was utilized to produce a renin-expressing cell line (As4.1) (Sigmund et al., 1990). Gross, Sigmund and coworkers used As4.1 cells to characterize the promoter elements that direct renin gene expression in these cells (Petrovic et al., 1996). They identified a sequence located approximately 2.5 kb upstream of the $Ren1^c$ transcription initiation site that satisfies the criteria of a classical enhancer in that it stimulated promoter activity in a distance and orientation independent fashion (Figure 11.4). Moreover, they showed that a sequence element, located approximately 60 bp upstream of the renin gene transcription start site, was required for the stimulatory effect of the enhancer. This same proximal sequence had previously been shown by Catanzaro

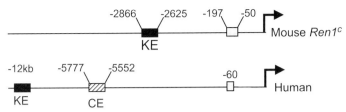

FIGURE 11.4 Relative location of the enhancer and proximal promoter regions characterized in the mouse $Ren1^c$ and human renin genes. KE, kidney enhancer; CE, chorionic enhancer. The gray boxes represent proximal promoter elements. The solid black arrow represents the start site of transcription of the renin genes. Drawing not to scale.

and coworkers to be required for human renin promoter activity in pituitary and placental cells (Sun et al., 1993, 1994; Catanzaro et al., 1994). The former group demonstrated that members of the POU family of transcription factors may be involved in binding to this site. More recently, Gross and coworkers have shown that a member of the HOX family of transcription factors present in As4.1 cells binds the −60 sequence (Pan et al., 2001).

Since enhancers are often cell specific in their activity, this prompted a search for a similar enhancer in the human renin gene. Because Southern blotting revealed no homologous sequences in existing renin clones that contained approximately 3 kb 5'-flanking DNA, Yan et al. isolated P1 human genomic clones that contained much longer renin flanking sequences. At approximately 12 kb upstream of the human renin gene, they identified a 242 bp region that was 40–80% identical to the mouse enhancer (Yan et al., 1997b). Although in As4.1 cells the transcriptional activity of the human enhancer was much lower than its mouse counterpart, these studies showed that the sequences were conserved between mouse and human genes, although in the human gene they were much further upstream (Figure 11.4). A number of transcription factors have been identified that bind these sequences and account for their activity in As4.1 cells (Pan and Gross, 2005).

Pinet and coworkers also reported a distinct enhancer sequence located approximately 5 kb upstream of the human renin coding sequence (Germain et al., 1998). This enhancer stimulated renin promoter activity about 60-fold in renin-expressing primary cultures of chorion laeve cells, but had relatively little effect in Calu-6 cells (a human cell line from an adenocarcinoma of the lung that expresses renin at low levels), or As4.1 cells. More recently, this group reported finding a single nucleotide polymorphism downstream of this enhancer element that had a modest effect on the transcriptional activity of constructs containing these sequences transfected in choriodecidual cell cultures (Fuchs et al., 2002). Although the function of this enhancer in vivo has yet to be determined, preliminary studies with transgenic mice suggest that it is dispensable for expression of the human renin gene (Zhou and Sigmund, 2008).

To overcome the lack of bona fide JG cell lines and to test for the in vivo relevance of renin gene control elements identified in tissue culture, several groups have resorted to transgenic mice. In particular, transgenic mice can be used to determine whether the transgene sequences are sufficient not only for expression at appropriate sites, but also sufficient to prevent expression at inappropriate sites. Most early studies indicated that relatively short mouse renin transgenes were sufficient to direct appropriate cell-specific expression (Fabian et al., 1989; Mullins and Ganten, 1990). Although at the time those studies were carried out the putative enhancer was unknown, it was fortuitously included in most of the transgenes tested. These studies took advantage of the fact that some strains of mice contain a duplication of the renin gene. Thus, mice were made transgenic for one of the renin genes it normally lacked and the renin mRNAs were distinguished by virtue of sequence differences between the genes. Generally, these studies demonstrated appropriate cell and tissue expression and regulation of the transgenes.

Transgenic animal lines were also made containing human renin transgenes that lacked the enhancer sequences (Fukamizu et al., 1989; Ganten et al., 1992; Sigmund et al., 1992). In each of these lines, human renin gene expression was detected in renal juxtaglomerular cells and in a variety of extrarenal sites that included normal sites of renin gene expression. However, other unusual sites of renin gene expression were also detected. Based on these observations, it was postulated that these unusual sites represented novel sites of renin gene expression that had previously gone undetected and that warranted further investigation to determine their physiological function (Seo et al., 1991). However, another interpretation of these observations is that the more distal sequences might not be required for expression in JG cells, but might instead prevent expression at extrarenal sites.

Yan et al. prepared three separate transgenic lines containing a 45 kb human renin transgene with approximately 25 kb 5'- and 8 kb 3'-flanking DNA plus all coding and intervening sequences (Yan et al., 1997a). These sequences included the enhancer region of the human renin gene. In each of these lines, human renin mRNA was detected in tissues which, with the exception of lung, represent normal sites of renin gene expression in the mouse (Yan et al., 1997a). Although these studies did not directly address the role of the enhancer, they showed that the extended flanking sequences were necessary to restrict expression of the transgene to physiologically appropriate sites. In a slightly different approach, Morris and colleagues (Adams et al., 2006) generated a germ line deletion of the putative mouse kidney enhancer by homologous recombination in C57BL mice. Consistent with the critical role played by this enhancer region, the investigators were unable to detect expression of the modified renin gene in either basal or

Bacteriophage and Plasmid-Derived Clones

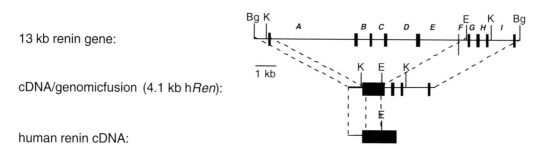

PAC and P1-Derived Clones

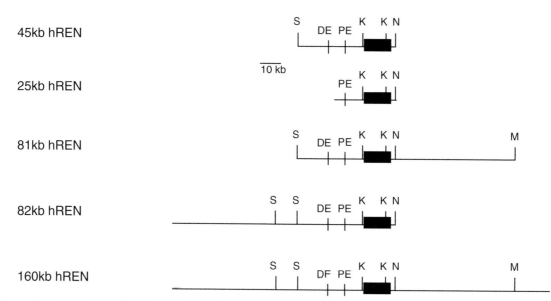

FIGURE 11.5 Genomic clones used in the generation of transgenic mice (see text). In the upper panel are shown bacteriophage and cDNA-derived clones. The 13 kb hREN gene was used to prepare transgenic mice (Sigmund et al., 1992). This clone was used to generate hybrids in which part (4.1 kb hREN) or all (human renin cDNA) of the genomic sequences were replaced by cDNA. In the lower panel are shown PAC and P1-derived sequences that were used to prepare transgenic with much longer human renin genomic sequences. Note differences in the scale bars. The 160 kb hREN is described in Sinn et al. (1999). Not shown is the 140 kb hREN sequence described in the same paper that contains 90 kb 3'-flanking DNA. Restriction enzyme cleavage sites present in thesE sequences are as follows: S—SalI, K—KpnI, N—NotI, M—MluI, E—EcoRI, Bg—BglII. DE and PE denote the distal and proximal enhancers, respectively.

low salt diet conditions. In contrast, Sigmund and colleagues (Sinn et al., 1999) made transgenic mice with a 160 kb transgene that encompassed the entire human renin gene locus including 75 kb 5'- and 70 kb 3'-flanking sequences. While this transgene exhibited the expected tissue-specific expression and was appropriately modulated by angiotensin II, mice generated with the same transgene in which the kidney enhancer had been specifically deleted still expressed and regulated the transgene appropriately, but at 3–10-fold lower expression levels (Zhou et al., 2006). Taken together, these data suggest that the so-called kidney 'enhancer' is not the major determinant of tissue-specificity of regulation of the renin gene, but that it may serve to boost its expression.

To determine whether other DNA sequences in the human renin gene might contribute to the control of its expression, Yan and coworkers (unpublished observations) also prepared a 4.1 kb renin genomic/cDNA fusion that deletes the introns between exons I–VI (Figure 11.5). This construct has the same 5'- and 3'-flanking DNA as the 13 kb hREN transgene that, although expressed at ectopic sites, appeared to be expressed in renal JG cells (Table 11.1). However, unlike the 13 kb hREN transgene, the 4.1 kb minigene was not expressed in renal JG cells or any other tissues of eight transgenic lines that were produced. These findings further implicate a critical role for intron sequences in directing hREN gene expression. In this respect, it is noteworthy that Zis, a nuclear factor that appears to be co-expressed in

TABLE 11.1 Tissue distribution of transgene expression among various hREN transgenic lines

Tissue	45 kb hREN mouse	140 kb hREN mouse	13 kb hREN mouse	15 kb hREN mouse	13 kb hREN rat	15 kb hREN rat
Aorta					+	
Adipose	−	−	+			
Adrenal	+		+			
Brain	+		−	+	+	−
Eye	+		+			
Heart	−	−	−	+	−	−
Intestines	−		+		+	−
Kidney	+	+	+	+	+	+
Liver	−	−	−	−	−	−
Lung	+	+	+	+	+	+
Ovary	+	−	+			+
Pancreas		−		−		−
Skeletal muscle	−		+			+
Spleen	−	−	+	+	+	+
Stomach	−		+	+		
SMG	−		−	−		−
Testis	−		+	+		
Thymus				+	+	
Thyroid					+	

Data are adapted from Fukamizu et al., 1989; Ganten et al., 1992; Sigmund et al., 1992; Takaori et al., 1993; Bohlender et al., 1997; Yan et al., 1998; Sinn et al., 1999.

renin-expressing cells (see above), may be involved in RNA splicing (Karginova et al., 1997).

Together, these observations suggested that the renin gene intervening sequences might be required for expression. The conserved sequence that was defined as an enhancer based on studies in As4.1 cells may not serve this role *in vivo*, or may require additional sequences to fulfill its biological role.

IV. RENIN GENE MUTATION AND DISEASE

A. Renin Gene Polymorphisms and Hypertension

Given the importance of renin levels in regulating the production of Ang II, it is natural that investigators have questioned whether mutations that affect the renin gene might control blood pressure and may perhaps even be responsible for some forms of hypertension. This hypothesis received some strong support with early reports that polymorphisms in the renin gene co-segregated with hypertension in the Dahl salt-sensitive rat (Rapp et al., 1989) and the spontaneously hypertensive rat (SHR) using cross-breeding experiments (Pravenec et al., 1991). In spite of this encouraging start and numerous linkage and association studies undertaken in humans (Naftilan et al., 1989; Soubrier et al., 1990; Zee et al., 1991; Jeunemaitre et al., 1992), there is no evidence to date that convincingly links variations in the renin gene to hypertension. While efforts continue in this regard using much higher density genomic markers currently available, it seems unlikely at this point that analysis of the renin gene will achieve sufficient predictive power to be used as a diagnostic tool in essential hypertension.

B. Inactivation of the Renin Gene in Mouse

Because of the availability of mouse embryonic stem cells in which homologous recombination can be used to modify (usually inactivate) a specific gene, mice have been particularly useful in investigating the role of RAAS genes. Until recently, embryonic stem cells were only available from two gene mice. Because the two renin genes are closely linked, they cannot be inactivated one at a time and then bred to produce double knockouts. Therefore, most of the early studies on inactivating RAAS genes targeted genes other than the renin gene. Although a full discussion of these studies is beyond the scope of the present chapter, the general findings of these studies was that inactivation of components of the RAAS (e.g. angiotensinogen, angiotensin-converting enzyme or the angiotensin II AT1 receptor) resulted in hypotension as well as a number of renal problems that became apparent just after birth including hydronephrosis and hypertrophy of interlobular arteries (Chen et al., 2004). Despite the complications presented by two closely linked renin genes present in some strains of mice, knockout studies of these mice provided valuable insights into the normal regulation of the renal RAAS. Several groups have used homologous recombination selectively to inactivate *Ren1^d* (Bertaux et al., 1997; Clark et al., 1997) or *Ren2* (Sharp et al., 1996). More recently, with the advent of embryonic stem cells from single gene mice, it has been possible to produce renin deficient mice (Yanai et al., 2000). These mice show a similar phenotype to mice in which other key genes of the RAAS were knocked out. However, they also suggested that renin may be dispensable in maintaining the integrity of the blood–brain barrier as previously suggested by studies of angiotensinogen knockout mice (Kakinuma et al., 1998). Although one strain mice with disruption of the $Ren1^d$ gene appeared to be normotensive (Bertaux et al., 1997), another group reported that renin-deficient mice were hypotensive and had degranulation of the JG cells (Clark et al., 1997). In a subsequent study, Gomez and coworkers showed that $Ren2$ can compensate for the inactivation of $Ren1^d$, although blood pressure was reduced and there was a reduction in juxtaglomerular renin secretory granules (Pentz et al., 2001). Moreover, they demonstrated that both $Ren1^d$ and $Ren2$ were subject to the same regulatory process involving recruitment of afferent arteriolar smooth muscle cells.

C. Inactivation of the Renin Gene in Humans

Considering the contribution of the RAAS to cardiovascular disease and the fact that patients tolerate pharmacological inhibition of the RAAS so well, it may seem surprising that mutations resulting in the inactivation of renin or other RAAS components have not been found in healthy humans. However, the disastrous results of early clinical trials testing the use of ACE inhibitors in pregnancy-induced hypertension (Fiocchi et al., 1984) and the characteristic renal defects seen when any of the components of the RAAS were inactivated in mice discussed above raise the possibility that the RAAS plays a critical role in renal development in humans. Taking this characteristic as a cue, Marie Claire Gubler and associates screened RAAS genes from individuals afflicted with autosomal recessive renal tubular dysgenesis (RTD). Of the 16 individuals tested with RTD, eight had mutations that inactivated the renin gene while the remaining eight had mutations in either angiotensinogen, ACE or the AT1 angiotensin II receptor (Gribouval et al., 2005). Other characteristics of this class of mutations were a scarcity of proximal tubules, pulmonary hypoplasia, early-onset oligohydramnios and some skull ossification defects. Taken together, these results confirm that the RAAS has two roles: to orchestrate some as yet poorly understood role in development and as a modulator of cardiovascular homeostasis in adult mammals. Surprisingly, a similar situation exists with the endothelin system which is required for craniofacial and intestinal development in the fetus while being primarily vasoconstrictive in the adult (Kurihara et al., 1994).

V. FUTURE PERSPECTIVES

Although it has been more than 100 years since renin activity was first described, the enzyme remains a focus of study relevant to clinical therapy in cardiovascular diseases. In particular, the process by which prorenin is converted to renin in JG cells is still poorly understood and the potential contribution of circulating prorenin to physiology and pathology remains largely a matter of conjecture. Likewise, in spite of much effort, we still do not have a clear understanding of the mechanisms that regulate expression of the renin gene either in terms of its tissue distribution or its regulation by physiological cues. A better understanding of any one of these processes could provide new pharmacologic targets for modulation of RAAS activity and an important complement to current therapy for cardiovascular diseases.

Acknowledgments

The authors would like to thank Chantal Mercure for the gift of previously unpublished images presented in Figures 11.1 and 11.2. This work was supported by an unrestricted grant from Merck-Frosst Canada Ltd. to TLR.

References

Adams, D. J., Head, G. A., Markus, M. A. et al. (2006). Renin enhancer is critical for control of renin gene expression and cardiovascular function. *J Biol Chem* **281**: 31753–31761.

Benjannet, S., Reudelhuber, T., Mercure, C., Rondeau, N., Chretien, M., and Seidah, N. G. (1992). Proprotein conversion is determined by a multiplicity of factors including convertase processing, substrate specificity, and intracellular environment. Cell type-specific processing of human prorenin by the convertase PC1. *J Biol Chem* **267**: 11417–11423.

Bertaux, F., Colledge, W. H., Smith, S. E., Evans, M., Samani, N. J., and Miller, C. C. (1997). Normotensive blood pressure in mice with a disrupted renin Ren-1d gene. *Transgenic Res* **6**: 191–196.

Bohlender, J., Fukamizu, A., Lippoldt, A. et al. (1997). High human renin hypertension in transgenic rats. *Hypertension* **29**: 428–434.

Brandt, C. R., Pumfery, A. M., Micales, B. et al. (1994). Renin mRNA is synthesized locally in rat ocular tissues. *Curr Eye Res* **13**: 755–763.

Buhler, F. R., Laragh, J. H., Baer, L., Vaughan, E. D., and Brunner, H. R. (1972). Propranolol inhibition of renin secretion. A specific approach to diagnosis and treatment of renin-dependent hypertensive diseases. *N Engl J Med* **287**: 1209–1214.

Campbell, D. J., Kladis, A., Skinner, S. L., and Whitworth, J. A. (1991a). Characterization of angiotensin peptides in plasma of anephric man. *J Hypertens* **9**: 265–274.

Campbell, D. J., Valentijn, A. J., and Condron, R. (1991b). Purification and amino-terminal sequence of rat kidney renin: evidence for a two-chain structure. *J Hypertens* **9**: 29–33.

Catanzaro, D. F., Mesterovic, N., and Morris, B. J. (1985). Studies of the regulation of mouse renin genes by measurement of renin messenger ribonucleic acid. *Endocrinology* **117**: 872–878.

Catanzaro, D. F., Sun, J., Gilbert, M. T. et al. (1994). A Pit-1 binding site in the human renin gene promoter stimulates activity in pituitary, placental and juxtaglomerular cells. *Kidney Int* **46**: 1513–1515.

Chen, Y., Lasaitiene, D., and Friberg, P. (2004). The renin-angiotensin system in kidney development. *Acta Physiol Scand* **181**: 529–535.

Clark, A. F., Sharp, M. G., Morley, S. D., Fleming, S., Peters, J., and Mullins, J. J. (1997). Renin-1 is essential for normal renal juxtaglomerular cell granulation and macula densa morphology. *J Biol Chem* **272**: 18185–18190.

Cohen-Haguenauer, O., Soubrier, F., Van, C. N. et al. (1989). Regional mapping of the human renin gene to 1q32 by in situ hybridization. *Ann Genet* **32**: 16–20.

Danser, A. H., Derkx, F. H., Schalekamp, M. A., Hense, H. W., Riegger, G. A., and Schunkert, H. (1998). Determinants of interindividual variation of renin and prorenin concentrations: evidence for a sexual dimorphism of (pro)renin levels in humans. *J Hypertens* **16**: 853–862.

Della, B. R., and Kurtz, A. (1995). Juxtaglomerular cells in culture. *Exp Nephrol* **3**: 219–222.

Derkx, F. H., Schalekamp, M. P., and Schalekamp, M. A. (1987). Two-step prorenin-renin conversion. Isolation of an intermediary form of activated prorenin. *J Biol Chem* **262**: 2472–2477.

Dickinson, D. P., Gross, K. W., Piccini, N., and Wilson, C. M. (1984). Evolution and variation of renin genes in mice. *Genetics* **108**: 651–667.

Do, Y. S., Shinagawa, T., Tam, H., Inagami, T., and Hsueh, W. A. (1987). Characterization of pure human renal renin. Evidence for a subunit structure. *J Biol Chem* **262**: 1037–1043.

Everett, A. D., Carey, R. M., Chevalier, R. L., Peach, M. J., and Gomez, R. A. (1990). Renin release and gene expression in intact rat kidney microvessels and single cells. *J Clin Invest* **86**: 169–175.

Fabian, J. R., Field, L. J., McGowan, R. A., Mullins, J. J., Sigmund, C. D., and Gross, K. W. (1989). Allele-specific expression of the murine Ren-1 genes. *J Biol Chem* **264**: 17589–17594.

Field, L. J., and Gross, K. W. (1985). Ren-1 and Ren-2 loci are expressed in mouse kidney. *Proc Natl Acad Sci USA* **82**: 6196–6200.

Fiocchi, R., Lijnen, P., Fagard, R. et al. (1984). Captopril during pregnancy. *Lancet* **2**: 1153.

Fuchs, S., Philippe, J., Germain, S. et al. (2002). Functionality of two new polymorphisms in the human renin gene enhancer region. *J Hypertens* **20**: 2391–2398.

Fukamizu, A., Nishi, K., Cho, T. et al. (1988). Structure of the rat renin gene. *J Mol Biol* **201**: 443–450.

Fukamizu, A., Seo, M. S., Hatae, T. et al. (1989). Tissue-specific expression of the human renin gene in transgenic mice. *Biochem Biophys Res Commun* **165**: 826–832.

Ganong, W. F., Deschepper, C. F., Steele, M. K., and Intebi, A. (1989). Renin-angiotensin system in the anterior pituitary of the rat. *Am J Hypertens* **2**: 320–322.

Ganten, D., Wagner, J., Zeh, K. et al. (1992). Species specificity of renin kinetics in transgenic rats harboring the human renin and angiotensinogen genes. *Proc Natl Acad Sci USA* **89**: 7806–7810.

Germain, S., Bonnet, F., Philippe, J., Fuchs, S., Corvol, P., and Pinet, F. (1998). A novel distal enhancer confers chorionic expression on the human renin gene. *J Biol Chem* **273**: 25292–25300.

Gomez, R. A., Chevalier, R. L., Everett, A. D. et al. (1990). Recruitment of renin gene-expressing cells in adult rat kidneys. *Am J Physiol* **259**: F660–F665.

Gribouval, O., Gonzales, M., Neuhaus, T. et al. (2005). Mutations in genes in the renin-angiotensin system are associated with autosomal recessive renal tubular dysgenesis. *Nat Genet* **37**: 964–968.

Griendling, K. K., Murphy, T. J., and Alexander, R. W. (1993). Molecular biology of the renin-angiotensin system. *Circulation* **87**: 1816–1828.

Hardman, J. A., Hort, Y. J., Catanzaro, D. F. et al. (1984). Primary structure of the human renin gene. *DNA* **3**: 457–468.

Heinrikson, R. L., Hui, J., Zurcher-Neely, H., and Poorman, R. A. (1989). A structural model to explain the partial catalytic activity of human prorenin. *Am J Hypertens* **2**: 367–380.

Hobart, P. M., Fogliano, M., O'Connor, B. A., Schaefer, I. M., and Chirgwin, J. M. (1984). Human renin gene: structure and sequence analysis. *Proc Natl Acad Sci USA* **81**: 5026–5030.

Holm, I., Ollo, R., Panthier, J. J., and Rougeon, F. (1984). Evolution of aspartyl proteases by gene duplication: the mouse renin gene is organized in two homologous clusters of four exons. *EMBO J* **3**: 557–562.

Hosoi, M., Kim, S., Tabata, T. et al. (1992). Evidence for the presence of differently glycosylated forms of prorenin in the plasma of anephric man. *J Clin Endocrinol Metab* **74**: 680–684.

Hsueh, W. A., and Baxter, J. D. (1991). Human prorenin. *Hypertension* **17**: 469–477.

Jeunemaitre, X., Rigat, B., Charru, A., Houot, A. M., Soubrier, F., and Corvol, P. (1992). Sib pair linkage analysis of renin gene haplotypes in human essential hypertension. *Hum Genet* **88**: 301–306.

Johns, D. W., Carey, R. M., Gomez, R. A. et al. (1987). Isolation of renin-rich rat kidney cells. *Hypertension* **10**: 488–496.

Jutras, I., and Reudelhuber, T. L. (1999). Prorenin processing by cathepsin B in vitro and in transfected cells. *FEBS Lett* **443**: 48–52.

Kakinuma, Y., Hama, H., Sugiyama, F. et al. (1998). Impaired blood-brain barrier function in angiotensinogen-deficient mice. *Nat Med* **4**: 1078–1080.

Karginova, E. A., Pentz, E. S., Kazakova, I. G., Norwood, V. F., Carey, R. M., and Gomez, R. A. (1997). Zis: a developmentally regulated gene expressed in juxtaglomerular cells. *Am J Physiol* **273**: F731–F738.

Kikkawa, Y., Yamanaka, N., Tada, J., Kanamori, N., Tsumura, K., and Hosoi, K. (1998). Prorenin processing and restricted endoproteolysis by mouse tissue kallikrein family enzymes (mK1, mK9, mK13, and mK22). *Biochim Biophys Acta* **1382**: 55–64.

Kim, H. S., Maeda, N., Oh, G. T., Fernandez, L. G., Gomez, R. A., and Smithies, O. (1999). Homeostasis in mice with genetically decreased angiotensinogen is primarily by an increased number of renin-producing cells. *J Biol Chem* **274**: 14210–14217.

Kim, S., Hosoi, M., Kikuchi, N., and Yamamoto, K. (1991). Amino-terminal amino acid sequence and heterogeneity in glycosylation of rat renal renin. *J Biol Chem* **266**: 7044–7050.

Kurihara, Y., Kurihara, H., Suzuki, H. et al. (1994). Elevated blood pressure and craniofacial abnormalities in mice deficient in endothelin-1. *Nature* **368**: 703–710.

Laragh, J. H. (2001). Abstract, closing summary, and table of contents for Laragh's 25 lessons in pathophysiology and 12 clinical pearls for treating hypertension. *Am J Hypertens* **14**: 1173–1177.

Matsuba, H., Watanabe, T., Watanabe, M. et al. (1989). Immunocytochemical localization of prorenin, renin, and cathepsins B, H, and L in juxtaglomerular cells of rat kidney. *J Histochem Cytochem* **37**: 1689–1697.

Mercure, C., Jutras, I., Day, R., Seidah, N. G., and Reudelhuber, T. L. (1996). Prohormone convertase PC5 is a candidate processing enzyme for prorenin in the human adrenal cortex. *Hypertension* **28**: 840–846.

Mercure, C., Ramla, D., Garcia, R., Thibault, G., Deschepper, C. F., and Reudelhuber, T. L. (1998). Evidence for intracellular generation of angiotensin II in rat juxtaglomerular cells. *FEBS Lett* **422**: 395–399.

Mesterovic, N., Catanzaro, D. F., and Morris, B. J. (1983). Detection of renin mRNA in mouse kidney and submandibular gland by hybridization with renin cDNA. *Endocrinology* **113**: 1179–1181.

Methot, D., Silversides, D. W., and Reudelhuber, T. L. (1999). In vivo enzymatic assay reveals catalytic activity of the human renin precursor in tissues. *Circ Res* **84**: 1067–1072.

Misono, K. S., Chang, J. J., and Inagami, T. (1982). Amino acid sequence of mouse submaxillary gland renin. *Proc Natl Acad Sci USA* **79**: 4858–4862.

Miyazaki, H., Fukamizu, A., Hirose, S. et al. (1984). Structure of the human renin gene. *Proc Natl Acad Sci USA* **81**: 5999–6003.

Mori, M., Ishizaki, K., Serikawa, T., and Yamada, J. (1992). Instability of the minisatellite sequence in the first intron of the rat renin gene and localization of the gene to chromosome 13q13 between FH and PEPC loci. *J Hered* **83**: 204–207.

Mullins, J. J., and Ganten, D. (1990). Transgenic animals: new approaches to hypertension research. *J Hypertens Suppl* **8**: S35–S37.

Mulrow, P. J. (1989). Adrenal renin: a possible local regulator of aldosterone production. *Yale J Biol Med* **62**: 503–510.

Nabi, A. H., Kageshima, A., Uddin, M. N., Nakagawa, T., Park, E. Y., and Suzuki, F. (2006). Binding properties of rat prorenin and renin to the recombinant rat renin/prorenin receptor prepared by a baculovirus expression system. *Int J Mol Med* **18**: 483–488.

Naftilan, A. J., Williams, R., Burt, D. et al. (1989). A lack of genetic linkage of renin gene restriction fragment length polymorphisms with human hypertension. *Hypertension* **14**: 614–618.

Nguyen, G., Delarue, F., Burckle, C., Bouzhir, L., Giller, T., and Sraer, J. D. (2002). Pivotal role of the renin/prorenin receptor in angiotensin II production and cellular responses to renin. *J Clin Invest* **109**: 1417–1427.

Pan, L., and Gross, K. W. (2005). Transcriptional regulation of renin: an update. *Hypertension* **45**: 3–8.

Pan, L., Xie, Y., Black, T. A., Jones, C. A., Pruitt, S. C., and Gross, K. W. (2001). An Abd-B class HOX.PBX recognition sequence is required for expression from the mouse Ren-1c gene. *J Biol Chem* **276**: 32489–32494.

Paul, M., Wagner, J., and Dzau, V. J. (1993). Gene expression of the renin-angiotensin system in human tissues. Quantitative analysis by the polymerase chain reaction. *J Clin Invest* **91**: 2058–2064.

Pentz, E. S., Lopez, M. L., Kim, H. S., Carretero, O., Smithies, O., and Gomez, R. A. (2001). Ren1d and Ren2 cooperate to preserve homeostasis: evidence from mice expressing GFP in place of Ren1d. *Physiol Genomics* **6**: 45–55.

Pentz, E. S., Sequeira Lopez, M. L., Cordaillat, M., and Gomez, R. A. (2007). Identity of the renin cell is mediated by cAMP and chromatin remodeling: an in vitro model for studying cell recruitment and plasticity. *Am J Physiol Heart Circ Physiol* **294**: H699–H707.

Petrovic, N., Black, T. A., Fabian, J. R. et al. (1996). Role of proximal promoter elements in regulation of renin gene transcription. *J Biol Chem* **271**: 22499–22505.

Pitarresi, T. M., Rubattu, S., Heinrikson, R., and Sealey, J. E. (1992). Reversible cryoactivation of recombinant human prorenin. *J Biol Chem* **267**: 11753–11759.

Pratt, R. E., Carleton, J. E., Roth, T. P., and Dzau, V. J. (1988). Evidence for two cellular pathways of renin secretion by the mouse submandibular gland. *Endocrinology* **123**: 1721–1727.

Pravenec, M., Simonet, L., Kren, V. et al. (1991). The rat renin gene: assignment to chromosome 13 and linkage to the regulation of blood pressure. *Genomics* **9**: 466–472.

Prescott, G., Silversides, D. W., and Reudelhuber, T. L. (2002). Tissue activity of circulating prorenin. *Am J Hypertens* **15**: 280–285.

Rapp, J. P., Wang, S. M., and Dene, H. (1989). A genetic polymorphism in the renin gene of Dahl rats cosegregates with blood pressure. *Science* **243**: 542–544.

Reddi, V., Zaglul, A., Pentz, E. S., and Gomez, R. A. (1998). Renin-expressing cells are associated with branching of the developing kidney vasculature. *J Am Soc Nephrol* **9**: 63–71.

Reudelhuber, T. L., Ramla, D., Chiu, L., Mercure, C., and Seidah, N. G. (1994). Proteolytic processing of human prorenin in renal and non-renal tissues. *Kidney Int* **46**: 1522–1524.

Sealey, J. E., and Rubattu, S. (1989). Prorenin and renin as separate mediators of tissue and circulating systems. *Am J Hypertens* **2**: 358–366.

Seo, M. S., Fukamizu, A., Saito, T., and Murakami, K. (1991). Identification of a previously unrecognized production site of human renin. *Biochim Biophys Acta* **1129**: 87–89.

Sequeira Lopez, M. L., Pentz, E. S., Nomasa, T., Smithies, O., and Gomez, R. A. (2004). Renin cells are precursors for multiple cell types that switch to the renin phenotype when homeostasis is threatened. *Dev Cell* **6**: 719–728.

Sequeira Lopez, M. L., Pentz, E. S., Robert, B., Abrahamson, D. R., and Gomez, R. A. (2001). Embryonic origin and lineage of juxtaglomerular cells. *Am J Physiol Renal Physiol* **281**: F345–F356.

Sharp, M. G., Fettes, D., Brooker, G. et al. (1996). Targeted inactivation of the Ren-2 gene in mice. *Hypertension* **28**: 1126–1131.

Sigmund, C. D., and Gross, K. W. (1991). Structure, expression, and regulation of the murine renin genes. *Hypertension* **18**: 446–457.

Sigmund, C. D., Jones, C. A., Kane, C. M., Wu, C., Lang, J. A., and Gross, K. W. (1992). Regulated tissue- and cell-specific expression of the human renin gene in transgenic mice. *Circ Res* **70**: 1070–1079.

Sigmund, C. D., Okuyama, K., Ingelfinger, J. et al. (1990). Isolation and characterization of renin-expressing cell lines from transgenic mice containing a renin-promoter viral oncogene fusion construct. *J Biol Chem* **265**: 19916–19922.

Sinn, P. L., Davis, D. R., and Sigmund, C. D. (1999). Highly regulated cell type-restricted expression of human renin in mice containing 140- or 160-kilobase pair P1 phage artificial chromosome transgenes. *J Biol Chem* **274**: 35785–35793.

Soubrier, F., Jeunemaitre, X., Rigat, B., Houot, A. M., Cambien, F., and Corvol, P. (1990). Similar frequencies of renin gene restriction fragment length polymorphisms in hypertensive and normotensive subjects. *Hypertension* **16**: 712–717.

Sun, J., Oddoux, C., Gilbert, M. T. et al. (1994). Pituitary-specific transcription factor (Pit-1) binding site in the human renin gene 5'-flanking DNA stimulates promoter activity in placental cell primary cultures and pituitary lactosomatotropic cell lines. *Circ Res* **75**: 624–629.

Sun, J., Oddoux, C., Lazarus, A., Gilbert, M. T., and Catanzaro, D. F. (1993). Promoter activity of human renin 5'-flanking DNA sequences is activated by the pituitary-specific transcription factor Pit-1. *J Biol Chem* **268**: 1505–1508.

Takaori, K., Kim, S., Fukamizu, A. et al. (1993). Biochemical characteristics of human renin expressed in transgenic mice. *Clin Sci (Lond)* **84**: 21–29.

Taugner, R., Buhrle, C. P., and Nobiling, R. (1984). Ultrastructural changes associated with renin secretion from the juxtaglomerular apparatus of mice. *Cell Tissue Res* **237**: 459–472.

Taugner, R., Kim, S. J., Murakami, K., and Waldherr, R. (1987). The fate of prorenin during granulopoiesis in epithelioid cells. Immunocytochemical experiments with antisera against renin and different portions of the renin prosegment. *Histochemistry* **86**: 249–253.

Taugner, R., Metz, R., and Rosivall, L. (1988). Macroautophagic phenomena in renin granules. *Cell Tissue Res* **251**: 229–231.

Wang, P. H., Do, Y. S., Macaulay, L. et al. (1991). Identification of renal cathepsin B as a human prorenin-processing enzyme. *J Biol Chem* **266**: 12633–12638.

Yan, Y., Chen, R., Pitarresi, T. et al. (1998). Kidney is the only source of human plasma renin in 45-kb human renin transgenic mice. *Circ Res* **83**: 1279–1288.

Yan, Y., Jones, C. A., Sigmund, C. D., Gross, K. W., and Catanzaro, D. F. (1997a). Conserved enhancer elements in human and mouse renin genes have different transcriptional effects in As4.1 cells. *Circ Res* **81**: 558–566.

Yan, Y., Jones, C. A., Sigmund, C. D., Gross, K. W., and Catanzaro, D. F. (1997b). Conserved enhancer elements in human and mouse renin genes have different transcriptional effects in As4.1 cells. *Circ Res* **81**: 558–566.

Yanai, K., Saito, T., Kakinuma, Y. et al. (2000). Renin-dependent cardiovascular functions and renin-independent blood-brain barrier functions revealed by renin-deficient mice. *J Biol Chem* **275**: 5–8.

Zee, R. Y., Ying, L. H., Morris, B. J., and Griffiths, L. R. (1991). Association and linkage analyses of restriction fragment length polymorphisms for the human renin and antithrombin III genes in essential hypertension. *J Hypertens* **9**: 825–830.

Zhou, X., Davis, D. R., and Sigmund, C. D. (2006). The human renin kidney enhancer is required to maintain base-line renin expression but is dispensable for tissue-specific, cell-specific, and regulated expression. *J Biol Chem* **281**: 35296–35304.

Zhou, X., and Sigmund, C. D. (2008). The chorionic enhancer is dispensable for regulated expression of the human renin gene. *Am J Physiol Regul Integr Comp Physiol* **294**(2): R279–R287.

CHAPTER **12**

Physiology and Regulation of the Renin–Angiotensin–Aldosterone System

ROBERT M. CAREY AND SHETAL H. PADIA
Division of Endocrinology and Metabolism, Department of Medicine, University of Virginia Health System, Charlottesville, VA, USA

Contents

I.	Introduction	147
II.	The classical circulating renin–angiotensin system (RAS)	147
III.	Renin biosynthesis and secretion	148
IV.	The renin receptor	149
V.	Angiotensin-converting enzyme (ACE)	150
VI.	The ACE-2/angiotensin (1–7)/*mas* receptor pathway	151
VII.	AT$_1$ receptors	151
VIII.	AT$_2$ receptors	152
IX.	Angiotensin receptor heterodimerization	153
X.	Tissue renin–angiotensin systems	153
XI.	Intrarenal renin–angiotensin system	153
XII.	Brain renin–angiotensin system	155
XIII.	Vascular tissue renin–angiotensin system	156
XIV.	Cardiac renin–angiotensin system	157
XV.	Aldosterone and mineralocorticoid receptors	157
XVI.	Clinical effects of the renin–angiotensin–aldosterone system (RAAS)	159
XVII.	Summary	161
	References	161

I. INTRODUCTION

The renin–angiotensin–aldosterone system (RAAS) is a major hormonal regulatory system in the control of blood pressure (BP) and hypertension (HT) (Carey and Siragy, 2003). Several new components and pathways of the RAAS have been described during the past 5 years. In this chapter, these new components and pathways will be described and their potential clinical significance discussed.

II. THE CLASSICAL CIRCULATING RENIN–ANGIOTENSIN SYSTEM (RAS)

The classical RAS (Figure 12.1) begins with the biosynthesis of the glycoprotein hormone, renin, by the juxtaglomerular (JG) cells of the renal afferent arteriole (Figure 12.2). Renin is encoded by a single gene and renin mRNA is translated into preprorenin, containing 401 amino acids (Peach, 1977; Griendling et al., 1993). In the JG cell endoplasmic reticulum, a 20-amino-acid signal peptide is cleaved from preprorenin, leaving prorenin, which is packaged into secretory granules in the Golgi apparatus, where it is further processed into 'active' renin by severance of a 46-amino-acid peptide from the N-terminal region of the molecule. Mature, 'active' renin is a glycosylated carboxypeptidase with a molecular weight of approximately 44 kDa. 'Active' renin is released from the JG cell by a process of exocytosis involving stimulus-secretion coupling. In contrast, 'inactive' prorenin is released constitutively across the cell membrane. Prorenin is converted to 'active' renin by a trypsin-like activation step (Hsueh and Baxter, 1991).

In the past, renin has been considered to have no intrinsic biological activity, serving solely as an enzyme catalytically cleaving angiotensinogen (Agt), the only known precursor of angiotensin peptides, to form the decapeptide angiotensin I (Ang I) (see Figure 12.1). Liver-derived Agt provides the majority of systemic circulating angiotensin (Ang) peptides, but Agt also is synthesized and constitutively released in other tissues, including heart, vasculature, kidney and adipose tissue. Angiotensin-converting enzyme (ACE), a glycoprotein (molecular weight 180 kDa) with two active carboxy-terminal enzymatic sites, hydrolyzes the inactive Ang I into biologically active Ang II (Soubrier et al., 1993) (see Figure 12.1). ACE exists in two molecular forms, soluble and particulate. ACE is localized on the plasma membranes of various cell types, including vascular endothelial cells, the apical brush border (microvilli) of epithelial cells (e.g. renal proximal tubule cells) and neuroepithelial cells. In addition to cleaving Ang I to Ang II, ACE metabolizes bradykinin (BK), an active vasodilator and natriuretic autacoid, to BK (1–7), an inactive metabolite (Erdos and Skidgel, 1997) (see Figure 12.1). ACE, therefore, increases the production of a potent vasoconstrictor, Ang II, while simultaneously degrading a vasodilator, BK. ACE also metabolizes substance P into inactive fragments.

Unlike renin and Agt, which have relatively long plasma half-lives, Ang II is degraded within seconds by peptidases,

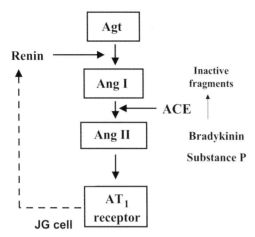

FIGURE 12.1 Schematic depiction of the classical renin–angiotensin system. Agt, angiotensinogen; Ang, angiotensin; ACE, angiotensin-converting enzyme; JG, juxtaglomerular. Dashed line: 'short-loop' negative feedback inhibition.

collectively termed angiotensinases, at different amino-acid sites, to form fragments, mainly des-aspartyl^1-Ang II (Ang III), Ang (1–7) and Ang (3–8) (Ang IV). Ang II is converted to Ang III by aminopeptidase A and Ang III is converted to Ang IV by aminopeptidase N (1). However, the expression levels and functional significance of these two critical enzymes, especially at the tissue level, are not completely understood at the present time and the functional role of the peptide fragments produced is largely unknown.

The vast majority of cardiovascular, renal and adrenal actions of Ang II are mediated by the Ang type-1 (AT_1) receptor, a seven transmembrane G-protein-coupled receptor that is widely distributed in these tissues, which is coupled positively to protein kinase C and is negatively coupled to adenylyl cyclase (de Gasparo et al., 2000). As shown in Table 12.1, AT_1 receptors mediate vascular smooth muscle cell contraction, aldosterone secretion, thirst, sympathetic nervous system stimulation, renal tubular Na^+ reabsorption and cardiac ionotropic and chronotropic responses. Ang II also binds to another cloned receptor, the Ang type-2 (AT_2) receptor but, until recently, the cell signaling mechanisms and functions of the AT_2 receptor were unknown (de Gasparo et al., 2000).

III. RENIN BIOSYNTHESIS AND SECRETION

Renin catalytic cleavage of Agt is the rate-limiting biochemical step in the RAS (2). The renal JG cell is thought to be the only source of circulating renin because, following bilateral nephrectomy, renin quantitatively disappears from the circulation (Sealey et al., 1977). However, nephrectomy does not alter circulating levels of prorenin, indicating that non-renal tissues both produce and secrete prorenin into the circulation. In addition, many organs, such as the heart, can take up renin from the circulation by uncertain mechanisms (Prescott et al., 2000, 2002) (see Chapter 14).

The primary means by which the RAAS contributes to acute changes in extracellular fluid volume and BP homeostasis is by varying the level of renin in the circulation. This process is mediated by active renin release from secretory granules of JG cells. A primary mechanism of renin release is the afferent arteriolar baroreceptor, which increases renin release when arterial (and renal) perfusion pressure decreases and vice versa. In addition, JG cells are innervated by sympathetic neurons, the activation of which stimulates norepinephrine release and subsequent stimulation of β_1-adrenergic receptors triggering renin release. Therefore, as shown in Figure 12.3, β_1-adrenergic receptor blockade suppresses renin release directly at JG cells. JG cells also express both AT_1 and AT_2 receptors and circulating Ang II

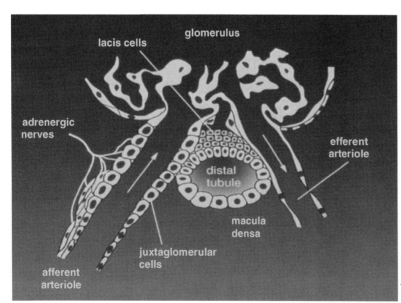

FIGURE 12.2 Schematic representation of the renal juxtaglomerular apparatus showing the various components. (See color plate section.)

TABLE 12.1 Effects of Ang II via AT$_1$ receptors

Vasoconstriction
Activation of the SNS
Aldosterone, vasopressin and endothelin secretion
Cardiac contractility
Renin inhibition
Sodium and water retention
Thrombosis:
- PAI-1 synthesis, platelet activation, aggregation and adhesion

Inflammation:
- activation of cytokine production by monocytes and macrophages

Cardiac and vascular remodeling
- vascular smooth muscle hypertrophy, migration, proliferation, growth and fibrosis

Endothelial dysfunction
- superoxide anion production
- NO destruction

Decreased vascular compliance
Tissue fibrosis
- collagen biosynthesis

SNS, sympathetic nervous system; PAI-1, plasminogen activator inhibitor-1; NO, nitric oxide.

participates in a short-loop negative feedback mechanism to inhibit renin release by binding to these two receptors (Siragy et al., 2005). Conversely, blockade of the RAS increases renin release and circulating renin levels (see Figure 12.4 for ACE inhibitors and Figure 12.5 for AT$_1$ receptor blockers). Indeed, chronic RAS blockade by AT$_1$ receptor antagonists or ACE inhibitors induces recruitment of new renin secreting cells in renal microvessels, further augmenting renin secretion (Gomez et al., 1988). Another renin secretory control mechanism is the *macula densa* segment of the early distal tubule, which relays a signal to the JG cell to increase renin release when a reduction in Na$^+$ and/or Cl$^-$ in the distal tubule is perceived.

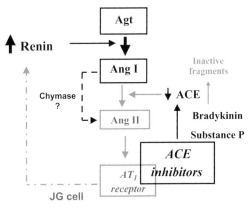

FIGURE 12.4 Schematic representation of changes in the renin–angiotensin system in response to ACE inhibition. Ang II formation and bradykinin and substance P degradation are simultaneously reduced (gray) while renin biosynthesis and secretion are markedly increased due to inhibition of Ang II interaction with the AT$_1$ receptor on JG cells (short-loop negative feedback).

IV. THE RENIN RECEPTOR

Although renin has been considered as the enzyme responsible for cleaving the decapeptide Ang I from substrate Agt and has been thought to have no direct biological actions, recent studies demonstrate that renin can bind to human glomerular mesangial cell membranes in culture and that binding causes cell hypertrophy and increased levels of plasminogen-activator inhibitor (Nguyen et al., 1996, 1998). The bound renin is not internalized or degraded. A renin receptor has now been cloned from mesangial cells and its functional significance is being clarified (Nguyen et al., 2002). The receptor is a 350-amino-acid protein with a single transmembrane domain that specifically binds

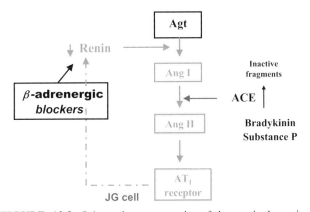

FIGURE 12.3 Schematic representation of changes in the renin–angiotensin system in response to β_1-adrenergic receptor blockade. Renin secretion and Ang peptide production are uniformly suppressed, as depicted in gray.

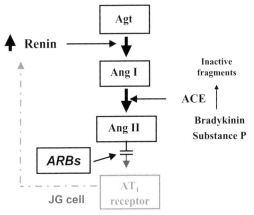

FIGURE 12.5 Schematic representation of changes in the renin–angiotensin system in response to angiotensin AT$_1$ receptor blockers (ARBs). Renin biosynthesis and secretion are driven to high levels by interruption of short-loop negative feedback (gray), leading to markedly increased Ang II levels.

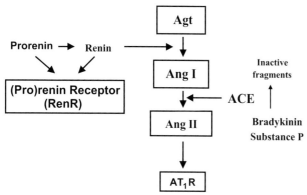

FIGURE 12.6 Schematic representation of the renin–angiotensin system depicting the interaction of prorenin and renin with a newly discovered and cloned (pro)renin receptor.

both renin and prorenin (Figure 12.6) (Nguyen et al., 2002). Binding induces the activation of the extracellular signal-related mitogen-activated protein kinases (ERK 1 and ERK 2) associated with serine and tyrosine phosphorylation and a fourfold increase in the catalytic conversion of Agt to Ang I (Figure 12.7). The receptor is localized on renal mesangial cell membranes and in the subendothelial layer of both coronary and renal arteries, associated with vascular smooth muscle cells, and co-localizes with renin (Nguyen et al., 2002). The receptor also is expressed in visceral adipocytes. In renal mesangial cells, the renin receptor mediates transforming growth factor-β production via MAP kinase phosphorylation (Figure 12.8). Although the possibility of a direct biological role of renin and prorenin via a renin/prorenin receptor exists, the functional importance of this receptor other than catalytic conversion of Agt to Ang I awaits further investigation.

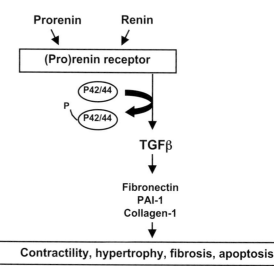

FIGURE 12.8 Schematic representation of potential Ang II-independent direct effects of renin and/or prorenin mediated by the recently discovered (pro)renin receptor. Receptor activation results in phosphorylation of MAP kinases (P42/44) which mediate increased production of transforming growth factor-β (TGFβ), resulting in fibronectin, PAI-1 and collagen-1 formation in renal mesangial cells. These changes lead to increased contractility, hypertrophy, fibrosis and apoptosis. Adapted from Huang et al., *Curr Hypertens Reports* 9, 133–139, 2007 with permission.

V. ANGIOTENSIN-CONVERTING ENZYME (ACE)

ACE inactivates two vasodilator peptides, BK and kallidin. BK is both a direct and an indirect vasodilator via stimulation of NO and cGMP and also by release of the vasodilator prostaglandins, PGE$_2$ and prostacyclin (Linz et al., 1995). Thus, when an ACE inhibitor is employed (see Figure 12.4),

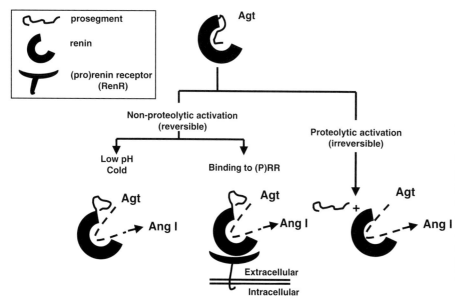

FIGURE 12.7 Schematic illustration of the interaction of renin with angiotensinogen to cleave the decapeptide angiotensin I. Exposure to low temperature, acidification or binding of renin to the (pro)renin receptor ((P)RR) displaces the prosegment peptide, allowing catalytic conversion of angiotensin I to occur in a reversible manner. Proteolytic cleavage of the prosegment peptide leads to irreversible catalytic cleavage. Adapted from Danser et al., *J Cardiovasc Pharmacol* 50, 105–111, 2007 with permission.

not only is the synthesis of Ang II inhibited but also the formation of BK, NO and prostaglandins is facilitated. ACE inhibition induces cross-talk between the BK B_2 receptor and ACE on the plasma membrane, abrogating B_2 receptor desensitization and potentiating both the levels of BK and the vasodilator action of BK at its B_2 receptor (Erdos and Marcic, 2001; Tschope et al., 2002). Also, in the presence of ACE inhibition, an alternative pathway of Ang II production via chymase may be activated (see Figure 12.4). The chymase pathway may serve as a major route of Ang II formation in the heart.

VI. THE ACE-2/ANGIOTENSIN (1–7)/*MAS* RECEPTOR PATHWAY

A second ACE has recently been discovered (Figure 12.9). ACE-2 is a zinc metalloproteinase consisting of 805 amino acids with significant sequence homology to ACE (Tipnis et al., 2000). Unlike ACE, however, ACE-2 functions as a carboxypeptidase rather than a dipeptidyl-carboxypeptidase. In contrast to ACE, ACE-2 hydrolyzes Ang I to Ang (1–9), but the major pathway is the conversion of Ang II to Ang (1–7) (Figure 12.9). ACE-2 also degrades BK to (des-Arg^9)-BK, an inactive metabolite. In marked contrast to ACE, ACE-2 does not convert Ang I to Ang II and its enzyme activity is not blocked with ACE inhibitors. Thus, ACE-2 is effectively an inhibitor of Ang II formation by stimulating alternate pathways for Ang I and, particularly, Ang II degradation. ACE-2 has been localized to the cell membranes of cardiac myocytes, renal endothelial and tubule cells and the testis. ACE-2 gene ablation does not alter BP but impairs cardiac contractility and induces increased Ang II levels, suggesting that ACE-2 may at least partially nullify the physiological actions of ACE (Crackower et al., 2002).

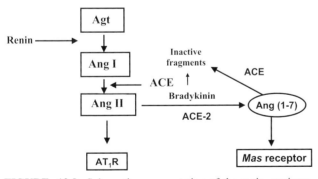

FIGURE 12.9 Schematic representation of the renin–angiotensin system depicting Ang II binding to both AT_1 and AT_2 receptors and the newly discovered ACE-2 pathway for conversion of Ang II directly to Ang(1–7), which interacts with the *mas* receptor to inhibit cell growth and stimulate vasodilation and natriuresis via prostaglandins and NO. Since Ang (1–7) is metabolized to inactive fragments by ACE, ACE inhibition results in increased Ang (1–7) levels.

The heptapeptide fragment of Ang II, Ang (1–7) (see Figure 12.9) has been discovered to have biological activity (Schiavone et al., 1988) (see Chapter 14 for additional details). Ang (1–7) can be formed directly from Ang I by a two-step process involving conversion to Ang (1–9) by ACE-2 followed by conversion to Ang (1–7) by endopeptidases. However, as stated above, the major pathway for Ang (1–7) formation is directly from Ang II by the action of ACE-2 (see Figure 12.9). Interestingly, the major catabolic pathway for inactivation of Ang (1–7) is by ACE (see Figure 12.9). Thus, ACE inhibitor administration markedly increases the level of Ang (1–7) (Chappell et al., 2001). The kidney is a major target organ for Ang (1–7). The peptide is formed in the kidney, where it has specific actions via a non-AT_1 or -AT_2 receptor. These actions include increased glomerular filtration rate (GFR), inhibition of Na/K/ATPase, vasorelaxation and downregulation of AT_1 receptors, all of which are blocked by the specific Ang (1–7) antagonist (D-Ala^7)-Ang (1–7) and are mediated at least in part by NO and prostacyclin (Diz et al., 2005). Most of these effects of Ang (1–7) oppose those of Ang II via the AT_1 receptor. Although a specific Ang (1–7) receptor has not been cloned, the peptide is an endogenous ligand for the *mas* oncogene, which mediates many of its actions (see Figure 12.9) (Santos et al., 2003).

VII. AT_1 RECEPTORS

Ang II, the major effector peptide of the RAS, binds to two major receptors, AT_1 and AT_2, that generally oppose each other (de Gasparo et al., 2000). The AT_1 receptor is widely distributed in the vasculature, heart and kidney (Harrison-Bernard et al., 1997; Matsubara et al., 1998; Miyata et al., 1999; Allen, 2000; Pounarat, 2002). Actions of Ang II mediated by the AT_1 receptor includes vasoconstriction, SNS activation, aldosterone, vasopressin and endothelin secretion, plasminogen activator inhibitor biosynthesis, platelet aggregation, thrombosis, cardiac contractility, superoxide formation, VSM growth and collagen formation (see Table 12.1). These actions are conducted by both G-protein-coupled and -independent pathways and involve phospholipases C, A_2 and D activation, increased intracellular Ca^{++} and inositol 1, 4, 5- trisphosphate, activation of MAP kinases, ERKs and the JAK/STAT pathway, enhanced protein phosphorylation and stimulation of early growth response genes (Murphy et al., 1991; Sasaki et al., 1991; Schmitz, 1997; Ishida et al., 1998; Lijnen and Petrov, 1999; Schmitz et al., 2001). Tyrosine phosphorylation and stimulation of MAP kinase phosphorylation are the major intracellular signaling pathways for the AT_1 receptor (Giasson et al., 1997). Ang II, via AT_1 receptors, activates c-SRC, generating reactive oxygen species via NADPH oxidase (Nox1). Many of the detrimental tissue effects of Ang II, including vascular smooth muscle contraction, hyperplasia/hypertrophy, fibrosis and inflammation, involve the actions of reactive oxygen species on these intracellular signaling pathways.

VIII. AT$_2$ RECEPTORS

The second major Ang II receptor is the AT$_2$ receptor (Figure 12.10). AT$_2$ receptors are highly expressed in fetal tissues but regress substantially in the postnatal period (Sadjadi et al., 2002; Crowley et al., 2005). However, the AT$_2$ receptor is still expressed at low copy in the adult vasculature, especially in the endothelium and renal vasculature, JG cells, glomeruli and tubules (van Kats et al., 1998). The AT$_2$ receptor acts via the third intracellular loop by a G$_i$ protein-mediated process involving stimulation of protein tyrosine phosphatases and reduction of ERK phosphorylation and activity (van Kats et al., 2001). The AT$_2$ receptor also induces sphingolipid and ceramide accumulation (van Kats et al., 2001). A major mechanism of action of AT$_2$ receptors is BK release (probably via kininogen activation through cellular acidification), with consequent NO and cGMP generation (Danser et al., 1994; van Kats et al., 1998, 2000) (Figure 12.11). AT$_2$ receptors also can stimulate NO directly without BK as an intermediate. The AT$_2$ receptor mediates vasodilation, natriuresis and inhibition of cell growth (Campbell et al., 1993; Danser et al., 1994; van Kats et al., 1998, 2000, 2001; Nussberger, 2000). Ang III, not Ang II, appears to be the preferred agonist for AT$_2$ receptor-mediated natriuresis. When the AT$_1$ receptor is blocked, augmentation of renin release by inhibition of short-loop negative feedback leads to increased Ang II formation (see Figure 12.11). Increased levels of Ang II, while inhibited from binding to the AT$_1$ receptor, are free to activate the

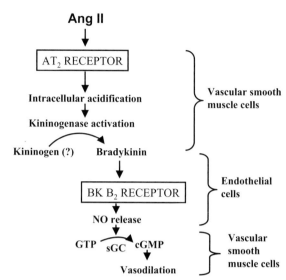

FIGURE 12.11 Schematic depiction of the paracrine vasodilator cascade elicited by activation of AT$_2$ receptors by Ang II. AT$_2$ receptor activation can stimulate NO production directly or can do so via increased levels of bradykinin (BK) via its B$_2$ receptor.

unblocked AT$_2$ receptor, potentially leading to vasodilation and/or natriuresis (see Figure 12.11). Acute studies in experimental animals from our laboratory have demonstrated that these beneficial effects of AT$_1$ receptor blockers are mediated, at least acutely, by activation of AT$_2$ receptors (Figure 12.12). Thus, clinically, it was anticipated that AT$_1$ receptor blockers would have a greater cardiovascular and renal protective effects than ACE inhibitors. This prediction has not been proven to be the case (The ONTARGET Investigators, 2008). However, equivalent clinical efficacy of ACE inhibitors and AT$_1$ receptor blockers does not test whether AT$_2$ receptor activation participates in the beneficial

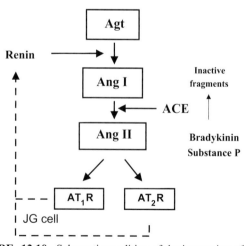

FIGURE 12.10 Schematic rendition of the interaction of angiotensin II with the AT$_1$ and AT$_2$ receptors, both of which mediate negative feedback inhibition of renin release (dashed lines) at renal juxtaglomerular cells.

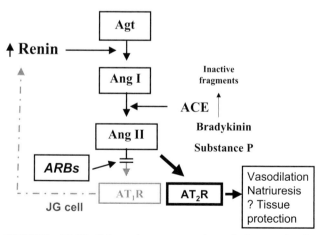

FIGURE 12.12 Schematic representation of changes in the renin–angiotensin system induced by AT$_1$ receptor blockade (ARBs). Ang II levels are increased by interruption of short-loop negative feedback on renin secretion (dashed gray line). Ang II is free to activate unblocked AT$_2$ receptors inducing vasodilation and natriuresis.

actions of AT_1 receptor blockade. AT_2 receptors, similar to AT_1 receptors, mediate short-loop negative feedback of renin biosynthesis and secretion at JG cells (see Figure 12.10).

IX. ANGIOTENSIN RECEPTOR HETERODIMERIZATION

Some of the actions of Ang II may be related to heterodimerization. If AT_1 and AT_2 receptors are expressed in the same cell, the physical association of these receptors on the cell membrane may inhibit the action of AT_1 receptors in a ligand-independent manner (Hilgers et al., 2001). Similarly, there is evidence for AT_1 receptor and BK B_2 receptor heterodimerization resulting in increased AT_1 receptor effects via G-protein activation and, as shown by our laboratory, AT_2-B_2 receptor heterodimerization resulting in increased cGMP formation (Re, 2004a; Chai and Danser, 2005).

X. TISSUE RENIN–ANGIOTENSIN SYSTEMS

As discussed above, the circulating RAS is only one component of the overall RAS. Tissue RASs that may operate completely independently of the circulating RAS have now been identified in brain, kidney, heart, blood vessels and the adrenal gland. Given the independence of tissue RASs, it is possible that they are activated even when the circulating RAS is normal or suppressed. However, it has been difficult to show the relative roles of the intrarenal RAS as opposed to the systemic RAS. This issue has been considered important because the long-term regulation of BP has been thought to involve the kidney and the ability to sustain a hypertensive process chronically has been regarded as requiring renal Na^+ retention. Indeed, the key sites that determine the level of BP could not be localized precisely utilizing ACE inhibitors or AT_1 receptor blockers which inhibit the RAS in all tissues or by conventional gene targeting experiments. Recent work, however, has helped clarify the tissue sites whereby the RAS regulates BP through a cross-transplantation approach in AT_{1A} receptor-deficient mice (Crowley et al., 2005). Rodents have two AT_1 receptors termed, AT_{1A} and AT_{1B} receptors. The AT_{1A} receptor is considered as the major AT receptor mediating the majority of actions of Ang II. In terms of expression, AT_{1A} receptors predominate in most organs except the adrenal gland and regions of the central nervous system, wherein AT_{1B} receptor expression is more predominant (Sadjadi et al., 2002). Absence of AT_{1A} receptors exclusively in the kidney, with normal receptors elsewhere, was sufficient to lower BP by about 20 mmHg (Crowley et al., 2005). Thus, renal AT_1 receptors were demonstrated to have a unique and non-redundant role in the control of BP homeostasis. As aldosterone levels were unaffected in these experiments, BP appears to be regulated by the direct action of AT_1 receptors upon kidney cells, independent of mineralocorticoids. However, in addition to the kidney, AT_1 receptors outside the kidney were demonstrated to make an equivalent, unique and non-redundant contribution to BP control: animals with a full complement of AT_{1A} receptors in the kidney, but without AT_{1A} receptors in extrarenal tissues, also had BP reductions of about 20 mmHg (Crowley et al., 2005). This finding also was independent of aldosterone levels. Taken altogether, this new evidence indicates that AT_1 receptors, either in the vasculature or the central nervous system, mediate the component of BP control that is independent of the kidney, at least in rodents. Since humans have only one form of the AT_1 receptor, it is uncertain if a similar scenario applies to them.

It is clear that Ang I and II are synthesized in tissue sites. Indeed, most, if not all, tissue Ang II is synthesized locally from tissue-derived Ang I (van Kats et al., 1998, 2001). In addition, the beneficial actions of RAS blockers are most likely due to interference with tissue Ang II rather than Ang II in the circulation (van Kats et al., 2000). Although it was originally thought that the renin required for local Ang I synthesis was synthesized locally, studies in nephrectomized animal models proved that this was not the case (Campbell et al., 1993; Danser et al., 1994; Nussberger, 2000; Hilgers et al., 2001). In many tissues, such as the heart and blood vessel wall, local Ang I synthesis depends on tissue uptake of kidney-derived renin (Chai and Danser, 2005), although currently this is controversial (see Chapter 14). In other tissues, such as the kidney, adrenal gland and brain, renin is synthesized locally (Re, 2004b). In addition, prorenin, the inactive precursor of renin, may contribute to Ang I generation at tissue sites (van den Eijnden et al., 2001; Saris et al., 2001). This would require activation of prorenin to renin following uptake from the systemic circulation, unless the recently described pro/renin receptor is involved.

XI. INTRARENAL RENIN–ANGIOTENSIN SYSTEM

The intrarenal RAS was first recognized in the 1970s and early 1980s when selective intrarenal inhibition of the RAS was demonstrated to increase GFR and renal Na^+ and water excretion (Kimbrough et al., 1977; Levens et al., 1983). Since that time, the intrarenal RAS has increasingly been recognized as a fundamental system in the regulation of Na^+ excretion and the long-term control of arterial BP (Navar et al., 2002, 2003). Indeed, there is growing recognition that inappropriate activation of the intrarenal RAS prevents the kidney from maintaining normal Na^+ balance at normal arterial pressures and is an important cause of hypertension (Wang et al., 2000; Navar et al., 2002; Adamczak et al., 2002). Several experimental models support an overactive intrarenal RAS in the development and maintenance of hypertension (Ploth, 1983; Mitchell and Navar, 1995; Frohlich,

1997; Inada et al., 1997; Navar et al., 2002, 2003). These include 2-kidney, 1-clip (2K1C) Goldblatt hypertension, Ang II-infused hypertension, transgenic rat (TGR) mRen2 hypertension with an extra renin gene, the remnant kidney hypertensive model and several mouse models overexpressing the renin or Agt gene (Navar et al., 2002, 2003). Indeed, there is increasing recognition that, in many forms of hypertension, the intrarenal RAS is inappropriately activated, limiting the ability of the kidney to maintain Na$^+$ balance when perfused at normal arterial pressure (Mitchell and Navar, 1995; Navar, 1997; Navar and Ham, 1999). In addition to Na$^+$ and fluid retention and progressive hypertension, other long-term consequences of an inappropriately activated intrarenal RAS include renal vascular, glomerular and tubulointerstitial injury and fibrosis (Adamczak et al., 2002; Wolf et al., 2002; Wolf, 2003).

Evidence for overactivation of the intrarenal RAS in hypertension has accumulated for many of the components of the intrarenal RAS. In Ang II-dependent hypertension, renal vascular and glomerular AT$_1$ receptors are downregulated but proximal tubule receptors are either upregulated or not significantly altered (Navar et al., 2002, 2003). In some forms of HT (e.g. 2K1C Goldblatt hypertension, Ang II-induced hypertension and TGR (mRen2) hypertension), net intrarenal Ang II content is increased due to the intrarenal production of Ang II as well as increased uptake of the peptide from the circulation via an AT$_1$ receptor-mediated process (Navar et al., 2002, 2003; Ingert et al., 2002). A sustained increase in circulating Ang II causes progressive accumulation of Ang II within the kidney in these models. A substantial fraction of the increase in intrarenal Ang II is due to AT$_1$ receptor-mediated endocytosis (Navar et al., 2002, 2003; Zhuo et al., 2002). In the 2K1C Goldblatt model, intrarenal Ang II was elevated in both the clipped and nonclipped kidneys both during the development phase (1 week) and up to 12 weeks following clipping (Sadjadi et al., 2002; Tokuyama et al., 2002). These increases in intrarenal Ang II were present even in the absence of increased plasma Ang II (Sadjadi et al., 2002). Renal interstitial Ang II levels also have been reported as elevated in several Ang II-dependent models of hypertension, including the 2-kidney, 1-wrap (2K1W) Grollman model and the Ang II-infused hypertensive model (Siragy and Carey, 1999; Nishiyama et al., 2002). However, measurements of renal tubular fluid Ang II have not demonstrated significant differences between control and hypertensive rats (Siragy et al., 1995; Cervenka et al., 1999). Because Ang II bound to the AT$_1$ receptor is internalized by receptor-mediated endocytosis, endosomal accumulation of Ang II in renal cells has been studied in Ang II-infused hypertension. Endosomal Ang II was increased and endosomal Ang II accumulation was blocked by AT$_1$ receptor blockade (Zhuo et al., 2002). At least some of the internalized Ang II remains intact and contributes to the increased total Ang II content as measured in renal cortical homogenates in this model (Chen et al., 2000; Zhuo et al., 2002). The internalized Ang II could be recycled and secreted, being available to act at plasma membrane AT$_1$ receptors, or may act at cytosolic receptors, as have been described for vascular smooth muscle cells (Haller et al., 1996; Navar et al., 2002, 2003). Another possibility is that Ang II could exert genomic effects in the nucleus as a part of an intracrine system (Haller et al., 1996; Chen et al., 2000; Re, 2003, 2004b; Re and Cook, 2006). Because Ang II may exert positive feedback stimulation of Agt in mRNA, intracellular Ang II may upregulate Agt or renin gene expression in renal proximal tubule cells (Kobori et al., 2004).

Agt is the only known precursor of Ang II and most of the intrarenal Agt mRNA and protein is localized to the proximal tubule cell (Figure 12.13), suggesting that intratubular Ang II is produced from locally formed Agt (Kobori et al., 2001a, b). Both Agt and its metabolite Ang II derived from proximal tubule cells are secreted directly into the tubule lumen (Rohrwasser et al., 1999). In response to a 2-week infusion of Ang II, intrarenal Agt mRNA and protein were upregulated (Kobori et al., 2001a, b). Therefore, an intrarenal positive feedback loop is probably present whereby increased Ang II stimulates its requisite precursor, leading to markedly increased Ang peptide levels in hypertension (Navar et al., 2002, 2003).

Renin is not only synthesized and secreted in the JG cells of the afferent arteriole but also by connecting tubules, indicating that renin is probably secreted into distal tubule fluid (Rohrwasser et al., 1999). Because intact Agt is present in urine, it is possible that some of the proximally formed Agt is converted to Ang II in the distal nephron (Rohrwasser et al., 1999; Kobori et al., 2003). Indeed, Ang II infusion significantly increased urinary Agt in a time- and dose-dependent manner associated with increased renal Ang II levels (Kobori et al., 2003, 2004). Furthermore, collecting duct renin is upregulated by Ang II via the AT$_1$ receptor

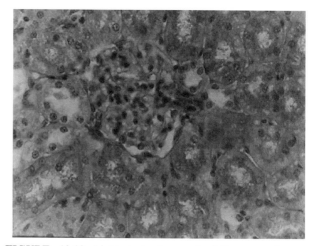

FIGURE 12.13 Light photomicrograph of the rat renal cortex demonstrating angiotensinogen protein (brown) by immunohistochemistry. Angiotensinogen within the kidney is synthesized largely in cortical proximal tubule cells. (See color plate section.)

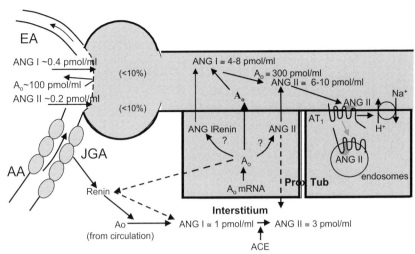

FIGURE 12.14 Schematic representation of the intrarenal renin–angiotensin system, the most fully characterized of the independent tissue renin–angiotensin systems. Adapted from Navar et al., *Hypertension* 39, 316–322, 2002 with permission.

(Prieto-Carrasquero et al., 2004, 2005). Therefore, several intrarenal mechanisms provide positive feedback control to enhance Ang II concentrations at both proximal and distal tubule sites, where Ang II has potent Na^+-retaining actions (Saccomani, 1990; Peti-Peterdi et al., 2002; Komlosi et al., 2003). A schematic depiction of the intrarenal RAS is shown in Figure 12.14.

In addition to aforementioned work suggesting the seminal role of kidney AT_1 receptors in the production of hypertension, recent studies have demonstrated that production of renin and Agt in the proximal tubule can increase BP independently of the circulating RAS (Rocha et al., 1998). Transgenic mice expressing human Agt selectively in the proximal tubule via the kidney androgen-regulated protein (KAP) promoter, when bred with mice expressing human renin systemically, had a 20 mmHg increase in BP despite having normal Ang II levels in plasma (Rocha et al., 1999). The increase in BP could be abolished with AT_1 receptor blockade (Rocha et al., 1999), indicating an Ang II-dependent HT. This was the first demonstration of systemic hypertension from isolated renal tissue activation of the RAS (Rocha et al., 1999). Furthermore, when purely proximal tubule overexpression of both human renin and Agt was achieved, hypertension also was present, supporting the concept that intrarenal tubular RAS activation could induce hypertension (Rocha et al., 1998). In these studies, it is unclear whether Ang I was first generated within the proximal tubule cell from intracellular cleavage of Agt by renin or whether the renin and Agt interaction occurred in the tubule lumen after secretion (Rocha et al., 1999). Also, whether the HT is due to increased proximal or distal Na^+ reabsorption, or both, remains unanswered.

Finally, it appears that the ability of Ang II to induce hypertension and cardiac hypertrophy resides exclusively in activation of AT_1 receptors within the kidney to reduce urinary Na^+ excretion. The basis for this principle is renal AT_{1A} receptor cross-transplantation studies demonstrating that, in animals with renal but not systemic AT_{1A} receptors, Ang II infusion was able to induce a hypertensive phenotype with cardiac hypertrophy. On the other hand, animals with systemic but not renal AT_{1A} receptor expression were unable to mount a hypertensive or cardiac hypertrophic response to Ang II. The ability of Ang II to induce hypertension was directly via renal AT_{1A} receptors and did not require an increase in aldosterone secretion. Therefore, the evidence strongly suggests that renal AT_1 receptors, in glomeruli, blood vessels and/or tubules, are critical for the development of Ang II-induced hypertension (Crowley et al., 2006). However, it is important to emphasize that these findings in experimental animals have been difficult to document in humans. Thus, the mechanisms by which Ang II induces hypertension in humans are uncertain.

XII. BRAIN RENIN–ANGIOTENSIN SYSTEM

Although many tissues express all of the RAS components necessary for the biosynthesis and action of Ang II, the ability of the tissues actually to produce Ang II and the specific role of locally generated Ang II has only been proven for the kidney (Rocha et al., 1999) and only in experimental animals. An independently functioning RAS in the brain remains controversial because the level of the rate-limiting component, renin, is extremely low and difficult to detect. There is no question that administration of exogenous Ang II centrally increases BP, sympathetic outflow, vasopressin release, drinking behavior and attenuation of baroreceptor reflux activity and that these effects are blocked with AT_1 receptor blockade (Printz et al., 2003). In addition, specific brain nuclei clearly mediate Ang II responses, including the ventrolateral medulla (VLM), nucleus tractus solitarii (NTS), paraventricular nucleus (PVN) and subfornical organ (SFO), among several others (Moulik et al., 2002). Although all of the RAS components are present in various regions of the brain, renin expression in very low levels has been detected in the pituitary and pineal glands, choroid plexus, hypothalamus, cerebellum and amygdala as well as other locations (Printz et al., 2003). At the cellular level, renin has recently been detected in both neuronal and glial tissue (Lavoie et al., 2004a, b). If a neuronal source of renin is coupled with a glial source of Agt, Ang II could derive from secreted precursors in the extracellular space. However, recent studies have demonstrated co-localization of Agt with a novel non-secreted form of renin, opening the door to intracellular Agt synthesis and possible action (Re, 2003; Lavoie et al., 2004b).

Because brain renin levels are low, investigators have searched for a renin-independent Ang II generating system in the brain. Many enzymes are present in brain which can generate Ang II either from Ang I or directly from Agt, including trypsin, tonin, elastase, cathepsin C, kallekrein, chymase and chemostatin-sensitive Ang II – generating enzyme (Sakai and Sigmund, 2005).

In addition to non-renin pathways, some investigators have suggested the involvement of non-Ang II peptides, including Ang III, Ang IV and Ang (1–7), as important regulators of BP (Ferrario and Chappell, 2004). In particular, Ang III appears to have a prominent role (Reaux-Le Goazigo et al., 2005). Aminopeptidase A (APA), which metabolizes Ang II to Ang III, is present in the brain (Wright et al., 2003). Ang II and Ang III are equally potent pressor substances when infused directly into the brain and the pressor action of Ang II was abolished by preadministration of an APA inhibitor, suggesting that conversion to Ang III may be required (Reaux et al., 1999; Fournie-Zaluski et al., 2004). Ang II and Ang III have equal affinity for the AT_1 receptor and both also are agonists at the AT_2 receptor. However, the Ang III mediated increase in BP appears to be mediated by the AT_1 receptor, as its action can be blocked with an AT_1 receptor blocker. In addition, inhibition of endogenous brain Ang III formation by intracerebroventricular, but not intravenous, APA inhibitor induced a large, dose-dependent reduction in BP in conscious SHR and deoxycorticosterone acetate (DOCA)-salt rat, a RAS-independent model of HT (Reaux et al., 1999; Fournie-Zaluski et al., 2004). On the other hand, administration of an inhibitor of aminopeptidase N (APN), which metabolizes Ang III to Ang IV, into the brain induces a pressor effect that is abolished with an AT_1 receptor blocker (Reaux et al., 1999). Thus, increasing endogenous brain Ang III levels increases BP via the AT_1 receptor. Moreover, the pressor action of an APN inhibitor could be blocked with an APA inhibitor, confirming the existence of an endogenous brain Ang III cascade in the control of BP (Reaux et al., 1999). Finally, work employing non-metabolizable analogs D-Asp^1-Ang II and D-Asp^1-Ang III demonstrated that Ang III is a centrally active agonist of the brain RAS (Wright et al., 2003).

There is also evidence that Ang IV may be an endogenous ligand of the brain RAS. When Ang IV is overexpressed specifically in the brain, these transgenic mice developed HT that was abolished by an AT_1 receptor antagonist (Lochard et al., 2004). The role of Ang (1–7) as a counter-regulatory peptide to the pressor actions of Ang II is the subject of current studies (Ferrario and Chappell, 2004; Sakai and Sigmund, 2005). These studies take on additional importance due to the recent discovery of ACE-2, which converts Ang II directly to Ang (1–7) and the identification of the *mas* oncogene as an Ang (1–7) receptor. Furthermore, the receptor for renin and prorenin, which probably converts Agt to Ang I and activates extracellular signal-related kinases (ERKs), has recently been cloned and is highly expressed in brain. It is, therefore, possible that the renin receptor may enhance the formation of Ang I at selective neuronal sites within the central nervous system. As stated above, because of the low levels of renin in most sites in the brain, the role of a local renin–angiotensin system in mediating physiologic or pathophysiologic effects is uncertain. Furthermore, as in the kidney, direct documentation of a substantial effect of a brain renin–angiotensin system in humans is lacking.

XIII. VASCULAR TISSUE RENIN–ANGIOTENSIN SYSTEM

Evidence for local RASs have been found in the heart, large blood vessels, adrenal, uterus, ovaries, testes, placenta and pancreas (Danser, 1996; Dostal and Baker, 1999). This evidence demonstrates that local systems may act independently from the circulating system to produce Ang II. Within the vasculature, molecular expression of Agt has been found within the walls of large vessels, including the saphenous and umbilical vein and aorta (Paul et al., 1993). Renin and Agt mRNA also has been detected within the aorta (Samani et al., 1988) and in isolated small resistance arteries in skeletal muscle (Agoudemos and Greene, 2005). Microvessels also express AT_1 and AT_2 mRNA and protein (Linderman and Greene, 2001). There is functional evidence of a local vascular RAS (Oliver and Sciacca, 1984; Vicaut and Hou, 1994; Boddi et al., 1998) and the concentration of Ang II within microvessels is much higher than in plasma (Agoudemos and Greene, 2005). Of note, renin or prorenin has been identified in fetal microvessels in the kidney (Gomez et al., 1989) and in renal vessels of adult animals after ACE inhibition (Gomez et al., 1990; Everett et al., 1990). Thus, recent evidence would support the concept that renin is synthesized in vessels, even though this has been a past topic of controversy (von Lutterotti et al., 1994). Since anephric patients and nephrectomized animals have undetectable plasma renin activities (PRAs), local microvascular production of renin does not contribute to circulating renin (Berman et al., 1972; Thurston and Swales, 1977). Thus, the local vascular RAS is likely to act within the vessels. Whether or not the local vascular RAS contributes to vascular tone is currently unknown. However, there seems to be an increase in the local microvessel concentration of renin and Ang II in SHR compared to Sprague-Dawley or WKY

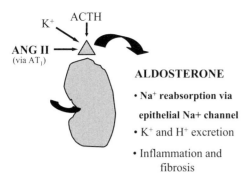

FIGURE 12.15 Schematic depiction of the control of aldosterone secretion from the *zona glomerulosa* cells of the adrenal cortex. The major long-term control mechanisms are angiotensin II (Ang II) and potassium (K^+). Adrenocorticotropic hormone (ACTH) stimulates aldosterone secretion in transient but not sustained fashion. Renal actions of aldosterone are shown.

control rats (Samani et al., 1988; Vicaut and Hou, 1994), suggesting the possibility of an overactive local vascular RAS that could play a pathophysiologic role in HT.

XIV. CARDIAC RENIN–ANGIOTENSIN SYSTEM

Renin and its mRNA were originally found in the heart in 1987 (Dzau and Re, 1987). Conclusive evidence now exists that all of the necessary RAS components for synthesis of Ang II are present in the heart and that peptide formation does indeed occur (Dostal and Baker, 1999; Dostal, 2000) (see Chapter 14 for additional details).

XV. ALDOSTERONE AND MINERALOCORTICOID RECEPTORS

(See Chapters 21–24 for additional details.)

Aldosterone is a mineralocorticoid hormone synthesized and secreted by the adrenal *zona golomerulosa* in response to Ang II, K^+ and ACTH, that interacts with mineralocorticoid receptors in the renal cortical collecting duct, colon, salivary and sweat glands to promote unidirectional Na^+ flux (Figure 12.15). Na^+ retention in response to aldosterone is accompanied by water retention such that inappropriately high aldosterone levels expand blood and extracellular fluid volumes leading to hypertension. Non-epithelial sites of physiological aldosterone action include the central nervous system, where it increases salt appetite, and the vascular wall, where it increases vascular contractility (Rossi et al., 2005). Aldosterone is responsible for a large number of additional effects, including myocardial remodeling, vascular inflammation and fibrosis, reduced vascular compliance, baroreceptor and endothelial dysfunction, potentiation of Ang II- and catecholamine-induced vasoconstriction, ventricular ectopy, progressive renal dysfunction and thrombosis, all of which contribute to cardiovascular and renal target-organ damage (Figure 12.16).

The involvement of aldosterone in the pathophysiology of hypertension is clearly delineated in primary aldosteronism in which there is a primary increase in aldosterone production by a unilateral adrenal aldosterone-producing adenoma (APA) or idiopathic bilateral adrenal hyperplasia (IHA). Patients with these disorders have Na^+ and water retention, hypokalenic alkalosis and hypertension that, in the case of APA, is reversible with removal of the adenoma. In the case of IHA causing primary aldosteronism, bilateral adrenalectomy does not cure the hypertension so that other pathophysiological factors must be involved in the hypertension of patients with this disorder. In recent years, there has been a revival of interest in the diagnosis of primary aldosteronism with the finding that this disorder may be present in 2–8% of hypertensive and 15–20% of treatment-resistant hypertensive patients (125 and Chapter 23). However, it has been much more difficult to demonstrate a role of aldosterone in primary hypertension. While it is clear that mineralocorticoid receptor antagonists such as spironolactone are effective in lowering BP of patients with resistant hypertension, these agents are only modestly effective in the early

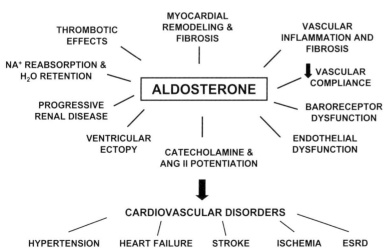

FIGURE 12.16 Cardiovascular and renal actions of aldosterone.

stages of hypertension (Laragh, 2001; Nishizaka et al., 2003). Interestingly, an analysis for the identification of genetic markers of blood pressure did not identify plasma aldosterone as a familial correlate of hypertension (Watt et al., 1992).

Experimentally, large doses of aldosterone that are associated with a threefold increase in plasma aldosterone levels and salt loading are required to induce arterial remodeling in humans (Virdis et al., 2002). However, these studies were performed only for a short time. What effect smaller increases for a longer time would have is unknown. The major action of aldosterone on arteries is stimulation of collagen turnover and induction of fibrosis and perivascular inflammation (Virdis et al., 2002). These changes are not typical of the eutropic inward remodeling in small resistance arteries in primary hypertension, although the data supporting this conclusion in all forms of primary hypertension are limited. In uninephrectomized rats on a low Na^+ diet, large doses of aldosterone, which result in marked elevation of plasma aldosterone, do not induce perivascular fibrosis in the myocardium (Weber et al., 1995).

Recently, several new concepts have emerged regarding the role of aldosterone and mineralocorticoids that may be of major importance:

1. extra-adrenal synthesis of aldosterone
2. aldosterone induction of vascular and cardiac inflammation and fibrosis
3. aldosterone induction of progressive cerebral and renal damage and cardiovascular disease
4. the action of glucocorticoid hormones at mineralocorticoid receptors.

The classic site of aldosterone biosynthesis is the *zona glomerulosa* of the adrenal cortex. Recent research indicates, however, that the necessary chemical machinery (e.g. aldosterone synthase) may be expressed in non-adrenal tissues, such as blood vessels, brain, myocardium and kidney (Takeda et al., 1997; Gomez-Sanchez et al., 1997; Silvestre et al., 1998; Xue and Siragy, 2005). In each of these non-adrenal tissues it has been argued that local aldosterone production occurs, albeit in small quantities, due to aldosterone synthase expression in the tissue and that aldosterone may serve as a paracrine substance to inflict tissue injury. On the other hand, non-adrenal tissue can take up aldosterone from circulating plasma and this is easily demonstrated for the heart (Gomez-Sanchez et al., 2004). Irrespective of its origin via local synthesis or uptake from plasma, aldosterone can be detected in cardiac homogenates at 17-fold the concentration in plasma and its level is markedly increased after 3 h of perfusion by Ang II or ACTH and is sensitive to dietary Na^+ and K^+. Recent work by a number of groups now demonstrates that the heart does not make aldosterone in appreciable quantities (Funder, 2004; Gomez-Sanchez et al., 2004; Ye et al., 2005). However, the door remains open for vascular and or renal aldosterone biosynthesis (Takeda, 2004; Xue and Siragy, 2005).

Aldosterone has been shown to induce fibrosis in the heart, as increased levels of the steroid in combination with high Na^+ intake induce myocardial interstitial fibrosis (Brilla and Weber, 1992). The fibrotic reaction induced by aldosterone is independent of elevated blood pressure and increased plasma aldosterone is not prerequisite for the steroid to mediate fibrosis (Brilla et al., 1993). Selective mineralocorticoid receptor blockade abolishes the cardiac effects of administered aldosterone (Brilla et al., 1993). In rats on a high Na^+ intake infused with Ang II and NOS inhibitor L-NAME, which do not have increased plasma aldosterone, administration of mineralocorticoid receptor antagonist eplerenone or adrenalectomy abolished coronary vascular injury without lowering BP (Rocha et al., 2000) (Figure 12.17). The protective effect of adrenalectomy was reversed by aldosterone replacement (Rocha et al., 2000). Thus, aldosterone in the presence of a high Na^+ environment induces coronary damage through blood pressure-independent mechanisms (Rocha et al., 2000). In uninephrectomized rats receiving a high Na^+ diet and exogenous aldosterone, severe HT developed after 2 weeks (Rocha et al., 2002). The coronary vasculature was markedly inflamed with medial fibrinoid necrosis and perivascular inflammation associated with increased medial expression vascular adhesion molecule-1 (VCAM-1), cyclo-oxygenase-2 (COX-2), osteopontin and monocyte chemoattractant protein (MCP-1) (Rocha et al., 2002). Mineralocorticoid receptor blockade with eplerenone reduced BP and vascular inflammation to control levels (Rocha et al., 2002). The fact that aldosterone itself is not toxic to tissues but requires a high Na^+ environment to induce these changes represents an intriguing unsolved problem in vascular biology (Rossi et al., 2005; Dluhy and Williams, 2004).

Aldosterone also has the capacity to induce progressive renal and cerebral vascular damage (Rocha et al., 1998, 1999). In the stroke-prone hypertensive rat, aldosterone appears to be a primary mediator of vascular inflammation in the brain and kidneys. Mineralocorticoid receptor antagonist administration or adrenalectomy afforded the same degree of tissue protection as did ACE inhibitors (Rocha et al., 1998, 1999). Thus, Ang II *per se* is insufficient to induce vascular and glomerular damage but aldosterone plays a central role in the pathology. Aldosterone causes microvascular damage, vascular inflammation, oxidative stress and endothelial dysfunction and mineralocorticoid receptor antagonists are efficacious in preventing these changes (Joffe and Adler, 2005).

Aldosterone acts by binding to the mineralocorticoid receptor, which was first characterized in classical Na^+-transporting epithelial cells and subsequently at a large number of non-epithelial sites, such as cardiomyocytes and blood vessels (Rousseau et al., 1972; Funder et al., 1973, 1989; Pearce and Funder, 1987). Glucocorticoids bind to

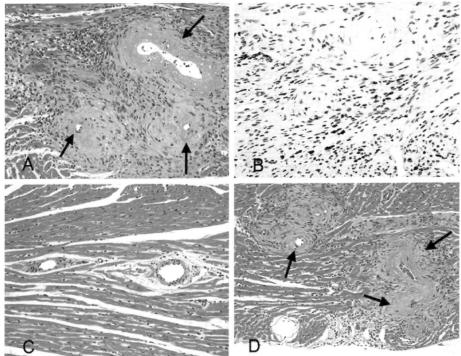

FIGURE 12.17 Effect of mineralocorticoid receptor blockade or adrenalectomy on myocardial and vascular damage. Photomicrographs of representative coronal sections of hearts from rats in different experimental groups. Focal lesions of medial fibrinoid necrosis were observed in response to Ang II/salt treatment (arrows), associated with a prominent perivascular inflammatory response (A). Macrophages were frequently found associated with coronary lesions and infiltrating the perivascular spare (B). Adrenalectomy or eplerenone treatment attenuated lesion development (C). Severe vascular inflammatory lesions were observed in all adrenalectomized animals with aldosterone treatment (D, arrows). From Rocha et al., *Endocrinology* 143, 4828–4836, 2002 with permission. (See color plate section.)

mineralocorticoid receptors with equal affinity as aldosterone, but cortisol circulates at plasma concentrations 1000-fold higher than aldosterone. In epithelia, blood vessels and certain brain areas, however, mineralocorticoid receptors are protected against activation by glucocorticoids by expression of the enzyme 11-β-hydroxysteroid dehydrogenase-type 2 (11-βOHSD2), which converts cortisol to receptor-inactive cortisone and the co-factor NAD to NADPH (Funder et al., 1988; Gordon et al., 2005) (Figure 12.18).

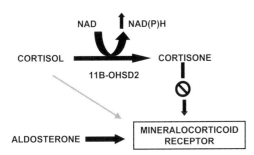

FIGURE 12.18 Schematic representation of interactions of aldosterone and cortisol at the mineralocorticoid receptor. Both steroids bind the mineralocorticoid receptor with approximately equal affinity. However, the receptor is protected from glucocorticoid exposure in tissues such as the kidney by 11-β-hydroxysteroid dehydrogenase type-2 (11-βOHSD2) enzymatic conversion to cortisone, which is inactive at the receptor. Aldosterone is not similarly metabolized because its C11-OH is protected by cyclization with the signature C18-CHO to give an 11, 18 hemiacetal. NAD is a co-factor for cortisol conversion to cortisone by this mechanism. The heart is devoid of 11-βOHSD2. Adapted from Funder et al., *Trends in Endocrinol Metab* 15, 139–142, 2004 with permission.

In epithelial tissues, such as the renal distal tubule, there is no question that mineralocorticoid receptors can be activated by cortisol. However, glucocorticoids can occupy both epithelial and non-epithelial mineralocorticoid receptors without activating the receptor but competitively excluding aldosterone (Funder, 2007) (Figure 12.18). The exact role of glucocorticoid activation of mineralocorticoid receptors is the subject of intense current investigation. Experimental evidence is currently being sought for the hypothesis that Ang II, inflammation or fibrosis, via activation of NAD(P)H oxidase, can reduce NAD(P)H, leading to reduction of 11-β OHSD2 activity and an increased supply of cortisol availability to mineralocorticoid receptors in tissues (Figure 12.19).

XVI. CLINICAL EFFECTS OF THE RENIN–ANGIOTENSIN–ALDOSTERONE SYSTEM (RAAS)

From the preceding discussion, it is apparent that the RAAS is a highly complex hormonal cascade with multiple interacting, redundant pathways and several newly discovered protein/peptide agonists, receptors and mechanisms. It is now clear that the RAS functions not only as a systemic hormonal system but also as several local tissue systems independently of systemic influence. The intrarenal RAS exerts a powerful influence on renal function, independent of circulating renin and Ang peptides (Schmitz, 1997; Schmitz et al., 2001). There is also considerable evidence that local tissue RASs are present in the brain, heart, blood

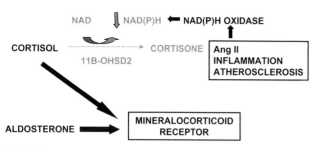

FIGURE 12.19 Schematic representation of the hypothesis that inflammation, fibrosis and/or angiotensin II (Ang II) increase the bioavailability of cortisol at the mineralocorticoid receptor by reducing cofactor NAD(P)H through activation of NAD(P)H oxidase, thus reducing 11βOHSD2 activity. Adapted from Funder et al., *Trends in Endocrinol Metab* 15, 139–142, 2004 with permission.

TABLE 12.2 List of clinical conditions benefited by renin–angiotensin system blockade with ACE inhibitor (ACEI) and/or AT_1 receptor blocker (ARB) therapy

Hypertension, high risk profile (ACEI)
Hypertension, LVH (ARB)
Stroke (ACEI)
Congestive heart failure (ACEI, ARB)
LV dysfunction (ACEI)
Post-myocardial infarction (ACEI, ARB)
Chronic stable coronary disease (ACEI)
Diabetes mellitus (ACEI)
Diabetes mellitus, renal failure (ACEI, ARB)
Diabetes mellitus, hypertension, microalbuminuria (ARB)
Chronic renal disease, diabetic and non-diabetic (ARB)

Studies were randomized, double-blinded, placebo-controlled trials. LVH, left ventricular hypertrophy.

vessels, adrenal glands and even in adipocytes. The precise regulatory role of cell-to-cell (paracrine), cell-to-same-cell (autocrine) and even intracellularly confined (intracrine) RAS activity and function await future investigation.

From the clinical standpoint, there is no question that the major effector mechanism of the RAS is Ang II action via the AT_1 receptor, which is highly expressed in most tissues. Ang II by this mechanism engenders antinatriuresis directly at the renal tubule (Ishida et al., 1998), aldosterone secretion, sympathetic nervous system stimulation, vasoconstriction and tissue damage via cell growth and proliferation, inflammation, thrombosis and fibrosis leading to vascular and cardiac remodeling and renal disease. Virtually all of the pharmaceutical effort thus far has justifiably been placed in developing blockers of the RAS. ACE inhibitors and AT_1 receptor antagonists have had a major clinical impact to improve/prevent progression of cardiovascular and renal disease. As shown in Table 12.2, clinical trials have demonstrated the efficacy of RAS blockade in hypertension, stroke, congestive heart failure, chronic stable coronary disease, prevention of cardiac remodeling post-myocardial infarction, diabetes mellitus with and without hypertension and/or renal failure and even in non-diabetic renal disease. Although the available clinical trials have been performed with either ACE inhibitor or Ang II receptor blocker, very few studies are available comparing responses to these two methods of RAS inhibition. The available studies generally demonstrate the equivalence of each approach. Since several studies have documented long-term escape of Ang II and aldosterone suppression with ACE inhibitors, several studies are presently in progress evaluating the efficacy of combination therapy with both ACE inhibitor and AT_1 receptor blocker. However, it is important to note that there is limited documentation that the mechanisms by which the renin–angiotensin system functions and/or produces cardiovascular and renal damage in experimental animals also applies in humans.

A novel approach to RAS blockade has been introduced in 2007 in the form of the first clinically available inhibitor of the enzymatic action of renin (Figure 12.20). Aliskiren is a new orally active renin inhibitor with similar antihypertensive efficacy when given alone to hydrochlorothiazide (Lijnen and Petrov, 1999). Interestingly, however, the antihypertensive actions of aliskiren are additive with those of hydrochlorothiazide and also with those of ACE inhibitors or AT_1 receptor blockers. As shown in Figure 12.20, renin inhibition blocks the catalytic action of renin to cleave the decapeptide Ang I from Agt. Under some circumstances, renin inhibition could theoretically be circumvented, at least in part, by enzymes such as cathepsin D or tonins. Renin inhibitors block the RAS at the highest point achieved so far in the hormonal cascade resulting in reductions in both Ang I and Ang II. Importantly, plasma renin activity is suppressed by renin inhibitors, providing a useful laboratory marker of clinical efficacy. However, renin inhibitors do not block the biosynthesis or secretion of renin, which is augmented by inhibition of the short-loop negative feedback loop of Ang II via JG cell AT_1 receptors. If the activation of the (pro)renin receptor is proven to induce tissue damage, this mechanism may limit cardiovascular and renal effectiveness of all RAS blockers, including renin inhibitors.

The RAAS is a hormonal cascade of major importance in the regulation of blood pressure and cardiovascular and renal function. Within the past decade, several new components, pathways and mechanisms have been discovered, including the AT_2 receptor, ACE-2, Ang (1–7), the *mas* receptor, the (pro)renin receptor, the local tissue RASs and the role of Ang II and aldosterone in target-organ damage. While Ang II activation of the AT_1 receptor remains an important pathway, leading to cardiovascular and renal damage, several counter-regulatory pathways, including the AT_2 receptor and the ACE-2–Ang (1–7)–*mas* receptor pathways, may limit the damaging tissue actions of Ang II. These pathways appear to have a potential beneficial role, especially in the presence of AT_1 receptor blocker therapy. The recent discovery of a direct action of prorenin and renin at their specific receptor

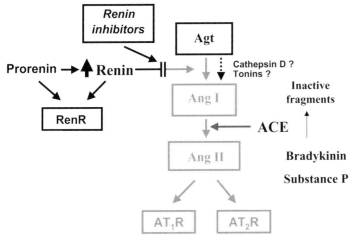

FIGURE 12.20 Schematic depiction of changes in the renin–angiotensin system in response to renin inhibitor therapy. Renin inhibitors result in blockade of the catalytic conversion of Agt to decapeptide Ang I. The entire Ang cascade is suppressed (gray), but increased renin biosynthesis and secretion still occur via interruption of short-loop negative feedback via AT_1 and AT_2 receptors on JG cells. Renin and prorenin may be free to act at (pro)renin receptors (RenR).

leading to cell signaling mechanisms which may be detrimental opens the door for an additional therapeutic target.

Similarly, the mineralocorticoid receptor antagonists, spironolactone and eplerenone, have been demonstrated as efficacious in the treatment of congestive heart failure, resistant hypertension and the prevention of progression to end-stage renal disease. It is now clear that both Ang II and aldosterone contribute to tissue damage in a variety of cardiovascular and renal disorders. Finally, the discovery of the role of 11-βOHSD2 in protecting the mineralocorticoid receptor from exposure to cortisol opens the door to the role of both glucocorticoids and mineralocorticoids in tissue damage.

XVII. SUMMARY

The renin–angiotensin–aldosterone system (RAAS) is a major hormonal cascade in the control of blood pressure (BP), hypertension and tissue damage. The primary means by which the RAAS contributes to acute changes in extracellular fluid volume and BP homeostasis is by adjusting the level of renin in the circulation. Angiotensin II, the major effector peptide of the renin–angiotensin system (RAS), binds to two major receptors, AT_1 and AT_2, that generally oppose each other. During the past five years, several new pathways in the RAS have been discovered and/or clarified, including a (pro)renin receptor, the functional properties of the AT_2 receptor and the ACE-2–angiotensin (1–7)–*mas* receptor pathway. The circulating RAS is only one component of the overall RAS and tissue renin–angiotensin systems (RASs) that operate independently of the circulating RAS have been identified and characterized, especially in the kidney. RAAS blockade has been central to our treatment of hypertension and its complications such as heart failure, atrial fibrillation, coronary artery disease and the prevention of progression to renal failure. Renin inhibition constitutes a promising new therapeutic target in RAAS blockade. These and other discoveries identify new potential therapeutic targets for prevention of morbidity and mortality due to cardiovascular and renal disease in the future.

References

Adamczak, M., Zeier, M., Dikow, R., and Ritz, E. (2002). Kidney and hypertension. *Kidney Int Sup*, pp. 62–67.

Agoudemos, M. M., and Greene, A. S. (2005). Localization of the renin–angiotensin system components to the skeletal muscle microcirculation. *Microcirculation* **12:** 627–636.

Allen, A. (2000). Localization and function of angiotensin AT1 receptors. *Am J Hypertens* **13:** 31S–38S.

Berman, L. B., Vertes, V., Mitra, S., and Gould, A. B. (1972). Renin–angiotensin system in anephric patients. *N Engl J Med* **286:** 58–61.

Boddi, M., Poggesi, L., Coppo, M. et al. (1998). Human vascular renin-angiotensin system and its functional changes in relation to different sodium intakes. *Hypertension* **31:** 836–842.

Brilla, C. G., and Weber, K. T. (1992). Reactive and reparative myocardial fibrosis in arterial hypertension in the rat. *Cardiovasc Res* **26:** 671–677.

Brilla, C. G., Matsubara, L. S., and Weber, K. T. (1993). Anti-aldosterone treatment and the prevention of myocardial fibrosis in primary and secondary hyperaldosteronism. *J Mol Cell Cardiol* **25:** 563–575.

Campbell, D. J., Kladis, A., and Duncan, A. M. (1993). Nephrectomy, converting enzyme inhibition, and angiotensin peptides. *Hypertension* **22:** 513–522.

Carey, R. M., and Siragy, H. M. (2003). Newly recognized components of the renin–angiotensin system: potential roles in cardiovascular and renal regulation. *Endocr Rev* **24:** 261–271.

Cervenka, L., Wang, C. T., Mitchell, K. D., and Navar, L. G. (1999). Proximal tubular angiotensin II levels and renal functional responses to AT1 receptor blockade in nonclipped kidneys of Goldblatt hypertensive rats. *Hypertension* **33:** 102–107.

Chai, W., and Danser, A. H. (2005). Is angiotensin II made inside or outside of the cell? *Curr Hypertens Rep* **7:** 124–127.

Chappell, M. C., Allred, A. J., and Ferrario, C. M. (2001). Pathways of angiotensin-(1–7) metabolism in the kidney. *Nephrol Dial Transplant* **16**(Suppl 1): 22–26.

Chen, R., Mukhin, Y. V., Garnovskaya, M. N. et al. (2000). A functional angiotensin II receptor-GFP fusion protein: evidence for agonist-dependent nuclear translocation. *Am J Physiol Renal Physiol* **279**: F440–F448.

Crackower, M. A., Sarao, R., Oudit, G. Y. et al. (2002). Angiotensin-converting enzyme 2 is an essential regulator of heart function. *Nature* **417**: 822–828.

Crowley, S. D., Gurley, S. B., Oliverio, M. I. et al. (2005). Distinct roles for the kidney and systemic tissues in blood pressure regulation by the renin-angiotensin system. *J Clin Invest* **115**: 1092–1099.

Crowley, S. D., Gurley, S. B., Herrera, M. J. et al. (2006). Angiotensin II causes hypertension and cardiac hypertrophy through its receptors in the kidney. *Proc Natl Acad Sci USA* **103**: 17985–17990.

Danser, A. H. (1996). Local renin-angiotensin systems. *Mol Cell Biochem* **157**: 211–216.

Danser, A. H., van Kats, J. P., Admiraal, P. J. et al. (1994). Cardiac renin and angiotensins. Uptake from plasma versus in situ synthesis. *Hypertension* **24**: 37–48.

de Gasparo, M., Catt, K. J., Inagami, T., Wright, J. W., and Unger, T. (2000). International union of pharmacology. XXIII. The angiotensin II receptors. *Pharmacol Rev* **52**: 415–472.

Diz, D. I., Chappell, M. C., Tallant, E. A., and Ferrario, C. M. (2005). Angiotensin (1–7). *Hypertension*, pp. 100–110.

Dluhy, R. G., and Williams, G. H. (2004). Aldosterone – villain or bystander? *N Engl J Med* **351**: 8–10.

Dostal, D. E. (2000). The cardiac renin–angiotensin system: novel signaling mechanisms related to cardiac growth and function. *Regul Pept* **91**: 1–11.

Dostal, D. E., and Baker, K. M. (1999). The cardiac renin–angiotensin system: conceptual, or a regulator of cardiac function? *Circ Res* **85**: 643–650.

Dzau, V. J., and Re, R. N. (1987). Evidence for the existence of renin in the heart. *Circulation* **75**: I134–I136.

Erdos, E.G., and Skidgel, R.A. (1997). Metabolism of bradykinin by peptidases in health and diesease. In Farmer S.C. (ed.). The Kinin system: handbook of immunopharmology. pp. 112–41.

Erdos, E. G., and Marcic, B. M. (2001). Kinins, receptors, kininases and inhibitors – where did they lead us? *Biol Chem* **382**: 43–47.

Everett, A. D., Carey, R. M., Chevalier, R. L., Peach, M. J., and Gomez, R. A. (1990). Renin release and gene expression in intact rat kidney microvessels and single cells. *J Clin Invest* **86**: 169–175.

Ferrario, C. M., and Chappell, M. C. (2004). Novel angiotensin peptides. *Cell Mol Life Sci* **61**: 2720–2727.

Fournie-Zaluski, M. C., Fassot, C., Valentin, B. et al. (2004). Brain renin–angiotensin system blockade by systemically active aminopeptidase A inhibitors: a potential treatment of salt-dependent hypertension. *Proc Natl Acad Sci USA* **101**: 7775–7780.

Frohlich, E. D. (1997). Arthus C. Corcoran Memorial Lecture. Influence of nitric oxide and angiotensin II on renal involvement in hypertension. *Hypertension* **29**: 188–193.

Funder, J. W. (2004). Cardiac synthesis of aldosterone: going, going, gone? *Endocrinology* **145**: 4793–4795.

Funder, J. W. (2007). Why are mineralocorticoid receptors so nonselective? *Curr Hypertens Rep* **9**: 112–116.

Funder, J. W., Feldman, D., and Edelman, I. S. (1973). The roles of plasma binding and receptor specificity in the mineralocorticoid action of aldosterone. *Endocrinology* **92**: 994–1004.

Funder, J. W., Pearce, P. T., Smith, R., and Campbell, J. (1989). Vascular type I aldosterone binding sites are physiological mineralocorticoid receptors. *Endocrinology* **125**: 2224–2226.

Funder, J. W., Pearce, P. T., Smith, R., and Smith, A. I. (1988). Mineralocorticoid action: target tissue specificity is enzyme, not receptor, mediated. *Science* **242**: 583–585.

Giasson, E., Servant, M. J., and Meloche, S. (1997). Cyclic AMP-mediated inhibition of angiotensin II-induced protein synthesis is associated with suppression of tyrosine phosphorylation signaling in vascular smooth muscle cells. *J Biol Chem* **272**: 26879–26886.

Gomez, R. A., Lynch, K. R., Chevalier, R. L. et al. (1988). Renin and angiotensinogen gene expression and intrarenal renin distribution during ACE inhibition. *Am J Physiol* **254**: F900–F906.

Gomez, R. A., Lynch, K. R., Sturgill, B. C. et al. (1989). Distribution of renin mRNA and its protein in the developing kidney. *Am J Physiol* **257**: F850–F858.

Gomez, R. A., Chevalier, R. L., Everett, A. D. et al. (1990). Recruitment of renin gene-expressing cells in adult rat kidneys. *Am J Physiol* **259**: F660–F665.

Gomez-Sanchez, C. E., Zhou, M. Y., Cozza, E. N., Morita, H., Foecking, M. F., and Gomez-Sanchez, E. P. (1997). Aldosterone biosynthesis in the rat brain. *Endocrinology* **138**: 3369–3373.

Gomez-Sanchez, E. P., Ahmad, N., Romero, D. G., and Gomez-Sanchez, C. E. (2004). Origin of aldosterone in the rat heart. *Endocrinology* **145**: 4796–4802.

Gordon, R. D., Laragh, J. H., and Funder, J. W. (2005). Low renin hypertensive states: perspectives, unsolved problems, future research. *Trends Endocrinol Metab* **16**: 108–113.

Griendling, K. K., Murphy, T. J., and Alexander, R. W. (1993). Molecular biology of the renin–angiotensin system. *Circulation* **87**: 1816–1828.

Haller, H., Lindschau, C., Erdmann, B., Quass, P., and Luft, F. C. (1996). Effects of intracellular angiotensin II in vascular smooth muscle cells. *Circ Res* **79**: 765–772.

Harrison-Bernard, L. M., Navar, L. G., Ho, M. M., Vinson, G. P., and el-Dahr, S. S. (1997). Immunohistochemical localization of ANG II AT1 receptor in adult rat kidney using a monoclonal antibody. *Am J Physiol* **273**: F170–F177.

Hilgers, K. F., Veelken, R., Muller, D. N. et al. (2001). Renin uptake by the endothelium mediates vascular angiotensin formation. *Hypertension* **38**: 243–248.

Hsueh, W. A., and Baxter, J. D. (1991). Human prorenin. *Hypertension* **17**: 469–477.

Inada, Y., Wada, T., Ojima, M. et al. (1997). Protective effects of candesartan cilexetil (TCV-116) against stroke, kidney dysfunction and cardiac hypertrophy in stroke-prone spontaneously hypertensive rats. *Clin Exp Hypertens* **19**: 1079–1099.

Ingert, C., Grima, M., Coquard, C., Barthelmebs, M., and Imbs, J. L. (2002). Contribution of angiotensin II internalization to intrarenal angiotensin II levels in rats. *Am J Physiol Renal Physiol* **283**: F1003–F1010.

Ishida, M., Ishida, T., Thomas, S. M., and Berk, B. C. (1998). Activation of extracellular signal-regulated kinases (ERK1/2) by angiotensin II is dependent on c-Src in vascular smooth muscle cells. *Circ Res* **82**: 7–12.

Joffe, H. V., and Adler, G. K. (2005). Effect of aldosterone and mineralocorticoid receptor blockade on vascular inflammation. *Heart Fail Rev* **10**: 31–37.

Kimbrough, H. M., Vaughan, E. D., Carey, R. M., and Ayers, C. R. (1977). Effect of intrarenal angiotensin II blockade on renal function in conscious dogs. *Circ Res* **40:** 174–178.

Kobori, H., Prieto-Carrasquero, M. C., Ozawa, Y., and Navar, L. G. (2004). AT1 receptor mediated augmentation of intrarenal angiotensinogen in angiotensin II-dependent hypertension. *Hypertension* **43:** 1126–1132.

Kobori, H., Harrison-Bernard, L. M., and Navar, L. G. (2001a). Enhancement of angiotensinogen expression in angiotensin II-dependent hypertension. *Hypertension* **37:** 1329–1335.

Kobori, H., Harrison-Bernard, L. M., and Navar, L. G. (2001b). Expression of angiotensinogen mRNA and protein in angiotensin II-dependent hypertension. *J Am Soc Nephrol* **12:** 431–439.

Kobori, H., Nishiyama, A., Harrison-Bernard, L. M., and Navar, L. G. (2003). Urinary angiotensinogen as an indicator of intrarenal angiotensin status in hypertension. *Hypertension* **41:** 42–49.

Komlosi, P., Fuson, A. L., Fintha, A. et al. (2003). Angiotensin I conversion to angiotensin II stimulates cortical collecting duct sodium transport. *Hypertension* **42:** 195–199.

Laragh, J. (2001). Laragh's lessons in pathophysiology and clinical pearls for treating hypertension. *Am J Hypertens* **14:** 837–854.

Lavoie, J. L., Cassell, M. D., Gross, K. W., and Sigmund, C. D. (2004a). Localization of renin expressing cells in the brain, by use of a REN-eGFP transgenic model. *Physiol Genomics* **16:** 240–246.

Lavoie, J. L., Cassell, M. D., Gross, K. W., and Sigmund, C. D. (2004b). Adjacent expression of renin and angiotensinogen in the rostral ventrolateral medulla using a dual-reporter transgenic model. *Hypertension* **43:** 1116–1119.

Levens, N. R., Freedlender, A. E., Peach, M. J., and Carey, R. M. (1983). Control of renal function by intrarenal angiotensin II. *Endocrinology* **112:** 43–49.

Lijnen, P., and Petrov, V. (1999). Renin–angiotensin system, hypertrophy and gene expression in cardiac myocytes. *J Mol Cell Cardiol* **31:** 949–970.

Linderman, J. R., and Greene, A. S. (2001). Distribution of angiotensin II receptor expression in the microcirculation of striated muscle. *Microcirculation* **8:** 275–281.

Linz, W., Wiemer, G., Gohlke, P., Unger, T., and Scholkens, B. A. (1995). Contribution of kinins to the cardiovascular actions of angiotensin-converting enzyme inhibitors. *Pharmacol Rev* **47:** 25–49.

Lochard, N., Thibault, G., Silversides, D. W., Touyz, R. M., and Reudelhuber, T. L. (2004). Chronic production of angiotensin IV in the brain leads to hypertension that is reversible with an angiotensin II AT1 receptor antagonist. *Circ Res* **94:** 1451–1457.

Matsubara, H., Sugaya, T., Murasawa, S. et al. (1998). Tissue-specific expression of human angiotensin II AT1 and AT2 receptors and cellular localization of subtype mRNAs in adult human renal cortex using in situ hybridization. *Nephron* **80:** 25–34.

Mitchell, K.D. and Navar, L.G. (1995). Intrarenal actions of angiotensin II in the pathogenesis of experimental hypertension. In Hypertension: pathophysiology, diagnosis and management, (Laragh, J.H., Brenner, B.M., eds). Lippincott, Williams and Wilkins, Philadelphia, PA, pp. 1437–1450.

Miyata, N., Park, F., Li, X. F., and Cowley, A. W. (1999). Distribution of angiotensin AT1 and AT2 receptor subtypes in the rat kidney. *Am J Physiol* **277:** F437–F446.

Moulik, S., Speth, R. C., Turner, B. B., and Rowe, B. P. (2002). Angiotensin II receptor subtype distribution in the rabbit brain. *Exp Brain Res* **142:** 275–283.

Murphy, T. J., Alexander, R. W., Griendling, K. K., Runge, M. S., and Bernstein, K. E. (1991). Isolation of a cDNA encoding the vascular type-1 angiotensin II receptor. *Nature* **351:** 233–236.

Navar, L., and Ham, L.L. (1999). The kidney in blood pressure regulation. In Wilcox, CS, ed. Atlas of Diseases of the Kidney, 1.1–1.2.

Navar, L. G. (1997). The kidney in blood pressure regulation and development of hypertension. *Med Clin North Am* **81:** 1165–1198.

Navar, L. G., Harrison-Bernard, L. M., Nishiyama, A., and Kobori, H. (2002). Regulation of intrarenal angiotensin II in hypertension. *Hypertension* **39:** 316–322.

Navar, L. G., Kobori, H., and Prieto-Carrasquero, M. (2003). Intrarenal angiotensin II and hypertension. *Curr Hypertens Rep* **5:** 135–143.

Nguyen, G., Delarue, F., Berrou, J., Rondeau, E., and Sraer, J. D. (1996). Specific receptor binding of renin on human mesangial cells in culture increases plasminogen activator inhibitor-1 antigen. *Kidney Int* **50:** 1897–1903.

Nguyen, G., Bouzhir, L., Delarue, F., Rondeau, E., and Sraer, J. D. (1998). Evidence of a renin receptor on human mesangial cells: effects on PAI1 and cGMP. *Nephrologie* **19:** 411–416.

Nguyen, G., Delarue, F., Burckle, C., Bouzhir, L., Giller, T., and Sraer, J. D. (2002). Pivotal role of the renin/prorenin receptor in angiotensin II production and cellular responses to renin. *J Clin Invest* **109:** 1417–1427.

Nishiyama, A., Seth, D. M., and Navar, L. G. (2002). Renal interstitial fluid concentrations of angiotensins I and II in anesthetized rats. *Hypertension* **39:** 129–134.

Nishizaka, M. K., Zaman, M. A., and Calhoun, D. A. (2003). Efficacy of low-dose spironolactone in subjects with resistant hypertension. *Am J Hypertens* **16:** 925–930.

Nussberger, J. (2000). Circulating versus tissue angiotensin II. Angiotensin II receptor antagonists. In: Epstein, M. and Brunner, H. (eds), Angiotensin II receptor antagonists. Hanley and Belfus Inc., Philadelphia, PA.

Oliver, J. A., and Sciacca, R. R. (1984). Local generation of angiotensin II as a mechanism of regulation of peripheral vascular tone in the rat. *J Clin Invest* **74:** 1247–1251.

Paul, M., Wagner, J., and Dzau, V. J. (1993). Gene expression of the renin-angiotensin system in human tissues. Quantitative analysis by the polymerase chain reaction. *J Clin Invest* **91:** 2058–2064.

Peach, M. J. (1977). Renin-angiotensin system: biochemistry and mechanisms of action. *Physiol Rev* **57:** 313–370.

Pearce, P., and Funder, J. W. (1987). High affinity aldosterone binding sites (type I receptors) in rat heart. *Clin Exp Pharmacol Physiol* **14:** 859–866.

Peti-Peterdi, J., Warnock, D. G., and Bell, P. D. (2002). Angiotensin II directly stimulates ENaC activity in the cortical collecting duct via AT(1) receptors. *J Am Soc Nephrol* **13:** 1131–1135.

Ploth, D. W. (1983). Angiotensin-dependent renal mechanisms in two-kidney, one-clip renal vascular hypertension. *Am J Physiol* **245:** F131–F141.

Pounarat, J. (2002). The luminal membrane of rat limb expresses AT1 receptor and aminopeptidase activities. *Kidney Int* **62:** 434–445.

Prescott, G., Silversides, D. W., Chiu, S. M., and Reudelhuber, T. L. (2000). Contribution of circulating renin to local synthesis of angiotensin peptides in the heart. *Physiol Genomics* **4**: 67–73.

Prescott, G., Silversides, D. W., and Reudelhuber, T. L. (2002). Tissue activity of circulating prorenin. *Am J Hypertens* **15**: 280–285.

Prieto-Carrasquero, M. C., Kobori, H., Ozawa, Y., Gutierrez, A., Seth, D., and Navar, L. G. (2005). AT1 receptor-mediated enhancement of collecting duct renin in angiotensin II-dependent hypertensive rats. *Am J Physiol Renal Physiol* **289**: F632–F637.

Prieto-Carrasquero, M. C., Harrison-Bernard, L. M., Kobori, H. et al. (2004). Enhancement of collecting duct renin in angiotensin II-dependent hypertensive rats. *Hypertension* **44**: 223–229.

Printz, M.P., Unger, T., and Phillips, M.I. (2003). The brain renin–angiotensin system. In: Ganten, D., Printz, M.P., Phillips, M.I. and Sholkens, B.A. (eds). The renin-angiotensin system in the brain: a model for synthesis of peptides in the brain. Springer-Verlag, Berlin.

Re, R. N. (2003). Intracellular renin and the nature of intracrine enzymes. *Hypertension* **42**: 117–122.

Re, R. N. (2004a). Mechanisms of disease: local renin–angiotensin–aldosterone systems and the pathogenesis and treatment of cardiovascular disease. *Nat Clin Pract Cardiovasc Med* **1**: 42–47.

Re, R. N. (2004b). Tissue renin angiotensin systems. *Med Clin North Am* **88**: 19–38.

Re, R. N., and Cook, J. L. (2006). The intracrine hypothesis: An update. *Regul Pept*, pp. 1–9.

Reaux, A., Fournie-Zaluski, M. C., David, C. et al. (1999). Aminopeptidase A inhibitors as potential central antihypertensive agents. *Proc Natl Acad Sci USA* **96**: 13415–13420.

Reaux-Le Goazigo, A., Iturrioz, X., Fassot, C., Claperon, C., Roques, B. P., and Llorens-Cortes, C. (2005). Role of angiotensin III in hypertension. *Curr Hypertens Rep* **7**: 128–134.

Rocha, R., Stier, C. T., Kifor, I. et al. (2000). Aldosterone: a mediator of myocardial necrosis and renal arteriopathy. *Endocrinology* **141**: 3871–3878.

Rocha, R., Rudolph, A. E., Frierdich, G. E. et al. (2002). Aldosterone induces a vascular inflammatory phenotype in the rat heart. *Am J Physiol Heart Circ Physiol* **283**: H1802–H1810.

Rocha, R., Chander, P. N., Zuckerman, A., and Stier, C. T. (1999). Role of aldosterone in renal vascular injury in stroke-prone hypertensive rats. *Hypertension* **33**: 232–237.

Rocha, R., Chander, P. N., Khanna, K., Zuckerman, A., and Stier, C. T. (1998). Mineralocorticoid blockade reduces vascular injury in stroke-prone hypertensive rats. *Hypertension* **31**: 451–458.

Rohrwasser, A., Morgan, T., Dillon, H. F. et al. (1999). Elements of a paracrine tubular renin-angiotensin system along the entire nephron. *Hypertension* **34**: 1265–1274.

Rossi, G., Boscaro, M., Ronconi, V., and Funder, J. W. (2005). Aldosterone as a cardiovascular risk factor. *Trends Endocrinol Metab* **16**: 104–107.

Rousseau, G., Baxter, J. D., Funder, J. W., Edelman, I. S., and Tomkins, G. M. (1972). Glucocorticoid and mineralocorticoid receptors for aldosterone. *J Steroid Biochem* **3**: 219–227.

Saccomani, G. (1990). Angiotensin II stimulation of Na(+)-H+ exchange in proximal tubule cells. *Am J Phynol Renal Physiol* **258**: F1188–F1193.

Sadjadi, J., Puttaparthi, K., Welborn, M. B. et al. (2002). Upregulation of autocrine-paracrine renin-angiotensin systems in chronic renovascular hypertension. *J Vasc Surg* **36**: 386–392.

Sakai, K., and Sigmund, C. D. (2005). Molecular evidence of tissue renin-angiotensin systems: a focus on the brain. *Curr Hypertens Rep* **7**: 135–140.

Samani, N. J., Swales, J. D., and Brammar, W. J. (1988). Expression of the renin gene in extra-renal tissues of the rat. *Biochem J* **253**: 907–910.

Santos, R. A., Simoes e Silva, A. C., Maric, C. et al. (2003). Angiotensin-(1–7) is an endogenous ligand for the G protein-coupled receptor Mas. *Proc Natl Acad Sci USA* **100**: 8258–8263.

Saris, J. J., Derkx, F. H., Lamers, J. M., Saxena, P. R., Schalekamp, M. A., and Danser, A. H. (2001). Cardiomyocytes bind and activate native human prorenin: role of soluble mannose 6-phosphate receptors. *Hypertension* **37**: 710–715.

Sasaki, K., Yamano, Y., Bardhan, S. et al. (1991). Cloning and expression of a complementary DNA encoding a bovine adrenal angiotensin II type-1 receptor. *Nature* **351**: 230–233.

Schiavone, M. T., Santos, R. A., Brosnihan, K. B., Khosla, M. C., and Ferrario, C. M. (1988). Release of vasopressin from the rat hypothalamo-neurohypophysial system by angiotensin-(1–7) heptapeptide. *Proc Natl Acad Sci USA* **85**: 4095–4098.

Schmitz, U. (1997). Angiotensin II signal transduction; stimulation of multiple mitogen-active protein kinase pathways. *Trends Endocrinol Metab* **8**: 261–266.

Schmitz, U., Thommes, K., Beier, I. et al. (2001). Angiotensin II-induced stimulation of p21-activated kinase and c-Jun NH2-terminal kinase is mediated by Rac1 and Nck. *J Biol Chem* **276**: 22003–22010.

Sealey, J. E., White, R. P., Laragh, J. H., and Rubin, A. L. (1977). Plasma prorenin and renin in anephric patients. *Circ Res* **41**: 17–21.

Silvestre, J. S., Robert, V., Heymes, C. et al. (1998). Myocardial production of aldosterone and corticosterone in the rat. Physiological regulation. *J Biol Chem* **273**: 4883–4891.

Siragy, H. M., and Carey, R. M. (1999). Protective role of the angiotensin AT2 receptor in a renal wrap hypertension model. *Hypertension* **33**: 1237–1242.

Siragy, H. M., Xue, C., Abadir, P., and Carey, R. M. (2005). Angiotensin subtype-2 receptors inhibit renin biosynthesis and angiotensin II formation. *Hypertension* **45**: 133–137.

Siragy, H. M., Howell, N. L., Ragsdale, N. V., and Carey, R. M. (1995). Renal interstitial fluid angiotensin. Modulation by anesthesia, epinephrine, sodium depletion, and renin inhibition. *Hypertension* **25**: 1021–1024.

Soubrier, F., Wei, L., Hubert, C., Clauser, E., Alhenc-Gelas, F., and Corvol, P. (1993). Molecular biology of the angiotensin I converting enzyme: II. Structure-function. Gene polymorphism and clinical implications. *J Hypertens* **11**: 599–604.

Takeda, Y. (2004). Vascular synthesis of aldosterone: role in hypertension. *Mol Cell Endocrinol* **217**: 75–79.

Takeda, Y., Miyamori, I., Inaba, S. et al. (1997). Vascular aldosterone in genetically hypertensive rats. *Hypertension* **29**: 45–48.

(2008). Telmisartan, ramipril or both in patients at high risk for vascular events. The ONTARGET Investigators. *N Engl J Med* **358**: 1547–1559.

Thurston, H., and Swales, J. D. (1977). Blood pressure response of nephrectomized hypertensive rats to converting enzyme inhibition: evidence for persistent vascular renin activity. *Clin Sci Mol Med* **52**: 299–304.

Tipnis, S. R., Hooper, N. M., Hyde, R., Karran, E., Christie, G., and Turner, A. J. (2000). A human homolog of angiotensin-converting enzyme. Cloning and functional expression as a captopril-insensitive carboxypeptidase. *J Biol Chem* **275**: 3323.

Tokuyama, H., Hayashi, K., Matsuda, H. et al. (2002). Differential regulation of elevated renal angiotensin II in chronic renal ischemia. *Hypertension* **40**: 34–40.

Tschope, C., Schultheiss, H. P., and Walther, T. (2002). Multiple interactions between the renin-angiotensin and the kallikrein-kinin systems: role of ACE inhibition and AT1 receptor blockade. *J Cardiovasc Pharmacol* **39**: 478–487.

van den Eijnden, M. M., Saris, J. J., de Bruin, R. J. et al. (2001). Prorenin accumulation and activation in human endothelial cells: importance of mannose 6-phosphate receptors. *Arterioscler Thromb Vasc Biol* **21**: 911–916.

van Kats, J. P., Danser, A. H., van Meegen, J. R., Sassen, L. M., Verdouw, P. D., and Schalekamp, M. A. (1998). Angiotensin production by the heart: a quantitative study in pigs with the use of radiolabeled angiotensin infusions. *Circulation* **98**: 73–81.

van Kats, J. P., Schalekamp, M. A., Verdouw, P. D., Duncker, D. J., and Danser, A. H. (2001). Intrarenal angiotensin II: interstitial and cellular levels and site of production. *Kidney Int* **60**: 2311–2317.

van Kats, J. P., Duncker, D. J., Haitsma, D. B. et al. (2000). Angiotensin-converting enzyme inhibition and angiotensin II type 1 receptor blockade prevent cardiac remodeling in pigs after myocardial infarction: role of tissue angiotensin II. *Circulation* **102**: 1556–1563.

Vicaut, E., and Hou, X. (1994). Local renin-angiotensin system in the microcirculation of spontaneously hypertensive rats. *Hypertension* **24**: 70–76.

Virdis, A., Neves, M. F., Amiri, F., Viel, E., Touyz, R. M., and Schiffrin, E. L. (2002). Spironolactone improves angiotensin-induced vascular changes and oxidative stress. *Hypertension* **40**: 504–510.

von Lutterotti, N., Catanzaro, D. F., Sealey, J. E., and Laragh, J. H. (1994). Renin is not synthesized by cardiac and extrarenal vascular tissues. A review of experimental evidence. *Circulation* **89**: 458–470.

Wang, C. T., Chin, S. Y., and Navar, L. G. (2000). Impairment of pressure-natriuresis and renal autoregulation in ANG II-infused hypertensive rats. *Am J Physiol Renal Physiol* **279**: F319–F325.

Watt, G. C., Harrap, S. B., Foy, C. J. et al. (1992). Abnormalities of glucocorticoid metabolism and the renin-angiotensin system: a four-corners approach to the identification of genetic determinants of blood pressure. *J Hypertens* **10**: 473–482.

Weber, K., Sun, Y. and Guarda, E. (1995). Nonclassical actions of angiotensin II and aldosterone in non-classic target tissue (the heart): relevance to hypertensive heart disease. In Hypertension: pathophysiology, diagnosis, 2nd edn. 2203–2223.

Wolf, G. (2003). The renin–angiotensin system and progression of renal diseases. *Nephron Physiol*. Lippincott, Williams and Wilkins, Philadelphia, PA, pp. P3–P13.

Wolf, G., Wenzel, U., Burns, K. D., Harris, R. C., Stahl, R. A., and Thaiss, F. (2002). Angiotensin II activates nuclear transcription factor-kappaB through AT1 and AT2 receptors. *Kidney Int* **61**: 1986–1995.

Wright, J. W., Tamura-Myers, E., Wilson, W. L. et al. (2003). Conversion of brain angiotensin II to angiotensin III is critical for pressor response in rats. *Am J Physiol Regul Integr Comp Physiol* **284**: R725–R733.

Xue, C., and Siragy, H. M. (2005). Local renal aldosterone system and its regulation by salt, diabetes, and angiotensin II type 1 receptor. *Hypertension* **46**: 584–590.

Ye, P., Kenyon, C. J., MacKenzie, S. M. et al. (2005). The aldosterone synthase (CYP11B2) and 11beta-hydroxylase (CYP11B1) genes are not expressed in the rat heart. *Endocrinology* **146**: 5287–5293.

Zhuo, J. L., Imig, J. D., Hammond, T. G., Orengo, S., Benes, E., and Navar, L. G. (2002). Ang II accumulation in rat renal endosomes during Ang II-induced hypertension: role of AT(1) receptor. *Hypertension* **39**: 116–121.

Further reading

Baker, K. M., Chernin, M. I., Schreiber, T. et al. (2004). Evidence of a novel intracrine mechanism in angiotensin II-induced cardiac hypertrophy. *Regul Pept* **120**: 5–13.

Balcells, E., Meng, Q. C., Johnson, W. H., Oparil, S., and Dell'Italia, L. J. (1997). Angiotensin II formation from ACE and chymase in human and animal hearts: methods and species considerations. *Am J Physiol* **273**: H1769–H1774.

Danser, A. H., van Kesteren, C. A., Bax, W. A. et al. (1997). Prorenin, rennin, angiotensinogen, and angiotensin-converting enzyme in normal and failing human hearts. Evidence for renin binding. *Circulation* **96**: 220–226.

Dell'Italia, L. J., Meng, Q. C., Balcells, E. et al. (1997). Compartmentalization of angiotensin II generation in the dog heart. Evidence for independent mechanisms in intravascular and interstitial spaces. *J Clin Invest* **100**: 253–258.

Mazzolai, L., Nussberger, J., Aubert, J. F. et al. (1998). Blood pressure-independent cardiac hypertrophy induced by locally activated renin–angiotensin system. *Hypertension* **31**: 1324–1330.

Neri Serneri, G. G., Boddi, M., Coppo, M. et al. (1996). Evidence for the existence of a functional cardiac renin-angiotensin system in humans. *Circulation* **94**: 1886–1893.

Peters, J., Farrenkopf, R., Clausmeyer, S. et al. (2002). Functional significance of prorenin internalization in the rat heart. *Circ Res* **90**: 1135–1141.

Sinn, P. L., and Sigmund, C. D. (2000). Identification of three human renin mRNA isoforms from alternative tissue-specific transcriptional initiation. *Physiol Genomics* **3**: 25–31.

Urata, H., Boehm, K. D., Philip, A. et al. (1993). Cellular localization and regional distribution of an angiotensin II-forming chymase in the heart. *J Clin Invest* **91**: 1269–1281.

CHAPTER 13

The Renin–Angiotensin–Aldosterone System and the Kidney

BENJAMIN KO AND GEORGE BAKRIS
University of Chicago, Department of Medicine, Sections of Nephrology and Endocrinology, Diabetes and Metabolism, Hypertensive Diseases Unit, Chicago IL, USA

Contents

I.	Introduction	167
II.	Historical background	167
III.	Overview of the RAS pathway	168
IV.	Physiologic effects of RAS	169
V.	Renin inhibitors	174
VI.	Conclusion	175
	References	175

I. INTRODUCTION

The relationship between the renin–angiotensin system (RAS) and its modulation of kidney function dates back to when renin was first described in the late 1800s and integrated into renal physiology with the experiments of Harry Goldblatt. Studies over the past four decades have further crystallized our knowledge of the cellular and molecular role of this system on kidney function. Acting both as an endocrine organ and a target organ, the kidney profoundly affects and is affected by the RAS, resulting in a variety of physiologic and pathophysiologic effects with wide-reaching treatment implications in diseases as diverse as diabetes and congenital renal abnormalities.

II. HISTORICAL BACKGROUND

The prominent vasopressor actions of this system first led to the discovery of the RAS. In the late nineteenth century, Tigerstedt and Bergman, having observed the association between kidney disease and hypertension, injected kidney extracts into rabbits and found elevated blood pressure in these animals (Tigerstedt, 1898). They theorized that a soluble protein contained within the kidney acted directly as a vasopressor and named this compound renin. Numerous unsuccessful models of renal-injury hypertension followed until 1934, when Harry Goldblatt developed his model of renovascular hypertension (Goldblatt et al., 1934). By placing silver clamps over canine renal arteries, Goldblatt produced a sustained increase in blood pressure. Clamps over other large, non-renal, arteries did not elicit a similar blood pressure response, and so, like Tigerstedt and Bergman, Goldblatt suggested that renal ischemia caused the secretion of a compound that triggers vasoconstriction.

The nature of renin became clear due to the work of many researchers, most prominently by two laboratories under the direction of Eduardo Braun-Menendez and Irvine Page. Both groups ultimately demonstrated that renin was secreted by the kidney but was, in fact, not directly responsible for vasoconstriction. Instead, a blood component was necessary that was enzymatically cleaved by renin to generate an active molecule (Page and Helmer, 1939; Braun-Menendez et al., 1940). This active peptide was named hypertensin by Braun-Menendez and angiotonin by Page before the groups ultimately compromised and dubbed it 'Angiotensin' (Braun-Menendez and Page, 1958). Leonard Skeggs et al. would later discover that this protein actually existed in two forms, with angiotensin-converting enzyme (ACE) converting angiotensin I into its active form, angiotensin II (Ang II) (Skeggs et al., 1954).

Ang II then acts on type I angiotensin receptors (AT_1R) in a variety of tissues including vascular smooth muscle cells, neuronal cells, adrenal glomerulosa cells and renal tubular cells, resulting in vasoconstriction, increased tubular sodium reabsorption, aldosterone secretion and sympathetic activation (Davis et al., 1962; Laverty, 1963; Quan and Baum, 1996; Wang and Giebisch, 1996; Miyata et al., 1999). Additionally, Ang II is thought to mediate thirst and inhibit renin release (Bunag et al., 1967; Epstein et al., 1969).

The familiar actions of the RAS, such as vasoconstriction, are mediated via this pathway, but recent studies have expanded this pathway and revealed a number of additional effectors in this system. Among the first of these identified, type II angiotensin receptors (AT_2) seem to oppose many actions of the RAS mediated via AT_1. Present in blood

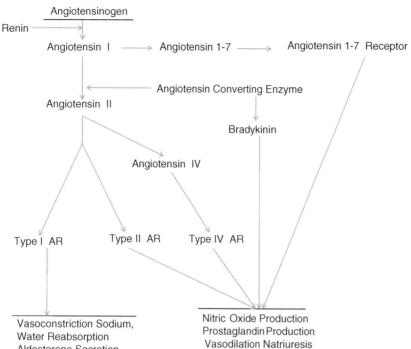

FIGURE 13.1 Schematic of the renin–angiotensin system.

vessels as well as renal epithelial, interstitial and tubular cells, AT₂ is expressed heavily in the fetal kidney, but remains to some degree in the adult, where expression increases in response to sodium depletion and various hormonal influences (Kambayashi et al., 1996; Ozono et al., 1997). Via stimulation of bradykinin and subsequently nitric oxide, AT₂ receptor activation triggers vasodilation (Ichiki et al., 1995; Siragy and Carey, 1997). Within the kidney, the role of AT₂ receptor activation is not clear, although it is likely that promotion of nitric oxide formation enhances the pressure natriuretic response (Tanaka et al., 1999).

While Ang II is the main effector in the RAS, many of the actions that result from the RAS cascade are mediated by other products and pathways. The most well-known example of this is the action of ACE on bradykinin. ACE hydrolyzes bradykinin, preventing its actions to stimulate nitric oxide and cGMP production (Dorer et al., 1974). Indeed, many of the beneficial effects of ACE inhibition may be just as due to increased bradykinin levels as to decreased Ang II formation.

III. OVERVIEW OF THE RAS PATHWAY

The combined efforts of these investigators defined the framework for the renin–angiotensin pathway. Subsequent investigators have built on the foundation of their work establishing a complex series of biochemical pathways and ligand-receptor interactions to form our current understanding of the RAS (Figure 13.1).

Forming the backbone of this system is the well-known cascade by which renin triggers angiotensinogen conversion into angiotensin I (Ang I) and ultimately angiotensin II (Ang II). In the rate-limiting step of this pathway, renin, an aspartyl protease, is secreted by the juxtaglomerular cells in the kidney and cleaves the glycoprotein, angiotensinogen, secreted by the liver into Ang I. Ang I is a biologically inactive decapeptide that is rapidly metabolized by angiotensin-converting enzyme (ACE), found in many tissues, but particularly in the lung, into the biologically active octapeptide Ang II.

Metabolites of angiotensinogen further downstream also have biologic activity. Cleaving the two terminal amino acids from Ang II creates angiotensin IV (Ang IV), which binds to type 4 angiotensin receptors (AT₄R). Ang IV activation of AT₄R promotes natriuresis, presumably mediated by decreased proximal and distal tubule sodium reabsorption (Hamilton et al., 2001).

The action of endopeptidases on either Ang I or Ang II can result in the formation of another active metabolite, angiotensin 1–7 (Ang1–7). Ang1–7 acts as a vasodilator in the vasculature via production of nitric oxide, prostaglandins or other chemokines (Paula et al., 1995; Gironacci et al., 2004). Concentrations of Ang1–7 are similar to Ang II levels in the kidney, but its role in the kidney is unclear with various studies demonstrating both diuretic and antidiuretic effects (DelliPizzi et al., 1994; Magaldi et al., 2003; Pendergrass et al., 2006).

Finally, the circulating levels of Ang II and other active metabolites may not be the key determinant in the renal response to the RAS. Instead, a local renal RAS appears to act in an autocrine and paracrine fashion. Renin, angiotensinogen and ACE have been found in the kidney, as well as various angiotensin receptors and Mas, the receptor for A1–7 (Terada et al., 1983; Bruneval et al., 1986; Gomez et al., 1988; Ingelfinger et al., 1990; Su et al., 2006). Proximal tubule cells appear to produce both renin and angiotensinogen and secrete them into the tubule lumen (Rohrwasser et al., 1999). ACE is present in the brush border of the proximal tubule and likely converts Ang I into Ang II, as Ang II levels in the renal interstitium are some 1000 times higher than the blood (Sibony et al., 1993; Nishiyama et al., 2002). In addition to its effects on renal hemodynamics and tubular transport, Ang II appears to stimulate proximal tubule production of angiotensinogen, thus potentially creating a positive feedback loop (Kobori et al., 2001).

IV. PHYSIOLOGIC EFFECTS OF RAS

The effectors and receptors that make up the renin–angiotensin system mediate a vast number of non-renal physiologic effects throughout the body, notably in the heart, vasculature, brain, lung and adrenal gland. Within the kidney, the RAS regulates renal hemodynamics, tubular transport, inflammation and cell growth and differentiation.

A. RAS and Renal Hemodynamics

Perhaps the most well-known effect of the RAS in the kidney is the action of Ang II on the renal vasculature. Ang II functions as a vasoconstrictor in the interlobular artery and both the afferent and efferent glomerular arterioles, but with greater potency in the efferent arteriole. As a result, Ang II's net effect on the glomerulus is to elevate the intraglomerular capillary pressure. Ang II also constricts the glomerular mesangium and enhances the afferent arteriole response to tubuloglomerular feedback. These actions have contrasting effects on the glomerular filtration rate (GFR). The preferential vasoconstriction of the efferent arteriole and resulting increase in glomerular hydrostatic pressure tends to preserve GFR under states of RAS activation, but mesangial, afferent and systemic vasoconstriction lower renal blood flow and therefore decrease GFR. The net result is generally an increase in filtration fraction, owing to a relatively smaller decrease in GFR compared to renal blood flow.

The differential vasoconstrictor activity responsible for this phenomenon is best explained by the interaction of Ang II-mediated vasoconstriction with various local mediators of vasodilation stimulated by Ang II (Figure 13.2). The rationale for this hypothesis comes from contrasting results between *in vivo* and *in vitro* vasoconstrictor studies. Microperfusion studies have demonstrated a 100-fold greater sensitivity for Ang II to elicit a vasoconstrictor response in the efferent versus afferent arteriole (Edwards, 1983). *In vitro* studies, however, failed to show a difference in vasoconstrictor sensitivity in these vascular beds, suggesting that the production of other local factors stimulated by Ang II may contribute to the differential response (Carmines et al., 1986). Among these are vasodilatory prostaglandins secreted by the glomerulus in response to Ang II. Experimental inhibition of prostaglandin synthesis with indomethacin enhances afferent and efferent arteriole vasoconstriction by Ang II, with afferent arteriolar response to Ang II increasing tenfold (Arima et al., 1994). In addition, AT_2R activation in the afferent arteriole may play a significant role. AT_2R blockade augments Ang II-induced vasoconstriction as well as inhibiting Ang II-mediated vasodilation in preconstricted vessels (Arima et al., 1997; Endo et al., 1997). As AT_2R stimulation is linked to nitric oxide production, it is likely that nitric oxide mediates this effect.

The phenomenon of renal autoregulation of blood flow is essential to normal kidney function and the RAS plays a prominent role in its maintenance. Via autoregulation, the

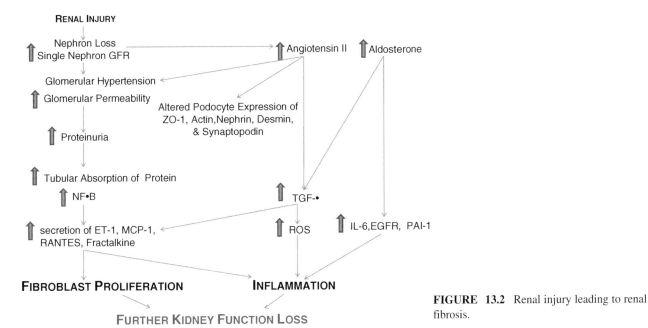

FIGURE 13.2 Renal injury leading to renal fibrosis.

kidney essentially maintains constant renal blood flow over a wide range of blood pressures. This effect is mediated by the myogenic reflex and by tubuloglomerular feedback (TGF). Micropuncture studies have demonstrated that Ang II affects TGF, making the afferent arteriole more sensitive to TGF-signaled vasoconstriction (Schnermann and Briggs, 1990). Vasoconstriction reduces GFR, promoting volume retention. Inhibiting the RAS via an ACE inhibitor or angiotensin receptor blocker (ARB) has the opposite effect, dampening the signals from TGF and promoting diuresis (Arendshorst and Finn, 1977).

B. RAS and Tubular Function

1. Proximal Tubule

The principal action of the proximal tubule is to reabsorb some 60–80% of the water and solute filtered at the glomerulus. Ang II seems to have a direct effect on this process. Micropuncture and microperfusion studies have shown increased proximal tubular reabsorption in response to Ang II. Thus, traditionally Ang II has been thought to stimulate proximal tubule sodium and water transport (Mitchell and Navar, 1987; Liu and Cogan, 1988). Recent studies have suggested that Ang II's impact on the proximal tubule is more complex, however. For example, sodium transport and bicabonate reclamation via luminal Na^+/H^+ and basolateral $Na^+/HCO3$ exchangers in the proximal tubule are regulated via Ang II levels. Sub-nanomolar concentrations of Ang II stimulate both of these exchangers, resulting in increased sodium and water transport, as well as an increased sensitivity for bicarbonate reabsorption in response to changes in blood carbon dioxide levels (Liu and Cogan, 1988; Wang and Chan, 1990; Ruiz et al., 1995; Wu et al., 1996; Romero et al., 1998; Zhou et al., 2006). In addition, Ang II stimulates brush border expression of H^+/ATPase, activity of the Na^+/K^+ ATPase and transport via the Na^+/glucose co-transporter: the net result being increased sodium, water, and bicarbonate absorption (Garvin, 1990; Wang and Chan, 1990; Wu et al., 1996; Wagner et al., 1998).

At high concentrations (10^{-9} M), however, Ang II appears to have an inhibitory effect on the Na^+/H^+ exchanger, basolateral $Na^+/HCO3^-$ exchanger, and Na^+/K^+ ATPase (Harris and Young, 1977). Both of these effects appear to be mediated via AT_1R activation and coupling with various G- or Gq-like proteins, with low concentrations reducing cAMP and activating protein kinase C and higher concentrations leading to elevated Ca^{2+}, nitric oxide and phospholipase A2 levels (Houillier et al., 1996; Kennedy and Burns, 2001).

The significance of this biphasic effect of Ang II on the proximal tubule is unclear, as are the relative contributions of the systemic and local RAS systems. Because of intrarenal Ang II synthesis, concentrations of Ang II in the proximal tubule are more likely to be at levels similar to those in which inhibitory effects were observed experimentally, but this seems contrary to the widely held dogma that Ang II enhances proximal tubule sodium uptake. However, controversy exists as to whether the intrarenal RAS displays the same responses to dietary salt loading as the systemic RAS, as some experimental data suggests that the local RAS may in fact increase Ang II production during salt loading (Thomson et al., 2006).

2. Loop of Henle

This biphasic effect of Ang II is also present within the loop of Henle. As in the proximal tubule, infusions of low concentrations of Ang II enhance sodium and bicarbonate reabsorption by stimulating the luminal Na^+/H^+ exchanger (Capasso et al., 1994). Importantly, low dose Ang II stimulates activity of the key transport mechanisms within the thick ascending limb, the electroneutral $Na^+/K^+/2Cl^-$ co-transporter and the ROMK channel (Lu et al., 1996; Amlal et al., 1998). The net effect of stimulating these transporters is to increased sodium reabsorption, leading to further distal volume reabsorption. Again with higher concentrations of Ang II, an inhibitory effect has been shown experimentally, both in total chloride absorption and by reduced activity of the Na^+/H^+ exchanger, the $Na^+/K^+/2Cl^-$ co-transporter and the ROMK channel (Lu et al., 1996; Amlal et al., 1998; Good et al., 1999; Lerolle et al., 2004).

3. Distal Convoluted Tubule

Traditionally, the bulk of RAS-mediated effects in the distal convoluted tubule (DCT) are believed to be mediated via RAS effects on aldosterone and aldosterone's subsequent upregulation of sodium chloride co-transporter (NCC) and epithelial sodium channel (eNaC) abundance (Palmer et al., 1982; Kim et al., 1998). AT_1Rs, however, are distributed throughout the DCT, suggesting a direct role for Ang II in the DCT (Mujais et al., 1986). While Ang II's effect on NCC function remains unclear, studies have clearly demonstrated that Ang II stimulates the Na^+/H^+ exchanger located in the early DCT and eNaC within the late DCT (Wang and Giebisch, 1996). Notably, these effects were demonstrated at Ang II concentrations of 10^{-11} M, the effect of Ang II at other concentrations has not been studied.

4. Collecting Duct

As in the DCT, while aldosterone in a major determinant of collecting duct activity, AT_1Rs have long been known to be present in the cortical collecting duct and a number of recent studies have explored the effects of Ang II on the collecting duct (Burns et al., 1996). B-type intercalated cells secrete increased bicarbonate and reabsorb increased chloride after Ang II stimulation of AT_1Rs in a pendrin-dependent mechanism (Weiner et al., 1995; Pech et al., 2006). Potassium secretion by ROMK is inhibited by Ang II infusion (Wang and Giebisch, 1996; Wei et al., 2007). In addition, Ang II stimulation of AT_1Rs appears to enhance vasopressin targeting of aquaporin-2 to the apical membrane (Lee et al., 2007).

C. RAS in Human Renal Disease

In addition to its role in maintaining GFR and ensuring salt and water homeostasis, the RAS effects cell signaling, inflammation, growth and differentiation. These actions play a significant role in the pathogenesis of a great many disease processes within the kidney and have shaped a number of treatment options for these diseases.

1. RAS in Chronic Renal Disease

Chronic kidney disease (CKD) is an ever-increasing entity that represents the effects of a number of different disease processes on the kidney. Whether the cause is diabetes mellitus, hypertension or glomerular disease, progressive CKD is characterized by tubular atrophy, interstitial fibrosis and glomerulosclerosis. These changes are, at least in part, triggered by glomerular hypertension, proteinuria-induced injury and renal fibrosis mediated by the TGF-β pathway. Ang II and other mediators of the RAS can directly influence these pathways and inhibiting these actions has been shown to be beneficial in both experimental models and clinical practice (see Figure 13.2).

With progressive kidney injury and nephron loss, the remaining nephrons adapt to maintain overall GFR with an increase in glomerular capillary pressure, resulting in increased single-nephron GFR (Hostetter et al., 1981). The consequence of this adaptive change is that hyperfiltration leads to increased glomerular permeability (Rovira-Halbach et al., 1986; Yoshioka et al., 1987). This allows for passage of potentially tubulo-toxic proteins into the ultrafiltrate, ultimately causing more nephron loss and further hyperfiltration by the surviving nephrons.

Typically, protein in the ultrafiltrate is reabsorbed by the proximal tubule (Scott et al., 1964). In cases of heavy proteinuria, this reabsorbed protein tends to accumulate in lysosomes, resulting in injury and cell death (Brenner, 1983; Haga et al., 1987; Prado et al., 1993; Thomas et al., 1999). Furthermore, animal models have demonstrated that proteinuria and albuminuria and their related accumulation in the proximal tubule increases the expression of a number of inflammatory and pro-fibrotic mediators secreted into the interstitium, possibly via increases in NF-κB (Eddy and Giachelli, 1995; Zoja et al., 1998). These mediators, including endothelin-1, monocyte chemoattractant protein 1, activate normal T-cell expression and secretion of RANTES, IL-8 and fractalkine, thereby triggering inflammation and fibroblast proliferation and, thus, interstitial fibrosis (Zoja et al., 1995, 1998; Wang et al., 1997; Tang et al., 2003; Donadelli et al., 2003).

In addition, the RAS profoundly affects the proteins which make up the podocyte. Ang II stimulation of AT$_1$Rs decreases nephrin and increasing desmin and synaptopodin levels (Yaoita et al., 1999; Reiser et al., 2000; Bonnet et al., 2001; Kelly et al., 2002; Davis et al., 2003; Blanco et al., 2005). Furthermore, AT$_1$R stimulation alters zonula occludens-1 and actin structure (Macconi et al., 2006; Rincon-Choles et al., 2006). These changes are associated with increased proteinuria and can be prevented via ACE inhibitors or angiotensin receptor blocker (ARB) treatment (Rincon-Choles et al., 2006).

The RAS also plays a role in another fibrogenic pathway, as increased Ang II secretion enhances transcription and release of active TGF-β (Wolf et al., 1995; Weigert et al., 2002; Naito et al., 2004; Zhou et al., 2006). TGF-β have been shown to play a key role in renal fibrogenesis, stimulating a wide array of cytokines, enzymes and growth factors in mesangial, endothelial and tubular cells (Border et al., 1990; Border and Noble, 1993; Wolf et al., 1995; Okada et al., 1997; Yokoi et al., 2001). It additionally serves as a chemoattractant for inflammatory cells such as monocytes and a regulator of cell cycle, survival and growth in lymphocytes (Allen et al., 1990; Gorelik and Flavell, 2000; Sung et al., 2003). Furthermore, TGF-β generates reactive oxygen species in the kidney, disrupting autoregulation, and exerts a vasoconstrictive effect in the periphery (Sharma et al., 2005; Zacchigna et al., 2006). TGF-β also enhances angiotensinogen expression in the proximal tubule, creating a positive feedback loop that further accelerates renal injury (Brezniceanu et al., 2006). Notably, aldosterone, also stimulated by Ang II, and renin itself act to enhance the effects of TGF-β (Huang et al., 2006).

These cascades can be ameliorated by reducing hyperfiltration, proteinuria, inflammation and fibrosis via RAS blockade.

ACE inhibitors and ARBs have been shown to reduce filtration pressure and resistive indices (Ogata et al., 2005). Both ACE inhibitors and ARBs reduce excretion of markers of oxidative stress in the urine (Dohi et al., 2003; Ogawa et al., 2006). These effects correlate with a reduction in proteinuria (Ogawa et al., 2006). Similarly, urinary TGF-β levels decrease under conditions of RAS blockade (Gomez-Garre et al., 2006; Song et al., 2006). Likely by this mechanism, treatment with ACE inhibitors decreases renal fibrosis in experimental models of urinary obstruction and chronic renal failure (Wu et al., 2006; Shirazi et al., 2007). In patients who have received a kidney transplant, ACE inhibitors have been shown to reduce allograft fibrosis and, notably, urinary markers of collagen formation decreased in diabetic patients treated with ARBs (Woo et al., 2006; Tylicki et al., 2007). An added benefit with both ACE inhibitors and ARBs has been demonstrated: the two agents in combination reduce renal Ang II levels lower than either agent alone (Komine et al., 2002).

The role of one of Ang II's key downstream targets, aldosterone secretion, in renal fibrosis and proteinuria is becoming increasingly important. Aldosterone is the principal mineralocorticoid secreted by the adrenal gland and has been thought to act primarily in the distal nephron to promote sodium and water reabsorption and potassium excretion. Mineralocorticoid receptors are, however, found throughout the kidney and considerable experimental evidence exists to support a role for aldosterone in kidney injury (Roland et al., 1995; Hirasawa et al., 1997). As with Ang II, aldosterone enhances TGF-β release and generation of reactive oxygen species, as well as other pro-fibrotic factors such as epidermal growth factor receptor and plasminogen activator inhibitor-1 (Brown et al., 2000a, b; Sun et al., 2000).

Interestingly, while RAS blockade should theoretically also result in decreased aldosterone levels, this effect is not sustained long term. Numerous animal and human studies have demonstrated reduced aldosterone levels in the first three months following ACE inhibitor or ARB administration, but these levels increase over time and can even exceed baseline levels at 12 months (Staessen et al., 1981; Biollaz et al., 1982; Sato and Saruta, 2003). Therefore, aldosterone blockade, in addition to RAS blockade, should provide added benefit. Numerous animal studies have supported this theory, showing decreased proteinuria, TGF-β, IL-6, fibrosis and tubulointerstitial damage with administration of the mineralocorticoid receptor blockers, spironolactone or eplerenone (Greene et al., 1996; Blasi et al., 2003; Feria et al., 2003; Zhou et al., 2004). In humans, a number of randomized, placebo control studies have demonstrated that the addition of aldosterone blockade to RAS blockade results in a further reduction in proteinuria (Rachmani et al., 2004; Rossing et al., 2005; Sato et al., 2005; Chrysostomou et al., 2006; Epstein et al., 2006).

2. RAS AND RAS BLOCKADE IN DIABETIC NEPHROPATHY

Based on data from *in vitro* and animal models, it appears that the RAS, largely via Ang II, contributes substantially to the development of chronic kidney disease. In diabetic nephropathy, Ang II's contribution may be even more significant. Animal models of diabetes mellitus (DM) have consistently demonstrated decreased renal expression of the AT_2R. Studies measuring nitric oxide production in these models have observed decreased nitric oxide, consistent with downregulation of the AT_2R (Brown et al., 1997; Wehbi et al., 2001; Awad et al., 2004). As AT_2R activation seems to oppose the actions of AT_1R activation, this observation suggests that, in DM, unopposed AT_1R stimulation by Ang II promotes accelerated hyperfiltration, proteinuria, tubule toxicity, fibrosis and inflammation by the previously described pathways.

Clinically, RAS blockade with either an ACE inhibitor or ARBs represents a cornerstone of treatment in patients with diabetes and diabetic nephropathy, although the studies are not entirely without controversy. In the early 1990s, the Collaborative Study Group first demonstrated the beneficial effects of ACE inhibitors in diabetic patients (Lewis et al., 1993). Four hundred and nine patients with type I diabetes mellitus, proteinuria (>500 mg/day) and mild renal insufficiency (serum creatinine $\leq$2.5 mg/dl) were randomized to captopril or placebo. Follow-up was 3 years on average and captopril treatment demonstrated a 43% decrease in doubling of serum creatinine compared to control. Additionally, the secondary endpoint of combined mortality, need for renal replacement therapy, including transplantation decreased by 50%. A 30% decrease in proteinuria was also observed. ARBs display similar effects, as evident by the Reduction of Endpoints in Non-Insulin-Dependent Diabetes Mellitus with the Angiotensin II Antagonist Losartan (RENAAL) and Irbesartan in Diabetic Nephropathy (IDNT) trials (Brenner et al., 2001; Lewis et al., 2001). Both studies followed a large number of patients with type II diabetes, renal insufficiency and proteinuria for approximately 3 years after randomizing them to either ARB or placebo (also versus amlodipine in the IDNT trial). The IDNT trial had a greater mean proteinuria compared to the RENAAL trial (2.9 g/day versus 1.2 g/day), but both studies showed a significant decrease in patients reaching the combined endpoint of doubling of serum creatinine, end-stage renal disease (ESRD) or death compared to placebo (16% reduction in RENAAL, 20% in IDNT versus placebo, 23% in IDNT versus amlodipine). Both studies showed that a 50% reduction in proteinuria resulted in a substantially greater decreased risk of reaching the combined endpoint. This was attributed to the anti-proteinuric effect of these medications.

A recent meta-analysis, however, has argued that RAS blockade with either ACE inhibitors or ARBs provide no additional reno-protective effect other than blood pressure control (Casas et al., 2005). This meta-analysis, however, suffered from a deficiency that afflicts all meta-analyses: what studies and therefore, subjects, were included in the analysis. In Casas et al.'s analyses, the results were heavily influenced by the inclusion of the Antihypertensive and Lipid-Lowering Treatment to Prevent Heart Attack (ALLHAT) trial that probably enrolled very few patients with diabetic nephropathy (Mann et al., 2006). Since the benefit of RAS blockade has been in studies with diabetic patients with renal insufficiency and proteinuria, the exclusion from ALLHAT of patients receiving ACE inhibitors for pre-existing renal disease or if they had a creatinine greater than 2.0 g/dl means that ALLHAT followed a very different patient population than the ones previously studied. The findings of this meta-analysis, therefore, may simply underscore that RAS blockade is of most benefit in diabetic patients with proteinuria and stage 3 or higher renal insufficiency. Alternatively, it may mean that renal protective effects of inhibitors of the RAS in subjects with minimal renal disease may not become evident by 3–5 years, the duration of follow-up in the ALLHAT study.

Of interest, animal studies have suggested that Ang II levels are related to insulin resistance. The presence of Ang II induces phosphorylation of insulin receptor substrate-1, reducing its ability to activate the phosphotidylinositol-3-kinase pathway in response to insulin (Velloso et al., 1996; Ogihara et al., 2002). The HOPE trial (Heart Outcomes

Prevention Evaluation) seemed to confirm these findings, as the use of ramipril, an ACE inhibitor, reduced diabetes in a high-risk population by 34%, as compared with placebo (Yusuf et al., 2001). Similar findings were observed in the ALLHAT trial (2002). However, the findings in these and other studies were either in *ad hoc* analysis or with self-reporting of diabetes. In the DREAM study (Diabetes Reduction Assessment with Ramipril and Rosiglitazone Medication), the use of an ACE inhibitor for 3 years in patients with impaired glucose tolerance or insulin resistance did not result in a reduction in diabetes or in mortality (Bosch et al., 2006). An improvement in glycemic control was observed, but many attribute this finding to improved insulin resistance because of improved blood pressure control rather than a specific ACE inhibitor effect (Bosch et al., 2006; Bangalore et al., 2007). A trend towards an improvement in the incidence of diabetes was observed towards the end of the 3-year study and further follow-up, as well as longer studies, are underway to assess for long-term benefit (Bosch et al., 2006). For now, while RAS blockade is a mainstay of treatment in patients with overt diabetes, there are no data available currently for or against a preventative role for ACE inhibitor or ARB in diabetes.

3. RAS BLOCKADE IN OTHER PROTEINURIC CHRONIC KIDNEY DISEASES

As proteinuria is common to many renal diseases other than diabetic nephropathy and certainly damaging to the kidney, the reno-protective effect of RAS blockade in non-diabetic chronic kidney disease has been an area of extensive interest. The Angiotensin-Converting-Enzyme Inhibition in Progressive Renal Insufficiency (AIPRI) study followed 583 non-diabetic patients with chronic kidney disease for 3 years after randomization to either ACE inhibitor or placebo (Maschio et al., 1996). These patients displayed a mean creatinine of 2.1 mg/dl and an average proteinuria of 1.8 g/day. The primary endpoint was either a doubling of serum creatinine or progression to ESRD. Treatment with an ACE inhibitor was associated with an overall risk reduction of 53% compared to placebo in reaching the primary endpoint. Similarly, the Ramipril Efficiency in Nephropathy trial (REIN) randomized non-diabetic patients with renal insufficiency and proteinuria (average proteinuria >3 g/day) to treatment with ACE inhibitor or placebo (GISEN, 1997). Even after correcting for changes in blood pressure, patients receiving ramipril displayed significant reductions in proteinuria and combined endpoint of doubling of serum creatinine or progression to ESRD. These findings have been replicated in a number of trials, with a meta-analysis concluding that ACE inhibitor treatment resulted in a 30% decrease in the combined endpoint of doubling serum creatinine or progression to ESRD (Jafar et al., 2001).

Like the data regarding diabetic nephropathy, the benefit of treatment with an ACE inhibitor or ARB being more than a reflection of their antihypertensive effect has been called into question. The evidence for this comes mainly from the ALLHAT trial (ALLHAT, 2002; Rahman et al., 2005). In ALLHAT, no differences in rates of ESRD were seen in either group with GFR to 60–89 ml/min/1.73 m^2 or <60 ml/min/1.73 m^2. This discrepancy is likely due to patient selection. The primary benefit for RAS blockade in short-term studies (3–5 years) seems to be with subjects that have heavy proteinuria or advanced renal disease, as evidenced by the REIN trial, whose patients had nephritic-range proteinuria (GISEN, 1997).

Urinary protein excretion was not measured in the ALLHAT trial. Exclusion criteria for ALLHAT included a serum creatinine greater than 2.0 mg/dl and treatment with an ACE inhibitor for underlying kidney disease. Given these limitations, no conclusion can be drawn from ALLHAT regarding ACE inhibitor efficacy in patients with proteinuria and more advanced renal insufficiency. Instead, given the evidence, these patients would likely benefit from RAS blockade. Indeed, the Combination Treatment of Angiotensin-II Receptor Blocker and Angiotensin-converting-enzyme Inhibitor in Non-diabetic Renal Disease trial (COOPERATE) prospectively demonstrated that combination doses of ACE inhibitor and ARB slowed progression of chronic renal insufficiency and that this effect was primarily generated by a reduction in proteinuria rather than further blood pressure reduction (Nakao et al., 2003). Most experts, therefore, consider a reduction of proteinuria more predictive of improved renal and cardiovascular outcomes than blood pressure control alone (Khosla and Bakris, 2006). While no outcome data exist, further proteinuria reduction via aldosterone blockade has been shown to reduce proteinuria in conjunction with RAS blockade in non-diabetic, proteinuric chronic renal failure (Sato et al., 2005; Chrysostomou et al., 2006). Therefore, this may further retard progression of renal disease and represents an emerging treatment option.

4. RAS IN LUPUS AND OTHER AUTOIMMUNE DISEASE

Ang II-mediated inflammation may play a significant role in systemic lupus erythematosus (SLE) and other autoimmune diseases in the kidney. Renin and ACE activity has long been known to be elevated in SLE and rheumatoid arthritis (RA) (Sheikh and Kaplan, 1987; Goto et al., 1990; Boers et al., 1990). These elevations are due to secretions of various inflammatory cells that results in elevated Ang II levels (Snyder et al., 1985; Kinoshita et al., 1991; Kitazono et al., 1995; Diet et al., 1996). As in chronic kidney disease, RAS activation causes an inflammatory cascade, acting as a positive feedback loop and triggering further inflammation and injury (Wang et al., 1997; Dzau, 2001; Tang et al., 2003).

Murine models of SLE treated with RAS blockade have demonstrated improvement in most clinical and laboratory parameters, including survival, lymphoid hyperplasia, hypertension and proteinuria (Herlitz et al., 1988a, b; Tarkowski et al., 1990). In addition, TGF-β levels sharply decreased when compared to control mice (De Albuquerque et al., 2004). Unfortunately, the role for ACE inhibitors and ARBs in the treatment of SLE in humans is not as well established. A few uncontrolled studies have demonstrated dramatic reductions in hypertension, proteinuria and renal function, but these studies cannot be taken as definitive evidence due to their non-randomized, uncontrolled design and other mitigating factors such as concurrent use of immuno-modulators (Herlitz et al., 1984; Shapira et al., 1990; Kanda et al., 2005). Although design of these studies precludes drawing conclusions from their data, these data are promising and suggest a greater role for RAS blockade in future treatment of SLE and other autoimmune disorders.

5. RAS IN ACUTE RENAL FAILURE

The role of the RAS in the pathogenesis of acute renal failure is unclear. An experimental model of ischemic renal failure demonstrated increases in renin activity. In a second model, the addition of Ang II enhanced tubular damage (Brezis et al., 1990; Honda et al., 1992). RAS blockade, however, has not consistently demonstrated a benefit in these models (Abdulkader et al., 1988; Koelz et al., 1988; Long et al., 1993).

In acute renal failure due to scleroderma renal crisis, however, ACE inhibition has become the standard of care. Scleroderma renal crisis, notable for its marked hypertension and pathologic changes consistent with malignant hypertension, develops due to hyperactivity of the RAS (Traub et al., 1983). The mainstay of treatment is blood pressure control with ACE inhibitors producing a greater degree of improved renal function and survival as compared to other antihypertensive medications (Traub et al., 1983; Beckett et al., 1985).

6. RAS IN PRE-ECLAMPSIA

7. RAS IN OBSTRUCTIVE UROPATHY

Acute urinary tract obstruction resulting in tubular atrophy and interstitial fibrosis is mediated primarily by the RAS. Within one hour of obstruction, renin secretion by the obstructed kidney rises, as well as renin mRNA expression throughout the kidney. In addition, angiotensinogen production increases and intrarenal vessels display greater ACE expression. Studies have demonstrated that RAS blockade is protective against fibrosis in models of unilateral ureteral obstruction (Ishidoya et al., 1995; Manucha et al., 2005). As in chronic kidney disease, this appears to be primarily due to decreased TGF-β levels, although decreased inflammation mediated by a reduction in NF-κB also plays a role.

8. RAS IN CONGENITAL ABNORMALITIES

The RAS appears to be intimately involved in fetal development of the kidney and collecting system. The first evidence for this role came about with the introduction of ACE inhibitors for hypertension. Fetuses carried by women taking ACE inhibitors, particularly in the second and third trimester, developed a series of congenital renal abnormalities including anuria and oligohydramnios (Quan, 2006). First trimester exposure to RAS blockade has also been shown to increase risk for renal dysplasia compared to other antihypertensive medications (Cooper et al., 2006). These findings suggested that an intact RAS was necessary for normal renal embryogenesis.

Additional evidence implicating a critical role for Ang II in renal development has come from human genetic studies. Patients with renal tubular dysgenesis, a disease characterized by absent proximal tubules and anuria, were found to have mutations in genes for renin, Ang II, ACE or AT$_1$R (Gribouval et al., 2005). In a study of patients with renal hypodysplasia, an ACE insertion/deletion polymorphism and an AT$_1$R single-nucleotide polymorphism (AT1R A1166C) were more frequent in patients whose hypodysplasia included obstructive uropathy due to posterior urethral valves (Peruzzi et al., 2005). The authors theorized that Ang II modulates ureteral peristalsis, preventing elevated pressures in the collecting system (Peruzzi et al., 2005).

Given that AT$_2$Rs are quite prominent in the fetal kidney, it is logical to assume that AT$_2$R activation also plays a role in development. Human studies have found that a single-nucleotide polymorphism in the AT$_2$R gene (A1332G) is associated with a host of congenital malformations including multicystic dysplastic kidney, ureteropelvic junction stenosis, posterior urethral valves, hypoplastic kidney and unilateral renal agenesis (Nishimura et al., 1999; Rigoli et al., 2004; Hahn et al., 2005). The mechanism by which these abnormalities occur is unclear and remains under investigation.

V. RENIN INHIBITORS

Renin inhibitors are a newly introduced class of antihypertensive agents that specifically inhibit renin activity and hence reduce the genesis of Ang II and to a lesser extent aldosterone, although the duration of the reported studies are too short to determine if the effect on aldosterone production differs from ACE inhibitors or ARBs (O'Brien, 2006; Oh et al., 2007). These agents have not been evaluated in treatment of unilateral renal arterial disease or other high renin conditions but show theoretical promise. They have been shown to add to the benefits of proteinuria reduction when added to an ARB (Azizi et al., 2004); the converse, however, has not been examined.

These agents are mentioned for completeness but it will be at least 3–5 years before we appreciate the impact on outcome in different kidney diseases. Ongoing clinical trials in diabetic nephropathy and systolic hypertension will be completed in 2010.

VI. CONCLUSION

The renin–angiotensin system traditionally is thought of as a system of circulating hormones that mediated vasoconstriction, regulated GFR, stimulated aldosterone secretion and exerted mild stimulatory effects on the renal tubule. In recent years, however, the RAS has been found to have a wide range of physiologic effects including altering cell expression, inflammation, fibrosis and fetal development, thereby impacting a whole host of diseases. The benefits of research into the mechanisms of action of the RAS are seen today in the widespread use of ACE inhibitors and ARBs in slowing progression of chronic kidney disease, but will likely expand to wider clinical use as the benefits of RAS blockade in other disease processes are fully elucidated. Furthermore, the potential clinical role of the RAS has focused on Ang II; the clinical relevance of many of the other mediators of the RAS such as AT_2R, Ang IV and even renin itself remain largely unexplored. The potential benefits of AT_2R activation, for instance, are many and AT_2R agonists may one day be used to reduce hypertension and further forestall renal disease. The RAS has proved to be a more complex system than Tigerstedt or Goldblatt imagined and research into those complex interactions may open up new treatment options for a vast array of human diseases.

References

Abdulkader, R. C. et al. (1988). Prolonged inhibition of angiotensin II attenuates glycerol-induced acute renal failure. *Braz J Med Biol Res* **21**: 233–239.

Allen, J. B. et al. (1990). Rapid onset synovial inflammation and hyperplasia induced by transforming growth factor beta. *J Exp Med* **171**: 231–247.

ALLHAT. (2002). Major outcomes in high-risk hypertensive patients randomized to angiotensin-converting enzyme inhibitor or calcium channel blocker vs diuretic: the antihypertensive and lipid-lowering treatment to prevent heart attack trial (ALLHAT). *J Am Med Assoc* **288**: 2981–2997.

Amlal, H. et al. (1998). Ang II controls Na+-K+(NH4+)-2Cl- cotransport via 20-HETE and PKC in medullary thick ascending limb. *Am J Physiol Cell Physiol* **274**: C1047–C1056.

Arendshorst, W. J., and Finn, W. F. (1977). Renal hemodynamics in the rat before and during inhibition of angiotensin II. *Am J Physiol Renal Physiol* **233**: F290–F297.

Arima, S. et al. (1994). Glomerular prostaglandins modulate vascular reactivity of the downstream efferent arterioles. *Kidney Int* **45**: 650–658.

Arima, S. et al. (1997). Possible role of P-450 metabolite of arachidonic acid in vasodilator mechanism of angiotensin II type 2 receptor in the isolated microperfused rabbit afferent arteriole. *J Clin Invest* **100**: 2816–2823.

Awad, A. S. et al. (2004). Renal nitric oxide production is decreased in diabetic rats and improved by AT1 receptor blockade. *J Hypertens* **22**: 1571–1577.

Azizi, M. et al. (2004). Pharmacologic demonstration of the synergistic effects of a combination of the renin inhibitor aliskiren and the AT1 receptor antagonist valsartan on the angiotensin II-renin feedback interruption. *J Am Soc Nephrol* **15**: 3126–3133.

Bangalore, S. et al. (2007). Effect of ramipril on the incidence of diabetes. *N Engl J Med* **356**: 522–524.

Beckett, V. L. et al. (1985). Use of captopril as early therapy for renal scleroderma: a prospective study. *Mayo Clin Proc* **60**: 763–771.

Biollaz, J. et al. (1982). Antihypertensive therapy with MK 421: angiotensin II-renin relationships to evaluate efficacy of converting enzyme blockade. *J Cardiovasc Pharmacol* **4**: 966–972.

Blanco, S. et al. (2005). ACE inhibitors improve nephrin expression in Zucker rats with glomerulosclerosis. *Kidney Int Suppl* **93**: S10–S14.

Blasi, E. R. et al. (2003). Aldosterone/salt induces renal inflammation and fibrosis in hypertensive rats. *Kidney Int* **63**: 1791–1800.

Boers, M. et al. (1990). Raised plasma renin and prorenin in rheumatoid vasculitis. *Ann Rheum Dis* **49**: 517–520.

Bonnet, F. et al. (2001). Irbesartan normalises the deficiency in glomerular nephrin expression in a model of diabetes and hypertension. *Diabetologia* **44**: 874–877.

Border, W. A., and Noble, N. A. (1993). Cytokines in kidney disease: the role of transforming growth factor-beta. *Am J Kidney Dis* **22**: 105–113.

Border, W. A. et al. (1990). Suppression of experimental glomerulonephritis by antiserum against transforming growth factor [beta]1. *Nature* **346**: 371–374.

Bosch, J. et al. (2006). Effect of ramipril on the incidence of diabetes. *N Engl J Med* **355**: 1551–1562.

Braun-Menendez, E., and Page, I. H. (1958). Suggested revision of nomenclature – angiotensin. *Science* **127**: 242.

Braun-Menendez, E. et al. (1940). The substance causing renal hypertension. *J Physiol* **98**: 283–298.

Brenner, B. M. (1983). Hemodynamically mediated glomerular injury and the progressive nature of kidney disease. *Kidney Int* **23**: 647–655.

Brenner, B. M. et al. (2001). Effects of losartan on renal and cardiovascular outcomes in patients with type 2 diabetes and nephropathy. *N Engl J Med* **345**: 861–869.

Brezis, M. et al. (1990). Angiotensin II augments medullary hypoxia and predisposes to acute renal failure. *Eur J Clin Invest* **20**: 199–207.

Brezniceanu, M. L. et al. (2006). Transforming growth factor-beta 1 stimulates angiotensinogen gene expression in kidney proximal tubular cells. *Kidney Int* **69**: 1977–1985.

Brown, L. et al. (1997). Tissue-specific changes in angiotensin II receptors in streptozotocin-diabetic rats. *J Endocrinol* **154**: 355–362.

Brown, N. J. et al. (2000a). Synergistic effect of adrenal steroids and angiotensin II on plasminogen activator inhibitor-1 production. *J Clin Endocrinol Metab* **85**: 336–344.

Brown, N. J. et al. (2000b). Aldosterone modulates plasminogen activator inhibitor-1 and glomerulosclerosis in vivo. *Kidney Int* **58**: 1219–1227.

Bruneval, P. et al. (1986). Angiotensin I converting enzyme in human intestine and kidney. Ultrastructural immunohistochemical localization. *Histochemistry* **85**: 73–80.

Bunag, R. D., Page, I. H., and McCubbin, J. W. (1967). Inhibition of renin release by vasopressin and angiotensin. *Cardiovasc Res* **1**: 67–73.

Burns, K. D. et al. (1996). Immortalized rabbit cortical collecting duct cells express AT1 angiotensin II receptors. *Am J Physiol Renal Physiol* **271**: F1147–F1157.

Capasso, G. et al. (1994). Bicarbonate transport along the loop of Henle. II. Effects of acid-base, dietary, and neurohumoral determinants. *J Clin Invest* **94**: 830–838.

Carmines, P. K., Morrison, T. K., and Navar, L. G. (1986). Angiotensin II effects on microvascular diameters of in vitro blood-perfused juxtamedullary nephrons. *Am J Physiol Renal Physiol* **251**: F610–F618.

Casas, J. P. et al. (2005). Effect of inhibitors of the renin-angiotensin system and other antihypertensive drugs on renal outcomes: systematic review and meta-analysis. *Lancet* **366**: 2026–2033.

Chrysostomou, A. et al. (2006). Double-blind, placebo-controlled study on the effect of the aldosterone receptor antagonist spironolactone in patients who have persistent proteinuria and are on long-term angiotensin-converting enzyme inhibitor therapy, with or without an angiotensin II receptor blocker. *Clin J Am Soc Nephrol* **1**: 256–262.

Cooper, W. O. et al. (2006). Major congenital malformations after first-trimester exposure to ACE inhibitors. *N Engl J Med* **354**: 2443–2451.

Davis, J. O. et al. (1962). The role of the renin–angiotensin system in the control of aldosterone secretion. *J Clin Invest* **41**: 378–389.

Davis, B. J. et al. (2003). Disparate effects of angiotensin II antagonists and calcium channel blockers on albuminuria in experimental diabetes and hypertension: potential role of nephrin. *J Hypertens* **21**: 209–216.

De Albuquerque, D. A. et al. (2004). An ACE inhibitor reduces Th2 cytokines and TGFbeta1 and TGF-beta2 isoforms in murine lupus nephritis. *Kidney Int* **65**: 846–859.

DelliPizzi, A. M., Hilchey, S. D., and Bell-Quilley, C. P. (1994). Natriuretic action of angiotensin(1-7). *Br J Pharmacol* **111**: 1–3.

Diet, F. et al. (1996). Increased accumulation of tissue ACE in human atherosclerotic coronary artery disease. *Circulation* **94**: 2756–2767.

Dohi, Y. et al. (2003). Candesartan reduces oxidative stress and inflammation in patients with essential hypertension. *Hypertens Res* **26**: 691–697.

Donadelli, R. et al. (2003). Protein overload induces fractalkine upregulation in proximal tubular cells through nuclear factor {kappa}b- and p38 mitogen-activated protein kinase-dependent pathways. *J Am Soc Nephrol* **14**: 2436–2446.

Dorer, F. E. et al. (1974). Hydrolysis of bradykinin by angiotensin-converting enzyme. *Circ Res* **34**: 824–827.

Dzau, V. J. (2001). Tissue angiotensin and pathobiology of vascular disease: a unifying hypothesis. *Hypertension* **37**: 1047–1052.

Eddy, A. A., and Giachelli, C. M. (1995). Renal expression of genes that promote interstitial inflammation and fibrosis in rats with protein-overload proteinuria. *Kidney Int* **47**: 1546–1557.

Edwards, R. M. (1983). Segmental effects of norepinephrine and angiotensin II on isolated renal microvessels. *Am J Physiol Renal Physiol* **244**: F526–F534.

Endo, Y. et al. (1997). Function of angiotensin II type 2 receptor in the postglomerular efferent arteriole. *Kidney Int Suppl* **63**: S205–S207.

Epstein, A. N., Fitzsimons, J. T., and Simons, B. J. (1969). Drinking caused by the intracranial injection of angiotensin into the rat. *J Physiol* **200**: 98–100.

Epstein, M. et al. (2006). Selective aldosterone blockade with eplerenone reduces albuminuria in patients with type 2 diabetes. *Clin J Am Soc Nephrol* **1**: 940–951.

Feria, I. et al. (2003). Therapeutic benefit of spironolactone in experimental chronic cyclosporine A nephrotoxicity. *Kidney Int* **63**: 43–52.

Garvin, J. L. (1990). Angiotensin stimulates glucose and fluid absorption by rat proximal straight tubules. *J Am Soc Nephrol* **1**: 272–277.

Gironacci, M. M. et al. (2004). Angiotensin-(1-7) inhibitory mechanism of norepinephrine release in hypertensive rats. *Hypertension* **44**: 783–787.

GISEN. (1997). Randomised placebo-controlled trial of effect of ramipril on decline in glomerular filtration rate and risk of terminal renal failure in proteinuric, nondiabetic nephropathy. The GISEN Group (Gruppo Italiano di Studi Epidemiologici in Nefrologia). *Lancet* **349**: 1857–1863.

Goldblatt, H. L. J., Hanzal, R. F., and Summerville, W. W. (1934). Studies on experimental hypertension. I: the production of experimental hypertension due to renal ischemia. *J Exp Med* **59**: 347–349.

Gomez, R. A. et al. (1988). Renin and angiotensinogen gene expression in maturing rat kidney. *Am J Physiol Renal Physiol* **254**: F582–F587.

Gomez-Garre, D. et al. (2006). Losartan improves resistance artery lesions and prevents CTGF and TGF-beta production in mild hypertensive patients. *Kidney Int* **69**: 1237–1244.

Good, D. W., George, T., and Wang, D. H. (1999). Angiotensin II inhibits HCO-3 absorption via a cytochrome P-450-dependent pathway in MTAL. *Am J Physiol Renal Physiol* **276**: F726–F736.

Gorelik, L., and Flavell, R. A. (2000). Abrogation of TGFbeta signaling in T cells leads to spontaneous T cell differentiation and autoimmune disease. *Immunity* **12**: 171–181.

Goto, M. et al. (1990). Spontaneous release of angiotensin converting enzyme and interleukin 1 beta from peripheral blood monocytes from patients with rheumatoid arthritis under a serum free condition. *Ann Rheum Dis* **49**: 172–176.

Greene, E. L., Kren, S., and Hostetter, T. H. (1996). Role of aldosterone in the remnant kidney model in the rat. *J Clin Invest* **98**: 1063–1068.

Gribouval, O. et al. (2005). Mutations in genes in the renin-angiotensin system are associated with autosomal recessive renal tubular dysgenesis. *Nat Genet* **37**: 964–968.

Haga, H. J. et al. (1987). Changes in lysosome populations in the rat kidney cortex induced by passive Heymann glomerulonephritis. *Ren Physiol* **10**: 249–260.

Hahn, H. et al. (2005). Implication of genetic variations in congenital obstructive nephropathy. *Pediatr Nephrol* **20**: 1541–1544.

Hamilton, T. A. et al. (2001). A role for the angiotensin IV/AT4 system in mediating natriuresis in the rat. *Peptides* **22**: 935–944.

Harris, P. J., and Young, J. A. (1977). Dose-dependent stimulation and inhibition of proximal tubular sodium reabsorption by angiotensin II in the rat kidney. *Pflugers Arch* **367**: 295–297.

Herlitz, H. et al. (1984). Captopril treatment of hypertension and renal failure in systemic lupus erythematosus. *Nephron* **38**: 253–256.

Herlitz, H. et al. (1988a). Beneficial effect of captopril on systemic lupus erythematosus-like disease in MRL lpr/lpr mice. *Int Arch Allergy Appl Immunol* **85**: 272–277.

Herlitz, H. et al. (1988b). Effect of captopril on murine systemic lupus erythematosus disease. *J Hypertens Suppl* **6**: S684–S686.

Hirasawa, G. et al. (1997). Colocalization of 11 beta-hydroxysteroid dehydrogenase type II and mineralocorticoid receptor in human epithelia. *J Clin Endocrinol Metab* **82**: 3859–3863.

Honda, N., Hishida, A., and Kato, A. (1992). Factors affecting severity of renal injury and recovery of function in acute renal failure. *Ren Fail* **14**: 337–340.

Hostetter, T. H. et al. (1981). Hyperfiltration in remnant nephrons: a potentially adverse response to renal ablation. *Am J Physiol Renal Physiol* **241**: F85–93.

Houillier, P. et al. (1996). Signaling pathways in the biphasic effect of angiotensin II on apical Na/H antiport activity in proximal tubule. *Kidney Int* **50**: 1496–1505.

Huang, Y. et al. (2006). Renin increases mesangial cell transforming growth factorbeta1 and matrix proteins through receptor-mediated, angiotensin II-independent mechanisms. *Kidney Int* **69**: 105–113.

Ichiki, T. et al. (1995). Effects on blood pressure and exploratory behaviour of mice lacking angiotensin II type-2 receptor. *Nature* **377**: 748–750.

Ingelfinger, J. R. et al. (1990). In situ hybridization evidence for angiotensinogen messenger RNA in the rat proximal tubule. An hypothesis for the intrarenal renin angiotensin system. *J Clin Invest* **85**: 417–423.

Ishidoya, S. et al. (1995). Angiotensin II receptor antagonist ameliorates renal tubulointerstitial fibrosis caused by unilateral ureteral obstruction. *Kidney Int* **47**: 1285–1294.

Jafar, T. H. et al. (2001). Angiotensin-converting enzyme inhibitors and progression of nondiabetic renal disease. A meta-analysis of patient-level data. *Ann Intern Med* **135**: 73–87.

Kambayashi, Y. et al. (1996). Insulin and insulin-like growth factors induce expression of angiotensin type-2 receptor in vascular-smooth-muscle cells. *Eur J Biochem* **239**: 558–565.

Kanda, H. et al. (2005). Antiproteinuric effect of ARB in lupus nephritis patients with persistent proteinuria despite immunosuppressive therapy. *Lupus* **14**: 288–292.

Kelly, D. J. et al. (2002). Expression of the slit-diaphragm protein, nephrin, in experimental diabetic nephropathy: differing effects of anti-proteinuric therapies. *Nephrol Dial Transplant* **17**: 1327–1332.

Kennedy, C. R., and Burns, K. D. (2001). Angiotensin II as a mediator of renal tubular transport. *Contrib Nephrol* **135**: 47–62.

Khosla, N., and Bakris, G. (2006). Lessons learned from recent hypertension trials about kidney disease. *Clin J Am Soc Nephrol* **1**: 229–235.

Kim, G.-H. et al. (1998). The thiazide-sensitive Na-Cl cotransporter is an aldosterone-induced protein. *Proc Natl Acad Sci USA* **95**: 14552–14557.

Kinoshita, A. et al. (1991). Multiple determinants for the high substrate specificity of an angiotensin II-forming chymase from the human heart. *J Biol Chem* **266**: 19192–19197.

Kitazono, T. et al. (1995). Evidence that angiotensin II is present in human monocytes. *Circulation* **91**: 1129–1134.

Kobori, H., Harrison-Bernard, L. M., and Navar, L. G. (2001). Enhancement of angiotensinogen expression in angiotensin II-dependent hypertension. *Hypertension* **37**: 1329–1335.

Koelz, A. M. et al. (1988). The angiotensin converting enzyme inhibitor enalapril in acute ischemic renal failure in rats. *Experientia* **44**: 172–175.

Komine, N. et al. (2002). Effect of combining an ACE inhibitor and an angiotensin II receptor blocker on plasma and kidney tissue angiotensin II levels. *Am J Kidney Dis* **39**: 159–164.

Laverty, R. (1963). A nervously mediated action of angiotensin in anesthetized rats. *J Pharm Pharmacol* **15**: 63.

Lee, Y.-J. et al. (2007). Increased AQP2 targeting in primary cultured IMCD cells in response to angiotensin II through AT1 receptor. *Am J Physiol Renal Physiol* **292**: F340–F350.

Lerolle, N. et al. (2004). Angiotensin II inhibits NaCl absorption in the rat medullary thick ascending limb. *Am J Physiol Renal Physiol* **287**: F404–F410.

Lewis, E. J. et al. (1993). The effect of angiotensin-converting-enzyme inhibition on diabetic nephropathy. The Collaborative Study Group. *N Engl J Med* **329**: 1456–1462.

Lewis, E. J. et al. (2001). Renoprotective effect of the angiotensin-receptor antagonist irbesartan in patients with nephropathy due to type 2 diabetes. *N Engl J Med* **345**: 851–860.

Liu, F. Y., and Cogan, M. G. (1988). Angiotensin II stimulation of hydrogen ion secretion in the rat early proximal tubule. Modes of action, mechanism, and kinetics. *J Clin Invest* **82**: 601–607.

Long, G. W. et al. (1993). Protective effects of enalaprilat against postischemic renal failure. *J Surg Res* **54**: 254–257.

Lu, M. et al. (1996). Effect of angiotensin II on the apical K+ channel in the thick ascending limb of the rat kidney. *J Gen Physiol* **108**: 537–547.

Macconi, D. et al. (2006). Permselective dysfunction of podocyte-podocyte contact upon angiotensin II unravels the molecular target for renoprotective intervention. *Am J Pathol* **168**: 1073–1085.

Magaldi, A. J. et al. (2003). Angiotensin-(1-7) stimulates water transport in rat inner medullary collecting duct: evidence for involvement of vasopressin V2 receptors. *Pflugers Arch* **447**: 223–230.

Mann, J. F. et al. (2006). Progression of renal disease – can we forget about inhibition of the renin-angiotensin system? *Nephrol Dial Transplant* **21**: 2348–2351 discussion 2352-3.

Manucha, W. et al. (2005). Angiotensin II type I antagonist on oxidative stress and heat shock protein 70 (HSP 70) expression in obstructive nephropathy. *Cell Mol Biol (Noisy-le-grand)* **51**: 547–555.

Maschio, G. et al. (1996). Effect of the angiotensin-converting-enzyme inhibitor benazepril on the progression of chronic renal insufficiency. The Angiotensin-Converting-Enzyme Inhibition in Progressive Renal Insufficiency Study Group. *N Engl J Med* **334**: 939–945.

Mitchell, K. D., and Navar, L. G. (1987). Superficial nephron responses to peritubular capillary infusions of angiotensins I and II. *Am J Physiol Renal Physiol* **252**: F818–F824.

Miyata, N. et al. (1999). Distribution of angiotensin AT1 and AT2 receptor subtypes in the rat kidney. *Am J Physiol Renal Physiol* **277**: F437–F446.

Mujais, S. K., Kauffman, S., and Katz, A. I. (1986). Angiotensin II binding sites in individual segments of the rat nephron. *J Clin Invest* **77**: 315–318.

Naito, T. et al. (2004). Angiotensin II induces thrombospondin-1 production in human mesangial cells via p38 MAPK and JNK: a mechanism for activation of latent TGF-beta1. *Am J Physiol Renal Physiol* **286**: F278–F287.

Nakao, N. et al. (2003). Combination treatment of angiotensin-II receptor blocker and angiotensin-converting-enzyme inhibitor in non-diabetic renal disease (COOPERATE): a randomised controlled trial. *Lancet* **361**: 117–124.

Nishimura, H. et al. (1999). Role of the angiotensin type 2 receptor gene in congenital anomalies of the kidney and urinary tract, CAKUT, of mice and men. *Mol Cell* **3**: 1–10.

Nishiyama, A., Seth, D. M., and Navar, L. G. (2002). Renal interstitial fluid concentrations of angiotensins I and II in anesthetized rats. *Hypertension* **39**: 129–134.

O'Brien, E. (2006). Aliskiren: a renin inhibitor offering a new approach for the treatment of hypertension. *Expert Opin Investig Drugs* **15**: 1269–1277.

Ogata, C. et al. (2005). Evaluation of intrarenal hemodynamics by Doppler ultrasonography for renoprotective effect of angiotensin receptor blockade. *Clin Nephrol* **64**: 352–357.

Ogawa, S. et al. (2006). Angiotensin II type 1 receptor blockers reduce urinary oxidative stress markers in hypertensive diabetic nephropathy. *Hypertension* **47**: 699–705.

Ogihara, T. et al. (2002). Angiotensin II-induced insulin resistance is associated with enhanced insulin signaling. *Hypertension* **40**: 872–879.

Oh, B. H., Mitchell, J., Herron, J. R., Chung, J., Khan, M., and Keefe, D. (2007). Aliskiren, an oral renin inhibitor, provides dose-dependent efficacy and sustained 24-hour blood pressure control in patients with hypertension. *J Am Coll.Cardiol* **49**: 1157.

Okada, H. et al. (1997). Early role of Fsp1 in epithelial-mesenchymal transformation. *Am J Physiol Renal Physiol* **273**: F563–F574.

Ozono, R. et al. (1997). Expression of the aubtype 2 angiotensin (AT2) receptor protein in rat kidney. *Hypertension* **30**: 1238–1246.

Page, I. H., and Helmer, O. M. (1939). A crystalline pressor substance, Angiotonin. *Proc Center Soc. Clin Invest* **12**: 17.

Palmer, L. G. et al. (1982). Aldosterone control of the density of sodium channels in the toad urinary bladder. *J Membr Biol* **64**: 91–102.

Paula, R. D. et al. (1995). Angiotensin-(1-7) potentiates the hypotensive effect of bradykinin in conscious rats. *Hypertension* **26**: 1154–1159.

Pech, V. et al. (2006). Angiotensin II increases chloride absorption in the cortical collecting duct in mice through a pendrin-dependent mechanism. *Am J Physiol Renal Physiol* 00361.2006.

Pendergrass, K. D. et al. (2006). Differential expression of nuclear AT1 receptors and angiotensin II within the kidney of the male congenic mRen2. Lewis rat. *Am J Physiol Renal Physiol* **290**: F1497–F1506.

Peruzzi, L. et al. (2005). Low renin-angiotensin system activity gene polymorphism and dysplasia associated with posterior urethral valves. *J Urol* **174**: 713–717.

Prado, M. J. et al. (1993). Nephrotoxicity of human Bence Jones protein in rats: proteinuria and enzymuria profile. *Braz J Med Biol Res* **26**: 633–638.

Quan, A. (2006). Fetopathy associated with exposure to angiotensin converting enzyme inhibitors and angiotensin receptor antagonists. *Early Hum Dev* **82**: 23–28.

Quan, A., and Baum, M. (1996). Endogenous production of angiotensin II modulates rat proximal tubule transport. *J Clin Invest* **97**: 2878–2882.

Rachmani, R. et al. (2004). The effect of spironolactone, cilazapril and their combination on albuminuria in patients with hypertension and diabetic nephropathy is independent of blood pressure reduction: a randomized controlled study. *Diabet Med* **21**: 471–475.

Rahman, M. et al. (2005). Renal outcomes in high-risk hypertensive patients treated with an angiotensin-converting enzyme inhibitor or a calcium channel blocker vs a diuretic: a report from the Antihypertensive and Lipid-Lowering Treatment to Prevent Heart Attack Trial (ALLHAT). *Arch Intern Med* **165**: 936–946.

Reiser, J. et al. (2000). Regulation of mouse podocyte process dynamics by protein tyrosine phosphatases rapid communication. *Kidney Int* **57**: 2035–2042.

Rigoli, L. et al. (2004). Angiotensin-converting enzyme and angiotensin type 2 receptor gene genotype distributions in Italian children with congenital uropathies. *Pediatr Res* **56**: 988–993.

Rincon-Choles, H. et al. (2006). ZO-1 expression and phosphorylation in diabetic nephropathy. *Diabetes* **55**: 894–900.

Rohrwasser, A. et al. (1999). Elements of a paracrine tubular renin-angiotensin system along the entire nephron. *Hypertension* **34**: 1265–1274.

Roland, B. L., Krozowski, Z. S., and Funder, J. W. (1995). Glucocorticoid receptor, mineralocorticoid receptors, 11 beta-hydroxysteroid dehydrogenase-1 and -2 expression in rat brain and kidney: in situ studies. *Mol Cell Endocrinol* **111**: R1–7.

Romero, M. F. et al. (1998). Cloning and functional expression of rNBC, an electrogenic Na+-HCO-cotransporter from rat kidney. *Am J Physiol Renal Physiol* **274**: F425–F432.

Rossing, K. et al. (2005). Beneficial effects of adding spironolactone to recommended antihypertensive treatment in diabetic nephropathy: a randomized, double-masked, cross-over study. *Diabetes Care* **28**: 2106–2112.

Rovira-Halbach, G. et al. (1986). Single nephron hyperfiltration and proteinuria in a newly selected rat strain with superficial glomeruli. *Ren Physiol* **9**: 317–325.

Ruiz, O. S. et al. (1995). Regulation of the renal Na-HCO3 cotransporter: IV. Mechanisms of the stimulatory effect of angiotensin II. *J Am Soc Nephrol* **6**: 1202–1208.

Sato, A., and Saruta, T. (2003). Aldosterone breakthrough during angiotensin-converting enzyme inhibitor therapy. *Am J Hypertens* **16**: 781–788.

Sato, A., Hayashi, K., and Saruta, T. (2005). Antiproteinuric effects of mineralocorticoid receptor blockade in patients with chronic renal disease. *Am J Hypertens* **18**: 44–49.

Schnermann, J., and Briggs, J. P. (1990). Effect of angiotensin and other pressor agents on tubuloglomerular feedback responses. *Kidney Int Suppl* **30**: S77–80.

Scott, W. N. et al. (1964). Inulin and albumin absorption from the proximal tubule in necturus kidney. *Science* **146**: 1588–1590.

Shapira, Y. et al. (1990). Antiproteinuric effect of captopril in a patient with lupus nephritis and intractable nephrotic syndrome. *Ann Rheum Dis* **49**: 725–727.

Sharma, K. et al. (2005). TGF-beta impairs renal autoregulation via generation of ROS. *Am J Physiol Renal Physiol* **288**: F1069–F1077.

Sheikh, I. A., and Kaplan, A. P. (1987). Assessment of kininases in rheumatic diseases and the effect of therapeutic agents. *Arthritis Rheum* **30**: 138–145.

Shirazi, M. et al. (2007). Captopril reduces interstitial renal fibrosis and preserves more normal renal tubules in neonatal dogs with partial urethral obstruction: a preliminary study. *Urol Int* **8**: 173–177.

Sibony, M. et al. (1993). Gene expression and tissue localization of the two isoforms of angiotensin I converting enzyme. *Hypertension* **21**: 827–835.

Siragy, H. M., and Carey, R. M. (1997). The subtype 2 (AT2) angiotensin receptor mediates renal production of nitric oxide in conscious rats. *J Clin Invest* **100**: 264–269.

Skeggs, L. T. et al. (1954). The existence of two forms of hypertensin. *J Exp Med* **99**: 275–282.

Snyder, R. A., Kaempfer, C. E., and Wintroub, B. U. (1985). Chemistry of a human monocyte-derived cell line (U937): identification of the angiotensin I-converting activity as leukocyte cathepsin G. *Blood* **65**: 176–182.

Song, J. H. et al. (2006). Effect of low-dose dual blockade of renin-angiotensin system on urinary TGF-beta in type 2 diabetic patients with advanced kidney disease. *Nephrol Dial Transplant* **21**: 683–689.

Staessen, J. et al. (1981). Rise in plasma concentration of aldosterone during long-term angiotensin II suppression. *J Endocrinol* **91**: 457–465.

Su, Z., Zimpelmann, J., and Burns, K. D. (2006). Angiotensin-(1-7) inhibits angiotensin II-stimulated phosphorylation of MAP kinases in proximal tubular cells. *Kidney Int* **69**: 2212–2218.

Sun, Y. et al. (2000). Local angiotensin II and transforming growth factor-beta1 in renal fibrosis of rats. *Hypertension* **35**: 1078–1084.

Sung, J. L., Lin, J. T., and Gorham, J. D. (2003). CD28 co-stimulation regulates the effect of transforming growth factor-[beta]1 on the proliferation of naive CD4+ T cells. *Int Immunopharmacol* **3**: 233–245.

Tanaka, M. et al. (1999). Vascular response to sngiotensin II is exaggerated through an upregulation of AT1 receptor in AT2 knockout mice. *Biochem Biophysl Res Commun* **258**: 194–198.

Tang, S. et al. (2003). Albumin stimulates interleukin-8 expression in proximal tubular epithelial cells in vitro and in vivo. *J Clin Invest* **111**: 515–527.

Tarkowski, A. et al. (1990). Differential effects of captopril and enalapril, two angiotensin converting enzyme inhibitors, on immune reactivity in experimental lupus disease. *Agents Actions* **31**: 96–101.

Terada, Y., Nonoguchi, H., and Marumo, F. (1983). PCR localization of angiotensin II receptor and angiotensinogen mRNA in rat kidney. *Kidney Int* **43**: 1251–1259.

Thomas, M. E. et al. (1999). Proteinuria induces tubular cell turnover: a potential mechanism for tubular atrophy. *Kidney Int* **55**: 890–898.

Thomson, S. C. et al. (2006). An unexpected role for angiotensin II in the link between dietary salt and proximal reabsorption. *J Clin Invest* **116**: 1110–1116.

Tigerstedt, R. B. P. (1898). Niere und Kreislauf. *Skand Arch Physiol* **8**: 223–271.

Traub, Y. M. et al. (1983). Hypertension and renal failure (scleroderma renal crisis) in progressive systemic sclerosis. Review of a 25-year experience with 68 cases. *Medicine (Balt.)* **62**: 335–352.

Tylicki, L. et al. (2007). Renal allograft protection with angiotensin II type 1 receptor antagonists. *Am J Transplant* **7**: 243–248.

Velloso, L. A. et al. (1996). Cross-talk between the insulin and angiotensin signaling systems. *Proc Natl Acad Sci USA* **93**: 12490–12495.

Wagner, C. A. et al. (1998). Angiotensin II stimulates vesicular H+-ATPase in rat proximal tubular cells. *Proc Natl Acad Sci USA* **95**: 9665–9668.

Wang, T., and Chan, Y. L. (1990). Mechanism of angiotensin II action on proximal tubular transport. *J Pharmacol Exp Ther* **252**: 689–695.

Wang, Y. et al. (1997). Induction of monocyte chemoattractant protein-1 in proximal tubule cells by urinary protein. *J Am Soc Nephrol* **8**: 1537–1545.

Wehbi, G. J. et al. (2001). Early streptozotocin-diabetes mellitus downregulates rat kidney AT2 receptors. *Am J Physiol Renal Physiol* **280**: F254–F265.

Wei, Y. et al. (2007). Angiotensin II Inhibits the ROMK-like small conductance K channel in renal cortical collecting duct during dietary potassium restriction. *J Biol Chem* **282**: 6455–6462.

Weigert, C. et al. (2002). Angiotensin II induces human TGF-beta 1 promoter activation: similarity to hyperglycaemia. *Diabetologia* **45**: 890–898.

Weiner, I. D. et al. (1995). Regulation of luminal alkalinization and acidification in the cortical collecting duct by angiotensin II. *Am J Physiol Renal Physiol* **269**: F730–F738.

Wolf, G. et al. (1995). Angiotensin II-stimulated expression of transforming growth factor beta in renal proximal tubular cells: attenuation after stable transfection with the c-mas oncogene. *Kidney Int* **48**: 1818–1827.

Woo, V. et al. (2006). Effects of losartan on urinary secretion of extracellular matrix and their modulators in type 2 diabetes mellitus patients with microalbuminuria. *Clin Invest Med* **29**: 365–372.

Wu, M.-S. et al. (1996). Role of NHE3 in mediating renal brush border Na+-H+ exchange. Adaptation to metabolic acidosis. *J Biol Chem* **271**: 32749–32752.

Wu, K. et al. (2006). Valsartan inhibited the accumulation of dendritic cells in rat fibrotic renal tissue. *Cell Mol Immunol* **3**: 213–220.

Yaoita, E. et al. (1999). Identification of renal podocytes in multiple species: higher vertebrates are vimentin positive/lower vertebrates are desmin positive. *Histochem Cell Biol* **111**: 107–115.

Yokoi, H. et al. (2001). Role of connective tissue growth factor in profibrotic action of transforming growth factor-beta: a potential target for preventing renal fibrosis. *Am J Kidney Dis* **38**: S134–S138.

Yoshioka, T. et al. (1987). Role of abnormally high transmural pressure in the permselectivity defect of glomerular capillary wall: a study in early passive Heymann nephritis. *Circ Res* **61**: 531–538.

Yusuf, S. et al. (2001). Ramipril and the development of diabetes. *J Am Med Assoc* **286**: 1882–1885.

Zacchigna, L. et al. (2006). Emilin1 links TGF-beta maturation to blood pressure homeostasis. *Cell* **124**: 929–942.

Zhou, X. et al. (2004). Aldosterone antagonism ameliorates proteinuria and nephrosclerosis independent of glomerular dynamics in L-NAME/SHR model. *Am J Nephrol* **24**: 242–249.

Zoja, C. et al. (1995). Proximal tubular cell synthesis and secretion of endothelin-1 on challenge with albumin and other proteins. *Am J Kidney Dis* **26**: 934–941.

Zoja, C. et al. (1998). Protein overload stimulates RANTES production by proximal tubular cells depending on NF-[kgr]B activation. *Kidney Int* **53**: 1608–1615.

Further reading

AbdAlla, S. et al. (2001). Increased AT(1) receptor heterodimers in preeclampsia mediate enhanced angiotensin II responsiveness. *Nat Med* **7**: 1003–1009.

Gant, N. F. et al. (1973). A study of angiotensin II pressor response throughout primigravid pregnancy. *J Clin Invest* **52**: 2682–2689.

Haller, H. et al. (1989). Increased intracellular free calcium and sensitivity to angiotensin II in platelets of preeclamptic women. *Am J Hypertens* **2**: 238–243.

Hanssens, M. et al. (1991). Angiotensin II levels in hypertensive and normotensive pregnancies. *Br J Obstet Gynaecol* **98**: 155–161.

Juurlink, D. N. et al. (2004). Rates of hyperkalemia after publication of the Randomized Aldactone Evaluation Study. *N Engl J Med* **351**: 543–551.

Wallukat, G. et al. (1999). Patients with preeclampsia develop agonistic autoantibodies against the angiotensin AT1 receptor. *J Clin Invest* **103**: 945–952.

Wang, T., and Giebisch, G. (1996). Effects of angiotensin II on electrolyte transport in the early and late distal tubule in rat kidney. F143–149.

Wang, T., and Giebisch, G. (1996b). Effects of angiotensin II on electrolyte transport in the early and late distal tubule in rat kidney. *Am J Physiol Renal Physiol* **271**: F143–F149.

Zhou, Y., Bouyer, P., and Boron, W. F. (2006a). Effects of angiotensin II on the CO2 dependence of HCO3- reabsorption by the rabbit S2 renal proximal tubule. *Am J Physiol Renal Physiol* **290**: F666–F673.

Zhou, Y. et al. (2006b). Thrombospondin 1 mediates angiotensin II induction of TGF-beta activation by cardiac and renal cells under both high and low glucose conditions. *Biochem Biophys Res Commun* **339**: 633–641.

CHAPTER 14

The Renin–Angiotensin System and the Heart

AARON J. TRASK AND CARLOS M. FERRARIO

The Hypertension & Vascular Research Center, Department of Physiology & Pharmacology, Wake Forest University School of Medicine, Winston-Salem, North Carolina, 27157, USA

Contents

I.	Introduction	181
II.	Cardiac RAS: local versus endocrine origin	181
III.	RAS actions at the cellular level	182
IV.	RAS and the coronary circulation	184
V.	Significance of the RAS on cardiac function	184
VI.	Conclusions	185
	References	185

I. INTRODUCTION

For many years, the renin–angiotensin system (RAS) was thought to be mainly a traditional circulating hormonal system whereby renal renin-dependent production of angiotensin II (Ang II) occurred in response to a fall in macula densa sodium concentration, low arterial pressure or a decrease in circulating blood volume. Renin could then act upon its circulating substrate, angiotensinogen – primarily produced in the liver – to produce the inactive precursor decapeptide, angiotensin I (Ang I). This process served as the starting point for the RAS cascade in that Ang I could be acted upon by several different enzymes to produce the biologically active peptide hormones, Ang II and angiotensin-(1–7) (Ang-(1–7)). A diagrammatic figure of the current expanded view of the RAS as primarily characterized by our laboratories (Ferrario et al., 2005) is shown in Figure 14.1.

Recent advances over the last several decades showed that the RAS is not merely an endocrine system – body tissues harbor local renin–angiotensin systems which can alter physiologic processes by exerting autocrine/paracrine actions. Local renin–angiotensin systems (Lee et al., 1993; Paul et al., 2006) have been found to date in the brain, kidney, vasculature, pancreas, uterus, placenta, the intestine and the focus of this chapter, the heart. These local systems are thought to exert effects on the tissues in which they reside, independent of blood pressure alterations (Lee et al., 1993; Paul et al., 2006). The local cardiac RAS is no exception.

This chapter will encompass the origin of the cardiac RAS and summarize how it acts to regulate cardiac processes and coronary blood flow. Most of the experimental findings regarding the cardiac RAS reported in this chapter are derived from or performed in animal models, including rodents (see Paul et al., 2006 for review). With the advent and wide use of angiotensin-converting enzyme (ACE) inhibitors and angiotensin receptor blockers (ARBs) for the treatment of hypertension and heart failure also came clinical data that now begin to complement years of experimental findings.

II. CARDIAC RAS: LOCAL VERSUS ENDOCRINE ORIGIN

In order for an organ to have a complete RAS, it must possess all of the necessary components, including the genes leading to the expression of the precursor protein angiotensinogen, as well as all of the processing enzymes that determine which biologically-active peptides will be produced. Using these criteria, the heart does indeed possess a complete RAS. Angiotensinogen and renin, either synthesized locally or uptaken from the circulation, serve as the precursors to Ang I, which can be acted upon by either angiotensin-converting enzyme (ACE) or chymase (Urata et al., 1990) to yield the potent mitogenic vasopressor and growth-promoting hormone Ang II. Newer studies (Ferrario et al., 2005) now show that Ang II can then be hydrolyzed by angiotensin-converting enzyme 2 (ACE2) to produce Ang-(1–7) (Trask et al., 2007), an action that allows ACE2 to regulate the balance of the two biologically-active arms of the RAS. Ang I can also be acted upon by the endopeptidases prolyl oligopeptidase (POP), neprilysin (NEP) and thimet oligopeptidase (TOP) to produce Ang-(1–7) (Welches et al., 1993). Although not all of these activities have been shown to produce Ang-(1–7) directly in the heart, all of these enzymes have been shown to hydrolyze Ang I into Ang-(1–7) (Yamamoto et al., 1992; Chappell et al., 2000).

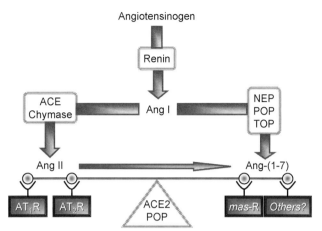

FIGURE 14.1 Current view of the cardiac renin angiotensin system's enzymes and peptides. ACE, angiotensin-converting enzyme (EC 3.4.15.1); ACE2, angiotensin-converting enzyme 2; NEP, neprilysin (EC 3.4.24.11); POP, prolyl oligopeptidase (EC 3.4.21.26); TOP, thimet oligopeptidase (EC 3.4.24.15) controlling angiotensin-(1–7) [Ang-(1–7)] production from angiotensin I or angiotensin II; Angiotensin receptors are the AT_1-R, the AT_2-R, and the *mas*-R. (Adapted from Trask and Ferrario, 2007.)

Since the heart contains all of these necessary components to produce Ang-(1–7) from Ang I (Cicilini et al., 1994; Fielitz et al., 2002; Linardi et al., 2004), it is plausible to consider that, given the availability of substrate, Ang-(1–7) may also be produced from Ang I in the heart. While this area of research still requires further investigation, early work from our laboratory showed the release of Ang-(1–7) in canine coronary sinus post induction of acute myocardial ischemia (Santos et al., 1990). In addition, newer studies showed the involvement of ACE2 in accounting for the increased formation of Ang-(1–7) in failing human heart ventricles (Zisman et al., 2003a, b).

As illustrated above, the existence of a RAS that is harbored within the heart is without question, but the origin of the local cardiac RAS is still somewhat of a controversy. In support of local production are the following data. First, cardiac myocytes and fibroblasts have the ability to produce the angiotensin precursor protein, angiotensinogen, as well as the rate-limiting enzyme that serves as the starting point of the RAS cascade, renin. Second, intriguing evidence suggests that cardiac renin may be synthesized by mast cells, while low in abundance in the normal heart, may be recruited into the cardiac tissue during pathological processes such as ischemia (Xiao and Bernstein, 2005; Francis and Tang, 2006; Le and Coffman, 2006; Mackins et al., 2006; Miyazaki et al., 2006; Reid et al., 2007). Third, some studies have detected ample amounts of angiotensinogen (Kunapuli and Kumar, 1987; Sawa et al., 1992) in the human heart. In a comparative study of angiotensinogen and renin, Dzau and colleagues (Dzau et al., 1987) found that angiotensinogen mRNA levels were in far excess of renin in rodent hearts, suggesting the possibility of excess substrate for the rate-limiting enzyme, renin. Fourth, angiotensinogen and renin may be localized only in the atria, but not in the ventricles (Chernin et al., 1990). These studies provide support that the heart contains the machinery by which to synthesize angiotensin precursors.

However, some studies have cast doubt on the *de novo* production of renin in the heart. First, there are data that renin is uptaken from the circulation into cardiac tissue for the processing of angiotensin peptides (Danser, 2003). Second, the finding that cardiac renin falls to undetectable levels after bilateral nephrectomy is indeed compelling (Danser et al., 1994), although recent identification of renin in cardiac mast cells may provide an alternative explanation (Xiao and Bernstein, 2005; Francis and Tang, 2006; Le and Coffman, 2006; Mackins et al., 2006; Miyazaki et al., 2006; Reid et al., 2007). Third, the transition from a circulating endocrine to a local autocrine/paracrine system was shown to be mediated by several different receptors. Both the Ang II AT_1 receptor (van Kats et al., 1997; de Lannoy et al., 1998) and the (pro) renin receptor (Nguyen et al., 2002, 2004; Nguyen and Burckle, 2004) have been reported as mediating their ligand's uptake into the heart, respectively. Fourth, Kalinyak et al. (Kalinyak and Perlman, 1987) found that angiotensinogen mRNA was undetectable in the heart. Fifth, perfused isolated rat hearts required the addition of angiotensinogen and renin into the perfusate in order to detect angiotensin I formation (de Lannoy et al., 1997). These data show that components of the renin–angiotensin system may likely be uptaken from the circulation for processing by cardiac proteases.

In summary, the origin of the cardiac RAS is likely a result of both local production and uptake from the circulation, both compartments communicating in endocrine, autocrine, paracrine and intracrine ways we do not yet understand (Figure 14.2). Studies that support either the synthesis or uptake of RAS components in the heart are limited in that both require the isolation of the heart and/or its components from the organism. While there is no doubt that a wealth of knowledge has stemmed from these types of studies, including studies from our laboratory, discovering ways in which to dissect how a particular organ and/or organ system operates and communicates within the organism is of the utmost importance. Additional insights into this question may provide a better understanding of the biological physiology of tissue RAS in general. As previously suggested by Re (Re, 2004), we are in agreement that these two sources of the RAS likely interact via some mechanism by which we do not yet understand.

III. RAS ACTIONS AT THE CELLULAR LEVEL

Two biologically-active peptides of the RAS exert their effects on cardiac dynamics and growth at the cellular level. It is now evident that both Ang II and Ang-(1–7) have

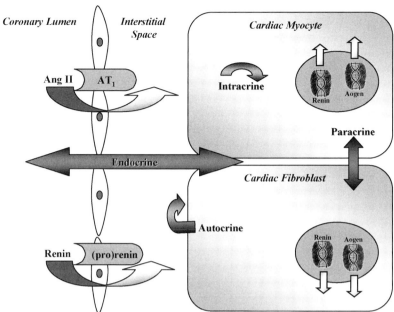

FIGURE 14.2 The two possible sources of the cardiac RAS are shown. Ang II and/or renin may be uptaken from the coronary circulation, or renin and the precursor angiotensin protein, angiotensinogen (aogen), may be synthesized in the nuclei of cardiac cells. These two sources of RAS components likely interact via endocrine, paracrine, autocrine and even intracrine mechanisms to produce and regulate the bioactive peptides of the RAS, Ang II and Ang-(1–7).

opposing actions on cardiac myocytes, fibroblasts and coronary endothelial cells. Ang II, via the AT_1 receptor, facilitates calcium (Ca^{2+}) handling in the cardiac cells (Freer et al., 1976; Kass and Blair, 1981; Peach, 1981; Baker et al., 1984, 1989), triggering enhanced cardiac contractility. This occurs via increases in cytosolic Ca^{2+} occurring both via increased uptake at the cellular membrane and by activation of inositol phosphates leading to Ca^{2+} release from sarcoplasmic reticulum (Baker et al., 1989). In both cardiac myocytes and fibroblasts, Ang II also exhibits growth-promoting effects by activating the mitogen-activated protein (MAP) kinase cascade, a pathway long recognized to be important in cellular growth (Booz et al., 1994; Sadoshima et al., 1995; Schorb et al., 1995; Yamazaki et al., 1995). Solid evidence for Ang II-mediated cardiac fibrosis was provided when Villarreal and colleagues (Villarreal et al., 1993) reported that Ang II could bind AT_1 receptors in cardiac fibroblasts. Further studies showed that Ang II may act through transforming growth factor beta (TGF-β) to induce increased collagen deposition (Campbell and Katwa, 1997). Proliferation of fibroblasts leads to an increased deposition of collagen in the myocardium which, when combined with myocyte hypertrophy, leads to left ventricular hypertrophy (LVH) – a major risk factor for hypertension and heart failure.

The above-mentioned effects of Ang II have been attributed to the AT_1 receptor, however, Ang II also binds to another receptor – the AT_2 receptor – with high affinity (de Gasparo et al., 2000). This receptor is thought to oppose the actions of Ang II-induced activation of the AT_1 receptor (Nakajima et al., 1995; Yamada et al., 1998; Carey, 2005). However, more recent evidence purports that cardiac AT_2 receptors can act as constitutive growth-promoting receptors that do not antagonize the hypertrophy-promoting consequences of AT_1-receptor-mediated activation by Ang II (D'Amore et al., 2005). As one can easily appreciate, the underlying void of a complete understanding of these two Ang II receptors and its implication in the modulation of cardiac function remains to be clarified.

In contrast to the cellular effects of Ang II in the heart, Ang-(1–7) was recently shown to inhibit the growth of cardiac myocytes via activation of the *mas* receptor (Tallant et al., 2005), which was previously shown to be a functional receptor for Ang-(1–7) (Santos et al., 2003). This finding followed several years of studies that showed Ang-(1–7) could inhibit cellular growth in vascular smooth muscle cells (Freeman et al., 1996; Strawn et al., 1999; Tallant and Clark, 2003). Moreover, Ang-(1–7) can also mitigate fibrosis under cell culture conditions (Iwata et al., 2005), as well as fibrosis associated with Ang II-driven (Grobe et al., 2007) and DOCA-salt-driven (Grobe et al., 2006) cardiac hypertrophy. Indeed, studies from our laboratory confirmed the existence of Ang-(1–7) in the heart and further showed an increase in Ang-(1–7) immunoreactivity in cardiac myocytes, but not in cardiac fibroblasts in response to coronary artery ligation (Averill et al., 2003). In two separate studies, Zisman also confirmed that Ang-(1–7) could be generated in the human heart (Zisman et al., 2003a, b). This local Ang-(1–7) may act on the cells of the heart to improve cardiac function, as will be discussed later in this chapter. Furthermore, Ang-(1–7) augments the threshold to ischemic-induced arrhythmias (Ferreira et al., 2001, 2002; Santos et al., 2006) as well as hyperpolarizing the ischemic heart fibers and re-establishing impulse propagation (De Mello, 2004). The beneficial effects of Ang-(1–7) are dose-dependent because, at higher concentrations

(10^{-7} M), the heptapeptide elicits an appreciable increase of action potential duration and early-after depolarizations (De Mello et al., 2007). In keeping with these findings, progressive conduction and rhythm disturbances with sustained ventricular tachycardia and terminal ventricular fibrillation occurred in transgenic mice with increased cardiac ACE2 expression (Donoghue et al., 2003).

The intracellular, or 'intracrine', RAS may mediate biological activity on its own. Very early studies showed that Ang II could be localized to nuclei of both smooth muscle and cardiac muscle (Robertson and Khairallah, 1971) and that the action of the octapeptide may stimulate intracellular changes in conductance and calcium handling. Moreover, significant support for intracellular actions of angiotensin peptides is accumulating. Baker and colleagues (Baker et al., 2004) found that exogenously-administered Ang II could cause cardiomyocyte hypertrophy, as well as stimulate protein synthesis, effects that could not be reversed by administration of an AT_1 receptor antagonist in the extracellular milieu. These data suggest that intracellular Ang II could promote cardiac hypertrophy independent of activation of the AT_1 receptors on the cellular membrane.

IV. RAS AND THE CORONARY CIRCULATION

The coronary circulation serves as the supply line to not only the heart, but largely to the whole organism because, if coronary blood flow is interrupted, heart function is depleted and systemic perfusion can decline. The coronary circulation also serves as a portal – a portal that allows the exchange of nutrients and hormones so that the heart can function properly. One of those hormonal systems that regulates the coronary circulation is the RAS. It is well accepted that blood flow in any tissue is regulated by both the autonomic nervous system and local effectors. The heart is no exception. Cardiac blood flow and coronary maintenance can be regulated locally by the production of adenosine, nitric oxide, Ang II and Ang-(1–7), just to name a few. Reduction of coronary blood flow by Ang II may either be direct or it may be mediated by the stimulatory effect of the peptide on the release of endothelin-1 (Schaefer et al., 2007). The relevance of the vasoconstrictor effects of Ang II on the coronary circulation is highlighted by the observation that enalaprilat improves coronary blood flow post-angioplasty (Schaefer et al., 2007). Furthermore, Zhang et al. (Zhang et al., 2005) showed that AT_1 receptor-mediated coronary constriction is augmented in the prediabetic metabolic syndrome and contributes to impaired control of coronary blood flow via increases in circulating Ang II and coronary arteriolar AT_1 receptor density.

Evidence for modulation of the coronary circulation by Ang-(1–7) was first discovered by Kumagai and colleagues (Kumagai et al., 1990) in the hamster heart. These investigators found that Ang-(1–7) produced a vasoconstrictor-like activity, likely due to the high doses used. Later studies found that Ang-(1–7) in fact acted upon the coronary endothelium to induce nitric oxide release, which produced a dose-dependent vasodilatory effect (Porsti et al., 1994; Brosnihan et al., 1996; Li et al., 1997; Almeida et al., 2000). These findings are consistent with the actions of Ang-(1–7) in other vascular beds (Oliveira et al., 1999; Feterik et al., 2000; Ren et al., 2002; Neves et al., 2003). Moreover, Ang-(1–7) can also modulate the distribution of blood flow via changes in systemic hemodynamics (Sampaio et al., 2003). Experimental evidence clearly and directly shows that Ang-(1–7) can act as a vasodilator in not only the coronary circulation, but also other systemic vascular beds. The heptapeptide may also play a role in the distribution of blood flow to various systemic vascular beds as physiological needs change.

V. SIGNIFICANCE OF THE RAS ON CARDIAC FUNCTION

As one can undeniably appreciate given the above-mentioned consequences of Ang II and Ang-(1–7) at the cellular level of the heart, these biologically-active peptides ultimately modulate myocardial performance. Although the effects of the local cardiac RAS at the cellular level have unveiled potential mediators of heart disease, the ultimate significance of the local RAS in the heart boils down to its effects on how the heart performs.

A. Pathophysiology of RAS in Cardiac Function

The development of angiotensin-converting enzyme (ACE) inhibitors and the later introduction of angiotensin receptor blockers (ARBs) drastically changed the medical approaches to the management of cardiac pathology, including heart failure. Both in patients and in experimental animal models of left ventricular dysfunction, these agents were proven to reverse cardiac hypertrophy (Dahlof, 1992; Dahlof et al., 1992, 2002), correct left ventricular systolic dysfunction (Kober et al., 1995; Buksa, 2000; Pfeffer et al., 2003; Braunwald et al., 2004; Young et al., 2004) and ameliorate progression of heart failure (Banerjee et al., 2003; Pfeffer et al., 2003; Hedrich et al., 2005; Levine and Levine, 2005; Voors and van Veldhuisen, 2005; Abdulla et al., 2006; Giles, 2007).

Cardioprotection mediated by blockade of increased cardiac expression of Ang II has been demonstrated in experimental models of induced cardiac pathology, while in humans, direct evidence for a local tissue mechanism is for the most part inferred from experiments in animals. However, patients with unstable angina produced Ang II in greater amount than in patients with stable angina; these changes were associated with increased expression of angiotensinogen, ACE and AT₁-receptor genes together with upregulation of tumor necrosis factor (TNF-α), interleukin-6 and iNOS genes (Neri Serneri et al., 2004). Human cardiac Ang II-forming activity is increased in autopsied hearts of patients with myocardial infarction (Ihara et al., 2000), a finding that correlates with studies in the rat (Ishiyama et al., 2004). Given that changes in cardiac Ang II in infarcted or remnant myocardium are very limited, it remains to be determined whether blockade of the octapeptide Ang II in cardiac tissue contributes to the beneficial effects of blockers of the RAS on cardiac remodeling post-myocardial infarction.

Contrasting with the functional aspects of Ang II on cardiac performance, some evidence exists as to whether the heptapeptide Ang-(1–7) may be a positive modulator of cardiac function, counterbalancing the hypertrophic and pro-fibrotic actions of Ang II. Initial studies to characterize the functional effects of Ang-(1–7) on the heart showed that its administration could improve both ischemia-induced functional impairments and cardiac arrhythmias (Ferreira et al., 2001, 2002). The latter effect may be due to activation of the sodium pump in cardiac muscle (De Mello, 2004) that may act to hyperpolarize the cell and increase conduction velocity (De Mello et al., 2007). Additionally, there is accumulating *in vivo* evidence of the positive effects of Ang-(1–7) in the heart. For example, administration of Ang-(1–7) after coronary artery ligation in rats attenuated the development of heart failure (Loot et al., 2002). Furthermore, plasma Ang-(1–7) was augmented in response to coronary artery ligation in rats, with a corresponding increase in cardiac ACE2 (Ishiyama et al., 2004). Taken together, these findings suggest that ACE2 may act to facilitate the conversion of Ang II into Ang-(1–7) as part of a feedforward mechanism as previously described by us, although in cardiac hypertrophy it appears that the levels of Ang-(1–7) are insufficient to counterbalance the deleterious effects of Ang II (Ferrario et al., 2005). Additional evidence for this hypothesis stems from severe cardiac functional impairments in both ACE2- and *mas*-receptor-knockout mice (Crackower et al., 2002; Santos et al., 2006) and the demonstration that ACE2 overexpression is associated with abrogation of experimentally-induced cardiac hypertrophy and fibrosis (Huentelman et al., 2005).

VI. CONCLUSIONS

As in all biological systems, the integration of the components of various hormonal systems is required for the proper assessment of physiological and pathophysiological function. The RAS is no exception. The regulation of the cardiac RAS is likely not independent of the circulation, although its various components can undoubtedly exert direct effects on the tissue itself. Nor possibly does the cardiac RAS operate independent of other tissues, as a very recent report showed that renal AT₁ receptors were required for the development of cardiac hypertrophy (Crowley et al., 2006). Our understanding of the complexity of the system continues to evolve. One thing is for sure – the RAS is not solely a circulating endocrine system. Increasing data, much of which is discussed in this chapter, have shown that the RAS also exerts autocrine, paracrine and intracrine actions that may work in concert within the organism to regulate physiological processes that, when out of balance, may induce pathology.

References

Abdulla, J., Pogue, J., Abildstrom, S. Z. et al. (2006). Effect of angiotensin-converting enzyme inhibition on functional class in patients with left ventricular systolic dysfunction – a meta-analysis. *Eur J Heart Fail* **8:** 90–96.

Almeida, A. P., Frabregas, B. C., Madureira, M. M., Santos, R. J., Campagnole-Santos, M. J., and Santos, R. A. (2000). Angiotensin-(1–7) potentiates the coronary vasodilatatory effect of bradykinin in the isolated rat heart. *Braz J Med Biol Res* **33:** 709–713.

Averill, D. B., Ishiyama, Y., Chappell, M. C., and Ferrario, C. M. (2003). Cardiac angiotensin-(1–7) in ischemic cardiomyopathy. *Circulation* **108:** 2141–2146.

Baker, K. M., Campanile, C. P., Trachte, G. J., and Peach, M. J. (1984). Identification and characterization of the rabbit angiotensin II myocardial receptor. *Circ Res* **54:** 286–293.

Baker, K. M., Chernin, M. I., Schreiber, T. et al. (2004). Evidence of a novel intracrine mechanism in angiotensin II-induced cardiac hypertrophy. *Regul Pept* **120:** 5–13.

Baker, K. M., Singer, H. A., and Aceto, J. F. (1989). Angiotensin II receptor-mediated stimulation of cytosolic-free calcium and inositol phosphates in chick myocytes. *J Pharmacol Exp Ther* **251:** 578–585.

Banerjee, A., Talreja, A., Sonnenblick, E. H., and LeJemtel, T. H. (2003). Evolving rationale for angiotensin-converting enzyme inhibition in chronic heart failure. *Mt Sinai J Med* **70:** 225–231.

Booz, G. W., Dostal, D. E., Singer, H. A., and Baker, K. M. (1994). Involvement of protein kianse C and Ca2+ in angiotensin II-induced mitogenesis of cardiac fibroblasts. *Am J Physiol* **267:** C1308–C1318.

Braunwald, E., Domanski, M. J., Fowler, S. E. et al. (2004). Angiotensin-converting-enzyme inhibition in stable coronary artery disease. *N Engl J Med* **351:** 2058–2068.

Brosnihan, K. B., Li, P., and Ferrario, C. M. (1996). Angiotensin-(1–7) dilates canine coronary arteries through kinins and nitric oxide. *Hypertension* **27:** 523–528.

Buksa, M. (2000). Trandolapril in the prevention of the sequelae of left ventricular systolic dysfunction after acute myocardial infarct. *Med Arh* **54:** 103–106.

Campbell, S. E., and Katwa, L. C. (1997). Angiotensin II stimulated expression of transforming growth factor-beta1 in cardiac fibroblasts and myofibroblasts. *J Mol Cell Cardiol* **29:** 1947–1958.

Carey, R. M. (2005). Cardiovascular and renal regulation by the angiotensin type 2 receptor: the AT2 receptor comes of age. *Hypertension* **45:** 840–844.

Chappell, M. C., Gomez, M. N., Pirro, N. T., and Ferrario, C. M. (2000). Release of angiotensin-(1–7) from the rat hindlimb: influence of angiotensin-converting enzyme inhibition. *Hypertension* **35**: 348–352.

Chernin, M. I., Candia, A. F., Stark, L. L., Aceto, J. F., and Baker, K. M. (1990). Fetal expression of renin, angiotensinogen, and atriopeptin genes in chick heart. *Clin Exp Hypertens A* **12**: 617–629.

Cicilini, M. A., Ramos, P. S., Vasquez, E. C., and Cabral, A. M. (1994). Heart prolyl endopeptidase activity in one-kidney, one clip hypertensive rats. *Braz J Med Biol Res* **27**: 2821–2830.

Crackower, M. A., Sarao, R., Oudit, G. Y. et al. (2002). Angiotensin-converting enzyme 2 is an essential regulator of heart function. *Nature* **417**: 822–828.

Crowley, S. D., Gurley, S. B., Herrera, M. J. et al. (2006). Angiotensin II causes hypertension and cardiac hypertrophy through its receptors in the kidney. *Proc Natl Acad Sci USA* **103**: 17985–17990.

D'Amore, A., Black, M. J., and Thomas, W. G. (2005). The angiotensin II type 2 receptor causes constitutive growth of cardiomyocytes and does not antagonize angiotensin II type 1 receptor-mediated hypertrophy. *Hypertension* **46**: 1347–1354.

Dahlof, B. (1992). Structural cardiovascular changes in essential hypertension. Studies on the effect of antihypertensive therapy. *Blood Press Suppl* **6**: 1–75.

Dahlof, B., Devereux, R. B., Kjeldsen, S. E. et al. (2002). Cardiovascular morbidity and mortality in the Losartan Intervention For Endpoint reduction in hypertension study (LIFE): a randomised trial against atenolol. *Lancet* **359**: 995–1003.

Dahlof, B., Pennert, K., and Hansson, L. (1992). Reversal of left ventricular hypertrophy in hypertensive patients. A metaanalysis of 109 treatment studies. *Am J Hypertens* **5**: 95–110.

Danser, A. H. (2003). Local renin-angiotensin systems: the unanswered questions. *Int J Biochem Cell Biol* **35**: 759–768.

Danser, A. H., van Kats, J. P., Admiraal, P. J. et al. (1994). Cardiac renin and angiotensins. Uptake from plasma versus in situ synthesis. *Hypertension* **24**: 37–48.

de Gasparo, M., Catt, K. J., Inagami, T., Wright, J. W., and Unger, T. (2000). International union of pharmacology. XXIII. The angiotensin II receptors. *Pharmacol Rev* **52**: 415–472.

de Lannoy, L. M., Danser, A. H., Bouhuizen, A. M., Saxena, P. R., and Schalekamp, M. A. (1998). Localization and production of angiotensin II in the isolated perfused rat heart. *Hypertension* **31**: 1111–1117.

de Lannoy, L. M., Danser, A. H., van Kats, J. P., Schoemaker, R. G., Saxena, P. R., and Schalekamp, M. A. (1997). Renin-angiotensin system components in the interstitial fluid of the isolated perfused rat heart. Local production of angiotensin I. *Hypertension* **29**: 1240–1251.

De Mello, W. C. (2004). Angiotensin (1–7) re-establishes impulse conduction in cardiac muscle during ischaemia-reperfusion. The role of the sodium pump. *J Renin Angiotensin Aldosterone Syst* **5**: 203–208.

De Mello, W. C., Ferrario, C. M., and Jessup, J. A. (2007). Beneficial versus harmful effects of angiotensin (1–7) on impulse propagation and cardiac arrhythmias in the failing heart. *J Renin Angiotensin Aldosterone Syst* **8**: 74–80.

Donoghue, M., Wakimoto, H., Maguire, C. T. et al. (2003). Heart block, ventricular tachycardia, and sudden death in ACE2 transgenic mice with downregulated connexins. *J Mol Cell Cardiol* **35**: 1043–1053.

Dzau, V. J., Ellison, K. E., Brody, T., Ingelfinger, J., and Pratt, R. E. (1987). A comparative study of the distributions of renin and angiotensinogen messenger ribonucleic acids in rat and mouse tissues. *Endocrinology* **120**: 2334–2338.

Ferrario, C. M., Trask, A. J., and Jessup, J. A. (2005). Advances in biochemical and functional roles of angiotensin-converting enzyme 2 and angiotensin-(1–7) in regulation of cardiovascular function. *Am J Physiol Heart Circ Physiol* **289**: H2281–H2290.

Ferreira, A. J., Santos, R. A., and Almeida, A. P. (2001). Angiotensin-(1–7): cardioprotective effect in myocardial ischemia/reperfusion. *Hypertension* **38**: 665–668.

Ferreira, A. J., Santos, R. A., and Almeida, A. P. (2002). Angiotensin-(1–7) improves the post-ischemic function in isolated perfused rat hearts. *Braz J Med Biol Res* **35**: 1083–1090.

Feterik, K., Smith, L., and Katusic, Z. S. (2000). Angiotensin-(1–7) causes endothelium-dependent relaxation in canine middle cerebral artery. *Brain Res* **873**: 75–82.

Fielitz, J., Dendorfer, A., Pregla, R. et al. (2002). Neutral endopeptidase is activated in cardiomyocytes in human aortic valve stenosis and heart failure. *Circulation* **105**: 286–289.

Francis, G. S., and Tang, W. H. (2006). Histamine, mast cells, and heart failure: is there a connection? *J Am Coll Cardiol* **48**: 1385–1386.

Freeman, E. J., Chisolm, G. M., Ferrario, C. M., and Tallant, E. A. (1996). Angiotensin-(1–7) inhibits vascular smooth muscle cell growth. *Hypertension* **28**: 104–108.

Freer, R. J., Pappano, A. J., Peach, M. et al. (1976). Mechanism for the positive inotropic effect of angiotensin II on isolated cardiac muscle. *Circ Res* **39**: 178–183.

Giles, T. D. (2007). Renin-angiotensin system modulation for treatment and prevention of cardiovascular diseases: toward an optimal therapeutic strategy. *Rev Cardiovasc Med* **8**(Suppl 2): S14–21.

Grobe, J. L., Mecca, A. P., Lingis, M. et al. (2007). Prevention of angiotensin II-induced cardiac remodeling by angiotensin-(1–7). *Am J Physiol Heart Circ Physiol* **292**: H736–H742.

Grobe, J. L., Mecca, A. P., Mao, H., and Katovich, M. J. (2006). Chronic angiotensin-(1–7) prevents cardiac fibrosis in DOCA-salt model of hypertension. *Am J Physiol Heart Circ Physiol* **290**: H2417–H2423.

Hedrich, O., Patten, R. D., and Denofrio, D. (2005). Current treatment options for chf management: focus on the renin-angiotensin-aldosterone system. *Curr Treat Options Cardiovasc Med* **7**: 3–13.

Huentelman, M. J., Grobe, J. L., Vazquez, J. et al. (2005). Protection from angiotensin II-induced cardiac hypertrophy and fibrosis by systemic lentiviral delivery of ACE2 in rats. *Exp Physiol* **90**: 783–790.

Ihara, M., Urata, H., Shirai, K. et al. (2000). High cardiac angiotensin-II-forming activity in infarcted and non-infarcted human myocardium. *Cardiology* **94**: 247–253.

Ishiyama, Y., Gallagher, P. E., Averill, D. B., Tallant, E. A., Brosnihan, K. B., and Ferrario, C. M. (2004). Upregulation of angiotensin-converting enzyme 2 after myocardial infarction by blockade of angiotensin II receptors. *Hypertension* **43**: 970–976.

Iwata, M., Cowling, R. T., Gurantz, D. et al. (2005). Angiotensin-(1–7) binds to specific receptors on cardiac fibroblasts to initiate antifibrotic and antitrophic effects. *Am J Physiol Heart Circ Physiol* **289**: H2356–H2363.

Kalinyak, J. E., and Perlman, A. J. (1987). Tissue-specific regulation of angiotensinogen mRNA accumulation by dexamethasone. *J Biol Chem* **262**: 460–464.

Kass, R. S., and Blair, M. L. (1981). Effects of angiotensin II on membrane current in cardiac Purkinje fibers. *J Mol Cell Cardiol* **13**: 797–809.

Kober, L., Torp-Pedersen, C., Carlsen, J. E. et al. (1995). A clinical trial of the angiotensin-converting-enzyme inhibitor trandolapril in patients with left ventricular dysfunction after myocardial infarction. Trandolapril Cardiac Evaluation (TRACE) Study Group. *N Engl J Med* **333**: 1670–1676.

Kumagai, H., Khosla, M., Ferrario, C., and Fouad-Tarazi, F. M. (1990). Biological activity of angiotensin-(1–7) heptapeptide in the hamster heart. *Hypertension* **15**: I29–33.

Kunapuli, S. P., and Kumar, A. (1987). Molecular cloning of human angiotensinogen cDNA and evidence for the presence of its mRNA in rat heart. *Circ Res* **60**: 786–790.

Le, T. H., and Coffman, T. M. (2006). A new cardiac MASTer switch for the renin-angiotensin system. *J Clin Invest* **116**: 866–869.

Lee, M. A., Bohm, M., Paul, M., and Ganten, D. (1993). Tissue renin-angiotensin systems. Their role in cardiovascular disease. *Circulation* **87**: IV7–13.

Levine, T. B., and Levine, A. B. (2005). Clinical update: the role of angiotensin II receptor blockers in patients with left ventricular dysfunction (Part II of II). *Clin Cardiol* **28**: 277–280.

Li, P., Chappell, M. C., Ferrario, C. M., and Brosnihan, K. B. (1997). Angiotensin-(1–7) augments bradykinin-induced vasodilation by competing with ACE and releasing nitric oxide. *Hypertension* **29**: 394–400.

Linardi, A., Panunto, P. C., Ferro, E. S., and Hyslop, S. (2004). Peptidase activities in rats treated chronically with N(omega)-nitro-L-arginine methyl ester (L-NAME). *Biochem Pharmacol* **68**: 205–214.

Loot, A. E., Roks, A. J., Henning, R. H. et al. (2002). Angiotensin-(1–7) attenuates the development of heart failure after myocardial infarction in rats. *Circulation* **105**: 1548–1550.

Mackins, C. J., Kano, S., Seyedi, N. et al. (2006). Cardiac mast cell-derived renin promotes local angiotensin formation, norepinephrine release, and arrhythmias in ischemia/reperfusion. *J Clin Invest* **116**: 1063–1070.

Miyazaki, M., Takai, S., Jin, D., and Muramatsu, M. (2006). Pathological roles of angiotensin II produced by mast cell chymase and the effects of chymase inhibition in animal models. *Pharmacol Ther* **112**: 668–676.

Nakajima, M., Hutchinson, H. G., Fujinaga, M. et al. (1995). The angiotensin II type 2 (AT2) receptor antagonizes the growth effects of the AT1 receptor: gain-of-function study using gene transfer. *Proc Natl Acad Sci USA* **92**: 10663–10667.

Neri Serneri, G. G., Boddi, M., Modesti, P. A. et al. (2004). Cardiac angiotensin II participates in coronary microvessel inflammation of unstable angina and strengthens the immunomediated component. *Circ Res* **94**: 1630–1637.

Neves, L. A., Averill, D. B., Ferrario, C. M. et al. (2003). Characterization of angiotensin-(1–7) receptor subtype in mesenteric arteries. *Peptides* **24**: 455–462.

Nguyen, G., and Burckle, C. A. (2004). [The (pro)renin receptor: biology and functional significance]. *Bull Acad Natl Med* **188**: 621–628.

Nguyen, G., Burckle, C. A., and Sraer, J. D. (2004). Renin/prorenin-receptor biochemistry and functional significance. *Curr Hypertens Rep* **6**: 129–132.

Nguyen, G., Delarue, F., Burckle, C., Bouzhir, L., Giller, T., and Sraer, J. D. (2002). Pivotal role of the renin/prorenin receptor in angiotensin II production and cellular responses to renin. *J Clin Invest* **109**: 1417–1427.

Oliveira, M. A., Fortes, Z. B., Santos, R. A., Kosla, M. C., and de Carvalho, M. H. (1999). Synergistic effect of angiotensin-(1–7) on bradykinin arteriolar dilation in vivo. *Peptides* **20**: 1195–1201.

Paul, M., Poyan, M. A., and Kreutz, R. (2006). Physiology of local renin-angiotensin systems. *Physiol Rev* **86**: 747–803.

Peach, M. J. (1981). Molecular actions of angiotensin. *Biochem Pharmacol* **30**: 2745–2751.

Pfeffer, M. A., McMurray, J. J., Velazquez, E. J. et al. (2003). Valsartan, captopril, or both in myocardial infarction complicated by heart failure, left ventricular dysfunction, or both. *N Engl J Med* **349**: 1893–1906.

Porsti, I., Bara, A. T., Busse, R., and Hecker, M. (1994). Release of nitric oxide by angiotensin-(1–7) from porcine coronary endothelium: implications for a novel angiotensin receptor. *Br J Pharmacol* **111**: 652–654.

Re, R. N. (2004). Mechanisms of disease: local renin-angiotensin-aldosterone systems and the pathogenesis and treatment of cardiovascular disease. *Nat Clin Pract Cardiovasc Med* **1**: 42–47.

Reid, A. C., Silver, R. B., and Levi, R. (2007). Renin: at the heart of the mast cell. *Immunol Rev* **217**: 123–140.

Ren, Y., Garvin, J. L., and Carretero, O. A. (2002). Vasodilator action of angiotensin-(1–7) on isolated rabbit afferent arterioles. *Hypertension* **39**: 799–802.

Robertson, A. L., and Khairallah, P. A. (1971). Angiotensin II: rapid localization in nuclei of smooth and cardiac muscle. *Science* **172**: 1138–1139.

Sadoshima, J., Qiu, Z., Morgan, J. P., and Izumo, S. (1995). Angiotensin II and other hypertrophic stimuli mediated by G protein-coupled receptors activate tyrosine kinase, mitogen-activated protein kinase, and 90-kD S6 kinase in cardiac myocytes. The critical role of Ca(2+)-dependent signalling. *Circ Res* **76**: 1–15.

Sampaio, W. O., Nascimento, A. A., and Santos, R. A. (2003). Systemic and regional hemodynamic effects of angiotensin-(1–7) in rats. *Am J Physiol Heart Circ Physiol* **284**: H1985–H1994.

Santos, R. A., Brum, J. M., Brosnihan, K. B., and Ferrario, C. M. (1990). The renin-angiotensin system during acute myocardial ischemia in dogs. *Hypertension* **15**: I121–I127.

Santos, R. A., Castro, C. H., Gava, E. et al. (2006). Impairment of in vitro and in vivo heart function in angiotensin-(1–7) receptor MAS knockout mice. *Hypertension* **47**: 996–1002.

Santos, R. A., Simoes e Silva, A. C., Maric, C. et al. (2003). Angiotensin-(1–7) is an endogenous ligand for the G protein-coupled receptor Mas. *Proc Natl Acad Sci USA* **100**: 8258–8263.

Sawa, H., Tokuchi, F., Mochizuki, N. et al. (1992). Expression of the angiotensinogen gene and localization of its protein in the human heart. *Circulation* **86**: 138–146.

Schaefer, U., Kurz, T., Bonnemeier, H. et al. (2007). Intracoronary enalaprilat during angioplasty for acute myocardial infarction: alleviation of postischaemic neurohumoral and inflammatory stress? *J Intern Med* **261**: 188–200.

Schorb, W., Conrad, K. M., Singer, H. A., Dostal, D. E., and Baker, K. M. (1995). Angiotensin II is a potent stimulator of MAP-kinase activity in neonatal rat cardiac fibroblasts. *J Mol Cell Cardiol* **27**: 1151–1160.

Strawn, W. B., Ferrario, C. M., and Tallant, E. A. (1999). Angiotensin-(1–7) reduces smooth muscle growth after vascular injury. *Hypertension* **33**: 207–211.

Tallant, E. A., and Clark, M. A. (2003). Molecular mechanisms of inhibition of vascular growth by angiotensin-(1–7). *Hypertension* **42**: 574–579.

Tallant, E. A., Ferrario, C. M., and Gallagher, P. E. (2005). Angiotensin-(1–7) inhibits growth of cardiac myocytes through activation of the mas receptor. *Am J Physiol Heart Circ Physiol* **289**: H1560–H1566.

Trask, A. J., Averill, D. B., Ganten, D., Chappell, M. C., and Ferrario, C. M. (2007). Primary role of angiotensin-converting enzyme-2 in cardiac production of angiotensin-(1–7) in transgenic Ren-2 hypertensive rats. *Am J Physiol Heart Circ Physiol* **292**: H3019–H3024.

Trask, A. J., and Ferrario, C. M. (2007). Angiotensin-(1–7): pharmacology and new perspectives in cardiovascular treatments. *Cardiovasc Drug Rev* **25**: 162–174.

Urata, H., Kinoshita, A., Misono, K. S., Bumpus, F. M., and Husain, A. (1990). Identification of a highly specific chymase as the major angiotensin II-forming enzyme in the human heart. *J Biol Chem* **265**: 22348–22357.

van Kats, J. P., de Lannoy, L. M., Danser, A. H. J., van Meegan, J. R., Verdouw, P. D., and Schalekamp, M. A. (1997). Angiotensin II type 1 (AT1) receptor-mediated accumulation of angiotensin II in tissues and its intracellular half-life in vivo. *Hypertension* **30**: 42–49.

Villarreal, F. J., Kim, N. N., Ungab, G. D., Printz, M. P., and Dillmann, W. H. (1993). Identification of functional angiotensin II receptors on rat cardiac fibroblasts. *Circulation* **88**: 2849–2861.

Voors, A. A., and van Veldhuisen, D. J. (2005). Pharmacological treatment of chronic heart failure according to the 2005 guidelines of the European Society of Cardiology. *Minerva Cardioangiol* **53**: 233–239.

Welches, W. R., Brosnihan, K. B., and Ferrario, C. M. (1993). A comparison of the properties and enzymatic activities of three angiotensin processing enzymes: angiotensin converting enzyme, prolyl endopeptidase and neutral endopeptidase 24.11. *Life Sci* **52**: 1461–1480.

Xiao, H. D., and Bernstein, K. E. (2005). Mast cells: the missing source of cardiac renin? *Mol Interv* **5**: 11–14.

Yamada, T., Akishita, M., Pollman, M. J., Gibbons, G. H., Dzau, V. J., and Horiuchi, M. (1998). Angiotensin II type 2 receptor mediates vascular smooth muscle cell apoptosis and antagonizes angiotensin II type 1 receptor action: an in vitro gene transfer study. *Life Sci* **63**: L289–L295.

Yamamoto, K., Chappell, M. C., Brosnihan, K. B., and Ferrario, C. M. (1992). In vivo metabolism of angiotensin I by neutral endopeptidase (EC 3.4.24.11) in spontaneously hypertensive rats. *Hypertension* **19**: 692–696.

Yamazaki, T., Komuro, I., Kudoh, S. et al. (1995). Angiotensin II partly mediates mechanical stress-induced cardiac hypertrophy. *Circ Res* **77**: 258–265.

Young, J. B., Dunlap, M. E., Pfeffer, M. A. et al. (2004). Mortality and morbidity reduction with Candesartan in patients with chronic heart failure and left ventricular systolic dysfunction: results of the CHARM low-left ventricular ejection fraction trials. *Circulation* **110**: 2618–2626.

Zhang, C., Knudson, J. D., Setty, S. et al. (2005). Coronary arteriolar vasoconstriction to angiotensin II is augmented in prediabetic metabolic syndrome via activation of AT1 receptors. *Am J Physiol Heart Circ Physiol* **288**: H2154–H2162.

Zisman, L. S., Keller, R. S., Weaver, B. et al. (2003a). Increased angiotensin-(1–7)-forming activity in failing human heart ventricles: evidence for upregulation of the angiotensin-converting enzyme Homologue ACE2. *Circulation* **108**: 1707–1712.

Zisman, L. S., Meixell, G. E., Bristow, M. R., and Canver, C. C. (2003b). Angiotensin-(1–7) formation in the intact human heart: in vivo dependence on angiotensin II as substrate. *Circulation* **108**: 1679–1681.

CHAPTER 15

Renin–Angiotensin Blockade: Therapeutic Agents

DOMENIC A. SICA
Clinical Pharmacology and Hypertension, Virginia Commonwealth University Health System, Richmond, Virginia 23298-0160, USA

Contents

I. Introduction 189
II. Therapeutic classes 189
III. Pharmacology 189
IV. ACE inhibitors and angiotensin receptor blockers with other agents 196
V. Select side-effects of ACE inhibitors and angiotensin-receptor blockers 197
VI. Summary 198
References 198

I. INTRODUCTION

Over the last 50 years, many treatment strategies have emerged for the treatment of hypertension and cardiovascular (CV) disease. Many of these therapeutic strategies saw their popularity gradually wane owing to burdensome side-effects and/or the absence of definitive outcomes data supporting their use. Such was the case for centrally-acting agents, direct vasodilators and, most recently, peripheral α-blockers. The renin–angiotensin system (RAS), however, has been observed to wield an important influence on hypertension and end-organ disease that has persisted as an attractive target for pharmacologic intervention. This chapter will broadly discuss the pharmacokinetics, pharmacodynamics, response and outcomes data for agents that interfere with RAS activity. The reader will be directed to sources that provide more comprehensive discussion on particular themes that cannot be discussed because of space constraints. Although aldosterone-receptor antagonists (ARAs) are discussed elsewhere in this book, where appropriate to their concomitant use with agents that interfere with RAS activity, they will be discussed.

II. THERAPEUTIC CLASSES

Several drug classes are known to interfere with RAS activity including angiotensin-converting enzyme (ACE) inhibitors, angiotensin-receptor blockers (ARBs) and direct renin inhibitors (DRIs) with the greatest treatment experience existing for ACE inhibitors (Sica and Gehr, 2005). The ACE inhibitor class has expanded to the degree that there are currently 10 such compounds available in the USA and several more in use on a global basis. Currently, there are seven ARBs and one DRI available with additional compounds in each of these classes in various stages of development.

III. PHARMACOLOGY

A. Angiotensin-Converting Enzyme Inhibitors

The first orally active ACE inhibitor was the drug captopril, which was commercially released in 1981. Captopril is a sulfhydryl-containing compound, with a rapid onset and not particularly prolonged duration of action. The more long-acting ACE inhibitor enalapril maleate became available shortly thereafter. Enalapril is a prodrug requiring *in vivo* hepatic and intestinal wall esterolysis to produce the active diacid inhibitor enalaprilat. All ACE inhibitors are administered in their prodrug forms with the exception of lisinopril and captopril (White, 1998). It was initially thought that prodrug conversion of an ACE inhibitor to an active diacid form would be hindered in the presence of hepatic impairment, such as in advanced heart failure (HF), however, this has proven not to be the case. The extent of absorption, the degree of hydrolysis and the bioavailability of the prodrug enalapril in patients with HF are comparable to those values observed in normal subjects other than the rates of absorption and ester hydrolysis being somewhat slower in HF (Dickstein, 1986).

All ACE inhibitors decrease the activity of ACE, however, ACE inhibitors are fundamentally heterogeneous on a structural basis. The chemical composition of the ACE binding ligand of an ACE inhibitor separates these drugs into three groupings. For example, the ACE binding ligand for captopril is a sulfhydryl moiety, for fosinopril a phosphinyl group and for virtually all other ACE inhibitors a carboxyl group. The side group on an ACE inhibitor has been proposed as the basis for differing pharmacological responses

TABLE 15.1 Pharmacokinetic parameters of angiotensin-converting-enzyme inhibitors

Drug	Onset/duration (h)	Peak hypotensive effect (h)	Protein binding* (%)	Plasma half-life (h)	Elimination**
Benazepril	1/24	2–4	>95	10–11	Renal/some biliary
Captopril	0.25/dose-related	1–1.5	25–30	<2	Renal as disulfides
Enalapril	1/24	4–6	50	11	Renal
Fosinopril	1/24	2–6	95	11	Renal = hepatic
Lisinopril	1/24	6	10	13	Renal
Moexipril	1/24	4–6	50	2–9	Renal/some biliary
Perindopril	1/24	3–7	10–20	3–10	Renal
Quinapril	1/24	2	97	2	Renal > hepatic
Ramipril	1–2/24	3–6	73	13–17	Renal
Trandolapril	2–4/24	6–8	80–94	16–24	Renal > hepatic

* Protein binding may vary for the prodrug and the active diacid of an ACE inhibitor.
** The concept of renal elimination of an ACE inhibitor takes into account both prodrug elimination and that of the active diacid where applicable.

among these compounds (Herman, 1992). For example, the sulfhydryl group on captopril is proposed to act as a recyclable free radical scavenger and, for this reason, captopril has been suggested differentially to retard atherogenesis and protect from myocardial infarction (MI) and diabetes development, however, this has yet to be clinically substantiated. In addition, captopril directly promotes prostaglandin synthesis, whereas other ACE inhibitors bring this about indirectly by increasing bradykinin activity (Zusman, 1987). Alternatively, the sulfhydryl side group found on captopril is believed to lead to a higher rate of maculopapular skin rashes and dysgeusia than what is seen with other ACE inhibitors (Chalmers et al., 1992). The phosphinyl group on fosinopril has been offered as the reason for its low incidence of cough (Sharif et al., 1994) and its ability selectively to improve diastolic dysfunction (Zusman et al., 1992). In the instance of the latter, the phosphinyl group may facilitate the myocardial penetration and/or retention of fosinopril and thereby differentially improve myocardial energetics.

Though ACE inhibitors can be distinguished by differences in absorption, protein binding, half-life and metabolic disposition, they behave quite similarly in how they lower BP (Table 15.1) (Reid, 1997; White, 1998; Sica and Gehr, 2005). Rarely, beyond the issue of frequency of dosing, should these pharmacologic differences influence the choice of an ACE inhibitor. This being said, two pharmacologic considerations for an ACE inhibitor, route of systemic elimination and tissue binding, have led to considerable recent debate and merit specific discussion (Hoyer et al., 1993; Dzau et al., 2001).

1. PHARMACOKINETICS

There is no evidence for systemic accumulation of the prodrugs ramipril, enalapril, fosinopril, trandolapril or benazepril in chronic kidney disease (CKD). This suggests that these compounds undergo intact biliary clearance or that the metabolic conversion of these drugs to their active diacid form is unaffected by renal failure (Ebihara and Fujimura, 1991). These pharmacokinetic findings have been offered as evidence for a dual route (hepatic/renal) of elimination for these particular compounds. A dual route of drug elimination in CKD is viewed as being advantageous in that dosage adjustment becomes unnecessary on a pharmacokinetic basis. For these compounds, however, this pharmacokinetic feature is immaterial to their dosing in CKD since these prodrugs exhibit marginal activity. True dual route of elimination ACE inhibitors are those whose active diacid form undergoes proportional hepatic and renal clearance. In that regard, the active diacids of only two ACE inhibitors, fosinoprilat and trandolaprilat, undergo a significant degree of hepatic (and renal) clearance (Sica et al., 1991; Danielson et al., 1994). For the remaining ACE inhibitors, systemic elimination is almost exclusively renal, occurring through both filtration and tubular secretion (Hoyer et al., 1993). The tubular secretion of an ACE inhibitor is compound specific and typically occurs via the organic anion secretory pathway (Noormohamed et al., 1990; Shionoiri, 1993). This property of combined renal and hepatic elimination minimizes accumulation of these compounds at steady state in CKD (Sica et al., 1991; Danielson et al., 1994). To date, a specific concentration-dependent adverse effect, attributable to ACE inhibitor accumulation in CKD, has not been identified, however, it is probable though that the longer ACE inhibitor concentrations remain elevated – once a blood pressure (BP) response had occurred – the more likely it is that BP will remain reduced. Thus, the major adverse consequence of ACE inhibitor accumulation may be that of prolonged hypotension and its organ-specific sequelae (Sica and Deedwania, 1997).

2. TISSUE-BINDING

The second controversial pharmacologic feature of the ACE inhibitors relates to the concept of binding to tissue ACE (Brown and Vaughn, 1998; Dzau et al., 2001).

The physicochemical differences among ACE inhibitors, including binding affinity, potency, lipophilicity and depot effect, allows for the arbitrary classification of ACE inhibitors according to their particular affinity for tissue-ACE (Dzau et al., 2001). The degree of *in vivo* functional inhibition of tissue ACE from administration of an ACE inhibitor corresponds to two properties of a compound: the inhibitor's binding affinity and the free inhibitor concentration within a specific tissue compartment. The free concentration of an ACE inhibitor, in turn, represents the state of dynamic equilibrium, which evolves from the transport of an ACE inhibitor to various tissues and its subsequent washout and return into the blood. Free inhibitor tissue concentrations are influenced by customary pharmacologic variables including dose frequency/amount, absolute bioavailability, plasma half-life, tissue penetration and, ultimately, the capacity for retention at the tissue level. The bioavailability and half-life of an ACE inhibitor in blood can be readily determined and are important elements in the selection of an ACE inhibitor dose (Sica and Gehr, 2005). When blood levels of an ACE inhibitor are high – typically in the first half of the dosing period – tissue retention of an ACE inhibitor is not a critical determinant of functional ACE inhibition. However, during the second-half of an ACE inhibitor dosing period as blood levels drop, two factors appear to be critical to extending functional ACE inhibition:

1. inhibitor binding affinity for ACE and
2. tissue retention, which will have a direct influence on the free inhibitor concentration in tissue (Dzau et al., 2001).

The order of potency for several ACE inhibitors has been ranked by competition analyses (Johnston et al., 1989; Fabris et al., 1990) and by direct binding of tritium-labeled ACE inhibitors to tissue ACE (Kinoshita et al., 1993). These studies show a hierarchal potency of: quinaprilat = benazeprilat > ramiprilat > perindoprilat > lisinopril > enalaprilat > fosinoprilat > captopril (Johnston et al., 1989; Fabris et al., 1990; Kinoshita et al., 1993). The process of tissue retention of ACE inhibitors has also been studied. Isolated organ bath studies examining the duration of ACE inhibition after the removal of ACE inhibitor from the bathing milieu show that functional inhibition of ACE lasts well beyond (2 to 5 times longer) the time predicted solely on the basis of inhibitor dissociation rates or binding affinity (Kinoshita et al., 1993). The ranking of tissue retention is quinaprilat > lisinopril > enalaprilat > captopril and this ordering reflects both the binding affinity and lipophilicity of these particular ACE inhibitors (Dzau et al., 2001).

The question arises as to whether an ACE inhibitor exhibits tissue protective effects independent of the degree to which they lower BP as suggested by the Heart Outcomes Prevention Evaluation (HOPE) Study (Yusuf et al., 2000). This is a different issue than whether there are differences in BP lowering efficacy between various ACE inhibitors (Sica and Gehr, 2005). ACE inhibitors reduce angiotensin-II levels, increase nitric oxide bioavailability and improve endothelial function, which may represent BP independent mechanisms by which these compounds confer vascular protection. The effects of ACE inhibitors on endothelium-dependent relaxation appear to differ among several reports and appear to be dependent on the agents used and the construct of the experimental design. It should be noted that consistent improvement in endothelial function is reported with those ACE inhibitors with higher tissue-ACE affinity, such as quinapril and ramipril. Despite the inherent appeal of such findings there have been few head-to-head trials, which directly compare highly tissue bound ACE inhibitors to those with more limited tissue binding. In circumstances where such comparisons have been undertaken, the results do not persuasively support the claim of overall superiority for lipophilic ACE inhibitors (Leonetti and Cuspidi, 1995).

3. Application of Pharmacologic Differences

Class effect is a concept often invoked to legitimatize use of a less costly ACE inhibitor (or an ARB) when a higher priced agent in the class has been the one specifically studied in a disease state, such as HF or diabetic nephropathy (Pfeffer et al., 1991; Lewis et al., 1993; Yusuf et al., 2000; Sica, 2001a). The concept of *class effect* may be best suited for application to the use of ACE inhibitors in the treatment of hypertension. Therein, little appears to distinguish one ACE inhibitor from another if equivalent doses are given (Sica, 2001a). Alternatively, it is less certain as to what represents true dose equivalence among ACE inhibitors when they are being used for specific outcomes benefits in the case of proteinuric renal disease or HF. In the treatment of proteinuric renal disease, the dose–response relationship for an ACE inhibitor and proteinuria reduction is incompletely explored, a situation made more complex by the observation that the higher the baseline urine protein excretion the greater the antiproteinuric effect of an ACE inhibitor (Jafar et al., 2003), however, since there are so few hard end-point studies with ACE inhibitors in nephropathic patients it would seem reasonable to suppose interchangeability among drugs in this class. In the case of HF, an ACE inhibitor is titrated to a presumed maximal tissue effect dose since any reduction in the morbidity and mortality with ACE inhibitors is dose-dependent. The utility of an ACE inhibitor in HF may derive from therapy-related neurohumoral and tissue-based changes and not solely from ACE-related inhibition of angiotensin-II production (Tang et al., 2002). Since not all ACE inhibitors have been thoroughly studied in HF or for that matter are clinically approved for HF use, particularly relative to secondary neurohumoral response parameters, it is less likely that specific interchangeable doses of different ACE inhibitors can be accurately identified in the treatment of HF (Kazi and Deswal, 2008).

TABLE 15.2 Pharmacokinetics of angiotensin-receptor blockers

Drug	Half-life (h)	Bioavailability (%)	Volume of distribution	Renal/hepatic clearance (%)
Candesartan cilexitil	9	15	0.13 l/kg	60/40
Eprosartan	5	6–29	13 l	30/70
Irbesartan	11–15	60–80	53–93 l	1/99
Losartan	2	33	34 l	10/90
E-3174	6–9	–	12 l	50/50
Olmesartan	10–15	28	17 l	45/55
Telmisartan	24	42–58	500 l	1/99
Valsartan	6	≈25	17 l	30/70

B. Angiotensin-Receptor Blockers

Angiotensin-receptor blockers work selectively at the angiotensin type-1 (AT_1) receptor subtype. The AT_1 receptor is fairly ubiquitous and its stimulation is the basis for virtually all of the known physiologic effects of angiotensin-II as relates to cardiovascular and cardiorenal homeostasis. Pharmacologic differences do exist for the various ARBs (Sica, 2001b), however, it is unlikely that any such differences have a meaningful effect on the head-to-head BP reducing ability of these compounds (Table 15.2).

1. Pharmacokinetics

a. Bioavailability. The bioavailability of the individual ARBs is quite variable. Three of the ARBs are administered in a prodrug form – losartan, candesartan cilexitil and olmesartan medoxomil – although, technically speaking, losartan is an active compound albeit one ultimately converted to its more potent E-3174 metabolite (Sica et al., 2005). The bioavailability of eprosartan is low (≈13%), which is not due to high first-pass elimination *per se* (Cox et al., 1996). Eprosartan absorption is to a degree saturable over the dose range of 100–800 mg, most probably due to the physicochemical properties of the drug (Chapelsky et al., 1998). Irbesartan demonstrates a bioavailability in the 60–80% range and is without a food effect (Vachharajani et al., 1998). Losartan has a moderate bioavailability (≈33%) with 14% of an administered dose being transformed to the E-3174 metabolite (Sica et al., 2005). Telmisartan has a saturable first-pass effect for its absorption with 42 and 58% bioavailability respectively at doses of 40 and 80 mg (Stangier et al., 2000a). Although the intrasubject variability in ARB absorption is seldom reported, it is probably in the order of 25–40%.

b. Dose Proportionality. The concept of dose proportionality becomes important when dose escalation of an antihypertensive agent is required to establish BP control. One form of dose proportionality is exhibited by irbesartan. In this regard, the results of two double-blind, placebo-controlled studies involving 88 healthy subjects show irbesartan to display linear, dose-related pharmacokinetics for its area-under-the-curve (AUC) with escalating doses over a dose range from 10 to 600 mg. The maximum plasma concentration (C_{max}) over this same dose range was related to the dose in a linear but less than dose-proportional manner. Increases in plasma AUC and C_{max} in subjects receiving 900 mg of irbesartan were smaller than predicted from dose proportionality (Gillis and Markham, 1997). Possible explanations for this phenomenon are that intestinal absorption is dose limited, perhaps due to saturation of a carrier system at high drug concentrations, or that the dissolution characteristics of this compound are dose-dependent (Ludden, 1991). In the instance of irbesartan, the observation that the terminal half-life of irbesartan is unchanged with higher irbesartan doses suggests that the intestinal absorption of irbesartan may saturate with increasing doses but that its metabolism and excretion are not so dose limited. Each of these proposed mechanisms may explain the lack of dose proportionality for eprosartan at doses above 400 mg (Chapelsky et al., 1998). It should be noted that the absence of dose proportionality for several of the ARBs, at doses which are seldom employed clinically, has limited relevance to the use of these compounds in the treatment of hypertension.

c. Volume of Distribution. Angiotensin-receptor blockers typically have a volume of distribution (V_D), which approximates extracellular fluid (ECF) volume, in part, in relationship to the extensive protein binding of these compounds. For example, the V_D for losartan and its E-3174 metabolite are 34 and 12 l (Lo et al., 1995), respectively, while the V_D for candesartan, olmesartan, valsartan and eprosartan are ≈10 (0.13 l/kg b.w.), 30 l, 17 l and 13 l, respectively (Sica, 2001b). Alternatively, telmisartan and irbesartan have the highest V_D of any of the ARBs with values of 500 and 53–93/l, respectively (Gillis and Markham, 1997; Stangier et al., 2000a). The fact that telmisartan has a V_D that is so high likely relates to a loose binding relationship with its predominant protein carrier, albumin (Stangier et al., 2000a). The V_D of the various ARBs in disease states, such as CKD, is unreported. To date, the clinical significance of ARBs having a high V_D remains uncertain. It has been suggested, however, that the higher the V_D for an ARB the more likely it is that extravascular AT_1-receptors are accessed and, therefore, at least in theory, the more significant the vasodepressor response.

d. Protein Binding. The protein binding of all ARBs is typically well in excess of 90% (Christ, 1995; Gillis and Markham, 1997; Martin et al., 1998; Stangier et al., 2000a). Irbesartan is the exception to this in that it has a plasma free fraction of approximately 10% (Gillis and Markham, 1997). In general, none of the ARBs binds to partitions in red blood cells in a meaningful fashion (Stangier et al., 2000a). Furthermore, the extent of protein binding for the ARBs remains constant over a wide concentration range. Typically, protein binding dictates the V_D for a compound and, in fact, irbesartan demonstrates a V_D somewhat higher than that of the other ARBs, with the exception of telmisartan (Stangier et al., 2000a). The significance of the level of protein binding for any of the ARBs remains to be determined.

e. Metabolism and Active Metabolite Generation. Metabolic conversion of an ARB can be viewed in two different ways. First, it may be a step required in order to produce an active metabolite, such is the case with losartan (Yun et al., 1995; Stearns et al., 1995), candesartan cilexitil (Hubner et al., 1997) and olmesartan medoxomil (von Bergmann et al., 2001). Alternatively, metabolic conversion may factor into the conversion of a compound to a physiologically inactive metabolite, as in the case of irbesartan (Hallberg et al., 2002). Losartan, an active substrate molecule, is converted via the P_{450} isozyme system (2C9 and 3A4) to its more active metabolite, E-3174 (Yun et al., 1995; Stearns et al., 1995), whereas candesartan cilexitil, a prodrug, is hydrolyzed to the active moiety candesartan in the course of absorption from the gastrointestinal tract (Hubner et al., 1997).

The metabolic conversion of candesartan cilexitil, an ester prodrug, seems not to be impacted to any degree by either disease states, genetic variation in metabolism or chronic dosing (Hubner et al., 1997). Variants of CYP2C9 have now been identified that influence the metabolic conversion of losartan to its active metabolite E-3174. The presence of certain of these variants (CYP2C9*3 allele) is associated with significant reductions in the conversion of losartan to its active E-3174 metabolite (McCrea et al., 1999). Likewise, the presence of the CYP2C9*2 allele appears to predict those individuals most likely to be responders to irbesartan (Hallberg et al., 2002). Less than 1% of the population of patients exposed to therapy with losartan has this abnormal genetic profile for the metabolism of losartan. Thus, it is unlikely that a metabolic polymorphism for losartan breakdown will ever be found in sufficient numbers of patients to matter clinically. Telmisartan is exclusively metabolized by conjugation to an inactive glucuronic acid metabolite (Stangier et al., 2000a). This lack of cytochrome CYP_{450} dependent metabolism distinguishes telmisartan from other ARBs.

It has also been suggested though that known inhibitors of the P_{450} system and, more specifically, inhibitors of the P_{450} 2C9 and 3A4 isozymes, such as fluconazole and ketoconazole might interfere with the conversion of losartan to its E-3174 metabolite (Sica et al., 2005b). In theory, such drugs might interfere either with the rate or the extent of losartan metabolism. As a consequence, BP control might then become more difficult to achieve and/or maintain if losartan-treated patients were simultaneously treated with such enzyme inhibitors. Although this hypothesis seemed attractive initially, it is not supported by the available data. Drug–drug interactions of this nature are difficult to predict in broad population bases thus, if a losartan-treated patient is simultaneously treated with inhibitors of either the P_{450} 2C9 and/or 3A4 isozymes, BP should be closely monitored (Kazierad et al., 1997). A final metabolic consideration with losartan is its degree of interaction with grapefruit juice. Although not formally tested as to its influencing the BP-lowering effect of losartan, grapefruit juice given together with losartan will reduce its conversion to E-3174 as well as activate P-glycoprotein, which, in sum significantly increases the $AUC_{losartan}/AUC_{E3174}$ ratio (Zaidenstein et al., 2001).

f. Route of Elimination. It is well recognized that the systemic clearance of a compound is dependent on the level of both renal and hepatic function. As a result, if renal and/or hepatic dysfunction exists in a patient, repeated dosing of an antihypertensive compound will ultimately lead to systemic drug accumulation and the need to dose adjust in order to lessen the chance of concentration-related side-effects. All ARBs undergo a significant degree of hepatic elimination with the exception of candesartan, olmesartan and the E-3174 metabolite of losartan, which are 40, 60 and 50% hepatically cleared, respectively (Sica et al., 1995; de Zeeuw et al., 1997; Schwocho and Masonson, 2001). Irbesartan and telmisartan undergo the greatest degree of hepatic elimination among the ARBs with each having >95% hepatic clearance (Sica et al., 1997; Stangier et al., 2000b). Valsartan and eprosartan each undergo about 70% hepatic clearance (Prasad et al., 1997; Kovacs et al., 1999). On the surface, the mode of elimination for an ARB may seem like a minor issue. In reality, it proves to be an important variable in the renally-compromised patient and may, in fact, dictate various elements of the change in renal function that occasionally occurs in the renal failure patient. In those who develop acute renal failure upon receipt of a hepatically cleared ARB, the duration of a renal failure episode will be moderated by the hepatic disposition of the compound, a process that does not occur when the compound is renally cleared (Sica and Deedwania, 1997).

g. Receptor Binding and Half-life. The half-life ($t_{1/2}$) of a compound is a pure pharmacokinetic term which often poorly correlates with a compound's duration of effect. This has typically been the case with antihypertensive medications including both ACE inhibitors and ARBs. The discrepancy between the pharmacokinetic and pharmacodynamic $t_{1/2}$ of a compound derives from the fact that a component of drug action derives from extravascular effects. Because of the inability to sample at these extravascular sites of action, the more meaningful tissue-based $t_{1/2}$ cannot be determined. This is particularly the case for the ARBs since AT_1-receptors are found in multiple extravascular locations and blocking these receptors may, in an as yet undefined fashion, influence the manner in which BP is reduced.

With the above in mind, the pharmacokinetic $t_{1/2}$ of an ARB will roughly approximate its duration of effect as long as the plasma concentration stays above the threshold for a BP-lowering effect. Several of the ARBs, such as candesartan, olmesartan, telmisartan and irbesartan, are considered once-daily compounds in pharmacokinetic terms. The true impact of pharmacologic $t_{1/2}$ for these compounds probably lies more so in the fact that drug is available for a longer period of time and thereby binds to additional AT_1-receptors as they are formed during the latter portion of a dosing interval. This phenomenon becomes obvious in evaluating both *ex vivo* and *in vivo* responses to an ARB across a range of doses, particularly when doses in the high-end of a dosing range are used (Maillard et al., 2002). Such data, however, at best provide guidelines for therapy since patient responses to this drug class, as is the case with ACE inhibitors, are typically highly individualized.

h. Application of Pharmacologic Differences/Receptor Affinity. Receptor affinity is just one of several factors that determine the response to an ARB. An ARB demonstrates *insurmountable* or *non-competitive* blockade if increasing concentrations of angiotensin II are unable to break through receptor blockade. The terms *surmountable*, *competitive*, *insurmountable* and *non-competitive* are often interchangeably used and frequently in an inconsistent fashion (McConnaughey et al., 1998). Surmountable antagonism implies that receptor blockade can eventually be overcome if high enough concentrations of angiotensin II are made available. Surmountable antagonists, such as losartan, shift concentration–response curves parallel and rightwards without reducing the maximal agonist response. In the case of competitive antagonism, as is the case with eprosartan, mass action kinetics exist and both agonists and antagonists individually compete for receptor binding. Non-competitive, irreversible antagonism is a phenomenon characterized by loss of receptor numbers, which occurs by a process of chemical modification.

Insurmountable antagonism mimics non-competitive antagonism. Insurmountable antagonists bind to their receptor in a semi-irreversible fashion, which differs from the permanent binding that occurs with non-competitive antagonists. An insurmountable antagonist releases from its receptor slowly, thus, the drug-receptor dissociation constant can be prolonged. Insurmountable antagonists elicit a parallel shift in agonist concentration–response curves with a depression in the maximal agonist response that is not overcome by increasing concentrations of the agonist. Valsartan, irbesartan, telmisartan and the E-3174 metabolite of losartan exhibit this form of antagonism. Insurmountable antagonists can also elicit non-parallel shifts of the agonist concentration–response curves, again depressing the maximal response to the agonist, a process that still is not overcome by increasing concentrations of the agonist. Candesartan demonstrates this form of insurmountable antagonism (McConnaughey et al., 1998).

To date, the specific mode of receptor occupancy and/or differential pharmacokinetic features of an ARB have not been clearly linked with differing BP responses to these drugs. If differing responses exist among these drugs on a pharmacologic basis, they would be most likely to be detected at low to mid-range doses. At high-end doses, the surfeit of medication available would most likely keep plasma concentrations above a threshold for effect so as to make between drugs efficacy comparisons a moot point (Maillard et al., 2002).

C. Direct Renin Inhibitors

There have been numerous attempts to bring renin inhibitors to market with most such efforts failing because of limited bioavailability and/or potency. Recently, the compound aliskiren has become available and is a drug whose BP-lowering efficacy is similar to or better than what is seen with standard therapeutic doses of ACE inhibitors or ARBs.

1. Pharmacokinetics

Aliskiren, an octanamide, is the first of a new class of completely non-peptide, low-molecular-weight, orally active transition-state renin inhibitors (Rahuel et al., 2000; Nussberger et al., 2003). This compound is a highly specific *in vitro* inhibitor of both human and primate renin with an IC_{50} of 0.6 nmol/l. This level of potency and specificity against human renin offsets the low absolute bioavailability of the drug which is in the order of 2–3%. There is a modest food effect with aliskiren when administered with a high fat meal (Azizi et al., 2006; Frampton and Curran, 2007).

Aliskiren has good water solubility and low level of lipophilicity and is resistant to biodegradation by peptidases found in the intestine, circulation and/or the liver (Nussberger et al., 2003). The plasma concentrations of aliskiren peak between 2 and 4 h following its administration and its mean steady-state $t_{1/2}$ is in the order of 23–36 h (Azizi et al., 2006). Aliskiren demonstrates dose-linear pharmacokinetics over the dose range 75–600 mg in healthy volunteers. Aliskiren is, however, subject to accumulation pharmacokinetics when dosed once daily to steady state with accumulation ratios ranging from 1.4 to 3.9; accumulation is more pronounced at higher doses (Azizi et al., 2006).

After the administration of a single 20 mg intravenous dose of aliskiren over 20 min in healthy male subjects, plasma clearance is $\approx$9 l/h and the volume of distribution at steady state ≈35 l. This latter figure points to there being extravascular distribution locations for aliskiren. The mean protein binding for aliskiren is 49.5% with concentration-independent binding in the range of 10–500 ng/ml (Azizi et al., 2006). This level of protein binding partially explains the large volume of distribution for aliskiren. Following oral administration, aliskiren undergoes minimal metabolism and is mainly eliminated as unchanged (mostly unabsorbed) compound in the feces. The pharmacokinetics of aliskiren are not effected by moderate to severe chronic kidney disease or hepatic impairment (Vaidyanathan et al., 2007a, b).

Aliskiren has a low potential for drug interactions relating to its modest protein binding, limited metabolism as well as its lack of effect on a wide range of CYP_{450} isozymes. One aliskiren drug–drug interaction of potential clinical relevance occurs when it is co-administered with furosemide. Co-administration of aliskiren 300 mg/day with furosemide 20 mg once daily to healthy volunteers did not effect aliskiren concentrations in a meaningful manner, whereas Cmax and AUC values were reduced by ≈50 and ≈30%, respectively (Dieterich et al., 2007). This is a finding of potential clinical significance to a range of patients, however, this drug–drug interaction requires further study before it can be viewed as definitive since furosemide absorption in and of itself is so highly variable.

The pharmacologic aspect of the direct renin inhibitor aliskiren with the potential for the greatest effect on response is its limited bioavailability, however, this seems not to influence unduly the therapeutic response to this compound in the dose range of 150–300 mg/day (Gradman et al., 2005; Sica et al., 2006). Low bioavailability of a compound, *per se*, will not necessarily impact drug action if the amount of drug absorbed still reaches and sustains a blood level adequate for the sought after pharmacologic effect. Moreover, aliskiren is retained in the kidney for a prolonged period of time. After a 3-week washout period, the renal concentration of aliskiren still exceeds its IC_{50} of 0.6 nmol/l by 100-fold (Feldman et al., 2006) with apparent localization in glomeruli and renal arteries/arterioles and possibly in juxtaglomerular cells. This partitioning of aliskiren in the kidney has been used as an explanation for the persistence of a BP and plasma renin activity lowering effect for days to weeks after its withdrawal (Sica et al., 2006; Oh et al., 2007).

D. Blood Pressure Lowering Effect

Increasingly ACE inhibitors, ARBs and DRIs are viewed as suitable first-step options in the treatment of hypertension. Considerable dosing flexibility exists with the available ACE inhibitors and ARBs. There is considerably more information available with the use of ACE inhibitors and ARBs in the treatment of hypertension, thus, the following comments will speak primarily to these two drug classes.

Where indicated, the specifics of the BP response to DRI attention will be addressed. The enthusiasm for the use of ACE inhibitors is not a matter of effectiveness since they have a pattern of efficacy comparable to (and no better than) most other drug classes, with response rates ranging from 40 to 70% in stage I or II hypertension (Materson et al., 1993). A similar range of response rates is observed with ARBs and DRIs.

In head-to-head BP trials comparing ACE inhibitors and ARBs there appears to be scant difference between the two drug classes (Elliott, 2000), likewise, comparison trials between a DRI and either an ACE inhibitor or an ARB show similar responses, which to a small degree favor DRI therapy (Gradman et al., 2005; Frampton et al., 2007; Andersen et al., 2008). Clinical trial results for BP reduction, obviously, do not reflect usage conditions in actual practice where the favorable side-effect profile of ACE inhibitors and ARBs and their highly touted end-organ protection features seem to dominate the thinking of many practitioners. In this regard, ACE inhibitors are extensively used since they are well tolerated and drugs that patients will continue to take over a long period (Caro et al., 1999). The same can be said for ARBs, a drug class with a tolerability profile which is better than that of ACE inhibitors (Mazzolai and Burnier, 1999).

The enthusiasm for these two drug classes must be put in proper perspective since in uncomplicated non-diabetic hypertensive patients, a number of drug classes given at low to mid-range doses can prove effective and well tolerated at a fraction of the cost of an ACE inhibitors or an ARB. Alternatively, evidence supports the favored use of ACE inhibitors and, more recently, ARBs (on a BP-independent basis) in the diabetic and/or at risk cardiac/renal patient with either established atherosclerotic disease or proteinuric renal disease (Yusuf et al., 2000; Brenner et al., 2001; Lewis et al., 2001; Halkin and Keren, 2002), however, there is a growing appreciation that the so-called BP-independent effects of these drugs are of less significance than first thought, particularly when BP changes are more carefully examined (Blood Pressure Lowering Treatment Trialists Collaboration, 2007).

There are hardly any predictors of the BP lowering response to ACE inhibitors. When hypertension is accompanied by significant activation of the RAA axis, such as in renal artery stenosis, the response to an ACE inhibitor or an ARB can be immediate and sometimes extreme (Maillard et al., 2001). In most other cases, any relationship between pre- and/or post-treatment plasma renin activity (PRA) and the vasodepressor response to an ACE inhibitor and/or an ARB is generally lacking. Certain patient phenotypes exhibit lower response rates to ACE inhibitor and ARB monotherapy including low-renin, salt-sensitive individuals such as the diabetic, African-American or elderly hypertensive (Flack et al., 2001), however, in the African-American this lesser response to ACE inhibition (or ARB therapy) is more so on a population basis with individual patients still having the capacity to be responders to therapy (Mokwe et al., 2004). This pattern of response suggests that eliminating even small amounts of RAS activity can result in significant BP reduction in a number (but not all) low-renin patients.

The low-renin state, characteristic of the elderly hypertensive is to be distinguished from other low-renin forms of hypertension in that it develops not in response to volume expansion but rather, because of senescence-related changes in the activity of this axis. The elderly generally respond well to ACE inhibitors at usual doses though senescence-related renal failure, which reduces the elimination of these drugs, makes interpretation of dose-specific treatment responses difficult. The elderly hypertensive patient, with systolic-predominant hypertension, also responds well to ARB monotherapy.

Results from a number of head-to-head trials support the comparable antihypertensive efficacy and tolerability of the various ACE inhibitors and ARBs. However, there are differences among the ACE inhibitors as to the time to onset of effect and/or the time to maximum BP reduction which may relate to the absorption and disposition characteristics of the various compounds. These differences, however, do not translate into different response rates *if* comparable doses of the individual ACE inhibitors or ARBs are given. Typical confounding variables, which confuse the interpretation of the findings in BP studies with ACE inhibitors, have included differences in study design/methodology, as well as dose frequency and/or amount. ACE inhibitors and ARBs labeled as 'once-daily', vary in a dose-dependent manner in their ability to reduce BP for a full 24 hours, as defined by a trough:peak ratio exceeding 50%. Unfortunately, the trough:peak ratio, as an index of duration of BP control, is oftentimes prone to misrepresent the true BP reduction seen with a compound (Zannad et al., 1996), thus, dosing instructions for many of these compounds include the proviso to administer a second-daily dose if the antihypertensive effect has dissipated at the end of the dosing interval.

The question is often raised as to what to do if an ACE inhibitor (or an ARB) falls short of normalizing BP. One approach is simply to raise the dose, however, the dose–response curve for ACE inhibitors, like most antihypertensive agents, is steep at the beginning doses and thereafter becomes shallow to flat (Izzo and Sica, 2008). Responders to ACE inhibitors typically do so at doses well below those necessary for 24-hour complete suppression of ACE. In addition, the maximal BP-lowering response to an ACE inhibitor does not occur until several weeks after beginning therapy, which may relate to an incremental effect on BP owing to vascular remodeling (Schiffrin, 2001). Thus, only with total absence of a response to an ACE inhibitor, should a switch to an alternative drug class be made. If a modest (partial) response occurs then therapy with an ACE inhibitor can be continued in anticipation of an additional drop in BP over the next several weeks. Alternatively, an additional compound such as a diuretic, calcium channel blocker (CCB) or peripheral α-blocker can be combined with an ACE inhibitor if circumstances dictate getting to goal BP more rapidly.

Virtually all of the prior comments directed to the pharmacologic response to ACE inhibitors apply to the ARB and DRI classes of drugs. This includes considerations of predictors of response, onset and duration of response, the makeup of their dose–response curves and their capacity for a late onset additional BP response. One exception with the DRI class relates to the duration of effect with the compound aliskiren, which can be quite prolonged with a very slow decay in its effect with drug discontinuation (Sica et al., 2006; Oh et al., 2007). A number of head-to-head studies have been conducted between different ARBs and would seem to favor several of the more recent additions to the ARB class over losartan. Head-to-head studies (other than with losartan) have not found significant differences among the drugs in the ARB class when full strength doses are studied (Giles et al., 2007).

IV. ACE INHIBITORS AND ANGIOTENSIN RECEPTOR BLOCKERS WITH OTHER AGENTS

To date, there appears to be little difference in the BP-lowering that occurs with ACE inhibitors, ARBs and DRIs given in combination with other drug classes including diuretics and CCBs. The BP-lowering effect of these compounds is notably enhanced with the co-administration of a diuretic, particularly in the African-American hypertensive. This pattern of response has spurred the development of a number of fixed-dose combination products, comprised of each of these RAS inhibitor classes and low to moderate doses of thiazide-type diuretics. The rationale for combining these two drug classes derives from the observation that the sodium depletion produced by a diuretic activates the RAA axis. Consequently, BP shifts to a more angiotensin II dependent mode and thus the additivity in response (Sica, 2002).

ACE inhibitors have been given together with β-blockers. The rationale behind this combination is that the β-blocker will presumably blunt the hyperreninemia caused by an ACE inhibitor. It was presumed that in so doing that the ACE inhibitor response might be more robust (Sica, 2002). Although this hypothesis originally seemed attractive, in practice, only a modest additional vasodepressor response occurs when these two drug classes are combined (Hansson, 1989). When BP meaningfully falls upon addition of a β-blocker to an ACE inhibitor, it is generally because of a favorable effect on systemic hemodynamics relating to pulse rate reduction. Alternatively, the addition of a peripheral α-antagonist, such as doxazosin, to an ACE inhibitor can be followed by a significant additional BP response (Black et al., 2000). The mechanism behind this additive response remains to be more completely elucidated. Finally, the BP-lowering effect of an ACE inhibitor is considerably enhanced by the co-administration of a CCB (88-89). This additive response occurs whether the CCB being given is a dihydropyridine (e.g. felodipine or amlodipine) (Chrysant et al., 2007) or a non-dihydropyridine, such as verapamil (DeQuattro and Lee, 1997). The potency of this combination has provided the practical basis for the development of a number of fixed-dose combination products comprised of an ACE inhibitor and a CCB and, more recently, an ARB together with a dihydropyridine CCB (Kereiakes et al., 2007). Adding an ACE inhibitor or an ARB to a CCB is also useful in that either of these drug classes appears to attenuate the peripheral edema that can follow from CCB therapy (Sica, 2003).

The efficacy of both ACE inhibitors and ARBs as antihypertensive agents is well documented. Quite logically, this has led to the belief that in combination these two drug classes may reduce BP better than if either were to be given alone. Of the available studies in hypertension where an ACE inhibitor has been given with an ARB, the observed results, however, have generally shown minimal additivity with typically inconsistent and poorly generalizable findings (Doulton et al., 2005). Alternatively, the studies of combination ACE inhibitor and ARB therapy in proteinuric CKD have shown more consistent findings, particularly as relates to providing an incremental benefit in reducing protein excretion (Wolf and Ritz, 2005). The same can be said for combination therapy in HF where the addition of an ARB to an ACE inhibitor affords symptomatic benefit in addition to improving outcomes beyond what is seen with ACE inhibitor therapy alone (Cohn and Tognoni, 2001; McMurray et al., 2003).

Finally, a number of studies have demonstrated the utility of ACE inhibitors in the management of hypertensive patients otherwise unresponsive to multiple drug combinations (Julien et al., 1990; Damasceno et al., 1997). Typically, such combinations have included a diuretic as well as either minoxidil, a CCB and/or a peripheral α-blocker. In addition, if an acute reduction in BP is desired it can be achieved with either oral or sublingual captopril in that its onset of action is as soon as 15 min after its administration (Gemici et al., 1999). An additional option for the management of hypertensive emergencies is that of parenteral therapy with enalaprilat (Hirschl et al., 1997). Compounds that interrupt RAA axis activity, such as ACE inhibitors, should be administered cautiously in patients suspected of a marked activation of the RAA axis (e.g. prior treatment with diuretics). In such subjects, sudden and extreme drops in BP have occasionally been observed with the first dose of an ACE inhibitor (Sica, 2001c).

V. SELECT SIDE-EFFECTS OF ACE INHIBITORS AND ANGIOTENSIN-RECEPTOR BLOCKERS

Shortly after the release of ACE inhibitors, a pattern of 'functional renal insufficiency' was noticed as a class effect. This phenomenon was initially reported in patients with renal artery stenosis and a solitary kidney or in the presence of bilateral renal artery stenosis. Predisposing conditions to this development include dehydration, HF, non-steroidal anti-inflammatory drug (NSAID) use and/or either micro- or macrovascular renal disease (Schoolwerth et al., 2001). The common theme in these conditions is a reduction in afferent arteriolar flow. When this occurs, glomerular filtration transiently declines. In response to this reduction in glomerular filtration, there is an increase in the local (renal) production of angiotensin II. Together with this increase in angiotensin II, the efferent or post-glomerular arteriole constricts, which restores hydrostatic pressures within the more proximal glomerular capillary bed (Sica and Gehr, 2005). The abrupt removal of angiotensin II, as occurs with an ACE inhibitor (or an ARB or a DRI), will rapidly open the efferent arteriole in concert with a reduction in systemic BP. In combination, these hemodynamic changes drop glomerular hydrostatic pressure and glomerular filtration drops. This type of 'functional renal insufficiency' is best treated by discontinuation of the responsible agent, careful volume expansion (if intravascular volume contraction is a contributing factor) and, if warranted on clinical grounds, evaluation for the presence of renal artery stenosis (Schoolwerth et al., 2001). The functional renal insufficiency with these compounds relates to drug concentration, thus, renally-cleared compounds will have a more prolonged negative effect on renal function than might be the case with compounds undergoing a significant degree of hepatic clearance (Schoolwerth et al., 2001).

Hyperkalemia is an additional ACE inhibitor (or ARB or DRI) associated side-effect. ACE inhibitor-related hyperkalemia occurs infrequently unless a specific predisposition to hyperkalemia exists – such as diabetes, HF with renal failure (receiving potassium-sparing diuretics or potassium supplements) (Textor et al., 1982; Juurlink et al., 2003; Cruz et al., 2003). The combination of an ACE inhibitor and an ARB or either of these classes together with a potassium-sparing diuretic (such as spironolactone) can cause significant hyperkalemia if being given in the setting of a reduced GFR as may occur in CKD and/or HF (Phillips et al., 2007).

Angioneurotic edema is a life-threatening complication of ACE inhibitors that is more common in blacks (Gibbs et al., 1999). Angioedema can be managed in the long term with simple discontinuation of the ACE inhibitor at fault (Cicardi et al., 2004). Angioedema can occasionally recur with ARB therapy in a patient having previously experienced it with an ACE inhibitor, but it is generally mild and not life threatening in its severity (Granger et al., 2003). Angioedema of the intestine (more common in women) can also occur with ACE inhibitor therapy. Its typical presentation is that of abdominal pain/diarrhea with or without facial and/or oropharyngeal swelling (Byrne et al., 2000).

A final side-effect consideration with ACE inhibitors (and ARBs) is that of anemia. ACE inhibitors suppress the production of erythropoietin in a dose-dependent manner which presents a particular problem when ACE inhibitors are administered in the presence of renal failure. Alternatively, this aspect of ACE inhibitor effect can be used therapeutically in the instance of post-transplant erythrocytosis and high-altitude polycythemia (Sica and Mannino, 2007). DRIs have not been studied in this regard but are likely to have a similar effect on red cell production.

VI. SUMMARY

The renin–angiotensin system has been and remains a notable target in the treatment of hypertension and select end-organ diseases. A number of drug classes reduce the activity of this axis including ACE inhibitors, ARBs and DRIs. Blood pressure reduction with each of these classes is comparable if equipotent doses are used. Blood pressure reduction is one factor in how these medications decrease event rate. Systolic forms of HF and proteinuric CKD seem to benefit in a BP-independent fashion from use of these drugs. Alternatively, question remains as to the degree of BP-independent protection afforded the high-risk coronary artery disease (CAD) patient treated with these medications.

References

Andersen, K., Weinberger, M. H., Egan, B. et al. (2008). Comparative efficacy and safety of aliskiren, an oral direct renin inhibitor, and ramipril in hypertension: a 6-month, randomized, double-blind trial. *J Hypertens* **26**: 589–599.

Azizi, M., Webb, R., Nussberger, J., and Hollenberg, N. K. (2006). Renin inhibition with aliskiren: where are we now and where are we going? *J Hypertens* **24**: 243–256.

Black, H. R., Sollins, J. S., and Garofalo, J. L. (2000). The addition of doxazosin to the therapeutic regimen of hypertensive patients inadequately controlled with other antihypertensive medications: a randomized, placebo-controlled study. *Am J Hypertens* **13**: 468–474.

Blood Pressure Lowering Treatment Trialists Collaboration, Turnbull, F., Neal, B., Pfeffer, M. et al. (2007). Blood pressure-dependent and independent effects of agents that inhibit the renin-angiotensin system. *J Hypertens* **25**: 951–958.

Brenner, B. M., Cooper, M. E., de Zeeuw, D. et al. (2001). Effects of losartan on renal and cardiovascular outcomes in patients with type 2 diabetes and nephropathy. *N Engl J Med* **345**: 861–869.

Brown, N. J., and Vaughn, D. E. (1998). Angiotensin-converting enzyme inhibitors. *Circulation* **97**: 1411–1420.

Byrne, T. J., Douglas, D. D., Landis, M. E., and Heppell, J. P. (2000). Isolated visceral angioedema: an underdiagnosed complication of ACE inhibitors? *Mayo Clin Proc* **75**: 1201–1204.

Caro, J. J., Speckman, J. L., Salas, M., Raggio, G., and Jackson, J. D. (1999). Effect of initial drug choice on persistence with antihypertensive therapy: the importance of actual practice data. *Can Med Assoc J* **160**: 41–46.

Chalmers, D., Whitehead, A., and Lawson, D. H. (1992). Postmarketing surveillance of captopril for hypertension. *Br J Clin Pharmacol* **34**: 215–223.

Chapelsky, M. C., Martin, D. E., Tenero, D. M. et al. (1998). A dose proportionality study of eprosartan in healthy male volunteers. *J Clin Pharmacol* **38**: 34–39.

Christ, D. D. (1995). Human plasma protein binding of the angiotensin II receptor antagonist losartan potassium (DuP 753/MK 954) and its pharmacologically active metabolite EXP3174. *J Clin Pharmacol* **35**: 515–520.

Chrysant, S. G., Sugimoto, D. H., Lefkowitz, M. et al. (2007). The effects of high-dose amlodipine/benazepril combination therapies on blood pressure reduction in patients not adequately controlled with amlodipine monotherapy. *Blood Press* (Suppl) **1**: 10–7.

Cicardi, M., Zingale, L. C., Bergamaschini, L., and Agostoni, A. (2004). Angioedema associated with angiotensin-converting enzyme inhibitor use: outcome after switching to a different treatment. *Arch Int Med* **164**: 910–913.

Cohn, J. N., and Tognoni, G. (2001). A randomized trial of the angiotensin-receptor blocker valsartan in chronic heart failure. Valsartan Heart Failure Trial Investigators. *N Engl J Med* **345**: 1667–1675.

Cox, P. J., Bush, B. D., and Gorycki, P. D. (1996). The metabolic fate of eprosartan in healthy volunteers. *Exp Toxicol Pathol* **48** (Supp II): 75–82.

Cruz, C. S., Cruz, A. A., and Marcilio de Souza, C. A. (2003). Hyperkalaemia in congestive heart failure patients using ACE inhibitors and spironolactone. *Nephrol Dial Transplant* **18**: 1814–1819.

Damasceno, A., Ferreira, B., Patel, S., Sevene, E., and Polonia, J. (1997). Efficacy of captopril and nifedipine in black and white patients with hypertensive crisis. *J Hum Hypertens* **11**: 471–476.

Danielson, B., Querin, S., LaRochelle, P. et al. (1994). Pharmacokinetics and pharmacodynamics of trandolapril after repeated administration of 2 mg to patients with chronic renal failure and healthy control subjects. *J Cardiovasc Pharmacol* **23**(Suppl 4): S50–S59.

DeQuattro, V., and Lee, D. (1997). Fixed-dose combination therapy with trandolapril and verapamil SR is effective in primary hypertension. *Amer J Hypertens* **10**(Suppl 2): 138S–145S.

de Zeeuw, D., Remuzzi, G., and Kirch, W. (1997). The pharmacokinetics of candesartan cilexitil in patients with renal or hepatic impairment. *J Hum Hypertens* **11**(Supp 2): S37–42.

Dickstein, K. (1986). Pharmacokinetics of enalapril in congestive heart failure. *Drugs* **32**(Suppl 5): 40–44.

Dieterich, H. A., Yeh, C., and Howard, D. (2007). Assessment of the pharmacokinetic interaction between the oral direct renin inhibitor aliskiren and furosemide: a study in healthy volunteers. *Clin Pharmacol Ther*, p. S110.

Doulton, T. W., He, F. J., and MacGregor, G. A. (2005). Systematic review of combined angiotensin-converting enzyme inhibition and angiotensin receptor blockade in hypertension. *Hypertension* **45**: 880–886.

Dzau, V. J., Bernstein, K., Celermajer, D. et al. (2001). The relevance of tissue angiotensin-converting enzyme: manifestations in mechanistic and endpoint data. *Am J Cardiol* **88**(Suppl 9): 1L–20L.

Ebihara, A., and Fujimura, A. (1991). Metabolites of antihypertensive drugs. An updated review of their clinical pharmacokinetic and therapeutic implications. *Clin Pharmacokinet* **21**: 33143.

Elliott, W. J. (2000). Therapeutic trials comparing angiotensin converting enzyme inhibitors and angiotensin II receptor blockers. *Curr Hypertens Rep* **2**: 402–411.

Fabris, B., Yamada, H., Cubela, R., Jackson, B., Mendelsohn, F. A., and Johnston, C. I. (1990). Characterization of cardiac angiotensin converting enzyme and in vivo inhibition following oral quinapril to rats. *Br J Pharmacol* **100**: 651–655.

Feldman, D. L., Persohn, E., and Schutz, H. (2006). Renal localization of the renin inhibitor aliskiren. *J Clin Hypertens* **8**(Suppl A): A80.

Flack, J. M., Saunders, E., Gradman, A. et al. (2001). Antihypertensive efficacy and safety of losartan alone and in combination with hydrochlorothiazide in adult African Americans with mild to moderate hypertension. *Clin Ther* **23**: 1193–1208.

Frampton, J., and Curran, M. (2007). Aliskiren: a review of its use in the management of hypertension. *Drugs* **67**: 1767–1792.

Gemici, K., Karakoç, Y., Ersoy, A., Baran, I. I., Güllüülü, S., and Cordan, J. (1999). A comparison of safety and efficacy of sublingual captopril with sublingual nifedipine in hypertensive crisis. *Int J Angiol* **8**: 147–149.

Gibbs, C. R., Lip, G. Y.H., and Beevers, D. G. (1999). Angioedema due to ACE inhibitors: increased risk in patients of African origin. *Br J Clin Pharmacol* **48**: 861–865.

Giles, T. D., Oparil, S., Silfani, T. N., Wang, A., and Walker, J. F. (2007). Comparison of increasing doses of olmesartan medoxomil, losartan potassium, and valsartan in patients with essential hypertension. *J Clin Hypertens* **9**: 187–195.

Gillis, J. C., and Markham, A. (1997). Irbesartan. A review of its pharmacodynamic and pharmacokinetic properties and therapeutic use in the management of hypertension. *Drugs* **54**: 885–902.

Gradman, A. H., Schmieder, R. E., Lins, R. L., Nussberger, J., Chiang, Y., and Bedigian, M. P. (2005). Aliskiren, a novel orally effective renin inhibitor, provides dose-dependent antihypertensive efficacy and placebo-like tolerability in hypertensive patients. *Circulation* **111**: 1012–1018.

Granger, C. B., McMurray, J. J., Yusuf, S. et al. (2003). Effects of candesartan in patients with chronic heart failure and reduced left ventricular systolic function intolerant to angiotensin-converting enzyme inhibitors: the CHARM-Alternative trial. *Lancet* **362**: 772–776.

Halkin, A., and Keren, G. (2002). Potential indications for angiotensin-converting enzyme inhibitors in atherosclerotic vascular disease. *Am J Med* **112**: 126–134.

Hallberg, P., Karlsson, J., Kurland, L. et al. (2002). The CYP2C9 genotype predicts the blood pressure response to irbesartan: results from the Swedish Irbesartan Left Ventricular Hypertrophy Investigation vs Atenolol (SILVHIA) trial. *J Hypertens* **20**: 2089–2093.

Hansson, L. (1989). Beta blockers with ACE inhibitors – a logical combination? *J Hum Hypertens* **3**: 97–100.

Herman, A. G. (1992). Differences in structure of angiotensin-converting enzyme inhibitors might predict differences in action. *Am J Cardiol* **70**: 102C–108C.

Hirschl, M. M., Binder, M., Bur, A. et al. (1997). Impact of the renin-angiotensin-aldosterone system on blood pressure response to intravenous enalaprilat in patients with hypertensive crises. *J Hum Hypertens* **11**: 177–183.

Hoyer, J., Schulte, K.-L., and Lenz, T. (1993). Clinical pharmacokinetics of angiotensin converting enzyme inhibitors in renal failure. *Clin Pharmacokinet* **24**: 230–254.

Hubner, R., Hogemann, A. M., Sunzel, M., and Riddell, J. G. (1997). Pharmacokinetics of candesartan after single and multiple doses of candesartan cilexitil in young and elderly healthy volunteers. *J Hum Hypertens* **11**(Suppl 2): S19–25.

Izzo, J. L., and Sica, D. A. (2008). Antihypertensive drugs: pharmacologic principles and dosing effect. In: *Hypertension Primer*, 4th edn, (Izzo, J.L., Sica, D.A., Black, H.R., eds) Lippincott, Williams & Wilkins, Baltimore, pp. 432–434.

Jafar, T. H., Stark, P. C., Schmid, C. H. et al. (2003). Progression of chronic kidney disease: the role of blood pressure control, proteinuria, and angiotensin-converting enzyme inhibition: a patient-level meta-analysis. *Ann Intern Med* **139**: 244–252.

Johnston, C. I., Fabris, B., Yamada, H. et al. (1989). Comparative studies of tissue inhibition by angiotensin converting enzyme inhibitors. *J Hypertens* **7**(Suppl): S11–S16.

Julien, J., Dufloux, M. A., Prasquier, R. et al. (1990). Effects of captopril and minoxidil on left ventricular hypertrophy in resistant hypertensive patients: a 6-month double-blind comparison. *J Amer Coll Cardiol* **16**: 137–142.

Juurlink, D. N., Mamdani, M., and Kopp, A. (2003). Drug–drug interactions among elderly patients hospitalized for drug toxicity. *J Am Med Assoc* **289**: 1652–1658.

Kazi, D., and Deswal, A. (2008). Role and optimal dosing of angiotensin-converting enzyme inhibitors in heart failure. *Cardiol Clin* **26**: 1–14.

Kazierad, D. J., Martin, D. E., Blum, R. A. et al. (1997). Effect of fluconazole on the pharmacokinetics of eprosartan and losartan in healthy male volunteers. *Clin Pharmacol Ther* **62**: 417–425.

Kereiakes, D. J., Neutel, J. M., Punzi, H. A., Xu, J., Lipka, L. J., and Dubiel, R. (2007). Efficacy and safety of olmesartan medoxomil and hydrochlorothiazide compared with benazepril and amlodipine besylate. *Am J Cardiovasc Drugs* **7**: 361–372.

Kinoshita, A., Urata, H., Bumpus, F. M., and Husain, A. (1993). Measurement of angiotensin I converting enzyme inhibition in the heart. *Circ Res* **73**: 51–60.

Kovacs, S. J., Tenero, D. M., Martin, D. E., Ilson, B., and Jorkasky, D. K. (1999). Pharmacokinetics and protein binding of eprosartan in hemodialysis-dependent patients with end-stage renal disease. *Pharmacotherapy* **19**: 612–619.

Leonetti, G., and Cuspidi, C. (1995). Choosing the right ACE inhibitor. A guide to selection. *Drugs* **49**: 516–535.

Lewis, E. J., Hunsicker, L. G., Clarke, W. R. et al. (2001). Renoprotective effect of the angiotensin-receptor antagonist irbesartan in patients with nephropathy due to type 2 diabetes. *N Engl J Med* **345**: 851–880.

Lewis, E. J., Hunsicker, L. G., Bain, R. P., and Rohde, R. D. (1993). The effect of angiotensin converting enzyme inhibition on diabetic nephropathy. The Collaborative Study Group. *N Engl J Med* **329**: 1456–1462.

Lo, M. W., Goldberg, M. R., McCrea, J. B., Lu, H., Furtek, C. I., and Bjornsson, T. D. (1995). Pharmacokinetics of losartan, an angiotensin II receptor antagonist, and its active metabolite, EXP3174 in humans. *Clin Pharmacol Ther* **58**: 641–649.

Ludden, T. M. (1991). Nonlinear pharmacokinetics: clinical implications. *Clin Pharmacokinet* **20**: 429–446.

Maillard, J. O., Descombes, E., Fellay, G., and Regamey, C. (2001). Repeated transient anuria following losartan administration in a patient with a solitary kidney. *Ren Fail* **23**: 143–147.

Maillard, M. P., Würzner, G., Nussberger, J., Centeno, C., Burnier, M., and Brunner, H. R. (2002). Comparative angiotensin II receptor blockade in healthy volunteers: the importance of dosing. *Clin Pharmacol Ther* **71:** 68–76.

Martin, D. E., Chapelsky, M. C., Ilson, B. et al. (1998). Pharmacokinetics and protein binding of eprosartan in healthy volunteers and in patients with varying degrees of renal impairment. *J Clin Pharmacol* **38:** 129–137.

Materson, B. J., Reda, D. J., Cushman, W. C. et al. (1993). Single-drug therapy for hypertension in men. A comparison of six antihypertensive agents with placebo. *N Engl J Med* **328:** 914–921.

Mazzolai, L., and Burnier, M. (1999). Comparative safety and tolerability of angiotensin II receptor antagonists. *Drug Saf* **21:** 23–33.

McConnaughey, M. M., McConnaughey, J. S., and Ingenito, A. J. (1998). Practical considerations of the pharmacology of angiotensin receptor blockers. *J Clin Pharmacol* **39:** 547–559.

McCrea, J. B., Cribb, A., Rushmore, T. et al. (1999). Phenotypic and genotypic investigations of a healthy volunteer deficient in the conversion of losartan to its active metabolite E-3174. *Clin Pharmacol Ther* **65:** 348–352.

McMurray, J. J., Ostergren, J., Swedberg, K. et al. (2003). Effects of candesartan in patients with chronic heart failure and reduced left-ventricular systolic function taking angiotensin-converting-enzyme inhibitors: the CHARM-Added trial. *Lancet* **362:** 767–771.

Mokwe, E., Ohmit, S. E., Nasser, S. A. et al. (2004). Determinants of blood pressure response to quinapril in black and white hypertensive patients: the Quinapril Titration Interval Management Evaluation trial. *Hypertension* **43:** 1202–1207.

Noormohamed, F. H., McNabb, W. R., and Lant, A. F. (1990). Pharmacokinetic and pharmacodynamic actions of enalapril in humans: effect of probenecid pretreatment. *J Pharmacol Exp Ther* **253:** 362–368.

Nussberger, J., Wuerzner, G., Jensen, C., and Brunner, H. R. (2003). Angiotensin II suppression in humans by the orally active renin inhibitor aliskiren (SPP100): comparison with enalapril. *Hypertension* **39:** E1–8.

Oh, B. H., Mitchell, J., Herron, J. R., Chung, J., Khan, M., and Keefe, D. L. (2007). Aliskiren, an oral renin inhibitor, provides dose-dependent efficacy and sustained 24-hour blood pressure control in patients with hypertension. *J Am Coll Cardiol* **49:** 1157–1163.

Pfeffer, M. A., Braunwald, E., Moye, L. A. et al. (1991). Effect of enalapril on survival in patients with reduced left ventricular ejection fractions and congestive heart failure. The SOLVD investigators. *N Engl J Med* **325:** 293–302.

Phillips, C. O., Kashani, A., Ko, D. K., Francis, G., and Krumholz, H. M. (2007). Adverse effects of combination angiotensin II receptor blockers plus angiotensin-converting enzyme inhibitors for left ventricular dysfunction: a quantitative review of data from randomized clinical trials. *Arch Intern Med* **167:** 1930–1936.

Prasad, P., Mangat, S., and Choi, L. (1997). Effect of renal function on the pharmacokinetics of valsartan. *Clin Drug Invest* **13:** 207–214.

Rahuel, J., Rasetti, V., Maibaum, J. et al. (2000). Structure-based drug design: the discovery of novel non peptide orally active inhibitors of human renin. *Chem Biol* **7:** 493–504.

Reid, J. L. (1997). Kinetics to dynamics: are there differences between ACE inhibitors? *Eur Heart J* **18**(Suppl E): E14–E18.

Schiffrin, E. L. (2001). Effects of antihypertensive drugs on vascular remodeling: do they predict outcome in response to antihypertensive therapy? *Curr Opin Nephrol Hypertens* **10:** 617–624.

Schoolwerth, A., Sica, D. A., Ballermann, B. J., and Wilcox, C. S. (2001). Renal considerations in angiotensin converting enzyme inhibitor therapy. A Statement for Healthcare Professionals from the Council on the Kidney in Cardiovascular Disease and the Council for High Blood Pressure Research of the American Heart Association. *Circulation* **104:** 1985–1991.

Schwocho, L. R., and Masonson, H. N. (2001). Pharmacokinetics of CS-866, a new angiotensin II receptor blocker, in healthy subjects. *J Clin Pharmacol* **41:** 515–527.

Sharif, M. N., Evans, B. L., and Pylypchuk, G. B. (1994). Cough induced by quinapril with resolution after changing to fosinopril. *Ann Pharmacother* **28:** 720–722.

Shionoiri, H. (1993). Pharmacokinetic drug interactions with ACE inhibitors. *Clin Pharmacokinet* **25:** 20–58.

Sica, D., and Gehr, T. W. (2005). Angiotensin converting enzyme inhibitors. In: *Hypertension, a companion to the kidney,* 2nd edn, (Oparil, S., Weber, M., eds) W. B. Saunders, Philadelphia, pp. 669–682.

Sica, D., Gradman, A., Lederballe, I., Meyers, M., Cai, J., and Zhang, J. (2006). Aliskiren, an oral renin inhibitor, provides long-term antihypertensive efficacy and safety in patients with hypertension. *Eur Heart J* **27**(Suppl): P797.

Sica, D. A. (2001a). The HOPE Study: ACE inhibitors – are their benefits a class effect or do individual agents differ? *Curr Opin Nephrol Hypertens* **10:** 597–601.

Sica, D. A. (2001b). Pharmacology and clinical efficacy of angiotensin-receptor blockers. *Amer J Hypertens* **14:** 242S–247S.

Sica, D. A. (2001c). Dosage considerations with perindopril for hypertension. *Am J Cardiol* **88**(Suppl 1): 13–18.

Sica, D. A. (2002). Rationale for fixed-dose combinations in the treatment of hypertension: the cycle repeats. *Drugs* **62:** 443–462.

Sica, D. A. (2003). Calcium-channel blocker edema: Can it be resolved? *J Clin Hypertens* **5:** 291–294.

Sica, D. A., and Deedwania, P. C. (1997). Renal considerations in the use of angiotensin-converting enzyme inhibitors in the treatment of congestive heart failure. *Cong Heart Fail* **3:** 54–59.

Sica, D. A., and Mannino, R. (2007). Antihypertensive medications and anemia. *J Clin Hypertens* **9:** 723–727.

Sica, D. A., Cutler, R. E., Parmer, R. J., and Ford, N. F. (1991). Comparison of the steady-state pharmacokinetics of fosinopril, lisinopril, and enalapril in patients with chronic renal insufficiency. *Clin Pharmacokinet* **20:** 420–427.

Sica, D. A., Gehr, T. W., and Ghosh, S. (2005). Clinical pharmacokinetics of losartan. *Clin Pharmacokinet* **44:** 797–814.

Sica, D. A., Marino, M. R., Hammett, J. L., Ferreira, I., Gehr, T. W., and Ford, N. F. (1997). The pharmacokinetics of irbesartan in renal failure and maintenance hemodialysis. *Clin Pharmacol Ther* **62:** 610–618.

Sica, D. A., Shaw, W. C., Lo, M. W. et al. (1995). The pharmacokinetics of losartan in renal insufficiency. *J Hypertens* **13**(Supp 1): S49–52.

Stangier, J., Schmid, J., Turck, D. et al. (2000a). Absorption, metabolism, and excretion of intravenously and orally administered (^{14}C) telmisartan in healthy volunteers. *J Clin Pharmacol* **40:** 1312–1322.

Stangier, J., Su, C. A., Brickl, R., and Franke, H. (2000b). Pharmacokinetics of single-dose telmisartan 120 mg given during and between hemodialysis in subjects with severe renal insufficiency: comparison with healthy volunteers. *J Clin Pharmacol* **40:** 1365–1372.

Stearns, R. A., Chakravarty, P. K., Chen, R., and Chiu, S. H. (1995). Biotransformation of losartan to its active carboxylic acid metabolite in human liver microsomes. Role of cytochrome P4502C and 3A subfamily members. *Drug Met Disp* **23:** 207–215.

Tang, W. H., Vagelos, R. H., Yee, Y. G. et al. (2002). Neurohormonal and clinical responses to high- versus low-dose enalapril therapy in chronic heart failure. *J Amer Coll Cardiol* **39:** 70–78.

Textor, S. C., Bravo, E. L., Fouad, F. M., and Tarazi, R. C. (1982). Hyperkalemia in azotemic patients during angiotensin-converting enzyme inhibition and aldosterone reduction with captopril. *Am J Med* **73:** 719–725.

Vachharajani, N. N., Shyu, W. C., Chando, T. J., Everett, D. W., Greene, D. S., and Barbhaiya, R. H. (1998). Oral bioavailability and disposition characteristics of irbesartan, an angiotensin antagonist, in healthy volunteers. *J Clin Pharmacol* **38:** 702–707.

Vaidyanathan, S., Bigler, H., Yeh, C. et al. (2007a). Pharmacokinetics of the oral direct renin inhibitor aliskiren alone and in combination with irbesartan in renal impairment. *Clin Pharmacokinet* **46:** 661–675.

Vaidyanathan, S., Warren, V., Yeh, C., Bizot, M. N., Dieterich, H. A., and Dole, W. P. (2007b). Pharmacokinetics, safety, and tolerability of the oral Renin inhibitor aliskiren in patients with hepatic impairment. *J Clin Pharmacol* **47:** 192–200.

von Bergmann, K., Laeis, P., Puchler, K., Sudhop, T., Schwocho, L. R., and Gonzales, L. (2001). Olmesartan medoxomil: influence of age, renal and hepatic function on the pharmacokinetics of olmesartan medoxomil. *J Hypertens* **19**(Suppl 1): S33–40.

White, C. M. (1998). Pharmacologic, pharmacokinetic, and therapeutic differences among ACE inhibitors. *Pharmacotherapy* **18:** 588–599.

Wolf, G., and Ritz, E. (2005). Combination therapy with ACE inhibitors and angiotensin II receptor blockers to halt progression of chronic renal disease: pathophysiology and indications. *Kidney Int* **67:** 799–812.

Yun, C. H., Lee, H. S., Lee, H., Rho, J. K., Jeong, H. G., and Gungerich, F. P. (1995). Oxidation of the angiotensin II receptor antagonist losartan (DuP 753) in human liver microsomes; role of cytochrome P4503A (4) in formation of the active metabolite EXP3174. *Drug Met Disp* **23:** 285–289.

Yusuf, S., Sleight, P., Pogue, J., Bosch, J., Davies, R., and Dagenais, G. (2000). Effects of an angiotensin-converting enzyme inhibitor, ramipril, on cardiovascular events in high-risk patients. The Heart Outcomes Prevention Evaluation Study Investigators. *N Engl J Med* **342:** 145–153.

Zaidenstein, R., Soback, S., Gips, M. et al. (2001). Effect of grapefruit juice on the pharmacokinetics of losartan and its active metabolite E3174 in healthy volunteers. *Ther Drug Monit* **23:** 369–373.

Zannad, F., Matzinger, A., and Larché, J. (1996). Trough/peak ratios of once daily angiotensin converting enzyme inhibitors and calcium antagonists. *Am J Hypertens* **9:** 633–643.

Zusman, R. M. (1987). Effects of converting-enzyme inhibitors on the renin-angiotensin-aldosterone, bradykinin, and arachidonic acid-prostaglandin systems: Correlation of chemical structure and biological activity. *Am J Kid Dis* **10**(Suppl 1): 13–23.

Zusman, R. M., Christensen, D. M., Higgins, J., and Boucher, C. A. (1992). Effects of fosinopril on cardiac function in patients with hypertension. Radionuclide assessment of left ventricular systolic and diastolic performance. *Am J Hypertens* **5:** 219–223.

PART II

The Kidney as an Hormonal Target

Section IV. Antidiuretic Hormone

CHAPTER **16**

Vasopressin in the Kidney: Historical Aspects

LYNN E. SCHLANGER AND JEFF M. SANDS

Emory University, School of Medicine, Atlanta, GA 30322, USA

Contents

I.	Introduction	205
II.	Hypothalamus	205
III.	Vasopressin receptors	206
IV.	Vasopressin regulated urea transport	210
V.	Aquaporins	212
VI.	Nephrogenic diabetes insipidus	215
VII.	Vaptans	216
	References	217

I. INTRODUCTION

A century ago, it was well established that the infusion of pituitary extract could cause antidiuresis and a concentrated urine. The structure of this neurohypophyseal hormone, mammalian vasopressin (AVP) or antidiuretic hormone (ADH) was discovered in 1954 by Acher and Chavet. Around the same time, a new syndrome was becoming apparent, the syndrome of inappropriate antidiuretic hormone or Schwartz-Bartter syndrome. In 1957, two patients were diagnosed with bronchogenic cancer and subsequently developed hyponatremia (Schwartz et al., 1957). These patients had elevated urine osmolality in the presence of a low serum osmolality, elevated excretion of sodium and weight gain. The authors concluded that the probable cause for the decline in serum sodium and osmolality was a sustained inappropriate release of antidiuretic hormone. The presence of both a low serum sodium and a hypertonic urine with normal renal function constituted *prima facie* evidence of the presence of antidiuretic hormone. Similar clinical presentations and findings of hyponatremia were reported in patients with diverse diseases: cerebral diseases (Carter et al., 1961), tuberculosis (Sims et al., 1950), idiopathic (Grumer et al., 1962) and bronchogenic carcinoma (Kaye, 1966).

Water homeostasis is maintained by the interaction between the distal tubule cells of the kidney and the hypothalamic hormone, AVP. The proximal tubule accounts for the reabsorption of a large percentage of the isotonic glomerular filtrate. The descending and ascending limbs of the loops of Henle are responsible for the generation and maintenance of the hypertonicity of the medullary interstitium. The collecting tubules are responsible for water reabsorption resulting in concentrated urine in the presence of vasopressin and a hypertonic interstitium. Arginine vasopressin, the mammalian antidiuretic hormone, is the sole hormone regulating electrolyte-free water reabsorption in the water permeable collecting tubules.

Over the past 50 years, there has been a vast amount of information inferred and discovered about vasopressin, vasopressin receptors, aquaporins, urea transporters and the treatment of hyponatremia and nephrogenic diabetes insipidus. We review some of this information in this chapter.

II. HYPOTHALAMUS

In 1947, Verney proposed that osmotic pressure resulted in secretion of vasopressin from the hypothalamus (Verney, 1947). A decade later, the osmoreceptors were identified in the anterior hypothalamus adjacent to the third ventricle (Jewell and Verney, 1957). However, the exact osmotic stimulus for vasopressin release, hypertonicity versus sodium salt, was uncertain. To answer this question, rhesus monkeys were studied to determine the effect of various hypertonic solutions infused into the anterior hypothalamus, third ventricle and carotids (Swaminathan, 1980). It was concluded that osmoreceptors were 'salt sensitive' with a greater release of ADH with exposure to NaCl than Na acetate, sucrose or mannitol. Besides this osmotic stimulus, hypovolemia was also known to cause a release of vasopressin through baroreceptors. In 1973, a radioimmunoassay was developed that allowed vasopressin levels in the serum to be measured accurately (Dunn et al., 1973). The contribution of the osmotic and volume effect on release of vasopressin from the hypothalamus was evaluated in Sprague-Dawley rats. A linear relationship between serum osmolality and vasopressin secretion was noted, while an exponential increase of

Textbook of Nephro-Endocrinology.
ISBN: 978-0-12-373870-7

vasopressin secretion was found with a decrease in blood volume of more than 7%. It appeared that the baroreceptors had a basal inhibitory effect on vasopressin and oxytocin secretion in addition to their role in hypovolemic stimulation (Morris and Alexander, 1989).

Arginine 8 vasopressin, whose gene is located on chromosome 20 (Simpson, 1988), is a mammalian cyclic nonapeptide hormone and is synthesized in the magnocelluar neurons of supraoptic and paraventricular nuclei of the hypothalamus in a precursor form composed of vasopressin, neurophysin II and glycopeptides known as prepro-AVP-NPII (Land et al., 1982; Schmale and Ritcher, 1984). The precursor form is packaged into vesicles and is transported down the axon to the posterior lobe of the pituitary gland where it is stored (Brownstein et al., 1980). During transport, the vesicles are converted to the active protein and await release into the circulation. The release of vasopressin is either the result of the direct activation of the osmoreceptors located in the third ventricle near the supraoptic and periventricular nuclei or indirectly by the venous baroreceptors or arterial stretch receptors that activate vagal and hypoglossophyseal nerves to signal the medulla and then the hypothalamus (Rai et al., 2006).

III. VASOPRESSIN RECEPTORS

Initially, vasopressin was named for its pressor effect but, as more information surfaced and its major role in water balance emerged, its name has been interchanged with antidiuretic hormone. Vasopressin receptors have diverse physiological actions on liver, smooth muscle, myocardium, platelets, brain and kidney (Ostrowski et al., 1993; Terada et al., 1993; Thibonnier et al., 1993; Goldsmith, 2005). Although it has been known since the 1960s that vasopressin activates adenylyl cyclase and increases cAMP in the rat inner medulla resulting in concentrated urine (Chase and Aurbach, 1968), the diversity of vasopressin receptors has only recently been appreciated. In 1979, it was proposed that there existed two types of vasopressin receptors, a V1 and V2 receptor, based on differences in signaling pathways and pharmacological studies with synthetic peptide agonists and antagonists (Michell et al., 1979; Gullion et al., 1980). This was confirmed by the discovery of different vasopressin receptors in the rat renal medulla and liver with similar molecular weights, 83 000 and 80 000, respectively, but possessing different functions (Gullion et al., 1980). Subsequently, the V1 receptors were further subdivided into V1a and V1b or V3 by different pharmacological profiles with similar signaling pathways (Antonio et al., 1984). For example, the V1b receptor is located primarily in the anterior pituitary gland and is associated with ACTH secretion.

Although the V1a and V1b receptors both stimulate the heterotrimeric G-protein subtype Gq resulting in activation of phospholipase C with increasing inositol phosphate production and intracellular calcium (Michell et al., 1979; Chabardes et al., 1980; Thibonnier et al., 1994; Briley et al., 1994), the V1a receptor also stimulates phospholipase A_2 and D (Briley et al., 1994). In contrast, the V2 receptor stimulates the heterotrimeric G-protein subtype Gs which activates adenylyl cyclase and, in turn, increases cAMP followed by activation of PKA. Thus, the V2 receptor plays a major role in water transport in the collecting duct and is responsible for concentrating the urine via aquaporin-2 (AQP2) water channels (Chase and Aurbach, 1968; Birnbaumer et al., 1992; Lolait et al., 1992).

With the use of *in situ* hybridization, the expression of rat mRNA for the V1a receptor and the V2 receptor was found at different embryologic and adult stages in various tissues (Ostrowski et al., 1993). The V1a receptor mRNA was found in hepatocytes surrounding the central veins at the time of birth, in the cortex of the kidney on the 16th day of gestation and localized in medulla, mesangial cells and vascular elements 1 day postpartum. In adult rats, the V1a receptor mRNA was found in the medullary vessels, a short segment of cortical distal tubules and the pelvis. The V2 receptor mRNA in rats was expressed on days 16 through 19 of gestation in the medullary and cortical collecting ducts and at birth in cells known to participate in water homeostasis in adult rats.

In addition, the V1a receptors have been localized by immunohistochemical studies and radioligand techniques in a variety of tissues and demonstrate different actions. The V1a receptors are found in vascular smooth muscle, vasa recta, mesangial cells, medullary thick ascending limbs (MTAL), adrenal cortex, brain, myocardium and platelets and mediate vasoconstriction, glycogen metabolism and platelet aggregation (Lolait et al., 1992; Ostrowski et al., 1993; Terada et al., 1993; Thibonnier et al., 1993; Goldsmith, 2005). Using a rat liver cDNA library, molecular cloning and expression of rat V1a receptors was performed in 1992 by Morel and colleagues (Morel et al., 1992). They noted that the rat V1a receptor cDNA encodes 394 amino acids with the translated protein having seven transmembrane spanning domains corresponding to the hydrophobic clusters connected by three intracellular and three extracellular loops with structure homologous to the family of G-protein-coupled receptors. In 1994, the human V1a vasopressin receptor was cloned (Thibonnier et al., 1994). The human V1a receptor encodes 418 amino acids with the putative seven transmembrane domains found in G-protein-coupled receptors. The homology of the human V1a receptor to rat liver V1a receptor, human and rat V2 receptors and human oxytocin receptor was 72, 36, 37 and 45%, respectively (Thibonnier et al., 1994).

In 1992, rat and human V2 receptors were cloned and characterized (Lolait et al., 1992; Birnbaumer et al., 1992). Two putative transmembrane domains derived from cDNA templates of the V1a receptors, were used for polymerase chain reaction (PCR) to identify the V2 receptors. Using the oligonucleotide complementary to one of the PCR clones, a rat kidney cDNA library was screened and a positive clone isolated (Lolait et al., 1992). Analysis of this translated protein revealed seven putative transmembrane domains that were characteristic of G-protein-coupled receptors. Northern blot analysis identified a 2.2 kb mRNA solely from the kidney. *In situ* hybridization showed high levels of mRNA in the collecting tubules, medullary thick ascending limbs and peri-glomerular tubules. Radioligand studies were characteristically consistent with the V2 receptor and not V1a or oxytocin receptors. Specifically, there was an increase in production of cAMP after exposure to AVP or dDAVP in transfected mouse L cells that was blocked by a V2 receptor antagonist (Lolait et al., 1992). Similar molecular structures and pharmacological analysis were found in human V2 receptors (Birnbaumer et al., 1992).

V1b receptors, also called V3, are located in the anterior pituitary gland and are important in ACTH secretion (de Keyzer et al., 1994). Two years after cloning V1a and V2 receptors, the human V1b receptor was cloned and characterized (de Keyzer et al., 1994). A V1a cDNA probe was used to screen a corticotrophin pituitary adenoma library and a 424 amino acid protein was found which was consistent with a G-protein-coupled receptor. Human V1b has a relatively high degree of amino acid homology with human V1a (45.5%), human oxytocin (44.8%), human V2 (37.3%), rat V1a (46.2%) and rat V2 (44.3%) (de Keyzer et al., 1994). Pharmacological studies revealed that this V1b receptor has distinct binding properties and is a new subtype different from V2 or V1a. The signaling pathway, determined using transfected COS cells, is coupled to the PLC pathway with a fivefold increase in inositol production after addition of 100 nM AVP; no cAMP production was found (de Keyzer et al., 1994). The V1b receptors were localized by reverse transcription polymerase chain reaction (RT-PCR) analysis to the pituitary and a faint band in the kidney while the V1a receptor was expressed in liver, adrenals, myometrium, brain, spleen, muscle, lung, foreskin fibroblast and kidney. A recent study suggested that there was a vasopressin V2-like receptor in the inner medullary collecting ducts which may be a V1b receptor rather than a vasopressin V2 receptor at high density (Saito et al., 2000).

A. Localization of Vasopressin Receptors in the Kidney

The presence of basolateral plasma membrane V2 receptors in the kidney and their role in water reabsorption in the principal cells of the collecting duct are well established. In 1993, the RT-PCR technique was a new method of measuring the relative levels of mRNA and was used to measure the V2 and V1a receptor mRNAs in rat microdissected tubules (Terada et al., 1993). The V2 receptor mRNA was found in the collecting tubules, medullary thick ascending limbs and peri-glomerular tubules. In the microdissected tubules, V1a receptor RT-PCR showed a single band of 598 bp in glomerulus, initial collecting duct, cortical collecting duct, outer medullary collecting duct (OMCD), inner medullary collecting duct and arcuate artery. There was a faint signal in the proximal convoluted and straight tubules, inner medullary thin limb and medullary thick ascending limb. The greatest signal was from the glomerulus. The V2 receptor RT-PCR showed a single band of 625 bp in the inner medullary collecting duct, outer medullary collecting duct, cortical collecting duct, initial collecting duct, medullary thick ascending limb and the inner medullary thin limb (Terada et al., 1993). Large signals from the cortical collecting duct, outer medullary collecting duct and inner medullary collecting duct correlated with the location and functionality studies demonstrated by others (Lolait et al., 1992; Birnbaumer et al., 1992; Ostrowski et al., 1993). Interestingly, there was no detection of a V2 receptor in the glomerulus, arcuate artery and proximal convoluted or straight tubules. In a dehydrated state following a 72-hour water restriction, V2 receptor expression decreased in the cortical collecting ducts and inner medullary collecting ducts while no change in V1a receptor expression occurred (Terada et al., 1993). This is consistent with downregulation of expression of V2 receptors in a chronic antidiuretic state in the presence of elevated vasopressin levels (Tian et al., 2000).

Hermossila and Schulein (2001) transfected the MDCK cell line with truncated forms of the cytoplasmic domains of the V2 receptor to determine which domain is responsible for sorting the receptor to the basolateral membrane. The second cytoplasmic loop appears to promote basolateral membrane expression and contains this sorting signal. The C-terminus is involved in the apical transport of the receptor and the third cytoplasmic loop leads to the retention of the receptor in the endoplasmic reticulum.

The localization and regulation of the V2 receptor by a V2 receptor agonist was studied in stably transfected polarized Madin-Darby canine kidney cells (Robben et al., 2004). The cells were stably transfected with a construct encoding human V2 receptor, C-terminally tagged with GFP and confirmed by immunoblot. The immunocytochemistry study showed that the V2 receptor mostly co-localized with the basolateral membrane marker E-cadherin with a small percent (11%) in the apical membrane (Robben et al., 2004). Once the agonist dDAVP and V2 receptor-GFP became bound, the structure was internalized and transported transiently to early lysosomes and then transported to the late endosomes where the acidic environment led to dissociation of the agonist and the receptor. The V2 receptor was shown to be degraded after internalization via the endosome/lysosomal pathway rather than recycled to the basolateral membrane (Robben et al., 2004).

B. Action of Vasopressin in the Toad Bladder, Collecting Tubules and Inner Medulla

The major action of vasopressin in the kidney of all mammals is to increase water permeability in the collecting tubules resulting in a concentrated urine. Toad bladder and Brattleboro rats have been studied to evaluate the effect of vasopressin on water permeability in epithelial cells. The homozygous Brattleboro rats, derived from parental Long-Evan rats, have central diabetes insipidus, produce large amount of urine and are unable to concentrate urine. They contain no functional vasopressin secondary to a single nucleotide mutation in the C-terminal region of the vasopressin precursor (Schmale and Ritcher, 1984). This results in a block in mRNA translation that does not allow production of the protein (Ivell et al., 1986).

Using horseradish perioxidase as a marker, Wade and colleagues (Wade and DiScala, 1971) studied the effect of an osmotic gradient and vasopressin on the movement of water through the urinary bladder of *Bufo marinus*. The movement of water into the intercellular space required the presence of an osmotic gradient and vasopressin. The exact path of water movement was not discerned. Using micropuncture, Woodhall and Tisher (1973) determined the histological and functional heterogeneity of the distal convoluted, connecting tubules and late distal tubules via micropuncture of Brattleboro rat. Morphological changes were found in the rats exposed to vasopressin under both light and electron microscopy. There was marked dilatation of lateral and basilar intercellular spaces of the entire late distal tubule, consistent with vasopressin-sensitive morphological changes as water entered the cells (Woodhall and Tisher, 1973). This was the first *in vivo* documentation of vasopressin 'responsiveness' by this segment of renal tubules.

The technique of freeze fracture electron microscopy, which results in membrane fracture within its hydrophobic interior, allows evaluation of membrane structure. Two fractured surfaces are produced: an inner and outer membrane face. In the urinary bladder of *Bufo marinus*, freeze fracture revealed that there was no aggregation of intramembranous particles in the absence of vasopressin but there was aggregation of particles with exposure to vasopressin (Kachadorian et al., 1977; Brown, 1989). A linear relationship resulted between the amount of aggregates and osmotic water flow following exposure to vasopressin, suggesting that vasopressin altered the distribution of intramembranous particles and induced changes in the water permeability of the membrane (Kachadorian et al., 1977). These aggrephores were thought to contain channels for water transport across the bilayer membrane. Biophysical analysis revealed that water transport across the toad bladder or tubules relied upon channels within the membrane rather than simple diffusion through bilayer membranes (Macey, 1984).

In the late 1970s and early 1980s, studies of isolated perfused collecting ducts revealed accumulation of intramembranous particle clusters induced by vasopressin. The aggrephores, containing vesicles, fused to plasma membranes by exocytosis and detached by endocytosis. The aggrephores were noted to contain vesicles coated with clathrin and were thought to be important for vesicle trafficking transport (Brown and Orci, 1983; Brown, 1989). This was consistent with the 'shuttle hypothesis' proposed by Wade in the early 1980s to describe the shuttling of intracellular vesicles to the apical membrane and detached from the membrane in the presence of vasopressin (Wade et al., 1981).

By the 1990s, it was established that, in vasopressin-sensitive epithelial cells, water traverses through the cell membrane via water channels. The first water channel was identified, CHIP28 (see section on aquaporins). Support of the 'shuttle' hypothesis was shown by Nielsen, in 1995, who perfused the rat inner medullary collecting ducts with 100 pM AVP (added to the bath) and noted an increase in osmotic water permeability (Pf) by fivefold and a decrease in Pf after wash out of AVP (Nielsen et al., 1995).

Immunocytochemistry confirmed exocytic insertion of intracellular vesicles containing AQP2 in the apical membrane and endocytosis following removal of vasopressin. This proved that vasopressin increases water permeability of collecting duct cells by inducing a reversible translocation of the aquaporin 2 (AQP2) water channel from intracellular vesicles to apical membrane by exocytosis (Nielsen et al., 1995).

After identification of the water channels, a study was performed on the urinary bladder of anurean *Xenopus laevis*. Because it is less responsive to vasopressin than other amphibian bladder, the effect of antiduiretic hormone on aggrephores and apical aggregates can be more easily discerned (Calamita et al., 1994). Moreover, the *Xenopus laevis* are aquatic and reabsorb water through the skin and excrete water via the bladder in minimally concentrated urine. They also exhibit a poor response to vasopressin (Calamita et al., 1994). Using the volumetric technique, net water flow in response to changes of hydrostatic or osmotic pressure was determined and only occurred with an osmotic change across the membrane (Calamita et al., 1994). This effect was enhanced by the presence of AVP (100 mU/ml). Morphological studies using freeze-fracture electron microscopy revealed aggrephores localized to the subapical membrane. In the presence of AVP, the intramembranous aggregates are inserted into the apical membrane in the granular cells. This process was reversed after a 60-minute washout period of the AVP and correlated with the finding on light and electron microscopy of an increase in fluid accumulation in the intercellular space and suggested that the apical borders are more permeable to water. This provided further support to the 'shuttle' theory proposed by Wade et al. (1981).

Hayashi and colleagues (1994) found an increase in AQP2 water channels in the apical membrane and subapical compartment in response to an increase of endogenous vasopressin in rat inner medulla. An infusion of a V2 receptor antagonist (OPC-31260) decreased the apical and the subapical AQP2 immunostaining, while a V1a receptor antagonist (OPC-21268) had no effect on AQP2 water channel distribution. Therefore, the study demonstrated that the location and distribution of AQP2 water channels is regulated by the V2 receptor. On the other hand, the numbers of apical AQP2 water channels increased under conditions of dehydration in response to a rise in endogenous vasopressin. This was confirmed by immunocytochemistry and immunoblot. There was no decrease in the amount of subapical vesicles, suggesting the change in distribution was not from relocation of these vesicles (Hayashi et al., 1994).

Vasopressin increases cAMP, however, it also stimulates phospholipase C and increased cytosolic Ca^{2+} in rat and rabbit collecting ducts and in LLC-PK1 cells (Burnatowska-Heiden and Spielman, 1987; Teitelbaum, 1991; Ando et al., 1991). The exact receptor responsible for this signaling pathway was uncertain with some studies suggesting V1 receptors while others suggesting the V2 receptors and oxytocin receptors (Burnatowska-Heiden and Spielman, 1989; Ando et al., 1991; Ando and Asano, 1993). In rat inner medullary collecting duct (IMCD), the V2 receptor agonist (dDAVP) and a selective oxytocin receptor (Thr^4, $Gly7$) agonist increased the IP_3 production, but a V1 receptor agonist (Ho1,Phe2,Orn8)VT had no effect on IP3 production (Teitelbaum, 1991). The AVP agonist [VDAVP] 10^{-7} M–10^{-13} M had a dose-dependent response on IP_3 production that was abolished with the addition of a V2 antagonist [d (CH2)5Tyr (Me) AVP 10^{-7} M]. Similar results were obtained with an oxytocin agonist and antagonist and no effect was seen by a V1 agonist or antagonist. This demonstrated that production of IP_3 by AVP or V2 agonist receptors may be through the oxytocin receptor (Teitelbaum, 1991).

The presence of apical vasopressin receptors in collecting tubules was discovered in addition to the well known basolateral V2 receptors in MTAL, cortical collecting duct (CCD) and IMCD (Ando et al., 1991; Nonoguchi et al., 1995). In 1991, Ando and colleagues revealed that the application of luminal AVP at nanomolar concentration in the presence of basolateral picomolar concentration of AVP in rabbit CCD cells decreased hydraulic conductivity by 35%. This effect was speculated to be secondary to stimulation of the V1a receptor since both V2 and V1a receptors have been localized to the kidney. The physiological concentration of AVP at the picomolar concentrations activates basolateral V2 receptors. Since nanomolar concentrations of AVP have been found in the urine and may bind to these luminal AVP receptors, this may result in self-inhibition of antidiuretic AVP in the CCD (Fressinaud et al., 1974; Pruszczynski et al., 1984).

Similarly, immunohistochemical studies of microdissected rat terminal IMCD stained both the basolateral and apical plasma membrane for V2 receptor and the apical plasma membrane to a lesser degree (Nonoguchi et al., 1995). The functional role that the apical plasma membrane played was determined by microperfusing the terminal IMCD and measuring osmotic water permeability (Pf), urea permeability (Pu) and cAMP production. With application of 10^{-9} M AVP to the luminal and basolateral baths, two different AVP receptors were discovered with differing pharmacologic properties and intracellular signaling properties. Luminal 10^{-9} M AVP showed a small increase in Pf, cAMP and Pu consistent with a V2 receptor. While luminal exposure to luminal 10^{-9} M AVP followed by basolateral 10^{-9} M AVP revealed a decreases in Pf (43.7%), Pu (19%) and cAMP. Moreover, V1a antagonists (OPC 21268 and PTME-AVP) and oxytocin antagonists applied to the lumen in the presence of luminal AVP abolished the inhibitory effect; luminal V2 antagonist showed no effect. These findings suggested that there is a luminal V1a or oxytocin receptor in the apical membrane of the terminal IMCD which possibly plays a negative feedback role on the stimulatory effect of basolateral vasopressin in maintaining water balance during chronic stimulation by vasopressin (Nonoguchi et al., 1995). The pathophysiological and clinical importance of the various locations of the V1, oxytocin and V2 receptors requires further investigation in their role in water homeostasis.

C. Vasopressin Effect in the Other Renal Tubules

The countercurrent multiplier is required for vasopressin to concentrate the urine and was proposed in the 1950s by Hargitay and Kuhn. They hypothesized that tubular urine is first osmotically concentrated in the descending limb of the loop of Henle and then diluted in the ascending limb of the loop of Henle before it is finally concentrated in the collecting tubules (Wirz et al., 1951). In the hamster kidney, micropuncture measurements of the fluid in the bend of the loop of Henle and collecting tubule at the same level showed similar osmolality while fluid in the more proximal collecting ducts was more dilute (Gottschalk and Mylle, 1958). Micropuncture study of the composition of the fluid at the hairpin loop of Henle in desert rodents and hamster confirmed this hypothesis (Lassiter et al., 1961; Gottschalk et al., 1962). These findings supported the countercurrent hypothesis with different concentrations of osmoles being measured in the loop of Henle versus the collecting tubules. Sodium and urea were the major active osmoles in the loop of Henle while urea was the major active osmole in the collecting tubules (see section on urea transport).

There have been numerous publications on the regulation of thick ascending limb transport by vasopressin (see Knepper et al., 1999). Vasopressin, V1a and V2, receptors have been localized to different segments of the tubules including the thin and thick ascending limb, collecting tubules, and proximal and distal convoluting tubules. In addition to the regulation of AQP2 water channels, V2 receptors regulate sodium reabsorption by the Na-K-2Cl co-transporters and the NaCl thiazide-sensitive transporter (Kim et al., 1999; Bertuccio et al., 2002). Microdissected rat tubules of the MTAL, CAL, OMCD, CCD, and IMCD were exposed to a physiological vasopressin dose, 10^{-12} M and a hyperosmolar environment to determine effect of vasopressin on the subunits of the epithelial sodium channel (ENAC) (Machida et al., 2003). The distribution of the three subunits of ENAC mRNA, α, β and γ, by RT-PCR showed expression of all three subunits in the medulla and expression only of the α subunit in the MTAL. There was an increase in expression of the α subunit of ENAC in the MTAL but none in the collecting duct with exposure to vasopressin and hyperosmolar environment. The significance of these findings needs to be further investigated.

In medullary thick ascending limb, there is evidence that vasopressin stimulates Na-K ATPase and the NKCC2 co-transporter (Kim et al., 1999; Bertuccio et al., 2002). The direct effect of endogenous vasopressin on the NKCC2 co-transporter and on Na-K ATPase in rats with normal renal function and with induced renal failure was studied (Bertuccio et al., 2002). The rats received an infusion of the V2 antagonist OPC-31260 for 3 days before decapitation. With the infusion, there was an increase in urine output and a decrease in urine osmolality in both treated groups. Western blot analysis of the outer medulla showed a decrease in the Na-K ATPase α subunit and the NKCC2 co-transporter in both treated groups as compared to their controls. Both the co-transporter and ATPase appear to be regulated by endogenous vasopressin. Similar findings were found in Sprague-Dawley and Brattleboro rats with an elevated vasopressin level induced by either dehydration in the former or dDVAP infusion in the latter for 7 days (Kim et al., 1999). Therefore, vasopressin appears to have a dual role in maintaining the concentration gradient in the medullary interstitium while allowing its antidiuretic effect downstream in the collecting ducts.

IV. VASOPRESSIN REGULATED UREA TRANSPORT

Kidney (and red blood cell) urea transporters play a pivotal role in the conservation of body water and the urine concentrating mechanism. In 1934, Gamble and colleagues wrote that it is 'an interesting instance of biological substances that the largest "waste product" in urine incidentally performs an important service to the organism' (Gamble et al., 1934; Fenton et al., 2006a). Several studies show that maximal urine concentrating ability is decreased in protein-deprived humans and animals and is restored by urea infusion (Epstein et al., 1957; Hendrikx and Epstein, 1958; Crawford et al., 1959; Levinsky and Berliner, 1959; Klahr and Alleyne, 1973; Pennell et al., 1975; Peil et al., 1990).

Recently, mice in which one of three different urea transporters was genetically knocked-out were each shown to have urine concentrating defects (Yang and Verkman, 2002; Yang et al., 2002; Fenton et al., 2004, 2005, 2006b; Klein et al., 2004; Uchida et al., 2005). Thus, any hypothesis regarding the mechanism by which the kidney concentrates urine needs to include some effect derived from urea.

In 1972, Kokko and Rector, and Stephenson, each proposed the passive equilibration model for urine concentration in the inner medulla. The passive mechanism hypothesis depends upon the delivery of large quantities of urea to the deepest portion of the inner medulla. The terminal inner medullary collecting duct is the only medullary collecting duct subsegment in which vasopressin stimulates urea reabsorption (Sands and Knepper, 1987; Sands et al., 1987), thereby limiting urea reabsorption to the deepest portion of the inner medulla. In addition, the asymmetry of red blood cell urea transport minimizes the loss of urea from the inner medulla by increasing the efficiency of urea's countercurrent exchange in vasa rectae (Macey and Yousef, 1988). Thus, the ability to produce concentrated urine and conserve body water depends upon urea transporters in both the kidney medulla and red blood cells.

The reader may wonder why the kidney or red blood cells need urea transporters when most textbooks state that urea is freely permeable across cell membranes and is not osmotically active. Urea ($OC(NH_2)_2$) is a small (mol. wt 60 Da) polar molecule and its permeability in artificial lipid bilayer membranes is quite low (Galluci et al., 1971). However, urea will diffuse across cell membranes and achieve equilibrium given enough time. In the kidney, the transit time for tubule fluid through inner medullary collecting ducts or red blood cells through the vasa rectae is too fast to allow urea concentrations to reach equilibrium by passive diffusion. In addition, generation of a concentrated urine, which is dependent upon concentrating urea within the inner medullary interstitium and maintenance of the urea gradient, suggests that a regulated urea transport process by vasopressin must be present (Kokko and Rector, 1972; Stephenson, 1972).

A. Red Blood Cells

Urea transport across red blood cell membranes has been studied since the 1930s, when the lysis rate of red blood cells in iso-osmotic solutions of different solutes was used as a rough measure to show that red blood cells have a high urea permeability (reviewed in Sands et al., 1997). In the 1950s and 1960s, the prevailing hypothesis was that urea permeated red blood cells through putative water channels (aquaporins had not yet been discovered). This hypothesis was discredited by studies showing that water and urea transport could be separated by inhibitors such as pCMBS (p-chloromercuribenzene sulfonate) and phloretin, which greatly inhibited urea flux but had no effect on water flux (Macey and Farmer,

1970; Wieth et al., 1974; Mayrand and Levitt, 1983); other inhibitors affected water flux but not urea flux. Thus, the stage was set for the future molecular identification of urea transporters and aquaporins as separate proteins with separate functions.

B. Inner Medullary Collecting Duct

Vasopressin was first shown to increase urea permeability in the collecting duct by Jaenike in 1961 using a combination of tissue slice and clearance studies in dogs (Jaenike, 1961). Morgan and colleagues showed in 1968 that vasopressin increased urea permeability in rat inner medullary collecting ducts using the isolated papilla technique (Morgan and Berliner, 1968; Morgan et al., 1968; Morgan, 1970). However, direct evidence that vasopressin could increase urea transport was not obtained until the 1980s, when three groups used tubular perfusion of rat terminal inner medullary collecting ducts to show that vasopressin stimulates urea permeability (Rocha and Kudo, 1982; Kondo and Imai, 1987; Sands et al., 1987).

These early studies did not examine the mechanism by which urea was transported across inner medullary collecting ducts. Sands and colleagues proposed in 1987 that a specific facilitated urea transport process was present in terminal inner medullary collecting ducts. The terminal inner medullary collecting duct is unique among medullary collecting duct subsegments in that it has a high basal urea permeability which is further stimulated by vasopressin (Knepper, 1983; Sands and Knepper, 1987; Sands et al., 1987). Even in the absence of vasopressin, the urea permeability measured in terminal inner medullary collecting ducts is 85 times greater than the calculated maximal urea permeability which could be achieved by simple lipid-phase diffusion and paracellular transport (Sands et al., 1987). This high urea permeability, its stimulation by vasopressin and the earlier studies in red blood cells showing that urea flux and water flux could be independently inhibited, raised the possibility that terminal inner medullary collecting ducts expressed a specific urea transporter protein (Sands et al., 1987).

Solvent drag was another mechanism that had been proposed to explain the high urea permeability across terminal inner medullary collecting ducts. However, Knepper and colleagues demonstrated that there is no solvent drag of urea in perfused terminal inner medullary collecting ducts (Knepper et al., 1989; Chou et al., 1990a, b). Thus, these functional studies in the late 1980s led to the conclusion that a facilitated urea transporter protein mediated the high urea permeability present in terminal inner medullary collecting ducts. They also provided a functional definition for a facilitated urea transporter, which was used by Hediger and colleagues to expression clone the first urea transporter, now named UT-A2, in 1993 from the rabbit inner medulla (You et al., 1993).

C. Urea Transport Proteins

Currently, six UT-A protein isoforms and the UT-A gene (*Slc14a2*) have been cloned (reviewed in Sands, 2003, 2004; Bagnasco, 2006; Smith and Fenton, 2006). UT-A1 is the best studied and largest UT-A protein. It is expressed in the apical membrane of inner medullary collecting ducts and is stimulated by vasopressin (Nielsen et al., 1996; Shayakul et al., 1996). By western blot, UT-A1 bands are detected at both 117 and 97 kDa; both bands represent glycosylated versions of a non-glycosylated 88 kDa UT-A1 protein (Bradford et al., 2001). UT-A2 is expressed in thin descending limbs but is not stimulated by vasopressin (Wade et al., 2000). UT-A3 is expressed in inner medullary collecting ducts, along with UT-A1, and is stimulated by vasopressin (Karakashian et al., 1999; Terris et al., 2001; Stewart et al., 2004; Lim et al., 2006). UT-A4 is expressed in the kidney medulla but its exact tubular location is unknown; it is stimulated by vasopressin (Karakashian et al., 1999). The other two UT-A urea transporters are not expressed in kidney: UT-A5 is in testis (Fenton et al., 2000) and UT-A6 is in colon (Smith et al., 2004) and the effect of vasopressin on these urea transporters has not been tested.

The red blood cell facilitated urea transporter, named UT-B, derives from a second gene, *Scl14a1*. It was initially cloned by Olives and colleagues in 1994 from a human erythropoietic cell line (Olives et al., 1994). In people, the UT-B protein is the Kidd antigen and several mutations of the UT-B/Kidd antigen (*Slc14a1*) gene have been reported (reviewed in Bagnasco, 2006; Sands and Bichet, 2006). Individuals who lack Kidd antigen also lack facilitated urea transport in their red blood cells (Fröhlich et al., 1991; Sands et al., 1992). These same individuals are unable to concentrate their urine above 800 mOsm/kg H_2O following overnight water deprivation (Sands et al., 1992). Recently, a UT-B knock-out mouse was shown to have a similar phenotype to people who lack Kidd antigen. UT-B is not regulated by vasopressin.

D. Acute Regulation of Urea Transport by Vasopressin

The primary method for investigating the rapid regulation of urea transport has been perfusion of rat inner medullary collecting ducts. This method provides physiologically relevant, functional data, although it cannot determine which urea transporter isoform is responsible for a specific functional effect since both UT-A1 and UT-A3 are expressed in this collecting duct subsegment. Vasopressin stimulates urea transport by increasing the number of functional urea transporters (V_{max}) without changing the affinity (K_m) to urea (Chou and Knepper, 1989). Functional studies show that phloretin-inhibitable urea transport is present in both the apical and basolateral membranes of rat terminal inner medullary collecting ducts, with the apical membrane being the rate-limiting barrier for vasopressin-stimulated urea transport (Star, 1990).

Vasopressin rapidly (within minutes) increases urea transport by binding to V2-receptors, stimulating adenylyl cyclase and increasing cAMP production (Sands et al., 1987; Star et al., 1988; Sands and Schrader, 1991). One mechanism for rapid regulation is that vasopressin increases UT-A1 phosphorylation (Zhang et al., 2002). Vasopressin increases UT-A1 phosphorylation within 2 min in rat inner medullary collecting duct suspensions, consistent with the time-course for vasopressin-stimulated urea transport in perfused rat terminal IMCDs (Star et al., 1988; Wall et al., 1992; Zhang et al., 2002). A selective V2-receptor agonist, dDAVP (desmopressin), also increases UT-A1 phosphorylation (Zhang et al., 2002). Recently, vasopressin was shown also to increase urea transport by increasing UT-A1 accumulation in the apical plasma membrane (Klein et al., 2006). Thus, vasopressin rapidly increases urea transport in inner medullary collecting ducts through cAMP-dependent increases in UT-A1 phosphorylation and UT-A1 accumulation in the plasma membrane.

In addition to vasopressin's direct effects on inner medullary collecting duct urea transport, vasopressin also increases inner medullary osmolality. In perfused terminal inner medullary collecting ducts, increasing osmolality by adding NaCl or mannitol, but not urea, increases urea permeability independently of vasopressin and also has an additive stimulatory effect to vasopressin to increase urea permeability (Chou et al., 1990a, b; Sands and Schrader, 1991; Gillin and Sands, 1992; Kudo et al., 1992).

Lithium is widely used to treat patients suffering from bipolar (manic-depressive) disorders, but can cause nephrogenic diabetes insipidus and an inability to concentrate urine (reviewed in Igarashi et al., 1996). While the mechanisms are not entirely understood, lithium-treated rats have a marked reduction in aquaporin-2 (Marples et al., 1995; Klein et al., 2002) and UT-A1 (Klein et al., 2002) protein abundances. In addition, vasopressin does not increase UT-A1 phosphorylation in inner medullary collecting ducts from lithium-fed rats (Klein et al., 2002), suggesting that lithium is interfering with the vasopressin signaling pathway.

Angiotensin II causes a further increase in vasopressin-stimulated urea permeability in rat terminal inner medullary collecting ducts, but has no effect in the absence of vasopressin (Kato et al., 2000). Angiotensin II increases UT-A1 phosphorylation through a protein kinase C-mediated pathway (Zhang et al., 2002). By augmenting vasopressin-stimulated permeability, angiotensin II may play a physiologic role in the urinary concentrating mechanism.

E. Long-Term Regulation of Urea Transport by Vasopressin

Vasopressin increases UT-A1 protein abundance indirectly by increasing inner medullary osmolality. Inducing water diuresis in rats for 2 weeks (to suppress endogenous vasopressin levels) decreases UT-A1 protein abundance in the inner medulla (Kim et al., 2004). Similarly, administering vasopressin for 12 days to Brattleboro rats (which lack vasopressin and have central diabetes insipidus) increases UT-A1 protein abundance (Kim et al., 2004). However, administering vasopressin for only 5 days to Brattleboro rats has no effect on UT-A1 protein abundance (Terris et al., 1998; Kim et al., 2004). This delayed increase in UT-A1 abundance is consistent with the time-course for the increase in inner medullary urea content in Brattleboro rats receiving vasopressin (Harrington and Valtin, 1968).

A molecular explanation for the delayed increase may be that the rat UT-A promoter I does not contain a cyclic AMP response element (CRE) and promoter activity is not increased by cyclic AMP (Nakayama et al., 2000, 2001). However, UT-A promoter I does contain a tonicity enhancer (TonE) element and promoter activity is increased by hyperosmolality (Nakayama et al., 2000, 2001). This suggests that vasopressin initially increases Na-K-2Cl co-transporter NKCC2/BSC1 transcription through its CRE (Igarashi et al., 1996; Yasui et al., 1997), thereby increasing inner medullary osmolality through countercurrent multiplication which, in turn, increases UT-A1 transcription through its TonE element.

V. AQUAPORINS

It had been known for years that the distal nephron and, in particular, the collecting tubules, are responsible for electrolyte, water and acid–base balance. Water is reabsorbed along different segments of the nephron with some segments under hormonal control while others are not. Water reabsorption can be by diffusion, although a majority is by transversion of the lipid bilayer membrane through water channels (Wade et al., 1981; Brown and Orci, 1983). Biophysical studies of water reabsorption showed that water reabsorption in the amphibian bladder or collecting tubules is consistent with the presence of water channels. The operational definition of a water channel in biological membranes is based on studies of human RBC by Macey and colleagues (1984) and includes:

1. a low activation energy E_a = 4–6 kcal/mol
2. a high water permeability (Pf) of 0.02 cm/s at 23 °C
3. inhibited by mercurial diuretics
4. high osmotic to diffusion permeability ratio Pf/Pd of 3.5.

If the flow of water occurred by diffusion through a bilayer, one would expect:

1. a high activation energy around 10–15 Cal/mol
2. insensitivity to mercurial reagents
3. a low ratio of osmotic to diffusion of water permeability Pf/Pd.

But the contrary was found following vasopressin stimulation of toad bladder and collecting tubule cells; a low Ea and a Pf inhibited by mercurial reagents, suggested water channels were responsible for water transport (Wade and DiScala, 1977; Nielsen et al., 1993; Inoue et al., 1999).

Aquaporins, a family of membrane water channel proteins, are responsible for water transport across the plasma membrane in kidney and other tissues. There are now 13 cloned water channels identified with six localized in the kidney (see Nielsen et al., 2002; Agre, 2006). Aquaporin 1 and 7 are located in the proximal convoluted tubule and descending limb of Henle. AQP2 is located in the apical membrane and AQP3 and AQP4 in the basolateral membrane of collecting duct principal cells. AQP6 is also located in the collecting duct and is associated with H^+-ATPase but not located in the plasma membrane.

The first aquaporin discovered was CHIP28, channel-like integral protein of 28 kDa, and was isolated from erythrocyte membranes and renal tubule cells by Denker and Agre (Denker et al., 1988; Preston and Agre, 1991). The 28 kDa protein was fortuitously found while isolating 32 kDa Rh polypeptides from human erythrocytes and was presumed to be a degradation product (Denker et al., 1988). The immunoblot of erythrocyte membranes revealed two bands, one at 28 kDa and a broad band between 35–50 kDa (HMW–28 kDa) which was demonstrated to be the N-glycosylated form of the 28 kDa protein. Besides being isolated from the rat and human erythrocyte membranes, it is present in the proximal tubules and thin descending limbs of the loops of Henle in the kidney (Denker et al., 1988).

In 1991, Preston and Agre isolated the cDNA template for CHIP28 from human fetal liver cDNA by three-step PCR with the third-step product used as a probe to isolate CHIP28 (AQP1) from a human bone marrow cDNA library. Northern blot analyses revealed a similar single transcript of 3.1 kb in human bone marrow and kidney mRNA. Computer analysis of this amino acid sequence demonstrated six hydrophobic regions corresponding to bilayer spanning domains with cytoplasmic hydrophilic NH_2 and COOH termini. The proposed topology of CHIP28 protein revealed five loops, exofacial loops, A, C and E, and endofacial loops B and D (Preston and Agre, 1991). The two potential N-glycosylated sites are located on loops A and E, while the loops B and D are hydrophobic (Preston and Agre, 1991). The protein has two internal repeats and each contain the conserved amino acid sequence Asn-Pro-Ala (NPA) (Agre, 2006). The B and E loops are equal in distance and symmetry and were found by crystallography to fold on themselves to form an hourglass shape (Jung et al., 1994). The isolated cDNA of CHIP28 was found to have homology (42%) with MIP26 (Smith and Agre, 1991; Preston et al., 1991). The N-termini of the CHIP28 and MIP26 are 42% identical when aligned, but the C-termini are identical in only four of 35 residues. MIP26, a 26 kDa major intrinsic protein of bovine lens fiber cells, is a prototype of a family of membrane proteins present in various species. When reconstituted in an artificial plasma bilayer, the MIP26 was found to form a tetramer channel regulating conductance and function as a channel (Smith and Agre, 1991; Preston et al., 1991). The channel-like characteristic was supported by the CHIP28's electron diffraction pattern which revealed an annular shape with a diameter of 31–34 Å, and a rim 7–8 Å wide with a central pore (Mitra et al., 1995).

In 1992, Xenopus oocytes were transfected with CHIP28 (now named AQP1) mRNA in order to assess the functionality of this protein (Preston et al., 1992). Immunoblotting confirmed expression of the protein, detecting a 28 kDa band and broad band at 35–45 kDa, representing non-glycosylated and glycosylated forms, respectively. The osmotic water permeability activity was present and greatly increased in the injected cells compared to the non-injected cells. Moreover, the osmotic water permeability was inhibited by mercuric chloride, an inhibitor of water channels, thus confirming the role of CHIP28 as a water channel. The structure of these channels in various membranes, proteoliposomes with reconstituted purified CHIP28, kidney proximal tubules and stably transfected CHO cells, was evaluated by rotatory shadowing freeze fracture microscopy (Verbavatz et al., 1993) and demonstrated that the water channels formed tetramers within the plasma membrane.

In 1993, the human AQP-CHIP gene was localized to chromosome 7p14 by *in situ* hybridization (Moon et al., 1993). The structure of the AQP-CHIP gene contains four exons corresponding to amino acids 1–128, 129–183, 184–210 and 211–269 separated by three introns of 9.6, 0.43 and 0.80 kb. The human kidney and bone marrow CHIP28 cDNA amino acid exon sequences were found to be identical, consistent with a single gene (Moon et al., 1993).

Using a rat kidney cDNA library, Fushimi cloned the WCH-CD in 1993 (renamed AQP-CD and AQP2) (Fushimi et al., 1993) and, in 1994, the human AQP-CD gene was isolated (Sasaki et al., 1994; Uchida et al., 1994). Rat AQP2 had conserved residues with CHIP28 and MIP and internal tandem repeats to suggest it belongs to the MIP family. Hydropathy analysis of rat AQP2 translated protein suggested it contains six transmembrane domains, similar to AQP1, and contains a putative site for glycosylation and one putative site for phosphorylation (serine 256) for cAMP-dependent protein kinase (Fushimi et al., 1993). Northern blot analysis revealed that expression is primarily in the medulla while immunocytochemistry of principal cells localized AQP2 to the apical and subapical compartment in the collecting tubules. The human AQP2 water channel contains 271 amino acids with molecular weight of 28 968 and 91% identity sequence to rat AQP-CD and 48% to the human AQP1 (Sasaki et al., 1994). Both northern and western blots showed that AQP2 expression was greater in the medullary of the kidney than the cortex. The western blots of the renal medullary membranes used antibodies to the C-terminus and revealed two bands, a broad band at 40–50 kDa and single band at 29 kDa (Sasaki et al., 1994). Similar findings with immunohistochemical studies localized AQP2 mainly in the apical membrane of the collecting duct. Chromosomal mapping revealed the AQP2 gene locus at chromosome 12q13 (Sasaki et al., 1994). *Xenopus* oocytes injected with cRNA of human AQP2 revealed properties consistent with a water channel.

The human AQP2 gene spanned 5.1 kb and consisted of four exons similar to the AQP1 and MIP26 (Uchida et al., 1994). The 5' flanking region was characterized by a TATA box, three E boxes and a cyclic AMP-responsive element (CRE) (Uchida et al., 1994).

LLC-PK1 cells, transfected with cDNA for AQP1 and AQP2, were incubated with anti-AQP1 serum antibodies or anti-c-Myc monoclonal antibodies (Katsura et al., 1995). AQP1 was mainly localized in the membranes with a small quantity in the intracellular compartment while AQP2 was predominantly localized in intracellular vesicles. AQP1 did not relocalize to the membrane following exposure to vasopressin. However, AQP2 relocated to the plasma membrane following a 10-minute exposure to forskolin. After a washout of vasopressin, there was an increase in endocytosis of AQP2 water channels followed by transport to intracellular vesicles. Water permeability, determined by fluorescence, increased in the cells transfected with AQP2 after exposure to vasopressin and diminished with $HgCl_2$ consistent with vasopressin-sensitive AQP2 channels but not AQP1 channels (Katsura et al., 1995).

That mecurial diuretics have been found to attenuate water permeability in most of the water channels, suggested that there is a cysteine residue at the entrance of the water channel. When *Xenopus* oocytes were transfected with mutant mRNA of four known cysteine sites in AQP1 water channels (87, 102, 152 or 189) with the cysteine sites replaced by serine (Preston et al., 1993), the Pf of the wild type and all the mutants except for C189S mutant were inhibited by 1 mM $HgCl_2$. Thus, it appeared that cysteine 189 is the Hg^{2+}-sensitive site and is important for water permeability. Interestingly, the substitution of a large side chain at C87 did not alter the Pf as compared to wildtype, however, substitution of the large side chain C189 caused substantial reduction in the Pf. Immunoblots of the transfected oocyte membranes from the various mutants were analyzed to determine if the expressed water channels exhibited negligible Pf. The immunoblots showed expression of the water channels at 28 kDa band, but loss of 35–45 kDa broad band and a new band slightly above the 28 kDa subunit band in those with C189 substitutions. These findings suggest that the C189 residue is important in the functionality and processing of the AQP1 water channel (Preston et al., 1993).

AQP2 was identified as the vasopressin-sensitive water channel by immunohistochemical studies of the distal nephron in apical and subapical areas and by water permeability characteristics following injection in *Xenopus* oocytes (Nielsen et al., 1995; Katsura et al., 1995). Nielsen and colleagues (1995) dehydrated Sprague-Dawley rats for 24 h before microdissecting collecting tubules. The tubules were microperfused with either vehicle, vasopressin for 40 min, or perfused with vasopressin followed by washout. The Pf increased in the vasopressin treated group and decreased in the washout group. Moreover, in the vasopressin treated group immunohistochemistry confirmed that there was an increase in apical membrane vesicles as a result of transfer from intracellular vesicles to the apical membrane by exocytosis. Following withdrawal of vasopressin, there was an increase of AQP2 in intracellular vesicles and a decrease in the apical membrane. These molecular and functionality findings were consistent with vasopressin stimulating an increase in the water permeability of the apical membrane through the insertion of water channels.

Two other water channels, AQP3 and AQP4, have been localized to the collecting ducts and play a role in water homeostasis. Both are located in the basolateral membrane of the collecting tubules and allow water to exit the cell. The human AQP3 gene has been isolated (Inase et al., 1995; Echervarria et al., 1996) and is comprised of six exons of 7 kb. The 5' flanking region has promoter activity which is upregulated by phorbol ester but not by cAMP. AQP3 is highly water permeable and upregulated by vasopressin and dehydration (Kim et al., 1999). To a lesser degree, it is permeable to glycerol and urea, placing it into the aquaglyceroporin group (Echervarria et al., 1996). Since there is no decrease in Pf with exposure to mercurial diuretic, AQP4 is known as the MIWC, mercurial insensitive water channel, and has been localized in the inner medullary collecting tubule cells, lung and brain (Hasegawa et al., 1994; Kim et al., 2001b).

A. Acute Regulation of Aquaporin Transport by Vasopressin

Vasopressin has a short-term and a long-term effect on AQP2. The short-term effect occurs within 5 min and results in a rise in osmotic water permeability from <50 um/s to >1000 um/s (Knepper, 1994). As vasopressin binds to the basolateral V2 receptor-G_s protein complex, adenylyl cyclase is activated and causes an increase in intracellular cAMP (Chabardes et al., 1980). cAMP further activates PKA and AQP2 is phosphorylated, which results in the shuttling of the intracellular vesicles to the apical plasma membrane (Christensen et al., 1998, 2000). Kuwahara and colleagues (1995) determined that the phosphorylation of AQP2 by cAMP-dependent phosphorylation is responsible for exocytosis of the water channels. When *Xenopus* oocytes were transfected with AQP2 mutants with various amino acid substitutions at the Ser-256 site (the proposed cAMP-dependent protein kinase residue) or with wild type AQP2, and incubated with cAMP plus forskolin or cAMP alone, Pf increased only with the wild type. No phosphorylation of the vesicles occurred in mutant proteins. The activation of PKA results in phosphorylation of serine 256 and is essential for movement of cytoplasmic vesicles to the apical membrane (Kuwahara et al., 1995). In addition to the phosphorylation of the vesicles, it appears that phosphorylation of at least three of the monomers of an AQP2 tetramer is required for apical relocation (Kamsteeg et al., 2000).

A more recent study evaluating the role of PKA and serine 256 phosphorylation on AQP2 trafficking and recycling in transfected MDCK-C7 cells used confocal laser microscopy to localize AQP2 (Nejsum et al., 2005). For exocytosis of the intracellular vesicles to occur, AQP2 must be phosphorylated in the presence of activated PKA. Endocytosis does not appear to require dephosphorylation. The exact mechanism of exocytosis secondary to phosphorylation of serine 256 is not clear but there is some evidence of an interaction between microtubules and the AQP2-containing vesicles since disruption of the microtubules or actin filaments results in loss of vesicle exocytosis (Brown et al., 1998). Over the past decade, much progress has been made in defining and understanding the AQP2 trafficking and the reader is referred to recent reviews (Brown et al., 1998; Valenti et al., 2005) and other sections in this textbook for more details.

B. Long-Term Effect on Aquaporin by Vasopressin

Long-term regulation by vasopressin has been shown to increase the expression of AQP2 protein and mRNA abundance (Nielsen et al., 1993; Hasler et al., 2002). Immunoblots from crude membranes of the inner medulla from rats thirsted for 24 h showed an increase in AQP2 protein expression and water loaded rats showed a decrease in AQP2 protein expression (Nielsen et al., 1993). Immunolabeling was consistent with these findings; there was an increased labeling of AQP2 in the apical membrane, subapical vesicles and multivesicular bodies in the principal cells of collecting ducts in the thirsted rats.

DiGiovanni and co-workers (1994) evaluated the long-term effect of AVP on AQP2 expression and function in Brattleboro rats. The expression AQP2 protein was shown to increase threefold in Brattleboro rats receiving a 5-day infusion of AVP. This corresponded to an increase in Pf and an increase in apical and subapical vesicles. Similar findings were seen in mpkCCDc14 cells, a mouse principal cell line, exposed to vasopressin at different time intervals (Hasler et al., 2002). The expressions of both AQP2 mRNA and protein were dose and time dependent and attenuated by the V2 receptor antagonist, SR121463B and withdrawal of AVP. The long-term response began at 3 h and gradually rose to a twofold increase in protein by 72 h after being exposed to basolateral 10^{-9} M AVP (Hasler et al., 2002). This upregulation of AQP2 mRNA expression appeared to be AVP-dependent as shown in thirsted Brattleboro rats that lacked an increase AQP2 mRNRA in the absence of vasopressin (Saito et al., 1999). The 5′ flanking region of the AQP2 gene contains the element that is responsive to cAMP (CRE) which moderates the long-term effect of vasopressin by increasing AQP2 mRNA and protein abundances (Matsumura et al., 1997). More recently, vasopressin-independent pathways regulating AQP2 expression have been proposed.

VI. NEPHROGENIC DIABETES INSIPIDUS

The identification of the single genes for the V2 receptor and the AQP2 water channel were known by the 1990s. There are two types of nephrogenic diabetes insipidus (NDI): acquired and genetic. Genetic NDI is rare and has various patterns of inheritance, X-linked and autosomal recessive or dominant (Sasaki, 2004). The X-linked pattern is the most common cause and is a result of mutations of the V2 receptor gene (Lolait et al., 1992). It may involve an incompetent receptor, misfolding of the vasopressin receptor in the endosomes or the instability of the transcribed receptor (Fujiwara and Bichet, 2005). There are at least 183 putative mutations at multiple sites (Fujiwara and Bichet, 2005). There is preponderance in Nova Scotia and New Brunswick thought to be due to the 'founder defect' (Fujiwara and Bichet, 2005). The autosomal pattern results from gene defects of AQP2, which is located in chromosome region 12q13; approximately 35 mutations have been identified (Bichet et al., 1997; Fujiwara and Bichet, 2005).

The genetic forms are diagnosed in infancy or childhood and clinical presentation varies according to severity and early treatment (Fujiwara and Bichet, 2005; Bernier et al., 2006). The main problems arise from the inability to concentrate urine, resulting in excessive electrolyte-free water loss. Signs and symptoms vary and include polyuria, hydronephrosis, bladder dilatation and dysfunction, renal insufficiency, polydipsia, dehydration, nausea and vomiting, fever, seizure and mental and physical retardation (Bichet et al., 1997). The X-linked and autosomal mutations are phenotypically similar except that in the autosomal genetic defect there is a normal coagulation and fibrinolytic cascade and a vasodilatory response to dDAVP (Bichet et al., 1997; Fujiwara and Bichet, 2005). Severe symptoms, such as mental retardation, may be avoided if the individuals are well hydrated, especially in infancy and early childhood (Bichet et al., 1997).

The treatment options are limited for genetic cases and include thiazides and non-steroidal anti-inflammatory drugs (NSAIDs). However, research with chemical or pharmacological chaperones may prove to be helpful (Tamarappoo and Verkman, 1998; Bouley et al., 2005; Bernier et al., 2006). The proteins produced within a cell that are misfolded or misassembled are often retained in the endoplasmic reticulum (ER) and targeted for degradation (Bonifacino and Lippincot-Schwartz, 1991). Changes in post-translational modification, folding and oligomerization have been the target of these various chaperones to allow the retained ER proteins to be transported to their appropriate destinations. Many of the mutations result in retention of the mutant vasopressin receptors or AQP2 vesicles in the endoplasmic recticulum and they do not get to the plasma membrane. In both human and animal studies, the use of the cell-permeable V1a receptor non-peptide antagonist, SR49059, rescues the V2 receptor mutant in the ER by binding to it and allowing it to reach the membrane except in nonsense mutations (Bernier et al., 2006). The five patients with various X-linked mutations who were treated showed an increase in urine osmolality, decrease in urinary volume and water intake and stable serum electrolytes. *In vitro* studies showed rescue and restoration of function of the mutant vasopressin V2 receptors when transiently or stably transfected into HEK and COS cells and exposed to SR29059 and YM087 (Bernier et al., 2006).

An increase in intracellular cGMP has been suggested to increase insertion of AQP2 into the membrane (Bouley et al., 2005) by phosphorylation of the COOH terminus, thereby bypassing the vasopressin cAMP-dependent pathway. To test this, Bouley and associates (2005) showed that phosphodiesterase-5 inhibitors, sidenafil and MBMQ, led to an increase in AQP2 insertion in the plasma membrane in LLC-PK_1 cells that were stably transfected with AQP2 and in principal cells in rat collecting ducts.

Beside pharmacochaperones, the use of chemical chaperones may be useful in the treatment of the X-linked form (Tamarappoo and Verkman, 1998). *Xenopus* oocytes were transfected with 5-single point mutations of the AQP2 gene or wild type to determine their intrinsic water permeability (Tamarappoo and Verkman, 1998). Pf was similar to wild type in the functional mutations L22V, T126M, A147T, but not in the R187C and C181W mutations. Transfection of these cDNAs in CHO and MDCK cells revealed an increase in degradation of T126M, A147T, R187C and C181W with retention in the endoplasmic recticulum of L22V compared with wild type; this was confirmed by immunoblotting of fractionated vesicles. However, incubation with the chemical chaperone, glycerol, resulted in an almost complete redistribution to membrane and endosomes. Similar results were obtained with the chemical chaperones, TMAO and DMSO.

Acquired causes of nephrogenic diabetes insipidus are more common than genetic causes and include lithium (Marples et al., 1995), hypokalemia (Marples et al., 1996), renal disease, hypercalcemia (Earm et al., 1998), hypomagnesemia, cisplatinum (Kim et al., 2001a) or bilateral ureteral obstruction (Frøkiaer et al., 1996). Immunoblotting and immunohistochemistry and immunocytochemistry studies of animal models of acquired nephrogenic diabetes insipidus reveal a downregulation of aquaporin 2 expression and a trafficking defect (Marples et al., 1996; Frøkiaer et al., 1996; Earm et al., 1998) and, in some cases, a downregulation of AQP1 and -3 as well (Kim et al., 2001a).

VII. VAPTANS

Serum osmolality is maintained between 275 and 290 mOsm/kg and daily changes in fluid intake are usually undetectable secondary to release of arginine vasopressin in response to changes of 1–2% in serum osmolality. The clinical criterion for diagnosis of the syndrome of inappropriate antidiuretic hormone secretion (SIADH) was described by Schwartz et al. in 1957. Changes in serum sodium out of the normal range, dysnatremias, are a result of an imbalance between water intake and excretion by the kidney. Hypernatremia is defined as serum sodium greater than 145 mEq/l and is equivalent to hypertonicity. Hypotonic hyponatremia is defined as serum sodium less than 135 and may represent various volume states – hypovolemic, euvolemic or hypervolemic; the euvolemic or hypervolemic forms are likely to benefit from vaptans.

The dysnatremias are common electrolyte abnormalities in the hospital, both at time of admission and in the hospital (Snyder et al., 1987; Palevsky et al., 1996) and are associated with high morbidity and mortality. Nursing home, elderly and female patients are more commonly associated with hyponatremia at the time of hospitalization, while younger, female and postoperative patients are characteristically associated with hyponatremia acquired during hospitalization (Ayus et al., 1992).

Hyponatremia is commonly caused by the syndrome of inappropriate antidiuretic hormone, congestive heart failure or cirrhosis. It has had limited treatment options other than fluid restriction, hypertonic fluid, demeclocycline and loop diuretics (Lien et al., 1990; Peter, 2001). In the early 1970s, Manning and Sawyer designed a vasopressin agonist, dVDAVP with potent antidiuretic and minimal pressor characteristics followed by the synthesis of the first selective and potent V1 vasopressin antagonist peptide by incorporating a dimethyl group at the β carbon in position 1, $d(CH_2)_5$VDAVP (Lowbridge et al., 1977; Verbalis, 2002). By the 1980s, there were 15 new vasopressin antagonist analogs selective for antidiuretic response by replacing the valine residue at position 4 with various amino acids. The majority had vasopressor antagonist effect as well (Manning et al., 1984). In 1991, new antagonist peptides tested strongly as antagonists in animals, but paradoxically had weak V2 receptor agonist qualities in humans, therefore making them less effective for clinical use. These peptide antagonists had poor oral bioavailability and a short half-life further limiting their use (Verbalis, 2002).

In 1993, Ohnishi and colleagues characterized the first non-peptide V2 receptor antagonist – OPC-31260. The non-peptide V2 receptor antagonist was shown to block the effect of excessive vasopressin at the V2 receptor in the collecting ducts of the kidney (Verbalis, 2002). These antagonists are called vaptans or aquarectics. The V2 receptor antagonists are tolvaptan (OPC-41061), lixivaptan (VPA-985), satavaptan (SR-121463A) and the combination of V1a/V2 antagonist receptor is called conivaptan (YM087) (Lemmons-Gruber and Kamyar, 2006). Conivaptan, the intravenous form, is the only vaptan that is FDA approved for euvolemic hyponatremia in hospitalized patients (Goldsmith, 2005). In one human study, conivaptan (20 or 40 mg) improved cardiac hemodynamics and urine output in a randomized, double-blinded, short-term study in patients with CHF (NYHA class III or IV) (Udelson et al., 2001). In an animal study, the effect of YM087 versus SR-121463A on cardiac function in ischemic congestive heart failure in rats was tested. There was a greater benefit from the dual antagonist, V1a and V2, compared with the sole V2 antagonist (Wada et al., 2004). A 5-day randomized, placebo-controlled trial with two doses of oral conivaptan (40 mg or 80 mg) in euvolemic or hypervolemic hyponatremic patients from diverse causes evaluated the efficacy and safety of oral conivaptan since substrate is an inhibitor of CYP3A4. Both doses showed a statistically significant increase in serum sodium compared with placebo without significant side effects (Ghali et al., 2006). The other vaptans are under investigation in various human clinical trials (Verbalis, 2002; Schrier et al., 2006; Konstam et al., 2007).

Tolvaptan has been involved recently in phase three trials. A small pilot study evaluated hospitalized patients with euvolemic or hypervolemic hyponatremia with oral tolvaptan titrated from 10 mg to 60 mg daily for 27 days with follow up for 65 days (Gheorghiade et al., 2006). The results showed an increase in serum sodium and suggested a benefit compared with standard treatment, but this was a small study with large withdrawal in both the control and treated groups. In a recent multi-center randomized controlled study, named SALT I/II, tolvaptan was evaluated in various disease states (cirrhosis, CHF and SIADH) with inappropriate vasopressin secretion with low serum sodium (Schrier et al., 2006). The trial was performed twice to test for reproducibility. The treatment lasted for 30 days and continued until day 37. The oral dose of tolvaptan was titrated to 60 mg per day on day four, depending on the serum sodium. The serum sodium rose in the treatment group compared to control ($P < 0.001$). Tolvaptan had minimal side effects and appeared to be effective in various diseases with excessive vasopressin secretion.

The EVEREST outcome trial, an event-driven, randomized, double-blinded placebo-controlled study, evaluated the effect of low dose tolvaptan (30 mg) on short- and long-term treatment and outcome in patients with severe CHF (Konstam et al., 2007). The primary endpoints of all-cause death, the composite of cardiovascular death or hospitalizations for CHF, were no different in the treated and placebo groups. Secondary endpoints, composite of cardiovascular mortality or cardiovascular hospitalization, incidence of cardiovascular death and incidence of worsening CHF symptoms or clinical visits, showed no significant differences in either group as well. A statistically significant difference in the two groups was observed in the short treatment course with a decrease in symptoms, increase in serum sodium, decrease in urine osmolality and decrease in weight. As the new vaptans become available for clinical use for various causes of hyponatremia, it is hoped that there will be a decrease in mortality and morbidity from these underlying conditions.

References

Acher, R., and Chavet, J. (1954). La structure de la vasopressine de boeuf. *Biochim Biophys Acta* **14:** 421–429.

Agre, P. (2006). The aquaporin water channels. *Proc Am Thorac Soc* **3:** 5–13.

Ando, Y., Tabei, K., and Asano, Y. (1991). Luminal vasopressin modulates transport in the rabbit cortical collecting duct. *J Clin Invest* **88:** 952–959.

Ando, Y., and Asano, Y. (1993). Functional evidence for an apical V1 receptor in rabbit cortical collecting duct. *Am J Physiol* **264:** F467–F471.

Antoni, F. A., Holmes, M. C., Makara, G. B., Karteszi, M., and Laszlo, F. A. (1984). Evidence that the effects of arginine-8-vasopressin (AVP) on pituitary corticotrophin (ACTH) release are mediated by a novel type of receptor. *Peptides* **5:** 519–522.

Ayus, J. C., Wheeler, J. M., and Arieff, A. I. (1992). Postoperative hyponatremic encephalopathy in menstruant women. *Ann Int Med* **117:** 891–897.

Bagnasco, S. M. (2006). The erythrocyte urea transporter UT-B. *J Mem* **212:** 133–138.

Bernier, V., Morello, J. P., Zarruck, A. et al. (2006). Pharmacologic chaperones as a potential treatment for X- linked nephrogenic diabetes insipidus. *Am J Soc Nephrol* **17:** 232–243.

Bertuccio, C. A., Ibarra, F. R., Toledo, J. E., Arrizurieta, E. E., and Martin, R. S. (2002). Endogenous vasopressin regulates Na-K-ATPase and Na^+- K^+-Cl^- cotransporter rbsc-1 in rat outer medulla. *Am J Physiol Renal Physiol* **282:** F265–F270.

Bichet, D. G., Oksche, A., and Rosenthal, W. (1997). Congenital nephrogenic diabetes insipidus. *J Am Soc Nephrol* **8:** 1951–1958.

Birnbaumer, M., Seibold, A., Gilbert, S. et al. (1992). Molecular cloning of the receptor for human antidiuretic hormone. *Nature* **357:** 333–335.

Bonifacino, J. S., and Lippincot-Schwartz, J. (1991). Degradation of proteins within the endoplasmic reticulum. *Curr Opin Cell Biol* **3:** 592–600.

Bouley, R., Pator-Soler, N., Cohen, O., Mclaughlin, M., Brenton, S., and Brown, D. (2005). Stimulation of AQP-2 membrane insertion in renal epithelial cells in vitro and in vivo by cGMP phosphodiesterase inhibitor sidenafil citrate (Viagra). *Am J Physiol Renal Physiol* **288:** F1103–F1112.

Bradford, A. D., Terris, J., Ecelbarger, C. A. et al. (2001). 97 and 117 kDa forms of the collecting duct urea transporter UT-A1 are due to different states of glycosylation. *Am J Renal Physiol* **281:** F133–F143.

Briley, E. M., Lolait, S. J., Axelrod, J., and Felder, C. C. (1994). The cloned vasopressin V1a receptor stimulates phospholipase A2, phospholipase C and phospholipase D through activation of receptor-operated calcium channels. *Neuropeptides* **27:** 63–74.

Brown, D. (1989). Vesicles recycling and cell-specific function in kidney epithelial cells. *Annu Rev Physiol* **51:** 771–784.

Brown, D., and Orci, L. (1983). Vasopressin stimulates formation of coated pits in rat kidney collecting ducts. *Nature* **302**: 253–255.

Brown, D., Katsura, T., and Gustafson, C. E. (1998). Cellular mechanisms of aquaporin trafficking. *Am J Physiol* **275**: F328–F331.

Brownstein, M. J., Russell, J. T., and Gainer, H. (1980). Synthesis, transport and release of posterior pituitary hormones. *Science* **207**: 373–378.

Burnatowska-Heiden, M. A., and Spielman, W. S. (1987). Vasopressin increases cystolic free calcium in LLC-PK$_1$ cells through V$_1$-receptor. *Am J Physiol* **253**: 328–332.

Burnatowska-Heiden, M. A., and Spielman, W. S. (1989). Vasopressin V$_1$ receptors on the principal cells of the rabbit collecting tubule. Stimulation of cytosolic free calcium and inositol phosphate production via coupling the pertussin toxin subtrate. *J Clin Invest* **83**: 84–89.

Calamita, G., Gounnon, P., Gobin, R., and Bourguet, S. (1994). Antidiuretic response in the urinary bladder of *Xenopuus laevis*: presence of subapical aggrephores apical aggregrates. *Biol Cell* **80**: 35–42.

Carter, N. W., Rector, F. C., and Seldin, D. W. (1961). Hyponatremia in cerebral; disease resulting from inappropriate secretion of antidiuretic hormone. *N Eng J Med* **264**: 67–72.

Chabardes, D., Gagnan-Brunette, M., Imbert-Teboul, M. et al. (1980). Adenylyl cyclase responsiveness to hormones in various portions of human nephron. *J Clin Invest* **65**: 439–448.

Chase, L. R., and Aurbach, G. D. (1968). Renal adenyl cyclase: anatomically separate sites for parathyroid hormone and vasopressin. *Science* **159**: 545–547.

Chou, C.-L., and Knepper, M. A. (1989). Inhibition of urea transport in inner medullary collecting duct by phloretin and urea analogues. *Am J Physiol* **257**: F359–F365.

Chou, C.-L., Sands, J. M., Nonoguchi, H., and Knepper, M. A. (1990a). Concentration dependence of urea and thiourea transport pathway in rat inner medullary collecting duct. *Am. J. Physiol* **258**: F486–F494.

Chou, C.-L., Sands, J. M., Nonoguchi, H., and Knepper, M. A. (1990b). Urea-gradient associated fluid absorption with s$_{urea}$ = 1 in rat terminal collecting duct. *Am J Physiol* **258**: F1173–F1180.

Christensen, B. M., Marples, D., Jensen, U. B. et al. (1998). Acute effects of vasopressin V2-receptor antagonist on kidney AQP-2 expression and subcellular distribution. *Am J Physiol* **275**: F285–F297.

Christensen, B. M., Zelenina, M., Aperia, A., and Nielsen, S. (2000). Localization and regulation of PKA-phosphorylated AQP2 in response to V2-receptors agonist/antagonist treatment. *Am J Physiol Renal Physiol* **278**: F29–F42.

Crawford, J. D., Doyle, A. P., and Probst, H. (1959). Service of urea in renal water conservation. *Am J Physiol* **196**: 545–548.

de Keyzer, Y., Auzan, C., Lenne, F. et al. (1994). Cloning and characterization of the human V3 pituitary vasopressin receptor. *FEBS Lett* **356**: 215–220.

Denker, B. M., Smith, B. L., Kuhajda, F. P., and Agre, P. (1988). Identification, purification, and partial characterization of a novel. Mr 28,000 integral membrane protein from erythrocytes and renal tubules. *J Biol Chem* **263**: 15634–15642.

DiGiovanni, S. R., Nielsen, S., Christensen, E. I., and Knepper, M. A. (1994). Regulation of collecting duct water channel expression by vasopressin in Brattleboro rat. *Proc Natl Acad Sci USA* **91**: 8984–8988.

Dunn, F. L., Brennan, T. J., Nelson, A. E., and Robertson, G. L. (1973). The role of blood osmolality and volume in regulating vasopressin secretion in the rat. *J Clin Invest* **52**: 3212–3219.

Earm, J. H., Christensen, B. M., Frøkiaer, J. et al. (1998). Decreased aquaporin-2 expression and apical plasma membrane delivery in kidney collecting ducts of polyuria hypercalcemic rats. *J Am Soc Nephrol* **9**: 2181–2193.

Echervarria, M., Windhanger, E. E., and Frindt, G. (1996). Selectivity of renal collecting duct water channel aquaporin-3. *J Biol Chem* **271**: 25079–25082.

Epstein, F. H., Kleeman, C. R., Pursel, S., and Hendrikx, A. (1957). The effect of feeding protein and urea on the renal concentrating process. *J Clin Invest* **36**: 635–641.

Fenton, R. A., Howorth, A., Cooper, G. J., Meccariello, R., Morris, I. D., and Smith, C. P. (2000). Molecular characterization of a novel UT-A urea transporter isoform (UT-A5) in testis. *Am J Physiol Cell Physiol* **279**: C1425–C1431.

Fenton, R. A., Chou, C.-L., Stewart, G. S., Smith, C. P., and Kneppe, M. A. (2004). Urinary concentrating defect in mice with selective deletion of phloretin-sensitive urea transporters in the renal collecting duct. *Proc Natl Acad Sci USA* **101**: 7469–7474.

Fenton, R. A., Flynn, A., Shodeinde, A., Smith, C. P., Schnermann, J., and Knepper, M. (2005). Renal phenotype of UT-A urea transporter knockout mice. *J Am Soc Nephrol* **16**: 1583–1592.

Fenton, R. A., Chou, C. L., Sowersby, H., Smith, C. P., and Knepper, M. A. (2006a). Gamble's 'economy of water' revisited: studies in urea transporter knockout mice. *Am J Physiol Renal Physiol* **291**: F148–F154.

Fenton, R. A., Smith, C. P., and Knepper, M. A. (2006b). Role of collecting duct urea transporters in the kidney – insights from mouse models. *J Membr Biol* **212**: 119–131.

Fressinaud, P., Corvol, P., and Menard, J. (1974). Radioimmunoassay of urinary antidiuretic hormone in man: stimulation-suppression tests. *Kidney Int* **6**: 184–190.

Fröhlich, O., Macey, R. I., Edwards-Moulds, J., Gargus, J. J., and Gunn, R. B. (1991). Urea transport deficiency in Jk(a-b-) erythrocytes. *Am J Physiol* **260**: C778–C783.

Frøkiaer, J., Marples, D., Knepper, M. A., and Nielsen, S. (1996). Bilateral ureteral obstruction downregulates expression vasopressin- sensitive AQP-2 water channel in rat kidney. *Am J Renal Physiol Renal Physiol* **270**: F657–F668.

Fujiwara, T. M., and Bichet, D. G. (2005). Molecular biology of hereditary diabetes insipidus. *Am J Soc Nephrol* **16**: 2836–2846.

Fushimi, K., Uchida, S., Hara, Y., Hirata, Y., Marumo, F., and Sasaki, S. (1993). Cloning and expression of apical membrane water channel of rat collecting tubule. *Nature* **361**: 549–552.

Galluci, E., Micelli, S., and Lippe, C. (1971). Non-electrolyte permeability across thin lipid membranes. *Arch Int Physiol Biochim* **79**: 881–887.

Gamble, J. L., McKhann, C. F., Butler, A. M., and Tuthill, E. (1934). An economy of water in renal function referable to urea. *Am J Physiol* **109**: 139–154.

Ghali, J. K., Koren, M. J., Taylor, J. R. et al. (2006). Efficacy and safety of oral conivaptan: a V$_{1A}$/V$_2$ vasopressin receptor antagonist, assessed in a randomized, placebo-controlled trial in patients with euvolemic or hypervolemic hyponatremia. *J Clin Endocrinol Metab* **91**: 2145–2152.

Gheorghiade, M., Gottlieb, S. S., Udelson, J. E. et al. (2006). Vasopressin V2 receptor blocker with Tolvaptan versus fluid restriction in the treatment of hypernatremia. *Am J Cardiol* **97**: 1064–1067.

Gillin, A. G., and Sands, J. M. (1992). Characteristics of osmolarity-stimulated urea transport in rat IMCD. *Am J Physiol* **262**: F1061–F1067.

Goldsmith, S. R. (2005). Current treatments and novel pharmacologic treatments for hyponatremia in congestive heart failure. *Am J Cardiol* **95**: 14b–23b.

Gottschalk, C. W., and Mylle, M. (1958). Evidence that mammalian nephron function as a countercurrent multiplier system. *Science* **128**: 594.

Gottschalk, C. W., Lassiter, W. E., Mylle, M. et al. (1962). Micropuncture study of composition of loop of Henle fluid in desert rodents. *Am J Physiol* **204**: 532–535.

Grumer, H. A., Derryberry, W., Dubin, A., and Waldstein, S. S. (1962). Idiopathic, episodic inappropriate secretion of antidiuretic hormone. *Am J Med* **32**: 954–963.

Gullion, G., Peirre-Olivier, C., Butlen, D., Cantau, B., and Serge, J. (1980). Size of vasopressin receptors from rat liver and kidney. *Eur J Biochem* **111**: 287–294.

Harrington, A. R., and Valtin, H. (1968). Impaired urinary concentration after vasopressin and its gradual correction in hypothalamic diabetes insipidus. *J Clin Invest* **47**: 502–510.

Hasegawa, H., Ma, T., Skach, E., Matthay, M., and Verkman, A. S. (1994). Molecular cloning of a mercurial-insensitive water channel expressed in selected water transporting tissues. *J Biol Chem* **269**: 5479–5550.

Hasler, U., Mordansini, D., Bens, M. et al. (2002). Long term regulation of aquaporins-2 expression in vasopressin-responsive renal collecting duct principal cells. *J Biol Chem* **277**: 10379–10386.

Hayashi, M., Sasaki, S., Tsuganezawa, H. et al. (1994). Expression and distribution of aquaporin of collecting duct are regulated by vasopressin V_2 receptor in rat kidney. *J Clin Invest* **94**: 1778–1783.

Hendrikx, A., and Epstein, F. H. (1958). Effect of feeding protein and urea on renal concentrating ability in the rat. *Am J Physiol* **195**: 539–542.

Hermossila, R., and Schulein, R. (2001). Sorting function of individual cytoplasmic domains of G protein-coupled vasopressin V2 receptor in Madin Darby Canine kidney epithelial cells. *Mol Pharmacol* **60**: 1031–1039.

Igarashi, P., Whyte, D. A., and Nagami, G. T. (1996). Cloning and kidney cell-specific activity of the promoter of the murine renal Na-K-Cl cotransporter gene. *J Biol Chem* **271**: 9666–9674.

Inase, N., Fushimi, K., Ishibashi, K. et al. (1995). Isolation of human aquaporin 3 gene. *J Biol Chem* **270**: 17913–17916.

Inoue, T., Terris, J., Ecelbarger, C. A., Chou, C.-L., Nielsen, S., and Knepper, M. K. (1999). Vasopressin regulates apical targeting of aquaporin-2 but not of UT1 urea transporter in renal colleting duct. *Am J Physiol Renal Physiol* **276**: F559–F566.

Ivell, R., Schmale, H., Krisch, B., Nahke, P., and Richter, D. (1986). Expression of a mutant vasopressin gene: different polyadenylation and read-through of mRNA 3' end in a frame-shift mutant. *EMBO J* **5**: 971–977.

Jaenike, J. R. (1961). The influence of vasopressin on the permeability of the mammalian collecting duct to urea. *J Clin Invest* **40**: 144–151.

Jewell, P. A., and Verney, E. B. (1957). An experimental attempt to determine the site of neurophyseal osmoreceptors in the dog. *Phil Trans Roy Soc* **240**: 197–324.

Jung, S. J., Preston, G. M., Smith, B. L., Guggino, W. B., and Agre, P. (1994). Molecular studies of the water channel through transport aquaporin CHIP. The hourglass model. *J Biol Chem* **269**: 14648–14654.

Kachadorian, W. A., Levine, S. D., Wade, J. B., DiScala, V. A., and Hays, R. M. (1977). Relationship of aggregated intramembranous particles to water permeability in vasopressin–treated toad urinary bladder. *J Clin Invest* **59**: 576–581.

Kamsteeg, E. J., Heijner, I., van Os, C. H., and Deen, P. M. T. (2000). The subcellular localization of an aquaporin-2 tetramer depends on the stoichiometry and phosphorylated and nonphosphorylated monomers. *J Cell Biol* **151**: 919–929.

Karakashian, A., Timmer, R. T., Klein, J. D., Gunn, R. B., Sands, J. M., and Bagnasco, S. M. (1999). Cloning and characterization of two new mRNA isoforms of the rat renal urea transporter: UT-A3 and UT-A4. *J Am Soc Nephrol* **10**: 230–237.

Kato, A., Klein, J. D., Zhang, C., and Sands, J. M. (2000). Angiotensin II increases vasopressin-stimulated facilitated urea permeability in rat terminal IMCDs. *Am J Physiol Renal Physiol* **279**: F835–F840.

Katsura, T., Verbavatz, J.-M., Farinas, J. et al. (1995). Constitutive and regulated membrane expression of aquaporin 1 and aquaporin 2 water channels in stably transfected LLC-PK_1 epithelial cells. *Proc Natl Acad Sci USA* **92**: 7212–7216.

Kaye, M. (1966). An investigation into cause of hyponatremia in the syndrome of inappropriate secretion of antidiuretic hormone. *Am J Med* **41**: 910–926.

Kim, D.-U., Sands, J. M., and Klein, J. D. (2004). Role of vasopressin in diabetes mellitus-induced changes in medullary transport proteins involved in urine concentration in Brattleboro rats. *Am J Physiol Renal Physiol* **286**: F760–F766.

Kim, G.-H., Ecelbarger, C. A., Mitchell, C., Packer, R. K., Wade, J. B., and Knepper, M. A. (1999). Vasopressin increases Na-K-Cl cotransporter expression in the thick ascending limb of Henle's loop. *Am J Physiol Renal Physiol* **276**: F96–103.

Kim, S.-W., Lee, J.-U., Nah, M.-Y. et al. (2001a). Cisplatin decreases the abundance of aquaporin water channels in rat kidney. *J Am Soc Nephrol* **12**: 875–882.

Kim, Y.-E., Earm, J.-H., Ma, T. et al. (2001b). Aquaporin-4 expression in adult and developing mouse and rat kidney. *J Am Soc Nephrol* **12**: 1795–1804.

Klahr, S., and Alleyne, G. A. O. (1973). Effects of chronic protein-calorie malnutrition on the kidney. *Kidney Int* **3**: 129–141.

Klein, J. D., Gunn, R. B., Roberts, B. R., and Sands, J. M. (2002). Down-regulation of urea transporters in the renal inner medulla of lithium-fed rats. *Kidney Int* **61**: 995–1002.

Klein, J. D., Sands, J. M., Qian, L., Wang, X., and Yang, B. (2004). Upregulation of urea transporter UT-A2 and water channels AQP2 and AQP3 in mice lacking urea transporter UT-B. *J Am Soc Nephrol* **15**: 1161–1167.

Klein, J. D., Froehlich, O., Blount, M. A., Martin, C. F., Smith, T. D., and Sands, J. M. (2006). Vasopressin increases plasma membrane accumulation of urea transporter UT-A1 in rat inner medullary collecting ducts. *J Am Soc Nephrol* **17**: 2680–2686.

Knepper, M. A. (1983). Urea transport in isolated thick ascending limbs and collecting ducts from rats. *Am J Physiol* **245**: 634–639.

Knepper, M. A. (1994). The aquaporin family of molecular water channels. *Proc Natl Acad Sci USA* **91**: 6255–6258.

Knepper, M. A., Sands, J. M., and Chou, C.-L. (1989). Independence of urea and water transport in rat inner medullary collecting duct. *Am J Physiol* **256**: F610–F621.

Knepper, M. A., Kim, G.-H., Fernandez-Llama, P., and Ecelbarger, C. A. (1999). Regulation of thick ascending limb transport by vasopressin. *J Am Soc Nephrol* **10**: 628–634.

Kokko, J. P., and Rector, F. C. (1972). Countercurrent multiplication system without active transport in inner medulla. *Kidney Int* **2**: 214–223.

Kondo, Y., and Imai, M. (1987). Effects of glutaraldehyde fixation on renal tubular function. I. Preservation of vasopressin-stimulated water and urea pathways in rat papillary collecting duct. *Pfluegers Arch* **408**: 479–483.

Konstam, M. A., Gheorghiade, M., Burnett, J. C. et al. (2007). Effects of oral tolvaptan in patients hospitalized for worsening heart failure – the EVEREST outcome trial. *J Am Med Assoc* **297**: 1319–1331.

Kudo, L. H., César, K. R., Ping, W. C., and Rocha, A. S. (1992). Effect of peritubular hypertonicity on water and urea transport of inner medullary collecting duct. *Am J Physiol Renal Fluid Electrolyte Physiol* **262**: F338–347.

Kuwahara, M., Fushimi, K., Terada, Y., Bai, L., Marumo, F., and Sasaki, S. (1995). cAMP-dependent phosphorylation stimulates water permeability of aquaporin-collecting duct water channel protein expressed in *Xenopus* oocytes. *J Biol Chem* **270**: 10384–10387.

Land, H., Schutz, G., Schmale, H., and Ritcher, D. (1982). Nucleotide sequence of cloned cDNA encoding bovine arginine vasopressin-neurophysin II precursor. *Nature* **295**: 299–303.

Lassiter, W. E., Gottschalk, C. W., and Mylle, M. (1961). Micropuncture of net transtubular movement of water and urea in nondiuretic mammalian kidney. *Am J Physiol* **200**: 1139–1147.

Lemmons-Gruber, R., and Kamyar, M. (2006). Vasopressin antagonists. *Cell Mol Life Sci* **63**: 1766–1779.

Levinsky, N. G., and Berliner, R. W. (1959). The role of urea in the urine concentrating mechanism. *J Clin Invest* **38**: 741–748.

Lien, Y.-H., Shapiro, J. I., and Chan, L. (1990). Effects of hypernatremia on organic osmoles. *J Clin Invest* **85**: 1427–1435.

Lim, S.-W., Han, K.-H., Jung, J.-Y. et al. (2006). Ultrastructural localization of UT-A and UT-B in rat kidneys with different hydration status. *Am J Physiol Regul Integr Comp Physiol* **290**: R479–R492.

Lolait, S. J., O'Carroll, A.-M., McBride, O. W., Konig, M., Morel, A., and Brownstein, M. J. (1992). Cloning and characterization of vasopressin V2 receptor and possible link to nephrogenic diabetes insipidus. *Nature* **357**: 336–339.

Macey, R. I. (1984). Transport of water and urea in red blood cells. *Am J Physiol* **246**: C195–C203.

Macey, R. I., and Framer, R. E. L. (1970). Inhibition of water and solute permeability in human red cells. *Biochim Biophys Acta* **211**: 104–106.

Macey, R. I., and Yousef, L. W. (1988). Osmotic stability of red cells in renal circulation requires rapid urea transport. *Am J Physiol* **254**: C669–C674.

Machida, K., Nonoguchi, H., Wakamatsu, S. et al. (2003). Acute regulation of epithelial sodium channel gene by vasopressin and hyperosmolality. *Hypertens Res* **26**: 624–629.

Manning, M., Nawrocka, E., Misicka, A. et al. (1984). Potent and selective antagonists of antidiuretic responses to arginine-vasopressin based on modification of [1-(β-mercapto-β, β-pentamethylenepropionic acid], 2-D-isoleucine,4-valine] arginine-vasopressin at positon 4. *J Med Chem* **27**: 423–429.

Marples, D., Christensen, S., Christensen, E. I., Ottosen, P. D., and Nielsen, S. (1995). Lithium-induced downregulation of aquaporin-2 water channel expression in rat kidney medulla. *J Clin Invest* **95**: 1838–1845.

Marples, D., Frøkiaer, J., Dorup, J., Knepper, M. A., and Nielsen, S. (1996). Hypokalemia-induced downregulation of aquaporin-2 water channel expression in rat kidney medulla and cortex. *J Clin Invest* **97**: 1960–1968.

Matsumura, Y., Uchida, S., Rai, T., Sasaki, S., and Marumo, F. (1997). Transcriptional regulation of aquaporin-2 water channel gene by cAMP. *J Am Soc Nephrol* **8**: 861–867.

Mayrand, R. R., and Levitt, D. G. (1983). Urea and ethylene glycol-facilitated transport systems in the human red cell membrane. *J Gen Physiol* **81**: 221–227.

Michell, R. H., Kirk, C. J., and Bilah, M. M. (1979). Hormonal stimulation of phosphatidylinositol breakdown, with particular reference to the hepatic effects of vasopressin. *Biochem Soc Trans* **7**: 861–865.

Mitra, A. K., van Hoek, A. N., Wiener, M. C., Verkman, A. S., and Yeager, M. (1995). The CHIP28 water channel visualized in ice by electron crystallography. *Nature Struct Biol* **2**: 726–729.

Moon, C., Preston, G. M., Griffin, C. A., Jabs, E. W., and Agre, P. (1993). The human aquaporin-CHIP gene: structure, organization, and chromosomal localization. *J Biol Chem* **268**: 15772–15778.

Morel, A., O'Carroll, A.-M., Brownstein, M. J., and Lolait, S. J. (1992). Molecular cloning and expression of a rat V1a arginine vasopressin receptor. *Nature* **356**: 523–526.

Morgan, T., and Berliner, R. W. (1968). Permeability of the loop of Henle, vasa recta, and collecting duct to water, urea, and sodium. *Am J Physiol* **215**: 108–115.

Morgan, T., Sakai, F., and Berliner, R. W. (1968). In vitro permeability of medullary collecting duct to water and urea. *Am J Physiol* **214**: 574–581.

Morgan, T. (1970). Permeability of the nephron to urea. In: (Schmidt-Nielsen, B., Kerr, D.W.S., eds) *Urea and the kidney*, Excerpta Medica Foundation, Amsterdam, pp. 186–192.

Morris, M., and Alexander, N. (1989). Baroreceptors influence on oxytocin and vasopressin secretion. *Hypertension* **13**: 110–114.

Nakayama, Y., Peng, T., Sands, J. M., and Bagnasco, S. M. (2000). The TonE/TonEBP pathway mediates tonicity-responsive regulation of UT-A urea transporter expression. *J Biol Chem* **275**: 38275–38280.

Nakayama, Y., Naruse, M., Karakashian, A., Peng, T., Sands, J. M., and Bagnasco, S. M. (2001). Cloning of the rat Slc14a2 gene and genomic organization of the UT-A urea transporter. *Biochim Biophys Acta* **1518**: 19–26.

Nejsum, L. N., Zelenina, M., Aperia, A., Frokiaer, J., and Nielsen, S. (2005). Bidirectional regulation of AQP2 trafficking and recycling:involvement of AQP2-S256 phosphorylation. *Am J Physiol Renal Physiol* **288**: F930–F938.

Nielsen, S., DiGiovanni, S. R., Christensen, E. I., Knepper, M. A., and Harris, H. W. (1993). Cellular and subcellular immunolocalization of vasopressin-regulated water channel in rat kidney. *Proc Natl Acad Sci USA* **90**: 1663–1667.

Nielsen, S., Chou, C.-L., Marples, D., Christensen, E. L., Kishore, B. K., and Knepper, M. A. (1995). Vasopressin increases water permeability of kidney collecting duct by inducing translocation of aquaporin-CD water channels to plasma membrane. *Proc Natl Acad Sci USA* **92**: 1013–1017.

Nielsen, S., Terris, J., Smith, C. P., Hediger, M. A., Ecelbarger, C. A., and Knepper, M. A. (1996). Cellular and subcellular localization of the vasopressin-regulated urea transporter in rat kidney. *Proc Natl Acad Sci USA* **93**: 5495–5500.

Nielsen, S., Frøkiaer, J., Marples, D., Kwon, T.-H., Agre, P., and Knepper, M. (2002). Aquaporins in the kidney: from molecules to medicine. *Physiol Rev* **2**: 205–244.

Nonoguchi, H., Owada, A., Kobayashi, N. et al. (1995). Immunohistochemical localization of V2 vasopressin receptor along the nephron and functional role of luminal V2 receptor in terminal inner medullary collecting ducts. *J Clin Invest* **96**: 1768–1778.

Ohnishi, A., Orita, Y., Okahara, R. et al. (1993). Potent aquaretic agent, a novel nonpeptide selective vasopressin 2 antagonist (OPC-31260) in men. *J Clin Invest* **92**: 2653–2659.

Olives, B., Neau, P., Bailly, P. et al. (1994). Cloning and functional expression of a urea transporter from human bone marrow cells. *J Biol Chem* **269**: 649–652.

Ostrowski, N. L., Young, W. S., Knepper, M. A., and Lolait, S. J. (1993). Expression of V1a and V2 receptor messenger ribonucleic acid in liver and kidney of embryonic, developing and adult rats. *Endocrinology* **133**: 1849–1859.

Palevsky, P. M., Bhagrath, R., and Greenberg, A. (1996). Hypernatremia in hospitalized patients. *Ann Intern Med* **124**: 197–203.

Peil, A. E., Stolte, H., and Schmidt-Nielsen, B. (1990). Uncoupling of glomerular and tubular regulations of urea excretion in rat. *Am J Physiol Renal Fluid Electrolyte Physiol* **258**: F1666–F1674.

Pennell, J. P., Sanjana, V., Frey, N. R., and Jamison, R. L. (1975). The effect of urea infusion on the urinary concentrating mechanism in protein-depleted rats. *J Clin Invest* **55**: 399–409.

Peter, G. (2001). Correction of hyponatremia. *Semin Nephrol* **21**: 269–272.

Preston, G. M., and Agre, P. (1991). Isolation of cDNA for erythrocytes intergral membrane protein of 28 kilodaltons: member of an ancient channel family. *Proc Natl Acad Sci USA* **88**: 11110–11114.

Preston, G. M., Carroll, T. P., Guggino, W. B., and Agre, P. (1992). Appearance of water channels in *Xenopus* oocytes expressing red cell CHIP28 protein. *Science* **256**: 385–387.

Preston, G. M., Jung, J. S., Guggino, W. B., and Agre, P. (1993). The mercury-sensitive residue at cysteine 189 in the CHIP 28 water channel. *J Biol Chem* **268**: 17–20.

Pruszczynski, W., Caillens, H., Drieu, L., Moulonguet-Doleris, L., and Ardaillou, R. (1984). Renal excretion of antidiuretic hormone in healthy subjects and patients with renal failure. *Clin Sci* **67**: 307–312.

Rai, A., Whaley-Connell, A., McFarlane, S., and Sowers, J. R. (2006). Hyponatremia, arginine vasopressin dysregulation, and vasopressin receptor antagonism. *Am J Nephrol* **26**: 579–589.

Robben, J. H., Knoers, N. V. A. M., and Deen, P. M. T. (2004). Regulation of the vasopressin V2 receptor by vasopressin in polarized renal collecting duct cells. *Mol Biol Cell* **15**: 5963–5969.

Rocha, A. S., and Kudo, L. H. (1982). Water, urea, sodium, chloride, and potassium transport in the in vitro perfused papillary collecting duct. *Kidney Int* **22**: 485–491.

Saito, M., Tahara, A., Sugimoto, T., Abe, K., and Furuichi, K. (2000). Evidence that atypical V2 receptor in inner medulla of kidney is V1b receptor. *Eur J Pharmacol* **401**: 289–296.

Saito, T., Ishikawa, S., Sasaki, S. et al. (1999). Lack of vasopressin-independent upregulation of AQP-2 gene expression in homozygous Brattleboro rats. *Am J Physiol* **277**: R427–R433.

Sands, J. M. (2003). Molecular mechanisms of urea transport. *J Membr Biol* **191**: 149–163.

Sands, J. M. (2004). Renal urea transporters. *Curr Opin Nephrol Hypertens* **13**: 525–532.

Sands, J. M., and Bichet, D. G. (2006). Nephrogenic diabetes insipidus. *Ann Intern Med* **144**: 186–194.

Sands, J. M., and Knepper, M. A. (1987). Urea permeability of mammalian inner medullary collecting duct system and papillary surface epithelium. *J Clin Invest* **79**: 138–147.

Sands, J. M., and Schrader, D. C. (1991). An independent effect of osmolality on urea transport in rat terminal IMCDs. *J Clin Invest* **88**: 137–142.

Sands, J. M., Nonoguchi, H., and Knepper, M. A. (1987). Vasopressin effects on urea and H_2O transport in inner medullary collecting duct subsegments. *Am J Physiol* **253**: F823–F832.

Sands, J. M., Gargus, J. J., Fröhlich, O., Gunn, R. B., and Kokko, J. P. (1992). Urinary concentrating ability in patients with Jk(a-b-) blood type who lack carrier-mediated urea transport. *J Am Soc Nephrol* **2**: 1689–1696.

Sands, J. M., Timmer, R. T., and Gunn, R. B. (1997). Urea transporters in kidney and erythrocytes. *Am J Physiol* **273**: F321–F339.

Sasaki, S. (2004). Nephrogenic diabetes insipidus: update of genetic and clinical aspects. *Nephrol Dial Transplant* **19**: 1351–1353.

Sasaki, S., Fushimi, K., Saito, H. et al. (1994). Cloning, characterization, and chromosomal mapping of human aquaporin of collecting duct. *J Clin Invest* **93**: 1250–1256.

Schmale, H., and Ritcher, D. (1984). Single base deletion in the vasopressin gene is cause of diabetes insipidus in Brattleboro rats. *Nature* **308**: 705–709.

Schrier, Gross, P., Gherghiade, M. et al. (2006). Tovalptan, a selective oral vasopressin V2-receptor antagonist, for hyponatremia. *N Engl J Med* **355**: 2099–2112.

Schwartz, W. B., Bennett, W., Curelop, S., and Bartter, F. C. (1957). A syndrome of renal sodium loss and hyponatremia probably resulting from inappropriate secretion of antidiuretic hormone. *Am J Med* **23**: 529–542.

Shayakul, C., Steel, A., and Hediger, M. A. (1996). Molecular cloning and characterization of the vasopressin-regulated urea transporter of rat kidney collecting ducts. *J Clin Invest* **98**: 2580–2587.

Simpson, N. E. (1988). The map of chromosome 20. *J Med Genet* **25**: 794–804.

Sims, E. A., Welt, L., Orloff, J. G., and Needham, J. W. (1950). Asymptomatic hyponatremia in pulmonary tuberculosis. *J Clin Invest* **29**: 846–847.

Smith, B. L., and Agre, P. (1991). Erythrocytes Mr 28,000 transmembrane protein exists as a multisubunit oligomer similar to channel proteins. *J Biol Chem* **266**: 6407–6415.

Smith, C. P., and Fenton, R. A. (2006). Genomic organization of the mammalian SLC14a2 urea transporter genes. *J Membr Biol* **212**: 109–117.

Smith, C. P., Potter, E. A., Fenton, R. A., and Stewart, G. S. (2004). Characterization of a human colonic cDNA encoding a structurally novel urea transporter, UT-A6. *Am J Physiol Cell Physiol* **287**: C1087–C1093.

Snyder, N. A., Fiegal, D. W., and Arieff, A. I. (1987). Hypernatremia in elderly patients: a heterogeneous, morbid, and iatrogenic entity. *Ann Int Med* **107**: 309–319.

Star, R. A. (1990). Apical membrane limits urea permeation across the rat inner medullary collecting duct. *J Clin Invest* **86**: 1172–1178.

Star, R. A., Nonoguchi, H., Balaban, R., and Knepper, M. A. (1988). Calcium and cyclic adenosine monophosphate as second messengers for vasopressin in the rat inner medullary collecting duct. *J Clin Invest* **81**: 1879–1888.

Stephenson, J. L. (1972). Concentration of urine in a central core model of the renal counterflow system. *Kidney Int* **2**: 85–94.

Stewart, G. S., Fenton, R. A., Wang, W. et al. (2004). The basolateral expression of mUT-A3 in the mouse kidney. *Am J Physiol Renal Physiol* **286**: F979–F987.

Swaminathan, S. (1980). Osmoreceptors or sodium receptors: investigation into ADH release in rhesus monkey. *J Physiol* **307**: 71–83.

Tamarappoo, B. K., and Verkman, A. S. (1998). Defective aquaporin −2 trafficking in nephrogenic diabetes insipidus and correction by chemical chaperones. *J Clin Invest* **101**: 2257–2267.

Teitelbaum, I. (1991). Vasopressin-stimulated phosphoinositide hydrolysis in cultured rat inner medullary collecting duct cells is mediated by the oxytocin receptor. *J Clin Invest* **87**: 2122–2126.

Terada, Y., Tomita, K., Nonoguchi, H., Yang, T., and Marumo, F. (1993). Different localization and regulation of two types of vasopressin receptor messenger RNA in microdissected rat nephron segments using reverse transcription polymerase chain reaction. *J Clin Invest* **92**: 2339–2345.

Terris, J., Ecelbarger, C. A., Sands, J. M., and Knepper, M. A. (1998). Long-term regulation of collecting duct urea transporter proteins in rat. *J Am Soc Nephrol* **9**: 729–736.

Terris, J. M., Knepper, M. A., and Wade, J. B. (2001). UT-A3: localization and characterization of an additional urea transporter isoform in the IMCD. *Am J Physiol Renal Physiol* **280**: F325–F332.

Thibonnier, M., Goraya, T., and Berti-Mattera, L. (1993). G protein coupling of human platelets V1 vascular vasopressor receptors. *Am J Physiol* **264**: C1336–C1344.

Thibonnier, M., Auzan, A., Madhun, Z., Wilkins, P., Berti-Mattera, L., and Clauser, E. (1994). Molecular cloning, sequencing, and functional expression of a cDNA encoding human V1a vasopressin receptor. *J Biol Chem* **269**: 3304–3310.

Tian, Y., Sandberg, K., Murase, T., Baker, E. A., Speth, R. C., and Verbalis, J. G. (2000). Vasopressin V2 receptor binding is downregulated during renal escape from vasopressin-induced antidiuresis. *Endocrinology* **141**: 307–314.

Uchida, S., Sasaki, S., Fushimi, K., and Marumo, F. (1994). Isolation of human aquaporin-CD gene. *J Biol Chem* **38**: 23451–23455.

Uchida, S., Sohara, E., Rai, T., Ikawa, M., Okabe, M., and Sasaki, S. (2005). Impaired urea accumulation in the inner medulla of mice lacking the urea transporter UT-A2. *Mol Cell Biol* **25**: 7357–7363.

Udelson, J. E., Smith, W. B., Hendrix, G. H. et al. (2001). Acute hemodynamic effect of conivaptan, or dual V(1a) and V(2) vasopressin receptor antagonist in patients with advanced heart failure. *Circulation* **104**: 2417–2423.

Valenti, G., Procino, G., Tamma, G., Carmosino, M., and Svelto, M. (2005). Minireveiw: aquaporin 2 trafficking. *Endocrinology* **146**: 5063–5070.

Verbalis, J. G. (2002). Receptor antagonists – vasopressin V2 receptor antagonists. *J Mol Endocrinol* **29**: 1–9.

Verbavatz, J.-M., Brown, D., Sabolic, I. et al. (1993). Tetrameric assembly of CHIP28 water channels in liposomes and cell membranes: a freeze-fracture study. *J Cell Biol* **123**: 605–618.

Verney, E. B. (1947). The antidiuretic hormone and the factors which determine its release. *Proc Roy Soc* **135**: 25–106.

Wada, K., Fujimori, A., Matsukawa, U. et al. (2004). Intravenous administration of conivaptan hydrochloride improves cardiac hemodynamics in rats with myocardial-induced congestive heart failure. *Eur J Pharmacol* **507**: 145–151.

Wade, J. B., and DiScala, V. A. (1971). The effect of osmotic flow on the distribution of horseradish peroxidase within the intercellular spaces of toad bladder epithelium. *J Cell Biol* **51**: 553–558.

Wade, J. B., Stetson, D. L., and Lewis, S. A. (1981). ADH action: evidence for membrane shuttle mechanism. *Ann NY Acad Sci* **372**: 106–117.

Wade, J. B., Lee, A. J., Liu, J. et al. (2000). UT-A2: a 55 kDa urea transporter protein in thin descending limb of Henle's loop whose abundance is regulated by vasopressin. *Am J Physiol* **278**: F52–62.

Wall, S. M., Suk, H. J., Chou, C.-L., and Knepper, M. A. (1992). Kinetics of urea and water permeability activation by vasopressin in rat terminal IMCD. *Am J Physiol Renal Fluid Electrolyte Physiol* **262**: F989–F998.

Wieth, J. O., Funder, J., Gunn, R. B., and Brahm, J. (1974). Passive transport pathways for chloride and urea through the red cell membrane. In: (Bolis, K., Bloch, K., Luria, S.E., Lynen, F., eds) *Comparative biochemistry and physiology of transport*, Elsevier/North-Holland, Amsterdam, pp. 317–337.

Wirz, H., Hargitay, B., and Kuhn, W. (1951). Localization of the concentration process in the kidney by direct kryoscopy. *Helv Physiol Pharmcol Acta* **9**: C26–C27.

Woodhall, P. B., and Tisher, C. C. (1973). Response of the distal tubule and cortical collecting duct to vasopressin in the rat. *J Clin Invest* **52**: 3095–3108.

Yang, B., Bankir, L., Gillespie, A., Epstein, C. J., and Verkman, A. S. (2002). Urea-selective concentrating defect in transgenic mice lacking urea transporter UT-B. *J Biol Chem* **277**: 10633–10637.

Yang, B., and Verkman, A. S. (2002). Analysis of double knockout mice lacking aquaporin-1 and urea transporter UT-B. *J Biol Chem* **277**: 36782–36786.

Yasui, M., Zelenin, S. M., Celsi, G., and Aperia, A. (1997). Adenylate cyclase-coupled vasopressin receptor activates AQP2 promoter via a dual effect on CRE and AP1 elements. *Am J Physiol Renal Physiol* **272**: F443–F450.

You, G., Smith, C. P., Kanai, Y., Lee, W.-S., Stelzner, M., and Hediger, M. A. (1993). Cloning and characterization of the vasopressin-regulated urea transporter. *Nature* **365**: 844–847.

Zhang, C., Sands, J. M., and Klein, J. (2002). Vasopressin rapidly increases the phosphorylation of the UT-A1 urea transporter activity in rat IMCDs through PKA. *Am J Physiol Renal Physiol* **282**: F85–90.

Further reading

Bondy, C., Chin, E., Smith, B. L., Preston, G. M., and Agre, P. (1993). Developmental gene expression and tissue distribution of CHIP28 water-channel protein. *Proc Natl Acad. Sci. USA* **90**: 4500–4504.

Fried, L. F., and Palevsky, P. M. (1997). Hyponatremia and hypernatremia. *Med Clin N Am* **81**: 585–606.

Gullion, G., Bulten, D., Cantau, B., Barth, T., and Jard, S. (1982). Kinetic and pharmacological characterization of vasopressin membrane receptors for human kidney medulla: relation to adenylate cyclase activation. *Eur J Pharmacol* **85**: 291–304.

Kahn, A., Brachet, E., and Blum, D. (1979). Controlled fall in natremia and risk of seizures in hypertonic dehydration. *Intens Care Med* **5**: 27–31.

Lowbridge, J., Manning, M., Jaya, H., and Sawyer, W. H. (1978). [1-(β-mercapto-β, β-cyclopenyamethylenepropionic acid), 4-valine-8-D-arginine]vasopressin, a potent and selective inhibitor of vasopressor response to arginine-vasopressin. *J Med Chem* **21**: 313–315.

Morel, A., Lolait, S. J., and Brownstein, M. J. (1993). Molecular cloning and expression of rat V1a and V2 arginine receptors. *Regulat Peptides* **45**: 53–59.

Nguyen, M. K., Nielsen, S., and Kurtz, I. (2003). Molecular pathogenesis of nephrogenic diabetes insipidus. *Clin Exp Nephrol* **7**: 9–17.

Stephenson, J. L. (1973). Concentrating engines and the kidney. II. Multisolute central core systems. *Biophys J* **13**: 546–567.

Timmer, R. T., and Sands, J. M. (1999). Lithium intoxication. *J Am Soc Nephrol* **10**: 666–674.

Van Balkom, B. W. M., Graat, M. P. J., van Raak, M., Hofman, E., van der Sluijs, P., and Deen, P. M. T. (2004). Role of cytoplasmic termini in sorting and shuttling of the aquaporin-2 water channel. *Am J Physiol Cell Physiol* **286**: C372–C379.

Van Den Ouweland, A., Dreesen, J., Verdijk, M. et al. (1992). Mutations in the vasopressin type 2 receptor gene (AVPR2) associated with nephrogenic diabetes insipidus. *Nature* **2**: 99–102.

Yamamura, Y., Ogawa, H., Yamashita, H. et al. (1992). Characterization of a novel aquaretic agent, OPC-31260, as an orally effective, nonapeptide vasopressin V2 receptor antagonist. *Br J Pharmacol* **105**: 787–791.

Yatsu, T., Tomura, Y., Tahara, A. et al. (1997). Pharmacological profile of YM087, novel nonpeptide dual vasopressin V1a and V2 receptor antagonist, in dogs. *Eur J Pharmacol* **321**: 225–230.

CHAPTER **17**

Molecular Biology and Gene Regulation of Vasopressin

SWASTI TIWARI AND CAROLYN A. ECELBARGER
Department of Medicine, Division of Endocrinology and Metabolism, Georgetown University, Washington DC, 20007, USA

Contents

I.	Introduction	225
II.	The antidiuretic hormone, vasopressin	225
III.	Vasopressin receptors	228
IV.	Cellular regulation of water, electrolyte and mineral reabsorption	234
V.	Vasopressin, renal hemodynamics and blood pressure	240
	References	242

I. INTRODUCTION

Vasopressin or, in humans and rodents, arginine vasopressin (AVP), is the most important hormone in the regulation of urine concentration. Without the ability to concentrate our urine, we would excrete approximately 173 l of urine per day given an average glomerular filtration rate of 120 ml/min. Other factors have some role, such as the renal sympathetic nervous system and possibly, to a small extent the renin–angiotensin–aldosterone system, but none of these other systems or regulatory-hormone cascades even approaches the importance of AVP. This is illustrated by the fact that mutations that either result in the loss of circulating AVP or affect AVP actions on renal tubules, such as mutations in aquaporin 2 water channel or the V2 receptor, always result in an overt disease state, i.e. diabetes insipidus (discussed further in Chapter 19). In this chapter, we discuss in detail: (1) the synthesis, release and regulation of AVP; (2) the various subclasses of AVP receptors, their molecular structures, differences in their patterns of expression, regulation and signaling; (3) the regulation of renal transport including water (greater detail in Chapter 16), sodium, potassium, chloride, bicarbonate, calcium, magnesium and urea and their specific transporters and channels by AVP; and (4) AVP actions on renal hemodynamics including its potential role in blood pressure regulation and underlying mechanisms.

II. THE ANTIDIURETIC HORMONE, VASOPRESSIN

Vasopressin is a peptide hormone secreted by the pituitary glands. The name 'vasopressin' symbolizes the fact that this hormone can cause vasoconstriction in arterioles and thus increase arterial blood pressure. This same hormone is also known as the 'antidiuretic hormone' (ADH) as it causes antidiuresis (decreases urinary water excretion) by increasing the water permeability of the renal tubules.

A. Structure, Synthesis and Secretion

Vasopressin or 'arginine vasopressin' (AVP) consists of nine amino acid residues. The amino acid sequence of arginine vasopressin is Cys-Tyr-Phe-Gln-Asn-Cys-Pro-Arg-Gly. The arginine (Arg) is replaced by cysteine (Cys) in the structure of cysteine vasopressin. It is synthesized in magnocellular neurons of para- and supraventricular nuclei of the hypothalamus (Swaab et al., 1975). Vasopressin is synthesized as a precursor prohormone in the cell body of the magnocellular neurons and then packed into membrane-bound neurosecretory granules (Sachs et al., 1967). This prohormone consists of vasopressin and a glycopeptide and vasopressin-associated carrier protein, neurophysin. Cleavage of this large prohormone occurs during its axonal transport in the pituitary stalk to the posterior pituitary. The cleavage results in release, followed by storage, of the final hormonal product, AVP in the axonal bulb of the pars nervosa. Upon stimulation for its secretion, AVP is released from these storage lobes. Various kinds of stimuli that result in AVP release are discussed in detail later in this chapter. These stimuli also enhance the biosynthesis (Takabatake and Sachs, 1964), transport and post-translational cleavage process (Russell et al., 1981) of AVP and thus help to maintain the amount of the final hormone product in the neural storage lobe. The entire process of the synthesis until storage of AVP (including the transport and post-translational cleavage of the prohormone) takes approximately 1–2 h

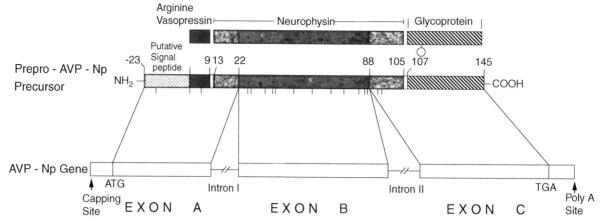

FIGURE 17.1 Schematic representation of the structural organization of the AVP gene and the deduced protein precursor. The protein precursor (prepro-AVP-Np) includes putative signal peptide (dotted bar); arginine vasopressin, AVP (black bar), neurophysin, Np (shaded bar); glycoprotein (hatched bar). The open bars in the prepro-AVP-Np precursor represent the amino acids involved in post-translational processing. The positions of significant amino acid residues are numbered; cysteine residues are marked by vertical lines below the protein precursor. Post-translational modification results in the formation of AVP-Np gene, which consists of the three principal functional domains, AVP, Np and glycoprotein on the three different exons (A, B and C in the figure). The processing amino acid sequence, Gly-Lys-Arg, essential for the release of hormone is located together with AVP on exon A. The sites of capping, translational initiation and termination and the poly (A) site are indicated in the AVP-Np gene structure. (Figure is from Schmale et al., 1983).

(Pickford, 1976). Figure 17.1 illustrates AVP gene structure and its precursor.

B. Half-life and Clearance

Normal plasma AVP levels in a healthy, hydrated human are approximately 1–2 pg/ml or about 1 pM. In the circulation, AVP circulates as an unbound peptide hormone and is rapidly metabolized by vasopressinase in the kidney and the liver, the two sites of AVP clearance (Baumann and Dingman, 1976). The half-life of AVP in rats is between 1 and 8 min, in humans between 10 and 35 min and, in dogs, between 4 and 8 min. Vasopressin is present in significant amounts in urine as there is little tubular reabsorption and degradation of AVP after glomerular filtration due to a disulfide bond between the two cysteine residues. However, urinary AVP levels should not be considered as a measure of plasma levels, as its excretion varies with glomerular filtration rate (GFR), tubular reabsorption and luminal degradation of the hormone.

C. Actions or Functions

Kidney and blood vessels are the two principal sites of vasopressin action. AVP modulates various aspects of renal function, which are described in greater detail in later sections of this chapter. Briefly, it acts on:

1. thick ascending limb (TAL), connecting tubule (CNT) and/or collecting duct (CD) to cause increased water reabsorption (not in TAL), sodium reabsorption and potassium secretion (in CNT and CD)
2. inner medullary CD cells to cause increased urea reabsorption
3. renal afferent and efferent arterioles, as well as epithelial medullary CD cells, to affect renal blood flow (RBF) and GFR.

In the vasculature, AVP binds with its receptor on smooth muscle cells of blood vessels and induces vasoconstriction, which increases arterial blood pressure. However, its vasoactive effects are negligible under normal physiological conditions. Nevertheless, under severe hypovolemic shock (decreased plasma volume), there is increased secretion of AVP, which increases systemic vascular resistance as a compensatory mechanism. Besides its effects on vasculature and renal tubules, AVP also acts on the central nervous system (CNS), where it regulates multiple physiologic activities including: learning and memory processes, blood pressure, adaptive sexual behavior and body temperature. For greater detail regarding CNS effects of AVP, readers should consult the review by de Wied et al. (de Wied et al., 1993). AVP also regulates blood glucose levels via its actions on the liver where it stimulates gluconeogenesis and glycogenolysis. In addition, it also stimulates insulin's and glucagon's secretion from the pancreas. Vasopressin also plays a physiological role via the V1a receptor in regulating both protein catabolism and glucose homeostasis (Hiroyama et al., 2007).

D. Similarity to Oxytocin

The structure of oxytocin, another nonapeptide hormone, is very similar to that of vasopressin with a substitution of two amino acids, i.e. isoleucine (Ile) in place of phenylalanine

(Phe) and leucine (Leu) in place of arginine (Arg). This may explain why oxytocin is slightly diuretic. Besides the similarities in their structure, oxytocin and vasopressin have other similarities too, such as the fact that they are located on the same chromosome (separated by 1500 genes). Like oxytocin, vasopressin can cause some uterine contractions.

E. Factors Regulating its Secretion

There are various factors that trigger vasopressin production and secretion into the bloodstream. Among them, the most important stimulus is a change in extracellular fluid osmolality, which is detected by the presence of 'osmoreceptors' located in the hypothalamus. Osmoreceptors sense changes in extracellular fluid osmolality caused by gain or loss of water and send a signal for the increase or decrease of AVP secretion. An increase in osmolality increases the rate of AVP secretion, while reduced osmolality inhibits AVP secretion.

The other 'important' factor regulating the secretion of AVP is a change in blood volume and pressure, which is detected by pressure-sensitive receptors, baroreceptors. A decrease in extracellular volume due to conditions such as hemorrhage or diarrhea is sensed by the baroreceptors present in the cardiac atria, aorta and carotid sinus. The signal is carried by neurons of the vagus and glossopharyngeal nerves to the central nervous system (CNS) resulting in the stimulation of AVP secretion. However, a decrease in plasma volume (hypovolemia) alone cannot cause a rise in plasma AVP until it is accompanied by a steep fall in blood pressure, in order to trigger arterial baroreceptors and AVP secretion. Thus, a decrease in cardiovascular pressure leads to an increase in AVP secretion and vice versa. The increased levels of circulating AVP result in water conservation through its antidiuretic actions on the kidney, thereby maintaining extracellular volume and hence blood pressure. Thus, it can be predicted that diseases associated with decreased input from the arterial baroreceptors, such as congestive heart failure, would have elevated levels of AVP secretion. During a severe decrease in extracellular volume and pressure, for example due to hemorrhage, strong signals sent by baroreceptors result in secretion of AVP which far exceeds that which is required for its antidiuretic actions on kidney. Under these circumstances, AVP directly acts on the vascular smooth muscle cells as a vasoconstrictor, which helps to maintain cardiovascular pressure independently of its slow antidiuretic action. Thus, excessive increases in AVP levels are an adaptive reflex in response to the prompt need for the blood pressure maintenance under a severe fall in extracellular volume and cardiovascular pressure.

In the past 20 or so years, the subfornical organs of the lamina terminalis, a part of the brain, has been recognized as a crucial site for the physiological regulation of vasopressin secretion. Circulating hormones such as angiotensin II and relaxin, as well as plasma hypertonicity, are among factors that have been suggested to exert their actions on the lamina terminus to regulate vasopressin secretion (McKinley et al., 2004).

Angiotensin II was the first blood-borne peptide hormone discovered to exert action on the lamina terminalis. It was shown that circulating angiotensin II could act on the neurons in the subfornical organ of the lamina terminalis to stimulate thirst in rats. In addition to driving thirst, systemically-infused angiotensin II was shown to stimulate an increase in vasopressin secretion in the dog and rat. This action of angiotensin II was attenuated when the efferent neural connections of the subfornical organ were disrupted, suggesting that the subfornical organ is the main site of action of circulating angiotensin II to induce vasopressin release.

Several studies have suggested that relaxin, a hormone secreted by the corpus luteum of the ovary during pregnancy in humans and rodents, may influence body fluid homeostasis. When injected systemically or directly into the brain, relaxin acts as both dipsogenic and a stimulus for vasopressin release. Moreover, plasma osmolality falls during the course of pregnancy in humans and rats and this effect can be mimicked by systemic infusion of relaxin over several days into ovariectomized rats, which is consistent with the proposal that there is a resetting of the osmoreceptor during pregnancy and that increased blood concentrations of relaxin may be the cause of such a resetting.

Sex steroids have also been shown to regulate the secretion of vasopressin. In this regard, estrogen and testosterone have both been shown to effect central neuropeptide transmission. In rodents, estrogen, in particular, has been shown to stimulate the expression of the AVP gene, resulting in a massive increase in AVP peptide in the bed nucleus of the stria terminalis. This estrogen or testosterone stimulated upregulation of vasopressin may play a key role in maintaining or regulating olfactory memory (Fink et al., 1996, 1998).

F. Pathology Associated with Dysregulation of AVP Secretion and Actions

The inappropriate or sustained secretion of AVP by the hypothalamic–pituitary system leads to a condition known as SIADH (syndrome of inappropriate antidiuretic hormone secretion), a relatively common cause of hyponatremia (Verbalis, 2003). Elevated AVP secretion leads to renal water retention and extracellular fluid expansion, which is compensated for by increased urinary Na^+ excretion. The combination of water retention and Na^+ excretion leads to hyponatremia. Bartter and Schwartz (Bartter and Schwartz, 1967) first characterized SIADH as 'hypotonic hyponatremia', urine osmolality greater than appropriate for the concomitant plasma osmolality, increased natriuresis, absence of edema or volume depletion and normal renal and adrenal function. There are various reasons/conditions responsible for inappropriately high levels of circulating AVP, including cancer that induces ectopic AVP production,

HIV/AIDS, pulmonary disease, endocrine disease, neurologic disease or trauma and surgery. This condition is expanded upon in Chapter 19.

Another potentially life-threatening disorder associated with dysregulation of AVP actions or secretion is exercise-associated hyponatremia (EAH). The primary cause of EAH is inappropriate body water retention leading to dilutional hyponatremia. EAH results from the combination of a high or excessively high fluid intake in the setting of a prolonged or sustained cardiovascular activity, which seems to be associated with modest elevations in plasma AVP levels (Siegel et al., 2007; Verbalis, 2007). The mechanism(s) responsible for non-osmotic stimulation of vasopressin secretion in EAH remain to be determined. However, possible factors are pain and hypotension, which commonly occur in marathon runners. In addition, the release of muscle-derived interleukin-6 during rhabdomyolysis, as shown in human research subjects, may also stimulate secretion of arginine vasopressin. Furthermore, rhabdomyolysis has been observed in other clinical conditions in which SIADH may accompany inflammatory stress (Siegel, 2006). Finally, use of non-steroidal prostaglandin inhibitors prophylactically as a means to alleviate pain or inflammation due to the activity may also be detrimental and lead to increased fluid salt and water retention (Page et al., 2007). Prostaglandins may be helpful in that they antagonize the actions of AVP with regard to cAMP generation and urine concentration (Murase et al., 2003).

III. VASOPRESSIN RECEPTORS

A. Receptor Subclasses

Essentially three main G-protein-coupled vasopressin receptors have been cloned and characterized. These are as follows: (1) type 1a (V1a); (2) type 1b (V1b); and (3) type 2 (V2) receptors. However, besides these three receptors, vasopressin also has equal affinity with oxytocin for oxytocin receptors (OT), a related G-protein-coupled receptor, and may exert some of its action via OT signaling. In addition to these classical receptors, a dual angiotensin II/vasopressin receptor has been isolated and characterized (Ruiz-Opazo, 1998). This dual receptor type has been localized in the outer medullary thick ascending limb and inner medullary collecting duct cells. Pharmacological characterization has revealed that this is an angiotensin II type I receptor (AT1)/V2 type receptor.

B. V1a, V1b and V2 Cloning

In 1992, Morel et al. (Morel et al., 1992) cloned the complementary DNA (cDNA) encoding the hepatic arginine vasopressin receptor type 1a (V1a) gene from the rat liver. V1a cDNA is 1354 base pairs in length and encodes a 394 amino acid protein with the molecular weight $(M_r) = 44\,202$.

Subsequently, the structural organization of the rat V1a receptor gene was reported in 1995; the gene spans 3.8 kb and consists of two coding exons separated by a non-coding intron of 1.8 kb (Murasawa et al., 1995). The first exon encodes six transmembrane domains and the second exon encodes the last transmembrane domain (i.e seventh). Two different transcription initiation sites were reported, i.e. at -243 and -237 base pairs upstream of the initiation codon (ATG). The region between -296 and -222 (relative to initiation codon) was found to exhibit promoter activity.

The human V1a receptor cDNA was cloned by Thibonnier and associates (Thibonnier et al., 1994). They reported a 1472 nucleotide sequence encoding a 418 amino acid protein $(M_r = 46\,745)$ deduced from the open-reading frame of the 1254 nucleotides spanning from 62 to 1315 of the cloned cDNA. Subsequently, they reported its structure, sequence and chromosomal localization (Thibonnier et al., 1996). Using northern blot analysis, they found a 5.5 kb mRNA transcript. The gene encodes two coding exons and a 2.2 kb intron located before the seventh transmembrane domain, as in the case of the rat V1a gene. For the human V1a receptor gene, the transcription initiation site is at 1973 bp upstream of the initiation codon. There are untranslated regions of 2 and 1 kb, respectively, at the 5' and 3' ends of the gene.

At approximately the same time, molecular cloning of the cDNA of human V1b (also known as V3) receptor was accomplished from pituitary (Sugimoto et al., 1994). The V1b receptor mRNA is a 5.2-kb transcript as detected by northern blot analysis. The deduced 424 amino acid sequence (calculated relative molecular mass is 47 034) has seven putative transmembrane domains. The human V1b receptor has quite high homology with the other vasopressin receptor subtypes, i.e. V1a (45%) and V2 (35%) and also with oxytocin receptors (45%) (Sugimoto et al., 1994; Saito et al., 1995).

The V1b receptor was also cloned and characterized from mouse and rat pituitary (Saito et al., 1995; Ventura et al., 1999). The rat V1b receptor gene contains three exons and two introns. The first short intron (161 bp) is located in the 5'-untranslated region from bp +124 to +285 (+1 was assigned to the proximal transcriptional start point). Similar to the other vasopressin receptor subtypes, the second intron was found to be present at the end of the sixth transmembrane domain. Two major putative transcription initiation sites have been mapped: (1) the proximal site, which is assigned +1 and (2) a distal site located at -31.

The human V2 receptor gene was cloned (Birnbaumer et al., 1992) and shown to be localized to the long arm of the X-chromosome at the q28-qter region. V2R cDNA predicts a protein with 341 amino acids with a calculated M_r of 40 285 in the absence of any post-transcriptional modification. Also in 1992, Morel, Lolait and colleagues (Lolait et al., 1992) cloned the rat V2 gene from kidney and reported that, in humans, the gene appeared on the X

TABLE 17.1 Actions of vasopressin through its receptor subtypes on the target cell. This table has been taken from (Holmes et al., 2001)

Receptors	Tissues	Principal Effects	Intracellular Signaling
V1aR	Vascular smooth muscle	Direct and indirect vasoconstriction	Phosphoinositide pathway (activate phospholipase C)
	Kidney (bladder, adipocytes, platelets, spleen, testis)		Increased intracellular Ca^{++}
V2R	Renal collecting duct	Increased permeability to water	Increased cAMP
	Endothelium	Vasodilation	NO mediated
V1bR	Pituitary	Neurotransmitter ACTH release	Increased cAMP
OTR	Uterus, mammary gland	Vasoconstriction	Phospholipase C
	Endothelium	Vasodilation	NO mediated

*V3R = V3 pituitary receptors.

chromosome near a locus for nephrogenic diabetes insipidus (NDI). Subsequently, Seibold et al. (Seibold et al., 1993) described the structure of the V2 receptor gene. The human V2 gene consists of three exons separated by two non-coding intronic sequences. The first intron is 360 bp long and interrupts a codon corresponding to the ninth amino acid of the receptor sequence. The second intron separates the sixth and seventh transmembrane domains. The human V2 receptor gene has 35% homology with the human $\beta1$ adrenergic receptor and its ligand-binding properties are similar to those of V1 receptors and oxytocin receptor (Birnbaumer et al., 1992).

C. Tissue/Cellular Localization

Vasopressin receptor subtypes are expressed in various tissue sites. Table 17.1 describes the location and the principal effects of these receptors. The V1a receptor is expressed in a variety of tissues including the brain, vasculature and kidney. In the brain, it is found in the septum, cerebral cortex, hippocampus and hypothalamus. V1a receptors are also localized on the smooth muscle cells of vessels of the systemic, splanchnic, renal and coronary circulation. In the kidney, V1a has been localized in renal tubules and vasculature of the kidney cortex and medulla (Park et al., 1997). With regard to tubular expression, Terada and associates (Terada et al., 1993) examined mRNA expression of V1a in microdissected segments and found high expression in the glomerulus, initial cortical CD, cortical CD, outer medullary CD, inner medullary CD and arcuate artery. Small but detectable signals were found in proximal convoluted and straight tubules, inner medullary thin limbs and medullary TAL. The presence of V1a in the TAL is somewhat controversial, however. At least a couple of laboratories (Ostrowski et al., 1992; Firsov et al., 1994) did not detect V1a message using quantitative RT-PCR (reverse transcriptase polymerase chain reaction) or *in situ* hybridization in rat TAL. Additional function studies did, however, suggest the presence of V1a receptors in TAL. Binding studies in isolated medullary TAL cells have suggested that a V1a-like receptor is present in this segment (there were sites bindable by a V1a antagonist) (Ammar et al., 1992). Furthermore, vasopressin has been demonstrated to trigger a transient increase in intracellular Ca^{2+}, possibly via a V1 receptor mediated signaling pathway, in the cortical TAL cells of the rabbit (Nitschke et al., 1991). In the collecting duct cells, V1a receptor protein has been localized at the luminal membrane in the rat kidney using immunobased approaches (Nonoguchi et al., 1995). Subsequently, Tashima et al. (Tashima et al., 2001) have demonstrated V1a receptor mRNA and protein in microdissected cortical CD. They reported that this receptor is present primarily in the principal and intercalated cells of the CD.

In contrast to the V1a receptor, the V1b receptor had a limited tissue distribution. It was initially defined as a pituitary-specific subtype in rat and humans, indeed, its mRNA was markedly abundant in pituitary glands. V1b receptors have also been localized in the anterior pituitary gland, adrenal glands, kidneys and pancreas (Jard et al., 1986; Lolait et al., 1995; Grazzini et al., 1996; Hurbin et al., 1998; Vaccari et al., 1998). However, it appears to have a much weaker expression in the kidney and the specific cell types which express V1b in kidney are not clearly defined.

The V2 receptor is expressed primarily in the kidney and mediates the vasopressin stimulated antidiuretic action on renal tubules. The localization of V2 receptor protein has been determined by several groups using immunohistochemistry. A decade ago, Nonoguchi and colleagues (Nonoguchi et al., 1995) demonstrated V2 receptor protein expression in the CD and in some TAL segments, but not all, using an antibody raised against the intracellular loop between the fifth and sixth transmembrane domain. Expression of the receptor was primarily basolateral, however, they did show luminal and basolateral membrane expression in the CD, especially in the terminal inner medullary CD. In contrast, Sarmiento and colleagues (Sarmiento et al., 2005), using an antibody directed against the second intracellular loop, reported that the V2 receptor is expressed in all thick ascending limbs, distal convoluted tubules, connecting

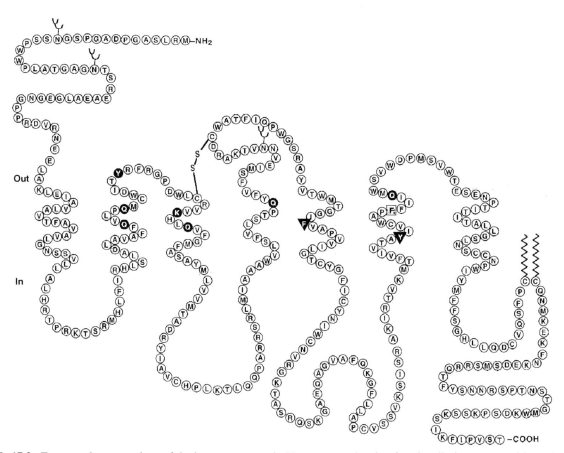

FIGURE 17.2 Transmembrane topology of the human vasopressin V1a receptor showing functionally important residues. Amino acids highlighted in black circles are critically involved in agonist binding; that in the grey square is possibly involved in antagonist binding; that in the grey circle is a conserved amino acid involved in activation of the receptor; those in black triangles modulate the process of receptor activation in oxytocin receptor. Potential glycosylation (on Asn 14, 27 and 196) and palmitoylation (on Cys365 and Cys366) sites are also indicated (taken from Barberis et al., 1998).

tubules and collecting ducts upon examination of rat tissue. Recently, Fenton et al. (Fenton et al., 2007) have generated and characterized a polyclonal antibody targeted against the N-terminus of the rat V2 receptor. Using their antibody, they have shown the cellular and subcellular localization of V2 receptor in the kidneys of rats and mice. They found predominantly intracellular expression of the V2 receptor with some apparent labeling in the basolateral membrane domains. Regarding the cellular localization, they demonstrated that V2 receptor was expressed abundantly in the collecting duct and was also apparent in the connecting tubule in both rat and mouse kidney. However, there was a complete absence of labeling in vascular structures and other renal tubules, including the TAL. However, they did find V2 receptor mRNA in microdissected rat TAL segments.

D. Molecular Structure

Molecular cloning has confirmed that vasopressin receptor subtypes are members of the G-protein-coupled receptor (GPCR) superfamily, consisting of seven hydrophobic transmembrane alpha-helices joined by alternating intracellular and extracellular loops, an extracellular N-terminal domain and a cytoplasmic C-terminal domain (Figure 17.2). The vasopressin receptor subtypes, including the oxytocin receptor, display a high degree of sequence identity, showing about 102 invariant amino acids among the 370–420 amino acids in the human receptors (Figure 17.3). As anticipated, these receptor subtypes exhibit structural features which are characteristic of most of the G-protein-coupled receptors such as the presence of a disulfide bridge between two highly conserved cysteine residues in the second and third extracellular domains, glycosylation on asparagine residues present in the extracellular domains and two relatively well-conserved cysteine residues within the C-terminal domain, which have been shown to be palmitoylated in other GPCRs (O'Dowd et al., 1989). The disulfide bond appears to be required for the correct folding of vasopressin receptors. The Cys124 and Cys205 of the rat and human platelet V1a receptor and also of bovine and porcine V2 receptors are probably involved in the tertiary structure of the receptor (Gopalakrishnan et al., 1988; Pavo and Fahrenholz, 1990).

Upon binding with AVP, they activate G-proteins and initiate a cascade of intracellular signaling events

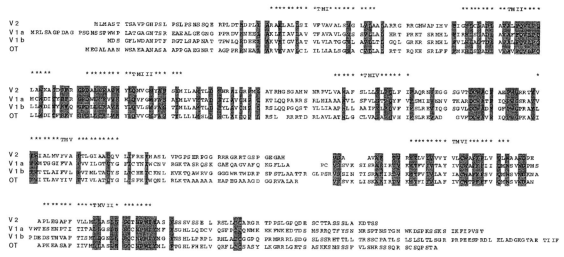

FIGURE 17.3 Primary sequence comparisons between human vasopressin/oxytocin receptors. Gaps have been introduced to maximize sequence identities. Asterisks denote predicted transmembrane regions I–VI. V2, amino acid sequence of the human V2 receptor; V1a, amino acid sequence of the human V1a receptor; V1b, amino acid sequence of the human V1b receptor; OT, amino acid sequence of the human oxytocin receptor (taken from Barberis et al., 1998).

(Jard, 1983). The ligand-binding site on the receptor is proposed to be in a pocket-like structure formed by the arrangement of the seven transmembrane domains (Mouillac et al., 1995a, b). Furthermore, mutagenesis experiments suggest that the agonist-binding site is located in a narrow cleft encircled by transmembrane (TM) domains (primarily TM II–VII), about 15 Å away from the extracellular surface.

Mutational analysis techniques have also revealed several domains which impart G-protein-coupling selectivity and are important for the activation and function of the receptors. The conserved aspartate (Asp^{97}) residue in transmembrane II is found to be functionally important for the activation of V1a vasopressin receptor. The conserved Pro^{322} situated on transmembrane VII in the human V2 receptor is probably also necessary to allow the relative movements within the helical bundle that are required for receptor activation. This highly conserved residue has also been found to have a key role in the vasopressin receptor activation process. The highly conserved triplet Asp-Arg-Tyr (Asp-Arg-His in human V2) located at the N-terminal of the intracellular loop 2 is also required for efficient G-protein activation. The importance of this conserved triplet is also true for various other receptors. In the human V2 receptor, an Arg137 to His137 mutation was found to abolish coupling to G-proteins and results in phenotypic nephrogenic diabetes insipidus. By analogous replacement of the intracellular loops between V1a and V2 receptors, it has been demonstrated that intracellular loop 3 of the V2 receptor has a key role in correct recognition and activation of Gs G-protein while intracellular loop 2 of the V1a receptor is critically involved in selective activation of Gq/G11 G-protein (Liu and Wess, 1996).

Sites for N-linked glycosylation are present in the extracellular segments of all the three receptor subtypes. In V1a, two N-linked glycosylation sites are present, one at the N-terminus and a second in the extracellular loop 2. Whereas both V1b and V2 receptor have only one N-linked glycosylation site present at the extracellular N-terminus (Innamorati et al., 1996). Glycosylation at the extracellular sites has been suggested to increase stability of the protein. Furthermore, both Cys residues at the C-terminus of the V2 receptor and equivalent sites (three) in the V1 receptors have been shown to undergo palmitoylation. The palmitoylation has been shown to impart stability to the V2 receptor at the cell membrane (Sadeghi et al., 1997). Furthermore, there are 8 threonine and 17 serine residues in the third cytoplasmic loop and C-terminal region of the human V1a receptor, which could be sites for regulatory phosphorylation.

Loss of function mutations in the gene encoding the V2 receptor causes a severe disturbance in water homeostasis known as X-linked nephrogenic diabetes insipidus (NDI) (Bichet et al., 1993; Bichet, 1996). In this disorder, patients are unable to concentrate their urine despite increased serum vasopressin levels. Over 180 gene mutations in the V2 receptor gene have been described which may disrupt the V2 receptor signaling and thus making the principal cell collecting duct insensitive to plasma vasopressin levels. The step at which the V2 receptor signaling is disrupted depends on the site and type of mutation. Based on the fate of the cellular process, Robben and associates (Robben et al., 2005) have recently classified the V2 receptor mutations into five different classes (Table 17.2).

In addition, functional mutations in the AQP2 gene, the major apical water channel of the collecting duct principal cell, can also a type of congenital NDI. This NDI can either be autosomal recessive or dominant. In addition to congenital forms of NDI, there are also acquired forms of NDI (due to electrolyte disturbances, urinary tract obstruction or lithium-ingestion). The mechanisms underlying these disorders are still not entirely clear, but likely involve some

TABLE 17.2 Vasopressin type 2 receptor mutations in nephrogenic diabetes insipidus

Nucleotide	Ammo Acid	Functionality	Conserved (Location)	Class
492T > C	L44P	F	Y (tmdl)	II
488T > A	146 K	F	N (tmdl)	II
548T > C	L62P	?	Y (tmdl)	II?
545-553 del	Δ62–64	G	N (tmdl)	II
574G > A	W71X		N (ICL 1)	I
612C > A	A84D	A	N (tmd2)	II
614G > A	D85N	F	Y (tmd2)	III
623G > A	V88M	?	Y (tmd2)	II
692T > C	W99R	A	Y (ECL1)	II, IV
671C > T	R104C	F	Y (ECL1)	II
674T > G	F105V	A	Y (ECL1)	IV
698C > T	R113W	F	Y (tmd3)	II, IV
749A > T	I130F	F	N (tmd3)	II
771G > A	R137H	G	Y (ICL2)	II, III, V
860T > A	S167T	F	N (tmd4)	II
861C > T	S167L	A	N (tmd4)	II
902C > T	R181C	A	N (tmd4)	IV
914G > T	G185C	A	N (ECL3)	IV
963G > A	G201D	F	Y (ECL3)	II, IV
965C > T	R202C	A	N (ECL3)	IV
966-967 del	AR202	A	N (ECL3)	IV
972C > A	T204N	A	N (ECL3)	II
975A > G	Y205C	A	Y (ECL3)	II
978T > A	V206D	A	N (tmd5)	II
1431C > T	P322S	F	Y (tmdT)	III, IV
1476C > T	R337X	A	N (C-tail)	I

F, Functional; A, disturbed AVP binding; G, disturbed G_s protein biding; conserved (Y) or not (N) between vasopressin receptors; tmd, transmembrane domain; ECL, extracellular loop; ICL, intracellular loop. This table has been abridged from the review article by Robben et al. (Robben et al., 2006).

disruption in the AVP/V2R/cAMP/AQP2 signaling cascade. The topic of NDI will be examined in greater detail in Chapter 19.

E. Signaling of Vasopressin Receptors

For most practical purposes, V1 and V2 receptors can be thought to signal through different pathways. V1 (V1a and V1b) receptors signal by modifying phospholipase activity, whereas V2 receptors regulate adenylyl cyclase (AC) activity. Figure 17.4 depicts the major components of V1 and V2 signaling pathways. The vasopressin V1a receptor is associated with Gq protein and phospholipase C and is primarily responsible for vasoconstriction of the vascular smooth muscle cells. Upon activation by AVP, there is an increase in intracellular Ca^{2+} via its coupling to phospholipase C (PLC) and protein kinase C (PKC). The pathways of intracellular Ca^{2+} elevation following activation of V1a receptors have been described in detail in Barrett et al. (2007). In addition to elevation of intracellular Ca^{2+}, V1a receptor activation may also sensitize the contractile apparatus to the effects of calcium via the inhibition of myosin light chain phosphatase by PKC. Similar to V1a receptors, V1b signaling is also coupled to Gq protein activation and phospholipase C signaling cascades.

The vasopressin type 2 receptor is primarily coupled to adenylate cyclase by the heterotrimeric G-protein, G_s. G_s is a GTP-binding protein comprising three subunits: -α, -β and -γ. Upon activation by AVP, the α-subunit releases GDP, binds to GTP and dissociates from the -β and -γ subunits. This Gα–GTP complex, in turn, activates adenylate cyclase which catalyzes the conversion of ATP to cAMP. Cyclic AMP then activates protein kinase A, initiates downstream signaling cascades resulting in increased urine concentration (as discussed below). Expression of Gsα can directly influence V2 signaling. Mice heterozygous for Gsα were demonstrated to have impaired urine concentrating ability in response to the V2 agonist, dDAVP (Ecelbarger et al., 1999).

In addition, there is evidence that the V2 receptor is coupled to calcium mobilization as well. Knepper and associates (Star et al., 1988; Ecelbarger et al., 1996a) demonstrated that vasopressin and the V2-select agonist, dDAVP induced calcium mobilization in microdissected inner medullary collecting ducts. Further studies by this group revealed that the Ca^{2+} mobilization appeared to be due to a mechanism other than activation of the phosphoinositide hydrolysis pathway and may involve ryanodine-sensitive Ca^{2+} stores (Chou et al., 1998, 2000).

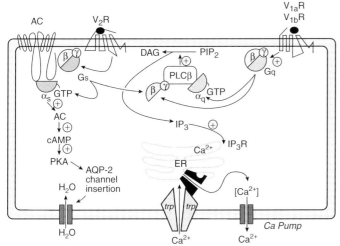

FIGURE 17.4 A hypothetical cell containing V1R (right side) and V2R (left side) signaling. The V2R signaling is shown in the collecting duct principal cells of the kidney (left portion): upon binding with the agonist, the V2 receptor promotes dissociation of heterotrimeric Gs–GTP into its α and $\beta\gamma$ subunits; the α subunit stimulates AC activity and the subsequent increase in cAMP activates PKA and the insertion of the water channel (AQP2) into the luminal surface of the collecting duct cell; the $\beta\gamma$ dimers dissociated from Gs might cause stimulation of PLCβ activity. The signaling by V1 receptors (right portion) promotes the dissociation of Gq/11–GTP. The α_q-subunit-induced stimulation of PLCβ activity promotes hydrolysis of PIP2, increasing the intracellular levels of DAG and IP3; IP3 in turn stimulates the IP3R that resides in the membrane of the ER and promotes Ca^{2+} release from intracellular stores. In addition to activating downstream signaling, emptying of the Ca^{2+} stores activates divalent cationic channels named *trp* that allow the flux of extracellular Ca^{2+} into the cell to replenish the stores. AC, adenylyl cyclase; AQP-2, aquaporin, AVP-regulated water channel; AVP, arginine vasopressin; DAG, diacylglycerol; ER, endoplasmic reticulum; IP3R, inositol (1,4,5)-trisphosphate receptor; PIP2, phosphatidylinositol (4,5)-bisphosphate; PKA, protein kinase A; PLCβ, phospholipase Cb. (From Birnbaumer, 2000.)

F. Receptor Regulation and Desensitization

Following agonist stimulation, G-protein-coupled receptors develop a reduction in responsiveness called 'desensitization'. Vasopressin-induced V1a and V2 receptor desensitization (homologous desensitization) has been described in various tissues. Desensitization involves phosphorylation of the receptors by GPCR kinases (GRKs). Receptor phosphorylation promotes the binding of arrestin which, in turn, uncouples the receptor from the G-protein and enhances sequestration (Innamorati et al., 1996, 1998a, 1999). The kinetics of receptor desensitization are relatively quick for the V1a receptor and rather slow for the V2 receptor (Innamorati et al., 1998a). Upon binding with vasopressin, both V1a and V2 receptors are phosphorylated. The receptor gets internalized in approximately 3 min following agonist binding. In the case of the V1a receptor, phosphatases dephosphorylate the receptor in approximately 6 min ($t_{1/2}$) and then it recycles back to the plasma membrane, which takes around 60 min. Unlike the V1a receptor, the phosphatases remain associated with the V2 receptor for about 3 h after the removal of vasopressin. This difference in sensitivity to dephosphorylation is likely due to the particular sites of phosphorylation on the V2 molecule (SER 362, 363 and 364) (Innamorati et al., 1998b). Thus, the V2 receptor fails to recycle back to the plasma membrane. V1a receptors are also subjected to heterogeneous desensitization by angiotensin II (Ancellin et al., 1999).

In addition, vasopressin receptors have been shown to be regulated by a variety of other physiologic states. In the CD, V1a receptor mRNA and protein has been demonstrated to increase under chronic metabolic acidosis (Tashima et al., 2001). However, dehydration leads to either a decrease or no change in its mRNA levels in the CD (Terada et al., 1993; Tashima et al., 2001).

With regard to the V1b receptor, the number of V1b receptors has been demonstrated to be regulated during stress in the anterior pituitary and is directly correlated to corticotroph responsiveness. Molecular regulation of pituitary V1b involves transcriptional and translational mechanisms. V1b receptor gene transcription depends on a number of responsive elements in the promoter region, of which the stretch of GA repeats near the transcription start point (GAGA box) is essential. The V1b receptor is mainly regulated at the translational level. The repressor effect of small open-reading frames present upstream of the main V1b receptor and an internal ribosome entry site, are the two potential mechanisms by which the 5' untranslated region of the receptor mediates negative and positive regulation of its translation.

In addition, the V2 receptor has been shown to be down-regulated by chronic acidosis and upregulated by dehydration (Machida et al., 2007). Its upregulation by dehydration has been shown to involve increased synthesis of prostaglandins in the renal medulla (Machida et al., 2007). The balance between synthesis and destruction of the V2 receptor in unstimulated cells is maintained by a slow degradative pathway. This slow pathway continuously degrades the V2 receptor by a ubiquitin-independent mechanism. However, when stimulated by an agonist, V2 receptors are degraded by a rapid ubiquitin-dependent mechanism (Martin et al., 2003).

G. Actions and Interactions of AVP Receptors

V1a receptors are primarily involved in vasoconstrictive responses in vascular tissue. Furthermore, V1a receptors also mediate cell proliferation, platelet aggregation, coagulation factor release and gluconeogenesis. V2 receptors in the kidney are primarily responsible for setting into motion a series of regulatory events aimed at conserving (reabsorbing) the filtered water load (this is detailed in the next two sections of this chapter). In addition, body fluid homeostasis may be regulated not only by basolateral AVP, but also by luminal AVP, possibly binding to V1a receptors. Recently, Izumi et al. (Izumi et al., 2007) used the LLC-PK1 cell line, derived from the proximal tubule of porcine kidney, to demonstrate an interaction between V2 promoter activity and the V1a receptor pathway. Their results suggested that intracellular cross-talk between the V1a receptor and V2 receptor pathways is important in the maintenance of body fluid homeostasis. They have shown that the V2 receptor is down-regulated via the V1a receptor pathway and upregulated via the V2 receptor pathway. Their results suggest that the opposite effects of PKC (downregulation of V2 receptor) and PKA (upregulation of V2 receptor) stimulation could be exerted on the same consensus sequences in the promoter region of V2 receptor.

Moreover, the V2 receptor may also mediate the vasodilatory effects of AVP (Cowley et al., 1995, 2003). AVP-induced vasorelaxation has been suggested to be mediated by cAMP-stimulated nitric oxide (NO) generation in endothelial cells. The localization of V2 receptor on the human endothelial cells further strengthens this possibility. More regarding the regulation of nitric oxide by vasopressin in the kidney is provided later in this chapter.

IV. CELLULAR REGULATION OF WATER, ELECTROLYTE AND MINERAL REABSORPTION

Vasopressin plays a major role in the reabsorption of a variety of molecules and elemental substances including sodium, chloride, potassium, magnesium, calcium, bicarbonate, urea and water. Most of this occurs via activated channels and transporters that are integral to the plasma membrane of the renal tubule cells, primarily in the thick ascending limb (TAL) through the collecting duct (CD). Vasopressin acting through the V2 receptor subtype is responsible for the majority of regulated transport.

A. Water

Vasopressin is clearly the pre-eminent hormone regulator of water balance. As discussed above, during periods of thirsting or volume depletion, vasopressin release from the posterior pituitary is stimulated. AVP binding to the V2 receptor, thought to be basolateral in most renal cell types, increases the production of cyclic AMP. Cyclic AMP stimulates the trafficking of the major water channel (aquaporin-2) into the collecting duct principal cell apical membrane. Greater detail surrounding the regulation of this channel and of other related aquaporins is provided in Chapter 16.

B. NaCl

Vasopressin also stimulates NaCl reabsorption in both the TAL and the CD. Vasopressin action to increase NaCl reabsorption in the CD and TAL is clearly relevant to water balance in that it facilitates the establishment of the corticomedullary gradient allowing for more efficient concentration of urine. Its relevance as a hormone system to overall sodium balance is less clear, given the fact that the renin–angiotensin–aldosterone system (RAAS) seems to play a much more dominant role in this regulation. Nonetheless, vasopressin has been demonstrated to affect sodium balance. Human patients administered the V2-selective agonist, dDAVP, had reduced urine flow rate and sodium excretion (Bankir et al., 2005). However, this effect was not apparent in patients with diabetes insipidus as the result of a mutation in the V2 receptor.

C. The CD and NaCl Uptake in Response to AVP

In the collecting duct, several investigators have shown, in acute studies, that *in vitro* application of vasopressin will increase sodium transport in perfused tubules obtained from the cortex of rat (Reif et al., 1984, 1986; Tomita et al., 1985; Schlatter and Schafer, 1987; Chen et al., 1990; Hawk et al., 1996) and rabbit (Frindt and Burg, 1972; Chen et al., 1990) kidneys, as well as in primary rabbit cortical CD suspensions (Canessa and Schafer, 1992) and in several cell lines such as A6 cells (cultured toad kidney cells) (Bindels et al., 1988; Wong and Chase, 1988) and M-1 cells (derived from mouse cortical collecting duct) (Nakhoul et al., 1998). Furthermore, in perfused tubules, the effect was additive to, and thus independent from, the increase in transport observed with aldosterone (Reif et al., 1986; Chen et al., 1990; Barbry and Hofman, 1997). Moreover, in several of the above perfused-tubule studies, it was shown that without pretreatment of animals with mineralocorticoids, sodium transport in response to vasopressin was nearly imperceptible in the cortical collecting ducts (Tomita et al., 1985; Reif et al., 1986). Chronically elevated vasopressin levels also increase the sodium reabsorptive capacity of the collecting duct. In 1997, Djelidi and colleagues (Djelidi et al., 1997) demonstrated that chronic exposure of rat CCD cells to vasopressin increased ^{22}Na influx.

D. Regulation of CD ENaC by AVP

Whether chronic or acute, increased sodium reabsorptive activity of the CD in response to AVP involves increased activity of the sole apical sodium entry route, the amiloride-sensitive epithelial sodium channel (ENaC). ENaC, in the kidney, is a hetero-multimer, either a tetramer (Firsov et al.,

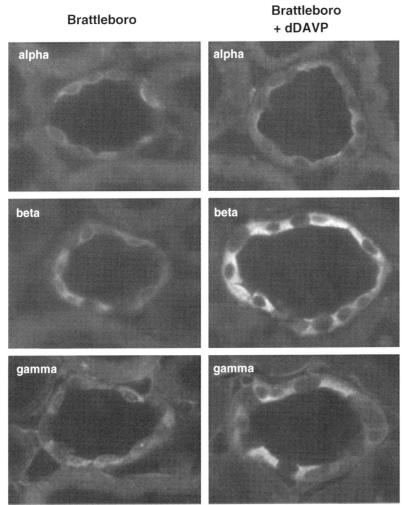

FIGURE 17.5 Immunohistochemical detection of ENaC subunits in representative CCD profiles of control and dDAVP-treated homozygous Brattleboro rats. Chronic dDAVP infusion to male Brattleboro rats markedly increased the cytoplasmic staining of beta- and gamma-, but not alpha-ENaC in collecting duct principal cells (Sauter et al., 2006).

1998; Dijkink et al., 2002) or a nonomer (Staruschenko et al., 2004) made up of three distinct subunits: α, β and γ. The three subunits were originally cloned by Canessa and colleagues (Canessa et al., 1993, 1994). The stoichiometry in kidney has been proposed to be either 2-αs, 1-β and 1-γ (Firsov et al., 1998; Dijkink et al., 2002) or three of each (Staruschenko et al., 2004). Recent studies have shown that acute ENaC stimulation by cyclic AMP in renal cell lines is the result of the trafficking of ENaC towards the apical membrane from a recycling subapical channel pool (Morris and Schafer, 2002; Butterworth et al., 2005). These studies suggest that AVP may regulate ENaC by changes in subcellular distribution (trafficking), similar to the other major hormones that activate ENaC, such as insulin (Tiwari et al., 2007) and aldosterone (Loffing et al., 2001). However, studies by Sauter and associates (Sauter et al., 2006) did not find differences in ENaC subunit subcellular distribution in response to chronically elevated dDAVP (by infusion) in Brattleboro (lack endogenous vasopressin) and Sprague-Dawley rats.

However, ENaC subunit expression does change in response to chronically elevated vasopressin. Chronic treatment with AVP or dDAVP has been demonstrated to increase the expression of protein and/or mRNA for the β- and γ-subunits, (but not α-) in a rat cortical CD (CCD) cell line (Djelidi et al., 1997) and in the rat kidney (Ecelbarger et al., 2000; Nicco et al., 2001; Sauter et al., 2006) (Figure 17.5). In our studies (Ecelbarger et al., 2000), we also showed that chronic water restriction of Sprague-Dawley rats resulted in a similar upregulation of β- and γ-, but not α- subunits. This effect was accompanied by significant increase in sodium transport in the CCD (Djelidi et al., 1997; Nicco et al., 2001) suggesting concomitant changes in functional ENaC membrane proteins, despite no measurable change in the abundance of α-ENaC.

The molecular mechanism(s) underlying the increase in mRNA and protein levels for β- and γ-ENaC in response to AVP are not known. However, chronic exposure to vasopressin is thought to lead to increased expression of aquaporin-2 protein via potentially multiple effects on regulatory motifs (or elements) in the 5'-flanking region of the gene known as cyclic AMP regulatory elements (CRE) (Matsumura et al., 1997). A CRE has also been found in the 5'-flanking region of γ-ENaC.

E. The TAL and NaCl Uptake in Response to AVP

Similar to the CD, NaCl is reabsorbed from the TAL along its entire length from medulla to cortex. Morel et al. (Imbert et al., 1975; Imbert-Teboul et al., 1978) were the first to demonstrate the presence of vasopressin-sensitive adenylyl cyclase in both the medullary TAL and cortical TAL of rodents and rabbits. This provided the first evidence that these segments were vasopressin sensitive. Subsequent studies demonstrated that vasopressin was able to increase the intracellular level of cAMP in microdissected thick ascending limb segments (Torikai et al., 1981; Torikai and Imai, 1984). Increased NaCl reabsorption in the TAL is critical for both concentration and dilution of urine. The involvement of the TAL or the loop of Henle, *per se*, in urine concentration was originally conceived by Swiss scientists, i.e. Werner Kuhn and colleagues, in a process described as 'countercurrent multiplication' (Knepper et al., 1999). In this process, a single act or 'effect', which requires a small input of energy (ATP) is compounded multiple times to amplify the overall effect. In this case, the development of a renal medulla more concentrated than plasma. This is assisted by the counterflow between the ascending and descending limbs of Henle's loop, increasing the tonicity of the interstitium as one progresses deeper into the medulla (Knepper et al., 1999). The countercurrent multiplier hypothesis became widely accepted after the classic paper from Gottschalk and Mylle (Gottschalk and Mylle, 1959) utilizing micropuncture to demonstrate increased tonicity of inner medullary tubule fluid relative to plasma.

With the advent of the perfused-tubule approach, it became possible to demonstrate directly the mechanism of luminal dilution (Burg and Green, 1973; Rocha and Kokko, 1973). NaCl is rapidly absorbed by active transport, lowering the luminal NaCl concentration and osmolality below levels in the peritubular fluid. Because the osmotic water permeability of the thick ascending limb is very low, water transport does not occur despite the high transepithelial osmotic gradient. This is essential for the establishment of the cortico-medullary osmotic gradient.

Greger and associates (Greger and Schlatter, 1983a, b; Greger et al., 1983) were instrumental in determining the stoichiometry underlying NaCl reabsorption in the TAL. NaCl transport across the apical plasma membrane occurs via a furosemide or bumetanide-inhibitable Na-K-2Cl co-transporter. This protein was cloned by a couple of groups and is referred to as either NKCC2 (Payne et al., 1995) or BSC1 (Delpire et al., 1994). The potassium ions that enter the cell on the co-transporter are largely recycled into the lumen via apical potassium channels. Patch-clamp studies have demonstrated that there are two such K^+ channels in the apical plasma membrane with unit conductances of 72 and 30 pS (Wang, 1994). Sodium ions are actively transported from the cell into the interstitium by the basolateral Na-K-ATPase. The chloride ions that enter the cell via NKCC2 exit across the basolateral plasma membrane via a chloride channel and/or by K-Cl co-transport. Overall, electroneutrality is maintained by paracellular sodium ion movement from the lumen driven by a lumen-positive voltage across the epithelium. The net process results in the transfer of equal numbers of sodium and chloride ions from lumen to interstitium.

F. AVP Increases the Activity of NKCC2

Vasopressin acts acutely (within seconds or minutes) to increase the rate of NaCl absorption in the medullary thick ascending limb of rodents (Hall and Varney, 1980; Sasaki and Imai, 1980; Hebert et al., 1981; de Rouffignac et al., 1983). Theoretically, vasopressin could act at any or all of the transporters expressed in this cell type. However, perfused tubule studies have clearly demonstrated that the NKCC2 is one protein with increased activity (Hebert et al., 1984; Molony et al., 1987; Sun et al., 1991). The mechanism of this action is as yet unknown, although it has been well established that a closely related bumetanide-sensitive Na-K-2Cl co-transporter (NKCC1 or BSC2), which is responsible for fluid secretion in various tissues, is regulated by phosphorylation (Tanimura et al., 1995; Lytle and Forbush, 1996). Immunoelectron microscopic studies have demonstrated that NKCC2 is abundant in both the apical plasma membrane of thick ascending limb cells and in intracellular vesicles of medullary TAL cells (Nielsen et al., 1998), raising the possibility that vasopressin could increase plasma membrane NKCC2 activity through regulation of trafficking. In addition, Knepper and associates (Kim et al., 1999) found that moderate water restriction of rats markedly increased the renal abundance of NKCC2, as did dDAVP infusion into Brattleboro rats. Thus, the long-term adaptation of transepithelial NaCl transport in the medullary TAL is apparently mediated in part by an increase in the number of co-transporters per cell. Increased abundance of NKCC2 in response to chronically elevated AVP or its analogs may be the results of increased cellular levels of cyclic AMP. Consistent with this possibility are findings of Igarashi et al. (Igarashi et al., 1996), who have recently cloned and sequenced a large portion of the 5' flanking region of the mouse NKCC2 gene. Analysis of the sequence revealed the presence of a cyclic AMP regulatory element that could potentially be a mediator of cyclic AMP-mediated transcriptional regulation. AVP binds to the V2 receptor which then couples to adenylyl cyclase through the heterotrimeric G-protein, Gs. The adenylyl cyclase isoform expressed in the thick ascending limb is AC-VI, a calcium-inhibitable isoform (Chabardes et al., 1996). Activation of adenylyl cyclase increases cyclic AMP in the cell which, in turn, activates protein kinase A. This in turn phosphorylates various proteins in the cell including transcription factors. Furthermore, mice which are heterozygous for the α-subunit of the heterotrimeric G-protein Gs were demonstrated to have 50%

reduced levels of cyclic AMP in their TAL and a 50% reduction in NKCC2 (Ecelbarger et al., 1999). Since the V2 receptor is expressed in the TAL and signals through increased cyclic AMP, these data further strengthen the conclusion that there is a causal link between V2-mediated signaling and NKCC2 expression.

G. Regulation of Na-K-ATPase by AVP

The 'sodium pump', sodium, potassium, adenosine triphosphatase (Na-K-ATPase) actively pumps sodium from the proximal tubule cells, as well as both TAL and CD cells into the interstitium. The pump is composed of α-, β- and γ-subunits. Activity of this pump in the TAL is vital for the establishment of the cortico-medullary gradient for urine concentration. In the distal tubule, it is also needed for sodium reabsorption against a gradient. Its actions lower intracellular sodium concentrations so that the apical entry of sodium becomes a rate-limiting step in sodium reabsorption. Na-K-ATPase has been reported to be activated by vasopressin in the TAL (Charlton and Baylis, 1990). Vasopressin and cyclic AMP, its second messenger, stimulate Na-K-ATPase activity within minutes through translocation of sodium pumps from a brefeldin A-sensitive intracellular pool to the plasma membrane. However, another group has reported the opposite effect, a decrease in Na-K-ATPase activity in the TAL in response to vasopressin (Aperia et al., 1994). Additional studies are needed to clarify this controversy. Likewise, in the distal tubule, including the collecting duct, AVP has been demonstrated to stimulate Na-K-ATPase activity (Feraille et al., 2003; Vinciguerra et al., 2005). In the TAL, hyperosmolality was also found to be a stimulus for increased Na-K-ATPase activity. However, this does not appear to be the case for the outer and inner medullary collecting cells (Takayama et al., 1999; Sakuma et al., 2005).

H. Potassium Conductance is Regulated by AVP

In the TAL, two groups of investigators have demonstrated an increase in apical potassium conductance in response to AVP (Reeves et al., 1989; Wang, 1994). They used the patch-clamp technique to demonstrate that the apical plasma membrane of the rat TAL contains a 30-pS ATP-sensitive K^+ channel whose activity is increased by application of cAMP or vasopressin. The cDNA for a renal medullary potassium channel (ROMK) responsible for the apical K^+ conductance in the TAL was cloned in 1993 (Ho et al., 1993). Subsequent studies have demonstrated that ROMK is a substrate for phosphorylation by protein kinase A (Xu et al., 1996). In 2001, Ecelbarger and associates (Ecelbarger et al., 2001) demonstrated that infusion of dDAVP for 7 days into Brattleboro rats resulted in dramatic increases in apical membrane labeling of ROMK in the TAL of the dDAVP-treated rats, as assessed by immunocytochemical analyses.

Using immunoblotting, a more than threefold increase in immunoreactive ROMK levels was observed in the outer medulla after dDAVP infusion. Restriction of water intake to increase vasopressin levels also significantly increased TAL ROMK immunolabeling and abundance in immunoblots.

In the collecting duct, AVP increases potassium secretion, rather than reabsorption. It is still not entirely clear which AVP receptor is responsible for this effect and what potassium channel. There are two apical potassium channels involved with secretion in the distal tubule: (1) ROMK, expressed solely in principal cells and (2) the more recently characterized (BK) or the 'big potassium' channel, expressed in principal and intercalated cells (Sansom and Welling, 2007). One group (Schafer et al., 1990) showed a cyclic AMP analog and forskolin increased potassium secretion, as did AVP in perfused tubules, suggesting a V2-receptor mediated mechanism. Amorim and colleagues (Amorim et al., 2004), using receptor-selective antagonists, have reported that apically-applied AVP stimulates potassium secretion via the V1 receptor and activation of the phospholipase C/protein kinase C signaling cascade. They further suggest AVP acts via the BK channel since secretion was blocked by a BK-selective inhibitor derived from scorpion venom, iberiotoxin.

Nonetheless, the physiological relevance of AVP-mediated increased potassium secretion is somewhat uncertain. Rieg and colleagues (Rieg et al., 2007) demonstrated that wild-type mice given a V2 antagonist have increased, rather than decreased potassium secretion. This effect was absent in mice with knockout of the BK receptor, which has been shown to be upregulated primarily by increased collecting duct flow, which is decreased in high AVP states and increased during diuresis. Flow may be the most important stimulus for net potassium flux in the distal tubule.

I. Regulation of Chloride (Cl^-) Conductance by AVP

Numerous studies using primarily micropuncture and perfused tubules have demonstrated that AVP regulates not only sodium, but also chloride conductance in the TAL through the CD. The net flux of chloride ions in response to AVP can be positive or negative depending on the cell type being regulated and specific conditions and the physiological state of the organism.

J. AVP increases chloride reabsorption in the TAL

Schlatter and Greger have provided evidence that acute vasopressin, acting through cyclic AMP, directly increases basolateral chloride conductance in the mouse medullary thick ascending limb (Schlatter and Greger, 1985). Studies conducted by Winters and associates (Winters et al., 1997) provide evidence that this effect may be mediated by the chloride channel (rbClC-Ka) localized to the basolateral plasma membrane of thick ascending limb. Increased chloride reabsorption in response to chronic, rather than acute, AVP administration has also been reported. Brattleboro rats infused with AVP for 10 to 21 days had an approximate doubling of their spontaneous rate of active chloride absorption in perfused isolated medullary TALs (Besseghir et al., 1986).

K. AVP can Elicit Both Reabsorption and Secretion of Chloride in the Distal Tubule

AVP is stimulatory of ENaC activity in the collecting duct principal cells to reabsorb sodium. However, how AVP regulates the counter ion chloride to result in overall net NaCl reabsorption is less clear. There has been described a chloride permeable paracellular pathway, as well as transcellular transport through the β-intercalated cells via an apically located Cl^-/HCO_3^- exchanger (SLC26A4, pendrin) and basolateral Cl^- conductance (Schuster and Stokes, 1987; Van Huyen et al., 2001). A number of studies using the short-circuit current (I_{sc}) method have demonstrated that amphibian A6 cells and cultured CCD and inner medullary collecting duct (IMCD) cells can be stimulated by vasotocin or vasopressin to secrete Cl^-. However, under more physiological conditions, i.e. open-circuit conditions, vasopressin (or vasotocin) preferentially increases net Cl^- absorption by the same cells (Chalfant et al., 1993; Verrey, 1994). Thus, net Cl^- absorption, in response to AVP, probably results from a balance between absorption and secretion (Simmons, 1993).

One physiological state in which the distal tubule exhibits Cl^- secretion is hyperkalemia (Wingo, 1990). A specific chloride channel which may have a role in this secretion is the cystic fibrosis transmembrane regulator (CFTR). CFTR mRNA has been detected by reverse transcriptase polymerase chain reaction (RT-PCR) all along the renal tubule in renal tissues and in cultured cells (Vandewalle, 2007). Also specific pharmacologic blockade of CFTR in DCT, CCD and IMCD cell types resulted in impaired chloride secretion (Van Huyen et al., 2001).

L. AVP and Chloride/Bicarbonate Exchange

Chloride/bicarbonate exchange in renal epithelial cells has been demonstrated to be enhanced by AVP. SLC26A4 (pendrin) is an Na^+-independent, $Cl^-/HCO_3^-/OH^-$ exchanger that is expressed in the apical regions of type B and non-A, non-B intercalated cells within the cortical collecting duct (CCD), the connecting tubule and the distal convoluted tubule where it mediates HCO_3^- secretion and Cl^- absorption. Amlal et al. (Amlal et al., 2006) recently showed increased SLC26A4 in the kidney of Brattleboro rats infused with dDAVP. Furthermore, increased activity of SLC26A4 is likely one means for increased chloride reabsorption in response to AVP. Northern hybridization and immunofluorescence labeling indicated a significant increase (+80%) in SLC26A4 mRNA expression along with a sharp increase in protein abundance in B-type intercalated cells in the cortical CD in the dDAVP-treated rats (Amlal et al., 2006). They also demonstrated reduced plasma bicarbonate levels in these rats suggestive of net bicarbonate secretion. Two additional major regulators of the activity and expression of SLC26A4 are aldosterone analogs and dietary Cl^- restriction. On the other hand, in the inner medullary CD, the abundance of the basolateral Cl^-/HCO_3^- exchanger (AE1) and apical H+-ATPase were significantly reduced in dDAVP-treated rats.

In addition to SLC26A4, another Cl^-/HCO_3^- exchanger, SLC26A7 has been demonstrated to be upregulated by AVP. Petrovic et al. (Petrovic et al., 2006) demonstrated the absence of expression of SLC26A7 in the OMCD of Brattleboro rats and its induction in response to dDAVP to a level similar to what was normally seen in the Sprague-Dawley rat. DDAVP did not affect the mRNA expression of SLC26A7 suggesting that this protein is upregulated at the post-transcriptional level. Activation of this exchanger should result in bicarbonate reabsorption rather than secretion. In the kidney, SLC26A7 localizes to the basolateral membrane of acid-secreting α-intercalated cells of the outer medullary collecting duct (OMCD), the portion of the CD with the highest rate of acid secretion. There is no evidence of SLC26A7 expression in the cortical CD. Secretion of acid in α-intercalated cells of the OMCD occurs via vacuolar H^+-ATPase in conjunction with H^+-K^+-ATPase and results in the generation of intracellular HCO_3^-, which is then reabsorbed across the basolateral membrane via SLC26A7. (Thus, this protein may also have a role in Cl^- secretion in response to AVP.) Co-localization of SLC26A7 with AE1 on the basolateral membrane of α-intercalated cells suggests an important role for SLC26A7 in acid secretion and HCO_3^- absorption in the OMCD. Studies in *Xenopus* oocytes showed that increasing the tonicity of the medium also increased the activity of SLC26A7 (Xu et al., 2006).

M. Urea Reabsorption is Stimulated by AVP

Urea reabsorption by the mammalian renal tubule occurs by two different mechanisms:

1. a constitutive process that occurs in the proximal tubule and accounts for reabsorption of nearly 40% of the filtered load of urea
2. a regulated process that occurs in the distal tubule and depends on the level of antidiuresis among other factors (Fenton and Knepper, 2007).

Regulation of urea reabsorption in the collecting duct has been the subject of intense scrutiny by several laboratories in

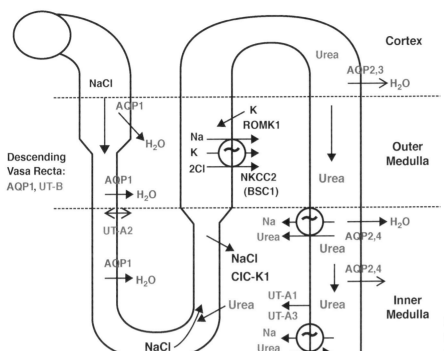

FIGURE 17.6 Diagram showing the location of major medullary transport proteins involved in the urine concentrating mechanism.

the last three to four decades. A model of urea recycling along the renal tubule is shown in Figure 17.6. The cortical collecting duct (CCD) has a very low urea permeability that is not increased by vasopressin (AVP) (Grantham and Burg, 1966). On the other hand, the terminal part of the inner medullary collecting duct (IMCD) possesses extraordinarily high urea permeability augmented by AVP (Sands and Knepper, 1987; Sands et al., 1987). Urea transport in the IMCD is inhibited by phloretin and urea analogs and is a saturable process, consistent with a transporter-mediated (facilitated) mechanism (Chou and Knepper, 1989; Chou et al., 1990).

N. Molecular Biology and UT-A1

The urea transporter(s) were cloned in the early 1990s (Smith et al., 1995). There are two distinct but closely related urea transporter genes: UT-A (*Slc14a2*) and UT-B (*Slc14a1*). Several urea transporter isoforms are derived from the UT-A gene via alternative splicing and alternative promotors (Fenton and Knepper, 2007). Multiple cDNAs that encode urea transporters have been isolated and characterized. UT-A1 is expressed exclusively in IMCD cells. On immunoblots, it is composed of two bands, one at approximately 97 and the other at 117 kDa as a result of different states of glycosylation (Bradford et al., 2001). DDAVP infusion or water restriction reduced the density of the 117 kDa form (glycosylated form) in rats, whereas treatments which resulted in diuresis, i.e. furosemide administration, increased the 117 kDa band of UT-A1 (Terris et al., 1998). Diuresis was also found to increase the degree of glycosylation of other transporter or channel proteins regulated by AVP including NKCC2 and aquaporin 2 (Ecelbarger et al., 1996b; Terris et al., 1996). Finally, unexpectedly, water restriction producing chronically elevated levels of AVP resulted in decreased basal and AVP-stimulated urea transport in terminal IMCD (Kato et al., 1998).

O. UT-A2 is Increased by AVP

UT-A2 is expressed in the inner stripe of the outer medulla, where it is localized to the lower portions of the thin descending limbs (tDL) of short loops of Henle and, under prolonged antidiuretic conditions, in the inner medulla where it is localized to the tDL of long loops of Henle (Fenton and Knepper, 2007). AVP increases UT-A2 abundance and recent studies have determined that UT-A2-mediated urea transport can be regulated acutely by cAMP. In a similar manner to UT-A1, expression of UT-A3 is restricted to the terminal IMCD where, in mouse, it is both intracellular and in the basolateral membrane domains. UT-A3 mRNA abundance can be upregulated by the prolonged action of AVP. In contrast to the multiple UT-A isoforms, the mouse UT-B gene encodes only a single protein that is expressed throughout the kidney medulla in the basolateral and apical regions of the descending vasa recta (DVR) endothelial cells (Fenton and Knepper, 2007). Long-term treatment with the type II vasopressin receptor agonist dDAVP causes downregulation of UT-B protein abundance.

P. Regulation of Calcium and Magnesium Reabsorption by AVP

Early studies by Costanzo and Windhager (Costanzo and Windhager, 1980) did not observe any change in calcium absorption in the microperfused rat distal convoluted tubule with administration of AVP. However, other studies suggest that AVP can cause the reabsorption of calcium and magnesium from the thick ascending limb (TAL), distal convoluted tubule (DCT) and collecting duct (CD) (Hoenderop et al., 1999a, 2000). In the distal tubule, AVP-regulated active calcium (Ca^{2+}) reabsorption proceeds against a transepithelial electrochemical gradient and involves several components, each of which forms a potential target of hormone action. First, a Ca^{2+} influx pathway, the transient receptor protein-vanilloid TRPV5 (also known as the epithelial calcium channel, ECaC), present in the luminal membrane of the cell, allows Ca^{2+} to enter into the cytosolic compartment down a steep electrochemical gradient (Hoenderop et al., 1999b, c). Next, a vitamin D-dependent Ca^{2+} binding protein (calbindin-D_{28K}) binds Ca^{2+} with high affinity and shuttles it to the basolateral membrane. Finally, a plasma membrane Ca^{2+}-ATPase and Na^+–Ca^{2+} exchanger transport Ca^{2+} out of the cytosol against a steep electrochemical gradient. Presently, it is unknown which component(s) of the Ca^{2+} reabsorptive machinery are increased in activity by AVP. Other studies provide evidence that AVP may indeed have a Ca^{2+}-sparing effect *in vivo* (Rodriguez-Soriano et al., 1989). They demonstrated that dDAVP administered as a nasal spray to a 7-year-old boy with hypercalciuria reduced urinary Ca^{2+}, as well as Mg^{2+} excretion. The physiological relevance of increased Ca^{2+} reabsorption with AVP is uncertain, but has been speculated as a means to decrease the risk of development of urolithiasis (Hoenderop et al., 2000).

AVP has also been shown to be an effective magnesium-conserving hormone in anesthetized and conscious hormone-deprived rats (Dai et al., 1998, 2001). Micropuncture studies suggested that the AVP was acting principally in TAL cells to conserve magnesium. Elalouf and colleagues (Elalouf et al., 1984) failed to discern any change in fractional magnesium (Mg^{2+}) absorption in the superficial distal tubule after physiological administration of AVP. In those studies, fractional calcium absorption significantly increased, whereas the change in fractional magnesium transport with AVP was not significant, although it increased marginally. AVP has also been shown to increase Mg^{2+} entry into mouse DCT cells in a concentration-dependent fashion that is sensitive to nifedipine (Dai et al., 2001). Other hormones have an additive effect on Mg^{2+} reabsorption in mouse DCT, such as glucagon. These observations suggest that AVP may play a role in control of both magnesium and calcium conservation in the TAL, DCT and CD.

Q. Chronic Adaptations in the TAL Due to AVP

In addition to the short-term actions of vasopressin on the thick ascending limb, the MTAL can undergo structural and functional changes as a result of long-term alterations in circulating vasopressin level or water intake. Bankir and associates (Bouby et al., 1985; Bankir et al., 1988) infused Brattleboro and normal rats for 5–6 weeks with vasopressin and measured medullary TAL morphometric characteristics. They found evidence for marked hypertrophy of the medullary TAL and widening of the inner stripe of the outer medulla, a region of the kidney highly enriched in TAL. In addition, long-term water restriction in Wistar rats resulted in hypertrophy of the TAL segment (Bouby and Bankir, 1988). The hypertrophy of the medullary TAL in vasopressin-treated Brattleboro rats was accompanied by increases in hormone-sensitive adenylyl cyclase activities and in Na-K-ATPase activity (Trinh-Trang-Tan et al., 1985).

V. VASOPRESSIN, RENAL HEMODYNAMICS AND BLOOD PRESSURE

AVP, via binding to both V1a and V2 receptors in the kidney, has the capacity to influence renal hemodynamics including: glomerular filtration rate (GFR), renal blood flow (RBF) and tubular glomerular feedback (TGF). Changes in these physical parameters can then augment direct transporter regulation in influencing, primarily urine concentration. However, inappropriately high levels of AVP, can result not only in the syndrome of inappropriate ADH (SIADH) and related hyponatremia (discussed in greater detail in Chapter 19), but also to increased blood pressure, via primarily changes in hemodynamics.

A. AVP and Renal Blood Flow

Cowley, Nakanishi, Mattson and associates (Mattson et al., 1993; Cowley et al., 1995, 2003; Nakanishi et al., 1995) have, throughout the years, developed a number of quite elegant approaches to the measurement of renal hemodynamic responses to AVP. Acutely implanted optical fibers allowed for the study of the influence of AVP on medullary blood flow (Nakanishi et al., 1995). These studies, performed in renal denervated rats, also determined the relative contribution of V1 and V2 receptors to the response to AVP. Infusion of a V1 receptor agonist into the renal medulla selectively reduced blood flow in the outer medulla by 15% and to the inner medulla by 35%. Equimolar doses of AVP also decreased outer medullary blood flow by 15%, but the fall in inner medullary flow (17%) was significantly less than that observed with the V1 agonist. Stimulation of V2 receptors by medullary interstitial infusion of dDAVP, or infusion of AVP in rats pretreated with a V1 receptor antagonist, increased medullary blood flow by 16% and 27%, respectively. These studies demonstrate that AVP has two diametrically opposed actions on medullary blood flow. V1 activity reduces blood flow and V2 activity increases it.

B. Glomerular Filtration rate (GFR)

Vasopressin has been reported to increase glomerular filtration rate (GFR) in normal rats and result in relative hyperfiltration (Bardoux et al., 1999; Bankir et al., 2001). Moreover, GFR has been demonstrated to correlate positively with urine osmolality in dDAVP-treated or water-restricted rats. The mechanism underlying this effect is not certain, but has been postulated to involve an alteration in tubuloglomerular feedback as a result of AVP's effects on urea recycling in the medulla (Bankir et al., 1993; Bouby et al., 1996). That is, AVP should increase urea in the lumen of the TAL via uptake in the loop of Henle (see Figure 17.6). Higher concentrations of urea at the macula densa (MD) may reduce water back flux into the lumen allowing for more efficient dilution of the urine. Low sodium levels in MC cells will thus feedback to the afferent arterioles as a signal to reduce vasoconstriction and increase filtration rate. Levels of AVP also are elevated in disease states such as diabetes mellitus (DM) and may contribute to pathogenesis. Brattleboro rats with DM exhibited no or markedly reduced hyperfiltration, albuminuria and renal hypertrophy as compared to Long-Evans rats (with endogenous AVP) with DM (Bardoux et al., 1999).

C. Vasopressin and Blood Pressure Control

The degree to which AVP participates in elevated blood pressure or to pathological hypertension is somewhat controversial. Whereas AVP is a potent vasoconstrictor *in vitro*, its ability to raise blood pressure after infusion *in vivo* has been less than impressive (Webb et al., 1986). In fact, dDAVP intraveneous infusion has been shown to result in at least a transient fall in blood pressure in patients with essential hypertension (Brink et al., 1994). One mechanism blunting the rise in blood pressure in response to AVP is baroreceptor-reflex buffering. This response is fairly unique for AVP. Cowley et al. (Cowley et al., 1974) demonstrated that sino-aortic denervation increased AVP-evoked pressor sensitivity 60–100-fold in dogs, whereas it only increased pressor sensitivity due to Ang II or norepinephrine 4–6-fold.

D. V2 Receptors are Coupled to Nitric Oxide Generation

Another aspect of blood pressure regulation by AVP is concomitant binding of V1 and V2 receptors. In the renal medulla, as mentioned above, AVP action via V1 receptors reduces blood flow, whereas AVP acting via V2 receptors appears to increase blood flow. Reduced medullary blood flow is associated with rises in blood pressure (Cowley et al.,

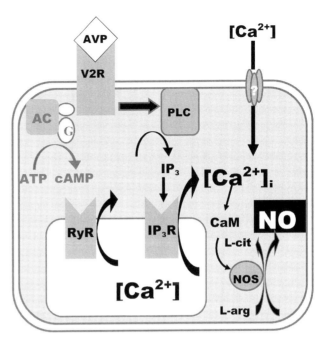

FIGURE 17.7 Diagrammatic representation of proposed pathway whereby AVP stimulates NO production in IMCD.

1995; Nakanishi et al., 1995). Furthermore, infusion of a V1 receptor select agonist of AVP has been demonstrated to increase blood pressure to a greater extent than AVP alone (Cowley et al., 1995). When renal medullary nitric oxide synthase (NOS) activity is reduced, however, even small elevations in circulating AVP produce sustained reductions of medullary blood flow and persistent hypertension (Szentivanyi et al., 2000). Tubule NOS activity is highest in inner medullary collecting duct segments (IMCD) (Wu et al., 1999) and the mRNA for V2 receptor is present only in tubules such as IMCD but not in the renal vasculature (Park et al., 1997), indicating that renal medullary NO production in response to AVP may be primarily mediated by IMCD (Mori et al., 2002). There is evidence that administration of the V2 receptor selective peptide agonist dDAVP alone results in increased medullary NO production while administration of the selective V1 receptor agonist [Phe^2, Ile^3, Orn^8]-vasopressin does not (Park et al., 1998). These observations indicate that AVP stimulates medullary NO production via activation of vasopressin V2-like receptors. Recently, O'Connor and Cowley (O'Connor and Cowley, 2007) demonstrated that signaling via phospholipase C (PLC) activation and calcium mobilization are involved in V2 activation of nitric oxide synthesis (Figure 17.7).

E. Hypertension may Correlate With Urinary Concentrating Ability

Nonetheless, there is evidence to suggest that AVP acting through V2 receptors may contribute to increased blood pressure. Recent work out of Bankir's laboratory

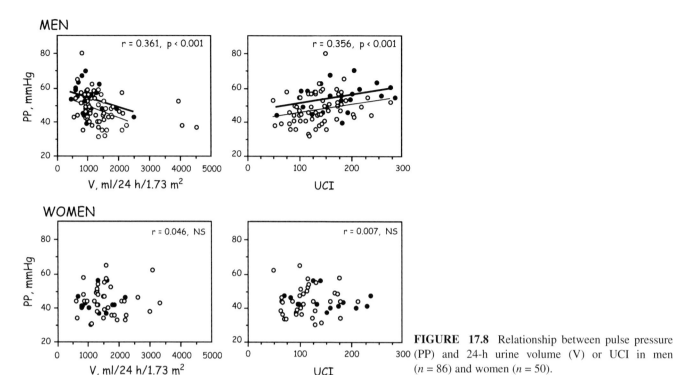

FIGURE 17.8 Relationship between pulse pressure (PP) and 24-h urine volume (V) or UCI in men ($n = 86$) and women ($n = 50$).

(Bankir et al., 2007) has shown that, in black men, urine concentrating ability is higher than in white men. This higher urine concentrating ability was significantly correlated with increased pulse pressure in these men (Figure 17.8). Other studies have reported higher basal levels of AVP in men of African origin as compared to Caucasians, which was related to higher incidence of hypertension (Bakris et al., 1997). Higher levels of AVP in the basal state may suggest reduced sensitivity of the thirst mechanism in these subjects and may have had some evolutionary advantage when water was limited. This also suggests a potentially therapeutic role for the use of V2-antagonists in the treatment of resistant forms of hypertension.

References

Amlal, H., Sheriff, S., Faroqui, S. et al. (2006). Regulation of acid-base transporters by vasopressin in the kidney collecting duct of Brattleboro rat. *Am J Nephrol* **26**: 194–205.

Ammar, A., Roseau, S., and Butlen, D. (1992). Pharmacological characterization of V1a vasopressin receptors in the rat cortical collecting duct. *Am J Physiol* **262**: F546–F553.

Amorim, J. B., Musa-Aziz, R., Mello-Aires, M., and Malnic, G. (2004). Signaling path of the action of AVP on distal K+ secretion. *Kidney Int* **66**: 696–704.

Ancellin, N., Preisser, L., Le Maout, S. et al. (1999). Homologous and heterologous phosphorylation of the vasopressin V1a receptor. *Cell Signal* **11**: 743–751.

Aperia, A., Holtback, U., Syren, M. L., Svensson, L. B., Fryckstedt, J., and Greengard, P. (1994). Activation/deactivation of renal Na^+, $K(+)$-ATPase: a final common pathway for regulation of natriuresis. *Faseb J* **8**: 436–439.

Bakris, G., Bursztyn, M., Gavras, I., Bresnahan, M., and Gavras, H. (1997). Role of vasopressin in essential hypertension: racial differences. *J Hypertens* **15**: 545–550.

Bankir, L., Ahloulay, M., Bouby, N. et al. (1993). Is the process of urinary urea concentration responsible for a high glomerular filtration rate?. *J Am Soc Nephrol* **4**: 1091–1103.

Bankir, L., Bardoux, P., and Ahloulay, M. (2001). Vasopressin and diabetes mellitus. *Nephron* **87**: 8–18.

Bankir, L., Fernandes, S., Bardoux, P., Bouby, N., and Bichet, D. G. (2005). Vasopressin-V2 receptor stimulation reduces sodium excretion in healthy humans. *J Am Soc Nephrol* **16**: 1920–1928.

Bankir, L., Fischer, C., Fischer, S., Jukkala, K., Specht, H. C., and Kriz, W. (1988). Adaptation of the rat kidney to altered water intake and urine concentration. *Pflugers Arch* **412**: 42–53.

Bankir, L., Perucca, J., and Weinberger, M. H. (2007). Ethnic differences in urine concentration: possible relationship to blood pressure. *Clin J Am Soc Nephrol* **2**: 304–312.

Barberis, C., Mouillac, B., and Durroux, T. (1998). Structural bases of vasopressin/oxytocin receptor function. *J Endocrinol* **156**: 223–229.

Barbry, P., and Hofman, P. (1997). Molecular biology of Na^+ absorption. *Am J Physiol* **273**: G571–G585.

Bardoux, P., Martin, H., Ahloulay, M. et al. (1999). Vasopressin contributes to hyperfiltration, albuminuria, and renal hypertrophy in diabetes mellitus: study in vasopressin-deficient Brattleboro rats. *Proc Natl Acad Sci USA* **96**: 10397–10402.

Barrett, L. K., Singer, M., and Clapp, L. H. (2007). Vasopressin: mechanisms of action on the vasculature in health and in septic shock. *Crit Care Med* **35**: 33–40.

Bartter, F. C., and Schwartz, W. B. (1967). The syndrome of inappropriate secretion of antidiuretic hormone. *Am J Med* **42**: 790–806.

Baumann, G., and Dingman, J. F. (1976). Distribution, blood transport, and degradation of antidiuretic hormone in man. *J Clin Invest* **57**: 1109–1116.

Besseghir, K., Trimble, M. E., and Stoner, L. (1986). Action of ADH on isolated medullary thick ascending limb of the Brattleboro rat. *Am J Physiol* **251**: F271–F277.

Bichet, D. G. (1996). Vasopressin receptors in health and disease. *Kidney Int* **49**: 1706–1711.

Bichet, D. G., Arthus, M. F., Lonergan, M. et al. (1993). X-linked nephrogenic diabetes insipidus mutations in North America and the Hopewell hypothesis. *J Clin Invest* **92**: 1262–1268.

Bindels, R. J., Schafer, J. A., and Reif, M. C. (1988). Stimulation of sodium transport by aldosterone and arginine vasotocin in A6 cells. *Biochim Biophys Acta* **972**: 320–330.

Birnbaumer, M. (2000). Vasopressin receptors. *Trends Endocrinol Metab* **11**: 406–410.

Birnbaumer, M., Seibold, A., Gilbert, S. et al. (1992). Molecular cloning of the receptor for human antidiuretic hormone. *Nature* **357**: 333–335.

Bouby, N., Ahloulay, M., Nsegbe, E., Dechaux, M., Schmitt, F., and Bankir, L. (1996). Vasopressin increases glomerular filtration rate in conscious rats through its antidiuretic action. *J Am Soc Nephrol* **7**: 842–851.

Bouby, N., and Bankir, L. (1988). Effect of high protein intake on sodium, potassium-dependent adenosine triphosphatase activity in the thick ascending limb of Henle's loop in the rat. *Clin Sci (Lond.)* **74**: 319–329.

Bouby, N., Bankir, L., Trinh-Trang-Tan, M. M., Minuth, W. W., and Kriz, W. (1985). Selective ADH-induced hypertrophy of the medullary thick ascending limb in Brattleboro rats. *Kidney Int* **28**: 456–466.

Bradford, A. D., Terris, J. M., Ecelbarger, C. A. et al. (2001). 97- and 117-kDa forms of collecting duct urea transporter UT-A1 are due to different states of glycosylation. *Am J Physiol Renal Physiol* **281**: F133–F143.

Brink, H. S., Derkx, F. H., Boomsma, F., and Schalekamp, M. A. (1994). Effects of DDAVP on renal hemodynamics and renin secretion in subjects with essential hypertension. *Clin Nephrol* **42**: 95–101.

Burg, M. B., and Green, N. (1973). Function of the thick ascending limb of Henle's loop. *Am J Physiol* **224**: 659–668.

Butterworth, M. B., Edinger, R. S., Johnson, J. P., and Frizzell, R. A. (2005). Acute ENaC stimulation by cAMP in a kidney cell line is mediated by exocytic insertion from a recycling channel pool. *J Gen Physiol* **125**: 81–101.

Canessa, C. M., Horisberger, J. D., and Rossier, B. C. (1993). Epithelial sodium channel related to proteins involved in neurodegeneration. *Nature* **361**: 467–470.

Canessa, C. M., and Schafer, J. A. (1992). AVP stimulates Na+ transport in primary cultures of rabbit cortical collecting duct cells. *Am J Physiol* **262**: F454–F461.

Canessa, C. M., Schild, L., Buell, G. et al. (1994). Amiloride-sensitive epithelial Na+ channel is made of three homologous subunits. *Nature* **367**: 463–467.

Chabardes, D., Firsov, D., Aarab, L. et al. (1996). Localization of mRNAs encoding Ca^{2+}-inhibitable adenylyl cyclases along the renal tubule. Functional consequences for regulation of the cAMP content. *J Biol Chem* **271**: 19264–19271.

Chalfant, M. L., Coupaye-Gerard, B., and Kleyman, T. R. (1993). Distinct regulation of Na+ reabsorption and Cl− secretion by arginine vasopressin in the amphibian cell line A6. *Am J Physiol* **264**: C1480–C1488.

Charlton, J. A., and Baylis, P. H. (1990). Stimulation of rat renal medullary Na+/K(+)-ATPase by arginine vasopressin is mediated by the V2 receptor. *J Endocrinol* **127**: 213–216.

Chen, L., Williams, S. K., and Schafer, J. A. (1990). Differences in synergistic actions of vasopressin and deoxycorticosterone in rat and rabbit CCD. *Am J Physiol* **259**: F147–F156.

Chou, C. L., and Knepper, M. A. (1989). Inhibition of urea transport in inner medullary collecting duct by phloretin and urea analogues. *Am J Physiol* **257**: F359–F365.

Chou, C. L., Rapko, S. I., and Knepper, M. A. (1998). Phosphoinositide signaling in rat inner medullary collecting duct. *Am J Physiol* **274**: F564–F572.

Chou, C. L., Sands, J. M., Nonoguchi, H., and Knepper, M. A. (1990). Concentration dependence of urea and thiourea transport in rat inner medullary collecting duct. *Am J Physiol* **258**: F486–F494.

Chou, C. L., Yip, K. P., Michea, L. et al. (2000). Regulation of aquaporin-2 trafficking by vasopressin in the renal collecting duct. Roles of ryanodine-sensitive Ca^{2+} stores and calmodulin. *J Biol Chem* **275**: 36839–36846.

Costanzo, L. S., and Windhager, E. E. (1980). Effects of PTH, ADH, and cyclic AMP on distal tubular Ca and Na reabsorption. *Am J Physiol* **239**: F478–F485.

Cowley, A. W., Mattson, D. L., Lu, S., and Roman, R. J. (1995). The renal medulla and hypertension. *Hypertension* **25**: 663–673.

Cowley, A. W., Monos, E., and Guyton, A. C. (1974). Interaction of vasopressin and the baroreceptor reflex system in the regulation of arterial blood pressure in the dog. *Circ Res* **34**: 505–514.

Cowley, A. W., Mori, T., Mattson, D., and Zou, A. P. (2003). Role of renal NO production in the regulation of medullary blood flow. *Am J Physiol Regul Integr Comp Physiol* **284**: R1355–R1369.

Dai, L. J., Bapty, B., Ritchie, G., and Quamme, G. A. (1998). Glucagon and arginine vasopressin stimulate Mg^{2+} uptake in mouse distal convoluted tubule cells. *Am J Physiol* **274**: F328–F335.

Dai, L. J., Ritchie, G., Kerstan, D., Kang, H. S., Cole, D. E., and Quamme, G. A. (2001). Magnesium transport in the renal distal convoluted tubule. *Physiol Rev* **81**: 51–84.

de Rouffignac, C., Corman, B., and Roinel, N. (1983). Stimulation by antidiuretic hormone of electrolyte tubular reabsorption in rat kidney. *Am J Physiol* **244**: F156–F164.

de Wied, D., Diamant, M., and Fodor, M. (1993). Central nervous system effects of the neurohypophyseal hormones and related peptides. *Front Neuroendocrinol* **14**: 251–302.

Delpire, E., Rauchman, M. I., Beier, D. R., Hebert, S. C., and Gullans, S. R. (1994). Molecular cloning and chromosome localization of a putative basolateral Na(+)-K(+)-2Cl− cotransporter from mouse inner medullary collecting duct (mIMCD-3) cells. *J Biol Chem* **269**: 25677–25683.

Dijkink, L., Hartog, A., van Os, C. H., and Bindels, R. J. (2002). The epithelial sodium channel (ENaC) is intracellularly located as a tetramer. *Pflugers Arch* **444**: 549–555.

Djelidi, S., Fay, M., Cluzeaud, F. et al. (1997). Transcriptional regulation of sodium transport by vasopressin in renal cells. *J Biol Chem* **272**: 32919–32924.

Ecelbarger, C. A., Chou, C. L., Lolait, S. J., Knepper, M. A., and DiGiovanni, S. R. (1996). Evidence for dual signaling pathways for V2 vasopressin receptor in rat inner medullary collecting duct. *Am J Physiol* **270**: F623–F633.

Ecelbarger, C. A., Kim, G. H., Knepper, M. A. et al. (2001). Regulation of potassium channel Kir 1.1 (ROMK) abundance in the thick ascending limb of Henle's loop. *J Am Soc Nephrol* **12**: 10–18.

Ecelbarger, C. A., Kim, G. H., Terris, J. et al. (2000). Vasopressin-mediated regulation of epithelial sodium channel abundance in rat kidney. *Am J Physiol Renal Physiol* **279**: F46–53.

Ecelbarger, C. A., Terris, J., Hoyer, J. R., Nielsen, S., Wade, J. B., and Knepper, M. A. (1996). Localization and regulation of the rat renal Na(+)-K(+)-2Cl⁻ cotransporter, BSC-1. *Am J Physiol* **271**: F619–F628.

Ecelbarger, C. A., Yu, S., Lee, A. J., Weinstein, L. S., and Knepper, M. A. (1999). Decreased renal Na-K-2Cl cotransporter abundance in mice with heterozygous disruption of the G(s)alpha gene. *Am J Physiol* **277**: F235–F244.

Elalouf, J. M., Roinel, N., and de Rouffignac, C. (1984). Effects of antidiuretic hormone on electrolyte reabsorption and secretion in distal tubules of rat kidney. *Pflugers Arch* **401**: 167–173.

Fenton, R. A., Brond, L., Nielsen, S., and Praetorius, J. (2007). Cellular and subcellular distribution of the type-2 vasopressin receptor in the kidney. *Am J Physiol Renal Physiol* **293**: F748–F760.

Fenton, R. A., and Knepper, M. A. (2007). Urea and renal function in the 21st century: insights from knockout mice. *J Am Soc Nephrol* **18**: 679–688.

Feraille, E., Mordasini, D., Gonin, S. et al. (2003). Mechanism of control of Na, K-ATPase in principal cells of the mammalian collecting duct. *Ann NY Acad Sci* **986**: 570–578.

Fink, G., Sumner, B. E., McQueen, J. K., Wilson, H., and Rosie, R. (1998). Sex steroid control of mood, mental state and memory. *Clin Exp Pharmacol Physiol* **25**: 764–775.

Fink, G., Sumner, B. E., Rosie, R., Grace, O., and Quinn, J. P. (1996). Estrogen control of central neurotransmission: effect on mood, mental state, and memory. *Cell Mol Neurobiol* **16**: 325–344.

Firsov, D., Gautschi, I., Merillat, A. M., Rossier, B. C., and Schild, L. (1998). The heterotetrameric architecture of the epithelial sodium channel (ENaC). *Embo J* **17**: 344–352.

Firsov, D., Mandon, B., Morel, A. et al. (1994). Molecular analysis of vasopressin receptors in the rat nephron. Evidence for alternative splicing of the V2 receptor. *Pflugers Arch* **429**: 79–89.

Frindt, G., and Burg, M. B. (1972). Effect of vasopressin on sodium transport in renal cortical collecting tubules. *Kidney Int* **1**: 224–231.

Gopalakrishnan, V., McNeill, J. R., Sulakhe, P. V., and Triggle, C. R. (1988). Hepatic vasopressin receptor: differential effects of divalent cations, guanine nucleotides, and N-ethylmaleimide on agonist and antagonist interactions with the V1 subtype receptor. *Endocrinology* **123**: 922–931.

Gottschalk, C. W., and Mylle, M. (1959). Micropuncture study of the mammalian urinary concentrating mechanism: evidence for the countercurrent hypothesis. *Am J Physiol* **196**: 927–936.

Grantham, J. J., and Burg, M. B. (1966). Effect of vasopressin and cyclic AMP on permeability of isolated collecting tubules. *Am J Physiol* **211**: 255–259.

Grazzini, E., Lodboerer, A. M., Perez-Martin, A., Joubert, D., and Guillon, G. (1996). Molecular and functional characterization of V1b vasopressin receptor in rat adrenal medulla. *Endocrinology* **137**: 3906–3914.

Greger, R., and Schlatter, E. (1983). Properties of the basolateral membrane of the cortical thick ascending limb of Henle's loop of rabbit kidney. A model for secondary active chloride transport. *Pflugers Arch* **396**: 325–334.

Greger, R., and Schlatter, E. (1983). Properties of the lumen membrane of the cortical thick ascending limb of Henle's loop of rabbit kidney. *Pflugers Arch* **396**: 315–324.

Greger, R., Schlatter, E., and Lang, F. (1983). Evidence for electroneutral sodium chloride cotransport in the cortical thick ascending limb of Henle's loop of rabbit kidney. *Pflugers Arch* **396**: 308–314.

Hall, D. A., and Varney, D. M. (1980). Effect of vasopressin on electrical potential difference and chloride transport in mouse medullary thick ascending limb of Henle's loop. *J Clin Invest* **66**: 792–802.

Hawk, C. T., Li, L., and Schafer, J. A. (1996). AVP and aldosterone at physiological concentrations have synergistic effects on Na⁺ transport in rat CCD. *Kidney Int Suppl* **57**: S35–41.

Hebert, S. C., Culpepper, R. M., and Andreoli, T. E. (1981). NaCl transport in mouse medullary thick ascending limbs. I. Functional nephron heterogeneity and ADH-stimulated NaCl cotransport. *Am J Physiol* **241**: F412–F431.

Hebert, S. C., Friedman, P. A., and Andreoli, T. E. (1984). Effects of antidiuretic hormone on cellular conductive pathways in mouse medullary thick ascending limbs of Henle: I. ADH increases transcellular conductance pathways. *J Membr Biol* **80**: 201–219.

Hiroyama, M., Aoyagi, T., Fujiwara, Y. et al. (2007). Hyperammonaemia in V1a vasopressin receptor knockout mice caused by the promoted proteolysis and reduced intrahepatic blood volume. *J Physiol* **581**: 1183–1192.

Ho, K., Nichols, C. G., Lederer, W. J. et al. (1993). Cloning and expression of an inwardly rectifying ATP-regulated potassium channel. *Nature* **362**: 31–38.

Hoenderop, J. G., De Pont, J. J., Bindels, R. J., and Willems, P. H. (1999). Hormone-stimulated Ca^{2+} reabsorption in rabbit kidney cortical collecting system is cAMP-independent and involves a phorbol ester-insensitive PKC isotype. *Kidney Int* **55**: 225–233.

Hoenderop, J. G., van der Kemp, A. W., Hartog, A. et al. (1999). Molecular identification of the apical Ca^{2+} channel in 1,25-dihydroxyvitamin D3-responsive epithelia. *J Biol Chem* **274**: 8375–8378.

Hoenderop, J. G., van der Kemp, A. W., Hartog, A. et al. (1999). The epithelial calcium channel, ECaC, is activated by hyperpolarization and regulated by cytosolic calcium. *Biochem Biophys Res Commun* **261**: 488–492.

Hoenderop, J. G., Willems, P. H., and Bindels, R. J. (2000). Toward a comprehensive molecular model of active calcium reabsorption. *Am J Physiol Renal Physiol* **278**: F352–F360.

Holmes, C. L., Patel, B. M., Russell, J. A., and Walley, K. R. (2001). Physiology of vasopressin relevant to management of septic shock. *Chest* **120**: 989–1002.

Hurbin, A., Boissin-Agasse, L., Orcel, H. et al. (1998). The V1a and V1b, but not V2, vasopressin receptor genes are expressed in the supraoptic nucleus of the rat hypothalamus, and the transcripts are essentially colocalized in the vasopressinergic magnocellular neurons. *Endocrinology* **139**: 4701–4707.

Igarashi, P., Whyte, D. A., Li, K., and Nagami, G. T. (1996). Cloning and kidney cell-specific activity of the promoter of the murine renal Na-K-Cl cotransporter gene. *J Biol Chem* **271**: 9666–9674.

Imbert, M., Chabardes, D., Montegut, M., Clique, A., and Morel, F. (1975). Vasopressin dependent adenylate cyclase in single segments of rabbit kidney tubule. *Pflugers Arch* **357**: 173–186.

Imbert-Teboul, M., Chabardes, D., Montegut, M., Clique, A., and Morel, F. (1978). Vasopressin-dependent adenylate cyclase activities in the rat kidney medulla: evidence for two separate sites of action. *Endocrinology* **102**: 1254–1261.

Innamorati, G., Sadeghi, H., and Birnbaumer, M. (1996). A fully active nonglycosylated V2 vasopressin receptor. *Mol Pharmacol* **50**: 467–473.

Innamorati, G., Sadeghi, H., and Birnbaumer, M. (1998). Transient phosphorylation of the V1a vasopressin receptor. *J Biol Chem* **273**: 7155–7161.

Innamorati, G., Sadeghi, H., and Birnbaumer, M. (1999). Phosphorylation and recycling kinetics of G protein-coupled receptors. *J Recept Signal Transduct Res* **19**: 315–326.

Innamorati, G., Sadeghi, H. M., Tran, N. T., and Birnbaumer, M. (1998). A serine cluster prevents recycling of the V2 vasopressin receptor. *Proc Natl Acad Sci USA* **95**: 2222–2226.

Izumi, Y., Nakayama, Y., Mori, T. et al. (2007). Downregulation of vasopressin V2 receptor promoter activity via V1a receptor pathway. *Am J Physiol Renal Physiol* **292**: F1418–F1426.

Jard, S. (1983). Vasopressin isoreceptors in mammals: relation to cAMP-dependent and cAMP-independent transduction mechanisms. In: Kleinzeller A, editor. *Current topics in membranes and transport* vol. 18: Academic Press, pp. 255–285.

Jard, S., Gaillard, R. C., Guillon, G. et al. (1986). Vasopressin antagonists allow demonstration of a novel type of vasopressin receptor in the rat adenohypophysis. *Mol Pharmacol* **30**: 171–177.

Kato, A., Naruse, M., Knepper, M. A., and Sands, J. M. (1998). Long-term regulation of inner medullary collecting duct urea transport in rat. *J Am Soc Nephrol* **9**: 737–745.

Kim, G. H., Ecelbarger, C. A., Mitchell, C. et al. (1999). Vasopressin increases Na-K-2Cl cotransporter expression in thick ascending limb of Henle's loop. *Am J Physiol* **276**: F96–F103.

Knepper, M. A., Kim, G. H., Fernandez-Llama, P., and Ecelbarger, C. A. (1999). Regulation of thick ascending limb transport by vasopressin. *J Am Soc Nephrol* **10**: 628–634.

Liu, J., and Wess, J. (1996). Different single receptor domains determine the distinct G protein coupling profiles of members of the vasopressin receptor family. *J Biol Chem* **271**: 8772–8778.

Loffing, J., Zecevic, M., Feraille, E. et al. (2001). Aldosterone induces rapid apical translocation of ENaC in early portion of renal collecting system: possible role of SGK. *Am J Physiol Renal Physiol* **280**: F675–F682.

Lolait, S. J., O'Carroll, A. M., and Brownstein, M. J. (1995). Molecular biology of vasopressin receptors. *Ann NY Acad Sci* **771**: 273–292.

Lolait, S. J., O'Carroll, A. M., McBride, O. W., Konig, M., Morel, A., and Brownstein, M. J. (1992). Cloning and characterization of a vasopressin V2 receptor and possible link to nephrogenic diabetes insipidus. *Nature* **357**: 336–339.

Lytle, C., and Forbush, B. (1996). Regulatory phosphorylation of the secretory Na-K-Cl cotransporter: modulation by cytoplasmic Cl. *Am J Physiol* **270**: C437–C448.

Machida, K., Wakamatsu, S., Izumi, Y. et al. (2007). Downregulation of the V2 vasopressin receptor in dehydration: mechanisms and role of renal prostaglandin synthesis. *Am J Physiol Renal Physiol* **292**: F1274–F1282.

Martin, N. P., Lefkowitz, R. J., and Shenoy, S. K. (2003). Regulation of V2 vasopressin receptor degradation by agonist-promoted ubiquitination. *J Biol Chem* **278**: 45954–45959.

Matsumura, Y., Uchida, S., Rai, T., Sasaki, S., and Marumo, F. (1997). Transcriptional regulation of aquaporin-2 water channel gene by cAMP. *J Am Soc Nephrol* **8**: 861–867.

Mattson, D. L., Lu, S., Roman, R. J., and Cowley, A. W. (1993). Relationship between renal perfusion pressure and blood flow in different regions of the kidney. *Am J Physiol* **264**: R578–R583.

McKinley, M. J., Mathai, M. L., McAllen, R. M. et al. (2004). Vasopressin secretion: osmotic and hormonal regulation by the lamina terminalis. *J Neuroendocrinol* **16**: 340–347.

Molony, D. A., Reeves, W. B., Hebert, S. C., and Andreoli, T. E. (1987). ADH increases apical Na^+, K^+, $2Cl^-$ entry in mouse medullary thick ascending limbs of Henle. *Am J Physiol* **252**: F177–F187.

Morel, A., O'Carroll, A. M., Brownstein, M. J., and Lolait, S. J. (1992). Molecular cloning and expression of a rat V1a arginine vasopressin receptor. *Nature* **356**: 523–526.

Mori, T., Dickhout, J. G., and Cowley, A. W. (2002). Vasopressin increases intracellular NO concentration via Ca(2+) signaling in inner medullary collecting duct. *Hypertension* **39**: 465–469.

Morris, R. G., and Schafer, J. A. (2002). cAMP increases density of ENaC subunits in the apical membrane of MDCK cells in direct proportion to amiloride-sensitive Na(+) transport. *J Gen Physiol* **120**: 71–85.

Mouillac, B., Chini, B., Balestre, M. N. et al. (1995). The binding site of neuropeptide vasopressin V1a receptor. Evidence for a major localization within transmembrane regions. *J Biol Chem* **270**: 25771–25777.

Mouillac, B., Chini, B., Balestre, M. N. et al. (1995). Identification of agonist binding sites of vasopressin and oxytocin receptors. *Adv Exp Med Biol* **395**: 301–310.

Murasawa, S., Matsubara, H., Kijima, K., Maruyama, K., Mori, Y., and Inada, M. (1995). Structure of the rat V1a vasopressin receptor gene and characterization of its promoter region and complete cDNA sequence of the 3'-end. *J Biol Chem* **270**: 20042–20050.

Murase, T., Tian, Y., Fang, X. Y., and Verbalis, J. G. (2003). Synergistic effects of nitric oxide and prostaglandins on renal escape from vasopressin-induced antidiuresis. *Am J Physiol Regul Integr Comp Physiol* **284**: R354–R362.

Nakanishi, K., Mattson, D. L., Gross, V., Roman, R. J., and Cowley, A. W. (1995). Control of renal medullary blood flow by vasopressin V1 and V2 receptors. *Am J Physiol* **269**: R193–200.

Nakhoul, N. L., Hering-Smith, K. S., Gambala, C. T., and Hamm, L. L. (1998). Regulation of sodium transport in M-1 cells. *Am J Physiol* **275**: F998–F1007.

Nicco, C., Wittner, M., DiStefano, A., Jounier, S., Bankir, L., and Bouby, N. (2001). Chronic exposure to vasopressin upregulates ENaC and sodium transport in the rat renal collecting duct and lung. *Hypertension* **38**: 1143–1149.

Nielsen, S., Maunsbach, A. B., Ecelbarger, C. A., and Knepper, M. A. (1998). Ultrastructural localization of Na-K-2Cl cotransporter in thick ascending limb and macula densa of rat kidney. *Am J Physiol* **275**: F885–F893.

Nitschke, R., Frobe, U., and Greger, R. (1991). Antidiuretic hormone acts via V1 receptors on intracellular calcium in the isolated perfused rabbit cortical thick ascending limb. *Pflugers Arch* **417**: 622–632.

Nonoguchi, H., Owada, A., Kobayashi, N. et al. (1995). Immunohistochemical localization of V2 vasopressin receptor along the nephron and functional role of luminal V2 receptor in terminal inner medullary collecting ducts. *J Clin Invest* **96**: 1768–1778.

O'Connor, P., and Cowley, A. W. (2007). Vasopressin-induced nitric oxide production in rat inner medullary collecting duct is dependent on V2 receptor activation of the phosphoinositide pathway. *Am J Physiol Renal Physiol* **293**: F526–F532.

O'Dowd, B. F., Hnatowich, M., Caron, M. G., Lefkowitz, R. J., and Bouvier, M. (1989). Palmitoylation of the human beta 2-adrenergic receptor. Mutation of Cys341 in the carboxyl tail leads to an uncoupled nonpalmitoylated form of the receptor. *J Biol Chem* **264**: 7564–7569.

Ostrowski, N. L., Lolait, S. J., Bradley, D. J., O'Carroll, A. M., Brownstein, M. J., and Young, W. S. (1992). Distribution of V1a and V2 vasopressin receptor messenger ribonucleic acids in rat liver, kidney, pituitary and brain. *Endocrinology* **131**: 533–535.

Page, A. J., Reid, S. A., Speedy, D. B., Mulligan, G. P., and Thompson, J. (2007). Exercise-associated hyponatremia, renal function, and nonsteroidal antiinflammatory drug use in an ultraendurance mountain run. *Clin J Sport Med* **17**: 43–48.

Park, F., Mattson, D. L., Skelton, M. M., and Cowley, A. W. (1997). Localization of the vasopressin V1a and V2 receptors within the renal cortical and medullary circulation. *Am J Physiol* **273**: R243–R251.

Park, F., Zou, A. P., and Cowley, A. W. (1998). Arginine vasopressin-mediated stimulation of nitric oxide within the rat renal medulla. *Hypertension* **32**: 896–901.

Pavo, I., and Fahrenholz, F. (1990). Differential inactivation of vasopressin receptor subtypes in isolated membranes and intact cells by N-ethylmaleimide. *FEBS Lett* **272**: 205–208.

Payne, J. A., Xu, J. C., Haas, M., Lytle, C. Y., Ward, D., and Forbush, B. (1995). Primary structure, functional expression, and chromosomal localization of the bumetanide-sensitive Na-K-Cl cotransporter in human colon. *J Biol Chem* **270**: 17977–17985.

Petrovic, S., Amlal, H., Sun, X., Karet, F., Barone, S., and Soleimani, M. (2006). Vasopressin induces expression of the Cl-/HCO3- exchanger SLC26A7 in kidney medullary collecting ducts of Brattleboro rats. *Am J Physiol Renal Physiol* **290**: F1194–F1201.

Pickford, M. (1976). Progress in the study of biosynthesis and transport in the neurohypophysial system. In Neurohypophysis, Ink ConJ Key Biscayne, Florida. Karger, Basel, pp. 30–42.

Reeves, W. B., McDonald, G. A., Mehta, P., and Andreoli, T. E. (1989). Activation of K^+ channels in renal medullary vesicles by cAMP-dependent protein kinase. *J Membr Biol* **109**: 65–72.

Reif, M. C., Troutman, S. L., and Schafer, J. A. (1984). Sustained response to vasopressin in isolated rat cortical collecting tubule. *Kidney Int* **26**: 725–732.

Reif, M. C., Troutman, S. L., and Schafer, J. A. (1986). Sodium transport by rat cortical collecting tubule. Effects of vasopressin and desoxycorticosterone. *J Clin Invest* **77**: 1291–1298.

Rieg, T., Vallon, V., Sausbier, M. et al. (2007). The role of the BK channel in potassium homeostasis and flow-induced renal potassium excretion. *Kidney Int* **72**: 566–573.

Robben, J. H., Knoers, N. V., and Deen, P. M. (2005). Characterization of vasopressin V2 receptor mutants in nephrogenic diabetes insipidus in a polarized cell model. *Am J Physiol Renal Physiol* **289**: F265–F272.

Robben, J. H., Knoers, N. V., and Deen, P. M. (2006). Cell biological aspects of the vasopressin type-2 receptor and aquaporin 2 water channel in nephrogenic diabetes insipidus. *Am J Physiol Renal Physiol* **291**: F257–F270.

Rocha, A. S., and Kokko, J. P. (1973). Sodium chloride and water transport in the medullary thick ascending limb of Henle. Evidence for active chloride transport. *J Clin Invest* **52**: 612–623.

Rodriguez-Soriano, J., Vallo, A., and Dominguez, M. J. (1989). 'Chloride-shunt' syndrome: an overlooked cause of renal hypercalciuria. *Pediatr Nephrol* **3**: 113–121.

Ruiz-Opazo, N. (1998). Identification of a novel dual angiotensin II/vasopressin receptor. *Nephrologie* **19**: 417–420.

Russell, J. T., Brownstein, M. J., and Gainer, H. (1981). Time course of appearance and release of [35S]cysteine labelled neurophysins and peptides in the neurohypophysis. *Brain Res* **205**: 299–311.

Sachs, H., Poryanova, R., Haller, E. W., and Share, L. (1967). Cellular processes concerned with vasopressin biosynthesis, storage and release. In: Stutinsky F, editor. *Neurosecretion*. Springer-Verlag, Berlin, pp. 146–154.

Sadeghi, H. M., Innamorati, G., Dagarag, M., and Birnbaumer, M. (1997). Palmitoylation of the V2 vasopressin receptor. *Mol Pharmacol* **52**: 21–29.

Saito, M., Sugimoto, T., Tahara, A., and Kawashima, H. (1995). Molecular cloning and characterization of rat V1b vasopressin receptor: evidence for its expression in extra-pituitary tissues. *Biochem Biophys Res Commun* **212**: 751–757.

Sakuma, Y., Nonoguchi, H., Takayama, M. et al. (2005). Differential effects of hyperosmolality on Na-K-ATPase and vasopressin-dependent cAMP generation in the medullary thick ascending limb and outer medullary collecting duct. *Hypertens Res* **28**: 671–679.

Sands, J. M., and Knepper, M. A. (1987). Urea permeability of mammalian inner medullary collecting duct system and papillary surface epithelium. *J Clin Invest* **79**: 138–147.

Sands, J. M., Nonoguchi, H., and Knepper, M. A. (1987). Vasopressin effects on urea and H_2O transport in inner medullary collecting duct subsegments. *Am J Physiol* **253**: F823–F832.

Sansom, S. C., and Welling, P. A. (2007). Two channels for one job. *Kidney Int* **72**: 529–530.

Sarmiento, J. M., Ehrenfeld, P., Anazco, C. C. et al. (2005). Differential distribution of the vasopressin V receptor along the rat nephron during renal ontogeny and maturation. *Kidney Int* **68**: 487–496.

Sasaki, S., and Imai, M. (1980). Effects of vasopressin on water and NaCl transport across the in vitro perfused medullary thick ascending limb of Henle's loop of mouse, rat, and rabbit kidneys. *Pflugers Arch* **383**: 215–221.

Sauter, D., Fernandes, S., Goncalves-Mendes, N. et al. (2006). Long-term effects of vasopressin on the subcellular localization of ENaC in the renal collecting system. *Kidney Int* **69**: 1024–1032.

Schafer, J. A., Troutman, S. L., and Schlatter, E. (1990). Vasopressin and mineralocorticoid increase apical membrane driving force for K^+ secretion in rat CCD. *Am J Physiol* **258**: F199–210.

Schlatter, E., and Greger, R. (1985). cAMP increases the basolateral Cl^--conductance in the isolated perfused medullary thick ascending limb of Henle's loop of the mouse. *Pflugers Arch* **405**: 367–376.

Schlatter, E., and Schafer, J. A. (1987). Electrophysiological studies in principal cells of rat cortical collecting tubules. ADH increases the apical membrane Na$^+$-conductance. *Pflugers Arch* **409**: 81–92.

Schmale, H., Heinsohn, S., and Richter, D. (1983). Structural organization of the rat gene for the arginine vasopressin-neurophysin precursor. *Embo J* **2**: 763–767.

Schuster, V. L., and Stokes, J. B. (1987). Chloride transport by the cortical and outer medullary collecting duct. *Am J Physiol* **253**: F203–F212.

Seibold, A., Rosenthal, W., Bichet, D. G., and Birnbaumer, M. (1993). The vasopressin type 2 receptor gene. Chromosomal localization and its role in nephrogenic diabetes insipidus. *Regul Pept* **45**: 67–71.

Siegel, A. J. (2006). Exercise-associated hyponatremia: role of cytokines. *Am J Med* **119**: S74–S78.

Siegel, A. J., Verbalis, J. G., Clement, S. et al. (2007). Hyponatremia in marathon runners due to inappropriate arginine vasopressin secretion. *Am J Med* **120**: 461e11–e17.

Simmons, N. L. (1993). Renal epithelial Cl$^-$ secretion. *Exp Physiol* **78**: 117–137.

Smith, C. P., Lee, W. S., Martial, S. et al. (1995). Cloning and regulation of expression of the rat kidney urea transporter (rUT2). *J Clin Invest* **96**: 1556–1563.

Star, R. A., Nonoguchi, H., Balaban, R., and Knepper, M. A. (1988). Calcium and cyclic adenosine monophosphate as second messengers for vasopressin in the rat inner medullary collecting duct. *J Clin Invest* **81**: 1879–1888.

Staruschenko, A., Medina, J. L., Patel, P., Shapiro, M. S., Booth, R. E., and Stockand, J. D. (2004). Fluorescence resonance energy transfer analysis of subunit stoichiometry of the epithelial Na$^+$ channel. *J Biol Chem* **279**: 27729–27734.

Sugimoto, T., Saito, M., Mochizuki, S., Watanabe, Y., Hashimoto, S., and Kawashima, H. (1994). Molecular cloning and functional expression of a cDNA encoding the human V1b vasopressin receptor. *J Biol Chem* **269**: 27088–27092.

Sun, A., Grossman, E. B., Lombardi, M., and Hebert, S. C. (1991). Vasopressin alters the mechanism of apical Cl$^-$ entry from Na$^+$: Cl$^-$ to Na$^+$:K$^+$:2Cl$^-$ cotransport in mouse medullary thick ascending limb. *J Membr Biol* **120**: 83–94.

Swaab, D. F., Nijveldt, F., and Pool, C. W. (1975). Distribution of oxytocin and vasopressin in the rat supraoptic and paraventricular nucleus. *J Endocrinol* **67**: 461–462.

Szentivanyi, M., Park, F., Maeda, C. Y., and Cowley, A. W. (2000). Nitric oxide in the renal medulla protects from vasopressin-induced hypertension. *Hypertension* **35**: 740–745.

Takabatake, Y., and Sachs, H. (1964). Vasopressin biosynthesis. 3. In vitro studies. *Endocrinology* **75**: 934–942.

Takayama, M., Nonoguchi, H., Yang, T. et al. (1999). Acute and chronic effects of hyperosmolality on mRNA and protein expression and the activity of Na-K-ATPase in the IMCD. *Exp Nephrol* **7**: 295–305.

Tanimura, A., Kurihara, K., Reshkin, S. J., and Turner, R. J. (1995). Involvement of direct phosphorylation in the regulation of the rat parotid Na(+)-K(+)-2Cl$^-$ cotransporter. *J Biol Chem* **270**: 25252–25258.

Tashima, Y., Kohda, Y., Nonoguchi, H. et al. (2001). Intranephron localization and regulation of the V1a vasopressin receptor during chronic metabolic acidosis and dehydration in rats. *Pflugers Arch* **442**: 652–661.

Terada, Y., Tomita, K., Nonoguchi, H., Yang, T., and Marumo, F. (1993). Different localization and regulation of two types of vasopressin receptor messenger RNA in microdissected rat nephron segments using reverse transcription polymerase chain reaction. *J Clin Invest* **92**: 2339–2345.

Terris, J., Ecelbarger, C. A., Nielsen, S., and Knepper, M. A. (1996). Long-term regulation of four renal aquaporins in rats. *Am J Physiol* **271**: F414–F422.

Terris, J., Ecelbarger, C. A., Sands, J. M., and Knepper, M. A. (1998). Long-term regulation of renal urea transporter protein expression in rat. *J Am Soc Nephrol* **9**: 729–736.

Thibonnier, M., Auzan, C., Madhun, Z., Wilkins, P., Berti-Mattera, L., and Clauser, E. (1994). Molecular cloning, sequencing, and functional expression of a cDNA encoding the human V1a vasopressin receptor. *J Biol Chem* **269**: 3304–3310.

Thibonnier, M., Graves, M. K., Wagner, M. S., Auzan, C., Clauser, E., and Willard, H. F. (1996). Structure, sequence, expression, and chromosomal localization of the human V1a vasopressin receptor gene. *Genomics* **31**: 327–334.

Tiwari, S., Nordquist, L., Halagappa, V. K., and Ecelbarger, C. A. (2007). Trafficking of ENaC subunits in response to acute insulin in mouse kidney. *Am J Physiol Renal Physiol* **793**: F178–F185.

Tomita, K., Pisano, J. J., and Knepper, M. A. (1985). Control of sodium and potassium transport in the cortical collecting duct of the rat. Effects of bradykinin, vasopressin, and deoxycorticosterone. *J Clin Invest* **76**: 132–136.

Torikai, S., and Imai, M. (1984). Effects of solute concentration on vasopressin stimulated cyclic AMP generation in the rat medullary thick ascending limbs of Henle's loop. *Pflugers Arch* **400**: 306–308.

Torikai, S., Wang, M. S., Klein, K. L., and Kurokawa, K. (1981). Adenylate cyclase and cell cyclic AMP of rat cortical thick ascending limb of Henle. *Kidney Int* **20**: 649–654.

Trinh-Trang-Tan, M. M., Bankir, L., Doucet, A. et al. (1985). Influence of chronic ADH treatment on adenylate cyclase and ATPase activity in distal nephron segments of diabetes insipidus Brattleboro rats. *Pflugers Arch* **405**: 216–222.

Vaccari, C., Lolait, S. J., and Ostrowski, N. L. (1998). Comparative distribution of vasopressin V1b and oxytocin receptor messenger ribonucleic acids in brain. *Endocrinology* **139**: 5015–5033.

Van Huyen, J. P., Bens, M., Teulon, J., and Vandewalle, A. (2001). Vasopressin-stimulated chloride transport in transimmortalized mouse cell lines derived from the distal convoluted tubule and cortical and inner medullary collecting ducts. *Nephrol Dial Transplant* **16**: 238–245.

Vandewalle, A. (2007). Expression and function of CLC and cystic fibrosis transmembrane conductance regulator chloride channels in renal epithelial tubule cells: pathophysiological implications. *Chang Gung Med J* **30**: 17–25.

Ventura, M. A., Rene, P., de Keyzer, Y., Bertagna, X., and Clauser, E. (1999). Gene and cDNA cloning and characterization of the mouse V3/V1b pituitary vasopressin receptor. *J Mol Endocrinol* **22**: 251–260.

Verbalis, J. G. (2003). Disorders of body water homeostasis. *Best Pract Res Clin Endocrinol Metab* **17**: 471–503.

Verbalis, J. G. (2007). Renal function and vasopressin during marathon running. *Sports Med* **37**: 455–458.

Verrey, F. (1994). Antidiuretic hormone action in A6 cells: effect on apical Cl and Na conductances and synergism with aldosterone for NaCl reabsorption. *J Membr Biol* **138**: 65–76.

Vinciguerra, M., Mordasini, D., Vandewalle, A., and Feraille, E. (2005). Hormonal and nonhormonal mechanisms of regulation of the NA, K-pump in collecting duct principal cells. *Semin Nephrol* **25**: 312–321.

Wang, W. H. (1994). Two types of K^+ channel in thick ascending limb of rat kidney. *Am J Physiol* **267**: F599–605.

Webb, R. L., Osborn, J. W., and Cowley, A. W. (1986). Cardiovascular actions of vasopressin: baroreflex modulation in the conscious rat. *Am J Physiol* **251**: H1244–H1251.

Wingo, C. S. (1990). Active and passive chloride transport by the rabbit cortical collecting duct. *Am J Physiol* **258**: F1388–F1393.

Winters, C. J., Zimniak, L., Reeves, W. B., and Andreoli, T. E. (1997). Cl^- channels in basolateral renal medullary membranes. XII. Anti-rbClC-Ka antibody blocks MTAL Cl^- channels. *Am J Physiol* **273**: F1030–F1038.

Wong, S. M., and Chase, H. S. (1988). Effect of vasopressin on intracellular [Ca] and Na transport in cultured toad bladder cells. *Am J Physiol* **255**: F1015–F1024.

Wu, F., Park, F., Cowley, A. W., and Mattson, D. L. (1999). Quantification of nitric oxide synthase activity in microdissected segments of the rat kidney. *Am J Physiol* **276**: F874–F881.

Xu, J., Worrell, R. T., Li, H. C. et al. (2006). Chloride/bicarbonate exchanger SLC26A7 is localized in endosomes in medullary collecting duct cells and is targeted to the basolateral membrane in hypertonicity and potassium depletion. *J Am Soc Nephrol* **17**: 956–967.

Xu, Z. C., Yang, Y., and Hebert, S. C. (1996). Phosphorylation of the ATP-sensitive, inwardly rectifying K^+ channel, ROMK, by cyclic AMP-dependent protein kinase. *J Biol Chem* **271**: 9313–9319.

CHAPTER **18**

Vasopressin Antagonists in Physiology and Disease

TOMAS BERL AND ROBERT W. SCHRIER

University of Colorado at Denver and Health Sciences Center, Division of Renal Diseases and Hypertension, 4200 East Ninth Avenue, BRB 423, Denver, CO 80262, USA

Contents

I.	Introduction	249
II.	Physiologic antagonists	249
III.	Vasopressin antagonists and their role in the treatment of water-retaining disorders	251
IV.	Are vasopressin antagonists safe?	256
V.	Summary and unanswered questions	257
	References	257

I. INTRODUCTION

Hyponatremia is one of the most common electrolyte abnormalities in hospitalized patients. It can present acutely or chronically. The treatment of hyponatremia has been reviewed elsewhere extensively but includes simple strategies such as free water restriction as well as more emergent measures such as the administration of hypertonic (3%) saline. Recently, a new class of agents has emerged – vasopressin V2-receptor antagonists. Additional targets, such as the V1A and V1B receptors have also generated interest. The purpose of this chapter is both to review the physiologic basis for vasopressin release and its inhibition and the role of vasopressin antagonists in water-retaining disorders.

II. PHYSIOLOGIC ANTAGONISTS

A. Inhibition of Vasopressin Release

The maintenance of serum tonicity, and therefore serum sodium concentration, within a very narrow physiologic range between 138 and 142 mEq/l is a reflection of the sensitive osmotic regulation of arginine vasopressin (AVP) secretion, a fact already recognized since the pioneering observations of Verney, more than 60 years ago (Verney, 1947). As such, vasopressin is a component of the intricate neuroendocrine system that controls body fluid homeostasis. The development of a radioimmunoassay for the hormone in the 1970s by Robertson and colleagues (Robertson et al., 1973) defined the high osmotic sensitivity of vasopressin release which translated into large changes in renal water excretion (Robertson et al., 1976). While the cells that secrete vasopressin are the magnocellular neurons of the supraoptic and paraventricular nuclei in the hypothalamus (Zimmerman et al., 1987), the sensing of osmolality is in close proximity but distinct from these neurons and probably resides in the anterior hypothalamus in the vicinity of the organum vasculosum of the lamina terminalis (OVLT) (Verbalis, 2007b).

The osmotic threshold, as well as the sensitivity for vasopressin release, is determined by genetic factors providing significant intra-individual variability (Zerbe et al., 1991). Nonetheless, this variability is modulated by a number of other factors, most of which such as pregnancy and aging either decrease the threshold (Lindheimer and Davison, 1995) or increase the sensitivity of the hormone's release (Helderman et al., 1978) respectively. In terms of inhibitory pathways, a decrement in plasma osmolality as small as 1%, with water intake, by causing swelling of osmoreceptor cells, is a powerful suppressor of vasopressin release. This process allows for dilution of the urine and the excretion of the water that was ingested. As such, diminished osmolality can be viewed as the most physiologic of all antagonists of vasopressin release. Also a number of neurotransmitters have been implicated in osmoregulatory function. In this regard, the supraoptic nuclei are innervated by numerous pathways, including catecholamines, opioids, γ aminobutyric acid (GABA), cholinergics, to name a few (Verbalis, 1993; Sladek and Kapoor, 2001), all of which probably interact with the osmotic input to provide an integrated response.

In addition to the osmotic pathways, the secretion of vasopressin is impacted by non-osmotic stimuli as well. Of these, changes in hemodynamics are probably the most important. While significant (5–7%) decrements in volume and perhaps as much as 20% decrements in pressure are required to activate vasopressin secretion, once activated these stimuli

TABLE 18.1 Drugs and hormones that inhibit vasopressin secretion

Norepinephrine
Fluphenazine
Haloperidol
Promethazine
Oxilorphan
Butorphanol
Opioid agonists
Morphine (low doses)
Ethanol
Glucocorticoids
Clonidine
Muscimol
Phencyclidine
?Phenytoin

Modified from Berl, T., Verbalis, J. (2004). Pathophysiology of water metabolism. In Brenner & Rector's The Kidney, 7th edn (Brenner, B., ed.). Saunders, Philadelphia, pp. 857–919.

TABLE 18.2 Drugs associated with nephrogenic diabetes insipidus

Alcohol
Phenytoin
Lithium
Demeclocycline
Acetohexamide
Tolazamide
Glyburide
Propoxyphene
Amphotericin
Fascarnet
Methoxyflurane
Norepinephrine
Vinblastine
Colchicine
Gentamicin
Methicillin
Isophosphamide
Angiographic dyes
Osmotic diuretics
Furosemide and ethacrynic acid

Modified from Bichet, D.G. (2007). Nephrogenic and central diabetes insipidus. In Disease of the kidney and urinary tract, 8th edn, (Schrier, R.W., ed.). Lippincott Williams & Wilkins, Philadelphia.

powerfully stimulate the release of the hormone in an exponential manner (Dunn et al., 1973). Although less well characterized, an increase in blood volume and/or blood pressure seems to have the opposite effect, namely to reduce vasopressin secretion (Goldsmith et al., 1984). These hemodynamic alterations appear to be mediated by changes in parasympathetic and sympathetic tone as the hypothalamus is richly innervated by these neural pathways. Thus, for example, the diuresis that accompanies acute left atrial distention is mediated by stimulation of the vagal afferents (Gauer and Henry, 1976). Vasopressin secretion is also suppressed by alpha adrenergic stimulation, but this is mediated by the increment in blood pressure rather than a direct effect on the hypothalamus (Berl et al., 1974). In addition, sensory afferents from the oropharynx via the glossopharyngeal nerve are felt to mediate the decrease in vasopressin associated with the act of drinking, which occurs independent of any decrease in tonicity (Verbalis, 1991; Geelen et al., 1996).

A number of hormones and drugs have been reported to inhibit vasopressin secretion. These are listed in Table 18.1. The ability of ethanol to suppress the secretion of the hormone has been known for a long time and served to establish bioassays (Kleeman et al., 1955). Some hormones, such as the aforementioned norepinephrine, do so by indirect mechanisms. In low doses, a number of opioid agonists, including morphine, met-enkephalin and kappa receptor agonists, inhibit both basal and stimulated vasopressin secretion (Miller, 1980; Oiso et al., 1988). The dopamine antagonists, such as promethazine and haloperidol, most likely act by suppressing the effects of the emetic center which, via the chemoreceptor trigger zone, is a powerful stimulating pathway for hormone release (Robertson, 1976). The effect of clonidine is most likely mediated by peripheral and central adrenoreceptors (Reid et al., 1984).

B. Inhibition of Vasopressin Action

The cellular response to vasopressin in the principal cells of the mammalian collecting duct is critical to the generation of a concentrated urine. The sequence of events that follows the binding of the hormone to its receptor in the basolateral membrane that ultimately culminate in the insertion of aquaporin 2 into the luminal membrane and render the tubule permeable to water have been well described (Nielson et al., 2007). Genetic disorders of the receptor and the water channel clearly result in resistance to the hormone (Bichet, 2007). There are other pathophysiologic and pharmacologic settings that also inhibit the hydro-osmotic response to the hormone resulting in renal concentrating defects. There is no identifiable counter-regulatory hormone to the action of vasopressin, but several autacoids, such as prostaglandins (Berl and Schrier, 1973; Berl et al., 1977a) and endothelin (Kohan and Hughes, 1993; Edwards et al., 1993), as well as alpha adrenergic stimulation (Schrier and Berl, 1973), have been shown to antagonize the hydro-osmotic properties of the hormone, thereby increasing water excretion. A number of pharmacologic agents that have been associated with this disorder are listed in Table 18.2. Not all the listed drugs cause the disorder by directly antagonizing vasopressin action, as other mechanisms such as failure to generate interstitial hypertonicity may be operant with some. Two of the pharmacologic agents deserve particular mention – lithium and demeclocycline.

By virtue of its widespread use in the treatment of affective disorders, lithium has emerged as perhaps the most common cause of nephrogenic diabetes insipidus, affecting as many as 50% of patients on the drug (Baton et al., 1987; Walker, 1993). There is no evidence that lithium impairs vasopressin release (Forrest et al., 1974; Baylis and Heath, 1978). In terms of the mechanism of its renal action, lithium does not interfere with accumulation of medullary solutes, thus, an intrinsic tubular defect is postulated. In this regard, lithium decreases vasopressin stimulated water transport in the perfused cortical collecting duct (Cogan and Abramow, 1986). An inhibition in adenylate cyclase and cAMP generation is observed in human tissue (Dousa, 1974) and cultured cells (Goldberg et al., 1988; Anger et al., 1990) exposed to the cation as well as animals chronically treated with lithium (Christensen et al., 1985). A downregulation of aquaporin-2 (AQP2) and aquaporin-3 (AQP3) has been described in lithium-treated rats (Marples et al., 1995). It is of interest that the aquaporin levels remained low after removal of lithium, in line with slow recovery of concentrating ability seen in humans (Baton et al., 1987). More recently, an effect of lithium on the epithelial sodium channel has also been described, which may explain the natriuresis that contributes to the polyuria (Nielsen et al., 2003).

Since it was first recognized as a cause of nephrogenic diabetes insipidus (Singer and Rotenberg, 1973), demeclocycline has become the drug of choice for the treatment of the syndrome of inappropriate antidiuretic hormone secretion (SIADH). It has yet to be determined if demeclocycline reduces AVP secretion. It is clear, however, that demeclocycline induces dose-dependent decreases in human renal medullary adenylate cyclase activity. Because the drug decreases not only vasopressin but also cAMP stimulated water flow, a post-cAMP defect may be operant (Singer and Rotenberg, 1973). The precise biochemical mechanism of demeclocycline, however, has eluded elucidation.

Hypokalemia has long been known to cause polyuria as a consequence of a vasopressin-resistant renal concentrating defect (Teitelbaum and Berl, 1984). Initially, the polyuria results from a primary effect of potassium depletion to stimulate water intake (Berl et al., 1977b). A renal concentrating defect that is independent of the high rate of water intake eventually supervenes. The defect is in part caused by a decrement in tonicity in the medullary interstitial, which most likely relates to a decrease in sodium chloride reabsorption in the thick ascending limb (Gutsche et al., 1984). The elaboration of vasopressin-resistant hypotonic urine suggests, however, a defect in the collecting duct's response to vasopressin, independent of the decreased medullary tonicity. A direct effect of hypokalemia on the collecting tubule is supported by studies in the toad bladder that show a decrease in cAMP and vasopressin-stimulated water flow when potassium is removed from the bathing solution (Finn et al., 1966). These findings suggest both a pre-cAMP and post-cAMP defect. The hypokalemia-induced resistance to vasopressin is associated with a decreased cAMP accumulation, apparently owing to decreased adenylate cyclase activity (Kim et al., 1984). A profound decrement in the vasopressin sensitive water channel (aquaporin-2) has been described in the hypokalemic rats (Marples et al., 1996). Hypokalemia from any cause (e.g. diarrhea, chronic diuretic use or primary aldosteronism) may be associated with a urinary concentrating defect. The defect generally is reversible but required a longer time (1–3 months) than would be expected from a purely functional defect (Relman and Schwartz, 1958).

Hypercalcemia is another well-recognized cause of impaired urinary concentrating ability (Teitelbaum and Berl, 1984). A decrement in medullary interstitial tonicity clearly is present with hypercalcemia (Levi et al., 1983), which may be related to diminished solute reabsorption in the thick ascending limb (Peterson, 1990). This defect is associated with a decrement in AVP-stimulated adenylate cyclase in this nephron segment (Berl, 1987). The concentrating defect, however, is multifactorial, because the elaboration of a vasopressin-resistant hypotonic urine implies an intrinsic defect in the collecting tubule. In this regard, studies in isolated toad bladders (Omachi et al., 1974), as well as papillary collecting ducts, revealed a decreased response to vasopressin in hypercalcemia. A similar inhibition of cAMP accumulation and hydro-osmotic response to AVP is seen with maneuvers that increase cell calcium (Teitelbaum and Berl, 1986; Ando et al., 1988). Two studies have examined the effect of hypercalcemia in AQP2 expression employing a vitamin D-treated rat. Both revealed a decrement in AQP2 (Earm et al., 1998; Sands et al., 1998).

III. VASOPRESSIN ANTAGONISTS AND THEIR ROLE IN THE TREATMENT OF WATER-RETAINING DISORDERS

Soon after Du Vigneaud's original report of the amino acid sequence of vasopressin (Du Vigneaud, 1956), several investigators undertook the development of both selective agonists and antagonists of the hormone (Manning and Sawyer, 1989). While the synthesis of dDAVP (desmopressin), a long-acting agonist proved particularly successful (Manning and Sawyer, 1989), efforts to develop peptide antagonists were met with significant challenges, as many of them proved to have some agonist action, and significant species differences in their effectiveness limited their application to humans (Kinter et al., 1991). Thereafter, attempts to develop peptide antagonists of the vasopressin V2 receptor were abandoned. Moreover, peptides have to be administered parenterally as they have poor oral bioavailability, further limiting their widespread applicability. Subsequently, oral non-peptide antagonists were synthesized employing a functional screening approach. These efforts culminated in the report of Ohnishi et al.

TABLE 18.3 Non-peptide vasopressin antagonists currently under commercial development

Compound	Receptor	Route	Manufacturer
Conivaptan (YM-087)	V1a + V2	I.V.	Astellas (Tokyo, Japan)
Lixivaptan (VPA-985)	V2	Oral	CardioKine (Philadelphia, PA)
Tolvaptan	V2	Oral	Otsuka (Tokyo, Japan)
Satavaptan	V2	Oral	Sanofi-Aventis (Paris, France)

Greenberg, A., Verbalis, J.G. (2006). Vasopressin receptor antagonists. *Kidney Int* 69, 2124–30.

demonstrating the aquaretic effect of one such oral agent in healthy antidiuretic human subjects (Ohnishi et al., 1993). The availability of these agents made it possible to define better the role of vasopressin in the pathogenesis of various disorders of water balance, previously possible only by correlations with measurements of radioimmunoassayable plasma vasopressin levels. At the same time, the non-peptide antagonists have been studied as therapeutic agents in the treatment of such disorders, clinically characterized by the presence of hyponatremia. Since this is the most common electrolyte disorder, occurring in as many as 15–30% of hospitalized patients (Anderson et al., 1985; Hoorn et al., 2006), the availability of a drug that can correct the disorder would appear to be most attractive. The non-peptide antagonists that are currently in development for commercial use are listed in Table 18.3. Of these, only conivaptan has so far been approved as an intravenous preparation for the treatment of hyponatremia (Greenberg and Verbalis, 2006). Structurally, they are all benzazepine or oxindole derivatives and, with the exception of conivaptan, that is both a V1a and V2 receptor antagonist, the others are selective for the V2 receptor. The agents displace radioactively labeled hormone from its receptor and thereby potently inhibit AVP stimulated adenylate cyclase stimulation (Yamamura et al., 1992; Tahara et al., 1997). More recently, molecular modeling has revealed that the binding sites for arginine vasopressin and the antagonists are only partially overlapping, whereas the native hormone binds on the extracellular surface of the receptor, the antagonists penetrate into the transmembrane region, as illustrated in Figure 18.1 (Macion-Dazard et al., 2006).

A. Vasopressin Antagonists in Euvolemic Hyponatremia

1. Hypothyroidism

Advanced hypothyroidism may be associated with myxedema and hyponatremia. Studies in hypothyroid patients demonstrated a failure to suppress plasma AVP during an acute water load (Schrier, 2006). Experimental hypothyroidism produced by aminotriazole treatment in rats has allowed for the better understanding of the impaired urinary dilution with diminished thyroid function (Chen et al., 2005a). This experimental model of hypothyroidism was associated with a decrease in cardiac output and a slower pulse rate as compared to euthyroid animals. These are systemic hemodynamic perturbations which occur in hypothyroid humans. Plasma AVP concentrations were elevated in spite of hypo-osmolality and impaired water excretion in these hypothyroid rats. This impaired water excretion was associated with increased medullary AQP2 expression. Thyroxin replacement normalized these abnormalities in water homeostasis. Moreover, a V2 receptor antagonist normalized the water excretion and the AQP2 expression (Chen et al., 2005a). Thus, the systemic hemodynamic alterations in hypothyroidism are associated with non-

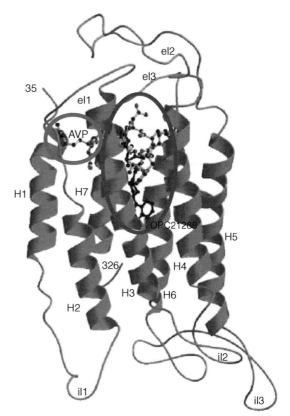

FIGURE 18.1 AVP versus AVP antagonist binding. The vasopressin receptor with its 7 transmembrane regions (H1–H7), extracellular (e) and intracellular domains (il). The site at which AVP binds at the surface of the receptor is circled in red. The site of the antagonist binding is deep in the transmembrane region and is circled in blue. The sites are distinct and partially overlap. The antagonist prevents the binding of AVP. Modified from Macion-Dazard, R. et al. (2006). *J Pharmacol Exp Ther* 316, 564–71. (See color plate section.)

osmotic, baroreceptor-mediated AVP stimulation with upregulation of medullary AQP2 expression, impaired water excretion and hyponatremia with reversal during V2 receptor antagonist administration. The use of V2 antagonists in the setting of patients with hyponatremia secondary to hypothyroidism has not been reported. In fact, hormone replacement is the most appropriate therapeutic approach to these patients.

2. Addison's Disease and Hypopituitarism

The impaired water excretion and hyponatremia associated with primary adrenal insufficiency (Addison's disease) involves both glucocorticoid and mineralocorticoid deficiency, whereas the hyponatremia of hypopituitarism involves only glucocorticoid deficiency since the renin–angiotensin–aldosterone axis is intact (Schrier, 2006). The separate effects of glucocorticoid and mineralocorticoid deficiency have been examined in adrenalectomized rats (Ishikawa and Schrier, 1982; Chen et al., 2005b). The control animals received physiological doses of both glucocorticoid and mineralocorticoid hormones. Then glucocorticoid deficiency was studied in adrenalectomized animals receiving only mineralocorticoid hormone, while animals with mineralocorticoid deficiency only received glucocorticoid hormone.

As mineralocorticoid deficiency results in negative sodium balance and decrements in extracellular fluid volume, it is truly a form of hypovolemic hyponatremia. In contrast, glucocorticoid deficiency which accompanies hypopituitarism is a euvolemic form of hyponatremia, as there is no negative sodium balance and, in fact, a positive balance may occur. There is, however, a decrement in cardiac output, a lower blood pressure and impaired systemic vascular resistance leading to baroreceptor mediated non-osmotic release of vasopressin (Schrier, 2006). This is associated with increased expression of the vasopressin dependent water channel AQP 2 (Saito et al., 2000; Chen et al., 2005b). In an experimental model of glucocorticoid deficiency, the administration of a vasopressin antagonist normalizes the response to a water load (Figure 18.2) and returns the expression of AQP2 to normal levels (Chen et al., 2005b). As with hypothyroidism, there are no well described patients with glucocorticoid deficiency who have received vasopressin antagonists. Since these disorders are best treated by hormone replacement, there is not a role for the novel antagonists in these causes of hyponatremia. Nonetheless, they have been helpful tools in delineating the mechanisms that underlie the diluting defect and the central role of vasopressin in their pathogenesis.

3. Syndrome of Inappropriate ADH secretion

Following the successful use of the first generation of oral vasopressin antagonists (OPC-31260) in an experimental model of SIADH in rats (Fujisawa et al., 1993), 11 patients

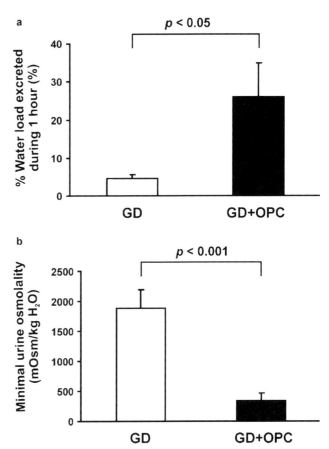

FIGURE 18.2 Treatment of GD rats with the vasopressin V2 receptor antagonist OPC-31260 (GD + OPC, $n = 8$) resulted in a marked increase in percent excretion of water load (a) and a significant decrease in minimal urinary osmolality (b) 1 h after a 40 ml/kg water load by oral gavage compared with treatment with vehicle (GD, $n = 8$). From Wang et al., 2006.

with the syndrome were treated with the drug (Saito et al., 1997). The authors describe a transient decrease in urinary osmolality and an increase in urine volume that was independent of changes in solute excretion. The agents listed in Table 18.3 have also been used in patients with SIADH. Specifically, in six patients given 50–100 mg bid of lixivaptan, the achieved serum sodium was 133 mmoles/l compared to 126 mmoles/l for those on placebo. This increment in serum sodium concentration was associated with a decrease in urinary sodium excretion, presumably as a consequence of the correction of volume expansion that accompanied the aquaretic response (Decaux, 2001a). As part of the North American Lixivaptan study, the four patients with SIADH who received the drug increased their serum sodium concentration from 127 to139 mmoles/l over a 7-day period (Wong et al., 2003). On the heels of these smaller studies, the largest study to date examined the effect of tolvaptan in a randomized double-blind, placebo-controlled study, named Study of Ascending Levels of Tolvaptan in hyponatremia in the United States (SALTII) as well as multiple international sites (SALT I) (Schrier et al., 2006). In these studies, 91 of

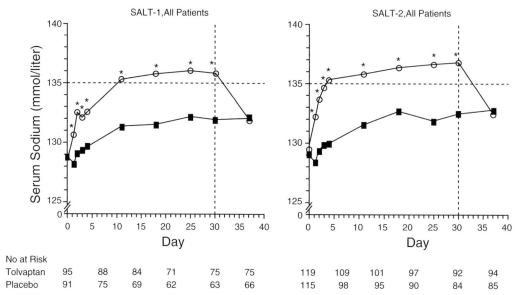

FIGURE 18.3 Effect of tolvaptan (15–60 mg) on serum sodium concentration in two trials. The serum sodium was significantly higher in tolvaptan treated patients (open circles) than on placebo (dark squares). From Schrier, R.W. et al. (2006). *N Engl J Med* 355, 2099.

the 214 patients who received the active drug had SIADH. The effect of tolvaptan and placebo on serum sodium concentration over the 30 days of the study and 7 days after discontinuation is shown in Figure 18.3. At all time points, the serum Na was higher in the tolvaptan treated subjects and the effect was reversible as the serum Na decreased to levels seen in subjects on placebo after 7 days of stopping tolvaptan. In this study, after 30 days of treatment, the mean increase in serum sodium in SIADH patients was 8.2 mmoles/l, an increase greater than that seen in patients with congestive heart failure or cirrhosis in the same study. The increment in serum Na was accompanied by an improvement in the mental component of the SF-12 survey (Schrier et al., 2006). Likewise, in 26 patients with SIADH, satavaptan increased serum Na from 125 to 136 mmoles/l over 5 days on 25 mg and to 140 with 50 mg of the drug. A longer term open label extension revealed persistent efficacy without escape from the drug's effects (Soupart et al., 2006). Finally, conivaptan was effective in raising serum sodium over several months in two patients with chronic SIADH who did not respond to water restriction (Decaux, 2001b). A recent report with this agent also confirms its efficacy over a 5-day period in a study that included 38 euvolemic patients, most of whom likely had SIADH (Ghali et al., 2006). However, conivaptan is a substrate and potent inhibitor of the microsomal enzyme cytochrome P450 (CYP) 3A4, thus limiting the concomitant use of many chemotherapeutic agents, calcium channel blockers, ketoconazole, itraconazole, clarithromycin, HMG CoA reductase inhibitors, benzodiazepines, ritonavir, indinavir and immunosuppressants (Walter, 2007). Thus, this drug is approved by the FDA for 4-day intravenous use only. This approval was obtained on the basis of a double-blind clinical trial in which 35 euvolemic patients (14 of whom had well documented SIADH) received a bolus of 20 mg of intravenous conivaptan followed by either 40 mg/day or 80 mg/day by continuous infusion. All patients were treated with a 2-liter water restriction. Patients treated with either dose of conivaptan had a predictable and significant ($P < 0.001$) rise in serum sodium. The greatest increase occurred in the first week but was sustained during the days of the infusion (Zeltser et al., 2007).

B. Vasopressin Antagonists in Hypervolemic Hyponatremia

1. Cardiac Failure

Activation of neurohumoral pathways, including vasopressin, is central to the disturbance in fluid homeostasis in heart failure (Schrier, 1988a, b). It is therefore not surprising that hyponatremia occurs in advanced heart failure and is, in fact, associated with decreased survival (Lee and Packer, 1986). With the advent of the sensitive radioimmunoassay for vasopressin, hyponatremic cardiac failure patients were shown not to suppress their plasma AVP concentrations, even though the level of plasma osmolality was sufficient to suppress plasma AVP in normal subjects (Szatalowicz et al., 1981). This observation was compatible with the non-osmotic, baroreceptor-mediated stimulation of AVP. Since AVP is known to modulate the short- and long-term regulation of AQP2 water channels, studies were undertaken to examine the role of AVP and AQP2 in experimental heart failure secondary to coronary artery ligation (Nielsen et al., 1997; Xu et al., 1997). In these studies, plasma AVP was increased in association with an upregulation of renal medullary AQP2 expression and trafficking. The AVP-mediated water retention, hyponatremia and increased expression and trafficking of AQP2 in these heart failure animals were reversed with the administration of a V2 vasopressin receptor antagonist (Xu et al., 1997).

On this background, several investigations were undertaken in patients with cardiac failure treated with V2 vasopressin antagonists. Studies in hyponatremic New York Heart Association Class II and III cardiac failure patients demonstrated an increase in electrolyte-free water excretion, correction of the hyponatremia and decreased urinary AQP2 excretion during the administration of the V2 receptor antagonist lixivaptan (Martin et al., 1999). This antagonist, as well as tolvaptan, was found to produce an aquaresis in patients with this degree of heart failure (Abraham et al., 2006; Costello-Boerrigter et al., 2006). Concomitantly, larger studies were undertaken in such patients. Gheorghiade reported a double-blind, placebo-controlled study to examine the effects of oral tolvaptan in 254 patients with class II or III congestive heart failure (CHF), 28% of whom were hyponatremic (Gheorghiade et al., 2003). Patients were treated with 30 mg, 45 mg or 60 mg tolvaptan daily for 25 days. They were maintained on furosemide, but were not water restricted. All patients on the drug lost weight and maintained their weight loss for the duration of the study. Further, their serum sodium rose by 3 mEq/l in the first 24 h and then drifted back to baseline. However, of the hyponatremic patients at baseline, 80% normalized their serum sodium concentration and maintained it (compared to 40% of those on placebo). The same group then examined the effect of tolvaptan for sicker, hospitalized patients with CHF (Gheorghiade et al., 2004); 319 patients with CHF and an ejection fraction of <40% were treated with 30 mg, 60 mg or 90 mg of tolvaptan. The drug treated group had a greater in-hospital weight loss than those on placebo.

These studies paved the way for a larger multicenter trial – The Efficacy of Vasopressin Antagonism in heart failure outcome Study with Tolvaptan (EVEREST) trial (Konstam et al., 2007). The study was designed to investigate the long-term effect of tolvaptan on morbidity and mortality in patients hospitalized with worsening CHF, accompanied by signs and symptoms. Here, 4733 patients hospitalized for CHF were randomized to either tolvaptan (30 mg/day) or placebo for a minimum of 60 days. There was no difference, either favorable or unfavorable, between the two study groups in all-cause mortality, cardiovascular mortality or hospitalization related to heart failure (Konstam et al., 2007). Tolvaptan, however, significantly improved secondary endpoints of perceived dyspnea, measured body weight and edema (Gheorghiade et al., 2007). It must be noted, however, that in this study of tolvaptan in CHF, less than 10% of patients were hyponatremic, a subgroup that is too small for meaningful independent analysis.

Some of the aforementioned studies support the possible use of vasopressin receptor antagonist as an adjunct to therapy for patients with varying degrees of CHF. Intravenous conivaptan is the only agent available at this time.

With its combined V1a and V2 receptor antagonism, conivaptan has been shown to have favorable hemodynamic effects in patients with advanced heart failure. Decreases in pulmonary capillary wedge pressure (PCWP) and right atrial pressure were accompanied by substantial increases in urine output, with no effects on systemic blood pressure, heart rate and electrolytes (Udelson et al., 2001). On the other hand, oral tolvaptan has not been shown to impact left ventricular volume, but in contrast to the above mentioned EVEREST trial, a significant favorable effect of tolvaptan was found on mortality and heart failure related hospitalization (Udelson et al., 2007). It must be noted that in this trial the finding is, as the authors admit, constrained by several factors as this was not a prespecified endpoint. The outcomes also were investigator reported rather than being adjudicated by a blinded committee (Udelson et al., 2007).

2. Cirrhosis

While a decrease in stroke volume causes arterial underfilling in low output heart failure, the arterial underfilling in cirrhosis is secondary to primary arterial vasodilation in the splanchnic circulation (Bosch and Garcia-Pagan, 2005). In hyponatremic cirrhotic patients, plasma vasopressin has been shown not to be suppressed (Bichet et al., 1982). Thus, there is non-osmotic, baroreceptor stimulation of vasopressin in cirrhosis as occurs in cardiac failure (Schrier, 1988a, b). Pretreatment hyponatremia in cirrhosis is an important risk factor in progression to hepatorenal syndrome and poor survival (Gines et al., 1993).

A considerable number of patients with cirrhosis have been exposed to vasopressin antagonists. Decaux described the effect of lixivaptan in five such patients whose mean serum sodium rose from 128 to 133 mmoles/l over a 72 h period on 50–100 mg bid of the drug (Decaux, 2001a). In contrast to the decrease in sodium excretion that accompanies the correction in SIADH, these patients sustained a mild natriuresis. Guyader et al. (2002) performed a pharmacodynamic study with ascending single doses of the drug from 25 to 300 mg and observed a dose-dependent aquaretic response and a small natriuretic response in 27 patients with cirrhosis. A similar dose-dependent effect was observed in the North American trial (Wong et al., 2003), as 33 patients with cirrhosis received either placebo, 25, 125 or 250 mg bid of lixivaptan. While the serum sodium was unchanged at 7 days in the placebo group, it rose by 3, 5 and 7 mmoles/l in the other three groups, respectively. Finally, in a larger multicenter European trial involving 60 patients on either placebo, 50 or 100 mg bid of lixivaptan, at 7 days those patients on placebo had no change in serum sodium while those on 100 mg bid of the drug had an increase of 6 mmoles/l. In fact, 50% of the patients in this group normalized their serum sodium as defined by a level >136 mmoles/l (Gerbes et al., 2003).

The precursor to tolvaptan, OPC31260, was described as causing a water diuresis in subjects with cirrhosis (Inoue et al., 1998) and, in the aforementioned SALT trial, approximately 30% of the 223 studied patients who received tolvaptan had cirrhosis. When compared to those on placebo, the patients with cirrhosis had a greater increase in serum sodium at 30 days, 1.5 versus 4.0 mmoles/l, an increase that is more modest than was seen in the SIADH group. In fact, 37% of the cirrhotic patients seemed resistant to the drug. Because conivaptan has V1a receptor blocker activity, it has not been used in patients with cirrhosis. This is because of concerns that the drug could cause hypotension and variceal bleeding secondary to increased portal pressure and sphlanchnic vasodilation. Thus, only V2 receptor antagonists may be appropriate for use in cirrhotic patients. Furthermore, the data obtained with lixivaptan and tolvaptan are not in full accordance and only further clinical experience will ascertain the role of these agents in the management of the hyponatremia in patients with cirrhosis.

C. Resistance to Vasopressin Antagonists

It has become evident in the course of clinical trials that the response to vasopressin antagonists is not uniform and that there are patients whose response is significantly blunted. In Table 18.4 are listed some of the potential causes for failure to respond to these agents. Patients with hyponatremia, be it of the euvolemic or hypervolemic forms, frequently have excess water intake. While the administration of vasopressin antagonists will clearly allow significant liberalization of water intake, the excessive intake of fluids may blunt their therapeutic effectiveness leading to small if any changes in serum sodium concentration.

While vasopressin plays a central role in the pathogenesis of most hyponatremic disorders, vasopressin-independent mechanisms can significantly contribute to the diminished maximal water excretion. Specifically, decrements in glomerular filtration rate, enhanced proximal reabsorption of filtrate and impaired function of the diluting segment of the nephron can limit solute-free water clearance. Such processes probably account for the more modest response referred to above in patients with cirrhosis and CHF when compared with patients with SIADH.

The measurement of plasma vasopressin in patients with SIADH undertaken by Robertson and his colleagues revealed that most had hormone levels in the 'normal' range, albeit at lower plasma osmolalities at which the levels should have been suppressed (Robertson et al., 1982). Such patients would be expected to respond to the vasopressin antagonists. There were, however, individuals who had very high levels of the hormone and it is conceivable that in those patients the drugs would be less effective. Of interest is that there were 10–15% of patients in whom the hormone was not measurable. The recent report of two male infants with hyponatremia who had gain of function mutations of the V2 receptor, located on the X chromosome (Feldman et al., 2005) has triggered the search for such mutation in the adult population as well. Recently, Decaux et al. (2007) reported on a family whose proband, a 74-year-old man with hyponatremia, was resistant to two different vasopressin receptor antagonists proved to have a mutation in the vasopressin receptor. Female carriers also had spontaneous episodes of hyponatremia. Such patients will be resistant to vasopressin antagonists.

IV. ARE VASOPRESSIN ANTAGONISTS SAFE?

As a class, the vaptans interact with the microsomal enzyme cytochrome P450 (CYP) 3A4 enzyme system, though this interaction has been noted to be the highest with conivaptan, accounting for its approval only in intravenous form. The other agents, tolvaptan, lixivaptan and satavaptan, are in various stages of development. As noted, conivaptan, with its combined V1a/V2 receptor antagonism, may be at least theoretically potentially detrimental in patients with cirrhosis and portal hypertension by blocking the vasoconstrictive effects of AVP in the splanchnic circulation. Therefore, until further data and experience accrue, the use of conivaptan should be avoided in cirrhotic patients (Verbalis, 2007). There were a few side effects reported common to the vaptans in all the studies. These include thirst, dry mouth and hypotension (Gheorghiade et al., 1984; Schrier et al., 2006; Ghali et al., 2006). The administration of the vaptans should be reserved for euvolemic and hypervolemic hyponatremia only. Patients with hypovolemic hyponatremia should be managed with volume repletion and appropriate solute-free water restriction (Thurman et al., 2003).

Osmotic demyelination from rapid correction of hyponatremia can lead to central pontine myelinolysis with devastating outcomes (Adrogue and Madias, 2000; Thurman et al., 2003). While this is a theoretical concern, no case occurred in any of the studies above or since conivaptan has been used clinically. A factor that possibly mitigates the rapid rise of serum sodium is ongoing water and fluid intake. Neither water nor fluid was absolutely restricted in any of the studies. None of the patients were concomitantly treated with hypertonic saline either.

TABLE 18.4 Potential causes for resistance to vasopressin antagonists

Excessive water intake
Vasopressin-independent diluting defect
Very high levels of vasopressin
Activating mutations of the vasopressin receptor

V. SUMMARY AND UNANSWERED QUESTIONS

The development of vasopressin antagonists (vaptans) and their advent to clinical practice clearly portend a new era in the treatment of hyponatremic disorders. By addressing the primary mechanism of the non-osmotic release of AVP that underlies these disorders, this is the most physiologic of approaches to the enhancement of free water excretion and the correction of the electrolyte abnormality. At the present, it is available in an IV form for euvolemic conditions such as the SIADH secretion, as well as for hypervolemic states. The aquaretic agents provide a reliable, predictable, yet not excessive, increase in serum sodium concentration without attendant significant changes in sodium or potassium excretion. The use of these drugs also appears to be free of significant adverse effects and to be extremely well tolerated allowing patients free water access. They therefore possess many of the qualities a clinician would be looking for in an ideal drug to treat these hyponatremic conditions and are superior to all presently available treatments. However, there remain a number of questions that require further assessment and data collection. At this time, it is unclear whether the treatment with intravenous conivaptan will suffice for patients with acute symptomatic hyponatremia and whether additional treatment with 3% saline will be needed to complement the drug. It is also not known whether the correction of the hyponatremia will impact significantly the patient's quality of life, cognitive function, rate of hospitalizations or overall mortality. Finally, an assessment of the value of sole V2 receptor blockade versus combined V1a/V2 blockade is of interest, particularly to assess whether V1a blockade would add to the effects of beta adrenergic, renin–angiotensin and aldosterone blockade in patients with advanced cardiac disease. Clearly, further clinical trials directed at providing answers to these unanswered questions are still needed.

References

Abraham, W. T., Shamshirsaz, A. A., McFann, K., Oren, R. M., and Schrier, R. W. (2006). Aquaretic effect of lixivaptan, an oral, non-peptide, selective V2 receptor vasopressin antagonist, in New York Heart Association functional class II and III chronic heart failure patients. *J Am Coll Cardiol* **47**: 1615–1621.

Adrogue, H. J., and Madias, N. E. (2000). Hyponatremia. *N Engl J Med* **342**: 1581–1589.

Anderson, R. J., Chung, H. M., Kluge, R., and Schrier, R. W. (1985). Hyponatremia: a prospective analysis of its epidemiology and the pathogenetic role of vasopressin. *Ann Intern Med* **102**: 164–168.

Ando, Y., Jacobson, H. R., and Breyer, M. D. (1988). Phorbol ester and A23187 have additive but mechanistly separate effects on vasopressin action in rabbit collecting tubule. *J Clin Invest* **81**: 1578–1584.

Anger, M. S., Shanley, P., Mansour, J., and Berl, T. (1990). Effects of lithium on cAMP generation in cultured rat inner medullary collecting tubule cells. *Kidney Int* **37**: 1211–1218.

Baton, T., Gavira, M., and Battle, D. (1987). Prevalance, pathogenesis and treatment of renal dysfunction associated with chronic lithium therapy. *Ann Intern Med* **79**: 679.

Baylis, P. H., and Heath, D. A. (1978). Water disturbances in patients treated with oral lithium carbonate. *Ann Intern Med* **88**: 607–609.

Berl, T. (1987). The cAMP system in vasopressin-sensitive nephron segments of the vitamin D-treated rat. *Kidney Int* **31**: 1065–1071.

Berl, T., and Schrier, R. W. (1973). Mechanism of effect of prostaglandin E 1 on renal water excretion. *J Clin Invest* **52**: 463–471.

Berl, T., Cadnapaphornchai, P., Harbottle, J. A., and Schrier, R. W. (1974). Mechanism of suppression of vasopressin during alpha-adrenergic stimulation with norepinephrine. *J Clin Invest* **53**: 219–227.

Berl, T., Raz, A., Wald, H., Horowitz, J., and Czaczkes, W. (1977a). Prostaglandin synthesis inhibition and the action of vasopressin: studies in man and rat. *Am J Physiol* **232**: F529–F537.

Berl, T., Linas, S. L., Aisenbrey, G. A., and Anderson, R. J. (1977b). On the mechanism of polyuria in potassium depletion. The role of polydipsia. *J Clin Invest* **60**: 620–625.

Bichet, D. G. (2007). Nephrogenic and central diabetes insipidus. In: *Diseases of the kidney and urinary tract,* 8th edn, (Schrier, R. W., eds) Lippincott Williams & Wilkins, Philadelphia.

Bichet, D., Szatalowicz, V., Chaimovitz, C., and Schrier, R. W. (1982). Role of vasopressin in abnormal water excretion in cirrhotic patients. *Ann Intern Med* **96**: 413–417.

Bosch, J., and Garcia-Pagan, J. (2005). The splanchnic circulation in cirrhosis. In: (Gines, P., Arroyo, V., Rodes, J., Schrier, R.W., eds) *Ascites and renal dysfunction in liver disease,* Blackwell Publishing, Oxford, pp. 156–163.

Chen, Y. C., Cadnapaphornchai, M. A., Yang, J. et al. (2005a). Nonosmotic release of vasopressin and renal aquaporins in impaired urinary dilution in hypothyroidism. *Am J Physiol* **289**: F672–F678.

Chen, Y. C., Cadnapaphornchai, M. A., Summer, S. N. et al. (2005b). Molecular mechanisms of impaired urinary concentrating ability in glucocorticoid-deficient rats. *J Am Soc Nephrol* **16**: 2864–2871.

Christensen, S., Kusano, E., Yusufi, A. N., Murayama, N., and Dousa, T. P. (1985). Pathogenesis of nephrogenic diabetes insipidus due to chronic administration of lithium in rats. *J Clin Invest* **75**: 1869–1879.

Cogan, E., and Abramow, M. (1986). Inhibition by lithium of the hydroosmotic action of vasopressin in the isolated perfused cortical collecting tubule of the rabbit. *J Clin Invest* **77**: 1507–1514.

Costello-Boerrigter, L. C., Smith, W. B., Boerrigter, G. et al. (2006). Vasopressin-2-receptor antagonism augments water excretion without changes in renal hemodynamics or sodium and potassium excretion in human heart failure. *Am J Physiol* **290**: F273–F278.

Decaux, G. (2001a). Difference in solute excretion during correction of hyponatremic patients with cirrhosis or syndrome of inappropriate secretion of antidiuretic hormone by oral vasopressin V2 receptor antagonist VPA-985. *J Lab Clin Med* **138**: 18–21.

Decaux, G. (2001b). Long-term treatment of patients with inappropriate secretion of antidiuretic hormone by the vasopressin receptor antagonist conivaptan, urea, or furosemide. *Am J Med* **110**: 582–584.

Decaux, G., Vandergheynst, F., Bouko, Y., Parma, J., Vassart, G., and Vilain, C. (2007). Nephrogenic syndrome of inappropriate antidiuresis in adults: high phenotypic variability in men and women from a large pedigree. *J Am Soc Nephrol* **18**: 606–612.

Dousa, T. P. (1974). Interaction of lithium with vasopressin-sensitive cyclic AMP system of human renal medulla. *Endocrinology* **95**: 1359–1366.

Du Vigneaud, V. (1956). Hormones of the posterior pituitary gland: oxytocin and vasopressin. In: (Du Vigneaud, V., Bing, R.J., Oncley, J.L., eds) *The Harvey Lectures,* Academic Press, New York, pp. 1954–1955.

Dunn, F. L., Brennan, T. J., Nelson, A. E., and Robertson, G. L. (1973). The role of blood osmolality and volume in regulating vasopressin secretion in the rat. *J Clin Invest* **52**: 3212–3219.

Earm, J. H., Christensen, B. M., Frokiaer, J. et al. (1998). Decreased aquaporin-2 expression and apical plasma membrane delivery in kidney collecting ducts of polyuric hypercalcemic rats. *J Am Soc Nephrol* **9**: 2181–2193.

Edwards, R. M., Stack, E. J., Pullen, M., and Nambi, P. (1993). Endothelin inhibits vasopressin action in rat inner medullary collecting duct via the ETB receptor. *J Pharmacol Exp Ther* **267**: 1028–1033.

Feldman, B. J., Rosenthal, S. M., Vargas, G. A. et al. (2005). Nephrogenic syndrome of inappropriate antidiuresis. *N Engl J Med* **352**: 1884–1890.

Finn, A. L., Handler, J. S., and Orloff, J. (1966). Relation between toad bladder potassium content and permeability response to vasopressin. *Am J Physiol* **210**: 1279–1284.

Forrest, J. N., Cohen, A. D., Torretti, J., Himmelhoch, J. M., and Epstein, F. H. (1974). On the mechanism of lithium-induced diabetes insipidus in man and the rat. *J Clin Invest* **53**: 1115–1123.

Fujisawa, G., Ishikawa, S., Tsuboi, Y., Okada, K., and Saito, T. (1993). Therapeutic efficacy of non-peptide ADH antagonist OPC-31260 in SIADH rats. *Kidney Int* **44**: 19–23.

Gauer, O. H., and Henry, J. P. (1976). Neurohormonal control of plasma volume. *Int Rev Physiol* **9**: 145–190.

Geelen, G., Greenleaf, J. E., and Keil, L. C. (1996). Drinking-induced plasma vasopressin and norepinephrine changes in dehydrated humans. *J Clin Endocrinol Metab* **81**: 2131–2135.

Gerbes, A. L., Gulberg, V., Gines, P. et al. (2003). Therapy of hyponatremia in cirrhosis with a vasopressin receptor antagonist: a randomized double-blind multicenter trial. *Gastroenterology* **124**: 933–939.

Ghali, J. K., Koren, M. J., Taylor, J. R. et al. (2006). Efficacy and safety of oral conivaptan: a V1A/V2 vasopressin receptor antagonist, assessed in a randomized, placebo-controlled trial in patients with euvolemic or hypervolemic hyponatremia. *J Clin Endocrinol Metab* **91**: 2145–2152.

Gheorghiade, M., Gattis, W. A., O'Connor, C. M. et al. (1984). Effects of tolvaptan, a vasopressin antagonist, in patients hospitalized with worsening heart failure: a randomized controlled trial. *J Am Med Assoc* **291**: 1963–1971.

Gheorghiade, M., Niazi, I., Ouyang, J. et al. (2003). Vasopressin V2-receptor blockade with tolvaptan in patients with chronic heart failure: results from a double-blind, randomized trial. *Circulation* **107**: 2690–2696.

Gheorghiade, M., Konstam, M. A., Burnett, J. C. et al. (2007). Short-term clinical effects of tolvaptan, an oral vasopressin antagonist, in patients hospitalized for heart failure: the EVEREST Clinical Status Trials. *J Am Med Assoc* **297**: 1332–1343.

Gines, A., Escorsell, A., Gines, P. et al. (1993). Incidence, predictive factors, and prognosis of the hepatorenal syndrome in cirrhosis with ascites. *Gastroenterology* **105**: 229–236.

Goldberg, H., Clayman, P., and Skorecki, K. (1988). Mechanism of Li inhibition of vasopressin-sensitive adenylate cyclase in cultured renal epithelial cells. *Am J Physiol* **255**: F995–1002.

Goldsmith, S. R., Cowley, A. W., Francis, G. S., and Cohn, J. N. (1984). Effect of increased intracardiac and arterial pressure on plasma vasopressin in humans. *Am J Physiol* **246**: H647–H651.

Greenberg, A., and Verbalis, J. G. (2006). Vasopressin receptor antagonists. *Kidney Int* **69**: 2124–2130.

Gutsche, H. U., Peterson, L. N., and Levine, D. Z. (1984). In vivo evidence of impaired solute transport by the thick ascending limb in potassium-depleted rats. *J Clin Invest* **73**: 908–916.

Guyader, D., Patat, A., Ellis-Grosse, E. J., and Orczyk, G. P. (2002). Pharmacodynamic effects of a nonpeptide antidiuretic hormone V2 antagonist in cirrhotic patients with ascites. *Hepatology* **36**: 1197–1205.

Helderman, J. H., Vestal, R. E., Rowe, J. W., Tobin, J. D., Andres, R., and Robertson, G. L. (1978). The response of arginine vasopressin to intravenous ethanol and hypertonic saline in man: the impact of aging. *J Gerontol* **33**: 39–47.

Hoorn, E. J., Lindemans, J., and Zietse, R. (2006). Development of severe hyponatraemia in hospitalized patients: treatment-related risk factors and inadequate management. *Nephrol Dial Transplant* **21**: 70–76.

Inoue, T., Ohnishi, A., Matsuo, A. et al. (1998). Therapeutic and diagnostic potential of a vasopressin-2 antagonist for impaired water handling in cirrhosis. *Clin Pharmacol Ther* **63**: 561–570.

Ishikawa, S., and Schrier, R. W. (1982). Effect of arginine vasopressin antagonist on renal water excretion in glucocorticoid and mineralocorticoid deficient rats. *Kidney Int* **22**: 587–593.

Kim, J. K., Summer, S. N., and Berl, T. (1984). The cyclic AMP system in the inner medullary collecting duct of the potassium-depleted rat. *Kidney Int* **26**: 384–391.

Kinter, L. B., Ilson, B. E., and Caltabianol, S. (1991). Antidiuretic hormone antagonists in humans: are there predictors? In: (Jard, S., eds) *Vasopressin Third International Vasopressin Conference,* John Libbey Eurotext, Paris, pp. 321–329.

Kleeman, C. R., Rubini, M. E., Lamdin, E., and Epstein, F. H. (1955). Studies on alcohol diuresis. II. The evaluation of ethyl alcohol as an inhibitor of the neurohypophysis. *J Clin Invest* **34**: 448–455.

Kohan, D. E., and Hughes, A. K. (1993). Autocrine role of endothelin in rat IMCD: inhibition of AVP-induced cAMP accumulation. *Am J Physiol* **265**: F126–F129.

Konstam, M. A., Gheorghiade, M., Burnett, J. C. et al. (2007). Effects of oral tolvaptan in patients hospitalized for worsening heart failure: the EVEREST Outcome Trial. *J Am Med Assoc* **297**: 1319–1331.

Lee, W. H., and Packer, M. (1986). Prognostic importance of serum sodium concentration and its modification by converting-enzyme inhibition in patients with severe chronic heart failure. *Circulation* **73**: 257–267.

Levi, M., Peterson, L., and Berl, T. (1983). Mechanism of concentrating defect in hypercalcemia. Role of polydipsia and prostaglandins. *Kidney Int* **23**: 489–497.

Lindheimer, M. D., and Davison, J. M. (1995). Osmoregulation, the secretion of arginine vasopressin and its metabolism during pregnancy. *Eur J Endocrinol* **132**: 133–143.

Macion-Dazard, R., Callahan, N., Xu, Z., Wu, N., Thibonnier, M., and Shoham, M. (2006). Mapping the binding site of six nonpeptide antagonists to the human V2-renal vasopressin receptor. *J Pharmacol Exp Ther* **316**: 564–571.

Manning, M., and Sawyer, W. H. (1989). Discovery, development, and some uses of vasopressin and oxytocin antagonists. *J Lab Clin Med* **114**: 617–632.

Marples, D., Christensen, S., Christensen, E. I., Ottosen, P. D., and Nielsen, S. (1995). Lithium-induced downregulation of aquaporin-2 water channel expression in rat kidney medulla. *J Clin Invest* **95**: 1838–1845.

Marples, D., Frokiaer, J., Dorup, J., Knepper, M. A., and Nielsen, S. (1996). Hypokalemia-induced downregulation of aquaporin-2 water channel expression in rat kidney medulla and cortex. *J Clin Invest* **97**: 1960–1968.

Martin, P. Y., Abraham, W. T., Lieming, X. et al. (1999). Selective V2-receptor vasopressin antagonism decreases urinary aquaporin-2 excretion in patients with chronic heart failure. *J Am Soc Nephrol* **10**: 2165–2170.

Miller, M. (1980). Role of endogenous opioids in neurohypophysial function of man. *J Clin Endocrinol Metab* **50**: 1016–1020.

Nielsen, S., Terris, J., Andersen, D. et al. (1997). Congestive heart failure in rats is associated with increased expression and targeting of aquaporin-2 water channel in collecting duct. *Proc Natl Acad Sci USA* **94**: 5450–5455.

Nielsen, J., Kwon, T. H., Praetorius, J. et al. (2003). Segment-specific ENaC downregulation in kidney of rats with lithium-induced NDI. *Am J Physiol* **285**: F1198–F1209.

Nielson, S., Knepper, M. A., Kwon, T., and Frokiaer, J. (2007). Regulation of water balance: urine concentration and dilution. In: *Diseases of the kidney and urinary tract*, 8th edn, (Schrier, R. W., eds) Lippincott Williams & Wilkins, Philadelphia.

Ohnishi, A., Orita, Y., Okahara, R. et al. (1993). Potent aquaretic agent. A novel nonpeptide selective vasopressin 2 antagonist (OPC-31260) in men. *J Clin Invest* **92**: 2653–2659.

Oiso, Y., Iwasaki, Y., Kondo, K., Takatsuki, K., and Tomita, A. (1988). Effect of the opioid kappa-receptor agonist U50488H on the secretion of arginine vasopressin. Study on the mechanism of U50488H-induced diuresis. *Neuroendocrinology* **48**: 658–662.

Omachi, R. S., Robbie, D. E., Handler, J. S., and Orloff, J. (1974). Effects of ADH and other agents on cyclic AMP accumulation in toad bladder epithelium. *Am J Physiol* **226**: 1152–1157.

Peterson, L. N. (1990). Vitamin D-induced chronic hypercalcemia inhibits thick ascending limb NaCl reabsorption in vivo. *Am J Physiol* **259**: F122–F129.

Reid, I. A., Ahn, J. N., Trinh, T., Shackelford, R., Weintraub, M., and Keil, L. C. (1984). Mechanism of suppression of vasopressin and adrenocorticotropic hormone secretion by clonidine in anesthetized dogs. *J Pharmacol Exp Ther* **229**: 1–8.

Relman, A. S., and Schwartz, W. B. (1958). The kidney in potassium depletion. *Am J Med* **24**: 764–773.

Robertson, G. L. (1976). The regulation of vasopressin function in health and disease. *Recent Prog Horm Res* **33**: 333–385.

Robertson, G. L., Mahr, E. A., Athar, S., and Sinha, T. (1973). Development and clinical application of a new method for the radioimmunoassay of arginine vasopressin in human plasma. *J Clin Invest* **52**: 2340–2352.

Robertson, G. L., Shelton, R. L., and Athar, S. (1976). The osmoregulation of vasopressin. *Kidney Int* **10**: 25–37.

Robertson, G. L., Aycinena, P., and Zerbe, R. L. (1982). Neurogenic disorders of osmoregulation. *Am J Med* **72**: 339–353.

Saito, T., Ishikawa, S., Abe, K. et al. (1997). Acute aquaresis by the nonpeptide arginine vasopressin (AVP) antagonist OPC-31260 improves hyponatremia in patients with syndrome of inappropriate secretion of antidiuretic hormone (SIADH). *J Clin Endocrinol Metab* **82**: 1054–1057.

Saito, T., Ishikawa, S. E., Ando, F., Higashiyama, M., Nagasaka, S., and Sasaki, S. (2000). Vasopressin-dependent upregulation of aquaporin-2 gene expression in glucocorticoid-deficient rats. *Am J Physiol* **279**: F502–F508.

Sands, J. M., Flores, F. X., Kato, A. et al. (1998). Vasopressin-elicited water and urea permeabilities are altered in IMCD in hypercalcemic rats. *Am J Physiol* **274**: F978–F985.

Schrier, R. W. (1988a). Pathogenesis of sodium and water retention in high-output and low-output cardiac failure, nephrotic syndrome, cirrhosis, and pregnancy (1). *N Engl J Med* **319**: 1065–1072.

Schrier, R. W. (1988b). Pathogenesis of sodium and water retention in high-output and low-output cardiac failure, nephrotic syndrome, cirrhosis, and pregnancy (2). *N Engl J Med* **319**: 1127–1134.

Schrier, R. W. (2006). Body water homeostasis: clinical disorders of urinary dilution and concentration. *J Am Soc Nephrol* **17**: 1820–1832.

Schrier, R. W., and Berl, T. (1973). Mechanism of effect of alpha adrenergic stimulation with norepinephrine on renal water excretion. *J Clin Invest* **52**: 502–511.

Schrier, R. W., Gross, P., Gheorghiade, M. et al. (2006). Tolvaptan, a selective oral vasopressin V2-receptor antagonist, for hyponatremia. *N Engl J Med* **355**: 2099–2112.

Singer, I., and Rotenberg, D. (1973). Demeclocycline-induced nephrogenic diabetes insipidus. In-vivo and in-vitro studies. *Ann Intern Med* **79**: 679–683.

Sladek, C. D., and Kapoor, J. R. (2001). Neurotransmitter/neuropeptide interactions in the regulation of neurohypophyseal hormone release. *Exp Neurol* **171**: 200–209.

Soupart, A., Gross, P., Legros, J. J. et al. (2006). Successful long-term treatment of hyponatremia in syndrome of inappropriate antidiuretic hormone secretion with satavaptan (SR121463B), an orally active nonpeptide vasopressin V2-receptor antagonist. *Clin J Am Soc Nephrol* **1**: 1154–1160.

Szatalowicz, V. L., Arnold, P. E., Chaimovitz, C., Bichet, D., Berl, T., and Schrier, R. W. (1981). Radioimmunoassay of plasma arginine vasopressin in hyponatremic patients with congestive heart failure. *N Engl J Med* **305**: 263–266.

Tahara, A., Tomura, Y., Wada, K. I. et al. (1997). Pharmacological profile of YM087, a novel potent nonpeptide vasopressin V1A and V2 receptor antagonist, in vitro and in vivo. *J Pharmacol Exp Ther* **282**: 301–308.

Teitelbaum, I., and Berl, T. (1984). Water metabolism in patients with electrolyte disorders. *Semin Nephrol* **4**: 354.

Teitelbaum, I., and Berl, T. (1986). Effects of calcium on vasopressin-mediated cyclic adenosine monophosphate formation in cultured rat inner medullary collecting tubule cells. Evidence for the role of intracellular calcium. *J Clin Invest* **77**: 1574–1583.

Thurman, J., Halterman, R., and Berl, T. (2003). Therapy of dysnatremic disorder. In: *Therapy in nephrology and hypertension,* 2nd edn, (Brady, H., Wilcox, C.S., eds) WB Saunders, Philadelphia, pp. 35.

Udelson, J. E., Smith, W. B., Hendrix, G. H. et al. (2001). Acute hemodynamic effects of conivaptan, a dual V(1A) and V(2) vasopressin receptor antagonist, in patients with advanced heart failure. *Circulation* **104**: 2417–2423.

Udelson, J. E., McGrew, F. A., Flores, E. et al. (2007). Multicenter, randomized, double-blind, placebo-controlled study on the effect of oral tolvaptan on left ventricular dilation and function in patients with heart failure and systolic dysfunction. *J Am Coll Cardiol* **49**: 2151–2159.

Verbalis, J. G. (1991). Inhibitory controls of drinking. In: (Ramsay, D.J., Booth, D.A., eds) *Thirst: physiological and psychological aspects,* Springer-Verlag, London.

Verbalis, J. G. (1993). Osmotic inhibition of neurohypophysial secretion. *Ann NY Acad Sci* **689**: 146–160.

Verbalis, J. G. (2007b). Vaptans for the treatment of hyponatremia: how, who, when and why. *Nephrol Self Assess Prog* **6**: 199–209.

Verney, E. B. (1947). The antidiuretic hormone and the factors which determine its release. *Proc R Soc Lond B Biol Sci* **135**: 25.

Walker, R. G. (1993). Lithium nephrotoxicity. *Kidney Int Suppl* **42**: S93–S98.

Walter, K. A. (2007). Conivaptan: new treatment for hyponatremia. *Am J Health Syst Pharm* **64**: 1385–1395.

Wang, W., Li, C., Summer, S. N. et al. (2006). Molecular analysis of impaired urinary diluting capacity in glucocorticoid deficiency. *Am J Physiol* **290**: F1135–F1142.

Wong, F., Blei, A. T., Blendis, L. M., and Thuluvath, P. J. (2003). A vasopressin receptor antagonist (VPA-985) improves serum sodium concentration in patients with hyponatremia: a multicenter, randomized, placebo-controlled trial. *Hepatology* **37**: 182–191.

Xu, D. L., Martin, P. Y., Ohara, M. et al. (1997). Upregulation of aquaporin-2 water channel expression in chronic heart failure rat. *J Clin Invest* **99**: 1500–1505.

Yamamura, Y., Ogawa, H., Yamashita, H. et al. (1992). Characterization of a novel aquaretic agent, OPC-31260, as an orally effective, nonpeptide vasopressin V2 receptor antagonist. *Br J Pharmacol* **105**: 787–791.

Zeltser, D., Rosansky, S., van Rensburg, H., Verbalis, J. G., and Smith, N. (2007). Assessment of the efficacy and safety of intravenous conivaptan in euvolemic and hypervolemic hyponatremia. *Am J Nephrol* **27**: 447–457.

Zerbe, R. L., Miller, J. Z., and Robertson, G. L. (1991). The reproducibility and heritability of individual differences in osmoregulatory function in normal human subjects. *J Lab Clin Med* **117**: 51–59.

Zimmerman, E. A., Ma, L. Y., and Nilaver, G. (1987). Anatomical basis of thirst and vasopressin secretion. *Kidney Int Suppl* **21**: S14–S19.

Further reading

Verbalis, J. G. (2007a). How does the brain sense osmolality? *J Am Soc Nephrol* **18**: 3056–3059.

Zaoral, M., Kolc, J., and Sorm, F. (1967). Amino acids and peptides. LXXII. Synthesis of 1-deamino-8-D-y-aminobutyrine-vasopressin, 1-deamino-8-D-lysine-vasopressin, 1-deamino-8-D-arginine-vasopressin. *Coll Czech Chem Commun* **32**: 1250–1257.

CHAPTER **19**

Diabetes Insipidus and SIADH

MICHAEL L. MORITZ[1] AND JUAN CARLOS AYUS[2]

[1]*Division of Nephrology, Department of Pediatrics, Children's Hospital of Pittsburgh of UPMC, The University of Pittsburgh School of Medicine, Pittsburgh, PA, USA*
[2]*Director of Clinical Research, Renal Consultants of Houston, Houston, TX, USA*

Contents

I. Introduction	261
II. Diabetes insipidus and SIADH	261
III. Hyponatremic encephalopathy	277
References	280

I. INTRODUCTION

Disorders in sodium and water homeostasis primarily result from perturbation in the release or response to arginine vasopressin (AVP). Impaired AVP secretion or response results in impaired renal concentration and is termed diabetes insipidus (DI). Hyponatremia that results from AVP production in the absence of an osmotic or hemodynamic stimulus is termed syndrome of inappropriate antidiuretic hormone secretion (SIADH). The goals of this chapter are to review both diabetes insipidus and SIADH, their causes and their treatment.

II. DIABETES INSIPIDUS AND SIADH

A. Arginine Vasopressin (AVP) and Water Homeostasis

The plasma osmolality is maintained within a narrow range, 275–290 mOsm/l. The hypothalamus senses alterations in osmolality which leads to changes in thirst and AVP release in order to return the osmolality to normal. Osmoreceptors are located in the anteroventral third ventricle of the hypothalamus and AVP is synthesized in the vasopressinergic neurons of supraoptic neuclei and pareventricular nuclei of the hypothalamus (Ishikawa and Schrier, 2003). AVP is stored and subsequently released from the posterior pituitary (Figure 19.1). The gene that encodes AVP is located on chromosome 20. The biosynthesis of AVP proceeds by way of a polypeptide precursor. This pro-hormone contains the AVP moiety, a protein known as neurophysin II (NPII) and a glycoprotein. This pro-hormone ultimately folds and dimerizes and is cleaved into AVP, where it is stored in the posterior pituitary until it is released into the cavernous sinus and superior vena cava (Robertson, 2001).

AVP is the primary determinant of free water excretion. Its major renal effect is to increase water permeability in the medullary and collecting tubules via the insertion of the aquaporin-2 (AQP2) water channel on the apical surface of the cortical collecting duct (Figure 19.2) (Oksche and Rosenthal, 1998). The action of AVP is mediated via the vasopressin V2 receptor that is primarily located on the basolateral surface of the principal cells. The V2 receptor is coupled to a guanine-nucleotide-binding protein, Gs. AVP binding of V2 receptors results in activation of adenylate cyclase with increases in cAMP and activation of c-AMP dependent protein kinase A (PKA). Activated PKA phosphorylates the serine residue at the C-terminus of the AQP2 protein which then leads to insertion of AQP2 on apical surface of the principal cells of the cortical collecting duct which markedly increases water permeability. Water movement across the basolateral membrane is facilitated by AQP3 and AQP4. Any defect along this pathway will prevent AQP2 expression and will result in a renal concentrating defect. Any process that will increase the expression of AQP2 will result in an impaired ability to excrete free water.

B. Diabetes Insipidus

Diabetes insipidus (DI) is a polyuric condition due to either inadequate vasopressin production, termed central diabetes insipidus, or the renal resistance to vasopressin, termed nephrogenic diabetes insipidus. An unusual form of DI, gestational diabetes insipidus, occurs during pregnancy from the degradation of plasma AVP by excessive vasopressinase produced in the placenta (Durr and Lindheimer, 1996). This condition responds to the administration of desmopressin (dDAVP) as it is not degraded by vassopressinase and remits following delivery. Polyuria, unrelated to a disorder in AVP production or response, can be due to excess fluid intake. This condition is referred to as primary polydypsia and can be seen in compulsive water drinking or psychogenic polydypsia. Primary polydypsia can be difficult to distinguish from DI at times. Polyuria is typically defined as the passage

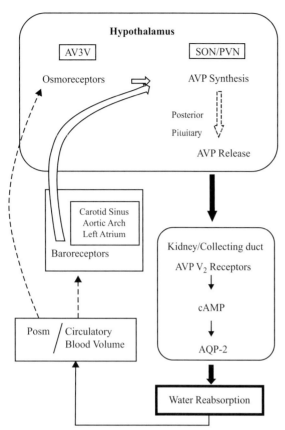

FIGURE 19.1 AVP and water homeostasis: anteroventral third ventricle (AV3V), supraoptic and paraventricular nuclei (SON/PVN). From Ishikawa, S.E. and Schrier, R.W. (2003). Pathophysiological roles of arginine vasopressin and aquaporin-2 in impaired water excretion. *Clin Endocrinol (Oxf.)* 58, 1–17.

of large volumes of hypotonic urine with an osmolality less than 300 mOsm/kg and in excess of 3 l/1.73 m^2/24 h.

1. Nephrogenic Diabetes Insipidus (Table 19.1)

Nephrogenic diabites insipidus (NDI) refers to polyuric states despite the adequate production of arginine vasopressin (AVP). NDI can be congenital or acquired, associated with severe polyuria or mild nocturia. Many pathophysiological mechanisms can result in a state of NDI:

1. resistance of AVP in the collecting tubule of the kidney
2. impairment in the medullary countercurrent mechanism
3. increase solute excretion by the kidney (Jamison and Oliver, 1982).

Acquired forms of NDI are the most common encountered in clinical practice, but the congenital forms are the most severe. The presenting features of NDI are usually polyuria or nocturia. Hypernatremia will only be a presenting feature if there is an impairment in access to water, such as during an acute illness or hospitalization. Numerous medications and disease states can result in an acquired NDI, so a high index of suspicion is required to make this diagnosis when a patient presents with or develops polyuria or hypernatremia.

2. Congenital Nephrogenic Diabetes Insipidus

Congenital nephrogenic diabetes insipidus (CNDI) is a rare hereditary disorder with an estimated prevalence of 1 in 250 000 males. There are two forms of inheritance, an X-linked recessive form, which affects the majority of individuals, and autosomal recessive and dominant forms, which affect the minority. Individuals can be equally affected with either form. In CNDI, there are varying degrees of resistance

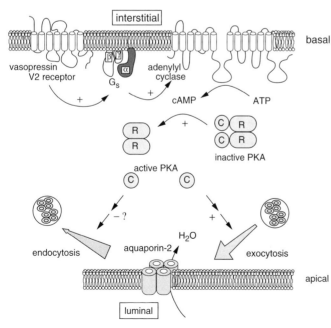

FIGURE 19.2 Action of AVP on the collecting duct: C: catalytic; R: regulatory. From Oksche, A. and Rosenthal, W. (1998). The molecular basis of nephrogenic diabetes insipidus. *J Mol Med* 76, 326–37.

CHAPTER 19 · Diabetes Insipidus and SIADH

TABLE 19.1 Causes of nephrogenic diabetes insipidus

Genetic
- X-linked recessive (V2- receptor gene)
- Autosomal recessive (aquaporin-2 gene)
- Autosomal dominant (aquaporin-2 gene)
- Renal disease
- Polycystic kidney disease
- Medullary cystic kidney disease
- Nephronophthisis
- Obstructive uropathy
- Renal dysplasia
- Reflux nephropathy
- Bartter syndrome
- Bardet–Biedl syndrome
- Chronic tubulointerstitial nephritis
- Anelgesic abuse
- Acute and chronic pyelonephritis
- Amyloidosis
- Sjögren's syndrome
- Advanced renal failure

Sickle cell disease
Electrolyte abnormalities
- Hypokalemia
- Hypercalcemia

Medications
- Lithium
- Demeclocycline
- Amphotericin B
- Dexamethasone
- Dopamine
- Ifosfamide
- Oflaxacin

to both exogenous and endogenous AVP in the collecting tubules of the kidney. There is an inability to appropriately reabsorb free water in the collecting duct, but urinary electrolyte handling is normal. First described in 1945, the cardinal features of this condition are polyuria, polydypsia, fever and episodes of hypernatremic dehydration (Waring, A. and Kajdi, L. 1945; Knoers et al., 1993). Pregnancies can be complicated by polyhadramnios, though this is not usually the case (Smith et al., 1990). Approximately 50% of affected males develop symptoms soon after birth. Affected infants have severe polyuria and polydypsia, a penchant for drinking cold beverages and frequently have constipation from dehydration. Complications reported in CNDI included mental retardation from repeated episodes of hypernatremia (Hoekstra et al., 1996), intracranial calcifications (Nozue et al., 1993), failure to thrive from inadequate caloric intake (van Lieburg et al., 1999) and non-obstructive hydronephrosis from the high urinary flow rate (Uribarri and Kaskas, 1993). Patients should have periodic renal sonograms as severe hydronephrosis has been reported. In general, patients who have been diagnosed early and received treatment with good supportive care achieve a good quality of life without serious long-term complications.

a. Genetics. CNDI can result from either loss-of-function mutations in the arginine vasopressin receptor 2 (AVP2) or the aquaporin 2 (AQP2) gene (Bichet, 2006; Sands and Bichet, 2006). The AVP2 gene is located in the region of chromosome Xq28. Defects in AVP2 gene are X-linked recessive and affect approximately 90% of patients with CNDI. There have been almost 200 putative gene mutations in AVPR2 resulting in X-linked CNDI. Males are severely affected and do not increase their urine osmolality in response to exogenous vasopressin. Males with X-linked CNDI typically have a maximal urine osmolality of less then 300 mOsm/kg/H_2O (van Lieburg et al., 1999). A severely affected male can produce over 10 liters of urine a day. Females have varying degrees of renal concentrating defects. Many females are asymptomatic. The AQP2 gene is located in the region chromosome 12q3. Mutations in the AQP2 gene can be either autosomal recessive or dominant and affect approximately 10% of families with CNDI, with 90% of these being autosomal recessive. There are approximately 30 putative AQP2 gene mutations. A diagnosis of CNDI can usually be made on history, as most children present in the first year of life, but molecular genetic testing and prenatal testing is clinically available (Sasaki, 2004).

b. Treatment. The primary management of CNDI is adequate administration of free water in order to replace urinary losses. Patients with CNDI should be on a low sodium diet in order to decrease the renal solute load. Infants should be on a low solute infant formula and offered fluid approximately every 2 hours. Infants typically prefer their formula at room temperature or slightly chilled while older children and adults crave ice water. A variety of medications have proven successful in the management of CNDI including hydrochlorothiazide, amiloride, the prostaglandin-synthetase inhibitor indomethicin and COX-2 inhibitors. A combination of hydrochlorothiazide and amiloride is most often recommended by experts in the field (van Lieburg et al., 1999).

Studies have demonstrated that thiazides can decrease urine volume by as much as 50% and almost double the urine osmolality in patients with DI (Earley and Orloff, 1962). It is not fully understood how thiazide diuretics induce this paradoxical effect. The most widely held view is that thiazides induce extracellular volume contraction via increased renal sodium excretion, which results in a slight decrease in GFR and increase proximal tubular sodium and water reabsorption. More recent data have emerged that thiazide diuretics directly increase the water permeability in the inner medullary collecting ducts and may upregulate the expression AQP2, Na-Cl co-transporter and the epithelial sodium channel (Cesar and Magaldi, 1999; Magaldi, 2000; Loffing, 2004; Christensen et al., 2004). The addition of amiloride to hydrochlorothiaze has an additive affect in reducing urine output and also helps prevent the thiazide-induced hypokalemia (Alon and Chan, 1985).

Non-steroidal anti-inflammaotry drugs (NSAIDs) and COX-2 inhibitors have been used successfully in patients with CNDI, but there clinical use has been limited by the undesirable side effects associated with these medications, such as GI bleeding, renal toxicity, hematopoietic effects and cardiac ischemia (Pattaragarn and Alon, 2003). The exact mechanism of prostaglandin inhibitors is uncertain. Prostaglandins are known to diminish the effect of AVP by inhibiting adenylate cyclase. For this reason, NSAIDs are believed to potentiate the effects of AVP. The effects of NSAIDs may also be due to reduced medullary blood flow, causing a further reduction in glomerular filtration rate (GFR) with increased proximal tubular reabsorption of sodium. Not all NSAIDs are effective in the treatment of CNDI. Indomethicin has been the most effective and ibuprofen has not been effective.

The exogenous administration of vasopressin is generally not effective in the treatment of CNDI. There may be a subpopulation of patients that partially respond and this could be of clinical benefit (van Lieburg et al., 1999). Surprisingly, the exogenous administration of dDAVP has been found useful in the treatment of nocturnal enuresis via extrarenal affects of vasopressin that convert enuresis to nocturia (Robben et al., 2007).

3. ACQUIRED NEPHROGENIC DIABETES INSIPIDUS

Acquired nephrogenic diabetes insipidus (ANDI) is much more common than congenital nephrogenic diabetes insipidus and is also less severe. In most forms of acquired nephrogenic diabetes insipidus, there is only a partial renal concentrating defect. Patients can be asymptomatic or have polyuria, polydypsia and nocturia. Hypernatremia is less common, but severe cases have been reported with intercurrent illnesses. ANDI can be either chronic in nature when occurring in the context of renal disease or reversible when due to medications or electrolyte abnormalities. ANDI should be suspected in:

1. patients with polyuria, urine output >3 l/1.73 m^2/day
2. a urine osmolality <300 mOsm/kg/H$_2$O or a urine specific gravity <1.015 on multiple occasions
3. mild hyperosmolality, serum osmolality >295 mOsm/kg/H$_2$O or a serum sodium >143 mEq/l, under normal conditions of fluid intake
4. a less then maximally concentrated urine <800 mOsm/kg/H$_2$O in the face of hypernatremia.

a. Renal Disease. Renal concentrating defects have been observed in a variety of renal diseases. There are many possible explanations for renal concentrating defects including decreased tubular responsiveness to AVP, damage to the renal medullary interstitium or impairment of the countercurrent mechanism and increased solute excretion in the remaining functioning nephrons (Osorio and Teitelbaum, 1997). The degree of renal concentrating impairment parallels the decline in GFR, with renal isosthenuria or hyposthenuria being an almost universal finding in advanced chronic renal failure (Tannen et al., 1969; Osorio and Teitelbaum, 1997). One of the primary mechanisms of renal concentrating impairment is vasopressin resistance, explaining the hyposthenuria in renal disease (Tannen et al., 1969). Chronic renal failure induced by 5/6 nephrectomy in rats results in polyuria due to decreased expression of AQP2 and AQP3 (Kwon et al., 1998). Polyuria is a common complication following ureteral obstruction. Rat models of bilateral ureteral obstruction have revealed markedly reduced expression of AQP2. AQP2 expression remains low after the ureteral obstruction has resolved and renal concentrating defects can still be documented as much as one month post-obstruction (Amlal et al., 2000). Polyuria is also a common feature in the recovery phase of ischemic acute renal failure. The mechanism behind this appears to be both defects in countercurrent multiplication and impaired water permeability in the collecting duct due to decreased expression of AQP2 and AQP3 (Ecelbarger et al., 2002).

b. Sickle Cell Disease. Renal concentrating defects are a universal finding in patients with sickle cell anemia and sickle cell trait. The extent of the renal concentrating defect is largely dependent on the percentage of hemoglobin S. The renal medulla is an environment conducive to hemoglobin S polymerization and red cell sickling as there is high rate of oxygen consumption with hypoxia, an acid environment and hypertonicity (Ataga and Orringer, 2000). Red cell sickling results in vaso-occlusion within the vasa recta with consequent inner medullary and papillary damage. Microradioangiographic studies have revealed almost complete loss of vasa recta in kidneys from patients with sickle cell disease (Statius van Eps et al., 1970). The renal concentrating defects primarily results from the loss of deep juxtaglomerular nephrons that are necessary for maximal renal concentration. The renal concentrating defect can be reversible early on with red cell transfusions, but in adulthood is irreversible. Exogenous vasopressin and indomethacin have little or no effect in improving renal concentration (De Jong et al., 1982). Patients with sickle cell disease usually have little difficulty maintaining water homeostasis unless there has been significant volume depletion or water deprivation.

c. Lithium. Lithium is the most common cause of drug-induced NDI causing a renal concentrating defect in approximately 55% of patients on chronic lithium therapy, with overt polyuria, >3 liters per day, occuring in 20% (Boton et al., 1987). Severe life-threatening hypernatremia has been reported to occur, especially in elderly patients with restricted access to water (Robben et al., 2007). Lithium is excreted mainly by the kidney and is reabsorbed in competition with sodium. NDI poses a risk to patients on lithium therapy as volume depletion can promote lithium retention via activation of the rennin–angiotensin–aldosterone system, leading to increased lithium absorption and acute lithium toxicity precipitating a viscous cycle of worsening NDI. Decreased dietary sodium alone can lead to lithium toxicity. The degree of renal impairment is dependent on both the duration of lithium therapy, average serum lithium ion level and the cumulative dose of lithium carbonate administered. Even in the well-controlled patient, urinary lithium levels are sufficiently high to impair renal concentration (Goldberg et al., 1988). The renal concentrating defect caused by lithium is not always reversible and permanent renal concentrating defects have been reported after lithium discontinuation. Amiloride has been used successfully in treating lithium-induced NDI. Amiloride appears to decrease the uptake of lithium in the principal cells of the collecting duct, thereby preventing the inhibitory of affect of lithium on water transport (Battle et al., 1985).

How lithium actually causes diabetes insipidus is complex and appears to involve multiple mechanisms. Initial animal studies revealed that lithium impaired water handling via decreased adenylate cyclase activity resulting in decreased generation and accumulation of cAMP in the response to AVP (Christensen et al., 1985). Lithium appears directly to inhibit the activation of vasopressin-sensitive adenylate cyclase in renal epithelia by competing with magnesium for activation of the GTP binding protein, Gs (Goldberg et al., 1988). Experiments in rat kidney medulla reveal that lithium caused marked downregulation of AQP2 expression that is only partially reversed by cessation of lithium therapy (Marples et al., 1995). In human studies, lithium exposure results in decreased urinary excretion of both AQP2 and cAMP (Neville et al., 2006). Recent experiments in mouse cortical collecting duct cell lines have revealed that short-term lithium exposure can decrease the expression of AQP2 via the AQP2 mRNA and independent of cAMP and adenyl cyclase (Li et al., 2006). Chronic lithium exposure can also affect sodium and water handling by changing the cellular profile in the collecting duct. Chronic lithium exposure decreases the percentage of principal cells in the rat cortical collecting duct, which are responsible for hormonally regulated sodium and water handling, and increase the percentage of intercalated cells that are involved in acid–base balance (Christensen et al., 2004). The cellular architecture can improve following the discontinuation of lithium. An additional mechanism for the polyuria seen with lithium administration is that of increased urinary sodium excretion. This appears to be due to a dysregulation of the epithelial sodium channel through a decreased responsiveness to vasopressin and aldosterone in the cortical collecting duct (Nielsen et al., 2006).

d. Hypokalemia. Total body potassium depletion can result in a significant renal concentrating defect in humans. Human subjects with a potassium deficit exceeding 300 mEq are only able to concentrate their urine to approximately 300 mOsm/l and some individuals develop hyposthenuria (Rubini, 1961). The renal concentrating defect is a consequence of chronic potassium depletion and not acute hypokalemia (Berl et al., 1977). It has been demonstrated that rats fed potassium-free diets manifest renal concentrating defects prior to the development of hypokalemia (Amlal et al., 2000). The primary cause of the concentrating defect appears to be decreased expression of AQP2, but the mechanism behind this is unclear (Marples et al., 1996). Chronic hypokalemia results in multiple morphological changes in the kidney, so there may be multiple factors that produce polyuria (Reungjui et al., 2008).

e. Hypercalcemia and Hypercalciuria. Impaired renal concentrating ability is a universal finding in patients with hypercalcemia and hypercalciura, regardless of the etiology (Gill and Bartter, 1961). A renal concentrating defect typically occurs when the serum calcium exceeds 11 mg/dl. Patients with serum calciums greater than 11 mg/dl are reported to have maximum urine osmolalities between 350 and 600 mOsm/l. Hypercalciura in the absence of hypercalcemia also results in impaired renal concentrating ability. Interestingly, patients with familial hypocalciuric hypercalcemia do not have a renal concentration defect, suggesting that the hypercalciuria is the main contributing factor. The renal concentrating defect is reversible with correction of hypercalcemia and hypercalciuria.

Animal studies have revealed that hypercalcemia leads to a renal concentrating defect with polyuria and polydypsia via downregulation of AQP2 and AQP3 (Earm et al., 1998; Wang et al., 2002). The mechanism behind this appears to be via the epithelial calcium-sensing receptor (CaSR). The CaSR is a G-protein that is expressed in renal epithelial cells that is activated by calcium. It has been demonstrated in a mouse cortical collecting duct cell line that luminal calcium activates the CaSR, resulting in reduced efficiency of coupling between the V2 receptor adenylate cyclase, leading to decreased cAMP levels and AQP2 expression (Bustamante et al., 2008). Basolateral calcium does not lead to a similar effect, explaining why hypercalciuria in the absence of hypercalcemia results in a concentrating defect in humans and why hypercalcemia in the absence of hypercalciuria does not. An additional mechanism, which may play a role in the polyuria associated with hypercalcemia, is the downregulation of sodium transporters presumably in response to activation of the CaSR (Wang et al., 2002).

4. CENTRAL DIABETES INSIPIDUS

Central diabetes insipidus (CDI) is defined as a urinary concentrating defect resulting from insufficient secretion of AVP in response to osmolar stimuli. CDI can be of varying severity depending on the etiology of disease. There are a number of defects that can affect the neurosecretory pathway for AVP secretion that can result in CDI. The cause of CDI can be broadly classified as either due to genetic mutations, where there is abnormal synthesis or packaging of AVP, or congenital or acquired damage to the AVP-producing magnocellular neurons or the supraopticohypophyseal tract. Approximately 50% of cases of CDI in children and young adults are idiopathic (Maghnie et al., 2000).

a. Genetic Mutations. Hereditary forms of CDI are rare. The most common is that of autosomal dominant neurohypophyseal diabetes insipidus, which is caused by mutations in the neurophysin-vasopressin II (AVP-NPII) gene (Robertson, 2001). Over 50 heterozygous mutations have been described (Siggaard et al., 2005). The symptoms of CDI are absent at birth, but develop months to years later and are progressive throughout life. It is believed that these mutations affect amino acid residues that play a critical role in removing the signal residue and/or the proper folding of the pro-hormone in the endoplasmic reticulum of the neuron. The accumulation of these mutant pro-hormones is believed to be cytotoxic to the AVP-producing magnocellular neurons located in the supraoptic and paraventricular nuclei, leading to progressive cell damage. Autosomal recessive neurohypophyseal diabetes insipidus has a similar clinical presentation as the autosomal dominant form, being absent at birth and presenting symptoms later in life which are progressive. The mechanism of DI is different for the autosomal recessive form than the dominant form. In the recessive form, there is a mutation in the pro-hormone that does not interfere with folding or transportation of the pro-hormone and does not result in destruction of magnocellular neurons. Rather, the mutation results in a mutant AVP hormone that is secreted but has little or no antidiuretic effect (Christensen et al., 2004). An X-linked recessive form of DI has been reported (Robertson, 2001). It affects males and results in a progressive loss of AVP production over time. Wolfram syndrome, also know as DIDMOAD syndrome, is a rare autosomal recessive neurodegenerative condition that typically presents in childhood and is characterized by diabetes insipidus, diabetes mellitus, optic atrophy and deafness (Domenech et al., 2006). Families with Wolfram syndrome have mutations in a gene encoding the WFS1 protein, located in 4p16.1 regions, a transmembrane protein in the endoplasmic reticulum expressed in numerous tissues (Fonseca et al., 2005).

b. Congenital Malformations. Congenital central nervous system malformations are the cause of CDI in approximately 5% of unselected children with CDI (Siggaard et al., 2005; De Buyst et al., 2007). There are a variety of congenital malformations which can result in CDI (Table 19.2), the most important of which involve midline defects and, in particular, septo-optic dysplasia. The pituitary gland develops at a very early stage of embryonogenesis and is closely linked to forebrain development. There is a strong association between developmental abnormalities of the pituitary and the forebrain. Septo-optic dysplasia is defined as the combination of optic nerve hypoplasia, midline neuroradiologic abnormalities and pituitary hypoplasia with hypopituitarism. There is great phenotypic variability and the diagnosis is usually made with the presence of two of the three features. A variety of gene defects have been implicated in this condition (Kelberman and Dattani, 2007). CDI is present in approximately 20% of patient with septo-optic dysplasia (Riedl et al., 2008).

TABLE 19.2 Cause of central diabetes insipidus

Genetic
- AVP neurophysin gene defects (chromosome 20). Autosomal dominant/recessive
- X-linked recessive (chromosome Xq28)
- DIDMOAD syndrome (diabetes insipidus, diabetes mellitus, optic atrophy and deafness)/Wolfram syndrome (chromosome 4p16, WFS1 gene)

Congenital
- Septo-optic dysplasia
- Midline craniofacial defects
- Holoprosencephaly syndromes
- Agenesis, hypogenesis of the pituitary

Acquired
- Trauma
- Neurosurgery, head injury
- Neoplasms

Craniopharyngioma, germinoma, meningioma, optic glioma, pituitary adenoma, CNS lymphoma, leukemia, metastases
- Granulomas

Langerhans cell histiocytosis, neurosarcoidosis, tuberculosis, Wegener's disease, xanthoma disseminatum
- Vascular

Internal carotid aneurysm, hypoxic/ischemic brain injury, Sheehan's syndrome, cerebral hemorrhage or infarction
- Infection

Chronic meningitis, viral encephalitis, congenital cytomegalovirus or toxoplasmosis
- Autoimmune

Lymphocytic infundibuloneurohypophysitis
Autoantibodies to AVP-secreting cells
- Toxins

Snake venom
- Idiopathic

c. Acquired Central Diabetes Insipidus. CDI is most commonly an acquired condition due to various causes. Thirty to 50% of cases of CDI are idiopathic. The most common identifiable causes of CDI are brain tumors, granulomatous diseases, autoimmune or vascular diseases and traumatic brain injury, subarachnoid hemorrhage or following brain surgery (Maghnie et al., 2000).

d. Idiopathic CDI and Autoimmunity. Autoimmunity has been associated with many cases of idiopathic CDI from autoantibodies to AVP-secreting cells (AVPcAb) (Pivonello et al., 2003). AVPcAb have been detected in approximately one-third of patients with idiopathic CDI. The likelihood of autoimmunity is 50% in patients less than 30 years and greater than 90% in patients with a history of autoimmune disease or thickening of the pituitary stalk on magnetic resonance imaging (MRI) (Pivonello et al., 2003). The presence of AVPcAb is not sufficient to make a diagnosis of autoimmunity as approximately two-thirds of patient with CDI and Langerhan's cell histiocytosis (LCH) are found to have AVPcAb (Scherbaum et al., 1986). It is felt that the presence of AVPcAb in patients with LCH or germinomas is a transient epiphenomena and not the primary cause (Maghnie et al., 2006). Abnormal blood supply to posterior pituitary due to vascular impairment of the inferior hypophyseal artery has also been implicated as a cause of idiopathic CDI (Maghnie et al., 2004).

e. Granulomatous Diseases. Granulomatous diseases such as Langerhan's cell histiocytosis (LCH), Wegener's granulomatosis, neuroscardosis and tuberculosis have all been associated with CDI. With the exception of LCH, it is unusual for granulomatous conditions to present with CDI without other extracranial manifestations. LCH is a rare condition that is responsible for CDI in approximately 15% of children. Ten percent of patients with LCH have CDI and, in half of these, CDI is the presenting feature (Prosch et al., 2004). LCH is due to a clonal proliferation of mononuclear phagocytic and dendritic cells, the etiology of which is uncertain. LCH typically presents in childhood between birth and 15 years of age with the peak incidence being 1–4 years of age. The most common associated features in patients presenting with CDI are infiltrative bony lesions, most often involving the cranial cavity, and skin involvement. A skeletal survey along with a detailed examination of the skin and mucous membranes is indicated in any child with idiopathic CDI who is suspected to have LCH. A lumbar puncture is indicated in patients without extracranial lesions to rule out germ cell tumor markers. If the pituitary stalk measures greater than 6.5 mm on MRI without extracranial evidence of LCH, then a biopsy of the hypothalamic–pituitary region may necessary to distinguish LCH from a tumor.

f. Neoplasms. Tumors arising in the hypothalamic–pituitary region and metastatic neoplasms affecting this region can result in CDI. Germinomas and craniopharyngiomas are the primary neoplasms that involve the hypothalamus and account for approximately 20% of the causes of CDI in children. Neoplasms affecting the pituitary rarely cause CDI as AVP is produced in the hypothalamus (Samarasinghe and Vokes, 2006). Germinomas can arise from any part of the neurohypophysis complex. Neurohypophyseal masses on MRI are strongly associated with germinomas. Patients with idiopathic CDI must be carefully evaluated for an occult germinoma as symptoms of CDI may precede radiologic evidence. Serial MRI and plasma and CSF measurement of human chorionic gonadotropin (hCG), a marker of germ-cell tumors, should be done in patients with idiopathic CDI. Elevated hCG level may precede radiologic evidence of a pituitary germinoma (Argyropoulou and Kiortsis, 2005). Cranipharyngiomas can occur anywhere from the hypothalamus to the pituitary gland with the primary associated endocrine abnormality being growth hormone deficiency, with CDI being less common. Panhypopituitarism with CDI can develop following surgical resection. Disorders in thirst are associated with craniopharyngiomas and/or their resection (Samarasinghe and Vokes, 2006).

g. Traumatic Brain Injury. Traumatic brain injury (TBI) is an important cause of CDI (Agha et al., 2005). In most cases, CDI is transient, but complete or partial CDI can be permanent. Recent studies have revealed that early CDI occurs in 26% of cases of TBI and permanent CDI in 6%. The severity of brain injury, as measured by the Glascow coma scale, appears to be predictive of early post-traumatic diabetes insipidus. CDI appears to result from inflammatory edema around the hypothalamus and pituitary, with resolution of CDI associated with recovery of edema. TBI can cause direct damage to the paraventricular and supraoptic hypothalamic neurons and damage to the posterior pituitary stalk. This can result in either transient or permanent CDI. Any patient who develops acute post-traumatic CDI should be evaluated for partial CDI in the convalescent phase as partial CDI could be easily overlooked.

h. Post-Surgical. Acute and transient CDI is a frequent complication of surgery involving the suprasellar hypothalamic area (Ecelbarger et al., 2002). Depending on the extent of the surgery and size of the tumor, permanent CDI can develop. Permanent DI following pituitary stalk resection typically follows a 'triphasic' pattern. The first phase involves DI which lasts several hours to days due to axonal shock and impaired function of damaged neurons. The second phase is the antidiuretic phase which can last from 2 to 14 days and results from the uncontrolled release of AVP from damaged neurons in the posterior pituitary which has been disconnected from the hypothalamus. The third phase is permanent CDI which occurs after depletion of AVP from the degenerating posterior pituitary.

5. MRI IN THE DIAGNOSIS CENTRAL DIABETES INSIPIDUS

MRI plays a key role in the diagnosis of CDI (Fujisawa, 2004). The posterior pituitary is seen as a hyperintense bright signal on sagittal T1-weighted MRI in the vast majority of normal individuals that identifies it distinctly from the anterior pituitary. The bright signal is thought to represent vasopressin-neurophysin storage in the posterior lobe of the pituitary. The signal intensity ratio in posterior lobe of the pituitary in comparison to the pons on T1-weighted imaging is reflective of vasopressin content in the posterior pituitary. The absence of this hyperintense signal is suggestive of CDI. Conditions other than CDI can also produce transient vasopressin depletion with absence of the bright signal, so absence of the bright signal is not necessarily diagnostic of CDI. This is of clinic significance because it is possible that patients with nephrogenic DI or hypernatremia could lose the bright signal from vasopressin depletion and it could be confused as a sign of CDI. The presence of a thickened pituitary stalk or infundibulum is suggestive of either a germinoma or an infiltrative disease such as Langerhans' cell histiocytosis and lymphocytic infundibuloneurohypophysitis (Maghnie et al., 2000). Follow-up MRI is indicated in patients with idiopathic CDI, as a germinoma or infiltrative process such as LCH may not be seen on MRI at the time of presentation, but could be detected in future evaluation.

6. DIAGNOSIS OF DI

Diabetes insipidus needs to be ruled out in two circumstances:

1. unexplained hypernatremia
2. polyuria and polydypsia.

Whenever a patient presents with hypernatremia (plasma sodium >145 mEq/l) out of proportion to the clinic history, a prompt evaluation of renal concentrating ability is indicated. Many patients who present with unexpected hypernatremia will have an underlying renal concentrating defect. Patients with hypernatremia should have a simultaneous measurement of plasma and urine osmolality. A urine osmolality that is less then maximally concentrated (<800 mOsm/kg) in the face of hypernatremia is evidence of a renal concentrating defect. A urine osmolality of <300 mOsm/kg is consistent with DI and urine osmolality between 300 and 800 mOsm/kg is consistent with a partial renal concentrating defect. When hypernatremia is associated with a renal concentrating defect, then NDI can be distinguished from CDI by:

1. measuring a plasma AVP level
2. assessing the renal response to the administration of dDAVP.

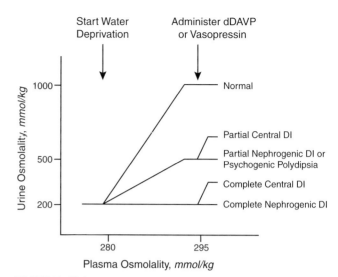

FIGURE 19.3 Response to a water deprivation test in individuals with and without diabetes insipidus (DI). From Sands, J.M. and Bichet, D.G. (2006). Nephrogenic diabetes insipidus. *Ann Intern Med* 144, 186–94.

An elevated AVP level and lack of response to exogenous vasopressin is consistent with NDI. A depressed AVP level and increase in urine osmolality of greater than 50% following dDAVP is consistent with CDI (Figure 19.3). A water deprivation test is contraindicated in the hypernatremic patient and should be considered dangerous (Moritz, 2005). Hypernatremia is a maximal stimulus for AVP release making a water deprivation test unnecessary.

Determining the etiology of polyuria and polydypsia can be a difficult one. A 24-hour urine should be sent for volume, osmolality and creatinine. A urine volume of greater than 3 l/1.73 m^2/day with a urine osmolality of <300 mOsm/kg is suspicious for DI and merits further evaluation. The first objective is to distinguish primary polydypsia from a form of DI. Patients with primary polydypsia generally will have a low normal serum osmolality, while patients with DI will have a high normal serum osmolality. In theory, measurement of an AVP level could be helpful in the evaluation but, in practice, a reliable and sensitive immunoassay can be difficult to obtain and false negative results are not uncommon. Distinguishing primary polydypsia from DI is best ascertained by performing a water deprivation test. Once a diagnosis of DI is established, differentiating CDI from NDI is done by measuring the renal response to dDAVP and a reliable plasma AVP level. Under certain circumstances, the administration of hypertonic saline is needed to determine if there is a partial DI.

Various protocols are available for performing a water deprivation test, but the principles are all similar (Samarasinghe and Vokes, 2006). The objective of the test is to produce an adequate hemodynamic or osmotic stimulus for AVP production in order to assess renal concentrating ability. Water restriction is instituted until there is either a

3–5% decrease in body weight or mild hypernatremia (plasma Na > 145 mEq/l). A urine concentration of <300 mOsm/kg is consistent with DI. The renal response to dDAVP is then assessed (see Figure 19.3). An increase in urine osmolality of >50% is consistent with CDI and a lack of response consistent with NDI. A water deprivation test and the renal response to dDAVP cannot always distinguish partial DI from primary polydypsia as in both cases the urine will concentrate to >300 mOsm/kg prior to the development of mild hypernatremia or significant dehydration. In this situation, hypertonic saline should be judiciously administered to produce mild hyernatremia and AVP levels subsequently measured. Elevated AVP levels are consistent with NDI or primary polydypsia and decreased levels with CDI. Primary polydypsia can be indistinguishable from partial NDI, as primary polydypsia causes a washout of the medullary concentration gradient and downregulates renal aquaporin-2 water channels with a suboptimal response to exogenous AVP (Verbalis, 2003).

7. Treatment of Central Diabetes Insipidus

The treatment of choice for CDI is the administration of dDAVP, a synthetic analog of AVP that has minimal pressor affects and a longer half-life than the native hormone (Samarasinghe and Vokes, 2006). dDAVP comes orally, intranasally or parenterally. The necessary dose will vary between patients and should be tailored to minimize polyuria. dDAVP can be used once or twice daily. In some cases, a large evening dose is sufficient. CDI rarely results in hypernatremia if there is unrestricted access to water with an intact thirst mechanism, therefore, excessively large doses of dDAVP should be avoided. dDAVP has been associated with life-threatening hyponatremic encephalopathy, so patients should be counseled not to drink excessively on dDAVP therapy. If hyponatremic encephalopathy does develop in a patient receiving dDAVP, the medication should not be abruptly discontinued as this can result in a brisk free water diuresis and a severe overcorrection of hyponatremia that could produce brain injury. It is far safer to continue the dDAVP and treat the hyponatremia with fluid restriction and hypertonic saline as needed to control the neurologic symptoms.

8. Treatment of Hypernatremia Due to Diabetes Insipidus

The management of hypernatremia in patients with diabetes insipidus can be challenging. These patients can have large ongoing free water losses and the amount of free water needed to correct the free water deficit and keep up with the ongoing free water losses can be staggering. In patients with congenital nephrogenic diabetes insipidus or complete central diabetes insipidus, standard hypotonic fluids such as 0.2–0.45% sodium chloride could actually worsen hypernatremia as these fluids are hypertonic in relationship to the urine. In patients with CDI with hypernatremia who are treated with dDAVP, a powerful antidiuresis will ensue that can result in the overcorrection of hypernatremia if hypotonic fluids are instituted. For these reasons, management of hypernatremia from CNDI and CDI will be discussed separately using case studies.

The cornerstone to managing hypernatremia is to provide adequate free water to correct the serum sodium. Hypernatremia is frequently accompanied by volume depletion, therefore, fluid resuscitation with normal saline or colloid should be instituted prior to correcting the free water deficit. Following initial volume expansion, the composition of parenteral fluid therapy largely depends on the etiology of the hypernatremia. Patients with CNDI will require a very hypotonic fluid, while patients with CDI can be managed with dDAVP and a less hypotonic fluid. Oral hydration should be instituted as soon as it can be safely tolerated. Plasma electrolytes should be checked every two hours until the patient is neurologically stable (Moritz and Ayus, 2005).

A simple way of estimating the minimum amount of free water necessary to correct the serum sodium is by the following equation which assumes the total body water to be 50% of body weight:

$$\text{Free water deficit (ml)} = 4\,\text{ml/kg} \times \text{lean body weight (kg)} \times [\text{Desired change in serum Na mEq/l}]$$

In general, 4 ml/kg of free water will decrease the serum sodium by 1 mEq/l, assuming no ongoing free water losses.

The rate of correction of hypernatremia is largely dependent on the severity of the hypernatremia and the etiology. Due to the brain's relative inability to extrude unmeasured organic substances, called idiogenic osmoles, rapid correction of hypernatremia can lead to cerebral edema (Ayus et al., 1996). In patients with hypernatremia and underlying brain injury such as head trauma, hypoxic ischemic injury or meningitis, even slow correction of hypernatremia could aggravate cerebral edema. While there are no definitive studies that document the optimal rate of correction that can be undertaken without developing cerebral edema, empirical data have shown that a rate of correction not exceeding 1 mEq/h or 15 mEq/24 h is reasonable (Banister et al., 1975; Rosenfeld et al., 1977; Pizarro et al., 1984). If there are other risk factors for cytotoxic or vasogenic cerebral edema present, a slower rate of correction is indicated. In severe hypernatremia (>170 mEq/l), serum sodium should not be corrected to below 150 mEq/l in the first 48–72 h (Rosenfeld et al., 1977). Seizures occurring during the correction of hypernatremia are not uncommon in children and may be a sign of cerebral edema (Hogan et al., 1969, 1984, 1985). They can usually be managed by slowing the rate of correction or by giving hypertonic saline to increase the serum sodium by a few milliequivalents. Seizures are usually self-limited and not a sign of long-term neurologic sequelae (Banister et al., 1975; Rosenfeld et al., 1977). Patients with acute hypernatremia, corrected by the oral route, can tolerate a more rapid rate of correction with a much lower incidence of seizures (Pizarro et al., 1983; Hogan et al., 1985).

a. Case 1 Nephrogenic Diabetes Insipidus (Moritz and Ayus, 2005). A 20-kg, 5-year-old child with nephrogenic diabetes insipidus has contracted a stomach flu and has not been able to keep down fluids for the past 12 h. He presents to the emergency department markedly dehydrated with sunken eyes, doughy skin and documented 2-kg weight loss. Serum biochemistries reveal serum sodium 172 mEq/l, potassium 4.5 mEq/l, blood urea nitrogen 20 mg/dl and creatinine 0.6 mg/dl. Urine biochemistries reveal sodium 25 mEq/l, potassium 15 mEq/l and osmolality 100 mOsm/kg/H_2O. On further history, his mother reports that he usually drinks between 4 and 5 liters of fluids a day.

This child's fluid deficit is primarily free water. To correct the serum sodium back to a normal value of 140 mEq/l would require approximately 2.5 l of free water (4 ml/kg × 20 kg × 30 mEq/l). Unfortunately, bolus therapy with free water will result in too rapid a fall in serum sodium. The child will initially have to be bolused with 40 ml/kg of 0.9 NaCl in order to restore the extracellular volume. Further fluid therapy would require maintenance fluids, which for him are about 4000 ml/day, plus the free water deficit to lower the serum sodium by 10 mEq/24 h which is 800 ml (4 ml/kg × 20 kg × 10 mEq/l) to give a total 24 h volume of 4800 ml or 200 ml/h. The sodium composition of the intravenous fluids will need to be more hypotonic than the urine in order to lower the serum sodium; 0.115% NaCl (Na 19 mEq/l) in 2.5% dextrose in water with 10 mEq of KCl per liter would be appropriate to start with. A higher dextrose concentration could result in hyperglycemia given high rate of fluid administration. Biochemistries would have to be monitored hourly initially as response to therapy can be unpredictable based on the urinary response. As soon as the child can keep down oral fluids, parenteral fluids should be tapered off.

b. Case #2 Central Diabetes Insipidus. A 45-year-old male weighing 70 kg was involved in a motor vehicle accident and suffered a closed head injury and was admitted to the intensive care unit. He was administered large amounts of 0.9% sodium chloride (NS) for fluid resucitation and was placed on continuous fluids with NS to prevent the development of hyponatremia. He developed raised intracranial pressure with evidence of cerebral edema. An extraventricular device was placed and he was started on an infusion of 3% sodium chloride to treat his cerebral edema. He then developed polyuria with urine output exceeding 500 ml/h. His serum sodium increased to 184 mEq/l. A spot urine osmolality was 120 mOsm/kg/H_2O with a combined urine sodium plus potassium of 50 mEq/l. A continuous dDAVP infusion was started and the urine osmolality increases to 800 mOsm/kg/H_2O.

This patient's hypernatremia is multifactorial, not only is there a brisk free water diuresis but the patient is receiving both NS and hypertonic saline. The excretion of free water plus retention of sodium has resulted in severe hypernatremia. The initiation of dDAVP has stopped the free water diuresis, so the hypernatremia should not worsen. Because the patient has a fixed inability to excrete free water while on dDAVP, even NS could, in theory, lead to a fall in serum sodium. NS should be administered at a restricted rate of 50 ml/h to maintain water and sodium homeostasis. The proper rate of correction of hypernatremia is difficult to determine. A serum sodium of >180 mEq/l could lead to brain injury, but a fall in the serum sodium could aggravate cerebral edema. In this situation, the rate of sodium correction should probably not exceed 10 mEq/l/24 h and 5 mEq/24 h should be initially attempted. If intracranial pressures increase with correction of hypernatremia, a 100 ml bolus of 3% sodium chloride should be administered to raise acutely the serum sodium and decrease the cerebral edema. The free water deficit to correct the serum sodium by 5 mEq is 1400 ml (4 ml/kg × 70 kg × 5 mEq/l) or 58 ml/h. In addition to the normal saline, 58 ml/h of 5% dextrose in water would be administered. Serum sodium would be checked every one to two hours. If the serum sodium is falling faster than anticipated or the intracranial pressure is rising, the rate of the free water infusion would be held or the rate decreased. The rate of correction may end up being greater than predicted if a natriuresis ensues.

C. Syndrome of Inappropriate Antidiuretic Hormone (SIADH) Secretion

1. Pathogenesis of Hyponatremia

Hyponatremia, defined as a serum sodium <135 mEq/l, is a common disorder that occurs in both inpatient and outpatient settings. The body's primary defense against developing hyponatremia is the kidney's ability to generate a dilute urine and excrete free water. Rarely is excess ingestion of free water alone the cause of hyponatremia, as an adult with normal renal function can typically excrete over 15 l of free water per day. It is also rare to develop hyponatremia from excess urinary sodium losses in the absence of free water ingestion. In order for hyponatremia to develop, it typically requires a relative excess of free water in conjunction with an underlying condition that impairs the kidney's ability to excrete free water (Table 19.3). Renal water handling is primarily under the control of AVP, which is produced in the hypothalamus and released from the posterior pituitary. AVP release impairs water diuresis by increasing the permeability to water in the collecting tubule. There are osmotic, hemodynamic and non-hemodynamic stimuli for AVP

TABLE 19.3 Disorders in impaired renal water excretion

Effective circulating volume depletion
1. Gastrointestinal losses: vomiting, diarrhea
2. Skin losses: cystic fibrosis
3. Renal losses: salt wasting nephropathy, diuretics, cerebral salt wasting, hypoaldosteronism
4. Edemetous states: heart failure, cirrhosis, nephrosis, hypoalbuminemia
5. Decreased peripheral vascular resistance: sepsis, hypothyroid

Thiazide diuretics

Renal failure
1. Acute
2. Chronic

Non-hypovolemic states of ADH excess
1. Syndrome of inappropriate secretion of antidiuretic hormone
2. Postoperative state
3. Nausea, vomiting
4. Pain, stress
5. Cortisol deficiency

release. In most cases of hyponatremia, there is a stimulus for vasopressin production which results in impaired free water excretion. The body will attempt to preserve the extracellular volume at the expense of the serum sodium, therefore, a hemodynamic stimulus for AVP production will override any inhibitory hypo-osmolar effect of hyponatremia (Dunn et al., 1973). There are numerous stimuli for AVP production (Table 19.3) that occur in hospitalized patients which can make virtually any hospitalized patient at risk for hyponatremia.

2. Pathogenesis of SIADH

SIADH was first reported in 1957 by Schwartz and Bartter in two patients with bronchogenic carcinoma and hyponatremia of unexplained etiology (Schwartz et al., 1957). These two patients displayed clinical features similar to that of human subjects who had received prolonged administration of a posterior pituitary extract, then called pitressin, which produced an antidiuresis (Leaf et al., 1953). Schwartz and Bartter postulated that there was inappropriate secretion of an antidiuretic hormone, hence the name SIADH (Bartter and Schwartz, 1967). It was later realized that this antidiuretic hormone was in fact AVP.

Human and animal studies have been done to elucidate the etiology of hyponatremia in SIADH, as the fall in serum sodium is not fully explained by free water retention alone (Jaenike and Waterhouse, 1961; Verbalis, 1984; Southgate et al., 1992). The prevailing view is that the fall in serum sodium is due to both a combination of free water retention and urinary solute excretion (Adler and Verbalis, 2006). AVP excess increases water permeability in the collecting duct leading to approximately 7–10% increase in total body water (Schrier et al., 2006). The volume expansion is subclincal and does not lead to overt signs of fluid overload such as hypertension and edema (Padfield et al., 1981). After the initial phase of volume expansion, hemodynamic regulatory mechanisms come into play that preserve the plasma volume at the expense of sodium and total body water remains relatively constant regardless of the amount of salt water administered. The hemodynamic stimulus that maintains plasma volume in SIADH is, in part, due to a pressure natriuresis and the release of natriuretic peptides. A pressure natriuresis, whereby expanded plasma volume increases renal perfusion and leads to a decrease in sodium reabsorption and increase in sodium excretion, plays a small role in the homeostatic response. It has been demonstrated that a natriuresis will continue to occur in SIADH even with partial clamping of the aorta and renal arteries (Bartter and Schwartz, 1967). The primary mechanism leading to a natriuresis in SIADH is believed to be the release of natriuretic peptides in response to volume expansion (Kaneko et al., 1987; Cogan et al., 1988). The natriuteic response is so powerful in SIADH that administration of normal saline has virtually no affect on correcting the hyponatremia or leading to additional volume expansion (Musch and Decaux, 1998). It is also postulated that a component of the hyponatremia results from the intracellular extrusion of electrolytes and osmolytes in order to preserve intracellular volume (Cooke et al., 1979).

An additional compensatory mechanism in SIADH is that of vasopressin escape. In the presence of either continuous endogenous or exogenous vasopressin, water excretion will increase over time with a fall in urine osmolality (Leaf et al., 1953; Cooke et al., 1979). Patients with chronic SIADH do not usually have urine osmolalities that exceed 600 mOsm/kg. Vasopressin escape appears to be due to both vasopressin dependent and vasopressin independent mechanisms, namely:

1. decreased V2 receptor binding capacity
2. decreased ability to generate cAMP in response to AVP
3. the downregulation of renal AQP2 protein and mRNA expression with decreased expression of AQ2 (Ecelbarger et al., 1997, 1998; Tian et al., 2000).

Hyponatremia will not develop in SIADH in the absence of free water administration. After approximately one week of constant sodium and water intake, a steady state will be achieved whereby urine and electrolyte output will match that of intake and serum sodium and total body water will remain constant. A natriuresis will not be evident unless there is additional volume expansion or increased sodium administration. Urine sodium concentration can be low in SIADH if there is either fluid or sodium restriction, which can confuse the diagnosis.

Hypouricemia is a prominent biochemical feature of SIADH that can aid in the diagnosis. The high uric acid clearance is not simply a factor of volume expansion. Experiments have revealed that healthy volunteers who are volume expanded with saline or patients with hyponatremia due to DDAVP do not have the same degree of hypouricemia and uric acid clearance as patients with SIADH (Decaux et al., 1996, 2000). In addition, uric acid excretion does not correlate with sodium excretion, as would be expected with volume expansion alone. It has been postulated that stimulation of V1 or V3 receptors by AVP may inhibit proximal tubular reabsorption of uric acid (Decaux et al., 1996).

3. AVP Regulation in SIADH (Figure 19.4)

AVP production in SIADH can arise from either the neurohypophysis or from ectopic production in the case of a malignancy. There are various patterns of AVP production seen in SIADH which are independent of the underlying disease. Four patterns of AVP production have been identified, types A–D, which helps explain the heterogeneity of clinic presentation and some of the difficulties in establishing a diagnosis and treatment (Figure 19.4) (Robertson et al., 1982; Robertson, 2006). Types A–C each affect approximately 30% of patients with SIADH. In type A, AVP production is erratic, constantly elevated and not influenced by serum osmolality. The urine osmolality remains relatively fixed and at the maximal achievable level for SIADH. In type B, AVP excretion increases in response to osmotic stimuli when serum osmolality is in the normal range, but AVP excretion cannot be maximally suppressed in the presence of hypo-osmolality. AVP levels will remain slightly elevated regardless of degree of hypo-osmolality. Type B patients will have a lower urine osmolality than type A, but will not be able maximally to dilute their urines. Type C is referred to as 'reset osmostat'. AVP secretion is under osmolar control but there is a downward resetting of the osmoregulatory system. With water loading, AVP secretion can be fully suppressed and a maximally dilute urine can be produced. If hypertonic saline is administered, there is an associated rise in AVP level and in urine osmolality, but it is inappropriate for the degree of hypo-osmolality. In some cases of reset osmostat, the thirst threshold is also reset. Treatment can be difficult as thirst will increase even in the hypo-osmolar range. Each type of SIADH remains constant in the same patient over time and changes in patterns of AVP production have not been reported.

4. Nephrogenic Syndrome of Inappropriate Antidiuresis (NSIAD)

Approximately 10% of patients with SIADH have non-detectable levels of AVP. Despite the lack of measurable AVP levels, there is an impaired ability to excrete free water. This had been referred to as the type D form of SIADH. It is now believed that many of these patients may in fact have a gain-of-function mutation of the AVP receptor 2 (AVPR2). Feldman et al. recently described two unrelated 3-month-old male infants with chronic SIADH and undetectable AVP levels who were found to have activating mutations of the V2R with a missense mutation in codon 137 resulting from arginine to cysteine or leucine (Feldman et al., 2005). The location of this mutation is the same as the loss-of-function mutation seen in X-linked congenital nephrogenic DI. The authors of this report have termed this condition nephrogenic syndrome of inappropriate antidiuresis (NSIAD). Since their initial report, this condition has since been observed in other children and adults (Bes et al., 2007; Decaux et al., 2007). Decaux et al. described a 74-year-old male with chronic SIADH who did not respond to vasopressin 2 antagonists and was on further study found to have a V2R mutation of R137C (Decaux et al., 2007). After investigating other family members, it was determined that this condition has phenotypic heterogeneity. Two additional males were identified, one presented with chronic hyponatremia at 7 months of age and the other was an adult who was asymptomatic throughout life, but was unable to dilute urine on a water loading test. Four heterozygous females were identified, three of whom displayed some degree of impaired diuresis with water loading. The prevalance of NSIAD is not known, but it should be suspected in idiopathic cases of chronic hyponatremia or in patients with unexplained recurrent hyponatremia.

5. Epidemiology of SIADH

SIADH is one of the most common causes of hyponatremia in the hospital setting and frequently leads to severe hyponatremia (plasma Na < 120 mEq/l) (Anderson et al., 1985). Following its initial description in 1957, it has become apparent that SIADH is a common occurrence in an ever

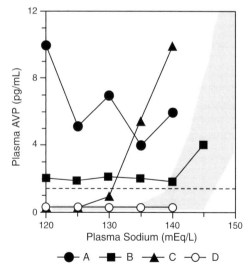

FIGURE 19.4 Osmoregulation of plasma arginine vasopressin (AVP) in patients with the syndrome of inappropriate antidiuresis is depicted for types A, B, C and D. 1 mEq/l = 1 mmol/l. From Robertson, G.L. (2006). Regulation of arginine vasopressin in the syndrome of inappropriate antidiuresis. *Am J Med* 119, S36–42.

CHAPTER 19 · *Diabetes Insipidus and SIADH* 273

TABLE 19.4 Causes of SIADH

Central nervous system disorders	Malignancies
Infection: meningitis, encephalitis	*Carcinomas*
Brain tumors	Bronchogenic
Vascular abnormalities	Oat cell of the lung
Psychosis	Stomach
Hydrocephalus	Duodenum
Congenital malformations	Pancreas
Post-pituitary surgery	Prostate
Head trauma	Ureter
Subarachnoid hemorrhage	Bladder
Cerebrovascular accident	Thymoma
Cavernous sinus thrombosis	Mesothelioma
Guillain–Barré syndrome	Endometrium
Multiple sclerosis	Neuroblastoma
Delerium tremens	Oropharynx
Amyotrophic lateral sclerosis	
Acute intermittent porphyria	Lymphomas
	Sarcoma
Pulmonary disorders	Medications
Pneumonia	Vincristine
Tuberculosis	Intravenous cytoxan
Aspergillosis	Carbamazepine
Asthma	Oxcarbazepine
Cystic fibrosis	Narcotics
Positive pressure ventilation	Seritonin reuptake inhibitors
Pneumothorax	Tricyclic antidepressants
	Nicotine
	3,4-methylenedioxymethamphetamine (ecstasy)
	Non-steroidal anti-inflammatory drugs

growing number of disease states (Table 19.4). SIADH most often occurs due to central nervous system disorders, pulmonary disorders, malignancies and medications (Zerbe et al., 1980). SIADH has been implicated in a variety of conditions that deserve special emphasis.

A condition similar to SIADH is seen in the postoperative setting, especially following orthopedic and head and neck surgery. Hospital acquired hyponatremia is commonly seen in the postoperative setting (Chung et al., 1986). Postoperative patients develop hyponatremia due to a combination of non-osmotic stimuli for ADH release, such as narcotics, subclinical volume depletion, pain, nausea, stress, edema-forming conditions and administration of hypotonic fluids. ADH levels are universally elevated postoperatively when compared to preoperative values (Wilson et al., 1988; Grant et al., 1991). Premenopausal females are most at risk for developing hyponatremic encephalopathy postoperatively (Ayus et al., 1992).

6. Exercise-Associated Hyponatremia

Hyponatremia is recognized as an important cause of race-related mortality in marathon runners. In a recent prospective trial, 13% of marathon runners developed a serum sodium of <135 mEq/l and 0.6% <20 mEq/l (Almond et al., 2005). Ayus et al. first reported hyponatremic encephalopathy as a cause of collapse in marathon runners (Ayus et al., 2000). These patients presented with nausea and non-cardiogenic pulmonary edema and were found to have cerebral edema on head CT. Patients treated with hypertonic saline had a good outcome, while those treated with normal saline under the mistaken belief that this is volume depleted state had a poor outcome. The majority of patients with neurologic morbidity are females. Marathon-associated hyponatremia appears to be due to elevated AVP levels, in addition to impaired free water excretion from non-steroidal anti-inflamatory drugs and excess free water intake. Patients with marathon-associated hyponatremia typically have demonstrated weight gain following the race with an increase in total body. It is speculated that with heavy exercise, blood flow is diverted from the GI tract to skeletal muscle and water pools in the GI tract. Following the cessation of exercise, water is rapidly absorbed through from the GI tract and acute hyponatremia develops. The most important measure to be taken to prevent hyponatremia is to avoid excessive hydration and to drink according to thirst. Any runner who exhibits signs of respiratory insufficiency, confusion, obtundation, nausea and vomiting should be evaluated for hyponatremia, with a rapid method on site in a medical tent. We have recommended that runners with symptomatic hyponatremia be treated with 100 ml of 3% sodium chloride solution for 10 min to raise the serum sodium concentration rapidly by 2 to 3 mEq/l and to decrease brain edema (Ayus et al., 2005).

7. 'Ecstasy' (3,4-Methylenediaxymethamphetamine [MDMA])

There have been many reported fatalities of hyponatremic encephalopathy with the recreational drug 'ecstasy' (MDMA). The metabolites of MDMA increase AVP release and results in SIADH (Fallon et al., 2002). MDMA is used at 'rave parties' where there is intense physical activity from dancing. MDMA users can drink excessive amounts of water to either prevent dehydration, due to thirst from dancing or from central polydypsia from the central nervous system effect of MDMA (Kalantar-Zadeh et al., 2006). The clinical presentation is similar to marathon-associated hyponatremia, primarily affecting young women who present with nausea, confusion and non-cardiogenic pulmonary edema.

8. HIV

Approximately 40–50% of hospitalized patients with AIDS have hyponatremia, with two-thirds of these patients having SIADH (Bevilacqua, 1994). The reasons for SIADH are not clear, but could be due to *Pneumocystis carinii* pneumonia, CNS infections, malignancies or medications. AIDS patients with hyponatremia are noted to have a higher mortality than those without.

9. Medications

A variety of medications have been associated with SIADH (see Table 19.4). The chemotherapeutic drugs vincristine and cytoxan, and the antiepileptic drug carbamazepine are among the most important. SIADH associated with vincristine usually occurs 4–10 days after the administration of the drug and is frequently coincides with vincristine neurotoxicity (Stuart et al., 1975). Recurrent hyponatremia has been noted with repeat drug administration. Excessive hydration and free water administration should be avoided. Fatal hyponatremic encephalopathy has been associated with hydration for IV cytoxan. Hyponatremic encephalopathy has not only been associated with high dose cytoxan for cancer treatment but also low dose cytoxan for rheumatologic conditions (Salido et al., 2003). It is not clear if cytoxan increases AVP levels or the renal sensitivity to AVP. The common practice of vigorous hydration with hypotonic fluid for cytoxan administration should be avoided and isotonic saline should be used. As many as 30% of patients treated with the antiepileptic drug carbamazepine and oxcarbazepine are noted to have hyponatremia (Van Amelsvoort et al., 1994). In most cases, it is mild but symptomatic hyponatremia that has been reported. Multiple mechanisms have been postulated for the cause of hyponatremia. As early as the 1960s, carbamazepine was noted to be useful in treating patients with CDI. There has been some evidence that carbamezapine increases AVP levels, increases the renal sensitivity to AVP, results in prolonged vasopressin half-life and can cause a reset osmostat. Patients receiving these drugs should be monitored for the development of hyponatremia.

10. dDAVP and Oxytocin

Hyponatremic encephalopathy is also a serious complication from the exogenous administration of antidiuretic hormone agonist dDAVP and oxytocin. Hyponatremic encephalopathy has been reported with the use of dDAVP for nocturnal enuresis and in the postoperative management of patients with hemophilia A or von Willebrand disease (Bernstein and Williford, 1997; Das et al., 2005). The FDA has requested manufacturers of intranasal desmopressin to stop marketing the formulation for the treatment of primary nocturnal enuresis (PNE) after receiving 61 post-marketing reports of hyponatremic-related seizures associated with use of the drug. The use of oxytocin to induce labor has been associated with hyponatremic encephalopathy in post-partum women and their infants (Schwartz and Jones, 1978). Oxytocin has a similar chemical structure to AVP and is also synthesized in the hypothalamus, stored in the posterior pituitary and increases renal water permeability via activation of the V2 receptor (Li et al., 2008). Hyponatremia, in large part, can be prevented by using 0.9% sodium chloride as the vehicle to administer the drug rather then 5% dextrose in water (Omigbodun et al., 1991).

11. Cerebral Salt Wasting (CSW)

Hyponatremia is frequently encountered in the neurosurgical setting and in patients with CNS injury. This has usually been attributed to SIADH, a condition whose hallmark is euvolemia, with the cornerstone of management being fluid restriction. More recently, it has become apparent that an increasing number of neurosurgical patients with hyponatremia have a distinct clinical entity called cerebral salt wasting (Wijdicks et al., 1985; Sivakumar et al., 1994), a condition whose hallmark is renal sodium loss leading to extracellular volume depletion, with the cornerstone of management being volume expansion and salt supplementation (Harrigan, 2001). Because these two diseases have many clinical similarities, it can be difficult to confirm a diagnosis of CSW. It is essential to be able to distinguish between these two conditions as their management is completely different and fluid restriction would be harmful in CSW.

The pathogenesis of CSW is not completely understood, but it appears to be due to the release of natriuretic peptides, such as atrial natriuretic peptide, brain natriuretic peptide and

TABLE 19.5 Diagnostic criteria of SIADH

1. Hyponatremia with hypo-osmolality
2. Urine osmolality less than maximally dilute
   - >100 mOsm/kg/H_2O
3. Urine sodium excretion increases with salt loading or water loading
4. Normal effective circulating volume
   - No edema-forming states
   - Normal blood pressure
5. Normal renal function
6. Normal adrenal function
7. Normal thyroid function

c-type natriuretic peptide (Harrigan, 2001). These peptides appear to lead to a natriuresis via a complex mechanism of:

1. hemodynamic effects leading to an increased GFR
2. inhibition of the renin angiotensin system and
3. inhibition of the secretion and action of AVP (Levin et al., 1998).

This complex mechanism can lead to biochemical features that are indistinguishable to SIADH with a low uric acid, plasma renin, aldosterone and vasopressin levels, despite volume depletion (Rabinstein and Wijdicks, 2003). The only distinguishing feature between CSW and SIADH is extracellular volume depletion. This can be particularly difficult to assess in CSW as the biochemistries may not be helpful. Central venous pressure or pulmonary capillary wedge pressures may be useful.

From a practical standpoint, the administration of normal saline should be an adequate prophylaxis against developing clinically significant hyponatremia, <130 mEq/l, in SIADH. If clinically significant hyponatremia develops in patients with a CNS disorder receiving only normal saline, then the diagnosis of CSW should be strongly considered. If there are no signs of extracellular volume depletion, then a brief period of fluid restriction could be tried. If there are signs of volume depletion or a lack of response to fluid restriction then the patient should be managed as CSW. Patients with CSW should be volume expanded with normal saline, followed by sufficient quantities of normal saline and 3% NaCl to main fluid balance and a normal serum sodium. The administration of fludrocortisone may be beneficial as aldosterone production is relatively decreased in CSW (Ishikawa et al., 1987).

12. Diagnosis of SIADH and Evaluation of Hyponatremia

SIADH is caused by elevated AVP secretion in the absence of an osmotic or hypovolemic stimulus (Bartter and Schwartz, 1967). SIADH is essentially a diagnosis of exclusion (Table 19.5, Figure 19.5). Before SIADH can be diagnosed, diseases causing decreased effective circulating volume, renal impairment, adrenal insufficiency and hypothyroidism must be excluded. Cortisol deficiency, in particular, should be ruled out as it can be clinically indistinguishable from SIADH. Glucocorticoid hormones exert an inhibitory affect on AVP synthesis, explaining why patients with glucocorticoid deficiency have markedly elevated AVP levels that are rapidly reversed by physiologic hydrocortisone replacement (Kim et al., 2001). The hallmarks of SIADH are: mild volume expansion with low to normal plasma concentrations of creatinine, urea, uric acid and potassium; impaired free water excretion with normal sodium excretion which reflects sodium intake (Cooke et al., 1979); and hyponatremia which is relatively unresponsive to sodium administration in the absence of fluid restriction.

The first step in the evaluation of hyponatremia is to confirm that hyponatremia is in fact associated with hypo-osmolality (see Figure 19.4). Hyponatremia can be associated with either a normal or an elevated serum osmolality. The most common reasons for this are hyperglycemia, severe hyperproteinemia or hyperlipidemia. Hyperglycemia results in hyperosmolality with a translocation of fluid from the intracellular space to the extracellular space, resulting in a 1.6 mEq/l fall in the serum sodium for every 100 mg/dl elevation in the serum glucose concentration above normal. Severe hyperlipidemia, hypercholesterolemia and hyperproteinemia can cause a displacement of plasma water, which will result in a decreased sodium concentration (pseudohyponatremia) with a normal serum osmolality (Turchin et al., 2003). Serum sodiums are currently measured by either direct or indirect reading ion-selective electrode potentiometry. The direct method will not result in pseudohyponatremia, as it measures the activity of sodium in the aqueous phase of serum only. The indirect method, on the other hand, can result in pseudohyponatremia as the specimen is diluted with a reagent prior to measurement (Weisberg, 1989). The indirect method is currently performed in approximately 60% of chemistry labs in the USA, therefore, pseudohyponatremia remains an entity that clinicians need to be aware of (Bruns et al., 2000). If hyponatremia is associated with hypo-osmolality (true hyponatremia), the next step is to measure the urinary osmolality to determine if there is an impaired ability to excrete free water (urine$_{osm}$ > 100$_{mosm/kg}$).

The information that is most useful in arriving at a correct diagnosis of hyponatremia is a detailed history of fluid balance, weight changes, medications (especially diuretics) and underlying medical illnesses. Hyponatremia is usually a multifactorial disorder and a detailed history will identify sources of salt and water losses, free water ingestion and underlying illnesses that cause a non-osmotic stimulus for vasopressin production. An assessment of the volume status on physical examination and the urinary electrolytes can be extremely helpful, but both can be misleading (Musch et al., 1995). In patients in whom hyponatremia is due to salt

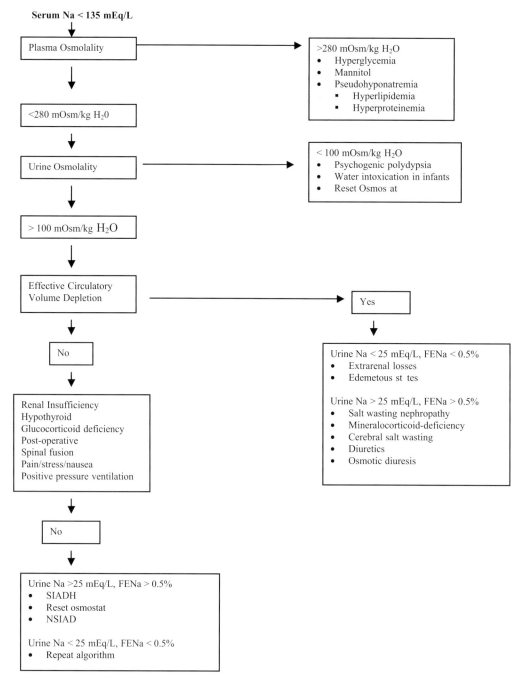

FIGURE 19.5 Diagnostic approach to hyponatremia.

losses, such as diuretics, signs of volume depletion may be absent on physical examination, as the volume deficit may be nearly corrected due to oral intake of hypotonic fluids if the thirst mechanism is intact.

In general, a urinary sodium concentration less than 25 mEq/l or a fractional excretion of sodium (FENa) less than 0.5% is consistent with effective circulating volume depletion, while a urine sodium greater than 25 mEq/l or a FENa greater than 0.5% is consistent with renal tubular dysfunction, use of diuretics or SIADH (Musch et al., 1995). Numerous factors can affect the urine sodium, mak-

ing interpretation difficult, therefore, the timing of the urinary measurements in relation to dosages of diuretics, intravenous fluid boluses or fluid and sodium restriction are also important. In some cases, assessment of volume status by the measurement of a central venous pressure may be helpful (Damaraju et al., 1997).

13. TREATMENT OF SIADH

SIADH is usually of short duration and resolves with treatment of the underlying disorder and discontinuation of the offending medication. Fluid restriction is the cornerstone to therapy, but is a slow method of correction and is frequently impractical in infants who receive most of their nutrition as liquids. All intravenous fluids should be of a tonicity of at least normal saline and, if this does not correct the plasma sodium, 3% sodium chloride may be given as needed. If a more rapid correction of hyponatremia is needed, the addition of a loop diuretic in combination with hypertonic saline is useful (Hantman et al., 1973; Perks et al., 1979).

Chronic SIADH can be difficult to manage and may be unresponsive to fluid restriction, increased sodium intake and loop diuretics. Demeclocycline, a tetracycline derivative that produces vasopressin-resistant polyuria, has been used in the past, but has nephrotoxicity as a side effect. The use of oral urea has been reported to be a safe and effective therapy for the managing chronic SIADH in both children and adults (Decaux and Genette, 1981; Huang et al., 2006); 30–60 g of oral urea administered once daily in adults and doses of up to 2 g/kg/day, divided in four doses, have been used successfully in children.

Vasopressin antagonists hold a promising yet undefined role in the management of hyponatremia in SIADH. The only currently FDA approved vasopressin antagonist is conivaptan (Vaprisol), which is a combined V1a and V2 receptor antagonist which is only approved for intravenous use. There are limited published data on the intravenous use of conivaptin in SIADH (Greenberg and Verbalis, 2006). Conivaptin is a potent inhibitor of the cytochrome P450 system, therefore its restricted use as an intravenous agent is to minimize adverse drug interactions. At this time, conivaptan should not be considered for the use of symptomatic hyponatremia, as it is not approved for this therapy and its efficacy has not been proven. In certain subsets of SIADH, such as reset osmostat and the NSIAD, V2 antagonists will be unsuccessful. Some patients have been diagnosed with NSIAD specifically because of lack of efficacy of V2 antagonists. In reset osmostat, AVP levels will increase significantly in response to V2 antagonist and thirst may increase limiting its efficacy. Agents which produce diabetes insipidus, such as demeclocyline, have been used in the past for chronic SIADH, but the nephrotoxicity association with it has limited persists for greater than a month and is unresponsive to fluid restriction, increased sodium intake and loop diuretics (Perks et al., 1979).

There are currently three oral selective V2 antagonists that are under investigation, tolvaptan, lixivaptan and satavaptan, but none are FDA approved at this time. A recent large multicenter, double-blinded, placebo-controlled trial has demonstrated that tolvaptan is safe and effective in the management of chronic euvolemic and hypervolemic hyponatremia. Many of these patients had SIADH. Serum sodium was 5 mEq/l higher after 30 days' treatment with tolvaptan than in controls (Schrier et al., 2006). Similar findings have been reported for satavaptan with one-year duration of therapy (Soupart et al., 2006). These data suggest that oral V2 receptor antagonists may be of benefit to a subset of patients with chronic SIADH who cannot be managed with moderate fluid restriction and salt supplementation.

III. HYPONATREMIC ENCEPHALOPATHY

A. Cerebral Edema and Brain Cell Volume Regulation

A major consequence of hyponatremia is cerebral edema. Cerebral edema causes increased intracranial pressure which can lead to brain ischemia, herniation and death. Hypo-osmolality results in cytotoxic cerebral edema with an influx of water into the intracellular space down a concentration gradient resulting in brain parenchymal swelling, without a disruption in the blood–brain barrier (BBB). This is different than vasogenic cerebral edema which results from the leakage of fluid across capillaries into the brain interstitium from a disruption in the BBB. Cerebral edema primarily results from cellular swelling of astrocytes, which are neuroglial cells that provide the supporting structure for the brain. Astrocytes are the primary site of AQP4 expression in the brain which makes them permeable to water (Papadopoulos and Verkman, 2007). It has been demonstrated that AQP4 null mice are relatively protected against cytotoxic cerebral edema from hyponatremia (Manley et al., 2000). It is therefore postulated that medications that inhibit AQP4 could protect the brain against cytotoxic cerebral edema.

A normal human has an intracranial volume of approximately 1.4 l, with 25% of this residing extracellularly in blood, CSF and interstitial fluid and 75% residing in the intracellular brain parenchyma. The brain's initial adaptation to cerebral edema occurs in minutes to hours and involves the bulk flow of fluid out of the extracellular space (Melton and Nattie, 1984). This is followed by an adaptive homeostatic mechanism, termed regulatory volume decrease (RVD). When astrocytes are subjected to hypo-osmolality, the RVD involves the immediate extrusion of sodium, followed by a slower extrusion of potassium and organic osmolytes (Melton et al., 1987; Law, 1999). Some of the organic osmolytes which are extruded by astrocytes are excitatory amino acids, such as glutamate and aspartate, which can produce seizure in the absence of clinically detectable cerebral edema (Kimelberg, 2004). The extrusion of intracellular electrolytes and organic osmolytes are energy dependent. It has been shown that inhibition of sodium potassium ATPase can impair the RVD and contribute to cerebral edema (Andersson et al., 2004).

B. Clinical Symptoms

The clinical manifestations of hyponatremia are primarily neurologic and related to cerebral edema caused by hypo-osmolality (Table 19.6). The symptoms of hyponatremic encephalopathy are quite variable between individuals with the only consistent symptoms being headache, nausea,

TABLE 19.6 Anatomic and biochemical changes and clinical symptoms of hyponatremic encephalopathy

Anatomic and biochemical changes	Clinical symptoms
Brain swelling	Headache
	Nausea
	Vomiting
Pressure on a rigid skull	Seizures
Excitatory amino acids	
Tentorial herniation	Respiratory arrest

TABLE 19.7 Risk factors for developing hyponatremic encephalopathy

Risk factor	Pathophysiologic mechanism
Children	Increase brain to intracranial volume ratio
Females	Sex steroids (estrogens) inhibit brain adaptation
	Increase vasopressin levels
	Cerebral vasoconstriction
	Hypoperfusion of brain tissue
Hypoxemia	Impaired brain adaptation

vomiting, emesis and weakness. As the cerebral edema worsens, patients then develop behavioral changes and impaired response to verbal and tactile stimuli. Advanced symptoms are signs of cerebral herniation, with seizures, respiratory arrest, dilated pupils and decorticate posturing. Hyponatremic encephalopathy is important to recognize early, as it accounts for one-third of the seizures encountered in the ICU setting (Wijdicks and Sharbrough, 1993). Not all patients have the usual progression in symptoms and advanced symptoms can present with sudden onset.

C. Neurogenic Pulmonary Edema

A common yet often unrecognized symptom of hyponatremic encephalopathy is neurogenic pulmonary edema (Ayus and Arieff, 1995, 2000; Nzerue et al., 2002). Neurogenic pulmonary edema is a well described yet under diagnosed condition that occurs as a complication of severe CNS injury (Fontes et al., 2003). There is typically a rapid onset of pulmonary edema following the development of cerebral edema. No cardiac etiology is found and there is a complete and rapid resolution of respiratory systems following appropriate treatment of hyponatremic encephalopathy with hypertonic saline. If not recognized early, the condition is almost universally fatal (Ayus and Arieff, 1995). The pathophysiologic mechanism of neurogenic pulmonary edema is unclear, but appears to be due to (1) increased microvascular permeability to proteins (McClellan et al., 1989) and (2) a sympathetic discharge resulting in pulmonary vasoconstriction with increased pulmonary hydrostatic pressure (Maron, 1990). The incidence of neurogenic pulmonary edema complicating hyponatremic encephalopathy is uncertain, but 15% of patients with severe hyponatremia do have radiographic evidence of pulmonary edema (Nzerue et al., 2002). Hyponatremic encephalopathy should be considered in any patient presenting with a non-cardiogenic pulmonary edema osmolality (McManus et al., 1995).

D. Risk Factors for Developing Hyponatremic Encephalopathy (Table 19.7)

Various factors can interfere with successful brain adaptation and may play a more important role than the absolute change in serum sodium in predicting whether a patient will suffer hyponatremic encephalopathy. Elevated AVP levels appear to be a contributing factor to the development of cerebral edema as AVP is known to increase brain water content in the absence of hyponatremia and impair brain regulatory volume mechanisms (Doczi et al., 1984; Vajda et al., 2001). The major factors that interfere with brain adaptation are physical factors related to age, hormonal factors related to gender and hypoxemia (Ayus and Arieff, 1993).

1. Age

Children under 16 years of age are at increased risk for developing hyponatremic encephalopathy due to their relatively larger brain to intracranial volume ratio as compared to adults (Arieff et al., 1992, 1995). A child's brain reaches adult size by 6 years of age, whereas the skull does not reach adult size until 16 years of age (Sgouros et al., 1999; Xenos et al., 2002). Consequently, children have less room available in their rigid skulls for brain expansion and are likely to develop brain herniation from hyponatremia at higher serum sodium concentrations than adults. Children will have a high morbidity from symptomatic hyponatremia unless appropriate therapy is instituted early (Arieff et al., 1992; Moritz and Ayus, 2001, 2003; Halberthal et al., 2001; McJunkin et al., 2001). After the third decade of life the brain begins to atrophy, with the steepest reduction in brain volume occurring after 50 years of age (Takeda and Matsuzawa, 1985; Courchesne et al., 2000). The brain volume of an 80 year old is approximately that of a young child. Consequently, the elderly are at the lowest risk of developing central nervous system manifestation of hyponatremia.

2. Gender

Recent epidemiological data have clearly shown that menstruant women are at substantially higher risk for developing permanent neurological sequelae or death from hyponatremic encephalopathy than men or post-menopausal females (Arieff, 1986; Ayus et al., 1992; Ayus and Arieff, 1996, 1999). The relative risk of death or permanent neurologic damage from hyponatremic encephalopathy is approximately 30 times greater for women compared to men and approximately 25 times greater for menstruant females than post-menopausal females (Ayus et al., 1992). Menstruant females can develop symptomatic hyponatremia at serum sodium values as high as 128 mEq/l (Arieff, 1986). Hyponatremic encephalopathy in menstruant females primarily occurs in healthy females following elective surgeries while receiving hypotonic fluids (Arieff, 1986; Ayus et al., 1992). Premenopausal women are at high risk for developing hyponatremic encephalopathy due to the inhibitory effects of sex hormones and the effects of vasopressin on the cerebral circulation which, in the female animal model as opposed to the male, are characterized by cerebral vasoconstriction and hypoperfusion to brain tissue (Fraser and Swanson, 1994; Arieff et al., 1995).

3. Hypoxia

Hypoxemia is a major risk factor for developing hyponatremic encephalopathy. The occurrence of a hypoxic event, such as respiratory insufficiency, is a major factor militating against survival without permanent brain damage in patients with hyponatremia (Arieff, 1986). The combination of systemic hypoxemia and hyponatremia is more deleterious than is either factor alone because hypoxemia impairs the ability of the brain to adapt to hyponatremia, leading to a vicious cycle of worsening hyponatremic encephalopathy (Vexler et al., 1994). Studies of hyponatremic animals have revealed that hypoxia impairs volume regulation of brain cells, decreases cerebral perfusion and increases the probability of neuronal lesions developing (Ayus et al., 2006). Patients with symptomatic hyponatremia can develop hypoxemia by at least two different mechanisms: neurogenic pulmonary edema or hypercapnic respiratory failure (Ayus and Arieff, 1995). Respiratory failure can be of very sudden onset in patients with symptomatic hyponatremia (Arieff, 1986; Ayus and Arieff, 2000). The majority of neurologic morbidity seen in patients with hyponatremia has occurred in patients who have had a respiratory arrest as a feature of hyponatremic encephalopathy (Arieff, 1986; Ayus et al., 1987, 1992; Arieff et al., 1992; Ayus and Arieff, 1999). Recent data have shown that hypoxia is the strongest predictor of mortality in patients with symptomatic hyponatremia (Nzerue et al., 2002).

E. 0.9% NaCl as Prophylaxis Against Hyponatremia

We have argued that administering isotonic saline in maintenance parenteral fluids is the most physiologic approach to preventing hospital-acquired hyponatremia (Moritz and Ayus, 2003, 2007). Several prospective studies in children and adults have revealed that administering 0.9% sodium chloride is effective prophylaxis against the development of hyponatremia (McFarlane and Lee, 1994; Scheingraber et al., 1999; Waters et al., 2001; Wilkes et al., 2001; Boldt et al., 2002; Takil et al., 2002; Aronson et al., 2002; Neville et al., 2006). Even in patients with SIADH and hyponatremia, the administration of normal saline does not aggravate the hyponatremia (Musch and Decaux, 1998). We have conducted a meta-analysis of 550 postoperative patients, 50 of whom were children, managed with either 0.9% NaCl or a more hypotonic fluid; 0.9% NaCl effectively prevented hyponatremia, while more hypotonic fluids, including Ringer's lactate, consistently resulted in a fall in serum sodium (Moritz et al., 2005). Ringer's lactate, which has a sodium concentration of 130 mmol/l, is hypotonic in relation to the plasma water and can produce hyponatremia (Steele et al., 1997). The avoidance of hypotonic fluids and the administration of 0.9% NaCl when parenteral fluids are required is the most physiologic approach to preventing hyponatremia. The administration of 0.9% NaCl is extremely safe, without there ever being a report of neurological complications from hyponatremia in non-neurosurgical patients. Neurosurgical patients can develop cerebral salt wasting, which is a condition where 0.9% NaCl may not be sufficient prophylaxis against hyponatremia and 3% NaCl may need to be administered to maintain a normal serum sodium.

F. Treatment of Hyponatremic Encephalopathy

Hyponatremic encephalopathy is a medical emergency that needs early recognition and treatment. Ayus and Arieff have previously shown that neurologic sequelae from hyponatremic encephalopathy result from insufficient therapy and not from rapid correction (Ayus et al., 1987; Ayus and Arieff, 1999). This has been confirmed by three recent studies in adults that have found a poor outcome to be associated with inadequate therapy (Nzerue et al., 2002; Hoorn et al., 2006; Huda et al., 2006). We have looked at risk factors for poor neurological outcome in hyponatremic encephalopathy in children, and have found lack of therapy to be the major factor leading to poor outcome (Moritz and Ayus, 2006).

Treatment of hyponatremia should be based on neurologic symptoms and not on the absolute serum sodium. Patients with symptomatic hyponatremia need aggressive management with 3% sodium chloride (513 mmol/l). Fluid restriction alone plays no role in the management of symptomatic hyponatremia. The treatment of hyponatremic encephalopathy should precede any neuroimaging studies to confirm cerebral edema and should occur in a monitored setting where the airway can be secured and serum sodiums measured every 2 h until the patient is stable. Patients with severe symptoms, including seizures, respiratory arrest or neurogenic pulmonary edema, should receive 100 ml of 3% NaCl as a bolus over 10 min rapidly to reverse brain edema (Ayus et al., 2005). This may need to be repeated 1–2 times until symptoms subside, with the remainder of therapy delivered via continuous infusion. Patients with lesser symptoms, such as headache, nausea, vomiting or lethargy, can be treated via an infusion pump to achieve a correction of 4–8 mmol/l in the first 4 h. In order to prevent complications from excessive therapy, 3% sodium chloride should be discontinued when symptoms subside, the rate of correction should not exceed 20 mmol/l in the first 48 h and correction should be to mildy hyponatremic values, avoiding normonatremia and hypernatremia in the first 48 hours. In general 1 ml/kg of 3% NaCl will result in about a 1 mmol/l rise in serum sodium. A continuous infusion of 3% NaCl at a rate of 50–100 ml/h administered over 4 hours is usually sufficient to reverse symptoms. Twelve ml/kg of 3% NaCl infused over 4 hours has been used in children with acute hyponatremic encephalopathy without apparent neurologic sequelae (Alam et al., 2006; Moritz, 2007). Much of the change in serum sodium will be related to the renal response to therapy, making formulas unreliable in predicting the change in serum sodium. Patients with SIADH are at low risk for overcorrection. Patient with hyponatremia from diuretics or psychogenic polydypsia will have a brisk free water diuresis during therapy and are prone to overcorrection (Moritz and Ayus, 2005). In these patients, active measures may be needed to prevent overcorrection of hyponatremia including a change to hypotonic fluids or dDAVP. The administration of dDAVP will stop the free water diuresis, and a controlled rate of sodium correction can be achieved with a combination of fluid restriction and 0.9% NaCl and 3% NaCl as needed. Recently, V2 receptor antagonists have received FDA approval for the treatment of hyponatremia (Greenberg and Verbalis, 2006). While preliminary data reveal that they may have a role in treatment of asymptomatic euvolemic or hypervolemic hyponatremia (Schrier et al., 2006), they currently have no role in the acute management of symptomatic hyponatremia.

G. Cerebral Demyelination in the Correction of Hyponatremia

Cerebral demyelination is a rare complication which has been associated with symptomatic hyponatremia (Norenberg et al., 1982). Animal data have shown that correction of hyponatremia by greater than 20–25 mEq/l can result in cerebral demyelination (Kleinschmidt-DeMasters and Norenberg, 1981). This has resulted in a mistaken belief that a rapid rate of correction is likely to result in cerebral demyelination (Sterns, 1994). Recent data have now shown that rate of correction has little to do with development of cerebral demyelinating lesions and that lesions seen in hyponatremic patients are more closely associated with other comorbid factors or extreme increases in serum sodium (Table 19.8) (Ayus et al., 1985, 1987; Moritz et al., 2001; Heng et al., 2002; Nzerue et al., 2002). Animals studies have shown that azotemia may decrease the risk of myelinolysis following the correction of hyponatremia (Soupart et al., 2000).

The lesions of cerebral demyelination can be pontine or extrapontine and typically develop many days after the correction of hyponatremia (Wright et al., 1979; Norenberg et al., 1982). Cerebral demyelination can be asymptomatic or can manifest in confusion, quadriplegia, pseudobulbar palsy and a pseudocoma with a 'locked-in stare'(Wright et al., 1979). The lesions of cerebral demyelination can be seen in the absence of any sodium abnormalities (Moritz et al., 2001). In fact, the primary cause of brain damage in patients with hyponatremia is not cerebral demyelination, but cerebral edema and herniation (Arieff, 1986; Ayus et al., 1987, 1992; Arieff et al., 1992; Ayus and Arieff, 1999). Most brain damage occurs in untreated patients and is not a consequence of therapy (Ayus and Arieff, 1999; Nzerue et al., 2002).

TABLE 19.8 Risk factors for developing cerebral demyelination in hyponatremic patients

1. Development of hypernatremia
2. Increase in serum sodium exceeding 25 mmol/l in 48 h
3. Hypoxemia
4. Severe liver disease
5. Alcoholism
6. Cancer
7. Severe burns
8. Malnutrition
9. Hypokalemia
10. Diabetes
11. Renal failure

In one prospective study it was observed that hyponatremic patients who develop demyelinating lesions had either:

1. been made hypernatremic inadvertently
2. had their plasma sodium levels corrected by greater then 25 mmol/l in 48 h
3. suffered a hypoxic event, or
4. had severe liver disease (see Table 19.3) (Ayus et al., 1987).

Others have cautioned that cerebral demyelination could develop with elevations in serum sodium of 12–15 mEq/l/24 h (Sterns et al., 1986). A recent prospective study evaluating the development of demyelinating lesions in hyponatremic patients found no association with a change in serum sodium (Heng et al., 2007). The only factor associated with demyelination was hypoxemia (Heng et al., 2007). We retrospectively reviewed our experience with cerebral demyelination seen on autopsy specimens in children over a 15-year period (Moritz et al., 2001). There was no association between change in serum sodium and demyelination when compared to a matched control group. The only predisposing factors identified were underlying liver disease or central nervous system radiation in children with cancer.

References

Adler, S. M., and Verbalis, J. G. (2006). Disorders of body water homeostasis in critical illness. *Endocrinol Metab Clin North Am* **35**: 873–894 xi.

Agha, A., Phillips, J., O'Kelly, P., Tormey, W., and Thompson, C. J. (2005). The natural history of post-traumatic hypopituitarism: implications for assessment and treatment. *Am J Med* **118:** 1416.

Alam, N. H., Yunus, M., Faruque, A. S. et al. (2006). Symptomatic hyponatremia during treatment of dehydrating diarrheal disease with reduced osmolarity oral rehydration solution. *J Am Med Assoc* **296:** 567–573.

Almond, C. S., Shin, A. Y., Fortescue, E. B. et al. (2005). Hyponatremia among runners in the Boston Marathon. *N Engl J Med* **352:** 1550–1556.

Alon, U., and Chan, J. C. (1985). Hydrochlorothiazide-amiloride in the treatment of congenital nephrogenic diabetes insipidus. *Am J Nephrol* **5:** 9–13.

Amlal, H., Krane, C. M., Chen, Q., and Soleimani, M. (2000). Early polyuria and urinary concentrating defect in potassium deprivation. *Am J Physiol Renal Physiol* **279:** F655–F663.

Anderson, R. J., Chung, H. M., Kluge, R., and Schrier, R. W. (1985). Hyponatremia: a prospective analysis of its epidemiology and the pathogenetic role of vasopressin. *Ann Intern Med* **102:** 164–168.

Andersson, R. M., Aizman, O., Aperia, A., and Brismar, H. (2004). Modulation of Na^+, K^+-ATPase activity is of importance for RVD. *Acta Physiol Scand* **180:** 329–334.

Argyropoulou, M. I., and Kiortsis, D. N. (2005). MRI of the hypothalamic-pituitary axis in children. *Pediatr Radiol* **35:** 1045–1055.

Arieff, A. I. (1986). Hyponatremia, convulsions, respiratory arrest, and permanent brain damage after elective surgery in healthy women. *N Engl J Med* **314:** 1529–1535.

Arieff, A. I., Ayus, J. C., and Fraser, C. L. (1992). Hyponatraemia and death or permanent brain damage in healthy children. *Br Med J* **304:** 1218–1222.

Arieff, A. I., Kozniewska, E., Roberts, T. P., Vexler, Z. S., Ayus, J. C., and Kucharczyk, J. (1995). Age, gender, and vasopressin affect survival and brain adaptation in rats with metabolic encephalopathy. *Am J Physiol* **268:** R1143–R1152.

Aronson, D., Dragu, R. E., Nakhoul, F. et al. (2002). Hyponatremia as a complication of cardiac catheterization: a prospective study. *Am J Kidney Dis* **40:** 940–946.

Ataga, K. I., and Orringer, E. P. (2000). Renal abnormalities in sickle cell disease. *Am J Hematol* **63:** 205–211.

Ayus, J. C., and Arieff, A. I. (1993). Pathogenesis and prevention of hyponatremic encephalopathy. *Endocrinol Metab Clin North Am* **22:** 425–446.

Ayus, J. C., and Arieff, A. I. (1995). Pulmonary complications of hyponatremic encephalopathy. Noncardiogenic pulmonary edema and hypercapnic respiratory failure. *Chest* **107:** 517–521.

Ayus, J. C., and Arieff, A. I. (1996). Brain damage and postoperative hyponatremia: the role of gender. *Neurology* **46:** 323–328.

Ayus, J. C., and Arieff, A. I. (1999). Chronic hyponatremic encephalopathy in postmenopausal women: association of therapies with morbidity and mortality. *J Am Med Assoc* **281:** 2299–2304.

Ayus, J. C., and Arieff, A. I. (2000). Noncardiogenic pulmonary edema in marathon runners. *Ann Intern Med* **133:** 1011.

Ayus, J. C., Krothapalli, R. K., and Armstrong, D. L. (1985). Rapid correction of severe hyponatremia in the rat: histopathological changes in the brain. *Am J Physiol* **248:** F711–F719.

Ayus, J. C., Krothapalli, R. K., and Arieff, A. I. (1987). Treatment of symptomatic hyponatremia and its relation to brain damage. A prospective study. *N Engl J Med* **317:** 1190–1195.

Ayus, J. C., Wheeler, J. M., and Arieff, A. I. (1992). Postoperative hyponatremic encephalopathy in menstruant women. *Ann Intern Med* **117:** 891–897.

Ayus, J. C., Armstrong, D. L., and Arieff, A. I. (1996). Effects of hypernatraemia in the central nervous system and its therapy in rats and rabbits. *J Physiol* **492:** 243–255.

Ayus, J. C., Varon, J., and Arieff, A. I. (2000). Hyponatremia, cerebral edema, and noncardiogenic pulmonary edema in marathon runners. *Ann Intern Med* **132:** 711–714.

Ayus, J. C., Arieff, A., and Moritz, M. L. (2005). Hyponatremia in marathon runners. *N Engl J Med* **353:** 427–428.

Ayus, J. C., Armstrong, D., and Arieff, A. I. (2006). Hyponatremia with hypoxia: effects on brain adaptation, perfusion, and histology in rodents. *Kidney Int* **69:** 1319–1325.

Banister, A., Matin-Siddiqi, S. A., and Hatcher, G. W. (1975). Treatment of hypernatraemic dehydration in infancy. *Arch Dis Child* **50:** 179–186.

Bartter, F. C., and Schwartz, W. B. (1967). The syndrome of inappropriate secretion of antidiuretic hormone. *Am J Med* **42:** 790–806.

Battle, D. C., von Riotte, A. B., Gaviria, M., and Grupp, M. (1985). Amelioration of polyuria by amiloride in patients receiving long-term lithium therapy. *N Engl J Med* **312:** 408–414.

Berl, T., Linas, S. L., Aisenbrey, G. A., and Anderson, R. J. (1977). On the mechanism of polyuria in potassium depletion. The role of polydipsia. *J Clin Invest* **60:** 620–625.

Bernstein, S. A., and Williford, S. L. (1997). Intranasal desmopressin-associated hyponatremia: a case report and literature review. *J Fam Pract* **44:** 203–208.

Bes, D. F., Mendilaharzu, H., Fenwick, R. G., and Arrizurieta, E. (2007). Hyponatremia resulting from arginine vasopressin receptor 2 gene mutation. *Pediatr Nephrol* **22:** 463–466.

Bevilacqua, M. (1994). Hyponatraemia in AIDS. *Baillieres Clin Endocrinol Metab* **8:** 837–848.

Bichet, D. G. (2006). Hereditary polyuric disorders: new concepts and differential diagnosis. *Semin Nephrol* **26:** 224–233.

Boldt, J., Haisch, G., Suttner, S., Kumle, B., and Schellhase, F. (2002). Are lactated Ringer's solution and normal saline solution equal with regard to coagulation? *Anesth Analg* **94:** 378–384 table of contents.

Boton, R., Gaviria, M., and Battle, D. C. (1987). Prevalence, pathogenesis, and treatment of renal dysfunction associated with chronic lithium therapy. *Am J Kidney Dis* **10:** 329–345.

Bruns, D. E., Ladenson, J. H., and Scott, M. G. (2000). Hyponatremia. *N Engl J Med* **343:** 886–887 author reply 8.

Bustamante, M., Hasler, U., Leroy, V. et al. (2008). Calcium-sensing receptor attenuates AVP-induced aquaporin-2 expression via a calmodulin-dependent mechanism. *J Am Soc Nephrol* **19:** 109–116.

Cesar, K. R., and Magaldi, A. J. (1999). Thiazide induces water absorption in the inner medullary collecting duct of normal and Brattleboro rats. *Am J Physiol* **277:** F756–F760.

Christensen, S., Kusano, E., Yusufi, A. N., Murayama, N., and Dousa, T. P. (1985). Pathogenesis of nephrogenic diabetes insipidus due to chronic administration of lithium in rats. *J Clin Invest* **75:** 1869–1879.

Christensen, B. M., Marples, D., Kim, Y. H., Wang, W., Frokiaer, J., and Nielsen, S. (2004). Changes in cellular composition of kidney collecting duct cells in rats with lithium-induced NDI. *Am J Physiol Cell Physiol* **286:** C952–C964.

Chung, H. M., Kluge, R., Schrier, R. W., and Anderson, R. J. (1986). Postoperative hyponatremia. A prospective study. *Arch Intern Med* **146:** 333–336.

Cogan, E., Debieve, M. F., Pepersack, T., and Abramow, M. (1988). Natriuresis and atrial natriuretic factor secretion during inappropriate antidiuresis. *Am J Med* **84:** 409–418.

Cooke, C. R., Turin, M. D., and Walker, W. G. (1979). The syndrome of inappropriate antidiuretic hormone secretion (SIADH): pathophysiologic mechanisms in solute and volume regulation. *Medicine (Balt.)* **58:** 240–251.

Courchesne, E., Chisum, H. J., Townsend, J. et al. (2000). Normal brain development and aging: quantitative analysis at in vivo MR imaging in healthy volunteers. *Radiology* **216:** 672–682.

Damaraju, S. C., Rajshekhar, V., and Chandy, M. J. (1997). Validation study of a central venous pressure-based protocol for the management of neurosurgical patients with hyponatremia and natriuresis. *Neurosurgery* **40:** 312–316 discussion 6–7.

Das, P., Carcao, M., and Hitzler, J. (2005). DDAVP-induced hyponatremia in young children. *J Pediatr Hematol Oncol* **27:** 330–332.

De Buyst, J., Massa, G., Christophe, C., Tenoutasse, S., and Heinrichs, C. (2007). Clinical, hormonal and imaging findings in 27 children with central diabetes insipidus. *Eur J Pediatr* **166:** 43–49.

De Jong, P. E., De Jong-van den Berg, L. T., De Zeeuw, D., Donker, A. J., Schouten, H., and Statius van Eps, L. W. (1982). The influence of indomethacin on renal concentrating and diluting capacity in sickle cell nephropathy. *Clin Sci (Lond.)* **63:** 53–58.

Decaux, G., and Genette, F. (1981). Urea for long-term treatment of syndrome of inappropriate secretion of antidiuretic hormone. *Br Med J* **283:** 1081–1083.

Decaux, G., Namias, B., Gulbis, B., and Soupart, A. (1996). Evidence in hyponatremia related to inappropriate secretion of ADH that V1 receptor stimulation contributes to the increase in renal uric acid clearance. *J Am Soc Nephrol* **7:** 805–810.

Decaux, G., Prospert, F., Soupart, A., and Musch, W. (2000). Evidence that chronicity of hyponatremia contributes to the high urate clearance observed in the syndrome of inappropriate antidiuretic hormone secretion. *Am J Kidney Dis* **36:** 745–751.

Decaux, G., Vandergheynst, F., Bouko, Y., Parma, J., Vassart, G., and Vilain, C. (2007). Nephrogenic syndrome of inappropriate antidiuresis in adults: high phenotypic variability in men and women from a large pedigree. *J Am Soc Nephrol* **18:** 606–612.

Doczi, T., Laszlo, F. A., Szerdahelyi, P., and Joo, F. (1984). Involvement of vasopressin in brain edema formation: further evidence obtained from the Brattleboro diabetes insipidus rat with experimental subarachnoid hemorrhage. *Neurosurgery* **14:** 436–441.

Domenech, E., Gomez-Zaera, M., and Nunes, V. (2006). Wolfram/DIDMOAD syndrome, a heterogenic and molecularly complex neurodegenerative disease. *Pediatr Endocrinol Rev* **3:** 249–257.

Dunn, F. L., Brennan, T. J., Nelson, A. E., and Robertson, G. L. (1973). The role of blood osmolality and volume in regulating vasopressin secretion in the rat. *J Clin Invest* **52:** 3212–3219.

Durr, J. A., and Lindheimer, M. D. (1996). Diagnosis and management of diabetes insipidus during pregnancy. *Endocr Pract* **2:** 353–361.

Earley, L. E., and Orloff, J. (1962). The mechanism of antidiuresis associated with the administration of hydrochlorothiazide to patients with vasopressin-resistant diabetes insipidus. *J Clin Invest* **41:** 1988–1997.

Earm, J. H., Christensen, B. M., Frokiaer, J. et al. (1998). Decreased aquaporin-2 expression and apical plasma membrane delivery in kidney collecting ducts of polyuric hypercalcemic rats. *J Am Soc Nephrol* **9:** 2181–2193.

Ecelbarger, C. A., Nielsen, S., Olson, B. R. et al. (1997). Role of renal aquaporins in escape from vasopressin-induced antidiuresis in rat. *J Clin Invest* **99:** 1852–1863.

Ecelbarger, C. A., Chou, C. L., Lee, A. J., DiGiovanni, S. R., Verbalis, J. G., and Knepper, M. A. (1998). Escape from vasopressin-induced antidiuresis: role of vasopressin resistance of the collecting duct. *Am J Physiol* **274:** F1161–F1166.

Ecelbarger, C. A., Murase, T., Tian, Y., Nielsen, S., Knepper, M. A., and Verbalis, J. G. (2002). Regulation of renal salt and water transporters during vasopressin escape. *Prog Brain Res* **139:** 75–84.

Fallon, J. K., Shah, D., Kicman, A. T. et al. (2002). Action of MDMA (ecstasy) and its metabolites on arginine vasopressin release. *Ann NY Acad Sci* **965:** 399–409.

Feldman, B. J., Rosenthal, S. M., Vargas, G. A. et al. (2005). Nephrogenic syndrome of inappropriate antidiuresis. *N Engl J Med* **352:** 1884–1890.

Fonseca, S. G., Fukuma, M., Lipson, K. L. et al. (2005). WFS1 is a novel component of the unfolded protein response and maintains homeostasis of the endoplasmic reticulum in pancreatic beta-cells. *J Biol Chem* **280:** 39609–39615.

Fontes, R. B., Aguiar, P. H., Zanetti, M. V., Andrade, F., Mandel, M., and Teixeira, M. J. (2003). Acute neurogenic pulmonary edema: case reports and literature review. *J Neurosurg Anesthesiol* **15:** 144–150.

Fraser, C. L., and Swanson, R. A. (1994). Female sex hormones inhibit volume regulation in rat brain astrocyte culture. *Am J Physiol* **267:** C909–C914.

Fujisawa, I. (2004). Magnetic resonance imaging of the hypothalamic-neurohypophyseal system. *J Neuroendocrinol* **16:** 297–302.

Gill, J. R., and Bartter, F. C. (1961). On the impairment of renal concentrating ability in prolonged hypercalcemia and hypercalciuria in man. *J Clin Invest* **40:** 716–722.

Goldberg, H., Clayman, P., and Skorecki, K. (1988). Mechanism of Li inhibition of vasopressin-sensitive adenylate cyclase in cultured renal epithelial cells. *Am J Physiol* **255:** F995–1002.

Grant, P. J., Hampton, K. K., Primrose, J., Davies, J. A., and Prentice, C. R. (1991). Vasopressin and haemostatic responses to inguinal hernia repair under local anaesthesia. *Blood Coagul Fibrinol* **2:** 647–650.

Greenberg, A., and Verbalis, J. G. (2006). Vasopressin receptor antagonists. *Kidney Int* **69:** 2124–2130.

Halberthal, M., Halperin, M. L., and Bohn, D. (2001). Lesson of the week: acute hyponatraemia in children admitted to hospital: retrospective analysis of factors contributing to its development and resolution. *Br Med J* **322:** 780–782.

Hantman, D., Rossier, B., Zohlman, R., and Schrier, R. (1973). Rapid correction of hyponatremia in the syndrome of inappropriate secretion of antidiuretic hormone. An alternative treatment to hypertonic saline. *Ann Intern Med* **78:** 870–875.

Harrigan, M. R. (2001). Cerebral salt wasting syndrome. *Crit Care Clin* **17:** 125–138.

Heng, A. E., Taillandier, A., Klisnick, A. et al. (2002). Determinants of osmotic demyelination syndrome following correction of hyponatremia: a prospective magnetic resonance imaging study. *J Am Soc Nephrol* **13**: SU–P0900.

Heng, A. E., Vacher, P., Aublet-Cuvelier, B. et al. (2007). Centropontine myelinolysis after correction of hyponatremia: role of associated hypokalemia. *Clin Nephrol* **67**: 345–351.

Hoekstra, J. A., van Lieburg, A. F., Monnens, L. A., Hulstijn-Dirkmaat, G. M., and Knoers, V. V. (1996). Cognitive and psychosocial functioning of patients with congenital nephrogenic diabetes insipidus. *Am J Med Genet* **61**: 81–88.

Hogan, G. R., Dodge, P. R., Gill, S. R., Master, S., and Sotos, J. F. (1969). Pathogenesis of seizures occurring during restoration of plasma tonicity to normal in animals previously chronically hypernatremic. *Pediatrics* **43**: 54–64.

Hogan, G. R., Dodge, P. R., Gill, S. R., Pickering, L. K., and Master, S. (1984). The incidence of seizures after rehydration of hypernatremic rabbits with intravenous or ad libitum oral fluids. *Pediatr Res* **18**: 340–345.

Hogan, G. R., Pickering, L. K., Dodge, P. R., Shepard, J. B., and Master, S. (1985). Incidence of seizures that follow rehydration of hypernatremic rabbits with intravenous glucose or fructose solutions. *Exp Neurol* **87**: 249–259.

Hoorn, E. J., Lindemans, J., and Zietse, R. (2006). Development of severe hyponatraemia in hospitalized patients: treatment-related risk factors and inadequate management. *Nephrol Dial Transplant* **21**: 70–76.

Huang, E. A., Feldman, B. J., Schwartz, I. D., Geller, D. H., Rosenthal, S. M., and Gitelman, S. E. (2006). Oral urea for the treatment of chronic syndrome of inappropriate antidiuresis in children. *J Pediatr* **148**: 128–131.

Huda, M. S., Boyd, A., Skagen, K. et al. (2006). Investigation and management of severe hyponatraemia in a hospital setting. *Postgrad Med J* **82**: 216–219.

Ishikawa, S. E., and Schrier, R. W. (2003). Pathophysiological roles of arginine vasopressin and aquaporin-2 in impaired water excretion. *Clin Endocrinol (Oxf.)* **58**: 1–17.

Ishikawa, S. E., Saito, T., Kaneko, K., Okada, K., and Kuzuya, T. (1987). Hyponatremia responsive to fludrocortisone acetate in elderly patients after head injury. *Ann Intern Med* **106**: 187–191.

Jaenike, J. R., and Waterhouse, C. (1961). The renal response to sustained administration of vasopressin and water in man. *J Clin Endocrinol Metab* **21**: 231–242.

Jamison, R. L., and Oliver, R. E. (1982). Disorders of urinary concentration and dilution. *Am J Med* **72**: 308–322.

Kalantar-Zadeh, K., Nguyen, M. K., Chang, R., and Kurtz, I. (2006). Fatal hyponatremia in a young woman after ecstasy ingestion. *Nat Clin Pract Nephrol* **2**: 283–288 quiz 9.

Kaneko, K., Okada, K., Ishikawa, S., Kuzuya, T., and Saito, T. (1987). Role of atrial natriuretic peptide in natriuresis in volume-expanded rats. *Am J Physiol* **253**: R877–R882.

Kelberman, D., and Dattani, M. T. (2007). Hypothalamic and pituitary development: novel insights into the aetiology. *Eur J Endocrinol* **157**(Suppl 1): S3–14.

Kim, J. K., Summer, S. N., Wood, W. M., and Schrier, R. W. (2001). Role of glucocorticoid hormones in arginine vasopressin gene regulation. *Biochem Biophys Res Commun* **289**: 1252–1256.

Kimelberg, H. K. (2004). Increased release of excitatory amino acids by the actions of ATP and peroxynitrite on volume-regulated anion channels (VRACs) in astrocytes. *Neurochem Int* **45**: 511–519.

Kleinschmidt-DeMasters, B. K., and Norenberg, M. D. (1981). Rapid correction of hyponatremia causes demyelination: relation to central pontine myelinolysis. *Science* **211**: 1068–1070.

Knoers, N., van den Ouweland, A., Dreesen, J., Verdijk, M., Monnens, L. A., and van Oost, B. A. (1993). Nephrogenic diabetes insipidus: identification of the genetic defect. *Pediatr Nephrol* **7**: 685–688.

Kwon, T. H., Frokiaer, J., Knepper, M. A., and Nielsen, S. (1998). Reduced AQP1, -2, and -3 levels in kidneys of rats with CRF induced by surgical reduction in renal mass. *Am J Physiol* **275**: F724–F741.

Law, R. O. (1999). Amino acid efflux and cell volume regulation in cerebrocortical minislices prepared from chronically hyponatraemic and hypernatraemic rats. *Neurochem Int* **35**: 423–430.

Leaf, A., Bartter, F. C., Santos, R. F., and Wrong, O. (1953). Evidence in man that urinary electrolyte loss induced by pitressin is a function of water retention. *J Clin Invest* **32**: 868–878.

Levin, E. R., Gardner, D. G., and Samson, W. K. (1998). Natriuretic peptides. *N Engl J Med* **339**: 321–328.

Li, C., Wang, W., Summer, S. N. et al. (2008). Molecular mechanisms of antidiuretic effect of oxytocin. *J Am Soc Nephrol* **19**: 225–232.

Loffing, J. (2004). Paradoxical antidiuretic effect of thiazides in diabetes insipidus: another piece in the puzzle. *J Am Soc Nephrol* **15**: 2948–2950.

Magaldi, A. J. (2000). New insights into the paradoxical effect of thiazides in diabetes insipidus therapy. *Nephrol Dial Transplant* **15**: 1903–1905.

Maghnie, M., Cosi, G., Genovese, E. et al. (2000). Central diabetes insipidus in children and young adults. *N Engl J Med* **343**: 998–1007.

Maghnie, M., Altobelli, M., Di Iorgi, N. et al. (2004). Idiopathic central diabetes insipidus is associated with abnormal blood supply to the posterior pituitary gland caused by vascular impairment of the inferior hypophyseal artery system. *J Clin Endocrinol Metab* **89**: 1891–1896.

Maghnie, M., Ghirardello, S., De Bellis, A. et al. (2006). Idiopathic central diabetes insipidus in children and young adults is commonly associated with vasopressin-cell antibodies and markers of autoimmunity. *Clin Endocrinol (Oxf.)* **65**: 470–478.

Manley, G. T., Fujimura, M., Ma, T. et al. (2000). Aquaporin-4 deletion in mice reduces brain edema after acute water intoxication and ischemic stroke. *Nat Med* **6**: 159–163.

Maron, M. B. (1990). Pulmonary vasoconstriction in a canine model of neurogenic pulmonary edema. *J Appl Physiol* **68**: 912–918.

Marples, D., Christensen, S., Christensen, E. I., Ottosen, P. D., and Nielsen, S. (1995). Lithium-induced downregulation of aquaporin-2 water channel expression in rat kidney medulla. *J Clin Invest* **95**: 1838–1845.

Marples, D., Frokiaer, J., Dorup, J., Knepper, M. A., and Nielsen, S. (1996). Hypokalemia-induced downregulation of aquaporin-2 water channel expression in rat kidney medulla and cortex. *J Clin Invest* **97**: 1960–1968.

McClellan, M. D., Dauber, I. M., and Weil, J. V. (1989). Elevated intracranial pressure increases pulmonary vascular permeability to protein. *J Appl Physiol* **67**: 1185–1191.

McFarlane, C., and Lee, A. (1994). A comparison of Plasmalyte 148 and 0.9% saline for intra-operative fluid replacement. *Anaesthesia* **49**: 779–781.

McJunkin, J. E., de los Reyes, E. C., Irazuzta, J. E. et al. (2001). La Crosse encephalitis in children. *N Engl J Med* **344**: 801–807.

McManus, M. L., Churchwell, K. B., and Strange, K. (1995). Regulation of cell volume in health and disease. *N Engl J Med* **333**: 1260–1266.

Melton, J. E., and Nattie, E. E. (1984). Intracranial volume adjustments and cerebrospinal fluid pressure in the osmotically swollen rat brain. *Am J Physiol* **246**: R533–R541.

Melton, J. E., Patlak, C. S., Pettigrew, K. D., and Cserr, H. F. (1987). Volume regulatory loss of Na, Cl, and K from rat brain during acute hyponatremia. *Am J Physiol* **252**: F661–F669.

Moritz, M. L. (2005). A water deprivation test is not indicated in the evaluation of hypernatremia. *Am J Kidney Dis* **46**: 1150–1211 author reply 1.

Moritz, M. L. (2007). Fluid replacement for severe hyponatremia. *J Am Med Assoc* **297**: 41–42.

Moritz, M. L., and Ayus, J. C. (2001). La Crosse encephalitis in children. *N Engl J Med* **345**: 148–149.

Moritz, M. L., and Ayus, J. C. (2003). Prevention of hospital-acquired hyponatremia: a case for using isotonic saline. *Pediatrics* **111**: 227–230.

Moritz, M. L., and Ayus, J. C. (2005). Preventing neurological complications from dysnatremias in children. *Pediatr Nephrol* **20**: 1687–1700.

Moritz, M. L., and Ayus, J. C. (2006). Risk factors for death or neurologic impairment from hyponatremic encephalopathy in children in the new millenium. *J Am Soc Nephrol* **17**: 38A.

Moritz, M. L., and Ayus, J. C. (2007). Hospital-acquired hyponatremia – why are hypotonic parenteral fluids still being used? *Nat Clin Pract Nephrol* **3**: 374–382.

Moritz, M. L., Ellis, D., Vats, A., and Ayus, J. C. (2001). Lack of relationship between changes in serum sodium and development of cerebral demyelination in children. *J Am Soc Nephrol* **12**: A0726.

Moritz, M. L., Potter, D. M., and Ayus, J. C. (2005). Post-operative hyponatremia: a meta-analysis. *J Am Soc Nephrol* **16**: 44A.

Musch, W., and Decaux, G. (1998). Treating the syndrome of inappropriate ADH secretion with isotonic saline. *Q J Med* **91**: 749–753.

Musch, W., Thimpont, J., Vandervelde, D., Verhaeverbeke, I., Berghmans, T., and Decaux, G. (1995). Combined fractional excretion of sodium and urea better predicts response to saline in hyponatremia than do usual clinical and biochemical parameters. *Am J Med* **99**: 348–355.

Neville, K. A., Verge, C. F., Rosenberg, A. R., O'Meara, M. W., and Walker, J. L. (2006). Isotonic is better than hypotonic saline for intravenous rehydration of children with gastroenteritis: a prospective randomised study. *Arch Dis Child* **91**: 226–232.

Nielsen, J., Kwon, T. H., Frokiaer, J., Knepper, M. A., and Nielsen, S. (2006). Lithium-induced NDI in rats is associated with loss of alpha-ENaC regulation by aldosterone in CCD. *Am J Physiol Renal Physiol* **290**: F1222–F1233.

Norenberg, M. D., Leslie, K. O., and Robertson, A. S. (1982). Association between rise in serum sodium and central pontine myelinolysis. *Ann Neurol* **11**: 128–135.

Nozue, T., Uemasu, F., Endoh, H., Sako, A., Takagi, Y., and Kobayashi, A. (1993). Intracranial calcifications associated with nephrogenic diabetes insipidus. *Pediatr Nephrol* **7**: 74–76.

Nzerue, C., Baffoe-Bonnie, H., and Dail, C. (2002). Predicters of mortality with severe hyponatremia. *J Am Soc Nephrol* **13**: A0728.

Oksche, A., and Rosenthal, W. (1998). The molecular basis of nephrogenic diabetes insipidus. *J Mol Med* **76**: 326–337.

Omigbodun, A. O., Fajimi, J. L., and Adeleye, J. A. (1991). Effects of using either saline or glucose as a vehicle for infusion in labour. *East Afr Med J* **68**: 88–92.

Osorio, F. V., and Teitelbaum, I. (1997). Mechanisms of defective hydroosmotic response in chronic renal failure. *J Nephrol* **10**: 232–237.

Padfield, P. L., Brown, J. J., Lever, A. F., Morton, J. J., and Robertson, J. I. (1981). Blood pressure in acute and chronic vasopressin excess: studies of malignant hypertension and the syndrome of inappropriate antidiuretic hormone secretion. *N Engl J Med* **304**: 1067–1070.

Papadopoulos, M. C., and Verkman, A. S. (2007). Aquaporin-4 and brain edema. *Pediatr Nephrol* **22**: 778–784.

Pattaragarn, A., and Alon, U. S. (2003). Treatment of congenital nephrogenic diabetes insipidus by hydrochlorothiazide and cyclooxygenase-2 inhibitor. *Pediatr Nephrol* **18**: 1073–1076.

Perks, W. H., Walters, E. H., Tams, I. P., and Prowse, K. (1979). Demeclocycline in the treatment of the syndrome of inappropriate secretion of antidiuretic hormone. *Thorax* **34**: 324–327.

Pivonello, R., De Bellis, A., Faggiano, A. et al. (2003). Central diabetes insipidus and autoimmunity: relationship between the occurrence of antibodies to arginine vasopressin-secreting cells and clinical, immunological, and radiological features in a large cohort of patients with central diabetes insipidus of known and unknown etiology. *J Clin Endocrinol Metab* **88**: 1629–1636.

Pizarro, D., Posada, G., Villavicencio, N., Mohs, E., and Levine, M. M. (1983). Oral rehydration in hypernatremic and hyponatremic diarrheal dehydration. *Am J Dis Child* **137**: 730–734.

Pizarro, D., Posada, G., and Levine, M. M. (1984). Hypernatremic diarrheal dehydration treated with 'slow' (12-hour) oral rehydration therapy: a preliminary report. *J Pediatr* **104**: 316–319.

Prosch, H., Grois, N., Prayer, D. et al. (2004). Central diabetes insipidus as presenting symptom of Langerhans cell histiocytosis. *Pediatr Blood Cancer* **43**: 594–599.

Rabinstein, A. A., and Wijdicks, E. F. (2003). Hyponatremia in critically ill neurological patients. *Neurology* **9**: 290–300.

Reungjui, S., Roncal, C. A., Sato, W. et al. (2008). Hypokalemic nephropathy is associated with impaired angiogenesis. *J Am Soc Nephrol* **19**: 125–134.

Riedl, S., Vosahlo, J., Battelino, T. et al. (2008). Refining clinical phenotypes in septo-optic dysplasia based on MRI findings. *Eur J Pediatr.*

Robben, J. H., Sze, M., Knoers, N. V., Eggert, P., Deen, P., and Muller, D. (2007). Relief of nocturnal enuresis by desmopressin is kidney and vasopressin type 2 receptor independent. *J Am Soc Nephrol* **18**: 1534–1539.

Robertson, G. L. (2001). Antidiuretic hormone. Normal and disordered function. *Endocrinol Metab Clin North Am* **30**: 671–694 vii.

Robertson, G. L. (2006). Regulation of arginine vasopressin in the syndrome of inappropriate antidiuresis. *Am J Med* **119**: S36–42.

Robertson, G. L., Aycinena, P., and Zerbe, R. L. (1982). Neurogenic disorders of osmoregulation. *Am J Med* **72**: 339–353.

Rosenfeld, W., deRomana, G. L., Kleinman, R., and Finberg, L. (1977). Improving the clinical management of hypernatremic dehydration. Observations from a study of 67 infants with this disorder. *Clin Pediatr (Phila.)* **16**: 411–417.

Rubini, M. E. (1961). Water excrtion in potassium-deficient man. *J Clin Invest* **40**: 2215–2224.

Salido, M., Macarron, P., Hernandez-Garcia, C., D'Cruz, D. P., Khamashta, M. A., and Hughes, G. R. (2003). Water intoxication induced by low-dose cyclophosphamide in two patients with systemic lupus erythematosus. *Lupus* **12**: 636–639.

Samarasinghe, S., and Vokes, T. (2006). Diabetes insipidus. *Expert Rev Anticancer Ther* **6**(Suppl 9): S63–S74.

Sands, J. M., and Bichet, D. G. (2006). Nephrogenic diabetes insipidus. *Ann Intern Med* **144**: 186–194.

Sasaki, S. (2004). Nephrogenic diabetes insipidus: update of genetic and clinical aspects. *Nephrol Dial Transplant* **19**: 1351–1353.

Scheingraber, S., Rehm, M., Sehmisch, C., and Finsterer, U. (1999). Rapid saline infusion produces hyperchloremic acidosis in patients undergoing gynecologic surgery. *Anesthesiology* **90**: 1265–1270.

Scherbaum, W. A., Wass, J. A., Besser, G. M., Bottazzo, G. F., and Doniach, D. (1986). Autoimmune cranial diabetes insipidus: its association with other endocrine diseases and with histiocytosis X. *Clin Endocrinol (Oxf.)* **25**: 411–420.

Schrier, R. W., Gross, P., Gheorghiade, M. et al. (2006). Tolvaptan, a selective oral vasopressin V2-receptor antagonist, for hyponatremia. *N Engl J Med* **355**: 2099–2112.

Schwartz, R. H., and Jones, R. W. (1978). Transplacental hyponatraemia due to oxytocin. *Br Med J* **1**: 152–153.

Schwartz, W. B., Bennet, W., Curelop, S., and Bartter, F. C. (1957). A syndrome of renal sodium loss and hyponatremia probably resulting from inappropriate secretion of antidiuretic hormone. *Am J Med* **23**: 529–542.

Sgouros, S., Goldin, J. H., Hockley, A. D., Wake, M. J., and Natarajan, K. (1999). Intracranial volume change in childhood. *J Neurosurg* **91**: 610–616.

Siggaard, C., Christensen, J. H., Corydon, T. J. et al. (2005). Expression of three different mutations in the arginine vasopressin gene suggests genotype-phenotype correlation in familial neurohypophyseal diabetes insipidus kindreds. *Clin Endocrinol (Oxf.)* **63**: 207–216.

Sivakumar, V., Rajshekhar, V., and Chandy, M. J. (1994). Management of neurosurgical patients with hyponatremia and natriuresis. *Neurosurgery* **34**: 269–274 discussion 74.

Smith, L. G., Kirshon, B., and Cotton, D. B. (1990). Indomethacin treatment for polyhydramnios and subsequent infantile nephrogenic diabetes insipidus. *Am J Obstet Gynecol* **163**: 98–99.

Soupart, A., Penninckx, R., Stenuit, A., and Decaux, G. (2000). Azotemia (48 h) decreases the risk of brain damage in rats after correction of chronic hyponatremia. *Brain Res* **852**: 167–172.

Soupart, A., Gross, P., Legros, J. J. et al. (2006). Successful long-term treatment of hyponatremia in syndrome of inappropriate antidiuretic hormone secretion with satavaptan (SR121463B), an orally active nonpeptide vasopressin V2-receptor antagonist. *Clin J Am Soc Nephrol* **1**: 1154–1160.

Southgate, H. J., Burke, B. J., and Walters, G. (1992). Body space measurements in the hyponatraemia of carcinoma of the bronchus: evidence for the chronic 'sick cell' syndrome? *Ann Clin Biochem* **29**: 90–95.

Statius van Eps, L. W., Pinedo-Veels, C., de Vries, G. H., and de Koning, J. (1970). Nature of concentrating defect in sickle-cell nephropathy. Microradioangiographic studies. *Lancet* **1**: 450–452.

Steele, A., Gowrishankar, M., Abrahamson, S., Mazer, C. D., Feldman, R. D., and Halperin, M. L. (1997). Postoperative hyponatremia despite near-isotonic saline infusion: a phenomenon of desalination. *Ann Intern Med* **126**: 20–25.

Sterns, R. H. (1994). Treating hyponatremia: why haste makes waste. *South Med J* **87**: 1283–1287.

Sterns, R. H., Riggs, J. E., and Schochet, S. S. (1986). Osmotic demyelination syndrome following correction of hyponatremia. *N Engl J Med* **314**: 1535–1542.

Stuart, M. J., Cuaso, C., Miller, M., and Oski, F. A. (1975). Syndrome of recurrent increased secretion of antidiuretic hormone following multiple doses of vincristine. *Blood* **45**: 315–320.

Takeda, S., and Matsuzawa, T. (1985). Age-related brain atrophy: a study with computed tomography. *J Gerontol* **40**: 159–163.

Takil, A., Eti, Z., Irmak, P., and Yilmaz Gogus, F. (2002). Early postoperative respiratory acidosis after large intravascular volume infusion of lactated ringer's solution during major spine surgery. *Anesth Analg* **95**: 294–298.

Tannen, R. L., Regal, E. M., Dunn, M. J., and Schrier, R. W. (1969). Vasopressin-resistant hyposthenuria in advanced chronic renal disease. *N Engl J Med* **280**: 1135–1141.

Tian, Y., Sandberg, K., Murase, T., Baker, E. A., Speth, R. C., and Verbalis, J. G. (2000). Vasopressin V2 receptor binding is down-regulated during renal escape from vasopressin-induced antidiuresis. *Endocrinology* **141**: 307–314.

Turchin, A., Seifter, J. L., and Seely, E. W. (2003). Clinical problem-solving. Mind the gap. *N Engl J Med* **349**: 1465–1469.

Uribarri, J., and Kaskas, M. (1993). Hereditary nephrogenic diabetes insipidus and bilateral nonobstructive hydronephrosis. *Nephron* **65**: 346–349.

Vajda, Z., Pedersen, M., Doczi, T. et al. (2001). Effects of centrally administered arginine vasopressin and atrial natriuretic peptide on the development of brain edema in hyponatremic rats. *Neurosurgery* **49**: 697–704 discussion 705.

Van Amelsvoort, T., Bakshi, R., Devaux, C. B., and Schwabe, S. (1994). Hyponatremia associated with carbamazepine and oxcarbazepine therapy: a review. *Epilepsia* **35**: 181–188.

van Lieburg, A. F., Knoers, N. V., and Monnens, L. A. (1999). Clinical presentation and follow-up of 30 patients with congenital nephrogenic diabetes insipidus. *J Am Soc Nephrol* **10**: 1958–1964.

Verbalis, J. G. (1984). An experimental model of syndrome of inappropriate antidiuretic hormone secretion in the rat. *Am J Physiol* **247**: E540–E553.

Verbalis, J. G. (2003). Disorders of body water homeostasis. *Best Pract Res Clin Endocrinol Metab* **17**: 471–503.

Vexler, Z. S., Ayus, J. C., Roberts, T. P., Fraser, C. L., Kucharczyk, J., and Arieff, A. I. (1994). Hypoxic and ischemic hypoxia exacerbate brain injury associated with metabolic encephalopathy in laboratory animals. *J Clin Invest* **93**: 256–264.

Wang, W., Li, C., Kwon, T. H., Knepper, M. A., Frokiaer, J., and Nielsen, S. (2002). AQP3, p-AQP2, and AQP2 expression is reduced in polyuric rats with hypercalcemia: prevention by cAMP-PDE inhibitors. *Am J Physiol Renal Physiol* **283**: F1313–F1325.

Waring, A., and Kajdi, L. (1945). A congenital defect of water metabolism. *Am J Dis Child* **69**: 323–324.

Waters, J. H., Gottlieb, A., Schoenwald, P., Popovich, M. J., Sprung, J., and Nelson, D. R. (2001). Normal saline versus lactated Ringer's solution for intraoperative fluid management in patients undergoing abdominal aortic aneurysm repair: an outcome study. *Anesth Analg* **93**: 817–822.

Weisberg, L. S. (1989). Pseudohyponatremia: a reappraisal. *Am J Med* **86**: 315–318.

Wijdicks, E. F., and Sharbrough, F. W. (1993). New-onset seizures in critically ill patients. *Neurology* **43**: 1042–1044.

Wijdicks, E. F., Vermeulen, M., ten Haaf, J. A., Hijdra, A., Bakker, W. H., and van Gijn, J. (1985). Volume depletion and natriuresis in patients with a ruptured intracranial aneurysm. *Ann Neurol* **18**: 211–216.

Wilkes, N. J., Woolf, R., Mutch, M. et al. (2001). The effects of balanced versus saline-based hetastarch and crystalloid solutions on acid-base and electrolyte status and gastric mucosal perfusion in elderly surgical patients. *Anesth Analg* **93**: 811–816.

Wilson, J., Grant, P. J., Davies, J. A., Boothby, M., Gaffney, P. J., and Prentice, C. R. (1988). The relationship between plasma vasopressin and changes in coagulation and fibrinolysis during hip surgery. *Thromb Res* **51**: 439–445.

Wright, D. G., Laureno, R., and Victor, M. (1979). Pontine and extrapontine myelinolysis. *Brain* **102**: 361–385.

Xenos, C., Sgouros, S., and Natarajan, K. (2002). Ventricular volume change in childhood. *J Neurosurg* **97**: 584–590.

Zerbe, R., Stropes, L., and Robertson, G. (1980). Vasopressin function in the syndrome of inappropriate antidiuresis. *Annu Rev Med* **31**: 315–327.

Further reading

Chung, H. M., Kluge, R., Schrier, R. W., and Anderson, R. J. (1987). Clinical assessment of extracellular fluid volume in hyponatremia. *Am J Med* **83**: 905–908.

Section V. The Atrial Natriuretic Peptides

CHAPTER **20**

ANP, BNP and CNP: Physiology and Pharmacology of the Cardiorenal Axis

CANDACE Y.W. LEE[1,2] AND JOHN C. BURNETT JR.[1,2]

[1]*Cardiorenal Research Laboratory, Division of Cardiovascular Diseases, Departments of Medicine and Physiology, Mayo Clinic and Mayo Clinic College of Medicine, Rochester, MN, USA*

[2]*Division of Clinical Pharmacology, Department of Molecular Pharmacology & Experimental Therapeutics, Mayo Clinic and Mayo Clinic College of Medicine, Rochester, MN, USA*

Contents

I.	Introduction	289
II.	The natriuretic peptides production, processing and release	289
III.	Natriuretic peptide particulate guanylyl cyclase receptors and physiological actions	290
IV.	Natriuretic peptide receptor and enzymatic pathways for clearance and metabolism	292
V.	Pathophysiologic implications in cardiorenal regulation	293
VI.	Pharmacology and therapeutics of native peptides	295
VII.	Novel delivery systems	299
VIII.	Novel chimeric and synthetic natriuretic peptides	299
IX.	Future directions	301
	References	301

I. INTRODUCTION

Over the past 25 year, there has been an explosion of knowledge in the biological, diagnostic and therapeutic aspects of the natriuretic peptides (NPs) which represent a humoral system of cardiac origin which links the heart and kidney in the control of cardiorenal homeostasis. Atrial natriuretic factor (ANF), also known as atrial natriuretic peptide or A-type natriuretic peptide (ANP) was first identified by de Bold et al. (1981) who observed marked natriuresis and diuresis upon injection of supernatants of atrial myocardial homogenates into anesthetized rats. Subsequently, other members of the NP family were identified (Figure 20.1). B-type natriuretic peptide (BNP) was discovered by Sudoh et al. (1988) in porcine brain and was later found to be produced also in cardiac myocytes and fibroblasts (Mukoyama et al., 1991; Tsuruda et al., 2002). In addition, C-type natriuretic peptide (CNP) was identified by Sudoh et al. (1990) in porcine brain but was also found in endothelial cells (Suga et al., 1992; Stingo et al., 1992), bone (Espiner et al., 2007), kidney (Nir et al., 1994) and other organs (Pagel-Langenickel et al., 2007). Urodilatin (URO) was identified in human urine (Schulz-Knappe et al., 1988). NPs are also found in snakes (Schweitz et al., 1992) and other non-mammalian vertebrates (Loretz and Pollina, 2000). In particular, *Dendroaspis* natriuretic peptide (DNP) was isolated from the venom of *Dendroaspis angusticeps* (the green mamba snake) (Schweitz et al., 1992) and its biological actions have been studied (Lisy et al., 1999, 2001; Chen et al., 2002; Best et al., 2002).

In this chapter, we will present our current understanding of the physiology and pharmacology of ANP, BNP and CNP. Selected aspects of DNP and URO, as well as novel designer NPs, will also be discussed. The unifying theme is that the NPs represent the central feature of a cardiorenal axis in which this family of humoral factors contribute to the overall regulation of optimal cardiorenal homeostasis and blood pressure regulation, which also provides therapeutic opportunities for cardiovascular and renal disease.

II. THE NATRIURETIC PEPTIDES PRODUCTION, PROCESSING AND RELEASE

The three mammalian NPs, ANP, BNP, CNP, are genetically distinct but share structural similarities (see Figure 20.1) (Lee and Burnett, 2007). Human ANP, BNP and CNP are synthesized as preprohormones, which are subsequently split into prohormones by proteolytic cleavage of an N-terminal signal peptide (Clerico et al., 2006). Human preproANP is a 151-amino acid (AA) peptide which is cleaved to the 126-AA proANP, whereas human preproBNP is a 134-AA peptide which is cleaved to the 108-AA proBNP (Potter et al., 2006). ProANP and

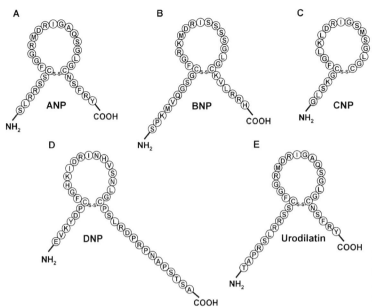

FIGURE 20.1 Structures and amino acid sequences of atrial natriuretic peptide (ANP), B-type natriuretic peptide (BNP), C-type natriuretic peptide (CNP), *Dendroaspis* natriuretic peptide (DNP) and urodilatin (URO).

proBNP are stored in secretory granules in atrial cardiomyocytes and are cleaved to form ANP and BNP, respectively, upon secretion (Clerico et al., 2006). The conversion of pro-ANP to ANP is mediated by corin, a transmembrane cardiac serine protease (Yan et al., 2000). Alternative processing of pro-ANP in the kidney generates URO, which shares the same AA sequence as ANP but also has an additional 4 AAs in the N-terminus (Potter et al., 2006; Lee and Burnett, 2007). Both human ANP, a 28-AA peptide, and human BNP, a 32-AA peptide, are released from the myocardium in response to various physiologic and pathophysiologic stimuli, such as myocardial wall stretch (Burnett et al., 2004).

Human proCNP consists of 103 AA residues and is processed by furin, an intracellular endoprotease, to the mature 53-AA CNP (Wu et al., 2003; Potter et al., 2006). CNP-53, which is found primarily in the brain, the heart and endothelial cells, may be further cleaved to CNP-22 (Potter et al., 2006). CNP-22 consists of a 17-AA ring structure, including a disulfide bond joining the two cysteine residues, and a single 5-AA extension in the N-terminal position, in contrast to the presence of two terminal extensions (both the N- and the C-termini) in other NPs (see Figure 20.1) (Lee and Burnett, 2007).

III. NATRIURETIC PEPTIDE PARTICULATE GUANYLYL CYCLASE RECEPTORS AND PHYSIOLOGICAL ACTIONS

A. NPR-A

Both ANP and BNP (as well as DNP and URO) are ligands for the natriuretic peptide receptor-A (NPR-A) (Figure 20.2), which is a member of the transmembrane guanylyl cyclase family and is widely distributed in the myocardium and other organ systems (Garbers et al., 2006). Based on competition binding experiments, Singh et al. (2006) reported the following rank order of potency for NPR-A: DNP > BNP ~ ANP ≫ CNP in a study that used a radioiodinated analog of DNP to evaluate the selectivity of DNP for NPR-A in the human myocardium. The ligand–receptor interactions result in activation of cyclic guanosine monophosphate (cGMP), which generates multiple important biological effects, including vasorelaxation, natriuresis, inhibition of renin and aldosterone, lusitropism, cytoprotection, anti-fibrosis, anti-hypertrophy, anti-apoptosis and anti-inflammation (Figure 20.2) (Furuya et al., 1991; Rautureau and Baxter, 2004; Chen et al., 2005; Lee and Burnett, 2007). In addition, the NPs have been demonstrated to be involved in the control of lipolysis in human adipose tissue (Sengenes et al., 2000).

In the nephron, the glomeruli and the inner medullary collecting duct (IMCD) cells are the areas where most NP receptors have been identified (Figure 20.3) (Candido et al., 2008). The main sites of action for ANP include the glomeruli and the IMCD, as well as the vascular system (Koike et al., 1993). Additional sites of action may also include the proximal convoluted tubules, medullary thick ascending limb and cortical collecting ducts (Koike et al., 1993). In cultures of human glomerular epithelial and mesangial cells, Ardaillou et al. (1986) demonstrated that ANP stimulated cGMP, the second messenger for the NPs. In rabbit IMCD cells, Zeidel et al. (1986) demonstrated that ANP inhibited sodium transport in medullary collecting duct and increased renal sodium excretion. In IMCD, receptor binding by ANP activated particulate guanylyl cyclase (pGC) which mediated cGMP

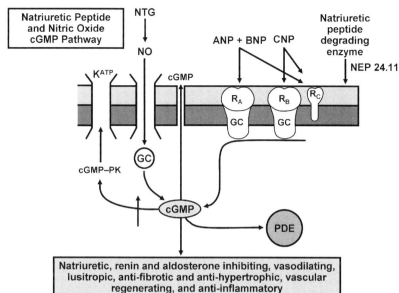

FIGURE 20.2 Signal transduction pathways of the particulate guanylyl cyclase and the soluble guanylyl cyclase systems. GC = guanylyl cyclase, NTG = nitroglycerin, NO = nitric oxide, cGMP = 3′,5′ cyclic guanosine monophosphate, PDE = phosphodiesterase type V.

production, resulted in natreuresis (Gunning et al., 1989). Moreover, in culture mesangial cells, the anti-proliferative effects of ANP have also been demonstrated (Haneda et al., 1993).

Previous studies have demonstrated that ANP increased natriuresis, diuresis and may also increase glomerular filtration rate (GFR) and filtration fraction (FF) (Richards et al., 1985; Hirsch et al., 2006; Candido et al., 2008). Potential mechanisms include ANP-induced relaxation of mesangial cells and an increase in ultrafiltration coefficient (Hirsch et al., 2006), augmentation of glomerular hydrostatic pressure (from afferent arteriolar dilatation and efferent arteriolar constriction) and direct tubular effects (Candido et al., 2008). In the proximal tubule, ANP has been demonstrated to inhibit angiotensin-mediated sodium and fluid reabsorption (Harris et al., 1987), as well as the Na^+/H^+ antiporter (Winaver et al., 1990). In addition, ANP exerts effects on other proteins involved in the transport of Na^+ and other ions (Hirsch et al., 2006). In the loop of Henle, ANP decreases reabsorption of chloride and inhibits Na^+/K^+-ATPase (Hirsch et al., 2006). In the IMCD, ANP exerts a dose-dependent inhibitory effect on oxygen consumption (Zeidel et al., 1986) and blocks the entry of Na^+ (Zeidel et al., 1988). ANP antagonizes the actions of vasopressin and of aldosterone (Candido et al., 2008). Moreover, Na^+ delivery to the

The Single Nephron and Inner Medullary Collecting Duct
Site of Sodium and Water Regulating Hormones

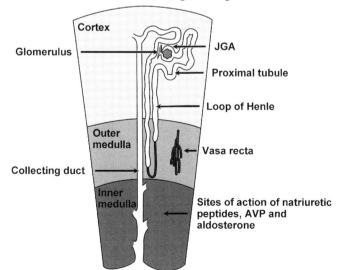

FIGURE 20.3 Sites of renal actions by the natriuretic peptides. JGA = juxtaglomerular apparatus, AVP = arginine vasopressin.

macula densa is enhanced by ANP, leading to indirect inhibition of renin secretion and subsequent angiotensin II and aldosterone secretion (Candido et al., 2008). ANP also directly suppresses aldosterone production by the adrenal gland which is underscored by the high density of NPR-A in zona glomerulosa cells (Kudo and Baird, 1984). BNP has been reported to also suppress aldosterone even when stimulated by potent diuretics such as furosemide (Cataliotti et al., 2004).

Urodilatin is synthesized in the renal distal tubules where it is secreted into the lumen (Hirsch et al., 2006). It binds to NPR-A on IMCD cells and plays an important role as a paracrine regulator of Na^+ excretion in the kidney (Goetz et al., 1990; Forssmann et al., 2001; Hirsch et al., 2006). Its cGMP-stimulating actions have also been detected in IMCD cells, glomeruli, proximal convoluted tubules and medullary thick ascending limbs, and its cGMP response is similar to that of ANP (Koike et al., 1993). It induces natriuresis and diuresis in healthy subjects (Carstens et al., 1998; Dorner et al., 1998) and mediates favorable hemodynamic effects in patients with decompensated heart failure (HF), such as significant reductions in pulmonary capillary wedge pressure (PCWP) (Mitrovic et al., 2005, 2006). In experimental studies, attenuation of myocardial ischemia-reperfusion injury (IRI) (Padilla et al., 2001) and possible anti-cancer effects have also been reported (Vesely et al., 2006).

B. NPR-B

Unlike the NPR-A agonists (ANP, BNP, DNP and URO), CNP binds preferentially to natriuretic peptide receptor-B (NPR-B), which shares topology with NPR-A and binds NPs in the following rank order: CNP ≫ ANP ≥ BNP (Potter et al., 2006). NPR-B has been reported in epithelial and mesangial cells of human glomeruli (Zhao et al., 1994), in addition to its wide distribution in the brain, chondrocytes, vascular smooth muscle cells, fibroblasts, myocardium, lung, kidney, adrenal, uterus and ovary (Canaan-Kuhl et al., 1992; Potter et al., 2006; Pagel-Langenickel et al., 2007). Importantly, the 17-AA core ring structure and the disulfide bond have been shown to be critical for NPR-B selectivity and for its cGMP-stimulating actions in vascular smooth muscle cells (Furuya et al., 1992).

C-type natriuretic peptide, an endothelial (Stingo et al., 1992) and renal (Nir et al., 1994) cell derived peptide consisting of 22 AA (Ahluwalia and Hobbs, 2005), is the most conserved NP across species and is thought to mediate its actions through a paracrine or autocrine mechanism (Potter et al., 2006; Del Ry et al., 2006; Pagel-Langenickel et al., 2007). Both renal production of CNP and NPR-B activation by CNP in the kidney have been demonstrated (Mattingly et al., 1994; Nir et al., 1994; Totsune et al., 1994; Terada et al., 1994; Dean et al., 1994, 1996; Millul et al., 1997; Cataliotti et al., 2002). Moreover, CNP exerts diverse biological actions and is involved in various regulatory processes, such as bone growth and cartilage homeostasis (Espiner et al., 2007; Pejchalova et al., 2008), control of vascular tone (Kelsall et al., 2006) and regulation of blood–testis–barrier dynamics in the rat (Xia et al., 2007). It exerts anti-proliferative and anti-inflammatory actions and has been shown to reduce cardiac preload (based upon the greater abundance of NPR-B in veins as compared to arteries), inhibit vascular smooth muscle proliferation, reduce leukocyte recruitment and leukocyte–platelet interactions, protect against cardiac IRI, prevent ventricular remodeling following myocardial infarction and attenuate cardiac dysfunction and inflammation in experimental acute myocarditis (Wei et al., 1993; Hobbs et al., 2004; Tokudome et al., 2004; Soeki et al., 2005; Scotland et al., 2005; Obata et al., 2007; Pagel-Langenickel et al., 2007; Wang et al., 2007). It exhibits greater anti-fibrotic properties than ANP or BNP (Horio et al., 2003; Garbers et al., 2006; Pagel-Langenickel et al., 2007). However, it exerts only limited natriuretic and diuretic actions (Stingo et al., 1992; Hunt et al., 1994; Pagel-Langenickel et al., 2007).

IV. NATRIURETIC PEPTIDE RECEPTOR AND ENZYMATIC PATHWAYS FOR CLEARANCE AND METABOLISM

A. NPR-C

The extracellular domain of NPR-C shares about 30% homology with NPR-A and NPR-B but it lacks guanylyl cyclase activity (van den Akker, 2001; Potter et al., 2006). It is the most abundant receptor subtype, with expression in most tissues, especially in the kidney, the vascular endothelium, smooth muscle cells and the heart (Potter et al., 2006; Candido et al., 2008). The rank order for NPR-C binding affinity for NPs (human and rats) is ANP ≥ CNP > BNP (Potter et al., 2006). Even though it was initially believed that NPR-C functions as a clearance receptor, accumulating evidence suggests that NPR-C may be involved in the regulation of cell function (Anand-Srivastava, 2005) and may play a role in mediating the anti-fibrotic effect of BNP in cardiac fibroblasts (Huntley et al., 2006).

B. Neutral Endopeptidase

Neutral endopeptidase (NEP) 24.11 is involved in the degradation of NPs and is distributed in the kidney, lung and vascular wall (Chen and Burnett, 2006). In the kidney, NEP is highly expressed in the brush borders of the proximal tubules, resulting in rapid degradation of ANP (Candido et al., 2008). NEP is often found to co-localize with angiotensin-converting enzyme (Burnett et al., 2004). The rank order of potency for hydrolysis by NEP *in vitro* has been reported to be CNP > ANP > BNP (Kenny et al., 1993; Schulz, 2005).

V. PATHOPHYSIOLOGIC IMPLICATIONS IN CARDIORENAL REGULATION

A. Heart Failure

Activation of NPs in heart failure (HF) serves as a compensatory mechanism to maintain sodium homeostasis and to suppress the renin–angiotensin–aldosterone system (RAAS) in HF (Burnett et al., 2004). In 1986, Burnett et al. identified that plasma ANP was elevated in human HF. Subsequently, elevation of BNP in human HF was reported by Mukoyama et al. (1991). In order to preserve and/or enhance renal function in HF, an understanding of intrarenal factors that regulate renal function may provide a direction on optimal use of current therapies and also lead to newer therapeutic strategies, especially as this relates to the NPs.

From an integrated cardiorenal physiological view, HF involves cardiac overload with release of ANP and BNP which activate the particulate guanylyl cyclase (pGC) NPR-A resulting in the generation of the second messenger cyclic cGMP and the effector protein kinase G (PKG) (Garbers et al., 2006). NPR-A is also the target of the intrarenal natriuretic peptide URO, as discussed above, that, like ANP and BNP, is being developed for the treatment of HF. Infusion of these three NP in animals and humans with HF results in natriuresis and diuresis and, at certain doses, an increase in GFR (Jensen et al., 1998; Chen et al., 2005). They possess other actions which include suppression of the RAAS, inhibition of fibrosis and cardiomyocyte hypertrophy and positive lusitropism. In severe experimental or human HF, a renal hyporesponsiveness may occur to the NPs due, in part, to excessive hypotension as well as upregulation of phosphodiesterase (PDE) V activity which degrades NP generated cGMP (Redfield et al., 1989; Margulies et al., 1991; Supaporn et al., 1996). The importance of the NPs and the NPR-A in renal regulation is underscored by studies of genetic and pharmacologic receptor disruption characterized by impaired renal sodium handling and, often, hypertension (John et al., 1996; Borgeson et al., 1998; Patel et al., 2005).

While cardiac volume overload, as in acute HF, may result in the release of ANP and BNP, the associated reduction of arterial pressure which may occur activates the intrarenal nitric oxide (NO) pathway in which NO stimulates soluble guanylyl cyclase (sGC) localized to the cell cytosol. Soluble guanylyl cyclase is a cGMP activator distinct from the ANP and BNP cGMP pathway, which involves activation of pGC.

From a clinical perspective in acute HF, both sodium nitroprusside (SNP) and nitroglycerin (NTG) are widely used for the treatment of acute HF and both are sGC activators and potent vasodilators. Again, this is in contrast to ANP and BNP that too are used for acute HF therapy but target pGC. To explore further cGMP mechanisms in HF, we recently employed a novel direct activator of sGC (BAY 58-2667) in a model of HF and observed potent renal vasodilation without natriuresis and diuresis or changes in GFR, although both produced a significant reduction in arterial pressure together with cardiac unloading (Boerrigter et al., 2007a). This is in contrast to pGC activation in the kidney by ANP or BNP which, as discussed above, may be hypotensive but may augment GFR, natriuresis and diuresis.

What is emerging in cGMP regulation of cardiorenal function is the concept of compartmentalization. Specifically, compartmentalization of cGMP signaling in cells has been advanced especially in the heart in which pGC and sGC have been demonstrated to have distinct roles in cardiomyocyte function (Fischmeister et al., 2006; Castro et al., 2006). Further, Airhart and co-workers have reported that the pGC agonist ANP, but not the sGC agonist S-nitroso-N-acetyl penicillamine, stimulates the translocation of PKG to the plasma membrane of renal cells augmenting the NPR-A receptor to which ANP, BNP and URO bind (Airhart et al., 2003). These observations strongly support *in vitro* distinct functional roles for pGC and sGC in the kidney, meaning a natriuretic peptide like ANP or BNP could have different renal actions than NTG or SNP despite both activating the second messenger cGMP.

Understanding the intrarenal roles of the NP/cGMP and NO/cGMP pathways *in vivo* in HF in the control of renal function would advance our knowledge of renal adaptations in this syndrome and help guide our therapeutic strategies. Therefore, we investigated *in vivo*, in a large animal model of HF, the physiological properties of the endogenous NP/cGMP and NO/cGMP pathways in the control of renal hemodynamic and excretory function (Martin et al., 2007). We hypothesized that each pathway would play specific roles in the maintenance of renal function in heart failure consistent with distinct GC enzymes for each system.

These studies revealed differential and complementary roles for these two endogenous cGMP activating systems in HF whereby the endogenous NPs appear to play a greater role in the preservation of GFR and sodium excretion while the endogenous NO system was more important in the control of renal blood flow. Thus, the preservation of renal function in experimental acute decompensated HF is mediated by dual cGMP systems which activate both pGC and sGC enzymes.

A paradox nonetheless does exist in HF which is that there is elevation of ANP and BNP yet the kidney is vasoconstricted and sodium retaining. The biological significance of this elevation in ANP and BNP has begun to unfold in recent years. Hawkridge et al. (2005) reported that, despite an elevation in BNP(1–32), as measured by point-of-care testing, in subjects with HF and New York Heart Association Class IV symptoms, an absence of BNP 1–32 was demonstrated by quantitative mass spectral analysis, suggesting that other molecular forms of BNP might have contributed to the results on point-of-care testing. Indeed, recent investigations have demonstrated reduced biological activities in other molecular forms of BNP, which contribute to the BNP immunoreactivity (Liang et al., 2007; Heublein et al., 2007; Boerrigter et al., 2007b).

From a therapeutic point of view, NP resistance is an important consideration in HF. Potential mechanisms for attenuated response to NP in patients with HF may include decreased receptor protein concentration (Bryan et al., 2007), increased renal neutral endopeptidase activity (Lindenfeld and Schrier, 2007), increased activity of PDEV (Forfia et al., 2007) and reduced delivery of sodium to distal tubules (Volpe et al., 1991). Moreover, HF has been shown to be a state of relative deficiency of NP (Redfield et al., 1989; Chen, 2007). It is likely that both an increased resistance to NP and a relative deficiency of the biologically active form of BNP are present in HF (Hawkridge et al., 2005; Chen, 2007).

Most recently, a key role for enhanced PDEV activity in the kidney has been highlighted in several studies (Margulies et al., 1991; Supaporn et al., 1996; Forfia et al., 2007). Chen et al. (2006) performed a study to elucidate the role of intrarenal PDEV in the renal hyporesponsiveness to BNP in HF. Here, the actions of PDEV inhibition (PDEVI) in experimental CHF were defined. They also assessed the response to acute subcutaneous BNP in the presence and absence of chronic PDEVI. It should be repeated that PDEV metabolizes cGMP and is abundant in the kidney and vasculature and has also been shown to be present in the heart. In renal disease states such as nephrotic syndrome, it has been demonstrated that PDEV contributes to renal impairment and reduced sensitivity to the natriuretic peptide system (Lee and Humphreys, 1996). In this recent study, there were two groups, one consisting of animals receiving PDEVI (Sildenafil for the 10 days of CHF) and the other one consisting of animals that received no PDEVI. Despite having higher cardiac output, there was no improvement of renal function in the PDEVI group after 10 days as compared with the control group. However, the PDEVI group had significantly higher plasma and urinary cGMP than the control group. When acute subcutaneous BNP was administered at day 11, the PDEVI group had a natriuretic and diuretic response associated with an increase in GFR that was not observed in the control group. Plasma BNP increased to a similar extent in both groups with subcutaneous BNP. In contrast, the PDEVI group had a greater urinary cGMP excretion than the control group. Even though chronic administration of PDEVI therapy did not enhance renal function, despite an improvement in cardiac output, PDEVI significantly enhanced the renal hemodynamic and excretory responses to exogenous BNP. This supports the idea that there is a role for PDEV as a contributing factor to renal maladaptation in the setting of experimental overt HF. Also, this study warrants new clinical trials that can address the possible strategy of maximizing the renal cGMP system by combined PDEVI and NP therapy in HF to improve renal function.

B. Myocardial Ischemia and Infarction

B-type natriuretic peptide and N-terminal pro-BNP have emerged as an important diagnostic and prognostic marker in myocardial ischemia and infarction, following its initial application in the setting of HF (Adams et al., 2006). Specifically, NT-proBNP and BNP have been shown to be powerful predictors of prognosis in patients who present with stable or unstable coronary artery disease (CAD), independent of their conventional risk factors for mortality (Adams et al., 2006; Bibbins-Domingo et al., 2007). During myocardial ischemia, circulating NPs increase rapidly from the release of stored form of NPs from the heart or from induction of synthesis.

From a therapeutic standpoint, previous experimental and clinical studies support the notion that endogenous natriuretic peptides (NPs), ANP, BNP, CNP and URO confer protection against myocardial IRI (Padilla et al., 2001; D'Souza et al., 2003; Baxter, 2004; Hobbs et al., 2004; Yang et al., 2006; Burley et al., 2007a). Potential mechanisms of protection include opening of mitochondrial K_{ATP} channels (D'Souza et al., 2003), stimulation of nitric oxide-soluble guanylyl cyclase pathway (D'Souza et al., 2003), activation of the reperfusion injury salvage kinase (RISK) pathway (Yang et al., 2006; Yellon and Hausenloy, 2007) modulation of Ca^{2+} homeostasis either by cGMP-dependent protein kinase (PKG) via modification of sarcoplasmic reticulum Ca^{2+} uptake or by PKG-independent mechanisms (Burley et al., 2007a), inhibition of mitochondrial permeability transition pore in early reperfusion (Burley et al., 2007a) and anti-apoptosis (Kato et al., 2005).

Importantly, it has been observed that the NPs conferred cardioprotection even when administered during early reperfusion (Burley et al., 2007b). This is of particular clinical relevance as most patients with ischemic symptoms present after the onset of ischemia or at the time of reperfusion.

A previous study by Maeda et al. (2000) reported that, during acute myocardial infarction (AMI), there was a relative insufficiency of ANP secretion. Subsequent investigations were in support of a role of ANP infusion in the setting of AMI. The Japan-Working Groups of Acute Myocardial Infarction for the Reduction of Necrotic Damage by ANP (J-WIND-ANP) Trial (Kitakaze et al., 2007) was a prospective, randomized, multicenter, single-blind, placebo-controlled study which assessed the effects of ANP on myocardial infarct size and cardiovascular outcomes in patients undergoing reperfusion therapy for AMI. Patients were randomized to ANP 0.025 μg/kg/min intravenously (i.v.) or placebo for 3 days and were followed for a mean duration of 2.5 years (Kitakaze et al., 2007). The primary endpoints were infarct size (as estimated by the area under the concentration versus time curve for creatine kinase) and left ventricular ejection fraction (LVEF) (as assessed by angiography of the LV) (Kitakaze et al., 2007). A total of 603 subjects were randomized to ANP or placebo. Infarct size was significantly reduced by ANP (by 14.7%), as compared to placebo, while EF remained unchanged (Kitakaze et al., 2007).

Recently, Sezai et al. (2007) conducted a randomized study evaluating the effects of continuous low-dose human ANP (hANP) versus placebo in 124 patients who underwent emergent coronary artery bypass grafting (CABG) for the acute coronary syndromes. Patients in the hANP group were observed to have significantly reduced peak creatine kinase-MB and peak creatinine levels and a lower incidence of arrhythmias in the postoperative period (Sezai et al., 2007). At 1 month following CABG, patients in the hANP group were noted to have significantly higher LVEF and lower LV end-diastolic pressure and LV end-diastolic volume index (Sezai et al., 2007). Moreover, hANP-treated patients had significantly lower postoperative plasma BNP levels for up to 1 year and a higher event-free rate for than those of the placebo group for up to 1 year following CABG (Sezai et al., 2007).

In another recent study, patients who presented with first AMI undergoing primary percutaneous coronary intervention were enrolled (Kasama et al., 2007). Exogenous ANP was shown to suppress sympathetic nerve activity and attenuated LV remodeling, as compared to isosorbide dinitrate (Kasama et al., 2007).

Recently, Singh et al. (2006) demonstrated significant downregulation of NPR-A receptor density, without change in binding affinity, in the LV and coronary arteries from patients with ischemic heart disease undergoing cardiac transplantation. The therapeutic implications of these findings warrant further investigation.

VI. PHARMACOLOGY AND THERAPEUTICS OF NATIVE PEPTIDES

A. Atrial Natriuretic Peptide

Atrial natriuretic peptide has a short half-life and a high total body clearance (Tan et al., 1993). Nakao et al. (1986) studied the pharmacokinetics of synthetic alpha-human ANP (α-hANP) 100 μg as an i.v. bolus in six healthy male subjects. The disappearance of α-hANP was fitted to a bi-exponential decay curve and the fast and slow half-times were reported to be 1.7 and 13.3 min, respectively (Nakao et al., 1986). The steady-state volume of distribution was 11.9 l and the mean plasma clearance was 1.52 l min^{-1} kg^{-1}. Eiskjaer and Pedersen (1993) studied the dose–response relationship of ANP as a bolus injection in healthy male subjects and observed dose-dependent increases in plasma and urinary cGMP. The increase in cGMP correlated with augmentation in urinary Na$^+$ excretion. Weidmann et al. (1986) evaluated synthetic α-hANP as an i.v. infusion (initial bolus 50 μg followed by infusion at 6.25 μg/min for 45 min) in 10 healthy normal subjects and observed an acute decrease in diastolic blood pressure, an increase in GFR (in the setting of unchanged or decreased total renal blood flow), natriuresis, diuresis, hemoconcentration and increased plasma norepinephrine. Tonolo et al. (1988) studied synthetic α-hANP 10 pmol/kg/min i.v. or placebo as a one-hour infusion in random order on two separate occasions immediately before hemodialysis in eight patients with end-stage renal disease (ESRD) and six normal volunteers. No significant changes in blood pressure (the normal subjects were on high-salt diet and were in semi-recumbent position during the study), heart rate, plasma renin concentration, electrolytes and serum creatinine were observed (Tonolo et al., 1988). The metabolic clearance rate of α-hANP in patients with ESRD and in normal subjects was 1.04 l/min and 2.6 l/min, respectively; the plasma half-life of α-hANP was 4 min 34 s and 3 min 30 s, respectively and the volume of distribution was 6.84 l and 11.1 l, respectively (Tonolo et al., 1988). A significant increase in microhematocrit was observed during α-hANP infusion and not during placebo infusion, in patients with ESRD, suggesting that plasma volume contraction induced by ANP was not dependent on renal function (Tonolo et al., 1988).

Kimura et al. (2007) studied the pharmacokinetics of synthetic α-human ANP (0.05 μg/kg/min) and human BNP (0.01 μg/kg/min) in 16 subjects with HF using a one-compartment model. The plasma half-life of ANP ($n = 8$) was reported to be 2.4 min and that of BNP ($n = 8$) was 12.1 min, which was significantly longer than the former (Kimura et al., 2007). The total body clearance of ANP was 48.2 ml/min/kg and that of BNP was 10.1 ml/min/kg, which was also significantly different from the former (Kimura et al., 2007). The volume of distribution was 153 and 181 ml/kg for ANP and BNP, respectively (Kimura et al., 2007).

In subjects with HF, Cody et al. (1986) reported that i.v. infusion of ANP induced natriuresis and diuresis, inhibited renin and aldosterone, decreased systemic blood pressure and pulmonary capillary wedge pressure (PCWP). Giles et al. (1991) studied the hemodynamic responses of 12 HF patients to human ANP. At 30 minutes following i.v. bolus injection (4.5 μg/kg/min), PCWP, right atrial pressure (RAP) and heart rate (HR) decreased significantly (Giles et al., 1991). Moreover, the efficacy and safety of carperitide were evaluated in a 6-year prospective open-label registry of 3777 patients with acute HF who were treated with a median dose of 0.085 μg/kg/min and a median duration of 65 hours (Suwa et al., 2005). Eighty-two percent of the patients improved clinically (Suwa et al., 2005). Notably, previous studies have also demonstrated bronchodilatory effects of ANP, as well as URO (Angus et al., 1993; Fluge et al., 1995).

ANP increases capillary hydrostatic permeability and mediates shifting of intravascular fluid to the extravascular compartment, thus contributing to a reduction in cardiac preload (Groban et al., 1990; Candido et al., 2008). Indeed, elegant studies in which NPR-A was genetically deleted from endothelial cells supported a key role for ANP mediated increases in capillary permeability in blood pressure regulation. Specifically, the phenotype of this novel murine model of tissue-specific NPR-A deletion was one of hypervolemia, hypertension and cardiac hypertrophy. These studies therefore underscored the important vascular action of ANP upon capillary permeability in blood pressure homeostasis (Sabrane et al., 2005).

ANP also increases venous capacitance, decreases peripheral vascular resistance, reduces plasma aldosterone and attenuates plasma aldosterone response to angiotensin II infusion (Hunt et al., 1996; Candido et al., 2008). Indeed, the therapeutic potential of ANP was evaluated in subjects with hypertension (Richards, 1989; Richards et al., 1985; Richards et al., 1989; Tonolo et al., 1989). It has been demonstrated that intravenous administration of ANP reduced systemic blood pressure, increased urinary sodium excretion and urine volume, decreased plasma norepinephrine and aldosterone (Richards et al., 1985) and suppressed plasma renin and angiotensin II in some investigations (Richards, 1989).

B. B-Type Natriuretic Peptide

Recombinant BNP (known as nesiritide in the USA) exhibits similar physiologic actions as native BNP (Keating and Goa, 2003; Sackner-Bernstein et al., 2006). The distribution half-life and the mean terminal elimination half-life of nesiritide are about 2 min and 18 min, respectively (Keating and Goa, 2003). The time to steady state is <90 min and the mean volume of distribution at steady state is 0.19 l/kg (Keating and Goa, 2003). Recombinant BNP is cleared via three mechanisms: binding to the NPR-C on cell surface; degradation by neutral endopeptidase; and renal filtration, although dose adjustment is not needed in patients with renal insufficiency (Keating and Goa, 2003; Sackner-Bernstein et al., 2006).

Nesiritide decreases cardiac filling pressure, systemic and pulmonary vascular resistance and increases cardiac output in a dose-dependent manner (Adams et al., 2006). At 0.01 µg/kg, nesiritide significantly reduces LV filling pressure (Adams et al., 2006). It induces natriuresis, with variable effects on GFR and renal blood flow (Jensen et al., 1999; Adams et al., 2006).

Colucci et al. (2000) studied the efficacy of nesiritide in hospitalized patients with decompensated HF, who were enrolled into either a double-blind efficacy trial or an open-label comparative trial. In the efficacy trial, 127 patients who had PCWP ≥18 mmHg and a cardiac index ≤2.7 l/min/m^2 were randomized to placebo or nesiritide (0.015 or 0.03 µg/kg/min infusion) for 6 h (Colucci et al., 2000). Nesiritide (at both doses) decreased PCWP versus an increase in PCWP with placebo and was associated with significant improvement in dyspnea, fatigue and global clinical status (Colucci et al., 2000). In the comparative trial, 305 patients were randomized to nesiritide (0.015 or 0.03 µg/kg/min) or standard therapy (Colucci et al., 2000). Nesiritide-treated patients had improvements in dyspnea, fatigue and overall clinical status, which were sustained for up to 7 days and were similar to those who received standard therapy (Colucci et al., 2000). The Vasodilation in the Management of Acute CHF (VMAC) study (Publication Committee for the VMAC Investigators, 2002) was a randomized, double-blind trial to compare the efficacy and safety of i.v. nesiritide, i.v. nitroglycerin (NTG) and placebo in 489 patients with decompensated HF and dyspnea at rest. Nesiritide, NTG, or placebo was administered for 3 h, followed by nesiritide or NTG for 24 h (Publication Committee for the VMAC Investigators, 2002). It was observed that nesiritide decreased mean PCWP to a greater extent than either NTG or placebo at 3 h and significantly more than NTG at 24 h (Publication Committee for the VMAC Investigators, 2002). Significant improvement in dyspnea at 3 h in the nesiritide-treated group over placebo was detected, although no difference was demonstrated between the nesiritide-treated group and the NTG-treated group (Publication Committee for the VMAC Investigators, 2002).

In subjects with mild-to-moderate essential hypertension, exogenous administration of human BNP (2 pmol/kg/min as a constant infusion) resulted in plasma cGMP activation, natriuresis and suppression of plasma aldosterone, without significant effects on systemic blood pressure (Richards et al., 1993). The safety of BNP and its effects on renal function have been controversial, with concerns raised by some studies (Sackner-Bernstein et al., 2005a, b) but not in others (Arora et al., 2006; Yancy et al., 2007) and are being further investigated.

Exogenous human ANP (carperitide) and human BNP (nesiritide) are currently available as intravenous therapies for the treatment of acute decompensated HF in Japan and the USA, respectively, as well as in other countries (Gheorghiade et al., 2005; Chen and Burnett, 2006). Increasing data demonstrate potential beneficial effects of ANP in cardioprotection against IRI in AMI (Kitakaze et al., 2006), cardiovascular surgery (Sezai et al., 2006a, b, 2007) and post-infarct remodeling in humans (Kasama et al., 2007). Renal protective effects of ANP have been demonstrated in the settings of heart failure, renal failure and perioperative states (Sward et al., 2001, 2004; Sato et al., 2006; Mentzer et al., 2007; Chen et al., 2007), despite disappointing results of previous clinical trials on anaritide, a synthetic form of ANP (102–106), in acute tubular necrosis (Allgren et al., 1997; Lewis et al., 2000; Candido et al., 2008). Moreover, the anti-cancer effects of ANP are being increasingly recognized (Saba and Vesely, 2006).

Similarly, cardiovascular and renal benefits of BNP have been demonstrated in a number of clinical studies.

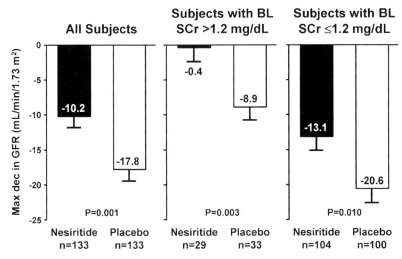

FIGURE 20.4 Maximal change in glomerular filtration rate by baseline renal function in the NAPA Trial. Max dec = maximum decrease, SCr = serum creatinine, BL = baseline. Modified from Mentzer, R.M. Jr et al. (2007). *J Am Coll Cardiol* 49, 716–26, with permission from the publisher, Elsevier and the corresponding author.

Accumulating evidence suggests potential benefits of exogenous BNP in the settings of perioperative renal protection (Zierer et al., 2006; Mentzer et al., 2007; Chen et al., 2007) and in AMI (Hillock et al., 2007).

The rationale for the therapeutic use of natriuretic peptides (NPs) during cardiac surgery includes reduction of neurohumoral activation, preservation of renal function, afterload reduction and optimization of end-organ perfusion (Moazami and Oz, 2005; Mentzer et al., 2007). Several retrospective (Beaver et al., 2006; Gordon et al., 2006) or small-scale prospective (Feldman et al., 2004; Salzberg et al., 2005) investigations in patients undergoing cardiac surgery reported improvement in renal function and/or cardiac hemodynamics with perioperative infusion of nesiritide. Experience with hANP has also been promising in patients who underwent CABG or aortic thoracic surgery (Sezai et al., 2000, 2006a, b, c, 2007). Salutary effects include increased diuresis, reduced peripheral vascular resistance, inhibitory effects on postoperative LV remodeling, suppression of the renin–angiotensin–aldosterone system and attenuation of ischemia-reperfusion injury (Sezai et al., 2000, 2006a, b, c, 2007).

Recently, the Nesiritide Administered Peri-Anesthesia in Patients Undergoing Cardiac Surgery (NAPA) Trial (Mentzer et al., 2007) was a prospective, randomized, double-blind, multicenter, placebo-controlled exploratory trial which was conducted in patients with known LV dysfunction with an EF of ≤40% and New York Heart Association functional class II to IV HF to evaluate the effects of nesiritide on postoperative renal function, hemodynamics, safety and clinical outcomes. Nesiritide was started as an i.v. infusion 0.01 μg/kg/min without a bolus after induction of anesthesia but prior to chest incision, and was continued for ≥24 h and up to a maximum duration of 96 h (Mentzer et al., 2007). Of the 303 patients who were randomized, 279 patients received the study drug and were followed for 6 months (Mentzer et al., 2007). In the nesiritide-treated group, as compared to placebo, significant augmentation in urine output and attenuations in both peak elevation in serum creatinine and decrease in GFR (during hospitalization or by study day 14, whichever came first) were observed (Mentzer et al., 2007). The nesiritide-treated group also had a significantly shorter duration of hospitalization and a lower mortality at 180 days (6.6% versus 14.7% in the placebo group, $P = 0.046$) (Mentzer et al., 2007). Importantly, in patients with pre-existing renal dysfunction (defined as baseline serum creatinine >1.2 mg/dl), nesiritide versus placebo was associated with greater attenuations in both peak increase in serum creatinine and fall in GFR (Figure 20.4), as compared to a similar analysis in patients without pre-existing renal dysfunction (Mentzer et al., 2007). Overall, adverse events occurred at similar rates in the nesiritide-treated group versus the placebo group (Mentzer et al., 2007). Despite the exploratory nature of the study, the NAPA trial has provided important data on the effects of fixed-dose nesiritide without a bolus on renal function in patients with LV dysfunction undergoing cardiac surgery.

More recently, Chen et al. (2007) conducted a double-blind, placebo-controlled pilot study in 40 patients with renal dysfunction (preoperative creatinine clearance <60 ml/min, as estimated by Cockcroft-Gault formula) undergoing CABG, who were randomized to nesiritide 0.005 μg/kg/min i.v. or placebo for 24 hours. Infusion of nesiritide or placebo was initiated after the induction of anesthesia but before cardiopulmonary bypass (Chen et al., 2007). It was observed that the nesiritide-treated group, but not the placebo group, had significant increases in plasma cGMP and BNP and a significant reduction in cystatin as compared to baseline (Figure 20.5) (Chen et al., 2007). Plasma aldosterone was significantly activated at the end of the 24-hour infusion in the placebo group but not in the nesiritide group (Chen et al., 2007). Low-dose nesiritide was well tolerated without inducing significant systemic hypotension (Chen et al., 2007). Overall, these studies are in support of the concept that exogenous ANP or BNP, especially

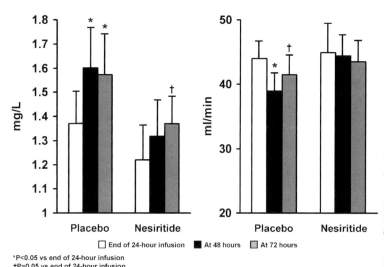

FIGURE 20.5 Plasma cystatin (left panel) and estimated creatinine clearance (right panel) at the end of 24-hour infusion, at 48 h and at 72 h in the placebo group and the nesiritide group. *$P < 0.05$ versus end of 24-hour infusion. $P = 0.05$ versus end of 24-hour infusion. Reproduced from Chen, H.H. et al. (2007). *Circulation* 116, I-134-8 with permission from the copyright owner, 2007, Lippincott Williams and Wilkins.

administered in low doses, preserves renal function in the perioperative period. Future studies are needed to evaluate if such therapy would confer simultaneous cardioprotection and to assess if it would impact on clinical outcomes.

Extended clinical application of BNP beyond the acute stage of HF is also being actively investigated, both as subcutaneous therapy (Chen et al., 2004) and as chronic serial infusions on an outpatient basis (Yancy et al., 2007).

Another property of BNP that may be applied clinically is its bronchodilating effect, such as for the treatment of asthma (Akerman et al., 2006). Moreover, recent evidence from population studies suggests a protective role of higher BNP levels against diabetes, possibly related to improved insulin sensitivity (Meirhaeghe et al., 2007). Further studies are needed to explore potential therapeutic applications of these findings.

C. C-Type Natriuretic Peptide

In experimental studies, CNP has been observed to exert predominantly venodilating effects (Wei et al., 1993) and is more potent in inducing smooth muscle relaxation as compared to ANP (Schulz, 2005). CNP immunoreactivity has been detected in human kidneys, specifically in proximal, distal and medullary collecting ducts (Mattingly et al., 1994; Cataliotti et al., 2002). Its renal effects have not been clearly defined, although it is believed that CNP is not natriuretic at physiologic concentrations (Potter et al., 2006). Charles et al. (1995) observed natriuretic response to CNP (both 1 and 10 pmol/kg/min) in conscious sheep, in which the plasma half-life of CNP was observed to be 1.6 ± 0.27 min. The metabolic clearance rates were reported to be 3.15 and 2.48 l/min, respectively, for the two doses studied (Charles et al., 1995). In anesthetized dogs, the natriuretic response to CNP appears to be related to the dose and the duration of infusion (Clavell et al., 1993; Lee et al., 2007). C-type natriuretic peptide is metabolized via internalization by NPR-C and via hydrolysis by neutral endopeptidase (Schulz, 2005).

In human studies, Hunt et al. (1994) evaluated the cardiorenal and hormonal effects of CNP in nine normal men who were given, in a random order, a 2-hour infusion of synthetic human CNP (5 pmol/kg/min) or placebo. In this study of single-blind design, baseline plasma CNP was undetectable in the subjects but it increased to a mean plateau level of 60 ± 6 pmol/l within 30 to 120 min (Hunt et al., 1994). The metabolic clearance rate and the plasma half-life of CNP were reported to be 4.8 ± 0.7 l/min and 2.6 min (Hunt et al., 1994). Infusion of CNP elicited significant increases in plasma cGMP and plasma ANP and a significant decrease in aldosterone, as compared to infusion of placebo (vehicle) (Hunt et al., 1994). No natriuresis or vasodepressor response was observed (Hunt et al., 1994). In contrast, Pham et al. (1997) reported dose-dependent natriuretic response to CNP in eight healthy subjects. Further studies are needed to clarify the renal effects of CNP in humans.

Exogenous CNP is being investigated as a novel therapy for achondroplasia (Nakao, 2007). Additional areas of investigation include prevention or/and treatment of vasospasm (Kelsall et al., 2005) in coronary artery bypass grafts, prevention of graft thrombosis or neointimal hyperplasia (Ohno et al., 2005), cardioprotection against myocardial ischemia-reperfusion injury (Hobbs et al., 2004), anti-remodeling following myocardial infarction (Soeki et al., 2005) and, most recently, valvular aortic stenosis (Peltonen et al., 2007). The latter report by Peltonen et al. (2007) documented that, in human aortic valve stenosis, there were downregulations of the gene for CNP expression, the processing enzyme (furin) for CNP, and both NPR-A and NPR-B, thus raising the possibility that CNP may have a therapeutic role in aortic valve stenosis. This warrants further investigation.

VII. NOVEL DELIVERY SYSTEMS

Oral delivery of peptides poses a challenge as a result of penetration and enzymatic barriers (Frokjaer and Otzen, 2005; Banga, 2006). Various methods for oral delivery of proteins and peptides have been under investigation (Banga, 2006). One developed technology is PEGylation, which entails the modification of a peptide or protein by linking polyethyleneglycol (PEG) chains (Harris and Chess, 2003; Pasut et al., 2006; Jorgensen et al., 2006). Potential advantages of pegylation include prolonged residence secondary to reduced renal clearance, reduced degradation by proteolytic enzymes secondary to shielding effect of the PEG chains and reduced immunogenicity (Harris and Chess, 2003; Pasut et al., 2006). More recently, proprietary technology has facilitated the development of oral BNP, to which short, amphiphilic oligomers were covalently linked, especially in the presence of lysine or serine residues which would enable linking of oligomers (Cataliotti et al., 2005, 2006, 2007; Miller et al., 2006). Importantly, cGMP-activating and anti-hypertensive actions have been demonstrated *in vivo,* including experimental hypertension (Cataliotti et al., 2005, 2006).

Currently, exogenous natriuretic peptides that are in clinical use, such as carperitide and nesiritide, in various countries are administered as i.v. infusions. An alternative strategy that has shown promise both in experimental (Clemens et al., 1998; Chen et al., 2000, 2006a) and clinical studies (Chen et al., 2004) is subcutaneous administration of BNP. Chen et al. (2004) have demonstrated the feasibility, short-term safety and efficacy of subcutaneous administration of BNP in human subjects with symptomatic HF. In addition, intra-renal administration of NP also appears promising in experimental (Chen et al., 2006b) and clinical (Heywood et al., 2006) studies.

VIII. NOVEL CHIMERIC AND SYNTHETIC NATRIURETIC PEPTIDES

The concept of novel peptide therapies with chimeric and synthetic NPs has been advanced by the Burnett Lab at the Mayo Clinic for the past 15 years (Wei et al., 1993). Selection and incorporation of isolated structural determinants from native NPs could result in novel peptides with specific cardiorenal activity profiles. It provides opportunities for creating peptides with unique biological actions that are not available in native NPs. These novel designer peptides could have enhanced pharmacologic profiles and may reduce undesirable effects of native peptides. One of the first peptides that was designed by our group was vasonatrin (VNP) (Wei et al., 1993). Vasonatrin was a 27-AA peptide which consisted of the entire 22-amino-acid (AA) sequence of CNP and the 5-AA C-terminus of ANP. It exhibited natriuretic, venodilating and vasorelaxant properties, but also arterial vasodilating actions which were not shared by ANP and CNP.

A. Dendroaspis Natriuretic Peptide and CD-NP

Snake venoms provide a diverse and potentially valuable source of peptides for new drug development. In 1992, Schweitz et al. purified and sequenced a 38-AA peptide in the venom of the green mamba snake (*Dendroaspis angusticeps*). This peptide shares structural similarities with ANP, BNP and CNP and also possesses a unique 15-amino-acid C-terminus beyond the second cysteine residue (see Figure 20.1). It exhibited picomolar potency in stimulating NPR-A (Johns et al., 2007). In pre-contracted rat aortic strips, DNP induced relaxation to a similar extent as ANP (Schweitz et al., 1992). In cultured rat aortic myocytes and bovine aortic endothelial cells, DNP elicited concentration-dependent increases in cGMP (Schweitz et al., 1992). Previous work from our laboratory has demonstrated the natriuretic and diuretic effects of DNP in normal dogs and in dogs with pacing-induced HF (Lisy et al., 1999, 2001). Moreover, DNP suppressed plasma renin activity, increased GFR and reduced cardiac filling pressures (Lisy et al., 2001). Intra-renal infusion of DNP (5 ng/kg/min) into normal anesthetized dogs resulted in significant increases in plasma and urinary cGMP, renal blood flow and GFR and Na^+ excretion, as well as a significant decrease in distal fraction Na^+ reabsorption (Chen et al., 2002). However, DNP induces systemic hypotension both in normal anesthetized dogs and in dogs with experimental HF (Lisy et al., 1999, 2001), which may be a limiting factor in its therapeutic application, as renal perfusion may be compromised in the presence of systemic hypotension (Lee and Burnett, 2007).

Recently, a novel chimeric natriuretic peptide, CD-NP (Figure 20.6), was designed and developed (Lisy and Burnett, 2003a). CD-NP consists of the 22-AA sequence of human CNP and the 15-AA C-terminus of *Dendroaspis* natriuretic peptide (Lisy and Burnett, 2003). The rationale for its design was to transform CNP, a venodilating peptide with diverse cardiovascular but limited renal actions, into a CNP-like peptide with augmented renal actions without inducing systemic hypotension.

Preclinical studies to date including experimental HF have demonstrated natriuretic, diuretic, GFR-enhancing, cardiac-unloading, anti-proliferative and renin-suppressing actions of CD-NP without significant effects on systemic blood pressure (Lisy and Burnett, 2003b; Lisy et al., 2006). Importantly, Lisy et al. (2006) demonstrated, in normal anesthetized dogs, that CD-NP 50 ng/kg/min i.v., when compared to an equimolar dose of BNP, was significantly less hypotensive and was associated with a significantly greater increase in GFR. When compared to an equimolar dose of CNP, Lee et al. (2007) demonstrated that CD-NP 50 ng/kg/min i.v. elicited greater increases in plasma cGMP, urinary cGMP excretion and net renal generation of cGMP,

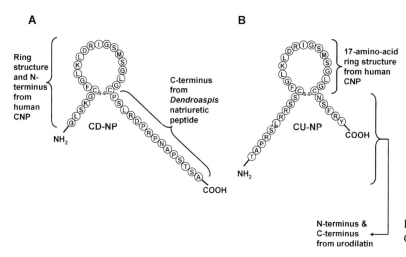

FIGURE 20.6 Structures and amino acid sequences of CD-NP and CU-NP.

which was associated with enhanced natriuresis. Moreover, CD-NP, not CNP, enhanced glomerular filtration rate and suppressed plasma renin activity and angiotensin II. This study demonstrates the successful transformation of CNP to a CNP-like peptide with enhanced renal and neurohumoral actions.

B. CU-NP

More recently, a novel peptide which consists of the 17-AA ring structure of human CNP and both the N- and the C-termini of URO was designed and synthesized (see Figure 20.6) (Lee and Burnett, 2007a). Preliminary canine studies demonstrate that this designer peptide, which consists of entirely human AA sequences, significantly increased plasma cGMP, urinary cGMP excretion and net renal generation of cGMP (Lee and Burnett, 2007a). It dose-dependently increased natriuresis, diuresis and GFR in normal anesthetized dogs (Lee and Burnett, 2007b). Moreover, CU-NP significantly decreased cardiac filling pressures without lowering systemic blood pressure and exhibited inhibitory actions on the RAAS (Lee and Burnett, 2007a).

Previous studies demonstrated the importance of NPR-A in mediating diuresis and natriuresis (Richards et al., 1985; Zeidel et al., 1987), although the potential significance of NPR-B, which is present in human kidneys (Canaan-Kuhl et al., 1992), has also been implicated (Totsune et al., 1994; Millul et al., 1997). Future studies are needed to delineate the renal mechanisms of action of designer NPs, which are in progress in our laboratory.

Another rationale for developing designer peptides as novel therapies for HF is that it may enable simultaneous activation of NPR-A, NPR-B and NPR-C signaling pathways, as compared to the current clinical strategy of NPR-A activation alone (with recombinant ANP or BNP) (Lee and Burnett, 2007). This new strategy is of potential clinical importance, as recent studies have demonstrated reduced NPR-A protein concentrations in experimental HF (Bryan et al., 2007) and downregulation of NPR-A receptor density in human HF in the presence of underlying ischemic heart disease (Singh et al., 2006), and the predominance of NPR-B (not NPR-A) in experimental HF (Dickey et al., 2007). Moreover, systemic hypotension, as observed in the clinical use of exogenous ANP and BNP (Publication Committee for the VMAC Investigators, 2002; Suwa et al., 2005), is a known consequence of arterial vasodilation induced by NPR-A activation and may be a potential reason for adverse renal effects observed in some clinical studies (Riter et al., 2006; Lee and Burnett, 2007).

C. Other Novel Peptides

AlbuBNP, which is a fusion of recombinant human BNP and human serum albumin with a prolonged half-life, was reported to activate plasma cGMP and lower blood pressure in spontaneous hypertensive rats (Wang et al., 2004). It also exhibits natriuretic, GFR-enhancing and aldosterone-inhibiting properties (Chen et al., 2006c). More recently, another novel peptide, ASBNP2.1, which is based on an alternatively spliced transcript for BNP in conjunction with a truncated C-terminus, was demonstrated by Chen et al. (2006d) to exert aquaretic and GFR-enhancing effects without inducing systemic hypotension in a canine model of experimental HF. Further studies are needed to define the mechanisms of actions of these novel peptides.

IX. FUTURE DIRECTIONS

Exogenous NPs that are in clinical use (ANP, BNP) or in development (URO) are all NPR-A agonists. There is an unmet need for novel therapies that either predominantly activate NPR-B or co-activate NPR-A and NPR-B. The currently available NPR-B agonist, CNP, exerts only limited natriuretic and diuretic actions which are important in sodium-retaining states, such as HF. Designer peptides that

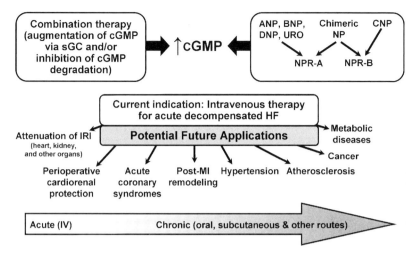

FIGURE 20.7 Natriuretic peptides: the evolving therapeutic paradigm. sGC = soluble guanylyl cyclase, IRI = ischemia-reperfusion injury, MI = myocardial infarction, NPR-A = natriuretic peptide receptor-A, NPR-B = natriuretic peptide receptor-B, HF = heart failure.

activate both NPR-A and NPR-B may hold potential as novel therapies for the treatment of HF. Moreover, whether NPR-C blockade, either alone or in combination with activation of NPR-A and/or NPR-B, would confer an additional therapeutic advantage remains to be evaluated.

With the advent of biotechnologies, chronic NP therapies via oral or other innovative routes hold promise as novel treatment strategies for chronic cardiovascular diseases, including systemic hypertension and chronic stable HF. The feasibility, efficacy and safety of chronic NP administration warrant further evaluation. Opportunities for combination therapy with co-activation of particulate and soluble guanylate cyclases or with simultaneous inhibition of NP or cGMP degradation deserve further evaluation. As the details on the mechanisms of NP degradation continue to unfold (Pankow et al., 2007), additional targets for pharmacologic intervention will likely become available.

The therapeutic applications of the NPs are an evolving paradigm (Figure 20.7). It is anticipated that the use of NPs will extend beyond acute HF to other clinical areas such as perioperative cardiorenal protection (including organ transplantation), cardiorenal protection in the acute coronary syndromes, amelioration of ventricular remodeling following myocardial infarction and prevention and/or treatment of metabolic diseases and cancer.

Acknowledgments

This work is supported by grants from the National Institutes of Health (HL PO1 HL76611; RO1 HL36634; and RO1 HL83231) and from the Mayo and Marriott Foundations to Dr Burnett. Dr Lee is supported by a 2007 Heart Failure Society of America Research Fellowship Award and is a recipient of the 2007 Young Investigator Award from the American Society for Clinical Pharmacology and Therapeutics.

References

Adams, K. F., Lindenfeld, J., Arnold, J. M. O. et al. (2006). HFSA 2006 comprehensive heart failure practice guidelines. *J Card Fail* **12**: e1–e122.

Ahluwalia, A., and Hobbs, A. J. (2005). Endothelium-derived C-type natriuretic peptide: more than just a hyperpolarizing factor. *Trends Pharmacol Sci* **26**: 162–167.

Airhart, N., Yang, Y. F., Roberts, C. T., and Silberbach, M. (2003). Atrial natriuretic peptide induces natriuretic peptide receptor-cGMP-dependent protein kinase interaction. *J Biol Chem* **278**: 38693–38698.

Akerman, M. J., Yaegashi, M., Khiangte, Z., Murugan, A. T., Abe, O., and Marmur, J. D. (2006). Bronchodilator effect of infused B-type natriuretic peptide in asthma. *Chest* **130**: 66–72.

Allgren, R. L., Marbury, T. C., Rahman, S. N. et al. (1997). Anaritide in acute tubular necrosis. *N Engl J Med* **336**: 828–834.

Anand-Srivastava, M. B. (2005). Natriuretic peptide receptor-C signaling and regulation. *Peptides* **26**: 1044–1059.

Angus, R. M., McCallum, M. J., Hulks, G., and Thomson, N. C. (1993). Bronchodilator, cardiovascular, and cyclic guanylyl monophosphate response to high-dose infused atrial natriuretic peptide in asthma. *Am Rev Respir Dis* **147**: 1122–1125.

Ardaillou, N., Nivez, M. P., and Ardaillou, R. (1986). Stimulation of cyclic GMP synthesis in human cultured glomerular cells by atrial natriuretic peptide. *FEBS Lett* **204**: 177–182.

Arora, R. R., Venkatesh, P. K., and Molnar, J. (2006). Short and long-term mortality with nesiritide. *Am Heart J* **152**: 1084–1090.

Banga, A.K. (2006). Oral delivery of peptide and protein drugs. In Therapeutic peptides and proteins: formulation, processing, and delivery systems, 2nd edn. CRC Press, Boca Raton, pp. 229–258.

Baxter, G. F. (2004). Natriuretic peptides and myocardial ischaemia. *Basic Res Cardiol* **99**: 90–93.

Beaver, T. M., Winterstein, A. G., Shuster, J. J. et al. (2006). Effectiveness of nesiritide on dialysis or all-cause mortality in patients undergoing cardiothoracic surgery. *Clin Cardiol* **29**: 18–24.

Best, P. J., Burnett, J. C., Wilson, S. H., Holmes, D. R., and Lerman, A. (2002). Dendroaspis natriuretic peptide relaxes isolated human arteries and veins. *Cardiovasc Res* **55**: 375–384.

Bibbins-Domingo, K., Gupta, R., Na, B., Wu, A. H., Schiller, N. B., and Whooley, M. A. (2007). N-terminal fragment of the prohormone brain-type natriuretic peptide (NT-proBNP), cardiovascular events, and mortality in patients with stable coronary heart disease. *J Am Med Assoc* **297**: 169–176.

Boerrigter, G., Costello-Boerrigter, L. C., Cataliotti, A., Lapp, H., Stasch, J.-P., and Burnett, J. C. (2007a). Targeting heme-oxidized soluble guanylate cyclase in experimental heart failure. *Hypertension* **49**: 1128–1133.

Boerrigter, G., Costello-Boerrigter, L. C., Harty, G. J., Lapp, H., and Burnett, J. C. (2007b). Des-Serine-Proline B-Type natriuretic peptide (BNP 3-32) in cardiorenal regulation. *Am J Physiol Renal Fluid Electrolyte Physiol* **292**: R897–R901.

Borgeson, D. D., Stevens, T. L., Heublein, D. M., Matsuda, Y., and Burnett, J. C. (1998). Activation of myocardial and renal natriuretic peptides during acute intravascular volume overload in dogs: functional cardiorenal responses to receptor antagonism. *Clin Sci (Lond.)* **95**: 195–202.

Bryan, P. M., Xu, X., Dickey, D. M., Chen, Y., and Potter, L. R. (2007). Renal hyporesponsiveness to atrial natriuretic peptide in congestive heart failure results from reduced atrial natriuretic peptide receptor concentrations. *Am J Physiol* **292**: F1636–F1644.

Burley, D. S., Ferdinandy, P., and Baxter, G. F. (2007a). Cyclic GMP and protein kinase-G in myocardial ischaemia-reperfusion: opportunities and obstacles for survival signaling. *Br J Pharmacol* **152**: 855–869.

Burley, D. S., Hamid, S. A., and Baxter, G. F. (2007b). Cardioprotective actions of peptide hormones in myocardial ischemia. *Heart Fail Rev*.

Burnett, J. C., Kao, P. C., Hu, D. C. et al. (1986). Atrial natriuretic peptide elevation in congestive heart failure in the human. *Science* **231**: 1145–1147.

Burnett, J. C., Costello-Boerrigter, L., and Boerrigter, G. (2004). Alterations in the kidney in heart failure: The cardiorenal axis in the regulation of sodium homeostasis. In: (Mann, D.L., eds) *Heart failure: a companion to Braunwald's heart disease,* Elsevier Inc., Philadelphia, p. 279–289.

Canaan-Kuhl, S., Jamison, R. L., Myers, B. D., and Pratt, R. E. (1992). Identification of 'B' receptor for natriuretic peptide in human kidney. *Endocrinology* **130**: 550–552.

Candido, R., Burrell, L. M., Jandeleit-Dahm, K. A. M., and Cooper, M. E. (2008). Vasoactive peptides and the kidney. In: *Brenner & Rector's the kidney,* 8th edn, (Brenner, B.M., eds) Saunders, Philadelphia, p. 333–362.

Carstens, J., Jensen, K. T., and Pedersen, E. B. (1998). Metabolism and action of urodilatin infusion in healthy volunteers. *Clin Pharmacol Ther* **64**: 73–86.

Castro, L. R., Verde, I., Cooper, D. M., and Fischmeister, R. (2006). Cyclic guanosine monophosphate compartmentation in rat cardiac myocytes. *Circulation* **113**: 2221–2228.

Cataliotti, A., Giordano, M., De Pascale, E. et al. (2002). CNP production in the kidney and effects of protein intake restriction in nephrotic syndrome. *Am J Physiol* **283**: F464–F472.

Cataliotti, A., Boerrigter, G., Costello-Boerrigter, L. C. et al. (2004). Brain natriuretic peptide enhances renal actions of furosemide and suppresses furosemide-induced aldosterone activation in experimental heart failure. *Circulation* **109**: 1680–1685.

Cataliotti, A., Schirger, J. A., Martin, F. L. et al. (2005). Oral human brain natriuretic peptide activates cyclic guanosine $3',5'$-monophosphate and decreases mean arterial pressure. *Circulation* **112**: 836–840.

Cataliotti, A., Heublein, D. M., James, K. D., and Burnett, J. C. (2006). Biological actions of a novel oral human BNP in an experimental model of acute hypertension (abst). *J Card Fail,* p. S30.

Cataliotti, A., Chen, H. H., James, K. D., and Burnett, J. J. C. (2007). Oral brain natriuretic peptide: a novel strategy for chronic protein therapy for cardiovascular disease. *Trends Cardiovasc Med* **17**: 10–14.

Charles, C. J., Espiner, E. A., Richards, A. M., Nicholls, M. G., and Yandle, T. G. (1995). Biological actions and pharmacokinetics of C-type natriuretic peptide in conscious sheep. *Am J Physiol* **268**: R201–R207.

Chen, H. H. (2007). Heart failure: a state of brain natriuretic peptide deficiency or resistance or both!. *J Am Coll Cardiol* **49**: 1089–1091.

Chen, H. H., and Burnett, J. C. (2006). Clinical application of the natriuretic peptides in heart failure. *Eur Heart J* **8**(Suppl E): E18–E25.

Chen, H. H., Grantham, J. A., Schirger, J. A., Jougasaki, M., Redfield, M. M., and Burnett, J. C. (2000). Subcutaneous administration of brain natriuretic peptide in experimental heart failure. *J Am Coll Cardiol* **36**: 1706–1712.

Chen, H. H., Lainchbury, J. G., and Burnett, J. C. (2002). Natriuretic peptide receptors and neutral endopeptidase in mediating the renal actions of a new therapeutic synthetic natriuretic peptide dendroaspis natriuretic peptide. *J Am Coll Cardiol* **40**: 1186–1191.

Chen, H. H., Redfield, M. M., Nordstrom, L. J., Horton, D. P., and Burnett, J. C. (2004). Subcutaneous administration of the cardiac hormone BNP in symptomatic human heart failure. *J Card Fail* **10**: 115–119.

Chen, H. H., Cataliotti, A., Schirger, J. A., Martin, F. L., and Burnett, J. C. (2005). Equimolar doses of atrial and brain natriuretic peptides and urodilatin have differential renal actions in overt experimental heart failure. *Am J Physiol Regul Integr Comp Physiol* **288**: R1093–R1097.

Chen, H. H., Huntley, B. K., Schirger, J. A., Cataliotti, A., and Burnett, J. C. (2006). Maximizing the renal cyclic 3(-5(-guanosine monophosphate system with type V phosphodiesterase inhibition and exogenous natriuretic peptide: a novel strategy to improve renal function in experimental overt heart failure. *J Am Soc Nephrol* **17**: 2742–2747.

Chen, H. H., Cataliotti, A., Martin, F. L., Schirger, J. A., and Burnett, J. C. (2006c). AlbuBNP, a recombinant human B-type natriuretic peptide and serum albumin fusion hormone has prolonged natriuretic, glomerular filtration rate enhancing and aldosterone inhibiting properties (abst). *J Card Fail* **12**(Suppl 1): S84.

Chen, H. H., Pan, S., Burnett, J. C., and Simari, R. D. (2006d). A novel designer natriuretic and diuretic peptide based upon an alternatively spliced BNP without vascular vasodilatory actions (abst). *Circulation* **114**(Suppl II): II-270.

Chen, H. H., Sundt, T. M., Cook, D. J., Heublein, D. M., and Burnett, J. C. (2007). Low dose nesiritide and the preservation of renal function in patients with renal dysfunction undergoing cardiopulmonary-bypass surgery: a double-blind placebo-controlled pilot study. *Circulation* **116**(suppl): I-134–I-138.

Clavell, A. L., Stingo, A. J., Wei, C. M., Heublein, D. M., and Burnett, J. C. (1993). C-type natriuretic peptide: a selective cardiovascular peptide. *Am J Physiol* **264:** R290–R295.

Clemens, L. E., Almirez, R. G., Baudouin, K. A., Mischak, R. P., Grossbard, E. B., and Protter, A. A. (1998). Pharmacokinetics and biological actions of subcutaneously administered human brain natriuretic peptide. *J Pharmacol Exp Ther* **287:** 67–71.

Clerico, A., Recchia, F. A., Passino, C., and Emdin, M. (2006). Cardiac endocrine function is an essential component of the homeostatic regulation network: Physiological and clinical implications. *Am J Physiol Heart Circ Physiol* **290:** H17–29.

Cody, R. J., Atlas, S. A., Laragh, J. H. et al. (1986). Atrial natriuretic factor in normal subjects and heart failure patients. Plasma levels and renal, hormonal, and hemodynamic responses to peptide infusion. *J Clin Invest* **78:** 1362–1374.

Colucci, W. S., Elkayam, U., Horton, D. P. et al. For the Nesiritide Study Group. (2000). Intravenous nesiritide, a natriuretic peptide, in the treatment of decompensated congestive heart failure. *N Engl J Med* **343:** 246–253.

de Bold, A. J., Borenstein, H. B., Veress, A. T., and Sonnenberg, H. (1981). A rapid and potent natriuretic response to intravenous injection of atrial myocardial extract in rats. *Life Sci* **28:** 89–94.

Dean, A. D., Vehaskari, V. M., and Greenwald, J. E. (1994). Synthesis and localization of C-type natriuretic peptide in mammalian kidney. *Am J Physiol* **266:** F491–F496.

Dean, A.D., Vehaskari, V.M., Ritter, D. and Greenwald, J.E. (1996). Distribution and regulation of guanylyl cyclase type B in the rat nephron. In F311–8.

Del Ry, S., Passino, C., Emdin, M., and Giannessi, D. (2006). C-type natriuretic peptide and heart failure. *Pharmacol Res* **54:** 326–333.

Dickey, D. M., Flora, D. R., Bryan, P. M. et al. (2007). Differential regulation of membrane guanylyl cyclases in congestive heart failure: NPR-B, not NPR-A, is the predominant natriuretic peptide receptor in the failing heart. *Endocrinology* **148:** 3518–3522.

Dorner, G. T., Selenko, N., Kral, T., Schmetterer, L., Eichler, H. G., and Wolzt, M. (1998). Hemodynamic effects of continuous urodilatin infusion: a dose-finding study. *Clin Pharmacol Ther* **64:** 322–330.

D'Souza, S. P., Yellon, D. M., Martin, C. et al. (2003). B-type natriuretic peptide limits infarct size in rat isolated hearts via KATP channel opening. *Am J Physiol Heart Circ Physiol* **284:** H1592–H1600.

Eiskjaer, H., and Pedersen, E. B. (1993). Dose-response study of atrial natriuretic peptide bolus injection in healthy man. *Eur J Clin Invest* **23:** 37–45.

Espiner, E. A., Prickett, T. C., Yandle, T. G. et al. (2007). ABCs of natriuretic peptides: growth. *Horm Res* **67**(Suppl 1): 81–90.

Feldman, D. S., Ikonomidis, J. S., Uber, W. E. et al. (2004). Human B-natriuretic peptide improves hemodynamics and renal function in heart transplant patients immediately after surgery. *J Card Fail* **10:** 292–296.

Fischmeister, R., Castro, L. R. V., Abi-Gerges, A. et al. (2006). Compartmentation of cyclic nucleotide signaling in the heart: the role of cyclic nucleotide phosphodiesterases. *Circ Res* **99:** 816–828.

Fluge, T., Fabel, H., Wagner, T. O., Schneider, B., and Forssmann, W. G. (1995). Urodilatin (ularitide, INN): a potent bronchodilator in asthmatic subjects. *Eur J Clin Invest* **25:** 728–736.

Forfia, P. R., Lee, M., Tunin, R. S., Mahmud, M., Champion, H. C., and Kass, D. A. (2007). Acute phosphodiesterase 5 inhibition mimics hemodynamic effects of B-type natriuretic peptide and potentiates B-type natriuretic peptide effects in failing but not normal canine heart. *J Am Coll Cardiol* **49:** 1079–1088.

Forssmann, W., Meyer, M., and Forssmann, K. (2001). The renal urodilatin system: clinical implications. *Cardiovasc Res* **51:** 450–462.

Frokjaer, S., and Otzen, D. E. (2005). Protein drug stability: a formulation challenge. *Nat Rev Drug Discov* **4:** 298–306.

Furuya, M., Ohnuma, N., Takehisa, M. et al. (1991). Pharmacological activities of brain natriuretic peptides of human, porcine and rat origin. *Eur J Pharmacol* **200:** 233–237.

Furuya, M., Tawaragi, Y., Minamitake, Y. et al. (1992). Structural requirements of C-type natriuretic peptide for elevation of cyclic GMP in cultured vascular smooth muscle cells. *Biochem Biophys Res Commun* **183:** 964–969.

Garbers, D. L., Chrisman, T. D., Wiegn, P. et al. (2006). Membrane guanylyl cyclase receptors: an update. *Trends Endocrinol Metab* **17:** 251–258.

Gheorghiade, M., Zannad, F., Sopko, G. et al. (2005). Acute heart failure syndromes: current state and framework for future research. *Circulation* **112:** 3958–3968.

Giles, T. D., Quiroz, A. C., Roffidal, L. E., Marder, H., and Sander, G. E. (1991). Prolonged hemodynamic benefits from a high-dose bolus injection of human atrial natriuretic factor in congestive heart failure. *Clin Pharmacol Ther* **50:** 557–563.

Goetz, K., Drummer, C., Zhu, J. L., Leadley, R., Fiedler, F., and Gerzer, R. (1990). Evidence that urodilatin, rather than ANP, regulates renal sodium excretion. *J Am Soc Nephrol* **1:** 867–874.

Gordon, G. R., Schumann, R., Rastegar, H., Khabbaz, K., and England, M. R. (2006). Nesiritide for treatment of perioperative low cardiac output syndromes in cardiac surgical patients: an initial experience. *J Anesth* **20:** 307–311.

Groban, L., Cowley, A. W., and Ebert, T. J. (1990). Atrial natriuretic peptide augments forearm capillary filtration in humans. *Am J Physiol* **259:** H258–H263.

Gunning, M., Silva, P., Brenner, B. M., and Zeidel, M. L. (1989). Characteristics of ANP-sensitive guanylate cyclase in inner medullary collecting duct cells. *Am J Physiol* **256:** F766–F775.

Haneda, M., Kikkawa, R., Koya, D. et al. (1993). Biological receptors mediate anti-proliferative action of atrial natriuretic peptide in cultured mesangial cells. *Biochem Biophys Res Commun* **192:** 642–648.

Harris, J. M., and Chess, R. B. (2003). Effect of pegylation on pharmaceuticals. *Nat Rev Drug Discov* **2:** 214–221.

Harris, P. J., Thomas, D., and Morgan, T. O. (1987). Atrial natriuretic peptide inhibits angiotensin-stimulated proximal tubular sodium and water reabsorption. *Nature* **326:** 697–698.

Hawkridge, A. M., Heublein, D. M., Bergen, H. R., Cataliotti, A., Burnett, J. C., and Muddiman, D. C. (2005). Quantitative mass spectral evidence for the absence of circulating brain natriuretic peptide (BNP-32) in severe human heart failure. *Proc Natl Acad Sci USA* **102:** 17442–17447.

Heublein, D. M., Huntley, B. K., Boerrigter, G. et al. (2007). Immunoreactivity and guanosine 3′,5′-cyclic monophosphate activating actions of various molecular forms of human B-type natriuretic peptide. *Hypertension* **49:** 1114–1119.

Heywood, J. T., Ho, A., and Mathur, V. (2006). Favorable renal hemodynamic effects of intra-renal nesiritide infusion in heart transplant patients (abst). *J Card Fail* **12** (Suppl 1): S3.

Hillock, R.J., Frampton, C.M., Yandle, T.G., Troughton, R.W., Lainchbury, J.G., and Richards, A.M. (2007). B-Type natriuretic peptide infusions in acute myocardial infarction. Heart published online July 16, 2007. Accessed July 17, 2007.

Hirsch, J. R., Meyer, M., and Forssmann, W. G. (2006). ANP and urodilatin: who is who in the kidney. *Eur J Med Res* **11**: 447–454.

Hobbs, A., Foster, P., Prescott, C., Scotland, R., and Ahluwalia, A. (2004). Natriuretic peptide receptor-C regulates coronary blood flow and prevents myocardial ischemia/reperfusion injury: Novel cardioprotective role for endothelium-derived C-type natriuretic peptide. *Circulation* **110**: 1231–1235.

Horio, T., Tokudome, T., Maki, T. et al. (2003). Gene expression, secretion, and autocrine action of C-type natriuretic peptide in cultured adult rat cardiac fibroblasts. *Endocrinology* **144**: 2279–2284.

Hunt, P. J., Richards, A. M., Espiner, E. A., Nicholls, M. G., and Yandle, T. G. (1994). Bioactivity and metabolism of C-type natriuretic peptide in normal man. *J Clin Endocrinol Metab* **78**: 1428–1435.

Hunt, P. J., Espiner, E. A., Nicholls, M. G., Richards, A. M., and Yandle, T. G. (1996). Differing biological effects of equimolar atrial and brain natriuretic peptide infusions in normal man. *J Clin Endocrinol Metab* **81**: 3871–3876.

Huntley, B. K., Sandberg, S. M., Noser, J. A. et al. (2006). BNP-induced activation of cGMP in human cardiac fibroblasts: Interactions with fibronectin and natriuretic peptide receptors. *J Cell Physiol* **209**: 943–949.

Jensen, K. T., Carstens, J., and Pedersen, E. B. (1998). Effect of BNP on renal hemodynamics, tubular function and vasoactive hormones in humans. *Am J Physiol* **274**: F63–72.

Jensen, K. T., Eiskjaer, H., Carstens, J., and Pedersen, E. B. (1999). Renal effects of brain natriuretic peptide in patients with congestive heart failure. *Clin Sci (Lond)* **96**: 5–15.

John, S. W., Veress, A. T., Honrath, U. et al. (1996). Blood pressure and fluid-electrolyte balance in mice with reduced or absent ANP. *Am J Physiol* **271**: R109–R114.

Johns, D. G., Ao, Z., Heidrich, B. J. et al. (2007). Dendroaspis natriuretic peptide binds to the natriuretic peptide clearance receptor. *Biochem Biophys Res Commun* **358**: 145–149.

Jorgensen, L., Moeller, E. H., van de Weert, M., Nielsen, H. M., and Frokjaer, S. (2006). Preparing and evaluating delivery systems for proteins. *Eur J Pharmaceut Sci* **29**: 174–182.

Kasama, S., Toyama, T., Hatori, T. et al. (2007). Effects of intravenous atrial natriuretic peptide on cardiac sympathetic nerve activity and left ventricular remodeling in patients with first anterior acute myocardial infarction. *J Am Coll Cardiol* **49**: 667–674.

Kato, T., Muraski, J., Chen, Y. et al. (2005). Atrial natriuretic peptide promotes cardiomyocyte survival by cGMP-dependent nuclear accumulation of zyxin and Akt. *J Clin Invest* **115**: 2716–2730.

Keating, G. M., and Goa, K. L. (2003). Nesiritide: a review of its use in acute decompensated heart failure. *Drugs* **63**: 47–70.

Kelsall, C. J., Chester, A. H., Amrani, M., and Singer, D. R. J. (2005). C-type natriuretic peptide relaxes human coronary artery bypass grafts preconstricted by endothelin-1. *Ann Thorac Surg* **80**: 1347–1351.

Kelsall, C. J., Chester, A. H., Sarathchandra, P., and Singer, D. R. J. (2006). Expression and localization of C-type natriuretic peptide in human vascular smooth muscle cells. *Vasc Pharmacol* **45**: 368–373.

Kenny, A. J., Bourne, A., and Ingram, J. (1993). Hydrolysis of human and pig brain natriuretic peptides, urodilatin, C-type natriuretic peptide and some C-receptor ligands by endopeptidase-24.11. *Biochem J* **291**: 83–88.

Kimura, K., Yamaguchi, Y., Horii, M. et al. (2007). ANP is cleared much faster than BNP in patients with congestive heart failure. *Eur J Clin Pharmacol*.

Kitakaze, M., Asakura, M., Shintani, Y. et al. On behalf of the J-WIND Investigators. (2006). Large-scale trial using atrial natriuretic peptide or nicorandil as an adjunct to percutaneous coronary intervention for ST-segment elevation acute myocardial infarction [abst]. *Circulation* **114**: 2425–2426.

Kitakaze, M., Asakura, M., Kim, J. et al. (2007). Human atrial natriuretic peptide and nicorandil as adjuncts to reperfusion treatment for acute myocardial infarction (J-WIND): two randomised trials. *Lancet* **370**: 1483–1493.

Koike, J., Nonoguchi, H., Terada, Y., Tomita, K., and Marumo, F. (1993). Effect of urodilatin on cGMP accumulation in the kidney. *J Am Soc Nephrol* **3**: 1705–1709.

Kudo, T., and Baird, A. (1984). Inhibition of aldosterone production in the adrenal glomerulosa by atrial natriuretic factor. *Nature* **312**: 756–757.

Lee, C. Y. W., and Burnett, J. C. (2007). Natriuretic peptides and therapeutic applications. *Heart Fail Rev* **12**: 131–142.

Lee, C. Y. W., and Burnett, J. C. (2007a). Discovery of a novel synthetic natriuretic peptide, CU-NP (abst). *J Card Fail* **13** (Suppl 2): S74.

Lee, C. Y., and Burnett, J. C. (2007b). Design, synthesis, and pharmacologic actions of a novel designer natriuretic peptide: CU-NP (abst). *Circulation* **116**(Suppl II): II 549–II 550.

Lee, E. Y., and Humphreys, M. H. (1996). Phosphodiesterase activity as a mediator of renal resistance to ANP in pathological salt retention. *Am J Physiol* **271**: F3–6.

Lee, C. Y., Boerrigter, G., Harty, G. J., Lisy, O., and Burnett, J. C. (2007). Pharmacodynamic profile of a novel chimeric natriuretic peptide, CD-NP, as compared to C-type natriuretic peptide (abst). *Circulation* **116**(Suppl II): II 550.

Lewis, J., Salem, M. M., Chertow, G. M. et al. (2000). Atrial natriuretic factor in oliguric acute renal failure. Anaritide Acute Renal Failure Study Group. *Am J Kidney Dis* **36**: 767–774.

Liang, F., O'Rear, J., Schellenberger, U. et al. (2007). Evidence for functional heterogeneity of circulating B-type natriuretic peptide. *J Am Coll Cardiol* **49**: 1071–1078.

Lindenfeld, J., and Schrier, R. W. (2007). The kidney in heart failure. In: *Congestive heart failure,* 3rd edn, (Hosenpud, J.D., Greenberg, B.H., eds) Lippincott Williams & Wilkins, Philadelphiap. 243–260.

Lisy, O., and Burnett, J. C. (2003a). The design, synthesis and cardiorenal actions of a new chimeric natriuretic peptide CD-NP (abst). *J Am Coll Cardiol*, p. 312A.

Lisy, O., and Burnett, J. C. (2003b). The new designer peptide CD-NP unloads the heart, suppresses renin and is natriuretic in vivo (abst). *J Card Fail* **9**(Suppl 1): S32.

Lisy, O., Jougasaki, M., Heublein, D. M. et al. (1999). Renal actions of synthetic dendroaspis natriuretic peptide. *Kidney Int* **56**: 502–508.

Lisy, O., Lainchbury, J. G., Leskinen, H., and Burnett, J. C. (2001). Therapeutic actions of a new synthetic vasoactive and natriuretic peptide, dendroaspis natriuretic peptide, in experimental severe congestive heart failure. *Hypertension* **37**: 1089–1094.

Lisy, O., Huntley, B. K., McCormick, D. J., Kurlansky, P. A., and Burnett, J. C. (2006). Design, synthesis and unique biological actions of CD-NP: a novel CNP-like chimeric natriuretic peptide (abst). *Circulation*, p. II-440.

Loretz, C. A., and Pollina, C. (2000). Natriuretic peptides in fish physiology. *Comp Biochem Physiol Part A: Mol Integ Physiol* **125**: 169–187.

Maeda, K., Tsutamoto, T., Wada, A. et al. (2000). Insufficient secretion of atrial natriuretic peptide at acute phase of myocardial infarction. *J Appl Physiol* **89**: 458–464.

Margulies, K. B., Heublein, D. M., Perrella, M. A., and Burnett, J. C. (1991). ANF-mediated renal cGMP generation in congestive heart failure. *Am J Physiol* **260**: F562–F568.

Martin, F. L., Supaporn, T., Chen, H. H. et al. (2007). Distinct roles for renal particulate and soluble guanylyl cyclases in preserving renal function in experimental acute heart failure. *Am J Physiol Regul Integr Comp Physiol* **293**: R1580–R1585.

Mattingly, M. T., Brandt, R. R., Heublein, D. M., Wei, C. M., Nir, A., and Burnett, J. C. (1994). Presence of C-type natriuretic peptide in human kidney and urine. *Kidney Int* **46**: 744–747.

Meirhaeghe, A., Sandhu, M. S., McCarthy, M. I. et al. (2007). Association between the T-381C polymorphism of the brain natriuretic peptide gene and risk of type 2 diabetes in human populations. *Hum Mol Genet* **16**: 1343–1350.

Mentzer, R. M., Oz, M. C., Sladen, R. N. et al. (2007). Effects of perioperative nesiritide in patients with left ventricular dysfunction undergoing cardiac surgery: The NAPA Trial. *J Am Coll Cardiol* **49**: 716–726.

Miller, M. A., Malkar, N. B., Severynse-Stevens, D. et al. (2006). Amphiphilic conjugates of human brain natriuretic peptide designed for oral delivery: In vitro activity screening. *Bioconjug Chem* **17**: 267–274.

Millul, V., Ardaillou, N., Placier, S., Baudouin, B., and Ronco, P. M. (1997). Receptors for natriuretic peptides in a human cortical collecting duct cell line. *Kidney Int* **51**: 281–287.

Mitrovic, V., Luss, H., Nitsche, K. et al. (2005). Effects of the renal natriuretic peptide urodilatin (ularitide) in patients with decompensated chronic heart failure: a double-blind, placebo-controlled, ascending-dose trial. *Am Heart J* **150**: 1239.e1–1239.e8.

Mitrovic, V., Seferovic, P. M., Simeunovic, D. et al. (2006). Haemodynamic and clinical effects of ularitide in decompensated heart failure. *Eur Heart J* **27**: 2823–2832.

Moazami, N., and Oz, M. C. (2005). Natriuretic peptides in the perioperative management of cardiac surgery patients. *Heart Surg Forum* **8**: E151–E157.

Mukoyama, M., Nakao, K., Hosoda, K. et al. (1991). Brain natriuretic peptide as a novel cardiac hormone in humans. Evidence for an exquisite dual natriuretic peptide system, atrial natriuretic peptide and brain natriuretic peptide. *J Clin Invest* **87**: 1402–1412.

Nakao, K. (2007). Translational research of natriuretic peptide family (abst). *BMC Pharmacol*, p. S32.

Nakao, K., Sugawara, A., Morii, N. et al. (1986). The pharmacokinetics of alpha-human atrial natriuretic polypeptide in healthy subjects. *Eur J Clin Pharmacol* **31**: 101–103.

Nir, A., Beers, K. W., Clavell, A. L. et al. (1994). CNP is present in canine renal tubular cells and secreted by cultured opossum kidney cells. *Am J Physiol* **267**: R1653–R1657.

Obata, H., Yanagawa, B., Tanaka, K. et al. (2007). CNP infusion attenuates cardiac dysfunction and inflammation in myocarditis. *Biochem Biophys Res Commun* **356**: 60–66.

Ohno, N., Itoh, H., Ikeda, T. et al. (2005). Accelerated reendothelialization with suppressed thrombogenic property and neointimal hyperplasia of rabbit jugular vein grafts by adenovirus-mediated gene transfer of C-type natriuretic peptide. *Circulation* **105**: 1623–1626.

Padilla, F., Garcia-Dorado, D., Agullo, L. et al. (2001). Intravenous administration of the natriuretic peptide urodilatin at low doses during coronary reperfusion limits infarct size in anesthetized pigs. *Cardiovasc Res* **51**: 592–600.

Pagel-Langenickel, I., Buttgereit, J., Bader, M., and Langenickel, T. H. (2007). Natriuretic peptide receptor B signaling in the cardiovascular system: Protection from cardiac hypertrophy. *J Mol Med* **85**: 797–810.

Pankow, K., Wang, Y., Gembardt, F. et al. (2007). Successive action of meprin A and neprilysin catabolizes B-type natriuretic peptide. *Circ Res* **101**: 875–882.

Pasut, G., Morpurgo, M., and Veronese, F. M. (2006). Basic strategies for PEGylation of peptid and protein drugs. In: (Torchilin, V.P., eds) *Delivery of protein and peptide drugs in cancer,* Imperial College Press, London, p. 53–84.

Patel, J. B., Valencik, M. L., Pritchett, A. M., Burnett, J. C., McDonald, J. A., and Redfield, M. M. (2005). Cardiac-specific attenuation of natriuretic peptide A receptor activity accentuates adverse cardiac remodeling and mortality in response to pressure overload. *Am J Physiol Heart Circ Physiol* **289**: H777–H784.

Pejchalova, K., Krejci, P. and Wilcox, W.R. (2008). C-natriuretic peptide: An important regulator of cartilage. Mol Genet Metab. In Press.

Peltonen, T. O., Taskinen, P., Soini, Y. et al. (2007). Distinct downregulation of C-type natriuretic peptide system in human aortic valve stenosis. *Circulation* **116**: 1283–1289.

Pham, I., Sediame, S., Maistre, G. et al. (1997). Renal and vascular effects of C-type and atrial natriuretic peptides in humans. *Am J Physiol Regul Integr Comp Physiol* **273**: R1457–R1464.

Potter, L. R., Abbey-Hosch, S., and Dickey, D. M. (2006). Natriuretic peptides, their receptors, and cyclic guanosine monophosphate-dependent signaling functions. *Endocr Rev* **27**: 47–72.

Publication Committee for the VMAC Investigators (Vasodilation in the Management of Acute CHF). (2002). Intravenous nesiritide vs nitroglycerin for treatment of decompensated congestive heart failure: a randomized controlled trial. *J Am Med Assoc* 287, 1531–40; Erratum, *J Am Med Assoc* 288, 577.

Rautureau, Y., and Baxter, G. F. (2004). Acute actions of natriuretic peptides in coronary vasculature and ischaemic myocardium. *Curr Pharm Des* **10**: 2477–2482.

Richards, A. M. (1989). Atrial natriuretic factor administered to humans: 1984–1988. *J Cardiovasc Pharmacol* **13**(Suppl 6): S69–74.

Richards, A. M., Espiner, E. A., Ikram, H., and Yandle, T. G. (1989). Atrial natriuretic factor in hypertension: bioactivity at normal plasma levels. *Hypertension* **14**: 261–268.

Richards, A. M., Crozier, I. G., Holmes, S. J., Espiner, E. A., Yandle, T. G., and Frampton, C. (1993). Brain natriuretic peptide: natriuretic and endocrine effects in essential hypertension. *J Hypertens* **11:** 163–170.

Riter, H. G., Redfield, M. M., Burnett, J. C., and Chen, H. H. (2006). Nonhypotensive low-dose nesiritide has differential renal effects compared with standard-dose nesiritide in patients with acute decompensated heart failure and renal dysfunction (letter). *J Am Coll Cardiol* **47:** 2334–2335.

Saba, S. R., and Vesely, D. L. (2006). Cardiac natriuretic peptides: hormones with anticancer effects that localize to nucleus, cytoplasm, endothelium, and fibroblasts of human cancers. *Histol Histopathol* **21:** 775–783.

Sabrane, K., Kruse, M. N., Fabritz, L. et al. (2005). Vascular endothelium is critically involved in the hypotensive and hypovolemic actions of atrial natriuretic peptide. *J Clin Invest* **115:** 1666–1674.

Sackner-Bernstein, J. D., Skopicki, H. A., and Aaronson, K. D. (2005a). Risk of worsening renal function with nesiritide in patients with acutely decompensated heart failure. *Circulation* **111:** 1487–1491.

Sackner-Bernstein, J. D., Kowalski, M., Fox, M., and Aaronson, K. (2005b). Short-term risk of death after treatment with nesiritide for decompensated heart failure: a pooled analysis of randomized controlled trials. *J Am Med Assoc* **293:** 1900–1905.

Sackner-Bernstein, J. D., Skopicki, H., and Aaronson, K. D. (2006). Natriuretic peptides for the treatment of heart failure. In: (Feldman, A., eds) *Heart failure: pharmacologic management*, Blackwell Publishing, Malden, p. 154–171.

Salzberg, S. P., Filsoufi, F., Anyanwu, A. et al. (2005). High-risk mitral valve surgery: perioperative hemodynamic optimization with nesiritide (BNP). *Ann Thorac Surg* **80:** 502–506.

Sato, K., Sekiguchi, S., Kawagishi, N. et al. (2006). Continuous low-dose human atrial natriuretic peptide promotes diuresis in oliguric patients after living donor liver transplantation. *Transplant Proc* **38:** 3591–3593.

Schulz, S. (2005). C-type natriuretic peptide and guanylyl cyclase B receptor. *Peptides* **26:** 1024–1034.

Schulz-Knappe, P., Forssmann, K., Herbst, F., Hock, D., Pipkorn, R., and Forssmann, W. G. (1988). Isolation and structural analysis of 'urodilatin', a new peptide of the cardiodilatin-(ANP)-family, extracted from human urine. *Klin Wochenschr* **66:** 752–759.

Schweitz, H., Vigne, P., Moinier, D., Frelin, C., and Lazdunski, M. (1992). A new member of the natriuretic peptide family is present in the venom of the green mamba (*Dendroaspis angusticeps*). *J Biol Chem* **267:** 13928–13932.

Scotland, R. S., Cohen, M., Foster, P. et al. (2005). C-type natriuretic peptide inhibits leukocyte recruitment and platelet-leukocyte interactions via suppression of P-selectin expression. *Proc Natl Acad Sci USA* **102:** 14452–14457.

Sengenes, C., Berlan, M., De Glisezinski, I., Lafontan, M., and Galitzky, J. (2000). Natriuretic peptides: a new lipolytic pathway in human adipocytes. *Faseb J* **14:** 1345–1351.

Sezai, A., Shiono, M., Orime, Y. et al. (2000). Low-dose continuous infusion of human atrial natriuretic peptide during and after cardiac surgery. *Ann Thorac Surg* **69:** 732–738.

Sezai, A., Hata, M., Wakui, S. et al. (2006a). Efficacy of low-dose continuous infusion of alpha-human atrial natriuretic peptide (hANP) during cardiac surgery: possibility of postoperative left ventricular remodeling effect. *Circ J* **70:** 1426–1431.

Sezai, A., Shiono, M., Hata, M. et al. (2006b). Efficacy of continuous low-dose human atrial natriuretic peptide given from the beginning of cardiopulmonary bypass for thoracic aortic surgery. *Surg Today* **36:** 508–514.

Sezai, A., Hata, M., Wakui, S. et al. (2006c). Efficacy of low-dose continuous infusion of alpha-human atrial natriuretic peptide (hANP) during cardiac surgery. *Circ J* **70:** 1426–1431.

Sezai, A., Hata, M., Wakui, S. et al. (2007). Efficacy of continuous low-dose hANP administration in patients undergoing emergent coronary artery bypass grafting for acute coronary syndrome. *Circ J* **71:** 1401–1407.

Singh, G., Kuc, R. E., Maguire, J. J., Fidock, M., and Davenport, A. P. (2006). Novel snake venom ligand Dendroaspis natriuretic peptide is selective for natriuretic peptide receptor-A in human heart: Downregulation of natriuretic peptide receptor-A in heart failure. *Circ Res* **99:** 183–190.

Soeki, T., Kishimoto, I., Okumura, H. et al. (2005). C-type natriuretic peptide, a novel antifibrotic and antihypertrophic agent, prevents cardiac remodeling after myocardial infarction. *J Am Coll Cardiol* **45:** 608–616.

Sudoh, T., Kangawa, K., Minamino, N., and Matsuo, H. (1988). A new natriuretic peptide in porcine brain. *Nature* **332:** 78–81.

Sudoh, T., Minamino, N., Kangawa, K., and Matsuo, H. (1990). C-type natriuretic peptide (CNP): a new member of natriuretic peptide family identified in porcine brain. *Biochem Biophys Res Commun* **168:** 863–870.

Suga, S., Nakao, K., Itoh, H. et al. (1992). Endothelial production of C-type natriuretic peptide and its marked augmentation by transforming growth factor-beta. Possible existence of 'vascular natriuretic peptide system'. *J Clin Invest* **90:** 1145–1149.

Supaporn, T., Sandberg, S. M., Borgeson, D. D. et al. (1996). Blunted cGMP response to agonists and enhanced glomerular cyclic $3',5'$-nucleotide phosphodiesterase activities in experimental congestive heart failure. *Kidney Int* **50:** 1718–1725.

Suwa, M., Seino, Y., Nomachi, Y., Matsuki, S., and Funahashi, K. (2005). Multicenter prospective investigation on efficacy and safety of carperitide for acute heart failure in the 'real world' of therapy. *Circ J* **69:** 283–290.

Sward, K., Valson, F., and Ricksten, S. E. (2001). Long-term infusion of atrial natriuretic peptide (ANP) improves renal blood flow and glomerular filtration rate in clinical acute renal failure. *Acta Anaesthesiol Scand* **45:** 536–542.

Sward, K., Valsson, F., Odencrants, P., Samuelsson, O., and Ricksten, S. E. (2004). Recombinant human atrial natriuretic peptide in ischemic acute renal failure: a randomized placebo-controlled trial. *Crit Care Med* **32:** 1310–1315.

Tan, A. C., Russel, F. G., Thien, T., and Benraad, T. J. (1993). Atrial natriuretic peptide. An overview of clinical pharmacology and pharmacokinetics. *Clin Pharmacokinet* **24:** 28–45.

Terada, Y., Tomita, K., Nonoguchi, H., Yang, T., and Marumo, F. (1994). PCR localization of C-type natriuretic peptide and B-type receptor mRNAs in rat nephron segments. *Am J Physiol* **267:** F215–F222.

Tokudome, T., Horio, T., Soeki, T. et al. (2004). Inhibitory effect of C-type natriuretic peptide (CNP) on cultured cardiac myocyte hypertrophy: Interference between CNP and endothelin-1 signaling pathways. *Endocrinology* **145:** 2131–2140.

Tonolo, G., McMillan, M., Polonia, J. et al. (1988). Plasma clearance and effects of alpha-hANP infused in patients with end-stage renal failure. *Am J Physiol* **254**: F895–F899.

Tonolo, G., Richards, A. M., Manunta, P. et al. (1989). Low-dose infusion of atrial natriuretic factor in mild essential hypertension. *Circulation* **80**: 893–902.

Totsune, K., Takahashi, K., Murakami, O. et al. (1994). Natriuretic peptides in the human kidney. *Hypertension* **24**: 758–762.

Tsuruda, T., Boerrigter, G., Huntley, B. K. et al. (2002). Brain natriuretic peptide is produced in cardiac fibroblasts and induces matrix metalloproteinases. *Circ Res* **91**: 1127–1134.

van den Akker, F. (2001). Structural insights into the ligand binding domains of membrane bound guanylyl cyclases and natriuretic peptide receptors. *J Mol Biol* **311**: 923–937.

Vesely, B. A., Eichelbaum, E. J., Alli, A. A., Sun, Y., Gower, W. R., and Vesely, D. L. (2006). Urodilatin and four cardiac hormones decrease human renal carcinoma cell numbers. *Eur J Clin Invest* **36**: 810–819.

Volpe, M., Tritto, C., De Luca, N. et al. (1991). Failure of atrial natriuretic factor to increase with saline load in patients with dilated cardiomyopathy and mild heart failure. *J Clin Invest* **88**: 1481–1489.

Wang, W., Ou, Y., and Shi, Y. (2004). AlbuBNP, a recombinant B-type natriuretic peptide and human serum albumin fusion hormone, as a long-term therapy of congestive heart failure. *Pharm Res* **21**: 2105–2111.

Wang, Y., de Waard, M. C., Sterner-Kock, A. et al. (2007). Cardiomyocyte-restricted over-expression of C-type natriuretic peptide prevents cardiac hypertrophy induced by myocardial infarction in mice. *Eur J Heart Fail* **9**: 548–557.

Weidmann, P., Hasler, L., Gnadinger, M. P. et al. (1986). Blood levels and renal effects of atrial natriuretic peptide in normal man. *J Clin Invest* **77**: 734–742.

Winaver, J., Burnett, J. C., Tyce, G. M., and Dousa, T. P. (1990). ANP inhibits Na^+-H^+ antiport in proximal tubular brush border membrane: role of dopamine. *Kidney Int* **38**: 1133–1140.

Wu, C., Wu, F., Pan, J., Morser, J., and Wu, Q. (2003). Furin-mediated processing of pro-C-type natriuretic peptide. *J Biol Chem* **278**: 25847–25852.

Xia, W., Mruk, D.D. and Cheng, C.Y. (2007). C-type natriuretic peptide regulates blood-testis barrier dynamics in adult rat testes. In 3841–6.

Yan, W., Wu, F., Morser, J., and Wu, Q. (2000). Corin, a transmembrane cardiac serine protease, acts as a pro-atrial natriuretic peptide-converting enzyme. *Proc Natl Acad Sci USA* **97**: 8525–8529.

Yancy, C., Massie, B., Krum, H., Silver, M., Stevenson, L., and Mills, R. (2007). Chronic serial infusion of nesiritide is not associated with worsening renal function in chronic decompensated heart failure patients with renal insufficiency: An analysis from the FUSION-II Trial (abst). *J Card Fail* **13**: S136.

Yang, X. M., Philipp, S., Downey, J. M., and Cohen, M. V. (2006). Atrial natriuretic peptide administered just prior to reperfusion limits infarction in rabbit hearts. *Basic Res Cardiol* **101**: 311–318.

Yellon, D. M., and Hausenloy, D. J. (2007). Myocardial reperfusion injury. *N Engl J Med* **357**: 1121–1135.

Zeidel, M. L., Seifter, J. L., Lear, S., Brenner, B. M., and Silva, P. (1986). Atrial peptides inhibit oxygen consumption in kidney medullary collecting duct cells. *Am J Physiol* **251**: F379–F383.

Zeidel, M. L., Silva, P., Brenner, B. M., and Seifter, J. L. (1987). cGMP mediates effects of atrial peptides on medullary collecting duct cells. *Am J Physiol* **252**: F551–F559.

Zeidel, M. L., Kikeri, D., Silva, P., Burrowes, M., and Brenner, B. M. (1988). Atrial natriuretic peptides inhibit conductive sodium uptake by rabbit inner medullary collecting duct cells. *J Clin Invest* **82**: 1067–1074.

Zhao, J., Ardaillou, N., Lu, C. Y. et al. (1994). Characterization of C-type natriuretic peptide receptors in human mesangial cells. *Kidney Int* **46**: 717–725.

Zierer, A., Voeller, R. K., Melby, S. J. et al. (2006). Potential renal protective benefits of intra-operative BNP infusion during cardiac transplantation. *Transplant Proc* **38**: 3680–3684.

Further reading

Chen, H. H., Schirger, J. A., Cataliotti, A., and Burnett, J. C. (2006a). Intact acute cardiorenal and humoral responsiveness following chronic subcutaneous administration of the cardiac peptide BNP in experimental heart failure. *Eur J Heart Fail* **8**: 681–686.

Chen, H. H., Schirger, J. A., Alessandro, C., Martin, F. L., and Burnett, J. C. (2006b). Intra-renal infusion of BNP in experimental heart failure: a novel strategy to maximize the renal enhancing actions of BNP while minimizing arterial hypotension (abst). *J Card Fail*, p. S32.

Redfield, M. M., Edwards, B. S., Heublein, D. M., and Burnett, J. C. (1989a). Restoration of renal response to atrial natriuretic factor in experimental low-output heart failure. *Am J Physiol* **257**: R917–R923.

Redfield, M. M., Edwards, B. S., McGoon, M. D., Heublein, D. M., Aarhus, L. L., and Burnett, J. C. (1989b). Failure of atrial natriuretic factor to increase with volume expansion in acute and chronic congestive heart failure in the dog. *Circulation* **80**: 651–657.

Richards, A. M., Nicholls, M. G., Ikram, H., Webster, M. W., Yandle, T. G., and Espiner, E. A. (1985a). Renal, haemodynamic, and hormonal effects of human alpha atrial natriuretic peptide in healthy volunteers. *Lancet* **1**: 545–549.

Richards, A. M., Nicholls, M. G., Espiner, E. A. et al. (1985b). Effects of alpha-human atrial natriuretic peptide in essential hypertension. *Hypertension* **7**: 812–817.

Stingo, A. J., Clavell, A. L., Heublein, D. M., Wei, C. M., Pittelkow, M. R., and Burnett, J. C. (1992a). Presence of C-type natriuretic peptide in cultured human endothelial cells and plasma. *Am J Physiol* **263**: H1318–H1321.

Stingo, A. J., Clavell, A. L., Aarhus, L., and Burnett, J. C. (1992b). Cardiovascular and renal actions of C-type natriuretic peptide. *Am J Physiol* **262**: H308–H312.

Wei, C. M., Aarhus, L. L., Miller, V. M., and Burnett, J. C. (1993a). Action of C-type natriuretic peptide in isolated canine arteries and veins. *Am J Physiol* **264**: H71–H73.

Wei, C. M., Kim, C. H., Miller, V. M., and Burnett, J. C. (1993b). Vasonatrin peptide: a unique synthetic natriuretic and vasorelaxing peptide. *J Clin Invest* **92**: 2048–2052.

Section VI. Aldosterone

CHAPTER 21

Aldosterone: History and Introduction

JOHN COGHLAN[1] AND JAMES F. TAIT[2]
[1]210 Clarendon St, East Melbourne Victoria 3002, Australia
[2]Granby Court, Granby Road, Harrogate, N. Yorkshire, UK

Contents

I. Early history of aldosterone 311
II. Post-discovery progress to approximately 1970 318
III. Blockers of aldosterone action 322
References 324

I. EARLY HISTORY OF ALDOSTERONE

A. Earliest Work on Adrenal Extract

1. ACTIVITY OF THE AMORPHOUS FRACTION

The first serious attempts to isolate the compound from the adrenal cortex responsible for maintaining the life of adrenalectomized animals started in about 1934. These followed the preparation of biologically active extracts in the early 1930s free of contaminating compounds from the adrenal medulla, such as epinephrine. The major groups involved were those of (a) Winstersteiner and Pfiffner (1936), mainly working at Columbia University, who retired from the experimental field after a few years after making significant contributions, (b) Kendall and Mason (Mason et al., 1936) at the Mayo Clinic and (c) Reichstein (1936) and his coworkers, particularly von Euw, in Basel. Mason and Reichstein were still involved in the race for the isolation and identification of aldosterone in 1953. Research efforts in the field were also intense in the 1930s and many other groups were involved, such as those of Grollman (1937) and Hartman (1940). Kuizenga and Cartland (1939) also made notable contributions and a switch to their earlier methods of adrenal extraction was one of the factors vital to the final success of Reichstein in isolating relatively large amounts of aldosterone (Kuizenga, 1944; Simpson et al., 1953a, b; Tait and Tait, 1998). In the early work, it was realized that the life maintenance tests for adrenalectomized animals under low stress conditions favored so-called mineralocorticoids, such as deoxycorticosterone, but similar tests with the animals under stressful conditions (e.g. the Selye Schenker test using adrenalectomized rats kept in the cold) favored so-called glucocorticoids, such as cortisol (Swingle and Remington, 1944). The Ingle work test (Ingle, 1940), used particularly by Kendall, measured chronic muscle fatigue and favored glucocorticoids. On the other hand, the Everse–de Femery assay (Everse and de Femery, 1932), particularly used by Reichstein, measured the immediate response of muscle and favored mineralocorticoids. Nevertheless, this division of corticosteroids into these classifications, although useful, was arbitrary. The glucocorticoids exhibited some mineralocorticoid activity and vice versa. When, in the 1950s, glucocorticoids such as cortisone, were used clinically in relatively large amounts, it became evident that they had marked effects on electrolyte metabolism. Also, although aldosterone is unlikely to have a significant effect on carbohydrate metabolism in humans compared with that of cortisol because of the relative amounts secreted, on an equi-weight basis, it is one-third to one-quarter as effective as cortisone in the mouse eosinophil assay of Speirs (Speirs et al., 1954). Results using this micromethod correlate well with other macrotests for glucorticoid activity, such as the liver glycogen test. Nevertheless, as a result of the application of such glucocorticoid and mineralocorticoid assay methods, many corticosteroids, of both types, were crystallized and identified before 1940, particularly by the teams of Kendall-Mason and Reichstein, and their biological properties thoroughly studied (Swingle and Remington, 1944).

It was realized by nearly all the early investigators in the field that the amorphous fraction remaining after the crystallization of these steroids, contained significant biological activity. Kuizenga (1944) stated that the amorphous fraction contained considerable activity, particularly when measured by mineralocorticoid assays. Glucocorticoid assays did not show such a high proportion of the total original activity. Also, Kuizenga (1944) concluded that the active component in the mineralocorticoid assays was not due to the most active known mineralocorticoid, deoxycorticosterone. From the properties of the compound responsible for most of the biological activity in these early studies, it seems clear that nearly all the early investigators had preparations containing substantial amounts of aldosterone (Winstersteiner and Pfiffner, 1936; Mason et al., 1936; Kuizenga and Cartland, 1939; Kuizenga, 1944). It was reported that, in contrast to deoxycorticosterone, the active compound was as polar (in partitioning between water and benzene) as the O_5 steroids, such as cortisone. It also lost its biological activity after acetylation and was very sensitive to acid and alkaline conditions (Mason, 1939; Kuizenga, 1944). All these properties reported by the early workers for the amorphous fraction are also exhibited by aldosterone.

Therefore, it seems highly likely that most of the biological activity found by the early workers in the amorphous fraction was due to aldosterone. The data of Kuizenga (1944) on its specific biological activity suggest that the best preparations contained nearly 20% aldosterone by weight. If that were so, then even the least efficient of the mild methods of purification available at the present time would have easily yielded pure material from such preparations. Crystalline aldosterone was actually not obtained until 1953, some 18 years after the serious start of the search for the active compound in adrenal extracts. The delay is partially because the Second World War intervened and, although there was still research activity in the adrenal field, this was mainly directed toward the possible use of glucocorticoids to prevent pilot fatigue. Immediately after the war came the success of the glucocorticoids in treating certain disease conditions, such as rheumatoid arthritis, and the award of the Nobel Prize for this work to the clinician Hench and the adrenal pioneers, Reichstein and Kendall.

2. LOSS OF FAITH IN SIGNIFICANCE OF THE AMORPHOUS FRACTION

There was then, it seems, a certain loss of faith in the significance of the amorphous fraction. It seems that only the Mayo Clinic team, now headed by H. Mason, retained a substantial belief in its importance. The use of steroids, such as cortisone (and cortisol), in clinical applications had made investigators realize that these glucocorticoids also had effects on electrolyte metabolism, although these were regarded as undesirable side effects. It was also being realized from the content of adrenal venous blood in animals, as found by Vogt, Bush and Nelson (Vogt, 1943; Bush, 1953a, b; Nelson et al., 1950) and from the total quantity of the appropriate urinary metabolites, that cortisol was secreted in amounts similar to those used clinically.

Therefore, for all these reasons, influential groups such as those of Albright and Bartter and Conn and co-workers (Fourman et al., 1950; Conn et al., 1951) proposed that cortisol was *the* single adrenal hormone, which could control both carbohydrate and electrolyte metabolism. Ironically, some of the co-authors of these papers, such as Bartter and Conn, later made outstanding contributions to the aldosterone field. Even Bush (1985), who had investigated the amorphous fraction originally prepared by Wintersteiner, stated, 'I had almost convinced myself to join the Unitarians'. We should emphasize that the unitarian views of all these workers were reasonably based on the evidence available at the time. On the other hand, the acceptance of the existence of aldosterone was certainly not a smooth development, as represented in some historical accounts.

However, we now know, as emphasized by Ulick (Ulick et al., 1979; Ulick, 1996), that cortisol can be a ubiquitous single adrenal hormone when there is excessive systemic or local production of cortisol, as in Cushing's syndrome or in the apparent mineralocorticoid excess (AME) syndrome. Even in normal subjects, cortisol probably contributes appreciably to the total mineralocorticoid activity (Tait and Tait, 1979). There is, however, an important difference in the physiological role of aldosterone and cortisol. The secretion of aldosterone (unlike that of cortisol) is physiologically controlled by such factors as the intake of electrolytes, so leading usually to a stable control system as the level of aldosterone in turn affects the excretion and distribution of these electrolytes. Cortisol is obviously part of a different, mainly adrenocorticotrophic hormone (ACTH), control system. In physiological situations, at the present time, it seems that aldosterone is the dominant hormone as regards the adrenal control of sodium and potassium metabolism.

This was certainly not the view of most people in the field in the early 1950s as regards the role of an active compound in the amorphous fraction. Apart from the views of Albright and Conn and co-workers, previously mentioned, Sayers (1950), an influential physiologist, stated that the activity of the amorphous fraction could be due to several compounds acting synergistically and there was no evidence for the secretion of a single active compound.

Also, the efforts of the very skilled chemical workers in the field, such as Kendall and Reichstein and their collaborators, had failed to isolate an appropriate active compound. It now seems that the relevant failure of the steroid chemists was probably due to an unfortunate juxtaposition of events. The Kendall group used very mild methods of separation, such as partition countercurrent distribution and these would have preserved aldosterone. However, the Ingle work test (Ingle, 1940), which they used, was most sensitive to the effects of glucocorticoids. Although it was later found that aldosterone had some effect on carbohydrate metabolism, it would probably have been impossible to follow aldosterone in fractionation work using the Ingle test. On the other hand, Reichstein used the Everse–de Femery assay (Everse and de Femery, 1932), which although also cumbersome and insensitive, measured mainly mineralocorticoid activity and aldosterone would have been very active in this assay. Unfortunately, for the aim of isolating aldosterone, Reichstein and von Euw used alumina columns for isolating steroids. This highly efficient procedure was responsible for their success in isolating many steroids from adrenal extracts. However, using this method, it was necessary first to acetylate steroids with an α ketol side chain to prevent their destruction. At the time, all the known 21-acetyl derivatives of steroids were found to be nearly as biologically active as the corresponding free compound, so this was not usually a problem. Unfortunately, under the same conditions of acetylation, aldosterone diacetate is formed. This compound is relatively inactive in the usual mineralocorticoids bioassays, such as the ^{24}Na/^{42}K assay and probably also in the Everse–de Femery test used then by Reichstein (Simpson and Tait, 1952). The efficient method of Reichstein and Shoppee (1943) of regenerating the free aldosterone followed by bioassay would probably have been successful in detecting aldosterone diacetate but was not considered to be necessary in the early work. The biological inactivity of the 18-acetyl derivatives of aldosterone was unfortunate, as this was not expected from previous experience in the field with the 21-acetates. However, it probably did delay by some years the isolation and identification of the hormone by Reichstein.

B. Direct Bioassays for Mineralocorticoid Activity

At about this time, various groups, such as Dorfman et al. (1947), Spencer (1950), Deming and Luetscher (1950), Simpson and Tait (1952) and Singer and Venning (1953), were devising methods that directly measured the action of steroid hormones on electrolyte metabolism. Particularly noteworthy were the pioneering studies of Dorfman et al. (1947), which initiated the work of Simpson and Tait (1952), and the early success of Deming and Luetscher (1950) in devising a practical method. All these methods had greater convenience and sensitivity than the traditional bioassay methods, such as the Everse–de Femery test. The direct microassays were mostly applied to human urine, as in the work of the Luetscher and Venning groups, but Grundy et al. (1952b) at the Middlesex Hospital Medical School (MHMS) investigated a beef adrenal extract prepared by Allen and Hanbury, UK. Following the publication by Dorfman et al. (1947), Simpson and Tait (1952) had devised a bioassay that measured the effects of steroids on the urinary ^{24}Na/^{42}K of adrenalectomized rats after injection of tracer amounts of the radioactive isotopes. Using such tiny amounts meant that the animals were not loaded with electrolytes that otherwise would have been necessary with the insensitive flame photometers then available to these investigators. After loading with electrolytes, it was found that steroids, such as cortisol, caused an increase in the urinary ^{24}Na/^{42}K ratio whereas deoxycorticosterone and aldosterone always lowered this ratio. However, with an injection of tracer quantities of electrolytes, all steroids tested acted unidirectionally and this simplified interpretations of the results of the effects of biological extracts particularly with assaying mixtures of hydrocortisone and aldosterone. This phenomenon could be due to the action of cortisol (and possibly other glucocorticoids) in increasing the glomerular filtration rate of adrenalectomized animals (cf. however, Gaunt et al., 1949) and emphasizes again the crude nature of the division of steroids into mineralocorticoids and glucocorticoids.

Later, other bioassays used the urinary Na/K ratio to indicate mineralocorticoid activity but measured the non-isotopic ratio by flame photometry (Maddox et al., 1953b; Kagawa, 1960). The relative steroid potencies were found to be similar using either the radioactive or non-isotopic ratio under the conditions employed.

C. Studies on Adrenal Extract Using the ^{24}Na/^{42}K Assay and Paper Chromatography

The ^{24}Na/^{42}K assay showed the very high activity of the Allen and Hanbury's adrenal extract that could not possibly be due simply to the content of deoxycorticosterone or any other known steroid (Tait et al., 1952; Grundy et al., 1952a). However, this fact did not advance the conclusions of the earlier workers on the activity of adrenal extract and could have been explained by synergistic actions between the known steroids (Kuizenga, 1944; Sayers, 1950). The extract was therefore fractionated by paper chromatography by the Zaffaroni method that used propylene glycol–toluene as solvents (Burton et al., 1951).

Much to the initial surprise of the investigators, in seeming confirmation of the prevailing unitarian ideas of the time, the compound responsible for the biological activity ran at exactly the same speed as cortisone, at least with the usual running times of the paper chromatographic procedure (Tait et al., 1952; Grundy et al., 1952a, b). Because of its low potency in the ^{24}Na/^{42}K assay, cortisone alone could not directly account for the activity but a synergistic effect involving cortisone could not be ruled out.

More direct evidence arose from overrunning the Zaffaroni system, for 7 days, which resulted in the complete separation of the active compound from cortisone (Grundy et al., 1952a, b). Later, the use of the Bush B5 paper chromatographic system (aqueous methanol:benzene) rapidly separated the active compound from cortisone and cortisol (Bush, 1952; Simpson and Tait, 1953). These results definitely eliminated the possibility that the biological activity could be due to synergism involving known steroids and strongly indicated that it was due to a single compound (then named electrocortin, later aldosterone).

D. Discovery of Electrocortin (Aldosterone) as a Hormone

It still remained to be demonstrated that electrocortin (aldosterone) was secreted by mammalian adrenal glands. At the time, near the MHMS in London where Simpson and Tait and collaborators were obtaining their results on the activity of adrenal extracts, Bush was working at the MRC Laboratories, Mill Hill. He had devised his paper chromatographic system for the microanalysis of steroids (Bush, 1952) and had made preparations of perfused mammalian adrenal glands (similar to those of Martha Vogt 1943). Using these methods, Bush had already achieved several important findings in the field such as the marked species difference in the ratio of secreted cortisol to corticosterone by the adrenal gland (Bush, 1953a, b). He was also interested in the components in the amorphous fraction of the adrenal extracts of the earlier workers. However, Bush was unable to identify the proposed active unknown compound from a Wintersteiner preparation using only paper chromatography. Collaboration between the two London groups then occurred and it was rapidly demonstrated that electrocortin was secreted by dog and monkey adrenal glands (Simpson et al., 1952; Bush, 1953b).

The active material in the ^{24}Na/^{42}K assay from the adrenal blood extracts ran at the same speed on paper chromatography in the Bush B5 system as the electrocortin obtained from the extracts of beef adrenal glands. As another important unique characteristic, the activity in the beef adrenal extract disappeared after acetylation but could be regenerated by alkali hydrolysis (with low but consistent yields by the inefficient methods used by Simpson and Tait at the time) from the non-polar regions of Bush paper chromatograms, indicating that the acetyl derivative contained more than one carboxyl group (Simpson et al., 1952; Bush, 1952). The active compound in the dog adrenal blood behaved in the same manner. These properties were unique and constituted proof that electrocortin (aldosterone) was secreted by mammalian adrenal glands and was not only present artefactually in adrenal extract. Later, this finding was confirmed for dog adrenal (Farrell et al., 1953) and for rat adrenals by Singer and Stack-Dunne (1955).

Soon afterwards, John Luetscher and co-workers, working in San Francisco, USA, obtained salt-retaining activity with similar properties from the more difficult starting material of human urine from patients with edema (Luetscher et al., 1954, 1955, 1956; Luetscher, 1955). In the earliest work of Luetscher et al. (1954), it was not known that the active compound originated from the adrenal gland. However, it became clear that the active compound in this material and electrocortin were similar and, after Simpson and Tait sent a sample of electrocortin to Luetscher, it was evident that they were identical. Electrocortin (at the time, then named aldosterone) was later crystallized from the urine of patients with the nephrotic syndrome (Luetscher et al., 1955). Relevant to the results of the RALES clinical trials (Pitt et al., 1999; Weber, 1999), Luetscher and co-workers also found that some patients with congestive heart failure were found to excrete increased amounts of the active compound, which was crystallized and identified as aldosterone (Luetscher et al., 1956). Therefore, these patients (and also normal humans), secreted aldosterone. However, it was established as a general mammalian hormone in 1952 and this could be considered, according to the usual convention in endocrinology, to be the year of the discovery of electrocortin (later named aldosterone) as a hormone.

E. Purification of Electrocortin (Aldosterone), Preliminary Results on Structure and Isolation in Crystalline Form

1. Purification and Early Structural Studies

After the discovery of electrocortin, the interest in the field then turned to its isolation in crystalline form in reasonable amounts, chemical identification and synthesis. Simpson and Tait (1953), with the help of various colleagues in London, applied various micromethods to the analysis of 1 mg sample of pure material prepared by partition column chromatography and found that electrocortin had a Δ^4 3 oxo group and an alpha ketol side chain. Using ^{14}C-labeled acetic anhydride, they also found that electrocortin had another acylable hydroxyl group but, according to infrared analysis, no other non-acylable hydroxyl groups. There was also some evidence from infrared analysis that the unknown carboxyl group of the diacetate was in proximity to that on the c-21 position. On this basis, Simpson and Tait suggested that one possibility for the position of the unknown carboxyl group was at c-16. This suggestion, although tentative, seems to have stimulated a great deal of work of some synthetic steroid chemists at the time. The identification of aldosterone, mainly by Reichstein, showed that the structure did not have a c-16 hydroxyl, although he seriously considered this possibility at quite a late stage (Tait and Tait, 1998). Ironically, it was eventually found that the substitution of steroids at c-16, e.g. in triamcinolone, led to the loss of the salt-retaining activity of the parent molecule and this was useful for clinical therapy using glucocorticoids.

2. Isolation in Crystalline Form

At this stage, there was clearly a need for larger quantities of pure material to elucidate the structure of electrocortin by more classical chemical means. Simpson and Tait then started full collaboration with T. Reichstein and J. von Euw (and the Ciba group, including Drs A. Wettstein, R. Neher and J. Schmidlin) and supplied them with all their information on the properties of electrocortin. Reichstein and von Euw then started the purification of the extract (mainly by the procedure of Cartland and Kuizenga) of 500 kg of beef adrenals prepared by Organon Ltd, Holland. Reichstein used partition chromatography on large kieselguhr columns with aqueous methanol:benzene solvent systems and eventually 21 mg of electrocortin was crystallized with the critical use, in effect, of the ^{24}Na/^{42}K bioassay in London to control the separations (Tait and Tait, 1998; Tait and Tait, 2004).

The 21 mg of crystalline electrocortin was sufficient in quantity for degradative studies, at least in the hands of a steroid chemist as skilled as Reichstein. Shortly afterwards, smaller quantities were also crystallized in the laboratories of the Ciba team and the Taits. Also, the Mayo group crystallized the hormone within a few weeks of Reichstein (Maddox et al., 1953a, b). In the following year, the Merck

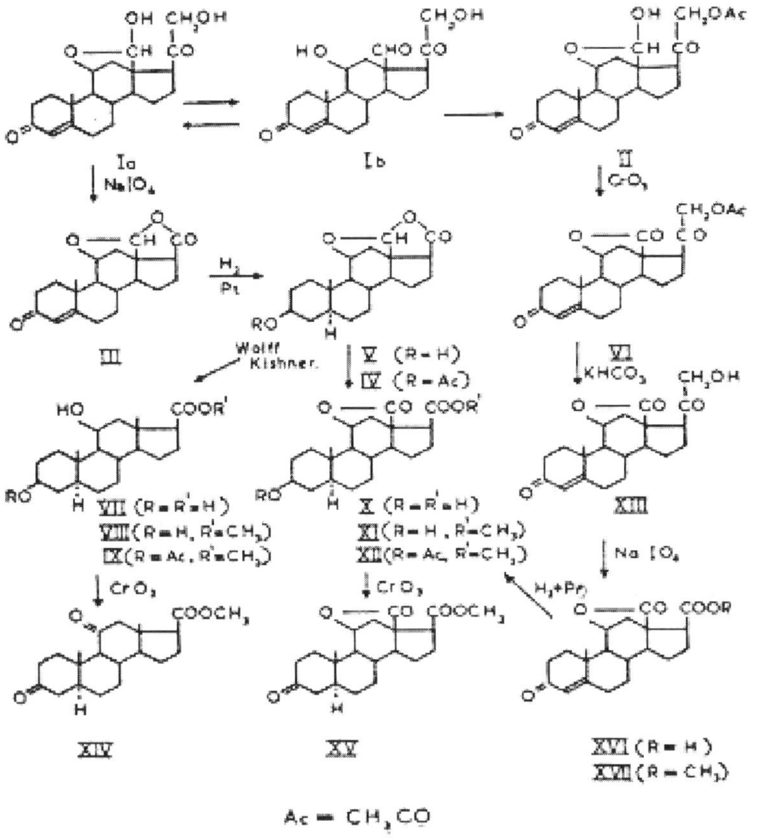

FIGURE 21.1 Degradation scheme used by Reichstein. Key use of lactone (III). From Simpson et al. (1953b). *Experientia* 10, 132.

group also succeeded in obtaining crystals of the hormone (Harman et al., 1954).

F. Identification and Synthesis of Aldosterone (Electrocortin)

Apart from in the original publications (Simpson et al., 1953a, b), the degradative studies used by Reichstein have been described by Wettstein (1954), Neher (1979) and Fieser and Fieser (1959), although these accounts differ slightly. Probably the most reliable full published account is based on the direct correspondence of Reichstein with the Taits at the time (Tait and Tait, 1998) (Figure 21.1).

The crucial compound was the oxidation product of the free compound (a γ lactone, VI, Figure 21.2) which sublimed and crystallized easily and could be purified directly from crude preparations (Wettstein, 1954). Reichstein concluded that there was an aldehyde group at the

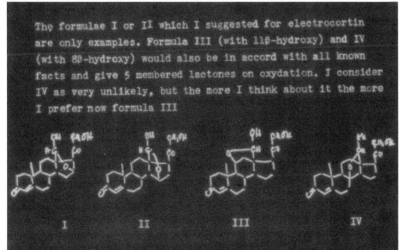

FIGURE 21.2 Reichstein intuition on structure of aldosterone. From Original correspondence of S.A.S. Simpson and J.F. Tait with T. Reichstein, Welcome Museum for the History of Science, London.

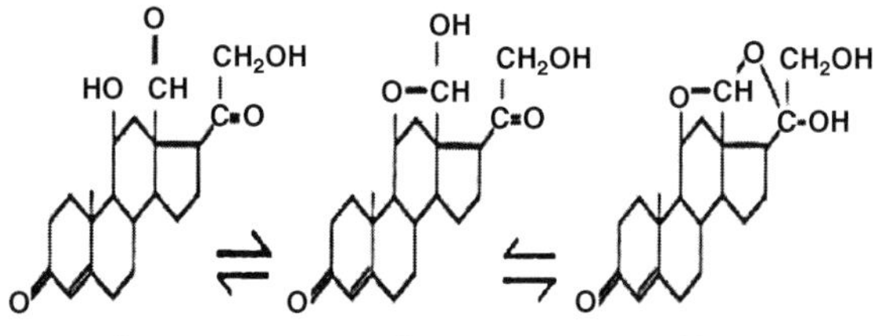

FIGURE 21.3 Tautomers of aldosterone. From Neher, R. (1979). *J Endocrinol* 81, 25P–35P.

c-18 position, mainly from the properties of this oxidation product. It was also concluded that there was a hemi-acetal formed from the c-18 aldehyde (e.g. after oxidation of the aldosterone 21 monoacetate). As previously discussed, Simpson and Tait had earlier found, using ^{14}C acetic anhydride, that the fully acetylated derivative was a diacetate (rather than a triacetate that had been suggested by preliminary results in the Ciba laboratories in Basel). This was a critical issue at one stage of the work as the diacetate did not readily crystallize. If there were two acylable hydroxyl groups (one presumably at position c-21) and no remaining hydroxyl groups (as indicated by infrared analysis), then a hemi-acetal was probably formed with a hydroxyl group at carbon positions 8, 11, 15 or 16 (Simpson and Tait, 1953; Wettstein, 1954; Fieser and Fieser, 1959; Neher, 1979; Tait and Tait, 1998).

Reichstein, who had first favored the c-15 or c-16 positions, suddenly and somewhat intuitively, chose the c-11β position (Figure 21.3). He was proven to be correct after further classical chemical studies including degradation from aldosterone to a known steroid (Simpson et al., 1954a, b; Tait and Tait, 1998). This proved, for the first time, that aldosterone was a steroid.

The c-11β hydroxyl structure in aldosterone might explain the significant glucocorticoid potency (on a stoichimetric basis) of the hormone, as previously discussed. This activity of aldosterone was discovered at about the same time Reichstein first considered that aldosterone might have a c-11β hydroxyl group (Tait and Tait, 1998). The function of the glucocorticoid activity of aldosterone (if any) has still to be elucidated.

Finally then, in 1954, the structure of aldosterone was elucidated as 11β, 21-dihydroxy-18-oxo-pregn-4-ene-3, 20 dione by Reichstein and co-workers (Simpson et al., 1953b). In solution, it can exist in three tautomeric forms, as shown in Figure 21.3 (Neher, 1979). This tautomerism explains the lack of the aldehyde peak and the weak 20-oxo peak in the IR absorption spectra, which Dr A. Kellie (of the MHMS) first observed (Simpson and Tait, 1953) and was confirmed with the crystalline material (Simpson et al., 1954a). Clearly, Reichstein was the key scientist involved in the final identification of aldosterone and he appropriately, with the agreement of Simpson and Tait, renamed electrocortin, aldosterone. Soon afterwards, both the Mayo Clinic (Maddox, 1955) and Merck (Ham et al., 1955) groups confirmed this structure. Aldosterone was first synthesized by the Ciba group in Basel (Schmidlin et al., 1955; Vischer et al., 1956), subsequently by Reichstein and chemists at Organon (Fieser and Fieser, 1959; Neher, 1979). Later, an original method of synthesis by Barton involved photolysis of corticosterone acetate nitrite to aldosterone acetate oxime, which could then be readily converted to aldosterone 21 monoacetate (Barton et al., 1961).

The history of the discovery, isolation and identification of aldosterone has also been presented elsewhere (Wettstein, 1954; Simpson and Tait, 1955; Tait and Tait, 1979, 1988, 1998; Tait et al., 2004).

G. Adrenal Site of Production of Steroids

As regards the classical question as to the purpose (if any) of the division of the adrenal into the zona glomerulosa and fasciculata (and zona reticularis) (Figure 21.4), there had been various theories advanced. One (the functional zonation theory) suggested that the various zones produced different steroids; another (the cellular migration theory) suggested that these were regions of different mitotic activity. Although these theories are not necessarily mutually exclusive, the choice between them was once the subject of much discussion.

The discovery of aldosterone meant that these theories could be investigated more directly. Giroud et al. (1956), in Montreal, demonstrated that the zona glomerulosa of the rat exclusively produced aldosterone, so confirming the classical but more indirect studies of Swann (1940) and Deane et al. (1948). At about the same time, the MHMS London group also found exclusive production of

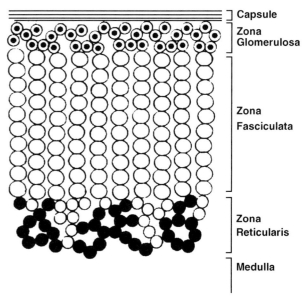

FIGURE 21.4 Histology of adrenal cortex. From Tait, S.A., Tait, J.F., and Coghlan, J.P. (2004). *Mol Cell Endocrinol* 217, 1–21.

aldosterone in both the rat and beef zona glomerulosa of adrenal glands and of cortisol in the zona fasciculata in the beef adrenal (Ayres et al., 1956). Both zones in the two species produced corticosterone.

The MHMS group also found preferential production of aldosterone by the zona glomerulosa and cortisol by the zona fasciculata of human adrenal glands, although, because of the more convoluted nature of the glands, this is more difficult to demonstrate than in rat and beef glands (Ayres et al., 1958). On the other hand, one adenoma from a case of primary aldosteronism produced the same steroids, including cortisol and aldosterone and also substantial amounts of corticosterone, for all slices from the outer to inner layers (Ayres et al., 1958).

H. Metabolism of Aldosterone

This structure of aldosterone led to some equally unique biochemical properties, apart from its high mineralocorticoid activity. Aldosterone is metabolized, as is cortisol, by reduction of its Δ4 3 oxo structure to A ring reduced metabolites such as 5β, 3α tetrahydroaldosterone (Kelly et al., 1962). However, the c-18 hydroxyl group in the hemi-acetal structure is also available for metabolism in humans as in the formation of aldosterone 18 glucuronide (Figure 21.5). This is the urinary metabolite of aldosterone, first used as an indication of aldosterone secretion by Luetscher, and later found by his group to be formed to a major extent in the kidney (Cheville et al., 1966). Underwood and Tait (1961) first indicated that it was the 18 glucuronide of aldosterone. This finding was later confirmed more rigorously by Carpenter and Maddox (1976), who also showed that the glycosidic linkage has the β configuration. This conjugate is hydrolyzed more readily in acid conditions than other glucuronides, such as those on the c-3 position, but not so readily by β glucuronidase.

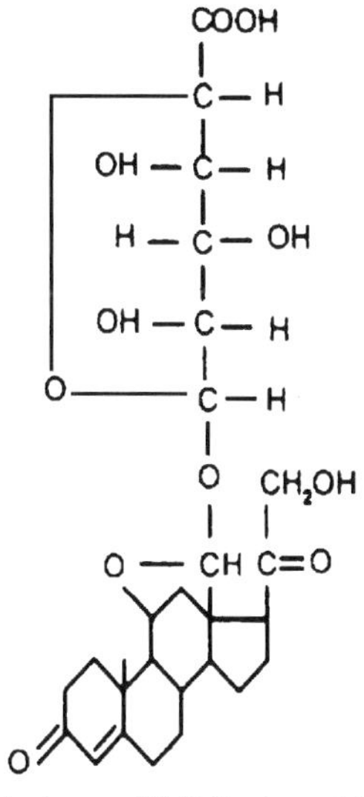

Aldosterone in the free form in urine, as for cortisol, is a small proportion of the secreted hormone. However, when free aldosterone is measured after mild acid (pH1) treatment of the urine, it reflects the excretion of aldosterone 18 glucuronide that is a major metabolite of the hormone. This glucuronide is formed directly from the hormone unlike the A ring glucuronide reduced metabolites of aldosterone and cortisol. Therefore the formation of the 18 glucuronide has a direct effect to increase the metabolic clearance rate (MCR) of aldosterone.

Also, as previously mentioned, acetylation of the c-18 hydroxyl group led to the formation of aldosterone 18 monoacetate (or the 18, 21 diacetate) that, unlike the 21 monoacetates, are biologically inactive. The lack of biological activity of the c-18 monoacetate in Na/K assays is probably because it is not hydrolyzed readily *in vivo* enzymatically, e.g. by cholinesterases or acetylesterases. There is also a relatively low rate of enzymatic hydrolysis of aldosterone 18 glucuronide, as previously discussed.

Because of the unique structure of aldosterone with the c-11 hemi-acetal group, that presumably protects the 11β hydroxyl, a conversion product with a c-11 oxo group of aldosterone is not formed. It was found that the interconversion of cortisol and cortisone was a significant factor in controlling the mineralocorticoid activity of cortisol as cortisone itself is biologically inactive (Ulick et al., 1979). However, with aldosterone, because the corresponding 11 oxo compound is not formed, this process cannot occur. In clinical cases of apparent mineralocorticoid excess (AME), or after administration of glycyrrhizinic acid, there are signs of increased mineralocorticoid activity (Ulick et al., 1979). At first, it was thought that aldosterone might be responsible. However, it was found that the 11β hydroxysteroid dehydrogenase activity is inhibited in these situations. Cortisol conversion to the inactive cortisone is then reduced and cortisol becomes the dominant mineralocorticoid. Because of the inhibition of control mechanisms by this dominant effect of cortisol, aldosterone secretion actually is normal or reduced.

The simplest explanation of the ready availability of the c-18 hydroxyl group for conjugation with glucuronic acid is that aldosterone exists *in vivo* as the hemi-acetal or acetal form (Neher, 1979). This is largely confirmed by the lack of a c-11 oxo metabolic product suggesting that the 11β hydroxyl is not available for conversion by 11β hydroxysteroid dehydrogenase, as previously discussed. The low binding of aldosterone to plasma corticosteroid binding globulin (CBG) also suggests that the 11β hydroxyl of aldosterone is not exposed, as other 11β hydroxyl Δ4- 3 oxo steroids, such as cortisol and corticosterone, (but not the 11 oxo cortisone) bind strongly to CBG (Tait and Burstein, 1964). It remains to be seen whether the high mineralocorticoid and appreciable glucocorticoid activities of aldosterone are also due to

FIGURE 21.5 Aldosterone 18 β-D-glucosiduronic acid (aldosterone 18 glucuronide). From Tait, S.A., Tait, J.F., and Coghlan, J.P. (2004). *Mol Cell Endocrinol* 217, 1–21.

binding of a preferred hemi-acetal or acetal structure to the appropriate receptors. Unfortunately, the structural range of aldosterone antagonists so far synthesized is too limited for them to be very helpful in this regard.

II. POST-DISCOVERY PROGRESS TO APPROXIMATELY 1970

A. Primary Aldosteronism

1. Definition

The general aim of clinical researchers from the discovery of a hormone or factor was to speed it as fast as possible to the bedside. A great advocate of this approach was Professor George Thorn. Aldosterone from its discovery in the early fifties was in the clinic in a record short time largely due to the recognition of the disease entity: primary hyperaldosteronism.

Long before Dr Jerome Conn of the University of Michigan Ann Arbor described the syndrome that bears his name, medical investigators had treated many patients with symptoms of what was to be also named primary aldosteronism. However, the basic pathology in those early days was thought to be renal defects (Conn et al., 1951). The initial case was of a single large, 1 cm, unilateral adrenal adenoma (Conn, 1955a). Jerome Conn was uniquely placed to make this discovery as his war time (WWII) research was to study 'acclimatization'. He had found that military personnel who had acclimatized to a hot climate had low sodium sweat levels. After aldosterone was discovered, he reasoned that individuals with high aldosterone levels would have low sodium sweat levels. Indeed, the first patient with an aldosteronoma of the adrenal cortex had a low sodium sweat level, thus documenting the utility of sweat sodium as an onboard bioassay. The Howard Florey Institute group used saliva sodium/potassium ratio as a similar bioassay in their extensive studies in sheep. Thus, the transfer from discovery to clinic appeared to be achieved with Conn's recognition of what is now named Conn's syndrome in 1954, or primary aldosteronism (Conn, 1955a, b).

The start of research on therapeutic antagonists began as early as 1957. However, the approach to the extensive physiology and pathophysiology of aldosterone and an activated mineralocorticoid receptor was limited to its action in the kidney for about 40 years. This 'mineralocorticoid' sobriquet for aldosterone dominated and confined thinking until recently. It is difficult to imagine the developments that would have taken place if the biomarker used during fractionation had been vascular inflammation (see Chapter 25). That mineralocorticoids acting via their receptors are involved in many more processes than epithelial sodium transport, including fast membrane action, is now very clear. More insights may have been gleaned earlier if more attention had been paid to the studies of the Robert Gaunt and Hans Selye groups (Gaunt and Eversole, 1949; Selye and Horava, 1953).

In Conn's day, he wrote that the incidence of Conn's syndrome, primary hyperaldosteronism, among essential hypertensives was still open to debate. Conn's work on acclimatization in tropical zones appeared later again among his prolific publications (Conn, 1963). In 1969, a review of the field (Coghlan et al., 1971b) raised considerations about the definition of the types of Conn's syndrome that was confusing at the time. How many essential hypertensives should be included was contentious then, and still is. Conn suggested up to 20% of primary aldosteronism might be found among essential hypertensives, an estimate based primarily on the frequency of low plasma renin activity levels in the essential hypertensive population. However, by 1970, the accepted figures ranged between 0.5% and 10%, but many clinicians believed that 5% was the more realistic absolute upper limit. The incidence may depend on the patients in a particular group, on genetics and especially on the environment. Tropical zones and high sodium intake and possibly other factors may be important. Before the 1970s, it seemed that bilateral hyperplasia was less common than unilateral adenomas but recent work has cast doubt on this (Lim et al., 2000; Stowasser and Gordon, 2004). As early as 1965, Conn had the perception of definitional change in the offing with a paper 'Normokalemic primary aldosteronism a detectable cause of curable essential hypertension'. Was this manifestation a slightly different form of the disease or merely early diagnosis before the total body potassium was significantly reduced (Conn et al., 1965)?

2. Diagnostic Criteria and Approaches

It could be confusing for us to leave the question of incidence among the essential hypertensive population as it was before 1970. The incidence of primary hyperaldosteronism in contemporary literature can be as high as 10% with about 55% of these having bilateral hyperplasia. New tools became available; in 1972, a procedure of adrenal venous sampling was introduced to lateralize unilateral adrenal malfunction utilizing the suppression of aldosterone in the contralateral adrenal (good adrenal), the tumor being on the opposite side that otherwise could only be recognized at surgery and not always even then (Scoggins et al., 1972). Although the left adrenal vein drained into the left inferior phrenic that was sampled and, in those days, the idiosyncratic disposition of the right adrenal vein made it too difficult. Improved techniques now allow this to be performed routinely. A large series using this approach demonstrated its value (Ma et al., 1986). Lateralization fell into disuse as claims were made that imaging was equivalent in specificity and sensitivity with much less morbidity. However, this claim has not proved to be true and adrenal venous cannulation is considered the gold standard for diagnosing an aldosteronoma if, in relation to patient sodium status, the aldosterone is found inappropriately raised (Mosso et al., 2003; Mulatero et al., 2004).

A number of screening tests for primary aldosterone has been proposed. Earlier, an unexplained low serum potassium was used. In the last decade, the most common screen test employed has been the ratio of aldosterone to renin activity in the blood with a high ratio being suggestive of primary aldosteronism. The rationale for this test is that patients with primary aldosteronism will have increased sodium retention, volume expansion and therefore suppression of plasma renin activity (PRA). Thus, the ratio of aldosterone to PRA will be high. However, its reliability is limited by the level of PRA. Very low PRA with normal or even low aldosterone levels often produces a false positive ratio. Several drugs and diseases other than primary aldosteronism are associated with very low PRA levels. Thus, for the test to be valid, most investigators now require the serum aldosterone levels to be elevated with the PRA measurable and above 0.1 ng/ml/h. The definitive test for primary hyperaldosteronism is based on the suppression of aldosterone levels either in the urine or blood. Two forms of suppression are used: salt loading either intravenously or with diet, or administration of synthetic mineralocorticoids.

3. INCIDENCE

Recognition that primary aldosteronism is the underlying problem in essential hypertension allows a surgical cure in the case of adenomas or an improved therapeutic approach in bilateral hyperplasia. Mosso et al. (2003), Mulatero et al. (2004) and Williams et al. (2006), however, report an incidence of only 3.2% in a normokalemic moderately hypertensive group. These studies for various reasons, strictly speaking, cannot be compared.

Nevertheless, much has been made of this issue in the recent clinical literature and will no doubt be addressed in full detail in the chapters that follow. Possibly, a new definition would help. It should be borne in mind that unilateral adenomas, single or multiple are frequent, although multiple adenomas are less common but varying with the group reporting. The single adenomas (microadenomas) are often quite small and hard to see at surgery. The incidence of bilateral hyperplasia is subject to international debate, but is more than 50% as set out above.

Allowing the number of variables involved between groups and the difference in hypertensive screening procedures, there is actually generally good agreement concerning the incidence of primary aldosteronism. Thus, the constellation of stigmata that were defining parameters in the original Conn case are not all present and this may be affected by many factors: different populations, different patient watershed, different hypertensive screening, earlier diagnosis or even better control of sodium intake and postural change (Table 21.1) (Gordon et al., 2005; Funder, 2005). In many respects, the entity, primary aldosteronism, like essential hypertension itself, has evolved from a 'disease' to a 'syndrome' with several independent mechanisms leading to the final common feature – dysregulated aldosterone production.

TABLE 21.1

Primary aldosteronism
1. Unilateral single adenoma (Conn's syndrome)
2. Unilateral multiple adenomas
3. Bilateral hyperplasia
4. Bilateral nodular (rare?)

Secondary hyperaldosteronism
1. Pregnancy?
2. Liver disease
3. Cardiac failure
4. Renal disease
5. Idiopathic edema?

B. Measurement

1. BIOASSAY

The bioassay of aldosterone is dealt with extensively in the history above; as always, an excellent bioassay being sensitive and reproducible, was critical for the isolation of the new hormone. This isotopic bioassay was not readily applicable for clinical screening although, as mentioned above, Conn used sweat sodium as a diagnostic aid.

Extraction, chromatographic purification and ^{24}Na/^{42}K bioassay was not routinely used as it was considered far too tedious and complex but, in fact, no more so than the double isotope dilution derivative assays that were subsequently used.

The Na/K ratio of parotid saliva was used extensively in the sheep, although about 90 minutes out of phase with the secretion change. The parotid salivary assay, of course, put a benchmark in place that had to be met by the emerging chemical assays.

C. Na Fluorescence

There was no chemical reaction that was unique to aldosterone so that it could be easily measured. Observing the yellow fluorescence on paper chromatograms after treatment with sodium hydroxide (NaOH) was devised by Bush (1952) as a specific test for Δ4, 3 oxo steroids and it was very successful. In the structural work on aldosterone, this method was used to correct the erroneous tentative conclusion using UV spectra. However, for the test to be specific for aldosterone, other Δ4, 3 oxo steroids had to be separated out usually by Bush paper chromatography (Bush, 1952; Bush and Willougby, 1957). At first, the method was only semi-quantitative but was improved to be reasonably quantitative by the method of Ayres et al. (1957). Nevertheless, although the overall method could be made specific, it lacked sensitivity compared with radioactive-labeled reagent methods and was soon replaced by them in most laboratories.

1. ISOTOPIC METHODS

a. *In vivo* Isotopic Procedures, Isotope Dilution. The metabolic profile of isotopic-labeled substances, especially ^{3}H-labeled, was common practice of this era. Pearlman et al. (1954) studied estrogen metabolism using deuterium-labeled material. Peterson (1959a, b) did pioneering studies with ^{3}H-labeled cortisol and corticosterone showing the effect of strong plasma protein binding on the metabolism and the consequence of saturating these binding sites. Similar studies using randomly tritium-labeled aldosterone were unfortunately flawed. As the aldosterone was racemic, dl aldosterone and the l isomer were not biologically active nor metabolized in the same way as the d isomer.

The metabolic clearance rate (MCR) can be calculated by integration of the area under the disappearance curve or more easily and perhaps more accurately from the plateaued value following venous infusion. MCR is defined as the volume of blood irreversibly cleared of a substance in time and, in the case of aldosterone, is usually about 1500 l per day. The use of *in vivo* isotope dilution procedures to calculate the production rate could be and was used together with the MCR to calculate the systemic blood levels even when this could not be measured by current methods.

Both at the MHMS and Worcester Foundation for Experimental Biology (WFEB), the Tait group, using ^{3}H-labeled aldosterone, found that the volume of distribution and MCR of the hormone was high compared to that of other adrenal steroids, such as cortisol and corticosterone (Tait and Burstein, 1964). This was found to be mainly because of the weak binding of aldosterone to circulating binding high affinity proteins, such as cortisol binding globulin (CBG), and the resultant high hepatic extraction (over 90%) (Tait and Burstein, 1964). There was also the contribution to the MCR from a significant renal clearance (due to the formation of the c-18 glucuronide) as previously discussed (Cheville et al., 1966). The high total MCR of aldosterone (about equal to the hepatic blood flow) means that the inertia of its metabolic system is relatively low for example, compared to cortisol. Thus, changes in aldosterone's blood concentrations closely follow its secretion. This may be relevant to the recent findings that the intrinsic rate of its action (both non-genomic and even genomic) can be fairly rapid (Lösel et al., 2004). Changes in the MCR of aldosterone could theoretically play a significant role in the physiological control of the plasma concentration of the hormone and this was shown with changes in posture when there are marked alterations in hepatic blood flow (Balikian et al., 1968). Presumably, the nearly complete hepatic extraction is maintained at higher blood flows. However, variation in MCR is probably not a significant factor in the effect of changes in electrolyte intake on aldosterone plasma levels. There is usually a change in the MCR of aldosterone in patients with severe congestive heart failure due to lowering of hepatic extraction and/or hepatic blood flow (Tait et al., 1965; Cheville et al., 1966; Coghlan et al., 1971a), but it remains to be seen if these changes are clinically important (Table 21.2).

b. Double Isotope Assays. The essentials of the double isotope derivate procedures were laid down even before aldosterone was discovered and were an essential part of the discovery process:

1. Extraction
2. Serial chromatographic separation
3. Radioactive derivative formation
4. Further derivative formation
5. Liquid scintillation spectrometer counting.

None of these steps were especially new but were cleverly put together by Kliman and Peterson (1960), albeit in a tedious but necessary procedure. Extraction to obtain a semi-purified starting material was a common first step. Bush variously published on multiple chromatographic steps and derivative formation to assist in the separation of steroid 'families' and multiple derivative formation was used to enhance these procedures (Bush, 1960, 1961). Modification of the Bush paper chromatography system by Peterson's laboratory (Kliman and Peterson, 1960) made these more suitable for the less polar acetate derivatives and generally more user friendly. The further derivative formation by oxidation to the 11–18 lactone of aldosterone by Peterson group was an important step. Isotope labeling via formation of the acetates had been used with great success in the original purification and structural analysis of

TABLE 21.2

PR = i × MCR	
Where PR is the secretion or production rate	μg/day
i is the systemic (peripheral) concentration	ng/100 ml
MCR is the metabolic clearance rate	l/day
Aldosterone values	
Aldosterone excretion rate (aldo 11–18 glucuronide)	5–20 μg/day
Aldosterone secretion rate (isotope dilution)	70–250 μg/day
Plasma aldosterone (calculated now measured)	5–15 ng/100 ml

The assumptions and conditions that must be met for valid secretion rate determination by isotope dilution are covered extensively in Tait, 1963, Tait and Burstein (1964), Coghlan and Blairwest (1967).

aldosterone. Labeling as a technique was not novel (Keston and Udenfriend, 1950; Bojesen, 1956; Avivi et al., 1954). Of course, the ready availability of 'hotter' ^{3}H-labeled acetic anhydride was important.

A substantial advance occurred with the commercial introduction of the liquid scintillation spectrometer. This instrument allowed, for the first time, efficient counting and electronic separation simultaneously of beta emitting isotopes from the high background produced by the amplification required.

Several other problems had to be addressed before reliable, reproducible results could be obtained. For example, during the acetylation reaction not only was aldosterone labeled but also other interfering steroids were and 'polymers' or 'polymeric substances' were produced. With paper chromatography these were 'steroid mimics' and exceedingly difficult to eliminate from the appropriately labeled steroid derivative. Second, the local presence of radiation could contribute to other artefacts. High specific activity ^{3}H-labeled acetic anhydride quickly discolored while unlabeled acetic anhydride remained colorless indefinitely. Third, in similar fashion, concentrated solutions of labeled steroids were subject to serious radiation damage leading to erroneous results. Finally, the original Kliman and Peterson method (Kliman and Peterson, 1960) did not have a recovery marker until after the acetylation step (after extraction, defatting, partitioning and acetylation) thereby compromising accuracy and reproducibility. These losses could occur by use of unsiliconized soda glass tubes by formation of c-17 iso-aldosterone (Schmidlin et al., 1957) or by traces of copper in copper distilled water by formation of the c-21 glyoxal (Lewbart and Mattox, 1959). Nonetheless, this assay was effective for clinical use and the assay was applied most widely to measuring aldosterone released from the 11–18 glucuronide after acid hydrolysis of urine. The specific activity of the aldosterone released by acid hydrolysis was used extensively as the endpoint in secretion rate analysis (Ayres et al., 1957). Others used enzymatically liberated urinary tetrahydo metabolite of aldosterone (Ulick, 1958).

c. Double Isotope Dilution Derivative Assay (DIDDA). This assay was developed to address the issue of losses throughout the assay procedure. It required that aldosterone labeled with ^{14}C or ^{3}H be added to the sample at the outset. ^{3}H aldosterone was available but ^{14}C-labeled acetic anhydride approaching theoretical specific activity 100% isotopic abundance at a single site was prohibitively expensive (see Peterson, 1964). Thus, assays used ^{3}H acetic anhydride and ^{14}C-labeled aldosterone biosynthesized using ^{14}C-labeled corticosterone incubated with bovine adrenal glands. This dilution marker of near theoretical specific activity was used by the Howard Florey institute group as early as 1962. Later NIH made a high specific activity 4-^{14}C aldosterone available through the efforts of Seymour Leiberman at Columbia and Gregory Pincus, Jim and Sylvia Tait and Marcel Gut at the WFEB. Dr Gut was very skillful and generous and made 7-^{3}H-aldosterone 1, 2-^{3}H-aldosterone, 4-^{14}C-progesterone, 4-^{14}C-corticosterone. These steroids were provided internationally as gifts from him or from NIH study sections. The availability of this material eventually culminated in DIDDA for aldosterone in peripheral plasma as a routine procedure (Coghlan and Scoggins, 1967; Brodie et al., 1967). Both these assays employed a reverse phase chromatographic step – mesitylene:methanol; water – that was of critical importance in removing unwanted material and was first used by Peterson (Peterson, 1964). Both assays required the repeated microvacuum distillation of the radioactive reagent, about 100 μl of acetic anhydride. Many laboratories found this procedure technically too demanding and formidable; Coghlan and Scoggins (1967) included an extensive statistical analysis of their method to allay scepticism about the low count rates above background. The real issue was specificity but this could be established by further derivative formation (Coghlan et al., 1979).

d. Radioimmunoassay. Effective radioimmunoassay has substantially transformed the ability to assay aldosterone both in clinical and experimental studies as documented by the thousands of reports that have used this approach since the first assay was reported by Haning et al. (see Ito et al., 1972; Coghlan et al., 1979 for reviews).

D. Enduring Concepts and Findings

During the period 1953–1970/75 there were about 5500 refereed papers published as well as chapters and monographs. In the historical overview after the discovery of aldosterone, as discussed above, we have made an emphasis on methodology as this was the preoccupation of the time. The lack of good reliable accurate assays inhibited development of the field, especially the physiological control of its secretion. With the development of reliable assays, particularly radioimmunoassays (RIA), of aldosterone, substantial progress began to be made in its regulation and mechanism of action in health and disease. At the time, release and secretion were often used to mean the same thing, but there is no stored aldosterone and increased secretion results from *de novo* synthesis (Coghlan and Blairwest, 1967).

Before 1967 and even after, until RIA procedures had been well established, the availability of methods of measurement limited both clinical and experimental studies.

In 1972, a procedure was introduced to use adrenal vein sampling to locate unilateral adenomas (Scoggins et al., 1972). This procedure is still in use though improved (see above).

One of the earliest experimental findings was that the response to aldosterone via its receptors had greatly increased sensitivity when sodium status was reduced (Blairwest et al., 1963a). Meanwhile, the aldosterone (mineralocorticoid) receptor was described (Feldman et al., 1973). Change in receptors' properties could not be shown to explain the increased sensitivity of response with negative sodium status (Butkus et al., 1976). The sensitivity of the aldosterone response to sodium deficit was shown likewise to be increased in the rat in extremely careful and well controlled studies (Singer and Stack-Dunne, 1955).

Many experimental studies were compromised by the use of anesthetized animals that often were also surgically traumatized. Adrenal vein blood was obtained by surgical procedures that made the results difficult or impossible to interpret, at best. This was only realized slowly across the field. The Melbourne group overcame this drawback by studies in sheep with adrenal gland auto transplants using the stimulus of sodium loss via parotid saliva or sodium restricted diet. The control of aldosterone secretion had a broad focus of inquiry but this was only advanced by incremental creep, as species differences demanded the most conservative interpretations.

The J.O. Davis group made the crucial observation that the kidney was required for aldosterone secretion. The Davis group documented that the responsible renal factor was renin and the immediately acting principal was angiotensin II (Davis et al., 1961, 1962). Clinical work, confirming these findings in human subjects, quickly followed.

Studies in sheep with transplanted adrenals documented a direct action of potassium and the adrenal's exquisite sensitivity to it (Funder et al., 1969). Using the same model, Blairwest et al. (1963b, 1971) documented that both angiotensin II and angiotensin III are equipotent on stimulating aldosterone but very different in producing vasoconstriction with angiotensin III having minimal vasoconstrictor activity. Much of the clinical thrust of this period was to determine the frequency of Conn's syndrome and its relationship to essential hypertension especially to 'low' renin hypertension. Was low renin hypertension a lesser form of Conn's syndrome? Was a modification of the definition of Conn's syndrome required? This debate still goes on.

So while the stalwarts of the aldosterone field pressed on – Davis, Bigleri, Thorn, Ganong, Mulrow, Farrell, Muller, Wright and Liddle – the luminaries of the hypertension field came to the fray – Page, Bumpus, Dustan, Peart, Tobian, Laragh, Doyle and Skinner. And, of course, there was a legion of other contributors in both categories. There was a huge number of papers and meetings about low renin hypertension as a subclass of essential hypertension. The great benefits were in the therapeutic control of blood pressure.

Many attempts to unravel aldosterone secretory control used *in vitro* rat adrenal approaches of one type or another. These studies made a contribution to the understanding of the interaction between the known proximate stimuli and sodium deficiency (Muller, 1971) and also made the interesting observation that investigators seeking a urinary factor in aldosterone control were misled by the ammonium ion that is a potent stimulus.

Two critical but tangential developments that were imperatives of the aldosterone and hypertension fields were the measurement of renin and angiotensin II in peripheral blood/plasma. Skinner and others at the Cleveland Clinic and Sealy and her colleagues in New York made monumental contributions to renin assay and defining the roles of active versus inactive renin (Skinner, 1967; Sealy et al., 2005). The Howard Florey Group were the first to measure angiotensin II by radioimmunoassay (Catt et al., 1971) and, in a long series of benchmark studies extending over 10 years, documented that angiotensin II was produced not just in the lung but in many vascular beds (Fei et al., 1980). This paper was very important in that it demonstrated the ability to raise specific antibodies against non-antigenic small peptides, indeed, against small molecules in general and, in the context of this chapter, steroid hormones in particular. The method used coupling of the small molecule to a larger antigenic molecule using the method of Goodfriend et al. (1964) of carbodiimide condensation.

Later the Howard Florey Institute group were able to show, with the development of appropriate assays for the known components modulating aldosterone secretion, a careful assessment of their roles in physiologic manipulations of them could then be undertaken. For example, in sheep, aldosterone could be induced to rise and fall normally with angiotensin II clamped in the normal range or kept far outside the normal by intravenous infusion. However, factors other than angiotensin II and potassium had to be involved both in regulating an increase and decrease in aldosterone secretion (Coghlan et al., 1971b; Boon et al., 1996, 1998). Aldosterone start up may involve other factors beyond plasma angiotensin II and plasma [K] but the turn off mechanism does not involve either. Thus, in clinical situations where the aldosterone level is inappropriately high for the level of sodium and potassium intake, the cause could be because one of the known proximate stimuli is raised or because the turn off of aldosterone secretion is defective. The control of aldosterone secretion, as it is not stored, is multifactorial and several factors are necessary but no simple algorithm has been devised completely to explain regulation of secretion.

III. BLOCKERS OF ALDOSTERONE ACTION

A. Spironolactone

In 1957, Kagawa and co-workers at the GD Searle Co. laboratories in Chicago, USA, used a bioassay to measure the anti-mineralocorticoid potency of compounds by assessing their ability to block the effect of deoxycorticosterone and

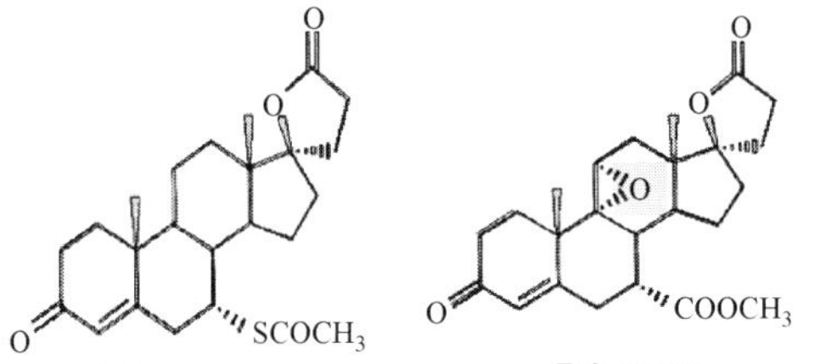

FIGURE 21.6 Chemical structure of aldosterone blockers spironolactone and eplerenone. From Garthwaite, S.A., and Mahon, E.G. (2004). *Mol Cell Endocrinol* 217, 27–31.

aldosterone on the urinary Na/K ratio of rats (Kagawa et al., 1957; Cella et al., 1959; Cella and Tweit, 1959; Kagawa, 1960). Using this bioassay, it was found that two steroids 3-(3-oxo-17β – hydroxyl-4-androsten-17α –yl) propionic acid γ -lactone and its c-19 nor analog were active as antimineralocorticoids. However, they did not have an effect on electrolyte metabolism when administered without aldosterone or deoxycorticosterone (Kagawa et al., 1957). Subsequently, the alkanethioloc acid adducts of these compounds (the 7α–acetyl thio derivatives of the 19 methyl and 19 nor steroids), subsequently named spironolactones, were found to be active orally as aldosterone blockers and were used extensively experimentally and clinically (Cella and Tweit, 1959) (Figure 21.6).

According to Garthwaite and McMahon (2004), Kagawa and co-workers (1957) synthesized these steroids with their particular structures because progesterone blocked the action of aldosterone and digitoxin. In the chemical design of these steroids, there does not seem to be any attempt to occupy the aldosterone receptor from knowledge of the unique structure of aldosterone, e.g. the 11–18 hemi-acetal group. Therefore, it would not be expected that their effects would lead to the identification of compounds that would specifically block aldosterone. Indeed, it was soon discovered that spironolactone also blocked the androgen and stimulated the progesterone receptors thereby giving rise to side effects in clinical anti-mineralocorticoid therapy, such as gynecomastia in men and menstrual irregularities in premenstrual women. Also Selye et al. (1969) found that spironolactone reduced the toxicity of digitalis-like compounds in rats, suggesting an interaction of spironolactone with the digitalis receptor. However, this would not have led to a cardiotonic effect, as hoped by Kagawa and co-workers according to Garthwaite and McMahon (2004). Nevertheless, spironolactone proved to be very useful as a research tool to reveal specific actions of aldosterone (Young et al., 1994). Also, it was effective in the treatment of congestive heart failure, as in the RALES trial (Pitt et al., 1999) and other clinical conditions presumably caused by excess mineralocorticoid effects, as described in Chapters 22–24.

B. Eplerenone

After the discovery of spironolactone, scientists at Ciba-Geigy, Roussel Uclef and Searle attempted to synthesize more specific antagonists (Garthwaite and McMahon, 2004; Ménard, 2004). Finally, Grob and co-workers (de Gasparo et al., 1987) at the Ciba-Geigy laboratories in Basle, Switzerland, did synthesize steroids which successfully took into account the unique structure of aldosterone. These were the various 9α, 11α epoxy derivatives of spironolactone which differed from one another only in the B ring structure. *In vitro*, these were found to be similar in binding affinity to the mineralocorticoid receptor compared to spironolactone but the binding to the androgen and progesterone receptors decreased between 10- and 50-fold respectively. *In vivo*, there were parallel results with a 3- to 10-fold decrease in the antiandrogenic and progestogenic effect compared to spironolactone in both the rat and rabbit. The direct derivative of spironolactone (see Figure 21.6) was subsequently named eplerenone. Therefore the Ciba laboratory in Basle, which took part in the elucidation of the structure of aldosterone, but did not benefit from the clinical applications of spironolactone, appropriately produced the first designer drug based on the unique structure of aldosterone. Unfortunately, they did not continue to apply their initial results and sold the rights to Searle Co. Eplerenone subsequently proved to be a specific anti-mineralocorticoid with many applications in basic research and also in clinical applications as shown in the EPHESUS trials (Pitt et al., 2003; Pitt, 2004). No significant anti-androgenic and progestational effects have been reported in animal laboratory or clinical investigations. Another advantage of eplerenone is the relative simplicity of its metabolism. Spironolactone has several active metabolites; a property not shared by eplerenone. This makes the duration and mechanism of action of eplerenone more predictable (Garthwaite and McMahon, 2004). The only serious side effect in clinical applications after administration of eplerenone (also shown with spironolactone) has been hyperkalemia in some patients. This is due to the inhibition of the physiological function of aldosterone to regulate potassium metabolism and might be considered to be difficult to overcome by chemical design of the structure of the compound. However, one of the surprising features of the results of the RALES and EPHESEUS trials was the low doses of spironolactone or eplerenone that were able to exhibit the beneficial effects. These doses of the antimineralocorticoids would not be expected to affect gross features of electrolyte metabolism such as potassium retention. Therefore, it would seem that adjustment of the dose administered and perhaps modification of the chemical structure of the anti-mineralocorticoid might further improve the risk benefit ratio of blockade of the mineralocorticoid receptor.

References

Avivi, P., Simpson, S. A., Tait, J. F., and Whitehead, J. F. (1954). The use of ^{3}H and ^{14}C labelled acetic anhydride as analytical reagents in micro chemistry. In: Johnson, J. P. (ed.) *Proceedings of the Second Radioisotope Conference,* Butterworth, London, pp. 313–324.

Ayres, P. J., Gould, P., Simpson, S. A. S., and Tait, J. F. (1956). The in vitro demonstration of differential corticosteroid production within the ox adrenal gland. *Biochem J.* **63:** 19.

Ayres, P. J., Garrod, O., Tait, S. A. S., Tait, J. F., Walker, G., and Pearlman, W. H. (1957). The use of 16-3H aldosterone in studies on human peripheral blood. *Ciba Foundation Colloq Endocrinol.* **11:** 309–327.

Ayres, P. J., Barlow, J., Garrod, O., Tait, S. A. S., and Tait, J. F. (1958). Primary aldosteronism (Conn's syndrome). In: Muller, A. (ed.) *Symposium on aldosterone,* J & A Churchill, London, pp. 143–153.

Balikian, H. M., Brodie, A. H., Dale, S. L., Melby, J. C., and Tait, J. F. (1968). Effect of posture on the metabolic clearance rate, plasma concentration and blood production rate in man. *J Clin Endocrinol Metab.* **28:** 1630–1640.

Barton, D. H. R., Beaton, J. M., Geller, L. E., and Pechet, M. M. (1961). A synthesis of aldosterone acetate. *J Am Chem Soc.* **83:** 4083–4089.

Blairwest, J. R., Coghlan, J. P., Denton, D. A., Goding, J. R., and Wright, R. D. (1963a). The effect of aldosterone, cortisol and corticosterone upon the sodium and potassium content of sheep parotid saliva. *J Clin Invest.* **4122:** 484–496.

Blairwest, J. R., Coghlan, J. P., Denton, D. A., Goding, J. R., Wintour, E. M., and Wright, R. D. (1963b). The control of aldosterone secretion. *Recent Prog Hormone Res.* **19:** 311–383.

Blairwest, J. R., Coghlan, J. P., Denton, D. A., Funder, J. W., Scoggins, B. A., and Wright, R. D. (1971). The effect of the heptpeptide (2–8) and hexapeptide (3–8) fragments of angiotensin II on aldosterone secretion. *J Clin Endocrinol.* **32:** 575–578.

Bojesen, E. (1956). Determination of cortisol in plasma using ^{35}S pipsan. *Scand J Clin Lab Invest.* **8:** 55–66.

Boon, W. C., McDougall, J. G., and Coghlan, J. P. (1996). Control of aldosterone secretion. 'Towards the molecular idiom'. In: Vinson, V.P. Anderson, D.C. (ed.) *Adrenal glands, vascular systems and hypertension,* Journal of Endocrinology Ltd, Bristol, pp. 159–185.

Boon, W. C., McDougall, J. G., and Coghlan, J. P. (1998). Hypothesis: aldosterone is synthesized by an alternate pathway during severe sodium depletion. 'New wine in an old bottle'. *Clin Exp Pharmacol Physiol.* **25:** 369–378.

Brodie, A. H., Shimizu, N., Tait, S. A. S., and Tait, J. F. (1967). A method for the measurement of aldosterone in peripheral plasma. *J Clin Endcrinol Metabol.* **27:** 997–1011.

Burton, R. B., Zaffaroni, A., and Keutmann, E. H. (1951). Paper chromatography of steroids. II. Corticosteroids and related compounds. *J Biol Chem.* **188:** 763–771.

Bush, I. E. (1952). Methods of paper chromatography of steroids applicable to the study of steroids in mammalian blood and tissues. *Biochem J.* **50:** 370–398.

Bush, I. E. (1953a). Species differences in adrenocortical secretion. *J Endocrinol.* **9:** 95–101.

Bush, I. E. (1953b). Species differences and other factors influencing adrenocortical secretion. In: Klyne, W. Wolstenholme, G. Cameron, M.P. (ed.) *Ciba F. Symp. Vol. VII. Synthesis and metabolism of adrenocortical steroids,* J and A Churchill Ltd, London.

Bush, I. E. (1960). In biosynthesis and secretion of adrenocortical steroids, (Clark, F., Grant, J.K., eds). Cambridge University Press, Cambridge.

Bush, I. E. (1961). *The chromatography of steroids.* Pergamon Press, Oxford.

Bush, I. E., and Willougby, J. (1985). Breaking the lipid barrier in partition choromatography. A steroid memoir. *Steroids.* **45:** 479–496.

Bush, I. E., and Willougby, J. (1957). The secretion of allotetrahydrocortisol in human urine. *Biochem J.* **67:** 689–692.

Butkus, A., Coghlan, J. P., Paterson, R., Scoggins, B. A., Robinson, and Funder, J. W. (1976). Mineralocorticoid receptors in sheep kidney and parotid: studies in Na replete and Na deplete states. *Clin Exp Pharmacol Physiol.* **3:** 557–565.

Carpenter, P. C., and Maddox, V. R. (1976). Isolation, determination of structure and synthesis of the acid-labile conjugate of aldosterone. *Biochem J.* **157:** 1–14.

Catt, K. J., Cran, E., Zimmett, P. J., Best, J. B., and Coghlan, J. P. (1971). Angiotensin II blood levels in human hypertension. *Lancet.* **1:** 459–464.

Cella, J., Brown, E. A., and Burtner, R. R. (1959). Steroid aldosterone blockers I. *J Org Chem.* **24:** 743–748.

Cella, J., and Tweit, R. (1959). Steroid aldosterone blockers. II. *J Org Chem.* **24:** 1109–1110.

Cheville, R. A., Luetscher, J. A., Hancock, E. W., Dowdy, A. J., and Nokes, G. W. (1966). Distribution, conjugation, and excretion of labelled aldosterone in congestive heart failure and in controls with normal circulation: development and testing of a model with an analog computer. *J Clin Invest.* **45:** 1302–1316.

Coghlan, J. P., and Blairwest, J. R. (1967). Aldosterone. In Hormones in blood, Vol. 2 (Gray, C.H., Bacharach, A.L., eds). Academic Press, London and New York, pp. 391–471.

Coghlan, J. P., and Scoggins, B. A. (1967). The measurement of aldosterone in peripheral blood. *J Clin Endocrinol Metab.* **27:** 1470–1486.

Coghlan, J. P., Scoggins, B. A., Stockigt, J. R., Meerkin, M. and Hudson, B. (1971a). Aldosterone in congestive heart failure. Suppl. Bull. Post-graduate Committee in Medicine. University of Sydney, **26:** 17–31.

Coghlan, J. P., Blairwest, J. R., Denton, D. A., Scoggins, B. A., and Wright, R. D. (1971b). Perspectives in aldosterone and renin control. *Aust NZ J Med.* **2:** 178–197.

Coghlan, J. P., Scoggins, B. A., and Wintour, E. M. (1979). Aldosterone. Gray, C. H., and James, V. T. H. (eds) *Hormones in blood.* **Vol. III:** Academic Press, New York, pp. 493–609.

Conn, J. (1955a). Presidential Address. Part I. Painting background; Part II. Primary aldosteronism, a new clinical syndrome. *J Lab Clin Med.* **45:** 3–17.

Conn, J. (1955b). Primary aldosteronism. *J Lab Clin Med.* **45:** 661–664.

Conn, J. W., Lewis, I. H., and Fajans, S. S. (1951). The probability of compound F (17 hydroxycorticosterone) is the hormone produced by the normal human adrenal cortex. *Science.* **113:** 713–714.

Conn, J. W. (1963). Aldosteronism in man. Some clinical and climatological aspects. *J Am Med Assoc.* **183:** 871–878.

Conn, J. W., Cohen, E. L., Rovner, D. R., and Nesbit, R. M. (1965). Normokalemic primary aldosteronism. *J Am Med Assoc.* **193:** 200–206.

Davis, J. O., Carpenter, C. C., Ayers, C. R., Holman, J. F., and Bahn, R. C. (1961). Evidence for a secretion of an aldosterone stimulating hormone by the kidney. *J Clin Invest.* **40:** 684–696.

Davis, J. O., Hartroft, P. M., Titus, P. O., Carpenter, C. C., Ayers, C. R., and Spiegal, H. E. (1962). The role of the renin angiotensin system in the control of aldosterone secretion. *J Clin Invest.* **41:** 378–389.

Deane, H. W., Shaw, J. H., and Greep, R. O. (1948). The effect of altered sodium and potassium intake on the width and cytochemistry of the cat's adrenal cortex. *Endocrinology.* **43:** 133–153.

de Gasparo, M., Joss, U., Ramjoué, H. P. et al. (1987). Three new epoxy-spironolactone derivatives: characterization in vivo and in vitro. *J Pharmacol Exp Ther.* **240:** 650–656.

Deming, Q. B., and Luetscher, J. A. (1950). Bioassay of deoxycorticosterone-like material in urine. *Proc Soc Exper Biol Med.* **73:** 171–175.

Dorfman, R. I., Potts, A. M., and Feil, M. L. (1947). Studies on the bioassay of hormones. The use of radiosodium for the detection of small quantities of deoxycorticosterone. *Endocrinology.* **41:** 464–469.

Everse, J. W. E., and de Femery, F. (1932). On a method for measuring fatigue in rats and its application for testing of the suprarenal cortical hormone. *Acta Brev Neerland.* **2:** 152–154.

Farrell, G. L., Royce, P. C., Rauskholb, E. W., and Hirschmann, H. (1953). Isolation and identification of aldosterone from adrenal venous blood. *Proc Soc Exper Biol Med.* **87:** 141–143.

Fei, D. T. W., Coghlan, J. P., Fernley, R. T., Scoggins, B. A., and Tregear, G. W. (1980). Peripheral production of angiotensin II and III in sheep. *Circ Res.* **1:** 135–137.

Feldman, D., Funder, J. W., and Edelman, I. S. (1973). Evidence for a new class of corticosterone receptors in the rat kidney. *Endocrinology.* **920:** 1429–1441.

Fieser, L. F., and Fieser, M. (eds) (1959). *Steroids,* Reinhold, Chapman and Hall, London, pp. 713–720.

Fourman, P., Bartter, F. C., Albright, F., Dempsey, E., Carroll, E., and Alexander, J. (1950). Effect of 17-hydroxycorticosterone (compound F) in man. *J Clin Invest.* **19:** 1462–1473.

Funder, J. W. (2005). Endocrine hypertension – review. *Trends Endocrinol Metab.* **16:** 79–80.

Funder, J. W., Blairwest, J. R., Coghlan, J. P., Denton, D. A., Scoggins, B. A., and Wright, R. D. (1969). Effect of plasma K on the secretion of aldosterone. *Endocrinology.* **85:** 381–384.

Garthwaite, S.M., and McMahon, E.G. (2004). The evolution of aldosterone anatgonists. Proceedings of the 2003 International Symposium on aldosterone, (Coghlan, J.P., Vinson, G.P., eds). *Mol Cell Endocrinol* **217:** 27–32.

Gaunt, R., and Eversole, W. J. (1949). Notes on the history of the adrenal cortical problem. *Ann NY Acad Sci.* **50:** 511–521.

Gaunt, R., Birnie, J. H., and Eversole, W. J. (1949). Adrenal cortex and water metabolism. *Physiol Rev.* **29:** 281–310.

Giroud, C. J. P., Stachenko, J., and Venning, E. H. (1956). Secretion of aldosterone by the zona glomerulosa of rat adrenal in vitro. *Proc Soc Exper Med.* **92:** 154–158.

Goodfriend, T. L., Levine, L., and Fasman, G. D. (1964). Antibodies to bradykinin and angiotensin: a use of carbodiimides in immunology. *Science.* **144:** 1344–1346.

Gordon, R. D., Laragh, J. H., and Funder, J. W. (2005). Low renin hypertensive states: perspective,unsolved problems and future research. *Trends Endocrinol Metab.* **16:** 108–113.

Grollman, A. (1937). Physiological and chemical studies on the adrenal cortical hormone. *Cold Spring Harbor Symp Quant Biol.* **5:** 313–322.

Grundy, H. M., Simpson, S. A., Tait, J. F., and Woodford, M. (1952a). Further studies on the properties of a highly active mineralocorticoid. *Acta Endocrinol (Copenh).* **11:** 199–220.

Grundy, H. M., Simpson, S. A. S., and Tait, J. F. (1952b). Isolation of a highly active mineralocorticoid from beef adrenal extract. *Nature.* **169:** 795–797.

Ham, E. A., Harman, R. E., DeYoung, J. J., Brink, N. G., and Sarrett, L. H. (1955). Studies on the chemistry of aldosterone. *Am Soc.* **77:** 1637–1641.

Harman, R. E., Ham, E. A., DeYoung, J. J., Brink, N. G., and Sarrettt, L. H. (1954). Isolation of aldosterone (electrocortin). *Am Soc.* **76:** 5035–5036.

Hartman, C. G. (1940). The usefulness of biological extracts. *Science.* **91:** 142.

Ingle, D. J. (1940). The work performance of adrenalectomized rats treated with corticosterone and chemically related compounds. *Endocrinology.* **26:** 472–477.

Ito, T., Woo, J., Hanig, R., and Horton, R. (1972). Radioimmunoassay for aldosterone with a comparison of alternative techniques. *J Clin Endocrinol Metab.* **34:** 106–112.

Kagawa, C. M. (1960). Blocking the renal electrolyte effects of mineralocorticoids with an orally active steroidal spironolactone. *Endocrinology.* **67:** 125–132.

Kagawa, C. M., Cella, J. A., and van Harman, C. G. (1957). Action of new steroids blocking effects of deoxycorticosterone and aldosterone. *Science.* **126:** 1015–1016.

Kelly, W. G., Bandi, L., Shoolery, J. N., and Lieberman, S. (1962). Isolation and characterization of of aldosterone metabolites from human urine; two metabolites bearing a bicyclic acetal structure. *Biochemistry.* **1:** 172–181.

Keston, A. S., and Udenfriend, S. (1950). The application of isotope derivative methods to analysis of proteins. *Cold Spring Harbour Symp Quant Biol.* **14:** 92–96.

Kliman, B., and Peterson, R. E. (1960). Double isotope derivative assay of aldosterone. *J Biol Chem.* **235:** 1639–1648.

Kuizenga, M. H. (1944). The isolation and chemistry of the adrenal hormones. *Chem Physiol Horm,* pp. 57–68.

Kuizenga, M. H., and Cartland, G. F. (1939). Fractionation studies on adrenal cortex extract with notes on the distribution of biological activity among the crystalline and amorphous fractions. *Endocrinology.* **24:** 526–535.

Lewbart, M. L., and Mattox, V. R. (1959). Destruction of cortisone and related steroids by traces of copper during purification procedures. *Nature.* **183:** 820–821.

Lim, P. O., Dow, E., Brennan, G., Jung, R. T., and MacDonald, T. M. (2000). High prevalence of primary aldosteronism in Tayside hypertension clinic. *J Hum Hyper.* **14:** 311–315.

Lösel, R., Schultz., A. and Wehling, M. (2004). A quick glance at rapid aldosterone action. Proceedings of the 2003 International Symposium on aldosterone, (Coghlan, J.P., Vinson, G.P., eds). *Mol Cell Endocrinol* **217:** 137–42.

Luetscher, J. A. (1955). Discussion. Recent progress in the methods of isolation, chemistry and physiology of aldosterone. *Recent Prog Horm Res.* **11:** 214–216.

Luetscher, J. A., Johnson, B. B., Dowdy, A., Harvey, J., Lew, W., and Poo, L. J. (1954). Chromatographic separation of the sodium-retaining corticoid from the urine of children with nephrosis compared with observations on normal children. *J Clin Invest*. **33**: 276–286.

Luetscher, J. A., Dowdy, A., Harvey, J., Neher, R., and Wettstein, A. (1955). Isolation of aldosterone from the urine of aldosterone from the urine of a child with nephrotic syndrome. *J Biol Chem*. **217**: 505–512.

Luetscher, J. A., Neher, R., and Wettstein, A. (1956). Isolation of crystalline aldosterone from the urine of patients with congestive heart failure. *Experientia*. **12**: 1–3.

Ma, J. T. C., Wang, C., Lam, K. S. L. et al. (1986). Fifty cases of primary aldosteronism in Hong Kong Chinese. Evaluation of techniques for tumour location. *Q J Med*. **235**: 1021–1037.

Maddox, V. R. (1955). Discussion. Recent progress in the methods of isolation, chemistry and physiology of aldosterone. *Recent Prog Horm Res*. **11**: 217–218.

Maddox, V. R., Mason, H. L., and Albert, A. (1953a). Isolation of a salt retaining substance from beef adrenal extract. *Mayo Clin Proc*. **28**: 569–576.

Maddox, V. R., Mason, H. L., Albert, A., and Code, J. C. (1953b). Properties of a sodium-retaining principle from beef adrenal extract. *J Am Chem Soc*. **75**: 4869–4870.

Mason, H. (1939). Chemistry of the adrenal cortical hormone. *J Endocrinol*. **25**: 405–412.

Mason, H., Myers, C. S., and Kendall, E. C. (1936). The chemistry of crystalline substances isolated from the suprarenal gland. *J Biol Chem*. **114**: 613–631.

Ménard, J. (2004). The 45-year story of the development of an anti-aldosterone more specific than spironolactone. Proceedings of the 2003 International Symposium on Aldosterone, (Coghlan, J. P., Vinson, G.P., eds). *Mol Cell Endocrinol* **217**: 45–52.

Mosso, L., Carajal, C., Gonzalez, A. et al. (2003). Primary aldosteronism and hypertensive disease. *Hypertension*. **42**: 161–168.

Mulatero, P., Stowasser, M., Loh, K. C. et al. (2004). Increased diagnosis of primary aldosteronism, including surgically correctable forms in centers from five continents. *J Clin Endocrinol Metab*. **89**: 1045–1050.

Muller, J. (ed.) (1971). *Regulation of aldosterone biosynthesis, Vol 5. Monographs in endocrinology,* Springer Verlag, Berlin.

Neher, R. (1979). Aldosterone: chemical aspects and related enzymology. *J Endocrinol*. **81**: 25–35.

Nelson, D. H., Reich, H., and Samuels, L. T. (1950). Isolation of a steroid hormone from the adrenal vein blood of dogs. *Science*. **111**: 578–589.

Pearlman, W. H., Pearlman, M. R. J., and Rakoff, A. E. (1954). Estrogen metabolism in pregnancy: a study with the aid of deuterium. *J Biol Chem*. **209**: 803–812.

Peterson, R. E. (1959a). Metabolism of adrencortical steroids in man. *Ann NY Acad Sci*. **81**: 846–853.

Peterson, R. E. (1959b). The miscible pool and turnover rates of adrenocortical steroids. *Recent Prog Hormone Res*. **15**: 231–241.

Peterson, R. E. (1964). Aldosterone measurement. In: Baulieu, E.E. Robel, P. (ed.) *Aldosterone*, Blackwell Scientific, Oxford.

Pitt, B. (2004). Effect of aldosterone blockade in patients with systolic left ventricular dysfunction: implications of the RALES and EPHESUS studies. Proceedings of the 2003 International Symposium on aldosterone, (Coghlan, J.P., Vinson, G.P., eds). *Mol Cell Endocrinol* **217**: 53–58.

Pitt, B., Zannad, F., Remme, W. J. et al. for the Randomized Aldactone Evaluation Study Investigators. (1999). The effect of spironolactone on morbidity and mortality in patients with severe heart failure. *N Engl J Med*. **341**: 709–717.

Pitt, B., Remme, W., Zannad, F. et al. (2003). Eplerenone, a selective aldosterone blocker in patients with left ventricular dysfunction after myocardial infarction. *N Engl J Med*. **348**: 1309–1382.

Reichstein, T. (1936). Uber Cortin, da Hormon der Nebbennieren Rinde (X) I Mitteilung. *Helv Chim Acta*. **19**: 29–63.

Reichstein, T., and Shoppee, C. W. (1943). The hormones of the adrenal cortex. *Vit Hormones*. **1**: 345–413.

Sayers, G. (1950). The adrenal cortex and homeostasis. *Physiol Rev*. **30**: 241–320.

Schmidlin, J., Anner, G., Billeter, J. R., and Wettstein, A. (1955). Uber synthesen in der Aldosterons-Reihe. *Experientia*. **40**: 365–368.

Schmidlin, J., Anner, G., Billiter, J. R. et al. (1957). Totalsynthese des Aldosterons. C. Recemisches Aldosteron und die beidon Enantiomeren des entsprechenden (18–11) Lactons. *Helv Chim Acta*. **40**: 2291.

Scoggins, B. A., Oddie, C. J., Hare, W. S. C., and Coghlan, J. P. (1972). Preoperative lateralization of aldosterone producing tumours in primary aldosteronism. *Ann Intern Med*. **76**: 891–897.

Sealy, J. E., Gordon, R. D., and Matero, F. (2005). Plasma renin and aldosterone measurements in low renin hypertensive states. *Trends Endocrinol Metab*. **16**: 86–91.

Selye, H., Mécs, I., and Tamura, T. (1969). Effect of spironolactone and norbolethone on the toxicity of digitalis compounds in the rat. *Br J Pharmacol*. **37**: 485–488.

Selye, H. and Horava, A. (eds). (1953). The stress concept in 1953. Third Annual Report on Stress. Acta Inc, Montreal, pp. 17–65.

Simpson, S. A., and Tait, J. F. (1952). A quantitative method for the bioassay of the effect of adrenal cortical hormones on mineral metabolism. *Endocrinology*. **50**: 150–161.

Simpson, S. A., and Tait, J. F. (1953). Physico-chemical methods of detection a previously unidentified adrenal hormone. *Mem Soc Endocrinol*. **2**: 9–24.

Simpson, S. A. S., and Tait, J. F. (1955). Recent progress in the methods of isolation, chemistry and physiology of aldosterone. *Recent Prog Horm Res*. **11**: 183–219.

Simpson, S. A., Tait, J. F., and Bush, I. E. (1952). Secretion of a salt-retaining hormone by the mammalian adrenal cortex. *Lancet*. **1**: 226–232.

Simpson, S. A., Tait, J. F., Wettstein, A. et al. (1953a). Isolierung eines neuen krystallisierten Hormons aus Nebennieren mit besonders hoher Wirksamkeit auf den Mineralstoffwechsel. *Experientia*. **9**: 333–335.

Simpson, S. A., Tait, J. F., Wettstein, A. et al. (1953b). Konstitution des Aldosterons, des neuen Mineralocorticoids. *Experientia*. **10**: 132–133.

Simpson, S. A., Tait, J. F., Wettstein, A. et al. (1954a). Aldosterons, Isolierung und Eigenschaften Uber Bestandteile der Nebbenierenrinde und verwandte Stoffe. 91 Mitteilung. *Helv Chim Acta*. **37**: 1163–1200s.

Simpson, S. A., Tait, J. F., Wettstein, A. et al. (1954b). Die Konstitut des Aldosterons. Uber Bestandteile der Nebbenierenrinde und verwandte Stoffe. 92 Mitteilung. *Helv Chim Acta*. **37**: 1200–1223.

Singer, B., and Venning, E. H. (1953). Method of assay of a sodium retaining hormone in human urine. *Endocrinology*. **52:** 623–633.

Singer, B., and Stack-Dunne, M. P. (1955). Secretion of aldosterone and corticosterone by rat adrenal. *J Endocrinol*. **12:** 130–145.

Skinner, S. L. (1967). Improved assay methods for renin 'concentration' and 'activity' in human plasma. *Methods using selective denaturation of renin substrate. Circ Res*. **4:** 391–402.

Speirs, R. S., Simpson, S. A., and Tait, J. F. (1954). Certain biological activities of crystalline electrocortin. *Endocrinology*. **55:** 233–236.

Spencer, A. G. (1950). Biological assay of small quantities of deoxycorticosterone. *Nature*. **166:** 32–33.

Stowasser, M., and Gordon, R.D. (2004). Primary aldosteronism – careful investigation is essential and rewarding. Proceedings of the 2003 International Symposium on aldosterone, (Coghlan, J. P., Vinson, G.P., eds). *Mol Cell Endocrinol* **217:** 33–40.

Swann, H. G. (1940). The pituitary-adrenocortical relationship. *Physiol Rev*. **20:** 493–521.

Swingle, W. W., and Remington, J. W. (1944). The role of the adrenal cortex in physiological processes. *Physiol Rev*. **24:** 89–127.

Tait, J. F., and Burstein, S. (1964). In vivo studies of steroid dynamics. Pincus, G., Thimann, K., and Astwood, E. B. (eds). *The hormones*. **vol. V:** Academic Press, New York, pp. 441–557.

Tait, J. F., and Tait, S. A. S. (1979). Recent perspectives on the history of the adrenal cortex. The Sir Henry Dale Lecture for 1979. *J Endocrinol*. **83:** 1P–24P.

Tait, J. F., and Tait, S. A. S. (1988). A decade (or more) of electrocortin. *Steroids*. **51:** 213–250.

Tait, J. F., and Tait, S. A. S. (1998). Personal history. The correspondence of S. A. S. Tait and J. F. Tait with T. Reichstein during their collaborative work on the isolation and elucidation of the structure of electrocortin (later aldosterone). *Steroids*. **63:** 440–453.

Tait, J. F., Bougas, J., Little, B., Tait, S. A. S., and Flood, C. (1965). Splanchnic extraction and clearance of aldosterone in subjects with minimal and marked cardiac dysfunction. *J Clin Endocrinol Metab*. **25:** 219–228.

Tait, J. F., Simpson, S. A. S., and Grundy, H. (1952). The effect of adrenal extract on mineral metabolism. *Lancet*. **1:** 122–129.

Tait, J.F., Tait, S.A.S., and Coghlan, J. (2004). The discovery, isolation and identification of aldosterone: reflections on emerging regulation and function. Proceedings of the 2003 International Symposium on aldosterone, (Coghlan, J.P., Vinson, G.P., eds). *Mol Cell Endocrinol* **217:** 1–21.

Tait, S. A. S., and Tait, J. F. (2004). *A quartet of unlikely discoveries*. Athena Press, London.

Ulick, S. (1958). The isolation of a urinary metabolite of aldosterone and its use to measure the rate of secretion. *Rauh Trans Assoc Am Physicians*. **71:** 225–235.

Ulick, S. (1996). Editorial: cortisol as mineralocorticoid. *J Clin Endocrinol Metab*. **81:** 1307–1308.

Ulick, S., Levine, L. S., Gunczler, P. et al. (1979). A syndrome of apparent mineralocorticoid excess associated with defects in the peripheral metabolism of cortisol. *J Clin Endocrinol Metab*. **49:** 757–764.

Underwood, R. H., and Tait, J. F. (1961). Purification, partial characterization and metabolism of an acid labile conjugate of aldosterone. *J Clin Endocrinol Metab*. **24:** 1110–1124.

Vischer, E., Schmidlin, J., and Wettstein, A. (1956). Mikrobiologische Spaltung razemischer Steroide. Synthese von d-Aldosterone. *Experientia*. **12:** 50–52.

Vogt, M. (1943). The output of cortical hormones by the mammalian suprarenal. *J Physiol*. **102:** 341–356.

Weber, K. T. (1999). Editorial. Aldosterone and spironolactone in heart failure. *N Engl J Med*. **341:** 753–755.

Wettstein, A. (1954). Advances in the field of adrenal cortical hormones. *Experientia*. **10:** 397–416.

Williams, J. S., Williams, G. H., Raji, A. et al. (2006). Prevalence of primary hyperaldosteronism in mild to moderate hypertension without hypocalcaemia. *J Hum Hypertens*. **20:** 129–136.

Wintersteiner, O., and Pfiffner, J. J. (1936). Chemical studies on the adrenal cortex. III Isolation of two new physiologically inactive crystalline compounds from adrenal extracts. *Biology*. **116:** 291–305.

Young, M., Fullerton, M., Dilley, R., and Funder, J. (1994). Mineralocorticoids, hypertension and cardiac fibrosis. *J Clin Invest*. **93:** 2578–2583.

Further reading

Coghlan, J. P., and Vinson, G. P. (2004). Proceedings of the 2003 International Symposium on aldosterone. *Mol Cell Endocrinol* **217:** 1–270.

CHAPTER 22

Aldosterone Receptors and Their Renal Effects: Molecular Biology and Gene Regulation

CELSO E. GOMEZ-SANCHEZ[1,2], ELISE P. GOMEZ-SANCHEZ[1,2] AND MARIO GALIGNIANA[3]
[1]*Endocrinology, G.V. Sonny Montgomery VA Medical Center, Jackson, MS 39216, USA*
[2]*University of Mississippi Medical Center, Jackson, MS 39216, USA*
[3]*Fundacion Leloir, Buenos Aires, Argentina*

Contents

I. Introduction 329
II. Aldosterone-binding sites and the mineralocorticoid receptor (MR) 329
III. Molecular biology of the MR 330
IV. Distribution of the mineralocorticoid receptor in the nephron 337
V. Proteins induced by aldosterone in transport epithelia 337
VI. Non-genomic effects of aldosterone in the kidney 340
References 342

I. INTRODUCTION

The kidney plays the primary role in salt and water homeostasis, maintaining osmolarity and volume in the extracellular space within a very narrow range despite wide variations in fluid and salt intake. Aldosterone plays a significant role in the maintenance of mammalian sodium, potassium, water and acid–base balance, primarily through effects on renal electrolyte excretion. Aldosterone promotes stimulation of Na^+ absorption, potassium and hydrogen secretion by tight epithelia that display high transepithelial electrical resistance and amiloride-sensitive sodium transport. This epithelium is found in distal segments of the nephron, bladder, distal parts of the colon and rectum and in the ducts of exocrine glands (salivary, sweat glands) (Verrey et al., 2000; Pearce et al., 2003). Many of the concepts of sodium transport stimulated by aldosterone were established using amphibian epithelial models such as the toad bladder and the *Xenopus laevis* kidney A6 cell line (Chen et al., 1998; Verrey, 1999).

Sodium transport across epithelia is driven by an electrochemical potential difference across the apical membrane allowing for passive movement of ions and water and by an active transport of ions across the basolateral membrane. The apical-membrane step is mediated by the opening of the amiloride-sensitive sodium channels that are sodium-selective. The basolateral membrane extrusion of sodium is mediated by activation of the ouabain-sensitive sodium potassium ATPase (Verrey et al., 2000).

II. ALDOSTERONE-BINDING SITES AND THE MINERALOCORTICOID RECEPTOR (MR)

Steroid hormones bind to and activate intracellular receptors which act as transcription factors that induce the synthesis of specific proteins within the target cells (Edelman et al., 1963; Rousseau et al., 1972). Aldosterone binds to high affinity cytoplasmic and nuclear receptors in the kidney and other mineralocorticoid target organs, inducing the synthesis of proteins that mediate the best understood effects of aldosterone (Funder et al., 1972; Fuller and Young, 2005; Pascual-Le Tallec and Lombes, 2005). For many years, despite very early evidence that the heart and vessels were also primary and direct targets for mineralocorticoid actions (Raab et al., 1950; Vanatta and Cottle, 1955; Jones and Hart, 1975), the effects of aldosterone were believed to be limited to the kidney and colon where they regulated Na^+ excretion. High affinity binding sites were also demonstrated in the brain (Anderson and Fanestil, 1976), heart (Pearce and Funder, 1987), skin (Kenouch et al., 1994) and other organs. In the early studies, aldosterone affinity for the renal receptor was found to be about 40 times greater than that of corticosterone, while the affinity for the receptor in the hippocampus was similar (Funder et al., 1973). When binding to corticosterone-binding globulin was taken into account, the binding of aldosterone and corticosterone to the hippocampus and the kidney were similar (Krozowski and Funder, 1983). The finding that specificity of the mineralocorticoid receptor to aldosterone is primarily exerted at the pre-receptor level explained how aldosterone could access the mineralocorticoid receptor in competition with glucocorticoids that exceed its concentration in the blood by two to three orders of magnitude (Stewart et al., 1987; Funder et al., 1988). Modern concepts of renal mineralocorticoid receptor specificity arose from these findings, as discussed below.

III. MOLECULAR BIOLOGY OF THE MR

The nuclear receptor superfamily is one of the most abundant classes of transcriptional regulators in metazoans. It comprises a large number of distantly related regulatory proteins that include receptors for hydrophobic molecules such as steroid hormones (estrogens, progesterone, androgens, glucocorticoids, mineralocorticoids, vitamin D, ecdysone, oxysterols, bile acids, etc.), retinoic acids (all-*trans* and 9-*cis* isoforms), thyroid hormones, dioxin, sterols, fatty acids, leukotrienes and prostaglandins. All the members of the steroid/thyroid/retinoic nuclear receptor family of ligand-dependent transcription factors consist of three principal domains: the N-terminal domain (A/B), the DNA-binding domain and the c-terminal or ligand binding domain (LBD) (Figure 22.1) (Arriza et al., 1987; Evans, 1988; Fuller and Young, 2005; Pascual-Le Tallec and Lombes, 2005).

The human mineralocorticoid receptor (MR) was cloned by Arriza and Evans by low-stringency hybridization with the glucocorticoid receptor cDNA from a human placental library and found to encode for a 107 kDa protein with 984 amino acids (Arriza et al., 1987). Its crystal structure resembles that of other receptors of the same family (Arriza et al., 1987; Evans, 1988; Fagart et al., 2005a). The DNA binding domain of the mineralocorticoid receptor comprises 66 amino acids corresponding to a highly conserved region among members of the nuclear receptor superfamily that has 94% homology with the glucocorticoid receptor and ≈90% homology with other nuclear receptors). Two groups of four cysteines form α-helices called 'zinc fingers', one of which lies in the major groove of the DNA facilitating specific contacts during transcription (Arriza et al., 1987). The zinc fingers contain a P box, the interacting surface with the half site of the inverse repeat of the hormone (glucocorticoid) response element (AGGTCANNNTGACCT), and a D box responsible for weak dimerization with the DNA (Liu et al., 1995, 1996). A specific mineralocorticoid response element has not been identified. An additional nuclear export signal is located between the two zinc fingers near the LBD.

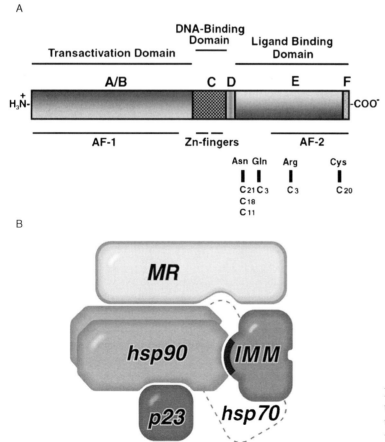

FIGURE 22.1 (A) Consensus structure of nuclear receptors. (B) Steroid receptor heterocomplex. Arrows indicate the position of key amino acids for the hMR that interact with the aldosterone molecule.

The hinge region is located between residues 671 and 732 and contains a proline stretch, which permits a twist of the DNA-binding domain relative to the ligand binding domain, positioning the receptor in contact with the general transcription machinery (Tsai and O'Malley, 1994). This region also possesses weak ligand-independent nuclear localization signal (NLS1) responsible for receptor subcellular translocation. In the glucocorticoid receptor, the hinge region is a potential link responsible for homodimerization, but not heterodimerization, and it may play a similar role in the mineralocorticoid receptor (Savory et al., 2001).

The crystal structure of the LBD closely resembles that of the glucocorticoid, androgen and progesterone receptors (Fagart et al., 2005a). The LBD of the human mineralocorticoid receptor has 251 amino acid residues with around 57% homology with the glucocorticoid receptor and more than 85% homology with mineralocorticoid receptors of other species (Arriza et al., 1987; Rogerson et al., 2004; Fuller and Young, 2005; Pascual-Le Tallec and Lombes, 2005; Bridgham et al., 2006). It is organized in 12 α-helices and one β-sheet forming three antiparallel layers (Auzou et al., 2000; Hellal-Levy et al., 2000b; Fagart et al., 2005a, b; Li et al., 2005). The N-terminal domains differ greatly between receptors of the superfamily; the human mineralocorticoid receptor has only 11% homology with the glucocorticoid receptor (Arriza et al., 1987; Evans, 1988).

The mineralocorticoid receptor has an activating function, AF2, within the LBD that becomes activated in a ligand-dependent manner after agonist binding in the hydrophobic pocket of the LBD. The AF2 is constituted by the H3, H4, H5 and H12 helices. Like that of the glucocorticoid receptor, the tau2 minimal domain of the mineralocorticoid receptor AF2 has 30 residues that are sufficient to recruit the general transcriptional machinery and activate transcription (Hollenberg and Evans, 1988).

The AF2 sequence is highly conserved for all SR functional domains and is located in H12 of the mineralocorticoid receptor. The positioning of aldosterone is ensured by the hydrophobic residues L938, F941, F946 and F956 of MR helices H11–12 and is stabilized by the interactions of the 3-ketone, 20-ketone, 21-hydroxyl and 18-hydroxyl groups of aldosterone to LBD polar residues Q776 and R817, C942 and N770 respectively (Couette et al., 1998; Hellal-Levy et al., 2000a, b; Fagart et al., 2005a, b; Li et al., 2005; Huyet et al., 2007). The Met852 residue acts as an organizer residue with two major roles. First, it allows steroids with no substituent at the C7 position to be accommodated within the ligand binding cavity and, second, it is involved in the steric hindrance that prevents C7-substituted spirolactones from folding the receptor in its active state (Fagart et al., 2005b). This results in the rotation of H12, closure of the pocket, rearrangement of helices H3, H5 and H11, which together expose the outside of the LBD, a hydrophobic groove that interacts with NR box of different co-activators defined by the LXXLL motif (where L is a leucine and X any residue) (Heery et al., 2001; Li et al., 2005). The L924, Q776R and L979 mutations in patients with type I pseudo-hypoaldosteronism impair aldosterone binding and function of the receptor (Sartorato et al., 2003, 2004a, b; Zennaro and Lombes, 2004).

The N-terminal domain of the MR, at 602 amino acids, is the longest and most highly variable among the steroid receptors. While it has only 15% homology with the N-terminus of the glucocorticoid receptor, the N-terminus has been conserved in evolution and is highly homologous among species. This domain contains multiple functional sites responsible for ligand-independent transactivation or transrepression.

Two possible evolutionary histories may have led to such a particular intra-molecular organization of nuclear receptors. Different domains may have had different origins, with those related to the regulation of metabolism eventually becoming fused to a DNA-binding motif to produce a transcription factor. Alternatively, a multidomain precursor that initially mediated a simple signal transduction mechanism may have acquired increasingly complex functions. In the case of steroid receptors (SRs), the archetypal ligand-activated transcription factors, how could selection drive the receptor's affinity if the hormone was not yet present? Conversely, without the receptor, what selection pressure could guide the evolution of the ligand?

The GR and MR share the highest homology among the members of the steroid receptor family and are derived from a common ancestor. Recent studies suggest that long before aldosterone evolved, the affinity for aldosterone was present in that ancestral receptor as a structural by-product of its partnership with chemically similar, more ancient ligands (Bridgham et al., 2006). Introducing two amino acid changes into the ancestral sequence recapitulates the evolution of present-day MR specificity. In other words, it appears that the sensitivity of corticoid receptors to aldosterone may have been more ancient than the hormone itself and that the ancient receptor was activated by a different ligand, perhaps deoxycorticosterone that is present in certain archaic fish that express MR, but not aldosterone (Bridgham et al., 2006).

On the other hand, the other characters of this plot, the ligands, share the general property of being small and hydrophobic molecules. Many of them are derived from common precursors that are also found in plants. However, there are no equivalent receptors between the animal and plant kingdoms (Clouse, 2002). This is interesting because it raises the question of how the same simple compounds have been assigned to shape the biological development and differentiation in metazoans. It may be possible that the information for hormonal regulation is written neither in the hormone nor in the receptor exclusively, but in both components of a complex *functional unit*, which also includes regulatory proteins such as the multimeric chaperone system associated to the transcription factor. In turn, this functional unit may be subject of other types of regulations, for example, differential recruitment of soluble factors, phosphorylation, acetylation, sumoylation, etc. (Galigniana and Piwien Pilipuk, 2004).

It is interesting to emphasize that ancient members of the steroid receptor superfamily such as the ER or PR are found primarily in the nucleus. However, the two receptors more recently evolved, GR and MR, in the absence of ligand are primarily cytoplasmic and rapidly move to the nucleus upon hormone binding ($t_{0.5}$ = 5 min). This property may provide the opportunity for more 'check-points' for the regulation of the molecular mechanisms of action before the receptor reaches its nuclear sites of action.

A. Genomic Structure and Organization

The MR (NR3C2 gene) is a member of the nuclear receptor superfamily of nuclear receptors and belongs to the steroid receptor subgroup of ligand-activated transcription factors (Arriza et al., 1987; Evans, 1988). The receptor is widely expressed including the kidney, brain, heart, colon and vasculature (Fuller and Young, 2005; Gomez-Sanchez et al., 2006).

The human MR gene is localized in chromosome 4 in the q31.1 region (Morrison et al., 1990), spans ≈450 kb and is composed of ten exons (Zennaro et al., 1995). The two first exons are referred as exon 1α and 1β and correspond to the 5' untranslated region in the human. They are followed by eight exons that code for the protein with exon 2 encoding the N-terminal domain (NTD or A/B region). Exons 3 and 4 code for the two zinc fingers of the DNA-binding domain (C region) and the last five exons code for the LBD (E) (Zennaro et al., 1995). Rats and mice have an additional 5' untranslated region (Kwak et al., 1993; Pascual-Le Tallec and Lombes, 2005). Alternative transcription for these 5'-untranslated exons generate different mRNA isoforms that are differentially expressed in aldosterone target tissues (Kwak et al., 1993; Zennaro et al., 1997). Mineralocorticoid receptor translation starts 2 bp downstream from the beginning of exon 2 and the translated protein is the same for the two (three in the rat) 5'-untranslated isoforms.

There are functional splice variants of the MR. A 12 bp insertion between the two zinc fingers results from the use of a cryptic splice site at the exon3/intron C splice junction (Bloem et al., 1995) creating a splice variant that is expressed in most tissues, but its transactivation activities are not significantly different from the wild type (Wickert et al., 1998, 2000; Wickert and Selbig, 2002). A 10-bp deletion in the rat and human MR leads to a truncation in the LBD at residue 807 from the rat sequence and unresponsiveness to aldosterone. It is expressed at low levels in the rat and human tissues and does not interfere with the wild-type mRNA activity (Zhou et al., 2000). An additional alternative splice variant skips exons 5 and/or 6, leading to the co-expression of the Δ5 or the Δ5,6 human mineralocorticoid receptor mRNA isoforms (Zennaro et al., 2001). This isoform retains the DNA-binding domain and can act in a ligand-independent manner (Zennaro et al., 2001).

The glucocorticoid receptor has been shown to express two translational variants (Yudt and Cidlowski, 2001). The mineralocorticoid receptor has two strong Kozak sequences suggesting the possibility of two different translation initiation sites. While the existence of the two variants has been demonstrated in a transcription-translation type of study (Pascual-Le Tallec et al., 2004), they have yet to be shown to be present *in vivo*. Use of a monoclonal antibody against amino acids 1–18 of the MR demonstrates that the long variant is expressed, but does not rule out the existence of the shorter one (Gomez-Sanchez et al., 2006). The transactivation properties of the short isoform that starts at amino acid 15 is greater than the classic form (Pascual-Le Tallec et al., 2004).

B. Post-Translational Modifications of the MR

The mineralocorticoid receptor is a phosphoprotein (Alnemri et al., 1991; Galigniana, 1998; Piwien-Pilipuk and Galigniana, 1998) with multiple consensus sites for phosphorylation (Figure 22.2). Rapid phosphorylation of serine and threonine residues occurs within minutes of exposure to aldosterone. These are mediated in part by protein kinase C alpha activation and might be involved in the rapid, non-genomic effects of aldosterone (Le Moellic et al., 2004). There is some evidence that phosphorylation by PKA enhances MR function, but this could be due to phosphorylation of an associated co-regulator rather than a direct effect (Massaad et al., 1999). Inhibition of serine/threonine phosphatases inhibits MR transformation and inhibits DNA binding (Piwien-Pilipuk and Galigniana, 1998). Mutation of one potential tyrosine phosphorylation site, position 73 of the N-terminal domain in the Brown Norway rat (Y73C), results in a robust gain of function for the MR (Marissal-Arvy et al., 2004) in comparison to the Fisher 344 rat, however, the original clone obtained from a Sprague Dawley rat has a similar tyrosine at position 73 as the Brown Norway rat (Patel et al., 1989). Except for the non-genomic effects, the role of phosphorylation remains unclear (Le Moellic et al., 2004; Mihailidou et al., 2004).

Sumoylation, modification by SUMO (small ubiquitin-related modifier), is a post-translational modification common to most steroid receptors (Poukka et al., 2000; Tian et al., 2002; Chauchereau et al., 2003). The MR has four sumoylation consensus motifs in the N-terminal end at positions K89, K399, K428 and one in the LBD at K953 of the human sequence (see Figure 22.2). The consensus motifs for sumoylation are named synergy control motifs and are defined by the sequence consensus ψKXE sequence, where X is any residue and the ψ is an aliphatic residue. These sites are highly conserved through evolution. A study of protein–protein interaction resulted in the identification of a protein inhibitor of activated signal transducer and activator of

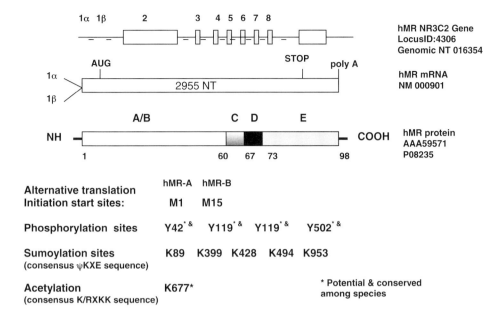

FIGURE 22.2 The human mineralocorticoid receptor gene, mRNA, protein functional domains and post-translational modifications. The intron, exon structure of the gene and the two different promoters are shown in the upper part of the figure. The middle part of the figure shows the mRNA and the lower part of the drawing shows the MR protein and its different domains. Alternative translation initiation sites, phosphorylation sites, sumoylation sites, potential acetylation and Genbank accession numbers are listed.

transcription (PIAS) and PIASxβ which is a SUMO E3-ligase and conjugated SUMO1 *in vivo* and *in vitro* and represses its ligand-dependent transcriptional activity (Tallec et al., 2003). Transient transfection of the SUMO1-conjugating enzyme Ubc9 increased MR transactivation of reporter constructs a mineralocorticoid-response element, ENAC or MMTV promoter in a ligand-sensitive manner (Yokota et al., 2007). Mutation of all four lysines of the sumoylation consensus eliminated sumoylation by Ubc9, but enhanced transactivation by Ubc9 of this mutant MR remained, indicating that the sumoylation activity is dispensable for the co-activation capacity of Ubc9 and the effects are distinct from the sumoylation of the receptor (Yokota et al., 2007). Other studies have shown that the transcriptional activity of the MR can be modulated by its sumoylation potential, as well as by the sumoylation of MR-interacting proteins, and requires the continuous function of the proteosome (Tirard et al., 2007). Acetylation of the receptor is also theoretically possible as it has a consensus sequence for acetylation, but it remains to be demonstrated (Pascual-Le Tallec and Lombes, 2005).

C. Oligomeric Structure of Steroid Receptors

In their mature form, SRs are associated to 90-kDa and 70-kDa heat shock proteins, the small acidic protein p23 and proteins that posses tetratricopeptide repeats (TPR), i.e. sequences of 34 amino acids repeated in tandem (see Figure 22.1b, *black crescent*) that are critical for protein–protein interactions (Lamb et al., 1995). In the SR heterocomplex, the TPR-acceptor site of hsp90 is normally occupied by either high molecular weight immunophilin (IMM) FKBP52, FKBP51, CyP40 or PP5 (Silverstein et al., 1999). IMMs are a family of intracellular receptors for immunosuppressant drugs also characterized for having peptidylprolyl isomerase (PPIase) enzymatic activity, which directs cis–trans isomerization of peptidylprolyl bonds (Galat, 2003). The oligomeric structure of untransformed SRs is also found in primarily nuclear untransformed receptors such as the progesterone or estrogen receptors.

The common pathway for complex assembly involves an initial ATP-dependent interaction of the client protein with the essential chaperone hsp70 and its non-essential co-chaperone hsp40 to form a client protein•hsp70 complex that is now 'primed' to bind hsp90 and the non-essential co-chaperone Hop (Pratt et al., 2004a, b). A second ATP-dependent reaction then occurs, producing a client protein•hsp90 complex in which the bound hsp90 is converted to its ATP-dependent conformation. The small, ubiquitous co-chaperone p23 then binds dynamically to the bound hsp90 to maintain it in the ATP-dependent conformation, thus stabilizing the client protein•hsp90 complex. This assembly machinery is ubiquitous and conserved among animal and plant cells, indicating that it performs essential functions (Pratt et al., 2001; Harrell et al., 2002). After their formation, the client protein•hsp90•p23 complexes diverge from the common pathway in that protein kinase•hsp90 complexes quite selectively bind p50^{cdc37}, whereas transcription factor•hsp90 complexes bind primarily the TPR domain immunophilins and immunophilin homologs. Both p50cdc37 and the TPR domain immunophilins bind directly to hsp90 but at different sites (Pratt et al., 1999). Steroid receptors form hsp90 heterocomplexes that contain a TPR domain immunophilin, the core of the TPR binding site on hsp90 being the MEEVD sequence of the chaperone (Scheufler et al., 2000; Brinker et al., 2002).

IMMs bind immunossuppressant drugs such as FK506, rapamycin and cyclosporine A. The common feature of all members of the IMM family is the presence of a peptidylprolyl isomerase (PPIase) domain, and they are divided into two classes: the FKBPs bind FK506 and rapamycin, and the cyclophilins (CyPs) bind cyclosporine A. The immunosuppressant drugs occupy the PPIase site on the IMM, blocking its ability to direct *cis–trans* isomerization of peptidyl–prolyl bonds. Three high molecular weight IMMs with TPR domains – FKBP52, FKBP51, CyP40 – have been found in steroid receptor•hsp90 complexes (Pratt et al., 2004b). A fourth SR•hsp90 complex protein, protein phosphatase 5 (PP5), is a protein-serine phosphatase with three TPRs and a PPIase homology domain. These TPR proteins can exchange for binding to hsp90, however, it has been shown that the IMMs exist in separate SR•hsp90 heterocomplexes. The relative amounts of FKBP52, FKBP51, CyP40 and PP5 may vary somewhat among the different SR heterocomplexes (Ratajczak et al., 2003) according to IMM interaction with the receptor itself.

D. Nuclear-Cytoplasmic Shuttling of Steroid Receptors

It is currently accepted that soluble signaling cascade factors such as SRs are not confined to the cytoplasm or the nucleus in a static manner, but are capable of shuttling dynamically through the nuclear pore (DeFranco, 2002; Vicent et al., 2002; Xu and Massague, 2004). Moreover, the reorganization of nuclear proteins is an essential step for them to acquire certain functions and/or repress others.

Protein transport across the nuclear envelope involves sequential steps: retrograde movement through the cytoplasm, recognition by nuclear import proteins, docking to the nuclear pore, translocation across the pore, movement through the nuclear compartment, anchorage to and release from the nuclear sites of action, recognition by nuclear export proteins, anterograde translocation across the nuclear pore and anterograde cytoplasmic movement. Several studies have been conducted to elucidate import and export mechanisms through the nuclear pore and, even though there is an uncountable number of questions that still deserve an answer, the information currently available in this field overwhelms the information we have on the molecular mechanism/s of protein movement within the cytoplasm and nucleus (Xu and Massague, 2004).

The possibility that this movement involves simple diffusion such that random collisions results in signaling proteins becoming trapped at their sites of action by protein–protein or protein–nucleic acid interactions does not suffice for a number of reasons. Diffusion does not explain how each protein exerts specific effects when a given cascade is activated, since the protein responsible for triggering the process would freely spread throughout one or more cell compartments. Cytoskeletal filamentous macromolecules have net negative charge and are able to interact with arginine and lysine groups of signaling proteins. Electrostatic interactions could restrict the motion of soluble solutes to the vicinity of filaments and provide the conditions for a 'directed' diffusion along the cytoskeleton constraining the molecules into more focused trajectories. Speed of signal transduction could be controlled by changes in the cytoskeleton structure of the cell. However, cryoelectron tomography shows that the cytoplasm is filled with large and highly packed assembles of filaments and macromolecules forming functional complexes rather than freely diffusing and colliding soluble complexes (Ellis and Minton, 2003) that would compromise the efficiency of free diffusion, particularly for large oligomeric macromolecules like SR•hsp90•IMM heterocomplexes (Seksek et al., 1997). In addition, it has been determined that diffusion in membrane-adjacent cytoplasm is significantly slower than that measured in the rest of the cytoplasm due to the high density of proteins near the cell membrane (Swaminathan et al., 1996). An extreme example is that of neurons, particularly their axons (Goldstein and Yang, 2000; Howe and Mobley, 2004). Even diffusion directed by electrostatic forces would not suffice for the delivery of protein solutes from the neural soma over long polarized axons, particularly those of animals the size of a human that may be greater than one meter in length with a volume that exceeds that of its cell body by three or more orders of magnitude. The more densely compacted nucleus would seem to present even greater obstacles to the movement of large particles than the cytoplasm. Alternative and more directed mechanisms for the movement of large oligomeric macromolecules like SR•hsp90•IMM heterocomplexes across cell compartments likely involve molecular motors and cytoskeletal or nucleoskeletal tracts in an active manner. This implies that specific interactions are required to determine the direction of signal protein movement.

E. MR Trafficking

Upon aldosterone binding, the MR moves rapidly, within minutes, towards the nucleus, whereas it cycles back to the cytoplasm upon ligand withdrawal much more slowly, more than 18 h. A classical model accepted *à bouche ouverte* for more than two decades posits that the ligand binding-dependent dissociation of the hsp90-based heterocomplex (a process frequently referred to as 'transformation') is a *sine qua non* for the nuclear translocation of SRs (Dahmer et al., 1984; Rousseau, 1984). This model was based on the assumption that the chaperone complex anchors SR to the cytoplasm, impairing its nuclear translocation. The constitutive nuclear localization of ER and PR occurs, it was assumed, because their nuclear localization signals (NLS) protrude more and are more exposed than those of primarily cytoplasmic SRs. This model was challenged in early publications in which it was postulated that both hsp90 (Koyasu et al., 1986; Nishida et al., 1986; Miyata and Yahara, 1991) and the hsp90-binding IMMs might be involved in protein trafficking (Gething and Sambrook, 1992), however, the transformation model, although unproven, prevailed until recent years.

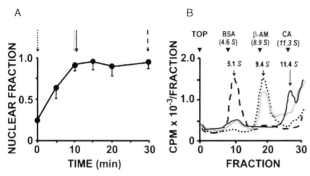

FIGURE 22.3 Transformation of MRs is not an early event. (A) Nuclear translocation rate of the MR after treatment of renal duct cells with 10 nM aldosterone. (B) Sucrose gradient of [³H]aldosterone-MR complexes in whole lysates before adding steroid (*dotted*); gradients of nuclear fractions after 10 min (*gray*) and 30 min (*dashed*) with steroid. The profile of cells treated for 10 min with steroid whose fractions were preincubated with an IgM antibody against hsp90 is shown in *black*. Calibration standards (bovine serum albumin, β-amylase and catalase) are shown on the top. Arrows in panel (A) mark the incubation times of those gradient profiles shown in panel (B).

A key discovery was that cytoplasmic dynein co-immunoprecipitates with FKBP52 (Galigniana et al., 2001). Dyneins are molecular motors that generate force towards the minus end of microtubules and are related to the retrograde movement of vesicles. Cargo attachment occurs via the dynein intermediate chain, whereas the ATP-hydrolytic domain responsible for the motor function is located in the heavy chains. The actual microtubule-binding site is a small globular unit that protrudes from heavy chains. It then became clear that the dynein-IMM interaction involves the PPIase domain of the IMMs (Galigniana et al., 2001; Pilipuk et al., 2007), a property that appears to be a common feature for most high molecular weight IMMs associated with SRs (Galigniana et al., 2002). Importantly, at least two of the most abundant IMMs found in SR●hsp90 complexes, FKBP52 and PP5, co-localize with microtubules.

These observations implied that an active transport system requiring hsp90, IMMs, dynein motor proteins and cytoskeletal tracts moves SRs within the cell towards the nucleus. If correct, one critical corollary of this hypothesis is that the hsp90●IMM complex associated with the untransformed receptor should not dissociate upon ligand binding because it would be required for the retrograde movement of the ligand–receptor complex. The results of the experiment described in Figure 22.3 clearly contradicts the unproven classical model: transformation is not an early event (Pilipuk et al., 2007). Figure 22.3a shows that, upon steroid binding, the MRs in renal collecting duct cell cytoplasm are rapidly translocated to the nucleus and that 100% of the MR population becomes entirely nuclear after 10 min of incubation with aldosterone. When the [³H]aldosterone–MR complexes in the nuclear compartment are analyzed by a continuous sucrose gradient (Figure 22.3b), more than 70% of the MR population still co-migrates with the 9.4S, untransformed MR form, even 10 min after the steroid was added to the medium, i.e. a time when the MR is totally nuclear (Figure 22.3a). This observation suggests that hsp90 is not immediately dissociated from SRs upon ligand binding. This notion is evidenced in a more convincing manner by resolving the gradient after a preincubation with an IgM antibody against hsp90, which switches the 9.4S peak to 11.4S (Figure 22.3b). After 15 min with aldosterone, the 9.4S peak is greatly reduced in favor of the transformed, 5.4S MR form, whereas after 20 min in the presence of hormone, 100% of the MR population is fully transformed and tightly bound to chromatin.

In agreement with the model, the cytoplasmic-nuclear movement of hsp90●IMM-chaperoned factors of SR is impaired or blocked by hsp90 inhibitors (i.e. geldanamycin or radicicol), by overexpression of the PPIase domain of FKBP52 that prevents dynein binding, by saturation of the hsp90●IMM binding site with the TPR peptide, and by overexpression of the dynactin complex subunit, p50/dynamitin, which disrupts the dynein-dynactin complex and dissociates cargoes from dynein (Figure 22.4). Thus, there is strong evidence that validates a model in which the retrograde movement of certain soluble factors occurs in an active manner via cytoskeletal tracts with the hsp90●IMM complex forming the bridge between the cargo and the motor protein responsible for the retrograde movement of SR and that this system can be uncoupled by some inhibitory agents.

The disruption of the molecular machinery of transport impairs, but does not block SR retrograde movement. For example, MR becomes entirely nuclear after 60 min of incubation with aldosterone (i.e. one order slower translocation rate than the normal mechanism) (Pilipuk et al., 2007).

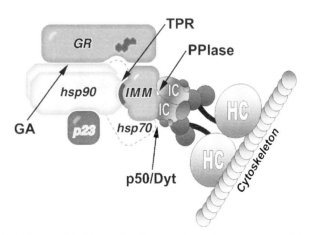

FIGURE 22.4 The molecular machinery for movement of the MR and the agents for selectively uncoupling the system. Arrows show the sites of uncoupling by hsp90-disrupting agents such as geldanamycin (*GA*), the TPR domain fragment of PP5, the PPIase domain fragment of FKBP52 and the p50/dynamitin (*Dyt*). Immunophilin TPR domain (*black crescent*). Dynein heavy chains (*HC*), intermediate chains (*IC*).

This implies the existence of a secondary, hsp90•IMM•dynenin-independent mechanism of movement, which is clearly less efficient than that mediated by the chaperone machinery. The nature of this secondary mechanism is unknown, although it may possibly be driven by simple diffusion forces.

The interaction between the hsp90-binding IMMs with dynein motor proteins seems to be conserved in nature. It has recently been described that plants possess two hsp90-binding IMMs, named wFKBP73 and wFKBP77, that are highly homologous with mammalian FKBP52 (Reddy et al., 1998). Interestingly, mGR•whsp90•wFKBPs•dyenin heterocomplexes can be assembled in wheat germ lysate (Harrell et al., 2002) and the resultant mGR-chaperone heterocomplex binds steroid with as high affinity as complexes assembled with mammalian proteins. The presence of the entire heterocomplex assembly machinery in plants implies that this complex plays a cardinal role in the biology of the eukaryotic cell, most likely protein trafficking via motor proteins. Thus, all of the conditions for ligand-regulated movement of SRs described in mammalian cells also exist in plants.

F. Mineralocorticoid Receptor Selectivity

The MR is not selective for aldosterone; it has similar affinity for glucocorticoids cortisol and corticosterone (Krozowski and Funder, 1983). Given the much greater abundance of the glucocorticoids compared to aldosterone in the circulation, it was difficult to explain the selectivity of the MR for aldosterone. Selectivity of the MR is accomplished at three different levels: pre-receptor, receptor and post-receptor selectivity (Funder and Myles, 1996; Farman and Rafestin-Oblin, 2001b; Pascual-Le Tallec and Lombes, 2005).

At the pre-receptor level, the 11β-hydroxysteroid dehydrogenase 2 enzyme (11β-HSD2) is expressed in aldosterone target cells such as that of the kidney connecting tubule and collecting tubule epithelia (Edwards et al., 1988; Funder et al., 1988; Edwards and Stewart, 1991). The 11β-HSD2 confers pre-receptor selection by catalyzing the unidirectional conversion of cortisol to cortisone using NAD$^+$ as a coenzyme of corticosterone very efficiently in these cells. When tubular cells were grown on semipermeable membranes and incubated with ^{3}H-corticosterone on the luminal side, only ^{3}H-11-dehydrocorticosterone was detected on the other side of the filter (basolateral side) (Naray-Fejes-Toth et al., 1991). The 11β-HSD2 is located within the endoplasmic reticulum (ER) with its N terminus protruding into the ER lumen and the C terminus in the cytoplasm (Odermatt et al., 1999). When HEK-293 cells were transfected with the cDNA of the MR by itself, the MR protein was distributed in the cytosol and nuclei. When co-expressed with the 11β-HSD2, the MR displayed a reticular distribution pattern, suggesting association with the 11β-HSD2 at the endoplasmic reticulum membrane (Odermatt et al., 2001). Aldosterone induced rapid nuclear translocation of the MR, whereas moderate concentrations of cortisol (10–200 nM) did not activate the receptor, due to oxidation of cortisol to cortisone. Cortisone, the product of the conversion of cortisol by the 11β-HSD2 blocked the aldosterone-induced MR activation by an 11β-HSD2-dependent mechanism (Odermatt et al., 2001). The MR also displayed exclusively reticular localization when co-expressed with the 11β-HSD2-deficient mutants from patients suffering from apparent mineralocorticoid excess in the absence of corticosteroids, further suggesting an interaction between the enzyme and the MR. In cells co-transfected with the MR and the defective 11β-HSD2 enzyme cDNA, low doses of cortisol produced a rapid translocation of the MR to the nucleus (Odermatt et al., 2001). Congenital or acquired deficiency (produced by licorice or its active compounds) of the 11β-HSD2 results in severe mineralocorticoid hypertension by allowing cortisol to bind and activate the mineralocorticoid receptor (Mune et al., 1995; Stewart et al., 1996).

At the receptor level, in aldosterone target tissues other than the kidney and even areas of the kidney where the 11β-HSD2 enzyme is not expressed, selectivity of the MR for aldosterone must be conferred by an alternate mechanism. NAD$^+$ has also been postulated to determine MR selectivity as, in the absence of the 11β-hydroxysteroid dehydrogenase 2 (congenital or acquired), accumulation of NAD$^+$ allows cortisol to become a full agonist (Funder, 2007).

Although aldosterone and cortisol bind to the MR with similar affinities, the dissociation kinetics are different (Lombes et al., 1994; Hellal-Levy et al., 2000a). Aldosterone dissociates from the receptor much more slowly than cortisol, indicating that the aldosterone-receptor complex is more stable and efficiently stabilizes the helix 12 active position (Hellal-Levy et al., 2000a, b). Binding of a ligand confers specific conformation resulting in agonist or antagonistic activity. Another possible mechanism of MR selectivity for aldosterone is the interactions between the N-terminal and LBDs that appear to be stronger in the presence of aldosterone than cortisol (Rogerson and Fuller, 2003; Rogerson et al., 2007). An S810L mutation of the MR has been found in a family of hypertensive patients whose hypertension is exacerbated during pregnancy. This mutated MR receptor is moderately constitutively activated and fully activated by progesterone, cortisone and the mineralocorticoid antagonist spironolactone (Geller et al., 2000; Rafestin-Oblin et al., 2003).

Post-receptor steroid selectivity occurs through binding of transcriptional regulators that either activate or repress receptor function. Some of these regulators are fairly generic in their action, affecting several nuclear receptors, while others are quite specific, either because they are specific for a given receptor or are present only in certain cells. Binding to the NTD allows the transcription regulators to confer specificity since this region is not conserved among receptors. One large family of co-activators that interact with all steroid receptors is the steroid receptor co-activator 1 (SRC-1) (McKenna et al., 1999). SRC-1 interacts with the steroid receptor bound to DNA to initiate transcription by the sequential recruitment of a series of proteins involved in chromatin remodeling (Kitagawa et al., 2002; McKenna and O'Malley, 2002; Pascual-Le Tallec and Lombes, 2005). MR has been shown to interact with different SRC-1 variants, mainly through interactions at the AF2 domain of the LBD (Auzou et al., 2000; Fuse et al., 2000; Hellal-Levy et al., 2000b; Meijer et al., 2000, 2005; Kitagawa et al., 2002). Other co-activators interacting with the MR include TIF2 and CBP/p300 (Fuse et al., 2000; Kitagawa et al., 2002), receptor-interacting protein 140 (Zennaro et al., 2001; Pascual-Le Tallec and Lombes, 2005) and the peroxisome proliferator-activated receptor γ co-activator 1 (Kressler et al., 2002). The elongation factor ELL (eleven-nineteen lysine-rich leukemia) is a particularly strong selective co-activator for the MR. The ELL increases RNA polymerase II processivity and elongation rate by suppressing the termination of mRNA synthesis and resuming transient pausing (Shilatifard et al., 1996; Sims et al., 2004; Kong et al., 2005; Pascual-Le Tallec et al., 2005). ELL interacts with the NTD of the MR and exerts an exclusive AF1b-dependent co-activation. It behaves as a transcriptional selector because it represses GR transactivation with no effect on the androgen or progesterone receptor (Pascual-Le Tallec et al., 2005). ELL may exert an important role in determining MR-positive versus GR-negative effects in epithelial cells and in the brain where both the MR and GR are frequently co-expressed.

Co-repressors participate in extinguishing MR activated-transcription. The death-associated protein (DAXX) is a co-repressor for both the MR and the GR (Obradovic et al., 2004). The NTD of the MR interacts with PIAS1, PIASxβ and UBC9, members of the PIAS family of corepressors (Tallec et al., 2003; Tirard et al., 2007; Yokota et al., 2007). These proteins behave as SUMO-E3 ligases that sumoylate the MR *in vitro* and *in vivo* (Tallec et al., 2003; Tirard et al., 2007; Yokota et al., 2007). PIAS1 is a specific co-repressor of the MR, that has no effect on GR transactivation (Tallec et al., 2003). Thus, MR selectivity is exerted by a combination of the pre-receptor, receptor and post-receptor steps that vary according to the various cells.

IV. DISTRIBUTION OF THE MINERALOCORTICOID RECEPTOR IN THE NEPHRON

Identification of target cells within the kidney has been intensively studied using ligand binding and functional methods (Marver and Schwartz, 1980; Gnionsahe et al., 1989; Farman et al., 1991). The results for many nephron segments have been contradictory and difficult to interpret, due in part of the substantial overlap of ligand affinities and to the fact that the corticoids bind more than one receptor.

The renal cell distribution of the MR has been studied using competitive PCR to quantify relative levels of the MR and GR mRNA from isolated nephron segments (Todd-Turla et al., 1993). While mRNA for the MR was detected in all segments of the nephron, levels of MR mRNA in the cortical collecting duct, outer stripe of outer medullary collecting duct, inner stripe of the outer medullary collecting duct and inner medullary collecting duct were about 10 times higher than other areas of the nephron (Todd-Turla et al., 1993). Immunolocalization of the MR in the rabbit kidney using a monoclonal auto anti-idiotypic antibody against the MR demonstrated presence of the MR in all segments of the distal tubules and collecting duct including its cortical, medullary and papillary portions (Lombes et al., 1990; Farman et al., 1991). Approximately 15% of cells of the cortical collecting tubule (intercalated cells) were devoid of MR staining. Weak staining was demonstrated in the loop of Henle, but not in the glomerulus or proximal tubules (Farman et al., 1991). Similar results were obtained using monoclonal antibodies against MR peptides, with additional staining demonstrated in mesangial cells of the glomerulus (Nishiyama et al., 2005; Gomez-Sanchez et al., 2006) (Figure 22.5). The distribution of the 11β-HSD2 was similar to that of the MR, except no 11β-HSD2 immunoreactivity was found in the papillary portion of the collecting duct where the MR is clearly detected (Figure 22.6).

V. PROTEINS INDUCED BY ALDOSTERONE IN TRANSPORT EPITHELIA

Vectorial sodium transfer induced by aldosterone occurs mainly in the distal nephron and distal colon. Sodium entry into the cell at the apical membrane is mediated by the amiloride-sensitive epithelial sodium channel (ENaC). Efflux of sodium from the epithelial cell at the basolateral membrane is energy-dependent and is mediated by the sodium-potassium ATPase (Na^+-K^+ATPase) (Figure 22.7). Aldosterone stimulates Na^+ transport after a short lag period of 20–60 min first by activating pre-existing sodium channels (Verrey, 1999). This increase in open probability of active channels may depend in part upon methyl esterification reactions (Kemendy et al., 1993; Eaton et al., 2001).

The epithelial sodium channel is composed of three subunits (Rossier et al., 1994), α, β, γ, which are regulated by corticosteroids in a tissue-specific manner, however, this is not the principal mechanism by which aldosterone regulates ENaC activity (Muller et al., 2003). A more important effect on ENaC activity by aldosterone is its ability to alter ENaC subunit turnover and increase the number of channels inserted in the plasma membrane. Study of the gain-of-function mutations in Liddle's syndrome that result in an increase in the number of ENaCs at the cell surface helped

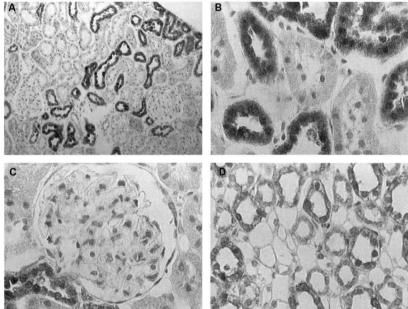

FIGURE 22.5 Photomicrographs of rat kidney immunostained with a monoclonal antibody against the mineralocorticoid receptor. (A) Cytosolic and nuclear binding in distal tubular epithelial cells of the rat renal cortex. (B) Higher magnification of distal tubular epithelial cells. (C) Kidney glomerulus showing immunolabeling of the distal tubule and glomerular mesangial cells. (D) Renal papilla showing immunolabeling of the collecting duct. (See color plate section.)

elucidate the mechanism of the aldosterone effect (Snyder, 2002). Turnover of ENaC is mediated by the ubiquitin protein ligase, Nedd4-2 (Snyder, 2002; Bhalla et al., 2005). Mutations of the C terminal of the β and γ subunits result in the identification of a sequence, PPPxYxxL, that plays an important role in controlling the surface expression of ENaC. Mutations of the PPPxYxxL motif or deletion of the area containing the motif results in Liddle syndrome (Schild et al., 1996; Snyder et al., 1995; Snyder, 2002). Nedd4 was identified as an interacting protein with the PPPxYxxL motif through the use of yeast two-hybrid technique (Staub et al., 1996). Nedd4 co-expression with ENaC decreased its surface expression in part by increasing channel degradation. Nedd4 has WW domains of approximately 38 amino acids that bind PY motifs of ENaC. It also contains a ubiquitin ligase domain that results in ubiquinilation of the ENaC and acceleration of its degradation within the proteosome (Snyder, 2002). The Nedd4-2 is the Nedd4 isoform that alters the function of ENaC. Aldosterone also increases the expression of 14-3-3β, a protein which interacts with phospho-Nedd4-2 and blocks its interaction with ENaC, thus increasing apical membrane ENaC density and enhancing Na^+ absorption (Liang et al., 2006; Nagaki et al., 2006).

Under Na^+ replete conditions, when aldosterone levels are low, ENaC is mainly in an intracellular location as shown by a diffuse punctuate immunostaining in the principal cells of the collecting duct (Masilamani et al., 1999; Nielsen et al., 2007). Na^+ absorption in response to aldosterone increases before any changes in mRNA for the subunits are observed (Garty and Palmer, 1997) suggesting that aldosterone first induces the transcription of proteins that modulate ENaC trafficking or function. One aldosterone-induced protein is SGK, a member of the serine-threonine kinase family. Although originally described as a kinase that increased with

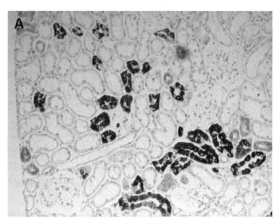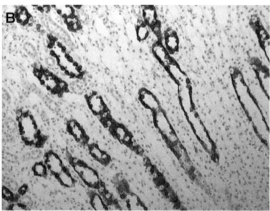

FIGURE 22.6 Immunostaining of rat kidney 11β-hydroxysteroid dehydrogenase 2 enzyme. (A) Distal nephron tubules in the cortex extending, (B) to the inner medullary collecting tubules and stopping when the tubules penetrate the papillae. (See color plate section.)

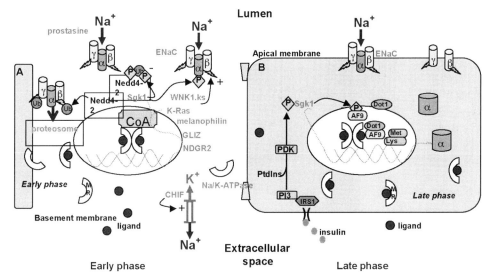

FIGURE 22.7 Mechanism of action of the mineralocorticoid receptor in polarized epithelial cells of the renal distal nephron. Cartoon cell A demonstrates early mechanisms of aldosterone regulation of ENaC and Na$^+$.K$^+$-ATPase. Aldosterone stimulates Sgk1 which phosphorylates Nedd4-2, preventing Nedd4-2 from ubiquinilating the ENaC β and γ subunits. Ubiquinylation accelerates removal of the channel from the membrane; Nedd4-2 phosphorylation allows a greater permanence of the ENaC subunits in the membrane and an increase in channel activity. Cartoon cell B demonstrates the late effects of aldosterone. At low aldosterone levels ENaCα gene transcription is low due to tonic inhibition by couples of Dot1a and AF9 which methylate Lys79 on histone H3. Aldosterone stimulates the transcription and translation of Sgk1 which phosphorylates multiple proteins including AF9 resulting in inhibition of the Dot1a-AF9 complex and diminished H3 methylation, thus enhancing MR access to chromatin and initiation of transcription. (See color plate section.)

glucocorticoids, it was later identified as an aldosterone-induced protein that affected ENaC (Webster et al., 1993; Chen et al., 1999; Naray-Fejes-Toth et al., 1999). Sgk1 overexpressed in *Xenopus* oocytes produced a large increase in Na$^+$ current, but had no effect on K$^+$ channel activity, due to an increase in the number of ENaC channels at the cell surface (Alvarez de la Rosa et al., 1999). Sgk1 phosphorylates Nedd4-2, decreasing the binding of Nedd4-2 to ENaC and resulting in an increase in ENaC surface expression (Snyder et al., 2001). Sgk can also regulate ENaC single-channel activity through direct phosphorylation of its α subunit, ENaCα (Diakov and Korbmacher, 2004).

Aldosterone stimulation results in a rapid increase in Sgk1 which peaks 1–2 h after exposure (Naray-Fejes-Toth et al., 1999). ENaCα is also stimulated by aldosterone, but peak expression takes at least 12 h (Mick et al., 2001). Although regulation of ENaCα gene transcription can be directly stimulated by the MR, it appears that it is not the only mechanism and Sgk1 might be implicated (Boyd and Naray-Fejes-Toth, 2005). To express kinase activity, Sgk1 needs to be phosphorylated at two residues (Thr^{256} and Ser^{422}). Inhibitors of the PI3K pathway abolish the aldosterone-induced increase in Na$^+$ current in A6 cells (Park et al., 1999), but the role in mammalian sodium absorption is unclear (Snyder, 2002).

Another potential mediator of early aldosterone response is K-Ras2, a small G protein. It was first shown in A6 cells that aldosterone increased K-Ras2 mRNA and protein within 2.5–4 h (Spindler and Verrey, 1999). K-Ras activates human ENaC reconstituted in CHO cells in a GTP-dependent manner, most likely by affecting their open probability. Inhibition of phosphoinositide-3-OH kinase (PI3K) abolished K-Ras actions on ENaC (Staruschenko et al., 2004).

The late phase of aldosterone regulation of ENaC involves the increase in the expression of αENaC mRNA and protein in the kidney with no changes in the β or γ subunits (Escoubet et al., 1997; Stokes and Sigmund, 1998). This transcriptional and translational regulation is tissue specific, as in the rat colon aldosterone produces large increases in mRNA for the β and γ subunits, but a small increase in the α subunit (Lingueglia et al., 1994; Asher et al., 1996; Stokes and Sigmund, 1998).

Sgk1 also phosphorylates the Kir1.1 (ROMK) subtype of inward rectifier K$^+$ channel, increasing channel density and contributing to the kaliuretic actions of aldosterone (Yoo et al., 2003). The Sgk1 knockout mouse has a mild phenotype that is only induced by severe sodium depletion, suggesting that additional or alternative mechanisms of sodium absorption must also be operative (Wulff et al., 2002; Huang et al., 2004).

The mechanism by which aldosterone modifies chromatin structure and thus transcription is unclear. A novel aldosterone signaling network governing ENaCα transcription has been demonstrated in the mouse inner medullary collecting duct cell line, mIMCD3 (Zhang et al., 2006) (see Figure 22.7). Basal repression of ENaCα was shown by a complex containing the histone H3Lys79 methyltransferase *d*isruptor *o*f *t*elomeric silencing alternative splice variant a (Dot1a) and the putative transcription factor ALL1-fused gene from chromosome 9(Af9) (Zhang et al., 2006, 2007). The nuclear repressor complex tonically associates with chromatin and hypermethylates histone H3Lys79 associated with specific regions of the ENaCα promoter, thus repressing transcription under basal conditions. Aldosterone inhibits Dot1a and Af9 expression, leading to hypomethylation of targeted histone H3Lys79 and release of the basal transcriptional repression of ENaCα. Aldosterone appears to signal to the complex through Sgk1. Af9 is a physiological target of Sgk1 and negatively regulates the Dot1a-Af9 repressor complex that controls transcription of ENaCα and very likely other aldosterone-induced genes (Pearce and Kleyman, 2007; Zhang et al., 2007). The negative regulation of the Dot1a–Af9 complex results in a decrease in hypermethylation of H3 and chromatin remodeling and allows the MR to bind to the ENaCα promoter to stimulate transcription.

The kidney-specific WNK1 isoform is induced by aldosterone and stimulates epithelial sodium channel-mediated Na^+ transport. The WNK1 (without lysine K) mutations have been associated with an autosomal dominant form of hypertension, pseudohypoaldosteronism type II or Gordon's syndrome (Wilson et al., 2001). A similar syndrome is produced by mutations of the WNK4 gene. The WNK4 inhibits the surface expression of several renal transporters including the secretory K^+ channel, ROMK and the Cl/base exchanger SLC26A6 (Kahle et al., 2005). Mutations of the WNK4 result in an increase in transporters. Hypertension-causing mutations of the WNK1 are mainly large intronic deletions resulting in increased expression of the WNK1 and stimulating Na^+ retention (Wilson et al., 2001, 2003; Choate et al., 2003). In addition to the classical long form of WNK1, there is a short form that is kidney-specific (Naray-Fejes-Toth et al., 2004) and is induced by aldosterone. Stable overexpression of the kidney-specific isoform increases transepithelial Na^+ transport and might participate in aldosterone-induced sodium retention (Naray-Fejes-Toth et al., 2004).

Several other genes have been shown to be regulated by aldosterone in transporting epithelia. The corticosteroid hormone-induced factor (CHIF) is rapidly induced by a primary transcriptional mechanism (Rogerson and Fuller, 2000). Melanophilin is a protein involved in vesicular trafficking in melanocytes that is also found in the mouse cortical collecting duct cells. Its mRNA is rapidly increased by aldosterone in cortical collecting duct cells. Transfection of melanophilin in CCD cells led to a modest increase in amiloride sensitive Na^+ current suggesting that it might be involved in ENaC trafficking (Martel et al., 2007). Stimulation of isolated renal cortical collecting duct from rats with aldosterone resulted in the identification of an early responsive cDNA highly homologous to human and murine NDRG2 (n-myc downstream regulated gene 2). NDRG2 consists of four isoforms belonging to a family of differentiation-related genes. It is expressed in classical aldosterone target epithelia in the kidney and is regulated by physiological amounts of aldosterone. Its function is unknown (Boulkroun et al., 2002).

VI. NON-GENOMIC EFFECTS OF ALDOSTERONE IN THE KIDNEY

Aldosterone injected intra-arterially or into the renal artery resulted in a decrease in sodium and increased potassium excretion with a delay of 5–30 min (Ganong and Mulrow, 1958) with no consistent effect in glomerular filtration rate. Acute subcutaneous administration of aldosterone increases sodium efflux from rat tail artery smooth muscle due to a specific action on MR (Moura and Worcel, 1984). Aldosterone has two types of action on sodium transport: a very rapid increase of passive sodium efflux, insensitive to actinomycin D (inhibitor of transcription), and a delayed stimulation of ouabain-dependent sodium efflux and an ouabain-independent sodium efflux that are completely blocked by actinomycin D. Both the early and late effects were blocked by the spirolactone RU 28318 (Moura and Worcel, 1984). These studies indicated that aldosterone had an early, non-genomic effect on sodium transport that was mediated by the MR and a late or classic genomic effect mediated by the MR (Funder, 2005). Non-genomic effects of aldosterone which occur too rapidly to be mediated by gene transcription, insensitive to inhibitors of transcription (actinomycin D) or translation (cycloheximide) have been postulated to be mediated by a membrane receptor different from the soluble MR (Boldyreff and Wehling, 2003; Koppel et al., 2003; Losel et al., 2003; Mihailidou and Funder, 2005). Some membrane MR effects are inhibited, but others are not inhibited by spirolactone derivatives (Alzamora et al., 2003; Wehling, 2005). Aldosterone induced a rapid and dose-dependent phosphorylation of ERK1/2 and c-Jun NH_2-terminal kinase (JNK) in Chinese hamster ovary cells (CHO cells) and human embryonic kidney cells transfected with the human MR cDNA, which was inhibited by the MR-antagonist spironolactone (Grossmann et al., 2005). Aldosterone also increased cytosolic Ca^{2+} in mock and MR-transfected CHO cells that were not inhibited by spironolactone (Grossmann et al., 2005) indicating that aldosterone exerts genomic and some non-genomic effects by the MR and some by an alternative, as yet unknown, mechanism not involving the MR (Figure 22.8). Most studies of non-genomic effects of aldosterone in the kidney have been done with renal cells in

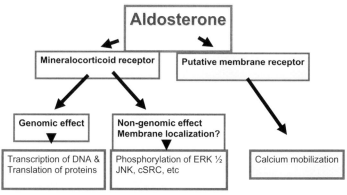

FIGURE 22.8 Aldosterone receptors and action: binding to the mineralocorticoid receptor to produce genomic and non-genomic effects and to a putative aldosterone membrane receptor that is not the classical mineralocorticoid receptor.

culture, however, a series of studies using isolated perfused tubules have also been done that also suggested the existence of a non-genomic effect of aldosterone (Good, 2007).

Non-genomic effects of aldosterone on ion transport proteins have been studied. Aldosterone has been shown to regulate the Na^+/K^+ exchange isoforms (NHE) NHE1 and NHE3 (Oberleithner et al., 1987, 1988; Gekle et al., 1996; Good et al., 2006; Watts et al., 2006). The Madin-Darby canine kidney cell has properties corresponding to collecting duct intercalated cells and demonstrates both genomic and non-genomic responses to aldosterone (Oberleithner et al., 1987, 1988; Gekle et al., 1996, 1997, 2001a, b). At physiological concentrations, aldosterone had rapid (1–2 min) non-genomic stimulatory effects on N^+/H^+ exchange in these cells as determined by increases in intracellular pH and Na^+ concentration (Gekle et al., 1996, 1997, 2001a, b). The rapid effects were not prevented by actinomycin D, cycloheximide or spironolactone and resulted from an increase in the affinity of the exchanger for intracellular H^+ (Gekle et al., 1996, 1997, 1998). Aldosterone also stimulates the Na^+/H^+ exchange activity in other epithelial and non-epithelial cells, probably through stimulation of the NHE1 isoform (Maguire et al., 1999; Gekle et al., 2001b). The stimulation of NHE1 in the MDCK-C11 cells by aldosterone requires the parallel activation of membrane H^+ conductance (Gekle et al., 1996). The rapid stimulation of Na^+/H^+ exchange by aldosterone depends on an increase in intracellular Ca^{2+}, as well as rapid phosphorylation of the extracellular signal-regulated kinase (ERK1/2) (Gekle et al., 1996, 2001b).

While NHE1 is expressed in all segments of the renal tubule (Good, 2007), it is unknown if this action of aldosterone through the NHE1 is of significance for renal tubular function, as there are no studies addressing the effect of aldosterone on the basolateral Na^+/H^+ exchange in the mammalian nephron (Good, 2007). NHE1 functions as a signaling molecule that regulates multiple cell functions including epithelial transport of ions and water, by controlling the organization of the actin cytoskeleton (Denker and Barber, 2002; Putney et al., 2002; Watts et al., 2005). Aldosterone could regulate multiple transporters by a non-genomic mechanism through NE1-induced cytoskeletal reorganization. Transport proteins that are regulated through cytoskeletal interactions include ENac, renal outer medullary K^+ channel (ROMK) and the Na^+-K^+-$2Cl^-$ co-transporter NKCC2 (Denker and Barber, 2002; Putney et al., 2002; Watts et al., 2005; Good, 2007).

Most of NaCl, $NaHCO_3$ and fluid absorption by the renal proximal tubule and most of the $NaHCO_3$ in the MTAL is mediated by the apical NHE3. Thus, regulation of the NHE3 is critical for maintenance of extracellular fluid volume, acid–base balance and blood pressure (Schultheis et al., 1998; Ledoussal et al., 2001). Aldosterone in physiological concentrations induces a rapid inhibition of $NaHCO_3$ in MTAL as a result of decrease in NHE3 activity (Good et al., 2006; Watts et al., 2006) that is not blocked by inhibitors of transcription or translation and not reproduced by glucocorticoids (Watts et al., 2006). The inhibition by aldosterone is mediated by rapid activation of the ERK 1/2 signaling pathway (Watts et al., 2006).

Aldosterone has genomic and non-genomic effects on proton secretion via vacuolar H^+-ATPases in the collecting duct from type A intercalated cells (Winter et al., 2004). The mechanism appears to be due to a rapid action of aldosterone producing increased trafficking of the H^+-ATPases to the apical membrane (Winter et al., 2004).

The epithelial sodium channels (ENaC) are located in the apical membrane of principal cells in the connecting tubule and collecting duct (Loffing et al., 2001a; Verrey, 2001). Aldosterone increases Na^+ absorption by increasing ENaC activity through MR-mediated transcriptional effects resulting in the induction of regulatory proteins, including Sgk1, that leads to an increase in the activity and number of functional channels in the membrane (Bhalla et al., 2005, 2006; McCormick et al., 2005; Grahammer et al., 2006; Pearce and Kleyman, 2007). In addition to the transcriptional effects, there is some evidence for rapid activation of ENaC by aldosterone, but only at supra-physiological concentrations (Zhou and Bubien, 2001).

Aldosterone increases the Na^+-K^+ATPase activity in collecting tubule principal cells through transcriptional regulation (Loffing et al., 2001a, b; Feraille et al., 2003; Verrey et al., 2003; Summa et al., 2004), but it also increased Na^+

pump activity in MDCK cells within 15 min, an effect that was unaffected by cycloheximide, but blocked by colchicine (Shahedi et al., 1993). Thus, it is likely that aldosterone has genomic and non-genomic effects on the Na^+-K^+ATPase.

The medullary thick ascending limb (MTAL), collecting duct and proximal tubules have also been shown to be sites of non-genomic regulation by aldosterone. Aldosterone induces rapid inhibition of apical NHE3 and HCO_3^- absorption in the MTAL. Changes in NHE3 activity in the MTAL for acid–base homeostasis, although it would appear to counterintuitive for the physiological action of aldosterone to promote Na^+ retention. In the MTAL, NHE3 mediates $NaHCO_3$ absorption, but absorption of NaCl is mediated by the NKCC2 which accounts for the majority of Na^+ absorption (Good, 1993; Jeck et al., 2005). Activation of the renin–angiotensin–aldosterone system by sodium and volume depletion promotes Na^+ retention, but also induces multiple transport effects that increase net renal acid excretion and promote metabolic alkalosis. Metabolic alkalosis stimulates HCO_3^- absorption by angiotensin II in the proximal and distal tubules, stimulates ammonium secretion in the proximal tubules and stimulates H^+ secretion in the collecting tubule (Good et al., 2002). The inhibitory effect of aldosterone on H^+ secretion and HCO_3^- absorption in the MTAL would oppose these changes in the proximal and distal tubules. The non-genomic regulation of NHE3 by aldosterone in the MTAL would minimize changes in H^+ excretion and maintain acid–base balance while permitting regulated changes in Na^+ excretion and contributing to the regulation of extracellular fluid volume and blood pressure (Good, 2007). As evidence for this, the rapid infusion of aldosterone in rats induced a significant change in Na^+ excretion with no effect on urinary pH or acid excretion (Rad et al., 2005). Aldosterone rapidly regulates the ERK 1/2 pathway in the MTAL (Watts et al., 2006) which, in turn, might regulate proteins involved in the absorption of NaCl by the MTAL, including NKCC2, apical K^+ recycling via ROMK and basolateral Cl^- efflux via the ClC-Kb Cl^- channels (Lifton et al., 2001; Jeck et al., 2005). Aldosterone also induces a rapid increase in cAMP in a human distal cell line and could stimulate NaCl absorption via cAMP by a non-genomic pathway (Koppel et al., 2003).

The cortical and outer and inner medullary collecting ducts have also been shown to be targets for the non-genomic regulation by aldosterone. Aldosterone increases H^+-ATPase activity in mouse outer medullary collecting duct and in principal cells from the rabbit CCD (Zhou and Bubien, 2001; Winter et al., 2004). Aldosterone produces an early increase in short-circuit current in the CCD cell line $RCCD_2$ that is not blocked by inhibitors of transcription or translation (Le Moellic et al., 2004).

The nature of the receptors involved in the non-genomic effects of aldosterone is not well known. These non-genomic effects are reproduced using aldosterone covalently conjugated to albumin (aldo-BSA) that cannot penetrate cells (Le Moellic et al., 2004). The possibility that steroid conjugated to BSA could be hydrolyzed has been raised and alternative aldosterone conjugates have been suggested, but these have not yet been tested (Harrington et al., 2006). Some of the effects of aldo-BSA could be mediated through the classic MR, as has been shown in endothelial cells or CHO cells transfected with the cDNA for the human MR, but others do not appear to be mediated by the MR (Grossmann et al., 2005), however, the putative receptor has not been isolated.

References

Alnemri, E. S. et al. (1991). Overexpression and characterization of the human mineralocorticoid receptor. *J Biol Chem* **266**: 18072–18081.

Alvarez de la Rosa, D. et al. (1999). The serum and glucocorticoid kinase sgk increases the abundance of epithelial sodium channels in the plasma membrane of Xenopus oocytes. *J Biol Chem* **274**: 37834–37839.

Alzamora, R. et al. (2003). Nongenomic effect of aldosterone on Na^+, K^+-adenosine triposphatase in arterial vessels. *Endocrinology* **144**: 1266–1272.

Anderson, N. D., and Fanestil, D. D. (1976). Corticoid receptors in rat brain: evidence for an aldosterone receptor. *Endocrinology* **98**: 676–684.

Arriza, J. W. et al. (1987). Cloning of human mineralocorticoid receptor complementary DNA: structural and functional kinship with the glucocorticoid receptor. *Science* **237**: 268–275.

Asher, C. et al. (1996). Aldosterone-induced increase in the abundance of Na^+ channel subunits. *Am J Physiol* **271**: C605–C611.

Auzou, G. et al. (2000). A single amino acid mutation of ala-773 in the mineralocorticoid receptor confers agonist properties to 11beta-substituted spirolactones. *Mol Pharmacol* **58**: 684–691.

Bhalla, V. et al. (2005). Serum- and glucocorticoid-regulated kinase 1 regulates ubiquitin ligase neural precursor cell-expressed, developmentally down-regulated protein 4-2 by inducing interaction with 14-3-3. *Mol Endocrinol* **19**: 3073–3084.

Bhalla, V. et al. (2006). Disinhibitory pathways for control of sodium transport: regulation of ENaC by SGK1 and GILZ. *Am J Physiol Renal Physiol* **291**: F714–F721.

Bloem, L. J. et al. (1995). Identification of a splice variant of the rat and human mineralocorticoid receptor genes. *J Steroid Biochem Mol Biol* **55**: 159–162.

Boldyreff, B., and Wehling, M. (2003). Non-genomic actions of aldosterone: mechanisms and consequences in kidney cells. *Nephrol Dial Transplant* **18**: 1693–1695.

Boulkroun, S. et al. (2002). Characterization of rat NDRG2 (N-Myc downstream regulated gene 2), a novel early mineralocorticoid-specific induced gene. *J Biol Chem* **277**: 31506–31515.

Boyd, C., and Naray-Fejes-Toth, A. (2005). Gene regulation of ENaC subunits by serum- and glucocorticoid-inducible kinase-1. *Am J Physiol Renal Physiol* **288**: F505–F512.

Bridgham, J. T. et al. (2006). Evolution of hormone-receptor complexity by molecular exploitation. *Science* **312**: 97–101.

Brinker, A. et al. (2002). Ligand discrimination by TPR domains. Relevance and selectivity of EEVD-recognition in Hsp70 x Hop x Hsp90 complexes. *J Biol Chem* **277**: 19265–19275.

Chauchereau, A. et al. (2003). Sumoylation of the progesterone receptor and of the steroid receptor coactivator SRC-1. *J Biol Chem* **278**: 12335–12343.

Chen, S. Y. et al. (1998). Aldosterone responsiveness of A6 cells is restored by cloned rat mineralocorticoid receptor. *Am J Physiol* **274**: C39–46.

Chen, S. Y. et al. (1999). Epithelial sodium channel regulated by aldosterone-induced protein sgk. *Proc Natl Acad Sci USA* **96**: 2514–2519.

Choate, K. A. et al. (2003). WNK1, a kinase mutated in inherited hypertension with hyperkalemia, localizes to diverse Cl^--transporting epithelia. *Proc Natl Acad Sci USA* **100**: 663–668.

Clouse, S. D. (2002). Brassinosteroids. Plant counterparts to animal steroid hormones? *Vitam Horm* **65**: 195–223.

Couette, B. et al. (1998). Folding requirements of the ligand-binding domain of the human mineralocorticoid receptor. *Mol Endocrinol* **12**: 855–863.

Dahmer, M. K. et al. (1984). Effects of molybdate and endogenous inhibitors on steroid-receptor inactivation, transformation, and translocation. *Annu Rev Physiol* **46**: 67–81.

DeFranco, D. B. (2002). Navigating steroid hormone receptors through the nuclear compartment. *Mol Endocrinol* **16**: 1449–1455.

Denker, S. P., and Barber, D. L. (2002). Ion transport proteins anchor and regulate the cytoskeleton. *Curr Opin Cell Biol* **14**: 214–220.

Diakov, A., and Korbmacher, C. (2004). A novel pathway of epithelial sodium channel activation involves a serum- and glucocorticoid-inducible kinase consensus motif in the C terminus of the channel's alpha-subunit. *J Biol Chem* **279**: 38134 38142.

Eaton, D. C. et al. (2001). Mechanisms of aldosterone's action on epithelial Na + transport. *J Membr Biol* **184**: 313–319.

Edelman, I. S. et al. (1963). On the mechanism of action of aldosterone on sodium transport: the role of protein synthesis. *Proc Natl Acad Sci USA* **50**: 1169–1177.

Edwards, C. R. W., and Stewart, P. M. (1991). The cortisol-cortisone shuttle and the apparent specificity of glucocorticoid and mineralocorticoid receptors. *J Steroid Biochem Molec Biol* **39**: 859–865.

Edwards, C. R. W. et al. (1988). Localisation of 11β-hydroxysteroid dehydrogenase-tissue specific protector of the mineralocorticoid receptor. *Lancet* **ii**: 986–989.

Ellis, R. J., and Minton, A. P. (2003). Cell biology: join the crowd. *Nature* **425**: 27–28.

Escoubet, B. et al. (1997). Noncoordinate regulation of epithelial Na channel and Na pump subunit mRNAs in kidney and colon by aldosterone. *Am J Physiol* **272**: C1482–C1491.

Evans, R. M. (1988). The steroid and thyroid hormone receptor superfamily. *Science* **240**: 889–895.

Fagart, J. et al. (2005a). Crystal structure of a mutant mineralocorticoid receptor responsible for hypertension. *Nat Struct Mol Biol* **12**: 554–555.

Fagart, J. et al. (2005b). The Met852 residue is a key organizer of the ligand-binding cavity of the human mineralocorticoid receptor. *Mol Pharmacol* **67**: 1714–1722.

Farman, N. et al. (1991). Immunolocalization of gluco- and mineralocorticoid receptors in rabbit kidney. *Am J Physiol* **260**: C226–C233.

Farman, N., and Rafestin-Oblin, M. E. (2001b). Multiple aspects of mineralocorticoid selectivity. *Am J Physiol Renal Physiol* **280**: F181–F192.

Feraille, E. et al. (2003). Mechanism of control of Na, K-ATPase in principal cells of the mammalian collecting duct. *Ann NY Acad Sci* **986**: 570–578.

Fuller, P. J., and Young, M. J. (2005). Mechanisms of mineralocorticoid action. *Hypertension* **46**: 1227–1235.

Funder, J., and Myles, K. (1996). Exclusion of corticosterone from epithelial mineralocorticoid receptors is insufficient for selectivity of aldosterone action: *in vivo* binding studies. *Endocrinology* **137**: 5264–5268.

Funder, J. W. (2005). The nongenomic actions of aldosterone. *Endocr Rev* **26**: 313–321.

Funder, J. W. (2007). Why are mineralocorticoid receptors so nonselective? *Curr Hypertens Rep* **9**: 112–116.

Funder, J. W. et al. (1972). Specific aldosterone binding in rat kidney and parotid. *J Steroid Biochem* **3**: 209–218.

Funder, J. W. et al. (1973). The roles of plasma binding and receptor specificity in the mineralocorticoid action of aldosterone. *Endocrinology* **92**: 994–1004.

Funder, J. W. et al. (1988). Mineralocorticoid action: target tissue specificity is enzyme, not receptor, mediated. *Science* **242**: 583–585.

Fuse, H. et al. (2000). Characterization of transactivational property and coactivator mediation of rat mineralocorticoid receptor activation function-1 (AF-1). *Mol Endocrinol* **14**: 889–899.

Galat, A. (2003). Peptidylprolyl cis/trans isomerases (immunophilins): biological diversity – targets – functions. *Curr Top Med Chem* **3**: 1315–1347.

Galigniana, M. D. (1998). Native rat kidney mineralocorticoid receptor is a phosphoprotein whose transformation to a DNA-binding form is induced by phosphatases. *Biochem J* **333**: 555–563.

Galigniana, M. D., and Piwien Pilipuk, G. (2004). Activation of the ligand-mineralocorticoid receptor functional unit by ancient, classical, and novel ligands. Structure-activity relationship. *Vitam Horm* **69**: 31–68.

Galigniana, M. D. et al. (2001). Evidence that the peptidylprolyl isomerase domain of the hsp90-binding immunophilin FKBP52 is involved in both dynein interaction and glucocorticoid receptor movement to the nucleus. *J Biol Chem* **276**: 14884–14889.

Galigniana, M. D. et al. (2002). Binding of hsp90-associated immunophilins to cytoplasmic dynein: direct binding and in vivo evidence that the peptidylprolyl isomerase domain is a dynein interaction domain. *Biochemistry* **41**: 13602–13610.

Ganong, W. F., and Mulrow, P. J. (1958). Rate of change in sodium and potassium excretion after injection of aldosterone into the aorta and renal artery of the dog. *Am J Physiol* **195**: 337–342.

Garty, H., and Palmer, L. G. (1997). Epithelial sodium channels: function, structure, and regulation. *Physiol Rev* **77**: 359–396.

Gekle, M. et al. (1996). Rapid activation of Na^+/H^+ exchange by aldosterone in renal epithelial cells requires Ca^{2+} and stimulation of a plasma membrane proton conductance. *Proc Natl Acad Sci USA* **93**: 10500–10504.

Gekle, M. et al. (1997). The mineralocorticoid aldosterone activates a proton conductance in cultured kidney cells. *Am J Physiol* **273**: C1673–C1678.

Gekle, M. et al. (1998). Non-genomic action of the mineralocorticoid aldosterone on cytosolic sodium in cultured kidney cells. *J Physiol* **511**(Pt 1): 255–263.

Gekle, M. et al. (2001a). Determination of basolateral Na(+)/H(+) exchange activity in MDCK cells using a multiwell-multilabel reader. *Anal Biochem* **296**: 174–178.

Gekle, M. et al. (2001b). Rapid activation of Na^+/H^+-exchange in MDCK cells by aldosterone involves MAP-kinase ERK1/2. *Pflugers Arch* **441**: 781–786.

Geller, D. S. et al. (2000). Activating mineralocorticoid receptor mutation in hypertension exacerbated by pregnancy. *Science* **289**: 119–123.

Gething, M. J., and Sambrook, J. (1992). Protein folding in the cell. *Nature* **355**: 33–45.

Gnionsahe, A. et al. (1989). Aldosterone binding sites along nephron of Xenopus and rabbit. *Am J Physiol* **257**: R87–95.

Goldstein, L. S., and Yang, Z. (2000). Microtubule-based transport systems in neurons: the roles of kinesins and dyneins. *Annu Rev Neurosci* **23**: 39–71.

Gomez-Sanchez, C. E. et al. (2006). Development of a panel of monoclonal antibodies against the mineralocorticoid receptor. *Endocrinology* **147**: 1343–1348.

Good, D. W. (1993). The thick ascending limb as a site of renal bicarbonate reabsorption. *Semin Nephrol* **13**: 225–235.

Good, D. W. (2007). Nongenomic actions of aldosterone on the renal tubule. *Hypertension* **49**: 728–739.

Good, D. W. et al. (2002). Aldosterone inhibits HCO absorption via a nongenomic pathway in medullary thick ascending limb. *Am J Physiol Renal Physiol* **283**: F699–706.

Good, D. W. et al. (2006). Nongenomic regulation by aldosterone of the epithelial NHE3 Na(+)/H(+) exchanger. *Am J Physiol Cell Physiol* **290**: C757–C763.

Grahammer, F. et al. (2006). Renal function of gene-targeted mice lacking both SGK1 and SGK3. *Am J Physiol Regul Integr Comp Physiol* **290**: R945–R950.

Grossmann, C. et al. (2005). Human mineralocorticoid receptor expression renders cells responsive for nongenotropic aldosterone actions. *Mol Endocrinol* **19**: 1697–1710.

Harrell, J. M. et al. (2002). All of the protein interactions that link steroid receptor.hsp90.immunophilin heterocomplexes to cytoplasmic dynein are common to plant and animal cells. *Biochemistry* **41**: 5581–5587.

Harrington, W. R. et al. (2006). Estrogen dendrimer conjugates that preferentially activate extranuclear, nongenomic versus genomic pathways of estrogen action. *Mol Endocrinol* **20**: 491–502.

Heery, D. M. et al. (2001). Core LXXLL motif sequences in CREB-binding protein, SRC1, and RIP140 define affinity and selectivity for steroid and retinoid receptors. *J Biol Chem* **276**: 6695–6702.

Hellal-Levy, C. et al. (2000a). Mechanistic aspects of mineralocorticoid receptor activation. *Kidney Int* **57**: 1250–1255.

Hellal-Levy, C. et al. (2000b). Crucial role of the H11-H12 loop in stabilizing the active conformation of the human mineralocorticoid receptor. *Mol Endocrinol* **14**: 1210–1221.

Hollenberg, S. M., and Evans, R. M. (1988). Multiple and cooperative trans-activation domains of the human glucocorticoid receptor. *Cell* **55**: 899–906.

Howe, C. L., and Mobley, W. C. (2004). Signaling endosome hypothesis: a cellular mechanism for long distance communication. *J Neurobiol* **58**: 207–216.

Huang, D. Y. et al. (2004). Impaired regulation of renal K^+ elimination in the sgk1-knockout mouse. *J Am Soc Nephrol* **15**: 885–891.

Huyet, J. et al. (2007). Structural basis of spirolactone recognition by the mineralocorticoid receptor. *Mol Pharmacol* **72**: 563–571.

Jeck, N. et al. (2005). Salt handling in the distal nephron: lessons learned from inherited human disorders. *Am J Physiol Regul Integr Comp Physiol* **288**: R782–R795.

Jones, A. W., and Hart, R. G. (1975). Altered ion transport in aortic smooth muscle during deoxycorticosterone acetate hypertension in rats. *Circ Res* **37**: 333–341.

Kahle, K. T. et al. (2005). Regulation of diverse ion transport pathways by WNK4 kinase: a novel molecular switch. *Trends Endocrinol Metab* **16**: 98–103.

Kemendy, A. E. et al. (1993). Aldosterone alters the open probability of amiloride-blockable sodium channels in A6 epithelia. *Am J Physiol* **263**: C825–C837.

Kenouch, S. et al. (1994). Human skin as target for aldosterone: coexpression of mineralocorticoid receptors and 11β-hydroxysteroid dehydrogenase. *J Clin Endocrinol Metab* **79**: 1334–1341.

Kitagawa, H. et al. (2002). Ligand-selective potentiation of rat mineralocorticoid receptor activation function 1 by a CBP-containing histone acetyltransferase complex. *Mol Cell Biol* **22**: 3698–3706.

Kong, S. E. et al. (2005). ELL-associated factors 1 and 2 are positive regulators of RNA polymerase II elongation factor ELL. *Proc Natl Acad Sci USA* **102**: 10094–10098.

Koppel, H. et al. (2003). Nongenomic effects of aldosterone on human renal cells. *J Clin Endocrinol Metab* **88**: 1297–1302.

Koyasu, S. et al. (1986). Two mammalian heat shock proteins, HSP90 and HSP100, are actin-binding proteins. *Proc Natl Acad Sci USA* **83**: 8054–8058.

Kressler, D. et al. (2002). The PGC-1-related protein PERC is a selective coactivator of estrogen receptor alpha. *J Biol Chem* **277**: 13918–13925.

Krozowski, Z. S., and Funder, J. W. (1983). Renal mineralocorticoid receptors and hippocampal corticosterone-binding species have identical intrinsic steroid specificity. *Proc Natl Acad Sci USA* **80**: 6056–6060.

Kwak, S. P. et al. (1993). 5'-Heterogeneity of the mineralocorticoid receptor messenger ribonucleic acid: differential expression and regulation of splice variants within the rat hippocampus. *Endocrinology* **133**: 2344–2350.

Lamb, J. R. et al. (1995). Tetratrico peptide repeat interactions: to TPR or not to TPR? *Trends Biochem Sci* **20**: 257–259.

Le Moellic, C. et al. (2004). Early nongenomic events in aldosterone action in renal collecting duct cells: PKCalpha activation, mineralocorticoid receptor phosphorylation, and cross-talk with the genomic response. *J Am Soc Nephrol* **15**: 1145–1160.

Ledoussal, C. et al. (2001). Renal salt wasting in mice lacking $NHE3\ Na^+/H^+$ exchanger but not in mice lacking NHE2. *Am J Physiol Renal Physiol* **281**: F718–F727.

Li, Y. et al. (2005). Structural and biochemical mechanisms for the specificity of hormone binding and coactivator assembly by mineralocorticoid receptor. *Mol Cell* **19**: 367–380.

Liang, X. et al. (2006). 14-3-3 isoforms are induced by aldosterone and participate in its regulation of epithelial sodium channels. *J Biol Chem* **281**: 16323–16332.

Lifton, R. P. et al. (2001). Molecular mechanisms of human hypertension. *Cell* **104**: 545–556.

Lingueglia, E. et al. (1994). Different homologous subunits of the amiloride-sensitive Na⁺ channel are differently regulated by aldosterone. *J Biol Chem* **269**: 13736–13739.

Liu, W. et al. (1995). Steroid receptor heterodimerization demonstrated in vitro and in vivo. *Proc Natl Acad Sci USA* **92**: 12480–12484.

Liu, W. et al. (1996). Steroid receptor transcriptional synergy is potentiated by disruption of the DNA-binding domain dimer interface. *Mol Endocrinol* **10**: 1399–1406.

Loffing, J. et al. (2001a). Mediators of aldosterone action in the renal tubule. *Curr Opin Nephrol Hypertens* **10**: 667–675.

Loffing, J. et al. (2001b). Aldosterone induces rapid apical translocation of ENaC in early portion of renal collecting system: possible role of SGK. *Am J Physiol Renal Fluid Elect Physiol* **280**: 675–682.

Lombes, M. et al. (1990). Immunohistochemical localization of renal mineralocorticoid receptor by using an anti-idiotypic antibody that is an internal image of aldosterone. *Proc Natl Acad Sci* **87**: 1086–1088.

Lombes, M. et al. (1994). The mineralocorticoid receptor discriminates aldosterone from glucocorticoids independently of the 11β-hydroxysteroid dehydrogenase. *Endocrinology* **135**: 834–840.

Losel, R. M. et al. (2003). Nongenomic steroid action: controversies, questions, and answers. *Physiol Rev* **83**: 965–1016.

Maguire, D. et al. (1999). Rapid responses to aldosterone in human distal colon. *Steroids* **64**: 51–63.

Marissal-Arvy, N. et al. (2004). Gain of function mutation in the mineralocorticoid receptor of the Brown Norway rat. *J Biol Chem* **279**: 39232–39239.

Martel, J. A. et al. (2007). Melanophilin, a novel aldosterone-induced gene in mouse cortical collecting duct cells. *Am J Physiol Renal Physiol* **293**: F904–F913.

Marver, D., and Schwartz, M. J. (1980). Identification of mineralocorticoid target sites in the isolated rabbit cortical nephron. *Proc Natl Acad Sci USA* **77**: 3672–3676.

Masilamani, S. et al. (1999). Aldosterone-mediated regulation of ENaC alpha, beta, and gamma subunit proteins in rat kidney. *J Clin Invest* **104**: R19–23.

Massaad, C. et al. (1999). Modulation of human mineralocorticoid receptor function by protein kinase A. *Mol Endocrinol* **13**: 57–65.

McCormick, J. A. et al. (2005). SGK1: a rapid aldosterone-induced regulator of renal sodium reabsorption. *Physiology (Bethesda)* **20**: 134–139.

McKenna, N. J., and O'Malley, B. W. (2002). Minireview: nuclear receptor coactivators – an update. *Endocrinology* **143**: 2461–2465.

McKenna, N. J. et al. (1999). Nuclear receptor coactivators: multiple enzymes, multiple complexes, multiple functions. *J Steroid Biochem Mol Biol* **69**: 3–12.

Meijer, O. C. et al. (2000). Differential expression and regional distribution of steroid receptor coactivators SRC-1 and SRC-2 in brain and pituitary. *Endocrinology* **141**: 2192–2199.

Meijer, O. C. et al. (2005). Steroid receptor coactivator-1 splice variants differentially affect corticosteroid receptor signaling. *Endocrinology* **146**: 1438–1448.

Mick, V. E. et al. (2001). The alpha-subunit of the epithelial sodium channel is an aldosterone-induced transcript in mammalian collecting ducts, and this transcriptional response is mediated via distinct cis-elements in the 5′-flanking region of the gene. *Mol Endocrinol* **15**: 575–588.

Mihailidou, A. S., and Funder, J. W. (2005). Nongenomic effects of mineralocorticoid receptor activation in the cardiovascular system. *Steroids* **70**: 347–351.

Mihailidou, A. S. et al. (2004). Rapid, nongenomic effects of aldosterone in the heart mediated by epsilon protein kinase C. *Endocrinology* **145**: 773–780.

Miyata, Y., and Yahara, I. (1991). Cytoplasmic 8 S glucocorticoid receptor binds to actin filaments through the 90-kDa heat shock protein moiety. *J Biol Chem* **266**: 8779–8783.

Morrison, N. et al. (1990). Regional chromosomal assignment of the human mineralocorticoid receptor gene to 4q31.1. *Hum Genet* **85**: 130–132.

Moura, A.-M., and Worcel, M. (1984). Direct action of aldosterone on transmembrane 22Na efflux from arterial smooth muscle: rapid and delayed effects. *Hypertension* **6**: 425–430.

Muller, O. G. et al. (2003). Mineralocorticoid effects in the kidney: correlation between alphaENaC, GILZ, and Sgk-1 mRNA expression and urinary excretion of Na⁺ and K⁺. *J Am Soc Nephrol* **14**: 1107–1115.

Mune, T. et al. (1995). Human hypertension caused by mutations in the kidney isozyme of 11β-hydroxysteroid dehydrogenase. *Nat Genet* **10**: 394–399.

Nagaki, K. et al. (2006). 14-3-3 Mediates phosphorylation-dependent inhibition of the interaction between the ubiquitin E3 ligase Nedd4-2 and epithelial Na⁺ channels. *Biochemistry* **45**: 6733–6740.

Naray-Fejes-Toth, A. et al. (1991). 11β-Hydroxysteroid dehydrogenase activity in the renal target cells of aldosterone. *Endocrinology* **129**: 17–21.

Naray-Fejes-Toth, A. et al. (1999). sgk is an aldosterone-induced kinase in the renal collecting duct. Effects on epithelial Na⁺ channels. *J Biol Chem* **274**: 16973–16978.

Naray-Fejes-Toth, A. et al. (2004). The kidney-specific WNK1 isoform is induced by aldosterone and stimulates epithelial sodium channel-mediated Na⁺ transport. *Proc Natl Acad Sci USA* **101**: 17434–17439.

Nielsen, J. et al. (2007). Maintained ENaC trafficking in aldosterone-infused rats during mineralocorticoid and glucocorticoid receptor blockade. *Am J Physiol Renal Physiol* **292**: F382–F394.

Nishida, E. et al. (1986). Calmodulin-regulated binding of the 90-kDa heat shock protein to actin filaments. *J Biol Chem* **261**: 16033–16036.

Nishiyama, A. et al. (2005). Involvement of aldosterone and mineralocorticoid receptors in rat mesangial cell proliferation and deformability. *Hypertension* **45**: 710–716.

Oberleithner, H. et al. (1987). Aldosterone activates Na⁺/H⁺ exchange and raises cytoplasmic pH in target cells of the amphibian kidney. *Proc Natl Acad Sci USA* **84**: 1464–1468.

Oberleithner, H. et al. (1988). Aldosterone-controlled linkage between Na⁺/H⁺ exchange and K+ channels in fused renal epithelial cells. *Ciba Found Symp* **139**: 201–219.

Obradovic, D. et al. (2004). DAXX, FLASH, and FAF-1 modulate mineralocorticoid and glucocorticoid receptor-mediated transcription in hippocampal cells – toward a basis for the opposite actions elicited by two nuclear receptors? *Mol Pharmacol* **65**: 761–769.

Odermatt, A. et al. (1999). The N-terminal anchor sequences of 11beta-hydroxysteroid dehydrogenases determine their orientation in the endoplasmic reticulum membrane. *J Biol Chem* **274**: 28762–28770.

Odermatt, A. et al. (2001). The intracellular localization of the mineralocorticoid receptor is regulated by 11beta-hydroxysteroid dehydrogenase type 2. *J Biol Chem* **276**: 28484–28492.

Park, J. et al. (1999). Serum and glucocorticoid-inducible kinase (SGK) is a target of the PI 3-kinase-stimulated signaling pathway. *Embo J* **18**: 3024–3033.

Pascual-Le Tallec, L., and Lombes, M. (2005). The mineralocorticoid receptor: a journey exploring its diversity and specificity of action. *Mol Endocrinol* **19**: 2211–2221.

Pascual-Le Tallec, L. et al. (2004). Human mineralocorticoid receptor A and B protein forms produced by alternative translation sites display different transcriptional activities. *Eur J Endocrinol* **150**: 585–590.

Pascual-Le Tallec, L. et al. (2005). The elongation factor ELL (eleven-nineteen lysine-rich leukemia) is a selective coregulator for steroid receptor functions. *Mol Endocrinol* **19**: 1158–1169.

Patel, P. D. et al. (1989). Molecular cloning of a mineralocorticoid (type I) receptor complementary DNA from rat hippocampus. *Mol Endocrinol* **3**: 1877–1885.

Pearce, D. et al. (2003). Aldosterone: its receptor, target genes, and actions. *Vitam Horm* **66**: 29–76.

Pearce, D., and Kleyman, T. R. (2007). Salt, sodium channels, and SGK1. *J Clin Invest* **117**: 592–595.

Pearce, P., and Funder, J. W. (1987). High affinity aldosterone binding sites (type I receptors) in rat heart. *Clin Exp Pharmacol Physiol* **14**: 859–866.

Pilipuk, G. P. et al. (2007). Evidence for NL1-independent nuclear translocation of the mineralocorticoid receptor. *Biochemistry* **46**: 1389–1397.

Piwien-Pilipuk, G., and Galigniana, M. D. (1998). Tautomycin inhibits phosphatase-dependent transformation of the rat kidney mineralocorticoid receptor. *Mol Cell Endocrinol* **144**: 119–130.

Poukka, H. et al. (2000). Covalent modification of the androgen receptor by small ubiquitin-like modifier 1 (SUMO-1). *Proc Natl Acad Sci USA* **97**: 14145–14150.

Pratt, W. B. et al. (1999). A model for the cytoplasmic trafficking of signalling proteins involving the hsp90-binding immunophilins and p50cdc37. *Cell Signal* **11**: 839–851.

Pratt, W. B. et al. (2001). Hsp90-binding immunophilins in plants: the protein movers. *Trends Plant Sci* **6**: 54–58.

Pratt, W. B. et al. (2004a). Role of hsp90 and the hsp90-binding immunophilins in signalling protein movement. *Cell Signal* **16**: 857–872.

Pratt, W. B. et al. (2004b). Role of molecular chaperones in steroid receptor action. *Essays Biochem* **40**: 41–58.

Putney, L. K. et al. (2002). The changing face of the Na^+/H^+ exchanger, NHE1: structure, regulation, and cellular actions. *Annu Rev Pharmacol Toxicol* **42**: 527–552.

Raab, W. et al. (1950). Potentiation of pressor effects of nor-epinephrine and epinephrine in man by desoxycorticosterone acetate. *J Clin Invest* **29**: 1397–1404.

Rad, A. K. et al. (2005). Rapid natriuretic action of aldosterone in the rat. *J Appl Physiol* **98**: 423–428.

Rafestin-Oblin, M. E. et al. (2003). The severe form of hypertension caused by the activating S810L mutation in the mineralocorticoid receptor is cortisone related. *Endocrinology* **144**: 528–533.

Ratajczak, T. et al. (2003). Immunophilin chaperones in steroid receptor signalling. *Curr Top Med Chem* **3**: 1348–1357.

Reddy, R. K. et al. (1998). High-molecular-weight FK506-binding proteins are components of heat-shock protein 90 heterocomplexes in wheat germ lysate. *Plant Physiol* **118**: 1395–1401.

Rogerson, F. M., and Fuller, P. J. (2000). Mineralocorticoid action. *Steroids* **65**: 61–73.

Rogerson, F. M., and Fuller, P. J. (2003). Interdomain interactions in the mineralocorticoid receptor. *Mol Cell Endocrinol* **200**: 45–55.

Rogerson, F. M. et al. (2004). Mineralocorticoid receptor binding, structure and function. *Mol Cell Endocrinol* **217**: 203–212.

Rogerson, F. M. et al. (2007). A critical region in the mineralocorticoid receptor for aldosterone binding and activation by cortisol: evidence for a common mechanism governing ligand binding specificity in steroid hormone receptors. *Mol Endocrinol* **21**: 817–828.

Rossier, B. C. et al. (1994). Epithelial sodium channels. *Curr Opin Neph Hypertens* **3**: 487–496.

Rousseau, G. et al. (1972). Glucocorticoid and mineralocorticoid receptors for aldosterone. *J Steroid Biochem* **3**: 219–227.

Rousseau, G. G. (1984). Structure and regulation of the glucocorticoid hormone receptor. *Mol Cell Endocrinol* **38**: 1–11.

Sartorato, P. et al. (2003). Different inactivating mutations of the mineralocorticoid receptor in fourteen families affected by type I pseudohypoaldosteronism. *J Clin Endocrinol Metab* **88**: 2508–2517.

Sartorato, P. et al. (2004a). New naturally occurring missense mutations of the human mineralocorticoid receptor disclose important residues involved in dynamic interactions with DNA, intracellular trafficking and ligand binding. *Mol Endocrinol* **18** (9): 2151–2165.

Sartorato, P. et al. (2004b). Inactivating mutations of the mineralocorticoid receptor in type I pseudohypoaldosteronism. *Mol Cell Endocrinol* **217**: 119–125.

Savory, J. G. et al. (2001). Glucocorticoid receptor homodimers and glucocorticoid-mineralocorticoid receptor heterodimers form in the cytoplasm through alternative dimerization interfaces. *Mol Cell Biol* **21**: 781–793.

Scheufler, C. et al. (2000). Structure of TPR domain-peptide complexes: critical elements in the assembly of the Hsp70-Hsp90 multichaperone machine. *Cell* **101**: 199–210.

Schild, L. et al. (1996). Identification of a PY motif in the epithelial Na channel subunits as a target sequence for mutations causing channel activation found in Liddle syndrome. *EMBO J* **15**: 2381–2387.

Schultheis, P. J. et al. (1998). Renal and intestinal absorptive defects in mice lacking the NHE3 Na^+/H^+ exchanger. *Nat Genet* **19**: 282–285.

Seksek, O. et al. (1997). Translational diffusion of macromolecule-sized solutes in cytoplasm and nucleus. *J Cell Biol* **138**: 131–142.

Shahedi, M. et al. (1993). Acute and early effects of aldosterone on Na-K-ATPase activity in Madin-Darby canine kidney epithelial cells. *Am J Physiol* **264**: F1021–F1026.

Shilatifard, A. et al. (1996). An RNA polymerase II elongation factor encoded by the human ELL gene. *Science* **271**: 1873–1876.

Silverstein, A. M. et al. (1999). Different regions of the immunophilin FKBP52 determine its association with the glucocorticoid receptor, hsp90, and cytoplasmic dynein. *J Biol Chem* **274**: 36980–36986.

Sims, R. J. et al. (2004). Elongation by RNA polymerase II: the short and long of it. *Genes Dev* **18**: 2437–2468.

Snyder, P. M. (2002). The epithelial na(+) channel: cell surface insertion and retrieval in na(+) homeostasis and hypertension. *Endocr Rev* **23**: 258–275.

Snyder, P. M. et al. (1995). Mechanism by which Liddle's syndrome mutations increase activity of a human epithelial Na^+ channel. *Cell* **83**: 969–978.

Snyder, P. M. et al. (2001). SGK modulates Nedd4-2-mediated inhibition of ENaC. *J Biol Chem* **5**: 5.

Spindler, B., and Verrey, F. (1999). Aldosterone action: induction of p21(ras) and fra-2 and transcription-independent decrease in myc, jun, and fos. *Am J Physiol* **276**: 1154–1161.

Staruschenko, A. et al. (2004). Ras activates the epithelial Na(+) channel through phosphoinositide 3-OH kinase signaling. *J Biol Chem* **279**: 37771–37778.

Staub, O. et al. (1996). WW domains of Nedd4 bind to the proline-rich PY motifs in the epithelial Na^+ channel deleted in Liddle's syndrome. *EMBO J* **15**: 2371–2380.

Stewart, P. M. et al. (1987). Mineralocorticoid activity of liquorice: 11-Beta-hydroxysteroid dehydrogenase deficiency comes of age. *Lancet* **ii**: 821–824.

Stewart, P. M. et al. (1996). Hypertension in the syndrome of apparent mineralocorticoid excess due to mutation of the 11 beta-hydroxysteroid dehydrogenase type 2 gene. *Lancet* **347**: 88–91.

Stokes, J. B., and Sigmund, R. D. (1998). Regulation of rENaC mRNA by dietary NaCl and steroids: organ, tissue, and steroid heterogeneity. *Am J Physiol* **274**: C1699–C1707.

Summa, V. et al. (2004). Isoform specificity of human Na(+), K(+)-ATPase localization and aldosterone regulation in mouse kidney cells. *J Physiol* **555**: 355–364.

Swaminathan, R. et al. (1996). Cytoplasmic viscosity near the cell plasma membrane: translational diffusion of a small fluorescent solute measured by total internal reflection-fluorescence photobleaching recovery. *Biophys J* **71**: 1140–1151.

Tallec, L. P. et al. (2003). Protein inhibitor of activated signal transducer and activator of transcription 1 interacts with the N-terminal domain of mineralocorticoid receptor and represses its transcriptional activity: implication of small ubiquitin-related modifier 1 modification. *Mol Endocrinol* **17**: 2529–2542.

Tian, S. et al. (2002). Small ubiquitin-related modifier-1 (SUMO-1) modification of the glucocorticoid receptor. *Biochem J* **367**: 907–911.

Tirard, M. et al. (2007). Sumoylation and proteasomal activity determine the transactivation properties of the mineralocorticoid receptor. *Mol Cell Endocrinol* **268**: 20–29.

Todd-Turla, K. M. et al. (1993). Distribution of mineralocorticoid and glucocorticoid receptor mRNA along the nephron. *Am J Physiol* **264**: F781–F791.

Tsai, M. J., and O'Malley, B. W. (1994). Molecular mechanisms of action of steroid/thyroid receptor superfamily members. *Annu Rev Biochem* **63**: 451–486.

Vanatta, J. C., and Cottle, K. E. (1955). Effect of desoxycorticosterone acetate on the peripheral vascular reactivity of dogs. *Am J Physiol* **151**: 119–122.

Verrey, F. (1999). Early aldosterone action: toward filling the gap between transcription and transport. *Am J Physiol* **277**: 319–327.

Verrey, F. (2001). Sodium reabsorption in aldosterone-sensitive distal nephron: news and contributions from genetically engineered animals. *Curr Opin Nephrol Hypertens* **10**: 39–47.

Verrey, F. et al. (2000). Control of Na^+ transport by aldosterone. Seldin, D. and Giebisch, G. editors. *The kidney. Physiology & pathophysiology* **vol. 1**: Lippincott Williams & Wilkins, New York, pp. 1441–1471.

Verrey, F. et al. (2003). Short-term aldosterone action on Na, K-ATPase surface expression: role of aldosterone-induced SGK1? *Ann NY Acad Sci* **986**: 554–561.

Vicent, G. P. et al. (2002). Differences in nuclear retention characteristics of agonist-activated glucocorticoid receptor may determine specific responses. *Exp Cell Res* **276**: 142–154.

Watts, B. A. et al. (2005). The basolateral NHE1 Na^+/H^+ exchanger regulates transepithelial HCO_3^- absorption through actin cytoskeleton remodeling in renal thick ascending limb. *J Biol Chem* **280**: 11439–11447.

Watts, B. A. et al. (2006). Aldosterone inhibits apical NHE3 and HCO_3^- absorption via a nongenomic ERK-dependent pathway in medullary thick ascending limb. *Am J Physiol Renal Physiol* **291**: F1005–F1013.

Webster, M. K. et al. (1993). Characterization of sgk, a novel member of the serine/threonine protein kinase gene family which is transcriptionally induced by glucocorticoids and serum. *Mol Cell Biol* **13**: 2031–2040.

Wehling, M. (2005). Rapid effects of aldosterone: relevant in cardiac ischemia? *Hypertension* **46**: 27–28.

Wickert, L., and Selbig, J. (2002). Structural analysis of the DNA-binding domain of alternatively spliced steroid receptors. *J Endocrinol* **173**: 429–436.

Wickert, L. et al. (1998). Mineralocorticoid receptor splice variants in different human tissues. *Eur J Endocrinol* **138**: 702–704.

Wickert, L. et al. (2000). Differential mRNA expression of the two mineralocorticoid receptor splice variants within the human brain: structure analysis of their different DNA binding domains. *J Neuroendocrinol* **12**: 867–873.

Wilson, F. H. et al. (2001). Human hypertension caused by mutations in WNK kinases. *Science* **293**: 1107–1112.

Wilson, F. H. et al. (2003). Molecular pathogenesis of inherited hypertension with hyperkalemia: the Na-Cl cotransporter is inhibited by wild-type but not mutant WNK4. *Proc Natl Acad Sci USA* **100**: 680–684.

Winter, C. et al. (2004). Nongenomic stimulation of vacuolar H+-ATPases in intercalated renal tubule cells by aldosterone. *Proc Natl Acad Sci USA* **101**: 2636–2641.

Wulff, P. et al. (2002). Impaired renal Na(+) retention in the sgk1-knockout mouse. *J Clin Invest* **110**: 1263–1268.

Xu, L., and Massague, J. (2004). Nucleocytoplasmic shuttling of signal transducers. *Nat Rev Mol Cell Biol* **5**: 209–219.

Yokota, K. et al. (2007). Coactivation of the N-terminal transactivation of mineralocorticoid receptor by Ubc9. *J Biol Chem* **282**: 1998–2010s.

Yoo, D. et al. (2003). Cell surface expression of the ROMK (Kir 1.1) channel is regulated by the aldosterone-induced kinase, SGK-1, and protein kinase A. *J Biol Chem* **278**: 23066–23075.

Yudt, M. R., and Cidlowski, J. A. (2001). Molecular identification and characterization of a and b forms of the glucocorticoid receptor. *Mol Endocrinol* **15**: 1093–1103.

Zennaro, M. C., and Lombes, M. (2004). Mineralocorticoid resistance. *Trends Endocrinol Metab* **15:** 264–270.

Zennaro, M. C. et al. (1995). Human mineralocorticoid receptor genomic structure and identification of expressed isoforms. *J Biol Chem* **270:** 21016–21020.

Zennaro, M. C. et al. (1997). Tissue-specific expression of alpha and beta messenger ribonucleic acid isoforms of the human mineralocorticoid receptor in normal and pathological states. *J Clin Endocrinol Metab* **82:** 1345–1352.

Zennaro, M. C. et al. (2001). A new human mr splice variant is a ligand-independent transactivator modulating corticosteroid action. *Mol Endocrinol* **15:** 1586–1598.

Zhang, W. et al. (2006). Dot1a-AF9 complex mediates histone H3 Lys-79 hypermethylation and repression of ENaCalpha in an aldosterone-sensitive manner. *J Biol Chem* **281:** 18059–18068.

Zhang, W. et al. (2007). Aldosterone-induced Sgk1 relieves Dot1a-Af9-mediated transcriptional repression of epithelial Na^+ channel alpha. *J Clin Invest* **117:** 773–783.

Zhou, M.-Y. et al. (2000). An alternatively spliced mineralocorticoid receptor mRNA causing truncation of the steroid binding domain. *Mol Cell Endocrinol* **159:** 125–131.

Zhou, Z. H., and Bubien, J. K. (2001). Nongenomic regulation of ENaC by aldosterone. *Am J Physiol Cell Physiol* **281:** C1118–C1130.

CHAPTER 23

Aldosterone and its Cardiovascular Effects

RAJESH GARG AND GAIL K. ADLER

Division of Endocrinology, Diabetes and Hypertension, Brigham and Women's Hospital, Harvard Medical School, Boston, MA 02115, USA

Contents

I.	Introduction	349
II.	Aldosterone and the heart	349
III.	Aldosterone and stroke	351
IV.	Aldosterone and renal disease	351
V.	Potential mechanisms mediating the adverse cardiovascular effects of aldosterone (Figure 23.1)	352
VI.	Therapeutic considerations	355
VII.	Conclusions	355
	References	356

I. INTRODUCTION

In recent years, our understanding of aldosterone has changed from considering it to be a hormone mainly responsible for fluid and electrolyte balance to a hormone with widespread cardiovascular and metabolic effects. A large body of literature demonstrates that activation of the mineralocorticoid receptor by aldosterone increases oxidative stress, inflammation, insulin resistance and vascular dysfunction, leading to renovascular and cardiovascular injury and stroke. Further, clinical studies using either the selective mineralocorticoid receptor antagonist, eplerenone, or the nonselective antagonist, spironolactone, have demonstrated beneficial cardiovascular and renovascular effects in patients with heart failure, diabetes and hypertension. The adverse cardiovascular actions of mineralocorticoid receptor involve both genomic and non-genomic mechanisms. This chapter discusses these evolving concepts of aldosterone action.

II. ALDOSTERONE AND THE HEART

A. Clinical Studies Supporting a Role for Aldosterone in the Pathophysiology of Cardiac Disease

At the turn of the century, approximately 50 years after the discovery of aldosterone, several large-scale clinical studies revealed potent beneficial effects of aldosterone receptor antagonists on the heart. The Randomized Aldactone Evaluation Study (RALES) was designed to test the effect of a mineralocorticoid receptor antagonist in addition to standard therapy, including angiotensin-converting enzyme (ACE) inhibitors, on mortality in patients with severe heart failure secondary to systolic left ventricular dysfunction with ejection fraction ≤35% (Pitt et al., 1999). One thousand six hundred and sixty three patients were enrolled in a double-blind fashion to 25–50 mg spironolactone or placebo given once daily. The follow-up was initially planned for 3 years but the trial was discontinued after a mean follow-up period of 2 years because of clear benefits with spironolactone (30% reduction in relative risk of death). Hospitalization rate for worsening heart failure was 35% lower in the spironolactone group than in the placebo group. In addition, spironolactone caused a significant improvement in the symptoms of heart failure. Those patients with an elevation in procollagen type III amino-terminal peptide, a marker of fibrosis, were the most likely to benefit from spironolactone (Zannad et al., 2000). The most feared complication of spironolactone when used with ACE inhibitors, i.e. hyperkalemia, was uncommon in this trial. Further, the beneficial effect of spironolactone was observed without significant effects on blood pressure.

The benefit of mineralocorticoid receptor blockade in heart disease was confirmed in the Eplerenone Post-Acute Myocardial Infarction Heart Failure Efficacy and Survival Study (EPHESUS). This study was designed to test the effect of eplerenone, a selective mineralocorticoid receptor antagonist, on mortality and hospitalization rates in patients with left ventricular dysfunction and heart failure after acute myocardial infarction (Pitt et al., 2003b). These patients also received standard medical therapy, including ACE inhibitors or angiotensin receptor blocker (ARB) therapy. This double-blind, placebo-controlled trial involved 6642 patients who were randomly assigned to eplerenone (25 mg per day initially, titrated to a maximum of 50 mg per day) or placebo. During a mean follow-up of 16 months, there was a 15% relative risk reduction in overall mortality and 17% relative risk reduction in cardiovascular mortality. Hospitalization rates were also

Textbook of Nephro-Endocrinology.
ISBN: 978-0-12-373870-7

significantly lower in the eplerenone group. The rate of serious hyperkalemia was 5.5% in the eplerenone group and 3.9% in the placebo group.

Clinical studies also demonstrated a role for the mineralocorticoid receptor in the pathophysiology of hypertensive heart disease. In a 9-month, double-blind study, 202 patients with left ventricular hypertrophy and hypertension were randomized to receive eplerenone 200 mg, enalapril 40 mg, or eplerenone 200 mg plus enalapril 10 mg daily and the effect on left ventricular mass was assessed (Pitt et al., 2003a). Open-label hydrochlorothiazide and/or amlodipine were added to achieve systolic blood pressure less than 180 mmHg and diastolic blood pressure less than 90 mmHg. Both eplerenone and enalapril caused similar reductions in left ventricular mass, however, the decrease in cardiac hypertrophy was almost double with combination therapy and this greater benefit appeared to be independent of changes in blood pressure. Thus, this study demonstrated the direct role of aldosterone in the pathophysiology of left ventricular hypertrophy. Consistent with the result of this study, patients with primary hyperaldosteronism had increased left ventricular hypertrophy and evidence of diastolic dysfunction compared to patients with essential hypertension (Rossi et al., 1997). Surgical removal of the aldosterone producing adenoma led to regression of hypertrophy (Rossi et al., 1996).

Two small clinical studies suggest a beneficial effect of mineralocorticoid receptor blockade on diastolic dysfunction. Treatment for 6 months with the mineralocorticoid receptor antagonist canrenone (50 mg/day) improved diastolic function, as compared to placebo treatment, in patients with hypertension and left ventricular diastolic dysfunction on ACE inhibitor and calcium channel blockade therapy (Grandi et al., 2002). The improvements in diastolic function with canrenone were not mediated by changes in blood pressure or left ventricular mass, suggesting that mineralocorticoid blockade has a direct beneficial effect on diastolic function. In another study in patients with dilated cardiomyopathy, 12-months treatment with spironolactone was demonstrated to decrease myocardial collagen content, decrease myocardial stiffness and ameliorate diastolic dysfunction (Izawa et al., 2005). The Aldosterone Antagonist Therapy for Adults with Heart Failure and Preserved Systolic Function (TOPCAT) trial is a large ongoing clinical trial that will recruit 4500 patients to determine the effect of spironolactone on mortality and morbidity in patients with diastolic heart failure (New England Research Institutes et al., 2008). The trial started in the year 2006 and is likely to end in 2011.

Thus, available clinical data demonstrate benefits of mineralocorticoid receptor blockade in the treatment of heart failure, left ventricular hypertrophy and diastolic dysfunction. The potential mechanisms for these beneficial effects have been investigated using cell culture systems and animals models of cardiovascular injuries.

B. Aldosterone and Animal Models of Heart Disease

Animal models of cardiac injury have clearly demonstrated an adverse effect of mineralocorticoids on the heart and vasculature. Activation of the mineralocorticoid receptor for 8 weeks through administration of aldosterone or deoxycorticosterone led to hypertension and cardiac fibrosis in uninephrectomized rats on a moderately high sodium diet (1% NaCl in the drinking fluid) and mineralocorticoid receptor blockade prevented this damage (Brilla and Weber, 1992; Young et al., 1994; Rickard et al., 2006). In rat post-infarct heart models, treatment with eplerenone reduced cardiac expression of monocyte chemoattractant protein-1 (MCP-1), plasminogen activator inhibitor 1 (PAI-1) and collagen types I and III, decreased interstitial fibrosis and attenuated systolic and diastolic dysfunction (Enomoto et al., 2005). Mineralocorticoid receptor blockade also decreased oxidative stress and inflammation in mice with chronic pressure overload and prevented heart failure (Kuster et al., 2005).

In addition, blockade or reduction of aldosterone had beneficial effects in models of cardiovascular injury associated with an activated renin–angiotensin–aldosterone system (Rocha et al., 2000; Oestreicher et al., 2003; Fiebeler et al., 2005). Treatment with an aldosterone synthase inhibitor or adrenalectomy reduced mortality and cardiac hypertrophy in double transgenic rats overexpressing the human renin and angiotensinogen genes (Fiebeler et al., 2005). In a short-term (14-day) rodent model of hypertension and cardiovascular injury, rats receiving a nitric oxide synthase inhibitor, angiotensin II and a moderately high sodium diet developed coronary artery injury, myocardial necrosis and inflammation in both right and left ventricles (Rocha et al., 2000; Oestreicher et al., 2003). Vascular injury was characterized by increased expression of PAI-1, inflammation, intimal thickening and vascular wall necrosis with surrounding granulation tissue. Treatment with eplerenone or spironolactone to block the actions of aldosterone or adrenalectomy to reduce circulating aldosterone levels prevented the injury, while injury recurred when adrenalectomized animals were infused with aldosterone. In this model, there was minimal cardiac fibrosis suggesting that vascular injury is an early event in aldosterone-mediated cardiovascular injury leading secondarily to myocardial ischemia and necrosis followed by repair and fibrosis.

Several investigators demonstrated that activation of the mineralocorticoid receptor by administration of aldosterone, deoxycorticosterone acetate or an inhibitor of 11β-hydroxysteroid dehydrogenase type 2, which allows corticosterone to activate the mineralocorticoid receptor, increased coronary vascular expression of pro-inflammatory molecules, cyclo-oxygenase 2 (COX-2), MCP-1 and osteopontin in uninephrectomized rats on a moderately high sodium diet (Rocha et al., 2002b; Young et al., 2003). These vascular changes preceded the development of cardiac fibrosis. Similarly, mineralocorticoid receptor blockade or adrenalectomy reduced coronary vascular injury and inflammation in rats treated with angiotensin II and a moderately high sodium diet (Rocha et al., 2002a). Treatment of adrenalectomized, angiotensin II-infused rats with aldosterone restored vascular injury, which was characterized by arterial fibrinoid necrosis, perivascular inflammation and increased vascular expression of COX-2 and osteopontin. Finally, aldosterone also may have a role in atherosclerosis as aldosterone administration increased (Keidar et al., 2003) and mineralocorticoid receptor blockade reduced atherosclerotic lesion development and macrophage superoxide anion formation in apolipoprotein E-deficient mice (Keidar et al., 2004).

In these preclinical studies of cardiac injury, the beneficial effects of blocking the mineralocorticoid receptor or reducing aldosterone via adrenalectomy appeared to be independent of effects on volume homeostasis and blood pressure as vascular injury was ameliorated without reductions in blood pressure (Rocha et al., 1998, 2000, 2002a; Fiebeler et al., 2005). One study that directly assessed the role of blood pressure control contrasted the anti-inflammatory effects of eplerenone versus triple anti-hypertensive therapy with hydralazine, hydrochlorothiazide and reserpine in spontaneously hypertensive rats. Both therapies decreased aortic expression of interleukin-1β (IL-1β), interleukin-6 (IL-6) and tumor necrosis factor-α (TNF-α), however, the magnitude of the effect was significantly greater in the eplerenone treated group despite equivalent reductions in blood pressure and only eplerenone reduced aortic nuclear factor-κB (NF-κB) expression (Sanz-Rosa et al., 2005). Further supporting the concept of dissociation between aldosterone's effects on blood pressure and at least some of its effects on cardiovascular injury, peripheral infusion of aldosterone caused cardiac injury even when the rise in blood pressure was prevented through intracerebroventricular administration of a mineralocorticoid receptor antagonist (Young et al., 1995).

Thus, vascular injury and dysfunction appeared to play a key role in the pathophysiology of aldosterone-induced cardiac injury. However, aldosterone's adverse vascular effects are not limited to the heart but, as discussed below, extend to brain, kidney and peripheral vasculature.

III. ALDOSTERONE AND STROKE

Several animal studies suggest that mineralocorticoid receptor blockade has beneficial effects in stroke. In stroke-prone spontaneously hypertensive rats, spironolactone reduced vascular injury, stroke and death (Rocha et al., 1998) and decreased cerebral infarct size after occlusion of the middle cerebral artery (Dorrance et al., 2001). Treatment with deoxycorticosterone acetate to activate the mineralocorticoid receptor increased cerebral infarct size in Wistar rats that had undergone middle cerebral artery occlusion (Dorrance et al., 2006).

IV. ALDOSTERONE AND RENAL DISEASE

A. Clinical Studies Supporting a Role for Aldosterone in the Pathophysiology of Renal Disease

Aldosterone blockade has been shown to reduce albuminuria in hypertensive patients independent of an effect on blood pressure itself (White et al., 2003b; Williams et al., 2004). In a study involving 499 patients with stage 1 or 2 hypertension randomized to receive eplerenone or enalapril, eplerenone caused an equivalent reduction in blood pressure but a significantly greater reduction in urinary albumin excretion (-61.5% versus -25.7%; $P = 0.01$) (Williams et al., 2004). Similarly, in older patients with isolated systolic hypertension, eplerenone reduced microalbuminuria to a much greater extent than amlodipine despite similar reductions in blood pressure (White et al., 2003b). One mechanism for the apparent increase in mineralocorticoid-mediated renal injury may involve increased renal expression of mineralocorticoid receptor. Renal biopsies of patients with renal disease demonstrated a marked increase in renal expression of the mineralocorticoid receptor in individuals with albuminuria greater than 2 g/day, compared to those with lesser amounts of albuminuria (Quinkler et al., 2005).

Studies have shown a beneficial effect of mineralocorticoid receptor blockade when added to ACE inhibitor or ARB therapy on proteinuria in patients with diabetic nephropathy (Sato et al., 2003; Rachmani et al., 2004; Rossing et al., 2005; Epstein et al., 2006; van den Meiracker et al., 2006). In one randomized double-blind controlled trial, 268 albuminuric patients with type 2 diabetes on enalapril 20 mg/day were randomized to add-on therapy for 12 weeks with placebo, eplerenone 50 mg and eplerenone 100 mg. Amlodipine was added to decrease systolic/diastolic blood pressure to less than 120/80 mmHg. Eplerenone 50 mg and eplerenone 100 mg led to significant reductions in albuminuria of 41% and 48%, respectively, whereas placebo did not reduce albuminuria. Reductions in albuminuria were seen as early as week 4 and were independent of the reductions in blood pressure (Epstein et al., 2006).

B. Aldosterone and Animal Models of Diabetic and Non-Diabetic Renal Injury

Mineralocorticoid receptor blockade also reduces renal injury in non-diabetic animal models. In rodents on a moderately high sodium diet, administration of a nitric oxide synthase inhibitor and angiotensin II caused proteinuria and renal arteriopathy that was prevented by administration of eplerenone or by adrenalectomy (Rocha et al., 2000). The injury was again present when adrenalectomized rats were infused with aldosterone, demonstrating that aldosterone is required for the induction of angiotensin II renal injury in this model. Similarly, blockade of the mineralocorticoid receptor or adrenalectomy reduced proteinuria and renal arteriopathy in other hypertensive rodent models of angiotensin II-mediated renal injury, the stroke-prone spontaneously hypertensive rat (Rocha et al., 1998) and the double transgenic rat overexpressing the human renin and angiotensinogen genes (Fiebeler et al., 2005). In these models, the beneficial effects of mineralocorticoid blockade on the kidney were independent of improvements in blood pressure (Rocha et al., 1998, 2000; Fiebeler et al., 2005). Further, spironolactone treatment reduced renal fibrosis in mice with complete unilateral ureteral obstruction (Trachtman et al., 2004) and reduced arteriolopathy and tubulointerstitial fibrosis in cyclosporine A nephrotoxicity (Feria et al., 2003). Consistent with the beneficial renal effects of mineralocorticoid receptor antagonists, infusion of aldosterone into uninephrectomized rats consuming a high sodium diet caused podocyte injury, albuminuria, renal inflammation and increased renal expression of pro-inflammatory factors (Blasi et al., 2003; Shibata et al., 2007).

Blockade of the mineralocorticoid receptor also reduced renal injury in animal models of type 1 diabetic renal injury, the streptozotocin-treated rat and type 2 diabetes, the obese *db/db* mouse and the Otsuka Long-Evans Tokushima Fatty (OLETF) rat (Miric et al. 2001; Guo et al., 2006; Han et al., 2006a, b). In the streptozotocin-treated uninephrectomized rat and the obese, diabetic *db/db* mouse models, treatment with spironolactone or eplerenone reduced albuminuria and histopathological evidence of renal injury. Specifically, mineralocorticoid receptor blockade decreased glomerular hypertrophy, mesangial matrix expansion, glomerular inflammation and interstitial inflammation (Guo et al., 2006). Mineralocorticoid receptor blockade also reduced renal expression of transforming growth factor-β_1 (TGF-β_1), osteopontin, MCP-1, macrophage migration inhibitory factor and connective tissue growth factor in diabetic rodents (Guo et al., 2006; Han et al., 2006a, b). In cultured mesangial and proximal tubule cells, aldosterone increased expression of connective tissue growth factor and MCP-1, but not TGF-β_1 (Han et al., 2006a, b). Pre-treatment with a NF-κB inhibitor prevented the increase in MCP-1, suggesting that aldosterone promotes renal inflammation through NF-κB activation (Han et al., 2006b). Further, in renal mesangial cells there was a synergistic increase in osteopontin by treating with aldosterone and either IL-1β or TNF-α (Gauer et al., 2008).

The increased activity of the mineralocorticoid receptor in diabetic nephropathy could be mediated by multiple mechanisms. Renal cortical expression of mineralocorticoid receptor was increased in rodent models of type 1 and type 2 diabetic nephropathy (Guo et al., 2006). Obese diabetic mice also had elevated urinary aldosterone levels suggesting increased adrenal production of aldosterone (Guo et al., 2008). Further, local renal aldosterone production was increased in diabetic adrenalectomized rats and inhibition of aldosterone synthase in these rats reduced renal aldosterone content, albuminuria, renal inflammation and matrix formation (Siragy and Xue, 2008). Additionally, it is possible that decreases in the activity of 11-β hydroxysteroid dehydrogenase type 2 could lead to increased activation of the mineralocorticoid receptor by glucocorticoids or that post-receptor mechanisms could augment aldosterone's adverse effects in diabetic kidneys.

V. POTENTIAL MECHANISMS MEDIATING THE ADVERSE CARDIOVASCULAR EFFECTS OF ALDOSTERONE (FIGURE 23.1)

A. Effect of Aldosterone on Intracellular Signaling Pathways

In addition to aldosterone's well-characterized effects on gene transcription, aldosterone has rapid non-genomic effects that appear to involve cross-talk between the mineralocorticoid receptor and other signaling cascades, many

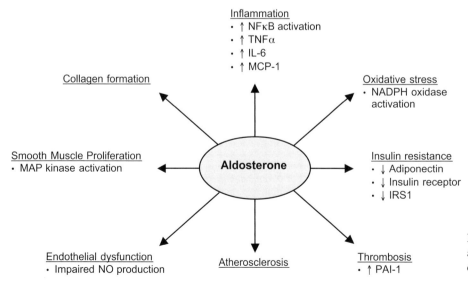

FIGURE 23.1 Potential mechanisms of aldosterone-mediated adverse cardiovascular effects.

of which are associated with cardiovascular injury. Grossman and colleagues demonstrated that aldosterone increased phosphorylation of extracellular signal-regulated kinase (ERK) through a process dependent on aldosterone-mediated increases in c-SRC phosphorylation and transactivation of epidermal growth factor receptor (EGFR) (Grossmann et al., 2005, 2007). Further, they identified the c-terminal EF domain of the mineralocorticoid receptor, which includes the ligand binding region, as the region which mediates these non-genomic actions of aldosterone on ERK activation (Grossmann et al., 2008). Other investigators reported that aldosterone interacts synergistically with angiotensin II to increase ERK activation (Min et al., 2005). Aldosterone-mediated increases in phosphorylation of c-Src were also shown to increase vascular p38 mitogen-activated protein (MAP) kinases and nicotinamide adenine dinucleotide phosphate (NADPH) oxidase activation in cultured rat vascular smooth muscle cells (Callera et al., 2005). In endothelial cells, aldosterone increased superoxide generation via activation of Src, NADPH oxidase and the small GTP-binding protein Rac1 (Iwashima et al., 2008). Thus, aldosterone has effects on multiple intracellular processes including activity of EGFR and angiotensin II receptor, oxidative stress and activity of ERK, c-SRC and MAP kinase signaling pathways.

Activation of these pathways may mediate some of the adverse effects of aldosterone. For example, aldosterone stimulation of MAP kinase 1 led to increased vascular smooth muscle proliferation (Ishizawa et al., 2005). Aldosterone increased collagen III abundance in the presence of low H_2O_2 concentrations through a non-genomic process involving EGFR (Grossmann et al., 2008). These rapid non-genomic actions of aldosterone may also lead to genomic effects. For example, aldosterone increased ACE gene expression via a JAK2-dependent pathway in cultured rat aortic endothelial cells (Sugiyama et al., 2005). Finally, angiotensin II activates gene transcription mediated by the mineralocorticoid receptor (Jaffe and Mendelsohn, 2005). Thus, there appear to be multiple interactions between angiotensin II and aldosterone affecting both genomic and non-genomic actions of these hormones. These interactions may modulate some of aldosterone's actions.

B. Aldosterone and Vascular Function

Aldosterone affects vasoconstriction and vasodilation. In mesenteric resistance vessels, aldosterone potentiated phenylephrine-mediated constriction through a non-genomic process involving phosphatidylinositol 3-kinase and protein kinase C, but not ERK activation (Michea et al., 2005). In rabbit preglomerular arterioles, aldosterone promoted vasoconstriction through these same pathways, but also stimulated vasodilation through increases in endothelial-derived nitric oxide (Arima et al., 2004). In intact aortic rings, the vasodilatory effects of aldosterone appeared to dominate, as aldosterone reduced phenylephrine-mediated vasoconstriction. However, when the endothelium was removed from the aortic ring, aldosterone increased phenylephrine-mediated vasoconstriction, indicating dominance of aldosterone's vasoconstrictive properties under these conditions (Liu et al., 2003).

Additional studies suggest that aldosterone may reduce endothelial nitric oxide. In human umbilical vein endothelial cells, mineralocorticoid receptor activation increased generation of reactive oxygen species, which can inactivate nitric oxide, and decreased endothelial nitric oxide synthase expression (Nagata et al., 2006). Further, aldosterone was shown to decrease endothelial glucose-6-phosphate dehydrogenase expression and activity leading to increased oxidant stress, decreased nitric oxide and impaired vascular reactivity (Leopold et al., 2007). Thus, while aldosterone has both vasodilatory and vasoconstrictive actions, most preclinical studies supported an adverse effect of aldosterone on vascular function.

Studies in healthy subjects have failed to show a consistent effect of aldosterone on vascular function. Specifically, aldosterone administration to healthy subjects has been shown to impair (Farquharson and Struthers, 2002), enhance (Nietlispach et al., 2007) and have no effect (Schmidt et al., 2003) on the nitric oxide-dependent forearm vasodilatory response to acetylcholine and to increase (Schmidt et al., 2003) or have no effect (Farquharson and Struthers, 2002) on the forearm vasoconstrictive response to noradrenaline or phenylephrine. Thus, it is likely that, in healthy subjects, aldosterone's vasodilatory and vasoconstrictive effects are balanced.

In disease states, such as heart failure, the adverse vascular effects of aldosterone appear dominant. In patients with heart failure, blockade of the aldosterone receptor with spironolactone significantly increased the forearm blood flow response to acetylcholine (Farquharson and Struthers, 2000). The effect of aldosterone on vascular function in diabetes may be more complex. Mineralocorticoid receptor blockade had either no effect on (Joffe et al., 2007) or impaired (Davies et al., 2004) nitric oxide-dependent forearm vasodilation in subjects with diabetes. However, mineralocorticoid receptor blockade improved coronary circulatory function, as compared with hydrochlorothiazide treatment, in subjects with diabetes on ACE inhibitor therapy (Joffe et al., 2007). Thus, the vascular effects of aldosterone may vary with the vascular bed under study and with the underlying health of the individual.

C. Aldosterone and Inflammation

Studies in animal models of cardiac and renal injury clearly demonstrated the pro-inflammatory effects of aldosterone and the anti-inflammatory effects of mineralocorticoid receptor blockade not only in target organs, but also in vessels and adipose tissue (Rocha et al., 2002a; Oestreicher et al.,

2003; Joffe and Adler, 2005; Kuster et al., 2005; Guo et al., 2006; Han et al., 2006b; Kang et al., 2006). *In vitro* studies, suggest that these inflammatory effects are due in part to direct pro-inflammatory actions of aldosterone. In human coronary artery smooth muscle cells, aldosterone increased gene expression of pro-inflammatory molecules as well as those involved in fibrosis and calcification (Jaffe and Mendelsohn, 2005). Aldosterone increased N-κB transcriptional activity and MCP-1 expression in cultured mesangial and proximal tubule cells (Han et al., 2006a, b) and increased TNF-α, IL-6 and MCP-1 in adipocyte cultures (Kraus et al., 2005; Guo et al., 2008). Aldosterone treatment also increased superoxide anion formation by mouse peritoneal macrophages in apolipoprotein E deficient mice (Keidar et al., 2004). In human peripheral blood mononuclear cells, spironolactone inhibited lipopolysaccharide or phytohemagglutinin-P stimulated production of pro-inflammatory cytokines (Hansen et al., 2004). Aldosterone acted synergistically with angiotensin II to increase PAI-1 (Brown et al., 2000) and mineralocorticoid receptor blockade decreased PAI-1 expression in animal models of cardiovascular injury (Oestreicher et al., 2003; Enomoto et al., 2005), suggesting that aldosterone may have pro-thrombotic effects.

These pro-inflammatory effects of aldosterone are observed in humans. An aldosterone infusion increased circulating IL-6 levels and spironolactone blocked the increase in IL-6 seen with the administration angiotensin II (Luther et al., 2006), suggesting that the pro-inflammatory effect of angiotensin II was mediated through stimulation of aldosterone (Luther et al., 2006).

D. Aldosterone and Insulin Sensitivity

Obesity is often associated with hypertension and one of the mechanisms of obesity-induced hypertension involves increased mineralocorticoid activity (Hiramatsu et al., 1981; Dustan, 1983). Studies that controlled for the effect of dietary sodium intake on aldosterone secretion demonstrated increased production of aldosterone in obese subjects (Goodfriend et al., 1995, 1998; Bentley-Lewis et al., 2007). Increased aldosterone levels were associated with an increase in insulin resistance (Goodfriend et al., 1995, 1998, 1999; Andronico et al., 2001; Bentley-Lewis et al., 2007). Moreover, weight loss reduced insulin sensitivity, as well as lowered blood pressure and aldosterone levels (Tuck et al., 1981; Rocchini et al., 1986; Goodfriend et al., 1999; Engeli et al., 2005). However, the strongest evidence for the effect of aldosterone on insulin resistance comes from studying patients with primary hyperaldosteronism. Insulin resistance was present in patients with tumoral and idiopathic hyperaldosteronism and treatment with surgery or mineralocorticoid receptor antagonists improved sensitivity to insulin (Haluzik et al., 2002; Catena et al., 2006). Finally, spironolactone reduced insulin resistance in patients with polycystic ovarian syndrome (Moghetti et al., 1996; Ganie et al., 2004; Zulian et al., 2005).

The underlying mechanism for aldosterone overproduction in obesity is not clear. Some studies linked increased production of aldosterone to mineralocorticoid releasing factors from the adipocytes (Ehrhart-Bornstein et al., 2003, 2004). Another hypothesis suggests the role of hepatic intermediaries in the stimulation of adrenal aldosterone by fatty acids produced by visceral adipocytes (Goodfriend et al., 1999).

The mechanism of aldosterone-induced insulin resistance may involve effects of aldosterone on adipose tissue. Mineralocorticoid receptor has been shown to mediate corticosteroid-induced adipocyte differentiation (Caprio et al., 2007). Further, aldosterone increased expression of pro-inflammatory factors, decreased expression of adiponectin and peroxisome proliferator-activated receptor-γ (PPAR-γ) and reduced insulin-stimulated glucose uptake in cultured adipocytes (Kraus et al., 2005; Guo et al., 2008). *In vivo* studies in the *db/db* mouse model of obesity and diabetes demonstrated that blockade of the mineralocorticoid receptor reduced adipose tissue inflammation, decreased adipose tissue expression of pro-inflammatory factors, increased adipose tissue expression of adiponectin and PPAR-γ and improved measures of insulin sensitivity (Guo et al., 2008). Also, aldosterone decreased protein levels of insulin receptor and insulin receptor substrate 1 and reduced insulin-induced AKT phosphorylation; all effects that may impair insulin action (Campion et al., 2002; Calle et al., 2003; Hitomi et al., 2007).

E. Dietary Sodium and Aldosterone-Mediated Cardiovascular Injury

In animal models, dietary sodium intake has been shown to have a profound effect on aldosterone-mediated cardiovascular injury. A low dietary sodium intake prevented the development of cardiac and renal injury in uninephrectomized rats infused with the mineralocorticoid deoxycorticosterone acetate for 8 weeks, whereas these animals developed profound cardiac and perivascular fibrosis when consuming a high sodium diet (Hitomi et al., 2007). Further, rats receiving the nitric oxide synthase inhibitor and angiotensin II developed vascular and cardiac injury when on a moderately high sodium diet. However, a low sodium diet prevented this injury despite markedly elevated blood levels of aldosterone (Brilla and Weber, 1992). These studies suggest that increases in circulating aldosterone that are appropriate for the level of dietary sodium consumption, e.g. elevated aldosterone levels in individuals consuming a low sodium diet, may not result in cardiovascular injury. Rather, cardiovascular injury results when aldosterone is elevated relative to dietary sodium intake (Martinez et al., 2002). The cellular and molecular mechanisms for this deleterious effect of increased dietary sodium and aldosterone remain to be determined. Further studies are needed to determine whether dietary sodium restriction has a beneficial effect on some or all aspects of aldosterone-mediated cardiovascular and renovascular injury.

In contrast to dietary sodium, modulation of dietary potassium does not have a major effect on aldosterone-mediated cardiovascular injury. In models of aldosterone-mediated cardiac injury, increased consumption of dietary potassium did not modulate cardiac injury, whereas injury was reduced with mineralocorticoid receptor blockade or dietary sodium restriction (Young et al., 1995; Weber, 2001).

F. Activation of Mineralocorticoid Receptor by Glucocorticoids

In the vasculature and many other tissues, the enzyme 11-β hydroxysteroid dehydrogenase type 2 prevents glucocorticoids from activating the mineralocorticoid receptor. However, if this enzyme is inactivated or absent, as occurs in the cardiomyocyte, cortisol becomes an effective mineralocorticoid receptor ligand (Martinez et al., 2002). In the uninephrectomized rat on a moderately high sodium diet, activation of the mineralocorticoid receptor by either glucocorticoids or mineralocorticoids led to coronary vascular inflammation (Funder, 2005). Thus, in humans, adverse effects of mineralocorticoid receptor activation may be mediated through cortisol as well as aldosterone.

VI. THERAPEUTIC CONSIDERATIONS

Spironolactone (25 mg daily) and eplerenone (50 mg daily) are effective antihypertensive agents, lowering blood pressure by $\approx$7 mmHg in patients with essential hypertension (White et al., 2003a). Mineralocorticoid receptor antagonists are particularly efficacious in patients with resistant hypertension (Calhoun et al., 2008). In two studies of individuals with poorly controlled hypertension despite treatment with two to four antihypertensive medications including, in most instances, an ACE inhibitor or ARB, addition of spironolactone lowered systolic blood pressure an additional $\approx$25 mmHg (Ouzan et al., 2002; Nishizaka et al., 2003). Plasma or 24-hour urinary aldosterone levels did not predict the blood pressure response to spironolactone. Further, recent studies indicate that primary hyperaldosteronism is a more common cause of hypertension than previously thought, occurring in $\approx$6% of all patients with hypertension and $\approx$20% of those with resistant hypertension (Calhoun et al., 2008). Mineralocorticoid receptor blockade is the treatment of choice for patients with hyperaldosteronism due to bilateral adrenal disease and for individuals with unilateral aldosteronoma who are not surgical candidates.

As described in this chapter, randomized controlled trials in patients with heart failure show a marked beneficial effect of mineralocorticoid receptor blockade on morbidity and mortality (Pitt et al., 1999, 2003b). Thus, spironolactone 25 mg or eplerenone 50 mg per day is indicated as add-on to standard therapy for stable patients with a left ventricular ejection fraction $\leq$40% and for patients with heart failure after a myocardial infarction. ACE inhibitors and mineralocorticoid receptor antagonists may have additive effects since they interfere with the activity of the renin–angiotensin–aldosterone system at two different steps. Further, aldosterone levels do not remain suppressed after $\approx$3 months of treatment with an ACE inhibitor or ARB. Due to the risk of hyperkalemia, mineralocorticoid receptor antagonists should not be started in patients with a serum potassium >5.5 mEq/l or a creatinine clearance $\leq$30 ml/min (<50 ml/min for hypertensive patients). In addition, the non-selective mineralocorticoid receptor antagonist spironolactone can cause gynecomastia in men because of its antiandrogen effects. Eplerenone is a specific antagonist for the mineralocorticoid receptor and does not have this side effect. It is well tolerated, but more expensive than spironolactone.

Currently, the use of eplerenone is contraindicated in patients with type 2 diabetes and microalbuminuria due to the risk of hyperkalemia. However, preclinical and clinical data strongly suggest a benefit for mineralocorticoid receptor blockade in diabetic nephropathy and this benefit is additive to that of ACE inhibition. In one study, 268 patients with type 2 diabetes and microalbuminuria on ACE inhibitor therapy were randomized to receive placebo, eplerenone 50 mg daily or eplerenone 100 mg daily (Epstein et al., 2006). Eplerenone significantly reduced albuminuria compared to placebo and the two doses of eplerenone were equally effective. Neither dose of eplerenone increased the risk of hyperkalemia above that of placebo. However, these were carefully selected patients who were $\leq$66 years of age, had hemoglobin A1c $\leq$8.5%, a serum potassium $\leq$5.0 mmol/l and a creatinine clearance $\geq$79 ml/min. Further clinical trials are needed to determine if mineralocorticoid receptor blockade reduces progression of renal injury, onset of renal failure or mortality in patients with type 2 diabetes on optimal medical therapy.

VII. CONCLUSIONS

Clinical and preclinical data support an important role for mineralocorticoid receptor in the pathophysiology of cardiovascular, renovascular and cerebrovascular disease. Multiple mechanisms may be involved in these adverse effects of aldosterone, including increased oxidative stress, inflammation, insulin resistance, vascular dysfunction and volume expansion leading to hypertension. Clinical studies in patients with diabetic nephropathy or heart failure demonstrated beneficial cardiovascular and renovascular effects of mineralocorticoid receptor blockade that are independent from its antihypertensive effects and are additive to benefits induced by ACE inhibition. These added benefits of ACE inhibition and mineralocorticoid receptor blockade may be due to the lack of prolonged suppression of serum aldosterone with ACE inhibition and/or to blockade of the interactions between aldosterone and angiotensin II on gene transcription and intracellular signaling pathways. Since combined mineralocorticoid receptor blockade and ACE inhibition can lead to hyperkalemia, additional clinical trials are needed to define the specific patient populations that will benefit from combination therapy. Further, there are many remaining questions regarding the cardiovascular effects of aldosterone, including how dietary sodium restriction abrogates aldosterone's adverse cardiovascular effects. Thus, more studies are needed to define the role of aldosterone in cardiovascular diseases.

References

Andronico, G., Cottone, S., Mangano, M. T. et al. (2001). Insulin, renin-aldosterone system and blood pressure in obese people. *Int J Obes Relat Metab Disord* **25**: 239–242.

Arima, S., Kohagura, K., Xu, H. L. et al. (2004). Endothelium-derived nitric oxide modulates vascular action of aldosterone in renal arteriole. *Hypertension* **43**: 352–357.

Bentley-Lewis, R., Adler, G. K., Perlstein, T. et al. (2007). Body mass index predicts aldosterone production in normotensive adults on a high-salt diet. *J Clin Endocrinol Metab* **92**: 4472–4475.

Blasi, E. R., Rocha, R., Rudolph, A. E., Blomme, E. A., Polly, M. L., and McMahon, E. G. (2003). Aldosterone/salt induces renal inflammation and fibrosis in hypertensive rats. *Kidney Int* **63**: 1791–1800.

Brilla, C. G., and Weber, K. T. (1992). Mineralocorticoid excess, dietary sodium, and myocardial fibrosis. *J Lab Clin Med* **120**: 893–901.

Brown, N. J., Kim, K. S., Chen, Y. Q., Blevins, L. S., Nadeau, J. H., Meranze, S. G., and Vaughan, D. E. (2000). Synergistic effect of adrenal steroids and angiotensin II on plasminogen activator inhibitor-1 production. *J Clin Endocrinol Metab* **85**: 336–344.

Calhoun, D. A., Jones, D., Textor, S. et al. (2008). Resistant hypertension: diagnosis, evaluation, and treatment. a scientific statement from the American Heart Association Professional Education Committee of the Council for High Blood Pressure Research. *Circulation* **117**: e510–e526.

Calle, C., Campion, J., Garcia-Arencibia, M., Maestro, B., and Davila, N. (2003). Transcriptional inhibition of the human insulin receptor gene by aldosterone. *J Steroid Biochem Mol Biol* **84**: 543–553.

Callera, G. E., Touyz, R. M., Tostes, R. C. et al. (2005). Aldosterone activates vascular p38MAP kinase and NADPH oxidase via c-Src. *Hypertension* **45**: 773–779.

Campion, J., Maestro, B., Molero, S., Davila, N., Carranza, M. C., and Calle, C. (2002). Aldosterone impairs insulin responsiveness in U-937 human promonocytic cells via the downregulation of its own receptor. *Cell Biochem Funct* **20**: 237–245.

Caprio, M., Feve, B., Claes, A., Viengchareun, S., Lombes, M., and Zennaro, M. C. (2007). Pivotal role of the mineralocorticoid receptor in corticosteroid-induced adipogenesis. *FASEB J* **21**: 2185–2194.

Catena, C., Lapenna, R., Baroselli, S. et al. (2006). Insulin sensitivity in patients with primary aldosteronism: a follow-up study. *J Clin Endocrinol Metab* **91**: 3457–3463.

Davies, J. I., Band, M., Morris, A., and Struthers, A. D. (2004). Spironolactone impairs endothelial function and heart rate variability in patients with type 2 diabetes. *Diabetologia* **47**: 1687–1694.

Dorrance, A. M., Osborn, H. L., Grekin, R., and Webb, R. C. (2001). Spironolactone reduces cerebral infarct size and EGF-receptor mRNA in stroke-prone rats. *Am J Physiol Regul Integr Comp Physiol* **281**: R944–R950.

Dorrance, A. M., Rupp, N. C., and Nogueira, E. F. (2006). Mineralocorticoid receptor activation causes cerebral vessel remodeling and exacerbates the damage caused by cerebral ischemia. *Hypertension* **47**: 590–595.

Dustan, H. P. (1983). Mechanisms of hypertension associated with obesity. *Ann Intern Med* **98**: 860–864.

Ehrhart-Bornstein, M., Lamounier-Zepter, V., Schraven, A. et al. (2003). Human adipocytes secrete mineralocorticoid-releasing factors. *Proc Natl Acad Sci USA* **100**: 14211–14216.

Ehrhart-Bornstein, M., Arakelyan, K., Krug, A. W., Scherbaum, W. A., and Bornstein, S. R. (2004). Fat cells may be the obesity-hypertension link: human adipogenic factors stimulate aldosterone secretion from adrenocortical cells. *Endocr Res* **30**: 865–870.

Engeli, S., Bohnke, J., Gorzelniak, K. et al. (2005). Weight loss and the renin-angiotensin-aldosterone system. *Hypertension* **45**: 356–362.

Enomoto, S., Yoshiyama, M., Omura, T. et al. (2005). Effects of eplerenone on transcriptional factors and mRNA expression related to cardiac remodelling after myocardial infarction. *Heart* **91**: 1595–1600.

Epstein, M., Williams, G. H., Weinberger, M. et al. (2006). Selective aldosterone blockade with eplerenone reduces albuminuria in patients with type 2 diabetes. *Clin J Am Soc Nephol* **1**: 940–951.

Farquharson, C. A., and Struthers, A. D. (2000). Spironolactone increases nitric oxide bioactivity, improves endothelial vasodilator dysfunction, and suppresses vascular angiotensin I/angiotensin II conversion in patients with chronic heart failure. *Circulation* **101**: 594–597.

Farquharson, C. A., and Struthers, A. D. (2002). Aldosterone induces acute endothelial dysfunction in vivo in humans: evidence for an aldosterone-induced vasculopathy. *Clin Sci (Lond.)* **103**: 425–431.

Feria, I., Pichardo, I., Juarez, P. et al. (2003). Therapeutic benefit of spironolactone in experimental chronic cyclosporine A nephrotoxicity. *Kidney Int* **63**: 43–52.

Fiebeler, A., Nussberger, J., Shagdarsuren, E. et al. (2005). Aldosterone synthase inhibitor ameliorates angiotensin II-induced organ damage. *Circulation* **111**: 3087–3094.

Funder, J. W. (2005). Mineralocorticoid receptors: distribution and activation. *Heart Fail Rev* **10**: 15–22.

Ganie, M. A., Khurana, M. L., Eunice, M. et al. (2004). Comparison of efficacy of spironolactone with metformin in the management of polycystic ovary syndrome: an open-labeled study. *J Clin Endocrinol Metab* **89**: 2756–2762.

Gauer, S., Hauser, I. A., Obermuller, N., Holzmann, Y., Geiger, H., and Goppelt-Struebe, M. (2008). Synergistic induction of osteopontin by aldosterone and inflammatory cytokines in mesangial cells. *J Cell Biochem* **103**: 615–623.

Goodfriend, T. L., Egan, B., Stepniakowski, K., and Ball, D. L. (1995). Relationships among plasma aldosterone, high-density lipoprotein cholesterol, and insulin in humans. *Hypertension* **25**: 30–36.

Goodfriend, T. L., Egan, B. M., and Kelley, D. E. (1998). Aldosterone in obesity. *Endocr Res* **24**: 789–796.

Grandi, A. M., Imperiale, D., Santillo, R. et al. (2002). Aldosterone antagonist improves diastolic function in essential hypertension. *Hypertension* **40**: 647–652.

Grossmann, C., Benesic, A., Krug, A. W. et al. (2005). Human mineralocorticoid receptor expression renders cells responsive for nongenotropic aldosterone actions. *Mol Endocrinol* **19**: 1697–1710.

Grossmann, C., Krug, A. W., Freudinger, R., Mildenberger, S., Voelker, K., and Gekle, M. (2007). Aldosterone-induced EGFR expression: interaction between the human mineralocorticoid receptor and the human EGFR promoter. *Am J Physiol Endocrinol Metab* **292**: E1790–E1800.

Grossmann, C., Freudinger, R., Mildenberger, S., Husse, B., and Gekle, M. (2008). EF-domains are sufficient for nongenomic mineralocorticoid receptor actions. *J Biol Chem* **283**: 7109–7116.

Guo, C., Martinez-Vasquez, D., Mendez, G. P. et al. (2006). Mineralocorticoid receptor antagonist reduces renal injury in rodent models of types 1 and 2 diabetes mellitus. *Endocrinology* **147**: 5363–5373.

Guo, C., Ricchiuti, V., Lian, B. Q. et al. (2008). Mineralocorticoid receptor blockade reverses obesity-related changes in expression of adiponectin, PPAR-gamma and pro-inflammatory adipokines. *Circulation* **117**: 2253–2261.

Haluzik, M., Sindelka, G., Widimsky, J., Prazny, M., Zelinka, T., and Skrha, J. (2002). Serum leptin levels in patients with primary hyperaldosteronism before and after treatment: relationships to insulin sensitivity. *J Hum Hypertens* **16**: 41–45.

Han, K. H., Kang, Y. S., Han, S. Y. et al. (2006a). Spironolactone ameliorates renal injury and connective tissue growth factor expression in type II diabetic rats. *Kidney Int* **70**: 111–120.

Han, S. Y., Kim, C. H., Kim, H. S. et al. (2006b). Spironolactone prevents diabetic nephropathy through an anti-inflammatory mechanism in type 2 diabetic rats. *J Am Soc Nephrol* **17**: 1362–1372.

Hansen, P. R., Rieneck, K., and Bendtzen, K. (2004). Spironolactone inhibits production of proinflammatory cytokines by human mononuclear cells. *Immunol Lett* **91**: 87–91.

Hiramatsu, K., Yamada, T., Ichikawa, K., Izumiyama, T., and Nagata, H. (1981). Changes in endocrine activities relative to obesity in patients with essential hypertension. *J Am Geriatr Soc* **29**: 25–30.

Hitomi, H., Kiyomoto, H., Nishiyama, A. et al. (2007). Aldosterone suppresses insulin signaling via the downregulation of insulin receptor substrate-1 in vascular smooth muscle cells. *Hypertension* **50**: 750–755.

Ishizawa, K., Izawa, Y., Ito, H. et al. (2005). Aldosterone stimulates vascular smooth muscle cell proliferation via big mitogen-activated protein kinase 1 activation. *Hypertension* **46**: 1046–1052.

Iwashima, F., Yoshimoto, T., Minami, I., Sakurada, M., Hirono, Y., and Hirata, Y. (2008). Aldosterone induces superoxide generation via rac1 activation in endothelial cells. *Endocrinology* **149**: 1009–1014.

Izawa, H., Murohara, T., Nagata, K. et al. (2005). Mineralocorticoid receptor antagonism ameliorates left ventricular diastolic dysfunction and myocardial fibrosis in mildly symptomatic patients with idiopathic dilated cardiomyopathy: a pilot study. *Circulation* **112**: 2940–2945.

Jaffe, I. Z., and Mendelsohn, M. E. (2005). Angiotensin II and aldosterone regulate gene transcription via functional mineralocorticoid receptors in human coronary artery smooth muscle cells. *Circ Res* **96**: 643–650.

Joffe, H. V., and Adler, G. K. (2005). Effect of aldosterone and mineralocorticoid receptor blockade on vascular inflammation. *Heart Fail Rev* **10**: 31–37.

Joffe, H. V., Kwong, R. Y., Gerhard-Herman, M. D., Rice, C., Feldman, K., and Adler, G. K. (2007). Beneficial effects of eplerenone versus hydrochlorothiazide on coronary circulatory function in patients with diabetes mellitus. *J Clin Endocrinol Metab* **92**: 2552–2558.

Kang, Y. M., Zhang, Z. H., Johnson, R. F. et al. (2006). Novel effect of mineralocorticoid receptor antagonism to reduce proinflammatory cytokines and hypothalamic activation in rats with ischemia-induced heart failure. *Circ Res* **99**: 758–766.

Keidar, S., Hayek, T., Kaplan, M. et al. (2003). Effect of eplerenone, a selective aldosterone blocker, on blood pressure, serum and macrophage oxidative stress, and atherosclerosis in apolipoprotein E-deficient mice. *J Cardiovasc Pharmacol* **41**: 955–963.

Keidar, S., Kaplan, M., Pavlotzky, E. et al. (2004). Aldosterone administration to mice stimulates macrophage NADPH oxidase and increases atherosclerosis development: a possible role for angiotensin-converting enzyme and the receptors for angiotensin II and aldosterone. *Circulation* **109**: 2213–2220.

Kraus, D., Jager, J., Meier, B., Fasshauer, M., and Klein, J. (2005). Aldosterone inhibits uncoupling protein-1, induces insulin resistance, and stimulates proinflammatory adipokines in adipocytes. *Horm Metab Res* **37**: 455–459.

Kuster, G. M., Kotlyar, E., Rude, M. K. et al. (2005). Mineralocorticoid receptor inhibition ameliorates the transition to myocardial failure and decreases oxidative stress and inflammation in mice with chronic pressure overload. *Circulation* **111**: 420–427.

Leopold, J. A., Dam, A., Maron, B. A. et al. (2007). Aldosterone impairs vascular reactivity by decreasing glucose-6-phosphate dehydrogenase activity. *Nat Med* **13**: 189–197.

Liu, S. L., Schmuck, S., Chorazcyzewski, J. Z., Gros, R., and Feldman, R. D. (2003). Aldosterone regulates vascular reactivity: short-term effects mediated by phosphatidylinositol 3-kinase-dependent nitric oxide synthase activation. *Circulation* **108**: 2400–2406.

Luther, J. M., Gainer, J. V., Murphey, L. J. et al. (2006). Angiotensin II induces interleukin-6 in humans through a mineralocorticoid receptor-dependent mechanism. *Hypertension* **48**: 1050–1057.

Martinez, D. V., Rocha, R., Matsumura, M. et al. (2002). Cardiac damage prevention by eplerenone: comparison with low sodium diet or potassium loading. *Hypertension* **39**: 614–618.

Michea, L., Delpiano, A. M., Hitschfeld, C., Lobos, L., Lavandero, S., and Marusic, E. T. (2005). Eplerenone blocks nongenomic effects of aldosterone on the Na^+/H^+ exchanger, intracellular Ca^{2+} levels, and vasoconstriction in mesenteric resistance vessels. *Endocrinology* **146**: 973–980.

Min, L. J., Mogi, M., Li, J. M., Iwanami, J., Iwai, M., and Horiuchi, M. (2005). Aldosterone and angiotensin II synergistically induce mitogenic response in vascular smooth muscle cells. *Circ Res* **97**: 434–442.

Miric, G., Dallemagne, C., Endre, Z., Margolin, S., Taylor, S. M., and Brown, L. (2001). Reversal of cardiac and renal fibrosis by pirfenidone and spironolactone in streptozotocin-diabetic rats. *Br J Pharmacol* **133**: 687–694.

Moghetti, P., Tosi, F., Castello, R. et al. (1996). The insulin resistance in women with hyperandrogenism is partially reversed by antiandrogen treatment: evidence that androgens impair insulin action in women. *J Clin Endocrinol Metab* **81**: 952–960.

Nagata, D., Takahashi, M., Sawai, K. et al. (2006). Molecular mechanism of the inhibitory effect of aldosterone on endothelial NO synthase activity. *Hypertension* **48**: 165–171.

New England Research Institutes and National Heart Lung and Blood Institute of the National Institutes of Health. (2008). TOPCAT Study. http://www.topcatstudy.com/.

Nietlispach, F., Julius, B., Schindler, R. et al. (2007). Influence of acute and chronic mineralocorticoid excess on endothelial function in healthy men. *Hypertension* **50**: 82–88.

Nishizaka, M. K., Zaman, M. A., and Calhoun, D. A. (2003). Efficacy of low-dose spironolactone in subjects with resistant hypertension. *Am J Hypertens* **16**: 925–930.

Oestreicher, E. M., Martinez-Vasquez, D., Stone, J. R. et al. (2003). Aldosterone and not plasminogen activator inhibitor-1 is a critical mediator of early angiotensin II/NG-nitro-L-arginine methyl ester-induced myocardial injury. *Circulation* **108**: 2517–2523.

Ouzan, J., Perault, C., Lincoff, A. M., Carre, E., and Mertes, M. (2002). The role of spironolactone in the treatment of patients with refractory hypertension. *Am J Hypertens* **15**: 333–339.

Pitt, B., Zannad, F., Remme, W. J. et al. (1999). The effect of spironolactone on morbidity and mortality in patients with severe heart failure. Randomized Aldactone Evaluation Study Investigators. *N Engl J Med* **341**: 709–717.

Pitt, B., Reichek, N., Willenbrock, R. et al. (2003a). Effects of eplerenone, enalapril, and eplerenone/enalapril in patients with essential hypertension and left ventricular hypertrophy: the 4E-left ventricular hypertrophy study. *Circulation* **108**: 1831–1838.

Pitt, B., Remme, W., Zannad, F. et al. (2003b). Eplerenone, a selective aldosterone blocker, in patients with left ventricular dysfunction after myocardial infarction. *N Engl J Med* **348**: 1309–1321.

Quinkler, M., Zehnder, D., Eardley, K. S. et al. (2005). Increased expression of mineralocorticoid effector mechanisms in kidney biopsies of patients with heavy proteinuria. *Circulation* **112**: 1435–1443.

Rachmani, R., Slavachevsky, I., Amit, M. et al. (2004). The effect of spironolactone, cilazapril and their combination on albuminuria in patients with hypertension and diabetic nephropathy is independent of blood pressure reduction: a randomized controlled study. *Diabet Med* **21**: 471–475.

Rickard, A. J., Funder, J. W., Fuller, P. J., and Young, M. J. (2006). The role of the glucocorticoid receptor in mineralocorticoid/salt-mediated cardiac fibrosis. *Endocrinology* **147**: 5901–5906.

Rocchini, A. P., Katch, V. L., Grekin, R., Moorehead, C., and Anderson, J. (1986). Role for aldosterone in blood pressure regulation of obese adolescents. *Am J Cardiol* **57**: 613–618.

Rocha, R., Chander, P. N., Khanna, K., Zuckerman, A., and Stier, C. T. (1998). Mineralocorticoid blockade reduces vascular injury in stroke-prone hypertensive rats. *Hypertension* **31**: 451–458.

Rocha, R., Stier, C. T., Kifor, I. et al. (2000). Aldosterone: a mediator of myocardial necrosis and renal arteriopathy. *Endocrinology* **141**: 3871–3878.

Rocha, R., Martin-Berger, C. L., Yang, P., Scherrer, R., Delyani, J., and McMahon, E. (2002a). Selective aldosterone blockade prevents angiotensin II/salt-induced vascular inflammation in the rat heart. *Endocrinology* **143**: 4828–4836.

Rocha, R., Rudolph, A. E., Frierdich, G. E. et al. (2002b). Aldosterone induces a vascular inflammatory phenotype in the rat heart. *Am J Physiol Heart Circ Physiol* **283**: H1802–H1810.

Rossi, G. P., Sacchetto, A., Visentin, P. et al. (1996). Changes in left ventricular anatomy and function in hypertension and primary aldosteronism. *Hypertension* **27**: 1039–1045.

Rossi, G. P., Sacchetto, A., Pavan, E. et al. (1997). Remodeling of the left ventricle in primary aldosteronism due to Conn's adenoma. *Circulation* **95**: 1471–1478.

Rossing, K., Schjoedt, K. J., Smidt, U. M., Boomsma, F., and Parving, H. H. (2005). Beneficial effects of adding spironolactone to recommended antihypertensive treatment in diabetic nephropathy: a randomized, double-masked, cross-over study. *Diabetes Care* **28**: 2106–2112.

Sanz-Rosa, D., Cediel, E., de las Heras, N. et al. (2005). Participation of aldosterone in the vascular inflammatory response of spontaneously hypertensive rats: role of the NFkappaB/IkappaB system. *J Hypertens* **23**: 1167–1172.

Sato, A., Hayashi, K., Naruse, M., and Saruta, T. (2003). Effectiveness of aldosterone blockade in patients with diabetic nephropathy. *Hypertension* **41**: 64–68.

Schmidt, B. M., Oehmer, S., Delles, C. et al. (2003). Rapid nongenomic effects of aldosterone on human forearm vasculature. *Hypertension* **42**: 156–160.

Shibata, S., Nagase, M., Yoshida, S., Kawachi, H., and Fujita, T. (2007). Podocyte as the target for aldosterone: roles of oxidative stress and Sgk1. *Hypertension* **49**: 355–364.

Siragy, H. M., and Xue, C. (2008). Local renal aldosterone production induces inflammation and matrix formation in Kidneys of diabetic rats. *Exp Physiol* **93**: 817–824.

Sugiyama, T., Yoshimoto, T., Tsuchiya, K. et al. (2005). Aldosterone induces angiotensin converting enzyme gene expression via a JAK2-dependent pathway in rat endothelial cells. *Endocrinology* **146**: 3900–3906.

Trachtman, H., Weiser, A. C., Valderrama, E., Morgado, M., and Palmer, L. S. (2004). Prevention of renal fibrosis by spironolactone in mice with complete unilateral ureteral obstruction. *J Urol* **172**: 1590–1594.

Tuck, M. L., Sowers, J., Dornfeld, L., Kledzik, G., and Maxwell, M. (1981). The effect of weight reduction on blood pressure, plasma renin activity, and plasma aldosterone levels in obese patients. *N Engl J Med* **304**: 930–933.

van den Meiracker, A. H., Baggen, R. G., Pauli, S. et al. (2006). Spironolactone in type 2 diabetic nephropathy: Effects on proteinuria, blood pressure and renal function. *J Hypertens* **24**: 2285–2292.

Weber, K. T. (2001). Aldosterone in congestive heart failure. *N Engl J Med* **345**: 1689–1697.

White, W. B., Carr, A. A., Krause, S., Jordan, R., Roniker, B., and Oigman, W. (2003a). Assessment of the novel selective aldosterone blocker eplerenone using ambulatory and clinical blood pressure in patients with systemic hypertension. *Am J Cardiol* **92**: 38–42.

White, W. B., Duprez, D., St Hillaire, R. et al. (2003b). Effects of the selective aldosterone blocker eplerenone versus the calcium antagonist amlodipine in systolic hypertension. *Hypertension* **41**: 1021–1026.

Williams, G. H., Burgess, E., Kolloch, R. E. et al. (2004). Efficacy of eplerenone versus enalapril as monotherapy in systemic hypertension. *Am J Cardiol* **93**: 990–996.

Young, M., Fullerton, M., Dilley, R., and Funder, J. (1994). Mineralocorticoids, hypertension, and cardiac fibrosis. *J Clin Invest* **93**: 2578–2583.

Young, M., Head, G., and Funder, J. (1995). Determinants of cardiac fibrosis in experimental hypermineralocorticoid states. *Am J Physiol* **269**: E657–E662.

Young, M. J., Moussa, L., Dilley, R., and Funder, J. W. (2003). Early inflammatory responses in experimental cardiac hypertrophy and fibrosis: effects of 11 beta-hydroxysteroid dehydrogenase inactivation. *Endocrinology* **144**: 1121–1125.

Zannad, F., Alla, F., Dousset, B., Perez, A., and Pitt, B. (2000). Limitation of excessive extracellular matrix turnover may contribute to survival benefit of spironolactone therapy in patients with congestive heart failure: insights from the randomized aldactone evaluation study (RALES). *Circulation* **102**: 2700–2706.

Zulian, E., Sartorato, P., Benedini, S. et al. (2005). Spironolactone in the treatment of polycystic ovary syndrome: effects on clinical features, insulin sensitivity and lipid profile. *J Endocrinol Invest* **28**: 49–53.

Further reading

Goodfriend, T. L., Egan, B. M., and Kelley, D. E. (1999a). Plasma aldosterone, plasma lipoproteins, obesity and insulin resistance in humans. *Prostaglandins Leukot Essent Fatty Acids* **60**: 401–405.

Goodfriend, T. L., Kelley, D. E., Goodpaster, B. H., and Winters, S. J. (1999b). Visceral obesity and insulin resistance are associated with plasma aldosterone levels in women. *Obes Res* **7**: 355–362.

CHAPTER 24

Regulation of Aldosterone Production

WILLIAM E. RAINEY[1], WENDY B. BOLLAG[2] AND CARLOS M. ISALES[2]

[1]Medical College of Georgia, Department of Physiology, 1120 15th Street, Augusta, Georgia 30912, USA
[2]Medical College of Georgia, The Institute of Molecular Medicine and Genetics, 1120 15th Street, Augusta, Georgia 30912, USA

Contents

I. Introduction	361
II. Aldosterone biosynthesis	361
III. Factors regulating aldosterone production	362
IV. Diseases of aldosterone production	371
V. Summary	375
References	375

I. INTRODUCTION

Aldosterone represents the primary mineralocorticoid produced by the adrenal gland and specifically within the outer adrenocortical cells of the glomerulosa layer. The need for the structural zonation of the adrenal cortex as a mechanism to control aldosterone production is apparent when one considers that the amount of aldosterone needed to control salt balance is 100–1000-fold less than the amount of cortisol needed to control carbohydrate metabolism. The unique histology of the three concentric zones of the mammalian adrenal cortex led to the names zona glomerulosa, zona fasciculata and zona reticularis (Arnold, 1866). It is now accepted that these zones have functionally distinct roles that allow the production of three major classes of steroid hormone, namely, zona glomerulosa-derived mineralocorticoids, zona fasciculata-derived glucocorticoids and zona reticularis-derived adrenal androgens. The mechanisms leading to adrenal zonation remain poorly defined but the gland's unique centripetal blood flow is believed to play a key role. Arterial blood enters the outer adrenal cortex, flows through fenestrated capillaries that start in the glomerulosa layer, progresses between the cords of fasciculata cells and drains inwardly into venules in the medulla. The directional flow of blood insures that the outer zona glomerulosa is unable to utilize fasciculata cell-produced precursor for the synthesis of aldosterone. An understanding of the steroid pathways used by the glomerulosa and fasciculata is helpful in making this connection.

II. ALDOSTERONE BIOSYNTHESIS

Aldosterone biosynthesis shares many of the same enzymatic reactions with those of cortisol production, however, zonal expression of key enzymes leads to the variation in glomerulosa versus fasciculata steroid products (Payne and Hales, 2004). Aldosterone is synthesized in the glomerulosa from cholesterol through the successive actions of four enzymes (Figure 24.1). Cholesterol side-chain cleavage (CYP11A1), 21-hydroxylase (CYP21) and aldosterone synthase (CYP11B2) are members of the cytochrome P450 family of enzymes. CYP11A1 and CYP11B2 are localized to the inner mitochondrial membrane, while CYP21 is found in the endoplasmic reticulum. These enzymes are cytochrome P450 heme-containing proteins that accept electrons from NADPH via accessory proteins and utilize molecular oxygen to perform hydroxylations (CYP21 and CYP11B2) or other oxidative conversions (CYP11A1). The fourth enzyme, type 2 3β-hydroxysteroid dehydrogenase (HSD3B2), is a member of the short-chain dehydrogenase family and is localized to endoplasmic reticulum. The first reactions of aldosterone biosynthesis (cholesterol to progesterone) are the same as occur for cortisol biosynthesis in the zona fasciculata (and for steroid hormone synthesis in general). Cholesterol is converted to pregnenolone by mitochondrial CYP11A1. This represents the rate-limiting reaction for all steroid-producing tissues and requires the transport of cholesterol from the cytoplasm to the mitochondrial outer membrane, followed by movement from the outer to the inner mitochondrial membrane where CYP11A1 is located (Capponi, 2004). This step is acutely regulated through the expression and phosphorylation of steroidogenic acute regulatory protein (StAR). The product of this reaction, pregnenolone is more soluble than cholesterol and can move via passive diffusion to the endoplasmic reticulum for conversion to progesterone by HSD3B2. Progesterone is hydroxylated to deoxycorticosterone by CYP21. Deoxycorticosterone can be converted to aldosterone by three successive oxidation reactions (11β- and 18-hydroxylation, followed by 18-oxidation) which, in humans, can be mediated by a single enzyme, aldosterone synthase CYP11B2. It should be noted that

Textbook of Nephro-Endocrinology.
ISBN: 978-0-12-373870-7

Copyright 2009, Elsevier Inc.
All rights reserved.

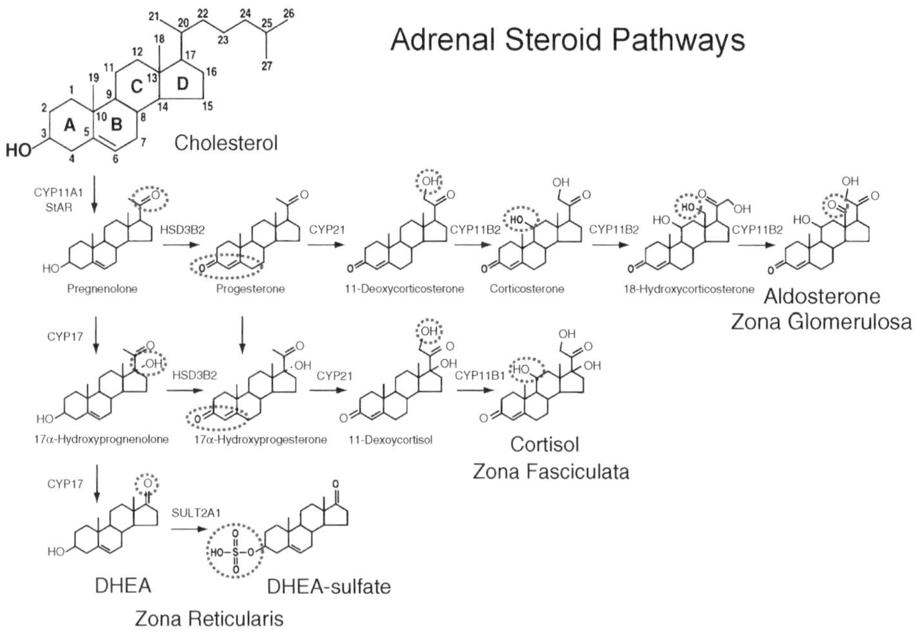

FIGURE 24.1 Adrenal steroid pathways leading to mineralocorticoid, glucocorticoids and adrenal androgens. Enzymes involved include cholesterol side-chain cleavage (CYP11A1), 3β-hydroxysteroid dehydrogenase type 2 (HSD3B2), 17α-hydroxylase, 17,20 lyase (CYP17), 21-hydroxylase (CYP21), 11β-hydroxylase (CYP11B1) and aldosterone synthase (CYP11B2). Steroidogenic acute regulatory (StAR) protein is needed for the rate-limiting movement of cholesterol to CYP11A1 in the inner mitochondrial membrane.

cortisol biosynthesis also requires 11β-hydroxylation to produce cortisol from 11-deoxycortisol and this is accomplished by 11β-hydroxylase (CYP11B1). However, this isozyme poorly catalyzes 18-hydroxylation and it does not catalyze 18-oxidation, thus preventing synthesis of aldosterone by the zona fasciculata.

In humans, functional zonation relies in part on the localized expression of two cytochrome P450 enzymes, specifically CYP11B2 and 17α-hydroxylase (CYP17). Expression of CYP11B2 is limited to the glomerulosa and this effectively prevents production of aldosterone in the other adrenocortical zones (Domalik et al., 1991; Ogishima et al., 1992; LeHoux et al., 1995; Pascoe et al., 1995). On the other hand, CYP17 diverts pregnenolone and progesterone away from the pathway leading to aldosterone and into that leading to cortisol (see Figure 24.1), explaining the reason for the lack of expression of CYP17 by the glomerulosa (Narasaka et al., 2001). Even in the presence of CYP17, the fasciculata produces large amounts of aldosterone precursors (in particular, progesterone and deoxycorticosterone) at levels that would lead to mineralocorticoid excess if they were even partially converted to aldosterone. Limiting expression of CYP11B2 to the outer zone in a situation where centripetal blood flow prevents these precursors from accessing CYP11B2 further controls the relative production rate of mineralocorticoids.

Historically, the regulation of aldosterone biosynthesis has been divided into two main phases (Clark et al., 1992; Muller, 1998). Acutely (minutes after a stimulus), aldosterone production is controlled by rapid signaling pathways that increase the movement of cholesterol into the mitochondria. This has been called the 'early regulatory step' and, as noted above, is mediated by increased expression and phosphorylation of StAR protein (Figure 24.2) (Cherradi et al., 1998). Chronically (hours to days), aldosterone production is regulated at the level of expression of the enzymes involved in the synthesis of aldosterone (Figure 24.3) (Bassett et al., 2004b). This has been called the 'late regulatory step' and is particularly dependent on increased transcription and expression of CYP11B2. Studies over the past decade have greatly increased our understanding of the signaling pathways that mediate early and late regulatory steps of aldosterone biosynthesis.

III. FACTORS REGULATING ALDOSTERONE PRODUCTION

A. The Renin–Angiotensin–Aldosterone System (RAAS)

1. Physiology of the RAAS (see Chapter 12 for Detailed Discussion)

Because the major function of aldosterone is to control body fluid volume by increasing sodium reabsorption by the kidneys, it is appropriate that the major regulator for aldosterone synthesis and secretion arises in the kidneys. Thus, the kidneys play the controlling role in the renin–angiotensin–aldosterone feedback system (Figure 24.4). Renin is a protease produced and stored in the juxtaglomerular cells that surround the glomerular afferent arterioles. Renin release is controlled by at least three mechanisms. First, release is activated by a decrease in the perfusion pressure of blood traversing the renal afferent arterioles, which is sensed by the juxtaglomerular apparatus functioning as a baroreceptor. This reduction

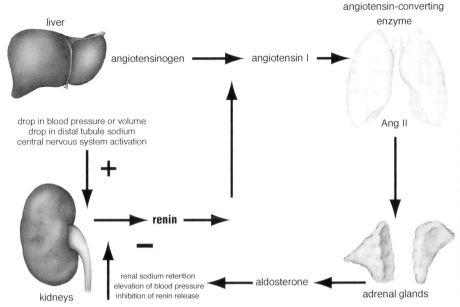

FIGURE 24.2 Aldosterone production is primarily regulated through a feedback loop that focuses on kidney production of renin. Once released, renin cleaves angiotensinogen to angiotensin I through the action of angiotensin-converting enzyme to produce Ang II. Ang II is the primary hormonal regulator of adrenal aldosterone production. Aldosterone indirectly exerts negative feedback to decrease renin release through renal sodium retention and elevations in blood pressure.

in perfusion pressure occurs as a result of a decrease in either systemic blood volume or blood pressure. Second, renin secretion can be stimulated by secretions from the macula densa as a result of a drop in sodium concentration in the distal tubule. Third, a drop in blood pressure will cause sympathetic stimulation of juxtaglomerular cells to stimulate both renin release and afferent arteriole constriction.

Acting in the blood, renin mediates the rate-limiting step in the production of angiotensin II (Ang II), cleaving the circulating precursor angiotensinogen to release the 10-amino acid peptide, angiotensin I (see Figure 24.2). Thereafter, inactive angiotensin I is converted rapidly to the potent octapeptide hormone Ang II by the action of angiotensin-converting enzyme, which is found in the plasma membrane of vascular endothelial cells throughout the body.

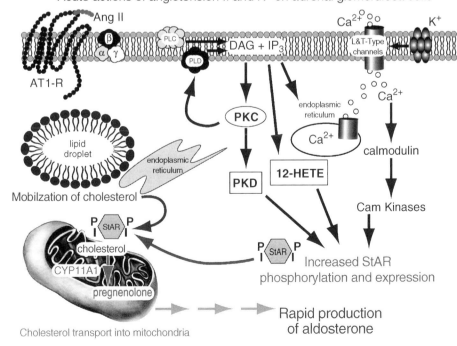

FIGURE 24.3 The acute regulation of aldosterone production is primarily regulated by Ang II and extracellular potassium (K^+). Ang II binds type 1 Ang II receptors (AT1-R) activating phospholipase C (PLC) to release diacylglycerol (DAG) and inositol 1,4,5-trisphosphate (IP_3). PLC activation also increases cellular levels of 12-HETE. DAG activates protein kinase C (PKC) and protein kinase D (PKD) and phospholipidase D (PLD). In turn, IP_3 causes release of intracellular calcium stores. Calcium activated kinases (CaM kinases), 12-HETE and PKD increase StAR protein levels and phosphorylation leading to increased cholesterol movement into the mitochondria. Within the mitochondria cholesterol is converted to pregnenolone by P-450 side-chain cleavage (CYP11A1) which is then metabolized to aldosterone.

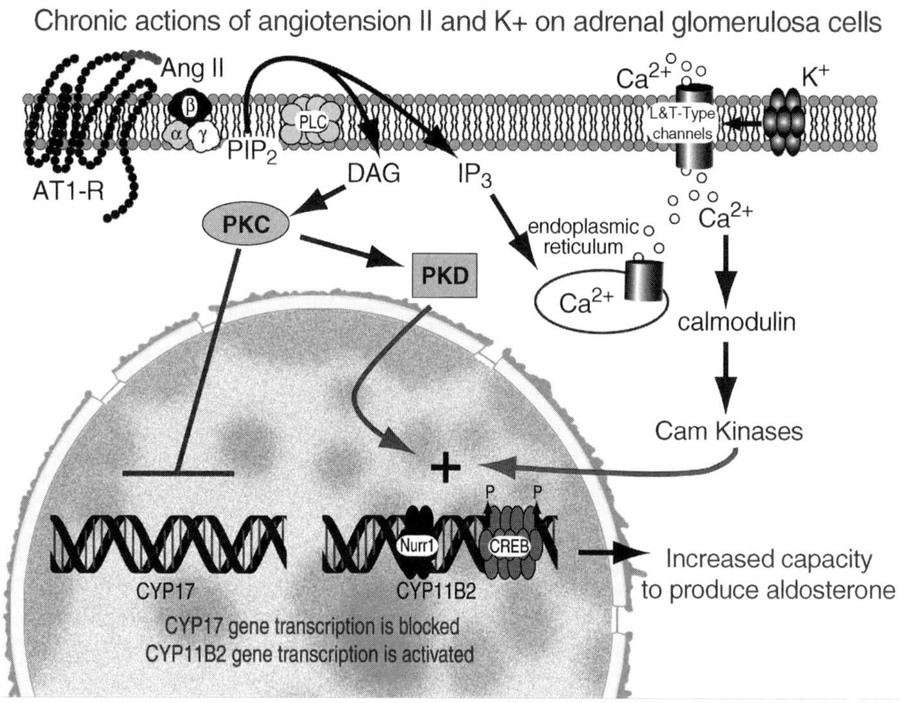

FIGURE 24.4 The chronic regulation of aldosterone production is regulated by Ang II and potassium (K+). Ang II binds type 1 Ang II receptors (AT1-R) to activate phospholipase C (PLC) activity, which releases diacylglycerol (DAG) and inositol 1,4,5-trisphosphate (IP$_3$). DAG activates protein kinase C (PKC) and protein kinase D (PKD), and IP$_3$ causes intracellular calcium release. PKC activation inhibits the transcription of 17α-hydroxylase (CYP17), while calcium and PKD increase transcription of aldosterone synthase (CYP11B2). This occurs through increased expression and phosphorylation of specific transcription factors that include Nurr1 and CREB. The increase in CYP11B2 increases the capacity to produce aldosterone.

Circulating Ang II is arguably the most important regulator of adrenal glomerulosa aldosterone production.

2. Ang II-Regulated Intracellular Glomerulosa Cell Signaling Pathways

In humans, Ang II has two G-protein-coupled receptors, type 1 (AT$_1$) and type 2 (AT$_2$), through which this hormone can elicit intracellular responses. In the glomerulosa, Ang II works primarily through AT$_1$ receptors to regulate aldosterone production (see Figure 24.2). The expression of AT$_1$ receptors is highest in the glomerulosa, which localizes the action of this hormone to aldosterone-producing cells. AT$_1$ receptors activate a variety of signaling pathways including phosphoinositide-specific phospholipase C (PLC), which hydrolyzes phosphatidylinositol 4,5-bisphosphate (PIP$_2$) to generate the two second messengers, inositol 1,4,5-trisphosphate (IP$_3$) and diacylglycerol (DAG) (Barrett et al., 1989; Ganguly and Davis, 1994). IP$_3$ is thought to initiate aldosterone secretion by eliciting a transient increase in the cytosolic calcium concentration and activating calcium/calmodulin-dependent protein kinases (CaM kinase), whereas DAG increases protein kinase C (PKC) activity. PKC activity has been suggested to underlie sustained aldosterone secretion from glomerulosa cells (Barrett et al., 1989). This idea is supported by the ability of synthetic ligands for PKC to stimulate aldosterone production and inhibitors of PKC to inhibit aldosterone production (Kojima et al., 1984, 1985a; Bollag et al., 1990; Kapas et al., 1995; Betancourt-Calle et al., 1999).

Ang II also increases calcium influx in glomerulosa cells. Influx of extracellular calcium acts to increase PKC activity, enhance PKC-stimulated steroidogenesis and maintain aldosterone production (Kojima et al., 1984, 1985a; Barrett et al., 1989; Bollag et al., 1990; Ganguly and Davis, 1994; Rasmussen et al., 1995; Betancourt-Calle et al., 1999). This increase in calcium influx is likely brought about by multiple mechanisms including Ang II depolarization of glomerulosa cells via effects on potassium channels (Spat and Hunyady, 2004) and voltage-gated calcium channels through G$_i$ (Lu et al., 1996). In addition, Ang II activates store-operated calcium channels (Aksoy et al., 1993) and/or calcium release-activated calcium (CRAC) channels (Aptel et al., 1999), which also presumably play a part in Ang II-induced calcium influx. Thapsigargin is a compound that inhibits endoplasmic reticulum calcium ATPase, leading to emptying of the intracellular calcium stores (Burnay et al., 1994). Thapsigargin has been found to increase steroidogenesis (Hajnoczky et al., 1991; Burnay et al., 1994) and to enhance Ang II- and potassium-stimulated aldosterone biosynthesis (Hajnoczky et al., 1991; Burnay et al., 1994).

Calcium influx is a critical component for sustained aldosterone production, but this signal alone does not appear to be sufficient to elicit maximal steroidogenesis. The other important signal for sustained aldosterone production is the PKC pathway. Indeed, activating PKC and calcium influx essentially reproduces the Ang II-stimulated aldosterone secretion (Kojima et al., 1984, 1985b), as well as the protein phosphorylation, pattern (Barrett et al., 1986). While generation of the DAG/PKC signal is initiated by activation of PLC, the phospholipase D (PLD) signaling system is also activated following Ang II treatment of glomerulosa cells (Bollag et al., 1990, 2002). PLD is activated in a sustained manner by Ang II, suggesting the possibility that PLD may mediate prolonged generation of DAG and stimulation of the PKC pathway. PLD hydrolyzes primarily phosphatidylcholine (Bollag et al., 2005), which has been proposed as a reservoir for sustained DAG production (Exton, 1998, 2000). In adrenal glomerulosa cells, Ang II stimulates the production of myristate-containing DAG (Bollag et al., 1991); since myristate is incorporated primarily into phosphatidylcholine (Bollag et al., 1991), this result suggests the possibility that both PLC and PLD mediate Ang II-induced DAG generation. Indeed, Hunyady et al. (1990) demonstrated a biphasic increase in DAG in glomerulosa cells, with an initial peak within minutes followed by a sustained plateau. Because in adrenal glomerulosa cells PLD is activated by Ang II in a sustained fashion, this result suggests the possibility that the early increase is the result of PLC-mediated PIP_2 hydrolysis and the second phase is due to PLD-mediated phosphatidylcholine breakdown, as has been suggested in other systems (Lee and Severson, 1994).

In addition to activating PKC, the DAG produced by Ang II-stimulated phospholipid hydrolysis also serves as a precursor for other signals regulating aldosterone secretion. Thus, arachidonic acid can be released by DAG lipase from arachidonic acid-containing DAG and is then metabolized by 12-lipoxygenase to generate 12-hydroxyeicosatetraenoic acid (12-HETE) (Natarajan et al., 1988, 1990). 12-HETE appears to play an important role in mediating Ang II-induced aldosterone secretion since blocking either arachidonic acid release with a DAG lipase inhibitor or its metabolism with a lipoxygenase inhibitor reduces Ang II-stimulated steroidogenesis (Nadler et al., 1987; Natarajan et al., 1988; Gu et al., 2003). In addition, production of this signal appears to be regulated by the PKC pathway (Shibata and Kojima, 1991), suggesting possible cross-talk between these two signals. This arachidonate metabolite also seems to be involved in Ang II-elicited increases cytosolic calcium levels (Stern et al., 1993) and this effect is presumably at least part of the mechanism by which Ang II-induced 12-HETE generation contributes to aldosterone production. Interestingly, 5-HETE has been reported to inhibit aldosterone secretion (Stern et al., 1989).

DAG metabolism to 12-HETE and its activation of PKC appear to play important roles in aldosterone production. However, there are several downstream pathways activated by PKC and their role in acute aldosterone production is not fully defined. In some systems, PKC activates the mitogen-activated protein kinase (MAPK) pathway culminating in extracellular signal-regulated kinase-1 and –2 (ERK-1/2) (Sugden, 2001). Ang II is known to activate ERK-1/2 in adrenal glomerulosa cell models and this activation occurs via PKC (Tian et al., 1998; Natarajan et al., 2002; Otis and Gallo-Payet, 2006). ERK-1/2 activation stimulates early growth response pathways, that should ultimately lead to proliferation (Tian et al., 1995) or hypertrophy (Otis et al., 2005). PKC can also activate protein kinase D (PKD), an enzyme that has some similarities to PKC in its ability to bind DAG and phorbol esters, but has been classified as a separate kinase family (Bollag et al., 2004). PKD is activated by Ang II in the H295R human adrenal cell model (Romero et al., 2006b) and in human aldosterone-producing adenomas (Chang et al., 2007). PKD is also known to activate ERK-1/2 and, in fact, increases the duration of activation of this MAPK in fibroblasts (Sinnett-Smith et al., 2004). Since ERK-1/2 is known to activate cholesterol ester hydrolase (CEH) (Cherradi et al., 2003), which releases cholesterol from cholesterol esters stored in lipid droplets for use in steroid synthesis, PKD may influence acute aldosterone production or, alternatively, modulate longer-term glomerulosa growth responses.

Some other signal transduction events initiated by Ang II in glomerulosa cell models include: p38 MAPK (Natarajan et al., 2002; Startchik et al., 2002), the Janus kinase (JAK)-signal transducer and activator of transcription (STAT) pathway (Li et al., 2003) and small GTPases, such as Rho, that modulate the cytoskeleton (Otis and Gallo-Payet, 2006; Gambaryan et al., 2006). The stress-activated kinase p38 is reported to alter glomerulosa cell calcium metabolism (Startchik et al., 2002) and 12-HETE production (Natarajan et al., 2002), while the JAK-STAT pathway increases StAR expression (Li et al., 2003). The small GTPase Rho appears to mediate aldosterone secretion, although this area is still under active investigation. Finally, Ang II in some systems is known to transactivate the epidermal growth factor receptor and exert some of its effects through this receptor tyrosine kinase (Shah et al., 2003). Although endothelin-1 has been shown to transactivate the epidermal growth factor receptor in adrenal glomerulosa cells (Shah et al., 2003), at present it remains unclear whether or not Ang II is also capable of doing so in these cells.

Activation of the above signaling pathways initiates acute aldosterone biosynthesis, which requires two distinct processes to occur. As discussed above, cleavage of cholesterol's side-chain is the rate-limiting reaction in steroidogenesis and this occurs inside the mitochondria. Therefore, cholesterol must first be mobilized to the mitochondria from storage sites in cytosolic lipid droplets. This first step is thought to be mediated by cytoskeletal rearrangements that allow juxtaposition of the mitochondria and lipid droplets (Hall, 1997). One protein that may be involved in this event may be the myristoylated alanine-rich C kinase substrate (MARCKS) protein. MARCKS binds actin in a manner which is regulated by both calcium/calmodulin binding and PKC phosphorylation and secretagogue-induced phosphorylation of this protein has been proposed to play a role in the cytoskeletal rearrangements associated with aldosterone production (Betancourt-Calle et al., 1999). Moreover, small GTPases like Rho can regulate the cytoskeleton and the role of these signaling molecules in glomerulosa cell function are currently under study (Otis and Gallo-Payet, 2006; Gambaryan et al., 2006). In addition to alignment of mitochondria with lipid droplets, cholesterol ester hydrolase must be activated to release free cholesterol from stored cholesterol esters (Cherradi et al., 2003). Ang II-induced CEH activation appears to be mediated by the MAPK, ERK-1/2 (Cherradi et al., 2003). Finally, the cholesterol used to maintain aldosterone production can also arise from *de novo* synthesis or from plasma levels of low density or high density lipoproteins (Capponi, 2004).

The second and rate-limiting step of steroidogenesis involves transfer of the mobilized cholesterol from the outer mitochondrial membrane to the inner mitochondrial membrane, where CYP11A1 is localized and initiates steroid production (Stocco and Clark, 1996). StAR possesses the characteristic properties of the labile acute regulator of steroidogenesis described in numerous studies from the laboratories of Orme-Johnson (Krueger and Orme-Johnson, 1983; Pon and Orme-Johnson, 1984; Pon et al., 1986) and Stocco (Stocco, 1992; Stocco and Ascoli, 1993). These criteria include:

1. the synthesis of StAR is increased upon hormone stimulation and is cycloheximide inhibited with a half-life of about 5 min
2. StAR is localized to the mitochondria where it increases the transfer of cholesterol from the outer to the inner membrane
3. StAR is specifically expressed in steroidogenic cells
4. increasing steroidogenic cell expression of StAR proteins using transgenes increases steroidogenesis
5. mutations in the StAR gene have been demonstrated in the human disease lipoid congenital adrenal hyperplasia, which is characterized by a complete inability to synthesize adrenal and gonadal steroid hormones
6. a mutant mouse generated to be deficient for the StAR gene demonstrated a phenotype similar to patients with lipoid congenital adrenal hyperplasia (Caron et al., 1997).

Thus, the StAR protein is thought to play a critical role in steroidogenesis, although other proteins are likely involved, as well (e.g. the steroidogenesis activator polypeptide, the sterol carrier protein-2 and the peripheral benzodiazepine receptor and its ligand, the diazepam binding inhibitor (Papadopoulos et al., 2007; Miller, 2007)).

The StAR gene encodes a 37 kDa protein with a mitochondrial targeting sequence, which does not appear to be required for its role in the transfer of cholesterol into the mitochondria (Arakane et al., 1998; Kallen et al., 1998; Wang et al., 2000). This 37 kDa protein is processed to several mitochondrial proteins that range in size (30–32 kDa) and isoelectric point (6.1 and 6.7) (Stocco and Clark, 1996). The more acidic products of StAR proteolysis represent modifications by phosphorylation (Stocco and Clark, 1996). Indeed, it is thought that phosphorylation is important to StAR's function, presumably yielding the active form(s) of the protein (Christenson and Strauss, 2001). Thus, steroidogenesis is decreased when cells are incubated with amino acid analogs that cannot be phosphorylated or when cells are transfected with mutant StAR protein with alanine substituted for serine 194 (hamster) or 195 (human) that prevents phosphorylation (Stocco and Clark, 1993; Arakane et al., 1997; Fleury et al., 2004). However, the amino acid residue mutated in this case represents a PKA rather than a PKC consensus phosphorylation site (Arakane et al., 1997).

On the other hand, StAR proteins bearing mutations of multiple phosphorylation sites for PKA, PKC and casein kinase 2 decreased steroidogenesis by almost 80% (Fleury et al., 2004). Because Ang II, potassium and adrenocorticotrophic hormone (ACTH) differentially alter StAR phosphorylation, it is likely that both PKC- and PKA-activating agonists are capable of regulating StAR activity via phosphorylation (Betancourt-Calle et al., 2001b). Nevertheless, a recent study suggests that PKC alone increases StAR protein expression without triggering its steroidogenic activity (Jo et al., 2005). In this case, the inclusion of low doses of cAMP analogs, which by themselves were without effect on StAR or steroidogenesis, resulted in an enhancement of StAR levels and steroid production (Jo et al., 2005). Another study examining the effect of sodium restriction *in vivo*, a manipulation that increases endogenous Ang II, concluded that subtle changes in StAR processing (i.e. proteolysis and phosphorylation), rather than large changes in StAR protein expression, mediate acute steroidogenesis (Lehoux et al., 1999). Indeed, all of the StAR forms (different sizes and isoelectric points) can be produced in cells transfected with a single StAR cDNA expression vector, indicating that these forms possess identical primary amino acid sequences but differ in some unknown post-translational modification (other than phosphorylation) (Clark et al., 1994).

The mechanism by which StAR induces cholesterol movement from the outer to the inner mitochondrial membrane is not clear. Several hypotheses have been proposed, none of which seem entirely satisfactory. The molten globule model suggests that near the mitochondrial membrane, a high concentration of protons converts a structured tertiary conformation of the StAR carboxy terminus into a semiordered molten globule, which has secondary but not tertiary structure (Miller, 2007). However, whether the pH is sufficiently low to trigger this transition or whether such a relatively unstructured domain could efficiently transport cholesterol is unclear. Other models also suffer from several shortcomings, such that the mechanism of StAR's action remains unclear (Miller, 2007). Nevertheless, StAR is clearly required for the translocation of cholesterol from the outer to the inner mitochondrial membrane and this rate-limiting step is controlled by signaling pathways (e.g. PKA, PKC, casein kinase 2) that are activated by aldosterone secretagogues.

3. CHRONIC EFFECTS OF ANG II

Chronically, Ang II increases adrenal aldosterone production through two major actions. First, Ang II increases expression of the enzymes needed to produce aldosterone, particularly

CYP11B2. Second, Ang II causes hypertrophy and hyperplasia of the adrenal glomerulosa, thus increasing the number of aldosterone-producing cells. *In vivo* studies have provided strong evidence that sodium restriction increases renin/Ang II levels causing an induction of glomerulosa CYP11B2 expression (Tremblay et al., 1992; Adler et al., 1993; Holland and Carr, 1993). Under these conditions adrenal expression of fasciculata CYP11B1 is not affected, suggesting that Ang II specifically increases the capacity to produce aldosterone but not glucocorticoid. However, unknown factors associated with sodium restriction, rather than Ang II, have not been excluded as mediators of this effect. The ability of low-sodium diets to increase aldosterone production and CYP11B2 expression can be largely (but not completely) inhibited by angiotensinogen inhibitors or antagonists of the AT_1 receptor (Kakiki et al., 1997a, b). These studies, along with the observation that Ang II receptors are higher in the glomerulosa than in the fasciculata layer, suggest an important role for Ang II in regulating glomerulosa expression of CYP11B2 and the long-term production of aldosterone (Breault et al., 1996).

In vitro studies have focused on defining the intracellular signals involved in Ang II-directed CYP11B2 expression. As noted above, adrenal AT_1 receptors couple to several signaling pathways. The most characterized of these is the activation of PLC, which increases intracellular calcium and DAG. These second messengers activate calmodulin (and CaM kinases) and PKC respectively. PKC activation with phorbol ester does not increase transcript levels of CYP11B2 but can act as an inhibitor for the expression of the fasciculata enzyme, CYP17 (McAllister and Hornsby, 1988; Bird et al., 1996) (see Figure 24.3). Therefore, PKC may play an important role in the zonation of the adrenal by blocking glomerulosa cell expression of CYP17. Several lines of evidence suggest that calcium signaling acts to increase glomerulosa cell expression of CYP11B2. For example, calcium signaling appears to be the major pathway used by extracellular potassium to increase CYP11B2 levels based on the ability of inhibitors of calcium influx, calmodulin or CaM kinase to block completely the ion's ability to induce CYP11B2 mRNA levels. However, Ang II-mediated increases in CYP11B2 mRNA are only partially sensitive to the inhibition of calmodulin or CaM kinase (Pezzi et al., 1997). Thus, Ang II may induce CYP11B2 expression through calcium and other signaling pathways (see Figure 24.3).

The increase in CYP11B2 expression appears to result from increased transcription of the gene (Clyne et al., 1997). Activation of transcription appears to rely on the activation of transcription factors that bind to a cyclic AMP response element (CRE) found in the proximal region of the CYP11B2 promoter (Bassett et al., 2000). In addition, both Ang II and potassium rapidly induce the expression of the nuclear hormone receptor NR4A3, which also binds the promoter and activates CYP11B2 transcription (Bassett et al., 2004a). This factor's expression is also increased in adrenal aldosterone-producing tumors and may play a role in tumor CYP11B2 levels (Lu et al., 2004).

Ang II regulation of long-term aldosterone relies on other signaling pathways. Ang II treatment also activates adrenal cell PKD and this activation is associated with increased CYP11B2 expression, suggesting that this pathway may be involved in CYP11B2 regulation. Further, Ang II-induced PKD activation is dependent upon PKC (Romero et al., 2006a; Chang et al., 2007). However, PKD is known to phosphorylate and stimulate transcriptional activity of the cAMP response element binding (CREB) protein transcription factor (Johannessen et al., 2007). As noted above, the promoter region of CYP11B2 is highly dependent on CREB response elements (Clyne et al., 1997), suggesting that this PKD-mediated CREB activation may be important in regulating chronic aldosterone synthetic capacity.

B. Potassium

1. POTASSIUM EFFECTS ON ALDOSTERONE PRODUCTION

The role of potassium in the regulation of aldosterone production is often underestimated. Infusion of potassium will cause an acute increase in aldosterone production and a high-potassium diet will increase aldosterone levels as well as the capacity of the adrenal to produce aldosterone. The development of angiotensinogen-deficient mice is perhaps the most convincing evidence for the role of potassium in the regulation of the adrenal glomerulosa. These animals are devoid of Ang II but maintain an adrenal zona glomerulosa and aldosterone production which appears regulated predominantly by circulating potassium. This potassium effect is appropriate, as aldosterone increases renal excretion of potassium. If the levels of plasma potassium are suddenly raised, as after a large meal of potassium-rich foods, aldosterone is secreted and acts on the kidney to excrete the excess potassium load. As described below, however, the effects of aldosterone to increase renal sodium retention and potassium excretion require an hour or more; thus, there must be more rapid ways to remove high levels of potassium entering the circulation, because a significant increase would be life threatening. The rapid transfer of potassium from extracellular fluid into cells is accomplished by a combination of insulin and epinephrine effects on potassium transport across cell membranes. Renal clearance through both acute and chronic mechanisms will also act to regulate potassium levels.

2. POTASSIUM-REGULATED INTRACELLULAR SIGNALING PATHWAYS

The mechanism by which potassium regulates aldosterone production relies on the extreme sensitivity of the glomerulosa cell membrane to small increases in potassium concentrations. Indeed, small increases in potassium stimulate calcium influx, via depolarization of the plasma membrane and activation of voltage-dependent calcium channels. As with Ang II stimulation, this influx is also thought to activate CaM kinase (Ganguly and Davis, 1994) and this influx is required for elevated potassium-induced aldosterone secretion, since inhibition of calcium influx abolishes the elevated potassium-stimulated secretory response (Barrett et al., 1989; Ganguly and Davis, 1994). DAG and PKC are generally not thought to play a role in potassium stimulation of aldosterone secretion. This idea was based on studies showing that, as opposed to Ang II treatment, potassium does not induce glomerulosa cell phosphoinositide hydrolysis, increases in DAG content or PKC translocation to the plasma membrane (Kojima et al., 1985b; Nadler et al., 1987; Hunyady et al., 1990; Ganguly et al., 1992). However, a previous report presented indirect evidence using a nonspecific inhibitor of PKC for a possible effect of increased potassium on the activity of this enzyme (Hajnoczky et al., 1992). In addition, it has been shown recently that small elevations in potassium trigger the phosphorylation of MARCKS (Betancourt-Calle et al., 2001a), an endogenous PKC substrate whose phosphorylation is thought to be a marker for PKC activation in intact cells (Nairn and Aderem, 1992). Furthermore, these elevations in potassium induce PLD activation (Betancourt-Calle et al., 2001b), suggesting a possible mechanism for the observed MARCKS phosphorylation. Thus, PKC activity may contribute to the calcium-signaling pathways activated by elevated potassium.

3. Chronic Effects of Potassium

It is accepted that small changes in circulating potassium can act directly on the adrenal glomerulosa to stimulate aldosterone biosynthesis, however, the direct influence of circulating potassium on glomerulosa production of aldosterone has not been studied as extensively as the influence of sodium balance. Indeed, rats on high-potassium diets exhibit increased circulating aldosterone levels and adrenal expression of CYP11B2. This stimulatory effect of potassium continues in the presence of angiotensin-converting enzyme (ACE) inhibition (LeHoux et al., 1994). In addition, studies using mice with targeted deletion of genes in the renin–angiotensin system have demonstrated that potassium can substitute for angiotensin and increase CYP11B2 expression and aldosterone production in the adrenal (Chen et al., 1997; Okubo and Ichikawa, 1997). These *in vivo* data support *in vitro* studies demonstrating that potassium stimulates expression of CYP11B2, both in primary cultures of rat glomerulosa and in human glomerulosa cell models (Bird et al., 1995b; Denner et al., 1996; Yagci and Muller, 1996; Yagci et al., 1996). Potassium signaling in glomerulosa cells involves a depolarization of the membrane leading to influx of calcium through T- and L-type channels. Consistent with this, pharmacologic elevation of cytosolic calcium increases expression of CYP11B2 mRNA in adrenal cell models (Clyne et al., 1997; Pezzi et al., 1997). Moreover, calcium channel blockers, such as nifedipine, block potassium induction of CYP11B2 (Denner et al., 1996; Yagci and Muller, 1996). Thus, a role for calcium in potassium induction of chronic aldosterone production and CYP11B2 expression appears likely.

Intracellular calcium signaling often occurs through the action of the calcium-binding protein, calmodulin. Calmodulin is a widely expressed protein that, in its calcium-bound form, activates a variety of enzymes and kinases. The tissue-specific expression of these kinases appears to be an important mechanism controlling the responses of different tissues to this signaling pathway. Of the various CaM kinases described to date, the multi-functional family of CaM kinases (types I, II and IV) are most likely to be involved in Ang II and potassium induction of aldosterone production. These kinases phosphorylate a wide variety of substrates and thus are distinct from the dedicated calcium-activated kinases such as myosin light chain kinase or phosphorylase kinase. Antagonists of calmodulin and CaM kinases completely inhibit potassium induction of CYP11B2 mRNA (Pezzi et al., 1997) as well as CYP11B2 transcription (Condon et al., 2002). In addition, the level of CaM kinase type I is elevated in the outer zones of the adrenal cortex. CaM kinase-mediated activation of transcription appears to rely on the activation of transcription factors that bind to a cyclic AMP response element (CRE) found in the proximal region of the CYP11B2 promoter (Bassett et al., 2000). Potassium-activated expression of the nuclear hormone receptor NURR1 (NR4A2) also increases CYP11B2 transcription (Bassett et al., 2004a). Within the adrenal cortex, the expression of NR4A2 is highest in the adrenal glomerulosa and its expression is elevated in aldosterone-producing adrenal tumors (Lu et al., 2004). These findings suggest that potassium stimulation of chronic aldosterone secretion relies on a relatively straightforward pathway involving an increased cytosolic calcium signal that activates CaM kinases, increasing expression of CYP11B2 (see Figure 24.3).

It should be noted that extracellular potassium levels also appear to regulate the response of glomerulosa cells to Ang II, at least *in vitro*. Thus, decreasing the potassium concentration to 2 mM blocks Ang II-elicited calcium influx and converts the sustained aldosterone secretory response to a transient one in bovine adrenal glomerulosa cells (Kojima et al., 1985b). This potassium modulation of aldosterone secretion likely serves as a fail-safe mechanism to prevent the Ang II-stimulated production of aldosterone, which could potentially decrease the serum potassium level to a lethal value. Similarly, in bovine adrenal glomerulosa cells *in vitro*, slight elevations in the extracellular potassium concentration can enhance the aldosterone secretory response to Ang II (Chen et al., 1999). Again, this suggests that potassium may serve to fine tune the production of aldosterone to allow precise regulation of sodium retention/excretion and blood volume/pressure, both acutely and chronically.

C. ACTH

1. Physiology of ACTH-Regulated Aldosterone

The effect of ACTH on aldosterone biosynthesis is powerful but short-lived. Infusion of ACTH will cause a rapid increase in serum aldosterone levels, which returns to basal levels within days. In addition, ACTH does not play an obligatory role in the production of aldosterone. Patients with familial glucocorticoid deficiency (lack of adrenal ACTH responsiveness) have normal aldosterone levels. In addition, humans or animals that have undergone hypophysectomy are able to respond to sodium restriction with a chronic increase in aldosterone production. On the other hand, ACTH excess may have an inhibitory effect on circulating aldosterone levels. Long-term infusion of ACTH (unless given in a pulsatile manner) or stresses that elevate ACTH will both lead to a drop in circulating aldosterone levels. However, these studies must be interpreted cautiously since other factors, mainly potassium, often are modified under these conditions. The mechanism for ACTH inhibition of aldosterone could be through direct or indirect effects on the glomerulosa cells. ACTH has been shown to have a direct action on isolated glomerulosa cells chronically to cause them to differentiate into fasciculata-like cells (Hornsby et al., 1974). In addition, elevated levels of fasciculata-derived weak mineralocorticoids, such as dexoycorticosterone, as well as elevated glucocorticoids, such as corticosterone and cortisol, can activate mineralocorticoid receptors in the kidney and feed back to inhibit the RAAS.

2. ACTH-Regulated Intracellular Signalling

The adrenal cortex, including the zona glomerulosa, expresses high levels of the melanocortin receptor 2 (MC2R). ACTH binds to MC2R and activates adenylate cyclase via the heterotrimeric G-protein, G_s. Adenylate cyclase produces cAMP which, in turn, stimulates the activity of cAMP-dependent protein kinase, also known as protein kinase A (PKA). PKA induces the expression and phosphorylation of StAR leading to increased cholesterol mobilization and mitochondrial pregnenolone production (Clark et al., 1995; Betancourt-Calle et al., 2001a). In addition, transcriptional activity of the CREB transcription factor can be regulated by PKA (Johannessen and Moens, 2007). Since CREB activity can stimulate expression of CYP11B2 and StAR, this PKA-mediated effect may be important in regulating aldosterone production. Finally, ACTH may also regulate aldosterone production through a PKA-independent pathway involving a guanine nucleotide exchange protein that is directly activated by cAMP (Gambaryan et al., 2006).

3. Chronic Action of ACTH on Aldosterone Production

It is apparent that there must be unique signaling pathways for the synthesis of aldosterone and cortisol, otherwise both hormones would simply be regulated by ACTH. ACTH plays an essential role in regulating CYP11B1 expression as well as genes encoding other enzymes involved in the synthesis of cortisol (Waterman and Bischof, 1996; Sewer and Waterman, 2003). However, the physiologic relevance of ACTH in aldosterone production and CYP11B2 transcription is less clear. Recent studies in mice with targeted ablation of pituitary corticotropes demonstrate that the adrenal can maintain CYP11B2, but not CYP11B1 in animals with very low circulating ACTH (Allen et al., 1995), a finding that confirms observations in humans with secondary adrenal insufficiency. Moreover, ACTH, while causing an acute increase in the production of aldosterone both *in vivo* and in isolated cells, acts chronically to decrease plasma aldosterone levels in humans as well as adrenal expression of CYP11B2 in animal models (Fuchs-Hammoser et al., 1980; Aguilera et al., 1981; Holland and Carr, 1993). *In vitro* studies have further shown that treatment with cAMP analogs preferentially increases CYP11B1 mRNA expression over that of CYP11B2 (Curnow et al., 1991; Denner et al., 1996). Whereas the chronic effects of ACTH on the induction of steroid hydroxylases mainly occur through increases in intracellular cAMP levels, the mechanisms through which ACTH inhibits expression of CYP11B2 are not clear. Cyclic AMP signaling has a negative effect on the expression of Ang II receptors in adrenocortical cells (Andoka et al., 1984; Yoshida et al., 1991; Bird et al., 1995a), thereby desensitizing adrenal cells to Ang II. As discussed above, ACTH may also decrease aldosterone synthesis through the induction of CYP11B1 and CYP17, enzymes that will effectively remove precursors from the pathway leading to mineralocorticoids and utilize them for cortisol production (McAllister and Hornsby, 1988; Bird et al., 1996).

The question remains as to how the glomerulosa maintains CYP11B2 expression in the presence of normal circulating levels of ACTH. While receptors for ACTH are present in both the glomerulosa and fasciculata layers (Xia and Wikberg, 1996), available data suggest that the glomerulosa has at least two mechanisms to control intracellular levels of cAMP generated by ACTH. First, in glomerulosa cells but not fasciculata cells, Ang II (through coupling to Gi) will inhibit ACTH-stimulated cAMP production (Hausdorff et al., 1987; Begeot et al., 1988). Second, a recent report has shown that the adrenal glomerulosa (but not the fasciculata) expresses adenylyl cyclases (types 5 and 6) that are inhibited by intracellular calcium, which is involved in signaling by both Ang II and potassium (Shen et al., 1997).

D. Other Aldosterone Secretagogues

Several other agents have also been shown to increase aldosterone secretion. One such agonist is parathyroid hormone (PTH), which has been shown to stimulate aldosterone secretion and to enhance the ability of Ang II to elicit steroidogenesis in adrenal glomerulosa cells (Isales et al., 1991; Hanley et al., 1993). PTH has only a small effect on cytosolic calcium itself but treatment with Ang II or elevated potassium potentiates the subsequent response to PTH. In addition, PTH increases cAMP levels. In rat glomerulosa cells, PTH enhances the ability of extracellular calcium to induce aldosterone production and these authors have suggested a ionophore-like effect of PTH (Olgaard et al., 1994).

Endothelin-1 is also reported to stimulate aldosterone secretion, via binding to both endothelin type A (ETA) and ETB receptors identified in adrenal glomerulosa cells. These two receptors seem to function through slightly different mechanisms, with ETA activating PLC, phospholipid hydrolysis and the calcium and DAG/PKC pathways (Andreis et al., 2001, 2002), presumably through the heterotrimeric G-protein, G_q (Shah et al., 2006). ETB activates PLC (Andreis et al., 2001, 2002), cyclo-oxygenase (Andreis et al., 2002) and a G_i-mediated cascade resulting in activation of ERK-1/2 (Shah et al., 2006). However, studies using glomerulosa cells suggest that endothelin-1 stimulates aldosterone secretion less efficiently, although in a manner similar to Ang II.

There is considerable evidence that glomerulosa cells respond to serotonin (5-hydroxytryptamine) and that these cells express serotonin receptors (Delarue et al., 2001). Serotonin has several receptors but it appears to be the 5-HT type 4 and 7 receptors that play the main role in regulating adrenal glomerulosa aldosterone production (Lenglet et al., 2002). Serotonin stimulates cAMP production and raises cytoplasmic calcium levels and these are believed to be the primary signaling pathways involved in activation of aldosterone biosynthesis (Davies et al., 1991; Larcher et al., 1992; Lenglet et al., 2002). The physiologic role of serotonin in the regulation of aldosterone production is still not clear. However, infusion of serotonin analogs increases aldosterone levels in human and animal models (Maestri et al., 1988; Davies et al., 1992). The observation that adrenal aldosterone-producing tumors have elevated expression of serotonin receptors and the ability of isolated tumor cells to respond to serotonin with increased aldosterone production (Lefebvre et al., 2000) has increased interest in this hormone's role in adrenal physiology and pathology.

E. Inhibitors of Aldosterone Biosynthesis

A number of compounds have been described that can inhibit aldosterone secretion *in vitro* including: adrenomedullin, dopamine and somatostatin, although the most physiologically relevant aldosterone inhibitor appears to be atrial natriuretic peptide (ANP). The existence of a 'natriuretic' factor had been proposed for some time; however, it was not until 1981, when de Bold and colleagues (de Bold et al., 1981) demonstrated that injection of an atrial extract into rats caused a natriuresis, that ANP was ultimately purified and characterized. Since then a family of related peptides has been identified. Brain natriuretic peptide is a 132 amino acid prepro-peptide initially identified in porcine brain but now known to be present in the heart (atria and ventricles). The biologically active peptide is the carboxy-terminal fragment and is 32 amino acids long. BNP is a vasorelaxant and has natriuretic properties. Type C natriuretic peptide (CNP) is initially synthesized as a 103 amino acid pro-peptide and is most highly expressed in endothelial cells and brain. The most biologically active fragment appears to be CNP22. CNP appears to have an important role in skeletal development. ANP is synthesized as a 151 amino acid prepro-hormone and processed to a biologically active 27 amino acid active fragment. It is expressed predominantly in cardiac atria and has both potent natriuretic and vasorelaxant properties. In addition, ANP has been shown to be a potent inhibitor of aldosterone secretion (see Chapter 20 for a detailed discussion).

Natriuretic peptide action is mediated through one of three receptors: natriuretic peptide receptor-A (NPR-A), NPR-B and NPR-C. Both NPR-A and -B are guanylate cyclase receptors. NPR-A binds ANP and BNP, NPR-B binds CNP. NPR-C binds all three natriuretic peptides and is thus considered to serve mainly as a 'clearance' receptor.

ANP is synthesized and released from cardiac atria upon stretching during states of volume overload. Thus, the ensuing natriuresis is a compensatory mechanism for the increased intravascular volume. ANP also appears to be the most potent inhibitor of aldosterone secretion. However, even though ANP works through NPR-A, a guanylate cyclase receptor, changes in cGMP do not appear to account fully for the inhibitory action of ANP on aldosterone secretion (Barrett and Isales, 1988). ANP is able to inhibit aldosterone secretion stimulated by any of the usual agonists, ACTH, potassium and Ang II, suggesting that ANP acts on an early step in aldosterone synthesis. Several potential mechanisms have been proposed including an inhibition of T-type calcium channels (McCarthy et al., 1990). ANP has also been shown to modulate potassium- and Ang II-stimulated phosphorylation of the PKC substrate MARCKS, as well as StAR, suggesting that ANP can modulate cholesterol transport (Calle et al., 2001).

Proadrenomedullin and adrenomedullin are peptides made in the adrenal cortex that can affect vascular tone. *In vitro*, both adrenomedullin and proadrenomedullin, acting through specific receptors on the adrenal cortex, have been shown to inhibit potassium- and Ang II-stimulated aldosterone secretion (Andreis et al., 1998). Interestingly, these peptides have also been shown to inhibit aldosterone secretion from adrenal adenomas removed from patients with primary hyperaldosteronism, suggesting that they may have a local paracrine role in regulating aldosterone secretion (Andreis et al., 1998).

The adrenal cortex expresses two subtypes of the dopamine receptor: D2 and D4. Dopamine has been known for some time to have an effect on aldosterone secretion since metoclopramide (a dopamine antagonist) is known to increase aldosterone secretion (Wu et al., 2002). Recently, Chang et al. (Chang et al., 2007) have shown that D2 receptors inhibit PKD activity and thus inhibit aldosterone secretion. They have also shown that the D2 receptor is downregulated in adrenal adenomas suggesting that this may be involved in the pathophysiology of APA aldosterone hypersecretion.

Somatostatin is a peptide known to inhibit the secretion of a wide variety of hormones. There are five somatostatin receptor subtypes (stt1–5) and all are expressed in the adrenal cortex and similarly in adrenal aldosterone-producing adenomas (Ueberberg et al., 2005). Somatostatin infusion in rats results in an inhibition of aldosterone secretion (Rebuffat et al., 1994), and it is possible that somatostatin is acting as a paracrine hormone released from the medulla to regulate aldosterone secretion; however, there are few data on the physiological role of this hormone (Aguilera and Catt, 1983).

Finally, it is important to point out three facts that should lead to a cautious interpretation of the above cited data.

1. *in vitro* results may not reflect the *in vivo* relationships
2. even *in vivo* studies may be difficult to interpret given the profound effect of environmental factors (e.g. sodium intake)
3. chronic *in vivo* treatment with Ang II leads to a variety of secondary modifications in addition to an increase in aldosterone that can confound the proper interpretation of the results.

IV. DISEASES OF ALDOSTERONE PRODUCTION

A. Genetic Defects Causing Aldosterone Deficiency

1. CYP21

Patients with congenital adrenal hyperplasia (CAH) have inherited abnormalities in the enzymes involved in cortisol biosynthesis in the adrenal gland. A key step in the synthesis of aldosterone and cortisol is the hydroxylation by 21-hydroxylase (or CYP21) of progesterone to deoxycorticosterone or 17-hydroxyprogesterone to 11-deoxycortisol, respectively. CYP21 deficiency thus results in either female newborns with ambiguous genitalia (due to virilization as the result of increased adrenal androgen synthesis) or normal appearing male newborns that present with a life-threatening 'Addisonian' type crisis ('salt wasters'). CYP21 deficiency is the most common cause of congenital adrenal hyperplasia (90–95% of CAH cases), although these patients can present as either classic salt wasters (presenting at birth and accounting for 75% of cases), classic simple virilizers (presenting between 2 and 4 years of age) or non-classical (presenting as adults) (White and Speiser, 2000). This condition is inherited in an autosomal recessive manner and the incidence is about 1:15 000 births, although the incidence varies greatly according to ethnic group.

a. Diagnosis. In the classical form of CAH, the diagnosis of the female newborn is relatively straightforward since the evaluation will be initiated by the findings of ambiguous genitalia. In male patients, the diagnosis requires a higher index of suspicion since the evaluation will not usually be initiated until 1–4 weeks after birth, because of symptoms of vomiting, failure to thrive, hyponatremia, hyperkalemia and hypotension or shock (White and Speiser, 2000). If CAH is suspected, prenatal diagnosis is possible and *in utero* treatment results in decreased virilization in the female newborn (Nimkarn and New, 2006).

In the classical but simple virilizing form of CYP21 deficiency, the diagnosis is usually made later in the child's life. The female child may present with ambiguous genitalia but no salt-wasting crisis. Rather, these patients may be taken to the pediatrician because of initial rapid growth followed by early closure of the epiphyses and short height, if not treated (White and Speiser, 2000). In the non-classical form of CYP21 deficiency, the male patient is usually asymptomatic, whereas female patients may present with increased acne, abnormal menstrual periods, polycystic ovaries or hirsutism (White, 2004; New, 2006). Because of the block of adrenal steroids at the 21-hydroxylation step, steroid metabolites prior to this step accumulate and are diagnostic of this condition. The most frequently measured metabolite for this condition is 17-hydroxyprogesterone, although androstenedione, progesterone and androgens are all significantly elevated (New, 2003). Typical 17-hydroxyprogesterone values for patients with the classic salt-wasting form of CAH are in excess of 20 000 ng/dl; between 10–20 000 ng/dl in patients with the simple virilizing form and greater than 1000 ng/dl after ACTH stimulation in patients with the non-classical form (White and Speiser, 2000). As mentioned previously, StAR mutations cause lipoid congenital adrenal hyperplasia and result in an inhibition of steroid hormone production, including aldosterone.

b. Treatment. CAH is treated by glucocorticoid replacement. Multiple steroid preparations are available, with the natural form, hydrocortisone (10–20 mg/m^2/day in divided doses) the steroid of choice because this steroid has less of a suppressive effect on growth in children. In patients with the salt-wasting form of CAH, mineralocorticoids should also be replaced (fludrocortisone 0.1–0.2 mg/day). In patients with ambiguous genitalia, sex assignment is generally made on the basis of genetics and corrective surgery performed early in the child's life (White and Speiser, 2000).

2. CYP11B2

An even rarer condition is that associated with isolated aldosterone deficiency with normal cortisol and sex hormone levels (White, 2004). These patients have inactivating mutations in the gene for CYP11B2. Aldosterone deficiency results in elevated serum potassium levels with a drop in serum sodium concentration in infants, although these values tend to normalize with age. Because this genetic abnormality affects only the final step of the aldosterone biosynthetic pathway, other metabolites with mineralocorticoid properties accumulate and tend to account for a less severe clinical presentation than that seen with CYP21 deficiency, for example. Treatment consists of increased dietary sodium and, if necessary, mineralocorticoid replacement with fludrocortisone.

B. Defects Causing Aldosterone Excess

1. Primary Hyperaldosteronism

Primary hyperaldosteronism (PA) is probably the most common endocrine cause of secondary hypertension. Because of its key role in maintaining the body's fluid volume/blood pressure, it is perhaps not surprising that the renin–angiotensin–aldosterone (RAA) axis contains some of the more potent vasoconstrictive (Ang II) and natriophilic (aldosterone) hormones in the body. Recall that blood pressure equals cardiac output (modulated by aldosterone) multiplied by the total peripheral resistance (which is modulated by Ang II). Thus, there are multiple ways in which abnormalities in the RAA axis, i.e. primary or secondary hyperaldosteronism, can result in arterial hypertension. This fact is also reflected by the clinical effectiveness of agents that block components of this axis [angiotensin-converting enzyme (ACE) inhibitors or angiotensin receptor blockers (ARBs) or aldosterone antagonists (aldactone or elprenolone)] in treating hypertension.

Jerome W. Conn, MD, was the first to describe aldosterone-producing adenomas (APA) as a cause of hypertension (Conn and Louis, 1955) in a patient whose condition improved after surgical removal of the adenoma. Dr Conn himself felt that adrenal adenomas overproducing aldosterone could be pathogenic in up to 20% of patients with arterial hypertension (see Chapter 23 for additional details.) Because of the potential clinical benefits of diagnosing and treating this surgically curable form of hypertension, many of Dr Conn's fellow physicians undertook the search for adrenal adenomas in hypertensive patients and were disappointed when the prevalence of 'Conn's disease' appeared to be much lower than expected. Thus, the prevailing medical opinion changed and primary hyperaldosteronism was felt to be a very rare disease, accounting for less than 1% of hypertensive patients (Bech and Hilden, 1975; Sinclair et al., 1987). Among these patients with primary hyperaldosteronism, the largest subgroup was felt to be those with an aldosterone-producing adrenal adenoma (Conn's disease or APA) accounting for 65–75% of patients, followed by patients with a bilateral symmetrical enlargement of the adrenal glands (idiopathic hyperaldosteronism or IHA) accounting for 25–35% of patients (Gomez-Sanchez et al., 1995).

More recently, the estimates of the prevalence of PA have increased dramatically (Schirpenbach and Reincke, 2006). In addition, the breakdown for causes of PA continues to evolve (Table 24.1). In a recent study where the prevalence of primary hyperaldosteronism was calculated to be about 11.2% in a hypertensive population, the prevalence of IHA was found to be greater than that of APA, 57% versus 43% of the total (Rossi et al., 2006). Since IHA is part of the spectrum of patients with low-renin hypertension, the increasing prevalence of IHA may reflect our improved diagnostic tools for classifying patients with essential hypertension.

a. Pathophysiology. Since elevations in blood pressure must be related to either changes in cardiac output or peripheral resistance, the expectation would be that patients with PA would be hypertensive secondary to an elevated blood volume. However, this does not seem to be the case; in fact up to 25% of patients with PA actually have a low blood volume compared to patients with essential hypertension (Bravo et al., 1983). The fact that patients with PA do not develop marked fluid retention and edema is, at least in part, related to the fact that these patients undergo an 'escape phenomenon'. In the setting of mineralocorticoid-induced hypertension, initial fluid retention is followed about 2

TABLE 24.1 Causes of hyperaldosteronism

Condition	Prevalence (%)
Idiopathic hyperaldosteronism (IHA)	55–65
Aldosterone producing adenoma (APA)	25–35
Bilateral cortical nodular hyperplasia	1
Gluocorticoid remediable hyperaldosteronism (GRA)	<1
Adrenal carcinoma	<1

weeks later by a significant natriuresis, resulting in reduced plasma volume (Fields et al., 1990). This natriuresis is due in part to elevated serum levels of atrial natriuretic peptide (ANP).

In fact, patients with PA have been shown to have elevations in both ANP and brain natriuretic peptide (BNP) (Kato et al., 2005). At pharmacologic concentrations, BNP is a vasodilator and reduces preload; in contrast, ANP reduces afterload and inhibits aldosterone secretion (Isales et al., 1989; Houben et al., 2005). Both BNP and ANP levels decrease in patients with APA after surgery and removal of the adenoma. Even though ANP receptors are known to be downregulated in PA (Shionoiri et al., 1988), an infusion of ANP will still result in increased urinary sodium excretion, although aldosterone levels do not drop (Pedrinelli et al., 1989). In most tissues, cGMP appears to be the main second messenger involved in ANP action. Studies using isolated adrenal glomerulosa cells suggest that ANP signaling pathways in the adrenal are intrinsically different from those in other tissues and may account for this dissociation in physiological responses to ANP (Barrett and Isales, 1988; Isales et al., 1989; Pedrinelli et al., 1989).

It is likely that aldosterone also has a direct effect on total peripheral resistance by modulating vasoconstriction (Yamakado et al., 1988). The mechanism for this effect is not clear but may relate to non-genomic aldosterone effects on vascular smooth muscle (Arima et al., 2003).

b. Diagnosis. Considerations as to the true prevalence of primary hyperaldosteronism and the best way to screen for or diagnose this condition are perhaps the most controversial. Differences in the reported prevalence of PA clearly reflect differences in the study population. For instance, the prevalence of PA appears to increase according to the severity of the hypertension. A study by Mosso et al. (Mosso et al., 2003) demonstrated that the prevalence of primary hyperaldosteronism increased from 1.99% in those patients with stage I hypertension (Joint National Committee on Prevention, Detection, Evaluation, and Treatment of High Blood Pressure, Stage 1 = 140/90–159/99 mmHg) to 8.02% in patients with stage II hypertension (160/100–179/109 mmHg) and to 13.2% in patients with stage III hypertension ($\geq$180/110 mmHg). In those patients requiring more than three medications for blood pressure control (resistant hypertension), the prevalence of PA is approximately 20% (Pimenta and Calhoun, 2006). Other findings which increase the likelihood of PA include spontaneous hypokalemia and the onset of hypertension at a young age (<40 years of age but in particular in those <20 years of age) (Young, 2007).

Another area of controversy relates to the best screening test for PA, as well as which subsequent confirmatory tests should be performed. Most researchers agree that a simultaneous morning measurement of plasma aldosterone (ng/dl) and plasma renin activity (ng/ml/h) while the patient is sitting is an appropriate screening test (Aldo/PRA). There are, however, several caveats associated with this test:

1. Hypokalemia should be corrected prior to testing.
2. The sensitivity and specificity of this screening test varies greatly depending on what cutoff values are used. Published studies have employed widely varying values but most use a ratio greater than 20 ng/dl of aldosterone per ng/ml/h of renin activity (Young, 2007). Using a cutoff value of 30 ng/dl per ng/ml/h, one study estimated that this screening test had a sensitivity of 90% and a specificity of 91% for detection of adrenal adenomas (Weinberger and Fineberg, 1993; Young, 2007).
3. There is an absolute requirement that the patient have stopped aldosterone antagonist medication (spironolactone, eplerenone, and probably amiloride) for at least 6 weeks prior to testing. In addition, ideally the patient should not take most other antihypertensive medications, such as angiotensin-converting enzyme (ACE) inhibitors, angiotensin receptor blockers (ARB), beta-blockers, calcium channel blockers and diuretics, for at least 2 weeks prior to the test. However, if the patient exhibits a suppressed PRA measurement despite being on an ACE or ARB this result is highly suggestive of PA. On the other hand, alpha-blockers and verapamil do not interfere with this measurement.
4. Because of inter-laboratory variability in determining PRA, it has been suggested that this measurement be replaced by an evaluation of plasma renin concentration. Initial results are encouraging but it is still too early to ascertain how reliable this measurement will be in replacing PRA measurements (Nicod et al., 2004).
5. The Aldo/PRA ratio is greatly influenced by the value used for PRA (e.g. some groups will report PRA as <0.1 ng/ml/h), thus it has been suggested that a minimum value of 0.5–0.7 ng/ml/h be used for PRA instead and that the measured aldosterone value be at least as high as 12–15 ng/dl.

The best confirmatory test to follow-up an elevated Aldo/PRA measurement is also controversial. The discriminatory accuracy of numerous tests has been examined. These include:

1. a 24-hour urinary aldosterone collection with or without oral sodium loading
2. an aldosterone suppression test with fludrocortisone
3. aldosterone suppression with either captopril (ACE inhibitor) or losartan (ARB)
4. aldosterone response to saline infusion.

A fludrocortisone (florinef) suppression test has been considered by many to be the most reliable test; however, it can be costly and labor intensive. This test requires administration of the fludrocortisone 0.1 mg every 6 h to a hospitalized patient together with sodium and potassium supplements. A study by Mulatero et al. (Mulatero et al., 2006) compared the sensitivity and specificity of the fludrocortisone loading test to a much shorter test involving infusion of 2 l of normal saline over 4 h to patients with suspected PA. These authors found that the saline loading test was a cheaper and reasonable alternative to the fludrocortisone loading test with a sensitivity of 90% and a specificity of 84%. Thus, even though the fludrocortisone loading test was a 'better' test for confirming the diagnosis of PA, the saline infusion test was a more practical confirmatory test for most medical centers. A subsequent study by Rossi et al. (Rossi et al., 2007) confirmed that the saline infusion test exhibited a high negative predictive value for excluding a diagnosis of PA, although it was not useful in distinguishing between APA and IHA. Thus, results from available studies support the use of the Aldo/PRA ratio as a screening tool followed by the saline infusion test as a confirmatory test.

Once the biochemical diagnosis of PA is confirmed, it then becomes important to differentiate whether the patient has IHA or APA. This is an important question since surgery is curative in those patients with APA but not in those with IHA. Imaging studies, such as computed tomography (CT) scanning or magnetic resonance imaging (MRI), are of limited value in distinguishing between APA and IHA. Most aldosterone-producing adenomas are small (usually <2 cm) and thus dedicated adrenal imaging must be undertaken. CT scans and MRI appear to have similar sensitivity and specificity. Because of the high prevalence of 'incidental adrenal masses' in the general population, diagnosing a patient with APA versus IHA with a CT scan is difficult. It is estimated that the prevalence of adrenal nodules >1 cm in patients undergoing abdominal CT scanning for other indications ranges between 0.35 and 4.36% (Kloos et al., 1995). This result is in contrast to autopsy series where the prevalence of incidentally discovered adrenal masses ranges between 1.0 and 32% (Kloos et al., 1995). Although it varies depending on the series, the likelihood of hormonal hypersecretion in an incidental adrenal mass is low (probably less than 10%) and among these secretory adrenal nodules the likelihood of it being an aldosterone-producing adenoma is even lower (less than 10% of the secretory adenomas) (Kloos et al., 1995; Zarco-Gonzalez and Herrera, 2004).

Further to complicate matters, Young et al. (Young et al., 2004) demonstrated that following lateralization of aldosterone secretion by adrenal vein sampling (AVS), almost 22% of these patients had no CT-detectable tumor in the aldosterone-producing adrenal. In addition, nearly 8% of those patients in whom the CT scan demonstrated an adrenal nodule on one side, in fact had lateralization to the opposite side. Thus, almost 50% of patients would have been misclassified for surgery based on the CT scan results.

AVS remains the most reliable method for accurately distinguishing between APA and IHA. The main concern about the use of the AVS relates to the technical difficulties involved in catheterizing the right adrenal vein, and thus the use of AVS is frequently restricted to those larger medical centers where this expertise is available. Aldosterone values obtained by AVS are divided by the respective cortisol value (A/C ratio) and, for the test to be diagnostic, generally a greater than fourfold difference between the adrenal samples is observed. Many centers also administer synthetic ACTH at the time of catheterization to stimulate cortisol secretion and document adequate sampling of the adrenal veins (Young, 2007).

c. Treatment. Treatment options for PA depend on whether the diagnostic workup is consistent with APA or IHA.

IHA
Treatment of IHA is limited to medical therapy since surgery does not result in a cure in these patients, as they will remain hypertensive. Currently, there are three medical alternatives: the aldosterone antagonists spironolactone (Aldactone), the newer eplerenone (Inspra) or the diuretic amiloride (Midamor). Spironolactone is the drug most studied for treatment of PA, although it is a non-selective aldosterone antagonist. It is available in 25, 50 or 100 mg strengths and the dosage should be titrated up to control hypokalemia and blood pressure. This medication can have some serious side effects including erectile dysfunction and gynecomastia in men. Eplerenone is a newer selective aldosterone antagonist which has been used in patients with PA (Young, 2003). Inspra is available in 25 or 50 mg strengths and its dose can also be titrated as needed. Amiloride is a diuretic rather than an aldosterone antagonist but is effective in the treatment of the hypokalemia associated with PA. It is available in 5 mg strengths with doses in excess of 20 mg sometimes required for prevention of hypokalemia in patients with PA.

APA
Unless there is a medical contraindication, surgery is the treatment of choice for patients with APA since all patients have corrections of their hypokalemia and improvement or correction of their arterial hypertension (in up to 60–70% of patients). Outcomes are less favorable in older patients (Obara et al., 1992). The preferred surgical approach is laparoscopic unilateral adrenalectomy.

2. SECONDARY HYPERALDOSTERONISM (SEE CHAPTER 23 FOR ADDITIONAL DETAILS)

Because of aldosterone's role in maintaining normal fluid volume, any condition in which there is fluid loss will result in secondary elevations in aldosterone. However, in these conditions of fluid loss, there is no associated hypertension. There are, however, a number of conditions, primarily involving perceived decreased fluid volume in which elevations in aldosterone result in increased blood pressure. These conditions usually primarily involve the kidney and include renal artery stenosis and renal ischemia. It is also possible to have a renin-secreting tumor and secondary elevations in blood pressure.

Renin-secreting tumors are rare and generally involve tumors of the juxtaglomerular apparatus of the kidney, although extra-renal rennin-secreting tumors have been reported (Pursell and Quinlan, 2003). Patients with renin-secreting tumors tend to be young (mean age 22.3) and have marked hypertension and hypokalemia and marked elevations in renin and aldosterone levels (Corvol et al., 1994). The best way of diagnosing these tumors is by CT scan and, if detected, tumors are treated by surgical resection of the tumor, which corrects the hypertension and hypokalemia (Haab et al., 1995). MRI has also been reported to be useful in the detection of these tumors (Wang et al., 1998).

Renovascular disease is the most common secondary cause of hypertension. A reduction in renal artery perfusion will result in secondary elevations in renin, Ang II and aldosterone. The term ischemic nephropathy has been used to classify patients with reductions in renal function associated with renovascular occlusive disease (Garcia-Donaire and Alcazar, 2005). The prevalence of 'ischemic nephropathy' increases with age with an incidence of 18% in younger (64–75 years of age) versus 42% in elderly (>75) hypertensive patients (Garcia-Donaire and Alcazar, 2005). Among older patients, atherosclerotic renal artery stenosis is the most common cause (90% of cases). There is some controversy as to whether these patients should be managed medically (with antihypertensive medication including ACE inhibitors and ARBs) or surgically (generally by renal artery angioplasty with stent placement) (Balk et al., 2006; Levin et al., 2007), although blood pressure control seems to be better after angioplasty. In younger patients with no history of hypertension, there may be a genetic variant of renal artery stenosis, fibromuscular dysplasia, that also benefits from angioplasty (Prisant et al., 2006).

C. Effects of Relative Aldosterone Excess (see Chapter 23 for Additional Details)

A landmark study by Pitt et al. (Pitt et al., 1999), demonstrated that low doses of the aldosterone antagonist spironolactone (25 mg daily) added to traditional hypertensive therapies was able to decrease mortality by 30% in patients with severe congestive heart failure (CHF) (ejection fractions <35%). In addition, spironolactone decreased hospitalizations for heart failure by 35%, despite the fact that these patients were already on an ACE inhibitor, a diuretic and usually digoxin. The findings from this study are consistent with those of other studies, which suggest that aldosterone itself has direct deleterious effects on the cardiovascular system. Thus, Pitt et al. (Pitt et al., 1999) suggest that aldosterone antagonists have a direct cardioprotective effect since elevated levels of aldosterone have been shown to promote vascular fibrosis and inhibit catecholamine uptake by the myocardium (Pitt, 2004). In fact, a related study by Zannad et al. (Zannad et al., 2000) demonstrated that spironolactone lowered mortality only in those patient with congestive heart failure in whom the markers of collagen turnover (i.e. cardiovascular fibrosis) improved.

Consistent with the findings that spironolactone had beneficial effects on cardiovascular mortality in patients with CHF, it was subsequently found that the more selective aldosterone antagonist eplerenone (50 mg daily) was also effective in lowering mortality among those patients who had a myocardial infarction with subsequent left ventricular failure. These investigators demonstrated that eplerenone reduced coronary artery inflammation and fibrosis, and improved endothelial dysfunction and vascular remodeling (Pitt et al., 2003). This beneficial effect on mortality by such low doses of mineralocorticoid receptor antagonists in the presence of near normal levels of aldosterone is unexpected. However, it is possible that ischemia and tissue damage, perhaps through the generation of reactive oxygen species, activate the mineralocorticoid receptor and thus promote fibrosis (Funder, 2005a, b; Rossi et al., 2005).

V. SUMMARY

Aldosterone is a key regulator of blood volume and pressure such that abnormalities in its production contribute to human diseases including hypertension, congestive heart failure and salt-wasting forms of congenital adrenal hyperplasia. The secretion of this steroid hormone is regulated primarily by angiotensin II and serum potassium levels, which increase aldosterone production both acutely, through stimulation of multiple signaling events, and chronically by regulating the expression of steroidogenic enzymes, in particular, aldosterone synthase (CYP11B2). A variety of cell culture model systems have been used to help define the mechanisms that regulate aldosterone biosynthesis. These models arise from different species and have been studied under diverse *in vitro* conditions; each of which may contribute to discrepancies in research findings discussed in this chapter. The importance of aldosterone in maintaining electrolyte balance and as a common cause of hypertension support the need for continued research to help define the detailed mechanisms controlling both normal and pathologic synthesis of aldosterone.

References

Adler, G. K., Chen, R., Menachery, A. I., Braley, L. M., and Williams, G. H. (1993). Sodium restriction increases aldosterone biosynthesis by increasing late pathway, but not early pathway, messenger ribonucleic acid levels and enzyme activity in normotensive rats. *Endocrinology* **133:** 2235–2240.

Aguilera, G., and Catt, K. J. (1983). Regulation of aldosterone secretion during altered sodium intake. *J Steroid Biochem* **19:** 525–530.

Aguilera, G., Fujita, K., and Catt, K. J. (1981). Mechanisms of inhibition of aldosterone secretion by adrenocorticotropin. *Endocrinology* **108:** 522–528.

Aksoy, I. A., Sochorova, V., and Weinshilboum, R. M. (1993). Human liver dehydroepiandrosterone sulfotransferase: nature and extent of individual variation. *Clin Pharmacol Ther* **54**: 498–506.

Allen, R. G., Carey, C., Parker, J. D., Mortrud, M. T., Mellon, S. H., and Low, M. J. (1995). Targeted ablation of pituitary pre-proopiomelanocortin cells by herpes simplex virus-1 thymidine kinase differentially regulates mRNAs encoding the adrenocorticotropin receptor and aldosterone synthase in the mouse adrenal gland. *Mol Endocrinol* **9**: 1005–1016.

Andoka, G., Chauvin, M. A., Marie, J., Saez, J. M., and Morera, A. M. (1984). Adrenocorticotropin regulates angiotensin II receptors in bovine adrenal cells in vitro. *Biochem Biophys Res Commun* **121**: 441–447.

Andreis, P. G., Tortorella, C., Mazzocchi, G., and Nussdorfer, G. G. (1998). Proadrenomedullin N-terminal 20 peptide inhibits aldosterone secretion of human adrenocortical and Conn's adenoma cells: comparison with adrenomedullin effect. *J Clin Endocrinol Metab* **83**: 253–257.

Andreis, P. G., Tortorella, C., Malendowicz, L. K., and Nussdorfer, G. G. (2001). Endothelins stimulate aldosterone secretion from dispersed rat adrenal zona glomerulosa cells, acting through ETB receptors coupled with the phospholipase C-dependent signaling pathway. *Peptides* **22**: 117–122.

Andreis, P. G., Neri, G., Tortorella, C., Aragona, F., Rossi, G. P., and Nussdorfer, G. G. (2002). Mechanisms transducing the aldosterone secretagogue signal of endothelins in the human adrenal cortex. *Peptides* **23**: 561–566.

Aptel, H. B., Burnay, M. M., Rossier, M. F., and Capponi, A. M. (1999). The role of tyrosine kinases in capacitative calcium influx-mediated aldosterone production in bovine adrenal zona glomerulosa cells. *J Endocrinol* **163**: 131–138.

Arakane, F., King, S. R., Du, Y. et al. (1997). Phosphorylation of steroidogenic acute regulatory protein (StAR) modulates its steroidogenic activity. *J Biol Chem* **272**: 32656–32662.

Arakane, F., Kallen, C. B., Watari, H. et al. (1998). The mechanism of action of steroidogenic acute regulatory protein (StAR). StAR acts on the outside of mitochondria to stimulate steroidogenesis. *J Biol Chem* **273**: 16339–16345.

Arima, S., Kohagura, K., Xu, H. L. et al. (2003). Nongenomic vascular action of aldosterone in the glomerular microcirculation. *J Am Soc Nephrol* **14**: 2255–2263.

Arnold, J. (1866). Ein Beitrag zu der feiner Struktur und dem Chemismus der Nebennieren. *Virchows Arch* **35**: 64–107.

Balk, E., Raman, G., Chung, M. et al. (2006). Effectiveness of management strategies for renal artery stenosis: a systematic review. *Ann Intern Med* **145**: 901–912.

Barrett, P. Q., and Isales, C. M. (1988). The role of cyclic nucleotides in atrial natriuretic peptide-mediated inhibition of aldosterone secretion. *Endocrinology* **122**: 799–808.

Barrett, P. Q., Kojima, I., Kojima, K., Zawalich, K., Isales, C. M., and Rasmussen, H. (1986). Temporal patterns of protein phosphorylation after angiotensin II, A23187 and/or 12-O-tetradecanoylphorbol 13-acetate in adrenal glomerulosa cells. *Biochem J* **238**: 893–903.

Barrett, P. Q., Bollag, W. B., Isales, C. M., McCarthy, R. T., and Rasmussen, H. (1989). Role of calcium in angiotensin II-mediated aldosterone secretion. *Endocr Rev* **10**: 496–518.

Bassett, M. H., Zhang, Y., White, P. C., and Rainey, W. E. (2000). Regulation of human CYP11B2 and CYP11B1: comparing the role of the common CRE/Ad1 element. *Endocr Res* **26**: 941–951.

Bassett, M. H., Suzuki, T., Sasano, H., White, P. C., and Rainey, W. E. (2004a). The orphan nuclear receptors NURR1 and NGFIB regulate adrenal aldosterone production. *Mol Endocrinol* **18**: 279–290.

Bassett, M. H., White, P. C., and Rainey, W. E. (2004b). The regulation of aldosterone synthase expression. *Mol Cell Endocrinol* **217**: 67–74.

Bech, K., and Hilden, T. (1975). The frequency of secondary hypertension. *Acta Med Scand* **197**: 65–69.

Begeot, M., Langlois, D., Penhoat, A., and Saez, J. M. (1988). Variations in guanine-binding proteins (Gs, Gi) in cultured bovine adrenal cells. Consequences on the effects of phorbol ester and angiotensin II on adrenocorticotropin-induced and choleratoxin-induced cAMP production. *Eur J Biochem* **174**: 317–321.

Betancourt-Calle, S., Bollag, W. B., Jung, E. M., Calle, R. A., and Rasmussen, H. (1999). Effects of angiotensin II and adrenocorticotropic hormone on myristoylated alanine-rich C-kinase substrate phosphorylation in glomerulosa cells. *Mol Cell Endocrinol* **154**: 1–9.

Betancourt-Calle, S., Calle, R. A., Isales, C. M., White, S., Rasmussen, H., and Bollag, W. B. (2001a). Differential effects of agonists of aldosterone secretion on steroidogenic acute regulatory phosphorylation. *Mol Cell Endocrinol* **173**: 87–94.

Betancourt-Calle, S., Jung, E. M., White, S. et al. (2001b). Elevated K(+) induces myristoylated alanine-rich C-kinase substrate phosphorylation and phospholipase D activation in glomerulosa cells. *Mol Cell Endocrinol* **184**: 65–76.

Bird, I. M., Mason, J. I., and Rainey, W. E. (1995a). Hormonal regulation of angiotensin II type 1 receptor expression and AT1-R mRNA levels in human adrenocortical cells. *Endocr Res* **21**: 169–182.

Bird, I. M., Mathis, J. M., Mason, J. I., and Rainey, W. E. (1995b). Ca(2+)-regulated expression of steroid hydroxylases in H295R human adrenocortical cells. *Endocrinology* **136**: 5677–5684.

Bird, I. M., Pasquarette, M. M., Rainey, W. E., and Mason, J. I. (1996). Differential control of 17 alpha-hydroxylase and 3 beta-hydroxysteroid dehydrogenase expression in human adrenocortical H295R cells. *J Clin Endocrinol Metab* **81**: 2171–2178.

Bollag, W. B., Barrett, P. Q., Isales, C. M., Liscovitch, M., and Rasmussen, H. (1990). A potential role for phospholipase-D in the angiotensin-II-induced stimulation of aldosterone secretion from bovine adrenal glomerulosa cells. *Endocrinology* **127**: 1436–1443.

Bollag, W. B., Barrett, P. Q., Isales, C. M., and Rasmussen, H. (1991). Angiotensin-II-induced changes in diacylglycerol levels and their potential role in modulating the steroidogenic response. *Endocrinology* **128**: 231–241.

Bollag, W. B., Dodd, M. E., and Shapiro, B. A. (2004). Protein kinase D and keratinocyte proliferation. *Drug News Perspect* **17**: 117–126.

Bollag, W. B., Jung, E., and Calle, R. A. (2002). Mechanism of angiotensin II-induced phospholipase D activation in bovine adrenal glomerulosa cells. *Mol Cell Endocrinol* **192**: 7–16.

Bollag, W. B., Zhong, X., Dodd, M. E., Hardy, D. M., Zheng, X., and Allred, W. T. (2005). Phospholipase d signaling and extracellular signal-regulated kinase-1 and -2 phosphorylation (activation) are required for maximal phorbol ester-induced transglutaminase activity, a marker of keratinocyte differentiation. *J Pharmacol Exp Ther* **312:** 1223–1231.

Bravo, E. L., Tarazi, R. C., Dustan, H. P. et al. (1983). The changing clinical spectrum of primary aldosteronism. *Am J Med* **74:** 641–651.

Breault, L., LeHoux, J. G., and Gallo-Payet, N. (1996). Angiotensin II receptors in the human adrenal gland. *Endocr Res* **22:** 355–361.

Burnay, M. M., Python, C. P., Vallotton, M. B., Capponi, A. M., and Rossier, M. F. (1994). Role of the capacitative calcium influx in the activation of steroidogenesis by angiotensin-II in adrenal glomerulosa cells. *Endocrinology* **135:** 751–758.

Calle, R. A., Bollag, W. B., White, S., Betancourt-Calle, S., and Kent, P. (2001). ANPs effect on MARCKS and StAR phosphorylation in agonist-stimulated glomerulosa cells. *Mol Cell Endocrinol* **177:** 71–79.

Capponi, A. M. (2004). The control by angiotensin II of cholesterol supply for aldosterone biosynthesis. *Mol Cell Endocrinol* **217:** 113–118.

Caron, K. M., Soo, S. C., Wetsel, W. C., Stocco, D. M., Clark, B. J., and Parker, K. L. (1997). Targeted disruption of the mouse gene encoding steroidogenic acute regulatory protein provides insights into congenital lipoid adrenal hyperplasia. *Proc Natl Acad Sci USA* **94:** 11540–11545.

Chang, H. W., Chu, T. S., Huang, H. Y. et al. (2007). Down-regulation of D2 dopamine receptor and increased protein kinase Cmu phosphorylation in aldosterone-producing adenoma play roles in aldosterone overproduction. *J Clin Endocrinol Metab* **92:** 1863–1870.

Chen, X. M., Li, W. G., Yoshid, H. et al. (1997). Targeting deletion of angiotensin type 1B receptor gene in the mouse. *Am J Physiol Renal Physiol* **41:** F299–F304.

Chen, X. L., Bayliss, D. A., Fern, R. J., and Barrett, P. Q. (1999). A role for T-type Ca2+ channels in the synergistic control of aldosterone production by ANG II and K+. *Am J Physiol* **276:** F674–F683.

Cherradi, N., Brandenburger, Y., and Capponi, A. M. (1998). Mitochondrial regulation of mineralocorticoid biosynthesis by calcium and the StAR protein. *Eur J Endocrinol* **139:** 249–256.

Cherradi, N., Pardo, B., Greenberg, A. S., Kraemer, F. B., and Capponi, A. M. (2003). Angiotensin II activates cholesterol ester hydrolase in bovine adrenal glomerulosa cells through phosphorylation mediated by p42/p44 mitogen-activated protein kinase. *Endocrinology* **144:** 4905–4915.

Christenson, L. K., and Strauss, J. F. (2001). Steroidogenic acute regulatory protein: an update on its regulation and mechanism of action. *Arch Med Res* **32:** 576–586.

Clark, A. J., Balla, T., Jones, M. R., and Catt, K. J. (1992). Stimulation of early gene expression by angiotensin II in bovine adrenal glomerulosa cells: roles of calcium and protein kinase C. *Mol Endocrinol* **6:** 1889–1898.

Clark, B. J., Wells, J., King, S. R., and Stocco, D. M. (1994). The purification, cloning, and expression of a novel luteinizing hormone-induced mitochondrial protein in MA-10 mouse Leydig tumor cells. Characterization of the steroidogenic acute regulatory protein (StAR). *J Biol Chem* **269:** 28314–28322.

Clark, B. J., Pezzi, V., Stocco, D. M., and Rainey, W. E. (1995). The steroidogenic acute regulatory protein is induced by angiotensin II and K+ in H295R adrenocortical cells. *Mol Cell Endocrinol* **115:** 215–219.

Clyne, C. D., Zhang, Y., Slutsker, L., Mathis, J. M., White, P. C., and Rainey, W. E. (1997). Angiotensin II and potassium regulate human CYP11B2 transcription through common cis-elements. *Mol Endocrinol* **11:** 638–649.

Condon, J. C., Pezzi, V., Drummond, B. M., Yin, S., and Rainey, W. E. (2002). Calmodulin-dependent kinase I regulates adrenal cell expression of aldosterone synthase. *Endocrinology* **143:** 3651–3657.

Conn, J. W., and Louis, L. H. (1955). Primary aldosteronism: a new clinical entity. *Trans Assoc Am Phys* **68:** 215–231 discussion, 231–233.

Corvol, P., Pinet, F., Plouin, P. F., Bruneval, P., and Menard, J. (1994). Renin-secreting tumors. *Endocrinol Metab Clin N Am* **23:** 255–270.

Curnow, K. M., Tusie-Luna, M. T., Pascoe, L. et al. (1991). The product of the CYP11B2 gene is required for aldosterone biosynthesis in the human adrenal cortex. *Mol Endocrinol* **5:** 1513–1522.

Davies, E., Edwards, C. R., and Williams, B. C. (1991). Serotonin stimulates calcium influx in isolated rat adrenal zona glomerulosa cells. *Biochem Biophys Res Commun* **179:** 979–984.

Davies, E., Rossiter, S., Edwards, C. R., and Williams, B. C. (1992). Serotoninergic stimulation of aldosterone secretion in vivo: role of the hypothalamo-pituitary adrenal axis. *J Steroid Biochem Mol Biol* **42:** 29–36.

de Bold, A. J., Borenstein, H. B., Veress, A. T., and Sonnenberg, H. (1981). A rapid and potent natriuretic response to intravenous injection of atrial myocardial extract in rats. *Life Sci* **28:** 89–94.

Delarue, C., Contesse, V., Lenglet, S. et al. (2001). Role of neurotransmitters and neuropeptides in the regulation of the adrenal cortex. *Rev Endocr Metab Disord* **2:** 253–267.

Denner, K., Rainey, W. E., Pezzi, V., Bird, I. M., Bernhardt, R., and Mathis, J. M. (1996). Differential regulation of 11 beta-hydroxylase and aldosterone synthase in human adrenocortical H295R cells. *Mol Cell Endocrinol* **121:** 87–91.

Domalik, L. J., Chaplin, D. D., Kirkman, M. S. et al. (1991). Different isozymes of mouse 11 beta-hydroxylase produce mineralocorticoids and glucocorticoids. *Mol Endocrinol* **5:** 1853–1861.

Exton, J. H. (1998). Phospholipase D. *Biochim Biophys Acta* **1436:** 105–115.

Exton, J. H. (2000). Phospholipase D. *Ann NY Acad Sci* **905:** 61–68.

Fields, C. L., Ossorio, M. A., Roy, T. M., Denny, D. M., and Varga, D. W. (1990). Thallium-201 scintigraphy in the diagnosis and management of myocardial sarcoidosis. *South Med J* **83:** 339–342.

Fleury, A., Mathieu, A. P., Ducharme, L., Hales, D. B., and LeHoux, J. G. (2004). Phosphorylation and function of the hamster adrenal steroidogenic acute regulatory protein (StAR). *J Steroid Biochem Mol Biol* **91:** 259–271.

Fuchs-Hammoser, R., Schweigwe, M., and Oelkers, W. (1980). The effect of chronic low-dose infusion of ACTH (1-24) on renin, renin substrate, aldosterone and other corticosteroids in sodium replete and deplete man. *Acta Endocrinol* **95:** 198–206.

Funder, J. W. (2005a). The nongenomic actions of aldosterone. *Endocr Rev* **26:** 313–321.

Funder, J. W. (2005b). RALES, EPHESUS and redox. *J Steroid Biochem Mol Biol* **93:** 121–125.

Gambaryan, S., Butt, E., Tas, P., Smolenski, A., Allolio, B., and Walter, U. (2006). Regulation of aldosterone production from zona glomerulosa cells by ANG II and cAMP: evidence for PKA-independent activation of CaMK by cAMP. *Am J Physiol Endocrinol Metab* **290:** E423–E433.

Ganguly, A., and Davis, J. S. (1994). Role of calcium and other mediators in aldosterone secretion from the adrenal glomerulosa cells. *Pharmacol Rev* **46:** 417–447.

Ganguly, A., Chiou, S., Fineberg, N. S., and Davis, J. S. (1992). Greater importance of Ca2+-calmodulin in maintenance of ANG II- and K+-mediated aldosterone secretion: lesser role of protein kinase C. *Biochem Biophys Res Commun* **182:** 254–261.

Garcia-Donaire, J. A., and Alcazar, J. M. (2005). Ischemic nephropathy: detection and therapeutic intervention. *Kidney Int Suppl*, pp. S131–S136.

Gomez-Sanchez, C. E., Gomez-Sanchez, E. P., and Yamakita, N. (1995). Endocrine causes of hypertension. *Semin Nephrol* **15:** 106–115.

Gu, J., Wen, Y., Mison, A., and Nadler, J. L. (2003). 12-lipoxygenase pathway increases aldosterone production, 3′,5′-cyclic adenosine monophosphate response element-binding protein phosphorylation, and p38 mitogen-activated protein kinase activation in H295R human adrenocortical cells. *Endocrinology* **144:** 534–543.

Haab, F., Duclos, J. M., Guyenne, T., Plouin, P. F., and Corvol, P. (1995). Renin secreting tumors: diagnosis, conservative surgical approach and long-term results. *J Urol* **153:** 1781–1784.

Hajnoczky, G., Varnai, P., Hollo, Z. et al. (1991). Thapsigargin-induced increase in cytoplasmic Ca^{2+} concentration and aldosterone production in rat adrenal glomerulosa cells: interaction with potassium and angiotensin-II. *Endocrinology* **128:** 2639–2644.

Hajnoczky, G., Varnai, P., Buday, L., Farago, A., and Spat, A. (1992). The role of protein kinase-C in control of aldosterone production by rat adrenal glomerulosa cells: activation of protein kinase-C by stimulation with potassium. *Endocrinology* **130:** 2230–2236.

Hall, P. F. (1997). The roles of calmodulin, actin, and vimentin in steroid synthesis by adrenal cells. *Steroids* **62:** 185–189.

Hanley, N. A., Wester, R. M., Carr, B. R., and Rainey, W. E. (1993). Parathyroid hormone and parathyroid hormone-related peptide stimulate aldosterone production in the human adrenocortical cell line, NCI-H295. *Endocr J* **1:** 447–450.

Hausdorff, W. P., Sekura, R. D., Aguilera, G., and Catt, K. J. (1987). Control of aldosterone production by angiotensin II is mediated by two guanine nucleotide regulatory proteins. *Endocrinology* **120:** 1668–1678.

Holland, O. B., and Carr, B. (1993). Modulation of aldosterone synthase messenger ribonucleic acid levels by dietary sodium and potassium and by adrenocorticotropin. *Endocrinology* **132:** 2666–2673.

Hornsby, P. J., O'Hare, M. J., and Neville, A. M. (1974). Functional and morphological observations on rat adrenal zona glomerulosa cells in monolayer culture. *Endocrinology* **95:** 1240–1251.

Houben, A. J., van der Zander, K., and de Leeuw, P. W. (2005). Vascular and renal actions of brain natriuretic peptide in man: physiology and pharmacology. *Fundam Clin Pharmacol* **19:** 411–419.

Hunyady, L., Baukal, A. J., Bor, M., Ely, J. A., and Catt, K. J. (1990). Regulation of 1,2-diacylglycerol production by angiotensin-II in bovine adrenal glomerulosa cells. *Endocrinology* **126:** 1001–1008.

Isales, C. M., Barrett, P. Q., Brines, M., Bollag, W., and Rasmussen, H. (1991). Parathyroid hormone modulates angiotensin II-induced aldosterone secretion from the adrenal glomerulosa cell. *Endocrinology* **129:** 489–495.

Isales, C. M., Bollag, W. B., Kiernan, L. C., and Barrett, P. Q. (1989). Effect of ANP on sustained aldosterone secretion stimulated by angiotensin II. *Am J Physiol* **256:** C89–C95.

Jo, Y., King, S. R., Khan, S. A., and Stocco, D. M. (2005). Involvement of protein kinase C and cyclic adenosine 3′,5′-monophosphate-dependent kinase in steroidogenic acute regulatory protein expression and steroid biosynthesis in Leydig cells. *Biol Reprod* **73:** 244–255.

Johannessen, M., and Moens, U. (2007). Multisite phosphorylation of the cAMP response element-binding protein (CREB) by a diversity of protein kinases. *Front Biosci* **12:** 1814–1832.

Johannessen, M., Delghandi, M. P., Rykx, A. et al. (2007). Protein kinase D induces transcription through direct phosphorylation of the cAMP-response element-binding protein. *J Biol Chem* **282:** 14777–14787.

Kakiki, M., Morohashi, K., Nomura, M., Omura, T., and Horie, T. (1997a). Expression of aldosterone synthase cytochrome P450 (P450aldo) mRNA in rat adrenal glomerulosa cells by angiotensin II type 1 receptor. *Endocr Res* **23:** 277–295.

Kakiki, M., Morohashi, K., Nomura, M., Omura, T., and Horie, T. (1997b). Regulation of aldosterone synthase cytochrome P450 (CYP11B2) and 11 beta-hydroxylase cytochrome P450 (CYP11B1) expression in rat adrenal zona glomerulosa cells by low sodium diet and angiotensin II receptor antagonists. *Biol Pharm Bull* **20:** 962–968.

Kallen, C. B., Arakane, F., Christenson, L. K., Watari, H., Devoto, L., and Strauss, J. F. (1998). Unveiling the mechanism of action and regulation of the steroidogenic acute regulatory protein. *Mol Cell Endocrinol* **145:** 39–45.

Kapas, S., Purbrick, A., and Hinson, J. P. (1995). Role of tyrosine kinase and protein kinase C in the steroidogenic actions of angiotensin II, alpha-melanocyte-stimulating hormone and corticotropin in the rat adrenal cortex. *Biochem J* **305**(Pt 2): 433–438.

Kato, J., Etoh, T., Kitamura, K., and Eto, T. (2005). Atrial and brain natriuretic peptides as markers of cardiac load and volume retention in primary aldosteronism. *Am J Hypertens* **18:** 354–357.

Kloos, R. T., Gross, M. D., Francis, I. R., Korobkin, M., and Shapiro, B. (1995). Incidentally discovered adrenal masses. *Endocr Rev* **16:** 460–484.

Kojima, I., Kojima, K., Kreutter, D., and Rasmussen, H. (1984). The temporal integration of the aldosterone secretory response to angiotensin occurs via two intracellular pathways. *J Biol Chem* **259:** 14448–14457.

Kojima, I., Kojima, K., and Rasmussen, H. (1985a). Effects of ANG II and K+ on Ca efflux and aldosterone production in adrenal glomerulosa cells. *Am J Physiol* **248:** E36–E43.

Kojima, I., Kojima, K., and Rasmussen, H. (1985b). Role of calcium fluxes in the sustained phase of angiotensin II-mediated aldosterone secretion from adrenal glomerulosa cells. *J Biol Chem* **260:** 9177–9184.

Krueger, R. J., and Orme-Johnson, N. R. (1983). Acute adrenocorticotropic hormone stimulation of adrenal corticosteroidogenesis. Discovery of a rapidly induced protein. *J Biol Chem* **258:** 10159–10167.

Larcher, A., Lamacz, M., Delarue, C., and Vaudry, H. (1992). Effect of vasotocin on cytosolic free calcium concentrations in frog adrenocortical cells in primary culture. *Endocrinology* **131**: 1087–1093.

Lee, M. W., and Severson, D. L. (1994). Signal transduction in vascular smooth muscle: diacylglycerol second messengers and PKC action. *Am J Physiol* **267**: C659–C678.

Lefebvre, H., Cartier, D., Duparc, C. et al. (2000). Effect of serotonin4 (5-HT4) receptor agonists on aldosterone secretion in idiopathic hyperaldosteronism. *Endocr Res* **26**: 583–587.

LeHoux, J. G., Bird, I. M., Rainey, W. E., Tremblay, A., and Ducharme, L. (1994). Both low sodium and high potassium intake increase the level of adrenal angiotensin-II receptor type 1, but not that of adrenocorticotropin receptor. *Endocrinology* **134**: 776–782.

LeHoux, J. G., Martel, D., LeHoux, J., Ducharme, L., Lefebvre, A., and Briere, N. (1995). P450aldo in hamster adrenal cortex: immunofluorescent and immuno-gold electron microscopic studies. *Endocr Res* **21**: 275–280.

LeHoux, J. G., Hales, D. B., Fleury, A., Briere, N., Martel, D., and Ducharme, L. (1999). The in vivo effects of adrenocorticotropin and sodium restriction on the formation of the different species of steroidogenic acute regulatory protein in rat adrenal. *Endocrinology* **140**: 5154–5164.

Lenglet, S., Louiset, E., Delarue, C., Vaudry, H., and Contesse, V. (2002). Activation of 5-HT(7) receptor in rat glomerulosa cells is associated with an increase in adenylyl cyclase activity and calcium influx through T-type calcium channels. *Endocrinology* **143**: 1748–1760.

Levin, A., Linas, S., Luft, F. C., Chapman, A. B., and Textor, S. (2007). Controversies in renal artery stenosis: a review by the American Society of Nephrology Advisory Group on Hypertension. *Am J Nephrol* **27**: 212–220.

Li, J., Feltzer, R. E., Dawson, K. L., Hudson, E. A., and Clark, B. J. (2003). Janus kinase 2 and calcium are required for angiotensin II-dependent activation of steroidogenic acute regulatory protein transcription in H295R human adrenocortical cells. *J Biol Chem* **278**: 52355–52362.

Lu, H. K., Fern, R. J., Luthin, D. et al. (1996). Angiotensin II stimulates T-type Ca^{2+} channel currents via activation of a G protein, Gi. *Am J Physiol* **271**: C1340–C1349.

Lu, L., Suzuki, T., Yoshikawa, Y. et al. (2004). Nur-related factor 1 and nerve growth factor-induced clone B in human adrenal cortex and its disorders. *J Clin Endocrinol Metab* **89**: 4113–4118.

Maestri, E., Camellini, L., Rossi, G. et al. (1988). Serotonin regulation of aldosterone secretion. *Horm Metab Res* **20**: 457–459.

McAllister, J. M., and Hornsby, P. J. (1988). Dual regulation of 3 beta-hydroxysteroid dehydrogenase, 17 alpha- hydroxylase, and dehydroepiandrosterone sulfotransferase by adenosine 3',5'-monophosphate and activators of protein kinase C in cultured human adrenocortical cells. *Endocrinology* **122**: 2012–2018.

McCarthy, R. T., Isales, C. M., Bollag, W. B., Rasmussen, H., and Barrett, P. Q. (1990). Atrial natriuretic peptide differentially modulates T- and L-type calcium channels. *Am J Physiol* **258**: F473–F478.

Miller, W. L. (2007). StAR search – what we know about how the steroidogenic acute regulatory protein mediates mitochondrial cholesterol import. *Mol Endocrinol* **21**: 589–601.

Mosso, L., Carvajal, C., Gonzalez, A. et al. (2003). Primary aldosteronism and hypertensive disease. *Hypertension* **42**: 161–165.

Mulatero, P., Milan, A., Fallo, F. et al. (2006). Comparison of confirmatory tests for the diagnosis of primary aldosteronism. *J Clin Endocrinol Metab* **91**: 2618–2623.

Muller, J. (1998). Regulation of aldosterone biosynthesis: the end of the road? *Clin Exp Pharmacol Physiol Suppl* **25**: S79–S85.

Nadler, J. L., Natarajan, R., and Stern, N. (1987). Specific action of the lipoxygenase pathway in mediating angiotensin II-induced aldosterone synthesis in isolated adrenal glomerulosa cells. *J Clin Invest* **80**: 1763–1769.

Nairn, A. C., and Aderem, A. (1992). Calmodulin and protein kinase C cross-talk: the MARCKS protein is an actin filament and plasma membrane cross-linking protein regulated by protein kinase C phosphorylation and by calmodulin. *Ciba Found Symp* **164**: 145–154 discussion 154–161.

Narasaka, T., Suzuki, T., Moriya, T., and Sasano, H. (2001). Temporal and spatial distribution of corticosteroidogenic enzymes immunoreactivity in developing human adrenal. *Mol Cell Endocrinol* **174**: 111–120.

Natarajan, R., Stern, N., Hsueh, W., Do, Y., and Nadler, J. (1988). Role of the lipoxygenase pathway in angiotensin II-mediated aldosterone biosynthesis in human adrenal glomerulosa cells. *J Clin Endocrinol Metab* **67**: 584–591.

Natarajan, R., Dunn, W. D., Stern, N., and Nadler, J. (1990). Key role of diacylglycerol-mediated 12-lipoxygenase product formation in angiotensin II-induced aldosterone synthesis. *Mol Cell Endocrinol* **72**: 73–80.

Natarajan, R., Yang, D. C., Lanting, L., and Nadler, J. L. (2002). Key role of P38 mitogen-activated protein kinase and the lipoxygenase pathway in angiotensin II actions in H295R adrenocortical cells. *Endocrine* **18**: 295–301.

New, M. I. (2003). Inborn errors of adrenal steroidogenesis. *Mol Cell Endocrinol* **211**: 75–83.

New, M. I. (2006). Extensive clinical experience: nonclassical 21-hydroxylase deficiency. *J Clin Endocrinol Metab* **91**: 4205–4214.

Nicod, J., Dick, B., Frey, F. J., and Ferrari, P. (2004). Mutation analysis of CYP11B1 and CYP11B2 in patients with increased 18-hydroxycortisol production. *Mol Cell Endocrinol* **214**: 167–174.

Nimkarn, S., and New, M. I. (2006). Prenatal diagnosis and treatment of congenital adrenal hyperplasia. *Pediatr Endocrinol Rev* **4**: 99–105.

Obara, T., Ito, Y., and Kanaji, Y. (1992). [Diagnosis and therapy of aldosterone-producing adrenal gland adenoma]. *Nippon Naika Gakkai Zasshi* **81**: 512–516.

Ogishima, T., Suzuki, H., Hata, J., Mitani, F., and Ishimura, Y. (1992). Zone-specific expression of aldosterone synthase cytochrome P-450 and cytochrome P-45011 beta in rat adrenal cortex: histochemical basis for the functional zonation. *Endocrinology* **130**: 2971–2977.

Okubo, S., and Ichikawa, I. (1997). Role of angiotensin: insight from gene targeting studies. *Kidney Int Suppl* **63**: S7–S9.

Olgaard, K., Lewin, E., Bro, S., Daugaard, H., Egfjord, M., and Pless, V. (1994). Enhancement of the stimulatory effect of calcium on aldosterone secretion by parathyroid hormone. *Miner Electrolyte Metab* **20**: 309–314.

Otis, M., and Gallo-Payet, N. (2006). Differential involvement of cytoskeleton and rho-guanosine 5'-triphosphatases in growth-promoting effects of angiotensin II in rat adrenal glomerulosa cells. *Endocrinology* **147**: 5460–5469.

Otis, M., Campbell, S., Payet, M. D., and Gallo-Payet, N. (2005). Angiotensin II stimulates protein synthesis and inhibits proliferation in primary cultures of rat adrenal glomerulosa cells. *Endocrinology* **146**: 633–642.

Papadopoulos, V., Liu, J., and Culty, M. (2007). Is there a mitochondrial signaling complex facilitating cholesterol import? *Mol Cell Endocrinol* **265–266**: 59–64.

Pascoe, L., Jeunemaitre, X., Lebrethon, M. C. et al. (1995). Glucocorticoid-suppressible hyperaldosteronism and adrenal tumors occurring in a single French pedigree. *J Clin Invest* **96**: 2236–2246.

Payne, A. H., and Hales, D. B. (2004). Overview of steroidogenic enzymes in the pathway from cholesterol to active steroid hormones. *Endocr Rev* **25**: 947–970.

Pedrinelli, R., Panarace, G., Spessot, M. et al. (1989). Low dose atrial natriuretic factor in primary aldosteronism: renal, hemodynamic, and vascular effects. *Hypertension* **14**: 156–163.

Pezzi, V., Clyne, C. D., Ando, S., Mathis, J. M., and Rainey, W. E. (1997). Ca(2+)-regulated expression of aldosterone synthase is mediated by calmodulin and calmodulin-dependent protein kinases. *Endocrinology* **138**: 835–838.

Pimenta, E., and Calhoun, D. A. (2006). Primary aldosteronism: diagnosis and treatment. *J Clin Hypertens (Greenwich)* **8**: 887–893.

Pitt, B. (2004). A new HOPE for aldosterone blockade? *Circulation* **110**: 1714–1716.

Pitt, B., Zannad, F., Remme, W. J., Cody, R. et al. (1999). The effect of spironolactone on morbidity and mortality in patients with severe heart failure. Randomized Aldactone Evaluation Study Investigators. *N Engl J Med* **341**: 709–717.

Pitt, B., Remme, W., Zannad, F. et al. (2003). Eplerenone, a selective aldosterone blocker, in patients with left ventricular dysfunction after myocardial infarction. *N Engl J Med* **348**: 1309–1321.

Pon, L. A., and Orme-Johnson, N. R. (1984). Protein synthesis requirement for acute ACTH stimulation of adrenal corticosteroidogenesis. *Endocr Res* **10**: 585–590.

Pon, L. A., Hartigan, J. A., and Orme-Johnson, N. R. (1986). Acute ACTH regulation of adrenal corticosteroid biosynthesis. Rapid accumulation of a phosphoprotein. *J Biol Chem* **261**: 13309–13316.

Prisant, L. M., Szerlip, H. M., and Mulloy, L. L. (2006). Fibromuscular dysplasia: an uncommon cause of secondary hypertension. *J Clin Hypertens (Greenwich)* **8**: 894–898.

Pursell, R. N., and Quinlan, P. M. (2003). Secondary hypertension due to a renin-producing teratoma. *Am J Hypertens* **16**: 592–595.

Rasmussen, H., Isales, C. M., Calle, R. et al. (1995). Diacylglycerol production, Ca2+ influx, and protein kinase C activation in sustained cellular responses. *Endocr Rev* **16**: 649–681.

Rebuffat, P., Belloni, A. S., Musajo, F. G. et al. (1994). Evidence that endogenous somatostatin (SRIF) exerts an inhibitory control on the function and growth of rat adrenal zona glomerulosa. The possible involvement of zona medullaris as a source of endogenous SRIF. *J Steroid Biochem Mol Biol* **48**: 353–360.

Romero, D. G., Plonczynski, M. W., Gomez-Sanchez, E. P., Yanes, L. L., and Gomez-Sanchez, C. E. (2006a). RGS2 is regulated by angiotensin II and functions as a negative feedback of aldosterone production in H295R human adrenocortical cells. *Endocrinology* **147**: 3889–3897.

Romero, D. G., Welsh, B. L., Gomez-Sanchez, E. P., Yanes, L. L., Rilli, S., and Gomez-Sanchez, C. E. (2006b). Angiotensin II-mediated protein kinase D activation stimulates aldosterone and cortisol secretion in H295R human adrenocortical cells. *Endocrinology* **147**: 6046–6055.

Rossi, G., Boscaro, M., Ronconi, V., and Funder, J. W. (2005). Aldosterone as a cardiovascular risk factor. *Trends Endocrinol Metab* **16**: 104–107.

Rossi, G. P., Bernini, G., Caliumi, C. et al. (2006). A prospective study of the prevalence of primary aldosteronism in 1,125 hypertensive patients. *J Am Coll Cardiol* **48**: 2293–2300.

Rossi, G. P., Belfiore, A., Bernini, G. et al. (2007). Comparison of the captopril and the saline infusion test for excluding aldosterone-producing adenoma. *Hypertension* **50**: 424–431.

Schirpenbach, C., and Reincke, M. (2006). Screening for primary aldosteronism. *Best Pract Res Clin Endocrinol Metab* **20**: 369–384.

Sewer, M. B., and Waterman, M. R. (2003). ACTH modulation of transcription factors responsible for steroid hydroxylase gene expression in the adrenal cortex. *Microsc Res Tech* **61**: 300–307.

Shah, B. H., Soh, J. W., and Catt, K. J. (2003). Dependence of gonadotropin-releasing hormone-induced neuronal MAPK signaling on epidermal growth factor receptor transactivation. *J Biol Chem* **278**: 2866–2875.

Shah, B. H., Baukal, A. J., Chen, H. D., Shah, A. B., and Catt, K. J. (2006). Mechanisms of endothelin-1-induced MAP kinase activation in adrenal glomerulosa cells. *J Steroid Biochem Mol Biol* **102**: 79–88.

Shen, T., Suzuki, Y., Poyard, M., Best-Belpomme, M., Defer, N., and Hanoune, J. (1997). Localization and differential expression of adenylyl cyclase messenger ribonucleic acids in rat adrenal gland determined by in situ hybridization. *Endocrinology* **138**: 4591–4598.

Shibata, H., and Kojima, I. (1991). Involvement of protein kinase C in angiotensin II-mediated release of 12-hydroxyeicosatetraenoic acid in bovine adrenal glomerulosa cells. *Endocrinol Jpn* **38**: 611–617.

Shionoiri, H., Hirawa, N., Takasaki, I. et al. (1988). Lack of atrial natriuretic peptide receptors in human aldosteronoma. *Biochem Biophys Res Commun* **152**: 37–43.

Sinclair, A. M., Isles, C. G., Brown, I., Cameron, H., Murray, G. D., and Robertson, J. W. (1987). Secondary hypertension in a blood pressure clinic. *Arch Intern Med* **147**: 1289–1293.

Sinnett-Smith, J., Zhukova, E., Hsieh, N., Jiang, X., and Rozengurt, E. (2004). Protein kinase D potentiates DNA synthesis induced by Gq-coupled receptors by increasing the duration of ERK signaling in swiss 3T3 cells. *J Biol Chem* **279**: 16883–16893.

Spat, A., and Hunyady, L. (2004). Control of aldosterone secretion: a model for convergence in cellular signaling pathways. *Physiol Rev* **84**: 489–539.

Startchik, I., Morabito, D., Lang, U., and Rossier, M. F. (2002). Control of calcium homeostasis by angiotensin II in adrenal glomerulosa cells through activation of p38 MAPK. *J Biol Chem* **277**: 24265–24273.

Stern, N., Natarajan, R., Tuck, M. L., Laird, E., and Nadler, J. L. (1989). Selective inhibition of angiotensin-II-mediated aldosterone secretion by 5-hydroxyeicosatetraenoic acid. *Endocrinology* **125**: 3090–3095.

Stern, N., Yanagawa, N., Saito, F. et al. (1993). Potential role of 12 hydroxyeicosatetraenoic acid in angiotensin II-induced calcium signal in rat glomerulosa cells. *Endocrinology* **133:** 843–847.

Stocco, D. M. (1992). Further evidence that the mitochondrial proteins induced by hormone stimulation in MA-10 mouse Leydig tumor cells are involved in the acute regulation of steroidogenesis. *J Steroid Biochem Mol Biol* **43:** 319–333.

Stocco, D. M., and Ascoli, M. (1993). The use of genetic manipulation of MA-10 Leydig tumor cells to demonstrate the role of mitochondrial proteins in the acute regulation of steroidogenesis. *Endocrinology* **132:** 959–967.

Stocco, D. M., and Clark, B. J. (1993). The requirement of phosphorylation on a threonine residue in the acute regulation of steroidogenesis in MA-10 mouse Leydig cells. *J Steroid Biochem Mol Biol* **46:** 337–347.

Stocco, D. M., and Clark, B. J. (1996). Regulation of the acute production of steroids in steroidogenic cells. *Endocr Rev* **17:** 221–244.

Sugden, P. H. (2001). Signalling pathways in cardiac myocyte hypertrophy. *Ann Med* **33:** 611–622.

Tian, Y., Balla, T., Baukal, A. J., and Catt, K. J. (1995). Growth responses to angiotensin II in bovine adrenal glomerulosa cells. *Am J Physiol* **268:** E135–E144.

Tian, Y., Smith, R. D., Balla, T., and Catt, K. J. (1998). Angiotensin II activates mitogen-activated protein kinase via protein kinase C and Ras/Raf-1 kinase in bovine adrenal glomerulosa cells. *Endocrinology* **139:** 1801–1809.

Tremblay, A., Parker, K. L., and LeHoux, J. G. (1992). Dietary potassium supplementation and sodium restriction stimulate aldosterone synthase but not 11 beta-hydroxylase P-450 messenger ribonucleic acid accumulation in rat adrenals and require angiotensin II production. *Endocrinology* **130:** 3152–3158.

Ueberberg, B., Tourne, H., Redman, A. et al. (2005). Differential expression of the human somatostatin receptor subtypes sst1 to sst5 in various adrenal tumors and normal adrenal gland. *Horm Metab Res* **37:** 722–728.

Wang, X. L., Bassett, M., Zhang, Y. et al. (2000). Transcriptional regulation of human 11beta-hydroxylase (hCYP11B1). *Endocrinology* **141:** 3587–3594.

Wang, Y. X., Chen, C. R., He, G. X., and Tang, A. R. (1998). CT findings of adrenal glands in patients with tuberculous Addison's disease. *J Belge Radiol* **81:** 226–228.

Waterman, M. R., and Bischof, L. J. (1996). Mechanisms of ACTH (cAMP)-dependent transcription of adrenal steroid hydroxylases. *Endocr Res* **22:** 615–620.

Weinberger, M. H., and Fineberg, N. S. (1993). The diagnosis of primary aldosteronism and separation of two major subtypes. *Arch Intern Med* **153:** 2125–2129.

White, P. C. (2004). Aldosterone synthase deficiency and related disorders. *Mol Cell Endocrinol* **217:** 81–87.

White, P. C., and Speiser, P. W. (2000). Congenital adrenal hyperplasia due to 21-hydroxylase deficiency. *Endocr Rev* **21:** 245–291.

Wu, W., Graves, L. M., Gill, G. N., Parsons, S. J., and Samet, J. M. (2002). Src-dependent phosphorylation of the epidermal growth factor receptor on tyrosine 845 is required for zinc-induced Ras activation. *J Biol Chem* **277:** 24252.

Xia, Y., and Wikberg, J. E. (1996). Localization of ACTH receptor mRNA by in situ hybridization in mouse adrenal gland. *Cell Tissue Res* **286:** 63–68.

Yagci, A., and Muller, J. (1996). Induction of steroidogenic enzymes by potassium in cultured rat zona glomerulosa cells depends on calcium influx and intact protein synthesis. *Endocrinology* **137:** 4331–4338.

Yagci, A., Oertle, M., Seiler, H., Schmid, D., Campofranco, C., and Muller, J. (1996). Potassium induces multiple steroidogenic enzymes in cultured rat zona glomerulosa cells. *Endocrinology* **137:** 2406–2414.

Yamakado, M., Nagano, M., Umezu, M., Tagawa, H., Kiyose, H., and Tanaka, S. (1988). Extrarenal role of aldosterone in the regulation of blood pressure. *Am J Hypertens* **1:** 276–279.

Yoshida, A., Nishidawa, T., Tamura, Y., and Yoshida, S. (1991). ACTH-induced inhibition of the action of angiotensin II in bovine zona glomerulosa cells. A modulatory effect of cyclic AMP on the angiotensin II receptor. *J Biol Chem* **266:** 4288–4294.

Young, W. F. (2007). Primary aldosteronism: renaissance of a syndrome. *Clin Endocrinol (Oxf.)* **66:** 607–618.

Young, W. F., Stanson, A. W., Thompson, G. B., Grant, C. S., Farley, D. R., and van Heerden, J. A. (2004). Role for adrenal venous sampling in primary aldosteronism. *Surgery* **136:** 1227–1235.

Young, W. F. (2003). Minireview: primary aldosteronism–changing concepts in diagnosis and treatment. *Endocrinology* **144:** 2208–2213.

Zannad, F., Alla, F., Dousset, B., Perez, A., and Pitt, B. (2000). Limitation of excessive extracellular matrix turnover may contribute to survival benefit of spironolactone therapy in patients with congestive heart failure: insights from the randomized aldactone evaluation study (RALES). Rales Investigators. *Circulation* **102:** 2700–2706.

Zarco-Gonzalez, J. A., and Herrera, M. F. (2004). Adrenal incidentaloma. *Scand J Surg* **93:** 298–301.

Section VII. Endocrine Disorders in Renal Failure

CHAPTER **25**

Insulin Resistance and Diabetes in Chronic Renal Disease

DONALD C. SIMONSON

Division of Endocrinology, Diabetes and Hypertension, Department of Medicine, Brigham and Women's Hospital, Harvard Medical School, Boston, MA, USA

Contents

I.	Introduction	385
II.	Historical perspective	385
III.	Cellular mechanisms of insulin secretion and action	386
IV.	Clinical physiology of insulin resistance	387
V.	Measurement of insulin resistance	388
VI.	Metabolic syndrome	389
VII.	Pathogenesis of insulin resistance in chronic kidney disease	398
VIII.	Regulation of renal glucose production	399
IX.	Syndromes of severe insulin resistance	399
X.	Treatment	400
XI.	Management of diabetes in chronic kidney disease	402
XII.	Hyperglycemia associated with renal transplantation	403
XIII.	Conclusions	403
	References	404

I. INTRODUCTION

Insulin resistance can broadly be defined as an impairment or defect in the ability of insulin to produce its normal biological, physiological or clinical effects. Despite the simplicity of this concept and the relative ease with which it can be demonstrated *in vitro* and *in vivo*, we only recently have begun to appreciate the true complexity and multiple clinical manifestations of insulin resistance. In addition to its well-established effects on carbohydrate, lipid and protein metabolism, insulin has a multitude of actions involving cell growth and differentiation, regulation of vascular tone and blood pressure, ion transport and inflammation.

Recently, there has been an increased appreciation of the role of the kidney in both the pathogenesis and consequences of insulin resistant states. Diabetes and hypertension – two common diseases characterized by insulin resistance – are also the leading causes of chronic kidney disease. Conversely, essentially all patients who develop end-stage renal disease, regardless of etiology, ultimately become insulin resistant. This chapter will provide an overview of the manifestations and implications of insulin resistance as they impact the kidney, with particular emphasis on the physiological and clinical aspects of the disorder.

II. HISTORICAL PERSPECTIVE

The concept of insulin resistance was first introduced in the 1930s when Himsworth (1936) reported that the ability of injected insulin to lower the blood glucose level was greater in patients with diabetes who were young and lean compared with patients who were older and obese. He designated these two groups as 'insulin-sensitive' and 'insulin-insensitive' forms of diabetes, respectively – a distinction that we recognize today as the difference between type 1 and type 2 diabetes mellitus.

Our understanding of insulin resistance took its next major step forward in the 1960s when Yalow and Berson (1960) developed the technique of radioimmunoassay that enabled the measurement of insulin concentrations in plasma. During the next decade, several studies revealed that plasma levels of insulin were actually elevated in patients with obesity, mild to moderate type 2 diabetes and other disorders characterized by elevated plasma glucose levels (Steinke et al., 1963; Karam et al., 1965; Yalow et al., 1965). This observation initially was considered paradoxical given the known ability of insulin to lower the plasma glucose concentration and led to a revision of the prevailing wisdom that all forms of diabetes or other states of glucose intolerance were caused by insulin deficiency. It began to be appreciated that 'resistance' to the biological effect of the hormone, and the body's compensatory increase in insulin secretion to overcome this defect, were cardinal features of many glucose intolerant states.

By the 1970s, investigators began to understand that insulin, like other peptide hormones, induced its metabolic effects and other actions by binding to a receptor on the cell surface, which then transmitted the hormonal signal to the interior of the cell. Several studies reported that abnormalities in insulin receptor binding, either due to a reduction in receptor number or binding affinity, were observed in type 2 diabetes, obesity and other hyperinsulinemic states (Kahn et al., 1972; Olefsky and Reaven, 1974; Amatruda et al., 1975; DeMeyts et al., 1976; Olefsky, 1976; Kahn, 1980). However, the impairment in insulin action could not be fully explained by receptor abnormalities, so attention turned to the post-receptor pathways involved in insulin signaling. The search for either genetic or acquired defects in the insulin signaling pathway still remains one of the most active areas of investigation into the mechanisms underlying insulin resistant states.

During the 1980s, a major conceptual advance occurred when investigators began to recognize that many diseases not classically associated with overt abnormalities in glucose tolerance were, in fact, characterized by insulin resistance. Although it had been recognized since the mid-1900s that diabetes, obesity, dyslipidemia, hypertension and cardiovascular disease commonly coexisted in the same individual, Reaven (1988) is generally acknowledged to be the first to propose the hypothesis that the frequent concurrence of these diseases was due to a common underlying defect in insulin action that was manifest to varying degrees in different individuals. He called this association 'syndrome X', although currently the terms 'insulin resistance syndrome' or 'metabolic syndrome' are more widely used. This concept forms the basis of our current physiological and clinical understanding of insulin resistance, which has been greatly expanded to include other abnormalities such as endothelial dysfunction, chronic inflammation, hypercoagulability, proteinuria and increased sympathetic nervous system activity. We also now appreciate that while hyperinsulinemia is present in essentially all insulin resistant states, unless there is a concurrent defect in insulin secretion, the varying manifestations of the syndrome often are due to tissue-specific differences in the relative sensitivity or resistance to insulin action.

III. CELLULAR MECHANISMS OF INSULIN SECRETION AND ACTION

Insulin is a 5800 kDa peptide consisting of a 21 amino acid A chain and a 30 amino acid B chain linked by two disulfide bridges. It is synthesized in the beta cells of the pancreatic islets as a single chain precursor, proinsulin, from which a connecting peptide (C-peptide) is removed to yield the active insulin molecule. Insulin and C-peptide, along with small amounts of proinsulin and inactive insulin split products, are stored in secretory granules where they are released into the circulation in response to nutrient ingestion. Glucose, the primary stimulus to insulin secretion, enters the beta cell via the GLUT2 transporter and is metabolized to generate adenosine triphosphate (ATP). In response to the increased intracellular levels of ATP, an ATP-sensitive K^+ channel on the cell surface closes, leading to depolarization of the cell membrane. In response to this depolarization, voltage sensitive Ca^{++} channels open, causing a rise in intracellular calcium levels and release of the insulin from the secretory vesicles (Figure 25.1). The release of insulin typically occurs in a biphasic pattern, with a first phase of secretion consisting primarily of rapid release of pre-stored insulin and a more prolonged second phase that primarily involves synthesis of new insulin in response to ongoing hyperglycemia (Aguilar-Bryan et al., 2001; Bratanova-Tochkova et al., 2002; Rhodes and White, 2002; Ahren and Taborsky, 2003; Henquin, 2005).

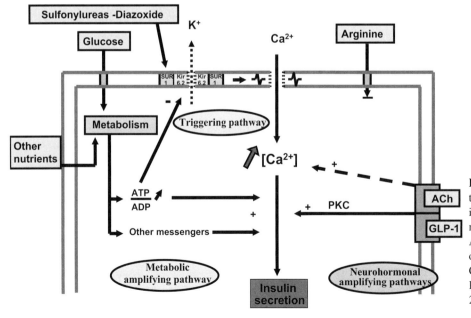

FIGURE 25.1 Schematic representation of the pathways of stimulation of insulin secretion by glucose and other nutrients. +, stimulation; −, inhibition; ATP, adenosine triphosphate; ADP, adenosine diphosphate; PKC, protein kinase C; Ach, acetylcholine; GLP-1, glucagon-like peptide-1. Adapted from Henquin, 2005.

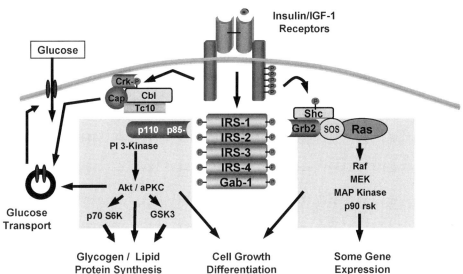

FIGURE 25.2 The insulin and insulin-like growth factor-1 (IGF-1) signaling network. The two major pathways are the phosphatidylinositol 3-kinase (PI 3-kinase) pathway and the Ras-mitogen-activated protein (MAP) kinase pathway. Adapted from Kahn and Saltiel, 2005.

Although glucose is considered to be the primary stimulus to the beta cell, insulin secretion also is induced by other nutrients including amino acids and, to a lesser degree, fatty acids. Various other hormones may potentiate the insulin response to nutrients (including gastric inhibitory polypeptide (GIP) and glucagon-like peptide-1 (GLP-1)), directly stimulate insulin secretion (cholecystokinin (CCK)), or inhibit insulin secretion (somatostatin, epinephrine, norepinephrine). Several different classes of pharmacologic agents, including sulfonylureas, meglitinides and GLP-1 agonists, also stimulate or potentiate insulin secretion and are widely used in the treatment of type 2 diabetes.

After its release from the beta cells into the circulation, insulin binds to receptors on the cell surface to initiate its diverse effects on metabolism and cell growth. The insulin receptor is a tetrameric protein consisting of two extracellular α-subunits attached to two transmembrane β-subunits (Figure 25.2). It belongs to the family of tyrosine kinase receptors and is closely related to the insulin-like growth factor-1 (IGF-1) receptor through which some of the growth-promoting effects of insulin are mediated. When insulin binds to the α-subunits, it induces a conformational change in the β-subunits, activating their tyrosine kinase activity. The β-subunits initially autophosphorylate, further increasing their kinase activity, and begin a cascade of phosphorylation of tyrosine residues on a series of intracellular proteins including insulin receptor substrates (IRS) 1-4, Gab-1 and Shc. These phosphorylated proteins function as docking sites for other intracellular proteins that contain Src-homology 2 (SH2) domains, initiating a complex series of enzymatic effects that mediate the diverse effects of the hormone through two major pathways. The phosphatidylinositol 3-kinase (PI-3 kinase) pathway primarily mediates the metabolic effects of insulin, including stimulation of glucose transport and glycogen synthesis, inhibition of lipolysis, stimulation of protein synthesis and, to a lesser extent, cell growth. The Ras-mitogen-activated protein kinase (Ras-MAP kinase) pathway is primarily involved in the effects of insulin on gene expression and cell growth and differentiation (Patti and Kahn, 1998; Avruch et al., 2001; Kanzaki and Pessin, 2001; Kido et al., 2001; White, 2002; Kahn and Saltiel, 2005). Insulin receptors are extensively distributed in almost all tissues of the body – including muscle, liver, adipose tissue, kidney, brain, the vasculature and many others – and the physiological effects of insulin vary widely depending upon the specific tissue type. The implications of these diverse effects and their role in the clinical manifestations of insulin resistant states are discussed in greater detail below.

IV. CLINICAL PHYSIOLOGY OF INSULIN RESISTANCE

In healthy individuals, the plasma glucose concentration typically ranges between 70 and 100 mg/dl after an overnight fast and usually does not exceed 120–140 mg/dl after ingestion of a meal. In the postabsorptive state, glucose is released into the circulation primarily from the liver, either by the breakdown of previously stored glycogen or by the process of gluconeogenesis – the synthesis of new glucose from precursors such as lactate, pyruvate, glycerol and the gluconeogenic amino acids, predominantly alanine and glutamine. Insulin is the primary hormone inhibiting the release of glucose from the liver and, after an overnight fast, circulates at concentrations ranging from about 3–10 μU/ml, depending on the age, weight and other demographic characteristics of the individual. Approximately half of all glucose produced in the fasting state is utilized by the brain, which does not require insulin for glucose uptake or metabolism (Shulman et al., 2003; Ferrannini and Mari, 2004) (Figure 25.3).

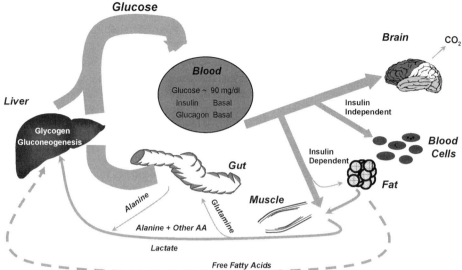

FIGURE 25.3 Whole-body fuel metabolism in the postabsorptive state. Adapted from Shulman et al., 2003.

After the ingestion of glucose or a mixed meal, nutrients are absorbed from the intestinal tract and stimulate the release of insulin from the pancreatic beta cells into the portal vein. Insulin initially reaches the liver where its major physiologic effects include stimulation of glycogen and triglyceride synthesis, inhibition of gluconeogenesis and suppression of lipolysis and ketogenesis (Figure 25.3). Approximately 50–60% of insulin secreted by the pancreas is removed by the liver, primarily by receptor-mediated clearance, before reaching the systemic circulation. Systemic levels of insulin vary widely after nutrient ingestion and may range from 20 to over 100 μU/ml, depending on the content and size of the meal as well as the metabolic characteristics of the individual. In response to the systemic hyperinsulinemia:

1. glucose is removed from the circulation, primarily by muscle, and either stored as glycogen or oxidized to meet ongoing energy requirements
2. lipolysis is inhibited and fatty acids are re-esterified into triglycerides
3. amino acids are taken up by a wide variety of tissues for tissue growth and repair.

After about 3–5 hours, most of the nutrients have been cleared from the circulation and the basal state is restored.

In patients with insulin resistance, several defects in this process are evident. In the fasting or postabsorptive state, insulin is no longer able fully to restrain hepatic glucose production, resulting in a small rise in the fasting plasma glucose concentration. This induces a compensatory increase in insulin secretion so that a new steady state is achieved with mild fasting hyperglycemia and hyperinsulinemia – the classic finding in insulin resistant states. In the post-prandial state, insulin also is not able fully to stimulate glucose uptake into muscle so there is an excessive rise in the plasma glucose concentration leading to even greater hyperinsulinemia (Bergman et al., 2002; LeRoith, 2002; Kahn, 2003).

Similar abnormalities in lipid metabolism also occur, with elevated fasting and post-prandial levels of free fatty acids and triglycerides. Although the pancreas initially is able to compensate for the higher levels of glucose, the continued exposure to hyperglycemia eventually leads to beta-cell failure – a process commonly known as glucose toxicity (Porte and Kahn, 2001; Robertson et al., 2004; Weir and Bonner-Weir, 2004). The first phase of insulin secretion is particularly sensitive to this process and typically becomes impaired when the fasting plasma glucose concentration exceeds 100 mg/dl, leading to more prolonged post-prandial hyperglycemia and even greater demand on the beta cell. Once the fasting plasma glucose concentration is consistently above 120 mg/dl, the second phase of insulin secretion also begins to fail, and the patient often progresses to overt type 2 diabetes shortly thereafter (DeFronzo, 2004) (Figure 25.4).

V. MEASUREMENT OF INSULIN RESISTANCE

Several clinical research techniques have been developed to measure insulin resistance. The most widely used method is the euglycemic hyperinsulinemic clamp technique, which is generally considered to be the gold standard for measuring insulin sensitivity *in vivo* (DeFronzo et al., 1979; Bergman et al., 1989). With this technique, a primed-continuous infusion of insulin is administered to raise the plasma insulin concentration to a predetermined physiological or pharmacological level. The plasma glucose concentration is then measured at 5-minute intervals and a variable infusion of exogenous glucose is administered to maintain the plasma glucose concentration constant at the fasting level. Since the plasma glucose concentration remains unchanged, the amount of exogenous glucose infused must equal the amount of glucose utilized in response to the hyperinsulinemia and, thus, provides a direct measure of whole-body sensitivity to

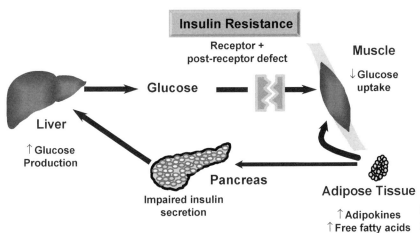

FIGURE 25.4 Pathophysiology of insulin resistance.

insulin. The insulin clamp is frequently combined with infusions of small amounts of stable or radioisotopically labeled glucose to measure hepatic glucose production or with labeled amino acids or fatty acids to measure protein or lipid metabolism. It also can be combined with indirect calorimetry to measure rates of glucose or lipid oxidation and non-oxidative glucose disposal, hemodynamic measures to assess the effects of insulin on vascular function or with imaging techniques to measure rates of glucose metabolism in specific tissues or organs.

The other most widely used method for assessing insulin sensitivity is the frequently sampled intravenous glucose tolerance test (FSIVGTT), also known as the minimal model technique (Bergman, 1989). With this procedure, a bolus of glucose, typically 300 mg/kg, is rapidly administered intravenously and plasma glucose and insulin levels are measured frequently for the next 3 hours. The resulting curves defined by the changes in plasma glucose and insulin are then fit using a non-linear modeling algorithm to derive indices reflecting the rate of change in plasma glucose level in response to the ambient insulin level (S_I, or insulin sensitivity), the rate of change in glucose independent of the level of insulin (S_G, or glucose effectiveness) and first and second phases of insulin secretion (Φ_1 and Φ_2). Although this technique is somewhat simpler to perform than the insulin clamp, it does not provide a steady state of glucose metabolism in response to a known level of insulin and thus typically cannot be combined with the other techniques used to assess intermediary metabolism, substrate oxidation, hemodynamic measures or imaging techniques. Many other techniques for measuring insulin sensitivity also have been developed, including the insulin suppression test (Greenfield et al., 1981), isolated organ or limb perfusion techniques and insulin measurements during an oral glucose tolerance test.

Because the aforementioned techniques are labor intensive and not suited for large population-based studies or routine clinical use, several simpler measures have been proposed for these purposes. The most widely used is the homeostasis model assessment of insulin resistance (HOMA-IR), which is calculated as [fasting plasma insulin level (in μU/ml) × fasting plasma glucose level (in mmol/l)]/ 22.5 (Matthews et al., 1985; Wallace et al., 2004). A value of 1.00 is considered normal and higher values indicate progressively severe states of insulin resistance. A similar measure known as the quantitative insulin sensitivity check index (QUICKI) is used by some investigators and is calculated as 1/[log fasting plasma glucose (in mg/dL) + log fasting plasma insulin (in μU/ml)] (Katz et al., 2000). Both of these measures are much more variable than either the insulin clamp or FSIVGTT, primarily due to the wide range of 'normal' values for fasting plasma insulin. Also, they cannot be used in patients with diabetes where the normal homeostatic relationship between plasma glucose and insulin levels no longer exists and they have theoretical limitations based on the fact that they attempt to measure insulin sensitivity in the fasting state when the majority of glucose uptake is independent of insulin.

VI. METABOLIC SYNDROME

A. Overview

The metabolic syndrome is a collection of clinical and laboratory abnormalities that are associated with insulin resistance and an increased risk of several diseases, predominantly atherosclerotic cardiovascular disease (ASCVD) and type 2 diabetes (Reaven, 1988, 2004; DeFronzo and Ferrannini, 1991; Rett, 1999; Grundy, 2006; Haffner, 2006; Pi-Sunyer, 2007). Although at different stages in its evolution it has been known as 'syndrome X', 'Reaven's syndrome' or the 'deadly quartet', the terms 'metabolic syndrome' and 'insulin resistance syndrome' are currently the most widely used in the medical literature. In medical practice, it is also known as 'dysmetabolic syndrome x', the term used to designate the clinical diagnosis of the disorder (ICD-9-CM diagnosis 277.7).

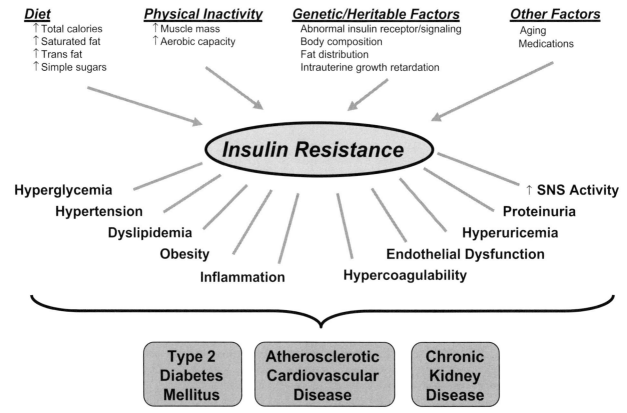

FIGURE 25.5 Features of the metabolic syndrome.

The core factors that define the syndrome are obesity (particularly intra-abdominal or visceral obesity), hyperglycemia, dyslipidemia, hypertension and atherosclerosis. Other associated abnormalities include elevated levels of inflammatory markers, a procoagulant state, increased sympathetic nervous system activity, endothelial dysfunction, hyperuricemia and proteinuria (Figure 25.5). Although chronic kidney disease traditionally has not been included as one of the primary components of the metabolic syndrome, it has become increasingly evident that renal disease plays a central role as both a cause and a consequence of the syndrome (Reaven, 1997; Chen et al., 2003, 2004; Shen et al., 2005; Lastra et al., 2006a; Peralta et al., 2006; Ritz, 2006, 2008; Sarafidis and Ruilope, 2006). Because all of these risk factors and their clinical consequences are associated with insulin resistance, many investigators believe that insulin resistance is the primary physiologic abnormality underlying the syndrome. While this may be true, based on our current understanding, it is preferable to consider this to be a true syndrome, i.e. the common association of pathological states that may have more than one etiology.

B. Definition

During the past decade, several attempts have been made to define the metabolic syndrome for both clinical and research purposes. The first definition was proposed by the World Health Organization (WHO) in 1998 and emphasized the central role of insulin resistance (defined as the presence of either impaired glucose tolerance, impaired fasting glucose, type 2 diabetes or as measured by the hyperinsulinemic clamp technique), plus two of the following criteria: obesity, hypertension, hypertriglyceridemia, decreased high density lipoprotein (HDL)-cholesterol or microalbuminuria (Alberti and Zimmet, 1998; World Health Organization, 1999). The following year, the European Group for the Study of Insulin Resistance (EGIR) modified the WHO criteria by requiring hyperinsulinemia as the primary laboratory abnormality, plus two of the following: abdominal obesity, hyperglycemia, hypertension, hypertriglyceridemia and low HDL-cholesterol (Balkau and Charles, 1999).

In 2001, the National Cholesterol Education Program Adult Treatment Panel III (NCEP-ATP III) put forth a definition that:

1. focused on the increased risk of ASCVD rather than insulin resistance
2. required no single defining criteria
3. recognized that certain ethnic groups, particularly South Asians, may have increased abdominal fat without an increase in body mass index (BMI), thus requiring a different threshold for waist circumference (Expert Panel on Detection, Evaluation, and Treatment of High Blood Cholesterol in Adults, 2001).

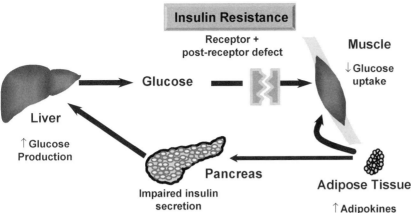

FIGURE 25.4 Pathophysiology of insulin resistance.

insulin. The insulin clamp is frequently combined with infusions of small amounts of stable or radioisotopically labeled glucose to measure hepatic glucose production or with labeled amino acids or fatty acids to measure protein or lipid metabolism. It also can be combined with indirect calorimetry to measure rates of glucose or lipid oxidation and non-oxidative glucose disposal, hemodynamic measures to assess the effects of insulin on vascular function or with imaging techniques to measure rates of glucose metabolism in specific tissues or organs.

The other most widely used method for assessing insulin sensitivity is the frequently sampled intravenous glucose tolerance test (FSIVGTT), also known as the minimal model technique (Bergman, 1989). With this procedure, a bolus of glucose, typically 300 mg/kg, is rapidly administered intravenously and plasma glucose and insulin levels are measured frequently for the next 3 hours. The resulting curves defined by the changes in plasma glucose and insulin are then fit using a non-linear modeling algorithm to derive indices reflecting the rate of change in plasma glucose level in response to the ambient insulin level (S_I, or insulin sensitivity), the rate of change in glucose independent of the level of insulin (S_G, or glucose effectiveness) and first and second phases of insulin secretion (Φ_1 and Φ_2). Although this technique is somewhat simpler to perform than the insulin clamp, it does not provide a steady state of glucose metabolism in response to a known level of insulin and thus typically cannot be combined with the other techniques used to assess intermediary metabolism, substrate oxidation, hemodynamic measures or imaging techniques. Many other techniques for measuring insulin sensitivity also have been developed, including the insulin suppression test (Greenfield et al., 1981), isolated organ or limb perfusion techniques and insulin measurements during an oral glucose tolerance test.

Because the aforementioned techniques are labor intensive and not suited for large population-based studies or routine clinical use, several simpler measures have been proposed for these purposes. The most widely used is the homeostasis model assessment of insulin resistance (HOMA-IR), which is calculated as [fasting plasma insulin level (in μU/ml) $\times$ fasting plasma glucose level (in mmol/l)]/22.5 (Matthews et al., 1985; Wallace et al., 2004). A value of 1.00 is considered normal and higher values indicate progressively severe states of insulin resistance. A similar measure known as the quantitative insulin sensitivity check index (QUICKI) is used by some investigators and is calculated as 1/[log fasting plasma glucose (in mg/dL) + log fasting plasma insulin (in μU/ml)] (Katz et al., 2000). Both of these measures are much more variable than either the insulin clamp or FSIVGTT, primarily due to the wide range of 'normal' values for fasting plasma insulin. Also, they cannot be used in patients with diabetes where the normal homeostatic relationship between plasma glucose and insulin levels no longer exists and they have theoretical limitations based on the fact that they attempt to measure insulin sensitivity in the fasting state when the majority of glucose uptake is independent of insulin.

VI. METABOLIC SYNDROME

A. Overview

The metabolic syndrome is a collection of clinical and laboratory abnormalities that are associated with insulin resistance and an increased risk of several diseases, predominantly atherosclerotic cardiovascular disease (ASCVD) and type 2 diabetes (Reaven, 1988, 2004; DeFronzo and Ferrannini, 1991; Rett, 1999; Grundy, 2006; Haffner, 2006; Pi-Sunyer, 2007). Although at different stages in its evolution it has been known as 'syndrome X', 'Reaven's syndrome' or the 'deadly quartet', the terms 'metabolic syndrome' and 'insulin resistance syndrome' are currently the most widely used in the medical literature. In medical practice, it is also known as 'dysmetabolic syndrome x', the term used to designate the clinical diagnosis of the disorder (ICD-9-CM diagnosis 277.7).

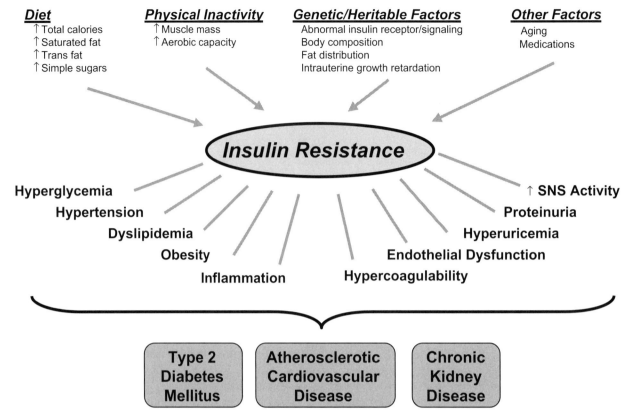

FIGURE 25.5 Features of the metabolic syndrome.

The core factors that define the syndrome are obesity (particularly intra-abdominal or visceral obesity), hyperglycemia, dyslipidemia, hypertension and atherosclerosis. Other associated abnormalities include elevated levels of inflammatory markers, a procoagulant state, increased sympathetic nervous system activity, endothelial dysfunction, hyperuricemia and proteinuria (Figure 25.5). Although chronic kidney disease traditionally has not been included as one of the primary components of the metabolic syndrome, it has become increasingly evident that renal disease plays a central role as both a cause and a consequence of the syndrome (Reaven, 1997; Chen et al., 2003, 2004; Shen et al., 2005; Lastra et al., 2006a; Peralta et al., 2006; Ritz, 2006, 2008; Sarafidis and Ruilope, 2006). Because all of these risk factors and their clinical consequences are associated with insulin resistance, many investigators believe that insulin resistance is the primary physiologic abnormality underlying the syndrome. While this may be true, based on our current understanding, it is preferable to consider this to be a true syndrome, i.e. the common association of pathological states that may have more than one etiology.

B. Definition

During the past decade, several attempts have been made to define the metabolic syndrome for both clinical and research purposes. The first definition was proposed by the World Health Organization (WHO) in 1998 and emphasized the central role of insulin resistance (defined as the presence of either impaired glucose tolerance, impaired fasting glucose, type 2 diabetes or as measured by the hyperinsulinemic clamp technique), plus two of the following criteria: obesity, hypertension, hypertriglyceridemia, decreased high density lipoprotein (HDL)-cholesterol or microalbuminuria (Alberti and Zimmet, 1998; World Health Organization, 1999). The following year, the European Group for the Study of Insulin Resistance (EGIR) modified the WHO criteria by requiring hyperinsulinemia as the primary laboratory abnormality, plus two of the following: abdominal obesity, hyperglycemia, hypertension, hypertriglyceridemia and low HDL-cholesterol (Balkau and Charles, 1999).

In 2001, the National Cholesterol Education Program Adult Treatment Panel III (NCEP-ATP III) put forth a definition that:

1. focused on the increased risk of ASCVD rather than insulin resistance
2. required no single defining criteria
3. recognized that certain ethnic groups, particularly South Asians, may have increased abdominal fat without an increase in body mass index (BMI), thus requiring a different threshold for waist circumference (Expert Panel on Detection, Evaluation, and Treatment of High Blood Cholesterol in Adults, 2001).

TABLE 25.1 Comparison of two most widely used criteria for diagnosis of metabolic syndrome

Measure	AHA/NHLBI criteria*	IDF criteria
Waist circumference	≥102 cm (40 in.) in men ≥88 cm (35 in.) in women	European, African ≥94 cm in men ≥80 cm in women South Asian, Chinese, Japanese ≥90 cm in men ≥80 cm in women
Triglycerides	≥150 mg/dl, or on drug treatment for elevated triglycerides	≥150 mg/dl, or on drug treatment for elevated triglycerides
HDL-cholesterol	<40 mg/dl in men <50 mg/dl in women, or on drug treatment for reduced HDL-cholesterol	<40 mg/dl in men <50 mg/dl in women, or on drug treatment for reduced HDL-cholesterol
Blood pressure	≥130/85 mmHg, or on drug treatment for hypertension	≥130/85 mmHg, or on drug treatment for hypertension
Fasting glucose	≥100 mg/dl, or on drug treatment for elevated glucose	≥100 mg/dl, or on drug treatment for elevated glucose

From data presented in Alberti et al. and IDF Epidemiology Task Force Consensus Group, 2005 and Grundy et al.; American Heart Association; National Heart, Lung, and Blood Institute, 2005.

*The American Heart Association/National Heart Lung and Blood Institute (AHA/NHLBI) requires any three of the five measures for a diagnosis of metabolic syndrome. The International Diabetes Federation (IDF) requires the presence of an elevated waist circumference plus any two of the remaining four measures for a diagnosis of metabolic syndrome.

This group proposed that the metabolic syndrome be defined by the presence of any three of the following five criteria: abdominal obesity, hypertension, hyperglycemia, hypertriglyceridemia or low HDL-cholesterol. Other consensus panels and professional groups have proposed definitions that, in addition to the criteria discussed above, may include hyperuricemia, polycystic ovary syndrome or family history of ASCVD (Einhorn et al., 2003).

Currently, the two most widely used definitions of the metabolic syndrome are the International Diabetes Federation (IDF) criteria (Alberti et al., 2005, 2006) and the American Heart Association/National Heart Lung and Blood Institute (AHA/NHLBI) modification of the NCEP-ATP III definition (Grundy et al., 2005) (Table 25.1). Both groups sought to produce a set of criteria that encompassed the key elements of the syndrome, had a strong basis in previous clinical and epidemiological research and could be easily measured in clinical practice. Although the two definitions are very similar, the IDF version requires the presence of abdominal obesity and contains different criteria for waist circumference in European versus Asian populations.

The metabolic syndrome has a high prevalence in the general population of the USA. The National Health and Nutrition Examination Survey-III (NHANES-III) estimated an overall prevalence of 23.7% (Ford et al., 2002) (Figure 25.6), while the Atherosclerosis Risk in the Community (ARIC) study found a prevalence of 23% (McNeill et al., 2005). In most studies, the prevalence is higher in the elderly and in most racial and ethnic minority groups, including African Americans, Hispanics, Asian Americans and Native Americans. Multiple epidemiological studies from the USA (NHANES-III, ARIC, San Antonio Heart

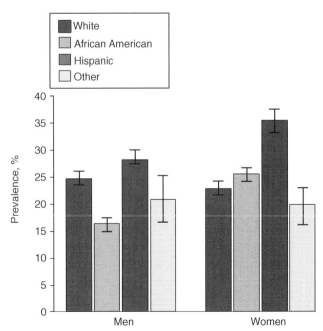

FIGURE 25.6 Age-adjusted prevalence of the metabolic syndrome among 8814 US adults aged at least 20 years, by sex and race or ethnicity, National Health and Nutrition Examination Survey-III, 1988–1994. Reprinted from Ford et al., 2002.

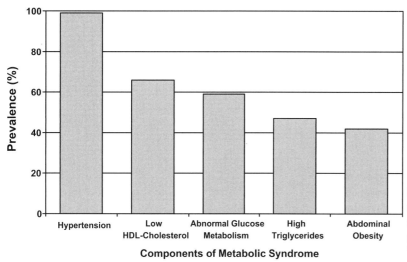

FIGURE 25.7 Prevalence of the individual components of the metabolic syndrome in patients on hemodialysis or peritoneal dialysis. Overall prevalence of metabolic syndrome is 69%. From data presented in Young et al., 2007.

Study, Framingham Heart Study) and Europe (Botnia Study, DECODE Study, Kuopio Study) have shown that the metabolic syndrome confers a three- to sixfold increase in risk of coronary heart disease, type 2 diabetes and cardiovascular mortality (Haffner, 2000; Isomaa et al., 2001; Lakka et al., 2002; Meigs et al., 2003; Hu et al., 2004).

There also is a strong association between the presence of the metabolic syndrome and the subsequent development of chronic kidney disease. In NHANES-III, patients with metabolic syndrome had a 2.6-fold increased odds of having a glomerular filtration rate (GFR) <60 ml/min/1.73 m^2 (Muntner et al., 2004; Beddhu et al., 2005). The risk increased from 2.2 when two components of the metabolic syndrome were present to 5.9 when all five components were present. Similarly, in the ARIC study, the odds ratio for GFR <60 ml/min/1.73 m^2 was 1.4 when the metabolic syndrome was present and this risk was substantially increased in persons over age 60 years (Muntner et al., 2005). Among patients receiving dialysis, the prevalence of metabolic syndrome is approximately 70%, predominantly due to very high rates of hypertension, low HDL-cholesterol and abnormal glucose metabolism (Young et al., 2007) (Figure 25.7).

C. Obesity

Obesity is a key component of the metabolic syndrome and is well established as a risk factor for cardiovascular disease and type 2 diabetes (Rexrode et al., 1998; Wilson et al., 2005; Meigs, 2008). Although any significant excess in body fat increases this risk, several studies have established that the risk is substantially greater for intra-abdominal or visceral fat versus subcutaneous fat (Carr and Brunzell, 2004; Després and Lemieux, 2006; Jensen, 2006; Haffner, 2007) (Figure 25.8). When compared with subcutaneous fat, visceral fat exhibits greater resistance to the antilipolytic effects of insulin and an enhanced response to lipolytic hormones such as epinephrine or cortisol (Bergman et al., 2006). Moreover, the free fatty acids and glycerol that are released from intra-abdominal fat are delivered directly to the liver where they provide increased substrate for gluconeogenesis and may be re-esterified to triglycerides, thereby producing a

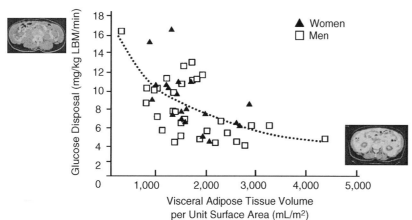

FIGURE 25.8 Relation between visceral adipose tissue and insulin action. LBM, lean body mass. Reprinted from Haffner, 2007.

fatty liver and further worsening insulin resistance. There are substantial racial and ethnic differences in regional fat distribution, with Asian populations often exhibiting increased visceral fat accumulation despite a relatively normal waist circumference and body habitus (Banerji et al., 1997; Raji et al., 2001a; Gerstein et al., 2003). Thus, criteria using waist/hip ratio or waist circumference as markers for visceral obesity must be adjusted to account for these differences.

Several physiologic mechanisms are responsible for the presence of insulin resistance in obesity. With greater adipose tissue mass, there is increased release of free fatty acids into the circulation. High levels of free fatty acids inhibit glucose oxidation and transport into cells, leading to insulin resistance and hyperglycemia. In addition, excess fatty acids are often deposited in ectopic locations, particularly muscle, further antagonizing insulin action. Excess adipose tissue mass also is accompanied by increased levels of many cytokines, including interleukin (IL)-1, IL-6 and tumor necrosis factor (TNF)-α. These mediators of inflammation can directly induce insulin resistance in skeletal muscle, liver and vascular endothelium and enhance the release of other inflammatory cardiovascular markers such as C-reactive protein (CRP) (Zoccali et al., 2005).

Adipose tissue also is the source of many adipokines that directly regulate fuel homeostasis. Leptin is produced in adipocytes and primarily acts on the hypothalamus to inhibit appetite and stimulate energy expenditure. Obese individuals are typically resistant to the effects of leptin and circulating levels are increased. Leptin also stimulates sympathetic nervous system activity and increases catecholamine turnover in the vasculature, brown adipose tissue and kidneys, thus raising blood pressure (Hall et al., 2001; Wolf et al., 2002). Leptin is independently associated with increased risk of coronary artery disease, hypertension and cardiomyopathy associated with obesity, increased production of reactive oxygen species and increased risk of thrombotic events (Correia and Haynes, 2004). Other adipokines, including resistin and visfatin, also are increased in obesity and can interfere with insulin action. Adiponectin is unique among the adipokines in that it appears to enhance insulin sensitivity and reduce the risk of developing type 2 diabetes and ASCVD. Adiponectin levels are typically low in obese patients and levels rise with weight loss.

Obesity also is well recognized as an independent risk factor for kidney disease (Adelman, 2002; Kincaid-Smith, 2004; Hsu et al., 2006; Kramer, 2006; Lastra et al., 2006b) (Figure 25.9). Obese individuals have an increase in glomerular filtration rate and renal plasma flow, enlarged glomeruli and up to 80% exhibit focal and segmental glomerulosclerosis in biopsy samples. These physiologic abnormalities are generally more severe in individuals who have been obese since childhood (Ejerblad et al., 2006; Hsu et al., 2006). The increased levels of free fatty acids and triglycerides and decreased HDL-cholesterol associated with increased fat mass are further exacerbated by renal insufficiency and fur-

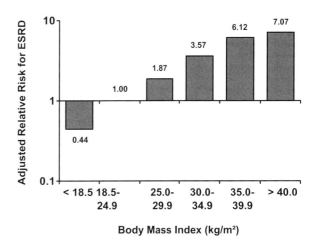

FIGURE 25.9 Association between body mass index and relative risk for development of end-stage renal disease (ESRD). Model adjusted for Multiphasic Health Checkup period, age, sex, race, education level, smoking status, history of myocardial infarction, serum cholesterol level, proteinuria, hematuria, and serum creatinine level. Adapted from Hsu et al., 2006.

ther increase the risk of subsequent cardiovascular disease (Chalmers et al., 2006). Moreover, chronic kidney disease is associated with elevated levels of inflammatory cytokines and other metabolically active hormones produced in adipose tissue (Zoccali et al., 2003; Stenvinkel, 2005; Kwan and Beddhu, 2007). Increased levels of IL-6, IL-1β, nuclear factor (NF)-κB and TNF-α commonly are observed in patients with obesity and kidney disease and are believed to contribute to the high rate of cardiovascular disease in this population (Axelsson et al., 2006a, b; Axelsson and Stenvinkel, 2008). Leptin levels also are increased in uremia, predominantly due to impaired renal clearance and may contribute to the anorexia and wasting that accompanies advanced renal disease. Although obesity is clearly a risk factor for development of kidney disease, once hemodialysis is initiated, overweight individuals frequently have increased survival, possibly due to better overall nutritional status or the increased muscle mass that accompanies obesity.

D. Hyperglycemia

Elevations in the plasma glucose concentration are the hallmark of insulin resistant states and are a fundamental component of the metabolic syndrome. The abnormalities in glucose homeostasis may include impaired fasting glucose (IFG; fasting plasma glucose of 100–125 mg/dl), impaired glucose tolerance (IGT; glucose level of 140–199 mg/dl 2 hours after a 75-g oral glucose tolerance test), or overt type 2 diabetes. The cellular and clinical features of insulin resistance and the natural history of the progression of insulin resistance to type 2 diabetes are discussed earlier in this chapter.

Diabetes is the leading cause of end-stage renal disease in the USA, accounting for 40–45% of all new cases annually. Approximately 15–20% of patients with type 1 diabetes will

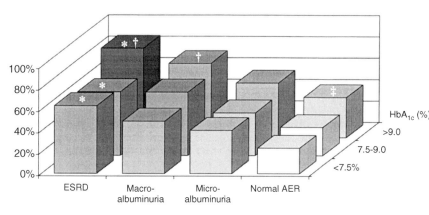

FIGURE 25.10 Prevalence of the metabolic syndrome in 2415 patients with type 1 diabetes according to glycemic control and different stages of albuminuria. *$P<0.001$ within the HbA_{1c} groups. †$P<0.001$ and ‡$P<0.05$ within the albuminuria groups. Reprinted from Thorn et al., 2005.

ultimately develop end-stage renal disease requiring dialysis or transplantation, although this proportion has been steadily decreasing over the past two decades due primarily to improved glycemic control, early treatment of microalbuminuria with angiotensin-converting enzyme (ACE) inhibitors or angiotensin receptor blockers (ARBs) and aggressive treatment of hypertension, dyslipidemia and other exacerbating factors. Among patients with type 1 diabetes, the prevalence of metabolic syndrome is higher in those with worse glycemic control and more severe renal insufficiency (Thorn et al., 2005) (Figure 25.10). The ability of improved glucose control to reduce the risk of development of nephropathy in those patients with normal renal function or reduce the progression in those with more advanced stages of proteinuria is well established (Ritz et al., 1997; DeFronzo, 2003; Williams and Stanton, 2005).

Estimates of the prevalence of end-stage renal disease among type 2 diabetic patients are much more variable, ranging from 10% to as high as 40% (Wolf and Ritz, 2003). The higher rates of renal disease in type 2 diabetes are thought to be due to the greater prevalence of several exacerbating factors associated with insulin resistance, including hypertension, dyslipidemia, obesity, atherosclerosis, endothelial dysfunction and increased levels of multiple vascular inflammatory markers (Endemann and Schiffrin, 2004). The subsequent development of kidney disease is closely related to the degree of impairment in glucose tolerance (Fox et al., 2005) (Figure 25.11). Hyperfiltration due to the presence of hyperglycemia and disordered regulation of intraglomerular hemodynamics also exacerbates the progression of renal disease (Castellino et al., 1990; Cooper, 2001). Molecular mechanisms involved in the development and progression of diabetic nephropathy include the accumulation of advanced glycosylation end products, increased activity of the polyol pathway, increased activity of diacyl glycerol-protein kinase C, increased levels of reactive oxygen species and increased activity of growth factors. Aggressive control of blood glucose, blood pressure and lipid levels are instrumental in slowing the decline in renal function and reducing the development of end-stage renal disease in this population (Gaede, 2006; Lewis, 2007).

E. Hypertension

The association between insulin resistance and hypertension is now well established (Hall et al., 1995b; Ferrannini et al., 1999; Corry and Tuck, 2001; Manrique et al., 2005). Although initial reports focused on patients with more severe elevations in blood pressure, the magnitude of the insulin resistance was generally less than that observed in obesity or

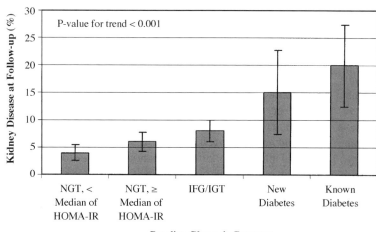

FIGURE 25.11 Prevalence of kidney disease at follow-up by baseline glycemic status among 2398 participants in the Framingham Heart Study. Bars represent 95% CI. Reprinted from Fox et al., 2005.

type 2 diabetes. Subsequent studies have shown that while patients with essential hypertension are insulin resistant as a group, as many as 50% have insulin sensitivity in the normal range. Insulin resistance also is less frequent in patients with secondary forms of hypertension compared to essential hypertension. More recently, insulin resistant subgroups within the essential hypertension population have been identified, particularly patients who have increased activity of the renin–angiotensin system and salt sensitivity.

Several factors contribute to the mechanisms by which insulin resistance and hyperinsulinemia may lead to hypertension (Hall, 1994; Aneja et al., 2004; El-Atat et al., 2004). Studies in animals using the isolated perfused kidney and chronic intrarenal insulin infusion have demonstrated that insulin directly enhances renal tubular sodium reabsorption independent of effects on GFR or renal plasma flow. In humans, DeFronzo et al. (1975), using the euglycemic hyperinsulinemic clamp technique, were the first to demonstrate that hyperinsulinemia decreased urinary sodium excretion by approximately 50% without a change in filtered load of glucose, GFR, renal plasma flow or aldosterone levels. Urinary potassium and phosphate excretion also both decreased during insulin administration although, unlike sodium, these may have been partially due to decreases in their respective plasma concentration (DeFronzo, 1981). Subsequent studies in humans have shown that the effect of insulin is dose-dependent up to circulating insulin levels of approximately 70 μU/ml – well within the physiologic range. Although initial studies concluded that the antinatriuretic effect of insulin was probably due to enhanced sodium reabsorption in the distal diluting segment, more recent reports have shown that insulin also promotes sodium retention in the loop of Henle and the proximal tubule. A primary distal tubular effect is supported by the fact that the greatest concentration of insulin receptor binding in the kidney occurs in the thick ascending loop and distal convoluted tubule. However, other studies have shown that the insulin enhances activity of sodium-proton exchanger and sodium-potassium ATPase, both of which are concentrated in the proximal tubule (Féraille and Doucet, 2001).

An important feature of this proposed mechanism is that the ability of insulin to enhance sodium retention by the kidney is not impaired in insulin resistant states, including obesity and type 2 diabetes (Hall et al., 1992; Hall, 1997). Thus, in states of insulin resistance accompanied by chronic hyperinsulinemia, there is a mild increase in total body exchangeable sodium and extracellular fluid volume. If peripheral vascular resistance does not change, then an increase in blood pressure would be expected. In healthy individuals, the potential adverse effects of this physiologic sequence are mitigated by the fact that insulin is also a peripheral vasodilator, so the net effect would be expected to be minimal. However, if vascular compliance is decreased due to the presence of atherosclerosis, or if the endothelial mechanisms mediating vascular tone are impaired – both of which are characteristic of insulin resistant states – then chronic elevation of blood pressure would more likely be the result (Ditzel et al., 1989; Hall et al., 1995a; Sarafidis and Bakris, 2007) (Figure 25.12).

States of insulin resistance also are frequently associated with increased activity of the renin–angiotensin system (Trevisan et al., 1998; Cooper, 2004). Angiotensin II is a potent vasoconstrictor that also induces the formation of reactive oxygen species that antagonize the effects of nitric oxide in vascular smooth muscle, promotes inflammation and stimulates vascular growth and fibrosis (Komers and Anderson, 2003; Cooper et al., 2007) (see Chapters 11–15 for additional details). In healthy individuals, these effects are mitigated by the ability of insulin and other hormones to induce vascular relaxation through their effects on the vascular endothelium and smooth muscle either directly or by modifying the production of endothelial nitric oxide. However, when insulin resistance is present, these effects are attenuated, leading to enhanced vasoconstriction and atherogenesis.

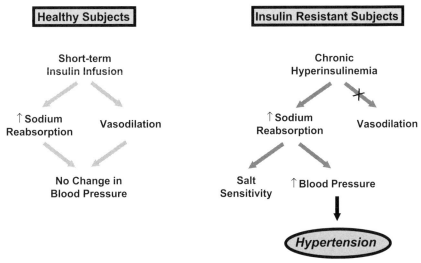

FIGURE 25.12 Effects of transient experimental hyperinsulinemia in healthy individuals and chronic hyperinsulinemia in subjects with insulin resistance. Although in both cases the antinatriuretic effect is similar, in the second case the continuous action of insulin, along with the absence of the compensating vasodilating effect, would result in blood pressure elevation. Reprinted from Sarafidis and Bakris, 2007.

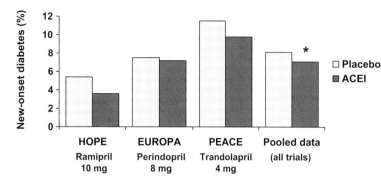

FIGURE 25.13 New-onset diabetes during the HOPE, EUROPA and PEACE studies comparing ACE inhibitors to placebo. In a pooled analysis of the 23 340 participants without diabetes mellitus at entry, 7.1% in the ACE inhibitor group and 8.1% in the placebo group developed diabetes during treatment (RR = 0.86; 95% CI 0.78, 0.95; $P = 0.0023$). From data presented in Dagenais et al., 2006.

Many studies have demonstrated that successful lowering of blood pressure is a highly effective means of slowing the progression from renal insufficiency to end-stage renal disease (Thomas and Atkins, 2006; Lewis, 2007). Because of the unique role of the renin–angiotensin system in regulating intrarenal hemodynamics, the angiotensin converting-enzyme (ACE) inhibitors and angiotensin receptor blockers (ARBs) are particularly effective in this group of patients (Mann et al., 2002; Appel et al., 2003). However, it is also known that both of these classes of antihypertensive medications also improve insulin sensitivity, which may provide additional benefits to the patients with renal insufficiency and insulin resistance (McFarlane et al., 2003). In a combined data analysis of three large clinical trials involving comparisons of ACE inhibitors with other antihypertensive agents, there was a significant reduction in new-onset diabetes in patients receiving the ACE inhibitors (Dagenais et al., 2006) (Figure 25.13). These effects may be mediated through direct interactions with insulin at the receptor and post-receptor level in both the intrarenal and peripheral vasculature, increased production of bradykinin or their partial agonist activity on the peroxisome proliferator activated receptor (PPAR)-γ system, similar to that seen with the thiazolidinediones (Damas et al., 2004).

Certain subgroups of patients with hypertension have been identified that may have a particularly high association with insulin resistance. One such group includes individuals with normal or high renin levels who are unable appropriately to modulate their blood pressure in response to changes in dietary sodium intake (Luft and Weinberger, 1997; Sechi, 1999; Rocchini, 2000; Raji et al., 2001b). Patients with this phenotype exhibit insulin resistance as measured either by HOMA-IR or the insulin clamp, an increased prevalence of dyslipidemia, increased risk of cardiovascular disease and a significantly greater family history of hypertension and other disorders associated with the metabolic syndrome (Gaboury et al., 1994, 1995; Ferri et al., 1999; Raji et al., 2006). They also have greater improvement in insulin sensitivity and renal hemodynamics during the administration of ACE inhibitors or ARBs, supporting the important role of the interaction between the renin–angiotensin system, insulin and sodium intake in the pathogenesis of hypertension in insulin resistant states (Perlstein et al., 2007).

F. Dyslipidemia

The classic dyslipidemia of the metabolic syndrome consists of elevated levels of triglycerides and apolipoprotein-B and a decrease in HDL-cholesterol. Although low density lipoprotein (LDL)-cholesterol levels are frequently normal, there is an increase in the small dense LDL phenotype as well as oxidized LDL, both of which have increased atherogenicity.

Chronic kidney disease is associated with many abnormalities in lipid metabolism that mimic the abnormalities seen in the metabolic syndrome (Saland and Ginsberg, 2007) (Table 25.2). In patients with proteinuria and mild reductions in GFR, LDL synthesis is increased, maturation of HDL-cholesterol particles is impaired and the clearance of chylomicrons and very low density lipoprotein (VLDL)-triglycerides is reduced due to decreased lipoprotein lipase activity (Kaysen, 2007a, b). Although total levels of LDL-cholesterol often are not severely elevated, there is an increase in the

TABLE 25.2 Lipid abnormalities in patients with kidney disease

	Nephrotic syndrome	CKD stage 1–4	CKD stage 5	Hemodialysis	Peritoneal dialysis	Post renal transplant
Total cholesterol	↑↑	↔	↔	↔	↑↑	↑
Triglyceride	↑↑	↔ or ↑	↑	↑	↑↑↑	↔ or ↑↑
LDL-cholesterol	↑	↔ or ↑ (↑small dense particles)	↔ or ↓ (↑small dense particles)	↔ or ↓ (↑small dense particles)	↑	↔ or ↑
HDL-cholesterol	↓	↓ or ↔	↓	↓	↓	↔

Adapted from Saland and Ginsberg, 2007.

small dense phenotype and lipoprotein (a) concentrations, both of which also increase the risk of cardiovascular disease. As the decrease in GFR progresses, there are further increases in triglycerides, apolipoprotein-B containing particles, remnant intermediate density lipoprotein (IDL)-cholesterol and VLDL-cholesterol, lipoprotein (a) and oxidized LDL-cholesterol and a decrease in HDL-cholesterol – all of which increase the risk of cardiovascular disease. Although LDL-cholesterol is not as strongly associated with cardiovascular events as in patients without chronic renal failure, the risk from low HDL-cholesterol and increased IDL appears to be greater.

Most of these abnormalities persist with only moderate improvements when hemodialysis is initiated. It is noteworthy that peritoneal dialysis typically exacerbates the dyslipidemia due to the large concentration of glucose and insulin in the dialysate, which stimulates hepatic lipoprotein synthesis. Treatment with HMG-CoA reductase inhibitors (statins) at low to moderate doses has been shown significantly to reduce cardiovascular mortality in patients on either hemodialysis or peritoneal dialysis by up to 50% – a greater reduction than is typically seen in the non-chronic kidney disease population (Shurraw and Tonelli, 2006). In addition to their specific effects on improving the lipid profile, statins exhibit many other beneficial effects in the patient with insulin resistance and kidney disease, including:

1. increased vascular and intrarenal nitric oxide production
2. decreased C-reactive protein (CRP), IL-6 and TNF-α
3. decreased responsiveness to angiotensin II, endothelin and other pressors
4. decreased thrombosis
5. decreased albuminuria
6. decreased risk of progression to type 2 diabetes (McFarlane et al., 2002).

G. Proteinuria

Increased urinary protein excretion is commonly found in patients with metabolic syndrome. Microalbuminuria, typically defined as urinary protein excretion of 30–300 mg/day, is an early manifestation of most types of renal disease. It also is independently associated with many of the components of the metabolic syndrome, including diabetes, hypertension, obesity and dyslipidemia, and is an independent predictor of risk of major cardiovascular events and death (Chen et al., 2004) (Figure 25.14). In NHANES-III, the odds ratio for the association between microalbuminuria and metabolic syndrome ranged from 1.9 to 4.1, depending on age, gender and the presence of diabetes. Moreover, among all participants, patients with metabolic syndrome had significantly increased rates of urinary albumin excretion (17.7 versus 11.5 mg per 24 h) (Palaniappan et al., 2003). Diabetes is the leading cause of end-stage renal disease in the USA and approximately 40–50% of all patients with type 2 diabetes will develop microalbuminuria during the course of their disease. Once microalbuminuria develops, about 5–10% of diabetic patients per year will progress to overt diabetic nephropathy.

Albuminuria is considered to be a marker of increased vascular permeability reflecting either endothelial dysfunction or vascular damage from atherosclerosis (Weinstock-Brown and Keane, 2001). It is frequently associated with increased glomerular filtration rate, impaired pressure natriuresis, chronic inflammation, abnormal autoregulation of renal blood flow due to endothelial dysfunction (particularly nitric oxide) and glomerular damage from hyperglycemia. Since albuminuria frequently results from increased activity of the renin–angiotensin system and intraglomerular pressure, medications like the angiotensin-converting enzyme inhibitors and angiotensin receptor blockers are frequently

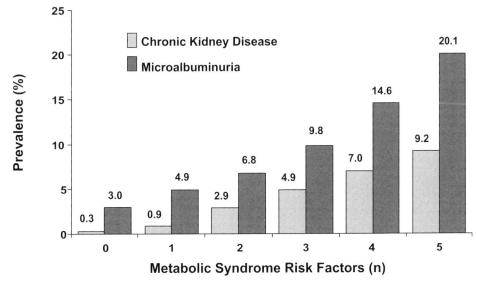

FIGURE 25.14 Prevalence of chronic kidney disease and microalbuminuria by number of the metabolic syndrome components. Adapted from Chen et al., 2004.

used to reduce albumin excretion. Several studies in both diabetic and hypertensive patients have shown that these classes of agents can slow the progression to more advanced stages of kidney disease and are specifically superior to other classes of antihypertensive drugs that provide similar degrees of reduction in blood pressure (Williams, 2005).

H. Hyperuricemia

Hyperuricemia frequently has been noted as a component of the metabolic syndrome and is an independent risk factor for the development of kidney disease, hypertension, atherosclerosis, cardiovascular disease, stroke and diabetes (Cirillo et al., 2006). Elevated levels of uric acid lead to renal vasoconstriction, intraglomerular hypertension, glomerular hypertrophy and interstitial damage and can hasten the progression to end-stage renal disease in experimental models of renal insufficiency. Although the precise mechanisms responsible for these effects are not known, uric acid is known to produce vascular dysfunction through inhibition of nitric oxide synthase and stimulation of the renin–angiotensin system. Moreover, studies in humans with moderate renal insufficiency have shown that lowering uric acid levels can lower blood pressure and reduce the rate of progression of renal disease.

VII. PATHOGENESIS OF INSULIN RESISTANCE IN CHRONIC KIDNEY DISEASE

Insulin resistance is present in almost all patients with chronic kidney disease. Because diabetes and hypertension are the two leading causes of renal disease, these patients already manifest insulin resistance before any clinical evidence of renal insufficiency has occurred. However, it is also clear that renal disease of any etiology, including polycystic kidney disease, glomerulonephritis, IgA nephropathy, or 5/6 nephrectomy models in animals also lead to insulin resistance and hyperinsulinemia once defects in renal function are evident, suggesting that some component of the uremic milieu is responsible for the defects in insulin action (Rigalleau and Gin, 2005).

Many studies in humans have demonstrated that impairments in insulin action typically occur when the GFR declines below 50–60 ml/min/1.73 m^2 (Fliser et al., 1998; Becker et al., 2005). Using the insulin clamp technique, it can be shown that the primary defect is in non-oxidative glucose disposal (i.e. glycolysis or glycogen synthesis), whereas glucose oxidation remains relatively normal until very advanced renal failure (DeFronzo et al., 1981; Schmitz, 1991; Castellino et al., 1992). Basal hepatic glucose production and its suppression by insulin are generally normal in uremic patients unless hyperglycemia or overt diabetes is also present (Capaldo et al., 1990). The defect in insulin stimulated glucose uptake is evident as both a reduction in the ability of maximal levels of insulin to stimulate normal glucose uptake (decreased responsiveness) as well as a rightward shift in the insulin dose–response curve (decreased sensitivity). Most studies have reported that insulin receptor binding is either normal or minimally reduced, suggesting that the defect lies in the insulin signaling pathway or glucose transport system (Jacobs et al., 1989; Friedman et al., 1991).

Although more than 30 years have passed since the first demonstration that uremia is associated with insulin resistance, there is still considerable uncertainty in our understanding of the uremic toxin(s) responsible for this effect. Recently, there has been interest in the role of asymmetric dimethyl arginine (ADMA), an endogenous inhibitor of nitric oxide production, as a potential mediator of the cardiovascular disease and insulin resistance in patients with kidney disease (Kielstein and Zoccali, 2005). ADMA is produced in increased amounts in renal insufficiency due to accelerated proteolysis and its clearance by the kidney is reduced. Since ADMA is a competitive inhibitor of arginine as the substrate for NO synthase, NO production is reduced and the ability of many vasodilators, including insulin, to produce vascular relaxation is impaired (Wilcken et al., 2007). Increased levels of ADMA are associated with other cardiovascular risk factors, particularly increased levels of homocysteine, as well as insulin resistance, obesity, carotid intima-media thickness and left ventricular hypertrophy.

The chronic metabolic acidosis that accompanies progressive renal insufficiency also significantly contributes to the insulin resistance (Kopple et al., 2005; Mitch, 2006a). Metabolic acidosis is present to some degree in almost all patients with renal disease when the GFR decreases below 30 ml/min/1.73 m^2. Induction of experimental metabolic acidosis in healthy volunteers by ammonium chloride administration produces a 15% reduction in insulin stimulated glucose uptake, but has minimal effects on basal hepatic glucose production or its suppression by insulin, thus mimicking the abnormalities seen in patients with renal failure (DeFronzo and Beckles, 1979). In patients receiving dialysis, aggressive treatment with sodium bicarbonate to maintain plasma bicarbonate levels above 22 mEq/l can significantly improve insulin sensitivity and secretion, but does not restore them completely to normal (Mak, 1998). Acidosis is also associated with accelerated rates of protein catabolism and decreased protein synthesis through activation of the ubiquitin-proteosome pathway, which contributes to chronic inflammation and impaired tissue growth and repair, thus further exacerbating the effects of insulin resistance (Kalantar-Zadeh et al., 2004).

Many other factors are believed to contribute to the insulin resistance of uremia, including anemia, hyperosmolality, elevated levels of carbamoylated amino acids, hyperparathyroidism and vitamin D deficiency (Mak, 1989). Although glucagon and growth hormone concentrations are frequently elevated in uremic patients, levels typically remain elevated after the initiation of hemodialysis even though insulin sensitivity improves. Insulin resistance also can be improved by low protein diets, increased physical activity, correction of hyperparathyroidism or initiation of dialysis. Insulin secretion is typically increased in non-diabetic patients with chronic kidney disease, representing an appropriate beta-cell response to the presence of insulin resistance. However, when compared to non-uremic patients with similar degrees of insulin resistance, it is apparent that a relative defect in insulin secretion is also present (Alvestrand et al., 1989).

The kidney also plays an important role in the clearance of insulin from the systemic circulation (Rabkin et al., 1984). Insulin is filtered by the glomerulus, reabsorbed in the proximal tubule by endocytosis and subsequently degraded. Insulin also undergoes receptor mediated clearance after binding to several sites in both the proximal and distal tubules, loop of Henle and collecting ducts where it regulates cation transport, or to cortical regions involved in gluconeogenesis. Insulin clearance begins to decline slightly as GFR decreases below 40 ml/min/1.73 m^2 and then decreases markedly when GFR falls below 15–20 ml/min/1.73 m^2 due to a significant decrease in tubular reuptake. Because of the important role of the kidney in removing insulin from the circulation and the presence of insulin resistance, patients with kidney disease often have elevated circulating levels of insulin, particularly in the postabsorptive state. The reduction in insulin clearance may also pose challenges for management of the insulin-treated diabetic patient, as insulin requirements often decrease and the pharmacokinetic properties of injected insulin become markedly altered with the progression of kidney disease.

VIII. REGULATION OF RENAL GLUCOSE PRODUCTION

It has long been recognized that the liver and the kidney are the two major organs providing glucose to the systemic circulation during periods of fasting. Although many tissues are capable of either synthesizing glucose for their metabolic needs or storing glycogen derived from uptake of circulating glucose, the liver and the kidney are unique in possessing glucose-6-phosphatase which enables them to release glucose into the circulation. In the postabsorptive state, the liver is the primary source of endogenous glucose production. Approximately half of the glucose released from the liver is derived from glycogenolysis and half is synthesized through the process of gluconeogenesis. In contrast, essentially all of the glucose released from the kidney is produced via gluconeogenesis, primarily in the renal cortex, since the kidney has very limited glycogen stores (Stumvoll et al., 1997, 1999; Battezzati et al., 2004).

After an overnight fast, the kidney contributes approximately 25% to total systemic glucose production, but utilizes approximately 20% of all glucose metabolized by the body. Thus, it contributes only about 5–10% to net systemic glucose appearance. After prolonged fasting, however, the kidney may contribute as much as 25–50% to net systemic glucose appearance and thus plays a critical role in maintaining whole-body glucose homeostasis during periods of severe nutrient restriction (Cersosimo et al., 2000).

Renal gluconeogenesis is inhibited by insulin, although some studies suggest that it is less responsive than hepatic gluconeogenesis to the inhibitory effects of the hormone (Cersosimo et al., 1994, 1999). Renal glucose production also can be greatly enhanced by epinephrine, but is relatively unresponsive to glucagon (Meyer et al., 2003; Gustavson et al., 2004). In humans and animal models of diabetes, renal gluconeogenesis is increased up to twofold due to a combination of insulin resistance, increased availability of gluconeogenic precursors and metabolic acidosis. Up to 40% of excess glucose production in patients with diabetes may be due to renal gluconeogenesis (Meyer et al., 2004).

IX. SYNDROMES OF SEVERE INSULIN RESISTANCE

Uncommon genetic diseases associated with severe insulin resistance, including the lipodystrophies, mutations in the insulin receptor or signaling pathways and autoantibodies to the insulin receptor, provide important insights into the relationship between insulin resistance, hyperinsulinemia and kidney disease. Although patients with these syndromes have marked hyperinsulinemia throughout life and frequently develop hyperglycemia or overt diabetes as children or adults, they are clearly different from the overwhelming majority of patients in whom the common variety of insulin resistance is present.

The congenital and acquired lipodystrophies are characterized by either partial or complete absence of adipose tissue. These patients have extreme hypertriglyceridemia, diabetes and ectopic deposition of fat in muscle, but are not obese and rarely have hypertension. Although more than half of the patients studied have gross proteinuria and over 90% have hyperfiltration, the most common renal lesions observed include focal segmental glomerulosclerosis and membranoproliferative glomerulonephritis, not diabetic glomerulosclerosis. Patients with mutations in the insulin receptor (type A insulin resistance) typically have the most severe hyperinsulinemia and the highest incidence of diabetes and also are the most likely of these rare syndromes to develop diabetic nephropathy. Patients with autoantibodies to the insulin receptor (type B insulin resistance) are also severely hyperinsulinemic, variably hyperglycemic and frequently have gross proteinuria, but renal lesions typically resemble autoimmune, not diabetic, nephropathy (Musso et al., 2006). Thus, despite the presence of severe insulin resistance and hyperinsulinemia, most of these patients do not manifest the classic features characteristic of the metabolic syndrome, further highlighting our current limited understanding of the genetic, cellular and physiologic etiology of the more common forms of insulin resistance.

X. TREATMENT

The cornerstone of treatment for patients with insulin resistance and most of the clinical disorders associated with the metabolic syndrome is diet and exercise. Although medications that directly improve insulin sensitivity (e.g. metformin or thiazolidinediones) have been developed for treatment of type 2 diabetes, none have been approved for the treatment of metabolic syndrome *per se* in the absence of the overt hyperglycemia that characterizes the diabetic state. Nevertheless, several clinical trials using thiazolidinediones have demonstrated improvements in many of the components of the metabolic syndrome in patients with chronic kidney disease. The treatment of other clinically overt manifestations of insulin resistance, particularly hypertension, cardiovascular disease and dyslipidemia, should be directed at the optimum pharmacologic approach for each disorder.

A. Dietary Management

Most patients with insulin resistance are obese, thus, the primary approach to treatment should focus on caloric restriction with the goal of achieving desirable body weight or body mass index. In addition, the content of the diet should be modified to reduce intake of saturated fats, *trans* fats, cholesterol, sodium and simple sugars and increase the consumption of monounsaturated fats, whole grains, fruits, vegetables, fish and poultry (Table 25.3). In the patient with chronic kidney disease, however, many additional dietary factors must be considered. Specifically, these patients present a unique combination of insulin resistance and the metabolic syndrome combined with general malnutrition and protein wasting. The problem is compounded by the fact that reduction in protein intake is generally recommended to slow the progression of renal disease and ameliorate the symptoms of uremia (Pedrini et al., 1996; Ikizler, 2004; Mandayam and Mitch, 2006).

Patients with advanced renal disease or receiving dialysis are in a chronic catabolic state characterized by anorexia, cachexia, chronic acidosis, increased metabolic rate, decreased skeletal muscle mass and general malnutrition – a condition also known as uremic wasting (Kopple, 1999; Mak et al., 2005). Recently, it has been appreciated that cytokines, including TNF-α, IL-1, IL-6 and leptin, increased levels of general markers of inflammation such as CRP, or increased activity of the ubiquitin-proteosome pathway may mediate much of the cachexia, increased metabolic rate and chronic malnutrition in end-stage renal disease, either by their direct effects on inflammation or through their effects on appetite (Mitch, 1998; Mitch and Du, 2004; Mak and Cheung, 2006). Since one of the major effects of insulin is to inhibit proteolysis and, when combined with increased levels of amino acids, promote net protein synthesis, this problem is even more difficult to manage in the diabetic patient with renal failure who may have more severe insulin resistance and/or insulin deficiency. Insulin-like growth factor-1 (IGF-1), a key hormone that mediates the effects of growth hormone on tissue growth and other anabolic processes, is also reduced in patients with chronic kidney disease. Metabolic balance studies have shown that the protein wasting is due to both a reduction in net protein synthesis in muscle and other lean tissues and an accelerated rate of proteolysis (Rennie and Wilkes, 2005; Mitch, 2006b). Although many abnormalities in protein metabolism have been characterized, supplementation of the branched chain amino acids (leucine, isoleucine and valine) or their keto-acids has been

TABLE 25.3 Nutrient composition of the therapeutic lifestyle changes diet

Nutrient	Recommended Intake
Saturated fat[*]	<7% of total calories
Polyunsaturated fat	Up to 10% of total calories
Monounsaturated fat	Up to 20% of total calories
Total fat	25–35% of total calories
Carbohydrate	50–60% of total calories
Fiber	20–39 g/day
Protein	Approximately 15% of total calories
Cholesterol	<200 mg/day
Total calories[‡]	Balance energy intake and expenditure to maintain desirable weight and prevent weight gain

Adapted from the Expert Panel on Detection, Evaluation, and Treatment of High Blood Cholesterol in Adults, 2001.
[*] *Trans* fatty acids are another LDL-raising fat that should be limited in the diet.
Carbohydrates should be derived predominantly from foods rich in complex carbohydrates, including grains, especially whole grains, fruits and vegetables.
[‡] Daily energy expenditure should include at least moderate physical activity (contributing ~200 kcal/day).

shown to improve insulin sensitivity and help decrease protein catabolism (Rigalleau et al., 1998). In addition, both peritoneal dialysis and hemodialysis produce chronic loss of amino acids and some proteins in the dialysate (Pupim et al., 2004).

Malnutrition due to uremia is a strong predictor of increased rates of hospitalization and mortality in this population. This problem is compounded by the delicate balance between reducing protein intake to slow the progression of renal disease and symptoms of uremia, typically to 0.5–0.8 g/kg/day, while maintaining sufficient amounts of high quality protein to prevent further wasting. This relationship may help explain the results from some studies in patients on dialysis demonstrating that higher body mass index and cholesterol levels are predictive of better survival, primarily because they reflect better nutritional status that outweighs any small increased risk of cardiovascular disease that they may present in this population.

Since diabetes is the leading cause of end-stage renal disease in the USA and the chronic insulin resistance and/or insulin deficiency that antedates the onset of renal insufficiency is already associated with increased catabolism and impairments in protein metabolism, the problem is particularly severe (Ikizler, 2007). The best approach is to optimize control of hyperglycemia, particularly before renal insufficiency has commenced, to prevent or delay the progression to more advanced stages of kidney disease. Also, if protein intake is restricted to slow the progression of the underlying renal disease and fat intake is maintained or mildly restricted to prevent exacerbation of the dyslipidemia, then the proportion of the diet composed of carbohydrates must be increased.

B. Exercise

The benefits of exercise for patients with insulin resistance and its associated clinical disorders are well documented. Many studies have demonstrated that regular moderate aerobic exercise of as little as 30 minutes per day for 5 or more days per week can significantly improve insulin sensitivity, improve glycemic control, raise HDL-cholesterol levels, lower blood pressure, help promote weight loss, preserve lean body mass and reduce the incidence of subsequent development of obesity, cardiovascular disease and diabetes.

In patients with renal disease, the benefits of moderate aerobic exercise or mild resistance training are even more striking. In patients who are not yet on dialysis, aerobic exercise improves insulin sensitivity, reduces oxidative stress, lowers the rate of progression of albuminuria and improves glomerular filtration rate. When combined with moderate resistance training, additional specific benefits are observed in the maintenance of lean body mass and improved strength and exercise capacity (Moinuddin and Leehey, 2008). In patients who have progressed to hemodialysis, aerobic exercise also improves insulin sensitivity and has favorable effects on improving dyslipidemia, lowering blood pressure, increasing hemoglobin levels and enhancing quality of life. The addition of resistance training further improves strength, preserves muscle mass, increases levels of IGF-1 and may improve the efficiency of dialysis.

C. Insulin Sensitizers

To the extent that insulin resistance contributes to the development of kidney disease in patients with the diabetes, hypertension and other components of the insulin resistance syndrome, then treatment with medications that improve insulin sensitivity may reduce the risk of progression of renal disease. There are two major classes of drugs used in the treatment of type 2 diabetes that improve insulin sensitivity – metformin and the thiazolidinediones (TZDs). Metformin should not be used in patients with even mild renal insufficiency (serum creatinine ≥ 1.4 mg/dl in women or ≥ 1.5 mg/dl in men) due to increased risk of lactic acidosis.

The thiazolidinediones, however, are primarily metabolized in the liver and can be used in patients with chronic kidney disease. These drugs, which include pioglitazone and rosiglitazone, directly improve insulin sensitivity by activating a nuclear receptor known as the peroxisome proliferator-activated receptor-gamma (PPAR-γ). These receptors are present primarily in adipocytes, but also are expressed in vascular smooth muscle and endothelial cells, macrophages and renal tubular and glomerular cells (Hsueh and Nicholas, 2002). In addition to lowering glucose levels in patients with diabetes, these medications also lower triglyceride levels, raise HDL-cholesterol levels, reduce visceral adiposity, lower blood pressure and decrease the levels of several markers of cardiovascular risk, including CRP and PAI-I (Dumasia et al., 2005; Schiffrin, 2005). Importantly, recent evidence in animals and humans indicates that they reduce proteinuria and preserve renal function in both diabetic and non-diabetic forms of kidney disease (Guan and Breyer, 2001; Guan, 2004).

Several small clinical studies using troglitazone, pioglitazone or rosiglitazone have shown short-term reduction of up to 60% in both microalbuminuria and gross proteinuria. However, the five largest studies using pioglitazone and rosiglitazone have shown 10–23% reduction in urinary albumin excretion over a 6–12 month period. In diabetic patients, this effect is independent of their glucose lowering effects, as similar improvements in renal function were not seen in trial comparing the TZDs with metformin, sulfonylureas or insulin despite similar improvements in HbA1c (Sarafidis and Bakris, 2006). Several possible mechanisms are thought to contribute to this effect, including lowering of plasma insulin levels, reduction in blood pressure, enhanced endothelial responsiveness, antiproliferative effects, anti-inflammatory effects, downregulation of the renin–angiotensin system and

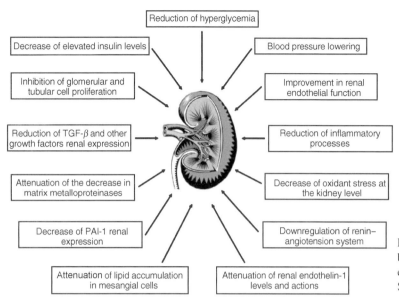

FIGURE 25.15 Actions of thiazolidinediones possibly contributing to the reduction of urinary albumin excretion and attenuation of renal injury. Reprinted from Sarafidis and Bakris, 2006.

reductions in circulating and intracellular lipid levels (Figure 25.15).

XI. MANAGEMENT OF DIABETES IN CHRONIC KIDNEY DISEASE

Because diabetes is the leading cause of end-stage renal disease in the USA and almost all other forms of kidney disease are associated with insulin resistance and defects in insulin secretion, many patients with chronic kidney disease require some form of pharmacologic treatment. Most patients with diabetes who develop end-stage renal disease either have type 1 or long-standing type 2 diabetes, hence, most are already being treated with insulin. Despite the fact that uremia is an insulin-resistant state, insulin requirements often decrease with the onset of renal insufficiency. This is primarily thought to result from the decrease in appetite, weight loss and cachexia that accompany the onset of renal insufficiency. Also, the kidney is a principal site of insulin clearance from the body, so the progressive loss of renal mass is accompanied by a prolongation of the effect of either exogenously administered or endogenously secreted insulin (Mak, 2000). If the patient also has diabetic autonomic neuropathy, particularly with gastroparesis, then precise matching of insulin dose with nutrient absorption is often very difficult and may be better managed with an insulin infusion pump. As with all insulin-treated patients, the specific doses and timing of insulin delivery should be guided by careful individual monitoring of fasting, pre-prandial and post-prandial glucose levels throughout the day and avoidance of frequent episodes of hypoglycemia.

For patients with type 2 diabetes who develop renal insufficiency before requiring insulin therapy, the doses of many oral antidiabetic medications need to be adjusted and others may need to be discontinued (Snyder and Berns, 2004). Metformin is excreted almost exclusively by the kidney and can accumulate in patients with renal insufficiency to levels that increase the risk of lactic acidosis. The use of metformin is contraindicated in patients with even mild renal insufficiency (serum creatinine ≥ 1.5 mg/dl in men or ≥ 1.4 mg/dl in women). Sulfonylureas (glipizide, glyburide, glimepiride) are extensively metabolized in the liver to intermediary compounds that have varying degrees of hypoglycemic activity. The metabolites plus a small percentage (typically 1–10%) of the parent drug are then excreted by the kidney, with the remainder of the metabolites being excreted in the biliary system. The accumulation of both the parent drug and partially active metabolites may occur in patients with severe reduction in GFR, so these drugs should be used with caution and at reduced doses in patients with kidney disease to avoid the risk of severe or prolonged hypoglycemia. The meglitinides (repaglinide and nateglinide) are extensively metabolized in the liver to inactive metabolites that are eliminated by both the liver and kidney. The guidelines for use of meglitinides in renal disease are similar to the sulfonylureas, i.e. use with caution and in reduced doses. Among the alpha glucosidase inhibitors, acarbose acts within the intestines and is not absorbed, but its metabolites are absorbed may accumulate in plasma in patients with renal failure. It is generally not used in patients with moderate to advanced renal disease. Miglitol, however, is extensively absorbed and excreted almost exclusively in the kidney, so its use in not recommended in patients with kidney disease. Thiazolidinediones (pioglitazone or rosiglitazone) are metabolized and excreted almost exclusively by the liver, so their use in patients with even moderately advanced renal insufficiency is frequently employed.

XII. HYPERGLYCEMIA ASSOCIATED WITH RENAL TRANSPLANTATION

Given the close interrelationship between insulin resistance and progressive renal failure, one might anticipate that kidney transplantation would improve insulin sensitivity and the accompanying glucose intolerance or diabetes. However, transplantation is frequently followed by a worsening state of glucose intolerance, a condition variably known as 'transplant-associated hyperglycemia', or 'post-transplant diabetes mellitus'. A meta-analysis of 19 prospective studies of new-onset diabetes noted that the incidence was as high as 50% during the first year after transplantation, although the two largest studies using the United States Medicare database found an incidence of 14–16% in the first year and 4–6% per year in subsequent years (Montori et al., 2002; Crutchlow and Bloom, 2007).

Several factors are believed to contribute to this phenomenon, particularly the immunosuppressive agents used to reduce the risk of transplant rejection. The hyperglycemia typically emerges within the first week after transplantation and subsides during the next few months as doses of immunosuppressive agents are reduced (Cosio et al., 2005) (Figure 25.16). Corticosteroids are well known for their ability to induce insulin resistance and exacerbate pre-existing diabetes. Although most current maintenance regimens use lower doses of steroids than in the past, or none at all, high doses are still frequently employed in the immediate post-transplant periods and/or during periods of threatened graft rejection. Both physiological and clinical studies have shown that reduction in steroid dose leads to an improvement in insulin sensitivity and overall glucose tolerance (van Hooff et al., 2004). The calcineurin inhibitors, predominantly tacrolimus and cyclosporine A, also are significantly associated with deterioration in glucose tolerance, although the mechanism is primarily related to their ability to inhibit insulin secretion rather than an effect on insulin sensitivity (Hjelmesaeth et al., 2002). This effect is particularly striking for tacrolimus, which increases the risk of new-onset diabetes by more than 60% compared with cyclosporine A (Webster et al., 2005). The risks of diabetes associated with sirolimus (rapamycin) are somewhat lower than with the calcineurin inhibitors, although sirolimus has been shown to increase triglyceride levels, ectopic fat deposition and also impair insulin secretion. Other factors thought to contribute to post-transplant diabetes include weight gain, improved insulin clearance by the newly functioning kidney and the risk of hepatitis C infection, which is a significant risk factor for development of type 2 diabetes. All patients undergoing renal transplant should be carefully evaluated for the glucose intolerance before and after the procedure and the principles of management should include appropriate diet and regular moderate aerobic exercise, use of the lowest possible doses of immunosuppressive medications, treatment of hyperlipidemia if indicated and use of antidiabetic medications if needed.

XIII. CONCLUSIONS

Over the past two decades, we have begun to appreciate that insulin resistance is not merely a pathophysiologic finding of interest to the endocrinologist or basic research scientist, but is an exceedingly common metabolic abnormality associated with many of the most frequent diseases seen in contemporary medical practice – obesity, type 2 diabetes, hypertension, dyslipidemia, coronary heart disease and chronic kidney disease. While great progress has been made in understanding the cellular and clinical physiology of insulin action, we still are not certain whether insulin resistance is truly the underlying cause of these diseases or simply a closely associated pathophysiologic abnormality. An even greater challenge, however, will be to develop new medical therapies designed to improve insulin sensitivity and, most importantly, to implement the lifestyle changes that are clearly so beneficial in ameliorating insulin resistance and

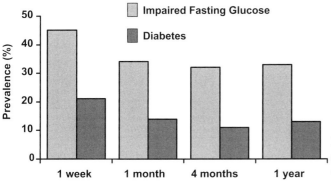

FIGURE 25.16 Prevalence of impaired fasting glucose and diabetes during the first year after kidney transplantation. Adapted from Cosio et al., 2005.

reducing the excessive morbidity and mortality of its associated disease states.

References

Adelman, R. D. (2002). Obesity and renal disease. *Curr Opin Nephrol Hypertens* **11**: 331–335.

Aguilar-Bryan, L., Bryan, J., and Nakazaki, M. (2001). Of mice and men: K(ATP) channels and insulin secretion. *Recent Prog Horm Res* **56**: 47–68.

Ahren, B., and Taborsky, G. J. (2003). Beta-cell function and insulin secretion. In: *Ellenberg & Rifkin's diabetes mellitus,* 6th edn, (Porte, D., Sherwin, R.S., Baron, A., eds) McGraw-Hill, New York, pp. 43–65.

Alberti, K. G., and Zimmet, P. Z. (1998). Definition, diagnosis and classification of diabetes mellitus and its complications. Part 1: diagnosis and classification of diabetes mellitus provisional report of a WHO consultation. *Diabet Med* **15**: 539–553.

Alberti, K. G., Zimmet, P., and Shaw, J. (2005). The metabolic syndrome – a new worldwide definition. IDF Epidemiology Task Force Consensus Group. *Lancet* **366**: 1059–1062.

Alberti, K. G., Zimmet, P., and Shaw, J. (2006). Metabolic syndrome – a new world-wide definition. A Consensus Statement from the International Diabetes Federation. *Diabet Med* **23**: 469–480.

Alvestrand, A., Mujagic, M., Wajngot, A., and Efendic, S. (1989). Glucose intolerance in uremic patients: the relative contributions of impaired beta-cell function and insulin resistance. *Clin Nephrol* **31**: 175–183.

Amatruda, J. M., Livingston, J. N., and Lockwood, D. H. (1975). Insulin receptor: role in the resistance of human obesity to insulin. *Science* **188**: 264–266.

Aneja, A., El-Atat, F., McFarlane, S. I., and Sowers, J. R. (2004). Hypertension and obesity. *Recent Prog Horm Res* **59**: 169–205.

Appel, G. B., Radhakrishnan, J., Avram, M. M. et al. (2003). Analysis of metabolic parameters as predictors of risk in the RENAAL study. *Diabetes Care* **26**: 1402–1407.

Avruch, J., Khokhlatchev, A., Kyriakis, J. M. et al. (2001). Ras activation of the Raf kinase: tyrosine kinase recruitment of the MAP kinase cascade. *Recent Prog Horm Res* **56**: 127–155.

Axelsson, J., and Stenvinkel, P. (2008). Role of fat mass and adipokines in chronic kidney disease. *Curr Opin Nephrol Hypertens* **17**: 25–31.

Axelsson, J., Carrero, J. J., Avesani, C. M., Heimbürger, O., Lindholm, B., and Stenvinkel, P. (2006a). Adipokine signaling in the peritoneal dialysis patient. *Contrib Nephrol* **150**: 166–173.

Axelsson, J., Heimbürger, O., and Stenvinkel, P. (2006b). Adipose tissue and inflammation in chronic kidney disease. *Contrib Nephrol* **151**: 165–174.

Balkau, B., and Charles, M. A. (1999). Comment on the provisional report from the WHO consultation. European Group for the Study of Insulin Resistance (EGIR). *Diabet Med* **16**: 442–443.

Banerji, M. A., Lebowitz, J., Chaiken, R. L., Gordon, D., Kral, J. G., and Lebovitz, H. E. (1997). Relationship of visceral adipose tissue and glucose disposal is independent of sex in black NIDDM subjects. *Am J Physiol* **273**: E425–E432.

Battezzati, A., Caumo, A., Martino, F. et al. (2004). Nonhepatic glucose production in humans. *Am J Physiol Endocrinol Metab* **286**: E129–E135.

Becker, B., Kronenberg, F., Kielstein, J. T. et al. (2005). Renal insulin resistance syndrome, adiponectin and cardiovascular events in patients with kidney disease: the mild and moderate kidney disease study. *J Am Soc Nephrol* **16**: 1091–1098.

Beddhu, S., Kimmel, P. L., Ramkumar, N., and Cheung, A. K. (2005). Associations of metabolic syndrome with inflammation in CKD: results From the Third National Health and Nutrition Examination Survey (NHANES III). *Am J Kidney Dis* **46**: 577–586.

Bergman, R. N. (1989). Lilly lecture 1989. Toward physiological understanding of glucose tolerance. Minimal-model approach. *Diabetes* **38**: 1512–1527.

Bergman, R. N., Hope, I. D., Yang, Y. J. et al. (1989). Assessment of insulin sensitivity in vivo: a critical review. *Diabetes Metab Rev* **5**: 411–429.

Bergman, R. N., Finegood, D. T., and Kahn, S. E. (2002). The evolution of beta-cell dysfunction and insulin resistance in type 2 diabetes. *Eur J Clin Invest* **32**: 35–45.

Bergman, R. N., Kim, S. P., Catalano, K. J. et al. (2006). Why visceral fat is bad: mechanisms of the metabolic syndrome. *Obesity* **14**: 16S–19S.

Bratanova-Tochkova, T. K., Cheng, H., Daniel, S. et al. (2002). Triggering and augmentation mechanisms, granule pools, and biphasic insulin secretion. *Diabetes* **51**: S83.

Capaldo, B., Cianciaruso, B., Napoli, R., Andreucci, V., Kopple, J. D., and Saccà, L. (1990). Role of the splanchnic tissues in the pathogenesis of altered carbohydrate metabolism in patients with chronic renal failure. *J Clin Endocrinol Metab* **70**: 127–133.

Carr, M. C., and Brunzell, J. D. (2004). Abdominal obesity and dyslipidemia in the metabolic syndrome: importance of type 2 diabetes and familial combined hyperlipidemia in coronary artery disease risk. *J Clin Endocrinol Metab* **89**: 2601–2607.

Castellino, P., Shohat, J., and DeFronzo, R. A. (1990). Hyperfiltration and diabetic nephropathy: is it the beginning? Or is it the end? *Semin Nephrol* **10**: 228–241.

Castellino, P., Solini, A., Luzi, L. et al. (1992). Glucose and amino acid metabolism in chronic renal failure: effect of insulin and amino acids. *Am J Physiol* **262**: F168–F176.

Cersosimo, E., Judd, R. L., and Miles, J. M. (1994). Insulin regulation of renal glucose metabolism in conscious dogs. *J Clin Invest* **93**: 2584–2589.

Cersosimo, E., Garlick, P., and Ferretti, J. (1999). Insulin regulation of renal glucose metabolism in humans. *Am J Physiol* **276**: E78–E84.

Cersosimo, E., Garlick, P., and Ferretti, J. (2000). Renal substrate metabolism and gluconeogenesis during hypoglycemia in humans. *Diabetes* **49**: 1186–1193.

Chalmers, L., Kaskel, F. J., and Bamgbola, O. (2006). The role of obesity and its bioclinical correlates in the progression of chronic kidney disease. *Adv Chronic Kidney Dis* **13**: 352–364.

Chen, J., Muntner, P., Hamm, L. L. et al. (2003). Insulin resistance and risk of chronic kidney disease in nondiabetic US adults. *J Am Soc Nephrol* **14**: 469–477.

Chen, J., Muntner, P., Hamm, L. L. et al. (2004). The metabolic syndrome and chronic kidney disease in US adults. *Ann Intern Med* **140**: 167–174.

Cirillo, P., Sato, W., Reungjui, S. et al. (2006). Uric acid, the metabolic syndrome, and renal disease. *J Am Soc Nephrol* **17**: S165–S168.

Cooper, M. E. (2001). Interaction of metabolic and haemodynamic factors in mediating experimental diabetic nephropathy. *Diabetologia* **44**: 1957–1972.

Cooper, M. E. (2004). The role of the renin-angiotensin-aldosterone system in diabetes and its vascular complications. *Am J Hypertens* **17**: 16S–20S.

Cooper, S. A., Whaley-Connell, A., Habibi, J. et al. (2007). Renin-angiotensin-aldosterone system and oxidative stress in cardiovascular insulin resistance. *Am J Physiol Heart Circ Physiol* **293**: H2009–H2023.

Correia, M. L., and Haynes, W. G. (2004). Leptin, obesity and cardiovascular disease. *Curr Opin Nephrol Hypertens* **13**: 215–223.

Corry, D. B., and Tuck, M. L. (2001). Selective aspects of the insulin resistance syndrome. *Curr Opin Nephrol Hypertens* **10**: 507–514.

Cosio, F. G., Kudva, Y., van der Velde, M. et al. (2005). New onset hyperglycemia and diabetes are associated with increased cardiovascular risk after kidney transplantation. *Kidney Int* **67**: 2415–2421.

Crutchlow, M. F., and Bloom, R. D. (2007). Transplant-associated hyperglycemia: a new look at an old problem. *Clin J Am Soc Nephrol* **2**: 343–355.

Dagenais, G. R., Pogue, J., Fox, K., Simoons, M. L., and Yusuf, S. (2006). Angiotensin-converting-enzyme inhibitors in stable vascular disease without left ventricular systolic dysfunction or heart failure: a combined analysis of three trials. *Lancet* **368**: 581–588.

Damas, J., Garbacki, N., and Lefèbvre, P. J. (2004). The kallikrein-kinin system, angiotensin converting enzyme inhibitors and insulin sensitivity. *Diabetes Metab Res Rev* **20**: 288–297.

DeFronzo, R. A. (1981). The effect of insulin on renal sodium metabolism. A review with clinical implications. *Diabetologia* **21**: 165–171.

DeFronzo, R. A. (2003). Diabetic nephropathy. In: *Ellenberg & Rifkin's diabetes mellitus*, 6th edn, (Porte, D., Sherwin, R.S., Baron, A., eds) McGraw-Hill, New York, pp. 723–745.

DeFronzo, R. A. (2004). Pathogenesis of type 2 diabetes mellitus. *Med Clin North Am* **88**: 787–835.

DeFronzo, R. A., and Beckles, A. D. (1979). Glucose intolerance following chronic metabolic acidosis in man. *Am J Physiol* **236**: E328–E334.

DeFronzo, R. A., and Ferrannini, E. (1991). Insulin resistance. A multifaceted syndrome responsible for NIDDM, obesity, hypertension, dyslipidemia, and atherosclerotic cardiovascular disease. *Diabetes Care* **14**: 173–194.

DeFronzo, R. A., Cooke, C. R., Andres, R., Faloona, G. R., and Davis, P. J. (1975). The effect of insulin on renal handling of sodium, potassium, calcium, and phosphate in man. *J Clin Invest* **55**: 845–855.

DeFronzo, R. A., Tobin, J. D., and Andres, R. (1979). Glucose clamp technique: a method for quantifying insulin secretion and resistance. *Am J Physiol* **237**: E214–E223.

DeFronzo, R. A., Alvestrand, A., Smith, D., Hendler, R., Hendler, E., and Wahren, J. (1981). Insulin resistance in uremia. *J Clin Invest* **67**: 563–568.

DeMeyts, P., Kahn, C. R., Roth, J., and Bar, R. S. (1976). Hormonal regulation of the affinity and concentration of hormone receptors in target cells. *Metabolism* **25**(Suppl 1): 1365–1370.

Després, J. P., and Lemieux, I. (2006). Abdominal obesity and metabolic syndrome. *Nature* **444**: 881–887.

Ditzel, J., Lervang, H. H., and Brøchner-Mortensen, J. (1989). Renal sodium metabolism in relation to hypertension in diabetes. *Diabete Metab* **15**: 292–295.

Dumasia, R., Eagle, K. A., Kline-Rogers, E., May, N., Cho, L., and Mukherjee, D. (2005). Role of PPAR- gamma agonist thiazolidinediones in treatment of pre-diabetic and diabetic individuals: a cardiovascular perspective. *Curr Drug Targets Cardiovasc Haematol Disord* **5**: 377–386.

Einhorn, D., Reaven, G. M., Cobin, R. H. et al. (2003). American College of Endocrinology position statement on the insulin resistance syndrome. *Endocr Pract* **9**: 237–252.

Ejerblad, E., Fored, C. M., Lindblad, P., Fryzek, J., McLaughlin, J. K., and Nyrén, O. (2006). Obesity and risk for chronic renal failure. *J Am Soc Nephrol* **17**: 1695–1702.

El-Atat, F. A., Stas, S. N., McFarlane, S. I., and Sowers, J. R. (2004). The relationship between hyperinsulinemia, hypertension and progressive renal disease. *J Am Soc Nephrol* **15**: 2816–2827.

Endemann, D. H., and Schiffrin, E. L. (2004). Endothelial dysfunction. *J Am Soc Nephrol* **15**: 1983–1992.

Expert Panel on Detection, Evaluation, and Treatment of High Blood Cholesterol in Adults. (2001). Executive summary of the third report of the national cholesterol education program (NCEP) expert panel on detection, evaluation, and treatment of high blood cholesterol in adults (Adult Treatment Panel III). *J Am Med Assoc* 285, 2486-97.

Féraille, E., and Doucet, A. (2001). Sodium-potassium-adenosine-triphosphatase-dependent sodium transport in the kidney: hormonal control. *Physiol Rev* **81**: 345–418.

Ferrannini, E., and Mari, A. (2004). Beta cell function and its relation to insulin action in humans: a critical appraisal. *Diabetologia* **47**: 943–956.

Ferrannini, E., Galvan, A. Q., Gastaldelli, A. et al. (1999). Insulin: new roles for an ancient hormone. *Eur J Clin Invest* **29**: 842–852.

Ferri, C., Bellini, C., Desideri, G. et al. (1999). Relationship between insulin resistance and nonmodulating hypertension: linkage of metabolic abnormalities and cardiovascular risk. *Diabetes* **48**: 1623–1630.

Fliser, D., Pacini, G., Engelleiter, R. et al. (1998). Insulin resistance and hyperinsulinemia are already present in patients with incipient renal disease. *Kidney Int* **53**: 1343–1347.

Ford, E. S., Giles, W. H., and Dietz, W. H. (2002). Prevalence of the metabolic syndrome among US adults: findings from the third National Health and Nutrition Examination Survey. *J Am Med Assoc* **287**: 356–359.

Fox, C. S., Larson, M. G., Leip, E. P., Meigs, J. B., Wilson, P. W., and Levy, D. (2005). Glycemic status and development of kidney disease: the Framingham Heart Study. *Diabetes Care* **28**: 2436–2440.

Friedman, J. E., Dohm, G. L., Elton, C. W. et al. (1991). Muscle insulin resistance in uremic humans: glucose transport, glucose transporters, and insulin receptors. *Am J Physiol* **261**: E87–E94.

Gaboury, C. L., Simonson, D. C., Seely, E. W., Hollenberg, N. K., and Williams, G. H. (1994). Relation of pressor responsiveness to angiotensin II and insulin resistance in hypertension. *J Clin Invest* **94**: 2295–2300.

Gaboury, C. L., Hollenberg, N. K., Hopkins, P. N., Williams, R., and Williams, G. H. (1995). Metabolic derangements in nonmodulating hypertension. *Am J Hypertens* **8:** 870–875.

Gaede, P. H. (2006). Intensified multifactorial intervention in patients with type 2 diabetes and microalbuminuria: rationale and effect on late-diabetic complications. *Dan Med Bull* **53:** 258–284.

Gerstein, H. C., Anand, S., Yi, Q. L., Vuksan, V., Lonn, E., and Teo, K. (2003). The relationship between dysglycemia and atherosclerosis in South Asian, Chinese, and European individuals in Canada: a randomly sampled cross-sectional study. *Diabetes Care* **26:** 144–149.

Greenfield, M. S., Doberne, L., Kraemer, F., Tobey, T., and Reaven, G. (1981). Assessment of insulin resistance with the insulin suppression test and the euglycemic clamp. *Diabetes* **30:** 387–392.

Grundy, S. M. (2006). Metabolic syndrome: connecting and reconciling cardiovascular and diabetes worlds. *J Am Coll Cardiol* **47:** 1093–1100.

Grundy, S. M., Cleeman, J. I., Daniels, S. R. et al. (2005). Diagnosis and management of the metabolic syndrome: an American Heart Association/National Heart, Lung, and Blood Institute scientific statement. American Heart Association; National Heart, Lung, and Blood Institute. *Circulation* **112:** 2735–2752.

Guan, Y. (2004). Peroxisome proliferator-activated receptor family and its relationship to renal complications of the metabolic syndrome. *J Am Soc Nephrol* **15:** 2801–2815.

Guan, Y., and Breyer, M. D. (2001). Peroxisome proliferator-activated receptors (PPARs): novel therapeutic targets in renal disease. *Kidney Int* **60:** 14–30.

Gustavson, S. M., Chu, C. A., Nishizawa, M. et al. (2004). Effects of hyperglycemia, glucagon, and epinephrine on renal glucose release in the conscious dog. *Metabolism* **53:** 933–941.

Haffner, S. M. (2000). Obesity and the metabolic syndrome: the San Antonio Heart Study. *Br J Nutr* **83**(Suppl 1): S67–S70.

Haffner, S. M. (2006). Risk constellations in patients with the metabolic syndrome: epidemiology, diagnosis, and treatment patterns. *Am J Med* **119**(Suppl 1): S3–S9.

Haffner, S. M. (2007). Abdominal adiposity and cardiometabolic risk: do we have all the answers? *Am J Med* **120**(9 Suppl 1): S10–16S.

Hall, J. E. (1994). Louis K. Dahl Memorial Lecture. Renal and cardiovascular mechanisms of hypertension in obesity. *Hypertension* **23:** 381–394.

Hall, J. E. (1997). Mechanisms of abnormal renal sodium handling in obesity hypertension. *Am J Hypertens* **10:** 49S–55S.

Hall, J. E., Brands, M. W., Hildebrandt, D. A., and Mizelle, H. L. (1992). Obesity-associated hypertension. Hyperinsulinemia and renal mechanisms. *Hypertension* **19**(Suppl): I45–I55.

Hall, J. E., Brands, M. W., Zappe, D. H., and Alonso-Galicia, M. (1995a). Cardiovascular actions of insulin: are they important in long-term blood pressure regulation? *Clin Exp Pharmacol Physiol* **22:** 689–700.

Hall, J. E., Brands, M. W., Zappe, D. H., and Alonso-Galicia, M. (1995b). Insulin resistance, hyperinsulinemia, and hypertension: causes, consequences, or merely correlations? *Proc Soc Exp Biol Med* **208:** 317–329.

Hall, J. E., Hildebrandt, D. A., and Kuo, J. (2001). Obesity hypertension: role of leptin and sympathetic nervous system. *Am J Hypertens* **14:** 103S–115S.

Henquin, J. C. (2005). Cell biology of insulin secretion. In: *Joslin's Diabetes Mellitus,* 14th edn, (Kahn, C.R., Weir, G.C., King, G. L., Moses, A.C., Smith, R.J., eds) Lippincott Williams & Wilkins, New York, pp. 83–107.

Himsworth, H. (1936). Diabetes mellitus: a differentiation into insulin-sensitive and insulin-insensitive types. *Lancet* **1:** 127–130.

Hjelmesaeth, J., Hagen, M., Hartmann, A., Midtvedt, K., Egeland, T., and Jenssen, T. (2002). The impact of impaired insulin release and insulin resistance on glucose intolerance after renal transplantation. *Clin Transplant* **16:** 389–396.

Hsu, C. Y., McCulloch, C. E., Iribarren, C., Darbinian, J., and Go, A. S. (2006). Body mass index and risk for end-stage renal disease. *Ann Intern Med* **144:** 21–28.

Hsueh, W. A., and Nicholas, S. B. (2002). Peroxisome proliferator-activated receptor-gamma in the renal mesangium. *Curr Opin Nephrol Hypertens* **11:** 191–195.

Hu, G., Qiao, Q., Tuomilehto, J., Balkau, B., Borch-Johnsen, K., and Pyorala, K. (2004). Prevalence of the metabolic syndrome and its relation to all-cause and cardiovascular mortality in nondiabetic European men and women. DECODE Study Group. *Arch Intern Med* **164:** 1066–1076.

Ikizler, T. A. (2004). Protein and energy: recommended intake and nutrient supplementation in chronic dialysis patients. *Semin Dial* **17:** 471–478.

Ikizler, T. A. (2007). Effects of glucose homeostasis on protein metabolism in patients with advanced chronic kidney disease. *J Ren Nutr* **17:** 13–16.

Isomaa, B., Almgren, P., Tuomi, T. et al. (2001). Cardiovascular morbidity and mortality associated with the metabolic syndrome. *Diabetes Care* **24:** 683–689.

Jacobs, D. B., Hayes, G. R., Truglia, J. A., and Lockwood, D. H. (1989). Alterations of glucose transporter systems in insulin-resistant uremic rats. *Am J Physiol* **257:** E193–E197.

Jensen, M. D. (2006). Is visceral fat involved in the pathogenesis of the metabolic syndrome? Human model. *Obesity* **14**(Suppl 1): 20S–24S.

Kahn, C. R. (1980). Role of insulin receptors in insulin-resistant states. *Metabolism* **29:** 455–466.

Kahn, C. R., and Saltiel, A. R. (2005). The molecular mechanism of insulin action and the regulation of glucose and lipid metabolism. In: *Joslin's Diabetes Mellitus,* 14th edn, (Kahn, C.R., Weir, G.C., King, G.L., Jacobson, A.M., Moses, A.C., Smith, R.J., eds) Lippincott Williams & Wilkins, New York, pp. 145–168.

Kahn, C. R., Neville, D. M., Gorden, P., Freychet, P., and Roth, J. (1972). Insulin receptor defect in insulin resistance: studies in the obese-hyperglycemic mouse. *Biochem Biophys Res Commun* **48:** 135–142.

Kahn, S. E. (2003). The relative contributions of insulin resistance and beta-cell dysfunction to the pathophysiology of type 2 diabetes. *Diabetologia* **46:** 3–19.

Kalantar-Zadeh, K., Mehrotra, R., Fouque, D., and Kopple, J. D. (2004). Metabolic acidosis and malnutrition-inflammation complex syndrome in chronic renal failure. *Semin Dial* **17:** 455–465.

Kanzaki, M., and Pessin, J. E. (2001). Signal integration and the specificity of insulin action. *Cell Biochem Biophys* **35:** 191–209.

Karam, J. H., Grodsky, G. M., Pavlatos, F. C., and Forsham, P. H. (1965). Critical factors in excessive serum-insulin response to glucose. Obesity in maturity-onset diabetes and growth hormone in acromegaly. *Lancet* **1:** 286–289.

Katz, A., Nambi, S. S., Mather, K. et al. (2000). Quantitative insulin sensitivity check index: a simple, accurate method for assessing insulin sensitivity in humans. *J Clin Endocrinol Metab* **85**: 2402–2410.

Kaysen, G. A. (2007a). Disorders in high-density metabolism with insulin resistance and chronic kidney disease. *J Ren Nutr* **17**: 4–8.

Kaysen, G. A. (2007b). Hyperlipidemia in chronic kidney disease. *Int J Artif Organs* **30**: 987–992.

Kido, Y., Nakae, J., and Accili, D. (2001). Clinical review 125: the insulin receptor and its cellular targets. *J Clin Endocrinol Metab* **86**: 972–979.

Kielstein, J. T., and Zoccali, C. (2005). Asymmetric dimethylarginine: a cardiovascular risk factor and a uremic toxin coming of age? *Am J Kidney Dis* **46**: 186–202.

Kincaid-Smith, P. (2004). Hypothesis: obesity and the insulin resistance syndrome play a major role in end-stage renal failure attributed to hypertension and labelled 'hypertensive nephrosclerosis'. *J Hypertens* **22**: 1051–1055.

Komers, R., and Anderson, S. (2003). Paradoxes of nitric oxide in the diabetic kidney. *Am J Physiol Renal Physiol* **284**: F1121–F1137.

Kopple, J. D. (1999). Pathophysiology of protein-energy wasting in chronic renal failure. *J Nutr* **129**(1S Suppl): 247S–251S.

Kopple, J. D., Kalantar-Zadeh, K., and Mehrotra, R. (2005). Risks of chronic metabolic acidosis in patients with chronic kidney disease. *Kidney Int (Suppl)* **95**: S21–S27.

Kramer, H. (2006). Obesity and chronic kidney disease. *Contrib Nephrol* **151**: 1–18.

Kwan, B. C., and Beddhu, S. (2007). A story half untold: adiposity, adipokines and outcomes in dialysis population. *Semin Dial* **20**: 493–497.

Lakka, H. M., Laaksonen, D. E., Lakka, T. A. et al. (2002). The metabolic syndrome and total and cardiovascular disease mortality in middle-aged men. *J Am Med Assoc* **288**: 2709–2716.

Lastra, G., Manrique, C., McFarlane, S. I., and Sowers, J. R. (2006a). Cardiometabolic syndrome and chronic kidney disease. *Curr Diab Rep* **6**: 207–212.

Lastra, G., Manrique, C., and Sowers, J. R. (2006b). Obesity, cardiometabolic syndrome, and chronic kidney disease: the weight of the evidence. *Adv Chronic Kidney Dis* **13**: 365–373.

LeRoith, D. (2002). Beta-cell dysfunction and insulin resistance in type 2 diabetes: role of metabolic and genetic abnormalities. *Am J Med* **113**(Suppl 6A): 3S–11S.

Lewis, E. J. (2007). Treating hypertension in the patient with overt diabetic nephropathy. *Semin Nephrol* **27**: 182–194.

Luft, F. C., and Weinberger, M. H. (1997). Heterogeneous responses to changes in dietary salt intake: the salt-sensitivity paradigm. *Am J Clin Nutr* **65**(2 Suppl): 612S–617S.

McFarlane, S. I., Muniyappa, R., Francisco, R., and Sowers, J. R. (2002). Clinical review 145: pleiotropic effects of statins: lipid reduction and beyond. *J Clin Endocrinol Metab* **87**: 1451–1458.

McFarlane, S. I., Kumar, A., and Sowers, J. R. (2003). Mechanisms by which angiotensin-converting enzyme inhibitors prevent diabetes and cardiovascular disease. *Am J Cardiol* **91**: 30H–37H.

McNeill, A. M., Rosamond, W. D., Girman, C. J. et al. (2005). The metabolic syndrome and 11-year risk of incident cardiovascular disease in the atherosclerosis risk in communities study. *Diabetes Care* **28**: 385–390.

Mak, R. H. (1989). Insulin secretion in uremia: effect of parathyroid hormone and vitamin D metabolites. *Kidney Int (Suppl)* **27**: S227–S230.

Mak, R. H. (1998). Effect of metabolic acidosis on insulin action and secretion in uremia. *Kidney Int* **54**: 603–607.

Mak, R. H. (2000). Impact of end-stage renal disease and dialysis on glycemic control. *Semin Dial* **13**: 4–8.

Mak, R. H., and Cheung, W. (2006). Energy homeostasis and cachexia in chronic kidney disease. *Pediatr Nephrol* **21**: 1807–1814.

Mak, R. H., Cheung, W., Cone, R. D., and Marks, D. L. (2005). Orexigenic and anorexigenic mechanisms in the control of nutrition in chronic kidney disease. *Pediatr Nephrol* **20**: 427–431.

Mandayam, S., and Mitch, W. E. (2006). Dietary protein restriction benefits patients with chronic kidney disease. *Nephrology* **11**: 53–57.

Mann, J. F., Gerstein, H. C., Pogue, J., Lonn, E., and Yusuf, S. (2002). Cardiovascular risk in patients with early renal insufficiency: implications for the use of ACE inhibitors. *Am J Cardiovasc Drugs* **2**: 157–162.

Manrique, C., Lastra, G., Whaley-Connell, A., and Sowers, J. R. (2005). Hypertension and the cardiometabolic syndrome. *J Clin Hypertens* **7**: 471–476.

Matthews, D. R., Hosker, J. P., Rudenski, A. S., Naylor, B. A., Treacher, D. F., and Turner, R. C. (1985). Homeostasis model assessment: insulin resistance and beta-cell function from fasting plasma glucose and insulin concentrations in man. *Diabetologia* **28**: 412–419.

Meigs, J. B. (2008). The role of obesity in insulin resistance. In: (Hansen, B.C., Bray, G.A., eds) *The metabolic syndrome. Epidemiology, clinical treatment, and underlying mechanisms*, Humana Press, Totowa, pp. 37–55.

Meigs, J. B., Wilson, P. W., Nathan, D. M., D'Agostino, R. B., Williams, K., and Haffner, S. M. (2003). Prevalence and characteristics of the metabolic syndrome in the San Antonio Heart and Framingham Offspring Studies. *Diabetes* **52**: 2160–2167.

Meyer, C., Stumvoll, M., Welle, S., Woerle, H. J., Haymond, M., and Gerich, J. (2003). Relative importance of liver, kidney, and substrates in epinephrine-induced increased gluconeogenesis in humans. *Am J Physiol Endocrinol Metab* **285**: E819–E826.

Meyer, C., Woerle, H. J., Dostou, J. M., Welle, S. L., and Gerich, J. E. (2004). Abnormal renal, hepatic, and muscle glucose metabolism following glucose ingestion in type 2 diabetes. *Am J Physiol Endocrinol Metab* **287**: E1049–E1056.

Mitch, W. E. (1998). Robert H Herman Memorial Award in Clinical Nutrition Lecture, 1997. Mechanisms causing loss of lean body mass in kidney disease. *Am J Clin Nutr* **67**: 359–366.

Mitch, W. E. (2006a). Metabolic and clinical consequences of metabolic acidosis. *J Nephrol* **19**(Suppl 9): S70–S75.

Mitch, W. E. (2006b). Proteolytic mechanisms, not malnutrition, cause loss of muscle mass in kidney failure. *J Ren Nutr* **16**: 208–211.

Mitch, W. E., and Du, J. (2004). Cellular mechanisms causing loss of muscle mass in kidney disease. *Semin Nephrol* **24**: 484–487.

Moinuddin, I., and Leehey, D. J. (2008). A comparison of aerobic exercise and resistance training in patients with and without chronic kidney disease. *Adv Chronic Kidney Dis* **15**: 83–96.

Montori, V. M., Basu, A., Erwin, P. J., Velosa, J. A., Gabriel, S. E., and Kudva, Y. C. (2002). Posttransplantation diabetes: a systematic review of the literature. *Diabetes Care* **25**: 583–592.

Muntner, P., He, J., Chen, J., Fonseca, V., and Whelton, P. K. (2004). Prevalence of non-traditional cardiovascular disease risk factors among persons with impaired fasting glucose, impaired glucose tolerance, diabetes, and the metabolic syndrome: analysis of the Third National Health and Nutrition Examination Survey (NHANES III). *Ann Epidemiol* **14**: 686–695.

Muntner, P., He, J., Astor, B. C., Folsom, A. R., and Coresh, J. (2005). Traditional and nontraditional risk factors predict coronary heart disease in chronic kidney disease: results from the atherosclerosis risk in communities study. *J Am Soc Nephrol* **16**: 529–538.

Musso, C., Javor, E., Cochran, E., Balow, J. E., and Gorden, P. (2006). Spectrum of renal diseases associated with extreme forms of insulin resistance. *Clin J Am Soc Nephrol* **1**: 616–622.

Olefsky, J. M. (1976). The insulin receptor: its role in insulin resistance of obesity and diabetes. *Diabetes* **25**: 1154–1162.

Olefsky, J. M., and Reaven, G. M. (1974). Decreased insulin binding to lymphocytes from diabetic subjects. *J Clin Invest* **54**: 1323–1328.

Palaniappan, L., Carnethon, M., and Fortmann, S. P. (2003). Association between microalbuminuria and the metabolic syndrome: NHANES III. *Am J Hypertens* **16**: 952–958.

Patti, M. E., and Kahn, C. R. (1998). The insulin receptor – a critical link in glucose homeostasis and insulin action. *J Basic Clin Physiol Pharmacol* **9**: 89–109.

Pedrini, M. T., Levey, A. S., Lau, J., Chalmers, T. C., and Wang, P. H. (1996). The effect of dietary protein restriction on the progression of diabetic and nondiabetic renal diseases: a meta-analysis. *Ann Intern Med* **124**: 627–632.

Peralta, C. A., Kurella, M., Lo, J. C., and Chertow, G. M. (2006). The metabolic syndrome and chronic kidney disease. *Curr Opin Nephrol Hypertens* **15**: 361–365.

Perlstein, T. S., Gerhard-Herman, M., Hollenberg, N. K., Williams, G. H., and Thomas, A. (2007). Insulin induces renal vasodilation, increases plasma renin activity, and sensitizes the renal vasculature to angiotensin receptor blockade in healthy subjects. *J Am Soc Nephrol* **18**: 944–951.

Pi-Sunyer, X. (2007). The metabolic syndrome: how to approach differing definitions. *Med Clin North Am* **91**: 1025–1040.

Porte, D., and Kahn, S. E. (2001). Beta-cell dysfunction and failure in type 2 diabetes: potential mechanisms. *Diabetes* **50**(Suppl 1): S160–S163.

Pupim, L. B., Flakoll, P. J., and Ikizler, T. A. (2004). Protein homeostasis in chronic hemodialysis patients. *Curr Opin Clin Nutr Metab Care* **7**: 89–95.

Rabkin, R., Ryan, M. P., and Duckworth, W. C. (1984). The renal metabolism of insulin. *Diabetologia* **27**: 351–357.

Raji, A., Seely, E. W., Arky, R. A., and Simonson, D. C. (2001a). Body fat distribution and insulin resistance in healthy Asian Indians and Caucasians. *J Clin Endocrinol Metab* **86**: 5366–5371.

Raji, A., Williams, G. H., Jeunemaitre, X. et al. (2001b). Insulin resistance in hypertensives: effect of salt sensitivity, renin status and sodium intake. *J Hypertens* **19**: 99–105.

Raji, A., Williams, J. S., Hopkins, P. N., Simonson, D. C., and Williams, G. H. (2006). Familial aggregation of insulin resistance and cardiovascular risk factors in hypertension. *J Clin Hypertens* **8**: 791–796.

Reaven, G. (2004). The metabolic syndrome or the insulin resistance syndrome? Different names, different concepts, and different goals *Endocrinol Metab Clin N Am* **33**: 283–303.

Reaven, G. M. (1988). Banting lecture 1988. Role of insulin resistance in human disease. *Diabetes* **37**: 1595–1607.

Reaven, G. M. (1997). The kidney: an unwilling accomplice in syndrome X. *Am J Kidney Dis* **30**: 928–931.

Rennie, M. J., and Wilkes, E. A. (2005). Maintenance of the musculoskeletal mass by control of protein turnover: the concept of anabolic resistance and its relevance to the transplant recipient. *Ann Transplant* **10**: 31–34.

Rett, K. (1999). The relation between insulin resistance and cardiovascular complications of the insulin resistance syndrome. *Diabetes Obes Metab* **1**(Suppl 1): S8–S16.

Rexrode, K. M., Carey, V. J., Hennekens, C. H. et al. (1998). Abdominal adiposity and coronary heart disease in women. *J Am Med Assoc* **280**: 1843–1848.

Rhodes, C. J., and White, M. F. (2002). Molecular insights into insulin action and secretion. *Eur J Clin Invest* **32**(Suppl 3): 3–13.

Rigalleau, V., and Gin, H. (2005). Carbohydrate metabolism in uraemia. *Curr Opin Clin Nutr Metab Care* **8**: 463–469.

Rigalleau, V., Aparicio, M., and Gin, H. (1998). Effects of low-protein diet on carbohydrate metabolism and energy expenditure. *J Ren Nutr* **8**: 175–178.

Ritz, E. (2006). Heart and kidney: fatal twins? *Am J Med* **119**(Suppl 1): S31–S39.

Ritz, E. (2008). Metabolic syndrome and kidney disease. *Blood Purif* **26**: 59–62.

Ritz, E., Keller, C., Bergis, K., and Strojek, K. (1997). Pathogenesis and course of renal disease in IDDM/NIDDM: differences and similarities. *Am J Hypertens* **10**: 202S–207S.

Robertson, R. P., Harmon, J., Tran, P. O., and Poitout, V. (2004). Beta-cell glucose toxicity, lipotoxicity, and chronic oxidative stress in type 2 diabetes. *Diabetes* **53**(Suppl 1): S119–S124.

Rocchini, A. P. (2000). Obesity hypertension, salt sensitivity and insulin resistance. *Nutr Metab Cardiovasc Dis* **10**: 281–294.

Saland, J. M., and Ginsberg, H. N. (2007). Lipoprotein metabolism in chronic renal insufficiency. *Pediatr Nephrol* **22**: 1095–1112.

Sarafidis, P. A., and Bakris, G. L. (2006). Protection of the kidney by thiazolidinediones: an assessment from bench to bedside. *Kidney Int* **70**: 1223–1233.

Sarafidis, P. A., and Bakris, G. L. (2007). The antinatriuretic effect of insulin: an unappreciated mechanism for hypertension associated with insulin resistance? *Am J Nephrol* **27**: 44–54.

Sarafidis, P. A., and Ruilope, L. M. (2006). Insulin resistance, hyperinsulinemia, and renal injury: mechanisms and implications. *Am J Nephrol* **26**: 232–244.

Schiffrin, E. L. (2005). Peroxisome proliferator-activated receptors and cardiovascular remodeling. *Am J Physiol Heart Circ Physiol* **288**: H1037–H1043.

Schmitz, O. (1991). Glucose metabolism in non-diabetic and insulin-dependent diabetic subjects with end-stage renal failure. *Dan Med Bull* **38**: 36–52.

Sechi, L. A. (1999). Mechanisms of insulin resistance in rat models of hypertension and their relationships with salt sensitivity. *J Hypertens* **17**: 1229–1237.

Shen, Y., Peake, P. W., and Kelly, J. J. (2005). Should we quantify insulin resistance in patients with renal disease? *Nephrology* **10**: 599–605.

Shulman, G. I., Barrett, E. J., and Sherwin, R. S. (2003). Integrated fuel metabolism. In: *Ellenberg & Rifkin's diabetes mellitus,* 6th edn, (Porte, D., Sherwin, R.S., Baron, A., eds) McGraw-Hill, New York, pp. 1–13.

Shurraw, S., and Tonelli, M. (2006). Statins for treatment of dyslipidemia in chronic kidney disease. *Perit Dial Int* **26**: 523–539.

Snyder, R. W., and Berns, J. S. (2004). Use of insulin and oral hypoglycemic medications in patients with diabetes mellitus and advanced kidney disease. *Semin Dial* **17**: 365–370.

Steinke, J., Soeldner, J. S., Camerini-Davalos, R. A., and Renold, A. E. (1963). Studies on serum insulin-like activity (ILA) in prediabetes and early overt diabetes. *Diabetes* **12**: 502–507.

Stenvinkel, P. (2005). Inflammation in end-stage renal disease – a fire that burns within. *Contrib Nephrol* **149**: 185–199.

Stumvoll, M., Meyer, C., Mitrakou, A., Nadkarni, V., and Gerich, J. E. (1997). Renal glucose production and utilization: new aspects in humans. *Diabetologia* **40**: 749–757.

Stumvoll, M., Meyer, C., Mitrakou, A., and Gerich, J. E. (1999). Important role of the kidney in human carbohydrate metabolism. *Med Hypotheses* **52**: 363–366.

Thomas, M. C., and Atkins, R. C. (2006). Blood pressure lowering for the prevention and treatment of diabetic kidney disease. *Drugs* **66**: 2213–2234.

Thorn, L. M., Forsblom, C., Fagerudd, J. et al. (2005). Metabolic syndrome in type 1 diabetes: association with diabetic nephropathy and glycemic control (the FinnDiane study). *Diabetes Care* **28**: 2019–2024.

Trevisan, R., Bruttomesso, D., Vedovato, M. et al. (1998). Enhanced responsiveness of blood pressure to sodium intake and to angiotensin II is associated with insulin resistance in IDDM patients with microalbuminuria. *Diabetes* **47**: 1347–1353.

van Hooff, J. P., Christiaans, M. H., and van Duijnhoven, E. M. (2004). Evaluating mechanisms of post-transplant diabetes mellitus. *Nephrol Dial Transplant* **19**(Suppl 6): vi8–vi12.

Wallace, T. M., Levy, J. C., and Matthews, D. R. (2004). Use and abuse of HOMA modeling. *Diabetes Care* **27**: 1487–1495.

Webster, A. C., Woodroffe, R. C., Taylor, R. S., Chapman, J. R., and Craig, J. C. (2005). Tacrolimus versus ciclosporin as primary immunosuppression for kidney transplant recipients: meta-analysis and meta-regression of randomized trial data. *Br Med J* **331**: 810–820.

Weinstock-Brown, W., and Keane, W. F. (2001). Proteinuria and cardiovascular disease. *Am J Kidney Dis* **38**(4 Suppl 1): S8–S13.

Weir, G. C., and Bonner-Weir, S. (2004). Five stages of evolving beta-cell dysfunction during progression to diabetes. *Diabetes* **53**(Suppl 3): S16–S21.

White, M. F. (2002). IRS proteins and the common path to diabetes. *Am J Physiol Endocrinol Metab* **283**: E413–E422.

Wilcken, D. E., Sim, A. S., Wang, J., and Wang, X. L. (2007). Asymmetric dimethylarginine (ADMA) in vascular, renal and hepatic disease and the regulatory role of L-arginine on its metabolism. *Mol Genet Metab* **91**: 309–317.

Williams, M. E. (2005). Diabetic nephropathy: the proteinuria hypothesis. *Am J Nephrol* **25**: 77–94.

Williams, M. E., and Stanton, R. C. (2005). Management of diabetic kidney disease. In: *Joslin's diabetes mellitus,* 14th edn, (Kahn, C.R., Weir, G.C., King, G.L., Jacobson, A.M., Moses, A.C., Smith, R.J., eds) Lippincott Williams & Wilkins, New York, pp. 925–950.

Wilson, P. W., D'Agostino, R. B., Parise, H., Sullivan, L., and Meigs, J. B. (2005). Metabolic syndrome as a precursor of cardiovascular disease and type 2 diabetes mellitus. *Circulation* **112**: 3066–3072.

Wolf, G., and Ritz, E. (2003). Diabetic nephropathy in type 2 diabetes prevention and patient management. *J Am Soc Nephrol* **14**: 1396–1405.

Wolf, G., Chen, S., Han, D. C., and Ziyadeh, F. N. (2002). Leptin and renal disease. *Am J Kidney Dis* **39**: 1–11.

World Health Organization. (1999). Definition, Diagnosis and Classification of Diabetes Mellitus and Its Complications: Report of a WHO Consultation. World Health Organization, Geneva.

Yalow, R. S., and Berson, S. A. (1960). Immunoassay of endogenous plasma insulin in man. *J Clin Invest* **39**: 1157–1175.

Yalow, R. S., Glick, S. M., Roth, J., and Berson, S. A. (1965). Plasma insulin and growth hormone levels in obesity and diabetes. *Ann NY Acad Sci* **131**: 357–373.

Young, D. O., Lund, R. J., Haynatzki, G., and Dunlay, R. W. (2007). Prevalence of the metabolic syndrome in an incident dialysis population. *Hemodial Int* **11**: 86–95.

Zoccali, C., Mallamaci, F., and Tripepi, G. (2003). Adipose tissue as a source of inflammatory cytokines in health and disease: focus on end-stage renal disease. *Kidney Int Suppl* **84**: S65–S68.

Zoccali, C., Tripepi, G., Cambareri, F. et al. (2005). Adipose tissue cytokines, insulin sensitivity, inflammation, and cardiovascular outcomes in end-stage renal disease patients. *J Ren Nutr* **15**: 125–130.

CHAPTER **26**

Growth Hormone

JOHN D. MAHAN
Children's Hospital, 700 Children's Drive, Columbus, OH 43205, USA

Contents

I.	Growth hormone and insulin-like growth factor-1 in renal failure	411
II.	Pediatric implications: growth failure and the GH/IGF-I axis	416
III.	Adult Implications: myriad effects of disturbed GH/IGF-I axis in CKD	418
IV.	Effects of recombinant growth hormone treatment in renal failure	419
V.	The horizon for improving growth and anabolism in renal failure	423
VI.	Summary	424
	References	424

I. GROWTH HORMONE AND INSULIN-LIKE GROWTH FACTOR-1 IN RENAL FAILURE

A. Disturbance of the Normal GH/IGF-I Axis

1. INTRODUCTION

Growth hormone (GH) primarily exerts its somatotropic and metabolic effects by stimulating production of insulin-like growth factor-1 (IGF-I) (Roelfsema and Clark, 2001; Pombo et al., 2001). Circulating IGF-I is released into the bloodstream as a result of liver production and functions as a classic endocrine hormone. At sites of local production, IGF-I provides paracrine/autocrine effects in specific tissues (Daughaday and Rotwein, 1989). The production of GH and IGF-I and the important interactions of these two proteins are quite complex and are typically altered in individuals with renal insufficiency, with the degree of alteration variably affected by level of renal impairment, nutritional status, altered systemic hormones and sex steroids (estrogen, testosterone) (Pombo et al., 2001).

Renal insufficiency in children and adults is marked by a number of significant metabolic and hormonal derangements, including alterations in the GH/IGF-I axis (Tonshoff et al., 2005; Mahesh et al., 2008). GH/IGF-I abnormalities may develop at any stage of renal insufficiency and are more likely to be clinically relevant with longer duration of renal dysfunction. In individuals with renal insufficiency, other factors than the level of renal function also impact the GH/IGF-I axis since the extent of the abnormalities is not linearly related to the level of renal insufficiency (Mahan and Warady, 2006). The GH/IGF-I axis is most importantly a regulator of growth and metabolism, but abnormalities in this axis may lead to a variety of other systemic effects. In children with renal disease, GH and IGF-I abnormalities may most notably result in impaired linear growth (Tonshoff and Mehls, 1995; Kaskel, 2003). Other potential consequences, such as decreased muscle mass, decreased bone mass, impaired neurocognitive development and disordered plasma lipids, may also develop and cause significant morbidity (Mahesh et al., 2008). In adults with renal insufficiency, GH and IGF-I abnormalities have also been identified and may lead to decreased muscle mass, decreased bone mass, altered plasma lipids and altered metabolism (Hirschberg and Adler, 1998). Altered GH/IGF-I effects are now recognized as a complication of chronic renal failure, referred to as chronic kidney disease (CKD), based on the terminology defined by the National Kidney Foundation, and may also be important in acute renal insufficiency (NKF Disease Outcomes Quality Initiative, 2002). Moreover, proper recognition of the clinical consequences and appropriate management of GH/IGF-I abnormalities can lead to decreased morbidity in children and adults with CKD. This represents an important new framework in the care of individuals with CKD.

B. Normal GH/IGF-I Axis

The complexities of the GH–IGF-I axis and the effects of these peptides have been well delineated (Roelfsema et al., 2001; Tonshoff et al., 2005; Mahesh et al., 2008). The GH–IGF-I axis involves:

1. the autocrine and paracrine effector peptides GH (191 amino acids, 22 000 D MW), IGF-I (70 amino acids, 7649 D MW) and IGF-2 and
2. a regulatory feedback process and modulation by important regulators, such as the hypothalamic hormones GH-releasing hormone (GHRH) and somatostatin (SRIF) and systemic peptides, like ghrelin (Pombo et al., 2001).

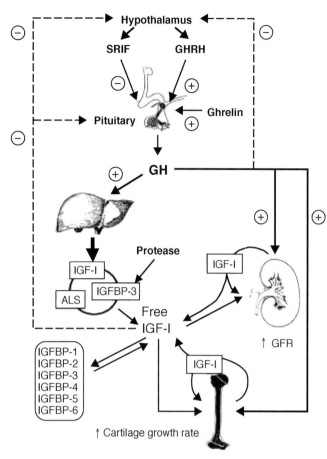

FIGURE 26.1 The somatotropic axis. The synthesis and release of growth hormone (GH) from the pituitary are controlled by the hypothalamic hormones GH-releasing hormone (GHRH) and somatostatin (SRIF) which, in turn, are regulated by feedback (dashed lines) from blood GH and insulin-like growth factor-I (IGF-I) concentrations. An endogenous GH-releasing peptide, called ghrelin, also stimulates GH release. Circulating GH acts directly on many organs to stimulate IGF-I production, with IGF-I production in the liver providing the main source of blood IGF-I. Most of the IGF-I in the circulation is bound to IGF-binding protein-3 (IGFBP-3) in a ternary complex with acid-labile subunit (ALS); a smaller fraction is bound to the five other IGFBP. A small fraction of the total IGF-I in blood is in a bioactive-free fraction. In the kidney, IGF-I increases renal plasma flow and GFR, whereas on bone it acts on the epiphysial plate, which leads to longitudinal bone growth. GH also has direct effects on many organs, including kidney and cartilage, which can be independent of IGF-I action.

As shown in Figure 26.1 and Table 26.1, normal pulsatile GH production and release by the anterior pituitary is stimulated by GHRH and inhibited by SRIF. Circulating GH and IGF-I levels provide negative feedback to the hypothalamus. Release of GH is also stimulated by the ghrelin, which is secreted by the stomach and hypothalamus and conveys nutritional regulation of the GH–IGF-I axis (Kojima et al., 1999; Pombo et al., 2001). Circulating GH stimulates the production and release of IGF-I, primarily from the liver. GH receptors are present in many tissues of epithelial and mesenchymal origin; additional IGF-I production by these tissues is thought to contribute little to circulating levels of IGF-I, but may have important and crucial local paracrine/autocrine effects in specific tissues, such as bone (Roelfsema et al., 2001).

The importance of local paracrine/autocrine effects of IGF-I is highlighted by studies in IGF-I knockout mice that demonstrated that IGF-I deficient mice that survive are profoundly growth retarded, while liver specific IGF-I knockout mice have low levels of circulating IGF-I but normal growth. GH supplementation of IGF-I deficient knockout mice does not improve growth (Liu et al., 1993; Sjogren et al., 1999). These findings indicate that, although the ability to generate IGF-I is essential for normal postnatal growth, liver generated circulating IGF-I is not required for normal growth, underscoring the importance of local GH effects and/or locally produced IGF-I on bone growth.

More than 97% of circulating IGF-I is bound to six IGF-binding proteins (IGFBP-1 to IGFBP-6) with the greatest amount complexed to IGFBP-3 and acid-labile subunit (ALS). Less than 1% of circulating IGF-I occurs in the free or bioactive form. The 150 kD complex of IGF-I, IGFBP-3 and ALS represents a storage form of IGF-I in the circulation and has a half-life of several hours. The binding of IGF-I to the IGFBPs limits the bioactivity of IGF-I since only free IGF-I appears to be able to activate the IGF-I receptor. Free IGF-I mediates many of the biologic effects of GH, including stimulation of longitudinal bone growth and regulation of renal hemodynamics. GH also appears to have a direct effect on several tissues, including bone. However, local production of IGF-I may mediate some or all of these important effects, particularly in growth cartilage (Isaakson et al., 1987). According to the dual effector theory (Green et al., 1985), GH and IGF-I act on different bone cell types to stimulate longitudinal growth. GH induces differentiation of epiphyseal growth plate precursor cells toward chondrocytes. These GH-stimulated chondrocytes then become responsive to IGF-I and concomitantly express IGF-I mRNA. IGF-I stimulates the clonal expansion of differentiated chondrocytes, thus leading to

TABLE 26.1 Outline normal GH/IGF axis; outline chronic kidney disease GH/IGF-I axis

Normal renal function	Decreased renal function
GH release from pituitary controlled by GHRH and SRIF and growth hormone secretagogues (e.g. ghrelin)	GH release from pituitary controlled by GHRH and SRIF and growth hormone secretagogues (e.g. ghrelin)
GHRH and SRIF controlled by serum GH and IGF-I	GHRH and SRIF controlled by serum GH and IGF-I
Serum GH binds to GH receptors on many cells	Serum GH levels are normal to elevated due to diminished renal clearance
GH receptor activation in liver leads to IGF-I production and most of serum IGF-I level	Serum GH binds to GH receptors on many cells – cellular GH receptor density is diminished
GH receptor activation in cells generates IGF-I independent effects	GH receptor activation in liver leads to less efficient IGF-I production and serum IGF-I levels are not increased
Most serum IGF-I is bound to IGFBP-3 in ternary complex with ALS	GH receptor activation in cells generates IGF-I independent effects
Unbound or free IGF-I activates IGF-I receptors to generate cellular responses:	More serum IGF-I is bound to increased levels of IGFBP1, 2 and 4 leading to less IGF-I bound to IGFBP-3 in ternary complex with ALS
Longitudinal bone growth	Increased proteolysis of ALS further diminishes IGF-I complexed to IGFBP3 and ALS
Increased renal plasma flow and GFR	Less unbound or free IGF-I available to activate IGF-I receptors therefore generating less cellular responses
Widespread cellular effects	

longitudinal bone growth (Isaakson et al., 1987; Kaplan and Cohen, 2007).

The GH/IGF-I axis is present in the kidney and is important to kidney structure and function. GH receptors are present in the proximal tubule and thick ascending limb (Hirschberg and Adler, 1998). IGF-I receptors are primarily present in the glomerulus (glomerular mesangial, endothelial and mesangial cells), proximal and distal tubules (Ernst et al., 2001; Berfield et al., 2006). Both circulating GH and IGF-I increase renal plasma flow and glomerular filtration rate (GFR). The GFR response to GH appears to be primarily related to the IGF-I level, since the response to the rise in IGF-I is quick while the response to the rise in GH is only seen after IGF-I levels rise. In IGF-I knockout mice, renal development is normal, suggesting that IGF-I is not essential to normal renal development (Liu et al., 1993). The effects of supraphysiologic levels of GH and IGF-I on the kidney can be quite different. For example, mice transgenic for IGF-I develop glomerular hypertrophy while mice that are transgenic for GH develop glomerular sclerosis (Doi et al., 1988).

C. Pathophysiology of the Disordered GH/IGF-I Axis in Renal Insufficiency

1. Overview

In renal insufficiency the GH/IGF-I axis is markedly deranged (Roelfsema et al., 2001) (Figure 26.2 and see Table 26.1). Circulating levels of GH are typically increased as a result of the combination of increased pulsatile release by the pituitary and reduced renal GH clearance (Haffner et al., 1994). Total IGF-I levels are normal, not elevated as might be expected in relation to circulating GH levels, and the biologic effectiveness of endogenous GH and IGF-I is reduced (Powell et al., 1998). Free, bioactive IGF-I levels are reduced as a result of increased levels of most circulating IGFBPs, especially the high affinity forms IGFBP-1, 2 and 6, in relation to the decline in renal function (Powell et al., 1996). The increased levels of IGFBP-1 and 2 appear to be most responsible for sequestering IGF from IGF-I receptors. Increased proteolysis of IGFBP-3 leads to less IGF-I in association with the IGFBP-3 and ALS complex. The end result is less IGF-I receptor activation at the cellular level.

Despite normal to elevated levels of circulating total GH and IGF-I, CKD is also marked by evidence of GH and IGF-I resistance (Powell et al., 1998). Undoubtedly, this resistance plays an important role in the reduction of linear bone growth observed in CKD and, therefore, is a significant factor in the growth impairment seen in children with CKD. The mechanisms responsible for IGF-I resistance in CKD are not completely understood but appear to involve a defect in the post-receptor GH-activated Janus kinase 2 (JAK2) signal transducer and activator of the transcription (STAT) pathway (Rabkin et al., 2005) (Figure 26.3). This signaling pathway is integral to the process of stimulated IGF-I gene expression which is the key cellular response to GH receptor activation. A decrease in cellular GH receptors may also contribute to the blunted IGF-I response to GH seen in CKD (Tonshoff et al., 1997).

D. Growth Hormone and Growth Hormone Resistance in CKD

The GH abnormalities in CKD are complex and include:

1. abnormal GH secretion
2. decreased serum growth hormone binding protein (GHBP) and
3. decreased renal clearance of GH.

The sum of these disturbances can be quite variable in uremic individuals but the most common pattern is a slight elevation in serum GH with blunted cellular responses to

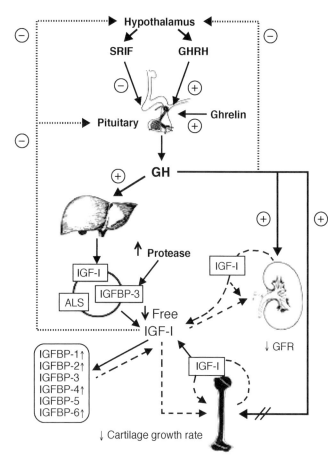

FIGURE 26.2 Deranged somatotropic axis in chronic renal failure. The GH/IGF-I axis in chronic kidney disease (CKD) is changed markedly compared with the normal axis, shown in Figure 26.1. In CKD, the total concentrations of the hormones in the GH/IGF-I axis are not reduced, but there is reduced effectiveness of endogenous GH and IGF-I, which probably plays a major role in reducing linear bone growth. The reduced effectiveness of endogenous IGF-I likely is due to decreased levels of free, bioactive IGF-I as levels of circulating inhibitory IGFBP are increased. In addition, less IGF-I is circulating in the complex with ALS and IGFBP-3 as a result of increased proteolysis of IGFBP-3. Together, there is decreased IGF-I receptor activation and decreased feedback to the hypothalamus and pituitary. Low free IGF-I and high IGFBP-1 and -2 levels may contribute to reduced renal function and lead to a reduced stature. The direct effects of GH on bone, which are poorly understood, also are blunted.

GH. The interplay of uremia and a number of additional factors modulate the extent and final outcome of these abnormalities.

In pre-pubertal children with CKD, increased pulsatile release of GH by the pituitary is noted. In pubertal children with CKD, the typical increase in GH secretion seen in adolescence appears to be impaired (Schaefer et al., 1994a). Schaefer and colleagues analyzed the rates of GH secretion and elimination in 43 peri-pubertal boys with CKD (Schaefer et al., 1994b). The estimated plasma GH half-life was significantly increased in children with CKD compared to control children. In the pre- and early pubertal CKD boys, the calculated GH secretion rate was low normal or reduced when expressed in absolute numbers or normalized per unit distribution volume or body surface. In late puberty, whereas body surface-corrected GH secretion was double the pre-pubertal value seen in normal boys, it did not differ significantly from the pre-pubertal rate of GH secretion seen in CKD boys. In adolescents with CKD, the lower than expected GH secretion resulted from a decrease in the amount of GH released within each burst, with burst frequency unchanged.

Since the major metabolic pathway for GH clearance is renal excretion, renal insufficiency delays clearance of GH, thereby contributing further to the tendency towards high serum GH levels (Tonshoff et al., 1995b). Balancing this is the decrease in serum growth hormone binding protein (GHBP) seen in CKD, which may allow relatively more renal excretion of free GH for the level of renal function (Postel-Vinay et al., 1991). The overall result is typically normal or mildly increased levels of serum GH (see Figure 26.1). In CKD, nutritional status may also contribute to GH levels since gastrointestinal hormones such as ghrelin, a 28 amino acid peptide with GH releasing properties (as a growth hormone secretagogue) secretion, are elevated in patients with CKD (Kojima et al., 1999).

In the face of an elevated serum GH, the presence of GH resistance in CKD is suggested from the absence of elevated circulating levels of IGF-I. The cellular basis for GH resistance in CKD has been explained, in part, by a decrease in cellular GH receptor number (Tonshoff et al., 1997). In support of a decrease in GH receptor number, based on observations in humans, has been the decrease in serum GH binding protein (GHBP), a product of proteolytic cleavage of the GH receptor that is quantitatively related to cellular GH receptor number (Maheshwari et al., 1992). In fact, in individuals with CKD, decreased levels of circulating GHBP are directly related to the level of glomerular filtration rate. In contrast, Greenstein and colleagues demonstrated that monocytes from individuals with CKD have normal levels of GH

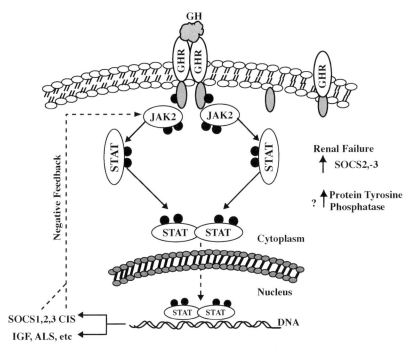

FIGURE 26.3 Growth hormone-mediated JAK/STAT signal transduction. GH activates several signaling pathways via JAK2, including the JAK/STAT pathway. Binding of GH to its receptor (GHR) activates JAK2, which then self-phosphorylates. This is followed by phosphorylation of the GHR and, subsequently, STAT1a, STAT3, STAT5a, and STAT5b, members of a larger family of cytoplasmic transcription factors. These phosphorylated STATs form dimers that enter the nucleus, where they bind to specific DNA sequences and activate their target genes, IGF-1 and some suppressors of cytokine signaling (SOCS). Deletion of STAT5 expression leads to retarded body growth and STAT5b is required for GH-mediated IGF-1 gene expression. In renal failure, phosphorylation of JAK2 and the downstream signaling molecules STAT5, STAT3, and STAT1 is impaired, as are the nuclear levels of phosphorylated STAT proteins. This important cause of uremic GH resistance may result, in part, from upregulation of SOCS2 and SOCS3 expression with suppressed GH signaling and also from increased protein tyrosine phosphatase activity, with enhanced dephosphorylation and deactivation of the signaling proteins.

receptors (Greenstein et al., 2006). Thus, not all tissues in individuals with CKD may participate in this mechanism for GH resistance.

GH resistance in CKD may also be mediated through inhibition of transduction signaling. As stated above, activation of the GH receptor phosphorylates JAK2, which in turn activates cytoplasmic STAT. Following translocation to the nucleus, STATs form homodimers to activate nuclear receptors thereby stimulating transcription of a number of proteins, including IGF-I (Rabkin et al., 2005). GH resistance in uremia may be caused by suppressors of cytokine signaling (SOCS) which are able to inhibit the JAK2/STAT pathway (Rabkin et al., 2005). In experimental uremia, increased SOCS expression has been demonstrated (Schaefer et al., 2001). This mechanism may occur via upregulation of SOCS2 and SOCS3 expression and/or from increased protein tyrosine phosphatase activity with enhanced dephosphorylation and deactivation of the signaling proteins (Wormald and Hilton, 2004) (see Figure 26.3). Whether increased SOCS expression is secondary to the CKD *per se* or to the often accompanying chronic inflammation is uncertain (Kaysen, 2001).

E. Insulin-Like Growth Factor-I and Insulin-Like Growth Factor-I Abnormalities in CKD

The presence of IGF-I resistance in CKD is at least partially due to increased levels of circulating IGFBP-1, -2, -4 and -6, which lead to a reduction in the concentration of bioavailable IGF-I (Roelfsema et al., 2001) (see Figure 26.2). IGFBP-1 and -2 appear to be most responsible for reducing IGF-I bioavailability. In addition, increased proteolysis of IGFBP-3 leads to a decrease in IGF-I available for the formation of IGF-I–ALS–IGFBP-3 complexes. Taken together, these events impair both the direct and indirect effects of GH by affecting IGF-I activity in children with CKD (Greenstein et al., 2006).

The mechanism for the increase in serum IGFBP-1 and IGFBP-2 (low molecular weight IGF binding proteins) seen in CKD is not clear but may be related to either a decrease in their renal excretion or increased hepatic production (Powell et al., 1993, 1996; Tonshoff et al., 2005). While the cause may be multifactorial, the elevated IGFBPs invariably lead to a low serum free IGF-I concentration (see Figure 26.2). In the circulation, IGF-I is mainly bound to the IGFBP-3 isoform, its free serum concentration representing under 0.5% of serum total IGF-I. In children with CKD, height correlates positively with serum IGF-I levels but inversely with serum IGFBP-2, which may be regarded as a marker for bioavailable IGF-I. The bioavailability of IGF-I is also affected by activity of IGFBP proteases, while malnutrition, a feature of CKD, influences serum IGF-I independent of the IGFBP mechanism (Haspolat et al., 2007).

As with GH resistance, IGF-I resistance may also involve the inhibition of transduction signaling pathways. Chronic inflammation, mediated by inflammatory cytokines such as IL-1 and tumor necrosis factor α (TNFα), is an important feature of CKD (Mak and Cheung, 2006). TNFα inhibits the phosphorylation of the IGF receptor docking molecules, the insulin receptor substrates, IRS-1 and IRS-2, an early step in the IGF-I transduction pathway (Broussard et al., 2003). IGF-I transduction signaling may also be inhibited downstream by the metabolic acidosis of CKD which can independently decrease phosphatidylinositol-3-kinase (PI3 kinase) activity and subsequent gene expression (Bailey et al., 2006).

II. PEDIATRIC IMPLICATIONS: GROWTH FAILURE AND THE GH/IGF-I AXIS

Growth failure is a common and significant clinical problem in children with chronic renal insufficiency. Children with CKD who experience growth failure exhibit a range of potentially serious medical and psychological complications, as well as increased mortality (Mahan and Warady, 2006). The etiology of growth failure in this population is multifactorial, reflecting both abnormalities in the GH/IGF-I axis and a variety of nutritional and metabolic problems (Mahan and Warady, 2006). Specific management of these abnormalities is required to improve growth in children with CKD. It is now clear that the component of abnormal GH/IGF-I activity characteristic of children with CKD is amenable to exogenous GH therapy. The safety and efficacy of recombinant human GH (rGH) therapy in promoting growth in children with CKD have now been well established (Fine et al., 1994).

A. Impact of Growth Failure in Children with CKD

The strong link between growth impairment in children requiring dialysis and increased risk of mortality and morbidity was clearly demonstrated by Furth and colleagues in two landmark studies. The first study involved United States Renal Data System (USRDS) data on 1112 children on dialysis followed from 1990 to 1995 (Furth et al., 2002a). In this group, 5-year death rates were higher in children with severe growth failure (height velocity SDS ≤ 3.0) or moderate growth failure (height velocity SDS ≤ 2.0 and ≤ 3.0) during the first year of observation compared to patients with normal growth (height velocity SDS ≤ 2.0) (16.2%, 11.5% and 5.6%, respectively). There was an almost threefold increase in mortality in the children on dialysis with severe growth impairment. These short children also had increased morbidity, particularly increased rates of hospitalization, primarily for the treatment of infection.

The second study evaluated 2306 patients from the NAPRTCS registry who began dialysis between 1992 and 2001 (Furth et al., 2002b). Based on height at initiation of dialysis, death rates were higher in shorter patients (height SDS ≤ 2.5) compared with rates in their taller counterparts (90.5 versus 39.6 deaths per 1000 patient-years, respectively). Although younger age was independently associated with higher mortality, poor growth was a consistent risk factor for death in both very young children (<1 year) and older children. Patients with short stature again had increased morbidity including more days per month in hospital. Although it is impossible simply to relate these adverse outcomes to poor height or growth alone, it is likely that short stature and poor growth are markers for the type of complex clinical course and poor nutrition that frequently affects some but not all children with CKD.

Although there are limited data available on the subject of altered psychosocial development in growth impaired children with CKD, clinical experience in children with growth hormone deficiency (GHD) has demonstrated that growth failure can have important effects on psychosocial development, psychological development and quality of life (QOL) (Stabler et al., 1998). Several studies have shown that short children are often perceived by adults to be younger than their true age, leading to 'juvenilization' or lowered expectations. These children are frequently not given age-appropriate responsibility or respect and may be overprotected and given excessive attention. As a result, they may become dependent and remain immature, leading to a dysfunctional position for the child in the family. Affected children may adopt 'mascot' or 'clown-like' behaviors to gain temporary acceptance by peers, however, these behaviors tend to reinforce the child's perceived immaturity and further impair self-esteem and QOL. In children with GHD, co-morbidities that affect psychological development and QOL, such as attention deficit hyperactivity disorder, learning disability and mood disorders, are increased in incidence. Interestingly, in a study by Stabler of 196 children with GHD or idiopathic short stature (ISS), growth improvement over a 3-year period was associated with a significant and sustained reduction in problem behavior as measured by the Child Behavior Checklist (Stabler et al., 1994). The NIH prospective study of children with CKD, Chronic Kidney Disease in Children (CKiD), is addressing the relationship of kidney function, growth and psychosocial development over time in over 500 children with CKD and should provide better understanding of these issues in children with CKD.

It is now clear that children with poor growth associated with CKD often become very short adults. In Hokken-Koelega's retrospective study of 52 children who underwent renal transplantation before the age of 15 years and did not receive rGH therapy, the median height SDS was below the third percentile for age at the time of first dialysis, decreased significantly during dialysis and did not improve following renal transplantation (Hokken-Koelega et al., 1994). The final adult height of these patients remained below the third percentile for age for 77% of the males and 71% of the females. Reports of children with CKD from North America and Europe (Hokken-Koelega et al., 1994; Fine and Stablein, 2005) have confirmed this conclusion: final adult height remains decreased following successful renal transplantation in children who were short at the time of transplantation.

Many adults who suffered significant CKD in childhood report that the adverse effects from their childhood renal disease and its complications continue in adulthood.

However, distinguishing the effects of CKD from the effects of immune suppressive therapy, particularly steroid therapy, is difficult. Busschbach evaluated QOL in five groups of short adults, including 17 male patients with childhood-onset renal failure (Busschbach et al., 1998). QOL was significantly lower than normal heighted controls for all groups (GH deficiency, CKD, Turner's syndrome, referred idiopathic short stature and non-referred idiopathic short stature) as measured by the Nottingham Health Profile. In all groups, the subjects perceived their chance of having a partner as low, except those individuals with idiopathic short stature. Only 40% of the CKD adults had a partner; however, this was attributed primarily to the constant risk of transplant rejection rather than to short stature. Of note, 60% of the CKD adults reported they would like to be taller and 33% were prepared to sacrifice an average of 4% of expected life-years to be taller. Time trade-offs for all symptoms associated with renal transplantation or dialysis (15% and 47% of expected life-years, respectively) were considerably larger than for height alone. This comparative study indicates that many children with short stature *per se*, regardless of etiology, have significant psychosocial effects that persist into adulthood.

Broyer investigated the social outcomes in 244 adults who underwent a renal transplant in childhood (Broyer et al., 2004). In this population, final adult height was significantly associated with marital status ($P < 0.0001$) and level of education ($P < 0.001$) and inversely related to level of employment ($P < 0.02$). Rosenkranz determined that 36% of young adults who had childhood CKD were dissatisfied with their adult height and that final adult height correlated with their subjective assessment of QOL ($P = 0.008$) (Rosenkranz et al., 2005).

B. Disordered Growth in Children with Renal Failure: Etiology of Growth Impairment

A number of other important factors can contribute to growth impairment in children with CKD, including age at onset of renal disease, type of renal disorder and presence of calorie-protein malnutrition, urine concentrating defect, metabolic acidosis, renal osteodystrophy (ROD) and anemia (Tonshoff and Mehls, 1995; Mahan and Warady, 2006) (Table 26.2). Because children typically reach one-third of their final adult height during the first 2 years of life, growth impairment during infancy has a greater impact on stature than onset of CKD later in childhood. Renal insufficiency limits normal growth patterns at every stage of childhood but is especially detrimental during puberty when delayed pubertal growth spurt and reduced pubertal height gain may be seen in affected children, resulting in a permanent loss in final adult height. Perturbations in serum GH binding proteins and the gonadotropic hormone axis may contribute to suboptimal pubertal growth and development in affected children.

TABLE 26.2 Factors affecting growth in children with CKD

Non-modifiable factors	Modifiable factors
Age of CKD onset	Excessive salt and water losses
Level of GFR	Metabolic acidosis
Type of renal disorder	Anemia
Corticosteroid therapy	Poor caloric intake
	Renal osteodystrophy
	GH/IGF-I resistance
	Gonadotropin abnormalities

1. Non-Modifiable Factors

The type of renal disorder also influences growth in children with CKD. For example, independent of the level of renal insufficiency, children with renal dysplasia tend to exhibit the most severe height deficits, whereas those with focal segmental glomerulosclerosis display less severe height deficits (Kaskel, 2003). This appears to be more than a manifestation of an early onset of subclinical CKD and may be related to the effects of tubular dysfunction and losses of renal substances important for growth (Mahesh and Kaskel, 2008).

2. Modifiable Factors

Caloric deficiency and abnormal protein metabolism may also play an important role in growth impairment, particularly in younger children. Energy intake correlates with growth velocity in infants with CKD. If energy intake exceeds 80–100% of recommended values, normal growth rates are observed (Graf et al., 2007). Reduced caloric intake in children with CKD may be the result of anorexia, emotional distress, altered taste sensation or nausea and vomiting. Inadequate protein intake can lead to malnutrition and poor growth; excessive protein intake may also be deleterious as it may lead to hyperfiltration and accelerated progression to end stage renal disease (ESRD) (although there are few data in children to support this concern). Increased catabolism in the face of decreased caloric intake further complicates the nutritional status of these children (Graf et al., 2007). There are multiple explanations for the increased catabolic state found in children with CKD, including the concept of 'cellular inefficiency' related to uremia (Graf et al., 2007) and the state of GH insensitivity that is related to diminished hepatic GH receptor expression (Tonshoff et al., 1994). Most importantly, multiple studies of children with CKD have demonstrated that nutritional supplementation improves growth velocity (Kari et al., 2000).

Metabolic acidosis is well recognized as a cause of impaired growth, as first clearly demonstrated in children with renal tubular acidosis (McSherry, 1978). Metabolic acidosis may contribute to derangements in the GH–IGF-I axis by reducing GH secretion and serum IGF-I levels (Maniar et al., 1996). Metabolic acidosis also appears to cause resistance to the anabolic actions of GH and has been shown to suppress albumin synthesis, promote calcium efflux from bone and accelerate protein degradation (Maniar et al., 1996). In children with CKD, a significant inverse correlation between plasma bicarbonate and the leucine rate of appearance, a measure of protein breakdown, suggests that protein breakdown, a factor that is associated with poor growth in a number of conditions, may be related to the degree of metabolic acidosis (Boirie et al., 2000). As a modifiable cause of growth failure, correction of metabolic acidosis may improve growth in children with CKD.

Growth may be further impaired by the effects of renal osteodystrophy (ROD). ROD represents a spectrum of disorders ranging from high-turnover bone disease (due to secondary hyperparathyroidism) to low-turnover osteomalacia and adynamic bone disease (Langman and Brooks, 2006). Secondary hyperparathyroidism appears to cause growth failure by altering genes involved in endochondral bone formation and disrupting the architecture of the growth plate (Salusky et al., 2004). In low turnover osteomalacia and adynamic bone disease, osteoblast activity and bone mineralization are diminished or absent leading to less bone formation and less growth over time. The clinical consequences of ROD can be quite profound in growing children and include reduced bone mineral density, increased bone fractures and bone deformities, in addition to growth impairment. There is good evidence that improvement in ROD can result in better growth in children with CKD (Mahesh and Kaskel, 2008; Wesseling et al., 2008).

Other modifiable causes of growth failure in children with CKD must also be considered. Children with salt-wasting disorders and urine concentrating defects may experience growth failure with inadequate replacement of excessive salt and water losses (Mahesh and Kaskel, 2008). Long-term steroid therapy may affect growth by several mechanisms, including depression of pulsatile GH secretion, inhibition of hepatic production of IGF-I and by interference with cartilage metabolism, bone formation, nitrogen retention and calcium metabolism (Hochberg et al., 2002). Anemia appears also to hinder growth in children with CKD. Correction of anemia has been associated with improved growth in such children (Boehm et al., 2007).

The multiple disturbances in the gonadotropic hormone axis that have been described in children with CKD may impair growth. The reduced renal clearance in CKD tends to elevate serum gonadotropin levels (Lane, 2005). Pituitary secretion of some compounds, such as luteinizing hormone (LH), is substantially reduced in children with CKD compared to normal adolescents. As Belgorosky identified, mean serum sex binding globulin (SBG) levels are elevated while non-sex binding globulin-bound testosterone and free testosterone are significantly lower in children with CKD (Belgorosky et al., 1991). The levels of serum non-SBG-bound testosterone and testosterone are similar to those seen in children with hypopituitarism and may contribute to the delay in puberty seen in children with CKD. Moreover, in children with CKD, serum testosterone is not positively correlated with the rate of GH secretion, as it is in normal children. GH secretion rates are lower than expected at each level of testosterone in boys with CKD (Schaefer et al., 1994b). The abnormal regulation of both GH/IGF-I and sex hormones in children with CKD may contribute to suboptimal pubertal growth and development in adolescents with CKD.

III. ADULT IMPLICATIONS: MYRIAD EFFECTS OF DISTURBED GH/IGF-I AXIS IN CKD

In adults with renal insufficiency, as in children, the GH/IGF-I axis is markedly deranged compared to that seen in normal individuals (Iglesias et al., 2004) (see Figure 26.2 and Table 26.1). Circulating levels of GH are typically increased while total IGF-I levels are normal, not elevated as might be expected in relation to circulating GH levels. There is good evidence that the biologic effectiveness of endogenous GH and IGF-I is reduced. Free, bioactive IGF-I levels are reduced as a result of increased levels of most circulating IGFBPs, in particular IGFBP-1 and 2, but not IGFBP-5, and the magnitude of the increase is related to the decline in renal function in adults as well (Frystyk et al., 1999). Increased proteolysis of IGFBP-3 leads to less IGF-I in association with the IGFBP-3 and ALS complex. The end result is less IGF-I receptor activation at the cellular level.

Despite the evidence of normal to elevated levels of circulating total GH and IGF-I in adults with CKD, there is also evidence of GH and IGF-I resistance. This resistance plays an important role in the abnormalities associated with the disturbed GH/IGF-I axis in adults with CKD. As in children, an important mechanism for IGF-I resistance in CKD presumably involves a defect in post-receptor GH-activated JAK2/STAT pathway (Rabkin et al., 2005). A decrease in GH receptors may also contribute to GH resistance and the diminished IGF-I response to GH in uremia.

While defects in growth are not an issue in adults with CKD, there are two areas where the disturbed GH/IGF-I axis may directly play an important role in the morbidity and complications seen in CKD in adults. There is now good evidence implicating these IGF-I abnormalities in the pathogenesis of cardiovascular disease (CVD) and in altering insulin sensitivity and augmenting end-organ damage in adults with CKD (Ferns et al., 1991; Laviades et al., 1997; Juul et al., 2002).

In adults with significant CKD (on dialysis), Abdulle recently highlighted the relationship of circulating IGF-I levels to conventional risk factors for CVD including blood pressure, body mass index (BMI) and age compared to that seen in healthy non-obese control subjects (Abdulle et al., 2007). Given the high rates of CVD and CVD morbidity and mortality in adults with CKD, this is a particularly important new area for investigation. Circulating levels of IGF-I were significantly lower in both male and female CKD patients compared to the control subjects and IGF-I was strongly and inversely correlated with both systolic blood pressure (SBP) and diastolic blood pressure (DBP) in the CKD group. When adjusted for age, the correlation was more significant, however, when adjusted for BMI no significant correlation was observed between IGF-I and blood pressure. IGF-I was inversely correlated with age and BMI in the control group, but not the patient group. In controls and patients with CKD, respectively, a positive correlation between leptin and BMI was observed. Although it is not clear whether low IGF-I levels are a cause or an effect of these cardiovascular risk factors, these interactions deserve further investigation to determine if interventions directed toward the GH/IGF-I axis may improve CVD in adults with CKD.

Another emerging area of interest in adults with CKD is the relationship of GH/IGF-I abnormalities to nutrition, appetite and serum leptin levels. The nutritional status of adults with CKD also impacts CKD mortality rates (Szczech et al., 2003). Leptin tends to be increased in CKD in adults and appears to be associated with malnutrition and anorexia in CKD patients (Fouque et al., 1998). Fouque demonstrated that leptin levels decrease during IGF-I treatment in adults with CKD but rise after administration of combination of GH and IGF-I therapy.

IV. EFFECTS OF RECOMBINANT GROWTH HORMONE TREATMENT IN RENAL FAILURE

A. Growth and the Pediatric Response to GH Therapy

Clinical trials have demonstrated the safety and efficacy of GH therapy in promoting linear growth in children with CKD (Vimalachandra et al., 2006). In a meta-analysis, four randomized, controlled trials were identified that evaluated the effect of GH versus placebo or no treatment on height SDS in children with CKD either predialysis or post-transplantation (Vimalachandra et al., 2006). After 1 year, GH treatment produced a significant increase in height SDS as measured by a weighted mean difference of 0.77 (95% CI, 0.51 to 1.04). One year of 28 IU/m^2/week rGH in children with CKD resulted in a 3.80 cm/year increase in height velocity above that of untreated patients. Moreover, long-term GH therapy in children with CKD results in catch-up growth and many patients achieve a final height within the normal range. Hokken-Koelega evaluated growth in 45 pre-pubertal children with CKD who received GH therapy for up to 8 years (Hokken-Koelega et al., 2000). Long-term GH therapy resulted in catch-up growth and a significant improvement in height relative to baseline ($P < 0.001$) with the mean SDS above the lower limit of normal (ht SDS > -2) after 3 years of therapy. The children nearly reached target height after 6 years of therapy. This improvement in growth was obtained without deleterious effects on GFR and bone maturation.

Haffner demonstrated that the administration of GH over a mean of 5 years in 38 children with CKD resulted in significantly greater pre-pubertal height gain compared to children who did not receive GH (boys, 18.6 ± 9.3 cm versus 9.9 ± 4.8, $P < 0.001$; girls, 16.6 ± 8.7 cm versus 9.1 ± 9.8, $P = 0.014$) (Haffner et al., 2000). The pubertal height gain and the duration of the pubertal growth spurt with GH therapy was still not equivalent to that seen in normal children, but two-thirds of treated children achieved a normal final adult height (Ht SDS > -1.88). Similar results have been demonstrated in adolescents treated with rGH following renal transplantation (Jabs et al., 1993).

Near final adult height with GH therapy in children with CKD was assessed recently by Nissel (Nissel et al., 2008). In a group of 240 children with CKD, mean height SDS increased continuously during rGH treatment for a total height SDS gain of +1.2 in boys and +1.6 in girls. Puberty had a significant impact on this response. The mean near adult final height differed significantly between pre-pubertal subjects with severely delayed puberty (-3.6) and those with normal onset of puberty (-2.0). The significant predictors of growth response to GH therapy were: the initial degree of impaired height; degree of bone age retardation; duration of rGH therapy; time spent on conservative treatment/dialysis; pubertal delay (>2SD); and gender and age at start of rGH treatment. Thus, long-term rGH therapy in pre-pubertal and pubertal children with CKD results in an increased adult height, but the response is less impressive in children on dialysis and/or with severely delayed puberty (Nissel et al., 2008).

Poor growth outcomes observed in children after renal transplantation are associated with a number of factors, including corticosteroid administration, decreased GFR and an abnormal GH/IGF-I axis. Guest demonstrated that rGH therapy significantly improved height velocity in a study of 90 pre- and early pubertal growth-impaired children after renal transplantation (Guest et al., 1998). In patients followed up for up to 4 years of rGH treatment, growth velocities declined following the first year of treatment but remained higher than baseline values. Data from a group of short renal transplant recipients in the NAPRTCS registry demonstrated a significant increase in height SDS with rGH treatment compared to untreated controls (Fine et al., 2002). Final adult height scores are significantly better in children who received rGH therapy post-renal transplant than in those who did not with no difference in incidence of graft rejection or loss.

Children with earlier pre-dialysis stages of CKD (CKD 2–4) typically experience better growth rates with rGH therapy than those who are on dialysis or post-renal transplantation. This observation has led to the recommendation to consider initiating rGH therapy at a young age and/or early in the evolution of CKD in order to achieve the greatest growth potential. 'Catch down' growth (substantial reduction in growth velocity) is seen in many children who discontinue rGH therapy after reaching target height. Restarting rGH can lead to additional improved growth (Mahan and Warady, 2006).

In addition to promoting growth, rGH therapy may provide other benefits to children with CKD. There are limited reports of beneficial effects of rGH treatment on selected parameters of neurodevelopmental status (Van Dyck et al., 2001a). GH can result in significant anabolic effects as evidenced by improvements in body weight, mid-arm circumference and mid-arm muscle circumference. Although data in patients with CKD are limited, rGH therapy may also improve QOL and parameters of bone metabolism. Several small series of children with CKD have demonstrated improved lumbar spine bone mineral content and bone mineral density with rGH therapy (Van Dyck et al., 2001b). While the effects of rGH therapy on cardiovascular health in children with CKD are poorly understood (Lilien et al., 2004), several studies have documented the benefits of rGH therapy in maintaining cardiovascular health in children and adolescents with GH deficiency and normal renal function.

B. Safety of GH Therapy

Multiple clinical studies now support the safety of rGH therapy in children with CKD (Van Es, 1991; Fine et al., 1994). Fine compared the frequency of rGH-related adverse events over a 6.5-year period in the NAPRTCS registry between rGH treated patients with CKD (pre-dialysis, dialysis and post-renal transplantation) and children with CKD who did not receive rGH therapy (Fine et al., 2003). Specifically, rGH treated children had no significant increase in the incidence of malignancy, slipped capital femoral epiphysis, avascular necrosis (AVN), benign intracranial hypertension, glucose intolerance, pancreatitis, progressive deterioration of renal function, acute allograft rejection or fluid retention (Fine et al., 2003).

Several studies have reported a significant elevation in insulin levels during the first year of treatment, which is consistent with the known activity of GH; typically, insulin levels return toward baseline with long-term treatment (Mahesh and Kaskel, 2008). Irreversible diabetes mellitus has not been observed in rGH treated children with CKD. Given the perturbations of carbohydrate metabolism typically seen in CKD (increased insulin and glucose levels), the potential for rGH therapy to alter glucose metabolism underscores the recommendation carefully to monitor glucose metabolism in these patients.

Although rGH therapy in children with CKD may accelerate bone abnormalities related to renal osteodystrophy and may appear to promote clinical manifestations of ROD in some children, a comprehensive study by Boechat of 205 children with CKD revealed no association between the incidence of avascular necrosis and the type or duration of renal disease or rGH therapy (Boechat et al., 2001). Children with poorly controlled ROD who experience increased growth may develop abnormal bone growth and deformities as part of their underlying bone abnormalities and ROD must be carefully assessed and controlled prior to and during rGH therapy.

In well-controlled studies, there has been no significant change in renal function observed in children with CKD pre- or post-renal transplantation who are treated with rGH when compared to children with CKD who do not receive GH therapy. Mentser, in an analysis of the NAPRTCS registry, found no significant increase in acute rejection episodes in children post-renal transplant who were treated with rGH compared to rejection rates prior to GH treatment or when compared to children with CKD who did not receive rGH therapy (Mentser et al., 1997).

C. Dosing Recommendations

Based on a number of clinical trials in children with CKD, the recommended dose of rGH to achieve the desired growth response is 0.35 mg/kg/week (28 IU/m^2/week) administered as a daily subcutaneous injection (Mahan and Warady, 2006). This dose is higher than that recommended for pre-pubertal children with idiopathic or acquired GH deficiency (0.30 mg/kg/week – 24 IU/m^2/week), also given on a daily basis.

Although the recommended rGH dose for children with CKD has been extensively evaluated in clinical trials, future research may allow further optimization of the therapeutic regimen in this population. Two approaches have been advocated: a fixed pubertal dosing regimen and dose modifications based on IGF-I levels. Studies in patients with GHD suggest that higher doses of rGH may be more effective during puberty for some adolescents, especially those with severe growth failure or a delayed diagnosis of growth impairment (Mauras et al., 2000). A study in adolescents with GHD showed that 'pubertal doses' of rGH (0.7 mg/kg/wk) resulted in greater near adult height than conventional rGH doses. Furthermore, these doses did not produce any advancement of bone age and were well tolerated (Mauras et al., 2000). However, additional research is required to determine the optimal pubertal rGH dosing regimen in patients with CKD.

Since GH therapy typically increases plasma IGF-I levels and many of the benefits of GH therapy are mediated by IGF-I, it has been suggested that plasma IGF-I may be a useful measure of efficacy and adherence to prescribed therapy in these individuals (Mahan and Warady, 2006). The relationship of serum concentration of IGF-I and the growth rate of children with CKD with rGH administration is unknown. Thus, it is not clear that there is any value to adjusting GH therapy to achieve a desired IGF-I response in children with CKD.

The 2006 Consensus Conference algorithm for evaluation and treatment of growth failure in children with CKD is shown in Figure 26.4.

D. Effects in Adults with Chronic Renal Failure Treated with rGH

There is limited information on the impact of endogenous rGH therapy in adults with chronic renal failure. Most of these investigations have involved adults with significant CKD (on dialysis) and there is much more work needed to define the response to rGH therapy and the risks/benefits of this pharmacologic intervention in adults with CKD.

Treatment with rGH may provide a significant anabolic impact in adults with CKD. Chu explored the response to rGH administration in stable malnourished elderly adults without significant CKD (Chu et al., 2001). Short-term rGH was associated with better total lean body mass (LBM), hemoglobin, serum albumin and walking speed.

Based on the clear evidence for beneficial effects of rGH in promoting protein synthesis and improving LBM in elderly persons with normal renal function, in persons with GH deficiency and in uremic children, endogenous rGH therapy has been explored in adults with significant CKD. In adults on hemodialysis treated with rhGH for 6 months, Hansen demonstrated improved LBM and reduced fat mass (Hansen et al., 2000). In a similar study in adults with CKD, Jensen documented increased left ventricular muscle mass but no improvements in ejection fraction or exercise capacity (Jensen et al., 2000).

These studies in adults with CKD have utilized rGH doses typically in the range of 2–4 IU/m^2/day (0.67–1.33 mg/m^2/day). No major side effects have been documented in these short-term studies (Hansen et al., 2000; Jensen et al., 2000).

The net protein catabolism seen in adults with advanced CKD may be related to acquired resistance to the anabolic actions of GH (Blum et al., 1991; Fouque et al., 1995). Johannsson showed, in a small group of adults on hemodialysis, that 6 months of rGH treatment increased serum albumin and muscle strength (Johannsson et al., 1999). Ericsson explored the relationship of underlying inflammation and the status of the anabolic GH–IGF-I axis in a double-blind, placebo-controlled study of rGH therapy given for 8 weeks in 35 adults on hemodialysis (Ericsson et al., 2004). There was a positive correlation between serum IGF-I response and normalized protein catabolic rate (PCRn) in this group of adults with rGH treatment. The improved PCRn was indicative of higher dietary protein intake. No differences were found in body weight, serum albumin or leptin between the two groups. The anabolic effect of rGH seemed to be abolished by subclinical inflammation, with the beneficial response only seen in the group with no evidence of inflammation.

Pupim demonstrated that short-term rGH administration induced a significant positive anabolic effect when added to additional nutrition (intradialytic parenteral nutrition) in a small group of seven chronic hemodialysis subjects (Pupim et al., 2005). Protein homeostasis was improved through an 18% increase in whole-body protein synthesis with less essential amino acid muscle loss during the GH treatment period. This effect was additive to that provided by the supplemental nutrition.

The impact of altered GH and IGF-I on the skeleton in adults with CKD has now received some investigation. Jehle studied 319 adults with CKD and demonstrated that patients with primary and secondary hyperparathyroidism had lower levels of IGFBP-5 but higher levels of IGFBP-1, -2, -3 and -6 than those without hyperparathyroidism, whereas total IGF-I and IGF-II levels were similar in the two groups (Jehle et al., 2000). A marked increase in serum levels of IGFBP-4 appeared to be characteristic for chronic renal failure in these adults. IGFBP-5 correlated with biochemical markers and histological indices of bone formation in ROD patients and was not influenced by renal function. These data suggest that IGFBP-5 may represent a useful serological marker for osteopenia and low bone turnover in adult patients on long-term dialysis.

The impact of rGH administration on bone metabolism in adults with CKD has been explored to a limited extent. Kotzmann studied the effects of a 12-month period of rGH therapy on bone parameters in a small group of 19 malnourished hemodialysis adult patients (Kotzmann et al., 2004). GH (0.25 IU/kg) was given subcutaneously 3×/week after each dialysis session. As expected, IGF-I concentrations rose significantly after 3 months. There was an impressive increase in bone turnover and PTH. There was a temporary reduction in BMD of the lumbar spine seen, which returned to baseline values after 12 months. In a shorter-term intervention of 12 weeks of rGH therapy in adults on hemodialysis, Kotzmann found that LBM remained stable during therapy (Kotzmann et al., 2001) while procollagen I carboxy terminal peptide, a marker of bone formation, increased significantly as parameters of bone resorption, like telopeptide ICTP, showed only a slight increase (Kotzmann et al., 2001). BMD at the lumbar spine decreased significantly after 3 months in the treatment group (0.8 ± 0.17 versus 0.77 ± 0.16 g/cm^2, $P < 0.01$), whereas BMD at the femoral neck remained stable in both groups. Polymorphonuclear (PMN) leukocyte phagocytic activity increased significantly after 3 months of therapy with rGH, whereas other parameters of PMN function were not affected by rGH. Interestingly, QOL was slightly improved in the rGH treated group, but decreased markedly in the placebo group.

Viidas further explored the response to rGH therapy in 20 elderly patients on chronic hemodialysis to investigate the effects on lipid profiles, blood pressure and bone

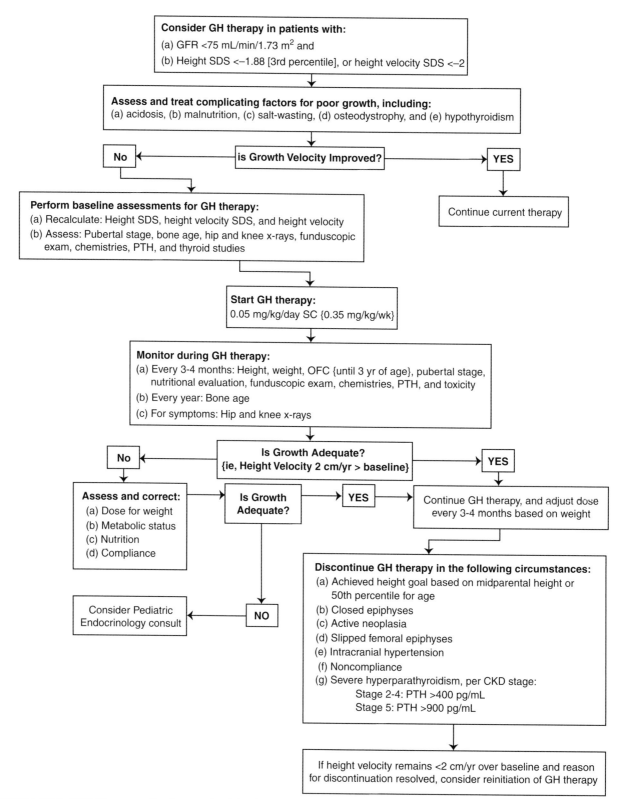

FIGURE 26.4 2006 Consensus Conference algorithm for evaluation and treatment of growth failure in children with CKD.

metabolism (Viidas et al., 2003). The typical lipid profile seen in CKD (increased triglycerides (TG), decreased high density lipoproteins, normal lipoprotein Apo-B, relatively low Apo-E values (Kaysen 2007)) was changed after rGH therapy. An unexpected decrease of TG and Apo-E with rGH therapy was noted. This, in fact, differs from the response seen in rGH treatment of non-uremic adults. Ambulatory 24-h blood pressures patterns were unchanged, while bone metabolism was increased in the rGH group as reflected by significant increases of osteocalcin (bone formation) and telopeptide of type I collagen (bone resorption) values.

E. Effects in Adults with Acute Renal Failure Treated with rGH

Exogenous rGH therapy has been investigated in adults with acute renal failure (ARF) (Saadeh et al., 2001). Saadeh performed an open label study of the response to rGH administered for 6 days in a small group of adults with ARF. All subjects had evidence of malnutrition at baseline and demonstrated improvements in negative nitrogen balance and a significant decrease in total nitrogen appearance rate. These changes corresponded to increases in serum growth hormone, IGF-I, IGF-I binding protein 3 and leptin levels after rGH treatment. More work is required better to define the effect of endogenous rGH on nutrition in this population. Based on experimental animal studies, there is no evidence that endogenous rGH treatment improves renal recovery (Fervenza et al., 1999) and there is no reason to expect a different response in adults with ARF.

F. Effects of Recombinant IGF-I Treatment in Chronic and Acute Renal Failure

In individuals with normal renal function, endogenous rIGF-I decreases serum glucose, facilitates the intracellular transport of glucose, suppresses lipolysis, protein degradation and ureagenesis, enhances protein synthesis, positive nitrogen balance and bone growth and increases GFR (Yakar et al., 2005). In renal failure, there is limited information about the effects of endogenous rIGF-I.

Because IGF-I appears to mediate most of the anabolic effects of GH, Fouque investigated the effects of endogenous IGF-I administration on nutritional status in malnourished adults on chronic peritonal dialysis (Fouque et al., 2000). Associated with a rise in serum IGF-I, nitrogen balance became strongly positive and serum phosphorus declined, consistent with an anabolic effect. Some patients experienced a reduction in blood pressure.

Basal plasma leptin and IGF-I levels are not correlated in adults with CKD or normal renal function. However, chronic administration of recombinant IGF-I has been associated with an early and sustained decrease in plasma leptin levels (Dagogo-Jack et al., 1998). Since lower serum leptin levels appear to be associated with better outcomes in patients with CKD and IGF-I may have an inhibitory effect on leptin secretion, this may be a therapeutic benefit for IGF-I therapy in patients with CKD.

In experimental models of acute renal failure (ARF), endogenous rIGF-I treatment increases the recovery of renal function, stimulating tubular cell proliferation, decreasing protein degradation and increasing protein synthesis. In studies where rIGF-I is administered before an ischemic event, less renal injury is seen. This effect appears to be mediated by nitric oxide generation (Noguchi et al., 1993). In these models of experimental ARF, rIGF-I administration is also associated with increased IGF-I receptor number and decreased IGFBP levels, resulting in increased bioavailability of the rIGF-I.

The results of clinical trials of rIGF-I therapy in adults with ARF, however, have been disappointing. In a study by Franklin, patients at risk of developing postoperative ARF as a complication of surgery involving the suprarenal aorta or renal arteries were administered rIGF-I (Franklin et al., 1997). The intervention group had no benefits in terms of length of hospital stay, dialysis or mortality compared to the placebo control group. Although a smaller proportion of patients administered rhIGF-I experienced a postoperative decline in renal function, this did not result in specific clinical benefits. In a larger multicenter study of 72 adults with ARF and significant co-morbidity, exogenous rIGF-I did not accelerate renal recovery (Hirschberg et al., 1999). This result may have been affected by a 6-day lag time before initiation of rIGF-I therapy. An additional clinical trial of acute tubular necrosis, involving delayed graft function following kidney transplantation, demonstrated a similar lack of response to rIGF-I even with prompt therapy (Hladunewich et al., 2003). In experimental ARF, it is now clear that IGF-I treatment can actually enhance the inflammatory response leading to cell toxicity due to neutrophil accumulation and this may explain the poor response to IGF-I therapy in this condition in adults (Fernández et al., 2001).

V. THE HORIZON FOR IMPROVING GROWTH AND ANABOLISM IN RENAL FAILURE

Recombinant human GH now is well established as a useful therapy for growth failure of children with CKD. The potential for rGH therapy to improve nutrition in adult patients on hemodialysis appears very promising. Recombinant GH therapy in both children and adults with CKD may have psychosocial and other benefits. Recombinant human IGF-I administration may be beneficial in the management of children and adults with CKD (Clark, 2005). Studies are now underway in Europe to assess the value of rIGF-I in children with CKD. Resistance to rIGF-I with continued administration has been a concern, however, since rIGF-I appears to suppress endogenous GH, IGFBP3 and ALS production, which may adversely affect growth.

One approach that may be worthy of further consideration is the combination of rGH and rIGF-I in treatment of growth delay and other abnormalities of the GH–IGF-I axis in uremia (Iglesias et al., 2004). In uremic rats, the combination of GH and IGF-I has additive effects on longitudinal growth and anabolism. Since there is evidence for resistance to endogenous GH and IGF-I in CKD, combination therapy may be the best method to correct the spectrum of abnormalities in CKD.

Modifications of rIGF-I may allow better realization of the therapeutic potential of this therapy in CKD. Mecasermin rinfabate, which combines rIGF-I with its major serum binding protein IGFBP-3, may prove to be an effective method to achieve the desired IGF-I effect while avoiding tolerance to therapy (Kemp et al., 2006). IGF-I displacers, which are peptides that displace endogenous IGF-I from its low molecular weight binding proteins (IGFBP-1/IGFBP-2) without affecting binding to IGFBP-3, have also been developed and may be a useful treatment strategy to circumvent resistance to exogenous rIGF-I administration (Roelfsema et al., 2000). Antagonists to GH and IGF-I may also have a therapeutic role in individuals with CKD. Animal studies suggest that pegvisomant, a GH antagonist, may slow the progression of chronic renal failure (Muller and van der Lely, 2004). IGF-I antagonists, however, which have the ability to decrease cell activity and have been developed for the treatment of cancer, have failed to inhibit renal fibrosis in a rat model of decreased renal function (Oldroyd et al., 2006).

VI. SUMMARY

GH, which primarily exerts its somatotropic and metabolic effects by stimulating production of IGF-I, initiates a wide range of cellular effects. The production of GH and IGF-I and the important interplay of these two proteins are quite complex and are altered in individuals with renal insufficiency. The alterations in the GH/IGF-I axis seen in renal failure can lead to profound effects. In children with renal disease, GH and IGF-I abnormalities may result in impaired linear growth, decreased muscle mass, decreased bone mass, impaired neurocognitive development and disordered plasma lipids. In adults with renal insufficiency, GH and IGF-I abnormalities have also been identified and may lead to decreased muscle mass, decreased bone mass, altered plasma lipids and altered metabolism.

Exogenous rGH therapy is safe and effective in the treatment of children with CKD who have growth failure. The benefit of rGH therapy for the other GH/IFGF-I mediated clinical problems seen in CKD is not proven but some of these may become important indications in the future. There are now tantalizing indications that rGH therapy may improve anabolism and other important functions in adults with CKD, but much work still remains to be done in this area. The impact of exogenous rIGF-I and other modifications of the IGF-I molecule, including IGF-I displacers, may have a role in the treatment of some renal disorders. Recognition of the clinical consequences and appropriate management of GH/IGF-I abnormalities in children and adults with CKD represents a valuable new framework for improving the lives of individuals with CKD.

References

Abdulle, A. M., Gillett, M. P., Abouchacra, S. et al. (2007). Low IGF-I levels are associated with cardiovascular risk factors in haemodialysis patients. *Mol Cell Biochem* **302**: 195–201.

Bailey, J. L., Zheng, B., Hu, Z., Price, S. R., and Mitch, W. E. (2006). Chronic kidney disease causes defects in signaling through the insulin receptor substrate/phosphatidylinositol 3-kinase/Akt pathway: implications for muscle atrophy. *J Am Soc Nephrol* **17**: 1388–1394.

Belgorosky, A., Ferraris, J. R., Ramirez, J. A., Jasper, H., and Rivarola, M. A. (1991). Serum sex hormone-binding globulin and serum nonsex hormone-binding globulin-bound testosterone fractions in prepubertal boys with chronic renal failure. *J Clin Endocrinol Metab* **73**: 107–110.

Berfield, A. K., Chait, A., Oram, J. F., Zager, R. A., Johnson, A. C., and Abrass, C. K. (2006). IGF-I induces rat glomerular mesangial cells to accumulate triglyceride. *Am J Physiol Renal Physiol* **290**: 38–47.

Blum, W. F., Ranke, M. B., Kietzmann, K., Tonshoff, B., and Mehls, O. (1991). Growth hormone resistance and inhibition of somatomedin activity by excess of insulin-like growth factor binding protein in uraemia. *Pediatr Nephrol* **5**: 539–544.

Boechat, M. I., Winters, W. D., Hogg, R. J., Fine, R. N., and Watkins, S. L. (2001). Avascular necrosis of the femoral head in children with chronic renal disease. *Radiology* **218**: 411–413.

Boehm, M., Riesenhuber, A., Winkelmayer, W. C., Arbeiter, K., Mueller, T., and Aufricht, C. (2007). Early erythropoietin therapy is associated with improved growth in children with chronic kidney disease. *Pediatr Nephrol* **22**: 1189–1193.

Boirie, Y., Broyer, M., Gagnadoux, M. F., Niaudet, P., and Bresson, J. L. (2000). Alterations of protein metabolism by metabolic acidosis in children with chronic renal failure. *Kidney Int* **58**: 236–241.

Broussard, S. R., McCusker, R. H., Novakofski, J. E., Strle, K., Shen, W. H., and Johnson, R. W. (2003). Cytokine-hormone interactions: tumor necrosis factor alpha impairs biologic activity and downstream activation signals of the insulin-like growth factor I receptor in myoblasts. *Endocrinology* **144**: 2988–2996.

Broyer, M., Le, B. C., Charbit, M. et al. (2004). Long-term social outcome of children after kidney transplantation. *Transplant* **77**: 1033–1037.

Busschbach, J. J., Rikken, B., Grobbee, D. E., De Charro, F. T., and Wit, J. M. (1998). Quality of life in short adults. *Horm Res* **49**: 32–38.

Chu, L. W., Lam, K. S., Tam, S. C. et al. (2001). A randomized controlled trial of low-dose recombinant human growth hormone in the treatment of malnourished elderly medical patients. *J Clin Endocrinol Metab* **86**: 1913–1920.

Clark, R. G. (2005). Recombinant insulin-like growth factor-1 as a therapy for IGF-I deficiency in renal failure. *Pediatr Nephrol* **20**: 290–294.

Dagogo-Jack, S., Franklin, S. C., Vijayan, A., Liu, J., Askari, H., and Miller, S. B. (1998). Recombinant human insulin-like growth factor-I (IGF-I) therapy decreases plasma leptin concentration in patients with chronic renal insufficiency. *Int J Obes Relat Metab Disord* **22**: 1110–1115.

Daughaday, W. H., and Rotwein, P. (1989). Insulin-like growth factors I and II. Peptide, messenger ribonucleic acid and gene structures, serum, and tissue concentrations. *Endoc Rev* **10**: 68–91.

Doi, T., Striker, L. J., Quaife, C. et al. (1988). Progressive glomerulosclerosis develops in transgenic mice chronically expressing growth hormone and growth hormone releasing factor but not in those expressing insulin-like growth factor-1. *Am J Path* **131**: 398–403.

Ericsson, F., Filho, J. C., and Lindgren, B. F. (2004). Growth hormone treatment in hemodialysis patients – a randomized, double-blind, placebo-controlled study. *Scand J Urol Nephrol* **38**: 340–347.

Ernst, F., Hetzel, S., Stracke, S. et al. (2001). Renal proximal tubular cell growth and differentiation are differentially modulated by renotropic growth factors and tyrosine kinase inhibitors. *Eur J Clin Invest* **31**: 1029–1039.

Fernández, M., Medina, A., Santos, F. et al. (2001). Exacerbated inflammatory response induced by insulin-like growth factor I treatment in rats with ischemic acute renal failure. *J Am Soc Nephrol* **12**: 1900–1907.

Ferns, G. A., Morani, A. S., and Anggard, E. E. (1991). The insulin-like growth factors: their putative role in atherogenesis. *Artery* **18**: 197–225.

Fervenza, F. C., Hsu, F. W., Tsao, T., Friedlaender, M. M., and Rabkin, R. (1999). Response to growth hormone therapy in experimental ischemic acute renal failure. *J Lab Clin Med* **133**: 434–439.

Fine, R. N., and Stablein, D. (2005). Long-term use of recombinant human growth hormone in pediatric allograft recipients: a report of the NAPRTCS Transplant Registry. *Pediatr Nephrol* **20**: 404–408.

Fine, R. N., Kohaut, E. C., Brown, D., and Perlman, A. J. (1994). Growth after recombinant human growth hormone treatment in children with chronic renal failure: report of a multicenter randomized double-blind placebo-controlled study. Genentech Co-operative Study Group. *J Pediatr* **124**: 374–382.

Fine, R. N., Stablein, D., Cohen, A. H., Tejani, A., and Kohaut, E. (2002). Recombinant human growth hormone post-renal transplantation in children: a randomized controlled study of the NAPRTCS. *Kidney Int* **62**: 688–696.

Fine, R. N., Ho, M., Tejani, A., and Blethen, S. (2003). Adverse events with rhGH treatment of patients with chronic renal insufficiency and end-stage renal disease. *J Pediatr* **142**: 539–545.

Fouque, D., Peng, S. C., and Kopple, J. D. (1995). Impaired metabolic response to recombinant insulin-like growth factor-1 in dialysis patients. *Kidney Int* **47**: 876–883.

Fouque, D., Juillard, L., Lasne, Y. et al. (1998). Acute leptin regulation in end-stage renal failure: the role of growth hormone and IGF-I. *Kidney Int* **54**: 932–937.

Fouque, D., Peng, S. C., Shamir, E., and Kopple, J. D. (2000). Recombinant human insulin-like growth factor-1 induces an anabolic response in malnourished CAPD patients. *Kidney Int* **57**: 646–654.

Franklin, S. C., Moulton, M., Sicard, G. A., Hammerman, M. R., and Miller, S. B. (1997). Insulin-like growth factor I preserves renal function postoperatively. *Am J Physiol* **272**: F257–F259.

Frystyk, J., Ivarsen, P., Skjaerbaek, C., Flyvbjerg, A., Pedersen, E. B., and Orskov, H. (1999). Serum-free insulin-like growth factor I correlates with clearance in patients with chronic renal failure. *Kidney Int* **56**: 2076–2084.

Furth, S. L., Hwang, W., Yang, C., Neu, A. M., Fivush, B. A., and Powe, N. R. (2002). Growth failure, risk of hospitalization and death for children with end-stage renal disease. *Pediatr Nephrol* **17**: 450–455.

Furth, S. L., Stablein, D., Fine, R. N., Powe, N. R., and Fivush, B. A. (2002). Adverse clinical outcomes associated with short stature at dialysis initiation: a report of the North American Pediatric Renal Transplant Cooperative Study. *Pediatrics* **109**: 909–913.

Graf, L., Candelaria, S., Doyle, M., and Kaskel, F. (2007). Nutrition assessment and hormonal influences on body composition in children with chronic kidney disease. *Adv Chronic Kidney Dis* **14**: 215–223.

Green, H., Morikawa, M., and Nixon, T. (1985). A dual effector theory of growth hormone action. *Differentiation* **29**: 195–198.

Greenstein, J., Guest, S., Tan, J. C., Tummala, P., Busque, C., and Rabkin, R. (2006). Circulating growth hormone binding protein levels and mononuclear cell growth hormone receptor expression in uremia. *J Ren Nutr* **16**: 141–149.

Guest, G., Berard, E., Crosnier, H., Chevallier, T., Rappaport, R., and Broyer, M. (1998). Effects of growth hormone in short children after renal transplantation. French Society of Pediatric Nephrology. *Pediatr Nephrol* **12**: 437–446.

Haffner, D., Schaefer, F., Girard, J., Ritz, E., and Mehls, O. (1994). Metabolic clearance of recombinant human growth hormone in health and chronic renal failure. *J Clin Invest* **93**: 1163–1171.

Haffner, D., Schaefer, F., Nissel, R., Wuhl, E., Tonshoff, B., and Mehls, O. (2000). Effect of growth hormone treatment on the adult height of children with chronic renal failure. German Study Group for Growth Hormone Treatment in Chronic Renal Failure. *N Engl J Med* **343**: 923–930.

Hansen, T. B., Gram, J., Jensen, P. B. et al. (2000). Influence of growth hormone on whole body and regional soft tissue composition in adult patients on hemodialysis. A double-blind, randomized, placebo-controlled study. *Clin Nephrol* **53**: 99–107.

Haspolat, K., Ece, A., Gurkan, F., Atamer, Y., Tutanc, M., and Yolbas, I. (2007). Relationships between leptin, insulin, IGF-I and IGFBP-3 in children with energy malnutrition. *Clin Biochem* **40**: 201–205.

Hirschberg, R., and Adler, S. (1998). Insulin-like growth factor system and the kidney: physiology, pathophysiology, and therapeutic implications. *Am J Kidney Dis* **31**: 901–919.

Hirschberg, R., Kopple, J., Lipsett, P. et al. (1999). Multicenter clinical trial of recombinant human insulin-like growth factor I in patients with acute renal failure. *Kidney Int* **55**: 2423–2432.

Hladunewich, M. A., Corrigan, G., Derby, G. C. et al. (2003). A randomized, placebo-controlled trial of IGF-I for delayed graft function: a human model to study postischemic ARF. *Kidney Int* **64**: 593–602.

Hokken-Koelega, A. C., van Zaal, M. A., van Bergen, W. et al. (1994). Final height and its predictive factors after renal transplantation in childhood. *Pediatr Res* **36**: 323–328.

Hokken-Koelega, A., Mulder, P., De Jong, R., Lilien, M., Donckerwolcke, R., and Groothof, J. (2000). Long-term effects of growth hormone treatment on growth and puberty in patients with chronic renal insufficiency. *Pediatr Nephrol* **14**: 701–706.

Iglesias, P., Díez, J. J., Fernández-Reyes, M. J. et al. (2004). Growth hormone, IGF-I and its binding proteins (IGFBP-1 and -3) in adult uraemic patients undergoing peritoneal dialysis and haemodialysis. *Clin Endocrinol* **60**: 741–749.

Isaakson, O. G., Lindahl, A., Nilsson, A., and Isgaard, J. (1987). Mechanism of the stimulatory effect of growth hormone on longitudinal bone growth. *Endoc Rev* **6:** 426–438.

Jabs, K., Van, D. C., and Harmon, W. E. (1993). Growth hormone treatment of growth failure among children with renal transplants. *Kidney Int* **43**(Suppl.): S71–S75.

Jehle, P. M., Ostertag, A., Schulten, K. et al. (2000). Insulin-like growth factor system components in hyperparathyroidism and renal osteodystrophy. *Kidney Int* **57:** 423–436.

Jensen, P. B., Ekelund, B., Nielsen, F. T., Baumbach, L., Pedersen, F. B., and Oxhøj, H. (2000). Changes in cardiac muscle mass and function in hemodialysis patients during growth hormone treatment. *Clin Nephrol* **53:** 25–32.

Johannsson, G., Bengtsson, B. A., and Ahlmén, J. (1999). Double-blind, placebo-controlled study of growth hormone treatment in elderly patients undergoing chronic hemodialysis: anabolic effect and functional improvement. *Am J Kidney Dis* **33:** 709–717.

Juul, A., Scheike, T., Davidsen, M., Gyllenborg, J., and Jorgensen, T. (2002). Low serum insulin-like growth factor-1 is associated with increased risk of ischemic heart disease: a population-based study case control. *Circulation* **106:** 939–944.

Kaplan, S. A., and Cohen, P. (2007). Review: the somatomedin hypothesis 2007: 50 years later. *J Clin Endocrinol Metab* **92:** 4529–4535.

Kari, J. A., Gonzalez, C., Ledermann, S. E., Shaw, V., and Rees, L. (2000). Outcome and growth of infants with severe chronic renal failure. *Kidney Int* **47:** 1681–1687.

Kaskel, F. (2003). Chronic kidney disease: a growing problem. *Kidney Int* **64:** 1141–1151.

Kaysen, G. A. (2001). The microinflammatory state in uremia: causes and potential consequences. *J Am Soc Nephrol* **12:** 1549–1557.

Kaysen, G. A. (2007). Hyperlipidemia in chronic kidney disease. *Int J Artif Organs* **30:** 987–992.

Kemp, S. F., Fowlkes, J. L., and Thrailkill, K. M. (2006). Efficacy and safety of mecasermin rinfabate. *Expert Opin Biol Ther* **6:** 533–538.

Kojima, M., Hosada, H., Date, Y., Nakazato, M., Matsuo, H., and Kanawa, K. (1999). Ghrelin is a growth hormone-releasing acylated peptide from stomach. *Nature* **402:** 656–660.

Kotzmann, H., Yilmaz, N., Lercher, P. et al. (2001). Differential effects of growth hormone therapy in malnourished hemodialysis patients. *Kidney Int* **60:** 1578–1585.

Kotzmann, H., Riedl, M., Pietschmann, P. et al. (2004). Effects of 12 months of recombinant growth hormone therapy on parameters of bone metabolism and bone mineral density in patients on chronic hemodialysis. *J Nephrol* **17:** 87–94.

Lane, P. H. (2005). Puberty and chronic kidney disease. *Adv Chronic Kidney Dis* **12:** 372–377.

Langman, C. B., and Brooks, E. R. (2006). Renal osteodystrophy in children: a systemic disease associated with cardiovascular manifestations. *Growth Horm IGF Res*, pp. S79–S83.

Laviades, C., Gil, M. J., Monreal, I., Gonzalez, A., and Diez, J. (1997). Is the tissue availability of circulating insulin-like growth factor involved in organ damage and glucose regulation in hypertension? *J Hypertens* **15:** 1159–1165.

Lilien, M. R., Schroder, C. H., Levtchenko, E. N., and Koomans, H. A. (2004). Growth hormone therapy influences endothelial function in children with renal failure. *Pediatr Nephrol* **19:** 785–789.

Liu, J. P., Baker, J., Perkins, A. S., Robrtson, E. J., and Efstrstiadia, A. (1993). Mice carrying null mutations of the genes encoding insulin-like growth factor I (IGF-I and type 1 EGF receptor (Igf1r). *Cell* **75:** 59–72.

Mahan, J. D., and Warady, B. A. for Consensus Committee (2006). Assessment and treatment of short stature in pediatric patients with chronic kidney disease: a consensus statement. *Pediatr Nephrol* **21:** 917–930.

Mahesh, S., and Kaskel, F. (2008). Growth hormone axis in chronic kidney disease. *Pediatr Nephrol* **23:** 41–48.

Maheshwari, H. G., Rifkin, I., Butler, J., and Norman, M. (1992). Growth hormone binding protein in patients with renal failure. *Acta Endocrinol* **127:** 485–488.

Mak, R. H., and Cheung, W. (2006). Energy homeostasis and cachexia in chronic kidney disease. *Pediatr Nephrol* **21:** 1807–1814.

Maniar, S., Kleinknecht, C., Zhou, X., Motel, V., Yvert, J. P., and Dechaux, M. (1996). Growth hormone action is blunted by acidosis in experimental uremia or acid load. *Clin Nephrol* **46:** 72–76.

Mauras, N., Attie, K. M, Reiter, E. O., Saenger, P., and Baptista, J. (2000). High dose recombinant human growth hormone (GH) treatment of GH-deficient patients in puberty increases near-final height: a randomized, multicenter trial. Genentech Inc Cooperative Study Group. *J Clin Endocrinol Metab* **85:** 3653–3660.

McSherry, E. (1978). Acidosis and growth in nonuremic renal disease. *Kidney Int* **14:** 349–354.

Mentser, M., Breen, T. J., Sullivan, E. K., and Fine, R. N. (1997). Growth hormone treatment of renal transplant recipients: the National Cooperative Growth Study experience – a report of the National Cooperative Growth Study and the North American Pediatric Renal Transplant Cooperative Study. *J Pediatr* **131:** S20–S24.

Muller, A. F., and van der Lely, A. J. (2004). Insights from growth hormone receptor blockade. *Curr Opin Investig Drugs* **5:** 1072–1079.

National Kidney Foundation Disease Outcomes Quality Initiative (2002). K/DOQI clinical practice guidelines for chronic kidney disease: evaluation, classification, and stratification. Part 4. Definition and classification of stages of chronic kidney disease. National Kidney Foundation. Available at http://www.kidney.org/professionals/kdoqi/guidelines_ckd/p4_class_g2.htm.

Nissel, R., Lindberg, A., Mehls, O., and Haffner, D. on behalf of the KIGS International Board (2008). Factors predicting the near-final height in growth hormone treated children and adolescents with chronic kidney disease. *J Clin Endocrinol Metab* **93:** 1359–1365.

Noguchi, S., Kashihara, Y., Ikegami, Y., Morimotoa, K., Miyamotao, M., and Nakao, K. (1993). Insulin-like growth factor-I ameliorates transient ischemia-induced acute renal failure in rats. *J Pharmacol Exp Ther* **267:** 919–926.

Oldroyd, S. D., Miyamoto, Y., Moir, A., Johnson, T. S., El Nahas, A. M., and Haylor, J. L. (2006). An IGF-I antagonist does not inhibit renal fibrosis in the rat following subtotal nephrectomy. *Am J Physiol Renal Physiol* **290:** F695–F702.

Pombo, M., Pombo, C. M., Garcia, A. et al. (2001). Hormonal control of growth hormone secretion. *Horm Res* **55**(Suppl 1): 11–16.

Postel-Vinay, M. C., Tar, A., Crosnier, H. et al. (1991). Plasma growth hormone-binding activity is low in uraemic children. *Pediatr Nephrol* **5:** 545–547.

Powell, D. R., Liu, F., Baker, B. K. et al. (1993). Characterization of insulin-like growth factor binding protein-3 in chronic renal failure serum. *Pediatr Res* **33:** 136–143.

Powell, D. R., Liu, F., Baker, B. K., Lee, P. D.K., and Hintz, R. L. (1996). Insulin-like growth factor binding proteins as growth inhibitors in children with chronic renal failure. *Pediatr Nephrol* **10:** 343–347.

Powell, D. R., Durham, S. K., Lu, F. et al. (1998). The insulin-like growth factor axis and growth in children with chronic renal failure: a report of the Southwest Pediatric Nephrology Study Group. *J Clin Endocrinol Metab* **83:** 1654–1661.

Pupim, L. B., Flakoll, P. J., Yu, C., and Ikizler, T. A. (2005). Recombinant human growth hormone improves muscle amino acid uptake and whole-body protein metabolism in chronic hemodialysis patients. *Am J Clin Nutr* **82:** 1235–1243.

Rabkin, R., Sun, D. F., Chen, Y., Tan, J., and Schafer, F. (2005). Growth hormone resistance in uremia, a role for impaired JAK/STAT signaling. *Pediatr Nephrol* **20:** 313–318.

Roelfsema, V., and Clark, R. S. (2001). The growth hormone and insulin-like growth factor axis: its manipulation for the benefit of growth disorders in renal failure. *J Am Nephrol* **12:** 1297–1306.

Roelfsema, V., Lane, M. H., and Clark, R. G. (2000). Insulin-like growth factor binding protein (IGFBP) displacers: relevance to the treatment of renal disease. *Pediatr Nephrol* **14:** 584–588.

Rosenkranz, J., Reichwald-Klugger, E., Oh, J., Turzer, M., Mehls, O., and Schaefer, F. (2005). Psychosocial rehabilitation and satisfaction with life in adults with childhood-onset of end-stage renal disease. *Pediatr Nephrol* **20:** 1288–1294.

Saadeh, E., Ikizler, T. A., Shyr, Y., Hakim, R. M., and Himmelfarb, J. (2001). Recombinant human growth hormone in patients with acute renal failure. *J Ren Nutr* **11:** 212–219.

Salusky, I. B., Kuizon, B. G., and Juppner, H. (2004). Special aspects of renal osteodystrophy in children. *Semin Nephrol* **24:** 69–77.

Schaefer, F., Veldhuis, J. D., Stanhoe, R., Jones, J., and Scharer, K. (1994). Alterations in growth hormone secretion and clearance in peripubertal boys with chronic renal failure and after renal transplantation. Cooperative Study Group on Pubertal Development in Chronic Renal Failure. *J Clin Endocrinol Metab* **78:** 1298–1306.

Schaefer, F., Veldhuis, J. D., Robertson, W. R., Dunger, D., and Scharer, K. (1994). Immunoreactive and bioactive luteinizing hormone in pubertal patients with chronic renal failure. Cooperative Study Group on Pubertal Development in Chronic Renal Failure. *Kidney Int* **45:** 1465–1476.

Schaefer, F., Chen, Y., Tsao, T., Nouri, P., and Rabkin, R. (2001). Impaired JAK-STAT signal transduction contributes to growth hormone resistance in chronic uremia. *J Clin Invest* **108:** 467–475.

Sjogren, K., Liu, J. L., Blad, K. et al. (1999). Liver-derived insulin-like growth factor I (IGF-I) is the principal source of IGF-I in blood but is not required for postnatal body growth in mice. *Proc Natl Acad Sci USA* **98:** 7088–7092.

Stabler, B., Clopper, R. R., Siegel, P. T., Stoppani, C., Compton, P. G., and Underwood, L. E. (1994). Academic achievement and psychological adjustment in short children. The National Cooperative Growth Study. *J Dev Behav Pediatr* **15:** 1–6.

Stabler, B., Siegel, P. T., Clopper, R. R., Stoppani, C. E., Compton, P. G., and Underwood, L. E. (1998). Behavior change after growth hormone treatment of children with short stature. *J Pediatr* **133:** 366–373.

Szczech, L. A., Reddan, D. N., Klassen, P. S. et al. (2003). Interactions between dialysis-related volume exposures, nutritional surrogates and mortality among ESRD patients. *Nephrol Dial Transplant* **18:** 1585–1591.

Tonshoff, B., and Mehls, O. (1995). Growth retardation in children with chronic renal insufficiency: current aspects of pathophysiology and treatment. *J Nephrol* **8:** 133–142.

Tonshoff, B., Eden, S., Weiser, E. et al. (1994). Reduced hepatic growth hormone (GH) receptor gene expression and increased GH binding protein in experimental uremia. *Kidney Int* **45:** 1085–1092.

Tonshoff, B., Veldhuis, J. D., Heinrich, U., and Mehls, O. (1995). Deconvolution analysis of spontaneous nocturnal growth hormone secretion in prepubertal children with preterminal renal failure and with end-stage renal disease. *Pediatr Res* **37:** 86–93.

Tonshoff, B., Cronin, M. J., Reichert, M. et al. (1997). Reduced concentration of serum growth hormone (GH)-binding protein in children with chronic renal failure: correlation with GH insensitivity. The European Study Group for Nutritional Treatment of Chronic Renal Failure in Childhood. The German Study for Growth Hormone Treatment in Chronic Renal Failure. *J Clin Endocrinol Metab* **82:** 1007–1013.

Tonshoff, B., Kiepe, D., and Ciarmatori, S. (2005). Growth hormone/insulin-like growth factor system in children with chronic renal failure. *Pediatr Nephrol* **20:** 279–289.

Van Dyck, M., Gyssels, A., Proesmans, W., Nijs, J., and Eeckels, R. (2001). Growth hormone treatment enhances bone mineralisation in children with chronic renal failure. *Eur J Pediatr* **160:** 359–363.

Van Es, A. (1991). Growth hormone treatment in short children with chronic renal failure and after renal transplantation: combined data from European clinical trials. The European Study Group. *Acta Paediatr Scand Suppl* **379:** 42–48.

Viidas, U., Johannsson, G., Mattson-Hulten, L., and Ahlmen, J. (2003). Lipids, blood pressure and bone metabolism after growth hormone therapy in elderly hemodialysis patients. *J Nephrol* **16:** 231–237.

Vimalachandra, D., Hodson, E. M., Willis, N. S., Craig, J. C., Cowell, C., and Knight, J. F. (2006). Growth hormone for children with chronic kidney disease. *Cochrane Database Syst Rev* **3:** CD003264.

Wesseling, K., Bakkaloglu, S., and Salusky, I. (2008). Chronic kidney disease mineral and bone disorder in children. *Pediatr Nephrol* **23:** 195–207.

Wormald, S., and Hilton, D. J. (2004). Inhibitors of cytokine signal transduction. *J Biol Chem* **279:** 821–824.

Yakar, S., Pennisi, P., Wu, Y., Zhao, H., and LeRoith, D. (2005). Clinical relevance of systemic and local IGF-I. *Endocr Dev* **9:** 11–16.

Further reading

Hochberg, Z. (2002). Mechanisms of steroid impairment of growth. *Horm Res* **58**(Suppl 1): 33–38.

Van Dyck, M., and Proesmans, W. (2001). Growth hormone therapy in chronic renal failure induces catch-up of head circumference. *Pediatr Nephrol* **16:** 631–636.

CHAPTER 27

Sexual Dysfunction in Men and Women with Chronic Kidney Disease

BIFF F. PALMER

Department of Medicine, Division of Nephrology, University of Texas Southwestern Medical Center, 5323 Harry Hines Blvd, Dallas Texas, 75390, USA

Contents

I.	Introduction	429
II.	Sexual dysfunction in uremic men	429
III.	Evaluation of sexual dysfunction in the uremic man	432
IV.	Treatment of sexual dysfunction in the uremic man	433
V.	Outcomes associated with hypogonadism and treatment	435
VI.	Sexual dysfunction in uremic women	437
VII.	Treatment	437
	References	439

I. INTRODUCTION

Disturbances in sexual function are common in patients with chronic kidney disease. Such disturbances include erectile dysfunction, decreased libido and marked declines in the frequency of intercourse (Procci et al., 1981; Toorians et al., 1997). Sexual dysfunction is multifactorial and primarily organic in origin. In addition to the uremic milieu, peripheral neuropathy, autonomic insufficiency, peripheral vascular disease and pharmacologic therapy all play an important role in the genesis of these problems. In addition, psychological and physical stresses are also commonly present in this setting.

II. SEXUAL DYSFUNCTION IN UREMIC MEN

Erectile dysfunction is one of the most common manifestations of sexual dysfunction in men with chronic kidney disease. The prevalence of this disorder has been reported to be as high as 70–80% and is similar between patients on hemodialysis and peritoneal dialysis (Rosas et al., 2001b; Turk et al., 2001). The high prevalence of this disorder is not surprising given that many of the diseases such as atherosclerosis, diabetes and hypertension that are associated with erectile dysfunction are commonly found in patients with chronic kidney disease. Normal male sexual function is achieved through the integrative response of the vascular, neurologic, endocrine and psychologic systems. Men with chronic kidney disease can exhibit abnormalities in any one or all of these systems (Table 27.1). Before discussing changes in sex hormones accompanying the loss of renal function, a brief discussion of other co-morbidities contributing to this disorder will be presented.

A. Vascular System

Normal erections require blood to flow from the hypogastric arterial system into specialized erectile chambers, including the paired corpora cavernosae flanking the penile urethra and the corpus spongiosum at the glans penis. As blood flow accelerates, the pressure within the intracavernosal spaces increases dramatically to choke off penile venous outflow from emissary veins. This combination of increased intracavernosal blood flow and reduced venous outflow allows a man to acquire and maintain a firm erection.

Studies in patients with chronic kidney disease have shown that vascular causes of erectile dysfunction are common (Kaufman et al., 1994). Decreased arterial inflow has been demonstrated to result from occlusive disease of the cavernosal artery. Erectile dysfunction can also result from occlusive disease in the more proximal ileac and pudendal arteries by producing a pelvic arterial steal syndrome (Michal et al., 1978). This syndrome results from blood being diverted from the corpus cavernosa in order to meet the oxygen requirements of the pelvis. In addition to reductions in arterial inflow many patients have evidence of veno-occlusive dysfunction. Such abnormalities lead to venous leakage and therefore inability to achieve or maintain an erection.

Erectile dysfunction in non-uremic men has been identified as a marker of atherosclerosis and endothelial

TABLE 27.1 Factors involved in the pathogenesis of erectile dysfunction in uremic men

Vascular system
    Occlusive arterial disease
    Veno-occlusive disease and venous leakage
Neurologic system
    Impaired autonomic function due to uremia and co-morbid conditions
Endocrine system
Psychologic system
Other factors
    Zinc deficiency
    Medications
    Anemia
    Secondary hyperparathyroidism

dysfunction in other vascular beds to include the coronary circulation (Billips, 2005). Erectile dysfunction is more common in men with evidence of coronary endothelial dysfunction as compared to those with normal function (Elesber et al., 2006). Such patients have increased circulating levels of the nitric oxide synthase inhibitor, asymmetric dimethylarginine (ADMA), suggesting a reduction of endogenous nitric oxide activity may be responsible for the abnormal vascular behavior. It is likely that a similar relationship exists between erectile dysfunction and vascular disease in uremic men given the high frequency of both disorders.

B. Neurologic System

Significant portions of chronic kidney disease patients display abnormalities in the function of the autonomic nervous system. Such derangements are due to co-morbid conditions, such as diabetes, but also result directly from uremic toxicity (Campese et al., 1982). Given the importance of the sympathetic and parasympathetic nervous system in normal sexual function, disturbances in the autonomic nervous system are likely to participate in the genesis of erectile dysfunction. Reductions in nocturnal penile tumescence and the frequency of sexual intercourse are positively correlated with disturbances in autonomic function as assessed by the Valsalva maneuver (Campese et al., 1982).

C. Endocrine System

While adequate penile blood flow and intact neural input are prerequisites for an erectile response, the endocrine system plays an integral role in normal male sexual function (Table 27.2). In men with chronic kidney disease, disturbances in the pituitary–gonadal axis can be detected with only moderate reductions in the glomerular filtration rate and progressively worsen as kidney failure progresses. These disorders rarely normalize with initiation of hemodialysis or peritoneal dialysis and, in fact, often progress. By comparison, a well-functioning kidney transplant is much more likely to restore normal sexual activity, although some features of reproductive function may remain impaired (Holdsworth et al., 1978; Diemont et al., 2000).

1. TESTICULAR FUNCTION

Chronic kidney disease is associated with impaired spermatogenesis and testicular damage, often leading to infertility (Prem et al., 1996). Semen analysis typically shows a decreased volume of ejaculate, either low or complete azoospermia and a low percentage of motility. These abnormalities are often apparent prior to the need for dialysis and then deteriorate further once dialytic therapy is initiated. Histologic changes in the testes show evidence of decreased spermatogenic activity with the greatest changes in the hormonally dependent later stages of spermatogenesis. The number of spermatocytes is reduced and there is little evidence of maturation to the stage of mature sperm. In most instances, the number of spermatogonia is normal but, on occasion, complete aplasia of germinal elements may also be present. Other findings include damage to the seminiferous tubules, interstitial fibrosis and calcifications. Interstitial fibrosis and calcification also develop in the epididymis and corpora cavernosa, particularly as the time on maintenance hemodialysis becomes prolonged (Guvel et al., 2004; Bellinghieri et al., 2004).

Unlike other causes of primary testicular disease, the Leydig and Sertoli cells show little evidence of hypertrophy or hyperplasia. This later finding suggests a defect in the hormonal regulation of the Leydig and Sertoli cells as might occur with gonadotropin deficiency or resistance, rather than a cytotoxic effect of uremia where spermatogonia would be most affected (Handelsman, 1985). The factors responsible for testicular damage in uremia are not well understood. It is possible that plasticizers in dialysis tubing, such as phthalate, may play a role in propagating the abnormalities once patients are initiated onto maintenance hemodialysis (Hauser et al., 2005).

TABLE 27.2 Endocrine factors involved in sexual dysfunction in uremic men and women

Men
↓ Gonadal function
    decreased production of testosterone
↓ Hypothalamic–pituitary function
    blunted increase in serum luteinizing hormone (LH) levels
    decreased amplitude of LH secretory burst
    variable increase in serum follicle stimulating hormone levels
    increased prolactin levels
Women
Anovulatory menstrual cycles
Lack of mid-cycle surge in LH
↑ Prolactin

2. SEX STEROIDS

In addition to impaired spermatogenesis, the testes also show evidence of impaired endocrine function. Total and free testosterone levels are typically reduced, although the binding capacity and concentration of sex hormone-binding globulin are normal (Lim and Fang, 1976; De Vries et al., 1984; Levitan et al., 1984; Albaaj et al., 2006). Acute stimulation of testosterone secretion with administration of human chorionic gonadotropin (HCG), a compound with luteinizing hormone-like actions, produces only a blunted response in uremic men (Stewart-Bentley et al., 1974). Lower free testosterone levels and impaired Leydig cell sensitivity to HCG are first detectable with only moderate reductions in the glomerular filtration rate and before basal levels of testosterone fall. The sluggish response of the Leydig cell to infusion of HCG may be due to the accumulation of a factor in uremic serum capable of blocking the luteinizing hormone receptor (Dunkel et al., 1997). This blocking activity is inversely correlated with glomerular filtration rate and largely disappears following transplantation. In comparison to testosterone, the total plasma estrogen concentration is often elevated in advanced kidney failure (Lim and Fang, 1976). However, the physiologically important estradiol levels are typically in the normal range. As with the lack of hypertrophy and hyperplasia of Leydig cells, normal levels of estradiol suggest a functional gonadotropin deficiency or resistance in uremia since increased luteinizing hormone levels should enhance the testicular secretion of estradiol (Handelsman, 1985).

3. HYPOTHALAMIC–PITUITARY FUNCTION

The plasma concentration of the pituitary gonadotropin, luteinizing hormone (LH), is elevated in uremic men (Lim and Fang, 1976). Elevated levels are found early in kidney insufficiency and progressively rise with deteriorating kidney function. The excess LH secretion in this setting is thought to result from the diminished release of testosterone from the Leydig cells, since testosterone normally leads to feedback inhibition of LH release. In addition, the metabolic clearance rate of LH is reduced as a result of decreased kidney clearance.

The increase in serum LH is variable and modest when compared to that observed in castrate non-uremic subjects. The lack of a more robust response of LH to low levels of circulating testosterone suggests a derangement in the central regulation of gonadotropin release. Infusion of gonadotropin-releasing hormone (GnRH) increases LH levels to the same degree as in normals, however, the peak value and return to baseline may be delayed (Schalch et al., 1975; LeRoith et al., 1980). Since the kidney contributes importantly to the clearance of GnRH and LH, decreased metabolism of these hormones may explain the observed variations. The abnormal LH response to GnRH precedes and is not corrected by dialytic therapy.

In addition to decreased metabolism, subtle disturbances in LH secretion have also been described. Under normal circumstances, LH is secreted in a pulsatile fashion. In uremic subjects, the number of secretory bursts remains normal but the amount of LH released per secretory burst is reduced (Schaefer et al., 1994a). It is not known whether this decrease in amplitude is the result of a change in the pattern of GnRH release from the hypothalamus or a change in the responsiveness of the pituitary. Under normal circumstances, GnRH release is pulsatile in nature. This pattern of release is critical for normal function of gonadotropin cells and reproductive capability. Uremia, with contributions from inadequate nutrient intake, stress and systemic illness, is associated with alterations in the pulsatile release of GnRH leading to a hypogonadal state (Ayub and Fletcher, 2000). The secretory pattern of GnRH and LH normalizes following the placement of a well functioning allograft.

Follicle stimulating hormone (FSH) secretion is also increased in men with chronic kidney failure, although to a more variable degree such that the LH/FSH ratio is typically increased. FSH release by the pituitary normally responds to feedback inhibition by a peptide product of the Sertoli cells called inhibin. The plasma FSH concentration tends to be highest in those uremic patients with the most severe damage to seminiferous tubules and presumably the lowest levels of inhibin. It has been suggested that increased FSH levels may portend a poor prognosis for recovery of spermatogenic function following kidney transplantation transplantation (Prem et al., 1996; Phocas et al., 1995).

Clomiphene is a compound that acts by competing with estrogen or testosterone for receptors at the level of the hypothalamus and prevents the negative feedback of gonadal steroids on the release of GnRh and subsequently the release of pituitary gonadotropins. When administered to chronic kidney failure patients, there is an appropriate rise in the levels of both LH and FSH suggesting that the negative feedback control of testosterone on the hypothalamus is intact and that the storage and release of gonadotropins by the pituitary is normal (Lim and Fang, 1976).

In summary, a number of observations suggest that gonadal failure is an important consequence of chronic kidney failure. The finding that LH levels are typically increased is consistent with the presence of testicular damage. However, the lack of Leydig cell hypertrophy and normal estradiol levels also raise the possibility of functional hypogonadism. The finding that LH levels are only modestly increased in chronic kidney failure suggests a diminished response of the hypothalamic–pituitary axis to lowered testosterone levels and impaired regulation of gonadotropin secretion. One explanation for the blunted rise in LH in response to low levels of testosterone is that the hypothalamic–pituitary axis in chronic kidney failure is reset in such a way that it is more sensitive to the negative feedback inhibition of testosterone. In this manner, the axis begins to assume a similar characteristic as seen in the pre-pubertal state where there is extreme sensitivity to the inhibitory effect of gonadal steroids (Handelsman, 1985).

4. Prolactin Metabolism

Elevated plasma prolactin levels are commonly found in dialyzed men (Gomez et al., 1980). Increased production is primarily responsible, since the kidney plays little, if any, role in the catabolism of this hormone. Prolactin release is normally under dopaminergic inhibitory control. Its secretion in chronic kidney disease, however, appears autonomous and resistant to stimulatory or suppressive maneuvers. As an example, dopamine infusion or the administration of oral L-dopa fails to decrease basal prolactin levels. On the other hand, procedures which normally increase prolactin secretion, such as arginine infusion, insulin-induced hypoglycemia or thyrotropin-releasing hormone infusion, either have no effect or only elicit a blunted response. These abnormalities resolve following a successful kidney transplant.

Increased prolactin secretion in chronic kidney disease may be related in part to the development of secondary hyperparathyroidism. An infusion of parathyroid hormone (PTH) in normal men enhances prolactin release, a response that can be suppressed by the administration of L-dopa (Isaac et al., 1978). Furthermore, partial inhibition of PTH release by the administration of calcitriol led, in one study, to an elevation in plasma testosterone levels, a reduction in plasma gonadotropin concentrations and improved sexual function (Massry et al., 1977). However, this benefit could not be confirmed in a controlled trial of calcitriol therapy (Blumberg et al., 1980). Depletion of total body zinc stores may also play an etiologic role in uremic hyperprolactinemia (Caticha et al., 1996).

The clinical significance of enhanced prolactin release in uremic men is incompletely understood. Extreme hyperprolactinemia is associated with infertility, loss of libido, low circulating testosterone levels and inappropriately low LH levels in men with normal kidney function. These observations led to the evaluation of therapy with bromocriptine, which reduces prolactin secretion. Although it can lower prolactin levels to near normal in men with advanced kidney disease, there has been an inconsistent effect on sexual potency and libido (Gomez et al., 1980). Use of the drug is complicated by a high incidence of side effects, particularly the development of hypotension.

5. Gynecomastia

Gynecomastia occurs in approximately 30% of men on maintenance hemodialysis. This problem most often develops during the initial months of dialysis and then tends to regress as dialysis continues. The pathogenesis of gynecomastia in this setting is unclear. Although elevated prolactin levels and an increased estrogen-to-androgen ratio seem attractive possibilities, most data fail to support a primary role for abnormal hormonal function. Alternatively, a mechanism similar to that responsible for gynecomastia following refeeding of malnourished patients may be involved.

D. Psychologic System

Primary depression may affect sexual function and lead to reduced libido and decreased frequency of intercourse. As a result, it has been suggested that depression may play a role in the genesis of sexual dysfunction in chronic kidney failure patients. Studies addressing this issue have produced conflicting results. Procci et al. found no association between the presence or absence of depression and measures of sexual function, such as frequency of intercourse or ability to develop an erection as determined by nocturnal penile tumescence testing (Procci et al., 1981). In a more recent study, Steele et al. surveyed a randomly selected group of 68 peritoneal dialysis patients that included both men and women and found that 63% of patients reported never having intercourse, 19% had intercourse less than or equal to two times per month and 18% had intercourse more than two times per month (Steele et al., 1996). In this study, standard psychologic questionnaires indicated that the patients who never had intercourse were more depressed and anxious and assessed their overall quality of life at a level that was significantly lower than that of the other two groups. Thus, subclinical depression may be underrated as a contributory cause of sexual dysfunction in both uremic men and women.

A negative body image is commonly present in patients receiving dialysis (Finkelstein et al., 2007). This concern arises from issues such as the necessity of having a catheter for dialysis access and the need to be connected to a machine. Sexual dysfunction can lead to marital discord between the patient and their partner further impairing the level of sexual function. It is also likely that mental symptoms of fatigue and listlessness associated with the treatment of end-stage kidney disease are also major factors in this problem (Toorians et al., 1997).

III. EVALUATION OF SEXUAL DYSFUNCTION IN THE UREMIC MAN

Sexual dysfunction can be manifest in several ways in the uremic man. Perhaps the most common complaint that the physician must address is that of impotence. In evaluating and ultimately treating the impotent kidney failure patient, one must not only consider disturbances in the hypothalamic–pituitary–gonadal axis discussed above, but also abnormalities in the sympathetic nervous system and derangements in the arterial supply or venous drainage of the penis. In addition, the psychological effects of a chronic illness and lifestyle limitations may negatively impact on sexual function.

A thorough history and physical can provide useful information during the initial evaluation of a patient with impotence. A history of normal erectile function prior to the development of kidney failure is suggestive of a secondary cause of impotence. Symptoms or physical findings of a neuropathy, as in a patient with a neurogenic bladder, would be particularly suggestive of a neurologic etiology. Similarly, symptoms or signs of peripheral vascular disease may be a clue to the presence of vascular obstruction to penile blood flow. One should look for the presence of secondary sexual characteristics, such as facial, axillary and pubic hair. The lack of these findings and the presence of small soft testicles suggest primary or secondary hypogonadism as the cause of the impotence. Neurogenic and vascular causes are more likely to be associated with normal sized testicles. Even when the history and physical examination point to a specific abnormality, one must also consider that an individual patient may have more than one factor responsible for the erectile dysfunction and other causes may need ultimately to be evaluated.

A review of the patient's medications may reveal a drug that could be potentially playing a role in impairing sexual function. Antihypertensive medications are common offenders with centrally acting agents and beta blockers being the most commonly implicated agents in causing impotence. The angiotensin-converting enzyme inhibitors or angiotensin receptor blockers are associated with a lower incidence of impotence and represent a useful alternative in kidney failure patients with hypertension. In one study of non-uremic hypertensive men, use of the angiotensin receptor blocker valsartan actually led to an increase in sexual activity as compared to subjects treated with carvedilol (Fogari et al., 2001). Other drugs commonly implicated include cimetidine, phenothiazines, tricyclic antidepressants, metoclopramide and peripheral alpha blockers (Bailie et al., 2007).

If the history and physical examination reveal no obvious cause, then a psychological cause of erectile dysfunction may need to be considered. Testing for the presence of nocturnal penile tumescence (NPT) has been utilized in some centers as a means to discriminate between a psychological and organic cause of impotence. The basis for this test is that during the rapid eye movement stage of sleep, males normally have an erection. The assumption is that a man with a psychological cause of impotence would still experience erections while asleep while the absence of an adequate erection would make an organic cause more likely. If a patient is found to have nocturnal erections then psychological testing and evaluation is indicated. It should be noted that NPT testing is not infallible and that if a patient has a normal test and no psychological cause is found, then evaluation for an organic cause should still be pursued.

There are tests that may aid in the discrimination between a neurogenic and vascular cause of impotence. Tests utilized to exclude a vascular etiology of impotence include Doppler studies to measure penile blood flow, measurement of penile blood pressure and penile pulse palpation. Neurogenic impotence is suggested by detecting a prolonged latency time of the bulbocavernous reflex or confirming the presence of a neurogenic bladder. With the availability of sildenafil (see below) to use as a therapeutic trial, such tests are generally reserved for non-responders who may eventually be considered for surgical placement of a penile prosthesis (Palmer, 2004).

As discussed previously, hormonal abnormalities are frequently detected in chronic kidney disease patients. Endocrine tests that are useful in the evaluation of an organic cause of impotence include measurement of serum LH, FSH, testosterone and prolactin levels. It should be noted that only a small percentage of uremic patients will have prolactin levels greater than 100 ng/ml. Imaging studies of the hypothalamic–pituitary region should be performed in patients with levels of greater magnitude in order to exclude the presence of a microadenoma or macroadenoma.

IV. TREATMENT OF SEXUAL DYSFUNCTION IN THE UREMIC MAN

The treatment of sexual dysfunction in the uremic man is initially of a general nature (Figure 27.1). One needs to ensure optimal delivery of dialysis and adequate nutritional intake. Administration of recombinant human erythropoietin has been shown to enhance sexual function in chronic kidney failure. It is likely that the associated improvement in well-being that comes with the correction of anemia probably plays an important role in this response. In some studies, erythropoietin therapy has been reported to cause normalization of the pituitary gonadal feedback mechanism with reduced plasma concentrations of LH and FSH and increases in plasma testosterone levels (Schaefer et al., 1994a; Kokot et al., 1995). Reductions in elevated plasma prolactin levels have also been noted (Schaefer et al., 1989). It is controversial as to whether these endocrinologic changes are solely the result of correction of the anemia or a direct effect of erythropoietin. As mentioned previously, controlling the degree of secondary hyperparathyroidism with 1,25 $(OH)_2$ vitamin D may be of benefit in lowering prolactin levels and improving sexual function in some patients.

One area that deserves further investigation is the impact of slow nocturnal hemodialysis on sexual function. In a pilot study of five patients undergoing dialysis 6 nights per week for 8 hours each night, serum testosterone levels increased in three patients over an 8-week period (O'Sullivan et al., 1998). Levels of LH and FSH remained unchanged. In a separate study, the percentage of patients who felt that sexual function was a problem declined from 80 to 29% after 3 months of nightly nocturnal hemodialysis (McPhatter et al., 1999).

Patients with normal NPT testing should be evaluated to determine if there is a psychologic component to the

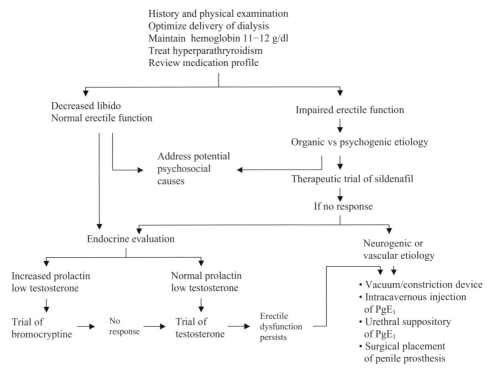

FIGURE 27.1 Approach to sexual dysfunction in uremic men.

impotence. If a problem is found then a trial of psychotherapy is warranted. The effectiveness of antidepressant medications and/or psychiatric counseling in chronic kidney failure patients with sexual dysfunction has not been well studied. Use of antidepressant medications can be problematic since many of these agents can cause sexual dysfunction (Kaplan, 2002).

With the recent approval of sildenafil for the treatment of men with erectile dysfunction, many physicians now utilize this agent as the first line therapy for patients with psychogenic, vascular or neurogenic causes of impotence. Since the majority of patients receiving renal replacement therapy in the USA are diabetic, it is noteworthy that a randomized controlled trial found that sildenafil was both effective and well tolerated in diabetic men with erectile dysfunction (Rendell et al., 1999). More than 95% of the patients in this study were felt to have either a neurogenic or vascular etiology of erectile dysfunction.

Several studies have examined the effectiveness of sildinafil in uremic men with chronic kidney disease (Turk et al., 2001; Chen et al., 2001; Rosas et al., 2001a; Sam and Patel, 2006). Each of these studies used the International Index of Erectile Function (IIEF) questionnaire as a means to gauge effectiveness of the therapy. The response rate ranged from 60 to 80%. Sildenafil was found to have similar efficacy in patients treated with either hemodialysis or peritoneal dialysis (Turk et al., 2001). There was no correlation between patients who failed to respond to therapy and etiology of erectile dysfunction including blood levels of testosterone and prolactin, etiology of kidney failure and patient age (Turk et al., 2001; Chen et al., 2001). In one study, patients not responding to sildenafil therapy had significantly lower penile blood flow (Turk et al., 2001). It is not uncommon for patients to require higher doses in order to demonstrate an adequate response (Dachille et al., 2006). Sildenafil remains an effective therapy in those patients with persistent erectile dysfunction after receiving a renal allograft (Sharma et al., 2006).

It is possible that chronic kidney disease patients may prove to be less responsive to the use of sildenafil due to alterations in nitric oxide metabolism that accompany the uremic state. The physiologic mechanism responsible for erection of the penis involves release of nitric oxide in the corpus cavernosum in response to sexual stimuli. Nitric oxide activates the enzyme guanylate cyclase which results in localized increased levels of cGMP thereby producing smooth muscle relaxation. Sildenafil is a potent and selective inhibitor of GMP-specific phosphodiesterase type 5, the predominant isoenzyme metabolizing cGMP in the cells of the corpus cavernosum. By inhibiting the breakdown of cGMP, sildenafil acts to facilitate flow of blood in the penis. Chronic kidney disease is associated with decreased nitric oxide production, in part, due to the accumulation of asymetric dimethylarginine (ADMA) (Ketteler and Ritz, 2000; Schmidt and Baylis, 2000). If deficient nitric oxide production is the limiting factor contributing to erectile dysfunction, then therapy designed to increase cGMP may not be sufficient to overcome the NO inhibitory properties of ADMA.

Decreased ability to synthesize nitric oxide also plays a role in the erectile dysfunction that occurs in the setting of testosterone deficiency and diabetic mellitus (Saenz de Tejada et al., 1989; Mills et al., 1992).

It has become common clinical practice first to administer sildenafil to patients who complain of erectile dysfunction and reserve further workup for only those patients who fail to achieve a therapeutic response (Bellinghieri et al., 2001). It should be emphasized that sildenafil is contraindicated in patients who are currently taking organic nitrates. Caution should also be exercised when prescribing this agent to patients with known coronary artery disease. To limit the possibility of hypotension among dialysis patients, some clinicians recommend use of sildenafil on non-dialysis days (Mohamed et al., 2000).

Patients found to have increased circulating levels of prolactin may benefit from a trial of bromocryptine. This agent is a dopaminergic agonist that has shown some efficacy in improving sexual function presumably by reducing elevated prolactin levels. However, its usefulness has been limited by a relatively high frequency of side effects. Other dopaminergic agonists, such as parlodel and lisuride, seem to be better tolerated but have only been used in small short-term studies.

Zinc deficiency has also been suggested as a cause of gonadal failure. Uremic patients are often deficient in zinc, probably due to reduced dietary intake, zinc malabsorption and/or possible leaching of zinc by dialysis equipment. In a controlled trial, supplemental zinc resulted in significant increases in the plasma testosterone concentration and sperm counts, as well as significant declines in LH and FSH levels as compared to a control group (Mahajan et al., 1982). Potency, libido and frequency of intercourse also improved in those patients given zinc. It is possible that normalization of total body zinc may also be effective in correcting uremic hyperprolactinemia (Caticha et al., 1996). Thus, the aggregate data suggest that the administration of zinc in a zinc-deficient man is a reasonable therapeutic option.

There are additional options for those patients with a neurogenic or vascular cause of impotence who have failed medical therapy to include a trial of sildenafil. One such therapy is a vacuum tumescence device. In a review of the experience of one kidney impotence clinic, vacuum tumescence devices were utilized in 26 impotent patients, all of whom had a normal pituitary–gonadal axis or hypogonadism corrected with testosterone replacement (Lawrence et al., 1998). The device completely corrected penile dysfunction in 19 individuals (73%).

Intraurethral administration of alprostadil (synthetic prostaglandin E_1) provides the delivery of prostaglandin to the corpus cavernosum resulting in an erection sufficient for intercourse. The drug is supplied in an applicator that is inserted in the urethra. Alprostadil can also be injected into the penis shaft, resulting in vasodilation and inhibition of platelet aggregation. The major side effects of intrapenile alprostadil therapy are penile pain, priapism and bleeding. Given the presence of platelet dysfunction with uremia, intracavernosal injections should be used with caution in patients with end-stage renal disease (Ayub and Fletcher, 2000). Surgical placement of a penile prosthesis is typically considered in patients who fail the less invasive first line treatments.

V. OUTCOMES ASSOCIATED WITH HYPOGONADISM AND TREATMENT

Testosterone deficiency also plays a role in sexual dysfunction in uremic men. The administration of androgens to non-uremic hypogonadal men generally results in an improvement in sexual function. By contrast, administration of testosterone to uremic men usually fails to restore libido or potency, despite increased testosterone levels and reduced release of LH and FSH (Barton et al., 1982; van Coevorden et al., 1986). In a recent report of 27 male dialysis patients with biochemically proven hypogonadism, administration of depot testosterone fully restored sexual function in only three patients (Lawrence et al., 1998). In a hypogonadal patient whose primary complaint is decreased libido, a trial of testosterone may be warranted. In very limited studies, administration of clomiphene citrate has also been reported to cause a normalization of plasma testosterone levels associated with some improvement in sexual function.

Testosterone deficiency in uremic men also plays a role in other clinical outcomes such as decreased bone mineralization, malnutrition, decreased muscle mass and anemia (Table 27.3). For example, androgens are known to play a significant role in bone metabolism. In hypogonadal non-uremic men, decreased bone mineral density is commonly present and is responsive to testosterone replacement therapy. Decreased bone mineral density is also seen in patients with chronic kidney disease. Decreases in bone mineral density can be detected with moderate reductions in the glomerular filtration rate and progressively worsen as renal function declines (Bianchi et al., 1992; Rix et al., 1999). Factors associated with this finding include secondary hyperparathyroidism, chronic metabolic acidosis, aluminum exposure,

TABLE 27.3 Clinical outcomes associated with hypogonadism in men with chronic kidney disease

---
Loss of libido and erectile dysfunction
Some regression of secondary sexual characteristics
Oligospermia or azoospermia
Mood changes
Fatigue
Decreased bone mineral density
Malnutrition
Loss of muscle mass and strength
Anemia

chronic use of heparin, prior exposure to glucocorticoids and deficiency of vitamin D.

In a recent prospective study of 88 dialysis patients, decreased bone mineral density was found in 48.9% of patients. A severe decrease was present in 19.3% of subjects (Taal et al., 1999). Similarly, a cross-sectional study of 250 dialysis patients found severe osteoporosis present in the lumbar spine, the femoral neck and distal radius in 8, 13 and 20% respectively (Stein et al., 1996). The presence of decreased bone mineral density is of major concern since it increases the risk of fractures by a factor of 3 to 4. Blacks with chronic kidney disease have higher bone mineral densities when compared to whites, but the percentage of loss per year is similar between the two groups (Stehman-Breen et al., 1999).

Given the role of testosterone deficiency in the genesis of decreased bone mineral density in subjects with normal kidney function, it follows that testosterone deficiency resulting from chronic kidney disease may also play an important role in this setting. To date, there are no studies completed that have specifically examined the effects of androgen replacement on changes in bone mineral density and fracture risk in men with chronic kidney disease. Such studies would be of considerable interest. However, given the mutifactorial nature of renal osteodystrophy, it is likely testosterone replacement therapy by itself may not offer significant benefit (Palmer, 2004). Rather, administration of testosterone may prove most useful as an adjunct to other agents designed to correct the deficit in bone mineral density. This is an area that deserves more investigation.

Testosterone deficiency has also been implicated in the development of malnutrition, decreased muscle mass and reduced levels of albumin in chronic kidney disease patients. The treatment of decreased muscle mass is currently of a general nature to include provision of appropriate amounts of protein and calories, correction of metabolic acidosis, intensifying the amount of delivered dialysis and aggressive management of co-morbid conditions. While such steps often lead to an improvement in muscle strength and nutrition, many men on hemodialysis continue to exhibit persistent residual abnormalities. As a result, additional treatment strategies are needed to correct fully the various manifestations of malnutrition in men on chronic hemodialysis.

To explore the role of testosterone therapy in the treatment of protein calorie malnutrition Johansen et al. conducted a randomized, double-blind, placebo-controlled trial in which nandralone decanoate was administered weekly by intramuscular injection for 16 months in 14 patients on either hemodialysis or peritoneal dialysis (Johansen et al., 1999). As compared to controls, the androgen treated patients showed a significant increase in lean body mass and an increase in serum creatinine concentration. In addition, there was evidence of functional improvement as evidenced by improvements in walking and stair climbing times.

While further study is needed, the encouraging results of this study suggest that use of androgenic steroids may prove useful in maximizing the nutritional status of men on dialysis. Transdermal delivery systems may prove to be the most effective way to administer the drug. In this regard, a transdermal preparation has recently been demonstrated to provide maintenance of serum concentrations of total and free testosterone in the mid-normal range in men on hemodialysis (Singh et al., 2001).

Anemia develops during the course of chronic kidney disease and is virtually present in all patients on dialysis. The genesis of anemia is multifactorial in origin, but erythropoietin deficiency is the primary cause of the abnormality. Before the availability of recombinant erythropoietin, transfusions and androgen therapy were the only viable treatment options for the management of anemia in dialysis patients (Hendler et al., 1974; Neff et al., 1981).

The beneficial effect of androgen therapy in treating anemia raises the possibility that hypogonadism in men on dialysis plays at least a contributory role in the genesis of this disorder. Prior clinical experience with testosterone as a treatment for anemia was that the response was generally modest and greatest in those men who were not anephric and not transfusion dependent (Johnson, 2000). The mechanism by which androgens improve anemia is thought to be increased production of endogenous erythropoietin. There is also evidence that androgens may increase the sensitivity of erythroid progenitors to the effects of erythropoietin (Teruel et al., 1995).

While recombinant erythropoietin is the primary means to treat anemia in patients on dialysis, there continues to be interest in using androgens to enhance the effect of erythropoietin. In a prospective, open label study of 20 hemodialysis patients, epoetin alpha therapy was withdrawn to permit the hematocrit to become $\leq 26\%$ (Gaughan et al., 1997). At this point, all patients were placed on epoetin alpha at a dose of 1500 units administered three times per week. Ten patients were also treated with nandrolone decanoate 100 mg intramuscularly each week. The patients who received both erythropoietin and androgen therapy had a significantly greater increase in the hematocrit consistent with a possible sensitizing effect of the androgen therapy.

While erythropoietin will remain the primary treatment of anemia in men with chronic kidney disease, androgens may prove useful as a sensitizing agent. Such a strategy may prove economically beneficial in that a given hematocrit can be maintained with less amount of the recombinant drug. The effectiveness of testosterone to enhance the response to erythropoietin stimulating therapy may depend on the dose and route of administration. In this regard, daily administration of 100 mg of topical 1% testosterone gel for 6 months did not have an impact on recombinant erythropoietin requirements in a study of 40 male hemodialysis patients (Brockenbrough et al., 2006). This is an area that deserves more investigation.

VI. SEXUAL DYSFUNCTION IN UREMIC WOMEN

Disturbances in menstruation and fertility are commonly encountered in women with chronic kidney disease, usually leading to amenorrhea by the time the patient reaches end-stage renal disease. The menstrual cycle typically remains irregular with scanty flow after the initiation of maintenance dialysis, although normal menses is restored in some women (Holley et al., 1997). In others, hypermenorrhagia develops, potentially leading to significant blood loss and increased transfusion requirements.

The major menstrual abnormality in uremic women is anovulation, with affected patients being infertile (Ginsburg and Owen, 1993). Women on chronic dialysis also tend to complain of decreased libido and reduced ability to reach orgasm (Steele et al., 1996; Toorians et al., 1997; Peng et al., 2005).

Pregnancy can rarely occur in advanced kidney failure, but fetal wastage is markedly increased. Some residual kidney function is usually present in the infrequent pregnancy that can be carried to term. The subject of pregnancy in chronic kidney insufficiency has recently been reviewed (Hou, 1999).

A. Normal Menstrual Cycle

The normal menstrual cycle is divided into a follicular or proliferative phase and a luteal or secretory phase. Normal follicular maturation and subsequent ovulation require appropriately timed secretion of the pituitary gonadotropins. Follicle stimulating hormone (FSH) secretion exhibits typical negative feedback, with hormone levels falling as the plasma estrogen concentration rises. In contrast, luteinizing hormone (LH) secretion is suppressed maximally by low concentrations of estrogen but exhibits positive feedback control in response to a rising and sustained elevation of estradiol. Thus, high levels of estradiol in the late follicular phase trigger a surging elevation in LH secretion which is responsible for ovulation. Following ovulation, progesterone levels increase due to production by the corpus luteum. Progesterone is responsible for the transformation of the endometrium into the luteal phase.

B. Hormonal Disturbances in Uremic Premenopausal Women

Indirect determination of ovulation suggests that anovulatory cycles are the rule in uremic women (Lim et al., 1980). For example, endometrial biopsies show an absence of progestational effects and there is a failure to increase basal body temperature at the time when ovulation would be expected. In addition, the preovulatory peak in LH and estradiol concentrations are frequently absent. The failure of LH to rise in part reflects a disturbance in the positive estradiol feedback pathway, since the administration of exogenous estrogen to mimic the preovulatory surge in estradiol fails to stimulate LH release (Lim et al., 1980). In contrast, feedback inhibition of gonadotropin release by low doses of estradiol remains intact. This can be illustrated by the ability of the antiestrogen clomiphene to enhance LH and FSH secretion.

It remains unclear whether the disturbances in cyclic gonadotropin production originate in the hypothalamus (via impaired production of gonadotropin-releasing hormone (GnRH)) or in the anterior pituitary (Ginsburg and Owen, 1993). It is possible, for example, that endorphins are involved. Circulating endorphin levels are increased in chronic kidney failure due primarily to reduced renal opioid clearance and endorphins can inhibit ovulation, perhaps by reducing the release of GnRH.

C. Prolactin and Galactorrhea

Women with chronic kidney disease commonly have elevated circulating prolactin levels. As in men, the hypersecretion of prolactin in this setting appears to be autonomous, as it is resistant to maneuvers designed to stimulate or inhibit its release.

It has been suggested that the elevated prolactin levels may impair hypothalamic–pituitary function and contribute to sexual dysfunction and galactorrhea in these patients. In this regard, non-uremic females with prolactin-producing pituitary tumors commonly present with amenorrhea, galactorrhea and low circulating gonadotropin levels. However, uremic women treated with bromocryptine rarely resume normal menses and continue to complain of galactorrhea (if present), despite normalization of the plasma prolactin concentration (Lim et al., 1980). Thus, factors other than hyperprolactinemia must be important in this setting.

D. Hormonal Disturbances in Uremic Postmenopausal Women

Postmenopausal uremic women have gonadotropin levels as high as those seen in non-uremic women of similar age (Lim et al., 1980; Zingraff et al., 1982). As mentioned above, the negative feedback of estrogen on LH and FSH release is intact in premenopausal uremic women. Presumably low estrogen levels in the postmenopausal state lead to the increased gonadotropin levels. The age at which menopause begins in chronic kidney disease tends to be decreased when compared to normal women.

VII. TREATMENT

The high frequency of anovulation leads sequentially to lack of formation of the corpus luteum and failure of progesterone secretion. Since progesterone is responsible for transforming the endometrium into the luteal phase, lack of progesterone

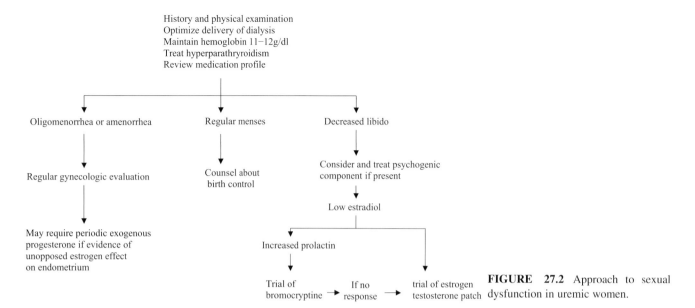

FIGURE 27.2 Approach to sexual dysfunction in uremic women.

is associated with amenorrhea. For patients who desire to resume menses, administration of a progestational agent during the final days of the monthly cycle will usually be successful. On the other hand, ongoing menses can contribute significantly to the anemia of chronic kidney disease, particularly in those patients with hypermenorrhagia. In this setting, administration of a progestational agent during the entire monthly cycle will terminate menstrual flow. Rarely, a patient may require hysterectomy for refractory uterine bleeding (Figure 27.2).

It is not known whether the usual absence of menses in women with chronic kidney disease predisposes to the development of endometrial hyperplasia and possible carcinoma. Since these patients are often anovulatory, there is no disruption of the proliferative effect of estrogen by the release of progesterone. It is therefore recommended that women be monitored closely by a gynecologist; it may be desirable in at least some cases to administer a progestational agent several times per year to interrupt the proliferation induced by unopposed estrogen release.

While pregnancy can rarely occur in women on chronic dialysis, restoration of fertility as a therapeutic goal should be discouraged. In comparison, the abnormalities in ovulation can usually be reversed and successful pregnancy achieved in women with a well-functioning kidney transplant. Uremic women who are menstruating normally should be encouraged to use birth control.

Studies addressing the therapy of decreased libido and sexual function in uremic women are lacking. Amenorrheic hemodialysis patients may have low estradiol levels that can secondarily lead to vaginal atrophy and dryness and result in discomfort during intercourse. Such patients may benefit from local estrogen therapy or vaginal lubricants. Low dose testosterone may be effective in increasing sexual desire, but is rarely used secondary to potential toxicity. Use of a testosterone patch has been shown effective in surgically menopausal women with hypoactive sexual disorder and deserves further investigation as to its use in uremic women (Simon et al., 2005). Bromocryptine therapy in hyperprolactinemic patients may help in restoring sexual function, but has not been well studied. Estrogen supplementation may improve sexual function in those patients with low circulating estradiol levels. Successful transplantation is clearly the most effective means to restore normal sexual desire in women with chronic kidney failure.

Amenorrheic women on hemodialysis may also be at increased risk for metabolic bone disease (Weisinger et al., 2000). In a recent study of 74 women on hemodialysis, trabecular bone mineral density was found to be lower in amenorrheic patients as compared to those with regular menses. Although the total serum estradiol concentration was normal in the amenorrheic women when compared with non-uremic women, the values were significantly lower than those in regularly menstruating women. Whether such patients would benefit from estrogen therapy deserves further study.

With regards to estrogen therapy, it has been noted that women on hemodialysis are often not treated in the same manner as non-uremic women. In three separate trials, only 4.8, 6 and 11.3% of postmenopausal females were noted to be receiving hormone replacement therapy (Cochrane and Regan, 1997; Holley et al., 1997; Weisinger et al., 2000). Given the potential benefits of estrogen therapy on bone disease and cardiovascular morbidity, it is likely that such therapy is being underutilized in this patient population (Holley and Schmidt, 2001).

References

Albaaj, F. et al. (2006). Prevalence of hypogonadism in male patients with renal failure. *Postgrad Med* **82**: 693–696.

Ayub, W., and Fletcher, S. (2000). End-stage kidney disease and erectile dysfunction: is there any hope? *Nephrol Dial Transplant* **15**: 1525–1528.

Bailie, G. et al. (2007). Sexual dysfunction in dialysis patients treated with antihypertensive or antidepressive medications: results from the DOPPS. *Nephrol Dial Transplant* Feb1 (Epub ahead of print).

Barton, C. H. et al. (1982). Effects of long-term testosterone administration on pituitary-testicular axis in end-stage kidney failure. *Nephron* **31**: 61–64.

Bellinghieri, G. et al. (2001). Erectile dysfunction in uremic dialysis patients: diagnostic evaluation in the sildenafil era. *Am J Kidney Dis* **38**(Suppl 1): S115–S117.

Bellinghieri, G. et al. (2004). Ultrastructural changes of corpora cavernosa in men with erectile dysfunction and chronic renal failure. *Semin Nephrol* **24**: 488–491.

Bianchi, M. et al. (1992). Bone mass status in different degrees of chronic renal failure. *Bone* **13**: 225–228.

Billips, K. L. (2005). Erectile dysfunction as a marker for vascular disease. *Curr Urol Rep* **6**: 439–444.

Blumberg, A. et al. (1980). Influence of 1,25 dihydroxycholecalciferol on sexual dysfunction and related endocrine parameters in patients on maintenance hemodialysis. *Clin Nephrol* **13**: 208–214.

Brockenbrough, A. et al. (2006). Transdermal androgen therapy to augment EPO in the treatment of anemia of chronic kidney disease. *Am J Kidney Dis* **47**: 251–262.

Campese, V. M. et al. (1982). Autonomic nervous system dysfunction and impotence in uremia. *Am J Nephrol* **2**: 140–143.

Caticha, O. et al. (1996). Total body zinc depletion and its relationship to the development of hyperprolactinemia in chronic kidney insufficiency. *J Endocrinol Invest* **19**: 441–448.

Chen, J. et al. (2001). Clinical efficacy of sildenafil in patients on chronic dialysis. *J Urol* **165**: 819–821.

Cochrane, R., and Regan, L. (1997). Undetected gynaecological disorders in women with kidney disease. *Hum Reprod* **12**: 667–670.

Dachille, G. et al. (2006). Sexual dysfunction in patients under diaytic treatment. *Minerva Urol Nefrol* **58**: 195–200.

De Vries, C. P. et al. (1984). Haemodialysis and testicular function. *Int J Andrology* **7**: 97–103.

Diemont, W. L. et al. (2000). Sexual dysfunction after kidney replacement therapy. *Am J Kidney Dis* **35**: 845–851.

Dunkel, L. et al. (1997). Circulating luteinizing hormone receptor inhibitor(s) in boys with chronic kidney failure. *Kidney Int* **51**: 777–784.

Elesber, A. et al. (2006). Coronary endothelial dysfunction is associated with erectile dysfunction and elevated asymmetric dimethylarginine in patients with early atherosclerosis. *Eur Heart J* **27**: 824–831.

Finkelstein, F. O. et al. (2007). Therapy insight:sexual dysfunction in patients with chronic kidney disease. *Nat Clin Pract Nephrol* **3**: 200–207.

Fogari, R. et al. (2001). Sexual activity in hypertensive men treated with valsartan or carvedilol: a crossover study. *Am J Hypertens* **14**: 27–31.

Gaughan, W. et al. (1997). A six month study of low dose recombinant human erythropoietin alone and in combination with androgens for the treatment of anemia in chronic hemodialysis patients. *Am J Kidney Dis* **30**: 495–500.

Ginsburg, E. S., and Owen, W. F. (1993). Reproductive endocrinology and pregnancy in women on hemodialysis. *Semin Dial* **6**: 105–116.

Gomez, F. et al. (1980). Endocrine abnormalities in patients undergoing long-term hemodialysis. The role of prolactin. *Am J Med* **68**: 522–530.

Guvel, S. et al. (2004). Calcification of the epididymis and the tunica albuginea of the corpora cavernosa in patients on maintenance hemodialysis. *J Androl* **25**: 752–756.

Handelsman, D. J. (1985). Hypothalamic-pituitary gonadal dysfunction in kidney failure, dialysis and kidney transplantation. *Endocr Rev* **6**: 151–182.

Hauser, R. et al. (2005). Evidence of interaction between polychlorinated biphenyls and phthalates in relation to human sperm motility. *Environ Hlth Perspect* **113**: 425–430.

Hendler, E. et al. (1974). Controlled study of androgen therapy in anemia of patients on maintenance hemodialysis. *N Engl J Med* **291**: 1046–1051.

Holdsworth, S. R. et al. (1978). A comparison of hemodialysis and transplantation in reversing the uremic disturbance of male reproductive function. *Clin Nephrol* **10**: 146–150.

Holley, J. L., and Schmidt, R. J. (2001). Hormone replacement therapy in postmenopausal women with end-stage kidney disease: a review of the issues. *Semin Dial* **14**: 146–149.

Holley, J. L. et al. (1997). Gynecologic and reproductive issues in women on dialysis. *Am J Kidney Dis* **29**: 685–690.

Hou, S. (1999). Pregnancy in chronic kidney insufficiency and end-stage kidney disease. *Am J Kidney Dis* **33**: 235–252.

Isaac, R. et al. (1978). Effect of parathyroid hormone on plasma prolactin in man. *J Clin Encrinol Metab* **47**: 18–23.

Johansen, K. et al. (1999). Anabolic effects of nandrolone decanoate in patients receiving dialysis: a randomized controlled trial. *J Am Med Assoc* **281**: 1275–1281.

Johnson, C. (2000). Use of androgens in patients with renal failure. *Semin Dialysis* **13**: 36–39.

Kaplan, M. J. (2002). Approaching sexual issues in primary care. *Prim Care* **29**: 113–124.

Kaufman, J. M. et al. (1994). Impotence and chronic kidney failure: a study of the hemodynamic pathophysiology. *J Urol* **151**: 612–618.

Ketteler, M., and Ritz, E. (2000). Kidney failure: a state of nitric oxide deficiency? *Kidney Int* **58**: 1356–1357.

Kokot, F. et al. (1995). Function of endocrine organs in hemodialyzed patients of long-term erythropoietin therapy. *Artif Org* **19**: 428–435.

Lawrence, I. G. et al. (1998). Correcting impotence in the male dialysis patient: experience with testosterone replacement and vacuum tumescence therapy. *Am J Kidney Dis* **31**: 313–319.

LeRoith, D. et al. (1980). Dissociation of pituitary glycoprotein response to releasing hormones in chronic kidney failure. *Acta Endocrinol* **93**: 277–282.

Levitan, D. et al. (1984). Disturbances in the hypothalamic-pituitary-gonadal axis in male patients with acute kidney failure. *Am J Nephrol* **4**: 99–106.

Lim, V. S., and Fang, V. S. (1976). Restoration of plasma testosterone levels in uremic men with clomiphene citrate. *J Clin Endocrinol Metab* **43**: 1370–1377.

Lim, V. S. et al. (1980). Ovarian function in chronic kidney failure: evidence suggesting hypothalamic anovulation. *Ann Intern Med* **93**: 21–27.

Mahajan, S. K. et al. (1982). Effect of oral zinc therapy on gonadal function in hemodialysis patients. *Ann Intern Med* **97**: 357–361.

Massry, S. G. et al. (1977). Impotence in patients with uremia: a possible role for parathyroid hormone. *Nephron* **19**: 305–310.

McPhatter, L. L. et al. (1999). Nightly home hemodialysis: improvement in nutrition and quality of life. *Adv Ren Replace Ther* **6**: 358–365.

Michal, V. et al. (1978). External iliac 'steal syndrome'. *J Cardiovasc Surg (Torino)* **19**: 355–357.

Mills, T. M. et al. (1992). Androgen maintenance of erectile function in the rat penis. *Biol Reprod* **46**: 342–348.

Mohamed, E. A. et al. (2000). Timing of sildenafil therapy in dialysis patients-lessons following an episode of hypotension. *Nephrol Dial Transplant* **15**: 926–927.

Neff, M. et al. (1981). A comparison of androgens for anemia in patients on hemodialysis. *N Engl J Med* **304**: 871–875.

O'Sullivan, D. A. et al. (1998). Improved biochemical variables, nutrient intake, and hormonal factors in slow nocturnal hemodialysis: a pilot study. *Mayo Clin Proc* **73**: 1035–1045.

Palmer, B. F. (2004). Outcomes associated with hypogonadism in men with chronic kidney disease. *Adv Chron Kidney Dis* **11**: 342–347.

Peng, Y. et al. (2005). Sexual dysfunction in female hemodialysis patients: a multicenter study. *Kidney Int* **68**: 760–765.

Phocas, I. et al. (1995). Serum α-immunoreactive inhibitin in males with kidney failure, under haemodialysis and after successful kidney transplantation. *Andrologia* **27**: 253–258.

Prem, A. R. et al. (1996). Male reproductive function in uraemia: efficacy of haemodialysis and kidney transplantation. *Br J Urol* **78**: 635–638.

Procci, W. R. et al. (1981). Sexual dysfunction in the male patient with uremia: a reappraisal. *Kidney Int* **19**: 317–323.

Rendell, M. S. et al. (1999). Sildenafil for treatment of erectile dysfunction in men with diabetes. A randomized controlled trial. *J Am Med Assoc* **281**: 421–426.

Rix, M. et al. (1999). Bone mineral density and biochemical markers of bone turnover in patients with predialysis chronic renal failure. *Kidney Int* **56**: 1084–1093.

Rosas, S. E. et al. (2001a). Preliminary observations of sildenafil treatment for erectile dysfunction in dialysis patients. *Am J Kidney Dis* **37**: 134–137.

Rosas, S. E. et al. (2001b). Prevalence and determinants of erectile dysfunction in hemodialysis patients. *Kidney Int* **59**: 2259–2266.

Saenz de Tejada, I. et al. (1989). Impaired neurogenic and endothelium-mediated relaxation of penile smooth muscle from diabetic men with impotence. *N Engl J Med* **320**: 1025–1030.

Sam, R., and Patel, P. (2006). Sildenafil in dialysis patients. *Int J Artif Organs* **29**: 264–268.

Schaefer, F. et al. (1994a). Changes in the kinetics and biopotency of luteinizing hormone in hemodialyzed men during treatment with recombinant human erythropoietin. *J Am Soc Nephrol* **5**: 1208–1215.

Schaefer, R. M. et al. (1989). Improved sexual function in hemodialysis patients on recombinant erythropoietin: a possible role for prolactin. *Clin Nephrol* **31**: 1–5.

Schalch, D. S. et al. (1975). Plasma gonadotropins after administration of LH-releasing hormone in patients with kidney or hepatic failure. *J Clin Endocrinol Metab* **41**: 921–925.

Schmidt, R. J., and Baylis, C. (2000). Total nitric oxide production is low in patients with chronic kidney disease. *Kidney Int* **58**: 1261–1266.

Sharma, R. et al. (2006). Treatment of erectile dysfunction with sildenafil citrate in renal allograft recipients: a randomized, double blind, placebo-controlled, crossover trial. *Am J Kidney Dis* **48**: 128–133.

Simon, J. et al. (2005). Testosterone patch increases sexual activity and desire in surgically menopausal women with hypoactive sexual desire disorder. *J Clin Endocrinol Metab* **90**: 5226–5233.

Singh, A. et al. (2001). Pharmacokinetics of a transdermal testosterone system in men with end stage renal disease receiving maintenance hemodialysis and healthy hypogonadal men. *J Clin Endocrinol Metab* **86**: 2437–2445.

Steele, T. E. et al. (1996). Sexual experience of the chronic peritoneal dialysis patient. *J Am So Nephrol* **7**: 1165–1168.

Stehman-Breen, C. et al. (1999). Racial differences in bone mineral density and bone loss among end-stage renal disease patients. *Am J Kidney Dis* **33**: 941–946.

Stein, M. et al. (1996). Prevalence and risk factors for osteopenia in dialysis patients. *Am J Kidney Dis* **28**: 515–522.

Stewart-Bentley, M. et al. (1974). Regulation of gonadal function in uremia. *Metabolism* **23**: 1065–1072.

Taal, M. et al. (1999). Risk factors for reduced bone density in haemodialysis patients. *Nephrol Dial Transplant* **14**: 1922–1928.

Teruel, J. et al. (1995). Evolution of serum erythropoietin after androgen administration to hemodialysis patients: a prospective study. *Nephron* **70**: 282–286.

Toorians, A. W. F. T. et al. (1997). Chronic kidney failure and sexual functioning: clinical status versus objectively assessed sexual response. *Nephrol Dial Transplant* **12**: 2654–2663.

Turk, S. et al. (2001). Erectile dysfunction and the effects of sildenafil treatment in patients on haemodialysis and continuous ambulatory peritoneal dialysis. *Nephrol Dial Transplant* **16**: 1818–1822.

van Coevorden, A. et al. (1986). Effect of chronic oral testosterone undecanoate administration on the pituitary-testicular axes of hemodialyzed male patients. *Clin Nephrol* **26**: 48–54.

Weisinger, J. R. et al. (2000). Role of persistent amenorrhea in bone mineral metabolism of young hemodialyzed women. *Kidney Int* **58**: 331–335.

Zingraff, J. et al. (1982). Pituitary and ovarian dysfunctions in women on haemodialysis. *Nephron* **30**: 149–153.

Further reading

Palmer, B. F. (1999). Editorial review: management of sexual dysfunction: responsibility of the primary care physician or the specialist. *Adv Kidney Replace Ther (Web-only editorial)* **6**: E4.

Schaefer, F. et al. (1994b). Immunoreactive and bioactive luteinizing hormone in pubertal patients with chronic kidney failure. *Kidney Int* **45**: 1465–1476.

CHAPTER 28

Thyroid Status in Chronic Renal Failure Patients – A Non-Thyroidal Illness Syndrome

VICTORIA S. LIM AND MANISH SUNEJA

Division of Nephrology, Department of Medicine, University of Iowa College of Medicine, Iowa City, Iowa 52242, USA

Contents

I.	Introduction	441
II.	Circulating thyroid hormone profile	441
III.	Thyroid hormone kinetics	443
IV.	Tissue T3 content and T4 uptake	445
V.	The hypothalamo-pituitary thyroid axis	448
VI.	Iodide retention, goiter, hypo- and hyperthyroidism	450
VII.	Effects of dialysis and transplantation	450
VIII.	Thyroid biology in chronic renal failure and other non-thyroidal illnesses	452
IX.	Should thyroid hormone be replaced in CRF and other non-thyroidal illness patients?	453
X.	Summary	454
	References	454

I. INTRODUCTION

Chronic renal failure (CRF) affects thyroid function in a multitude of ways including low circulating thyroid hormone concentration, altered peripheral thyroid hormone metabolism, inhibition of thyroid hormone binding to carrier proteins, decreased tissue thyroid hormone content, iodide retention and, possibly, impaired hypothalamic thyrotropin-releasing hormone (TRH) secretion. This chapter will review published work in the last three decades supporting the above statement and discuss its biologic significance. Due to the voluminous amount of literature on this issue and the limitation of space, the authors have selected what they perceive as the more pertinent and interesting findings. Most of the publications cited are derived from humans, but some important rodent studies also have been included. It should be emphasized that the above described abnormalities of thyroid function are also found in a variety of non-renal disease states such as malnutrition, heart failure, liver cirrhosis, severe infectious disorders, cancer and many others including most hospitalized patients, especially those in the intensive care units. Thyroid function in acute renal failure has not been studied systematically, but is included in many studies of acute critical illnesses. These will also be reviewed in this chapter.

The earliest and the most common thyroid function abnormality noted in CRF patients is a reduction in circulating T3, hence the original notation of this phenomenon as the 'low T3 syndrome'. Subsequently, as T4 was also noted to be reduced, the name was changed to 'sick euthyroid syndrome' believing that these patients are euthyroid. But arguments about the true thyroid status in these patients and questions regarding the need of replacement resulted in a more recent change of terminology to 'non-thyroidal illness' (NTI). In this review, we used mostly data from CRF patients, but also include papers derived from non-renal NTI populations.

II. CIRCULATING THYROID HORMONE PROFILE

In CRF patients, the most frequently observed deviation from normal is a reduction of serum tri-iodothyronine (T3), both total and free tri-iodothyronine (TT3 and FT3) concentrations are reduced (Spector et al., 1976; Wartofsky, 1994; Lim, 2001). Serum total thyroxine (TT4) is also reduced, but of a more minor magnitude and lesser frequency (Kaptein et al., 1987; Hershman et al., 1978; Kaptein, 1996). Serum free T4 index (FT4I) is likewise reduced. Serum free T4 ranged from subnormal to elevated levels depending on the assay method. Thyroxine binding globulin is generally normal or only mildly depressed (Lim et al., 1977). Serum total reverse T3 (rT3), which is never measured in clinical settings but assessed in research studies, is either normal or elevated. Serum free rT3 is high (Kaptein et al., 1983; Faber et al., 1983). Serum thyrotropin or thyroid-stimulating hormone (TSH) is not elevated, it is uniformly normal (Lim et al., 1977; Kaptein, 1996).

The low T3 is unequivocally a result of reduced production; the T4 to T3 conversion rate is markedly impaired at extrathyroidal sites (Lim et al., 1977). However, its metabolic clearance and degradation rates are actually reduced. The pathogenesis of the low T4 is more complex. T4 degradation rate and production rates, the latter equal to thyroidal T4 secretion rate, are both normal (Lim et al., 1977; Faber et al., 1983; Kaptein et al., 1987). The major contributing factor is a defect in T4 binding to circulating protein carriers (Neuhaus et al., 1975; Melmed et al., 1982; Oppenheimer et al., 1982; Munro et al., 1989). A concomitant defect in central hypothalamic thyrotropin-releasing hormone (TRH) secretion cannot be excluded, especially in those patients whose T4 and TSH are both severely reduced (Fliers et al., 1997). As T4 is >99% bound to carrier proteins, the low TT4 could be due to reduced binding molecules. Yet thyroxine binding globulin (TBG) is normal or only mildly depressed and transthyretin, another thyroid hormone binding protein, is also normal. Free thyroxine index (FT4I), a method used for correcting altered binding capacity, derived from the product of serum TT4 and ratio of labeled T3 resin uptake, is generally low as well. This is believed to be due to the presence of inhibitors of binding of T4 to TBG and, to varying extent, binding of T4 to the solid matrices used in *in-vitro* assays, including activated charcoal and resin. Furthermore, uremia decreases binding of drugs, such as phenytoin and salicylate to serum proteins, leaving higher free fractions of these drugs that then compete with T4 for TBG binding sites. In contrast to total T4, which is invariably low, serum free T4 varies from low to elevated values, depending on the assay method used. Currently, the most accurate method to measure free T4 in CRF patients is by equilibrium dialysis of undiluted serum followed by radioimmunoassay of the free T4 in the dialysate (Nelson and Tomel, 1988). With this procedure, free T4 is separated from the bound T4 by a semi-permeable membrane. Using this technique, free T4 in CRF patients ranges from normal to high value, but about 5–10% still have low free T4, perhaps secondary to small dialyzable inhibitors such as furan fatty acid and indoxyl sulfate (Lim et al., 1993a). Reduced binding to serum proteins appears not to be applicable to T3 as both total and free T3 are reduced proportionately. This could be explained by T4 binding more tightly to carrier proteins and is almost 50-fold higher in concentration than T3 in the circulation. Currently, there is no proof to support a hypothalamic deficit in TRH secretion causing low T4 secretion in the CRF population.

Despite low T3 and low or low normal T4, serum thyrotropin (TSH) is not elevated. Indeed, it is suppressed in patients who are severely ill. This lack of TSH elevation could be due to normal T3 content in the pituitary or a central defect in TRH secretion. In the nephrectomized uremic rat model, pituitary T3 content, unlike that in the liver, is not low (Lim et al., 1984). Evidence supporting a TRH secretory defect is the report of Fliers et al. showing a decreased TRH gene expression in the paraventricular nuclei of patients who died with NTI syndrome (Fliers et al., 1997). However, whether this also applies to individuals with uremia who are not acutely ill is unknown.

It should be emphasized that the abnormalities of circulating thyroid hormone profile described above for CRF patients have also been reported in a variety of non-renal, non-thyroidal illnesses (NTI), such as malnutrition, heart failure, liver cirrhosis, severe infectious disorders, cancer and many others including most hospitalized patients (Chopra, 1977; Docter et al., 1993; Nicoloff and LoPresti, 1996; DeGroot, 1999). There are two noted differences between CRF and other NTI, namely, normal serum total reverse T3 in CRF patients and reduced TSH in critically ill non-renal NTI. Serum total reverse T3, while uniformly elevated in all other NTI, is normal in CRF patients (Kaptein et al., 1983). Serum free reverse T3 is elevated in both groups of patients. Generally, when T3 is reduced, reverse T3 is increased. This relationship is found universally in all NTI patients due to the reciprocal activity of the iodothyronine outer ring 5′ and inner ring 5 deiodinase (Figure 28.1) activity (Bianco et al., 2002). While the former converts T4 to T3, the latter changed T4 to rT3. T3 is biologically many-fold more active than T4, while rT3 is void of metabolic activity. This regulation plays a critical role in energy conservation in all mammals. The non-elevated serum total reverse T3 in CRF patients is explained by a more rapid transit of rT3 from the serum to extravascular sites studied by a three-pool kinetic measurement. In the very seriously ill patients, including those with acute renal failure requiring dialysis, serum T4 and TSH are markedly reduced and their decrements appear temporally related (Wehmann et al., 1985; Peeters et al., 2005). It is such observation that led to the hypothesis of a concomitant central hypothalamic dysfunction.

Patients with acute renal failure have very similar thyroid profile as those with CRF. Bodziony et al. studied thyroid function in 36 acute renal failure patients and reported that during the oliguric phase, serum T4 and T3 levels were low and rT3 elevated. TSH was not different (Bodziony et al., 1981). This high rT3 is in contrast to that reported by Kaptein et al. (1983) in CRF patients whose total rT3 is normal.

Table 28.1 summarizes the serum thyroid profile in our study of the end-stage renal disease (ESRD) patients before and during maintenance hemodialysis as compared to renal transplantation recipients, whose serum creatinine is <2.0 mg/dl, and normal subjects. It showed that both groups of CRF patients have selective T3 deficiency as their TT4/TT3 ratio is markedly elevated. Serum TT4 and FT4 index, though lower, are not statistically different from the other two groups. Serum thyroxine binding globulin appears normal. Despite low circulating levels of T3 and low normal TT4 and FT4 index, serum thyroid stimulating hormone (TSH) is normal. Though these data were gathered decades ago, they are still valid today and have been confirmed by

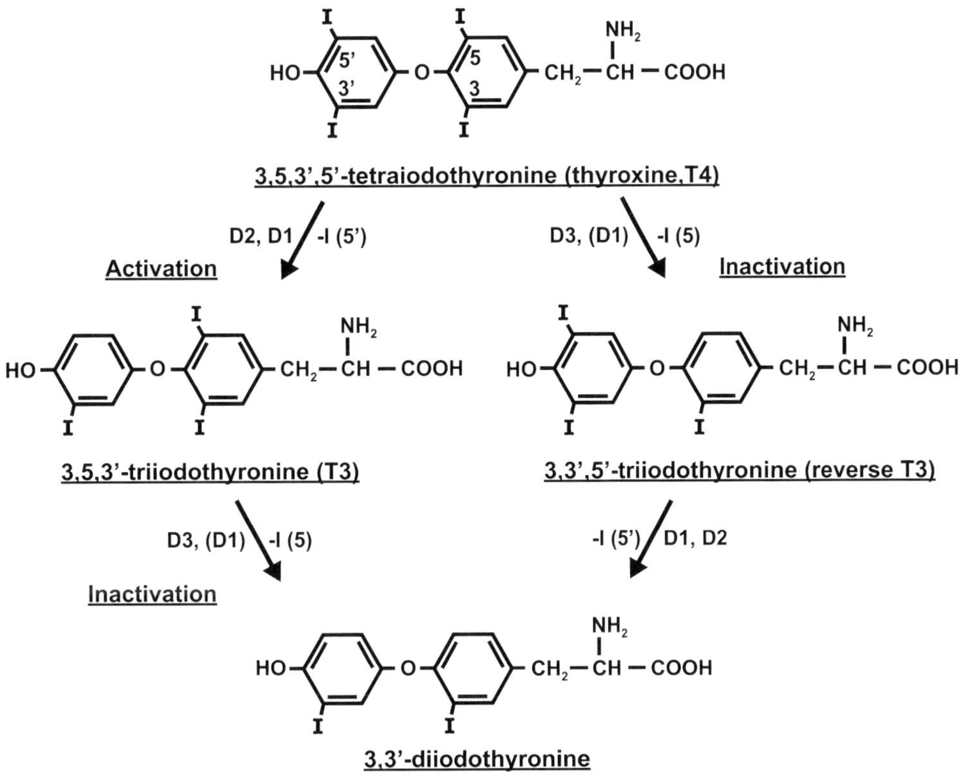

FIGURE 28.1 The iodothyronines. Outer ring 5′ deiodination of T4 by D1 and D2 enzymes and inner ring 5 deiodination of T4 by D3 enzyme produce, respectively, T3 and reverse T3. D1 and D2 enzymes deiodinate reverse T3 at outer ring 5′ position and D3 enzyme deiodinates T3 at inner ring 5 position, both give rise to di-iodothyronine. D1 is distributed mostly in the liver, D2 in hypothalamus, pituitary and D3 in both liver and skeletal muscle. In CRF and other NTI, D1 function is suppressed and D3 function upregulated. The net effect is a reduction in T3 and an increase in reverse T3.

TABLE 28.1 Serum thyroid hormone profile in normal subjects as compared to chronic renal failure patients before and after initiation of maintenance dialysis and renal transplantation

Group	TT_4 (μg/dl)	FT_4I	TBG capacity (μgT_4/dl)	TT_3 (ng/dl)	TT_4/TT_3	TSH (μU/ml)
Normal	6.6 ± 1.4	7.1 ± 1.0	18.3 ± 5.4	128 ± 25	55 ± 17	2.8 ± 1.2
pre-HD	6.2 ± 2.5	7.0 ± 2.4	16.4 ± 3.0	63 ± 17*	102 ± 43*	3.2 ± 1.5
HD	5.6 ± 1.2*	6.6 ± 1.5	16.5 ± 4.0	83 ± 22*	72 ± 28*	2.7 ± 2.5
RT	7.9 ± 1.6	7.8 ± 1.1	22.0 ± 6.4	134 ± 20	58 ± 13	3.3 ± 2.1
Anova: P	<0.01	ns	ns	<0.001	<0.001	ns

HD: hemodialysis; RT: renal transplant. FT_4I and TBG represent, respectively, free thyroxine index and thyroxine binding globulin. Values (mean ± SD) from all groups were tested by analysis of variance and the P values are listed in the bottom line. Significant differences between the means of each patient group and the normal controls, tested using mean square within value, are indicated by. (Reproduced with permission from Lim et al., 1977, *J Clin Invest.*)

many other studies. Table 28.2 lists the directional changes of the various parameters of thyroid function in the CRF population as compared to that obtained from patients of other NTI and hypothyroidism.

III. THYROID HORMONE KINETICS

To understand the pathophysiology leading to the above changes, we measured T4 and T3 kinetics after intravenous injections of ^{125}I-T4 and ^{131}I-T3 in four study groups: pre-hemodialysis (pre-HD), hemodialysis patients (HD), renal transplant recipients (RT) and normal controls (Lim et al., 1977). The daily turnover was estimated from the serum hormone concentrations and the disappearance curves of the injected labeled hormones using a non-compartmental technique. The study was performed under a condition in which the endogenous thyroid hormone secretion was suppressed by the administration of 0.2 mg of L-thyroxine per day for 7 days before and during the entire study period. Saturated solution of potassium iodide was administered 30 min before the isotope injection and repeated twice daily

TABLE 28.2 Directional changes in serum thyroid hormone profile in patients with chronic renal failure, non-thyroidal illness and hypothyroidism as compared to normal controls

	ESRD	Non-renal non-thyroidal illnesses	Hypothyroidism
TT_4	↓, N	↓, N	↓
FT_4I	↓, N	↓, N	↓
FT_4-RIA	↓, N	↓, N	↓
FT_4-ED	↓, N, ↑	N, ↑	↓
TT_3	↓	↓	↓
FT_3	↓	↓	↓
TrT_3	N	↑	↓
FrT_3	↓	↑	↓
TSH	N	N	↑
TRH test	↓, N	↓, N	↑

ESRD: end-stage renal disease ↓ = decreased, N = normal and ↑ = increased, FT_4-RIA and FT_4-ED represent, respectively, serum free T_4 measured by radioimmunoassay and equilibrium dialysis assay.

during the study period to prevent uptake and recirculation of the administered labeled iodide. With L-thyroxine replacement, serum TT4 increased equally in all study groups, indicating that thyroxine absorption in the CRF and transplant patients is normal and that the thyroxine pool can be easily increased in CRF patients as well as the other two groups of study subjects. Serum TT3 also increased from their respective baseline in all study groups (see Tables 28.1 and 28.3), but the increment is substantially reduced in the pre-HD and HD groups, and their serum concentrations remain subnormal. Figure 28.2 illustrates the changes in serum TT4 and TT3 before and during L-thyroxine replacement. While the rise in serum TT4 is of equal magnitude in all study groups and is unrelated to the pretreatment level, the increment in serum TT3 is proportional to the pretreatment level, greater in the control and the RT recipients and lower in the two CRF groups. These data suggest a T4 to T3 conversion defect. Similarly, Brendt and Hershman gave L-thyroxine at a dose of 1.5 μg/kg/day to patients with severe non-renal NTI and followed their serum hormone concentration for 14 days. Serum TT4 reached normal level within 3 days of treatment but serum TT3 remained subnormal (Brent and Hershman, 1986). Kinetic measurements of the injected labeled T4 and T3, as presented in Table 28.3, confirm the notion that there is a T4 to T3 conversion defect outside of the thyroid gland. T4 metabolic clearance and degradation rates are comparable in all four study groups. By contrast, T3 turnover is strikingly different in the pre-HD and HD patients and is characterized by reduced metabolic clearance and degradation and, more importantly, a profound reduction in T4 to T3 conversion rate. It should be mentioned that we could not measure thyroidal T4 and T3 secretion rates because the experiment was carried out during exogenous L-thyroxine replacement, but we estimated thyroidal

TABLE 28.3 T_4 and T_3 kinetics in chronic renal failure patients before and after initiation of maintenance hemodialysis treatment and after renal transplantation as compared to normal subjects

Group	Serum creatinine (mg/dl)	BW (kg)	T_4 kinetics				T_3 kinetics			$T_4 \rightarrow T_3$ conversion (%)
			TT_4 (μg/dl)	Absorption (%)	MCR (l/day)	D (μg/day)	TT_3 (ng/dl)	MCR (l/day)	D (μg/day)	
Normal	1.0 ± 0.2	65 ± 13	9.1 ± 1.2	54 + 6	1.2 ± 0.2	109 ± 12	184 ± 25	18.6 ± 3.6	33.8 ± 6.1	37.2 ± 5.8
pre-HD	10.0 ± 2.3*	65 ± 16	9.5 ± 0.8	52 + 1	1.1 ± 0.1	103 ± 2	94 ± 17*	14.6 ± 3.0	13.5 ± 2.6*	15.7 ± 3.1*
HD	12.2 ± 3.5*	75 ± 7	9.1 ± 2.9	61 + 15	1.4 ± 0.4	122 ± 31	105 ± 24*	12.5 ± 3.1*	12.9 ± 3.1*	12.8 ± 1.7*
RT	1.6 ± 0.6	75 ± 16	10.2 ± 1.5	58 + 17	1.2 ± 0.3	117 ± 34	168 ± 24	17.9 ± 2.4	30.0 ± 7.1	34.0 ± 14.7
Anova: P	<0.001	ns	ns	ns	ns	ns	<0.0001	<0.01	<0.0001	<0.0001

HD: hemodialysis; RT: renal transplant. Values (mean ± SD) from all four groups were tested by analysis of variance and the P values are listed in the bottom line. Significant difference between each patient group and the normal controls, tested using the mean square within values are indicated by *. For the kinetic data, MCR and D represent, respectively, metabolic clearance rate and daily degradation rate. TT_4 is elevated because the experiment was done during LT_4 replacement. (Data derived from Lim et al., 1977, with permission granted from *J Clin Invest*.)

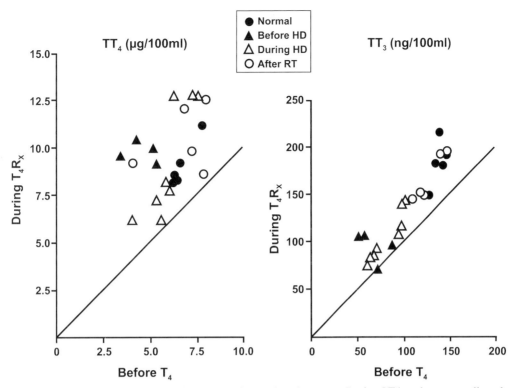

FIGURE 28.2 Serum thyroid hormone profile before and during LT4 replacement. During LT4 replacement, all study groups showed comparable increments of serum T4 compared to that of pretreatment value. By contrast, serum TT3 increment varied and correlated to pretreatment values, higher in normal controls and renal transplant recipients and significantly less in the two groups of renal failure patients. In the latter two groups of subjects, serum TT3 remains subnormal. (Data derived from Lim et al., 1977, with permission from *J Clin Invest*.)

T3 secretion indirectly by using the disappearance rates of the isotopes during the experiment and the serum TT4 and TT3 in the basal state before T4 supplementation and the results showed that thyroidal T3 secretion is not different among the four study groups.

Kaptein and Nicoloff studied T4 kinetics, using a three-pool model, plasma, a rapidly equilibrating pool and a slowly equilibrating pool. They found that the metabolic clearance and the appearance rates of thyroxine are comparable between the CRF patients and the healthy control subjects (Kaptein et al., 1987). In that same study, critically ill patients, however, behave differently in that T4 clearance is increased and T4 appearance rate is markedly reduced to about half of the control value. Assuming that appearance rate equals production rate, these data indicate that T4 production is impaired in the critically ill patients. The same investigators, in a similar study of reverse T3 kinetics, showed normal total rT3 clearance and normal rT3 production from T4, increased rT3 fractional transfer rate from serum to tissue sites, suggesting a shift of rT3 from vascular to the extravascular sites and thus accounting for the normal rT3 observed in the serum (Kaptein et al., 1983). Faber and associates measured the kinetics of T4, T3 and rT3 simultaneously and found that CRF patients, as compared to euthyroid normal controls, have normal T4 clearance and production rates. T3 clearance was slightly reduced and T3 production rate markedly low. As for rT3, the clearance rate was higher and the production rate normal (Faber et al., 1983). Increased clearance in the presence of a normal production rate of rT3 explains, in part, the absence of elevated rT3 in patients with CRF as opposed to other non-renal NTI.

IV. TISSUE T3 CONTENT AND T4 UPTAKE

There is virtually no information regarding thyroid hormone content in CRF patients, but there are data from a rodent model of uremia. In 5/6 nephrectomized (Nx) rats, serum blood urea nitrogen (BUN) rises to fivefold above that of the control littermates and there is a marked reduction of TT3, modest decrease in TT4 and normal TSH, strikingly similar to that found in CRF patients. In this uremic rat model, liver T3 content is markedly reduced. Additionally, the activity of two liver enzymes regulated by thyroid hormone, namely, mitochondrial α-glycerophosphate dehydrogenase (αGPD) and cytosol malate dehydrogenase (MDH) is also downregulated (Lim et al., 1980). Furthermore, liver nuclear T3 content and T3 receptor binding capacity are significantly reduced compared to the controls. T3 receptor binding affinity, however, is not different (Lim et al., 1984). More importantly, tri-iodothyronine replacement in the Nx rats results in an increase in the liver T3 content and restoration

TABLE 28.4 Serum thyroid hormone profile, liver and pituitary thyroid status in control, nephrectomized, thyroidectomized and nephrectomized-thyroidectomized rats

Rat Group	Serum				Liver					Pituitary
	BUN (mg/dl)	TT$_4$ (μg/dl)	TT$_3$ (ng/dl)	TSH (ng/ml)	T$_3$ (pg/mg) liver	(pg/mg) protein	αGPD (ΔOD/min/ mg) protein	MDH (units/mg) protein	Nuclear T$_3$ (pg/mg) liver	Cmax (pg/mg) liver	T$_3$ (pg/mg) pituitary
C	22 ± 1	4.7 ± 0.3	52 ± 6	703 ± 61	2.5 ± 0.2	19 ± 1	1.7 ± 0.1	0.037 ± 0.003	0.31 ± 0.05	0.36 ± 0.03	7.0 ± 1.5
Nx	112 ± 20*	3.3 ± 0.4*	30 ± 7*	441 ± 87*	1.5 ± 0.1*	12 ± 1*	1.1 ± 0.1*	0.024 ± 0.003*	0.16 ± 0.02*	0.23 ± 0.02*	13.0 ± 3.6
Tx	28 ± 2*	0.4 ± 0.1*	18 ± 2*	2249 ± 136*	0.5 ± 0.05*	4 ± 0.4*	0.6 ± 0.05*	0.019 ± 0.002*	0.04 ± 0.01*	0.24 ± 0.03*	3.1 ± 0.9*
NxTx	203 ± 24*	0.4 ± 0.2*	<10 ± 0*	2525 ± 292*	0.4 ± 0.04*	5 ± 0.4*	0.6 ± 0.12*	0.020 ± 0.001*	0.03 ± 0.01*	0.19 ± 0.03*	–
Anova: P	<0.0001	<0.0001	<0.0001	<0.0001	<0.0001	<0.0001	<0.0001	<0.0005	<0.0001	<0.0005	<0.05

BUN: blood urea nitrogen. Values (mean ± SEM) of the four groups of rats were tested by analysis of variance and the p values are listed in the last line of the table. Significant differences (*) between the means of each experimental rat group and the controls were assessed by the Student's *t* test. C, Nx, Tx and NxTx represent, respectively, control, nephrectomized, thyroidectomized and nephrectomized-thyroidectomized rats. αGPD and MDH indicate liver mitochondrial alpha-glycerophosphate dehydrogenase and cytosolic malate dehydrogenase. Cmax is T$_3$ binding capacity derived from Scatchard plot analysis. (Data derived from Lim et al., 1980, 1984, with permission granted from *J Clin Invest* and *Endocrinology*).

of the enzyme activity despite continued presence of uremia. These data, as listed in Table 28.4 and depicted in Figures 28.3 and 28.4, indicate that hypothyroidism is present in the liver of uremic rats. More importantly, this state of tissue hypothyroidism can be corrected by exogenous T3 treatment, if needed.

Arem et al. analyzed autopsied tissues from 12 patients who died of serious diseases with NTI and found very low T3 content in the cerebral cortex, hypothalamus, anterior pituitary, liver, kidney and lungs as compared to subjects who were previously healthy and died of acute trauma suddenly (Arem et al., 1993). Of great interest is the finding of a fivefold increase of T3 content in the skeletal muscle and the heart. The T3 content in the muscle, in this study, exceeded that of the circulating T3, suggesting that serum contamination is not likely the cause for the high value. Such high skeletal muscle T3 content, if confirmed, could explain the observation that, in CRF patients, while whole-body protein turnover is either normal or low, it is high in the skeletal muscle (Lim et al., 2005).

There are two possible explanations for the low tissue T3 content. First is an impairment of T4 transport from the circulation to the tissue and, second, a reduction in T4 to T3 conversion at tissue sites. T4 transport from the circulation to the tissue is a carrier-mediated, energy-dependent process. In a series of methodical experiments, Hennemann and his group studied T4 transport by measuring free iodide production from T4 by rat hepatocyte and came to the conclusion that serum from CRF and other non-renal NTI patients inhibits the transport of circulating T4 to tissue sites (Lim et al., 1993a, 1993b; Vos et al., 1995). This defect, independent of tissue deiodinase activity, may lead to the low T3 syndrome. Rat hepatocytes were cultured *in vitro* with ^{125}I-T4 added to the medium. At the end of incubation, aliquots of the medium were chromatographed to separate free iodide from conjugates and iodothyronines. Iodothyronines have to be conjugated with sulfates or glucuronides before they can be deiodinated. Using the conjugates/iodide ratios, together with T3 clearance from the medium, they were able to differentiate transport defect from deiodination suppression. When the transport system is intact and the deiodination process is defective, there would be an increase the conjugate/iodide ratio. If the deiodination is intact and transport is inhibited, the ratio should not change from baseline. T3 clearance rate is reduced in transport suppression and unchanged with deiodination defect. Using these techniques, they found that incubation of hepatocytes in the presence of 10% uremic serum resulted in reduced iodide production, unchanged conjugate/iodide ratio and inhibited T3 clearance consistent with a transport defect hypothesis. To investigate whether inhibition of T4 transport may contribute to the 'low T3 syndrome' in humans, van der Heyden and colleagues studied labeled T4 and T3 kinetics in obese subjects before and after 7 days of being on a daily intake

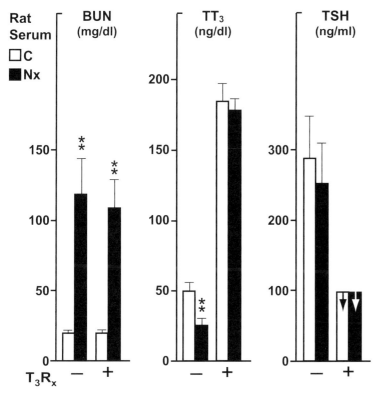

FIGURE 28.3 T3 replacement on BUN and serum thyroid profile in 5/6 nephrectomized rats. T3 treatment does not affect renal function, but it normalized serum T3 and suppresses serum TSH; C and Nx represent, respectively control and nephrectomized littermates.

of 240 Kcal diet. Kinetic analyses were performed according to a three-pool model. During caloric deprivation, T4 mass transfer rate from the plasma to the tissue is markedly reduced and the T4 pool, as a result, is decreased significantly (van der Heyden et al., 1986).

As for tissue deiodination, it is currently known that peripheral metabolism of thyroid hormone is regulated by three iodothyronine deiodinases, namely, D1, D2 and D3 (see Figure 28.1) (Bianco et al., 2002). While the function and distribution of these different deiodinases in different

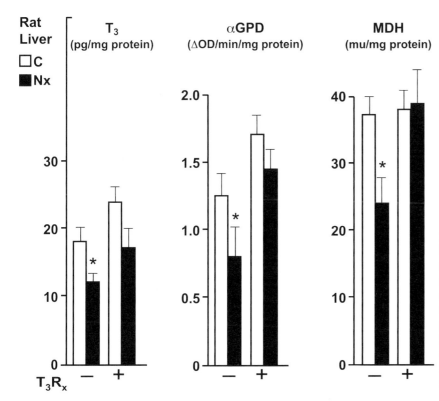

FIGURE 28.4 Liver thyroid hormone function in 5/6 nephrectomized rats. Liver T3 and the activity of two thyroid hormone-dependent enzymes are compared between 5/6 nephrectomized rats and their control littermates. αGPD and MDH represent, respectively, mitochondrial α glycerol phosphate dehydrogenase and cytosol malate dehydrogenase. T_3Rx- and T_3Rx+ indicate, respectively, data derived without and with T3 treatment. (Data derived from Lim et al., 1980, with permission from *J Clin Invest*.)

tissues is still being sorted out, the current data support the following patterns. Both D1 and D2 control outer ring 5′deiodinase resulting in the formation of T3. While D1 also has the ability to activate the inner ring 5 deiodinase, D2 acts strictly only on the outer ring. D1 is present in liver, kidney and the thyroid gland and is responsible for a major source of circulating T3. D2 is found mostly in the brain and the pituitary and, to a lesser extent, in the skeletal muscle. D3 exerts exclusive effect on the inner ring 5 deiodinase and produces rT3. It is found in the liver, skeletal muscle, CNS, uterus and placenta. Peeters et al. reported abnormal T4 deiodination patterns in humans with acute renal failure and non-renal NTI patients. In one study, they obtained serum, liver and skeletal muscle samples from 65 patients who died in the intensive care unit (Peeters et al., 2003). Liver and skeletal muscle biopsies were obtained within minutes after death. In the serum, T4, T3 and TSH were low and rT3 increased. D1 activity was downregulated in the liver and D3 activity was highly expressed in the liver and the skeletal muscle. Serum T3/rT3 ratio was positively correlated with liver D1 activity and negatively with liver D3 activity. Acute renal failure patients who required hemodialysis and were hemodynamically unstable behaved similarly to other non-renal NTI: liver D1 activity was reduced and liver and muscle D3 activity enhanced. It should be noted here that the finding of high D3 deiodinase in muscle suggest that muscle would not be able to produce T3 in the NTI patients, not consistent with the finding of increased muscle T3 content in Arem's autopsied specimens noted above. Thus, the issue of thyroid status in the muscle of CRF patients needs further investigation. In another study, the same investigators studied 451 critically ill patients hospitalized in the intensive care unit for at least 5 days. Seventy-one patients died, giving a crude mortality rate of 16%. They obtained serum samples on days 1, 5, 15 and last day of intensive care unit stay (Peeters et al., 2005). On day 1, serum thyroid hormone profile already had a prognostic value; serum rT3 was higher and T3/rT3 ratio lower in non-survivors. On day 5, serum TSH, T4 and T3 were lower in non-survivors. Serum TSH, T4 and T3 and T3/rT3 ratio increased over time in survivors, but decreased in non-survivors. Liver D1 deiodinase activity correlated positively with the last day serum T3/rT3 ratio and negatively with serum rT3. Additionally, liver and skeletal muscle D3 deiodinase activity correlated positively with the last day serum rT3. These data underscore the importance and the complexity of peripheral thyroid hormone metabolism in body energy homeostasis.

V. THE HYPOTHALAMO-PITUITARY THYROID AXIS

An enigma encountered in the study of thyroid abnormalities associated with uremia and other NTI is the absence of TSH elevation despite low circulating T3 and T4 levels. This phenomenon has been observed in humans with CRF and in uremic rat models. The absence of TSH elevation is not due to the inability of uremic pituitary to respond to low serum thyroid hormones because uremic patients who are truly hypothyroid could elicit a high TSH response. Moreover, nephrectomized rats that have low serum T4 and T3 have normal TSH, additional thyroidectomy leads to a generous fivefold rise in serum TSH, a magnitude equal to that seen in simple thyroidectomized rats (Table 28.4) (Lim et al., 1980). One could argue that the stress of hypothyroidism and thyroidectomy is so intense that heightened TSH secretion becomes obligatory. However, physiologic manipulation producing subtle changes of serum T3 concentration also elicits appropriate TSH responses in CRF patients as in the normal controls. Administration of slightly higher than physiologic dose of L-tri-iodothyronine, 50 μg per day, resulted in a rise in serum TT3 and suppression of TSH and serum TT4. By contrast, treatment with ipodate, a cholecystographic agent which inhibits hepatic T4 to T3 conversion,

TABLE 28.5 Serum thyroid hormone profile, nitrogen metabolism and leucine flux in hemodialysis patients and normal subjects in the basal state, during L- tri-iodothyronine replacement, and sodium ipodate administration

Subject/period	TT$_4$ (μg/dl)	TT$_3$ (ng/dl)	TSH (μU/dl)	Nin (g/24 h)	Nb (g/24 h)	Gurea (mg/min)	Q^{2}H$_3$ leu (μmol/kg/min)	Q^{15}N leu
Hemodialysis								
Basal	7.4 ± 0.4	79 ± 5	5.2 ± 1.6	9.2 ± 0.8	0.58 ± 0.3	4.6 ± 0.6	1.4 ± 0.1	2.5 ± 0.2
LT$_3$	5.5 ± 0.8*	172 ± 16*	1.5 ± 0.5	9.2 ± 0.8	−0.80 ± 0.4*	6.0 ± 0.5*	1.7 ± 0.1*	3.4 ± 0.2*
Ipodate	8.9 ± 0.7*	66 ± 9*	9.1v2.4*	8.7 ± 0.8	0.87 ± 0.4	4.7 ± 0.6	–	–
Normal								
Basal	7.4 ± 0.4	111 ± 6	2.2 ± 0.3	14.7 ± 1.1	0.02 ± 0.5	–	1.2 ± 0.1	2.1 ± 0.2
LT$_3$	5.3 ± 0.3*	214 ± 12*	1.0 ± 0.3	14.8 ± 1.1	0.22 ± 0.7	–	1.4 ± 0.1	2.5 ± 0.1
Ipodate	9.6 ± 1.0	74 ± 8	2.4 ± 0.5	14.6 ± 0.9	0.05 ± 0.6	–	–	–

Values (means ± SEM) during treatment that are significantly different from those of the basal period as assessed by paired t test are designated by *. Nin, Nb and Gurea represent, respectively nitrogen intake, nitrogen balance and urea generation rate. Q^{2}H$_3$ leu and Q^{15}N leu indicate leucine flux rates as measured using two stable isotopes. (Data derived from Lim et al., 1985, 1989, with permission granted, respectively, from *Kidney Int* and *Metabolism*.)

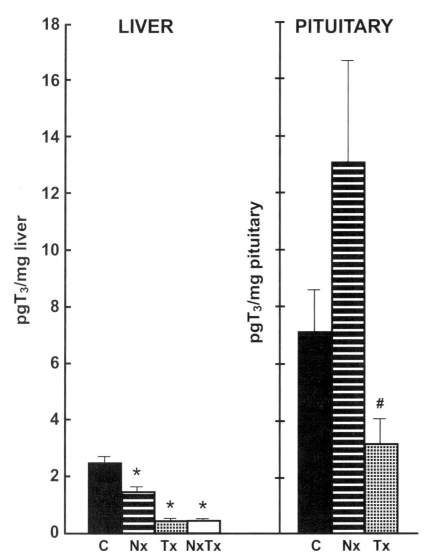

FIGURE 28.5 Liver and pituitary T3 contents. C, NX, Tx and NxTx represent, respectively, control littermates, 5/6 nephrectomized, thyroidectomized and nephrectomized-thyroidectomized rats. Pituitaries of NxTx rats were improperly handled and, therefore, the data are missing.

resulted in a mild lowering of serum TT3 and an unequivocal rise in TSH and TT4 (Lim et al., 1985). These data (Table 28.5) showed that the pituitary, and perhaps the hypothalamus as well, is sensitive to the negative feedback effect of serum T3 concentration. Why then is the TSH not elevated in the uremic and non-renal NTI patients? A possible answer to this question was found in the nephrectomized uremic rats. In these rats, pituitary T3 content, quantitated directly, is normal compared to the control littermates (see Table 28.4) (Lim et al., 1984). Thus, in uremia, the thyroid status of different tissues may vary widely. In the nephrectomized uremic rats, the liver is hypothyroid and the pituitary euthyroid (Figure 28.5). In an elegant *in-vitro* incubation study using labeled T4 and T3, Cheron et al. showed that, in the liver, T4 outer ring 5′deiodination is inhibited by both thiouracil and iopanoic acid; in the pituitary, thioruacil had no effect on T4 outer 5′deiodination. More importantly, fasting inhibits only the outer ring 5′deiodination in the liver but not in the pituitary (Cheron et al., 1979). In another experiment Silva and colleagues injected ^{125}I-T4 into euthyroid rats and measured nuclear to plasma ratio of ^{125}I-T3. The ratio is higher in the pituitary compared to the liver and the kidney (Silva et al., 1978). In this instance, nuclear ^{125}I-T3 is an index of *in-situ* T3 generation from T4. As it turns out, we now understand that propylthiouracil acts specifically on inhibiting D1 deiodinase in the liver and not the D2 deiodinase in the pituitary. Teleologically, this is the most logical adaptation for energy homeostasis. If the pituitary were solely dependent on circulating T3, then the purpose of protein/energy conservation of low thyroid function would be counteracted. This difference in pituitary T3 generation provides a mechanism for TSH regulation independent of circulating T3 concentration.

Nonetheless, the above findings do not exclude the concurrent presence of a hypothalamic defect in TRH secretion and there is evidence implicating central hypothalamo-pituitary dysfunction. Thyrotropin-releasing hormone (TRH) administration to CRF patients showed lesser and slower TSH increment compared to normal controls suggesting a blunted pituitary response (Lim et al., 1977). Bartalena et al.

showed that nocturnal TSH surge, found in normal subjects, is absent in CRF patients (Bartalena et al., 1990). While the morning serum TSH is identical in normal and CRF patients, peak TSH in the late evening hours is lower in the latter. Wehmann and colleagues studied thyroid function in 35 bone marrow transplant patients prospectively and longitudinally. All patients had normal thyroid hormone profile on admission. As they became severely ill with bone marrow transplantation and chemotherapy, not only T3 and T4 were reduced, but TSH as well. The fall in TSH usually preceded the decline in T4. During recovery, TSH and T4 increased concomitantly (Wehmann et al., 1985). Also, the decline in TSH bore little relationship to either glucocorticoid or dopamine administration. These data suggest that the low TSH was, at least, in part, responsible for the low T4 found in those patients. Moreover, Fliers et al. found decreased TRH messenger RNA by *in-situ* hybridization in the hypothalamus of 10 patients who died of a variety of acute and chronic illnesses. They had a set of thyroid function tests performed less than 24 h before death showing reduced T4, T3 and TSH (Fliers et al., 1997). Total TRH mRNA in the paraventricular nuclei showed a positive correlation with serum T3 and with log TSH, but not with T4. The puzzle in this study is the positive correlation of TRH mRNA with T3 and not T4. Because T4 is the precursor of tissue T3, it is difficult to explain the absence of correlation between serum T4 and hypothalamic TRH gene expression. The authors suggested that there might be direct transport of T3 from the circulation into the brain. They also hypothesized that, in the hypothalamus, T4 to T3 conversion may be accomplished via D2 deiodinase that, in contrast to the D1 deiodinase in the liver, is resistant to the inhibitory effect of thiouracil. Neither argument can account for the positive correlation between serum T3 and hypothalamic TRH mRNA. One plausible explanation is that whatever process inhibiting peripheral tissue outer ring 5' deiodination is also affecting hypothalamic TRH expression. Van den Berghe and colleagues showed that administration of TRH, 1 μg/kg/h, to patients with non-renal NTI lead to increased serum TSH, T4 and T3 concentrations (Van den Berghe et al., 1998). Thus, it appears that, if needed, thyroid function could be restored to normal state in NTI subjects, and perhaps CRF patients as well, by TRH administration.

The strongest argument for a central dysfunction is data derived from the rodent starvation model reported by Blake and colleagues (Blake et al., 1991). Three-month-old rats when food-deprived for 48 h showed a significant reduction of both serum T3 and portal blood TRH. Simultaneously, *in-situ* hybridization histochemistry revealed a significant decrease of pro-TRH mRNA in the paraventricular nuclei suggesting a reduction in hypothalamic TRH production. The thyroid function changes in fasting have been shown to link to leptin suppression as treatment with letpin in starved rats resulted in normalization of thyroid function despite continued fasting (Ahima et al., 1996; Flier et al., 2000). The difficulty in fitting the thyroid changes in CRF patients to leptin is the observation that their serum leptin is either normal or slightly elevated and not reduced (Heimburger et al., 1997; Deshmukh et al., 2005).

VI. IODIDE RETENTION, GOITER, HYPO- AND HYPERTHYROIDISM

Because of reduced urinary iodide excretion, both the serum inorganic iodide level and the thyroid gland iodine content, the latter determined by fluorescent thyroid scan, are increased in CRF patients. As a result, thyroid gland enlargement is frequently encountered. The incidence of goiter has been reported to be higher in CRF patients as compared to normal population, 37% and 43%, respectively, by Lim et al. and by Kaptein (Lim et al., 1977; Kaptein, 1996). Also, the frequency of goiter may increase further with the duration of hemodialysis, thought to be due to increased cutaneous iodide absorption of povidone iodine. Autoimmune thyroid disease is not more frequent in CRF patients: the incidence of positive thyroglobulin and thyroid microsomal antibodies is low. The frequency of hypothyroidism is slightly higher, which may be due, in part, to higher incidence of type 1 diabetes in the CRF population. Elevated TSH is a prerequisite for the diagnosis of hypothyroidism in CRF patients. Treatment for hypothyroidism is similar to that of the general population, i.e. l-thyroxine replacement. On rare occasions, T3 supplement may be needed as well. We have observed one patient whose TSH remained elevated even with large doses of l-thyroxine and came down only after addition of tri-iodothyronine to the treatment regimen. The frequency of hyperthyroidism is similar to that of the general population. To make a diagnosis of hyperthyroidism, a combination of elevated FT4 and suppressed TSH is needed. It is important to emphasize that to detect lower than normal serum TSH concentration, one should use a third generation assay method sufficiently sensitive to detect a TSH concentration of 0.02 mU/l. For treatment, the antithyroid drugs do not require dose adjustment. Treatment with ^{131}iodide could be done, but the dosage and the procedure need to be planned ahead with nuclear medicine colleagues. Also the dialysate for the next two to three sessions after radioactive iodine administration needs to be disposed of as any radioactive material.

VII. EFFECTS OF DIALYSIS AND TRANSPLANTATION

Hemodialysis, as currently practiced in the USA, does not normalize thyroid function in CRF patients. In a series of 306 ESRD patients, it was noted that thyroid hormone profile was not different between the non-dialyzed and those receiving

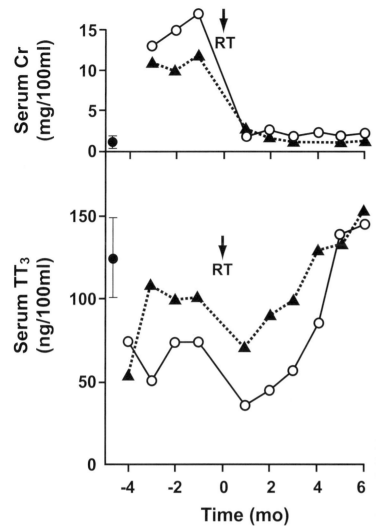

FIGURE 28.6 Changes in serum creatinine and TT_3 after renal transplantation. After renal transplantation (RT), serum creatinine (Cr) fell precipitously in less than one month. Serum TT3 is not normalized until 4 months later. In fact, serum TT3 declined to lower than baseline level shortly after transplantation.

9 hours of hemodialysis per week. We also did not find any significant changes in serum thyroid hormone profile before and 2–3 months after initiation of maintenance hemodialysis (Lim et al., 1977; Kaptein, 1996). An Australian study, on the other hand, found that serum TT4 and TT3 were higher in patients receiving 27 h per week of hemodialysis compared to those receiving 15–18 h of treatment (Savdre et al., 1978). The latter finding is not surprising as more adequate dialysis improves general health and thyroid hormone profile is a reflection of general health status. Single hemodialysis, however, can increase serum free T4, due to the effect of heparin in decreasing T4 binding to serum proteins. Neither T4 nor T3 is significantly dialyzed as they are, respectively, >99 and >95% bound to protein carriers. Serum thyroid hormone profile in peritoneal dialysis patients is similar to those on hemodialysis treatment, but thyroxine binding globulin (TBG) and albumin are lower (Robey et al., 1989). As the peritoneal membrane is more permeable than the artificial membranes used in hemodialysis, TBG, T4 and T3 are lost in the dialysis effluent. The documented losses, however, are quite small in amount, i.e. about 8 μg of T4 and less than 0.1 μg of T3 each day, representing less than 10% and less than 1%, respectively, for T4 and T3 production rates (Kerr et al., 1986). The daily production rates of T4 and T3 in ESRD patients are, respectively, approximately 120 and 13 μg (Chopra, 1976; Lim et al., 1977). Thus, no replacement is needed in peritoneal dialysis patients.

After renal transplantation, serum TT4 and TT3 increased to normal levels. As shown in Figure 28.6, with restoration of renal function, serum creatinine fell precipitously, while serum TT3 rose slowly, over a 3–4 month period. In fact, immediately after transplantation, serum TT3 transiently decreased to below the pre-transplant level. This suggests that low serum T3 may not be related to uremia, but to illness or inflammation. After transplantation, there is a transient period in which the general health status of patient declines related either to surgery and/or initiation of immunosuppressive therapy. The rise in TT4 is due to a combination of increased TBG and less binding inhibition of T4 to its protein carriers. The normalization of TT3, from our kinetic studies, is a consequence of increased T4 to T3 conversion in the extra-thyroidal tissues. Serum TSH response to TRH stimulation remains subnormal in renal transplant patients, likely related to glucocorticoid medication which has been shown to inhibit hypothalamic TRH output (Wilber and Utiger, 1969).

VIII. THYROID BIOLOGY IN CHRONIC RENAL FAILURE AND OTHER NON-THYROIDAL ILLNESSES

Thyroid hormone plays a critical role in fetal development, especially that of the central nervous system. With molecular biology techniques, we now know that, in the human fetus, the concentration of T3 varies in different tissues and at different stages of development (Burrow et al., 1994). Throughout life, thyroid hormone continues to play a dynamic role in body energy economy, particularly in the transition from energy-surplus to energy-deficit state. This is best illustrated in the rodent model of starvation. As energy deficit state threatens survival, there are complex systems developed through a long period of evolution to shut down unnecessary energy expenditure and preserve body mass. Among others, these include downregulation of reproductive and thyroid function. During starvation in the rats, serum TT3 and TT4 declined and TSH is normal. Humans with CRF and other NTI have an identical thyroid profile. Although this chapter only addresses thyroid function, it should be noted that, in CRF patients, gonadal function is likewise impaired due mostly to suppressed hypothalamic function in regulating gonadotropin-releasing hormone secretion (Lim, 1987). Are these patients starved? Perhaps, in some sense, they are: illnesses have put them in an energy-deficit state.

In contrast to other endocrine glands, thyroid function is regulated by two separate physiologic pathways, namely, the central hypothalamo-pituitary-thyroid axis and the peripheral metabolism of thyroid hormones. Thyrotropin-releasing hormone (TRH) controls the release of TSH which then stimulates the synthesis and secretion of thyroid hormones, T4 and T3. Beyond the central axis, there are factors affecting T4 transport from the circulation into the tissues and several iodothyronine deiodinases present in many tissues to metabolize T4 to T3, which is many-fold more active than T4, or to rT3 which is metabolically inert. While T4 is 100% secreted by the thyroid gland, almost 80% of T3 and 100% of rT3 are, respectively, derived from extra-thyroidal deiodination of T4. In an energy-deficit state, the body downregulates thyroid function by dual mechanisms: first, decreased T3 production at extrathyroidal sites due either to a reduction of T4 transport from the circulation to the tissue, reduced T4 outer ring $5'$ deiodination, or a combination of both processes (see section on Tissue T3 content and T4 uptake) and second, reduced secretion of TRH by the hypothalamus. The latter occurs in the most severe state of illnesses.

Evidence of diminished T3 production by extrathyroidal tissues is well documented. It is uniformly noted that liver T3 content is low in humans with NTI and in uremic rats. Pituitary T3 was found to be normal in uremic rats. The difference between the liver and the pituitary is most likely due to the presence of different deiodinases in the two tissues, D1 in the liver and D2 in the pituitary; the latter is less susceptible to inhibition during illnesses (Bianco et al., 2002). Peeters and colleagues found that in patients who died of critical illnesses in the intensive care units, liver D1 $5'$deiodinase activity correlates positively with serum T3 and serum T3/rT3 ratio. In the skeletal muscle, D3 inner ring 5 deiodinase activity correlates positively with serum rT3 (Peeters et al., 2003, 2005). These data suggest a diminished T3 production in the liver and an increased reverse T3 production in the muscle. Both would result in a reduction in total body T3 content. Support for inhibition of T4 transport from serum to tissues is derived mainly from *in-vitro* experiments. Serum from patients with uremia and other NTI inhibits hepatocytes T4 transport during incubation as evidenced by a reduction in iodide production (Lim et al., 1993b; Vos et al., 1995) (see section on Tissue T3 Content and T4 Uptake). As for reduced TRH secretion, evidence is scanty. In one study, Fliers and colleagues reported reduced TRH mRNA in paraventricular nuclei of patients dying with NTI (Fliers et al., 1997).

Although the abnormal thyroid hormone profile found in CRF and other NTI patients can be reproduced in rodent starvation model, NTI is more than starvation; it is an energy-deficient state plus inflammation. Recent research has produced data suggesting a major role of cytokines in suppressing tissue T3 generation. Stouthard et al. studied eight patients with metastatic renal cell carcinoma participating in a phase II clinical trial using recombinant human IL-6 (rhIL-6) (Stouthard et al., 1994). All participants were in good health status and their baseline thyroid profile was normal. They received intravenous infusion of IL-6 for 4 h and thyroid function was obtained before, during and for 2 h after drug administration. The results were compared to those obtained on a separate day of placebo normal saline infusion. IL-6 infusion did not produce any change in serum concentration of T4, free T4 or thyroxine hormone binding index, but it reduced the plasma concentration of TSH and T3 and increased that of reverse T3 to a degree significantly greater than that from placebo infusion. Crossmit and associates, in another project, studied eight healthy young men by giving them one subcutaneous injection of recombinant human interferon-α. Again, the interferon did not alter circulating T4, free T4, but produced a significant reduction in TSH and T3 and a rise in reverse T3 (Crossmit et al., 1995). Collectively, these data suggest that the cytokines may act simultaneously and independently on two separate sites that affect thyroid function, i.e. the peripheral tissue where T3 is generated from T4 and the hypothalamus where TRH is secreted.

In a cohort of 200 hemodialysis patients, Zoccali and associates reported that plasma free T3 is lower in the ESRD patients compared to healthy subjects and clinically euthyroid patients with normal renal function (Zoccali et al., 2005).

During the follow-up period, longest time 67 months, 102 patients died. Those who died had significantly lower plasma free T3 compared to those that survived. On multivariate Cox's regression analyses, adjusting for many traditional risk factors, patients who had higher plasma free T3 had significant reduction in the risk of death. Moreover, when the hemodialysis patients were subdivided into tertiles by the serum IL-6 levels, plasma free T3 was progressively lower across the tertile increments (Zoccali et al., 2006). Regression analyses showed strong and inverse association between plasma free T3 and IL-6, C-reactive protein and adhesion molecules like VCAM and ICAM. Similarly, in 41 CAPD patients, studied by the same group of investigators, plasma free T3 was lower in the patients as compared to healthy people (Enia et al., 2007). Free T3 correlated directly to serum albumin and inversely to IL-6 and C-reactive protein. During a mean follow-up period of 2.8 years, 27 patients died, plasma free T3 was lower in those who died compared to those who survived. Thus, plasma free T3 is an independent predictor of death in both hemodialysis and CAPD patients and this increase in the risk of death is related to increased levels of circulating inflammatory markers. Inflammation reduces T4 outer ring 5' deiodination and conversely increases inner ring 5-deiodination. The molecular mechanism underlying such a switch is yet undefined.

In rodents, starvation-induced hypothyroidism is believed to be related to low leptin. Fasting invariably reduced circulating leptin. A low leptin state in the hypothalamus down regulates α-melanocyte-stimulating hormone (α-MSH) production (Flier et al., 2000) and α-MSH, when administered intraventricularly in the central nervous system, stimulates TRH production (Kim et al., 2000). Using histochemistry technique, Fekete and associates found that α-MSH is contained in nerve terminals innervating the thyrotropin-releasing hormone-synthesizing neurons in the hypothalamic paraventricular nucleus (Fekete et al., 2000). Moreover, intracerebroventricular infusion of α-MSH prevented fasting-induced suppression of pro-TRH, hence starvation can lead to low thyroid state by inhibition of α-MSH via leptin suppression. It is not clear whether inflammation follows the same pathway.

The alteration in thyroid hormone profile described here, namely, reduced T3 and T4 and normal TSH, is not related to any specific organ dysfunction or specific toxic agent, but is seen in myriads of different disease states and illnesses. The common denominators responsible for the genesis of this syndrome include malnutrition, inflammation and poor general health status. There appears to be a negative correlation between thyroid function and gravity of illness. During sickness, the earliest change in thyroid function is a reduction in T3 production, followed by a reduced TRH secretion with consequent lowering of TSH and T4. The latter occurs in the more severely ill patients. Thus, at this moment, changes in thyroid hormone profile should be viewed as a non-specific adaptation to anything that threatens the well-being of mammals in general, including humans. In the history of medicine, comparable non-specific changes have been previously reported and these include the importance of the sympathetic nervous system in the 'fight-or-flight responses' described by Walter B Cannon and the role of adrenals in the 'general adaptation syndrome' studied by Hans Selye (Neylan, 1998).

IX. SHOULD THYROID HORMONE BE REPLACED IN CRF AND OTHER NON-THYROIDAL ILLNESS PATIENTS?

After three decades of research, the answer to the question of thyroid hormone replacement in NTI in general and in CRF patients in particular remains unsettled. Teleologically, there is reason to believe that downregulation of thyroid function with its consequent lowering of body metabolism is beneficial in terms of energy and protein conservation. In Table 28.5, we summarize the results of T3 replacement in CRF patient. LT3 treatment for 2 weeks, one week before the study and continued for another week during the experimental period, resulted in an increase in serum TT3 in both the CRF patients and the control subjects. In the CRF patients, urea generation rate increased and nitrogen balance became negative (Lim et al., 1985). Furthermore, amino acid turnover, measured by $^2H_3$ leucine and ^{15}N leucine fluxes, increased significantly during T3 replacement compared to the baseline period (Lim et al., 1989). Labeled leucine fluxes, in that experiment, were used as an index of body protein turnover. In the control subjects, identical replacement dose did not alter protein metabolism. Similarly, administration of replacement T3 dose to humans undergoing starvation restored serum T3 concentration, but urinary urea nitrogen and 3-methylhistidine excretion, an index of muscle catabolism, increased compared to starvation without T3 treatment (Vignati et al., 1978). Based on these data, it would not be prudent to correct the T3 deficiency in CRF patients. (Would it be more direct if we say 'it would be prudent not to correct T3 deficiency in CRF patients'?)

Lately, some studies have provided data suggesting that thyroid hormone replacement might be beneficial. Klemperer and associates reported that patients who underwent cardiac bypass surgery had a decline in serum T3 and that administration of LT3 intravenously altered some indices of cardiac function in a favorable manner (Klemperer et al., 1995). Premature neonates when given prophylactic T4 and T3 daily had a lower mortality than untreated infants (Schoenberger et al., 1979). Dogs subjected to hemorrhagic shock recovered cardiovascular function faster when given T3 intravenously compared to untreated animals (Shigematsu and Shatney, 1988). The premature neonate study is a different category from NTI because thyroid hormones are critical during fetal and early infant development. A study on severe burn injury patients with low serum T3, however, did not show survival benefits when given T3 supplementation (Becker et al., 1982). Since thyroid hormones, especially T3, regulate body metabolism and gene transcription, they affect function of virtually every organ/system. It is possible that thyroid hormone, when reduced to a critical extent, may threaten survival of the organism and, at that stage, replacement might be necessary. To what extent of decrement requires replacement therapy is unknown presently. In rats, surgical thyroidectomy performed on 5/6 nephrectomized rats increased the mortality of the animals tremendously compared to either simple thyroidectomy or 5/6 nephrectomy alone (Lim et al., 1980). However, this may simply be from the added stress of anesthesia and surgery and not necessarily due to a further decrement of thyroid hormones.

X. SUMMARY

Any acute and chronic illness not originating from the thyroid gland, hence non-thyroidal illness, evokes complex changes in thyroid function which follows a predictable pattern and has prognostic value for survival. The serum changes are characterized by, in its mildest form, reduced T3, increased reverse T3 and normal or mild T4 decrement. With increasing severity of illness, T3 is decreased and reverse T3 increased further and T4 becomes even lower. A markedly reduced TT4, <4 μg/dl, is associated with a high probability of death. Serum TSH starts out as normal in milder disease state and becomes low as the illness worsens. These changes are self-limiting; recovery from whatever illness is heralded by rising TSH and T4, followed by improvement in T3. On the other hand, if the illness remains grave, these abnormalities persist. There are precedences of hormone treatment with survival benefit in critically ill patients including control of hyperglycemia by insulin infusion and administration of glucocorticoid to hemodynamically unstable patients. The thyroid changes, however, are of the nature of energy conservation, nature's way of enhancing survival in an acute hostile environment. Even though physicians' temptation is to correct whatever abnormality is presented to them, our conclusion, based on reviewing the current literature, is not to intervene with nature's innate adaptation unless evidence emerges suggesting the contrary.

References

Ahima, R. S., Prabakaran, D., Mantzoros, C. et al. (1996). Role of leptin in the neuroendocrine response to fasting. *Nature* **382**: 250–252.

Arem, R., Wiener, G. J., Kaplan, S. G., Han-Seob, K., Reichlin, S., and Kaplan, M. M. (1993). Reduced tissue thyroid hormone levels in fatal illness. *Metabolism* **42**: 1102–1108.

Bartalena, L., Pacchiarotti, A., Palla, R. et al. (1990). Lack of nocturnal serum thyrotropin surge in patients with chronic renal failure undergoing regular maintenance hemofiltration: a case of central hypothyroidism. *Clin Nephrol* **34**: 30–34.

Becker, R. A., Vaughan, G. M., and Ziegler, M. G. (1982). Hypermetabolic low triiodothyronine syndrome on burn injury. *Cri Care Med* **10**: 870–875.

Bianco, A. C., Salvatore, D., Gereben, B., Berry, M. J., and Larsen, P. R. (2002). Biochemistry, cellular and molecular biology, and physiological roles of the iodothyronine selenodeiodinases. *Endocr Rev* **23**: 38–89.

Blake, N. G., Eckland, D. J., Foster, O. J., and Lightman, S. L. (1991). Inhibition of hypothalamic thyrotripin-releasing hormone messenger ribonucleric acid during food deprivation. *Endocrinology* **129**: 2714–2718.

Bodziony, D., Kokot, F., and Czekalski, S. (1981). Thyroid function in patients with acute renal failure. *Int Urol Nephrol* **13**: 81–88.

Brent, G. A., and Hershman, J. M. (1986). Thyroxine therapy in patients with severe nonthyroidal illnesses and lower serum thyroxine concentration. *J Clin Endocrinol Metab* **63**: 1–8.

Burrow, G. N., Fisher, D. A., and Larsen, P. R. (1994). Mechanisms of disease: maternal and fetal thyroid function. *N Engl J Med* **331**: 1072–1078.

Cheron, R. G., Kaplan, M. M., and Larsen, P. R. (1979). Physiological and pharmacological influences on thyroxine to 3,5,3'-triiodothyronine conversation and nuclear 3,5,3'-triiodothyronine binding in rat anterior pituitary. *J Clin Invest* **64**: 1402–1414.

Chopra, I. J. (1976). An assessment of daily production and significance of thyroidal secretion of 3,3'5'-triiodothyronine (reverse T$_3$) in man. *J Clin Invest* **58**: 32–40.

Crossmit, E. P. M., Heyligenberg, R., Endert, E., Sauerwein, H. P., and Romijn, J. A. (1995). Acute effects of interferon-*a* administration on thyroid hormone metabolism in healthy men. *J Clin Endocrinol Metab* **80**: 3140–3144.

DeGroot, L. J. (1999). Dangerous dogmas in medicine: the nonthyroidal illness syndrome. *J Clin Endocrinol Metab* **84**: 151–164.

Deshmukh, S., Phillips, B. G., O'Dorisio, T., Flanigan, M. J., and Lim, V. S. (2005). Hormonal responses to fasting and refeeding in chronic renal failure patients. *Am J Physiol Endocrinol Metab* **288**: E47–55.

Docter, R., Krenning, E. P., DeJong, M., and Hennemann, G. (1993). The sick euthyroid syndrome: changes in thyroid hormone serum parameters and hormone metabolism. *Clin Endocrinol (Oxf.)* **39**: 499–518.

Enia, G., Panuccio, V., Cutrupi, S. et al. (2007). Subclinical hypothyroidism is linked to micro-inflammation and predicts death in continuous ambulatory peritoneal dialysis. *Neph Dial Transplant* **22**: 538–544.

Faber, J., Heaf, J., Kirkegaard, C. et al. (1983). Simultaneous turnover studies of thyroxine, 3,5,3' and 3,3',5'-triiodothyronine, 3,5-, 3,3'-, and 3',5'-diiodothyronine, and 3'-monoiodothyronine in chronic renal failure. *J Clin Endocrinol Metab* **56**: 211–217.

Fekete, C., Légrádi, G., Mihály, E. et al. (2000). *a*-Melanocyte-stimulating hormone is contained in nerve terminals innervating thyrotropin-releasing hormone-synthesizing neurons in the hypothalamic paraventricular nucleus and prevents fasting-induced suppression of prothyrotropin-releasing hormone gene expression. *J Neurosci* **20**: 1550–1558.

Flier, J. S., Harris, M., and Hollenberg, A. N. (2000). Leptin, nutrition, and the thyroid: the why, the wherefore, and the wiring. *J Clin Invest* **150**: 859–860.

Fliers, E., Guldenaar, S. E. F., Wiersinga, W. M., and Swaab, D. F. (1997). Decreased hypothalamic thyrotropin-releasing hormone gene expression in patients with nonthyroidal illness. *J Clin Endocrinol Metab* **82**: 4032–4036.

Heimburger, O., Lönnqvist, F., Danielson, A., Nordenström, J., and Stenvinkel, P. (1997). Serum immunoreactive leptin concentration and its relation to body fat content in chronic renal failure. *J Am Soc Nephrol* **8**: 1423–1430.

Hershman, J. M., Krugman, L. G., Kopple, J. D., Reed, A. W., Azukizawa, M., and Shinaberger, J. H. (1978). Thyroid function in patients undergoing maintenance hemodialysis: unexplained low serum thyroxine concentration. *Metabolism* **27**: 755–759.

Kaptein, E. M. (1996). Thyroid hormone metabolism and thyroid diseases in chronic renal failure. *Endocr Rev* **17**: 45–63.

Kaptein, E. M., Feinstein, E. I., Nicoloff, J. T., and Massry, S. G. (1983). Serum reverse triiodothyronine and thyroxine kinetics in patients with chronic renal failure. *J Clin Endocrinol Metab* **57**: 181–189.

Kaptein, E. M., Kaptein, J. S., Chang, E. I., Egodage, P. M., Nicoloff, J. T., and Massry, S. G. (1987). Thyroxine transfer and distribution in critical nonthyroidal illnesses, chronic renal failure, and chronic ethanol abuse. *J Clin Endocrinol Metab* **65**: 606–616.

Kerr, D. J., Singh, V. K., Tsakiris, D., McConnell, K. N., Junor, B. J., and Alexander, W. D. (1986). Serum and peritoneal dialysate thyroid hormone levels in patients on continuous peritoneal dialysis. *Nephron* **43**: 164–168.

Kim, M. S., Small, C. J., Stanley, S. A. et al. (2000). The central melanocortin system affects the hypothalamo-pituitary thyroid axis and may mediate the effect of leptin. *J Clin Invest* **105**: 1005–1011.

Klemperer, J. D., Klein, I., and Gomez, M. (1995). Thyroid hormone treatment after coronary-artery bypass surgery. *N Engl J Med* **333**: 1522–1527.

Lim, C.-F., Bernard, B. F., DeJong, M., Doctor, R., Krenning, E. P., and Hennemann, G. (1993a). A furan fatty acid and indoxyl sulfate are the putative inhibitors of thyroxine hepatocyte transport in uremia. *J Clin Endocrinol Metab* **76**: 318–324.

Lim, C.-F., Docter, R., Visser, T. J. et al. (1993b). Inhibition of thyroxine transport into cultured rat hepatocytes by serum of nonuremic critically ill patients: effects of bilirubin and nonesterified fatty acids. *J Clin Endocrinol Metab* **76**: 1165–1172.

Lim, V. S. (1987). Reproductive function in patients with renal insufficiency. *Am J Kidney Dis* **9**: 363–367.

Lim, V. S. (2001). Thyroid function in patients with chronic renal failure. *Am J Kidney Dis* **38**: S80–S84.

Lim, V. S., Fang, V. S., Katz, A. I., and Refetoff, S. (1977). Thyroid dysfunction in chronic renal failure. *J Clin Invest* **60**: 522–534.

Lim, V. S., Henriquez, C., Seo, H., Refetoff, S., and Martino, E. (1980). Thyroid function in a uremic rat model. *J Clin Invest* **66**: 946–954.

Lim, V. S., Passo, C., Murata, Y., Ferrari, E., Nakamura, H., and Refetoff, S. (1984). Reduced triiodothyronine content in liver but not pituitary of the uremic rat model: demonstration of changes compatible with thyroid hormone deficiency in liver only. *Endocrinology* **114**: 280–286.

Lim, V. S., Flanigan, M. J., Zavala, D. C., and Freeman, R. M. (1985). Protective adaptation of low serum triiodothyronine in patients with chronic renal failure. *Kidney Int* **28**: 541–549.

Lim, V. S., Tsalikian, E., and Flanigan, M. J. (1989). Augmentation of protein degradation by L-triiodothyronine in uremia. *Metabolism* **38**: 1210–1215.

Lim, V. S., Ikizler, T. A., Raj, D. S. C., and Flanigan, M. J. (2005). Does hemodialysis increase protein breakdown? Dissociation between whole body amino acid turnover and regional muscle kinetics *J Am Soc Nephrol* **16**: 862–868.

Melmed, S., Geola, F. L., Reed, A. W., Pekary, A. E., and Hershman, J. M. (1982). A comparison of methods for assessing thyroid function in nonthyroidal illness. *J Clin Endocrinol Metab* **54**: 300–306.

Munro, S. L., Lim, C.-F., Hall, J. G. et al. (1989). Drug competition for thyroxine binding to transthyretin: Comparison with effects on thyroxin-binding globulin. *J Clin Endocrinol Metab* **68**: 1141–1147.

Nelson, J. C., and Tomel, R. T. (1988). Direct determination of free thyroxin in undiluted serum by equilibrium dialysis/radioimmunoassay. *Clin Chem* **34**: 1737–1744.

Neuhaus, K., Bauman, G., Walser, A., and Thoen, H. (1975). Serum thyroxine, thyroxine-binding proteins in chronic renal failure without nephrosis. *J Clin Endocrinol Metab* **41**: 395–398.

Neylan, T. C. (1998). Hans Seyle and the field of stress research. *Neuropsychiatr Classics* **10**: 230–231.

Nicoloff, J. T., and LoPresti, J. S. (1996). Non-thyroidal illness. In: (Braverman, L.E., Utiger, R.D., eds) *The thyroid,* Lippincott-Raven, Philadelphia, pp. 286–296.

Oppenheimer, J. H., Schwartz, H. L., Mariash, C. N., and Kaiser, F. E. (1982). Evidence for a factor in the sera of patients with nonthyroidal disease which inhibits iodothyronine binding by solid matrices, serum proteins, and rat hepatocytes. *J Clin Endocrinol Metab* **54**: 757–766.

Peeters, R. P., Wouters, P. J., Kaptein, E., van Toor, H., Visser, T. J., and Van den Berghe, G. (2003). Reduced activation and increased inactivation of thyroid hormone in tissues of critically ill patients. *J Clin Endocrinol Metab* **88**: 3202–3211.

Peeters, R. P., Wouters, P. J., van Toor, H., Kaptein, E., Visser, T. J., and Van den Berghe, G. (2005). Serum 3,3′,5′-triiodothyronine (rT_3) and 3,5,3′-triiodothyronine/rT_3 are prognostic markers in critically ill patients and are associated with post-mortem tissue deiodinase activities. *J Clin Endocrinol Metab* **90**: 4559–4565.

Robey, C., Shreedhar, K., and Batuman, V. (1989). Effect of chronic peritoneal dialysis on thyroid function tests. *Am J Kidney Dis* **13**: 99–103.

Savdre, E., Stewart, J. H., Mahoney, J. F., Hayes, J. M., Lazarus, L., and Simons, L. A. (1978). Circulating thyroid hormone and adequacy of dialysis. *Clin Nephrol* **9**: 68–72.

Schoenberger, W., Grimm, W., Emmrich, P., and Gempp, W. (1979). Thyroid administration lowers mortality in premature infants. *Lancet* **2**: 1181.

Shigematsu, H., and Shatney, C. H. (1988). The effects of triiodothyronine and reverse triiodothyronine on canine hemorrhagic shock. *Nippon Geka Gakhai Zasshi* **89**: 1587–1593.

Silva, J. E., Dick, T. E., and Larsen, P. R. (1978). The contribution of local tissue thyroxine monodeiodination to the nuclear 3,5,3′-triiodothyronine in pituitary, liver, and kidney of euthyroid rats. *Endocrinology* **103**: 1196–1206.

Stouthard, J. M. L., van der Poll, T., Endert, E. et al. (1994). Effects of acute and chronic interleukin-6 administration on thyroid hormone metabolism in humans. *J Clin Endocrinol Metab* **79:** 1342–1346.

Van den Berghe, G., DeZegher, F., and Baxter, R. C. (1998). Neuroendocrinology of prolonged critical illness: effects of exogenous thyrotropin-releasing hormone and its combination with growth hormone secretagogues. *J Clin Endocrinol Metab* **83:** 309–319.

van der Heyden, J. T. M., Docter, R., van Toor, H., Wilson, J. H. P., Hennemann, G., and Krenning, E. P. (1986). Effects of caloric deprivation on thyroid hormone tissue uptake and generation of low-T_3 syndrome. *Am J Physiol Endocrinol Metab* **251:** E156–E163.

Vignati, L., Finley, R. J., Hagg, S., and Aoki, T. T. (1978). Protein conservation during prolonged fast: a function of triiodothyronine levels. *Trans Assoc Am Phys* **91:** 169–179.

Vos, R. A., De Jong, M., Bernard, B. F., Docter, R., Krenning, E. P., and Hennemann, G. (1995). Impaired thyroxine and 3,5,3'-triiodothyronine handling by rat hepatocytes in the presence of serum of patients with nonthyroidal illness. *J Clin Endocrinol Metab* **80:** 2364–2370.

Wartofsky, L. (1994). The low T3 or 'sick euthyroid syndrome': update 1994. In Clinical and molecular aspects of diseases of the thyroid, (Braverman, L.E., Refetoff, S., eds). Endocr Rev Monographs 3, pp. 248–251.

Wehmann, R. E., Gregerman, R. I., Burns, W. H., Saral, R., and Santos, G. W. (1985). Suppression of thyrotropin in the low-thyroxine state of severe nonthyroidal illness. *N Engl J Med* **312:** 546–552.

Wilber, J. F., and Utiger, R. D. (1969). The effect of glucocorticoids on thyrotropin secretion. *J Clin Invest* **48:** 2096–2103.

Zoccali, C., Tripepi, G., Cutrupi, S., Pizzini, P., and Mallamaci, F. (2005). Low triiodothyronine: a new facet of inflammation in end-stage renal disease. *J Am Soc Nephrol* **16:** 2789–2795.

Zoccali, C., Mallamaci, F., Tripepi, G., Cutrupi, S., and Pizzini, P. (2006). Low triiodothyronine and survival in end-stage renal disease. *Kidney Int* **70:** 523–528.

Further reading

Chopra, I. J. (1997). Euthyroid sick syndrome: is it a misnomer? *J Clin Endocrinol Metab* **82:** 329–334.

Kaptein, E. M., Quion-Verde, H., Chooljian, C. J. et al. (1998). The thyroid in end-stage renal disease. *Medicine (Balt.)* **67:** 187–197.

Spector, D. A., Davis, P. J., Helderman, J. H., Bell, B., and Utiger, R. D. (1976). Thyroid function and metabolic state in chronic renal failure. *Ann Intern Med* **85:** 724–730.

CHAPTER **29**

Metabolic Acidosis of Chronic Kidney Disease

JEFFREY A. KRAUT AND GLENN T. NAGAMI

*UCLA Membrane Biology Laboratory, Division of Nephrology, Medical and Research Services, VA Greater Los Angeles Health Care System;
UCLA David Geffen School of Medicine, Los Angeles, California, 90073, USA*

Contents

I.	Introduction	457
II.	Regulation of acid–base balance with normal renal function and chronic kidney disease	457
III.	Acid–base production	457
IV.	Renal bicarbonate generation	459
V.	Cellular buffering	460
VI.	Renal tubular bicarbonate reabsorption	460
VII.	Hormonal regulation of acid–base balance with normal renal function and with CKD	461
VIII.	Clinical characteristics of the metabolic acidosis of chronic kidney disease	465
IX.	Clinical characteristics of acid–base parameters in dialysis patients	468
X.	Effects of metabolic acidosis of CKD on cellular function	470
XI.	Treatment of the metabolic acidosis of CKD	474
	References	475

I. INTRODUCTION

Approximately, 1 mEq/kg body weight of hydrogen ion is generated by metabolism of ingested foodstuffs (Kurtz et al., 1983) in normal adults each day. In addition, ≈4500 mEq/day (180 l × 25 mEq/l) of bicarbonate is filtered by the glomerulus. The bulk of filtered bicarbonate must be reabsorbed by the renal tubules and a quantity of bicarbonate equivalent to that produced by metabolism must be generated by renal tubules to enable blood pH and serum bicarbonate concentration to be maintained at normal levels of ≈7.38 ± 0.02 and 25.4 ± 0.09 mEq/l in males and ≈7.40 ± 0.02 and 24.4 ± 1.3 mEq/l in non-pregnant females, respectively (Moller, 1959). Chronic kidney disease (CKD) may cause impairment in one or both tubular processes, leading to the development of metabolic acidosis (Schwartz et al., 1959; Elkington, 1962; Kurtz et al., 1990).

Chronic kidney disease is the most common cause of chronic metabolic acidosis observed in the general population (Relman, 1964). The prevalence of the metabolic acidosis of CKD (often termed uremic acidosis) is expected to rise over the ensuing years (Clase et al., 2002; Garg et al., 2002), reflecting in part both aging of the population and the decline in glomerular filtration rate (GFR) which accompanies it (Frassetto and Sebastian, 1996). Importantly, this metabolic acidosis is not without adverse clinical consequences. Indeed, it has been implicated in producing dysfunction and disease of several organ systems and may even contribute to the progression of CKD (May et al., 1987; Alpern and Shakhaee, 1997; Sebastian et al., 2002; Kraut and Kurtz, 2005).

In the present chapter, we will review the mechanisms underlying development of metabolic acidosis with CKD, including the possible role of various hormones, the effect of metabolic acidosis on organ function and the impact of amelioration of the metabolic acidosis by the administration of base, and current recommendations for treatment.

II. REGULATION OF ACID–BASE BALANCE WITH NORMAL RENAL FUNCTION AND CHRONIC KIDNEY DISEASE

As indicated in Figure 29.1, regulation of acid–base balance in the healthy individual is the result of the interplay of several factors including: endogenous net acid production, buffering by cellular and extra cellular buffers and renal reclamation of filtered bicarbonate and generation of new bicarbonate (Kraut and Kurtz, 2005). Alterations in one or more of these processes can contribute to the development of metabolic acidosis with chronic kidney disease.

III. ACID–BASE PRODUCTION

A finite quantity of H^+ or base (bicarbonate) is produced from the hepatic metabolism of ingested food (Relman et al., 1961; Remer and Manz, 1994). Approximately 210 mEq of protons are generated daily from the metabolism of the neutral sulfur-containing amino acids, methionine and cysteine,

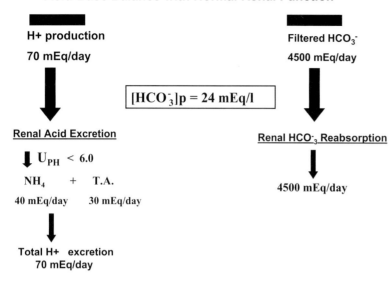

FIGURE 29.1 Acid–base balance in patients with normal renal function. Acid generation from metabolism of ingested foodstuffs is approximately 1 mEq/kg/day (70 mEq/day in 70 kg person). Sufficient HCO_3^- is generated as a consequence of net acid excretion to neutralize this acid. In addition, approximately 4500 mEq of HCO_3^- is filtered by the glomerulus each day which has to be recaptured by the renal tubules. Impairment of either tubule process can lead to development of metabolic acidosis.

which are converted to sulfate and H^+ ions, and the cationic amino acids, lysine, arginine and some histidine residues, which are converted into neutral products and H^+ (Halperin and Jungas, 1983). Approximately 160 mEq/day of bicarbonate is generated from the metabolism of the amino acids glutamate and aspartate and organic anions such as citrate, gluconate, malate, acetate and lactate (Halperin and Jungas, 1983). An additional 25–75 mEq of organic anions, half of which are metabolizable, are excreted in the urine (Relman et al., 1961; Lennon et al., 1966; Lawson et al., 1976). Therefore, net proton generation each day is ≈50 mEq. However, there is often great variability among individuals and therefore, net endogenous H^+ production can vary several fold (range 20–120 mEq/day) (Frassetto et al., 1998), reflecting primarily differences in dietary intake. The impact of dietary intake on acid–base balance in subjects with normal renal function is exemplified by the studies of Kurtz et al. (Kurtz et al., 1983). In normal subjects fed various diets designed to produce net acid excretion between 14 and 154 mEq/day, there was an inverse relationship between plasma bicarbonate concentration and endogenous acid load, the higher the acid load the lower the plasma bicarbonate concentration. This impact of dietary intake on plasma bicarbonate concentration was confirmed by Frassetto et al. (Frassetto et al., 1996; Frassetto et al., 1998) in a larger cohort of individuals. An increase in H^+ production of ≈100 mEq/day caused a fall in plasma HCO_3^- concentration of ≈1 mEq/l.

The magnitude of dietary net H^+ production in patients with CKD has been studied by several investigators. Relman et al. (Relman et al., 1961) studied acid–base balance in individuals with renal failure (GFR, 11–60) and metabolic acidosis ingesting a diet consisting of a purified formula which enabled them to estimate net H^+ production. Net H^+ production was not significantly different from healthy individuals with normal renal function (Goodman et al., 1965). Uribarri et al. (Uribarri et al., 1995b) measured dietary net H^+ production in patients:

1. with chronic renal failure (GFR 19–33 ml/min) and normal or reduced plasma HCO_3^- concentrations (Uribarri et al., 1995b)
2. maintained on chronic ambulatory peritoneal dialysis (CAPD) (Uribarri et al., 1995a) with normal plasma bicarbonate concentrations; and
3. maintained on thrice weekly hemodialysis (Uribarri et al., 1998) with mild pre-dialysis metabolic acidosis (plasma HCO_3^- concentrations of 21–23 mEq/l).

Net H^+ production in patients with CKD not yet on dialysis was either normal or slightly reduced. By contrast, in CAPD patients, in whom the plasma HCO_3^- concentration was in the normal range, net H^+ production was lower than that of individuals with normal renal function, but was similar to that of patients with CKD (Uribarri et al., 1995a). Similarly, the net H^+ production rate in stable chronic hemodialysis with mild pre-dialysis hypobicarbonatemia was reduced by ≈50%. The reduction in net H^+ production with both modalities of therapy was due to a decrease in sulfuric acid generation from cysteine and methionine-containing amino acids and the retention of metabolizable organic anions, potential sources of base. A fall in urinary excretion of metabolizable anions has also been described at earlier stages of CKD; whether this is due to the accompanying acidosis is not clear (Marangella et al., 1991).

Taken as a whole, the evidence indicates net endogenous H^+ production in CKD is either normal or reduced. These findings exclude an increment in net endogenous H^+ production as a contributory factor in the development of metabolic acidosis with CKD in most patients. Since acid production is correlated with dietary protein intake, however, any increase in protein intake over normal can contribute to the development or worsening of the metabolic acidosis, whereas, any reduction in protein intake below normal can lessen the severity of the metabolic acidosis.

IV. RENAL BICARBONATE GENERATION

The kidney is responsible for replenishment of bicarbonate lost in the process of buffering endogenous H^+ production. Bicarbonate is generated by the kidney by three processes:

1. secretion of H^+ and titration of filtered HPO_4
2. metabolism of α-ketoglutarate derived from glutamine
3. metabolism of filtered and reabsorbed organic anions such as lactate and citrate (Kurtz et al., 1990).

The secretion of protons into the tubule lumen generates intracellular bicarbonate because the secreted protons are derived from carbonic acid. Protons secreted into the tubule lumen can bind to inorganic (e.g. phosphate) and organic anions (e.g. citrate) depending on their respective pKs. Excretion of these protonated substances, so-called titratable acids (TAs), in the urine results in the intracellular generation of an equimolar amount of bicarbonate. In subjects with normal renal function, $\approx$10–30 mEq/day of new bicarbonate is generated by this mechanism each day (Halperin and Jungas, 1983; Kurtz et al., 1990). The dominant source of new bicarbonate ($\approx$60%) is derived from renal extraction and metabolism of glutamine: metabolism of 1 mole of glutamine generates 1 mole of α-ketoglutarate which yields 2 moles of HCO_3^- when converted into glucose during gluconeogenesis or oxidized in the Krebs cycle. Bicarbonate is then transported via a basolateral kNBC1 ultimately to the renal vein (Gross and Kurtz, 2002). Were all the NH_4^+ produced in the proximal tubule returned to the systemic circulation, new bicarbonate generated from α-ketoglutarate would be consumed within the liver in the urea cycle ($2NH_4^+ + 2HCO_3^- \rightarrow$ urea) resulting in no net bicarbonate generation by the kidney. However, out of the total of an estimated 54 mmoles of NH_4^+ and new bicarbonate per 1.73 m^2 produced each day from glutamine, 30 mmoles/1.73 m^2 of NH_4^+ are excreted daily in the urine (Tizianello et al., 1980). Since $\approx$54 mmoles/1.73 m^2 of new bicarbonate derived from glutamine is delivered to the renal vein, $\approx$24 mmoles/1.73 m^2 (54–30 = 24) is converted into urea, leaving $\approx$30 mmoles of bicarbonate/1.73 m^2 available to buffer the daily metabolic H^+ load.

In the early stages of metabolic acidosis, renal extraction of glutamine is not increased, however, by day 3–6, glutamine extraction increases substantially accounting for the vast majority of NH_4^+ and therefore new bicarbonate produced from α-ketoglutarate (Tizianello et al., 1982). Under these circumstances, glutamine extraction and metabolism per 100 ml GFR rises as much as sevenfold (Tizianello et al., 1980) and, concomitantly, there is a three- to fourfold increment in urinary NH_4 excretion (30 mEq, baseline; 90–120 mEq day with acidosis). The ability of the kidney to continue to generate bicarbonate at a heightened rate is supported by the increased renal delivery of glutamine resulting from increased production in other organs and tissues (Welbourne, 1988; Kurtz et al., 1990).

By contrast, in patients with severe CKD and metabolic acidosis (GFR < 20 ml/min), renal NH_4^+ excretion is decreased substantially to levels between 0.5–15 mEq/day and total renal NH_4^+ production is decreased by approximately 50% from 38 μmol/min/1.73 m^2 to 19 μmol/min/1.73 m^2 (Tizianello et al., 1980). On the other hand, when expressed per ml of GFR (an estimate of residual nephron function), both the NH_4^+ excretion and production rates are increased by three- to fourfold to levels seen with metabolic acidosis (89 μmol/min/100 ml GFR versus 116 μmol/min/100 ml GFR) indicating that the reduction in NH_4^+ production and new bicarbonate generation are primarily due to reduced functional renal mass (Tizianello et al., 1980).

The source of ammonia production is also altered in renal failure. Glutamine extraction is $\approx$one-tenth of that observed with ammonium chloride-induced metabolic acidosis in individuals with normal renal function (3 μmol/min compared to $\approx$31 μmol/min) (Tizianello et al., 1982). As a result, less than 35% of NH_4^+ produced can be accounted for by glutamine metabolism. The explanation for the impairment of glutamine metabolism is unclear, but it is not due to reduced delivery of glutamine to the kidney, as arterial levels of glutamine are not reduced in individuals with CKD and glutamine loading does not increase ammonia production or urinary ammonium excretion in patients with kidney disease (Welbourne et al., 1972; Tizianello et al., 1980). The remainder of the NH_4^+ is produced from the metabolism of other proteins and peptides generated within the kidney (Tizianello et al., 1980). Furthermore, the partitioning of NH_4^+ between the urine and renal vein is not as enhanced as might be expected for individuals with metabolic acidosis (Tizianello et al., 1980). Whether renal bicarbonate generation from organic anion metabolism is normal is not known.

The majority of patients with CKD can acidify their urine (Seldin et al., 1967) to less than 6.0 and closer to 5.0, but the urine pH achieved, is on average $\approx$1.5 pH units higher than that of individuals without kidney disease who have a similar degree of acidemia (Seldin et al., 1967). Also, it has been suggested hypobicarbonatemia and the attendant reduced distal bicarbonate delivery are essential for generation of a low urine pH (Schwartz et al., 1959; Seldin et al., 1967). This limitation on urinary acidification, *per se*, however, is not a major factor in the impaired net acid excretion, since it has little impact on TA formation or ammonium excretion given their respective pKs of 6.8 and 9.0. In a minority of patients with renal failure, urine pH remains above 6.0 and bicarbonaturia can be seen even though serum HCO_3^- concentration is substantially below normal (Schwartz et al., 1959; Lameire and Matthys, 1986). In these patients, not only is

TA reduced, but the elevation of urine pH would be predicted to alter the renal vein:urine NH_4^+ partitioning thereby reducing urinary NH_4^+ excretion while enhancing its delivery to the renal vein, similar to distal renal tubular acidosis (Kurtz et al., 1990).

The decrease in new bicarbonate generation from α-ketoglutarate discussed previously leaves the kidney more dependent on bicarbonate generation from TA excretion. TA excretion remains normal or only mildly decreased until the later stages of renal failure (GFR < 15 ml/min) when it is reduced to 2–10 mEq/day (Pitts, 1965). The preservation of TA excretion until later stages of renal failure primarily reflects the ability of the patient to reduce urine pH to less than 6.0 in association with excretion of normal quantities of phosphate, often promoted by elevated parathyroid hormone levels (PTH) (Muldowney et al., 1972). An increase in PTH levels if present can be the consequence of hypocalcemia and hyperphosphatemia or stimulation by metabolic acidosis (Bichara et al., 1990). Since urinary phosphate excretion is the dominant factor determining TA excretion, a decrease in TA excretion can be observed in patients prior to a significant decline in renal function if protein intake is markedly reduced or they are ingesting sufficient quantities of phosphate binders to reduce filtered phosphate load and consequently urinary phosphate excretion.

V. CELLULAR BUFFERING

Acid introduced in the body is neutralized by buffers in the extracellular and intracellular compartments (Lemann et al., 2003). A large fraction of the buffering done outside the extracellular compartment occurs within muscle (Bailey et al., 1996), however, it has been postulated that bone can also contribute significantly to this process (Lemann et al., 2003). The latter organ is primed for this purpose as it contains $\approx$35 000 mEq of exchangeable base in the form of carbonate (Burnell, 1971). Buffering of protons by bone is the result of both release of freely exchangeable base independent of cellular action (Lemann et al., 2003) and cell-dependent dissolution of bone (Kraut and Coburn, 1994; Lemann et al., 2003). Buffering of acid by bone has been postulated to begin very soon after acid is introduced into the body, but may be most important in the chronic buffering of acid. This process can be modulated by different hormones including PTH. The buffering of acid with short-term metabolic acidosis (few hours duration) has been examined by different investigators. Fraley et al. (Fraley and Adler, 1979) found that parathyroidectomized dogs and thyroidparathyroidectomized (TPTX) rats had a larger fall in plasma bicarbonate concentration in response to an acid load than animals with intact glands; results consistent with impaired buffering by non-extracellular buffers, presumably bone. By contrast, in other studies in TPTX rats, no impairment of buffering of an acute acid load was found (Madias et al., 1982). The reasons for the discrepant results are not clear.

Even if bone is not important for acute buffering of acid loads, studies in animals and man have suggested that it is important in the buffering of chronic acid loads. Administration of osteoclast-inhibitory drugs, thereby preventing release of base, produced a greater fall in serum bicarbonate in rats with CKD than normal controls (Goulding and Irving, 1974). Buffering of acid by bone was postulated to explain the stability of acid–base parameters in patients with chronic kidney disease and metabolic acidosis (Relman, 1964; Goodman et al., 1965). This inference was based on evidence that imposition of an acid load is associated with increased urinary calcium excretion with little change in intestinal calcium absorption (Lemann et al., 2003) implicating dissolution of bone in this process. Also, Goodman et al. (Goodman et al., 1965) studied a group of individuals with chronic renal failure (GFR 11–60 ml/min) and stable metabolic acidosis using a special diet in which hydrogen input could be determined. Over a period of several days, they demonstrated they were in positive H^+ balance by $\approx$12 mEq/day, although serum HCO_3^- concentration did not change. As will be discussed later, the validity of this assumption has been challenged recently. Studies showing decreased carbonate stores in dogs with metabolic acidosis (Burnell, 1971) and in patients with CKD with metabolic acidosis (Burnell et al., 1974) provide further support for an important role of bone buffering of acid loads with the chronic metabolic acidosis of CKD.

Whether depletion of carbonate stores or development of various types of bone disease affects the ability to buffer a chronic acid load has not been examined. However, Uribarri et al. (Uribarri et al., 1998) examined the bicarbonate space in a small group of stable chronic dialysis patients, some of whom presumably had bone disease. These investigators found bicarbonate space, an index of cellular buffering, to be similar to that of those with intact renal function previously reported i.e. $\approx$50% body weight. However, since the type and severity of bone disease in these patients was not examined, it is still conceivable that, in patients with very severe bone disease, buffering of acid loads by bone could be impaired.

VI. RENAL TUBULAR BICARBONATE REABSORPTION

Under normal conditions, 95% of filtered HCO_3^- is reabsorbed by the renal tubules and the urine is virtually bicarbonate free. The bicarbonate is predominately reabsorbed in the proximal tubule via a sodium–hydrogen exchange and proton translocating ATPase (H^+-ATPase) (Wagner, 2007). The bulk of bicarbonate reabsorption in this segment is mediated by the sodium–hydrogen exchanger. Studies using the

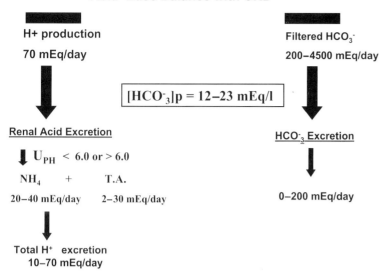

FIGURE 29.2 Acid–base balance in patients with chronic kidney disease. Acid generation from metabolism of ingested foodstuffs is similar to or slightly less than in patient with normal renal function, i.e. 1 mEq/kg/day. CKD can cause defective HCO_3^- reclamation and/or generation of HCO_3^- leading to a fall in serum HCO_3^- concentration and metabolic acidosis. The magnitude of the hypobicarbonatemia, bicarbonate excretion in the urine and net acid excretion varies among patients and also can vary in individual patients depending both on when in the course of CKD these parameters are measured. Values for net acid excretion and bicarbonate excretion are estimates.

remnant kidney model in dogs have suggested that the former transporter is upregulated with reduced GFR, thereby preventing substantial bicarbonate wasting (Cohn et al., 1982).

Studies in uremic rats and humans have found absolute bicarbonate reabsorptive capacity to be either increased or decreased (Slatopolsky et al., 1970; Lubowitz et al., 1971) depending on the model utilized. Despite the failure consistently to detect impaired bicarbonate reabsorptive capacity in experimental studies and the evidence that the sodium–hydrogen exchanger might be upregulated with renal failure, it is clear that some patients with chronic renal failure do have impaired bicarbonate absorption. Schwartz et al. (Schwartz et al., 1959) demonstrated that five of 12 subjects with renal failure continued to excrete bicarbonate in the urine when plasma bicarbonate concentrations fell below normal. Similarly, Lameire et al. (Lameire and Matthys, 1986) found that five of 17 patients with a GFR of 6–19 ml/min had fractional excretions of bicarbonate of 4.25% to 17.65% despite serum bicarbonate concentrations as low as 17 mEq/l. Other investigators also confirmed bicarbonate wasting can be seen with chronic renal failure (Muldowney et al., 1972). Defective bicarbonate absorption in renal failure could be ascribed to volume expansion, excessive PTH secretion (Muldowney et al., 1972; Schmidt, 1978), hyperfiltration in residual nephrons, an osmotic diuresis (Schmidt, 1978) or the presence of disorders that preferentially affects the proximal tubule.

In summary, endogenous acid production in chronic kidney disease is similar to or less than that with normal renal function, indicating this process is not a factor in the metabolic acidosis of CKD (Uribarri, 2000). Limited studies of bicarbonate space reflecting cellular buffering in humans have suggested that it is not different from that noted in patients with normal renal function (Uribarri et al., 1998). Bicarbonate reclamation can be impaired in some patients contributing to bicarbonate wasting, but the prevalence of this abnormality is not clear. Therefore, the major factor producing the metabolic acidosis in the majority of patients with CKD is a reduction in renal bicarbonate synthesis causing it to fall below acid production resulting in net positive hydrogen balance. Once hypobicarbonatemia has developed, it is not clear whether patients continue to remain in positive acid balance or are in neutral balance at that point. However, the persistence of the metabolic acidosis indicates the kidney is not capable of replenishing bicarbonate stores; therefore, it is unlikely that renal HCO_3^- generation is substantially greater than acid production. Figure 29.2 depicts acid–base balance in patients with CKD with metabolic acidosis. The magnitude of renal net acid excretion will vary depending upon the nature of acid excretory defect and the time in the course of CKD when this parameter is measured. Values given for net acid excretion and urinary bicarbonate excretion with CKD are estimates.

VII. HORMONAL REGULATION OF ACID–BASE BALANCE WITH NORMAL RENAL FUNCTION AND WITH CKD

The factors regulating renal acid excretion have been the subject of intense investigation. Although there is evidence that changes in blood pH, *per se*, can directly modulate renal acid excretion via various extrarenal and renal mechanisms, a significant role for various hormones in modulating this process has been demonstrated (Paillard and Bichara, 1989; Wesson, 2006). As summarized in Table 29.1, several hormones have been shown to affect acid–base balance by

TABLE 29.1 Interrelationship of acid–base balance and hormone function in normals and with CKD

Hormone	Effect of acidosis on serum level or response	Effect of hormone on acid–base balance	Role in acid–base regulation with CKD
Aldosterone	↑ Levels with metabolic acidosis	↑ Collecting duct H^+ secretion and net acid excretion	↑ Levels may help regulate acid–base balance ↓ Levels contribute to exacerbation of metabolic acidosis
Angiotensin II	↑ Levels with metabolic acidosis	↑ Collecting duct H^+ secretion and net acid excretion	↓ Levels may play a role in inducing hypoaldosteronism
Insulin	↓ Action due to impaired receptor binding and post-receptor signal transduction	↓ Net acid excretion no effect on acid–base parameters	Unclear
PTH	↑ Levels in some studies; attenuates end-organ response in others	↑May play role in cellular buffering ↓ Net acid secretion	Unknown role in acid–base regulation
Growth hormone	Blunted response to hormone	↑ Net acid excretion	Impaired action may contribute to development of metabolic acidosis
Glucagon	Unknown	↑ Net acid excretion	Unknown
Glucocorticoids		↑ Glutamine uptake with enhanced NH_4 production and possibly net acid excretion	Unknown
ADH	Unknown	↑ Net acid excretion	Unknown
Endothelin	Unknown	↑ Net acid excretion	Unknown

modulating cellular buffering of acid and/or renal net acid excretion.

A. Aldosterone

One of the major hormones modulating renal acid excretion is aldosterone (Henger et al., 2000; Wagner, 2007). Aldosterone modulates the reabsorption of sodium by the renal collecting duct (Schambelan et al., 1980). This process sets up the lumen negative potential that favors proton transport by the electrogenic H+-ATPase residing in the specialized intercalated cell (Henger et al., 2000). Aldosterone may also directly affect proton pumping by this transporter. It does not appear to modulate the activity of the electroneutral renal H^+-K+-ATPase, another proton transporter present in the kidney whose role in renal acid excretion remains unclear (Zies et al., 2007). Serum aldosterone levels rise with metabolic acidosis (Wagner, 2007) and this may be important in the kidney's ability to excrete an acid load. Indeed, states such as primary adrenal insufficiency (Henger et al., 2000) and hyporeninemic hypoaldosteronism (Schambelan et al., 1980) may be characterized by the development of a hyperchloremic metabolic acidosis, related in part to reduced aldosterone levels.

Serum aldosterone levels in some studies of patients with chronic kidney disease are in fact elevated (Leavey and Weitzel, 2002). However, in a subset of patients with diabetic renal disease, abnormalities in aldosterone production can develop either alone or in response to low renin levels (hyporeninemic hypoaldosteronism) (Schambelan et al., 1980) contributing to the development of metabolic acidosis. A potential side effect of the use of drugs that alter tubular response to aldosterone for the treatment of CKD is an elevation in serum potassium concentration and reduction in serum HCO_3^- concentration (Epstein, 2006). Therefore, both serum potassium and bicarbonate levels should be monitored carefully during treatment of these agents. Patients with pre-existing CKD may be particularly prone to develop more severe acidosis because they have a lower acid excretory reserve (Henger et al., 2000).

B. Angiotensin II

Angiotensin II has direct effects on acid–base transport and metabolism in the proximal tubule. It stimulates Na+-H+ exchange, Na+-bicarbonate co-transport and ammonia production and secretion (Liu and Cogan, 1987; Geibel et al., 1990; Chobanian and Julin, 1991; Bloch et al., 1992; Nagami, 1992, 1995; Eiam-Ong et al., 1993). Metabolic acidosis stimulates the renin–angiotensin–aldosterone system and this effect might be important in the adaptive response to an acid challenge (Perez et al., 1979; Gyorke et al., 1991; Wagner, 2007). Thus, studies in mice have demonstrated that blocking the type 1 angiotensin receptor prevented the adaptive increase in ammonia excretion that occurs after an acid challenge (Nagami, 2002). Receptor blockade had no effect on basal rates of urinary NH_4 excretion or proximal tubule ammonia production or secretion in non-acid-loaded mice. Also, in acid-loaded humans, Henger et al. (Henger et al., 2000) demonstrated that angiotensin-receptor blockade can reduce the renal response to an acid load by reducing ammonium and net acid excretion rates.

Exacerbation of metabolic acidosis caused by treatment with converting enzyme inhibitors and/or receptor blockers in patients with CKD is a potential risk of this therapy, although it is not common in the absence of an increased acid load. The administration of a reduced protein intake, thereby reducing the endogenous acid load, and administration of base early in the course of CKD, may explain in part the scarcity of cases in which treatment with these compounds worsened the metabolic acidosis.

C. Parathyroid Hormone

PTH may not only modulate release of bone buffers as described above, but can also affect acid–base balance by altering bicarbonate reabsorption and renal net acid excretion (Hulter, 1985; Paillard and Bichara, 1989). Although some studies had indicated that PTH administration inhibited bicarbonate reabsorption in the proximal tubule, thus inducing bicarbonaturia, this effect was rather shown to be due to changes in the filtered load of bicarbonate (Bichara et al., 1986; Paillard and Bichara, 1989). Studies by Bichara et al. (Bichara et al., 1990) not only showed that acute HCl infusion increased PTH levels, but it also enhances urinary phosphate, TA, ammonium and net acid excretion. Moreover, PTX rats had lower net acid excretion than intact rats both in the presence and absence of an acid load. Net acid excretion was restored to appropriate levels by administration of PTH. These data suggest that PTH contributes to renal regulation of acid–base balance under normal physiologic conditions and in response to acid loads. Chronically, PTH administration induces sustained elevations in serum bicarbonate concentration in several different species due to its effect on bone and kidney (Hulter, 1985). Theoretically, elevated PTH levels observed in many patients with CKD may also participate in the regulation of acid–base balance by its effect on bone buffering and renal acid excretion, however, the precise role of PTH in modulation of acid–base balance in CKD is unclear.

D. Glucocorticoids

Metabolic acidosis is associated with increased levels of adrenal corticosteroids (Sartorius et al., 1953; Perez et al., 1979). The resultant increase in corticosteroid levels stimulates both tubular bicarbonate reabsorption and generation, thereby reducing the severity of the acidosis. As a consequence, adrenalectomy is associated with reduced net acid excretion rates due to reduction in both TA and ammonium excretion. Glutamine delivery from peripheral tissues to the kidney and its extraction (Welbourne, 1988) and sodium–hydrogen exchange activity in the renal brush border (Kinsella et al., 1984), processes augmented with metabolic acidosis, are blunted in the absence of glucocorticoids. Further, glutaminase activity, a key ammonia generating enzyme, is upregulated in kidney cells exposed to dexamethasone, possibly a result of increased number of glucocorticoid receptors (Gowda et al., 1996). Taken as a whole, these data demonstrate that glucocorticoids, independent of aldosterone, contribute to renal acid excretion when stressed by acid load.

As mentioned previously, glutamine extraction is markedly blunted in individuals with CKD and this contributes to the reduced rates of ammonia production and excretion observed in patients with CKD (Tizianello et al., 1980). Whether changes in glucocorticoids levels or in the tubular response to glucocorticoids contribute to altered glutamine uptake and metabolism found with CKD is unclear. In individuals with CKD, existing data suggest that, except for a reduced degree of diurnal variation (lack of full morning suppression), cortisol levels are similar to normal individuals (Deshmukh et al., 2005). Therefore, the reduced renal extraction of glutamine in individuals with CKD cannot be attributed solely to a reduction in glucocorticoids levels.

Studies in peripheral blood lymphocytes from individuals with CKD indicate resistance to the effects of glucocorticoids does occur. This resistance is not mediated by changes in receptor number or affinity (Hirano et al., 1997), but occurs post receptor and has been attributed to impaired signal transduction (Takahashi et al., 2002). Therefore, it is possible cellular resistance in other organs could be present and this remains to be confirmed.

E. Andiuretic Hormone

In experimental studies in rats, antidiuretic hormone (ADH) acutely stimulates collecting duct proton transport and augments urinary net acid excretion (Bichara et al., 1987; Paillard and Bichara, 1989). However, its role in regulation of acid–base balance, both in individuals with normal renal function and in those with CKD, is unclear. Patients with inappropriate ADH excess have normal or near normal serum bicarbonate concentrations, despite dilutional reductions in the concentration of sodium and chloride (Cohen et al., 1976; Borensztein et al., 1993). Although the maintenance of a normal serum HCO_3^- concentration in this disorder has been attributed to changes in serum aldosterone concentration, it is also conceivable it is due part to excess ADH stimulation of renal bicarbonate synthesis (Cohen et al., 1976). ADH levels in patients with CKD have been found to be elevated, particularly when the renal function is markedly compromised (Nonoguchi et al., 1996). This elevation in ADH could contribute to an adaptive increase in tubular acid secretion with CKD, but this remains to be examined.

F. Glucagon

Acute glucagon infusion suppresses bicarbonate absorption in the ascending limb, distal tubule and collecting duct, thereby decreasing net acid excretion (Paillard and Bichara, 1989). Although these acute studies suggest glucagon could play a role in regulation of acid–base balance in patients with normal renal function and CKD, studies to examine these issues remain to be performed.

G. Endothelin

Endothelin is a potent vasoconstrictor hormone released from renal endothelial cells and other cells (Wesson, 2001). Endothelin 1 and 2 receptors are stimulated by the hormone whether it is released systemically or in a paracrine fashion by renal epithelial cells. Studies by Wesson (Wesson, 2006) have shown that endothelin stimulates renal net acid excretion by augmenting collecting duct proton transport. The stimulation of renal net acid excretion is observed whether metabolic acidosis is produced by administration of acid or increased endogenous acid production is augmented by increment in dietary protein intake, in the absence of detectable effect on acid–base parameters.

Both serum endothelin levels and the mRNA for endothelin receptors are increased with chronic renal failure (Richter, 2006). Since endothelin can enhance renal acid excretion, the increased levels of endothelin in CKD could contribute to preservation of renal acid excretion as renal function declines. Independent of its effect on renal acid–base balance, endothelin is pro-inflammatory and pro-fibrotic and therefore increase levels could contribute to progression of renal disease.

H. Insulin

Insulin is metabolized by the kidney. Therefore, progressive CKD reduces its metabolism explaining the improved glucose tolerance seen in some diabetic patients with CKD (Mak, 1998b). On the other hand, the metabolic acidosis found with CKD can reduce its action both by attenuating binding to its receptor and affecting the post-receptor signal transduction cascade (Mak, 1998a). The ultimate impact of CKD on the actions of insulin will depend on the competition between these two effects. Thus, any insulin resistance observed in patients with CKD might be due, in part, to the associated metabolic acidosis. In this regard, Kobayashi et al. (Kobayashi et al., 2005) demonstrated a strong inverse correlation between the serum HCO_3^- concentration and glucose disposal rate, a measure of insulin sensitivity. However, insulin resistance is not due purely to metabolic acidosis as correction of metabolic acidosis does not restore insulin sensitivity to normal as determined by the euglycemic clamp technique (Mak, 1998a). Cardiovascular disease is the major cause of morbidity in patients with CKD. Since it is more frequent and more severe in patients with the metabolic syndrome (characterized by insulin resistance), any interference with the actions of insulin which might contribute to development of this syndrome could be deleterious.

Experimental studies show insulin affects renal tubular sodium handling in animals and humans, but its role in regulation of sodium balance remains ill defined (Defronzo et al., 1975). Insulin also affects renal acid excretion: administration of insulin causes a fall in renal net acid excretion, primarily due to reduced ammonium excretion. Since insulin does not reduce glutamine extraction by the kidney (Gerich et al., 2000), any reduction in ammoniagenesis must occur at a step distinct from uptake of glutamine. Even though insulin decreases renal net acid excretion, there is no impact on acid–base parameters (Defronzo et al., 1975; Vaziri et al., 1987). The lack of change in acid–base status in the face of a fall in net acid excretion is unexplained but, theoretically, could result from a concomitant change in the buffering capacity of the body or reductions in acid production.

I. Growth Hormone and IGF-1

Growth hormone has diverse effects on renal function, including an increase in GFR rate as well as enhanced reabsorption of phosphate (Sicuro et al., 1998). Studies on hypophysectomized rats using the isolated kidney approach demonstrated a decrease in net acid excretion which was restored by administration of growth hormone. Further, the rats developed metabolic acidosis which was ameliorated by growth hormone administration (Welbourne and Cronin, 1991).

Administration of growth hormone to individuals receiving an acid load causes a rise in net acid excretion, primarily as a consequence of an increase in renal NH_4 excretion (Sicuro et al., 1998). This may be due to an effect of growth hormone on tubular ammonia production as, exposure of isolated proximal tubular segments obtained from animals with normal renal function and acid–base balance to growth hormone directly stimulates ammonia production, possibly by increasing glutamine transport into cells (Chobanian et al., 1992). A potential role for growth hormone in acid–base regulation with CKD is suggested by the presence of mild hypobicarbonatemia in children with growth hormone deficiency and its improvement with administration of growth hormone (Glaser et al., 1998). In patients with severe renal failure (GFR <15 ml/min), growth hormone levels are increased but IGF-1 levels are reduced. These findings suggest there is resistance to the effects of growth hormone with CKD. This resistance could arise from the presence in serum of IGF-1 binding proteins that sequester IGF-1 from its receptors (Tonshoff et al., 2005).

In summary, as noted above and detailed in Table 29.1, several hormones can alter renal acid–base excretion by the normal kidney and their interactions could contribute to the modulation of acid–base balance on a day-to-day basis. Dissecting how their effects are integrated remains a challenging problem. Also, the blood concentrations of and/or cellular responsiveness to several of these hormones are altered with CKD. How these changes are translated into changes in acid–base balance with CKD remain unclear since, the role of the individual hormones in modulating acid–base balance with CKD, with the exception of aldosterone and PTH, has not been examined in detail.

VIII. CLINICAL CHARACTERISTICS OF THE METABOLIC ACIDOSIS OF CHRONIC KIDNEY DISEASE

A. Onset, Prevalence and Magnitude

In the majority of patients with CKD, the onset of metabolic acidosis is correlated with the level of GFR, developing only when it falls below 20–25% of normal (Elkington, 1962; Simpson, 1971; Widmer et al., 1979; Hakim and Lazarus, 1988). This observation is supported by data from the third annual National Health and Nutrition Examination Survey (NHANES) (1988–1994) (Hsu and Chertow, 2002) in which a detectable fall in serum HCO_3^- concentration was not seen until the GFR was <20 ml/min. At this level of GFR, it is likely that ≈80% of patients will manifest some degree of metabolic acidosis (Wallia et al., 1986; Hakim and Lazarus, 1988; Caravaca et al., 1999). The metabolic acidosis and accompanying acidemia is generally mild to moderate in degree: serum HCO_3^- concentrations ranging between 12 mEq/l and 23 mEq/l and blood pH remaining greater than 7.2 (Schwartz et al., 1959; Elkington, 1962; Hakim and Lazarus, 1988).

An inverse correlation between the severity of the metabolic acidosis and the level of renal function usually exists, so the lower the GFR, the more severe the metabolic acidosis (Schwartz et al., 1959; Elkington, 1962; Widmer et al., 1979). Severe metabolic acidosis with serum HCO_3^- concentrations <12 mEq/l, however, are uncommon in the absence of an increased acid load or superimposed abnormality in renal tubular bicarbonate reabsorption and/or generation (Relman, 1964). Interestingly, there is variability in the serum HCO_3^- concentrations in patients with similar levels of GFR. The explanation for this variability is unknown, but could reflect inter-individual differences in tubular function, dietary acid production, and/or cellular buffering.

On the other hand, acid–base parameters can be within the normal range, even when GFR is ≤20 ml/min (Wallia et al., 1986). In two separate studies of patients with CKD and GFR less than 20 ml/min, as many as 20% of the patients had serum HCO_3^- concentrations close to or within the normal range (Wallia et al., 1986; Caravaca et al., 1999). The explanation for relatively normal acid–base parameters in the presence of severe CKD is unclear. However, it is also likely that if the dietary protein load was increased substantially, metabolic acidosis would ensue more rapidly than in patients with normal renal function, reflecting a reduced renal acid excretory reserve (Simpson, 1971; De Santo et al., 1997). Examination of acid excretory reserve in patients with all degrees of CKD remains to be accomplished.

Several factors other than the level of GFR can affect the onset and the severity of the metabolic acidosis in patients with CKD. Disorders that preferentially impair renal tubular function, including interstitial renal disease, sickle cell disease, two disorders that produce structural damage to the collecting duct, and hyporeninemic hypoaldosteronism, a disorder which is characterized by both reduced aldosterone levels and tubular resistance to the actions of aldosterone, have the potential to hasten the onset of metabolic acidosis or worsen a pre-existing acidosis (Schambelan et al., 1980; Ray et al., 1990). Graphs relating the degree of hypobicarbonatemia to the level of GFR in patients with CKD have been published (Elkington, 1962; Hakim and Lazarus, 1988). Based on these data, the presence of disorders preferentially affecting tubular function should be suspected if a fall in serum HCO_3^- is noted when GFR is above 25 ml/min, or if the serum HCO_3^- is lower than predicted based on the level of GFR. Since impaired collecting duct function can also reduce potassium secretion, the presence of hyperkalemia out of proportion to the level of GFR will be additional evidence for preferential tubular dysfunction.

Hyporeninemic hypoaldosteronism can be diagnosed by noting low basal and stimulated values for renin and aldosterone (Schambelan et al., 1980). If serum aldosterone values are not low, tubular damage can be suspected by the failure properly to acidify the urine when serum HCO_3^- is reduced below normal values. Some patients who, based on clinical, laboratory and/or pathological studies, have no evidence of abnormalities of aldosterone production or overt tubular disease appear to develop metabolic acidosis with a GFR > 25–35 ml/min (Widmer et al., 1979). In this regard, a retrospective analysis of 41 patients evaluated at a single institution revealed the presence of metabolic acidosis in patients with presumed glomerular disease when GFR was ≈30 ml/min or greater. However, in this study, the presence of tubular disease was assessed primarily on clinical findings (27/41) and the GFR rate was estimated based on serum creatinine concentration alone. The ability correctly to assess GFR based on serum creatinine concentration alone has been called into question by the results of the MDRD study (Coresh et al., 2003). Furthermore, many patients with clinical evidence of only glomerular disease will also be found to have tubulointerstitial disease on pathological examination (Ong and Fine, 1992). Taken as a whole, both findings suggest that the diagnosis of the type of renal disease present in the study of Widmer et al. (Widmer et al., 1979) may not be valid. Therefore, further elucidation of the factors that affect the onset and severity of metabolic acidosis in patients with CKD is necessary.

As noted above, one of the defining characteristics of the metabolic acidosis of CKD is that once the serum HCO_3^- has fallen, it remains relatively stable unless renal function declines further or an additional acid load is imposed, i.e. it is not progressive in nature (Relman, 1964; Goodman et al., 1965; Simpson, 1971). This conclusion has been based on early clinical observations and was supported by the seminal studies of Goodman et al. (Goodman et al., 1965) of patients with severe renal failure (GFR 11–60 ml/min). These investigators found that in patients with CKD and hypobicarbonatemia ingesting a fixed diet for a short period, plasma HCO_3^- remained stable, although careful measurements indicated they were in positive hydrogen balance.

The postulate that patients with CKD and metabolic acidosis are in positive hydrogen balance has been challenged. First, it has been suggested that the urinary parameters used to assess renal net bicarbonate generation may not be appropriate as calculation of renal net acid excretion (equivalent to new bicarbonate generation): $NH_4^+ + TA - HCO_3$ by Goodman et al. (Goodman et al., 1965) ignores the metabolizeable organic cations (which generate H^+) and metabolizable organic anions (which generate bicarbonate) which are also excreted in the urine. A more accurate assessment of new renal bicarbonate generation has been suggested by Kurtz et al. (Kurtz et al., 1990) as follows: $NH_4^+ + TA^+ + OC_M^+ - HCO_3^- - OA_M^-$, where OC_M^+ and OA_M^- refer to metabolizable organic cations and anions respectively. Studies using this more complete formulation have not yet been performed in patients with CKD to validate its usefulness.

Furthermore, Uribarri et al. (Uribarri et al., 1998) performed balance studies in patients with CKD, utilizing a modified method to assess net gastrointestinal and bone sources of alkali: net GI alkali absorption + internal bone alkali loss = (Na + K + Mg + Ca) − (Cl + 1.8P). At first examination, it appeared patients were in positive H^+ balance. However, when urinary electrolyte charge balance was assessed, total anions exceeded total cations (there was a cation gap), the magnitude of this difference was virtually identical to the positive hydrogen balance. Since a cation gap cannot be present in any body fluid, the authors inferred there was an error in the assessment of acid production or excretion and that the patients were not in positive proton balance. The discrepant results indicate the need for further studies to elucidate whether or not patients with CKD and hypobicarbonatemia are in continuous positive H^+ balance.

B. Serum Electrolyte Pattern

In a prior review by one of the authors, the concept of the serum anion gap, including its usefulness and limitations was discussed in depth (Kraut and Madias, 2007). The serum anion gap is defined as the sum of the Cl^- and HCO_3^- concentration subtracted from the serum Na concentration. This entity has proved useful in analysis of acid–base disorders, because various disorders producing metabolic acidosis can have different impacts on the serum electrolyte pattern (Kraut and Madias, 2007).

Disorders in which hydrochloric acid is retained (including those characterized by bicarbonate loss from the body) lead to the reduction in serum HCO_3^- and a reciprocal increment in serum chloride, but no change in the serum anion gap, hence the designation as a hyperchloremic or normal anion gap metabolic acidosis. By contrast, with accumulation of non Cl-containing acids such as lactic acid, the fall in serum HCO_3^- will be matched by an equivalent increment in the serum anion gap (at least initially). Hence, the designations of these forms of metabolic acidosis as high anion gap acidoses.

The metabolic acidosis of CKD has been considered to be a high anion gap form of metabolic acidosis (Relman, 1964). Bicarbonate reclamation and renal bicarbonate generation are tubular processes, while excretion of generated acid anions depends largely on GFR (although tubular reabsorption of anions contribute) (Dass and Kurtz, 1990). Therefore, it has been postulated that a proportionate reduction in both tubular and glomerular functions will result in equivalent retention of protons and generated anions. The high anion gap metabolic acidosis noted with CKD differed from the dominant hyperchloremic (normal anion gap) acidosis observed with the classic forms of renal tubular acidosis (RTA). In these disorders, tubular acidification is perturbed with little or no change in GFR. Based on analogy to the classic forms of RTA, it was inferred that patients with CKD who had a dominant hyperchloremic electrolyte pattern must have additional abnormalities in tubular acidification to those found in the usual patient with CKD (Johnson and Morgan, 1965; Morris et al., 1972).

Two observational studies have examined the serum electrolyte pattern in patients with CKD as renal failure progressed. Widmer et al. (Widmer et al., 1979) performed a retrospective analysis of a cohort of patients followed over several years in a renal clinic. They found that with mild to moderate renal failure (as defined by serum creatinine concentration of 2–4 mg/dl) the metabolic acidosis was primarily a normal anion gap type. However, as renal failure progressed, the decline in serum HCO_3^- was matched by a rise in the anion gap, i.e. a high anion gap metabolic acidosis developed. Hakim and Lazarus, in a retrospective analysis of over 900 patients followed over several years, also noted that with mild to moderate renal failure there was hyperchloremic metabolic acidosis present at early stages of renal failure, but serum anion gap rose as renal failure progressed (Hakim and Lazarus, 1988). On the other hand, in other studies, some degree of hyperchloremic acidosis remained present or developed even at very late stages of CKD (GFR $\leq$ 20 ml/min) (Enia et al., 1985; Wallia et al., 1986; Widmer et al., 1979; Caravaca et al., 1999).

The reasons for the serum anion gap to fail to rise in some patients with severe renal failure are unclear. Abnormalities in the gastrointestinal or renal absorption of organic/inorganic anions might account for the lower than expected anion gap in these patients. Reduced serum albumin levels that may occur in patients with proteinuric renal disease or those with malnutrition associated with severe renal failure may also mask a rise in the serum anion gap, although there was no evidence for this possibility in any of the studies. Although a hyperchloremic electrolyte pattern can be observed with CKD, a dominant hyperchloremic pattern should still indicate the possibility these patients might have interstitial renal disease, low serum aldosterone levels (frequently as a result of treatment with various medications that affect the renin–angiotensin system and/or tubular response to aldosterone such as (ACEI, ARB, NSAID) or disorders in which the tubules are preferentially damaged such as sickle cell disease (Battle et al., 1982) or obstructive uropathy (Battle et al., 1981). In this regard, treatment of the metabolic acidosis in these patients might include administration of mineralocorticoid or other measures.

C. Renal Tubular Bicarbonate Generation and Urinary Acidification

As noted above, elevation in net endogenous H^+ production is either unchanged or reduced with CKD (Goodman et al., 1965; Uribarri et al., 1995b, 1998) indicating a reduction in new renal bicarbonate generation and/or impairment in renal bicarbonate absorption (Tizianello et al., 1980; Dass and Kurtz, 1990) as the primary reason for the metabolic acidosis. The decrease in urine NH_4 excretion in patients with CKD is predominantly a reflection of decreased proximal renal tubular glutamine uptake and subsequent generation of ammonium and bicarbonate (from α-ketoglutatarate) (Tizianello et al., 1980). New renal bicarbonate generation from glutamine can be assessed indirectly by doubling the value of urine ammonium excretion. The latter calculation assumes that 50% of ammonium is delivered to the renal vein and 50% is excreted in the urine as in normal individuals.

The urine NH_4^+ excretion can be measured directly or assessed indirectly by determining the urine anion or osmolal gap (Kim et al., 1996). The urine anion gap, defined as the sum of urine $Na^+ + K^+ - Cl^-$, is generally negative by $\approx$30 mEq/l (Kim et al., 1996). In cases in which NH_4^+ excretion is low, it can be positive. The urine osmolal gap defined as the measured urine osmolality $-2 \times (Na^+ + K^+)$ + urea mg/dl/2.8 + glucose mg/dl/18 (Hakim and Lazarus, 1988). Dividing the urine osmolal gap by 2 gives an approximation of urine NH_4^+. Values measured directly or indirectly are usually below normal. Its use in CKD remains somewhat controversial, but studies by Kim et al. (Kim et al., 1996) have shown both calculations will be abnormal in patients with CKD, indicating either measurement may give a valid estimate of urine ammonium excretion in patients with CKD.

The decrease in bicarbonate generation from glutamine metabolism leaves the kidney more dependent on bicarbonate generation from TA excretion. In many patients with CKD, TA excretion will remain normal or only mildly decreased until the later stages of renal failure (GFR < 15 ml/min). At that point, it can be reduced to 2–10 mEq/day (Pitts, 1965; Simpson, 1971). TA excretion is preserved, because despite renal failure, urinary phosphate excretion remains normal during the course of CKD. The preserved urinary phosphate excretion occurs because an elevated serum phosphate concentration results in a higher filtered load of phosphate and elevated serum PTH levels inhibit phosphate reabsorption. A decrease in TA excretion can be found, even at higher levels of GFR in some patients, ingesting a very low protein intake, or in those receiving phosphate binders, both processes reducing serum phosphate concentrations and therefore urinary phosphate excretion.

Bicarbonaturia can also occur in patients with CKD (Schwartz et al., 1959; Muldowney et al., 1972; Lameire and Matthys, 1986). In one study of 17 patients, fractional excretions of bicarbonate ranged from 4.25% to 17.65%. Also, the bicarbonate threshold (serum HCO_3^- concentration at which bicarbonate is lost in the urine) was as low as 17 mEq/l (Lameire and Matthys, 1986). On the other hand, bicarbonate wasting was uncommon in an additional small study of patients with CKD (Ray et al., 1990). In all studies, the type of renal disease was not a predictor of the presence of bicarbonaturia. The actual prevalence of bicarbonate wasting in CKD is impossible to surmise from these studies since they involved small groups of patients. A small degree of impairment in bicarbonate reabsorption in many patients with CKD might be present, as it has been suggested that patients with CKD require a greater fall in serum HCO_3^- than in individuals with normal GFR before

TABLE 29.2 Characteristics of metabolic acidosis of chronic kidney disease

Serum
80% or more of patients with GFR < 20 ml to 30 ml/min will have serum HCO_3^- below normal
Serum bicarbonate range, 12–23 mEq/l; blood pH usually >7.2
Severity of hypobicarbonatemia correlated with level of GFR

[1]Low serum HCO_3^- with GFR > 20–30 ml/min:

- High dietary H^+ load
- Hyperkalemic DRTA due to hyporeninemic hypoaldosteronism or tubular resistance
- HCO_3^- wasting

Serum anion gap can be normal or high
Serum electrolyte pattern can be hyperchloremic, high anion gap or mixture of both patterns
Hyperchloremic pattern more frequent in presence of hyperkalemic distal tubular acidosis

Urine
Fractional excretion of bicarbonate usually low but can be increased in some patients as great as 17%
Urine pH < 5.5 when serum HCO_3^- below normal in majority of patients
Urine ammonium excretion low
Urine titratable acid excretion normal till GFR < 15 ml/min; can be low if patients taking phosphate binders

[1]Metabolic acidosis may appear in some patients with GFR > 20–30 ml/min for which there is no explanation. Studies to determine the mechanism of this hypobicarbonatemia are warranted.

they can reduce urine pH < 5.5 (Schwartz et al., 1959; Seldin et al., 1967).

The majority of patients with chronic renal failure are able to reduce their urine pH below 5.5 (Relman, 1964; Seldin et al., 1967) suggesting that there is not a gradient defect in renal acidification. However, the minimal urine pH achieved is higher than that of normal individuals with a similar degree of acidemia and, as noted, serum HCO_3^- concentration must be reduced to lower levels than in normals to generate the low urine pH (Schwartz et al., 1959; Seldin et al., 1967; Simpson, 1971). In addition to a mild degree of bicarbonate wasting, higher ammonium excretion rates or luminal flow rates per nephron may impede the fall in urine pH to levels observed in individuals without CKD.

Some patients with CKD, particularly diabetics, will have hyporeninemic hypoaldosteronism (Schambelan et al., 1980). Similar to other patients with CKD, patients with hyporeninemic hypoaldosteronism have low urine NH_4 excretion, but they can lower their urine pH < 5.5 in the face of hypobicarbonatemia. Therefore, based on tests of renal acidification, they may be difficult to distinguish from the usual patient with CKD. However, these patients usually have more severe hyperkalemia and hypobicarbonatemia than other patients with CKD at the same level of GFR and the metabolic acidosis is more likely to be hyperchloremic in nature. CKD patients with disorders primarily affecting the collecting ducts such as sickle cell disease, obstructive uropathy or amyloid (Battle, 1986), will not only manifest a reduced urinary NH_4 excretion, but will show impaired distal renal acidification as reflected by a urine pH above 6 in the face of hypobicarbonatemia.

In summary, the metabolic acidosis of CKD is mild to moderate in degree (serum HCO_3^- 12–23 mEq/l). Its onset and severity are related to the level of GFR, most frequently developing when GFR is <25 ml/min and being most severe the lower the GFR. Although the serum anion gap is often increased, electrolyte patterns characterized by hyperchloremia with or without an increase in the anion gap can be observed at all stages of CKD. Abnormalities in renal tubular bicarbonate reabsorption and generation are integral to the generation of the metabolic acidosis and thus urinary ammonium excretion will be low and in some cases bicarbonaturia will be present. Urine pH can be reduced appropriately below 5.5 in the absence of overt bicarbonaturia in most patients, differentiating it from classic forms of distal renal tubular acidoses. However, patients with CKD can also develop disorders that affect ammonia production or tubular acidification such as hyporeninemic hypoaldosteronism and/or damage to the collecting duct. The presence of one of both of these abnormalities can lead to more severe metabolic acidosis and a dominant hyperchloremic electrolyte pattern (Schambelan et al., 1980) (Table 29.2). Also, in patients with collecting duct damage urine pH will not be appropriately reduced.

IX. CLINICAL CHARACTERISTICS OF ACID–BASE PARAMETERS IN DIALYSIS PATIENTS

The characteristics of acid–base parameters in patients with end-stage renal disease (ESRD) maintained on chronic dialysis has been discussed in several excellent reviews (Mehrotra et al., 2003; Oh et al., 2004). The majority of patients ≈80% who reach end-stage renal failure (GFR < 15 ml/min) will have metabolic acidosis prior to beginning chronic maintenance dialysis (Wallia et al., 1986; Kraut and Kurtz, 2005).

With the loss of the ability of the kidney to generate bicarbonate, regulation of acid–base balance in most patients with ESRD depends on three major factors:

1. acid loads from endogenous acid production, largely the result of protein intake, or from certain medications, the most prominent of which is sevelamer (Brezina et al., 2004)
2. the space of distribution of generated acid (so-called bicarbonate or acid space)
3. the net delivery of base via dialysis (Gennari, 2000; Kraut, 2000).

As noted above, acid production is an important determinant of pre-dialysis acid–base parameters both in healthy individuals and those with CKD (Kurtz et al., 1983; Remer and Manz, 1994). In hemodialysis patients, plasma HCO_3^- rises from its pre-dialysis value as base is delivered to the patient. Within the first 1–2 h of completion of dialysis, plasma HCO_3^- may then fall, reflecting equilibration with cellular bicarbonate stores. Over the intervening days prior to the next dialysis, plasma HCO_3^- falls further, gradually returning to its pre-dialysis value (Gennari, 2000). The rate of fall as noted depends upon rate of acid production, cellular buffering and residual renal function, if any.

Since bone is a potential important source of chronic buffering of generated acid, there has been speculation that the presence of bone disease with depleted carbonate stores would alter the normal value for the bicarbonate space of $\approx 50\%$ body weight. However, Uribarri et al. (Uribarri et al., 1998) in a study of a small group of patients on chronic dialysis found bicarbonate space was also $\approx 50\%$ body weight. Since there was no information on the extent of bone disease in these patients, it is still possible that, in some patients with severe bone disease, the bicarbonate space might differ from this value.

Pre- and post-dialysis acid–base parameters will also depend upon the type of base used in the dialysate, the mode of dialytic treatment utilized and the duration of dialysis (Gennari, 2000). When dialysis was first begun in the early 1950s, bicarbonate was used as the source of base. However, it was supplanted by acetate because of the difficulty of delivering bicarbonate with central dialysis delivery systems and the risk of precipitation of added calcium. Acetate is metabolized in the liver resulting in the equimolar generation of bicarbonate. This remained the major buffer used for dialysis for more than 10 years. Since patients with CKD and metabolic acidosis were thought to have some depletion of bone bicarbonate stores, it was anticipated that these stores would need to be repleted before pre-dialysis plasma HCO_3^- concentrations would be raised substantially. Indeed, patients with metabolic acidosis prior to initiation of chronic hemodialysis can show a gradual increment in pre-dialysis plasma HCO_3^- concentrations after several months of dialysis (Gennari, 2000).

Examination of pre-dialysis acid–base parameters after at least a year of chronic hemodialysis with a dialysate containing 35 mEq/l of acetate failed to show complete resolution of the metabolic acidosis. In two separate studies, representative of the findings in many dialysis patients, mean pre-dialysis plasma HCO_3^- averaged 19 mEq/l (Gennari, 2000; Uribarri, 2000). Furthermore, in some patients plasma HCO_3^- bicarbonate concentrations actually fell during the dialysis treatment (Vinay et al., 1987). Net bicarbonate delivery during dialysis depends on the quantity of acetate delivered to the patient and its metabolic conversion to bicarbonate relative to the quantity of bicarbonate removed. The latter depends upon the blood-dialysate bicarbonate concentration gradient which is large since bicarbonate is absent from the dialysate. Careful studies indicated the failure of acetate dialysis to result in substantial net delivery of bicarbonate in some patients reflected the failure of the large quantities of acetate delivered to the patient to be rapidly metabolized to bicarbonate (acetate intolerance) (Vinay et al., 1987). As a consequence, depending on the severity of this acetate intolerance, little or no net bicarbonate delivery was accomplished, or there was even net bicarbonate removal. This so-called acetate intolerance was often associated with significant clinical findings including muscle cramps and hypotension.

The subsequent development of methods to deliver bicarbonate without complications enabled the substitution of bicarbonate for acetate. Initially, dialysate bicarbonate concentration was varied from 30 mEq/l to 35 mEq/l. Observational studies of large groups of patients revealed that even with dialysate bicarbonate concentration of 35 mEq/l, the average pre-dialysis plasma HCO_3^- approached 22 mEq/l, with a significant number of patients having values less than 19 mEq/l (Uribarri et al., 1999; Uribarri, 2000; Gennari, 2000). To raise plasma HCO_3^- concentration to more normal values, dialysate bicarbonate concentration was raised to 39–40 mEq/l. This change in dialysate bicarbonate concentration has resulted in the elevation of pre-dialysis plasma bicarbonate concentration above 23 mEq/l in the majority of patients (Harris et al., 1995). In some patients, despite this high dialysate bicarbonate concentration, plasma HCO_3^- concentration can remain subnormal. This has been attributed to a high endogenous acid production resulting from a high protein intake rather than inadequate base delivery or other factors (Uribarri et al., 1999). Of interest, recent studies have suggested that the acid generated from metabolism of a given protein intake might be actually less in hemodialysis patients than in those with normal renal function (Uribarri et al., 1998).

Dilution of plasma bicarbonate by fluid retained in the interdialytic period is an additional factor that can alter pre-dialysis HCO_3^- concentration (Feriani, 1998). The magnitude of this fall is correlated with the quantity of weight gained, but a decrement of $\approx 1-2$ mEq/l has been reported (Feriani, 1998). Indeed, in a cross-over study of 29 patients,

TABLE 29.3 Adverse effects of the metabolic acidosis of chronic kidney disease[1]

Effect	Comments
Muscle wasting	Can be seen with serum HCO_3^- only slightly below normal. Important factor in causing muscle breakdown in CKD
Reduced albumin synthesis	Acidosis one of several factors leading to hypoalbuminemia
Bone disease	Contributory rather than primary mechanism in producing bone disease
Impaired insulin sensitivity	Role unclear
β 2-microglobulin accumulation	Impact of acidosis unclear, much less important than renal failure or exposure to dialysis membrane
Accelerated progression of chronic kidney disease	Recent controlled study shows correction of acidosis slows progression and prevents ESRD
Impaired thyroid metabolism	May contribute to abnormalities of metabolic rate
Stunted growth in children	Correction of acidosis improves growth
Cardiac disease	Observation data suggest role in producing cardiac disease but more sophisticated studies needed
Increased inflammation	Could be important factor in progression or renal disease and in adverse effects of acidosis

KD, chronic kidney disease; [1]Potential adverse effects of metabolic acidosis from studies of normals and those with CKD; [2]syndrome X, syndrome characterized by dyslipidemia, hyperinsulinemia, hypertension, abdominal obesity, glucose intolerance and insulin resistance.

Fabris et al. reported that merely reducing the interdialytic weight gain by 1 kg caused pre-dialysis plasma HCO_3^- concentration to rise by ≈1.6 mEq/l (Fabris et al., 1988).

With bicarbonate in the dialysate, plasma HCO_3^- concentration rises predictably during dialysis. With a dialysate bicarbonate concentration of 35–39 mEq/l, plasma bicarbonate rose by ≈4–5 mEq/l at the completion of a 4-hour dialysis (Harris et al., 1995). Post-dialysis, values fall towards pre-dialysis levels, the rate and magnitude of this fall being dependent largely upon protein intake, cellular buffering, residual renal function and the duration of the interdialytic period as noted previously. In many of the patients dialyzed with the higher dialysate, base concentration of 39–40 mEq/l, blood pH rose above 7.40 transiently, although it returned to the normal range in the ensuing 48 hours (Kirschbaum, 2004).

Hemodiafiltration will also improve acid–base parameters, but plasma HCO_3^- concentrations may still be slightly below normal. In their group of patients treated with hemodiafiltration, Canaud et al. (Canaud et al., 1998) found an average pre-dialysis plasma HCO_3^- of 22.8 mEq/l. The average post-dialysis plasma HCO_3^- was 28 mEq/l. Given the efficiency of bicarbonate delivery with this technique, normalization of pr-dialysis plasma bicarbonate concentration should be possible in the majority of patients.

There has been a great deal of interest in daily dialysis as a potential alternative to thrice-weekly dialysis. Studies by Williams et al. (Williams et al., 2004) revealed that daily dialysis led to more stable pre-dialysis plasma HCO_3^- and less frequent alkalemia post-dialysis. This modality of dialysis treatment has additional benefits, but its impact on acid–base parameters might more accurately mimic the situation in patients with less severe renal failure. The impact of this treatment on the complications of metabolic acidosis noted with CKD remains to be explored.

In patients treated with continuous ambulatory peritoneal dialysis (CAPD) or automated peritoneal dialysis with dialysate lactate concentration of 35 mEq/l, 60–70% of patients have plasma HCO_3^- concentrations within the normal range (Mujais, 2003). A lesser number can have plasma HCO_3^- > 24 mEq/l (17–27%) and 10–12% can have values <22 mEq/l (Mujais, 2003). The failure completely to normalize plasma bicarbonate concentrations in some patients has been postulated to be due to a high endogenous acid load resulting from a high protein intake rather than a low delivery and/or generation of base from the organic anion precursor.

X. EFFECTS OF METABOLIC ACIDOSIS OF CKD ON CELLULAR FUNCTION

Table 29.3 summarizes the reported effects of metabolic acidosis on cellular function gleaned from studies done in animals and patients with normal renal function, animals and patients with renal failure not on dialysis and chronic hemodialysis and peritoneal dialysis patients (Lefebvre et al., 1989; Reaich et al., 1992; Ballmer et al., 1995; Kraut and Madias, 1996; Roberts et al., 1996, 2002; Brungger et al., 1997a; Graham et al., 1997a, b; Chauveau et al., 2000). Many of the studies are of short duration and it is not clear whether these results can be extrapolated to prolonged metabolic acidosis. It is also not clear if the effects of metabolic acidosis described in patients with normal renal function will be the same in patients with CKD, where many changes in the hormonal and electrolyte milieu can be present. These caveats aside, the abnormalities described below indicate metabolic acidosis can have a significant impact on the clinical condition of the patient.

A. Exacerbation or Production of Bone Disease

Bone disease is an important complication of CKD and contributes substantially to its morbidity. The bone disease is the consequence primarily of alterations in the secretion and/or biological action of parathyroid hormone, vitamin D, calcium and phosphorus and the exposure to various toxins such as aluminum (Bushinsky, 1999). However, experimental and clinical studies have shown that metabolic acidosis, *per se*, has diverse effects on the divalent ion system and bone and, therefore, can contribute to the development of or exacerbation of bone disease (Kraut and Coburn, 1994; Kraut, 2000; Lemann et al., 2003).

In vitro and *in vivo* studies have shown that metabolic acidosis acts directly on bone to stimulate bone resorption and inhibit bone formation (Kraut et al., 1986; Krieger et al., 1992), can inhibit vitamin D production (Chan et al., 1985) and affect the stimulation of parathyroid hormone or alter its end-organ responsiveness (Martin et al., 1980; Krieger et al., 1992; Coe et al., 1999; Kraut, 2000). In animals, metabolic acidosis is associated with the development of osteoporosis or exacerbation of parathyroid-induced bone disease (Barzel and Jowsey, 1969; Chan et al., 1985). Case reports of individual patients or reports of small numbers of patients have demonstrated that metabolic acidosis is associated with worsening of osteomalacia and osteitis fibrosa cystica (Kraut, 1995). This effect is roughly correlated with the severity of the metabolic acidosis (Cochran and Wilkinson, 1975; Coen et al., 1995, 1996).

Bone histomorphometric studies of children with hereditary distal renal tubular acidosis (DRTA) revealed that bone formation was low and mineralization was defective. One year after amelioration of the metabolic acidosis with base, there was substantial improvement in these bone parameters and in phosphate balance (Domrongkitchaiporn et al., 2002; Disthabanchong et al., 2003). Furthermore, improvement or normalization of acid–base parameters leads to acceleration of growth in children with renal tubular acidosis (McSherry and Morris, 1978).

In stable hemodialysis patients with metabolic acidosis who had either osteitis fibrosa or low bone turnover bone disease on bone biopsy, partial normalization of the metabolic acidosis attenuated the rise PTH, reduced bone resorption and improved bone formation (Lefebvre et al., 1989). Normalization of plasma bicarbonate concentration in dialysis patients with hyperparathyroidism enhanced the sensitivity of the parathyroid gland to changes in ionized calcium concentration, leading to better suppression of hyperparathyroidism (Graham et al., 1997a; Movilli et al., 2001).

Even in the absence of hypobicarbonatemia, the acid load provided by metabolism of ingested foodstuffs can worsen bone disease. In women with normal renal function and age-related osteoporosis who had acid–base parameters within the normal range, administration of sufficient base completely to neutralize the acid produced by metabolism improved bone metabolism, as assessed by various biochemical studies (Sebastian et al., 1994). Furthermore, administration of potassium citrate for one year to women with osteoporosis sufficient partially to neutralize endogenous acid production increased their bone mineral density (Jehle et al., 2006).

Collectively, these data show metabolic acidosis contributes to the development or exacerbation of bone disease in adults and to stunting of growth and bone disease of pediatric patients with CKD. The duration and severity of the metabolic acidosis is a critical determinant of the extent of the bone disease and so bone disease related to metabolic acidosis alone may take several years to become clinically overt. Concomitant abnormalities in divalent ion metabolism, parathyroid hormone secretion and vitamin D metabolism common with CKD will modify the impact of metabolic acidosis on bone.

B. Muscle Wasting

Muscle mass is reduced with progressive CKD (Coles, G.A., 1972). Since patients with metabolic acidosis and CKD have evidence of increased protein catabolism (Mitch and Price, 2001), it was inferred that metabolic acidosis, *per se*, might be an important factor in muscle wasting. Rats with renal failure and mild metabolic acidosis (mean serum HCO_3^- concentration of 20 mEq/l) had increased degradation of muscle protein without changes in protein synthesis (May et al., 1987; Mitch, 1998; Mitch and Price, 2003). Degradation of protein was accomplished via the ATP-dependent ubiquitin-proteosome system (Greiber and Mitch, 1992; Mitch et al., 1994; Bailey and Mitch, 2000): transcription of genes encoding proteins of this system was increased by metabolic acidosis.

Glucocorticoids can contribute to protein degradation with metabolic acidosis (May et al., 1987, 1996; Price et al., 1994). Since glucocorticoid synthesis in humans is stimulated by metabolic acidosis, the rise in glucocorticoids may act synergistically with metabolic acidosis to increase muscle breakdown (Price et al., 1994). Importantly, correction of metabolic acidosis attenuates protein degradation, reduces generation of urea, improves protein balance and increases muscle mass in individuals with severe renal failure both prior to and after initiation of chronic maintenance dialysis (Papadoyannakis et al., 1984; Williams et al., 1997; Lim et al., 1998; Pickering et al., 2002). In addition to the direct effects on muscle, metabolic acidosis has been shown to reduce expression of insulin-like growth factor and the abundance of growth hormone receptor (Challa et al., 1993); changes that themselves can lead to reduced muscle mass. With renal failure, high affinity binding proteins for IGF-1 may blunt the effect of circulating IGF-1 (Tonshoff et al., 2005).

As with bone disease, it appears that the normal endogenous acid loads can affect muscle metabolism despite having little impact on acid–base parameters: neutralization of the endogenous acid load with base reduced urinary nitrogen loss of women with post-menopausal osteoporosis (Frasetto et al., 1997). Therefore, taken as a whole, studies in both animal and human with either normal or impaired renal function implicate chronic metabolic acidosis as a potential factor in the development of muscle wasting.

C. Reduced Albumin Synthesis

A low serum albumin concentration is a strong predictor of poor survival in hospitalized patients in general and in those on chronic maintenance hemodialysis in particular (Owen et al., 1993). Subjects with normal renal function and ammonium chloride-induced metabolic acidosis had a reduction in albumin synthesis (Ballmer et al., 1995). A direct relationship between serum HCO_3^- and hypoalbuminemia was found in NHANES study (Eustace et al., 2004): the odds of developing hypoalbuminemia rose substantially when serum bicarbonate concentration was <22 mEq/l. Also, normalization of acid–base balance in patients with renal failure both prior to and after initiation of chronic maintenance dialysis resulted in an increase in serum albumin concentration and decrement in protein catabolic rate (Movilli et al., 1998; Verove et al., 2002). Whether any benefit will accrue from partial normalization of acid–base parameters is not clear since, in one study, serum albumin did not increase when serum bicarbonate concentration was not completely normalized (Brady and Hasbargen, 1998).

The mechanism(s) underlying the reduction in serum albumin with metabolic acidosis are not clear. A decrease in protein intake caused by metabolic acidosis did not appear to be a major contributor, as there was no change in protein intake in patients with CKD and metabolic acidosis after correction of the metabolic acidosis with base (Roberts et al., 1996). Rather, several studies had indicated that metabolic acidosis affects protein metabolism by reducing protein synthesis and increasing protein breakdown and amino acid oxidation (Reaich et al., 1993; Price et al., 1998). The importance of this effect can be surmised from studies that demonstrate that moderate metabolic acidosis, plasma $HCO_3^- \leq 15$ mEq/l, can result in the loss of $\approx$30 grams of protein per day (Ballmer et al., 1995).

D. Exacerbation of Renal Failure

Intraglomerular hypertension has been shown to be the major factor in the progression of renal failure (Tonelli et al., 2002; Klahr and Morrissey, 2003). Metabolic acidosis has been identified by some as an additional important factor (Nath et al., 1985; Gadola et al., 2004). Metabolic acidosis in rats causes increased proteinuria, exacerbation of tubulointerstitial injury and a more rapid decline in renal function acceleration of the progression of renal failure (Nath et al., 1985; Gadola et al., 2004). Correction of the metabolic acidosis by the administration of sodium bicarbonate or sodium citrate lessened the degree of injury (Nath et al., 1985; Gadola et al., 2004). Also, metabolic acidosis promoted cyst enlargement, interstitial fibrosis and exacerbation of renal insufficiency in animal models of polycystic kidney disease (Torres et al., 2001).

Two factors suggested as possible reasons for the worsening of renal injury include activation of the alternative complement pathway by ammonia with generation of inflammatory mediators (Nath et al., 1985) and alkalinization of the interstitium as bicarbonate exits the tubules and enters the interstitium possibly enhancing calcium phosphorus deposition (Halperin et al., 1989). Another possibility includes an augmenting effect of metabolic acidosis on the renal actions of angiotensin II. Nagami (Nagami, 2002) has demonstrated that stimulation of ammoniogenesis by angiotensin II, one of the effects of angiotensin II, is enhanced in the presence of an acid load.

Nevertheless, other studies in rats did not confirm the untoward effect of acidosis on the progression of kidney disease. Rats with renal failure, produced using a 5/6 nephrectomy model, manifested no aggravation of proteinuria, tubular interstitial injury or acceleration of the rate of progression of renal failure when compared to a similar group given base (Throssel et al., 1995). Furthermore, in rats with renal failure receiving a high-phosphate diet, the presence of metabolic acidosis actually lessened the rate of progression of renal failure (Jara et al., 2000), presumably by reducing the extent of calcium phosphorus precipitation in the kidney (Jara et al., 2004).

In humans, administration of oral bicarbonate to patients with mild to moderate renal failure led to reduced peptide catabolism and a fall in markers of tubular damage (Rustom et al., 1998), although there was no substantial impact on renal function. A recent randomized control study of 129 patients with stage 4 and 5 CKD and metabolic acidosis revealed that normalization of the acidosis by administration of base for 2 years slowed the progression of renal failure and prevented the development of end-stage renal failure (Ashurst et al., 2006). The mechanism(s) underlying this effect was not explored. Thus, despite conflicting evidence in animals and humans, there remains the possibility that metabolic acidosis could hasten the decline in renal function. This is important as administration of base could predispose to interstitial calcifications in the kidney and other tissues. Therefore, further study of this issue remains of great importance.

E. Exacerbation or Development of Cardiac Disease

Cardiac disease is the most important cause of death in chronic dialysis patients and prominent cardiac abnormalities are present at the time of inception of chronic maintenance dialysis in the majority of patients (McMahon et al., 2004). A retrospective analysis of laboratory data obtained from more than 12 000 hemodialysis patients showed an increased risk of death when plasma bicarbonate concentration was less than 15–17 mEq/l (Lowrie and Lew, 1990). Although not proven, the increased mortality was attributed to an increased presence of cardiovascular disease. A link between metabolic acidosis and cardiovascular disease in patients with CKD not on dialysis has not been explored.

Metabolic acidosis can induce apoptosis in cardiac cells (Thatte et al., 2004) and stimulate pro-inflammatory cytokines, both processes can potentially contribute to development of cardiomyopathy or ischemic heart disease. Elucidation of the pathophysiologic link between metabolic acidosis and cardiac disease is important, since normalization of acid–base parameters by administration of base has the potential actually to exacerbate cardiac disease (see below).

F. Impaired Glucose Homeostasis and Lipid Metabolism

Acute studies done *in vitro* and *in vivo* have demonstrated that an acidic milieu impairs insulin sensitivity (Cuthbert and Alberti, 1978; Whittaker et al., 1981, 1982), thereby impairing normal glucose regulatory mechanisms. The reduction in insulin sensitivity has been shown to be due to both reduced binding of insulin to its receptors and attenuation of post receptor activation of effector proteins (Whittaker et al., 1981). Further, normalization of plasma bicarbonate concentration improved insulin sensitivity of patients with CKD both prior to and after initiation of chronic maintenance dialysis (Reaich et al., 1995; Mak, 1998a). This impairment in insulin sensitivity could contribute to the development of vascular disease in dialysis patients and the insulin resistance syndrome.

Abnormalities in lipid metabolism and cholesterol metabolism are not uncommon in patients with CKD. This has been attributed in part to the high prevalence of diabetes as an underlying cause of the CKD. A possible role for metabolic acidosis in the hyperlipidemia was inferred from studies of nine patients with ESRD treated with bicarbonate to correct the acidosis (Mak, 1999). Correction of the acidosis led to a significant reduction in serum triglyceride concentrations without any alteration in cholesterol levels.

G. Leptin

Leptin is a hormone involved in regulation of appetite and could be involved in regulation of nutritional status in patients with CKD (Kalantar-Zadeh et al., 2004). In animals with renal failure, treatment with sodium bicarbonate causes a reduction in serum leptin levels (Teta et al., 1999). By contrast, in 25 patients with CKD, correction of metabolic acidosis caused an increase in serum leptin levels (Zheng et al., 2001). Larger interventional studies will be necessary to elucidate the interaction of metabolic acidosis and leptin and the impact of these changes on the nutritional status of patients with CKD.

H. Accumulation of β2-Microglobulin

One of the striking complications of CKD is the accumulation of β2-microglobulin in several tissues resulting in the development of amyloidosis. This amyloid accumulation may be the major cause of carpal tunnel syndrome, produce cysts in the bone and even cause collapse of cervical vertebrae (Sonikian et al., 1996). In rare cases, the amyloid material can be distributed in the heart producing various cardiac manifestations. The primary factors determining the accumulation of amyloid material appear to be the number of years the patient has been on dialysis and the type of dialyzer utilized (Sonikian et al., 1996).

Recently, metabolic acidosis has been suggested as an additional factor in promoting microglobulin accumulation based on studies showing an inverse correlation between plasma bicarbonate concentration and β2-microglobulin levels in renal failure (Sonikian et al., 1996) and higher β2-microglobulin levels in patients dialyzed with acetate than those dialyzed with bicarbonate. The higher values in the acetate group were attributed to lower plasma bicarbonate concentrations (Sonikian et al., 1996). However, it is possible that dialysis with acetate has other effects on β2-microglobulin levels. Thus, the actual contribution of metabolic acidosis to β2-microglobulin accumulation remains unclear.

I. Growth Hormone and Thyroid Function

Progressive renal failure is associated with alterations in growth and basal metabolic rate (Mak and Cheung, 2006). Initially, these abnormalities were ascribed to accumulation of uremic toxins, however, based on studies in patients with normal and impaired renal function, it has been suggested that metabolic acidosis might also contribute (Brungger et al., 1997b; Wiederkehr et al., 2004). Thus, production of metabolic acidosis by ammonium chloride administration causes reduced T3 and T4 levels and elevated thyroid stimulating hormone (TSH) levels, findings consistent with hypothyroidism (Brungger et al., 1997b). In patients with CKD or ESRD maintained on chronic hemodialysis, there are low plasma T3 and T4 levels with normal TSH and blunted response to TRH (Lim et al., 1977). Correction of the metabolic acidosis alone raised T3 levels (Wiederkehr et al., 2004) and presumably in basal metabolic rate.

Metabolic acidosis in rats lowers serum insulin-like growth factor 1 and 2 levels and IGF receptor mRNA levels (Challa et al., 1993; Kuemmerle et al., 1997). In patients with chronic metabolic acidosis and renal disease (IGF-1) levels are low (Brungger et al., 1997a) and appropriate growth hormone (GH)-induced increments in IGF-1 are blunted. Correction of metabolic acidosis by administration of sodium acetate for 4 weeks in hemodialysis patients improved the attenuated IGF-1 response to growth hormone administration (Wiederkehr et al., 2004). These data indicate that metabolic acidosis contributes to the abnormalities in the growth hormone axis noted with CKD.

J. Inflammatory Response

CKD is considered a chronic inflammatory state (Kalantar-Zadeh et al., 2004; Mak et al., 2006; Wu et al., 2006) and this contributes to muscle wasting, cachexia and, possibly, cardiovascular disease (Mak and Cheung, 2007). Metabolic acidosis may contribute to this inflammatory state, although there is some controversy about the extent of its role (Lin et al., 2002; Pickering et al., 2002; Wiederkehr et al., 2004; Kalantar-Zadeh et al., 2004). Exposure of macrophages to an acidic pH increased the production of TNFα and to upregulation of nitric oxide activity (Bellocq et al., 1998). In accord with this *in vitro* observation, correction of metabolic acidosis in chronic ambulatory peritoneal patients was associated with a reduction in tumor necrosis factor alpha levels (Pickering et al., 2002). On the other hand, no difference was found in markers of inflammation including C-reactive protein and interleukin-6 in hemodialysis patients with mild metabolic acidosis compared to those with normal acid–base values (Lin et al., 2002). Similarly, correction of metabolic acidosis with sodium citrate in hemodialysis patients had no significant effect on cytokines including interleukin 1-β and tumor necrosis factor nor acute phase reactants (Kalantar-Zadeh et al., 2004). These data involve small numbers of patients and larger studies are needed. If metabolic acidosis is a critical factor promoting inflammation, its correction will take on added importance, since chronic inflammation might be integral to many of the complications of CKD.

In summary, chronic metabolic acidosis is associated with several metabolic derangements and may contribute to the development of bone disease, muscle wasting, hypoalbuminemia, cachexia and, possibly, cardiovascular disease in CKD. It also might be a factor in inducing a chronic inflammatory state and in accelerating the progression of renal disease. Further studies to determine the magnitude and duration of the metabolic acidosis required for the full expression of these abnormalities and the interaction of metabolic acidosis with the other metabolic derangements accompanying the course of chronic progressive renal failure will be of value in delineating potential targets for therapy.

XI. TREATMENT OF THE METABOLIC ACIDOSIS OF CKD

The effect of improvement of acid–base parameters on cellular function has been examined in patients with CKD prior to and after initiation of chronic maintenance dialysis. The metabolic acidosis has been corrected by administration of base and/or increase in the dialysate base concentration. The impact of partial and complete normalization of acid–base balance has been determined. Taken collectively, the results of these studies indicate even partial normalization of acid–base parameters can improve bone disease in adults and pediatric patients, accelerate growth in children, reduce muscle wasting and improve nitrogen balance, improve glucose tolerance and normalize growth hormone and thyroid secretion (McSherry and Morris, 1978; Lefebvre et al., 1989; Reaich et al., 1995; Movilli et al., 1998; Mitch and Price, 2001; Wiederkehr et al., 2004). Controlled studies comparing partial and complete normalization have not been done, although in one study complete normalization was better than partial improvement (Brady and Hasbargen, 1998). Also, as noted previously, even the small acid load resulting from metabolism of ingested foodstuffs can have a deleterious impact on bone and muscle metabolism, even when plasma bicarbonate concentrations are within the normal range prior to administration of base (Sebastian et al., 1994; Frasetto et al., 1997). The results of these studies would argue for complete normalization of acid–base parameters in patients with CKD of any degree to prevent osteoporosis and muscle disease associated with aging.

On the other hand, complete normalization of acid–base parameters could increase the rate of calcifications in vessels and other tissues of the body. Theoretically, normalization of plasma bicarbonate concentration might promote metastatic calcification by decreasing the solubility of calcium phosphate, although studies done in dialysis patients in whom pH rose into the alkalemic range revealed no biochemical evidence of factors that would favor calcium deposition (Harris et al., 1995). Kirschbaum (Kirschbaum, 2004) also found no increase in the calculated concentration product ratio to favor calcifications, he suggested that if serum phosphorus concentration would rebound a few hours later, this ratio might be elevated enough to favor calcifications. Importantly, however, there have been no studies examining the prevalence of vascular or soft tissue calcification in patients with chronic renal failure stratified to various blood pH levels. Other potential complications of oral administration of base to patients include volume overload, congestive heart failure and exacerbation of pre-existing hypertension. These complications are related to sodium retention and can easily be avoided by either concomitant administration of diuretics or the use of calcium carbonate or citrate to provide the base. Moreover, sodium retention is less when the sodium is given as a non-chloride containing salt (Husted and Nolph, 1977). Oral sodium bicarbonate itself may not be well tolerated because of the generation of CO_2 in the gastrointestinal tract. Therefore, Shohl's solution (sodium citrate) is preferable since citrate is metabolized to form bicarbonate and CO_2 intracellularly. In using citrate, one must be cognizant of its ability to enhance aluminum absorption from the GI tract and cause acute and potentially irreversible aluminum toxicity (Drueke, 2004). Therefore, in CKD, citrate should not be given with aluminum-containing compounds such as antacids or aluminum-containing phosphate binders (Drueke, 2004). Since a portion of the carbonate or acetate administered along with calcium as a phosphate binder is absorbed, the quantity of exogenous base administered may be reduced in patients receiving calcium acetate or carbonate.

At present, the National Kidney Foundation Dialysis Outcome Quality Initiative (NKF/DOQI) recommend plasma bicarbonate be raised to 22 mEq/l, but not to completely normal values (Kraut and Kurtz, 2005). The CARI guidelines (Care of Australians with renal disease) has recommended raising plasma bicarbonate concentrations in patients with CKD and those on maintenance dialysis to 23–24 mEq/l (Baratt et al., 2004). As noted previously, pre-dialysis plasma bicarbonate concentration can be normalized in the majority of hemodialysis patients more effectively using a dialysate containing 39–40 mEq/l of base than one containing 35 mEq/l of base (Harris et al., 1995). In the small number of patients in which this is not possible, the addition of oral base can be effective.

In CAPD patients, acid–base parameters can be maintained within the normal range with the conventional 35 mEq/l lactate based dialysate in the majority of patients. However, some studies suggest better correction of acidosis with the 25 mEq/l bicarbonate/15 mEq/l lactate dialysate (Carrasco et al., 2001). As with hemodialysis patients, addition of oral base might be useful in maintaining normal acid–base parameters in those with persistent acidosis.

The recommendations by the National Kidney Foundation to limit daily calcium intake has resulted in a shift to greater use of non-calcium containing phosphate binders such as Renagel (sevelamer hydrochloride). A fall in serum bicarbonate concentration has been detected in studies of patients taking sevelamer which has been ascribed to an increased acid load and increased base loss due to the binding properties of sevelamer hydrochloride, but the precise mechanisms are under study (Brezina et al., 2004). Therefore, when this agent is utilized, base requirements might need to be modified.

As noted above, some patients with CKD may manifest bicarbonate wasting and its presence can substantially increase bicarbonate requirements. Therefore, if the plasma bicarbonate concentration does not rise as predicted after administration of base, urine pH should be determined to detect bicarbonate wasting and the quantity of base administered altered accordingly.

In summary, recognition of the metabolic acidosis associated with CKD is important, since this complication can have a large impact on cellular function and increase morbidity and possibly mortality. At present, there remains no consensus of whether partial or complete normalization of plasma bicarbonate concentration is necessary to eliminate completely the complications of metabolic acidosis. Furthermore, only a few studies have addressed the potential complications of base therapy and further information in this regard will be of value.

At present base therapy should still be given after assessment of the clinical characteristics of each patient. Factors that should be taken into consideration include but are not limited to:

1. the level of calcium and phosphorus control
2. presence of atherosclerotic vascular disease
3. protein intake
4. degree of muscle wasting
5. presence of co-morbid conditions
6. stage of renal failure.

Further studies to examine the risks and benefit of base therapy in various subsets of patients are clearly indicated to guide physicians in providing individualized treatment.

References

Alpern, R., and Shakhaee, K. (1997). The clinical spectrum of chronic metabolic acidosis:homeostatic mechanisms produce significant morbidity. *Am J Kid Dis* **29**: 291–302.

Ashurst, I., Varagunam, M., and Magdi, M. (2006). A randomized control trial to study the effect of bicarbonate supplementation on the rate of progression of renal failure and nutritional status in chronic kidney disease stage 4 & 5. *J Am Soc Nephrol* **42**: 221.

Bailey, J. L., and Mitch, W. E. (2000). Mechanisms of protein degradation: what do the rat studies tell us. *J Nephrol* **13**: 89–95.

Bailey, J. L., England, B. K., Long, R. C., and Mitch, W. E. (1996). Influence of acid loading, extracellular pH and uremia on intracellular pH in muscle. *Miner Electrol Metab* **22**: 66–68.

Ballmer, P. E., McNurlan, M. A., Hulter, H. N., Anderson, S. E., Garlick, P. J., and Krapf, R. (1995). Chronic metabolic acidosis decreases albumin synthesis and induces negative nitrogen balance in humans. *J Clin Invest* **95**: 39–45.

Baratt, L., Elder, G., Healy, H. et al. (2004). The Cari Guidelines for biochemical and haematological targets in chronic kidney disease. Acidosis: target bicarbonate levels. Available at http://www.kidney.org.au/cari/draft/biochem.html.

Barzel, U. S., and Jowsey, J. (1969). The effect of chronic acid and alkali administration on bone turnover in adult rats. *Clin Sci* **36**: 517–524.

Battle, D. C. (1986). Segmental characterization of defects in collecting tubule acidification. *Kidney Int* **30**: 546–553.

Battle, D. C., Arruda, J. A. L., and Kurtzman, N. A. (1981). Hyperkalemic distal renal tubular acidosis associated with obstructive uropathy. *N Engl J Med* **304**: 373–380.

Battle, D. C., Itsarayoungyuen, K., Arruda, J. A. L., and Kurtzman, N. A. (1982). Hyperkalemic hyperchloremic metabolic-acidosis in sickle-cell hemoglobinopathies. *Am J Med* **72**: 188–192.

Bellocq, A., Suberville, S., Philippe, C. et al. (1998). Low environmental pH is responsible for the induction of nitric-oxide synthase in macrophages – evidence for involvement of nuclear factor-kappa B activation. *J Biol Chem* **273**: 5086–5092.

Bichara, M., Mercier, O., Paillard, M., and Leviel, F. (1986). Effects of parathyroid hormone on urinary acidification. *Am J Physiol* **251**: F444–F453.

Bichara, M., Mercier, O., Houillier, P., Paillard, M., and Leviel, F. (1987). Effects of antidiuretic hormone on urinary acidification and on tubular handling of bicarbonate in the rat. *J Clin Invest* **80**: 621–630.

Bichara, M., Mercier, O., Borensztein, P., and Paillard, M. (1990). Acute metabolic acidosis enhances circulating parathyroid hormone, which contributes to the renal response against acidosis in the rat. *J Clin Invest* **86**: 430–443.

Bloch, R. D., Zikos, D., Fisher, K. A. et al. (1992). Activation of proximal tubular Na^+-H^+ exchange by angiotensin II. *Am J Physiol* **263**: F135–F142.

Borensztein, P., Juvin, P., Vernimmen, C., Poggioli, J., Paillard, M., and Bichara, M. (1993). Camp-dependent control of Na^+/H^+ antiport by Avp, Pth, and Pge2 in rat medullary thick ascending limb cells. *Am J Physiol* **264**: F354–F364.

Brady, J. P., and Hasbargen, J. A. (1998). Correction of metabolic acidosis and its effect on albumin in chronic hemodialysis patients. *Am J Kidney Dis* **31**: 35–40.

Brezina, B., Qunibi, W. Y., and Nolan, C. R. (2004). Acid loading during treatment with sevelamer hydrochloride: mechanisms and clinical implications. *Kidney Int* **66**: S39–45.

Brungger, M., Hulter, H. N., and Krapf, R. (1997a). Effect of chronic metabolic acidosis on the growth hormone IGF-1 endocrine axis: new cause of growth hormone insensitivity in humans. *Kidney Int* **51**: 216–221.

Brungger, M., Hulter, H. N., and Krapf, R. (1997b). Effect of chronic metabolic acidosis on thyroid hormone homeostasis in humans. *Am J Physiol* **272**: F648–F653.

Burnell, J. M. (1971). Changes in bone sodium and carbonate in metabolic acidosis and alkalosis in dog. *J Clin Invest* **50**: 327–331.

Burnell, J. M., Teubner, E., Wergedal, J. E., and Sherrard, D. J. (1974). Bone crystal maturation in renal osteodystrophy in humans. *J Clin Invest* **53**: 52–58.

Bushinsky, D. A. (1999). Nephrology forum: the contribution of acidosis to renal osteodystrophy. *Kidney Int* **47**: 1816–1832.

Canaud, B., Bose, J. Y., Leray, H. et al. (1998). On-line haemodiafiltration: state of the art. *Nephrol Dial Transplant* **13**: 3–11.

Caravaca, F., Arrobas, M., Pizarro, J. L., and Esparrago, J. F. (1999). Metabolic acidosis in advanced renal failure: differences between diabetic and nondiabetic patients. *Am J Kidney Dis* **33**: 892–898.

Carrasco, A. M., Rubio, M. A. B., Tomero, J. A. S. et al. (2001). Acidosis correction with a new 25 mmol/l bicarbonate/15 mmol/l lactate peritoneal dialysis solution. *Peritoneal Dial Int* **21**: 546–553.

Challa, A., Chan, W., Krieg, R. J. J. et al. (1993). Effect of metabolic acidosis on the expression of insulin-like growth factor and growth hormone receptor. *Kidney Int* **44**: 1224–1227.

Chan, Y. L., Sardie, E., Mason, R. S., and Posen, S. (1985). The effect of metabolic acidosis on vitamin D metabolism and bone histology in uremic rats. *Calcif Tiss Int* **37**: 158–164.

Chauveau, P., Fouque, D., Combe, C. et al. (2000). Acidosis and nutritional status in hemodialyzed patients. *Sem Dial* **13**: 241–246.

Chobanian, M. C., and Julin, C. M. (1991). Angiotensin II stimulates ammoniagenesis in canine renal proximal tubule cells. *Am J Physiol* **260**: F19–26.

Chobanian, M. C., Julin, C. M., Molteni, K. H., and Brazy, P. C. (1992). Growth hormone regulates ammoniagenesis in canine renal proximal tubule segments. *Am J Physiol* **262**: F878–F884.

Clase, C. M., Garg, A. X., and Kiberd, B. A. (2002). Prevalence of low glomerular filtration rate in nondiabetic Americans: Third National Health and Nutrition Examination Survey (NHANES III). *J Am Soc Nephrol* **13**: 1338–1349.

Cochran, M., and Wilkinson, R. (1975). Effect of correction of metabolic acidosis on bone mineralization rates in patients with renal osteomalacia. *Nephron* **15**: 98–110.

Coe, F. L., Firpo, D. J., Hollandsworth, L. et al. (1999). Effect of acute and chronic metabolic acidosis on serum immunoreactive parathyroid hormone in man. *Kidney Int* **8**: 262–273.

Coen, G., Manni, M., Addari, O. et al. (1995). Metabolic acidosis and osteodystrophic bone disease in predialysis chronic renal failure:effect of calcitriol treatment. *Miner Electrolyte Metab* **21**: 375–382.

Coen, G., Mazzaferro, S., and Ballanti, P. (1996). Renal bone disease in 76 patients with varying degrees of predialysis chronic renal failure:a cross sectional study. *Nephrol Dial Transplant* **11**: 813–819.

Cohen, J. J., Hulter, H. N., Smithline, N., Melby, J. C., and Schwartz, W. B. (1976). Critical role of adrenal gland in renal regulation of acid-base equilibrium during chronic hypotonic expansion. *J Clin Invest* **58**: 1201–1208.

Cohn, D. E., Hruska, K. A., Klahr, S., and Hammerman, M. R. (1982). Increased Na^+-H^+ exchange in brush border vesicles from dogs with renal failure. *Am J Physiol* **243**: F293–F299.

Coles, G. A. (1972). Body composition in chronic renal failure. *Q J Med* **41**: 25–47.

Coresh, J., Astor, B. C., Greene, T., Eknoyan, G., and Levey, A. S. (2003). Prevalence of chronic kidney disease and decreased kidney function in the adult US population: Third National Health and Nutrition Examination Survey. *Am J Kidney Dis* **41**: 1–12.

Cuthbert, C., and Alberti, K. G. (1978). Acidemia and insulin resistance in diabetic ketoacidotic rat. *Metabolism* **27**: 1903–1916.

Dass, P. D., and Kurtz, I. (1990). Renal ammonia and bicarbonate production in chronic renal failure. *Miner Electrol Metab* **16**: 308–314.

De Santo, N. G., Capasso, G., Malnic, G., Anastasio, P., Spitali, L., and D'Angelo, A. (1997). Effect of an acute oral protein load on renal acidification in healthy humans and in patients with chronic renal failure. *J Am Soc Nephrol* **8**: 784–792.

Defronzo, R. A., Cooke, C. R., Andres, R., Faloona, G. R., and Davis, P. J. (1975). Effect of insulin on renal handling of sodium, potassium, calcium, and phosphate in man. *J Clin Invest* **55**: 845–855.

Deshmukh, S., Phillips, B. G., O'Dorisio, T., Flanigan, M. J., and Lim, V. S. (2005). Hormonal responses to fasting and refeeding in chronic renal failure patients. *Am J Physiol Endocrinol Metab* **288**: E47–55.

Disthabanchong, S., Domrongkitchaiporn, S., Sirikulchayanont, V., and Rajatanvin, R. (2003). Effect of alkaline therapy on bone matrix protein composition in distal renal tubular acidosis. *J Am Soc Nephrol* **14**: 894A.

Domrongkitchaiporn, S., Pongskul, C., Sirikulchayanonta, V. et al. (2002). Bone histology and bone mineral density after correction of acidosis in distal renal tubular acidosis. *Kidney Int* **62**: 2160–2166.

Drueke, T. (2004). Intestinal absorption of aluminum in renal failure. *Nephrol Dial Transplant* **17**: 13–16.

Eiam-Ong, S., Hilden, S. A., Johns, C. A., and Madias, N. E. (1993). Stimulation of basolateral Na^+-HCO_3^- cotransporter by angiotensin II in rabbit renal cortex. *Am J Physiol* **265**: F195–203.

Elkington, J. R. (1962). Hydrogen ion turnover in health and disease. *Ann Intern Med* **57**: 660–684.

Enia, G., Catalano, C., Zoccali, C. et al. (1985). Hyperchloraemia: a non-specific finding in chronic renal failure. *Nephron* **41**: 189–192.

Epstein, M. (2006). Aldosterone blockade: an emerging strategy for abrogating progressive renal disease. *Am J Med* **119**: 912–919.

Eustace, J. A., Astor, B., Muntner, P. M., Ikizler, T. A., and Coresh, J. (2004). Prevalence of acidosis and inflammation and their association with low serum albumin in chronic kidney disease. *Kidney Int* **65**: 1031–1040.

Fabris, A., LaGreca, G., Chiaramonte, S. et al. (1988). The importance of ultrafiltration on acid-base status in a dialysis population. *ASAIO Transact* **34**: 200–201.

Feriani, M. (1998). Behavior of acid-base control with different dialysis schedules. *Nephrol Dial Transplant* **13**: 62–65.

Fraley, D. S., and Adler, S. (1979). Extra renal role for parathyroid hormone in the disposal of acute acid loads in rats and dogs. *J Clin Invest* **63**: 985–997.

Frassetto, L., and Sebastian, A. (1996). Age and systemic acid-base equilibrium: Analysis of published data. *J Gerontol* **51**: B91–B99.

Frassetto, L., Morris, R. C., and Sebastian, A. (1996). Effect of age on blood acid-base composition in adult humans. Role of age-related renal functional decline. *AmJPhysiol* **271**: 1114–1122.

Frassetto, L., Morris, R. C. J., and Sebastian, A. (1997). Potassium bicarbonate reduces urinary nitrogen excretion in postmenopausal women. *J Clin Endocrinol Metab* **82**: 254–259.

Frassetto, L. A., Todd, K. M., Morris, R. C., and Sebastian, A. (1998). Estimation of net endogenous noncarbonic acid production in humans from diet potassium and protein contents. *Am J Clin Nutr* **68**: 576–583.

Gadola, L., Noboa, O., Marquez, M. N. et al. (2004). Calcium citrate ameliorates the progression of chronic renal injury. *Kidney Int* **65**: 1224–1230.

Garg, A. X., Kiberd, B. A., Clark, W. F., Haynes, R. B., and Clase, C. M. (2002). Albuminuria and renal insufficiency prevalence guides population screening: results from the NHANES III. *Kidney Int* **61**: 2165–2175.

Geibel, J., Giebisch, G., and Boron, W. F. (1990). Angiotensin II stimulates both Na^+-H^+ exchange and Na^+/HCO_3^- cotransport in the rabbit proximal tubule. *Proc Natl Acad Sci* **87**: 7917–7920.

Gennari, F. J. (2000). Acid-base balance in dialysis patients. *Sem Dial* **13**: 235–239.

Gerich, J. E., Meyer, C., and Stumvoll, M. W. (2000). Hormonal control of renal and systemic glutamine metabolism. *J Nutrit* **130**: 995S–1001S.

Glaser, N. S., Shirali, A. C., Styne, D. M., and Jones, K. L. (1998). Acid-base homeostasis in children with growth hormone deficiency. *Pediatrics* **102**: 1407–1414.

Goodman, A. D., Lemann, J., Lennon, E. J., and Relman, A. S. (1965). Production, excretion, and net balance of fixed acid in patients with renal acidosis. *J Clin Invest* **44**: 495–506.

Goulding, A., and Irving, R. O. (1974). Acid base studies in chronic renal failure. *Prog Biochem Pharmacol* **9**: 196–205.

Gowda, B., Sar, M., Mu, X., Cidlowski, J., and Welbourne, T. (1996). Coordinate modulation of glucocorticoid receptor and glutaminase gene expression in LLC-PK1-F+ cells. *Am J Physiol* **270**: C825–C831.

Graham, K. A., Hoenich, N. A., Tarbit, M., Ward, M. K., and Goodship, T. H. (1997a). Correction of acidosis in hemodialysis patients increases the sensitivity of the parathyroid glands to calcium. *J Am Soc Nephrol* **8**: 627–631.

Graham, K. A., Reaich, D., Channon, S. M., Downie, S., and Goodship, T. H. (1997b). Correction of acidosis in hemodialysis decreases whole body protein degradation. *J Am Soc Nephrol* **8**: 632–637.

Greiber, S., and Mitch, W. E. (1992). Mechanisms for protein catabolism in uremia – metabolic-acidosis and activation of proteolytic pathways. *Miner Electrol Metab* **18**: 233–236.

Gross, E., and Kurtz, I. (2002). Structural determinants and significance of regulation of electrogenic Na(+)-HCO(3)(−) cotransporter stoichiometry. *Am J Physiol* **283**: 876–887.

Gyorke, Z. S., Sulyok, E., and Guignard, J. P. (1991). Ammonium chloride metabolic acidosis and the activity of renin-angiotensin-aldosterone system in children. *Eur J Pediatr* **150**: 547–549.

Hakim, R. M., and Lazarus, J. M. (1988). Biochemical parameters in chronic renal-failure. *Am J Kidney Dis* **11**: 238–247.

Halperin, M. L., Ethier, J. H., and Kamel, K. S. (1989). Ammonium excretion in chronic metabolic acidosis: benefits and risks. *Am J Kidney Dis* **14**: 267–271.

Halperin, M. L., and Jungas, R. L. (1983). Metabolic production and renal disposal of hydrogen ions. *Kidney Int* **24**: 709–713.

Harris, D. C., Yuill, E., and Chesher, D. W. (1995). Correcting acidosis in hemodialysis:effect on phosphate clearance and calcification risk. *J Am Soc Nephrol* **6**: 1607–1612.

Henger, A., Tutt, P., Riesen, W. F., Hulter, H. N., and Krapf, R. (2000). Acid-base and endocrine effects of aldosterone and angiotensin II inhibition in metabolic acidosis in human patients. *J Lab Clin Med* **136**: 379–389.

Hirano, T., Horigome, A., Oka, K. et al. (1997). Glucocorticoid-resistance in peripheral-blood lymphocytes does not correlate with number or affinity of glucocorticoid-receptors in chronic renal failure patients. *Immunopharmacology* **36**: 57–67.

Hsu, C. Y., and Chertow, G. M. (2002). Elevations of serum phosphorus and potassium in mild to moderate chronic renal insufficiency. *Nephrol Dial Transplant* **17**: 1419–1425.

Hulter, H. N. (1985). Effects and interrelationships of Pth, Ca-2+, vitamin-D, and pi in acid-base homeostasis. *Am J Physiol* **248**: F739–F752.

Husted, F. C., and Nolph, K. D. (1977). $NaHCO_3$ and NaCl tolerance in chronic renal failure. *Clin Nephrol* **7**: 21–25.

Jara, A., Felsenfeld, A. J., Bover, J., and Kleeman, C. R. (2000). Chronic metabolic acidosis in azotemic rats on a high phosphate diet halts the progression of renal disease. *Kidney Int* **58**: 1023–1032.

Jara, A., Chacon, C., Ibaceta, M., Valdivieso, A., and Felsenfeld, A. J. (2004). Effect of ammonium chloride and dietary phosphorus in the azotaemic rat. Part II – kidney hypertrophy and calcium deposition. *Nephrol Dial Transplant* **19**: 1993–1998.

Jehle, S., Zanetti, A., Muser, J., Hulter, H. N., and Krapf, R. (2006). Partial neutralization of the acidogenic western diet with potassium citrate increases bone mass in postmenopausal women with osteopenia. *J Am Soc Nephrol* **17**: 3213–3222.

Johnson, C. W., and Morgan, J. M. (1965). Acidosis: a clue to the etiology of renal failure. *South Med J* **58**: 1513–1516.

Kalantar-Zadeh, K., Mehrotra, R., Fouque, D., and Kopple, J. D. (2004). Metabolic acidosis and malnutrition-inflammation complex syndrome in chronic renal failure. *Sem Dial* **17**: 455–465.

Kim, G. H., Han, J. S., Kim, Y. S., Joo, K. W., Kim, S., and Lee, J. S. (1996). Evaluation of urine acidification by urine anion gap and urine osmolal gap in chronic metabolic acidosis. *Am J Kidney Dis* **27**: 42–47.

Kinsella, J., Cujdik, T., and Sacktor, B. (1984). Na^+-H^+ exchange in isolated renal brush-border membrane vesicles in response to metabolic acidosis. *J Biol Chem* **259**: 13224–13227.

Kirschbaum, B. (2004). Effect of high bicarbonate hemodialysis on ionized calcium and risk of metastatic calcification. *Clin Chim Acta* **343**: 231–236.

Klahr, S., and Morrissey, J. (2003). Progression of chronic renal disease. *Am J Kidney Dis* **41**: S3–7.

Kobayashi, S., Maesato, K., Moriya, H., Ohtake, T., and Ikeda, T. (2005). Insulin resistance in patients with chronic kidney disease. *Am J Kidney Dis* **45**: 275–280.

Kraut, J. A. (1995). The role of metabolic acidosis in the pathogenesis of renal osteodystrophy. *Adv Ren Replace Ther* **2**: 40–51.

Kraut, J. A. (2000). Disturbances of acid-base balance and bone disease in end-stage renal disease. *Sem Dial* **13**: 261–265.

Kraut, J. A., and Coburn, J. (1994). Bone, acid, and osteoporosis. *N Engl J Med* **330**: 1821–1822.

Kraut, J. A., and Kurtz, I. (2005). Metabolic acidosis of CKD: diagnosis, clinical characteristics, and treatment. *Am J Kidney Dis* **45**: 978–993.

Kraut, J. A., and Madias, N. (1996). Treatment of metabolic acidosis in end-stage renal failure: is dialysis with bicarbonate sufficient? *Sem Dial* **9**: 378–380.

Kraut, J. A., and Madias, N. E. (2007). Serum anion gap: its uses and limitations in clinical medicine. *Clin J Am Soc Nephrol* **2**: 162–174.

Kraut, J. A., Mishler, D. R., Singer, F. R., and Goodman, W. G. (1986). The effects of metabolic acidosis on bone formation and bone resorption in the rat. *Kidney Int* **30**: 694–700.

Krieger, N. S., Sessler, N. E., and Bushinsky, D. A. (1992). Acidosis inhibits osteoblastic and stimulates osteoclastic activity in vitro. *Am J Physiol* **262**: F442–F448.

Kuemmerle, N., Krieg, R. J., Latta, K., Challa, A., Hanna, J. D., and Chan, J. C. M. (1997). Growth hormone and insulin-like growth factor in non-uremic acidosis and uremic acidosis. *Kidney Int*, pp. S102–S105.

Kurtz, I., Maher, T., Hulter, H. N., Schambelan, M., and Sebastian, A. (1983). Effect of diet on plasma acid-base composition in normal humans. *Kidney Int* **24**: 670–680.

Kurtz, I., Dass, P. D., and Cramer, S. (1990). The importance of renal ammonia metabolism to whole-body acid-base balance – a reanalysis of the pathophysiology of renal tubular-acidosis. *Miner Electrol Metab* **16**: 331–340.

Lameire, N., and Matthys, E. (1986). Influence of progressive salt restriction on urinary bicarbonate wasting in uremic acidosis. *Am J Kidney Dis* **8**: 151–158.

Lawson, A. M., Chalmers, R. A., and Watts, R. W. E. (1976). Urinary organic-acids in man. 1. Normal patterns. *Clin Chem* **22**: 1283–1287.

Leavey, S. F., and Weitzel, W. F. (2002). Endocrine abnormalities in chronic renal failure. *Endocrinol Metab Clin N Am* **31**: 107–115.

Lefebvre, A., DeVernejoul, M. C., Gveris, J. et al. (1989). Optimal correction of acidosis changes progression of dialysis osteodystrophy. *Kidney Int* **36**: 1112–1118.

Lemann, J., Bushinsky, D. A., and Hamm, L. L. (2003). Bone buffering of acid and base in humans. *Am J Physiol* **285**: F811–F832.

Lennon, E., Lemann, J. J., and Litzow, J. R. (1966). The effects of diet and stool composition on the net external acid balance of normal subjec. *J Clin Invest* **45**: 1601–1607.

Lim, V. S., Fang, V. S., Katz, A. I., and Refetoff, S. (1977). Thyroid dysfunction in chronic rena failure. *J Clin Invest* **60**: 522–534.

Lim, V. S., Yarasheski, K. E., and Flanigan, M. J. (1998). The effect of uraemia, acidosis, and dialysis treatment on protein metabolism: a longitudinal leucine kinetic study. *Nephrol Dial Transplant* **13**: 1723–1730.

Lin, S. H., Lin, Y. F., Chin, H. M., and Wu, C. C. (2002). Must metabolic acidosis be associated with malnutrition in haemodialysed patients? *Nephrol Dial Transplant* **17**: 2006–2010.

Liu, F.-Y., and Cogan, M. G. (1987). Angiotensin II: a potent regulator of acidification in the rat early proximal convoluted tubule. *J Clin Invest* **80**: 272–275.

Lowrie, E. G., and Lew, N. L. (1990). Death risk in hemodialysis patients: the predictive value of commonly measured variables and an evaluation of death rate differences between facilities. *Am J Kidney Dis* **15**: 458–482.

Lubowitz, H., Purkerson, M. L., Rolf, D. B., Weisser, R., and Bricker, N. S. (1971). Effect of nephron loss on proximal tubular bicarbonate reabsorption in the rat. *Am J Physiol* **220**: 457–461.

Madias, N. E., Johns, C. A., and Homer, S. M. (1982). Independence of the acute acid buffering response from endogenous parathyroid hormone. *Am J Physiol* **243**: F141–F149.

Mak, R. H., and Cheung, W. (2006). Energy homeostasis and cachexia in chronic kidney disease. *Pediatr Nephrol* **21**: 1807–1814.

Mak, R. H., Cheung, W., Cone, R. D., and Marks, D. L. (2006). Mechanisms of disease: cytokine and adipokine signaling in uremic cachexia. *Nat Clin Prac Nephrol* **2**: 527–534.

Mak, R. H. K. (1998a). Effect of metabolic acidosis on insulin action and secretion in uremia. *Kidney Int* **54**: 603–607.

Mak, R. H. K. (1998b). Insulin, branched-chain amino acids, and growth failure in uremia. *Pediatr Nephrol* **12**: 637–642.

Mak, R. H. K. (1999). Effect of metabolic acidosis on hyperlipidemia in uremia. *Pediatr Nephrol* **13**: 891–893.

Mak, R. H. K., and Cheung, W. (2007). Cachexia in chronic kidney disease: role of inflammation and neuropeptide signaling. *Curr Opin Nephrol Hypert* **16**: 27–31.

Marangella, M., Vitale, C., Manganaro, M. et al. (1991). Renal handling of citrate in chronic renal insufficiency. *Nephron* **57**: 439–443.

Martin, K. J., Freitag, J. J., Bellorin-Font, E., Conrades, M. B., Klahr, S., and Slatopolsky, E. (1980). The effect of acute acidosis on the uptake of parathyroid hormone and the production of adenosine 3',5'-monophosphate by isolated perfused bone. *Endocrinology* **106**: 1607–1611.

May, R. C., Kelly, R. A., and Mitch, W. E. (1987). Mechanisms for defects in muscle protein metabolism in rats with chronic uremia: the influence of metabolic acidosis. *J Clin Invest* **79**: 1099–2003.

May, R. C., Bailey, J. L., Mitch, W. E., Masud, T., and England, B. K. (1996). Glucocorticoids and acidosis stimulate protein and amino acid catabolism in vivo. *Kidney Int* **49**: 679–683.

McMahon, L. P., Roger, S. D., and Levin, A. (2004). Development, prevention, and potential reversal of left ventricular hypertrophy in chronic kidney disease. *J Am Soc Nephrol* **15**: 1640–1647.

McSherry, E., and Morris, R. C. J. (1978). Attainment and maintenance of normal stature with alkali therapy in infants and children with classic renal tubular acidosis. *J Clin Invest* **61**: 509–527.

Mehrotra, R., Kopple, J. D., and Wolfson, M. (2003). Metabolic acidosis in maintenance dialysis patients: clinical considerations. *Kidney Int* **64**: S13–25.

Mitch, W. E. (1998). Mechanisms causing loss of lean body mass in kidney disease. *Am J Clin Nutr* **67**: 359–366.

Mitch, W. E., and Price, S. R. (2001). Mechanisms activated by kidney disease and the loss of muscle mass. *Am J Kidney Dis* **38**: 1337–1342.

Mitch, W. E., and Price, S. R. (2003). Mechanisms activating proteolysis to cause muscle atrophy in catabolic conditions. *J Renal Nutr* **13**: 149–152.

Mitch, W. E., Medina, R., Grieber, S. et al. (1994). Metabolic-acidosis stimulates muscle protein-degradation by activating the adenosine triphosphate-dependent pathway involving ubiquitin and proteasomes. *J Clin Invest* **93**: 2127–2133.

Moller, B. (1959). The hydrogen ion concentration in arterial blood. *Acta Med Scand* **348**: 1–5.

Morris, R. C. J., Sebastian, A., and McSherry, E. (1972). Renal acidosis. *Kidney Int* **1**: 322–340.

Movilli, E., Zani, R., Carli, O. et al. (1998). Correction of metabolic acidosis increases serum albumin concentrations and decreases kinetically evaluated protein intake in haemodialysis patients: a prospective study. *Nephrol Dial Transplant* **13**: 1719–1722.

Movilli, E., Zani, R., Carli, O. et al. (2001). Direct effect of the correction of acidosis on plasma parathyroid hormone concentrations, calcium and phosphate in hemodialysis patients: a prospective study. *Nephron* **87**: 257–262.

Mujais, S. (2003). Acid-base profile in patient on PD. *Kidney Int* **88**: S26–S36.

Muldowney, F. P., Dohohue, J. F., Carroll, D. V., Powell, D., and Freaney, R. (1972). Parathyroid acidosis in uremia. *Q J Med* **41**: 321–342.

Nagami, G. T. (1992). Effect of angiotensin II on ammonia production and secretion by mouse proximal tubules perfused in vitro. *J Clin Invest* **89**: 925–931.

Nagami, G. T. (1995). Effect of luminal angiotensin II on ammonia production and secretion by mouse proximal tubules. *Am J Physiol* **269**: F86–92.

Nagami, G. T. (2002). Enhanced ammonia secretion by proximal tubule segments from mice receiving NH_4Cl: role of angiotensin II. *Am J Physiol* **282**: F472–F477.

Nath, K. A., Hostetter, M. K., and Hostetter, T. H. (1985). Pathophysiology of chronic tubulointerstitial disease in rats. *J Clin Invest* **76**: 667–675.

Nonoguchi, H., Takayama, M., Owada, A. et al. (1996). Role of urinary arginine vasopressin in the sodium excretion in patients with chronic renal failure. *Am J Med Sci* **312**: 195–201.

Oh, M. S., Uribarri, J., Weinstein, J. et al. (2004). What unique acid-base considerations exist in dialysis patients? *Sem Dial* **17**: 351–354.

Ong, A., and Fine, L. G. (1992). Loss of glomerular function and tubulointerstitial fibrosis: cause or effect? *Kidney Int* **45**: 345–351.

Owen, W. F., Lew, N. L., Liu, Y., Lowrie, E. G., and Lazarus, J. M. (1993). The urea reduction ratio and serum-albumin concentration as predictors of mortality in patients undergoing hemodialysis. *N Engl J Med* **329**: 1001–1006.

Paillard, M., and Bichara, M. (1989). Peptide-hormone effects on urinary acidification and acid-base-balance – Pth, Adh, and glucagon. *Am J Physiol* **256**: F973–F985.

Papadoyannakis, N. J., Stefanidis, C. J., and McGeown, M. (1984). The effect of the correction of metabolic acidosis on nitrogen and protein balance of patients with chronic renal failure. *Am J Clin Nutr* **40**: 623–627.

Perez, G. O., Oster, J. R., Katz, F. H., and Vaamonde, C. A. (1979). The effect of acute metabolic acidosis on plasma control, renin activity and aldosteronea. *Horm Res* **11**: 12–21.

Pickering, W. P., Price, S. R., Bircher, G., Marinovic, A. C., Mitch, W. E., and Walls, J. (2002). Nutrition in CAPD: serum bicarbonate and the ubiquitin-proteasome system in muscle. *Kidney Int* **61**: 1286–1292.

Pitts, R. F. (1965). The renal regulation of acid-base balance with special reference to the mechanism of acidifying the urine. *Science* **102**: 81–85.

Price, S. R., England, B. K., Bailey, J. L., Vanvreede, K., and Mitch, W. E. (1994). Acidosis and glucocorticoids concomitantly increase ubiquitin and proteasome subunit messenger-Rnas in rat muscle. *Am J Physiol* **36**: C955–C960.

Price, S. R., Reaich, D., Marinovic, A. C. et al. (1998). Mechanisms contributing to muscle wasting in acute uremia: activation of amino acid catabolism. *J Am Soc Nephrol* **9**: 439–443.

Ray, S., Piraino, B., Chong, T. K., El-Shahawy, M., and Puschett, J. B. (1990). Acid excretion and serum electrolyte patterns in patients with advanced chronic renal failure. *Miner Electrolyte Metab* **16**: 355–361.

Reaich, D., Channon, S. M., Scrimgeour, C. M., Daley, S. E., Wilkinson, R., and Goodship, T. H. J. (1993). Correction of acidosis in humans with Crf decreases protein degradation and amino acid oxidation. *Am J Physiol* **265**: E230–E235.

Reaich, D., Channon, S. M., Scrimgeour, C. M., and Goodship, T. H. J. (1992). Ammonium chloride-induced acidosis increases protein breakdown and amino acid oxidation in humans. *Am J Physiol* **263**: E735–E739.

Reaich, D., Graham, K. A., Channon, S. M. et al. (1995). Insulin-mediated changes in PD and glucose uptake afte correction of acidosis in humans with CRF. *Am J Physiol* **268**: 121–126.

Relman, A. S. (1964). Renal acidosis and renal excretion of acid in health and disease. *Adv Intern Med* **12**: 295–347.

Relman, A. S., Lennon, E. J., and Lemann, J. J. (1961). Endogenous production of fixed acid and the measurement of the net balance of acid in normal subjects. *J Clin Invest* **40**: 1621.

Remer, T., and Manz, F. (1994). Estimation of the renal net acid excretion by adults consuming diets containing variable amounts of protein. *Am J Clin Nutr* **59**: 1356–1361.

Richter, C. (2006). Role of endothelin in chronic renal failure. *Rheumatology* **45**: 36–38.

Roberts, R. G., Gilmour, E. R., and Goodship, T. H. J. (1996). The correction of acidosis does not increase dietary protein intake in chronic renal failure patients. *Am J Kidney Dis* **28**: 350–353.

Roberts, R. G., Redfern, C. P. F., Graham, K. A., Bartlett, K., Wilkinson, R., and Goodship, T. H. J. (2002). Sodium bicarbonate treatment and ubiquitin gene expression in acidotic human subjects with chronic renal failure. *Eur J Clin Invest* **32**: 488–492.

Rustom, R., Grime, J. S., Costigan, M. et al. (1998). Oral sodium bicarbonate reduces proximal renal tubular peptide catabolism, ammoniogenesis, and tubular damage in renal patients. *Renal Fail* **20**: 371–382.

Sartorius, O. W., Calhoon, D., and Pitts, R. F. (1953). Studies on the interrelationships of the adrenal cortex and renal ammonia excretion by the rat. *Endocrinology* **52**: 256–265.

Schambelan, M., Sebastian, A., and Biglieri, E. G. (1980). Prevalence, pathogenesis, and functional significance of aldosterone deficiency in hyperkalemic patients with chronic renal insufficiency. *Kidney Int* **17**: 89–101.

Schmidt, R. W. (1978). Factors affecting HCO_3^- reabsorption in experimental renal insufficiency. *Am J Physiol* **234**: F472–F479.

Schwartz, W. B., Hall, P. W., Hays, R. M., and Relman, A. S. (1959). On the mechanism of acidosis in chronic renal disease. *J Clin Invest* **49**: 39–52.

Sebastian, A., Frassetto, L. A., Sellmeyer, D. E., Merriam, R. L., and Morris, R. C. (2002). Estimation of the net acid load of the diet of ancestral preagricultural Homo sapiens and their hominid ancestors. *Am J Clin Nutr* **76**: 1308–1316.

Sebastian, A., Harris, S. T., Ottaway, J. H., Todd, K. M., and Morris, R. C. J. (1994). Improved mineral balance and skeletal metabolism in postmenopausal women treated with potassium bicarbonate. *N Engl J Med* **330**: 1776–1781.

Seldin, D. W., Coleman, A. J., Carter, N. W., and Rector, F. C. J. (1967). The effect of Na_2SO_4 on urinary acidification in chronic renal disease. *J Lab Clin Med* **69**: 893–903.

Sicuro, A., Mahlbacher, K., Hulter, H. N., and Krapf, R. (1998). Effect of growth hormone on renal and systemic acid-base homeostasis in humans. *Am J Physiol* **43**: F650–F657.

Simpson, D. P. (1971). Control of hydrogen ion homeostasis and renal acidosis. *Medicine* **50**: 503–541.

Slatopolsky, E., Hoffsten, P., Puerkerson, M., and Bricker, N. S. (1970). On the influence of extracellular fluid volume expansion of uremia on bicarbonate reabsorption in man. *J Clin Invest* **49**: 988–998.

Sonikian, M., Gogusev, J., Zingraff, J. et al. (1996). Potential effect of metabolic acidosis on beta2-microglobulin generation: In vivo and in vitro studies. *J Am Soc Nephrol* **7**: 350–356.

Takahashi, E., Onda, K., Hirano, T. et al. (2002). Expression of c-fos, rather than c-jun or glucocorticoid-receptor mRNA, correlates with decreased glucocorticoid response of peripheral blood mononuclear cells in asthma. *Int Immunopharmacol* **2**: 1419–1427.

Teta, D., Bevington, A., Brown, J., Throssell, D., Harris, K. P. G., and Walls, J. (1999). Effects of acidosis on leptin secretion from 3T3-LI adipocytes and on serum leptin in the uraemic rat. *Clin Sci* **97**: 363–368.

Thatte, H. S., Rhee, J. H., Zagarins, S. E. et al. (2004). Acidosis-induced apoptosis in human and porcine heart. *Ann Thor Surg* **77**: 1376–1383.

Throssel, D., Brown, J., Harris, K. P., and Walls, J. (1995). Metabolic acidosis does not contribute to chronic renal injury in the rat. *Clin Sci* **89**: 643–650.

Tizianello, A., DeFerrari, G., Garibotto, G., Gurreri, G., and Robaudo, C. (1980). Renal metabolism of amino acids and ammonia in subjects with normal renal function and in patients with chronic renal insufficiency. *J Clin Invest* **65**: 1162–1173.

Tizianello, A., DeFerrari, G., Garibotto, G., Robaudo, C., Acquarone, N., and Ghiggeri, G. M. (1982). Renal ammoniagenesis in an early stage of metabolic acidosis in man. *J Clin Invest* **69**: 240–250.

Tonelli, M., Gill, J., Pandeya, S., Bohm, C., Levin, A., and Kiberd, B. A. (2002). Slowing the progression of chronic renal insufficiency. *Can Med Assoc J* **166**: 906–907.

Tonshoff, B., Kiepe, D., and Ciarmatori, S. (2005). Growth hormone/insulin-like growth factor system in children with chronic renal failure. *Pediatr Nephrol* **20**: 279–289.

Torres, V. E., Cowley, B. D. J., Branden, M. G., Yoshida, I., and Gattone, V. H. (2001). Long-term ammonium chloride or sodium bicarbonate treatment in two models of polycystic kidney disease. *Exp Nephrol* **9**: 171–180.

Uribarri, J. (2000). Acidosis in chronic renal insufficiency. *Sem Dial* **13**: 232–239.

Uribarri, J., Buquing, J., and Oh, M. S. (1995a). Acid-base balance in chronic peritoneal dialysis patients. *Kidney Int* **47**: 269–273.

Uribarri, J., Douyon, H., and Oh, M. S. (1995b). A re-evaluation of the urinary parameters of acid production and excretion in patients with chronic renal acidosis. *Kidney Int* **47**: 624–627.

Uribarri, J., Zia, M. M. J., Marcus, R. A., and Oh, M. S. (1998). Acid production in chronic hemodialysis patients. *J Am Soc Nephrol* **9**: 112–120.

Uribarri, J., Levin, N. W., Delmez, J. et al. (1999). Association of acidosis and nutritional parameters in hemodialysis patients. *Am J Kidney Dis* **34**: 493–499.

Vaziri, N. D., Byrne, C., Staten, M., and Charles, A. (1987). Effect of human insulin administration on urinary acidification in patients with insulin-dependent diabetes. *Gen Pharmacol* **18**: 441–443.

Verove, C., Maisonneuve, N., El Azouzi, A., Boldron, A., and Azar, R. (2002). Effect of the correction of metabolic acidosis on nutritional status in elderly patients with chronic renal failure. *J Renal Nutr* **12**: 224–228.

Vinay, P., Prud'Homme, M., Vinet, B. et al. (1987). Acetate metabolism and bicarbonate generation during hemodialysis: 10 years of observation. *Kidney Int* **31**: 1194–1204.

Wagner, C. A. (2007). Metabolic acidosis: new insights from mouse models. *Curr Opin Nephrol Hypert* **16**: 471–476.

Wallia, R., Greenberg, A., Piraino, B., Mitro, R., and Puschett, J. B. (1986). Serum electrolyte patterns in end-stage renal disease. *Am J Kidney Dis* **8**: 98–104.

Welbourne, T. (1988). Role of glucocorticoids in regulating interorgan glutamine flow during chronic metabolic acidosis. *Metabolism* **37**: 520–525.

Welbourne, T. C., and Cronin, M. J. (1991). Growth hormone accelerates tubular acid secretion. *Am J Physiol* **260**: R1036–R1042.

Welbourne, T. C., Weber, M., and Bank, N. (1972). Effect of glutamine administration on urinary ammonium excretion in normal subjects and patients with renal disease. *J Clin Invest* **51**: 1852–1860.

Wesson, D. E. (2001). Endogenous endothelins mediate increased acidification in remnant kidneys. *J Am Soc Nephrol* **12**: 1826–1835.

Wesson, D. E. (2006). Endothelin role in kidney acidification. *Sem Nephrol* **26:** 393–398.

Whittaker, J., Cuthbert, C., Hamond, V., and Alberti, K. G. M. (1981). Impaired insulin binding to isolated adipocytes in experimental diabetic ketoacidosis. *Diabetologia* **21:** 563–568.

Whittaker, J., Cuthbert, C., Hammond, V. A., and Alberti, K. G. M. M. (1982). The Effects of metabolic acidosis in vivo on insulin binding to isolated rat adipocytes. *Metabolism* **31:** 553–557.

Widmer, B., Gerhardt, R. E., Harrington, J. T., and Cohen, J. J. (1979). Serum electrolytes and acid base composition the influence of graded degrees of chronic renal failure. *Arch Intern Med* **139:** 1099–1102.

Wiederkehr, M. R., Kalogiros, J., and Krapf, R. (2004). Correction of metabolic acidosis improves thyroid and growth hormone axes in haemodialysis patients. *Nephrol Dial Transplant* **19:** 1190–1197.

Williams, A. J., Dittmer, I. D., McArley, A., and Clarke, J. (1997). High bicarbonate dialysate in haemodialysis patients:effects on acidosis and nutritional status. *Nephrol Dial Transplant* **12:** 2633–2637.

Williams, A. W., Chebrolu, S. B., Ing, T. S. et al. (2004). Early clinical, quality-of-life, and biochemical changes of 'daily hemodialysis' (6 dialyses per week). *Am J Kidney Dis* **43:** 90–102.

Wu, D. Y., Shinaberger, C. S., Regidor, D. L., McAllister, C. J., Kopple, J. D., and Kalantar-Zadeh, K. (2006). Association between serum bicarbonate and death in hemodialysis patients: Is it better to be acidotic or alkalotic? *Clin J Am Soc Nephrol* **1:** 70–78.

Zheng, F., Qiu, X. X., Yin, S. Y., and Li, Y. (2001). Changes in serum leptin levels in chronic renal failure patients with metabolic acidosis. *J Renal Nutr* **11:** 207–211.

Zies, D. L., Gumz, M. L., Wingo, C. S., and Cain, B. D. (2007). The renal H^+, K^+-ATPases as therapeutic targets. *Exp Opin Therap Targets* **11:** 881–890.

CHAPTER **30**

Pregnancy and the Kidney

CHUN LAM[1] AND S. ANANTH KARUMANCHI[2]
[1]*Merck Research Laboratories, Rahway, NJ, USA*
[2]*Beth Israel Deaconess Medical Center and Harvard Medical School, Boston, MA, USA*

Contents

I.	Introduction	483
II.	Normal pregnancy	483
III.	Pre-eclampsia and HELLP syndrome	485
IV.	Other hypertensive disorders of pregnancy	496
V.	Renal failure in pregnancy	498
	References	502

I. INTRODUCTION

The kidney and the placenta can both be viewed as endocrine organs. Pregnancy represents unique physiology which, in the context of kidney disease, presents challenges for the obstetrician, the nephrologist and the endocrinologist. This chapter will detail the major renal physiological changes during pregnancy, clinical features and pathogenesis of pre-eclampsia, other causes of hypertension in pregnancy and acute and chronic renal failure, as well as renal transplantation in pregnancy. The goal is to present relevant and recent developments in the understanding of mechanisms of renal disease and/or hypertension during pregnancy with the recognition of the multifaceted nature of these conditions.

II. NORMAL PREGNANCY

A. Renal Adaptation

Normal pregnancy is associated with increases in glomerular filtration rate (GFR) of 40–65% and renal plasma flow (RPF) of 50–85% above non-pregnant levels during the first half of gestation (Dunlop, 1981). These increases are a result of reductions in both afferent and efferent arteriolar resistances without glomerular pressure change that have been demonstrated in micropuncture studies (Baylis, 1980). Creatinine clearance has been found to be increased 25% by as early as the second week of gestation (Davison and Noble, 1981). In the final stages of pregnancy, RPF falls, while GFR is generally maintained throughout gestation (Conrad, 2004). The rise in GFR results in a corresponding reduction in plasma concentration of creatinine and blood urea nitrogen (BUN). Average plasma creatinine during pregnancy is about 0.5 mg/dl and BUN 9 mg/dl, respectively, compared to pre-pregnancy average levels of approximately 0.8 and 13 mg/dl, respectively (Sims and Krantz, 1958). This increase in GFR can also lead to the appearance of microalbuminuria in normal pregnancies, while women with pre-existing proteinuric renal disease can expect to have a dramatic increase in proteinuria after the first trimester (Higby et al., 1994; Gordon et al., 1996).

Accompanying the large rise in GFR is an increase in renal size by about 1 cm in length and 30% in volume, as well as dilation of the collecting systems (Bailey and Rolleston, 1971, Rasmussen and Nielsen, 1998). Physiologic hydronephrosis can be seen in up to 90% of pregnancies beginning in the first trimester and resolving after about 1 month postpartum (Peake et al., 1983). The etiology is attributed to the hormones that affect smooth muscle, such as progesterone, but primarily to mechanical compression by the gravid uterus, more pronounced on the right than the left because of dextrorotation of the uterus (Fainaru et al., 2002). Although typically asymptomatic, physiologic hydronephrosis can be a rare cause of acute renal failure in pregnancy that is characterized by abdominal pain and severe hydronephrosis resulting in obstruction (Khanna and Nguyen, 2001). More commonly, the physiologic dilation predisposes women to infection, which can range from asymptomatic bacteriuria to urosepsis (Puskar et al., 2001).

Renal handling of electrolytes is also affected by the rise of GFR. The increase in urate clearance results in a decline of serum uric acid to as low as 2–3 mg/dl by the second trimester (Lind et al., 1984). In the third trimester, increased renal tubular absorption of urate accounts for the return to pre-pregnant serum uric acid concentrations by delivery. By week 8 of gestation, calcium excretion is found to be increased, likely from elevated circulating 1,25 dihydroxyvitamin D_3, but this may be counteracted by a concomitant rise in excreted nephrocalcin, thought to be a crystalluria inhibitor (Heaney and Skillman, 1971; Davison et al., 1993). The overall incidence of nephrolithiasis is not increased compared to non-pregnant women of child-bearing age, which might be attributable to the presence of a urinary inhibitory protein, physiologic dilation and higher urine flow (Gorton and Whitfield, 1997).

Diminished serum bicarbonate levels in pregnant women are commonly seen. This is not caused by a metabolic acidosis, but rather represents compensation for respiratory alkalosis, which is also reflected by a decline in arterial pCO_2. Once compensation is achieved and steady state is again reached, urine pH returns to its usual acidity (Lyons, 1976). Progesterone both increases the sensitivity of the respiratory center to CO_2 and stimulates the respiratory drive directly. As a result, pCO_2 drops to approximately 27–32 mmHg. This allows for a high-normal pO2 to be sustained despite the 20–33% increase in oxygen consumption in pregnancy (Mason et al., 2005).

Women without diabetes may experience mild glycosuria and aminoaciduria in a normal pregnancy because of increased GFR and therefore load of glucose and amino acids, as well as the decreased glucose absorption capacity of the kidneys reabsorption (Buhling et al., 2004).

During the first trimester, plasma osmolality falls as a result of lowering of the osmotic threshold for thirst and secretion of antidiuretic hormone (ADH), shown to be induced by human chorionic gonadotropic hormone (Davison et al., 1990). The ability to excrete a water load is otherwise normal. As a result, mild hyponatremia is frequently observed at levels of 4–5 mEq/L below non-pregnant levels. Studies in rats have more recently suggested that upregulation of aquaporin 2 in the renal papillae may also play a role in the water retention of pregnancy (Ohara et al., 1998).

The kidney in pregnancy preserves the ability efficiently to excrete a sodium load (Chesley et al., 1958). However, sodium reabsorption is increased through activation of the renin–angiotensin–aldosterone system (RAAS), which offsets the additional renal loss of sodium from the rise in GFR and the natriuretic effect of other hormones, such as progesterone and atrial natriuretic peptide (ANP). This leads to a slow, gradual retention of sodium of about 900 mmol by term (Elsheikh et al., 2001). Total body water, distributed among the fetus and the maternal extracellular and interstitial space, increases by 6–8 liters (Davison, 1997). Because the increase in plasma volume is disproportionately greater than the increase in red cell mass, a physiologic fall in the hemoglobin concentration during pregnancy is commonly seen (Sifakis and Pharmakides, 2000).

B. Cardiac and Vascular Adaptation in Pregnancy

One of the earliest physiological adaptations in pregnancy is decreased vascular resistance and increased arterial compliance (Poppas et al., 1997). By as early as 6 weeks of gestation, even before placentation is complete, many of the major hemodynamic changes associated with pregnancy are already well underway: systemic vascular resistance drops, heart rate increases 15–20%, mean arterial blood pressure decreases by about 10 mmHg and cardiac output rises by approximately 20%, peaking at 50% above pre-conception level by the third trimester (Chapman et al., 1998; Desai et al., 2004). As delivery approaches, these changes begin to revert.

The mechanisms underlying the significant hemodynamic changes in early pregnancy are incompletely understood, however, the fall in systemic vascular resistance that precedes full development of the fetal–placental unit appears to stem from the increase in not only uterine blood flow, but also in renal and other extrauterine blood flow. By the end of the first trimester, diminished response to the vasopressors angiotensin II, norepinephrine and vasopressin, as well as increased production of vasodilators such as prostacyclin have been noted. Other hormones that also likely play a role in the systemic decrease in vascular tone include estrogen and progesterone (Gant et al., 1974; Fitzgerald et al., 1987; Baker et al., 1992).

In addition, during the first and second trimesters, as the placenta develops, 'pseudovasculogenesis' takes place, whereby cytotrophoblasts, transforming from an epithelial to endothelial phenotype in the process of invading uterine spiral arteries in order to remodel the pre-pregnancy high-resistance, low-capacitance maternal arteries into large-caliber, high-capacitance vessels in anticipation of the requirements of the developing fetus and placenta (Zhou et al., 2003). In the second and third trimesters, increasing uterine blood flow continues to play a part in persistent low systemic resistance.

Placental vascular development is complex and remains incompletely understood. The impact of angiogenic factors, such as vascular endothelial growth factor (VEGF), placental growth factor (PlGF), soluble fms-like tyrosine kinase 1 (sFlt1) and the angiopoietin receptors Tie-1 and Tie-2 (Kayisli et al., 2006), both systemically and on the development of the placenta have yet to be fully elucidated, but it seems clear that imbalance of these factors can lead to defective placental vascular development, most notably, pre-eclampsia.

In contrast to other conditions of peripheral vasodilation, such as sepsis, cirrhosis and high-output congestive heart failure that are not characterized by any change in renal plasma flow, pregnancy is marked by decreased renal vascular resistance and significantly increased renal plasma flow. The expectation that activation of the RAAS during pregnancy should lead to renal vasoconstriction therefore suggests the presence of a more dominant direct renal vasodilating factor. It has been proposed that the ovarian hormone relaxin plays this role in pregnancy. Relaxin is a 6 kDa peptide hormone isolated in the 1920s initially from pregnant serum and shown to relax the pelvic ligaments (Sherwood, 2005). It is secreted by the corpus luteum of the ovary and its release is stimulated by human chorionic gonadotropin (hCG) (Conrad et al., 2005). Studies in pregnant rats have indicated that relaxin mediates the renal vasodilation, glomerular hyperfiltration and fall in plasma osmolality (P_{osm}) of pregnancy (Novak et al., 2001). The mechanism is thought to be relaxin's upregulation of matrix metalloproteinase 2 which, in turn, promotes cleavage of big endothelin (ET) into ET1-32. This then leads to nitric oxide (NO)-mediated vasodilation via endothelial ET-B receptors (Smith et al., 2006). More recently, relaxin infusions over 5 hours in men and non-pregnant women resulted in a 50% increase in renal plasma flow without affecting GFR or peripheral blood pressure. No fall in P_{osm} was noted, however, scleroderma patients treated with relaxin over several weeks did exhibit a small, but significant drop in P_{osm}.

III. PRE-ECLAMPSIA AND HELLP SYNDROME

Pre-eclampsia is a pregnancy-associated hypertensive syndrome characterized by new-onset hypertension and proteinuria, often diagnosed in the third trimester, and is typically accompanied by edema and hyperuricemia. Clinical management still consists mainly of supportive measures, the only known definitive treatment being delivery of the fetus. The syndrome affects approximately 5% of pregnancies and continues to be a leading cause of maternal and neonatal morbidity and mortality, particularly in the developing world. The primary organs affected may be the liver (HELLP syndrome is the *h*emolysis, *e*levated *l*iver enzymes and *l*ow *p*latelet count syndrome), the brain (eclampsia) and the kidney (proteinuria and glomerular endotheliosis).

In the USA, pre-eclampsia and eclampsia represent 20% of pregnancy-related maternal mortality (MacKay et al., 2001). Worldwide, pre-eclampsia and eclampsia account for 10–15% of the roughly 500 000 women who die annually in childbirth (Duley, 1992; Hill et al., 2001). The most common causes of maternal death are eclampsia, cerebral hemorrhage, renal failure, hepatic failure and the HELLP syndrome. Worldwide, pre-eclampsia is associated with a perinatal and neonatal mortality rate of 10% (Altman et al., 2002). Neonatal death is usually caused by iatrogenic prematurity as a result of early delivery to preserve the health of the mother. The earlier pre-eclampsia presents in gestation, the higher the risk of neonatal mortality. Impaired uteroplacental blood flow or placental infarction can lead to fetal growth restriction. Less frequently seen are oligohydramnios and placental abruption. Despite significant recent advances in the understanding of its pathogenetic mechanisms, the pathophysiology of pre-eclampsia remains incompletely understood.

A. Epidemiology and Risk Factors

Pre-eclampsia has been described as a 'disease of first pregnancies' and its incidence is highest among nulliparous women, who account for roughly 75% of cases of pre-eclampsia (Sibai et al., 1995). Numerous risk factors for the development of pre-eclampsia have been identified. These include certain medical conditions, such as chronic hypertension, diabetes mellitus, renal disease and hypercoagulable states, as well as settings of increased placental mass, such as molar and multiple gestation pregnancies. Women with a prior history of pre-eclampsia are also at increased risk. Others that have been explored include genetic, nutritional and environmental risk factors. Although pre-eclampsia is more common in first pregnancies, multigravidas who are pregnant with a new partner are also at a similarly elevated risk (Tubbergen et al., 1999; Tuffnell et al., 2005). Once thought to be related to immunoprotection by exposure to paternal antigens, it is now recognized that interpregnancy interval is likely the determinant of increased risk in this group (Skjaerven et al., 2002).

Most cases occur without a family history; however, women who do have a first-degree relative with pre-eclampsia are at fourfold increased risk of severe pre-eclampsia, pointing to the influence of genetic factors on a woman's susceptibility to pre-eclampsia (Cincotta and Brennecke, 1998). Both men and women who were products of a pregnancy complicated by pre-eclampsia are also more likely to have a child whose *in-utero* course is complicated by pre-eclampsia (Esplin et al., 2001). Genome-wide scanning of Icelandic, Finnish and Dutch populations have revealed loci on chromosome 2p13, 2p25 and 12q, respectively, the last with a linkage to HELLP syndrome (Arngrimsson et al., 1999; Lachmeijer et al., 2001; Laivuori et al., 2003). A gene on chromosome 13 has also been suggested to raise susceptibility, since women with trisomy 13 fetuses have been found to have a higher incidence of pre-eclampsia (Tuohy and James, 1992). Two case-control studies have demonstrated higher incidence of pre-eclampsia in mothers who carry trisomy 13 fetuses, compared to other trisomies and control pregnant patients (Boyd et al., 1987; Tuohy and James, 1992). However, specific genetic mutations within these loci have not been identified.

Still other suspected risk factors that continue to be debated include infectious etiologies, racial/ethnic factors, thrombophilia and teenage pregnancy. In women with early-onset pre-eclampsia, anti-CMV and anti-chlamydia antibody (IgG) titers have been found to be increased relative to normal controls and women with late-onset pre-eclampsia (von Dadelszen et al., 2003). Parvovirus B19 infection has been reported in association with cases of pre-eclampsia (Yeh et al., 2004). Recently, it was suggested that perhaps periopathogenic bacteria may contribute to the pathogenesis of pre-eclampsia (Barak et al., 2007). On the other hand, certain other viruses have been associated with a lower incidence of pre-eclampsia (Trogstad et al., 2001). The true role of infectious agents in the pathogenesis of pre-eclampsia has not been established.

Some studies have seen a racial disparity in the incidence of pre-eclampsia, namely, a higher rate of pre-eclampsia among Hispanics and African-Americans, although the greater incidence in the latter group has not been borne out by studies conducted in healthy, nulliparous women (Siba et al., 1993; Levine et al., 1997; Wolf et al., 2004). As hypertension is a risk factor for pre-eclampsia, it has been suggested that the higher incidence of pre-eclampsia in black women seen in some studies stems from the increased rate of chronic hypertension in this subgroup (Samadi et al., 1996). In general, African-American women have also been noted to have a higher case-mortality rate, which could be attributed to more severe disease or inadequate prenatal care (MacKay et al., 2001). Interestingly, Hispanic ethnicity, while associated with an increased risk of pre-eclampsia, appeared to have a decreased risk of gestational hypertension (Wolf et al., 2004). Overall, it is difficult to determine racial differences in the incidence and severity of pre-eclampsia because of invariable confounding by socioeconomic and cultural factors.

Conflicting data currently exist on the association of pre-eclampsia with congenital or acquired thrombophilia (Roque et al., 2004; Rasmussen and Ravn, 2004; Stella et al., 2006). It is not surprising that data from the USA from 1979 to 1986 suggest that for every year past age 34, a woman's risk for pre-eclampsia increases, but some studies have also pointed to teenage pregnancy as a risk factor for pre-eclampsia, however, a subsequent meta-analysis and systematic review did not support the latter (Duckitt and Harrington, 2005).

B. Clinical Features and Pathophysiology

Pre-eclampsia is often a diagnostic challenge in the settings of proteinuric renal disease or chronic hypertension. It may be difficult to distinguish from other hypertensive disorders during pregnancy, such as gestational hypertension and chronic hypertension. Guidelines on the diagnosis of pre-eclampsia published by the American College of Obstetrics and Gynecology were last updated in 2002 (ACOG Practice Bulletin, 2002).

1. Hypertension

Hypertension, as one of the criteria for the diagnosis of pre-eclampsia, is defined by the American College of Obstetrics and Gynecology (ACOG) as a systolic blood pressure of 140 mmHg or higher or a diastolic blood pressure of 90 mmHg or higher in two separate measurements at least 2 h apart, in a previously normotensive woman after 20 weeks' gestation (ACOG Practice Bulletin, 2002). Blood pressure elevation in pre-eclampsia can vary considerably from mild, which may be treatable by bed rest alone, to severe, which may be resistant to multiple anti-hypertensive agents. When severe hypertension is accompanied by headache and visual disturbances as well, urgent delivery is indicated, as it may portend eclampsia.

In contrast to normal pregnancies, where peripheral vascular resistance and blood pressure are decreased, pre-eclampsia is marked by increased peripheral vascular resistance, which is the primary cause of the hypertension (Wallenberg, 1998). The rises in peripheral vascular resistance and blood pressure seen in pre-eclampsia are thought to be mediated by a substantial increase in sympathetic vasoconstrictor activity (Schobel et al., 1996). Results from a recent study suggested that this sympathetic over-activity may not be a secondary phenomenon of pre-eclampsia, but rather, a precursor of it (Fischer et al., 2004). An exaggerated response to angiotensin II and other hypertensive stimuli has also been found in pre-eclamptic women (Strauss, 1937; Gant et al., 1973; Gallery and Brown, 1987). In an animal model, an imbalance of expression of AT_1 relative to AT_2 receptors has been observed (Anguiano-Robledo et al., 2007). The hypertension seen in pre-eclampsia is distinctively characterized by suppression, rather than activation, of the renin–angiotensin–aldosterone system (August et al., 1990). Total plasma volume, however, is generally believed to be somewhat decreased (Redman, 1984). This perceived increase in effective circulating blood volume then leads to suppression of renin and aldosterone, as well as brain natriuretic peptide (Tapia et al., 1972; Okuno et al., 1999).

The perturbations of the balance of vasoactive substances are thought to reflect the contribution of endothelial dysfunction to the development of hypertension (reviewed later in the section on maternal endothelial function). Imbalances in these vasoactive substances that are predominantly synthesized by the vascular endothelium, including the vasoconstrictors norepinephrine, endothelin and potentially thromboxane and placental endothelin 1 (ET-1), as well as the vasodilators, such as prostacyclin and possibly nitric oxide, appear to be responsible for the prominent vasoconstriction seen in pre-eclampsia (Clark et al., 1992; Mills et al., 1999; Noris et al., 2004; Fiore et al., 2005). Prostaglandin I_2 (PGI_2, prostacyclin) is a circulating vasodilator produced chiefly by endothelial and smooth muscle cells and is increased in normal pregnancy (Goodman et al., 1982). In women with pre-eclampsia, but not pregnant women with chronic hypertension, production of PGI_2 is reduced before the appearance of hypertension and proteinuria (Fitzgerald et al., 1987; Moutquin et al., 1997; Mills et al., 1999). Thromboxane A_2 (TXA_2) is a potent vasoconstrictor produced by endothelial cells, activated platelets and macrophages. Its metabolites have been reported to be increased in the urine of pre-eclamptics in some studies, though not all (Fitzgerald et al., 1990; Paarlberg et al., 1998; Mills et al., 1999). TXA_2 synthesis was noted to be higher in patients with coagulopathy and marked platelet activation (Paarlberg et al., 1998). Studies examining the effect of aspirin on the incidence of pre-eclampsia via inhibition of platelet TXA_2 production have generally yielded conflicting data (CLASP Collaborative Group, 1994; Caritis et al., 1998).

Nitric oxide (NO) has generated interest because of its putative role in normal pregnancy as a vasodilator (Williams et al., 1996; Conrad et al., 1999). In pregnant rats, its inhibition by NG-nitro-L-arginine methyl ester (L-NAME), an exogenous nitric oxide synthase (NOS) inhibitor, induced some of the clinical characteristics of pre-eclampsia (Danielson et al., 1995). Supplementation of L-arginine reversed the hypertension and proteinuria caused by infusion of L-NAME and decreased the extent of glomerular injury (Helmbrecht et al., 1996). Data in humans have been inconsistent, with some studies demonstrating decreased nitric oxide production and others noting no change or an increase (Seligman et al., 1994; Davidge et al., 1996; Silver et al., 1996; Garmendia et al., 1997; Smarason et al., 1997; Ranta et al., 1999; Shaamash et al., 2000). Although levels of asymmetric dimethyl arginine, an endogenous inhibitor of NOS, are elevated in pre-eclampsia, its very low levels render it technically challenging to study and interpret. In addition, L-arginine supplementation has not been shown to be of significant benefit in pre-eclampsia (Holden et al., 1998; Staff et al., 2004).

Data on ET-1, a potent vasoconstrictor released by vascular endothelial cells in response to injury, have also been conflicting in pre-eclampsia (Clark et al., 1992; Schiff et al., 1992; Battistini and Dussault, 1998; Paarlberg et al., 1998; Slowinski et al., 2002). Some studies have reported increased circulating levels, while others have not. Current evidence suggests that endothelin alterations in pre-eclampsia are a secondary phenomenon (Greenberg et al., 1997; Scalera et al., 2001; Karumanchi et al., 2005).

2. Proteinuria

Although the urine dipstick method is frequently used to screen for proteinuria in routine prenatal monitoring, it has a high rate of false positives and false negatives in comparison to 24-hour urine protein measurement. In the non-obstetric population, the urine-protein-to-creatinine ratio (in units of mg protein per mg creatinine) is commonly used to estimate 24-hour protein excretion. While the urine-protein-to-creatinine ratio is not widely used among obstetricians, a number of studies have also supported use of the urine-protein-to-creatinine ratio in pregnant women, given that it closely approximates the 24-hour urinary protein excretion (Rodriguez-Thompson and Lieberman, 2001; Neithardt et al., 2002; Yamasmit et al., 2004). In these studies, spot urine protein-to-creatinine ratios >0.2 mg/mg are highly (>90%) sensitive for detection of significant (>300 mg) proteinuria by 24 hour collection in pregnant women in the third trimester.

Proteinuria in pre-eclampsia can vary widely, from minimal to nephrotic-range. Although proteinuria exceeding 5 g per day is defined as severe pre-eclampsia according to the ACOG practice guidelines, the degree of proteinuria is not considered an independent risk factor for adverse maternal or neonatal outcomes and, therefore, not by itself an indication for urgent delivery (Hall et al., 2002). The proteinuria of pre-eclampsia is 'non-selective' and thought to stem from loss of glomerular barrier charge selectivity (Moran et al., 2003). Postpartum, proteinuria generally resolves within 7–10 days, although it may persist for 3–6 months.

3. Edema

Edema is often present in normal pregnancies and is not specific to pre-eclampsia, however, sudden weight gain, particularly with edema of the feet, hands and face, is a common presenting symptom of pre-eclampsia. Women with pre-eclampsia given an intravenous saline load have been found to excrete a much smaller percentage than do normal pregnant women (Gallery and Brown, 1987). Unlike the edema of hepatic cirrhosis, congestive heart failure and nephrotic syndrome, in which low effective plasma volume results in activation of the RAAS and renal retention of salt, as mentioned, a primary renal retention of salt and water occurs in pre-eclampsia with suppression of the RAAS. The edema of pre-eclampsia is accompanied by a fall in GFR that is disproportionate to the decline in real plasma flow and, therefore, resembles the 'over-fill' edema of acute glomerulonephritis or of acute ischemic renal failure with volume overload (Schrier, 1998). Other contributing factors to the edema may be generalized increase in capillary permeability and hypoalbuminemia, but they are not unique to pre-eclampsia. Increased endothelial permeability is also seen in non-pregnant patients with heart failure and nephrotic syndrome (Galatius et al., 2000; Rostoker et al., 2000).

4. Uric Acid

A serum uric acid level greater than 5.5 mg/dl is a strong indicator of pre-eclampsia. The elevation of uric acid is attributed chiefly to decreased renal clearance and often precedes the onset of proteinuria and fall in GFR (Gallery and Gyory, 1979). In humans given infusions of vasoconstrictors, similar declines in uric acid clearance have been observed (Ferris and Gorden, 1968). Lowering serum uric acid with probenecid, however, does not appear to have any effect on the blood pressure in women with pre-eclampsia (Schackis, 2004). The serum level of uric acid rises as pre-eclampsia progresses and correlates with its severity, as well as with adverse pregnancy outcome; a level greater than 7.8 mg/dl is associated with significant maternal morbidity (Martin et al., 1999; Thadhani et al., 2005). The degree of uric acid rise also correlates with the severity of proteinuria and renal pathological changes and with fetal demise (Martin et al., 1999). Because the serum uric acid level in women with gestational hypertension similarly correlates with severity of disease and poor pregnancy outcomes, it is of limited clinical utility in distinguishing pre-eclampsia from other hypertensive disorders of pregnancy and/or as a clinical predictor of adverse outcomes (Lim et al., 1998; Lam et al., 2005; Roberts et al., 2005). A recent meta-analysis of data from 18 studies including nearly 4000 women concluded that serum uric acid is a weak predictor of maternal and fetal complications in women with pre-eclampsia (Thangaratinam et al., 2006). Nevertheless, pre-eclampsia superimposed on chronic renal disease may be a clinical setting where serum uric acid could be useful, as the diagnostic criteria of new-onset hypertension and proteinuria may be difficult to apply. In such cases, a serum uric acid level that exceeds 5.5 mg/dl and stable renal function may suggest the diagnosis of pre-eclampsia. Still debated is whether uric acid plays a direct role in the pathogenesis of pre-eclampsia by inducing endothelial dysfunction (Khosla et al., 2005).

5. RENAL CHANGES AND PATHOLOGY

In contrast to normal pregnancy, where GFR and renal plasma flow increase during early and mid-pregnancy, in pre-eclampsia, GFR and renal plasma flow are both decreased. Because of pregnancy's overall effect of augmenting GFR, BUN and serum creatinine often remain in the normal, non-pregnant range in pre-eclampsia despite the latter's significant GFR-lowering effect. Although acute renal failure can be seen in pre-eclampsia, it is more common for only proteinuria and renal sodium and water retention to manifest. Renal filtration fraction is lower in pre-eclamptics than in normal women in the 3rd trimester of pregnancy (Lafayette et al., 1998; Moran et al., 2003). The fall in renal blood flow (RBF) is a consequence of high renal vascular resistance, chiefly from increased afferent arteriolar resistance. GFR declines because of the decreases in RBF and in the ultrafiltration coefficient (Kf), which is attributed to endotheliosis in the glomerular capillary (Moran et al., 2003).

The urinary sediment of pre-eclamptics is usually 'bland', with no or few white and red blood cells, and cellular casts. Recent evidence reveals that 'podocyturia', the urinary excretion of glomerular visceral epithelial cells (podocytes) may be a novel way to distinguish women with pre-eclampsia from non-proteinuric, normotensive pregnant women (Garovic et al., 2007). As a diagnostic, though not necessarily a predictive, marker, this remains to be validated by larger studies.

The unique appearance of glomerular endothelial cells in pre-eclampsia is termed 'glomerular endotheliosis' and describes narrowed glomerular capillary lumen that are typically 'bloodless', enlarged glomeruli with generalized swelling and vacuolization of endothelial cells (Figure 30.1). Glomerular cellularity may be slightly increased and mesangial interposition may occur in severe cases or in the healing stages (Spargo et al., 1959; Pollak and Nettles, 1960). Immunofluorescence may reveal deposits of fibrin and fibrinogen within the endothelial cells, particularly in biopsies done within 2 weeks postpartum (Morris et al., 1964). Electron microscopy shows loss of glomerular endothelial fenestrae, but with relative preservation of the podocyte foot processes. Glomerular subendothelial and occasional

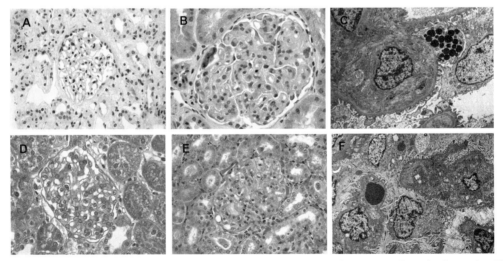

FIGURE 30.1 Glomerular endotheliosis. (A) Normal human glomerulus, H & E. (B) Human pre-eclamptic glomerulus, H & E – 33-year-old woman with twin gestation and severe pre-eclampsia at 26 weeks' gestation with urine protein/creatinine ratio of 26 at the time of biopsy. (C) Electron microscopy of glomerulus of the above patient described in (b). Note occlusion of capillary lumen cytoplasm and expansion of the subendothelial space with some electron-dense material. Podocyte cytoplasms show protein resorption droplets and relatively intact foot processes; original magnification 1500 ×. (D) Control rat glomerulus, H & E – note normal cellularity and open capillary loops. (E) sFlt-1 treated rat, H & E – note occlusion of capillary loops by swollen cytoplasm with minimal increase in cellularity. (F) Electon microscopy of sFlt-1 treated rat – note occlusion of capillary loops by swollen cytoplasm with relative preservation of podocyte foot processes; original magnification 2500 ×. All light micrographs taken at identical original magnification of 40×. This figure was reproduced with permission from Karumanchi et al. (2005). (See color plate section.)

mesangial electron-dense deposits can also be seen (Mautner et al., 1962; Lafayette et al., 1999). In contrast to other nephrotic diseases, where podocytes are damaged early in the disease, the primary focus of injury in pre-eclampsia is the endothelial cell. Other renal histological changes that have been described include atrophy of the macula densa and hyperplasia of the juxtaglomerular apparatus (Govan, 1954). It is now known that mild glomerular endotheliosis can be seen in non-pre-eclamptic pregnant women. Indeed, up to 50% of patients with non-proteinuric pregnancy-induced hypertension exhibit mild glomerular endotheliosis (Nochy et al., 1980; Fisher et al., 1981), perhaps suggesting that pregnancy-induced hypertension may represent an early or mild form of pre-eclampsia, or even a phenomenon that occurs at term in all pregnancies. The glomerular enlargement and endothelial swelling generally resolve within 8 weeks postpartum, along with resolution of the proteinuria and hypertension. However, persistent renal damage can follow pre-eclampsia in the form of focal segmental glomerulosclerosis (FSGS) in 50% or more of cases (Heaton and Turner, 1985; Gaber and Spargo, 1987).

6. Severe Pre-Eclampsia and Eclampsia

Severe pre-eclampsia should prompt a consideration to terminate pregnancy, given the potential life-threatening nature of maternal morbidity. The clinical and laboratory findings that indicate severe disease include: oliguria (less than 500 ml urine in 24 h, typically transient) and, uncommonly, acute renal failure. Pulmonary edema is seen in 2–3% of severe pre-eclampsia and can lead to respiratory failure (Tuffnell et al., 2005). Elevated liver enzymes can occur alone or as part of the HELLP syndrome and may be associated with epigastric or right upper quadrant pain. Persistent headache or visual disturbances can portend seizures (eclampsia), seen in roughly 2% of cases of pre-eclampsia in the USA (Saftlas et al., 1990). Typically, but not invariably, eclampsia occurs in the presence of hypertension and proteinuria. Late postpartum eclampsia is a diagnostic challenge, accounts for up to one-third of cases, occasionally days to weeks after delivery, and is frequently seen by non-obstetricians in the emergency room setting. Magnetic resonance imaging (MRI) or computed tomography (CT) of the head usually reveals vasogenic edema and infarctions in subcortical white matter and adjacent gray matter of the parieto-occipital lobes, however, radiological head imaging is not necessary if the diagnosis is otherwise clear (Sibai, 2005). The cerebral edema of eclampsia predominantly involves the posterior, parieto-occipital lobes and is similar to images described in reversible posterior leukoencephalopathy syndrome (RPLS), a syndrome characterized by headache, altered mental status, convulsions and cortical blindness (Hinchey et al., 1996). Cerebral edema is thought to result primarily from endothelial dysfunction rather than from hypertension, as it appears to parallel markers of endothelial damage, rather than severity of hypertension (Schwartz et al., 2000). This is confirmed by autopsy findings of cerebral edema and intracerebral parenchymal hemorrhage in women who have died from eclampsia. Cerebral vasospasm, excitation of brain receptors, a hyperactive sympathetic nervous system and hypertensive encephalopathy from cerebral overperfusion have all been associated with eclamptic seizures (Belfort et al., 2006). Most of these women have been found to have cerebral overperfusion rather than ischemia (Belfort et al., 2001). Interestingly, RPLS can be precipitated by acute blood pressure rises and treatment with antiangiogenic agents such as bevacizumab, a monoclonal antibody against VEGF and its receptors (Hinchey et al., 1996; Ozcan et al., 2006). This appears to align with the accumulating evidence on the role of antiangiogenic factors in the pathophysiology of pre-eclampsia/eclampsia.

7. HELLP Syndrome and Hematological Abnormalities

Although it can occur in the absence of proteinuria, the HELLP syndrome is generally considered to be a severe variant of pre-eclampsia. It develops in approximately 10–20% of women with severe pre-eclampsia. The HELLP syndrome can be complicated by eclampsia (6% of cases), placental abruption (10%), acute renal failure (5%), disseminated intravascular coagulation (8%) and pulmonary edema (10%). Rarely, hepatic hemorrhage and rupture can occur, even after delivery of the fetus; associated maternal and perinatal mortality rates of 59 and 42%, respectively, have been reported (Haddad et al., 2000; Dessole et al., 2007).

The HELLP syndrome is a consumptive coagulopathy and thrombotic microangiopathy that shares many clinical and biological features with thrombotic thrombocytopenic purpura (TTP) and hemolytic uremic syndrome (HUS) (Gando et al., 1995; Boehme et al., 1996; Rath et al., 2000; Mori et al., 2001). In normal pregnancy, coagulation is enhanced because the circulating levels of all coagulation factors are increased, including those made in the liver and in the vascular endothelium (Cadroy et al., 1993). In pre-eclampsia, however, only factors that are synthesized by the vascular endothelium are elevated (Friedman et al., 1995). These factors include prostacyclins (PGI_2), thrombomodulin, cellular fibronectin, PAI-1 and von Willebrand factor (vWF) (Estelles et al., 1989; Taylor et al., 1991; Hsu et al., 1993; Friedman et al., 1995). Plasma concentrations of cellular fibronectin are increased weeks before the onset of hypertension (Chavarria et al., 2002). Exposure of cultured endothelial cells to serum from pre-eclamptic women results in greater cellular fibronectin and thrombomodulin release compared to serum from normotensive pregnant women (Taylor et al., 1991; Roberts et al., 1992; Kobayashi et al., 1998). Recently, it was shown that the large vWF multimers (normally cleaved by the metalloproteinase ADAMTS13) that are highly reactive with platelets may be increased in women with the HELLP syndrome because of both endothelial activation and, similar to TTP, decreased ADAMTS13 levels (Hulstein et al., 2006). Other markers of endothelial injury that have also been reported will be reviewed in the section on maternal endothelial function.

C. Long-Term Cardiovascular and Renal Outcomes

Although the symptoms of pre-eclampsia appear to remit completely following delivery of the fetus, many of these women ultimately develop cardiovascular and renal morbidity later. Women with a history of pre-eclampsia continue to have impaired endothelial-dependent vasorelaxation, as measured by brachial artery flow-mediated dilatation up to 3 years postpartum, implying that changes in the maternal endothelium may be long-standing (Chambers et al., 2001; Agatisa et al., 2004). About 20% of women with pre-eclampsia are found to have hypertension or microalbuminuria on long-term follow-up (Nisell et al., 1995). One study observed that development of subsequent ischemic heart disease was increased by 1.7-fold and hypertension by 2.4-fold in women with pre-eclampsia, which was confirmed by two large subsequent European studies (Hannaford et al., 1997; Irgens et al., 2001; Smith et al., 2001). Women with pre-eclampsia and gestational hypertension have a roughly twofold increase in cardiovascular and cerebrovascular risk compared to age-matched controls (Irgens et al., 2001; Ray et al., 2005). The rates of heart disease among women with pre-eclampsia complicated by pre-eclampsia with preterm birth or intrauterine growth restriction and among those with severe or recurrent pre-eclampsia were increased by up to eightfold (Irgens et al., 2001; Smith et al., 2001). Risk of maternal renal morbidity requiring kidney biopsy has also been found to be increased in women with pre-eclampsia accompanied by low birth weight (Vikse et al., 2006).

Because pre-eclampsia and cardiovascular disease share many common risk factors, such as obesity, chronic hypertension, diabetes mellitus, renal disease and the metabolic syndrome, it may well be the reason their risks are so closely linked. However, the increase in long-term cardiovascular mortality appears to be present even for women who develop pre-eclampsia without clear cardiovascular risk factors, raising the possibility that perhaps pre-eclampsia itself causes vascular damage and persistent endothelial dysfunction. Children who were products of pregnancies complicated by low birth weight with or without pre-eclampsia have been observed to have a higher incidence of subsequent hypertension, diabetes, cardiovascular disease and chronic kidney disease (Zandi-Nejad et al., 2006).

D. Pathogenesis of Pre-Eclampsia (Figure 30.2)

1. Abnormal Placentation

Accumulating evidence has pointed to the placenta's pivotal role in the pathogenesis of pre-eclampsia. It is removal of the placenta, rather than the fetus, which leads to the amelioration of symptoms, as evident in cases of hyatidiform moles and extrauterine pregnancies, where delivery of the fetus is insufficient (Shembrey and Noble, 1995). Common pathological findings in pre-eclamptic placentas include atherosis, necrosis, sclerotic narrowing of arteries and arterioles, fibrin deposition, thrombosis, endothelial damage, atherosclerosis and infarcts, all consistent with placental hypoperfusion and ischemia, which appear to correlate to the severity of

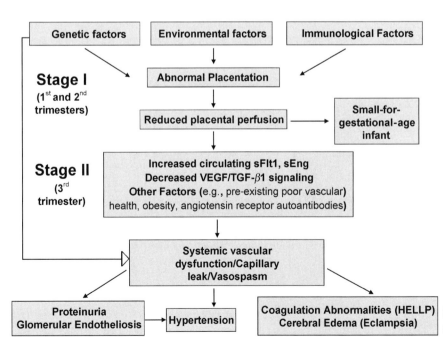

FIGURE 30.2 Summary of the pathogenesis of pre-eclampsia.

pre-eclampsia (Hertig, 1945; Salafia et al., 1998). Hypoxia has been implicated as a contributing factor because of the two- to four-fold increased incidence of pre-eclampsia in women living at high altitudes (Palmer et al., 1999). The finding of abnormal uterine artery Doppler ultrasound, suggesting decreased uteroplacental perfusion, typically precedes the onset of pre-eclampsia, although it is non-specific for pre-eclampsia (North et al., 1994; Lees et al., 2001; Frusca et al., 2003). In various species, constriction of uterine blood flow has reproduced hypertension, proteinuria and glomerular endotheliosis (Kumar, 1962; Combs et al., 1993; Podjarny et al., 1999). Finally, it has been suggested that a decrease in placental production of nitric oxide may contribute to placental ischemia (Noris et al., 2004).

Based on this evidence, a causative role for placental ischemia may be likely, but other data have challenged ischemia as the sole cause of pre-eclampsia. Animal models of uterine hypoperfusion have not been able to reproduce the full clinical syndrome seen in humans, such as seizures and the hallmark renal pathological finding of glomerular endotheliosis. Based on the frequent finding of placental ischemia and infarction in women with sickle cell disease, it could be theorized that they might have a higher incidence of pre-eclampsia, however, thus far, the data have been inconsistent (Larrabee and Monga, 1997; Stamilio et al., 2003). It appears that, while ischemia may be an important trigger, it is not present in all cases of pre-eclampsia and the maternal response to placental ischemia is variable.

2. Placental Vascular Development

In the course of normal placental development, cytotrophoblasts attach to the uterine decidua by means of anchoring villi; a small percentage of cytotrophoblasts in the anchoring villi migrate into the endometrium. The process whereby these extravillous cytotrophoblasts invade the uterine spiral arteries of the decidua and myometrium peaks around the 12th week of gestation. By 18–20 weeks, the cytotrophoblasts have lined the endometrial and superficial myometrial portion of the spiral arteries and have converted the arteries from small resistance vessels to high-caliber capacitance vessels (Brosens et al., 1972; De Wolf et al., 1980). Vascular remodeling permits an increase in uterine blood flow that is essential to nourish the developing fetus throughout pregnancy. In addition, the process by which the invasive cytotrophoblasts transform from an epithelial to endothelial phenotype, termed 'pseudovasculogenesis' is required to take place in order for normal placental development to ensue. This process describes the downregulation of the expression of adhesion molecules typical of epithelial cells and switch to expression of endothelial cell surface molecules by the invasive fetal cells (Zhou et al., 1997).

Defective placental vascular remodeling is thought to cause the placental ischemia seen in pre-eclampsia (Robertson et al., 1967). Pseudovasculogenesis fails to occur, which impedes normal invasion of the myometrial spiral arteries. This superficial invasion of the decidua results in narrow and undilated proximal segments of the spiral arteries, which ultimately leads to uterine hypoperfusion (Meekins et al., 1994; Zhou et al., 1997a, b). The abnormalities in cytotrophoblast differentiation found in the placentas of pre-eclamptics suggest that the mechanisms contributing to placental ischemia are set into motion very early in pregnancy, although their determinants have yet to be fully elucidated.

Placentas from women with pre-eclampsia have been found to over-express the hypoxia-inducible transcription factor proteins, HIF-1α and -2α (Rajakumar et al., 2003). The HIF-1α protein over-expressed in pre-eclamptic placentas has been shown to be capable of binding to the DNA hypoxia response element *in vitro* and target genes regulated by the HIF pathway appear to be altered in pre-eclamptic placentas *in vivo* (Rajakumar et al., 2003). Placental sFlt-1 expression has been observed to be increased by both physiologically and pathologically low levels of oxygen in women living at high altitudes and those with pre-eclampsia, respectively, which appears to be mediated by HIF-1 (Nevo et al., 2006). Other HIF-1 target genes, such as transforming growth factor beta-3 (TGF-β3) may block cytotrophoblast invasion (Caniggia et al., 2000). Expression of a number of angiogenic factors and receptors that are regulated by HIF on invasive cytotrophoblasts has been found to be altered in pre-eclampsia, including VEGF, PlGF and VEGFR-1 (Zhou et al., 2002). Epidermal growth factor (EGF) receptors are expressed in both extra-villous and villous cytotrophoblasts, as well as syncytiotrophoblasts of first trimester placentas. EGF was recently shown to promote differentiation of isolated term trophoblasts and regulate invasion of extravillous trophoblasts. It has also been demonstrated to be decreased in pre-eclamptic cytotrophoblasts, however, its exact role in pseudovasculogenesis it unclear (Moll et al., 2007). In non-human primates with uteroplacental hypoperfusion induced by aortic constriction, proteinuria and hypertension, but not defective placental cytotrophoblast invasion, were induced (Zhou et al., 1993). Genetic studies of the STOX1 gene, which encodes a putative DNA binding protein involved in trophoblast differentiation, have thus far yielded conflicting data on whether maternally-inherited mutations in this gene within the 10q22 locus might account for matrilineal pre-eclampsia in Dutch women (van Dijk et al., 2005; Berends et al., 2007). Despite these recent advances in our knowledge of the mechanisms underlying the abnormalities in pseudovasculogenesis, the defects of placental cytotrophoblast invasion remain incompletely understood.

3. Maternal Endothelial Function

Because pre-eclampsia appears to originate in defective placentation and progress to widespread endothelial dysfunction manifesting as vasoconstriction and end-organ damage, circulating factors derived from the placenta have been suspected as the culprits for the clinical syndrome (Roberts et al., 1989; Ferris, 1991). The search for these circulating factors has revealed numerous potential serum markers of endothelial dysfunction, such as von Willebrand antigen, cellular fibronectin, soluble tissue factor, soluble E-selectin, platelet-derived growth factor, endothelin, tumor necrosis factor-α (TNF-α), interleukin (IL)-6, IL-1, Fas ligand, neurokinin B, asymmetric dimethy L-arginine (ADMA), C-reactive protein and leptin (Roberts et al., 1989; Page et al., 2000; Roberts, 2000; Walker, 2000; Roberts and Cooper, 2001; Chappell et al., 2002; Savvidou et al., 2003; Qiu et al., 2004). Production of PGI_2, an endothelial-derived prostaglandin, is also diminished prior to the onset of clinical symptoms (Mills et al., 1999). Markers of endothelial injury, including the adhesion molecules soluble P-selectin, soluble E-selectin, and soluble vascular cell adhesion molecule-1 (VCAM-1) have been noted to be elevated in pre-eclampsia (Chaiworapongsa et al., 2002). That a number of these endothelial biomarkers of interest are altered before the onset of symptoms again points to the pivotal role of endothelial dysfunction in pre-eclampsia. Recent data suggest that increased syncytiotrophoblast microfragment shedding occurs as a result of placental apoptosis (Sargent et al., 2003). The syncytiotrophoblast microparticles have been found to circulate in bound (to monocytes) and unbound forms in pregnancy, but are substantially increased in the unbound form in the circulation of pre-eclamptics. This placental 'debris' appears to lead to greater stimulation of inflammatory cytokine production *in vitro* and has therefore been hypothesized to contribute to generalized endothelial dysfunction, although no *in vivo* data exist to support this (Cockell et al., 1997; Germain et al., 2007).

There has been recent interest in $p57^{Kip2}$, a cell-cycle inhibitor and paternally-imprinted gene in mice and humans. Pregnant mice that were heterozygous for $p57^{Kip2}$ deficiency appeared to exhibit the full pre-eclamptic syndrome, including abnormal placentation, hypertension, proteinuria, glomerular endotheliosis thrombocytopenia and increased endothelin levels in late pregnancy, in addition to trophoblastic hyperplasia (Kanayama et al., 2002). These findings, however, were not fully replicated in a subsequent study, which suggested that perhaps this is a model in which placental abnormalities cause pre-eclampsia only in the setting of specific environmental variables (Knox and Baker, 2007).

4. Immunologic Maladaptation

As mentioned previously, because pre-eclampsia is more common in first pregnancies, with a change in partners and with long inter-pregnancy interval, immunological intolerance to paternal fetal antigens has been invoked as a cause, although it remains unproven (Tubbergen et al., 1999; Skjaerven et al., 2002). In general, decreased exposure to a partner's sperm has been observed to increase the incidence of pre-eclampsia. These include contraceptive methods that reduce exposure to sperm and conception by intracytoplasmic sperm injection (ICSI) (Klonoff-Cohen et al., 1989; Wang et al., 2002). Prior exposure to paternal antigens, conversely, such as by co-habitation or oral tolerization, leads to lower risk of pre-eclampsia (Robillard et al., 1993; Koelman et al., 2000). Expression of HLA-G is typically found on invasive extravillous cytotrophoblasts and has been reported to be abnormally low or absent in pre-eclampsia, indicating a possible role in determining immune tolerance and possibly trophoblast invasion at the maternal–fetal interface.

Natural killer (NK) cells in the maternal–fetal interface, which are thought to play an important role in innate immunity, have been recently noted to play an important role during normal placental vascular remodeling. Moreover, genetic studies suggest that the susceptibility to pre-eclampsia may be influenced by polymorphic HLA-C ligands and the killer cell receptors (KIR) present on NK cells. Analysis of various human populations revealed a strong association between the lack of KIR AA and the presence of HLA-C2 haplotypes (that appear to favor unfavorable trophoblast invasion) and the occurrence of pre-eclampsia (Hiby et al., 2004). Based on these observations, a hypothesis was formulated indicating that, in normal pregnancies, dNK cells activation through interaction with HLA-C on extravilous trophoblasts would promote placental development and maternal decidual spiral artery modifications by extra-villous cytotrophoblasts. Insufficient dNK cell activation would halt this process prematurely resulting in poor decidual artery remodeling, increasing the risk of pre-eclampsia (Hiby et al., 2004; Parham, 2004).

5. Oxidative Stress

Accumulating evidence indicates that oxidative stress plays a role in human placental pathologies, such as pre-eclampsia and intrauterine growth restriction. Some studies have shown elevation of markers of oxidative stress in women with pre-eclampsia, although not all studies have confirmed this (Hubel, 1999; Regan et al., 2001). Superoxides and free radicals generated during maternal oxidative stress could attack cell membranes, proteins and nucleic acids, resulting in placental damage, and initiating maternal endothelial dysfunction and leukocyte activation (Hubel, 1999). The production of superoxide by neutrophils appears to differ for pregnant women with essential hypertension, in whom neutrophil activation remains unchanged postpartum, and for pre-eclamptics, in whom neutrophilic superoxide generation abates after delivery, prompting the suggestion of the presence of a neutrophil-activating serum factor in pre-eclampsia (Tsukimori et al., 2007). Some clinical and *in vitro* evidence has indicated that antioxidant supplementation may decrease the incidence of pre-eclampsia, including inhibition of monocyte adhesion by N-acetylcysteine, vitamin C and vitamin E to human umbilical vascular endothelial cells (HUVEC) in the presence of pre-eclamptic, but not normal pregnant serum (Ryu et al., 2007). Subsequent large trials, however, have not validated these findings (Chappell et al., 1999; Roberts and Speer, 2004; Poston et al., 2006; Rumbold et al., 2006). Given that dietary intake of food rich in antioxidants is more effective than vitamin E and C supplementation in preventing atherosclerotic vascular disease, it may well be that the focus in pre-eclampsia should be on different antioxidants. Nevertheless, this may also be consistent with the role of oxidative stress as a secondary phenomenon, rather than as a primary causative factor in the pathogenesis of pre-eclampsia, which remains to be proven.

6. Angiogenic Factors

Using gene expression profiling to identify the factor(s) produced by the placenta in pre-eclampsia, placental sFlt-1 mRNA was found to be upregulated (Maynard et al., 2003). sFlt-1 is a splice variant of vascular endothelial growth factor (VEGF) receptor Flt-1 that lacks the transmembrane and cytoplasmic domains and is produced and released by the placenta into maternal circulation (Kendall and Thomas, 1993; Clark et al., 1998; Zhou et al., 2002). It then binds to circulating VEGF and PlGF (Banks et al., 1998). In rats, sFlt-1 reproduced the maternal syndrome of hypertension, proteinuria and glomerular endotheliosis (see Figure 30.1). *In vitro* angiogenesis assays revealed that the anti-angiogenic effects of pre-eclamptic serum can be reversed by exogenous VEGF and PlGF (Maynard et al., 2003). In women with pre-eclampsia, circulating levels of sFlt-1 were found to be increased (Maynard et al., 2003; Koga et al., 2003; Tsatsaris et al., 2003; Chaiworapongsa et al., 2004). Correspondingly, levels of free VEGF and PlGF are decreased in pre-eclamptics, both before and during disease (Polliotti et al., 2003; Taylor et al., 2003). Beginning at about 5–6 weeks prior to the onset of the clinical syndrome, sFlt-1 concentrations rise dramatically and free VEGF and PlGF fall, correlating with earlier onset and severity of disease, as well as the risk of having a small-for-gestational-age infant (Hertig et al., 2004; Levine et al., 2004). In women who develop pre-eclampsia, a modest, but significant decrease in serum PlGF levels is seen as early as the 1st trimester, however, the concentration of free PlGF in plasma drops dramatically when sFlt-1 levels begin to rise in mid-pregnancy. Unbound PlGF is also freely filtered into the urine and, consistent with decreased serum concentrations of unbound PlGF, mid-gestation diminished levels of free PlGF in the urine are also predictive of development of pre-eclampsia (Buhimschi et al., 2005; Levine et al., 2005). These data suggest that sFlt-1 potentially plays a central role in the pathogenesis of pre-eclampsia.

VEGF promotes angiogenesis and plays a critical role in preserving glomerular endothelial cell health, as evidenced by proteinuria and glomerular endotheliosis in its absence (Ostendorf et al., 1999; Masuda et al., 2001; Eremina et al., 2003; Sugimoto et al., 2003). In the kidneys, it is constitutively expressed and serves to maintain the fenestrated endothelia; antagonism by sFlt-1 results in the expected glomerular injury that is seen in pre-eclampsia (Esser et al., 1998; Eremina et al., 2003). In addition, VEGF induces nitric oxide and vasodilatory prostacyclins in endothelial cells, which decrease vascular tone and blood pressure (He et al., 1999). Symptoms of headaches, hypertension, proteinuria and coagulopathy have been reported in reaction to neutralizing antibodies to VEGF used in oncology trials (Kuenen et al., 2002; Kabbinavar et al., 2003; Yang et al., 2003).

Current evidence points to the fact that angiogenic factors are very likely to be important in the regulation of placental vasculogenesis. The mechanism by which placental dysfunction influences sFlt-1 production has yet to be elucidated, however, evidence exists in *in vitro* primary cytotrophoblast cultures pointing to the possible role of placental hypoxia in upregulating sFlt-1 production (Nagamatsu et al., 2004). Recently, it was reported that isolated decidual cells express sFlt-1 mRNA, suggesting that they can synthesize sFlt-1 and that these mRNA levels in 1st trimester, but not term decidual cells were increased by thrombin. Based on this, it was hypothesized that thrombin could be influencing decidual cells to interfere with placental pseudovasculogenesis (Lockwood et al., 2007).

In vitro, sFlt-1 decreases cytotrophoblast invasiveness (Zhou et al., 2002). Placental tissue has high expression of VEGF ligands and receptors in the first trimester (Yancopoulos et al., 2000). Early in pregnancy, sFlt-1 levels are relatively low, but begin to rise in the 3rd trimester. The theory behind this timing is that placental vascular development relies on the balance between pro- and anti-angiogenic factors. An excess of the latter (e.g. sFlt-1) early in pregnancy might lead to defective cytotrophoblast invasion. In late gestation, high levels of sFlt-1 could be considered physiologic and represent completion of the vasculogenic phase of placental development and, therefore, a shift in angiogenic balance. Certain patterns of elevated circulating sFlt-1 have been observed in various groups: higher sFlt-1 levels have been found in first versus second pregnancies, in twin versus singleton pregnancies (Maynard et al., 2005; Wolf et al., 2005; Bdolah et al., 2006) and in non-smokers versus smokers, the latter of whom have also been noted to have a lower incidence of pre-eclampsia, possibly explained by the lower circulating sFlt-1 levels (Belgore et al., 2000; Lain et al., 2003; Levine et al., 2006).

Another placental anti-angiogenic protein, soluble endoglin (sEng), was recently reported to be high in pre-eclamptics. Endoglin (Eng or CD105) is an angiogenic receptor that is expressed on the surface of endothelial cells and placental syncytiotrophoblasts and acts as a co-receptor for TGF-β3, a potent proangiogenic molecule. Eng mRNA has been shown to be upregulated in the pre-eclamptic placenta (Venkatesha et al., 2006). Soluble endoglin (sEng) is the extracellular 65-kDa proteolytically cleaved form of endoglin that is released into the circulation and was recently found to be elevated in women with pre-eclampsia. In rats, sEng appeared to augment the vascular injury of sFlt-1, with a resultant severe pre-eclampsia-like syndrome similar to the HELLP syndrome. In explant cultures of trophoblasts from gestational week 5–8, mAb to Eng and antisense Eng oligonucleotides stimulated trophoblast outgrowth and migration (Caniggia et al., 1997). TGF-β1 and/or TGF-β3 inhibition of the migration and invasion appeared to be mediated by sEng, leading to the hypothesis that production of sEng by the placenta may be a compensatory mechanism to limit the effects of Eng. Subsequently, sEng was observed to be elevated in the serum of pre-eclamptic women 8–12 weeks before the clinical onset of the disease (Levine et al., 2006). Elevations in sEng were most marked and likely most useful for prediction in women who developed preterm pre-eclampsia or pre-eclampsia with a small-for-gestational-age fetus. Multivariate analysis indicated that sEng, sFlt-1 and PlGF were independently and significantly associated with pre-eclampsia, although the pattern of sEng concentration throughout pregnancy tended to parallel that of the sFlt-1/PlGF ratio. In this study, combining all three was even more strongly predictive of pre-eclampsia than the individual biomarkers. Recent in vitro data demonstrated that adenoviral overexpression of heme oxygenase-1 (HO-1) in endothelial cells resulted in decreased sFlt-1 and sEng production, while HO-1 small interfering RNA (siRNA) knockdown and pharmacological inhibition of HO-1 activity in placental villous explants potentiated endothelial sFlt-1 and sEng release (Cudmore et al., 2007). How exactly sEng might interact with sFlt-1 to produce this clinical picture in humans requires further investigation.

Other angiogenic molecules have also been identified. Endostatin is another antiangiogenic factor that is modestly raised in pre-eclampsia (Hirtenlehner et al., 2003). In pre-eclamptic placenta, the soluble form of Flk (VEGFR-2) has been found, although its function in placental vasculogenesis is unclear (Ebos et al., 2004).

Insulin resistance is common among women who develop pre-eclampsia and conversely, women with pre- or gestational diabetes mellitus are at greater risk of developing pre-eclampsia (Wolf et al., 2002; Duckitt and Harrington, 2005). Women with previous pre-eclampsia or gestational hypertension exhibit signs of altered angiogenesis and insulin resistance within the first decade postpartum. Therefore, insulin resistance and defective angiogenesis may represent the link between hypertensive diseases of pregnancy and long-term cardiovascular risk in this population (Thadhani et al., 2004; Wolf et al., 2004; Girouard et al., 2007).

7. Renin–Angiotensin System

As noted previously, vascular responsiveness to angiotensin II and other vasoconstrictive agents is heightened in pre-eclamptic compared to normal pregnancies. Plasma renin levels are also suppressed in pre-eclamptics compared to normotensive pregnant women as a response to systemic vasoconstriction and hypertension. Antiogensin-1 receptor (AT_1) agonistic autoantibodies have been identified in malignant renovascular hypertension and in transplant patients with vascular rejection, as well as in women with pre-eclampsia and have been hypothesized to account for the increased sensitivity to angiotensin II seen in pre-eclamptics (Wallukat et al., 1999; Fu et al., 2000). These AT_1 activating antibodies (AT_1-AA) were then shown to activate endothelial cells to produce tissue factor, an early marker of endothelial dysfunction (Dechend et al., 2000). Subsequently, AT_1-AAs were found to increase PAI-1 production and lower invasiveness of human trophoblasts in vitro, suggesting that they may contribute to defective placental pseudovasculogenesis (Xia et al., 2003). It was recently shown that significantly more women with a history of pre-eclampsia had persistently detectable AT_1-AAs even at 18 months' postpartum, compared to women who had normal pregnancies (Hubel, 1999). The autoantibody-positive women also had significantly increased sFlt-1, reduced free VEGF and higher insulin resistance homeostasis model assessment values compared with autoantibody-negative women. This raises the question of the long-term cardiovascular impact of AT_1-AAs. However, the autoantibodies may be non-specific for placental hypoperfusion, since they have also been found in women with abnormal second trimester uterine artery Doppler studies who later had normal pregnancies or intrauterine growth retardation (Walther et al., 2005). The finding of heterodimerization of AT_1 with bradykinin 2 receptors has been thought to explain sensitization to the vasopressor response of angiotensin II in pre-eclampsia (AbdAlla et al., 2001). In hypertensive rats, specific inhibition of these AT(1)/B-2 receptor heterodimers have revealed that they appear to mediate increased endothelin-1 secretion and enhanced angiotensin II-stimulated G alpha(q/11) activation in mesangial cells. Thus, AT(1)/B-2 receptor heterodimerization contributes to angiotensin II hyper-responsiveness of mesangial cells in experimental hypertension (AbdAlla et al., 2005).

E. Screening and Treatment

1. Screening

Despite the lack of effective treatment or prevention of pre-eclampsia, monitoring and supportive care is still beneficial to the patient and fetus, as evidenced by the poor outcomes seen with inadequate antenatal care (Abi-Said et al., 1995). Nulliparous women with any risk factors should be evaluated every 2–3 weeks after 20 weeks of gestation to monitor for the development of hypertension, proteinuria, headache, visual disturbances or epigastric pain. Screening for pre-eclampsia using uterine artery Doppler is not considered standard practice in the USA, however, the utility of combining uterine artery Doppler with serum markers remains to be seen (Parra et al., 2005). Biomarkers for pre-eclampsia that have shown promise have included cellular fibronectin (Chavarria et al., 2002), plasma neurokinin-B (Page et al., 2000) and, more recently, serum sFlt-1, PlGF and sEng, as well as urinary PlGF (Levine et al., 2004, 2005, 2006). Maternal sFlt-1 levels increase approximately 5–8 weeks before onset of pre-eclampsia. Serum PlGF concentrations are conversely lower starting in the 1st or early 2nd trimester and urinary PlGF may be significantly reduced starting in late 2nd trimester, although the latter is not borne out by all studies (Livingston et al., 2001). Recently, a placental protein-13, a member of the galectin family, was found to be depressed as early as the 1st trimester in patients who are at risk for developing pre-eclampsia (Chafetz et al., 2007).

Numerous approaches have been used in studies for the prevention of pre-eclampsia, including antiplatelet agents, calcium and antioxidants. Despite early trials showing aspirin's apparent benefit, it was not subsequently borne out in most trials to be effective in lowering the incidence of pre-eclampsia, except perhaps in those at highest risk, in whom the modest benefits would still need to be weighed against the risks (Salmela et al., 1993; Askie et al., 2007). In those at low risk for pre-eclampsia, it is thought that the benefit is minimal (Subtil et al., 2003). Calcium supplementation is recommended for women who have a low baseline dietary calcium intake (<600 mg/day) and for those who are at high risk of gestational hypertension. In a large, randomized, placebo-controlled trial, calcium did not lower the incidence of pre-eclampsia, but it did diminish the rate of eclampsia, gestational hypertension, complications of pre-eclampsia and neonatal mortality (Villar et al., 2006). As mentioned, antioxidants have been studied in pre-eclamptics, based on the theory that oxidative stress may play a central role in the pathogenesis of pre-eclampsia. Once again, large trials have failed to demonstrate that antioxidant supplementation reduces pre-eclampsia (Poston et al., 2006; Rumbold et al., 2006). Nutritional interventions, such as protein, salt and calorie restriction in obese pregnant women, have also not been shown to lower the incidence of pre-eclampsia (Villar et al., 2004). Although methods of prevention have been the focus of numerous studies, none has been identified to date to be of clear benefit.

2. Management and Treatment

Perinatal and neonatal mortality are extremely high (>80%) prior to 24–26 weeks of gestation and maternal complications are common. The presence of non-reassuring fetal testing, suspected abruption placentae, thrombocytopenia, rising liver enzymes and/or creatinine and symptoms suggestive of impending eclampsia or HELLP syndrome are generally considered indications for urgent delivery or possibly termination of pregnancy, particularly if the patient presents in the 2nd trimester.

Onset of pre-eclampsia from 24 to 34 weeks of gestation without signs and symptoms suspicious for severe pre-eclampsia may be managed by postponing delivery in the hope of improving neonatal outcome. However, the potential fetal benefit from delaying delivery must be balanced against the possibility of increased maternal morbidity. Most studies have supported the approach of postponing delivery to minimize fetal complications, provided that careful maternal and neonatal monitoring can be achieved (Odendaal et al., 1990; Sibai et al., 1994).

The goal of blood pressure management in the pre-eclamptic patient is to minimize maternal and fetal morbidity. Caution must be applied to avoid aggressive treatment of hypertension that compromises placental blood flow and fetal growth. Indeed, treatment of mild-to-moderate hypertension in pregnancy has not been shown to improve outcomes and has even been associated with an increased risk of small-for-gestational-age infants (Abalos et al., 2001; von Dadelszen et al., 2003). Acute and dramatic blood pressure reduction can potentially result in fetal distress or demise, particularly in the setting of pre-existing inadequate placental perfusion. Thus, the ACOG recommends that only blood pressure elevation greater than 150–160 mmHg systolic or 100–110 mmHg diastolic, where the risk of cerebral hemorrhage is significant, should be treated with anti-hypertensive agents (ACOG Practice Bulletin, 2002).

Magnesium has been shown to be superior to other agents for the management and prevention of eclampsia (Collaborative Eclampsia Trial, 1995; Lucas et al., 1995). It is generally given intravenously as a bolus, followed by a continuous infusion. Women receiving continuous infusions of magnesium must be monitored carefully for signs of neuromuscular toxicity, particularly in women with impaired magnesium excretion as a result of renal insufficiency.

3. Management of the HELLP Syndrome

For women in weeks 24–34 of gestation who have a reassuring fetal status and appear to be relatively stable, expectant management is an acceptable alternative to delivery. However, it should be noted that sudden and unpredictable deterioration is not uncommon in the clinical course of the HELLP syndrome and, therefore, some recommend urgent delivery upon confirmation of the diagnosis. Although intravenous steroids have been given based on retrospective and uncontrolled data, a recent randomized, controlled trial showed no benefit to high-dose dexamethasone treatment in HELLP syndrome (Fonseca et al., 2005).

Based on the recent advances in the understanding of the pathogenesis of pre-eclampsia, sFlt-1 may represent a promising target for therapeutic intervention (Li et al., 2007).

IV. OTHER HYPERTENSIVE DISORDERS OF PREGNANCY

A. Chronic Hypertension

Chronic essential hypertension in pregnancy is generally diagnosed prior to pregnancy or by the finding of a blood pressure >140/90 before 20 weeks of gestation. It is seen in 3–5% of pregnancies and is more prevalent with advanced maternal age, obesity and black race (Sibai, 1996; Magee et al., 1999). Pregnant women with chronic essential hypertension are at a higher risk of experiencing pre-eclampsia (21–25%), premature delivery (33–35%), IUGR (10–15%), placental abruption (1–3%) and perinatal mortality (4.5%) (Sibai et al., 1998; ACOG Practice Bulletin, 2001). The duration and the severity of hypertension correlate with neonatal morbidity and risk of pre-eclampsia (McCowan et al., 1996; August et al., 2004). Pre-existing proteinuria raises the risk of preterm delivery and IUGR, but not pre-eclampsia (Sibai et al., 1998). Most of these adverse outcomes, however, are seen in women with severe hypertension (diastolic blood pressure >110 mmHg) and pre-existing cardiovascular or renal disease. Pregnant women with mild, chronic hypertension can expect to have outcomes comparable to the general obstetric population (Sibai, 1996). The physiologic fall in blood pressure in the 2nd trimester also occurs in women with chronic essential hypertension and may mask underlying chronic hypertension. When this occurs, the chronic hypertension may be misdiagnosed as gestational hypertension when the blood pressure subsequently rises in the 3rd trimester.

In women without underlying proteinuric renal disease, new-onset proteinuria >300 mg/day accompanied by worsening hypertension likely indicates superimposed pre-eclampsia (ACOG Practice Bulletin, 2002). Diagnosing superimposed pre-eclampsia in women with chronic essential hypertension and pre-existing proteinuria, however, can be challenging. In this case, the diagnosis would require significant exacerbation of hypertension of at least 30 mmHg over baseline levels, along with other signs and symptoms suggestive of pre-eclampsia, such as headache, visual changes, epigastric pain and laboratory abnormalities of hemoconcentration, elevated liver enzymes and elevated uric acid. Pre-eclampsia can occasionally present before 20 weeks of gestation, thus, it should also be considered in women presenting with new hypertension and proteinuria in mid-gestation.

B. Gestational Hypertension

Gestational hypertension is new-onset hypertension without proteinuria after 20 weeks' gestation, which then resolves postpartum. Some women diagnosed with gestational hypertension in fact have pre-existing, undiagnosed essential hypertension. In these cases, a woman may be mistakenly presumed to be previously normotensive if she presents during the physiologic 2nd-trimester nadir in blood pressure. The diagnosis of chronic hypertension may not be made until after delivery, when blood pressure fails to normalize. In approximately 10–25% of cases, gestational hypertension progresses to pre-eclampsia (Saudan et al., 1998). In cases of severe gestational hypertension, the risk for adverse outcomes are similar to those of pre-eclampsia (Buchbinder et al., 2002). Interestingly, a renal biopsy study found a substantial percentage of gestational hypertensives to have renal glomerular endothelial damage, suggesting that gestational hypertension may share certain pathophysiologic mechanisms with pre-eclampsia (Fisher et al., 1981; Strevens et al., 2003). In another subset of women with gestational hypertension, it may be a transient unmasking of an underlying predisposition toward chronic hypertension. These women often have a strong family history of chronic hypertension and tend to develop hypertension in the 3rd trimester without hyperuricemia or proteinuria. Although gestational hypertension typically resolves after delivery, the women are at risk for development of hypertension later in life (Marin et al., 2000).

C. Secondary Causes of Hypertension in Pregnancy

The major forms of secondary hypertension seen in pregnant women are renal artery stenosis, primary hyperaldosteronism, Cushing's syndrome, pheochromocytoma and chronic renal failure from any cause. Hypertension that is severe and resistant to therapy in the 1st and 2nd trimesters of pregnancy should alert the clinician to the possibility of a secondary cause of hypertension.

The prevalence of renal artery stenosis varies from 0.5 to 2% of the total hypertensive population, with higher rates among patients with severe hypertension (Davis et al., 1979). Atherosclerosis is more common in men with late-onset hypertension, but 80% of patients with fibromuscular dysplasia are women of child-bearing age (Simon et al., 1972). The characteristic upper abdominal bruit of renal artery stenosis, even if present, will likely be masked in mid-gestation by the presence of a placental murmur. Angioplasty and stent placement in the 2nd and 3rd trimesters of pregnancy after diagnosis by magnetic resonance angiography (MRA) have been described (Hayashida et al., 2005).

Primary hyperaldosteronism is characterized by hypertension and hyperkalemic alkalosis. Plasma renin activity is low and plasma and urinary aldosterone levels are high. The prevalence is 0.5–1% of the hypertensive population (Gallery, 1999). Most reported cases have been hypokalemic, with suppressed plasma renin activity unresponsive to sodium deprivation (Hammond et al., 1982). Functional adrenal adenomas may be managed either medically or surgically, but as long as serum potassium and blood pressure are corrected, definitive therapy may be postponed until after delivery. Aldosterone antagonists, such as spironolactone, should be avoided, given the theoretical risks to the fetus (Gallery, 1999).

Cushing's syndrome may be diagnosed by measuring plasma cortisol and dexamethasone suppression testing. Fewer than 70 cases of Cushing's syndrome have been reported. Cortisol levels increase progressively throughout pregnancy and, by the 2nd trimester, reach levels that are two- to threefold greater than pre-pregnancy values, as a result of an increase in cortisol-binding globulin levels under the influence of estrogen. Low dose dexamethasone suppression tests can therefore give false negatives in pregnancy, but a high dose dexamethasone test that fails to suppress serum cortisol would be helpful in diagnosing an adrenal tumor causing Cushing's syndrome (Anjali et al., 2004). Cushing's syndrome in pregnancy may be complicated by hypertension in 54% and diabetes and pre-eclampsia in 13%. Adverse fetal outcomes include increased rates of spontaneous abortion and perinatal mortality (>10%) (Check et al., 1979). Pheochromocytoma is a catecholamine-secreting tumor that is a rare cause of hypertension (<0.5%), but can be life-threatening for both mother and fetus. It is occasionally unmasked during labor and delivery, when fatal hypertensive crisis is precipitated by vaginal delivery, uterine contractions, and anesthesia (Del Giudice et al., 1998). Screening of hypertensive pregnant women can be achieved by measurement of 24-hour urinary catecholamine excretion prior to conception, though it may also be done during gestation, since catecholamine excretion is unchanged in normal pregnancy (Schenker and Chowers, 1971). If at all possible, surgical intervention should be postponed until the postpartum period.

A rare form of hypertension in pregnant women is caused by an activating mineralocorticoid receptor mutation, which is exacerbated by pregnancy. The result is inappropriate receptor activation by progesterone, which leads to a marked exacerbation of hypertension in pregnancy, but no proteinuria or other features of pre-eclampsia (Geller et al., 2000).

D. Management of Chronic Hypertension in Pregnancy

Prior to conception, women with chronic hypertension should be counseled on the risks of adverse pregnancy outcomes and blood pressure management should be optimized. During pregnancy, they should be monitored closely for development of pre-eclampsia.

Anti-hypertensive therapy is indicated for the prevention of stroke and cardiovascular complications in severe hypertension (DBP >100 mmHg) (Report of the National High Blood Pressure Education Program Working Group on High Blood Pressure in Pregnancy, 2000). Treatment of less severe hypertension has not, however, shown clear maternal or fetal benefit. In three meta-analyses, no beneficial effect on the development of pre-eclampsia, neonatal death, preterm birth, small-for-gestational-age babies or other adverse outcomes was demonstrated with treatment of mild-to-moderate hypertension, although it did reduce the women's risk of developing severe hypertension (Magee et al., 1999; Ferrer et al., 2000; Abalos et al., 2001). Aggressive treatment of mild-to-moderate hypertension in pregnancy may impair fetal growth. Indeed, reduction of mean arterial pressure effected by anti-hypertensive treatment was associated with lower birth weight and fetal growth restriction, possible as a result of diminished uteroplacental perfusion (von Dadelszen et al., 2000). Therefore, practice guidelines recommend that anti-hypertensive treatment should be administered to newly-diagnosed chronic hypertensives only if evidence of end-organ injury (e.g. proteinuria or cardiomyopathy) is seen, or if SBP is greater than 150–160 mmHg, or DBP exceeds 100–110 mmHg. It is also advised that tapering or discontinuing therapy to meet these goal blood pressures be considered for women previously on anti-hypertensive treatment (ACOG Practice Bulletin, 2002).

Methyldopa, a centrally acting alpha-2 adrenergic agonist, is the first-line oral agent for the management of hypertension in pregnancy, although it is usually not used outside of pregnancy. Although it has the most extensive safety data and has no apparent adverse fetal effects, its main disadvantages are a short half-life, sedating effects and, rarely, elevated liver enzymes and hemolytic anemia. Less data are available on clonidine, but its mechanism and safety profile are largely comparable to those of methyldopa. With the exception of atenolol (fetal growth restriction), beta-adrenergic antagonists are non-teratogenic and used effectively and extensively in pregnancy (Magee and Duley, 2003).

Labetalol is also effective and widely used and may have the advantage of better preservation of uteroplacental blood flow due to its alpha inhibition (Podymow et al., 2004). Calcium channel blockers, including long-acting nifedipine and non-dihydropyridines, appear to be safe in pregnancy, although there has been somewhat less experience with this class of anti-hypertensives (Bortolus et al., 2000; Podymow et al., 2004). Diuretics may be appropriate and effective in the setting of pulmonary edema, however, they are not first-line agents, given that plasma volume is normal in pregnancy and low in pre-eclampsia, in which case they are avoided altogether. Nevertheless, there is no evidence that diuretics are associated with adverse fetal or maternal outcomes. ACE-inhibitors and angiotensin receptor blockers (ARBs) are contraindicated in the 2nd and 3rd trimesters of pregnancy because of their known adverse effects on fetal kidneys, including renal dysgenesis, fetal oliguria, oligohydramnios and neonatal anuric ARF leading to death (Podymow et al., 2004). The limited data on ARBs indicate that their impact on the fetus is similar to that of ACE-Is (Serreau et al., 2005).

Intravenous medications used for urgent control of severe hypertension are classified as pregnancy class C (no controlled studies in humans). However, there is broad clinical experience with several, which has thus far yielded no evidence of adverse effects. These include, in order of known safety, labetalol, calcium channel blockers such as nicardipine, and hydralazine. Nitroprusside is generally avoided, given that it can cause fetal cyanide poisoning if used for more than 4 hours.

V. RENAL FAILURE IN PREGNANCY

A. Acute Renal Failure

Over the past 40 years, the incidence of acute renal failure (ARF) in pregnancy in the developed world has decreased considerably (Selcuk et al., 1998). This decline is felt to stem from improved prenatal care and better availability of safe and legal abortion. In this series, the primary etiologies of ARF were abortion (30%), HELLP syndrome and pre-eclampsia (14%), pre-eclampsia or eclampsia (12%), postpartum hemorrhage (15%), fetal death (12%), abruptio placentae (6%) and placentae previa (1%). In general, the differential diagnosis of ARF in pregnancy includes: prerenal causes, such as hyperemesis gravidarum and uterine hemorrhage; infectious etiologies, such as acute pyelonephritis and septic abortion; renal cortical necrosis; those conditions unique to pregnancy, including severe pre-eclampsia, acute fatty liver of pregnancy and idiopathic postpartum acute renal failure; and obstructive causes, e.g. obstructive uropathy, rarely (Krane, 1988).

1. ACUTE TUBULAR NECROSIS

Acute tubular necrosis (ATN) in pregnancy can be the result of a number of factors. Hyperemesis gravidarum or uterine hemorrhage from placental abruption or previa, failure of the postpartum uterus to contract or uterine lacerations and perforations can lead to severe prerenal azotemia and subsequent ATN. ARF may also be precipitated by intra-amniotic saline administration, amniotic fluid embolism and other factors unrelated to pregnancy.

2. CORTICAL NECROSIS

Placental abruption and septic abortion are associated with bilateral cortical necrosis, which is often severe and irreversible. Septic abortion is infection, usually polymicrobial, of the uterus and surrounding tissues usually after a non-sterile (illegal) abortion, but can occur after any abortion. In countries where abortion is inaccessible or illegal, it remains a serious problem, although it has become rare in countries with more liberalized abortion laws. The typical presentation is vaginal bleeding, abdominal pain and fever, which may progress to shock, with renal failure occurring in up to 73% of cases, commonly accompanied by gross hematuria, flank pain and oligo-anuria (Finkielman et al., 2004).

Placenta abruption in the 3rd trimester of pregnancy can also cause bilateral patchy cortical necrosis characterized by oligo-anuria, although it can also affect the entire renal cortex and result in irreversible renal failure. Other than prompt administration of antibiotics, management of ATN in this setting is chiefly supportive, e.g. volume resuscitation, or possibly delivery if late enough in the pregnancy.

3. THROMBOTIC MICROANGIOPATHY

ARF in the context of thrombotic microangiopathy and pregnancy encompasses five pregnancy syndromes, which can be challenging to distinguish from each other: pre-eclampsia/HELLP, TTP/HUS, acute fatty liver of pregnancy, systemic lupus erythematosus (SLE) with the antiphospholipid antibody syndrome, and disseminated intravascular coagulation (DIC). ARF is rare in pre-eclampsia, but when present, requires urgent delivery and is often accompanied by coagulopathy, hepatic rupture, liver failure or is superimposed on pre-existing renal disease.

Acute fatty liver of pregnancy (AFLP) occurs after mid-gestation, most often near term (Fesenmeier et al., 2005) and is rare, affecting about 1 in 10 000 pregnancies, but potentially life-threatening, with a 10% case fatality rate (Pereira et al., 1997). It is characterized by jaundice and liver dysfunction and, in about half of cases, also complicated by pre-eclampsia. Hemolysis and thrombocytopenia, however, are usually absent, and their presence would be more indicative of HUS/TTP or the HELLP syndrome. The exact mechanism of renal failure is unknown, as renal lesions are both mild and non-specific and appear to be completely reversible. A defect in mitochondrial fatty acid oxidation due to mutations in the long-chain 3-hydroxyacyl CoA dehydrogenase deficiency has been proposed as a risk factor for the development of AFLP (Ibdah et al., 1999). Urgent delivery is necessary in the management of AFLP.

Pregnancy confers a higher risk of HUS/TTP, which is characterized by thrombocytopenia, hemolysis and ARF (or other organ dysfunction) (Vesely et al., 2004). TTP is associated more with the presence of neurological symptoms and HUS with renal failure. Occurrence of TTP is usually prior to 24 weeks and of HUS, typically near term or postpartum (George, 2003). Although relapse is a concern in women with a history of TTP, the data are too limited to determine risk of recurrence (Sibai, 2007). In non-pregnant states, the pathogenesis of TTP has been attributed to the deficiency of the von Willebrand factor cleaving protease, ADAMTS13, the levels of which decline in the 2nd and 3rd trimester. In distinguishing HUS/TTP from pre-eclampsia/HELLP syndrome, a history of preceding proteinuria, hypertension and severe liver injury is more suggestive of the HELLP syndrome, whereas the presence of renal failure and severe hemolytic anemia is more typical of HUS/TTP. The level of ADAMTS13 may also be helpful, as it is expected to remain detectable (albeit lower than in normal pregnant women) in HELLP syndrome, in contrast to its absence in TTP (Lattuada et al., 2003). Plasmapheresis has been reported in uncontrolled studies to be effective and safe (Sibai, 2007). In contrast to the HELLP syndrome, unless the neonate is compromised, the pregnancy may be continued in women with HUS/TTP.

4. Obstruction

In rare cases, ARF from ureteral obstruction by renal calculi can occur. Hypercalciuria during pregnancy is likely a result of a rise in circulating levels of 1, 25 dihydroxyvitamin D3 and higher intestinal calcium absorption (Kumar et al., 1980). However, the incidence of urolithiasis in pregnancy is no higher than that in non-pregnant women of child-bearing age, which is roughly 0.026–0.531%, or 1 in 200 to 1 in 2000 pregnancies (Coe et al., 1978; Drago et al., 1982; Gorton and Whitfield, 1997). The risk may be offset by increases in filtered citrate, magnesium and urinary glycosaminoglycans, which inhibit urinary lithogenesis, as well as possibly by an increase in urine flow and physiological dilation of the urinary tract (Kashyap et al., 2006). The vast majority of women who do develop urolithiasis in pregnancy do so after the 1st trimester (Horowitz and Schmidt, 1985). As in the general population, most stones formed during pregnancy are calcium oxalate and calcium phosphate. Infection risk is increased and severe pain from nephrolithiasis can contribute to premature labor (Hendricks et al., 1991). Treatment includes initial conservative management with hydration, analgesics and anti-emetics, with antibiotics in the case of complication by a urinary tract infection and ureteral catheterization and ureteral stent, if necessary. Lithotripsy, although contraindicated during pregnancy because of adverse fetal effects, has been reported to be used before 12 weeks of gestation without apparent adverse impact on the fetus (Cormier et al., 2006). Thiazide diuretics and allopurinol are contraindicated in pregnancy. If infection is present, antibiotics should be administered for 3–5 weeks, with continued suppressive treatment after delivery, since the stone may still serve as a nidus of infection.

Urinary tract infections (UTI) are the most frequent renal problem during pregnancy (Millar and Cox, 1997). The prevalence (2–10%) of asymptomatic bacteriuria is similar to that in non-pregnant populations, however, in pregnancy, it needs to be managed more aggressively for a number of reasons. Asymptomatic bacteriuria is associated with increased risk of premature delivery and low birth weight (Schieve et al., 1994). Pregnant women are susceptible to ascending pyelonephritis because of physiologic hydronephrosis. Untreated asymptomatic bacteriuria can progress to cystitis or acute pyelonephritis in up to 40% of pregnant women (Hill et al., 2005). Commonly presenting between 20 and 28 weeks of gestation as fevers, loin pain and dysuria, acute pyelonephritis can progress to sepsis, endotoxic shock, disseminated intravascular coagulation and ARF. Treatment of asymptomatic bacteriuria has been shown to reduce the risk of these complications and improve perinatal morbidity and mortality (Smaill, 2001). Pyelonephritis is frequently treated more aggressively with hospitalization and intravenous antibiotics.

B. Chronic Kidney Disease

Pregnancy in renal disease is recognized to be at higher risk for adverse maternal and fetal outcomes. Pre-eclampsia and maternal renal function deterioration, as well as prematurity, IUGR and neonatal death are common (Jones and Hayslett, 1996; Fischer et al., 2004). Compared to women without chronic renal disease, one study reported a near doubling of the frequency of fetal complications and a tripling of the rate of maternal adverse outcomes (Jones and Hayslett, 1996; Fischer et al., 2004). Chronic kidney disease may be exacerbated by the additional stress of increased renal blood flow during pregnancy, as evidenced by the frequent finding of worsening of pre-existing hypertension and proteinuria (Sanders and Lucas, 2001). Not surprisingly, maternal and fetal outcome correlates with the severity of pre-existing renal disease, proteinuria and hypertension. Most authors, however, agree that women with well-controlled blood pressure, serum creatinine 1.4 mg/dl, and no proteinuria can generally expect favorable maternal and neonatal outcomes and are at low risk for permanent renal function deterioration (Jungers et al., 1995; Hou, 1999). A >30% risk of accelerated progression to end-stage renal disease during and after pregnancy is seen when creatinine is >2.0 mg/dl (Jones and Hayslett, 1996). When serum creatinine exceeds 2.5 mg/dl, >70% of the pregnancies end with preterm delivery and >40% are complicated by pre-eclampsia (Jones and Hayslett, 1996; Sanders and Lucas, 2001). Several specific conditions, such as diabetic nephropathy and lupus nephritis may present additional challenges when encountered in the setting of pregnancy and chronic kidney disease.

1. Diabetic Nephropathy

Although diabetes mellitus is less common among women of child-bearing age, the number of diabetic women entering pregnancy is increasing, along with the general rise in incidence and prevalence of diabetes and diabetic nephropathy in the world (Dunne, 2005). In general, women with diabetes, with or without nephropathy are at an increased risk of adverse maternal and neonatal outcomes, relative to non-diabetics (Sibai et al., 2000; Dunne et al., 2003; Clausen et al., 2005). The risk of pre-eclampsia in women with pre-existing diabetes is more than twice that of non-diabetic women (Sibai et al., 2000). The presence of nephropathy further increases the risk of preterm delivery and pre-eclampsia. Suboptimal glycemic control has also been associated with higher risk of fetal malformations and pre-eclampsia (Hiilesmaa et al., 2000; Suhonen et al., 2000). Pregnancy alone does not appear to cause progression of renal disease, as long as kidney function is normal or only very mildly impaired (Hawthorne, 2005). Women with more severe renal impairment are more likely to have irreversible renal damage postpartum, which is evident as early as a few months after delivery (Biesenbach et al., 1999). Although aggressive treatment of hypertension before, throughout and after pregnancy in these women may mitigate the decline in renal function, disease progression is largely unavoidable (Rossing et al., 2002). As a result, women with moderate-to-severe renal impairment are often counseled to postpone pregnancy until after renal transplantation, given that it improves fertility and neonatal outcome and has no impact on kidney function as long as graft function is normal. It has been hypothesized that the children of diabetic mothers have inadequate nephrogenesis and lower nephron mass because of intrauterine diabetes exposure, leading to a higher risk of development of renal disease and hypertension later in life (Nelson et al., 1998; Biesenbach et al., 2000).

2. LUPUS NEPHRITIS

As with other types of chronic kidney disease, maternal renal disease is a strong predictor of adverse fetal outcome (Rahman et al., 1998). Women with SLE alone are already at higher risk for preterm birth, IUGR, pre-eclampsia and spontaneous abortions. With lupus nephritis, hypertension, or proteinuria, these risks are further increased (Khamashta and Hughes, 1996; Soubassi et al., 2004; Dhar et al., 2005; Moroni and Ponticelli, 2005).

Women with SLE and antiphospholipid antibodies are at highest risk of developing pre-eclampsia, thrombosis and experiencing fetal death (Ruiz-Irastorza and Khamashta, 2005). Those with proliferative (WHO class III or IV) or membranous lupus nephritis (WHO class V) incur the greatest risk for pre-eclampsia and low birth weight (Carmona et al., 2005). In addition, lupus flares have been reported to occur at a rate of 48–62% when conception has taken place during active lupus nephritis. Conception during remission corresponds to flare rates of 7–32% (Tandon et al., 2004). Prophylactic steroid therapy does not appear to prevent lupus flares during pregnancy (Khamashta et al., 1997). In general, women with lupus are advised to delay pregnancy until lupus activity is quiescent and immunosuppressants can be minimized (Moroni and Ponticelli, 2005).

A diagnostic challenge arises when a pregnant woman with a history of lupus nephritis presents with worsening proteinuria and hypertension. Distinguishing between pre-eclampsia and lupus flare, however, is critical for women presenting before 37 weeks of gestation because the treatments differ markedly. Lupus nephritis flares are managed by steroids and azathioprine and continuation of the pregnancy but, for pre-eclampsia, delivery is the only definitive treatment. Low complement levels and the presence of hematuria are typically not sensitive or specific enough to differentiate, although an active urinary sediment is often seen in lupus nephritis, while that of pre-eclampsia is bland.

Whether the measurement of angiogenic proteins such as sFlt-1 will help differentiate superimposed pre-eclampsia from lupus nephritis remains unknown (Williams et al., 2005). Nevertheless, a renal biopsy is frequently necessary, in the absence of diagnostic serologic tests, which have yet to be developed.

C. End-Stage Renal Disease and Dialysis

Pregnancy is rare in women undergoing hemodialysis (HD) or peritoneal dialysis. Most women with end-stage renal disease (ESRD) have severe hypothalamic-pituitary-gonadal abnormalities that result in menstrual irregularities, anovulation and infertility (Leavey and Weitzel, 2002), which is reversed by transplantation but not by dialysis. It is thought that uremia leads to abnormal neuroendocrine regulation of hypothalamic GnRH secretion resulting in diminished fertility (Handelsman and Dong, 1993). Nonetheless, successful pregnancy has been reported in women with ESRD on renal replacement. Data on fertility rates in women on chronic dialysis derive from surveys and large retrospective studies. These studies, including the RPDP study, revealed a pregnancy rate of roughly 2% in women of child-bearing age on chronic dialysis over 2–4 years (about 2% of the pregnancy rate in women of similar age in the general population), with approximately 40–60% of pregnancies producing a live infant, though the prematurity rate is extremely high (>80%) (Hou, 1994; Bagon et al., 1998; Okundaye et al., 1998). The most common adverse outcomes were spontaneous miscarriage, pregnancy-induced hypertension, preterm labor, premature rupture of membranes, polyhydramnios and IUGR. Increasing dialysis dose, rather than frequency, appears to improve fetal outcome. The current practice guidelines recommend augmenting the weekly dialysis dose to 20 h. In the RPDP study, 90% of the 10 surviving infants had a gestational age >32 weeks and 57% had birth weights >2.5 kg in the group of women dialyzed for >20 h/week. Among women receiving <14 h/week of dialysis, only 50% of the six surviving infants had a gestational age of >32 weeks and none had birth weight >2.5 kg (Okundaye et al., 1998). Data from a number of other studies have also provided evidence that increased dialysis correlates with greater birth weight and longer gestational age (Souqiyyeh et al., 1992; Bagon et al., 1998). Nocturnal hemodialysis (NHD) has been reported potentially to be the ideal modality to increase dialysis dose in pregnancy, given that it may allow for improved management of volume and blood pressure by minimizing fluctuations in extracellular fluid volume (Gangji et al., 2004). Peritoneal dialysis has also been used during pregnancy, however, increasing abdominal girth requires reductions in dwell volume and an increased number of exchanges to achieve adequate clearance, which may be difficult to maintain (Holley and Reddy, 2003). Finally, dosing of erythropoietin needs to be adjusted to maintain the physiologic anemia of pregnancy, as a high hematocrit can lead to adverse fetal outcomes.

D. Renal Transplant in Pregnancy

As mentioned, successful kidney transplantation normalizes hormonal function and can restore fertility within 6 months in about 90% of women of child-bearing age (Lessan-Pezeshki et al., 2004). Since 1958, more than 14 000 pregnancies in renal allograft recipients have been documented (McKay et al., 2005). Pregnancy following kidney transplantation is successful in >90%, with good maternal and fetal outcomes, however, these pregnancies require the involvement of and close monitoring by transplant nephrologists and obstetricians experienced in taking care of women with kidney disease (Hou, 2003; McKay and Josephson, 2006; Abbud-Filho et al., 2007).

1. Maternal, Fetal and Graft Outcomes

The outcome of the pregnancy is generally determined by maternal kidney function and blood pressure. Approximately 15–20% of pregnancies in kidney transplant patients end in elective termination and 10–15% in miscarriage. Women with serum creatinine <1.5 mg/dl are expected to deliver a live baby in >90% of pregnancies if it continues beyond the 1st trimester, while 75% of those with creatinine >1.5 mg/dl will have a live delivery (Abbud-Filho et al., 2007). There is, however, a high incidence of low birth weight (25–50%) and/or preterm delivery (30–50%) (Armenti et al., 2004; Gutierrez et al., 2005). Pregnancy does not appear adversely to affect graft function in transplant recipients, as long as baseline graft function is normal and significant hypertension is not present (Armenti et al., 2004). On a background of moderate renal insufficiency (creatinine >1.5–1.7 mg/dl), higher risks of developing pre-eclampsia and progressive renal impairment, and of having a small-for-gestational-age infant are incurred (Thompson et al., 2003; Galdo et al., 2005). The rejection rate (3–4%) is similar to that in non-pregnant controls when pregnancy occurs 1–2 years after transplantation (Armenti et al., 2004).

The most common adverse outcomes include hypertension (50–70%), pre-eclampsia (30%), intrauterine growth retardation (IUGR; 20%) and preterm birth (50%), in addition to anemia, urinary tract infection and diabetes (5–10%) (Abbud-Filho et al., 2007). Hypertension typically stems from underlying medical conditions and the use of calcineurin inhibitors. The American Society of Transplantation recommends aggressive treatment of hypertension in pregnant renal transplant recipients, with target blood pressure close to normal, which is lower than the goal blood pressure of pregnant, non-transplanted women with hypertension (Podymow et al., 2004; McKay et al., 2005). The most commonly used anti-hypertensive agents are methyldopa, non-selective beta-adrenergic antagonists (e.g. labetalol) and calcium channel blockers. ACE inhibitors are contraindicated in the 2nd and 3rd trimesters, but should generally be avoided in pregnancy. Transplant recipients are at risk for infections that may have implications for the fetus, including cytomegalovirus, herpes simplex and toxoplasmosis. The rate of bacterial urinary tract infections is also increased ($\approx$13–40%), but these are usually treatable and uncomplicated (Galdo et al., 2005).

Typically, women have been advised to wait 2 years after transplantation before conceiving. Pregnancy within the first 6–12 months after transplantation is not recommended because of a higher risk of acute rejection in this period, higher doses of immunosuppressive agents and an increased risk of infection (European best practice guidelines for renal transplantation, 2002; McKay et al., 2005). However, this needs to be weighed against the downside of age-related decreases in fertility, as many women who undergo renal transplantation are already of advanced maternal age. More recently, the American Society of Transplantation has suggested that in women on stable, low doses of immunosuppressive agents with normal renal function and with no prior rejection episodes, conception could be safely considered as early as 1 year post-transplant (McKay et al., 2005). Current data conflict on whether pregnancy decreases overall graft survival (Salmela et al., 1993; Sturgiss and Davison, 1995).

2. Immunosuppressive therapy

Cyclosporine or tacrolimus and steroids, with or without azathioprine, form the basis of immunosuppression during pregnancy. Low-dose corticosteroids are considered safe for use in pregnancy. Stress-dose steroids are needed at the time of and for 24–48 h after delivery. Azathioprine is safe at low doses, although doses >2 mg/kg/day should be avoided, since they have been associated with congenital anomalies and intrauterine growth retardation (European best practice guidelines for renal transplantation, 2002). Although high doses of cyclosporine and tacrolimus are associated with fetal resorption in animal studies, animal and human data indicate that lower doses of calcineurin inhibitors are safe in pregnancy (Danesi and Del Tacca, 2004; Garcia-Donaire et al., 2005). Thus far, clinical data have shown a possibly higher incidence of low birth weight, but not of congenital malformations (Danesi and Del Tacca, 2004). Cautious monitoring of cyclosporine and tacrolimus levels is necessary in pregnancy because of decreased GI absorption, increased volume of distribution and increased GFR, which can result in substantial fluctuations, with a concomitant risk of acute rejection (Garcia-Donaire et al., 2005). Sirolimus is teratogenic in rats at clinical doses and is therefore contraindicated in pregnancy. Mycophenolate mofetil (MMF), like sirolimus, is contraindicated in pregnancy, given its association with developmental toxicity, malformations and intrauterine death in animal studies at therapeutic dosages. Limited human data suggest it may also be associated with spontaneous abortions and major fetal malformations (Le Ray et al., 2004).

3. ACUTE REJECTION

Acute rejection during pregnancy should be considered if fever, oliguria, graft tenderness or deterioration in renal function is present. The incidence of acute rejection in pregnancy is similar to that in the non-pregnant population (European best practice guidelines for renal transplantation, 2002; Armenti et al., 2004). Confirmation by renal biopsy should be made before instituting treatment, which is administration of high-dose steroids, despite their association with fetal malformations (European best practice guidelines for renal transplantation, 2002; Mastrobattista and Katz, 2004). Limited data are available on the safety of OKT3, antithymocyte globulin, daclizumab and basiliximab in pregnancy (Danesi and Del Tacca, 2004).

References

Abalos, E., Duley, L., Steyn, D. W., and Henderson-Smart, D. J. (2001). Antihypertensive drug therapy for mild to moderate hypertension during pregnancy. *Cochrane Database Syst Rev* **2**: CD002252.

Abbud-Filho, M., Adams, P. L., Alberu, J. et al. (2007). A report of the Lisbon Conference on the care of the kidney transplant recipient. *Transplantation* **83**(Suppl): S1–22.

AbdAlla, S., Lother, H., el Massiery, A., and Quitterer, U. (2001). Increased AT(1) receptor heterodimers in pre-eclampsia mediate enhanced angiotensin II responsiveness. *Nat Med* **7**: 1003–1009.

AbdAlla, S., Abdel-Baset, A., Lother, H., el Massiery, A., and Quitterer, U. (2005). Mesangial AT1/B2 receptor heterodimers contribute to angiotensin II hyperresponsiveness in experimental hypertension. *J Mol Neurosci* **26**: 185–192.

Abi-Said, D., Annegers, J. F., Combs-Cantrell, D., Frankowski, R. F., and Willmore, L. J. (1995). Case-control study of the risk factors for eclampsia. *Am J Epidemiol* **142**: 437–441.

ACOG Practice Bulletin (2001). Chronic hypertension in pregnancy. ACOG Committee on Practice Bulletins. *Obstet Gynecol* **98** suppl, 177-85.

ACOG Practice Bulletin (2002). Diagnosis and management of pre-eclampsia and eclampsia. Number 33, January 2002. American College of Obstetricians and Gynecologists. *Int J Gynaecol Obstet* **77**: 67–75.

Agatisa, P. K., Ness, R. B., Roberts, J. M., Costantino, J. P., Kuller, L. H., and McLaughlin, M. K. (2004). Impairment of endothelial function in women with a history of preeclampsia: an indicator of cardiovascular risk. *Am J Physiol Heart Circ Physiol* **286**: H1389–H1393.

Altman, D., Carroli, G., Duley, L. et al. (2002). Do women with pre-eclampsia, and their babies, benefit from magnesium sulphate? The Magpie Trial: a randomised placebo-controlled trial. *Lancet* **359**: 1877–1890.

Anguiano-Robledo, L., Reyes-Melchor, P. A., Bobadilla-Lugo, R. A., Perez-Alvarez, V. M., and Lopez-Sanchez, P. (2007). Renal angiotensin-II receptors expression changes in a model of pre-eclampsia. *Hypertens Pregnancy* **26**: 151–161.

Anjali, B., Nair, A., Thomas, N., Rajaratnam, S., and Seshadri, M. S. (2004). Secondary hypertension in pregnancy due to an adrenocortical carcinoma. *Aust NZ J Obstet Gynaecol* **44**: 466–467.

Armenti, V. T., Radomski, J. S., Moritz, M. J. et al. (2004). Report from the National Transplantation Pregnancy Registry (NTPR): outcomes of pregnancy after transplantation. *Clin Transpl*, pp. 103–114.

Arngrimsson, R., Sigurdardottir, S., Frigge, M. L. et al. (1999). A genome-wide scan reveals a maternal susceptibility locus for pre-eclampsia on chromosome 2p13. *Hum Mol Genet* **8**: 1799–1805.

Askie, L. M., Duley, L., Henderson-Smart, D. J., Stewart, L. A., and Group, P. C. (2007). Antiplatelet agents for prevention of pre-eclampsia: a meta-analysis of individual patient data. *Lancet* **369**: 1791–1798.

August, P., Lenz, T., Ales, K. L. et al. (1990). Longitudinal study of the renin-angiotensin-aldosterone system in hypertensive pregnant women: deviations related to the development of superimposed preeclampsia. *Am J Obstet Gynecol* **163**: 1612–1621.

August, P., Helseth, G., Cook, E. F., and Sison, C. (2004). A prediction model for superimposed pre-eclampsia in women with chronic hypertension during pregnancy. *Am J Obstet Gynecol* **191**: 1666–1672.

Bagon, J. A., Vernaeve, H., De Muylder, X., Lafontaine, J. J., Martens, J., and Van Roost, G. (1998). Pregnancy and dialysis. *Am J Kidney Dis* **31**: 756–765.

Bailey, R. R., and Rolleston, G. L. (1971). Kidney length and ureteric dilatation in the puerperium. *J Obstet Gynaecol Br Commonw* **78**: 55–61.

Baker, P. N., Broughton Pipkin, F., and Symonds, E. M. (1992). Longitudinal study of platelet angiotensin II binding in human pregnancy. *Clin Sci (Lond)* **82**: 377–381.

Banks, R. E., Forbes, M. A., Searles, J. et al. (1998). Evidence for the existence of a novel pregnancy-associated soluble variant of the vascular endothelial growth factor receptor, Flt-1. *Mol Hum Reprod* **4**: 377–386.

Barak, S., Oettinger-Barak, O., Machtei, E. E., Sprecher, H., and Ohel, G. (2007). Evidence of periopathogenic microorganisms in placentas of women with preeclampsia. *J Periodontol* **78**: 670–676.

Battistini, B., and Dussault, P. (1998). The many aspects of endothelins in ischemia-reperfusion injury: emergence of a key mediator. *J Invest Surg* **11**: 297–313.

Baylis, C. (1980). The mechanism of the increase in glomerular filtration rate in the twelve-day pregnant rat. *J Physiol* **305**: 405–414.

Bdolah, Y., Lam, C., Rajakumar, A. et al. (2006). Pre-eclampsia in twin pregnancies: hypoxia or bigger placental mass? *J Soc Gynecol Investig* **12**: 285A-A 670 Suppl.

Belfort, M. A., Tooke-Miller, C., Allen, J. C. et al. (2001). Changes in flow velocity, resistance indices, and cerebral perfusion pressure in the maternal middle cerebral artery distribution during normal pregnancy. *Acta Obstet Gynecol Scand* **80**: 104–112.

Belfort, M. A., Clark, S. L., and Sibai, B. (2006). Cerebral hemodynamics in preeclampsia: cerebral perfusion and the rationale for an alternative to magnesium sulfate. *Obstet Gynecol Surv* **61**: 655–665.

Belgore, F. M., Lip, G. Y., and Blann, A. D. (2000). Vascular endothelial growth factor and its receptor, Flt-1, in smokers and non-smokers. *Br J Biomed Sci* **57**: 207–213.

Berends, A., Bertoli-Avella, A., de Groot, C., van Duijn, C., Oostra, B., and Steegers, E. (2007). STOX1 gene in pre-eclampsia and intrauterine growth restriction. *Br J Obstet Gynaecol* **114**: 1163–1167.

Biesenbach, G., Grafinger, P., Stoger, H., and Zazgornik, J. (1999). How pregnancy influences renal function in nephropathic type 1 diabetic women depends on their pre-conceptional creatinine clearance. *J Nephrol* **12**: 41–46.

Biesenbach, G., Grafinger, P., Zazgornik, J. et al. (2000). Perinatal complications and three-year follow up of infants of diabetic mothers with diabetic nephropathy stage IV. *Ren Fail* **22**: 573–580.

Boehme, M. W., Schmitt, W. H., Youinou, P., Stremmel, W. R., and Gross, W. L. (1996). Clinical relevance of elevated serum thrombomodulin and soluble E-selectin in patients with Wegener's granulomatosis and other systemic vasculitides. *Am J Med* **101**: 387–394.

Bortolus, R., Ricci, E., Chatenoud, L., and Parazzini, F. (2000). Nifedipine administered in pregnancy: effect on the development of children at 18 months. *Br J Obstet Gynaecol* **107**: 792–794.

Boyd, P. A., Lindenbaum, R. H., and Redman, C. (1987). Pre-eclampsia and trisomy 13: a possible association. *Lancet* **2**: 425–427.

Brosens, I. A., Robertson, W. B., and Dixon, H. G. (1972). The role of the spiral arteries in the pathogenesis of preeclampsia. *Obstet Gynecol Annu* **1**: 177–191.

Buchbinder, A., Sibai, B. M., Caritis, S. et al. (2002). Adverse perinatal outcomes are significantly higher in severe gestational hypertension than in mild preeclampsia. *Am J Obstet Gynecol* **186**: 66–71.

Buhimschi, C. S., Norwitz, E. R., Funai, E. et al. (2005). Urinary angiogenic factors cluster hypertensive disorders and identify women with severe preeclampsia. *Am J Obstet Gynecol* **192**: 734–741.

Buhling, K. J., Elze, L., Henrich, W. et al. (2004). The usefulness of glycosuria and the influence of maternal blood pressure in screening for gestational diabetes. *Eur J Obstet Gynecol Reprod Biol* **113**: 145–148.

Cadroy, Y., Grandjean, H., Pichon, J. et al. (1993). Evaluation of six markers of haemostatic system in normal pregnancy and pregnancy complicated by hypertension or pre-eclampsia. *Br J Obstet Gynaecol* **100**: 416–420.

Caniggia, I., Taylor, C. V., Ritchie, J. W., Lye, S. J., and Letarte, M. (1997). Endoglin regulates trophoblast differentiation along the invasive pathway in human placental villous explants. *Endocrinology* **138**: 4977–4988.

Caniggia, I., Mostachfi, H., Winter, J. et al. (2000). Hypoxia-inducible factor-1 mediates the biological effects of oxygen on human trophoblast differentiation through TGFbeta(3). *J Clin Invest* **105**: 577–587.

Caritis, S., Sibai, B., Hauth, J. et al. (1998). Low-dose aspirin to prevent pre-eclampsia in women at high risk. National Institute of Child Health and Human Development Network of Maternal-Fetal Medicine Units. *N Engl J Med* **338**: 701–705.

Carmona, F., Font, J., Moga, I. et al. (2005). Class III-IV proliferative lupus nephritis and pregnancy: a study of 42 cases. *Am J Reprod Immunol* **53**: 182–188.

Chafetz, I., Kuhnreich, I., Sammar, M. et al. (2007). First-trimester placental protein 13 screening for pre-eclampsia and intrauterine growth restriction. *Am J Obstet Gynecol* **197 35**: e1–7.

Chaiworapongsa, T., Romero, R., Yoshimatsu, J. et al. (2002). Soluble adhesion molecule profile in normal pregnancy and pre-eclampsia. *J Matern Fetal Neonatal Med* **12**: 19–27.

Chaiworapongsa, T., Romero, R., Espinoza, J. et al. (2004). Evidence supporting a role for blockade of the vascular endothelial growth factor system in the pathophysiology of preeclampsia. Young Investigator Award. *Am J Obstet Gynecol* **190**: 1541–1547; discussion 7–50.

Chambers, J. C., Fusi, L., Malik, I. S., Haskard, D. O., De Swiet, M., and Kooner, J. S. (2001). Association of maternal endothelial dysfunction with preeclampsia. *J Am Med Assoc* **285**: 1607–1612.

Chapman, A. B., Abraham, W. T., Zamudio, S. et al. (1998). Temporal relationships between hormonal and hemodynamic changes in early human pregnancy. *Kidney Int* **54**: 2056–2063.

Chappell, L. C., Seed, P. T., Briley, A. L. et al. (1999). Effect of antioxidants on the occurrence of pre-eclampsia in women at increased risk: a randomised trial. *Lancet* **354**: 810–816.

Chappell, L. C., Seed, P. T., Briley, A. et al. (2002). A longitudinal study of biochemical variables in women at risk of preeclampsia. *Am J Obstet Gynecol* **187**: 127–136.

Chavarria, M. E., Lara-Gonzalez, L., Gonzalez-Gleason, A., Sojo, I., and Reyes, A. (2002). Maternal plasma cellular fibronectin concentrations in normal and preeclamptic pregnancies: a longitudinal study for early prediction of preeclampsia. *Am J Obstet Gynecol* **187**: 595–601.

Check, J. H., Caro, J. F., Kendall, B., Peris, L. A., and Wellenbach, B. L. (1979). Cushing's syndrome in pregnancy: effect of associated diabetes on fetal and neonatal complications. *Am J Obstet Gynecol* **133**: 846.

Chesley, L. C., Valenti, C., and Rein, H. (1958). Excretion of sodium loads by nonpregnant and pregnant normal, hypertensive, and pre-eclamptic women. *Metabolism* **7**: 575–588.

Cincotta, R. B., and Brennecke, S. P. (1998). Family history of pre-eclampsia as a predictor for pre-eclampsia in primigravidas. *Int J Gynaecol Obstet* **60**: 23–27.

Clark, B. A., Halvorson, L., Sachs, B., and Epstein, F. H. (1992). Plasma endothelin levels in preeclampsia: elevation and correlation with uric acid levels and renal impairment. *Am J Obstet Gynecol* **166**: 962–968.

Clark, D. E., Smith, S. K., He, Y. et al. (1998). A vascular endothelial growth factor antagonist is produced by the human placenta and released into the maternal circulation. *Biol Reprod* **59**: 1540–1548.

CLASP Collaborative Group (1994). CLASP: a randomised trial of low-dose aspirin for the prevention and treatment of pre-eclampsia among 9364 pregnant women. CLASP (Collaborative Low-dose Aspirin Study in Pregnancy). *Lancet* **343**: 619–629.

Clausen, T. D., Mathiesen, E., Ekbom, P., Hellmuth, E., Mandrup-Poulsen, T., and Damm, P. (2005). Poor pregnancy outcome in women with type 2 diabetes. *Diabetes Care* **28**: 323–328.

Cockell, A. P., Learmont, J. G., Smarason, A. K., Redman, C. W., Sargent, I. L., and Poston, L. (1997). Human placental syncytiotrophoblast microvillous membranes impair maternal vascular endothelial function. *Br J Obstet Gynaecol* **104**: 235–240.

Coe, F. L., Parks, J. H., and Lindheimer, M. D. (1978). Nephrolithiasis during pregnancy. *N Engl J Med* **298**: 324–326.

Combs, C. A., Katz, M. A., Kitzmiller, J. L., and Brescia, R. J. (1993). Experimental pre-eclampsia produced by chronic constriction of the lower aorta: validation with longitudinal blood pressure measurements in conscious rhesus monkeys. *Am J Obstet Gynecol* **169**: 215–223.

Conrad, K. P. (2004). Mechanisms of renal vasodilation and hyperfiltration during pregnancy. *J Soc Gynecol Investig* **11**: 438–448.

Conrad, K. P., Kerchner, L. J., and Mosher, M. D. (1999). Plasma and 24-h NO(x) and cGMP during normal pregnancy and pre-eclampsia in women on a reduced NO(x) diet. *Am J Physiol* **277**: F48–57.

Cormier, C. M., Canzoneri, B. J., Lewis, D. F., Briery, C., Knoepp, L., and Mailhes, J. B. (2006). Urolithiasis in pregnancy: Current diagnosis, treatment, and pregnancy complications. *Obstet Gynecol Surv* **61**: 733–741.

Cudmore, M., Ahmad, S., Al-Ani, B. et al. (2007). Negative regulation of soluble Flt-1 and soluble endoglin release by heme oxygenase-1. *Circulation* **115**: 1789–1797.

Danesi, R., and Del Tacca, M. (2004). Teratogenesis and immunosuppressive treatment. *Transplant Proc* **36**: 705–707.

Davidge, S. T., Stranko, C. P., and Roberts, J. M. (1996). Urine but not plasma nitric oxide metabolites are decreased in women with preeclampsia. *Am J Obstet Gynecol* **174**: 1008–1013.

Davis, B. A., Crook, J. E., Vestal, R. E., and Oates, J. A. (1979). Prevalence of renovascular hypertension in patients with grade III or IV hypertensive retinopathy. *N Engl J Med* **301**: 1273–1276.

Davison, J. M. (1997). Edema in pregnancy. *Kidney Int Suppl* **59**: S90–S96.

Davison, J. M., and Noble, M. C. (1981). Serial changes in 24 hour creatinine clearance during normal menstrual cycles and the first trimester of pregnancy. *Br J Obstet Gynaecol* **88**: 10–17.

Davison, J. M., Shiells, E. A., Philips, P. R., and Lindheimer, M. D. (1990). Influence of humoral and volume factors on altered osmoregulation of normal human pregnancy. *Am J Physiol* **258**: F900–F907.

Davison, J. M., Nakagawa, Y., Coe, F. L., and Lindheimer, M. D. (1993). Increases in urinary inhibitor activity and excretion of an inhibitor of cyrstalluria in pregnancy-a defence against the hypercalciuria of normal gestation. *Hyperten Pregnancy* **12**: 25–35.

De Wolf, F., De Wolf-Peeters, C., Brosens, I., and Robertson, W. B. (1980). The human placental bed: electron microscopic study of trophoblastic invasion of spiral arteries. *Am J Obstet Gynecol* **137**: 58–70.

Dechend, R., Homuth, V., Wallukat, G. et al. (2000). AT(1) receptor agonistic antibodies from preeclamptic patients cause vascular cells to express tissue factor. *Circulation* **101**: 2382–2387.

Del Giudice, A., Bisceglia, M., D'Errico, M. et al. (1998). Extra-adrenal functional paraganglioma (phaeochromocytoma) associated with renal-artery stenosis in a pregnant woman. *Nephrol Dial Transplant* **13**: 2920–2923.

Desai, D. K., Moodley, J., and Naidoo, D. P. (2004). Echocardiographic assessment of cardiovascular hemodynamics in normal pregnancy. *Obstet Gynecol* **104**: 20–29.

Dessole, S., Capobianco, G., Virdis, P., Rubattu, G., Cosmi, E., and Porcu, A. (2007). Hepatic rupture after cesarean section in a patient with HELLP syndrome: a case report and review of the literature. *Arch Gynecol Obstet* **276**: 189–192.

Dhar, J. P., Essenmacher, L. M., Ager, J. W., and Sokol, R. J. (2005). Pregnancy outcomes before and after a diagnosis of systemic lupus erythematosus. *Am J Obstet Gynecol* **193**: 1444–1455.

Drago, J. R., Rohner, T. J., and Chez, R. A. (1982). Management of urinary calculi in pregnancy. *Urology* **20**: 578–581.

Duckitt, K., and Harrington, D. (2005). Risk factors for pre-eclampsia at antenatal booking: systematic review of controlled studies. *Br Med J* **330**: 565.

Duley, L. (1992). Maternal mortality associated with hypertensive disorders of pregnancy in Africa, Asia, Latin America and the Caribbean. *Br J Obstet Gynaecol* **99**: 547–553.

Dunlop, W. (1981). Serial changes in renal haemodynamics during normal human pregnancy. *Br J Obstet Gynaecol* **88**: 1–9.

Dunne, F. (2005). Type 2 diabetes and pregnancy. *Semin Fetal Neonatal Med* **10**: 333–339.

Dunne, F., Brydon, P., Smith, K., and Gee, H. (2003). Pregnancy in women with type 2 diabetes: 12 years outcome data 1990–2002. *Diabet Med* **20**: 734–738.

Ebos, J. M., Bocci, G., Man, S. et al. (2004). A naturally occurring soluble form of vascular endothelial growth factor receptor 2 detected in mouse and human plasma. *Mol Cancer Res* **2**: 315–326.

Elsheikh, A., Creatsas, G., Mastorakos, G., Milingos, S., Loutradis, D., and Michalas, S. (2001). The renin-aldosterone system during normal and hypertensive pregnancy. *Arch Gynecol Obstet* **264**: 182–185.

Eremina, V., Sood, M., Haigh, J. et al. (2003). Glomerular-specific alterations of VEGF-A expression lead to distinct congenital and acquired renal diseases. *J Clin Invest* **111**: 707–716.

Esplin, M. S., Fausett, M. B., Fraser, A. et al. (2001). Paternal and maternal components of the predisposition to preeclampsia. *N Engl J Med* **344**: 867–872.

Esser, S., Wolburg, K., Wolburg, H., Breier, G., Kurzchalia, T., and Risau, W. (1998). Vascular endothelial growth factor induces endothelial fenestrations in vitro. *J Cell Biol* **140**: 947–959.

Estelles, A., Gilabert, J., Aznar, J., Loskutoff, D. J., and Schleef, R. R. (1989). Changes in the plasma levels of type 1 and type 2 plasminogen activator inhibitors in normal pregnancy and in patients with severe preeclampsia. *Blood* **74**: 1332–1338.

European best practice guidelines for renal transplantation (2002). Section IV: Long-term management of the transplant recipient. IV.10. Pregnancy in renal transplant recipients. *Nephrol Dial Transplant* 17 Suppl 4, 50-5.

Fainaru, O., Almog, B., Gamzu, R., Lessing, J. B., and Kupferminc, M. (2002). The management of symptomatic hydronephrosis in pregnancy. *Br J Obstet Gynaecol* **109**: 1385–1387.

Ferrer, R. L., Sibai, B. M., Mulrow, C. D., Chiquette, E., Stevens, K. R., and Cornell, J. (2000). Management of mild chronic hypertension during pregnancy: a review. *Obstet Gynecol* **96**: 849–860.

Ferris, T. F. (1991). Pregnancy, preeclampsia, and the endothelial cell. *N Engl J Med* **325**: 1439–1440.

Ferris, T. F., and Gorden, P. (1968). Effect of angiotensin and norepinephrine upon urate clearance in man. *Am J Med* **44**: 359–365.

Fesenmeier, M. F., Coppage, K. H., Lambers, D. S., Barton, J. R., and Sibai, B. M. (2005). Acute fatty liver of pregnancy in 3 tertiary care centers. *Am J Obstet Gynecol* **192**: 1416–1419.

Finkielman, J. D., De Feo, F. D., Heller, P. G., and Afessa, B. (2004). The clinical course of patients with septic abortion admitted to an intensive care unit. *Intensive Care Med* **30**: 1097–1102.

Fiore, G., Florio, P., Micheli, L. et al. (2005). Endothelin-1 triggers placental oxidative stress pathways: putative role in preeclampsia. *J Clin Endocrinol Metab* **90**: 4205–4210.

Fisher, K. A., Luger, A., Spargo, B. H., and Lindheimer, M. D. (1981). Hypertension in pregnancy: clinical-pathological correlations and remote prognosis. *Medicine (Balt.)* **60**: 267–276.

Fitzgerald, D. J., Entman, S. S., Mulloy, K., and FitzGerald, G. A. (1987). Decreased prostacyclin biosynthesis preceding the clinical manifestation of pregnancy-induced hypertension. *Circulation* **75**: 956–963.

Fitzgerald, D. J., Rocki, W., Murray, R., Mayo, G., and FitzGerald, G. A. (1990). Thromboxane A2 synthesis in pregnancy-induced hypertension. *Lancet* **335**: 751–754.

Fonseca, J. E., Mendez, F., Catano, C., and Arias, F. (2005). Dexamethasone treatment does not improve the outcome of women with HELLP syndrome: a double-blind, placebo-controlled, randomized clinical trial. *Am J Obstet Gynecol* **193**: 1591–1598.

Friedman, S. A., Schiff, E., Emeis, J. J., Dekker, G. A., and Sibai, B. M. (1995). Biochemical corroboration of endothelial involvement in severe preeclampsia. *Am J Obstet Gynecol* **172**: 202–203.

Frusca, T., Soregaroli, M., Platto, C., Enterri, L., Lojacono, A., and Valcamonico, A. (2003). Uterine artery velocimetry in patients with gestational hypertension. *Obstet Gynecol* **102**: 136–140.

Fu, M. L., Herlitz, H., Schulze, W. et al. (2000). Autoantibodies against the angiotensin receptor (AT1) in patients with hypertension. *J Hypertens* **18**: 945–953.

Gaber, L. W., and Spargo, B. H. (1987). Pregnancy-induced nephropathy: the significance of focal segmental glomerulosclerosis. *Am J Kidney Dis* **9**: 317–323.

Galatius, S., Bent-Hansen, L., Wroblewski, H., Sorensen, V. B., Norgaard, T., and Kastrup, J. (2000). Plasma disappearance of albumin and impact of capillary thickness in idiopathic dilated cardiomyopathy and after heart transplantation. *Circulation* **102**: 319–325.

Galdo, T., Gonzalez, F., Espinoza, M. et al. (2005). Impact of pregnancy on the function of transplanted kidneys. *Transplant Proc* **37**: 1577–1579.

Gallery, E. D. (1999). Chronic essential and secondary hypertension in pregnancy. *Baillieres Best Pract Res Clin Obstet Gynaecol* **13**: 115–130.

Gallery, E. D., and Brown, M. A. (1987). Control of sodium excretion in human pregnancy. *Am J Kidney Dis* **9**: 290–295.

Gallery, E. D., and Gyory, A. Z. (1979). Glomerular and proximal renal tubular function in pregnancy-associated hypertension: a prospective study. *Eur J Obstet Gynecol Reprod Biol* **9**: 3–12.

Gando, S., Nakanishi, Y., Kameue, T., and Nanzaki, S. (1995). Soluble thrombomodulin increases in patients with disseminated intravascular coagulation and in those with multiple organ dysfunction syndrome after trauma: role of neutrophil elastase. *J Trauma* **39**: 660–664.

Gangji, A. S., Windrim, R., Gandhi, S., Silverman, J. A., and Chan, C. T. (2004). Successful pregnancy with nocturnal hemodialysis. *Am J Kidney Dis* **44**: 912–916.

Gant, N. F., Daley, G. L., Chand, S., Whalley, P. J., and MacDonald, P. C. (1973). A study of angiotensin II pressor response throughout primigravid pregnancy. *J Clin Invest* **52**: 2682–2689.

Gant, N. F., Chand, S., Whalley, P. J., and MacDonald, P. C. (1974). The nature of pressor responsiveness to angiotensin II in human pregnancy. *Obstet Gynecol* **43**: 854.

Garcia-Donaire, J. A., Acevedo, M., Gutierrez, M. J. et al. (2005). Tacrolimus as basic immunosuppression in pregnancy after renal transplantation. A single-center experience. *Transplant Proc* **37**: 3754–3755.

Garmendia, J. V., Gutierrez, Y., Blanca, I., Bianco, N. E., and De Sanctis, J. B. (1997). Nitric oxide in different types of hypertension during pregnancy. *Clin Sci (Lond.)* **93**: 413–421.

Garovic, V. D., Wagner, S. J., Turner, S. T. et al. (2007). Urinary podocyte excretion as a marker for preeclampsia. *Am J Obstet Gynecol* **196**(320): e1–e7.

Geller, D. S., Farhi, A., Pinkerton, N. et al. (2000). Activating mineralocorticoid receptor mutation in hypertension exacerbated by pregnancy. *Science* **289**: 119–123.

George, J. N. (2003). The association of pregnancy with thrombotic thrombocytopenic purpura-hemolytic uremic syndrome. *Curr Opin Hematol* **10**: 339–344.

Germain, S. J., Sacks, G. P., Soorana, S. R., Sargent, I. L., and Redman, C. W. (2007). Systemic inflammatory priming in normal pregnancy and preeclampsia: the role of circulating syncytiotrophoblast microparticles. *J Immunol* **178**: 5949–5956.

Girouard, J., Giguere, Y., Moutquin, J. M., and Forest, J. C. (2007). Previous hypertensive disease of pregnancy is associated with alterations of markers of insulin resistance. *Hypertension* **49**: 1056–1062.

Goodman, R. P., Killam, A. P., Brash, A. R., and Branch, R. A. (1982). Prostacyclin production during pregnancy: comparison of production during normal pregnancy and pregnancy complicated by hypertension. *Am J Obstet Gynecol* **142**: 817–822.

Gordon, M., Landon, M. B., Samuels, P., Hissrich, S., and Gabbe, S. G. (1996). Perinatal outcome and long-term follow-up associated with modern management of diabetic nephropathy. *Obstet Gynecol* **87**: 401–409.

Gorton, E., and Whitfield, H. N. (1997). Renal calculi in pregnancy. *Br J Urol* **80**(Suppl 1): 4–9.

Govan, A. D. (1954). Renal changes in eclampsia. *J Pathol Bacteriol* **67**: 311–322.

Greenberg, S. G., Baker, R. S., Yang, D., and Clark, K. E. (1997). Effects of continuous infusion of endothelin-1 in pregnant sheep. *Hypertension* **30**: 1585–1590.

Gutierrez, M. J., Acebedo-Ribo, M., Garcia-Donaire, J. A. et al. (2005). Pregnancy in renal transplant recipients. *Transplant Proc* **37**: 3721–3722.

Haddad, B., Barton, J. R., Livingston, J. C., Chahine, R., and Sibai, B. M. (2000). Risk factors for adverse maternal outcomes among women with HELLP (hemolysis, elevated liver enzymes, and low platelet count) syndrome. *Am J Obstet Gynecol* **183**: 444–448.

Hall, D. R., Odendaal, H. J., Steyn, D. W., and Grove, D. (2002). Urinary protein excretion and expectant management of early onset, severe pre-eclampsia. *Int J Gynaecol Obstet* **77**: 1–6.

Hammond, T. G., Buchanan, J. D., Scoggins, B. A., Thatcher, R., and Whitworth, J. A. (1982). Primary hyperaldosteronism in pregnancy. *Aust NZ J Med* **12**: 537–539.

Handelsman, D. J., and Dong, Q. (1993). Hypothalamo-pituitary gonadal axis in chronic renal failure. *Endocrinol Metab Clin North Am* **22**: 145–161.

Hannaford, P., Ferry, S., and Hirsch, S. (1997). Cardiovascular sequelae of toxaemia of pregnancy. *Heart* **77**: 154–158.

Hawthorne, G. (2005). Preconception care in diabetes. *Semin Fetal Neonatal Med* **10**: 325–332.

Hayashida, M., Watanabe, N., Imamura, H. et al. (2005). Congenital solitary kidney with renovascular hypertension diagnosed by means of captopril-enhanced renography and magnetic resonance angiography. *Int Heart J* **46:** 347–353.

He, H., Venema, V. J., Gu, X., Venema, R. C., Marrero, M. B., and Caldwell, R. B. (1999). Vascular endothelial growth factor signals endothelial cell production of nitric oxide and prostacyclin through flk-1/KDR activation of c-Src. *J Biol Chem* **274:** 25130–25135.

Heaney, R. P., and Skillman, T. G. (1971). Calcium metabolism in normal human pregnancy. *J Clin Endocrinol Metab* **33:** 661–670.

Heaton, J. M., and Turner, D. R. (1985). Persistent renal damage following pre-eclampsia: a renal biopsy study of 13 patients. *J Pathol* **147:** 121–126.

Helmbrecht, G. D., Farhat, M. Y., Lochbaum, L. et al. (1996). L-arginine reverses the adverse pregnancy changes induced by nitric oxide synthase inhibition in the rat. *Am J Obstet Gynecol* **175:** 800–805.

Hendricks, S. K., Ross, S. O., and Krieger, J. N. (1991). An algorithm for diagnosis and therapy of management and complications of urolithiasis during pregnancy. *Surg Gynecol Obstet* **172:** 49–54.

Hertig, A. T. (1945). Vascular pathology in the hypertensive albuminuric toxemias of pregnancy. *Clinics* **4:** 602–614.

Hertig, A., Berkane, N., Lefevre, G. et al. (2004). Maternal serum sFlt1 concentration is an early and reliable predictive marker of preeclampsia. *Clin Chem* **50:** 1702–1703.

Hiby, S. E., Walker, J. J., O'Shaughnessy, K. M. et al. (2004). Combinations of maternal KIR and fetal HLA-C genes influence the risk of pre-eclampsia and reproductive success. *J Exp Med* **200:** 957–965.

Higby, K., Suiter, C. R., Phelps, J. Y., Siler-Khodr, T., and Langer, O. (1994). Normal values of urinary albumin and total protein excretion during pregnancy. *Am J Obstet Gynecol* **171:** 984–989.

Hiilesmaa, V., Suhonen, L., and Teramo, K. (2000). Glycaemic control is associated with pre-eclampsia but not with pregnancy-induced hypertension in women with type I diabetes mellitus. *Diabetologia* **43:** 1534–1539.

Hill, K., AbouZhar, C., and Wardlaw, T. (2001). Estimates of maternal mortality for 1995. *Bull World Health Organ* **79:** 182–193.

Hill, J. B., Sheffield, J. S., McIntire, D. D., and Wendel, G. D. (2005). Acute pyelonephritis in pregnancy. *Obstet Gynecol* **105:** 18–23.

Hinchey, J., Chaves, C., Appignani, B. et al. (1996). A reversible posterior leukoencephalopathy syndrome. *N Engl J Med* **22** (334): 494–500.

Hirtenlehner, K., Pollheimer, J., Lichtenberger, C. et al. (2003). Elevated serum concentrations of the angiogenesis inhibitor endostatin in preeclamptic women. *J Soc Gynecol Investig* **10:** 412–417.

Holden, D. P., Fickling, S. A., Whitley, G. S., and Nussey, S. S. (1998). Plasma concentrations of asymmetric dimethylarginine, a natural inhibitor of nitric oxide synthase, in normal pregnancy and preeclampsia. *Am J Obstet Gynecol* **178:** 551–556.

Holley, J. L., and Reddy, S. S. (2003). Pregnancy in dialysis patients: a review of outcomes, complications, and management. *Semin Dial* **16:** 384–388.

Horowitz, E., and Schmidt, J. D. (1985). Renal calculi in pregnancy. *Clin Obstet Gynecol* **28:** 324–338.

Hou, S. H. (1994). Frequency and outcome of pregnancy in women on dialysis. *Am J Kidney Dis* **23:** 60–63.

Hou, S. (1999). Pregnancy in chronic renal insufficiency and end-stage renal disease. *Am J Kidney Dis* **33:** 235–252.

Hou, S. (2003). Pregnancy in renal transplant recipients. *Adv Ren Replace Ther* **10:** 40–47.

Hsu, C. D., Iriye, B., Johnson, T. R., Witter, F. R., Hong, S. F., and Chan, D. W. (1993). Elevated circulating thrombomodulin in severe preeclampsia. *Am J Obstet Gynecol* **169:** 148–149.

Hubel, C. A. (1999). Oxidative stress in the pathogenesis of preeclampsia. *Proc Soc Exp Biol Med* **222:** 222–235.

Hulstein, J. J., van Runnard Heimel, P. J., Franx, A. et al. (2006). Acute activation of the endothelium results in increased levels of active von Willebrand factor in hemolysis, elevated liver enzymes and low platelets (HELLP) syndrome. *J Thromb Haemost* **4:** 2569–2575.

Ibdah, J. A., Bennett, M. J., Rinaldo, P. et al. (1999). A fetal fatty-acid oxidation disorder as a cause of liver disease in pregnant women. *N Engl J Med* **340:** 1723–1731.

Irgens, H. U., Reisaeter, L., Irgens, L. M., and Lie, R. T. (2001). Long term mortality of mothers and fathers after pre-eclampsia: population based cohort study. *Br Med J* **323:** 1213–1217.

Jones, D. C., and Hayslett, J. P. (1996). Outcome of pregnancy in women with moderate or severe renal insufficiency. *N Engl J Med* **335:** 226–232.

Jungers, P., Houillier, P., Forget, D. et al. (1995). Influence of pregnancy on the course of primary chronic glomerulonephritis. *Lancet* **346:** 1122–1124.

Kabbinavar, F., Hurwitz, H. I., Fehrenbacher, L. et al. (2003). Phase II, randomized trial comparing bevacizumab plus fluorouracil (FU)/leucovorin (LV) with FU/LV alone in patients with metastatic colorectal cancer. *J Clin Oncol* **21:** 60–65.

Kanayama, N., Takahashi, K., Matsuura, T. et al. (2002). Deficiency in p57Kip2 expression induces preeclampsia-like symptoms in mice. *Mol Hum Reprod* **8:** 1129–1135.

Karumanchi, S. A., Maynard, S. E., Stillman, I. E., Epstein, F. H., and Sukhatme, V. P. (2005). Preeclampsia: a renal perspective. *Kidney Int* **67:** 2101–2113.

Kashyap, M. K., Saxena, S. V., Khullar, M., Sawhney, H., and Vasishta, K. (2006). Role of anion gap and different electrolytes in hypertension during pregnancy (preeclampsia). *Mol Cell Biochem* **282:** 157–167.

Kayisli, U. A., Cayli, S., Seval, Y., Tertemiz, F., Huppertz, B., and Demir, R. (2006). Spatial and temporal distribution of Tie-1 and Tie-2 during very early development of the human placenta. *Placenta* **27:** 648–659.

Kendall, R. L., and Thomas, K. A. (1993). Inhibition of vascular endothelial cell growth factor activity by an endogenously encoded soluble receptor. *Proc Natl Acad Sci USA* **90:** 10705–10709.

Khamashta, M. A., and Hughes, G. R. (1996). Pregnancy in systemic lupus erythematosus. *Curr Opin Rheumatol* **8:** 424–429.

Khamashta, M. A., Ruiz-Irastorza, G., and Hughes, G. R. (1997). Systemic lupus erythematosus flares during pregnancy. *Rheum Dis Clin North Am* **23:** 15–30.

Khanna, N., and Nguyen, H. (2001). Reversible acute renal failure in association with bilateral ureteral obstruction and hydronephrosis in pregnancy. *Am J Obstet Gynecol* **184:** 239–240.

Khosla, U. M., Zharikov, S., Finch, J. L. et al. (2005). Hyperuricemia induces endothelial dysfunction. *Kidney Int* **67:** 1739–1742.

Klonoff-Cohen, H. S., Savitz, D. A., Cefalo, R. C., and McCann, M. F. (1989). An epidemiologic study of contraception and pre-eclampsia. *J Am Med Assoc* **262**: 3143–3147.

Knox, K. S., and Baker, J. C. (2007). Genome-wide expression profiling of placentas in the p57Kip2 model of pre-eclampsia. *Mol Hum Reprod* **13**: 251–263.

Kobayashi, H., Sadakata, H., Suzuki, K., She, M. Y., Shibata, S., and Terao, T. (1998). Thrombomodulin release from umbilical endothelial cells initiated by pre-eclampsia plasma-induced neutrophil activation. *Obstet Gynecol* **92**: 425–430.

Koelman, C. A., Coumans, A. B., Nijman, H. W., Doxiadis, I. I., Dekker, G. A., and Claas, F. H. (2000). Correlation between oral sex and a low incidence of preeclampsia: a role for soluble HLA in seminal fluid? *J Reprod Immunol* **46**: 155–166.

Koga, K., Osuga, Y., Yoshino, O. et al. (2003). Elevated serum soluble vascular endothelial growth factor receptor 1 (sVEGFR-1) levels in women with preeclampsia. *J Clin Endocrinol Metab* **88**: 2348–2351.

Krane, N. K. (1988). Acute renal failure in pregnancy. *Arch Intern Med* **148**: 2347–2357.

Kuenen, B. C., Levi, M., Meijers, J. C. et al. (2002). Analysis of coagulation cascade and endothelial cell activation during inhibition of vascular endothelial growth factor/vascular endothelial growth factor receptor pathway in cancer patients. *Arterioscler Thromb Vasc Biol* **22**: 1500–1505.

Kumar, D. (1962). Chronic placental ischemia in relation to toxemias of pregnancy. A preliminary report. *Am J Obstet Gynecol* **84**: 1323–1329.

Kumar, R., Cohen, W. R., and Epstein, F. H. (1980). Vitamin D and calcium hormones in pregnancy. *N Engl J Med* **302**: 1143–1145.

Lachmeijer, A. M., Arngrimsson, R., Bastiaans, E. J. et al. (2001). A genome-wide scan for pre-eclampsia in the Netherlands. *Eur J Hum Genet* **9**: 758–764.

Lafayette, R. A., Druzin, M., Sibley, R. et al. (1998). Nature of glomerular dysfunction in pre-eclampsia. *Kidney Int* **54**: 1240–1249.

Lafayette, R. A., Malik, T., Druzin, M., Derby, G., and Myers, B. D. (1999). The dynamics of glomerular filtration after Caesarean section. *J Am Soc Nephrol* **10**: 1561–1565.

Lain, K. Y., Wilson, J. W., Crombleholme, W. R., Ness, R. B., and Roberts, J. M. (2003). Smoking during pregnancy is associated with alterations in markers of endothelial function. *Am J Obstet Gynecol* **189**: 1196–1201.

Laivuori, H., Lahermo, P., Ollikainen, V. et al. (2003). Susceptibility loci for pre-eclampsia on chromosomes 2p25 and 9p13 in Finnish families. *Am J Hum Genet* **72**: 168–177.

Lam, C., Lim, K. H., Kang, D. H., and Karumanchi, S. A. (2005). Uric acid and preeclampsia. *Semin Nephrol* **25**: 56–60.

Larrabee, K. D., and Monga, M. (1997). Women with sickle cell trait are at increased risk for preeclampsia. *Am J Obstet Gynecol* **177**: 425–428.

Lattuada, A., Rossi, E., Calzarossa, C., Candolfi, R., and Mannucci, P. M. (2003). Mild to moderate reduction of a von Willebrand factor cleaving protease (ADAMTS-13) in pregnant women with HELLP microangiopathic syndrome. *Haematologica* **88**: 1029–1034.

Le Ray, C., Coulomb, A., Elefant, E., Frydman, R., and Audibert, F. (2004). Mycophenolate mofetil in pregnancy after renal transplantation: a case of major fetal malformations. *Obstet Gynecol* **103**: 1091–1094.

Leavey, S. F., and Weitzel, W. F. (2002). Endocrine abnormalities in chronic renal failure. *Endocrinol Metab Clin North Am* **31**: 107–119.

Lees, C., Parra, M., Missfelder-Lobos, H., Morgans, A., Fletcher, O., and Nicolaides, K. H. (2001). Individualized risk assessment for adverse pregnancy outcome by uterine artery Doppler at 23 weeks. *Obstet Gynecol* **98**: 369–373.

Lessan-Pezeshki, M., Ghazizadeh, S., Khatami, M. R. et al. (2004). Fertility and contraceptive issues after kidney transplantation in women. *Transplant Proc* **36**: 1405–1406.

Levine, R. J., Hauth, J. C., Curet, L. B. et al. (1997). Trial of calcium to prevent preeclampsia. *N Engl J Med* **337**: 69–76.

Levine, R. J., Maynard, S. E., Qian, C. et al. (2004). Circulating angiogenic factors and the risk of preeclampsia. *N Engl J Med* **350**: 672–683.

Levine, R. J., Thadhani, R., Qian, C. et al. (2005). Urinary placental growth factor and risk of preeclampsia. *J Am Med Assoc* **293**: 77–85.

Levine, R. J., Lam, C., Qian, C. et al. (2006). Soluble endoglin and other circulating antiangiogenic factors in preeclampsia. *N Engl J Med* **355**: 992–1005.

Li, Z., Zhang, Y., Ying Ma, J. et al. (2007). Recombinant vascular endothelial growth factor 121 attenuates hypertension and improves kidney damage in a rat model of preeclampsia. *Hypertension* **50**: 686–692.

Lim, K. H., Friedman, S. A., Ecker, J. L., Kao, L., and Kilpatrick, S. J. (1998). The clinical utility of serum uric acid measurements in hypertensive diseases of pregnancy. *Am J Obstet Gynecol* **178**: 1067–1071.

Lind, T., Godfrey, K. A., Otun, H., and Philips, P. R. (1984). Changes in serum uric acid concentrations during normal pregnancy. *Br J Obstet Gynaecol* **91**: 128–132.

Livingston, J. C., Haddad, B., Gorski, L. A. et al. (2001). Placenta growth factor is not an early marker for the development of severe preeclampsia. *Am J Obstet Gynecol* **184**: 1218–1220.

Lockwood, C. J., Toti, P., Arcuri, F. et al. (2007). Thrombin regulates soluble fms-like tyrosine kinase-1 (sFlt-1) expression in first trimester decidua: implications for preeclampsia. *Am J Pathol* **170**: 1398–1405.

Lucas, M. J., Leveno, K. J., and Cunningham, F. G. (1995). A comparison of magnesium sulfate with phenytoin for the prevention of eclampsia. *N Engl J Med* **333**: 201–205.

Lyons, H. A. (1976). Centrally acting hormones and respiration. *Pharmacol Ther (B)* **2**: 743–751.

MacKay, A. P., Berg, C. J., and Atrash, H. K. (2001). Pregnancy-related mortality from pre-eclampsia and eclampsia. *Obstet Gynecol* **97**: 533–538.

Magee, L. A., and Duley, L. (2003). Oral beta-blockers for mild to moderate hypertension during pregnancy. *Cochrane Database Syst Rev* **3**: CD002863.

Magee, L. A., Ornstein, M. P., and von Dadelszen, P. (1999). *Fortnightly review: management of hypertension in pregnancy.* *Br Med J* **318**: 1332–1336.

Marin, R., Gorostidi, M., Portal, C. G., Sanchez, M., Sanchez, E., and Alvarez, J. (2000). Long-term prognosis of hypertension in pregnancy. *Hypertens Pregnancy* **19**: 199–209.

Martin, J. N., May, W. L., Magann, E. F., Terrone, D. A., Rinehart, B. K., and Blake, P. G. (1999). Early risk assessment of severe preeclampsia: admission battery of symptoms and laboratory tests to predict likelihood of subsequent significant maternal morbidity. *Am J Obstet Gynecol* **180**: 1407–1414.

Mason, R. J., Murray, J. F., Broaddus, V. C., and Nadel, J. A. (2005). *Murray and Nadel's Textbook of Respiratory Medicine*. Elsevier Inc., Philadelphia.

Mastrobattista, J. M., and Katz, A. R. (2004). Pregnancy after organ transplant. *Obstet Gynecol Clin North Am* **31**: 415–428 vii.

Masuda, Y., Shimizu, A., Mori, T. et al. (2001). Vascular endothelial growth factor enhances glomerular capillary repair and accelerates resolution of experimentally induced glomerulonephritis. *Am J Pathol* **159**: 599–608.

Mautner, W., Churg, J., Grishman, E., and Dachs, S. (1962). Preeclamptic nephropathy. An electron microscopic study. *Lab Invest* **11**: 518–530.

Maynard, S. E., Min, J. Y., Merchan, J. et al. (2003). Excess placental soluble fms-like tyrosine kinase 1 (sFlt1) may contribute to endothelial dysfunction, hypertension, and proteinuria in preeclampsia. *J Clin Invest* **111**: 649–658.

Maynard, S. E., Moore-Simas, T., and Solitro, M. (2005). Circulating soluble fms-like tyrosine kinase-1 (sFlt1) is increased in twin vs. singleton pregnancies (Abstract). Annual Meeting of the National Society for the Study of Hypertension in Pregnancy.

McCowan, L. M., Buist, R. G., North, R. A., and Gamble, G. (1996). Perinatal morbidity in chronic hypertension. *Br J Obstet Gynaecol* **103**: 123–129.

McKay, D. B., and Josephson, M. A. (2006). Pregnancy in recipients of solid organs – effects on mother and child. *N Engl J Med* **354**: 1281–1293.

McKay, D. B., Josephson, M. A., Armenti, V. T. et al. (2005). Reproduction and transplantation: report on the AST Consensus Conference on Reproductive Issues and Transplantation. *Am J Transplant* **5**: 1592–1599.

Meekins, J. W., Pijnenborg, R., Hanssens, M., McFadyen, I. R., and van Asshe, A. (1994). A study of placental bed spiral arteries and trophoblast invasion in normal and severe pre-eclamptic pregnancies. *Br J Obstet Gynaecol* **101**: 669–674.

Millar, L. K., and Cox, S. M. (1997). Urinary tract infections complicating pregnancy. *Infect Dis Clin North Am* **11**: 13–26.

Mills, J. L., DerSimonian, R., Raymond, E. et al. (1999). Prostacyclin and thromboxane changes predating clinical onset of preeclampsia: a multicenter prospective study. *J Am Med Assoc* **282**: 356–362.

Moll, S. J., Jones, C. J., Crocker, I. P., Baker, P. N., and Heazell, A. E. (2007). Epidermal growth factor rescues trophoblast apoptosis induced by reactive oxygen species. *Apoptosis* **12**: 1611–1622.

Moran, P., Baylis, P. H., Lindheimer, M. D., and Davison, J. M. (2003). Glomerular ultrafiltration in normal and preeclamptic pregnancy. *J Am Soc Nephrol* **14**: 648–652.

Mori, Y., Wada, H., Okugawa, Y. et al. (2001). Increased plasma thrombomodulin as a vascular endothelial cell marker in patients with thrombotic thrombocytopenic purpura and hemolytic uremic syndrome. *Clin Appl Thromb Hemost* **7**: 5–9.

Moroni, G., and Ponticelli, C. (2005). Pregnancy after lupus nephritis. *Lupus* **14**: 89–94.

Morris, R. H., Vassalli, P., Beller, F. K., and McCluskey, R. T. (1964). Immunofluorescent studies of renal biopsies in the diagnosis of toxemia of pregnancy. *Obstet Gynecol* **24**: 32–46.

Moutquin, J. M., Lindsay, C., Arial, N. et al. (1997). Do prostacyclin and thromboxane contribute to the 'protective effect' of pregnancies with chronic hypertension? A preliminary prospective longitudinal study *Am J Obstet Gynecol* **177**: 1483–1490.

Nagamatsu, T., Fujii, T., Kusumi, M. et al. (2004). Cytotrophoblasts up-regulate soluble fms-like tyrosine kinase-1 expression under reduced oxygen: an implication for the placental vascular development and the pathophysiology of preeclampsia. *Endocrinology* **145**: 4838–4845.

Neithardt, A. B., Dooley, S. L., and Borensztajn, J. (2002). Prediction of 24-hour protein excretion in pregnancy with a single voided urine protein-to-creatinine ratio. *Am J Obstet Gynecol* **186**: 883–886.

Nelson, R. G., Morgenstern, H., and Bennett, P. H. (1998). Intrauterine diabetes exposure and the risk of renal disease in diabetic Pima Indians. *Diabetes* **47**: 1489–1493.

Nevo, O., Soleymanlou, N., Wu, Y. et al. (2006). Increased expression of sFlt-1 in in vivo and in vitro models of human placental hypoxia is mediated by HIF-1. *Am J Physiol Regul Integr Comp Physiol* **29**: R1085–R1093.

Nisell, H., Lintu, H., Lunell, N. O., Mollerstrom, G., and Pettersson, E. (1995). Blood pressure and renal function seven years after pregnancy complicated by hypertension. *Br J Obstet Gynaecol* **102**: 876–881.

Nochy, D., Birembaut, P., Hinglais, N. et al. (1980). Renal lesions in the hypertensive syndromes of pregnancy: immunomorphological and ultrastructural studies in 114 cases. *Clin Nephrol* **13**: 155–162.

Noris, M., Todeschini, M., Cassis, P. et al. (2004). L-arginine depletion in pre-eclampsia orients nitric oxide synthase toward oxidant species. *Hypertension* **43**: 614–622.

North, R. A., Ferrier, C., Long, D., Townend, K., and Kincaid-Smith, P. (1994). Uterine artery Doppler flow velocity waveforms in the second trimester for the prediction of pre-eclampsia and fetal growth retardation. *Obstet Gynecol* **83**: 378–386.

Novak, J., Danielson, L. A., Kerchner, L. J. et al. (2001). Relaxin is essential for renal vasodilation during pregnancy in conscious rats. *J Clin Invest* **107**: 1469–1475.

Odendaal, H. J., Pattinson, R. C., Bam, R., Grove, D., and Kotze, T. J. (1990). Aggressive or expectant management for patients with severe pre-eclampsia between 28-34 weeks' gestation: a randomized controlled trial. *Obstet Gynecol* **76**: 1070–1075.

Ohara, M., Martin, P. Y., Xu, D. L. et al. (1998). Upregulation of aquaporin 2 water channel expression in pregnant rats. *J Clin Invest* **101**: 1076–1083.

Okundaye, I., Abrinko, P., and Hou, S. (1998). Registry of pregnancy in dialysis patients. *Am J Kidney Dis* **31**: 766–773.

Okuno, S., Hamada, H., Yasuoka, M. et al. (1999). Brain natriuretic peptide (BNP) and cyclic guanosine monophosphate (cGMP) levels in normal pregnancy and preeclampsia. *J Obstet Gynaecol Res* **25**: 407–410.

Ostendorf, T., Kunter, U., Eitner, F. et al. (1999). VEGF(165) mediates glomerular endothelial repair. *J Clin Invest* **104**: 913–923.

Ozcan, C., Wong, S. J., and Hari, P. (2006). Reversible posterior leukoencephalopathy syndrome and bevacizumab. *N Engl J Med* **354**: 980–982 discussion 982.

Paarlberg, K. M., de Jong, C. L., van Geijn, H. P., van Kamp, G. J., Heinen, A. G., and Dekker, G. A. (1998). Vasoactive mediators in pregnancy-induced hypertensive disorders: a longitudinal study. *Am J Obstet Gynecol* **179**: 1559–1564.

Page, N. M., Woods, R. J., Gardiner, S. M. et al. (2000). Excessive placental secretion of neurokinin B during the third trimester causes pre-eclampsia. *Nature* **405**: 797–800.

Palmer, S. K., Moore, L. G., Young, D., Cregger, B., Berman, J. C., and Zamudio, S. (1999). Altered blood pressure course during normal pregnancy and increased pre-eclampsia at high altitude (3100 meters) in Colorado. *Am J Obstet Gynecol* **180**: 1161–1168.

Parham, P. (2004). NK cells and trophoblasts: partners in pregnancy. *J Exp Med* **200**: 951–955.

Parra, M., Rodrigo, R., Barja, P. et al. (2005). Screening test for pre-eclampsia through assessment of uteroplacental blood flow and biochemical markers of oxidative stress and endothelial dysfunction. *Am J Obstet Gynecol* **193**: 1486–1491.

Peake, S. L., Roxburgh, H. B., and Langlois, S. L. (1983). Ultrasonic assessment of hydronephrosis of pregnancy. *Radiology* **146**: 167–170.

Pereira, S. P., O'Donohue, J., Wendon, J., and Williams, R. (1997). Maternal and perinatal outcome in severe pregnancy-related liver disease. *Hepatology* **26**: 1258–1262.

Podjarny, E., Baylis, C., and Losonczy, G. (1999). Animal models of preeclampsia. *Semin Perinatol* **23**: 2–13.

Podymow, T., August, P., and Umans, J. G. (2004). Antihypertensive therapy in pregnancy. *Semin Nephrol* **24**: 616–625.

Pollak, V. E., and Nettles, J. B. (1960). The kidney in toxemia of pregnancy: a clinical and pathologic study based on renal biopsies. *Medicine (Balt.)* **39**: 469–526.

Polliotti, B. M., Fry, A. G., Saller, D. N., Mooney, R. A., Cox, C., and Miller, R. K. (2003). Second-trimester maternal serum placental growth factor and vascular endothelial growth factor for predicting severe, early-onset preeclampsia. *Obstet Gynecol* **101**: 1266–1274.

Poppas, A., Shroff, S. G., Korcarz, C. E. et al. (1997). Serial assessment of the cardiovascular system in normal pregnancy. Role of arterial compliance and pulsatile arterial load. *Circulation* **95**: 2407–2415.

Poston, L., Briley, A. L., Seed, P. T., Kelly, F. J., and Shennan, A. H. (2006). Vitamin C and vitamin E in pregnant women at risk for pre-eclampsia (VIP trial): randomised placebo-controlled trial. *Lancet* **367**: 1145–1154.

Puskar, D., Balagovic, I., Filipovic, A. et al. (2001). Symptomatic physiologic hydronephrosis in pregnancy: incidence, complications and treatment. *Eur Urol* **39**: 260–263.

Qiu, C., Luthy, D. A., Zhang, C., Walsh, S. W., Leisenring, W. M., and Williams, M. A. (2004). A prospective study of maternal serum C-reactive protein concentrations and risk of preeclampsia. *Am J Hypertens* **17**: 154–160.

Rahman, P., Gladman, D. D., and Urowitz, M. B. (1998). Clinical predictors of fetal outcome in systemic lupus erythematosus. *J Rheumatol* **25**: 1526–1530.

Rajakumar, A., Doty, K., Daftary, A., Harge, R. G., and Conrad, K. P. (2003). Impaired oxygen-dependent reduction of HIF-1alpha and -2alpha proteins in pre-eclamptic placentae. *Placenta* **24**: 199–208.

Ranta, V., Viinikka, L., Halmesmaki, E., and Ylikorkala, O. (1999). Nitric oxide production with preeclampsia. *Obstet Gynecol* **93**: 442–445.

Rasmussen, P. E., and Nielsen, F. R. (1998). Hydronephrosis during pregnancy: a literature survey. *Eur J Obstet Gynecol Reprod Biol* **27**: 249–259.

Rasmussen, A., and Ravn, P. (2004). High frequency of congenital thrombophilia in women with pathological pregnancies? *Acta Obstet Gynecol Scand* **83**: 808–817.

Rath, W., Faridi, A., and Dudenhausen, J. W. (2000). HELLP syndrome. *J Perinat Med* **28**: 249–260.

Ray, J. G., Vermeulen, M. J., Schull, M. J., and Redelmeier, D. A. (2005). Cardiovascular health after maternal placental syndromes (CHAMPS): population-based retrospective cohort study. *Lancet* **366**: 1797–1803.

Redman, C. W. (1984). Maternal plasma volume and disorders of pregnancy. *Br Med J* **288**: 955–956.

Regan, C. L., Levine, R. J., Baird, D. D. et al. (2001). No evidence for lipid peroxidation in severe preeclampsia. *Am J Obstet Gynecol* **185**: 572–578.

Report of the National High Blood Pressure Education Program Working Group on High Blood Pressure in Pregnancy (2000). *Am J Obstet Gynecol* **183**: S1–S22.

Roberts, J. M. (2000). Preeclampsia: what we know and what we do not know. *Semin Perinatol* **24**: 24–28.

Roberts, J. M., and Cooper, D. W. (2001). Pathogenesis and genetics of pre-eclampsia. *Lancet* **357**: 53–56.

Roberts, J. M., and Speer, P. (2004). Antioxidant therapy to prevent preeclampsia. *Semin Nephrol* **24**: 557–564.

Roberts, J. M., Taylor, R. N., Musci, T. J., Rodgers, G. M., Hubel, C. A., and McLaughlin, M. K. (1989). Preeclampsia: an endothelial cell disorder. *Am J Obstet Gynecol* **161**: 1200–1204.

Roberts, J. M., Edep, M. E., Goldfien, A., and Taylor, R. N. (1992). Sera from preeclamptic women specifically activate human umbilical vein endothelial cells in vitro: morphological and biochemical evidence. *Am J Reprod Immunol* **27**: 101–108.

Roberts, J. M., Bodnar, L. M., Lain, K. Y. et al. (2005). Uric acid is as important as proteinuria in identifying fetal risk in women with gestational hypertension. *Hypertension* **46**: 1263–1269.

Robertson, W. B., Brosens, I., and Dixon, H. G. (1967). The pathological response of the vessels of the placental bed to hypertensive pregnancy. *J Pathol Bacteriol* **93**: 581–592.

Robillard, P. Y., Hulsey, T. C., Alexander, G. R., Keenan, A., de Caunes, F., and Papiernik, E. (1993). Paternity patterns and risk of pre-eclampsia in the last pregnancy in multiparae. *J Reprod Immunol* **24**: 1–12.

Rodriguez-Thompson, D., and Lieberman, E. S. (2001). Use of a random urinary protein-to-creatinine ratio for the diagnosis of significant proteinuria during pregnancy. *Am J Obstet Gynecol* **185**: 808–811.

Roque, H., Paidas, M. J., Funai, E. F., Kuczynski, E., and Lockwood, C. J. (2004). Maternal thrombophilias are not associated with early pregnancy loss. *Thromb Haemost* **91**: 290–295.

Rossing, K., Jacobsen, P., Hommel, E. et al. (2002). Pregnancy and progression of diabetic nephropathy. *Diabetologia* **45**: 36–41.

Rostoker, G., Behar, A., and Lagrue, G. (2000). Vascular hyperpermeability in nephrotic edema. *Nephron* **85**: 194–200.

Ruiz-Irastorza, G., and Khamashta, M. A. (2005). Management of thrombosis in antiphospholipid syndrome and systemic lupus erythematosus in pregnancy. *Ann NY Acad Sci* **1051**: 606–612.

Rumbold, A. R., Crowther, C. A., Haslam, R. R., Dekker, G. A., and Robinson, J. S. (2006). Vitamins C and E and the risks of pre-eclampsia and perinatal complications. *N Engl J Med* **354**: 1796–1806.

Ryu, S., Huppmann, A. R., Sambangi, N., Takacs, P., and Kauma, S. W. (2007). Increased leukocyte adhesion to vascular endothelium in pre-eclampsia is inhibited by antioxidants. *Am J Obstet Gynecol* **196**(400): e1–7; discussion e7–e8.

Saftlas, A. F., Olson, D. R., Franks, A. L., Atrash, H. K., and Pokras, R. (1990). Epidemiology of pre-eclampsia and eclampsia in the United States, 1979–1986. *Am J Obstet Gynecol* **163**: 460–465.

Salafia, C. M., Pezzullo, J. C., Ghidini, A., Lopez-Zeno, J. A., and Whittington, S. S. (1998). Clinical correlations of patterns of placental pathology in preterm pre-eclampsia. *Placenta* **19**: 67–72.

Salmela, K. T., Kyllonen, L. E., Holmberg, C., and Gronhagen-Riska, C. (1993). Impaired renal function after pregnancy in renal transplant recipients. *Transplantation* **56**: 1372–1375.

Samadi, A. R., Mayberry, R. M., Zaidi, A. A., Pleasant, J. C., McGhee, N., and Rice, R. J. (1996). Maternal hypertension and associated pregnancy complications among African-American and other women in the United States. *Obstet Gynecol* **87**: 557–563.

Sanders, C. L., and Lucas, M. J. (2001). Renal disease in pregnancy. *Obstet Gynecol Clin North Am* **28**: 593–600 vii.

Sargent, I. L., Germain, S. J., Sacks, G. P., Kumar, S., and Redman, C. W. (2003). Trophoblast deportation and the maternal inflammatory response in pre-eclampsia. *J Reprod Immunol* **59**: 153–160.

Saudan, P., Brown, M. A., Buddle, M. L., and Jones, M. (1998). Does gestational hypertension become pre-eclampsia? *Br J Obstet Gynaecol* **105**: 1177–1184.

Savvidou, M. D., Hingorani, A. D., Tsikas, D., Frolich, J. C., Vallance, P., and Nicolaides, K. H. (2003). Endothelial dysfunction and raised plasma concentrations of asymmetric dimethylarginine in pregnant women who subsequently develop pre-eclampsia. *Lancet* **361**: 1511–1517.

Scalera, F., Schlembach, D., and Beinder, E. (2001). Production of vasoactive substances by human umbilical vein endothelial cells after incubation with serum from preeclamptic patients. *Eur J Obstet Gynecol Reprod Biol* **99**: 172–178.

Schackis, R. C. (2004). Hyperuricaemia and preeclampsia: is there a pathogenic link? *Med Hypotheses* **63**: 239–244.

Schenker, J. G., and Chowers, I. (1971). Pheochromocytoma and pregnancy. *Review of 89 cases. Obstet Gynecol Surv* **26**: 739–747.

Schieve, L. A., Handler, A., Hershow, R., Persky, V., and Davis, F. (1994). Urinary tract infection during pregnancy: its association with maternal morbidity and perinatal outcome. *Am J Public Hlth* **84**: 405–410.

Schiff, E., Ben-Baruch, G., Peleg, E. et al. (1992). Immunoreactive circulating endothelin-1 in normal and hypertensive pregnancies. *Am J Obstet Gynecol* **166**: 624–628.

Schobel, H. P., Fischer, T., Heuszer, K., Geiger, H., and Schmieder, R. E. (1996). Pre-eclampsia – a state of sympathetic overactivity. *N Engl J Med* **335**: 1480–1485.

Schwartz, R. B., Feske, S. K., Polak, J. F. et al. (2000). Preeclampsia-eclampsia: clinical and neuroradiographic correlates and insights into the pathogenesis of hypertensive encephalopathy. *Radiology* **217**: 371–376.

Selcuk, N. Y., Tonbul, H. Z., San, A., and Odabas, A. R. (1998). Changes in frequency and etiology of acute renal failure in pregnancy (1980-1997). *Ren Fail* **20**: 513–517.

Seligman, S. P., Buyon, J. P., Clancy, R. M., Young, B. K., and Abramson, S. B. (1994). The role of nitric oxide in the pathogenesis of preeclampsia. *Am J Obstet Gynecol* **171**: 944–948.

Serreau, R., Luton, D., Macher, M. A., Delezoide, A. L., Garel, C., and Jacqz-Aigrain, E. (2005). Developmental toxicity of the angiotensin II type 1 receptor antagonists during human pregnancy: a report of 10 cases. *Br J Obstet Gynaecol* **112**: 710–712.

Shaamash, A. H., Elsnosy, E. D., Makhlouf, A. M., Zakhari, M. M., and Ibrahim, O. A. (2000). Maternal and fetal serum nitric oxide (NO) concentrations in normal pregnancy, pre-eclampsia and eclampsia. *Int J Gynaecol Obstet* **68**: 207–214.

Shembrey, M. A., and Noble, A. D. (1995). An instructive case of abdominal pregnancy. *Aust NZ J Obstet Gynaecol* **35**: 220–221.

Sherwood, O. D. (2005). An 'old hand's' perspective of relaxin 2004's place along the relaxin trail. *Ann NY Acad Sci* **1041**: xxix–xxix10.

Sibai, B. M. (1996). Treatment of hypertension in pregnant women. *N Engl J Med* **335**: 257–265.

Sibai, B. M. (2005). Diagnosis, prevention, and management of eclampsia. *Obstet Gynecol* **105**: 402–410.

Sibai, B. M. (2007). Imitators of severe preeclampsia. *Obstet Gynecol* **109**: 956–966.

Sibai, B. M., Mercer, B. M., Schiff, E., and Friedman, S. A. (1994). Aggressive versus expectant management of severe pre-eclampsia at 28 to 32 weeks' gestation: a randomized controlled trial. *Am J Obstet Gynecol* **171**: 818–822.

Sibai, B. M., Gordon, T., Thom, E. et al. (1995). Risk factors for pre-eclampsia in healthy nulliparous women: a prospective multicenter study. The National Institute of Child Health and Human Development Network of Maternal-Fetal Medicine Units. *Am J Obstet Gynecol* **172**: 642–648.

Sibai, B. M., Lindheimer, M., Hauth, J. et al. (1998). Risk factors for preeclampsia, abruptio placentae, and adverse neonatal outcomes among women with chronic hypertension. *National Institute of Child Health and Human Development Network of Maternal-Fetal Medicine Units. N Engl J Med* **339**: 667–671.

Sibai, B. M., Caritis, S., Hauth, J. et al. (2000). Risks of preeclampsia and adverse neonatal outcomes among women with pregestational diabetes mellitus. *National Institute of Child Health and Human Development Network of Maternal-Fetal Medicine Units. Am J Obstet Gynecol* **182**: 364–369.

Sifakis, S., and Pharmakides, G. (2000). Anemia in pregnancy. *Ann NY Acad Sci* **900**: 125–136.

Silver, R. K., Kupferminc, M. J., Russell, T. L., Adler, L., Mullen, T. A., and Caplan, M. S. (1996). Evaluation of nitric oxide as a mediator of severe preeclampsia. *Am J Obstet Gynecol* **175**: 1013–1017.

Simon, N., Franklin, S. S., Bleifer, K. H., and Maxwell, M. H. (1972). Clinical characteristics of renovascular hypertension. *J Am Med Assoc* **220**: 1209–1218.

Sims, E. A. H., and Krantz, K. E. (1958). Serial studies of renal function during pregnancy and the puerperium in normal women. *J Clin Invest* **37**: 1764–1774.

Skjaerven, R., Wilcox, A. J., and Lie, R. T. (2002). The interval between pregnancies and the risk of preeclampsia. *N Engl J Med* **346**: 33–38.

Slowinski, T., Neumayer, H. H., Stolze, T., Gossing, G., Halle, H., and Hocher, B. (2002). Endothelin system in normal and hypertensive pregnancy. *Clin Sci (Lond.)* **103**(Suppl 48): 446S–449S.

Smaill, F. (2001). Antibiotics for asymptomatic bacteriuria in pregnancy. *Cochrane Database Syst Rev* **2**: CD000490.

Smarason, A. K., Allman, K. G., Young, D., and Redman, C. W. (1997). Elevated levels of serum nitrate, a stable end product of nitric oxide, in women with pre-eclampsia. *Br J Obstet Gynaecol* **104**: 538–543.

Smith, G. C., Pell, J. P., and Walsh, D. (2001). Pregnancy complications and maternal risk of ischaemic heart disease: a retrospective cohort study of 129 290 births. *Lancet* **357**: 2002–2006.

Smith, M. C., Danielson, L. A., Conrad, K. P., and Davison, J. M. (2006). Influence of recombinant human relaxin on renal hemodynamics in healthy volunteers. *J Am Soc Nephrol* **17**: 3192–3197.

Soubassi, L., Haidopoulos, D., Sindos, M. et al. (2004). Pregnancy outcome in women with pre-existing lupus nephritis. *J Obstet Gynaecol* **24**: 630–634.

Souqiyyeh, M. Z., Huraib, S. O., Saleh, A. G., and Aswad, S. (1992). Pregnancy in chronic hemodialysis patients in the Kingdom of Saudi Arabia. *Am J Kidney Dis* **19**: 235–238.

Spargo, B., Mc, C. C., and Winemiller, R. (1959). Glomerular capillary endotheliosis in toxemia of pregnancy. *Arch Pathol* **68**: 593–599.

Staff, A. C., Berge, L., Haugen, G., Lorentzen, B., Mikkelsen, B., and Henriksen, T. (2004). Dietary supplementation with L-arginine or placebo in women with pre-eclampsia. *Acta Obstet Gynecol Scand* **83**: 103–107.

Stamilio, D. M., Sehdev, H. M., and Macones, G. A. (2003). Pregnant women with the sickle cell trait are not at increased risk for developing preeclampsia. *Am J Perinatol* **20**: 41–48.

Stella, C. L., How, H. Y., and Sibai, B. M. (2006). Thrombophilia and adverse maternal-perinatal outcome: controversies in screening and management. *Am J Perinatol* **23**: 499–506.

Strauss, M. (1937). Observations on the etiology of the toxemia of pregnancy-II. Production of acute exacerbations of toxemia by sodium salts in pregnant women with hypoproteinemia. *Am J Med* **6**: 772–783.

Strevens, H., Wide-Swensson, D., Hansen, A. et al. (2003). Glomerular endotheliosis in normal pregnancy and pre-eclampsia. *Br J Obstet Gynaecol* **110**: 831–836.

Sturgiss, S. N., and Davison, J. M. (1995). Effect of pregnancy on the long-term function of renal allografts: an update. *Am J Kidney Dis* **26**: 54–56.

Subtil, D., Goeusse, P., Puech, F. et al. (2003). Aspirin (100 mg) used for prevention of pre-eclampsia in nulliparous women: the Essai Regional Aspirine Mere-Enfant study (Part 1). *Br J Obstet Gynaecol* **110**: 475–484.

Sugimoto, H., Hamano, Y., Charytan, D. et al. (2003). Neutralization of circulating vascular endothelial growth factor (VEGF) by anti-VEGF antibodies and soluble VEGF receptor 1 (sFlt-1) induces proteinuria. *J Biol Chem* **278**: 12605–12608.

Suhonen, L., Hiilesmaa, V., and Teramo, K. (2000). Glycaemic control during early pregnancy and fetal malformations in women with type I diabetes mellitus. *Diabetologia* **43**: 79–82.

Tandon, A., Ibanez, D., Gladman, D. D., and Urowitz, M. B. (2004). The effect of pregnancy on lupus nephritis. *Arthritis Rheum* **50**: 3941–3946.

Tapia, H. R., Johnson, C. E., and Strong, C. G. (1972). Renin-angiotensin system in normal and in hypertensive disease of pregnancy. *Lancet* **2**: 847–850.

Taylor, R. N., Grimwood, J., Taylor, R. S., McMaster, M. T., Fisher, S. J., and North, R. A. (2003). Longitudinal serum concentrations of placental growth factor: evidence for abnormal placental angiogenesis in pathologic pregnancies. *Am J Obstet Gynecol* **188**: 177–182.

Thadhani, R., Ecker, J. L., Mutter, W. P. et al. (2004). Insulin resistance and alterations in angiogenesis: additive insults that may lead to preeclampsia. *Hypertension* **43**: 988–992.

Thadhani, R. I., Johnson, R. J., and Karumanchi, S. A. (2005). Hypertension during pregnancy: a disorder begging for pathophysiological support. *Hypertension* **46**: 1250–1251.

Thangaratinam, S., Ismail, K. M., Sharp, S., Coomarasamy, A., and Khan, K. S. (2006). Accuracy of serum uric acid in predicting complications of pre-eclampsia: a systematic review. *Br J Obstet Gynaecol* **113**: 369–378.

Thompson, B. C., Kingdon, E. J., Tuck, S. M., Fernando, O. N., and Sweny, P. (2003). Pregnancy in renal transplant recipients: the Royal Free Hospital experience. *Q J Med* **96**: 837–844.

Trogstad, L. I., Eskild, A., Bruu, A. L., Jeansson, S., and Jenum, P. A. (2001). Is pre-eclampsia an infectious disease? *Acta Obstet Gynecol Scand* **80**: 1036–1038.

Tsatsaris, V., Goffin, F., Munaut, C. et al. (2003). Overexpression of the soluble vascular endothelial growth factor receptor in pre-eclamptic patients: pathophysiological consequences. *J Clin Endocrinol Metab* **88**: 5555–5563.

Tsukimori, K., Nakano, H., and Wake, N. (2007). Difference in neutrophil superoxide generation during pregnancy between pre-eclampsia and essential hypertension. *Hypertension* **49**: 1436–1441.

Tubbergen, P., Lachmeijer, A. M., Althuisius, S. M., Vlak, M. E., van Geijn, H. P., and Dekker, G. A. (1999). Change in paternity: a risk factor for pre-eclampsia in multiparous women? *J Reprod Immunol* **45**: 81–88.

Tuffnell, D. J., Jankowicz, D., Lindow, S. W. et al. (2005). Outcomes of severe pre-eclampsia/eclampsia in Yorkshire 1999/2003. *Br J Obstet Gynaecol* **112**: 875–880.

Tuohy, J. F., and James, D. K. (1992). Pre-eclampsia and trisomy 13. *Br J Obstet Gynaecol* **99**: 891–894.

van Dijk, M., Mulders, J., Poutsma, A. et al. (2005). Maternal segregation of the Dutch pre-eclampsia locus at 10q22 with a new member of the winged helix gene family. *Nat Genet* **37**: 514–519.

Venkatesha, S., Toporsian, M., Lam, C. et al. (2006). Soluble endoglin contributes to the pathogenesis of preeclampsia. *Nat Med* **12**: 642–649.

Vesely, S. K., Li, X., McMinn, J. R., Terrell, D. R., and George, J. N. (2004). Pregnancy outcomes after recovery from thrombotic thrombocytopenic purpura-hemolytic uremic syndrome. *Transfusion* **44**: 1149–1158.

Vikse, B. E., Irgens, L. M., Bostad, L., and Iversen, B. M. (2006). Adverse perinatal outcome and later kidney biopsy in the mother. *J Am Soc Nephrol* **17**: 837–845.

Villar, J., Abalos, E., Nardin, J. M., Merialdi, M., and Carroli, G. (2004). Strategies to prevent and treat preeclampsia: evidence from randomized controlled trials. *Semin Nephrol* **24**: 607–615.

Villar, J., Abdel-Aleem, H., Merialdi, M. et al. (2006). World Health Organization randomized trial of calcium supplementation among low calcium intake pregnant women. *Am J Obstet Gynecol* **194**: 639–649.

von Dadelszen, P., Ornstein, M. P., Bull, S. B., Logan, A. G., Koren, G., and Magee, L. A. (2000). Fall in mean arterial pressure and fetal growth restriction in pregnancy hypertension: a meta-analysis. *Lancet* **355**: 87–92.

von Dadelszen, P., Magee, L. A., Krajden, M. et al. (2003). Levels of antibodies against cytomegalovirus and Chlamydophila pneumoniae are increased in early onset pre-eclampsia. *Br J Obstet Gynaecol* **110**: 725–730.

Walker, J. J. (2000). Pre-eclampsia. *Lancet* **356**: 1260–1265.

Wallenberg, HCS. (ed.) (1998). *Handbook of Hypertension*, Elsevier Science, New York.

Wallukat, G., Homuth, V., Fischer, T. et al. (1999). Patients with pre-eclampsia develop agonistic autoantibodies against the angiotensin AT1 receptor. *J Clin Invest* **103**: 945–952.

Walther, T., Wallukat, G., Jank, A. et al. (2005). Angiotensin II type 1 receptor agonistic antibodies reflect fundamental alterations in the uteroplacental vasculature. *Hypertension* **46**: 1275–1279.

Wang, J. X., Knottnerus, A. M., Schuit, G., Norman, R. J., Chan, A., and Dekker, G. A. (2002). Surgically obtained sperm, and risk of gestational hypertension and pre-eclampsia. *Lancet* **359**: 673–674.

Williams, D. J., Vallance, P. J., Neild, G. H., Spencer, J. A., and Imms, F. J. (1996). Nitric oxide-mediated vasodilation in human pregnancy. *Am J Physiol* **272**: H748–H752.

Williams, W. W., Ecker, J. L., Thadhani, R. I., and Rahemtullah, A. (2005). Case records of the Massachusetts General Hospital. *Case 38-2005. A 29-year-old pregnant woman with the nephrotic syndrome and hypertension. N Engl J Med* **353**: 2590–2600.

Wolf, M., Sandler, L., Munoz, K., Hsu, K., Ecker, J. L., and Thadhani, R. (2002). First trimester insulin resistance and subsequent preeclampsia: a prospective study. *J Clin Endocrinol Metab* **87**: 1563–1568.

Wolf, M., Shah, A., Jimenez-Kimble, R., Sauk, J., Ecker, J. L., and Thadhani, R. (2004). Differential risk of hypertensive disorders of pregnancy among Hispanic women. *J Am Soc Nephrol* **15**: 1330–1338.

Wolf, M., Shah, A., Lam, C. et al. (2005). Circulating levels of the antiangiogenic marker sFLT-1 are increased in first versus second pregnancies. *Am J Obstet Gynecol* **193**: 16–22.

Xia, Y., Wen, H., Bobst, S., Day, M. C., and Kellems, R. E. (2003). Maternal autoantibodies from preeclamptic patients activate angiotensin receptors on human trophoblast cells. *J Soc Gynecol Investig* **10**: 82–93.

Yamasmit, W., Chaithongwongwatthana, S., Charoenvidhya, D., Uerpairojkit, B., and Tolosa, J. (2004). Random urinary protein-to-creatinine ratio for prediction of significant proteinuria in women with preeclampsia. *J Matern Fetal Neonatal Med* **16**: 275–279.

Yancopoulos, G. D., Davis, S., Gale, N. W., Rudge, J. S., Wiegand, S. J., and Holash, J. (2000). Vascular-specific growth factors and blood vessel formation. *Nature* **407**: 242–248.

Yang, J. C., Haworth, L., Sherry, R. M. et al. (2003). A randomized trial of bevacizumab, an anti-vascular endothelial growth factor antibody, for metastatic renal cancer. *N Engl J Med* **349**: 427–434.

Yeh, S. P., Chiu, C. F., Lee, C. C., Peng, C. T., Kuan, C. Y., and Chow, K. C. (2004). Evidence of parvovirus B19 infection in patients of pre-eclampsia and eclampsia with dyserythropoietic anaemia. *Br J Haematol* **126**: 428–433.

Zandi-Nejad, K., Luyckx, V. A., and Brenner, B. M. (2006). Adult hypertension and kidney disease: the role of fetal programming. *Hypertension* **47**: 502–508.

Zhou, Y., Chiu, K., Brescia, R. J. et al. (1993). Increased depth of trophoblast invasion after chronic constriction of the lower aorta in rhesus monkeys. *Am J Obstet Gynecol* **169**: 224–229.

Zhou, Y., Fisher, S. J., Janatpour, M. et al. (1997). Human cytotrophoblasts adopt a vascular phenotype as they differentiate. A strategy for successful endovascular invasion? *J Clin Invest* **99**: 2139–2151.

Zhou, Y., Damsky, C. H., and Fisher, S. J. (1997). Pre-eclampsia is associated with failure of human cytotrophoblasts to mimic a vascular adhesion phenotype. One cause of defective endovascular invasion in this syndrome? *J Clin Invest* **99**: 2152–2164.

Zhou, Y., McMaster, M., Woo, K. et al. (2002). Vascular endothelial growth factor ligands and receptors that regulate human cytotrophoblast survival are dysregulated in severe pre-eclampsia and hemolysis, elevated liver enzymes, and low platelets syndrome. *Am J Pathol* **160**: 1405–1423.

Zhou, Y., Genbacev, O., and Fisher, S. J. (2003). The human placenta remodels the uterus by using a combination of molecules that govern vasculogenesis or leukocyte extravasation. *Ann NY Acad Sci* **995**: 73–83.

Further reading

Conrad, K. P., Jeyabalan, A., Danielson, L. A., Kerchner, L. J., and Novak, J. (2005). Role of relaxin in maternal renal vasodilation of pregnancy. *Ann NY Acad Sci* **1041**: 147–154.

Danielson, L. A., and Conrad, K. P. (1995). Acute blockade of nitric oxide synthase inhibits renal vasodilation and hyperfiltration during pregnancy in chronically instrumented conscious rats. *J Clin Invest* **96**: 482–490.

Fischer, M. J., Lehnerz, S. D., Hebert, J. R., and Parikh, C. R. (2004). Kidney disease is an independent risk factor for adverse fetal and maternal outcomes in pregnancy. *Am J Kidney Dis* **43**: 415–423.

Fischer, T., Schobel, H. P., Frank, H., Andreae, M., Schneider, K. T., and Heusser, K. (2004). Pregnancy-induced sympathetic overactivity: a precursor of preeclampsia. *Eur J Clin Invest* **34**: 443–448.

Rajakumar, A., Brandon, H. M., Daftary, A., Ness, R., and Conrad, K. P. (2004). Evidence for the functional activity of hypoxia-inducible transcription factors overexpressed in preeclamptic placentae. *Placenta* **25**: 763–769.

Schrier, R. W. (1988). Pathogenesis of sodium and water retention in high-output and low-output cardiac failure, nephrotic syndrome, cirrhosis, and pregnancy (2). *N Engl J Med* **319**: 1127–1134.

Sibai, B. M., Caritis, S. N., Thom, E. et al. (1993). Prevention of pre-eclampsia with low-dose aspirin in healthy, nulliparous pregnant women. The National Institute of Child Health and Human Development Network of Maternal-Fetal Medicine Units. *N Engl J Med* **329**: 1213–1218.

The Eclampsia Trial Collaborative Group (1995). Which anticonvulsant for women with eclampsia? Evidence from the Collaborative Eclampsia Trial *Lancet* **345**: 1455–1463.

Taylor, R. N., Crombleholme, W. R., Friedman, S. A., Jones, L. A., Casal, D. C., and Roberts, J. M. (1991). High plasma cellular fibronectin levels correlate with biochemical and clinical features of pre-eclampsia but cannot be attributed to hypertension alone. *Am J Obstet Gynecol* **165:** 895–901.

Taylor, R. N., Casal, D. C., Jones, L. A., Varma, M., Martin, J. N., and Roberts, J. M. (1991). Selective effects of preeclamptic sera on human endothelial cell procoagulant protein expression. *Am J Obstet Gynecol* **165:** 1705–1710.

Index

Acid–base balance, 100, 457
  cellular buffering, 460
  dialysis patients, 468–70
  hormonal regulation, 461–5
    aldosterone, 462–3
    angiotensin II, 463
    antidiuretic hormone, 464
    endothelin, 464
    glucagon, 464
    glucocorticoids, 463
    growth hormone, 464–5
    IGF-I, 464–5
    insulin, 464
    parathyroid hormone, 463
  renal tubular bicarbonate reabsorption, 460–1
  urinary acidification, 467–8 See also Metabolic acidosis
Acid–base production, 457–9
  renal bicarbonate generation, 459–60, 467
Acquired nephrogenic diabetes insipidus (ANDI), 264
Acute fatty liver of pregnancy (AFLP), 498
Acute renal failure, 174
  in pregnancy, 498–9
    acute tubular necrosis, 498
    cortical necrosis, 498
    obstruction, 499
    thrombotic microangiopathy, 498–9
  recombinant growth hormone treatment, 423
  recombinant IGF-I treatment, 423 See also Renal failure
Addison's disease, 253
Adiponectin, 393
Adrenal glands:
  amorphous fraction, 311–13
  steroid production site, 316–17
Adrenocorticotrophic hormone (ACTH):
  aldosterone regulation, 369
    chronic effects, 369–70
    intracellular signaling, 369
Adrenomedullin, 371
AIDS, 274
AKA O79, 110
AlbuBNP, 300
Albumin synthesis, in metabolic acidosis, 472
Albuminuria, 351, 352, 397–8
Aldosterone, 131, 157–9, 329
  acid–base balance regulation, 462–3
  binding sites, 329–30
  biosynthesis, 361–2
    inhibitors, 370–1
  blockers, 322–4
    eplerenone, 323–4
    spironolactone, 322–3, 340
  diseases of production, 371–5
    defects causing excess, 372–5
    effects of excess, 375
    genetic defects causing deficiency, 371–2
  heart disease and, 349–51
    animal models, 350–1
    dietary sodium relationships, 354–5
  history, 311–22
    adrenal extract amorphous fraction, 311–13
    adrenal site of steroid production, 316–17
    assay, 313, 319–21
    discovery, 313–14
    isolation in crystalline form, 314–15
    metabolic studies, 317–18
    mineralocorticoid activity bioassays, 313
    paper chromatography, 313
    secretion, 322
    synthesis, 315–16
  hypertension and, 157–8
  induced proteins in transport epithelia, 337–40
  inflammation and, 353–4
  insulin sensitivity and, 354
  intracellular signaling pathways and, 352–3
  non-genomic effects in the kidney, 340–2
  receptor, 322, 329–30 See also Mineralocorticoid receptor (MR)
  regulation of production, 362–11
    ACTH, 369–70
    endothelin, 1, 370
    inhibitors, 370–1
    parathyroid hormone, 370
    potassium, 367–9
    renin–angiotensin–aldosterone system (RAAS), 362–7
  renal disease and, 351–2
  stroke and, 351
  vascular function and, 353 See also Primary aldosteronism (PA) See also Renin–angiotensin–aldosterone system (RAAS)
Aldosterone-producing adrenal adenoma (APA), 372, 374
  treatment, 374
Aliskiren, 194–5
Alprostadil, 435
Altitude effects, 6–7
Ammoniagenesis, 101
Anemia, 436
  CERA effects, 42–3
  definition, 51
  in chronic kidney disease, 49, 50, 51–5, 436
    differential diagnosis, 51
    erythropoietin treatment, 55–6, 436
    ESA therapy effects on outcome, 54–5
    randomized controlled trials, 53–4
    target hemoglobin levels, 36, 52–3
    children, 55–6
Angiotensin (Ang), 147–8, 167, 181–2
  Ang I, 147, 167–9, 181
  Ang II, 147–8, 160–1, 167–9, 181–2
    acid–base balance regulation, 463
    chronic effects of, 366–7
    intracellular glomerulosa cell signaling pathway regulation, 364–6
    vasopressin regulation, 227 See also Renin–angiotensin system (RAS) See also Renin–angiotensin–aldosterone system (RAAS)
Angiotensin receptor blockers (ARBs), 152–3, 160, 171–3, 184–5, 189, 192
  blood pressure lowering effect, 195–6
  with other agents, 196–7
  pharmacokinetics, 192–4
    application of pharmacologic differences, 194
    bioavailability, 192
    dose proportionality, 192
    elimination route, 193
    metabolism, 193
    protein binding, 193
    receptor binding and half-life, 193–4
    volume of distribution, 192–3
  side-effects, 197–8
Angiotensin receptors:
  AT1 receptors, 151–2, 167
    pre-eclampsia and, 494
  AT2 receptors, 152, 167–8
  heterodimerization, 153 See also Angiotensin receptor blockers (ARBs) See also Renin–angiotensin system (RAS) See also Renin–angiotensin–aldosterone system (RAAS)

515

Index

Angiotensin-converting enzyme (ACE), 147, 150–1, 168, 181–2
  ACE inhibitors (angiotensin converting enzyme), 149, 160, 171–3, 184–5, 189–91
    application of pharmacologic differences, 191
    blood pressure lowering effect, 195–7, 396
    heart disease and, 349, 355
    order of potency, 191
    pharmacokinetics, 190
    side-effects, 197–8
    tissue-binding, 190–1
    with other agents, 196–7
  ACE-2/angiotensin (1-7)/*MAS* receptor pathway, 151
Angiotensinogen (Agt), 147, 181
Antidiuretic hormone (ADH)Vasopressin (AVP)
Apoptosis, erythropoietin effects, 29, 49–50
Aquaporins, 212–15
  acute regulation by vasopressin, 214–15
  aquaporin 1 (AQP1), 213–14
  aquaporin 2 (AQP2), 208–9, 213–16, 261
    in diabetes insipidus, 263, 264, 265
    in hyponatremia, 252–3, 254–5
  aquaporin 3 (AQ3), 214
  aquaporin 4 (AQ4), 214, 265, 277
  long-term effect of vasopressin, 215
Arginine vasopressin (AVP)Vasopressin (AVP)
Aryl hydrocarbon nuclear receptor translocator (ARNT), 10
ASBNP2.1, 300
Asymmetric dimethyl arginine (ADMA), 398, 430, 434–5
Atrial natriuretic peptide (ANP), 289
  actions, 290–2
    aldosterone regulation, 370–1
    heart failure and, 293, 295
    myocardial ischemia/infarction and, 294–5
  pharmacokinetics, 295
  primary hyperaldosteronism, 373
  production, 289–90
  therapeutics, 295–6 See also Natriuretic peptides (NPs)
Autoimmune disease, 173–4
  idiopathic central diabetes insipidus, 267
Autosomal dominant hypophosphatemic rickets (ADHR), 113–14
Autosomal recessive hypophosphatemia (ARHP), 114–16

B-type natriuretic peptide (BNP), 289, 370
  heart failure and, 293–4, 296
  pharmacokinetics, 296
  production, 289–90
  therapeutics, 296–8 See also Natriuretic peptides (NPs)

Bicarbonate, 457–9
  bicarbonaturia, 467–8
  dialysis patients, 469–70
  metabolic acidosis treatment, 474–5
  renal generation, 459–60, 467
  renal tubular reabsorption, 460–1 See also Acid–base balance

Blood pressure (BP) regulation, 153–5
  pre-eclampsia management, 495
  renin–angiotensin blockade effects, 195–6
    with other agents, 196–7
  vasopressin effect, 241–2 See also Hypertension
Blood vessels:
  erythropoietin receptor, 27–8
  neovascularization, 30
Bone, 83–4, 435–6
  acid–base balance and, 469
    metabolic acidosis, 471
  recombinant growth hormone treatment effects, 420, 421–3
Bone sialoprotein (BSP), 77
Bradykinin (BK), 147, 150–1, 152
Brain:
  erythropoietin receptor, 28
  renin–angiotensin system, 155–6
Brain natriuretic peptide (BNP), 373
Bromocriptine, 435, 438
Burst-forming unit erythroid (BFU-E), 6, 20, 49

C-peptide, 386
C-type natriuretic peptide (CNP), 289, 292, 370
  actions, 292
  pharmacokinetics, 298
  production, 289
  therapeutics, 298–9 See also Natriuretic peptides (NPs)
Calcineurin inhibitors, 403
Calcitriol, 69–70
  gene regulation, 77
  non-genomic actions, 78
  RNA translation regulation, 79–80
Calcitriol/VDR complex, 71, 73–4
  relevance in health and disease, 80–5
    bone, 83–4
    cardiovascular system, 85
    intestine, 80–1
    kidney, 84–5
    parathyroid glands, 81–3
Calcium:
  absorption, 80, 99
  reabsorption, 240
  regulation, 65, 99, 100, 240
Calcium sensing receptor (CaSR), 69, 81, 100
Calreticulin, 75
cAMP response element binding (CREB) protein, 367, 369
Captopril, 189–90
Cardiovascular system:
  adaptation in pregnancy, 484–5
  calcitriol/VDR complex relevance, 85
  erectile dysfunction and, 429–30
  metabolic acidosis and, 472–3
  renin–angiotensin system, 156–7 See also Heart
Cathepsin B, 137
Cbfa1, 77
CD-NP, 299
Central diabetes insipidus (CDI), 266
  case study, 270
  causes of, 266–70
    acquired CDI, 267
    congenital malformations, 266–7
    genetic mutations, 266
    granulomatous diseases, 267
    idiopathic, 267
    neoplasms, 267
    post-surgical, 267–8
    traumatic brain injury and, 267
  diagnosis, 268–9
    MRI, 268
  treatment, 269–70
    hypernatremia, 269–70
Cerebral demyelination, hyponatremia and, 280
Cerebral edema, 277
Cerebral salt wasting (CSW), 274–5
Children:
  anemia in chronic kidney disease, 55
    erythropoietin treatment, 55
    target hemoglobin levels, 55–6
  renal insufficiency, 411, 413, 414
    GH/IGF-1 axis disturbance, 416–18
    growth impairment and, 416–18, 423–4
    recombinant growth hormone treatment, 419–21
CHIP28, 213
Chloride regulation:
  chloride/bicarbonate exchange, 238
  conductance, 237
  reabsorption, 237–8
  secretion, 238
CHOIR study, 53–4
Chronic kidney disease (CKD), 36, 49, 119
  anemia of, 49, 50, 51–5, 436
    differential diagnosis, 51
    erythropoietin dose and adverse outcomes, 55
    erythropoietin treatment in children, 55–6
    ESA therapy effects on outcome, 54–5
    randomized controlled trials, 53–4
    target hemoglobin levels, 52–3, 55–6
  CERA effects, 42–3
  darbepoetin effects, 40–1
  diabetes management, 402–3
  drug clearance in, 190
  erythropoietin pathophysiology, 49–50
  GH/IGF-I axis disturbances, 413–19
    adult implications, 418–19
    growth impairment in children, 416–18, 423–4
    recombinant growth hormone treatment, 419–23
    recombinant IGF-I treatment, 423
  insulin resistance pathogenesis, 398–9
  metabolic acidosis of, 457
    clinical characteristics, 465–8
    effects on cellular function, 470–4
    treatment, 474–5 See also Acid–base balance
  parathyroid hormone dysregulation, 101–2
  pregnancy and, 499–500
  renin–angiotensin system in, 171–2
  sexual dysfunction andSexual dysfunction See also Renal failure See also Renal insufficiency
Chronic renal failure (CRF), 441
  goiter and, 450

thyroid function and, 441–54
 circulating hormone profile, 441–3
 dialysis effects, 450–51
 hypothalamo–pituitary–thyroid axis, 448–50
 kinetics, 443–5
 renal transplantation effects, 451–2
 tissue content and uptake, 445–8 See also Renal failure
Chuvash polycythemia, 16
Cinacalcet, 65–6
Cirrhosis, vasopressin antagonist role, 255–6
Class effect, 191
Clomiphene, 431
Col1a1 gene, 77
Colony-forming unit erythroid (CFU-E), 6, 20, 49
Common granulocyte myeloid precursor (CGMP), 19
Common lymphoid progenitor (CLP), 19
Common myeloid progenitor (CMP), 19
Congenital adrenal hyperplasia (CAH), 371
Congenital nephrogenic diabetes insipidus (CNDI), 262–4
 genetics, 263
 treatment, 263–4
Congestive heart failure (CHF), 375
Conivaptan, 217, 277 See also Vasopressin antagonists
Conn's syndrome, 318, 372
Continuous erythropoietin receptor activator (CERA), 37, 41–3
 experimental effects of, 41
 in healthy subjects, 41–2
 in patients with chronic kidney disease anemia, 42–3
 safety and tolerability, 43
Corticosteroids, 403
COX-2 inhibitors, 264
Craniopharyngioma, 267
CREATE study, 53–4
CTNO, 38
CU-NP, 300
CU-NO, 300
Cushing's syndrome, 497
Cyclic GMP (cGMP) pathway activation, 293, 295
Cyclosporine, 501
CYP11B2 deficiency, 372
CYP21 deficiency, 371
Cystic fibrosis transmembrane regulator (CFTR), 238

Darbepoetin alfa, 38–41
 hemodialysis patients, 39–40
 intravenous administration, 39–40
 pre-dialysis chronic kidney disease patients, 40–1
 subcutaneous administration, 40–1
Demeclocycline, 251, 277
Dendroaspis natriuretic peptide (DNP), 289, 299
Dentin matrix protein 1 (DMP1), 114–16
Desmopressin (dDAVP), 251, 261
 hyponatremia and, 274
Diabetes insipidus (DI), 261–70

 central (CDI), 266–70
  case study, 270
  causes of, 266–8
  diagnosis, 268–9
  treatment, 269–70
 hypercalcemia and hypercalciuria, 265–6
 hypokalemia, 265
 nephrogenic (NDI), 215–16, 262
  acquired (ANDI), 264
  case study, 270
  congenital (CNDI), 262–4
  lithium as cause, 251, 265
 renal disease, 264
 sickle cell disease, 264
Diabetic nephropathy, 172–4, 352
 pregnancy and, 499–500
Diacylglycerol (DAG), 364–5, 368
Dialysis:
 acid–base balance and, 468–70
 pregnancy and, 500–11
 thyroid function and, 450–1
DIDMOAD syndrome, 266
DIDMOAD syndrome, 266
1,25-Dihydroxyvitamin D/VDR complex, 70–85
 biological actions, 70–71
 gene expression regulation, 73–80
 gene targets, 70–1
 relevance in health and disease, 80–5
  bone, 83–4
  cardiovascular system, 85
  intestine, 80–1
  kidney, 84–5
  parathyroid glands, 81–3
 structure-function, 71–2 See also Vitamin D See also Vitamin D receptor (VDR)
Direct renin inhibitors (DRIs), 189, 194
 blood pressure lowering effect, 195–6
 pharmacokinetics, 194–5
Dopamine receptors, 371
Double isotope assays, 320–1
Double Isotope Dilution Derivative Assay (DIDDA), 321
Dyslipidemia, 396–7

Eclampsia, 489
'Ecstasy', 274
Edema:
 cerebral, 277
 neurogenic pulmonary, 278
 pre-eclampsia, 487
Electrocortin:
 discovery, 313–14
 isolation in crystalline form, 314–15
 purification, 314
 synthesis, 315–16 See also Aldosterone
Enalapril, 189, 350, 351
End-stage renal disease (ESRD), 173, 295, 468–9
 pregnancy and, 500
Endothelial function, pre-eclampsia and, 491–2
Endothelial progenitor cell (EPC) recruitment, 29–30
Endothelin, 464, 485
 endothelin-1, 370, 487
 type A (ETA) receptors, 370
 type B (ETB) receptors, 370

Epidermal growth factor receptor (EGFR), 76, 81–2, 84–5, 353
Epidermal nevus syndrome (ENS), 116
Epithelial sodium channel (ENaC) regulation, 234–5, 337–40
Eplerenone, 323–4, 350, 351, 355, 375
EPOErythropoietin (EPO)
EPO-mimetic peptide (EMP), 43
Epoetin therapy, 54
 epoetin delta, 37
 omega, 37
Erectile dysfunction, 429, 430
 neurologic system and, 430
 vascular system and, 429–30
Erythropoiesis, 19–20
 erythropoietin role, 21–2, 35
 growth factor involvement, 20
 historical overview, 4–5, 6–7
 hormonal regulation, 4–5
 iron role, 22–4
 regulation by hypoxia, 6–7
 gene regulation, 8–10
Erythropoiesis stimulating agents (ESAs), 43–4
 causes of poor response, 37
 in anemia of chronic kidney disease, 54
  hypertension and graft thrombosis risk, 54
 non-peptide-based ESAs, 44
 peptide-based ESAs, 43–4 See also Erythropoietin (EPO) analogs
Erythropoietin (EPO), 21, 27
 actions of, 29–30
  anti-apoptosis, 29
  anti-inflammation, 29
  endothelial progenitor cell recruitment, 29–30
  neovascularization, 30
 chronic kidney disease and, 49–50
  children, 55
  dose related to adverse outcomes, 55
  therapy impact on hypertension and graft thrombosis risk, 54
 EPO fusion proteins, 38
 historical overview, 5–6
  assays, 5
  effector mechanisms, 6
  isolation and characterization, 5–6
 intracellular signaling mediators of, 21–2
 mechanisms of action, 30–1
 receptorErythropoietin receptor (EPOR)
 regulation of synthesis, 50
 resistance, 52
 role in erythropoiesis, 21–2, 35
 site of production, 5, 50 See also Erythropoietin (EPO) analogs See also Recombinant human erythropoietin (EPO)
Erythropoietin (EPO) analogs, 36–43
 continuous erythropoietin receptor activator (CERA), 41–3
 darbepoetin alfa, 38–41
 strategies for creation of, 37–8 See also Erythropoiesis stimulating agents (ESAs)
Erythropoietin gene, 6
 regulation by hypoxia, 8–10
 regulatory elements, 7

Erythropoietin receptor (EPOR), 21, 27–8
  brain, 28
  expression regulation, 28
  heart and vasculature, 27–8
  kidney, 28
Estrogen therapy, 438
Exercise-associated hyponatremia (EAH), 228, 273–4
Extracellular signal-regulated kinase (ERK), 30–1, 109–10, 150, 353, 365

Fas ligand (FasL), 78
Fc–EPO fusion proteins, 38
Fibroblast growth factor 7 (FGF, 7), 111–12
Fibroblast growth factor 23 (FGF, 23), 69, 83–4
  elevated levels, 119
  mutations in tumoral calcinosis, 118
  phosphate regulation, 110–11
  vitamin D metabolism and, 113
Fibrous dysplasia (FD), 116–17
Focal segmental glomerulosclerosis (FSGS), 489
Follicle stimulating hormone (FSH), 431, 437
Fumarate hydratase (FH) mutation, 15

G-protein-coupled receptors (GPCRs), 230
  GPCR kinases (GRKs), 233
Galactorrhea, 437
Germinoma, 267
Gestational hypertension, 496
Glomerular endotheliosis, 488
Glomerular filtration rate (GFR), 169–70, 241
  pregnancy and, 483–4
    pre-eclampsia, 488
Glomerulosa cell signaling pathways, 364–6
Glucagon, 464
Glucocorticoids, 159
  acid–base balance regulation, 463
  deficiency, 253
  mineralocorticoid receptor activation, 355
  receptor (GR), 331, 332
Gluconeogenesis, 100–11
  renal, regulation, 399
Glucose, 386–8
  hyperglycemia, 393–4
  metabolic acidosis and, 473
  regulation of renal production, 399
Goiter, 450
Gonadotropin-releasing hormone (GnRH), 431, 437
Granulocyte-macrophage colony-stimulating factor (GMCSF), 77–8
Granulomatous diseases, 267
Growth hormone binding protein (GHBP), 414
Growth hormone (GH), 411
  acid–base balance regulation and, 464–5, 473–4
  deficiency (GHD), 416
  GH/IGF-I axis, 411–13
    disturbance, 411, 413–19
    growth disturbance in children, 416–18
  in chronic kidney disease, 413–15, 416–19
  recombinant GH treatment in renal failure, 419–23
    adults with chronic acute failure, 423
    adults with chronic renal failure, 421–3

bone metabolism effects, 420, 421–3
    dosing recommendations, 420–1
    pediatric response, 419–20
    safety, 420
  resistance, 415
Growth hormone-releasing hormone (GHRH), 411–12
Growth impairment with renal insufficiency, 416–18
  future approaches, 423–4
  modifiable factors, 417–18
  non-modifiable factors, 417
  recombinant growth hormone treatment, 419–20
    dosing recommendations, 420–1
    safety, 420
Gynecomastia, 432

Heart:
  aldosterone and, 349–51
    animal models, 350–1
    dietary sodium relationships, 354–5
  erythropoietin receptor, 27–8
  renin–angiotensin system, 157, 181–5
    actions at the cellular level, 182–4
    coronary circulation and, 184
    local versus endocrine origin, 181–2
    significance for cardiac function, 184–5
Heart failure:
  congestive (CHF), 375
  natriuretic peptides and, 293–4, 295, 296, 300
  vasopressin antagonist role, 254–5
HELLP, 484
HELLP syndrome, 485, 489–90, 498
  treatment, 495
Hematide, 43–4
Hematopoiesis, 19
  growth factor involvement, 20
Hematopoietic stem cell (HSC) precursors, 19
Heme, 23–4
Heme oxygenase-1 (HO-1), 494
Heme-regulated inhibitor (HRI), 24
Hemoglobin:
  target levels, 36, 52–3
    children, 55–6
Hemojuvelin (HJV), 23
Hepcidin, 22–3
Hereditary hypocalcemic vitamin D-resistant rickets (HVDRR), 71, 72
Hereditary hypophosphatemic rickets with hypercalciuria (HHRH), 117
HIV infection, 274
Human chorionic gonadotropin (HCG), 431
12-Hydroxyeicosatetraenoic acid (12-HETE), 365
Hyp mouse model, 113
Hyperaldosteronism:
  effects of, 375
  secondary, 374–5 See also Primary aldosteronism (PA)
Hypercalcemia, 251, 265–6
Hypercalciuria, 265–6
Hyperglycemia, 393–4
  renal transplantation and, 403
Hypernatremia, 216

treatment in diabetes insipidus, 269–70
Hyperphosphatemia, 117–19
  tumoral calcinosis (TC) with, 117–18
Hypertension:
  aldosterone involvement, 157–8
    blockade effects, 351, 355
  erythropoietin therapy and, 54
  in metabolic syndrome, 394–6
  in pregnancy, 496–8
    chronic hypertension, 496
    gestational hypertension, 496
    management, 497–8
    pre-eclampsia, 486–7
    secondary causes of, 496–7
  new monogenic forms, 132
  renin gene polymorphism and, 141
  renin–angiotensin system involvement, 153–5
    blockade effects, 195–7
  renovascular, 167
  urinary concentrating ability relationship, 241–2 See also Blood pressure (BP) regulation
Hyperthyroidism, 450
Hyperuricemia, 398
Hypoaldosteronism, 131
  hyporeninemic, 465
Hypokalemia, 251, 265
Hyponatremia, 216, 217
  cerebral demyelination and, 280
  cerebral salt wasting (CSW), 274–5
  drug-induced, 274
  evaluation, 275–7
  exercise-associated (EAH), 228, 273–4
  HIV and, 274
  pathogenesis, 270–71
  prophylaxis, 279
  treatment, 279–80
  vasopressin antagonist role, 252–6
    euvolemic hyponatremia, 252–4
    hypervolemic hyponatremia, 252–4 See also Syndrome of inappropriate antidiuretic hormone secretion (SIADH)
Hyponatremic encephalopathy, 277–80
  clinical symptoms, 278
  neurogenic pulmonary edema, 278
  risk factors, 278–9
    age, 278
    gender, 278–9
    hypoxia, 279
  treatment, 279–80
Hypoparathyroidism, 118–19
Hypophosphatemia, 113–17, 119
  autosomal dominant hypophosphatemic rickets (ADHR), 113–14
  autosomal recessive hypophosphatemia (ARHP), 114–16
  hereditary hypophosphatemic rickets with hypercalciuria (HHRH), 117
  X-linked hypophosphatemia (XLH), 114
Hypopituitarism, vasopressin antagonist role, 253
Hyporeninemic hypoaldosteronism, 465
Hypothalamic–pituitary function, 431–2

Hypothalamic–pituitary–thyroid axis, 448–50
Hypothalamus, 205–6
Hypothyroidism, 450, 453
  vasopressin antagonist role, 252–3
Hypouricemia, 272
Hypoxia:
  erythropoiesis regulation, 6–10
  hyponatremic encephalopathy and, 279
  tumor hypoxia, 14
Hypoxia-inducible factor-1 (HIF-1), 7, 10
  degradation of, 12–14
  erythropoietin gene regulation, 8–10, 50
  HIF-α subunits, 10, 12, 21
    regulation, 12–13, 15
  HIF-ß subunits, 10
  pharmacological manipulation, 16
  pre-eclampsia and, 491

Idiopathic central diabetes insipidus, 267
Idiopathic hyperaldosteronism (IHA), 372, 374
  treatment, 374
Immunophilin (IMM), 333–4, 335–6
Immunosuppressive therapy, in pregnancy, 501–2
Inflammation:
  aldosterone and, 353–4
  erythropoietin effects, 29
  metabolic acidosis and, 474
Inositol 1,4,5-trisphosphate (IP, 3), 364
Insulin:
  actions, 387
    acid–base balance regulation, 464
  clearance, 399
  receptor, 387
  secretion, 386–7
Insulin resistance, 354, 385
  clinical physiology of, 387–8
  historical perspective, 385–6
  measurement, 388–9
  pathogenesis in chronic kidney disease, 398–9
  syndromes of severe resistance, 399–400
  treatment, 400–2
    dietary management, 400–11
    exercise, 401
    insulin sensitizers, 401–2 See also
      Metabolic syndrome
Insulin-like growth factor-binding proteins
    (IGFBPs), 413, 415, 418, 423–4
Insulin-like growth factor-I (IGF-I), 411
  acid–base balance regulation, 464–5
  GH/IGF-I axis, 411–13
    disturbance, 411, 413–19
    growth disturbance in children, 416–18
  in chronic kidney disease, 415–19
  metabolic acidosis and, 473–4
  receptor, 387
  recombinant IGF-I treatment, 423
Interferon-γ (IFN-γ), 77
Iodine retention, 450
Iron:
  deficiency, 23, 50
  regulation of serum levels, 22–3
  role in erythropoiesis, 22–4
    regulation of hemoglobin production, 23–4
Ischemic nephropathy, 375

JAK2 kinases, 21
JAK2 kinases, role in erythropoiesis, 21–2
Juxtaglomerular (JG) cells, kidney, 135–8, 147, 148–9

K-Ras 2, 339
Kidney:
  aldosterone and, 351–2
  erythropoietin production, 5, 50
  erythropoietin receptor, 28
  failureRenal failure
  glomerular filtration rate (GFR), 169–70, 241
  glucose production regulation, 399
  oxygen sensing, 7
  phosphate transport, 107–12
    cellular and molecular aspects, 108–9
    regulation, 109–12
    tubular localization, 107–8
  renin production, 135–7
  renin–angiotensin system, 153–5
    fetal kidney development, 174
    in renal disease, 171–4
    renal hemodynamics and, 169–70
    tubular function and, 170–1
  renovascular hypertension, 167
  transplantation:
    hyperglycemia and, 403
    in pregnancy, 501–2
    thyroid function and, 451–2
  vasopressin function, 226
    receptor localization, 207, 229–30
    renal blood flow, 240–1
  vitamin D and, 84–5 See also Chronic kidney
    disease (CKD)

Labetalol, 497
Langerhan's cell histiocytosis (LCH), 267
Leptin, 393, 419, 423, 453
  metabolic acidosis and, 473
Linear nevus sebaceous syndrome (LNSS), 116
LIP, 82
Lipid metabolism, metabolic acidosis and, 473
Lithium, 250–1, 265
Lithocholic acid (LCA), 81
Loop of Henle, 170
Low density lipoprotein (LDL), 396–7
Lupus nephritis, 500
Luteinizing hormone (LH), 431–2, 437

McCune–Albright syndrome (MAS), 117
Magnesium:
  pre-eclampsia treatment, 495
  regulation, 99–100, 240
Malnutrition, 436
Megakaryoblast (MKP), 19
Megakaryocytic/erythrocytic progenitors
    (MEPs), 19, 20
Megalin, 84
Meglitinides, 402
Melanocyte-stimulating hormone (MSH), 453
Membrane-associated, rapid-response steroid-
    binding protein (1,25D-MARRS), 80
Menstrual cycle, 437
Metabolic acidosis of chronic kidney disease, 457
  dialysis patients, 468–70

  effects on cellular function, 470–4
    albumin synthesis, 472
    ß2-microglobulin, 473
    bone disease, 471
    cardiac disease, 472–3
    glucose homeostasis, 473
    growth hormone, 473–4
    inflammatory response, 474
    leptin, 473
    lipid metabolism, 473
    muscle wasting, 471–2
    renal failure, 472
    thyroid function, 473
  onset, 465–6
  prevalence, 465
  renal tubular bicarbonate generation, 467
  serum electrolyte pattern, 466–7
  severity, 465–6
  treatment, 474–5
  urinary acidification, 467–8 See also Acid–
    base balance
Metabolic syndrome, 389–98
  definition, 390 2
  dyslipidemia, 396–7
  hyperglycemia, 393–4
  hypertension, 394–6
  hyperuricemia, 398
  obesity, 392–3
  proteinuria, 397–8 See also Insulin resistance
Metformin, 401, 402
Methyldopa, 497
3,4-Methylenediaxymethamphetamine
    (MDMA), 274
ß2-Microglobulin, 473
Mineralocorticoid receptor (MR), 158–9, 329–30
  activation by glucocorticoids, 355
  blockade, 349–52, 355
    heart disease and, 349–51
    renal disease and, 351–2
  distribution in nephron, 337
  molecular biology, 330–7
    DNA binding domain, 330–1
    genomic structure, 332
    ligand binding domain (LBD), 331
    MR trafficking, 334–6
    N-terminal domain, 331
    post-translational modifications of, 332–3
    selectivity, 336–7
Mineralocorticoids:
  deficiency, 253
  early bioassays, 313
Mitogen-activated protein (MAP) kinase, 353, 365
Monocyte chemoattractant protein-1 (MCP-1), 350, 354
Muscle wasting, in metabolic acidosis, 471–3
Mycophenolate mofetil (MMF), 501–2
Myocardial ischemia/infarction, 294–5
Myristoylated alanine-rich C kinase substrate
    (MARCKS), 365

Na-K-ATPase regulation, 237
24Na/42K assay, 313, 319
Na/H exchange, 100

Na/H exchanger regulatory factor (NHERF), 95, 99, 102
NaCl regulation, 234, 236
NaPi-IIa mutations, 117
Natriuretic peptide receptors:
  type A (NPR-A), 84, 290–2
    in heart failure, 293
  type B (NPR-B), 292
  type C (NPR-C), 292
Natriuretic peptides (NPs), 289–90
  aldosterone regulation, 370–71
  chimeric/synthetic natriuretic peptides, 299–300
    CD-NP, 299–300
    CU-NP, 300
    Dendroaspis NP (DNP), 299
  delivery systems, 299
  future directions, 301
  pathophysiologic implications, 293–5
    heart failure, 293–4
    myocardial ischemia/infarction, 294–5
  production, 289–90
  use in cardiac surgery, 297–8 *See also* Natriuretic peptide receptors
Natural killer (NK) cells, pre-eclampsia and, 492
Neocytolysis, 50
Neovascularization, 30
Nephrogenic diabetes insipidus (NDI), 215–16, 262
  acquired (ANDI), 264
  case study, 270
  congenital (CNDI), 262–4
    genetics, 263
    treatment, 263–4
  lithium as cause, 251 *See also* Diabetes insipidus (DI)
Nephrogenic syndrome of inappropriate antidiuresis (NSIAD), 272
NesiritideB-type natriuretic peptide (BNP)
Neurogenic pulmonary edema, 278
Neurophysin II (NPII), 261
Neutral endopeptidase (NEP), 292–3
Nitric oxide (NO), 241, 353
  pre-eclampsia and, 487
Nitric oxide synthase (NOS), 241
NKCC 2, 236–7
Nocturnal penile tumescence (NPT), 433
Non-steroidal anti-inflammatory drugs (NSAIDs), 264
Non-thyroidal illness (NTI), 441, 442, 446–8, 452–3
Normal HCT Study, 53
Novel erythropoiesis stimulating protein (NESP), 38
NPT1, 2a, 2c, 84, 108–9
  gene disruption effect, 113
  regulation, 109–12
    dietary phosphate, 109
    fibroblast growth factor, 23, 110–11
    parathyroid hormone, 109–10

Obesity, 354, 392–3
Obstructive uropathy, 174
Osteoglophonic dysplasia (OGD), 116
Osteoprotegerin (OPG), 77

Oxidative stress:
  in pre-eclampsia, 492–3
  reactive oxygen species (ROS), 11–12, 29
Oxygen sensing, 7, 10–12
  disruption in cancer, 14–15
  disruption in hereditary polycythemia, 15–16
Oxytocin, 226–7
  hyponatremia and, 274
  receptors, 228

P13K, 30
Parathyroid glands, 81–3
Parathyroid hormone (PTH), 69–70, 95, 105
  actions, 99–111
    acid–base balance regulation, 100, 460, 463
    ammoniagenesis, 101
    gluconeogenesis, 100–11
    mineral ion homeostasis, 99–10, 109–10
    Na/H exchange, 100
    vitamin D synthesis regulation, 100
  aldosterone regulation, 370
  biosynthesis, 95
  chronic kidney disease and, 101–102
  circulating forms of, 97
  measurement, 97
  prolactin release and, 432
  receptors, 95, 97–8
    signaling, 98–9
    topology of PTH binding, 98
  regulation, 96–7, 100
  vitamin D receptor interactions, 71, 76–7
PediatricsChildren
PEGylation, 299
Pendrin, 238
Pheochromocytoma, 497
PHEX (phosphate-regulating enzyme), 77, 105
  X-linked hypophosphatemia, 114
Phosphate:
  homeostasis, 99, 106–7
  renal transport, 107–12
    cellular and molecular aspects, 108–9
    disorders with renal transport abnormalities, 113–19
    mouse models with defects, 113
    regulation, 109–12
    tubular localization, 107–8
  role in renal vitamin D metabolism, 112–13
  See also Hypophosphatemia
Phosphate response element (PRE), 109
Phosphodiesterase V (PDEV), 294
Phospholipase D (PLD), 364–5
PI3K, 30
Placental growth factor (PlGF), 484, 493, 495
Placentation:
  abnormalities in pre-eclampsia, 490–511
  placental vascular development, 491
Plasminogen activator inhibitor 1 (PAI-1), 350
Polycythemia, 15–16
Polyuria, 261–2
Potassium:
  conductance regulation, 237
  effects on aldosterone production, 367–8
    chronic effects, 368–9
    intracellular signaling pathways, 368
Pre-eclampsia, 485

outcomes, 490
  pathogenesis, 490–4
    angiogenic factors, 493–4
    immunologic maladaptation, 492
    maternal endothelial function, 491–2
    oxidative stress, 492–3
    placental vascular development, 491
    renin–angiotensin system, 494
  pathophysiology, 486–9
    edema, 487
    hypertension, 486–7
    proteinuria, 487
    renal changes and pathology, 488–9
    severe pre-eclampsia and eclampsia, 489
    uric acid, 487–8
  risk factors, 485–6
  screening, 494–5
  treatment, 495
Pregnancy, 483–5
  cardiovascular adaptation, 484–5
  hypertension in, 496–8
    chronic hypertension, 496
    gestational hypertension, 496
    management, 497–8
    secondary causes of, 496–7 *See also* HELLP syndrome *See also* Pre-eclampsia
  renal adaptation, 483–4
  renal failure in, 498–502
    acute renal failure, 498–9
    chronic kidney disease, 499–510
    end-stage renal disease, 500–1
    renal transplant, 501–2
Primary aldosteronism (PA), 318–19, 372–4, 496
  definition, 318
  diagnosis, 318–19, 373–4
  in pregnancy, 496–7
  incidence, 319
  pathophysiology, 372–3
  treatment, 374
Primary polydypsia, 261–2
Proadrenomedullin, 371
Progesterone, 437–8
Prolactin, 432, 435, 437
Prolyl hydroxylase domain (PHD) enzymes, 13–14, 50
Prorenin, 137–8, 147
Protein kinase A (PKA), 369
Protein kinase C (PKC), 364–5, 367, 368
Protein kinase D (PKD), 365, 367
Proteinuria, 171–73, 351, 397–8
  in pre-eclampsia, 487
Proximal tubule, 170
Pseudohypoparathyroidism, 118–19
PTHParathyroid hormone (PTH)
PTH-related polypeptide (PTH-rP), 77
Pyelonephritis, 499

Radioimmunoassay (RIA), aldosterone, 321
RANK, 76, 77, 83
RAP ditto, 110
Re1B, 78
Reactive oxygen species (ROS), 11–12, 29
Recombinant human erythropoietin (EPO), 436
  adverse effects, 36

benefits, 36, 49
  development, 35–6 See also Erythropoietin (EPO)
Regulatory volume decrease (RVD), 277
Relaxin, 227, 484–5
Renal artery stenosis, 496
Renal blood flow, 240–1
Renal cell carcinoma (RCC):
  von Hippel-Lindau syndrome-associated, 14–15
Renal failure:
  in pregnancyPregnancy
    metabolic acidosis and, 472
    recombinant growth hormone treatment, 419–23
      adults with acute renal failure, 423
      adults with chronic renal failure, 421–3
      dosing recommendations, 420–1
      pediatric response, 419–20
      safety, 420
    recombinant IGF-I treatment, 423
    vitamin D role, 65 See also Acute renal failure See also Chronic kidney disease (CKD) See also Chronic renal failure (CRF) See also Renal insufficiency
Renal injury, 351–2
Renal insufficiency:
  GH/IGF-I axis disturbances, 411, 413–19
    adult implications, 418–19
    growth impairment and, 416–18, 423–4
      modifiable factors, 417–18
      non-modifiable factors, 417 See also Chronic kidney disease (CKD) See also Renal failure
Renal osteodystrophy (ROD), 418, 420
Renal transplantation:
  hyperglycemia and, 403
  in pregnancy, 501–502
    acute rejection, 502
    immunosuppressive therapy, 501–2
    outcomes, 501
  thyroid function and, 451–2
Renin, 135, 147, 167
  activation mechanisms, 137–8
  biosynthesis and secretion, 147, 148–9, 154
  expression ontogeny and plasticity, 136–7
  future perspectives, 142
  gene mutation and disease, 141–2
    inactivation in humans, 142
    inactivation in mouse, 141–2
    polymorphisms and hypertension, 141
  gene structure and regulation, 138–41
    genomic structure, 138
    promoters and enhancers, 138–41
  inhibitors, 174–5, 189, 194
    blood pressure lowering effect, 195–6
    pharmacokinetics, 194–5
  physiological control of, 136
  receptor, 149–50
  tissue origin, 135–6 See also Renin–angiotensin system (RAS) See also Renin–angiotensin–aldosterone system (RAAS)
Renin–angiotensin system (RAS), 147–8, 167–74

blockade, 189
  blood pressure lowering effect, 195–6 See also Angiotensin receptor blockers (ARBs) See also Angiotensin-converting enzyme (ACE) See also Direct renin inhibitors (DRIs)
history, 129–33, 167–8
  20th century, 129–31
  21st century, 131–2
physiologic effects of, 169–74
  in renal disease, 171–4
  renal hemodynamics, 169–70
  tubular function, 170–1
pre-eclampsia and, 494
tissue RASs, 153, 181
  brain RAS, 155–6
  cardiac RAS, 157, 181–5
  hypertension and, 153–5
  intrarenal RAS, 153–5
  vascular tissue RAS, 156–7 See also Angiotensin (Ang) See also Renin See also Renin–angiotensin–aldosterone system (RAAS)
Renin–angiotensin–aldosterone system (RAAS), 131–2, 135, 141–2, 362–7
  clinical effects of, 159–61
  in pregnancy, 484 See also Aldosterone See also Angiotensin (Ang) See also Renin See also Renin–angiotensin system (RAS)
Renin-secreting tumors, 375
Retinoid X receptor (RXR), 71, 72
  VDR/RXR heterodimerization, 74
Reversible posterior leukoencephalopathy syndrome (RPLS), 489
Rickets:
  autosomal dominant hypophosphatemic rickets (ADHR), 113–14
  etiology, 63
    evidence for role of vitamin D, 63–4
  hereditary hypocalcemic vitamin D-resistant rickets (HVDRR), 71, 72
  hereditary hypophosphatemic rickets with hypercalciuria (HHRH), 117
  search for a remedy, 63
RNA translation regulation, 79–80
ROM, K potassium channel, 237

Serotonin, 370
Sexual dysfunction, 429
  uremic men, 429–36
    endocrine system and, 430
    evaluation, 432–3
    gynecomastia, 432
    hypothalamic–pituitary function and, 431–2
    neurologic system and, 430
    outcomes, 435–6
    prolactin metabolism, 432
    psychologic system and, 432
    sex steroids, 431
    testicular function, 430
    treatment, 433–5
    vascular system and, 429–30
  uremic women, 437–8

    postmenopausal hormonal disturbances, 437
    premenopausal hormonal disturbances, 437
    treatment, 437–8
SGK, 338
Syndrome of inappropriate antidiuretic hormone secretion (SIADH)
Sickle cell disease, 264
Sildenafil, 434–5
SLC26A4, 238
SLC26A7, 238
Somatostatin (SRIF), 371, 411–12
Spermatogenesis, 430
Spironolactone, 322–3, 340, 349–50, 352, 355
STAT5 transcription factor, 22, 31
Steroid receptors:
  nuclear-cytoplasmic shuttling of, 334
  oligomeric structure, 333–4
  translocation, 334–6
Steroidogenic acute regulatory protein (StAR), 366
Streptozotocin, 352
Stroke, 351
Succinate dehydrogenase deficiency (SDHD), 15
Syndrome of inappropriate antidiuretic hormone secretion (SIADH), 216, 227–8, 270–7
  causes of:
    'Ecstasy', 274
    HIV, 274
    medications, 274
  diagnosis, 275–7
  epidemiology, 273
  nephrogenic syndrome of inappropriate antidiuresis (NSIAD), 272
  pathogenesis, 271–2
  treatment, 251, 277
    vasopressin antagonist role, 253–4, 277
  vasopressin and, 256, 272
Syndrome X, 386 See also Metabolic syndrome
Systemic lupus erythematosus (SLE), 173–4

Tacrolimus, 403, 501
Targeted tumorigenesis, 138
Testicular function, 430
Testosterone, 431
  deficiency, 435–6
  therapy, 436, 438
Thapsigargin, 364
Thiazides, 263–4
Thiazolidinediones, 401, 403
Thrombotic microangiopathy, 498–9
Thromboxane A (TXA), 486
Thyroid hormone:
  in chronic renal failure, 441–54
    hypothalamo–pituitary–thyroid axis, 448–50
    kinetics, 443–5
    tissue content uptake, 445–8
  metabolic acidosis anf, 473
  replacement, 453–4
Thyroid-stimulating hormone (TSH), 441–2, 448–50, 452, 473
Thyrotropin-releasing hormone (TRH), 441, 442, 449–50, 452, 453
Thyroxine binding globulin (TBG), 442

Tolvaptan, 217
Transforming growth factor-α (TGF, α), 81–2
Traumatic brain injury (TBI), 267
TRPV5, 80, 84
TRPV6, 80, 84
Tuberose sclerosis, 15
Tumor hypoxia, 14
Tumor-induced osteomalacia (TIO), 119
Tumoral calcinosis (TC), 117–18

Ubiquitin-proteosomal pathway, 12–14
Urea:
   regulation, 210–12
      acute regulation by vasopressin, 211–12
      inner medullary collecting duct, 211
      long-term regulation by vasopressin, 212
      reabsorption, 238–9
      red blood cells, 210–1
      transport proteins, 211
   transport, 210, 239
      UT-A1, 239
      UT-A2, 239
Uremic acidosisMetabolic acidosis of chronic kidney disease
Ureteral obstruction, 499
Uric acid, 487–8
Urinary tract infections (UTI), 499
Urodilatin (URO), 289, 292
Urolithiasis, 499

Vaptans, 216–17
Vascular endothelial growth factor (VEGF), 30, 484, 493
Vasopressin antagonists, 251–6
   acid–base balance regulation, 464
   in euvolemic hyponatremia, 252–4
      Addison's disease, 253
      hypopituitarism, 253
      hypothyroidism, 252–3
      syndrome of inappropriate ADH secretion, 253–4, 277
   in hypervolemic hyponatremia, 254–6
      cardiac failure, 254–5
      cirrhosis, 255–6
   resistance to, 256
   safety of, 256
Vasopressin (AVP), 205, 225–8
   actions of, 208–10, 226
      aquaporin transport regulation, 214–15
      blood pressure control, 241–2
      calcium reabsorption regulation, 240
      chloride regulation, 237–8
      collecting duct (CD), 234–5
      inhibition of, 250–1
      magnesium reabsorption regulation, 240
      Na-K-ATPase regulation, 237
      NaCl regulation, 234, 236
      NKCC2 regulation, 236–7
      potassium conductance regulation, 237
      renal hemodynamics, 240–1
      thick ascending limb (TAL), 236, 237–8, 240
      urea regulation, 210–12, 238–9
      water regulation, 234, 261
   antagonistsVasopressin antagonists
   half-life and clearance, 226
   oxytocin similarity, 226–7
   pathology associated with dysregulation, 227–8, 272
   receptors, 206–10, 228–34
      actions and interactions of, 233–4
      antagonists, 216–17
      cloning, 228–9
      localization, 207, 229–30
      molecular structure, 230–2
      mutation in diabetes insipidus, 263
      nitric oxide generation and, 241
      regulation and desensitisation, 233
      signaling, 232
      subclasses, 228
   secretion, 205–6, 225–6
      regulation, 227, 249–50
   structure, 225
   synthesis, 225–6
Vitamin D, 63–5, 69–70
   calcium regulation role, 65
   deficiency, 65, 69
      parathyroid function and, 81
   FDA approval, 65
   phosphate role in metabolic regulation, 112–13
   role in disease states, 65
      kidney disease, 65
      rickets, 63–4
   structure elucidation, 64–5
   synthesis regulation, 100 *See also* 1,25-Dihydroxyvitamin D/VDR complex
Vitamin D receptor (VDR), 69, 70
   DNA binding, 75
   gene transactivation, 75–6
   gene transrepression, 76–8
   ligand/VDR complex formation, 73–4
   polymorphisms, 78–9
   structure-function, 71–2
      activation function 2 (AF, 2), 72
      DNA-binding domain (DBD), 71–2, 75
      ligand binding domain (LBD), 72
   VDR/RXR heterodimerization, 74, 75 *See also* Calcitriol/VDR complex *See also* 1,25-Dihydroxyvitamin D/VDR complex
Vitamin D-responsive elements (VDREs), 71, 75, 76
von Hippel-Lindau (VHL) syndrome, 12–13
   renal cell carcinoma association, 14–15

Water homeostasis, 205
   vasopressin role, 234, 261 *See also* Vasopressin (AVP)
Wolfram syndrome, 266

X-linked hypophosphatemia (XLH), 114

Zinc deficiency, 435

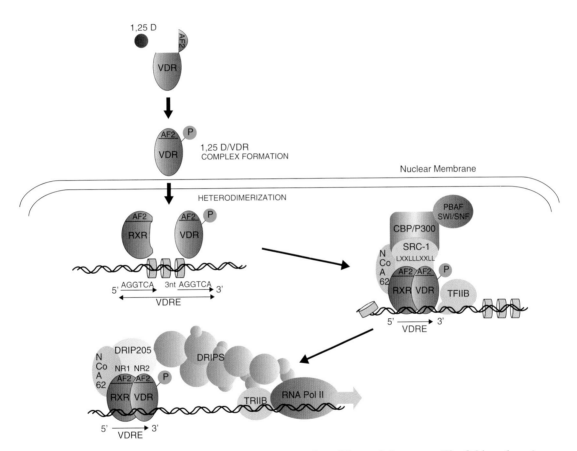

Plate 1 Calcitriol/VDR regulation of gene expression. (Figure 7.4 on page 73 of this volume)

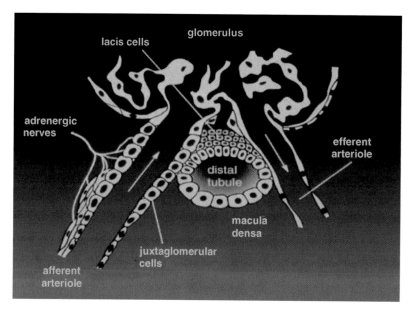

Plate 2 Schematic representation of the renal juxtaglomerular apparatus showing the various components. (Figure 12.2 on page 148 of this volume)

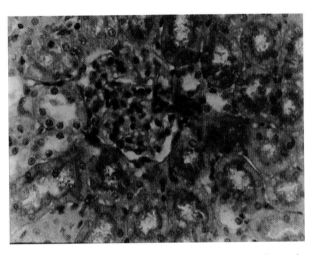

Plate 3 Light photomicrograph of the rat renal cortex demonstrating angiotensinogen protein (brown) by immunohistochemistry. Angiotensinogen within the kidney is synthesized largely in cortical proximal tubule cells. (Figure 12.13 on page 154 of this volume)

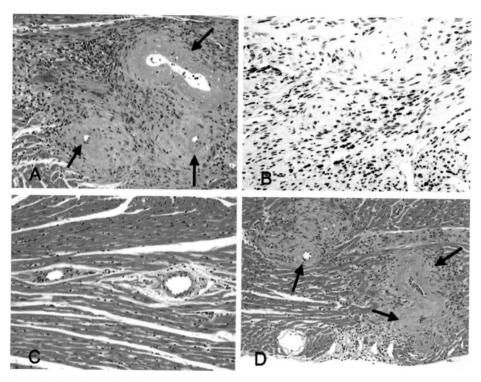

Plate 4 Effect of mineralocorticoid receptor blockade or adrenalectomy on myocardial and vascular damage. Photomicrographs of representative coronal sections of hearts from rats in different experimental groups. Focal lesions of medial fibrinoid necrosis were observed in response to Ang II/salt treatment (arrows), associated with a prominent perivascular inflammatory response (A). Macrophages were frequently found associated with coronary lesions and infiltrating the perivascular spare (B). Adrenalectomy or eplerenone treatment attenuated lesion development (C). Severe vascular inflammatory lesions were observed in all adrenalectomized animals with aldosterone treatment (D, arrows). From Rocha et al., *Endocrinology* 143, 4828–4836, 2002 with permission. (Figure 12.17 on page 159 of this volume)

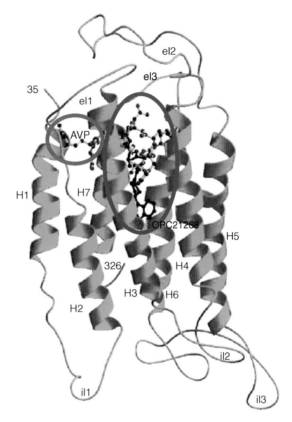

Plate 5 AVP versus AVP antagonist binding. The vasopressin receptor with its 7 transmembrane regions (H1–H7), extracellular (e) and intracellular domains (il). The site at which AVP binds at the surface of the receptor is circled in red. The site of the antagonist binding is deep in the transmembrane region and is circled in blue. The sites are distinct and partially overlap. The antagonist prevents the binding of AVP. Modified from Macion-Dazard, R. et al. (2006). *J Pharmacol Exp Ther* 316, 564–71.
(Figure 18.1 on page 252 of this volume)

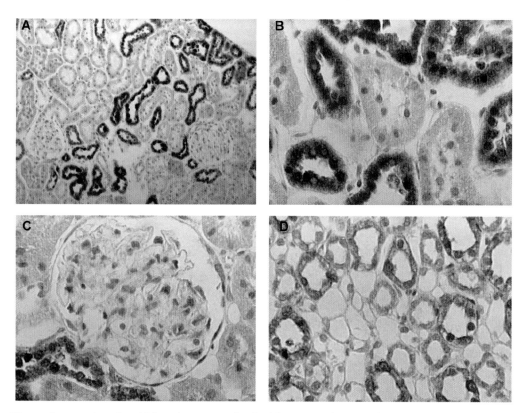

Plate 6 Photomicrographs of rat kidney immunostained with a monoclonal antibody against the mineralocorticoid receptor. (A) Cytosolic and nuclear binding in distal tubular epithelial cells of the rat renal cortex. (B) Higher magnification of distal tubular epithelial cells. (C) Kidney glomerulus showing immunolabeling of the distal tubule and glomerular mesangial cells. (D) Renal papilla showing immunolabeling of the collecting duct. (Figure 22.5 on page 338 of this volume)

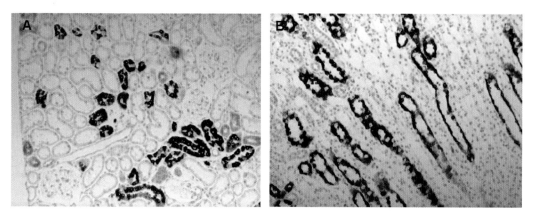

Plate 7 Immunostaining of rat kidney 11β-hydroxysteroid dehydrogenase 2 enzyme. (A) Distal nephron tubules in the cortex extending, (B) to the inner medullary collecting tubules and stopping when the tubules penetrate the papillae. (Figure 22.6 on page 338 of this volume)

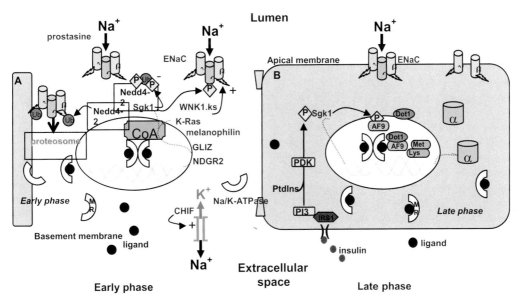

Plate 8 Mechanism of action of the mineralocorticoid receptor in polarized epithelial cells of the renal distal nephron. Cartoon cell A demonstrates early mechanisms of aldosterone regulation of ENaC and $Na^+.K^+$-ATPase. Aldosterone stimulates Sgk1 which phosphorylates Nedd4-2, preventing Nedd4-2 from ubiquinilating the ENaC β and γ subunits. Ubiquinylation accelerates removal of the channel from the membrane; Nedd4-2 phosphorylation allows a greater permanence of the ENaC subunits in the membrane and an increase in channel activity. Cartoon cell B demonstrates the late effects of aldosterone. At low aldosterone levels ENaCα gene transcription is low due to tonic inhibition by couples of Dot1a and AF9 which methylate Lys79 on histone H3. Aldosterone stimulates the transcription and translation of Sgk1 which phosphorylates multiple proteins including AF9 resulting in inhibition of the Dot1a-AF9 complex and diminished H3 methylation, thus enhancing MR access to chromatin and initiation of transcription. (Figure 22.7 on page 339 of this volume)

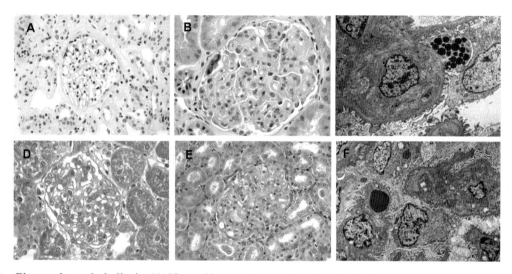

Plate 9 Glomerular endotheliosis. (A) Normal human glomerulus, H & E. (B) Human pre-eclamptic glomerulus, H & E – 33-year-old woman with twin gestation and severe pre-eclampsia at 26 weeks' gestation with urine protein/creatinine ratio of 26 at the time of biopsy. (C) Electron microscopy of glomerulus of the above patient described in (b). Note occlusion of capillary lumen cytoplasm and expansion of the subendothelial space with some electron-dense material. Podocyte cytoplasms show protein resorption droplets and relatively intact foot processes; original magnification 1500 ×. (D) Control rat glomerulus, H & E – note normal cellularity and open capillary loops. (E) sFlt-1 treated rat, H & E – note occlusion of capillary loops by swollen cytoplasm with minimal increase in cellularity. (F) Electon microscopy of sFlt-1 treated rat – note occlusion of capillary loops by swollen cytoplasm with relative preservation of podocyte foot processes; original magnification 2500 ×. All light micrographs taken at identical original magnification of 40×. This figure was reproduced with permission from Karumanchi et al. (2005). (Figure 30.1 on page 488 of this volume)